STEEL DESIGNERS' MANUAL

SIXTH EDITION

The Steel Construction Institute

Edited by

Buick Davison

Department of Civil & Structural Engineering, The University of Sheffield

Graham W. Owens

Director, The Steel Construction Institute

Blackwell
Science

© The Steel Construction Institute, 1992, 1994, 2003

Blackwell Science Ltd, a Blackwell Publishing company
Editorial offices:
Blackwell Science Ltd, 9600 Garsington Road, Oxford OX4 2DQ, UK
  Tel: +44 (0) 1865 776868
Blackwell Publishing Inc., 350 Main Street, Malden, MA 02148-5020, USA
  Tel: +1 781 388 8250
Blackwell Science Asia Pty Ltd, 550 Swanston Street, Carlton, Victoria 3053, Australia
  Tel: +61 (0)3 8359 1011

The right of the Author to be identified as the Author of this Work has been asserted in
accordance with the Copyright, Designs and Patents Act 1988.

First edition published Crosby Lockwood & Sons Ltd 1955
Second edition published 1960
Third edition published 1966
Fourth edition published by Granada Publishing Ltd in Crosby Lockwood Staples 1972
Fifth edition published by Blackwell Science Ltd 1992
Sixth edition published by Blackwell Science Ltd 2003
Reprinted 2004 (twice)
Reissued in paperback 2005

Library of Congress Cataloging-in-Publication Data is available

ISBN-10: 1-4051-3412-7
ISBN-13: 978-14051-3412-5

A catalogue record for this title is available from the British Library

Set in 10/12pt Times
by SNP Best-set Typesetters Ltd, Hong Kong
Printed and bound in India
by Gopsons Papers Ltd, New Delhi

The publisher's policy is to use permanent paper from mills that operate a sustainable forestry
policy, and which has been manufactured from pulp processed using acid-free and elementary
chlorine-free practices. Furthermore, the publisher ensures that the text paper and cover board
used have met acceptable environmental accreditation standards.

For further information on Blackwell Publishing, visit our website:
www.thatconstructionsite.com

Although care has been taken to ensure, to the best of our knowledge, that all data and
information contained herein are accurate to the extent that they relate to either matters of fact
or accepted practice or matters of opinion at the time of publication, the Steel Construction
Institute assumes no responsibility for any errors in or misinterpretations of such data and or
information or any loss or damage arising from or related to their use.

Extracts from the British Standards are reproduced with the permission of BSI. Complete
copies of the standards quoted can be obtained by post from BSI Sales, Linford Wood,
Milton Keynes, MK14 6LE.

Contents

SECTION 3: DESIGN THEORY

SECTION 4: ELEMENT DESIGN

Connection design

The Steel Construction Institute (SCI) promotes the proper and effective use of steel in construction.

Membership is open to organisations and individuals that are concerned with the use of steel in construction, and members include clients, designers, contractors, suppliers, fabricators, academics and government departments. SCI is financed by subscriptions from its members, revenue from research contracts, consultancy services and by the sale of publications.

SCI's work is initiated and guided through the involvement of its members on advisory groups and technical committees. A comprehensive advisory and consultancy service is available to members on the use of steel in construction.

Dr Graham Owens has 34 years' experience in designing, constructing, teaching and researching in structural steelwork. After six years' practical experience and 16 years at Imperial College, he joined the Steel Construction Institute at its formation in 1986. He was made director in 1992.

Dr Buick Davison is a Senior Lecturer in the Department of Civil and Structural Engineering at the University of Sheffield. In addition to his wide experience in teaching and research of steel structures, he is a Chartered Engineer and has worked in consultancies on the design of buildings and stadia.

Introduction to sixth edition

At the instigation of the Iron and Steel Federation, the late Bernard Godfrey began work in 1952 on the first edition of the *Steel Designers' Manual*. As principal author he worked on the manuscript almost continuously for a period of two years. On many Friday evenings he would meet with his co-authors, Charles Gray, Lewis Kent and W.E. Mitchell to review progress and resolve outstanding technical problems. A remarkable book emerged. Within approximately 900 pages it was possible for the steel designer to find everything necessary to carry out the detailed design of most conventional steelwork. Although not intended as an analytical treatise, the book contained the best summary of methods of analysis then available. The standard solutions, influence lines and formulae for frames could be used by the ingenious designer to disentangle the analysis of the most complex structure. Information on element design was intermingled with guidance on the design of both overall structures and connections. It was a book to dip into rather than read from cover to cover. However well one thought one knew its contents, it was amazing how often a further reading would give some useful insight into current problems. Readers forgave its idiosyncrasies, especially in the order of presentation. How could anyone justify slipping a detailed treatment of angle struts between a very general discussion of space frames and an overall presentation on engineering workshop design?

The book was very popular. It ran to four editions with numerous reprints in both hard and soft covers. Special versions were also produced for overseas markets. Each edition was updated by the introduction of new material from a variety of sources. However, the book gradually lost the coherence of its original authorship and it became clear in the 1980s that a more radical revision was required.

After 36 very successful years it was decided to rewrite and re-order the book, while retaining its special character. This decision coincided with the formation of the Steel Construction Institute and it was given the task of co-ordinating this activity.

A complete restructuring of the book was undertaken for the fifth edition, with more material on overall design and a new section on construction. The analytical material was condensed because it is now widely available elsewhere, but all the design data were retained in order to maintain the practical usefulness of the book as a day-to-day design manual. Allowable stress design concepts were replaced by limit state design encompassing BS 5950 for buildings and BS 5400 for bridges. Design examples are to the more appropriate of these two codes for each particular application.

The fifth edition was published in 1992 and proved to be a very worthy successor to its antecedents. It also ran to several printings in both hard and soft covers; an international edition was also printed and proved to be very popular in overseas markets.

This sixth edition maintains the broad structure introduced in 1992, reflecting its target readership of designers of structural steelwork of all kinds.

- Design synthesis
- Steel technology
- Design theory
- Element design
- Connection design
- Other elements
- Construction.

Design synthesis: Chapters 1–5

A description of the nature of the process by which design solutions are arrived at for a wide range of steel structures including:

- Single- and multi-storey buildings (Chapters 1 and 2)
- Heavy industrial frames (Chapter 3)
- Bridges (Chapter 4)
- Other diverse structures such as space frames, cable structures, towers and masts, atria and steel in housing (Chapter 5).

Steel technology: Chapters 6–8

Background material sufficient to inform designers of the important problems inherent in the production and use of steel, and methods of overcoming them in practical design.

- Applied metallurgy (Chapter 6)
- Fatigue and Fracture (Chapter 7)
- Sustainability and steel construction (Chapter 8).

Design theory: Chapters 9–12

A résumé of analytical methods for determining the forces and moments in structures subject to static or dynamic loads, both manual and computer-based.

Comprehensive tables for a wide variety of beams and frames are given in the Appendix.

- Manual and computer analysis (Chapter 9)
- Beam analysis (Chapter 10)
- Frame analysis (Chapter 11)
- Applicable dynamics (Chapter 12).

Element design: Chapters 13–22

A comprehensive treatment of the design of steel elements, singly, in combination or acting compositely with concrete.

- Local buckling and cross-section classification (Chapter 13)
- Tension members (Chapter 14)
- Columns and struts (Chapter 15)
- Beams (Chapter 16)
- Plate girders (Chapter 17)
- Members with compression and moments (Chapter 18)
- Trusses (Chapter 19)
- Composite floors (Chapter 20)
- Composite beams (Chapter 21)
- Composite columns (Chapter 22).

Connection design: Chapters 23–27

The general basis of design of connections is surveyed and amplified by consideration of specific connection methods.

- Bolts (Chapter 23)
- Welds and design for welding (Chapter 24)
- Plate and stiffener elements in connections (Chapter 25)
- Design of connections (Chapter 26)
- Foundations and holding-down systems (Chapter 27).

Other elements: Chapters 28–30

- Bearings and joints (Chapter 28)
- Piles (Chapter 29)
- Floors and orthotropic decks (Chapter 30).

Construction: Chapters 31–35

Important aspects of steel construction about which a designer must be informed if he is to produce structures which can be economically fabricated, and erected and which will have a long and safe life.

- Tolerances (Chapter 31)
- Fabrication (Chapter 32)
- Erection (Chapter 33)
- Fire protection and fire engineering (Chapter 34)
- Corrosion resistance (Chapter 35).

Finally, Chapter 36 summarizes the state of progress on the Eurocodes, which will begin to influence our design approaches from 2003 onwards.

A comprehensive collection of data of direct use to the practising designer is compiled into a series of appendices.

By kind permission of the British Standards Institution, references are made to British Standards throughout the manual. The tables of fabrication and erection tolerances in Chapter 31 are taken from the *National Structural Steelwork Specification,* second edition. Much of the text and illustrations for Chapter 33 are taken from *Steelwork Erection* by Harry Arch. Both these sources are used by kind permission of the British Constructional Steelwork Association, the publishers. These permissions are gratefully acknowledged.

Finally I would like to pay tribute both to the 38 authors who have contributed to the sixth edition and to my hard-working principal editor, Dr Buick Davison. All steelwork designers are indebted to their efforts in enabling this text book to be maintained as the most important single source of information on steel design.

Graham Owens

Contributors

Harry Arch

Harry Arch graduated from Manchester Faculty of Technology. For many years he worked for Sir William Arrol, where he became a director, responsible for all outside construction activities including major bridges, power stations and steelworks construction. In 1970 he joined Redpath Dorman Long International, working on offshore developments.

Mike Banfi

Mike Banfi joined Arup from Cambridge University in 1976. He has been involved in the design of various major projects, including: Cummins Engine Plant, Shotts; The Hong Kong and Shanghai Bank, Hong Kong; Usine L'Oreal, Paris; roofs for the TGV stations, Lille and Roissy; roofs for the Rad-und Schwimmsportshalle, Berlin; and various office blocks. He is now based in Arup Research & Development where he provides advice on projects; examples include: Wellcome Wing to the Science Museum, London; City Hall, London; T5, Heathrow. He is UK National Technical Contact for Eurocode 4 part 1.1 and was on the steering committee for the 4th edition of the NSSS. He is an Associate Director.

Hubert Barber

Hubert Barber joined Redpath Brown in 1948 and for five years gained a wide experience in steel construction. The remainder of his working life was spent in local government, first at Manchester and then in Yorkshire where he became chief structural engineer of West Yorkshire. He also lectured part-time for fourteen years at the University of Bradford.

Tony Biddle

Tony Biddle graduated in civil engineering from City University in 1966 and spent the early part of his career in contractors, designing in steel and reinforced concrete

before specializing in soil mechanics and foundation design in 1970. Between 1974 and 1993 he worked in the offshore industry, becoming a specialist in steel piling. He joined SCI in 1994 as manager for civil engineering and has developed the R&D research project work in steel piling related topics. He has been a drafting member for Eurocode 3 part 5, contributor to BS 8002 amendments, and author of several SCI publications.

Michael Burdekin

Michael Burdekin graduated from Cambridge University in 1959. After fifteen years of industrial research and design experience he went to UMIST, where he is now Professor of Civil and Structural Engineering. His specific expertise is the field of welded steel structures, particularly in the application of fracture mechanics to fracture and fatigue failure.

Brian Cheal

Brian Cheal graduated from Brighton Technical College in 1951 with an External Degree of the University of London. He was employed with W.S. Atkins and Partners from 1951 to 1986, becoming a technical director in 1979, and specialized in the analysis and design of steel-framed structures, including heavy structural framing for power stations and steelworks. He has written design guides and given lectures on various aspects of connection design and is co-author of *Structural Steelwork Connections*.

David Dibb-Fuller

David Dibb-Fuller started his career with the Cleveland Bridge and Engineering Company in London. His early bridge related work gave a strong emphasis to heavy fabrication; in later years he moved on to building structures. As technical director for Conder Southern in Winchester his strategy was to develop close links between design for strength and design for production. Currently he is a partner with Gifford and Partners in Southampton where he continues to exercise his skills in the design of steel structures.

Ian Duncan

Ian Duncan joined the London office of Ove Arup and Partners in 1966 after graduating from Surrey University. From 1975 he taught for four years at Univer-

sity College Cardiff before joining Buro Happold. He now runs his own practice in Bristol.

Michael Green

Michael Green graduated from Liverpool University in 1971. After an early career in general civil engineering, he joined Buro Happold, where he is now an executive partner. He has worked on a wide variety of building projects, developing a specialist expertise in atria and large-span structures.

Alan Hart

Alan Hart graduated from the University of Newcastle upon Tyne in 1968 and joined Ove Arup and Partners. During his career he has been involved in the design of a number of major award-winning buildings, including Carlsberg Brewery, Northampton; Cummins Engine Plant, Shotts, Lanarkshire; and the Hongkong and Shanghai Bank, Hong Kong. He is a project director of Ove Arup and Partners.

Alan Hayward

Alan Hayward is a bridge specialist and is principal consultant of Cass Hayward and Partners, who design and devise the erection methodology for all kinds of steel bridges, many built on a design:construct basis. Projects include London Docklands Light Railway viaducts, the M25/M4 interchange, the Centenary Lift bridge at Trafford Park and the Newark Dyke rail bridge reconstruction. Movable bridges and roll-on/roll-off linkspans are also a speciality. He is a former chief examiner for the Institution of Structural Engineers and was invited to become a Fellow of the Royal Academy of Engineers in 2001.

Eric Hindhaugh

Eric Hindhaugh trained as a structural engineer in design and constructional steelwork, timber and lightweight roll-formed sections. He then branched into promotional and marketing activities. He was a market development manager in construction for British Steel Strip Products, where he was involved in Colorcoat and the widening use of lightweight steel sections for structural steel products. He is now retired.

Roger Hudson

Roger Hudson studied metallurgy at Sheffield Polytechnic whilst employed by BISRA. He also has a Masters degree from the University of Sheffield. In 1968, he joined the United Steel Companies at Swinden Laboratories in Rotherham to work on the corrosion of stainless steels. The laboratories later became part of British Steel where he was responsible for the Corrosion Laboratory and several research projects. He is now principal technologist in the recently formed Corus company. He is a member of several technical and international standards committees, has written technical publications, and has lectured widely on the corrosion and protection of steel in structures. He is a long serving professional member of the Institute of Corrosion and is currently chairman of the Yorkshire branch and chairman of the Training and Certification Governing Board.

Ken Johnson

Ken Johnson was head of corrosion and coatings at British Steel's Swinden Laboratories. His early experience was in the paint industry but he then worked in steel for over twenty-five years, dealing with the corrosion and protection aspects of the whole range of British Steel's products, including plates, section, piling, strip products, tubes, stainless steels, etc. He represented the steel industry on several BSI and European Committees and was a council member of the Paint Research Association. He is now retired.

Alan Kwan

Alan Kwan graduated from the University of Sheffield and Cambridge University. He is a lecturer in structural engineering at Cardiff University, specializing in lightweight, deployable, tension and space structures, and numerical methods for their analysis.

Mark Lawson

A graduate of Imperial College, and the University of Salford, where he worked in the field of cold-formed steel, Mark Lawson spent his early career at Ove Arup and Partners and the Construction Industry Research and Information Association. In 1987 he joined the newly formed Steel Construction Institute as research manager for steel in buildings, with particular reference to composite construction, fire engineering and cold-formed steel. He is a member of the Eurocode 4 project team on fire-resistant design.

Ian Liddell

After leaving Cambridge, Ian Liddell joined Ove Arup and Partners to work on the roof of the Sydney Opera House and on the South Bank Art Centre. His early career encompassed a wide range of projects, with particular emphasis on shell structures and lightweight tension and fabric structures. Since 1976 he has been a partner of Buro Happold and has been responsible for a wide range of projects, many with special structural engineering features, including mosques, auditoriums, mobile and temporary structures, stadiums and retail atria.

Matthew Lovell

Matthew Lovell studied civil engineering at University College, London. After graduation Matthew worked for Arup on the Chur Station roof project. He is now senior associate at Buro Happold and has worked on many steel structures, including Thames Valley University LRC, the National Centre for Popular Music, and St David's RF Hotel. He has recently completed an MSc in Interdisciplinary Design at Cambridge University.

Stephen Matthews

Stephen Matthews graduated from the University of Nottingham in 1974 and completed postgraduate studies at Imperial College in 1976–77. His early professional experience was gained with Rendel Palmer and Tritton. During subsequent employment with Fairfield Mabey and Cass Hayward and Partners he worked on the design of several large composite bridges, including the Simon de Montfort Bridge Evesham, M25/M4 interchange, Poyle, and viaducts on the Docklands Light Railway. He is a director of WSP (Civils), where he has been manager of the Bridges Division since 1990. Work has included a number of major bridge repair schemes and drafting of the UK National Application Document for Eurocode 3 part 2 (steel bridges).

David Moore

David Moore graduated from the University of Bradford in 1981 and joined the Building Research Establishment (BRE) where he has completed over twenty years of research and specialist advisory work in the area of structural steelwork. He is the author of over 70 technical papers on a wide range of subjects. He has also made a significant contribution to a number of specialist steel and composite connection design guides, many of which are used daily by practising structural engineers and

steelwork fabricators. Currently he is the director of the Centre for Structural Engineering at BRE.

Rangachari Narayanan

Rangachari Narayanan graduated in civil engineering from Annamalai University (India) in 1951. In a varied professional career spanning over forty years, he has held senior academic positions at the Universities of Delhi, Manchester and Cardiff. He is the recipient of several awards including the Benjamin Baker Gold Medal and George Stephenson Gold Medal, both from the Institution of Civil Engineers. For many years he headed the Education and Publication Divisions at the Steel Construction Institute.

David Nethercot

Since graduating from the University of Wales, Cardiff, David Nethercot has completed thirty years of teaching, research and specialist advisory work in the area of structural steelwork. The author of over 300 technical papers, he has lectured frequently on post-experience courses; he is chairman of the BSI Committee responsible for BS 5950, and is a frequent contributor to technical initiatives associated with the structural steelwork industry. Since 1999 he has been head of the Department of Civil and Environmental Engineering at Imperial College.

Gerard Parke

Gerry Parke is a lecturer in structural engineering at the University of Surrey specializing in the analysis and design of steel structures. His particular interests lie in assessing the collapse behaviour of both steel industrial buildings and large-span steel space structures.

Phil Peacock

Phil Peacock is a member of the Corus Construction Centre. He started his career in 1965 at steelwork fabricators Ward Bros. Ltd., gained an HND at Teesside Polytechnic and moved to White Young Consulting Engineers in 1973 before joining British Steel (now Corus) in 1988. His experience covers the design management of a wide range of projects: heavy plant buildings and structures for the steel, petrochemical and coal industries, commercial offices, leisure and retail developments. He serves on several industry committees and is a past chairman of the Institution of Structural Engineers Scottish Branch.

Alan Pottage

Alan Pottage graduated from the University of Newcastle upon Tyne in 1976 and gained a Masters degree in structural steel design from Imperial College in 1984. He has gained experience in all forms of steel construction, particularly portal frame and multi-rise structures, and has contributed to various code committees, and SCI guides on composite design and connections.

Graham Raven

Graham Raven graduated from King's College, London in 1963 and joined Ove Arup and Partners. Following thirteen years with consulting engineers working on a variety of building structures he joined a software house pioneering work in structural steel design and detailing systems. In 1980 this experience took him to Ward Building Systems where he became technical director and was closely associated with the development of a range of building components and increased use of welded sections in buildings. Since 1991, with the exception of a year with a software house specialising in 3D detailing systems, he has been employed at the Steel Construction Institute, where he is the senior manager responsible for the Sustainability Group.

John Righiniotis

John Righiniotis graduated from the University of Thessalonika in 1987 and obtained an MSc in structural steel design from Imperial College in 1988. He worked at the Steel Construction Institute on a wide range of projects until June 1990 when he was required to return to Greece to carry out his military service.

John Roberts

John Roberts graduated from the University of Sheffield in 1969 and was awarded a PhD there in 1972 for research on the impact loading of steel structures. His professional career includes a period of site work with Alfred McAlpine, following which he has worked as a consulting engineer, since 1981 with Allott & Lomax/Babtie Group. He is a director of Babtie Group where he heads up the Structures and Buildings Teams that have designed many major steelwork structures. He was president of the Institution of Structural Engineers in 1999–2000 and is a council member of both the Steel Construction Institute and the BCSA.

Terry Roberts

Terry Roberts graduated in civil and structural engineering from the University of Wales Cardiff in 1967, and following three years of postgraduate study was awarded a PhD in 1971. His early professional experience was gained in bridge design and site investigation for several sections of the M4 motorway in Wales. He returned to academic life in 1975. He is the author of over 100 technical papers on various aspects of structural engineering, for which he received a DSc from the University of Wales and the Moisseiff Award from the Structural Engineering Institute of the American Society of Civil Engineers in 1997. Since 1996 he has been head of the Division of Structural Engineering in the Cardiff School of Engineering.

Jef Robinson

Jef Robinson graduated in metallurgy from Durham University in 1962. His early career in the steel industry included formulating high ductility steels for automotive applications and high-strength notch ductile steels for super tankers, drilling platforms and bridges. Later as market development manager for the structural division of British Steel (now Corus) he chaired the BSI committee that formulated BS 5950 Part 8: *Fire Resistant Design* for structural steelwork and served on a number of international fire related committees. He was appointed honorary professor at the University of Sheffield in 2000.

Alan Rogan

Alan Rogan is a leading consultant to the steel industry, working with prestigious clients such as Corus and Cleveland Bridge Engineering Group. Alan has been involved in the construction of many buildings, such as Canary Wharf, Gatwick Airport extension and many bridges from simple footbridges to complex multi-spans, in the UK and overseas.

Dick Stainsby

Dick Stainsby's career training started with an HNC and went on to include postgraduate studies at Imperial College London. His experience has encompassed steel structures of all kinds including bridgework. He was for many years chief designer with Redpath Dorman Long Middlesbrough. Since retiring from mainstream industry he has assisted the British Constructional Steelwork Association, the Steel Construction Institute and the Institution of Structural Engineers in the production of

technical publications relating to steelwork connections. He also compiled the National Structural Steelwork Specification for Building Construction, which is now in its 4th Edition.

Paul Tasou

Paul Tasou graduated from Queen Mary College, London in 1978 and subsequently obtained an MSc in structural steel design from Imperial College, London. He spent eleven years at Rendel Palmer and Tritton working on a wide range of bridge, building and civil engineering projects. He is now principal partner in Tasou Associates.

Colin Taylor

Colin Taylor graduated from Cambridge in 1959. He started his professional career in steel fabrication, initially in the West Midlands and subsequently in South India. After eleven years he moved into consultancy where, besides practical design, he became involved with graduate training, the use of computers for design and drafting, company technical standards and drafting work for British Standards and for Eurocodes as editorial secretary for Eurocode 3. Moving to the Steel Construction Institute on its formation as manager of the Codes and Advisory Division, he also became involved with the European standard for steel fabrication and erection *Execution of Steel Structures*.

John Tyrrell

John Tyrrell graduated from Aston University in 1965 and immediately joined Ove Arup and Partners. He has worked for them on a variety of projects in the UK, Australia and West Africa; he is now a project director. He has been responsible for the design of a wide range of towers and guyed masts. He currently leads the Industrial Structures Group covering diverse fields of engineering from telecommunications and broadcasting to the power industry.

Peter Wickens

Having graduated from Nottingham University in 1971, Peter Wickens spent much of his early career in civil engineering, designing bridges and Metro stations. In 1980,

he changed to the building structures field and was project engineer for the Billingsgate Development, one of the first of the new generation of steel composite buildings. He is currently manager of the Structural Division and head of discipline for Building Structures at Mott MacDonald.

Michael Willford

Michael Willford joined Arup in 1975, having graduated from Cambridge University. He has been a specialist in the design of structures subjected to dynamic actions for over twenty years. His design and analysis experience covers a wide variety of projects including buildings, bridges and offshore structures. He is currently a director of Arup and the leader of a team of specialists working in these fields based in London and San Francisco.

John Yates

John Yates was appointed to a personal chair in mechanical engineering at the University of Sheffield in 2000 after five years as a reader in the department. He graduated from Pembroke College, Cambridge in 1981 in metallurgy and materials science and then undertook research degrees at Cranfield and the University of Sheffield before several years of postdoctoral engineering and materials research. His particular interests are in developing structural integrity assessment tools based on the physical mechanisms of fatigue and fracture. He is the honorary editor of *Engineering Integrity* and an editor of the international journal *Fatigue and Fracture of Engineering Materials and Structures*.

Ralph Yeo

Ralph Yeo graduated in metallurgy at Cardiff and Birmingham and lectured at The University of the Witwatersrand. In the USA he worked on the development of weldable high-strength and alloy steels with International Nickel and US Steel and on industrial gases and the development of welding consumables and processes at Union Carbide's Linde Division. Commercial and general management activities in the UK, mainly with The Lincoln Electric Company, were followed by twelve years as a consultant and expert witness, with special interest in improved designs for welding.

Notation

Several different notations are adopted in steel design; different specializations frequently give different meanings to the same symbol. These differences have been maintained in this book. To do otherwise would be to separate this text both from other literature on a particular subject and from common practice. The principal definitions for symbols are given below. For conciseness, only the most commonly adopted subscripts are given; others are defined adjacent to their usage.

A		Area
	or	End area of pile
	or	Constant in fatigue equations
A_e		Effective area
A_g		Gross area
A_s		Shear area of a bolt
A_t		Tensile stress area of a bolt
A_v		Shear area of a section
a		Spacing of transverse stiffeners
	or	Effective throat size of weld
	or	Crack depth
	or	Distance from central line of bolt to edge of plate
	or	Shaft area of pile
B		Breadth
$\mathbf{B}$		Transformation matrix
b		Outstand
	or	Width of panel
	or	Distance from centreline of bolt to toe of fillet weld or to half of root radius as appropriate
b_e		Effective breadth or effective width
b_1		Stiff bearing length
C		Crack growth constant
$\mathbf{C}$		Transformation matrix
C_v		Charpy impact value
C_y		Damping coefficient
c		Bolt cross-centres
	or	Cohesion of clay soils

D		Depth of section
		Diameter of section or hole
$\mathbf{D}$		Elasticity matrix
D_r		Profile height for metal deck
D_s		Slab depth
d		Depth of web
	or	Nominal diameter of fastener
	or	Depth
d_e		Effective depth of slab
E		Modulus of elasticity of steel (Young's modulus)
e		End distance
e_y		Material yield strain
F_c		Compressive force due to axial load
F_s		Shear force (bolts)
F_t		Tensile force
F_v		Shear force (sections)
f		Flexibility coefficient
f_a		Longitudinal stress in flange
f_c		Compressive stress due to axial load
f_{cu}		Cube strength of concrete
f_m		Force per unit length on weld group from moment
f_r		Resultant force per unit length on weld group from applied concentric load
f_v		Force per unit length on weld group from shear
	or	Shear stress
G		Shear modulus of steel
g		Gravitational acceleration
H		Warping constant of section
	or	Horizontal reaction
h		Height
	or	Stud height
	or	Depth of overburden
I_o		Polar second moment of area of bolt group
I_{oo}		Polar second moment of area of weld group of unit throat about polar axis
I_x		Second moment of area about major axis
I_{xx}		Polar second moment of area of weld group of unit throat about xx axis
I_y		Second moment of area about minor axis
I_{yy}		Polar second moment of area of weld group of unit throat about yy axis
K		Degree of shear connection
	or	Stiffness
$\mathbf{K}$		Stiffness matrix
K_s		Curvature of composite section from shrinkage
	or	Constant in determining slip resistance of HSFG bolts
K_1, K_2, K_3		Empirical constants defining the strength of composite columns

k_a		Coefficient of active pressure
k_d		Empirical constant in composite slab design
k_p		Coefficient of passive resistance
L		Length of span or cable
L_y		Shear span length of composite slab
M		Moment
	or	Larger end moment
M_{ax}, M_{ay}		Maximum buckling moment about major or minor axis in presence of axial load
M_b		Buckling resistance moment (lateral – torsional)
M_E		Elastic critical moment
M_o		Mid-length moment on a simply-supported span equal to unrestrained length
M_{pc}		Plastic moment capacity of composite section
M_{rx}, M_{ry}		Reduced moment capacity of section about major or minor axis in the presence of axial load
M_x, M_y		Applied moment about major or minor axis
$\overline{M}_x, \overline{M}_y$		Equivalent uniform moment about major or minor axis
M_1, M_2		End moments for a span of a continuous composite beam
m		Equivalent uniform moment factor
	or	Empirical constant in fatigue equation
	or	Number of vertical rows of bolts
m_d		Empirical constant in composite slab design
N		Number of cycles to failure
N_c, N_q, N_γ		Constants in Terzaghi's equation for the bearing resistance of clay soils
n		Crack growth constant
	or	Number of shear studs per trough in metal deck
	or	Number of horizontal rows of bolts
	or	Distance from bolt centreline to plate edge
P		Force in structural analysis
	or	Load per unit surface area on cable net
	or	Crushing resistance of web
P_{bb}		Bearing capacity of a bolt
P_{bg}		Bearing capacity of parts connected by friction-grip fasteners
P_{bs}		Bearing capacity of parts connected by ordinary bolts
P_c		Compression resistance
P_{cx}, P_{cy}		Compression resistance considering buckling about major or minor axis only
P_o		Minimum shank tension for preloaded bolt
P_s		Shear capacity of a bolt
P_{sL}		Slip resistance provided by a friction-grip fastener
P_t		Tension capacity of a member or fastener
P_u		Compressive strength of stocky composite column
P_v		Shear capacity of a section

p		Ratio of cross-sectional area of profile to that of concrete in a composite slab
p_b		Bending strength
p_{bb}		Bearing strength of a bolt
p_{bg}		Bearing strength of parts connected by friction-grip fasteners
p_{bs}		Bearing strength of parts connected by ordinary bolts
p_c		Compressive strength
p_E		Euler strength
p_o		Minimum proof stress of a bolt
p_s		Shear strength of a bolt
p_t		Tension strength of a bolt
p_w		Design strength of a fillet weld
p_y		Design strength of steel
Q		Prying force
q		Ultimate bearing capacity
q_b		Basic shear strength of a web panel
q_{cr}		Critical shear strength of a web panel
q_e		Elastic critical shear strength of a web panel
q_f		Flange-dependent shear strength factor
R		Reaction
	or	Load applied to bolt group
	or	Radius of curvature
R_c		Compressive capacity of concrete section in composite construction
R_q		Capacity of shear connectors between point of contraflexure and point of maximum negative moment in composite construction
R_r		Tensile capacity in reinforcement in composite construction
R_s		Tensile capacity in steel section in composite construction
R_w		Compression in web section in composite construction
r		Root radius in rolled section
r_r		Reduction factor in composite construction
r_x, r_y		Radius of gyration of a member about its major or minor axis
S		Span of cable
S_R		Applied stress range
S_x, S_y		Plastic modulus about major or minor axis
s		Spacing
	or	Leg length of a fillet weld
T		Thickness of a flange or leg
	or	Tension in cable
t		Thickness of web
U		Elastic energy
U_s		Specified minimum ultimate tensile strength of steel
u		Buckling parameter of the section
V		Shear force
	or	Shear resistance per unit length of beam in composite construction
V_b		Shear buckling resistance of stiffened web utilizing tension field action

V_{cr}		Shear buckling resistance of stiffened or unstiffened web without utilizing tension field action
v		Slenderness factor for beam
W		Point load
	or	Foundation mass
	or	Load per unit length on a cable
	or	Energy required for crack growth
w		Lateral displacement
	or	Effective width of flange per bolt
	or	Uniformly distributed load on plate
X_e		Elastic neutral axis depth in composite section
x		Torsion index of section
x_p		Plastic neutral axis depth in composite section
Y		Correction factor in fracture mechanics
Y_s		Specified minimum yield stress of steel
Z_c		Elastic section modulus for compression
Z_{oo}		Elastic modulus for weld group of unit throat subject to torsional load
Z_x, Z_y		Elastic modulus about major or minor axis
z		Depth of foundation
α		Coefficient of linear thermal expansion
α_e		Modular ratio
β		Ratio of smaller to larger end moment
	or	Coefficient in determination of prying force
γ		Ratio M/M_o, i.e. ratio of larger end moment to mid-length moment on simply-supported span equal to unrestrained length
	or	Bulk density of soil
	or	Coefficient in determination of prying force
γ_f		Overall load factor
γ_m		Material strength factor
Δ		Displacements in vector
δ		Deflection
	or	Elongation
δ_c		Deflection of composite beam at serviceability limit state
δ_{ic}		Deflection of composite beam at serviceability limit state in presence of partial shear connection
δ_o		Deflection of steel beam at serviceability limit state
δ_{oo}		Deflection in continuous composite beam at serviceability limit state
ε		Constant $(275/p_y)^{1/2}$
	or	Strain
η		Load ratio for composite columns
λ		Slenderness, i.e. effective length divided by radius of gyration
λ_{cr}		Elastic critical load factor
λ_{LO}		Limiting equivalent slenderness
λ_{LT}		Equivalent slenderness

λ_o		Limiting slenderness
μ		Slip factor
μ_x, μ_y		Reduction factors on moment resistance considering axial load
σ		Stress
σ_ε		Tensile stress
ϕ		Diameter of composite column
	or	Angle of friction in granular soil

Chapter 1
Single-storey buildings

by GRAHAM RAVEN and ALAN POTTAGE

1.1 Range of building types

It is estimated that around 50% of the hot-rolled constructional steel used in the UK is fabricated into single-storey buildings, being some 40% of the total steel used in them. The remainder is light-gauge steel cold-formed into purlins, rails. cladding and accessories. Over 90% of single-storey non-domestic buildings have steel frames, demonstrating the dominance of steel construction for this class of building. These relatively light, long-span, durable structures are simply and quickly erected, and developments in steel cladding have enabled architects to design economical buildings of attractive appearance to suit a wide range of applications and budgets.

The traditional image was a dingy industrial shed, with a few exceptions such as aircraft hangars and exhibition halls. Changes in retailing and the replacement of traditional heavy industry with electronics-based products have led to a demand for increased architectural interest and enhancement.

Clients expect their buildings to have the potential for easy change of layout several times during the building's life. This is true for both institutional investors and owner users. The primary feature is therefore flexibility of planning, which, in general terms, means as few columns as possible consistent with economy. The ability to provide spans up to 60m, but most commonly around 30m, gives an extremely popular structural form for the supermarkets, do-it-yourself stores and the like which are now surrounding towns in the UK. The development of steel cladding in a wide variety of colours and shapes has enabled distinctive and attractive forms and house styles to be created.

Improved reliability of steel-intensive roofing systems has contributed to their acceptability in buildings used by the public and perhaps more importantly in 'high-tech' buildings requiring controlled environments. The structural form will vary according to span, aesthetics, integration with services, cost and suitability for the proposed activity. A cement manufacturing building will clearly have different requirements from a warehouse, food processing plant or computer factory.

The growth of the leisure industry has provided a challenge to designers, and buildings vary from the straightforward requirement of cover for bowls, tennis, etc., to an exciting environment which encourages people to spend days of their holidays indoors at water centres and similar controlled environments suitable for year round recreation.

1

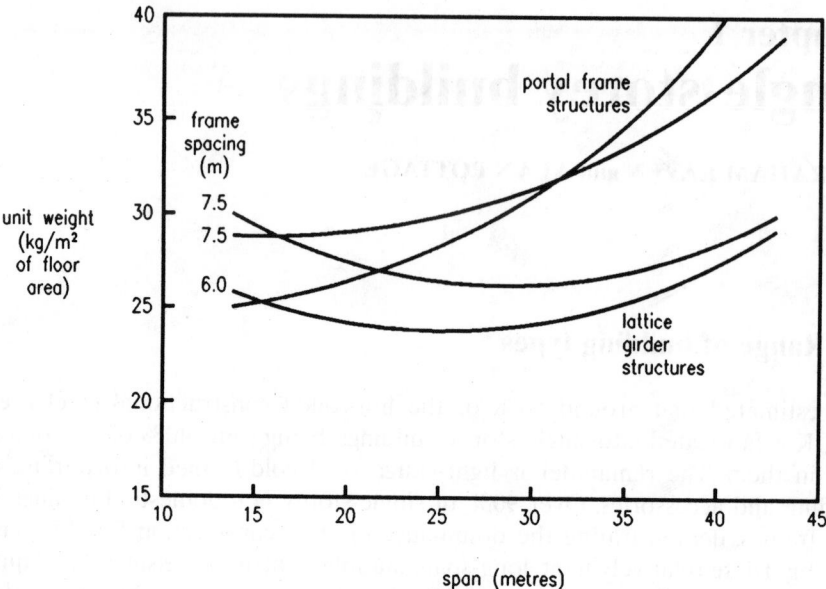

Fig. 1.1   Comparison of bare frame weights for portal and lattice structures

In all instances the requirement is to provide a covering to allow a particular activity to take place; the column spacing is selected to give as much freedom of use of the space consistent with economy. The normal span range will be from 12 m to 50 m, but larger spans are feasible for hangars and enclosed sports stadia.

Figure 1.1 shows how steel weight varies with structural form and span.[1]

1.2 Anatomy of structure

A typical single-storey building consisting of cladding, secondary steel and a frame structure is shown in Fig. 1.2.

1.2.1 Cladding

Cladding is required to be weathertight, to provide insulation, to have penetrations for daylight and access, to be aesthetically pleasing, and to last the maximum time with a minimum of maintenance consistent with the budget.

The requirements for the cladding to roofs and walls are somewhat different.

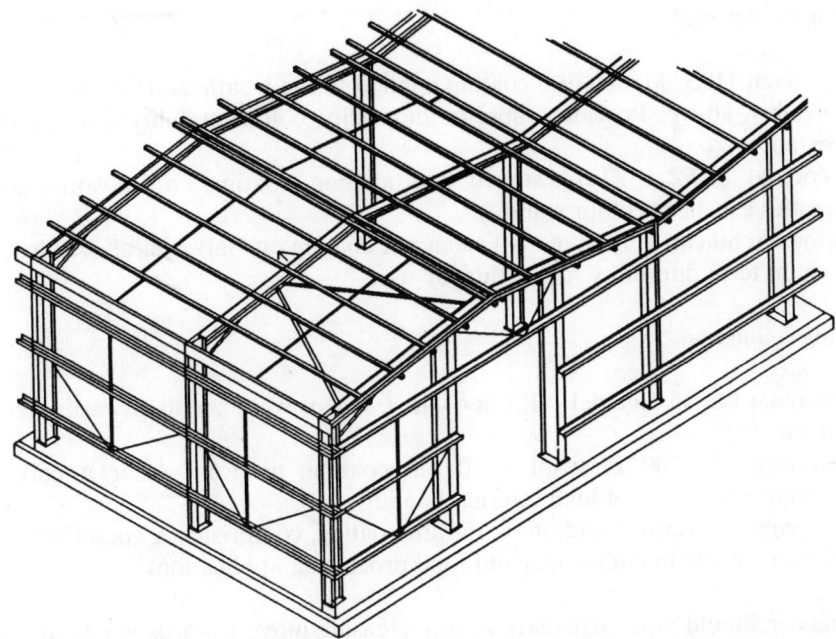

Fig. 1.2 Structural form for portal-frame building (some rafter bracing omitted for clarity)

The ability of the roof to remain weathertight is clearly of paramount importance, particularly as the demand for lower roof pitches increases, whereas aesthetic considerations tend to dictate the choice of walling.

Over the past 30 years, metal cladding has been the most popular choice for both roofs and walls, comprising a substrate of either steel or aluminium.

Cladding with a steel substrate tends to be more economical from a purely cost point of view and, coupled with a much lower coefficient of thermal expansion than its aluminium counterpart, has practical advantages. However, the integrity of the steel substrate is very much dependent on its coatings to maintain resistance to corrosion. In some 'sensitive' cases, the use of aluminium has been deemed to better serve the specification.

Typical external and internal coatings for steel substrates manufactured by Corus (formerly British Steel plc/Hoogovens) are detailed below.

Substrate – *steel*

- Galvatite, hot-dipped zinc coated steel to BS EN10147: 1992. Grade Fe E220G with a Z275 zinc coating.
- Galvalloy, hot-dipped alloy-coated steel substrate (95% zinc, 5% aluminium) to BS EN 10214. Grade S220 GD+ZA with a ZA255 alloy coating.

Coatings – *external*

- Colorcoat HPS200 – 200 μm coating applied to the weatherside of the sheet on Galvalloy, above. Provides superior durability, colour stability and corrosion resistance.
- Colorcoat PVF2 – 27 μm, stoved fluorocarbon, coating on Galvatite, above. Provides excellent colour stability.
- Colorcoat Silicon Polyester – An economic coating on Galvatite, above. Provides medium term durability for worldwide use.

Coatings – *internal*

- Colorcoat Lining Enamel – 22 μm coating, 'bright white', with an easily cleaned surface.
- Colorcoat HPS200 Plastisol – 200 μm coating, used in either a corrosive environment or one of high internal humidity.
- Colorcoat Stelvetite Foodsafe – 150 μm coating, comprising a chemically inert polymer for use in cold stores and food processing applications.

The reader should note that there is an increasing move towards whole life-cycle costing of buildings in general, on which the cladding element has a significant influence. A cheaper cladding solution at the outset of a project may result in a smaller initial outlay for the building owner. Over the life of the building, however, running costs could offset (and possibly negate) any savings that may have accrued at procurement stage. A higher cladding specification will reduce not only heating costs but also carbon dioxide (CO_2) emissions.

The construction of the external skin of a building can take several forms, the most prevalent being:

(1) single-skin trapezoidal
(2) double-skin trapezoidal shell
(3) standing seam with concealed fixings
(4) composite panels.

Further information on the above topics can be found by reference to the cladding manufacturers' technical literature, and section 1.4.8 below.

1.2.2 Secondary elements

In the normal single-storey building the cladding is supported on secondary members which transmit the loads back to main structural steel frames. The spacing of the frames, determined by the overall economy of the building, is normally in the range 5–8 m, with 6 m and 7.5 m as the most common spacings.

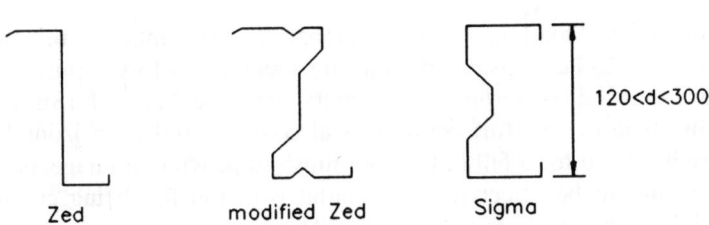

Fig. 1.3   Popular purlin and frame sections

A combination of cladding performance, erectability and the restraint require-ments for economically-designed main frames dictates that the purlin and rail spacing should be 1.5–2 m.

For this range the most economic solution has proved to be cold-formed light-gauge sections of proprietary shape and volume produced to order on computer numerically controlled (CNC) rolling machines. These have proved to be extremely efficient since the components are delivered to site pre-engineered to the exact requirements. which minimizes fabrication and erection times and eliminates mater-ial wastage. Because of the high volumes, manufacturers have been encouraged to develop and test all material-efficient sections. These fall into three main categories: Zed, modified Zed and Sigma sections. Figure 1.3 illustrates the range.

The Zed section was the first shape to be introduced. It is material-efficient but the major disadvantage is that the principal axes are inclined to the web. If subject to unrestrained bending in the plane of the web, out-of-plane displacements occur: if these are restrained, out-of-plane forces are generated.

More complicated shapes have to be rolled rather than press braked. This is a feature of the UK, where the market is supplied by relatively few manufacturers and the volumes produced by each allow the advanced manufacturing techniques to be employed, giving competitive products and service.

As roof pitches become lower, modified Zed sections have been developed with the inclination of the principal axis considerably reduced, so enhancing overall performance. Stiffening has been introduced, improving material efficiency.

The Sigma shape, in which the shear centre is approximately coincident with the load application line, has advantages. One manufacturer now produces, using rolling, a third-generation product of this configuration, which is economical.

1.2.3 Primary frames

The frame supports the cladding but, with increasing architectural and service demands, other factors are important. The basic structural form has developed against the background of achieving the lowest cost envelope by enclosing the minimum volume. Plastic design of portal frames brings limitations on the spacing of restraints of around 1.8–2 m. The cladding profiles are economic in this range:

they can support local loads and satisfy drainage requirements. The regime is therefore for the loads to be transferred from the sheeting on to the purlins and rails, which in turn must be supported on a primary structure. Figure 1.4 shows the simplest possible type of structure with vertical columns and a horizontal spanning beam. There is a need for a fall in the roof finish to provide drainage, but for small spans the beam can be effectively horizontal with the fall being created in the finishes or by a nominal slope in the beam. The minimum slope is also a function of weatherproofing requirements of the roof material.

The simple form shown would be a mechanism unless restraint to horizontal forces is provided. This is achieved either by the addition of bracing in both plan and vertical planes or by the provision of redundancies in the form of moment-resisting joints. The important point is that all loads must be transmitted to the foundations in a coherent fashion even in the simplest of buildings, whatever their size.

The range of frame forms is discussed in more detail in later sections but Fig. 1.5 shows the structural solutions commonly used. The most common is the portal shape with pinned bases, although this gives a slightly heavier frame than the fixed-base option. The overall economy, including foundations, is favourable. The portal form is both functional and economic with overall stability being derived from the provision of moment-resisting connections at eaves and apex.

The falls required to the roof are provided naturally with the cladding being carried on purlins, which in turn are supported by the main frame members. Architectural pressures have led to the use of flatter slopes compatible with weathertightness; the most common is around 6°, but slopes as low as 1° are used, which means deflection control is increasingly important.

Traditionally, portal frames have been fabricated from compact rolled sections and designed plastically. More recently the adoption of automated welding techniques has led to the introduction of welded tapered frames, which have been extensively used for many years in the USA. For economy these frames have deep slender sections and are designed elastically. In addition to material economies, the benefit is in the additional stiffness and reduced deflections.

Although the portal form is inherently pleasing to the eye, given a well-proportioned and detailed design, the industrial connotation, together with increased

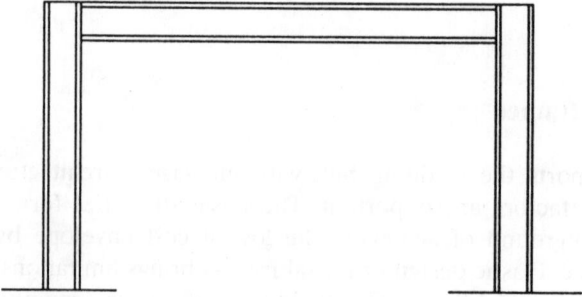

Fig. 1.4 Simplest single-storey structure

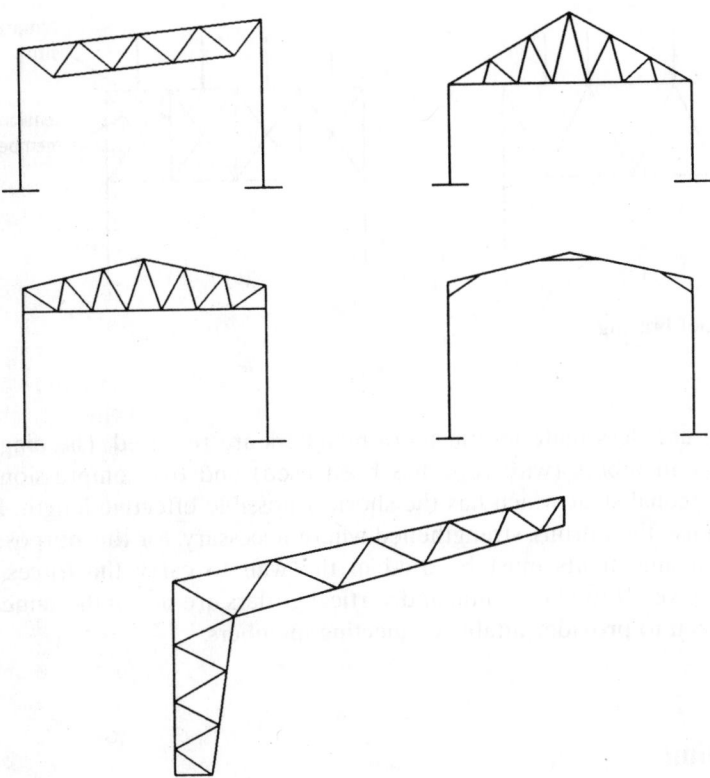

Fig. 1.5 A range of structural forms

service requirements, has encouraged the use of lattice trusses for the roof structure. They are used both in the simple forms with fixed column bases and as portal frames with moment-resisting connections between the tops of the columns for long-span structures such as aircraft hangars, exhibition halls and enclosed sports facilities.

1.2.4 Resistance to sway forces

Most of the common forms provide resistance to sidesway forces in the plane of the frame. It is essential also to provide resistance to out-of-plane forces; these are usually transmitted to the foundations with a combination of horizontal and vertical girders. The horizontal girder in the plane of the roof can be of two forms as shown in Fig. 1.6. Type (a) is formed from members, often tubes, capable of carrying tension or compression. One of the benefits is in the erection stage as the braced bay can be erected first and lined and levelled to provide a square and safe springboard for the erection of the remainder.

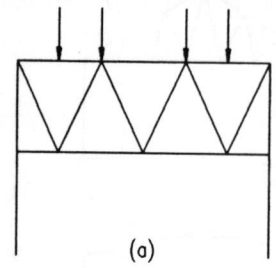
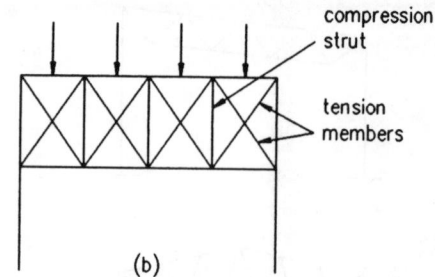

Fig. 1.6   Roof bracing

Type (b) uses less material but more members are required. The diagonals are tension-only members (wire rope has been used) and the compression is taken in the orthogonal strut which has the shortest possible effective length. It may be possible to use the purlins, strengthened where necessary, for this purpose.

Similar arrangements must be used in the wall to carry the forces down to foundation level. If the horizontal and vertical girders are not in the same bay, care must be taken to provide suitable connecting members.

1.3 Loading

1.3.1 External gravity loads

The dominant gravity loading is snow, with a general case being the application of a *minimum* basic uniform snow load of $0.60\,kN/m^2$ as an imposed load. However, there are certain geographical areas where this basic minimum will be unduly conservative, a fact that is recognised in BS 6399: Part 3.[2]

BS 6399: Part 3 contains a map showing values of 'basic snow load on the ground' in a form similar to a 'contour map'. In ascertaining the snow load for which the structure is to be designed, the designer must take cognisance of the latter value at the building's location *and* the altitude of the building above sea level. By substituting the latter values in a simple algorithm, the site snow load can be determined and used in building design. (It should be noted that any increase in the value of snow load on the ground only takes effect at altitudes greater than 100 m.)

The trend towards both curved and multi-span pitched and curved roof structures, with eaves and gable parapets, further adds to the number of load combinations that the designer must recognise.

BS 6399: Part 3 recognizes the possibility of drifting in the valleys of multi-span structures and adjacent to parapets, in addition to drifting at positions of abrupt changes in height. The process that must be followed by the designer in order to arrive at the relevant load case is illustrated by means of diagrams and associated

flow charts. In all instances, the drift is idealized as a varying, triangularly distributed load.

The drift condition must be allowed for not only in the design of the frame itself, but also in the design of the purlins that support the roof cladding, since the intensity of loading at the position of maximum drift is often far in excess of the minimum basic uniform snow load.

In practice, the designer will invariably design the purlins for the uniform load case, thereby arriving at a specific section depth and gauge. In the areas subject to drift, the designer will maintain that section and gauge by reducing the purlin spacing local to the greater loading in the area of maximum drift. (In some instances, however, it may be possible to maintain purlin depth but increase purlin gauge in the area of the drift. An increase in purlin gauge implies a stronger purlin, which in turn implies that the spacing of the purlins may be increased over that of a thinner gauge. However, there is the possibility on site that purlins which appear identical to the eye, but are of different gauge, may not be positioned in the location that the designer envisaged. As such, the practicality of the site operations should also be considered, thereby minimising the risk of construction errors.)

Over the years, the calculation of drift loading and associated purlin design has been made relatively straightforward by the major purlin manufacturers, a majority of whom offer state of the art software to facilitate rapid design, invariably free of charge.

1.3.2 Wind loads

A further significant change that must be accounted for in the design (in the UK) of structures in general (including the single-storey structures to which this chapter alludes) has been the inception of BS 6399: Part 2 – *Code of practice for wind loads* in lieu of CP3: Chapter V: Part 2, which has been declared obsolescent.

A cursory inspection of the former will show that BS6399: Part 2 addresses the calculation of the wind loading in a far more rigorous way than CP3: Chapter V: Part 2, and offers two alternative methods for determining the loads that the structure must withstand:

- Standard method – this method uses a simplified procedure to obtain a standard effective wind speed, which is used with standard pressure coefficients to determine the wind loads for *orthogonal* design cases
- Directional method – this method derives wind speeds and pressure coefficients for each wind direction, either orthogonal or oblique.

In both methods, the dynamic wind pressure, q_s, is calculated as follows:

$$q_s = 0.613\,V_e^2$$

$$V_e = V_s \times S_b$$

$$V_s = V_b \times S_a \times S_d \times S_s \times S_p$$

V_b = the basic wind speed
S_a = an altitude factor
S_d = a direction factor
S_s = a seasonal factor
S_p = a probability factor
V_s = the site wind speed
V_e = the effective wind speed
S_b = a terrain factor

The internal and external pressures that are applied to the structure are calculated from the generic equation:

$$p = q_s \times C_p \times C_a$$

p = either the internal or external applied pressure
C_p = either the internal or external pressure coefficient
C_a = the size effect factor for either internal or external pressures

BS 6399: Part 2 recognizes both site topography and location, in either town or country, the latter being influenced by the distance to the sea.

The reader will note that the size effect factor was not present in CP3: Chapter V: Part 2, and the calculation of this factor alone is worthy of further mention.

The size effect factor, C_a, is dependent upon a 'diagonal dimension – a', which varies for each loaded element. BS 6399: Part 2 recognizes the fact that elements with large diagonal dimensions can have the load to which they may be subject 'reduced' from the sum of the design loading of each of the elements that they support.

For example, the 'diagonal dimension – a' for the rafter of a typical portal frame shown in Fig. 1.7 below is greater than that for each purlin it supports. (This is consistent with the method of BS 6399: Part 1, which allows a percentage reduction in imposed load on a floor beam, say, depending on the tributary loading area of the beam.)

However, many purlin manufacturers have updated their design software to incorporate the requirements of this code of practice, and most of the somewhat complicated analysis is automatically executed.

BS 6399: Part 2 contains an abundance of tables and graphs from which external pressure coefficients for many types of building can be determined, and recognises the fact that certain areas of the structure (adjacent to the eaves, apex and corners, for example) must be designed to allow for high, local pressure coefficients. This fact in itself implies that the number of secondary elements such as purlins and side-rails may increase over those that were required under the criteria of CP3: Chapter V: Part 2.

As mentioned above, BS 6399: Part 2 offers a rigorous approach to the calculation of wind loads to structures, and a more detailed treatment of this topic is outside the scope of this chapter. It is hoped that the somewhat brief treatment above will induce the reader to study BS 6399: Part 2 in some depth, and perhaps calibrate any

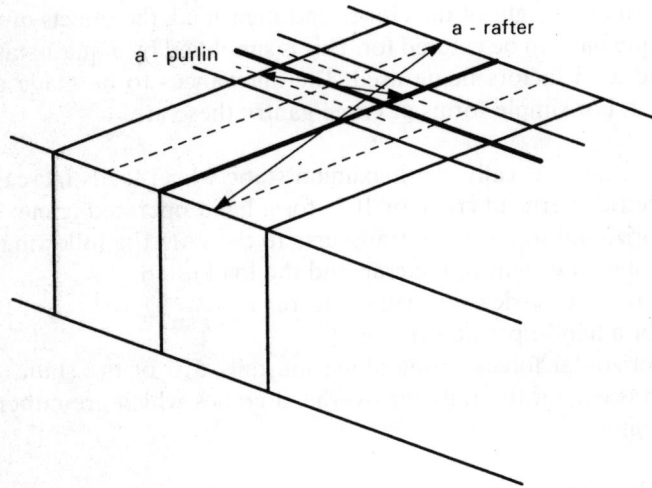

a - rafter

a - purlin

Fig. 1.7   Diagonal dimension 'a' for purlins and rafters

designs undertaken by hand calculation with those obtained from the manufacturers' software, to get a 'feel' for this new code of practice.

1.3.3 Internal gravity loads

Service loads for lighting, etc., are reasonably included in the global $0.6\,\text{kN/m}^2$. As service requirements have increased, it has become necessary to consider carefully the provision to be made.

Most purlin manufacturers can provide proprietary clips for hanging limited point loads to give flexibility of layout. Where services and sprinklers are required, it is normal to design the purlins for a global service load of $0.1–0.2\,\text{kN/m}^2$ with a reduced value for the main frames to take account of likely spread. Particular items of plant must be treated individually. The specifying engineer should make a realistic assessment of the need as the elements are sensitive, and while the loads may seem low, they represent a significant percentage of the total and affect design economy accordingly.

1.3.4 Cranes

The most common form of craneage is the overhead type running on beams supported by the columns. The beams are carried on cantilever brackets or in heavier cases by providing dual columns.

In addition to the weight of the cranes and their load, the effects of acceleration and deceleration have to be catered for. This is simplified by a quasi-static approach with enhanced load factors being used. The allowances to be made are given in BS 6399: Part 1. For simple forms of crane gantry these are:

(1) for loads acting vertically, the maximum static wheel loads increased by 25% for an electric overhead crane or 10% for a hand-operated crane;
(2) for the horizontal force acting transverse to the rails, the following percentage of the combined weight of the crab and the load lifted:
    (a) 10% for an electric overhead crane, or
    (b) 5% for a hand-operated crane;
(3) for the horizontal forces acting along the rails, 5% of the static wheel loads which can occur on the rails for overhead cranes which are either electric or hand-operated.

For heavy, high-speed or multiple cranes the allowances should be specially calculated with reference to the manufacturers.

The combination load factors for design are given in BS 5950: Part 1. The constant movement of a crane gives rise to a fatigue condition. This is, however, restricted to the local areas of support, i.e. the crane beam itself, the support bracket and its connection to the main column. It is not normal to design the whole frame for fatigue as the stress levels due to the crane travel are relatively low.

1.3.5 Notional horizontal forces

During the construction of any structure, there will always be the possibility that practical imperfections such as lack of verticality of the columns, for example, will be evident. To allow for this eventuality, BS 5950: Part 1 dictates that *when considering the strength of the structure*, all structures should be capable of resisting notional horizontal forces (NHFs) equivalent to 0.50% of the factored dead and imposed loads applied at the same level.

These NHFs should be assumed to act in any one direction at a time, and should be applied at each roof and floor level or their equivalent. They should be taken as acting simultaneously with the factored vertical dead and imposed loads, but need not be applied to a structure that contains crane loading, as the loads from the latter already contain significant horizontal loads.

The NHFs should not:

- Be applied when considering overturning
- Be applied when considering pattern loads
- Be combined with applied horizontal loads
- Be combined with temperature effects
- Be taken to contribute to the net reactions at the foundations.

In up-to-date analysis packages, the application of these NHFs is invariably allowed for when the load cases and load combinations are being derived.

In addition to the use of NHFs in the assessment of frame strength, they are also used in ascertaining the stiffness of frames relevant to in-plane stability checks. This latter consideration is covered in greater detail in 1.4.1, below.

1.4 Design of common structural forms

1.4.1 In-plane stability

Without doubt, one of the most significant changes in the recent amendment to BS 5950: Part 1 is the provision of in-plane stability checks to both multi-storey and moment-resisting portal frames.

The somewhat simple procedures of Section 2 of BS 5950: Part 1 cannot be utilized when considering the in-plane stability of portal frames since they do not consider axial compression within the rafters. Axial compression in the rafters has a significant effect on the stability of the frame as a whole.

In-plane stability checks are required to ensure that the load that would cause buckling of the frame as a whole is greater than the sum of the applied forces.

Unlike a beam and column structure, a single-storey, moment-resisting portal frame does not generally have any bracing in the plane of the frame. As such, restraint afforded to individual columns and rafters is a function of the stiffness of the members to which they connect. In simplistic terms, rafters rely on columns, which in turn rely on rafters. The stability check for the frame must therefore account for the stiffness of the frame itself.

When any structure is loaded it deflects. To this end, the deflected shape is different from the idealized representation of the frame when it is deemed unloaded or 'at rest'. If a frame is relatively stiff, such deflections are minimized. If, however, a frame is of such a 'small' stiffness as to induce significant deflections when loaded, 'second-order' effects impact on both the frame's stiffness and the individual members' ability to withstand the applied load.

Consider the example of a horizontal, axially loaded strut as shown in Fig. 1.8.

Prior to application of the axial force, the strut would deflect under its self-weight. If the strut was relatively stiff, the self-weight deflection, Δ, would be small. On application of an axial force, P, a bending moment equal to $P.\Delta$ would be induced at

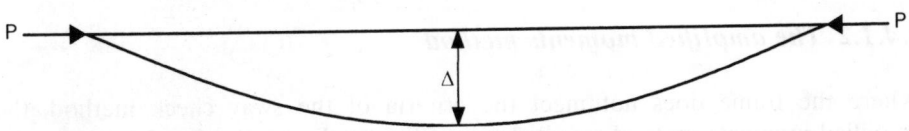

Fig. 1.8   Horizontal, axially loaded strut

mid-span. The strut would need to be designed for the combined effects of axial force and bending moment. In the case in question, axial load would have the greater influence on the design.

If, however, the strut were of a stiffness such that the initial self-weight deflection was relatively large, the induced second-order effect, $P.\Delta$, would significantly increase, and it is possible that the bending moment would play a greater part in the design of the member, since a greater deflection would induce a larger second-order bending moment, which in turn would induce a further deflection and a further second-order moment, etc.

In relation to the deflections of portal frames under load, the above phenomenon is more than sufficient to reduce the frame's ability to withstand the applied loads.

BS 5950: Part 1 addresses the problem of in-plane stability of portal frames by the use of one of the following methods:

(1) the sway check method with the snap-through check
(2) the amplified moments method
(3) second-order analysis.

1.4.1.1 The sway check method

This particular check is the simplest of those referred to above, in so far as elastic analysis can be used to determine the deflections at the top of the columns due to the application of the NHFs previously mentioned.

If, under the NHFs of the gravity load combination, the horizontal deflections at the tops of the columns are less than height/1000, the load factor for frame stability to be satisfied, λ_r, can be taken as 1.0.

The code also offers the possibility that for frames not subject to loads from either cranes, valley beams or other concentrated loads, the height/1000 criterion can be satisfied by reference to the L_b/D ratio, where L_b is the effective span of the frame, and D the cross-sectional depth of the rafter.

It should be noted, however, that this particular check can only be applied to frames that fall within certain geometric limits.

The reader is referred to both BS 5950: Part 1 Section 5.5.4.2 and reference [5] for a more in-depth treatment of the above.

1.4.1.2 The amplified moments method

Where the frame does not meet the criteria of the sway check method, the amplified moments method referred to in BS 5950: Part 1 Section 5.5.4.4 may be used.

The basis of this method revolves around the calculation of a parameter known as the lowest elastic critical load factor, λ_{cr}, for a particular load combination. (N.B. BS 5950: Part 1 does not give a method for calculating λ_{cr}, but refers to reference [5].)

A detailed treatment of the calculation of λ_{cr} is outside the scope of this particular chapter, and the reader is again referred to reference [5]. Fortunately, the industry software to carry out the analysis of portal frames is capable of calculating λ_{cr} to BS 5950: Part 1. Accordingly, the task is not as daunting as first appears. However, it is imperative that the reader understands the background to this particular methodology.

On determining λ_{cr}, the required load factor, λ_r, is calculated as follows:

$$\lambda_{cr} \geqslant 10 : \lambda_r = 1.0$$
$$10 > \lambda_{cr} \geqslant 4.6 : \lambda_r = 0.9\lambda_{cr}/(\lambda_{cr} - 1)$$

If $\lambda_{cr} < 4.6$, the amplified moments method cannot be used and a second-order analysis should be carried out.

The application of λ_r in the design process is as follows:

- For plastic design, ensure that the plastic collapse factor, $\lambda_p \geqslant \lambda_r$, and check the member capacities at this value of λ_r.
- For elastic design, if $\lambda_r > 1.0$, multiply the ultimate limit state moments and forces arrived at by elastic analysis by λ_r, and check the member capacities for these 'amplified' forces.

1.4.1.3 Second-order analysis

Second-order analysis, briefly referred to above, accounts for additional forces induced in the frame due to the axial forces acting eccentrically to the assumed member centroids as the frame deflects under load.

These secondary effects, often referred to as 'P-Delta' effects, can be best illustrated by reference to Fig. 1.9 of a simple cantilever.

As can be seen, the second-order effects comprise an additional moment of $P\Delta$ due to the movement of the top of the cantilever, Δ, induced by the horizontal force, H, in addition to a moment within the member of $P\delta$ due to deflection of the member itself between its end points. (It should be noted that, in certain instances, second-order effects can be beneficial. Should the force P, above, have been tensile, the bending moment at the base would have been *reduced* by $P\Delta$ kNm.)

In the case of a portal frame, there are joint deflections at each eaves and apex, together with member deflections in each column and rafter. In the case of the columns and rafters, it is standard practice to divide these members into several sub-elements between their start and end nodes to arrive at an accurate representation of the secondary effects due to member deflections.

As a consequence of these effects, the stiffness of the portal frame is reduced

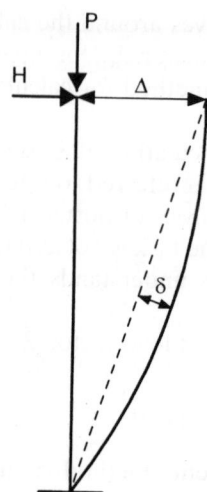

Fig. 1.9   Second-order effects in a vertical cantilever

below that arrived at from a first-order elastic analysis. Again, the industry software packages include modules for the rapid calculation of these second-order effects, and utilize either the matrix or energy methods of analysis to arrive at the required solution, in most cases employing an iterative solution.

1.4.1.4 Tied portals

The reader's attention is drawn to the fact that for tied portals, *it is mandatory to use a second-order analysis to check for in-plane stability.* In the case of the tied portal under gravity loading, the tie is subject to quite high tensile loads and, as such, would elongate under load. This in turn would cause the eaves to spread, with a subsequent downward deflection of the apex. This downward deflection of the apex reduces the lever arm between the rafter and the tie, and as such increases the tensile force in the tie and the compressive force in the rafter. A first-order elastic analysis would not truly represent what actually occurs in practice, and, subject to the dimensions of the tie, may be somewhat unsafe in arriving at the design forces to which the frame is subjected.

1.4.2 Beam and column

The cross-section shown in Fig. 1.4 is undoubtedly the simplest framing solution which can be used to provide structural integrity to single-storey buildings. Used

predominantly in spans of up to 10 m, where flat roof construction is acceptable, the frame comprises standard hot-rolled sections having simple or moment-resisting joints.

Flat roofs are notoriously difficult to weatherproof, since deflections of the horizontal cross-beam induce ponding of rainwater on the roof, which tends to penetrate the laps of traditional cladding profiles and, indeed, any weakness of the exterior roofing fabric. To counteract this, either the cross-member is cambered to provide the required fall across the roof, or the cladding itself is laid to a predetermined fall, again facilitating drainage of surface water off the roof.

Due to the need to control excessive deflections, the sections tend to be somewhat heavier than those required for strength purposes alone, particularly if the cross-beam is designed as simply-supported. In its simplest form, the cross-beam is designed as spanning between columns, which, for gravity loadings, are in direct compression apart from a small bending moment at the top of the column due to the eccentricity of the beam connection. The cross-beam acts in bending due to the applied gravity loads, the compression flange being restrained either by purlins, which support the roof sheet, or by a proprietary roof deck, which may span between the main frames and which must be adequately fastened. The columns are treated as vertical cantilevers for in-plane wind loads.

Resistance to lateral loads is achieved by the use of a longitudinal wind girder, usually situated within the depth of the cross-beam. This transmits load from the top of the columns to bracing in the vertical plane, and thence to the foundation. The bracing is generally designed as a pin-jointed frame, in keeping with the simple joints used in the main frame. Details are shown in Fig. 1.10.

Buildings which employ the use of beam-and-column construction often have brickwork cladding in the vertical plane. With careful detailing, the brickwork can be designed to provide the vertical sway bracing, acting in a similar manner to the shear walls of a multi-storey building.

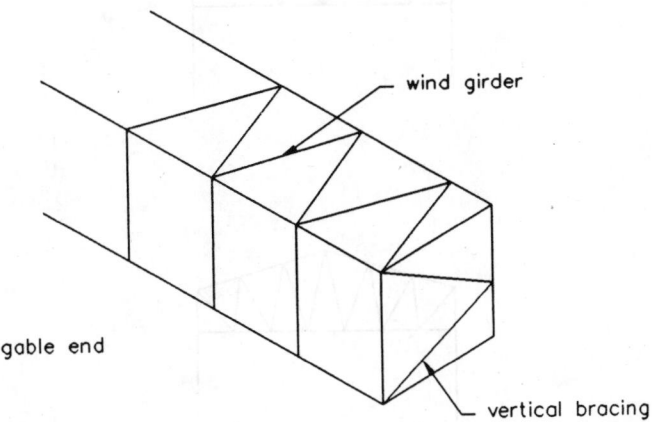

Fig. 1.10   Simple wind bracing system

Resistance to lateral loading can also be achieved either by the use of rigid connections at the column/beam joint or by designing the columns as fixed-base cantilevers. The latter point is covered in more detail in the following sub-section relating to the truss and stanchion framing system.

1.4.3 Truss and stanchion

The truss and stanchion system is essentially an extension of the beam-and-column solution, providing an economic means of increasing the useful span.

Typical truss shapes are shown in Fig. 1.11.

Members of lightly-loaded trusses are generally hot-rolled angles as the web elements, and either angles or structural tees as the boom and rafter members, the latter facilitating ease of connection without the use of gusset plates. More heavily loaded trusses comprise universal beam and column sections and hot-rolled channels, with connections invariably employing the use of heavy gusset plates.

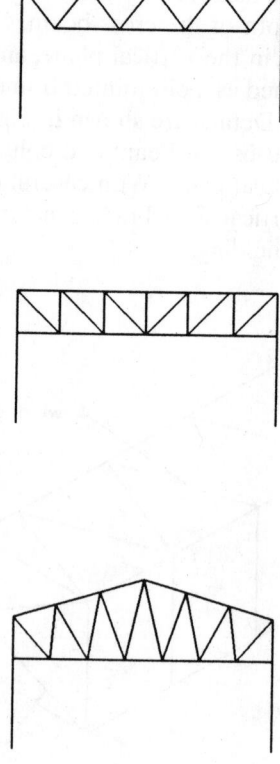

Fig. 1.11   Truss configurations

In some instances there may be a requirement for alternate columns to be omitted for planning requirements. In this instance load transmission to the foundations is effected by the use of long-span eaves beams carrying the gravity loads of the intermediate truss to the columns: lateral loading from the intermediate truss is transmitted to points of vertical bracing, or indeed vertical cantilevers by means of longitudinal bracing as detailed in Fig. 1.12. The adjacent frames must be designed for the additional loads.

Considering the truss and stanchion frame shown in Fig. 1.13, the initial assumption is that all joints are pinned, i.e. they have no capacity to resist bending moment. The frame is modelled in a structural analysis package or by hand calculation, and, for the load cases considered, applied loads are assumed to act at the node points. It is clear from Fig. 1.13 that the purlin positions and nodes are not coincident; consequently, due account must be taken of the bending moment induced in the rafter section. The rafter section is analysed as a continuous member from eaves to apex, the node points being assumed as the supports, and the purlin positions as the points of load application (Fig. 1.14).

The rafter is sized by accounting for bending moment and axial loads, the web members and bottom chord of the truss being initially sized on the basis of axial load alone.

Use of structural analysis packages allows the engineer to rapidly analyse any number of load combinations. Typically, dead load, live load and wind load cases are analysed separately, and their factored combinations are then investigated to determine the worst loading case for each individual member. Most software packages provide an envelope of forces on the truss for all load combinations, giving

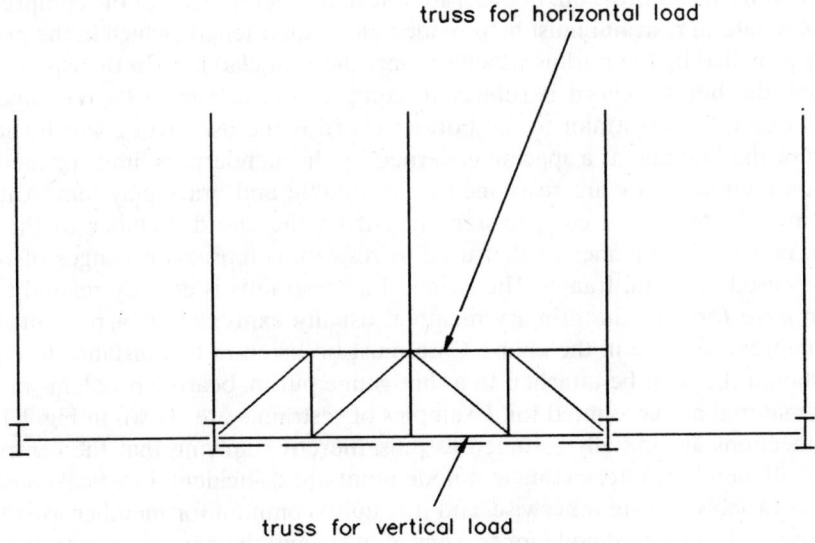

Fig. 1.12   Additional framing where edge column is omitted

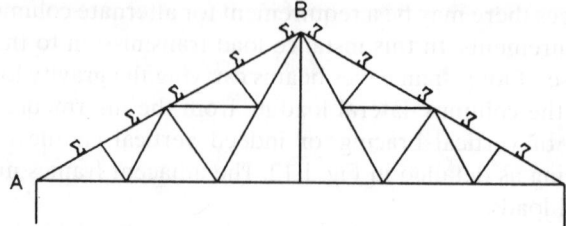

Fig. 1.13 Truss with purlins offset from nodes

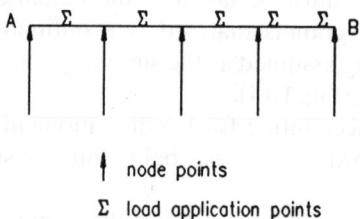

$\uparrow$ node points

Σ load application points

Fig. 1.14 Rafter analysed for secondary bending

maximum tensile and compressive forces in each individual member, thus facilitating rapid member design.

Under gravity loading the bottom chord of the truss will be in tension and the rafter chords in compression. In order to reduce the slenderness of the compression members, lateral restraint must be provided along their length, which in the present case is provided by the purlins which support the roof cladding. In the case of load reversal, the bottom chord is subject to compression and must be restrained. A typical example of restraint to the bottom chord is the use of ties, which run the length of the building at a spacing governed by the slenderness limits of the compression member; they are restrained by a suitable end bracing system. Another solution is to provide a compression strut from the chord member to the roof purlin, in a similar manner to that used to restrain compression flanges of rolled sections used in portal frames. The sizing of all restraints is directly related to the compressive force in the primary member, usually expressed as a percentage of the compressive force in the chord. Care must be taken in this instance to ensure that, should the strut be attached to a thin-gauge purlin, bearing problems in thin-gauge material are accounted for. Examples of restraints are shown in Fig. 1.15.

Connections are initially assumed as pins, thereby implying that the centroidal axes of all members intersecting at a node point are coincident. Practical considerations invariably dictate otherwise, and it is quite common for member axes to be eccentric to the assumed node for reasons of fit-up and the physical constraints that

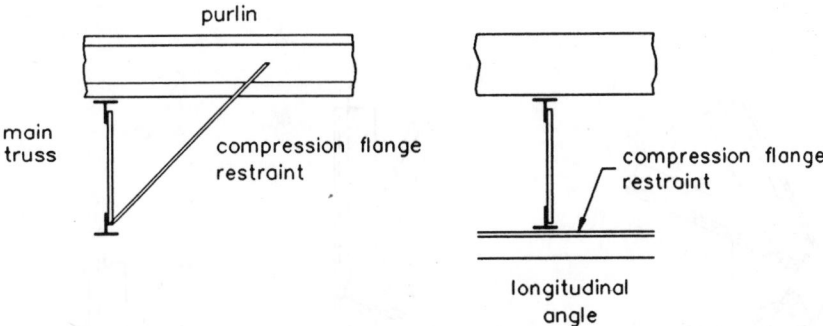

Fig. 1.15   Restraints to bottom chord members

are inherent in the truss structure. Such eccentricities induce secondary bending stresses of the node points, which must be accounted for not only by local bending and axial load checks at the ends of all constituent members, but also in connection design. Typical truss joints are detailed in Fig. 1.16.

It is customary to calculate the net bending moment at each node point due to any eccentricities, and proportion this moment to each member connected to the node in relation to member stiffness.

In heavily-loaded members secondary effects may be of such magnitude as to require member sizes to be increased quite markedly above those required when considering axial load effects alone. In such instances, consideration should be given to the use of gusset plates, which can be used to ensure that member centroids are coincident at node points, as shown in Fig. 1.17. Types of truss connections are very much dependent on member size and loadings. For lightly-loaded members, welding is most commonly used with bolted connections in the chords if the truss is to be transported to site in pieces and then erected. In heavily-loaded members, using gusset plates, either bolting, welding or a combination of the two may be used. However, the type of connection is generally based on the fabricator's own reference.

Where the roof truss has a small depth at the eaves, lateral loading is resisted either by longitudinal wind girders in the plane of the bottom boom and/or rafter, or by designing the columns as vertical fixed-base cantilevers, as shown in Fig. 1.18. Where the truss has a finite depth at the eaves, benefit can be obtained by developing a moment connection at this position with the booms designed for appropriate additional axial loads. This latter detail may allow the column base to be designed as a pin, rather than fixed, depending on the magnitude of the applied loading, and the serviceability requirements for deflection.

Longitudinal stability is provided by a wind girder in the plane of the truss boom and/or rafter at the gable wall, the load from the gable being transmitted to the foundations by vertical bracing as shown in Fig. 1.19.

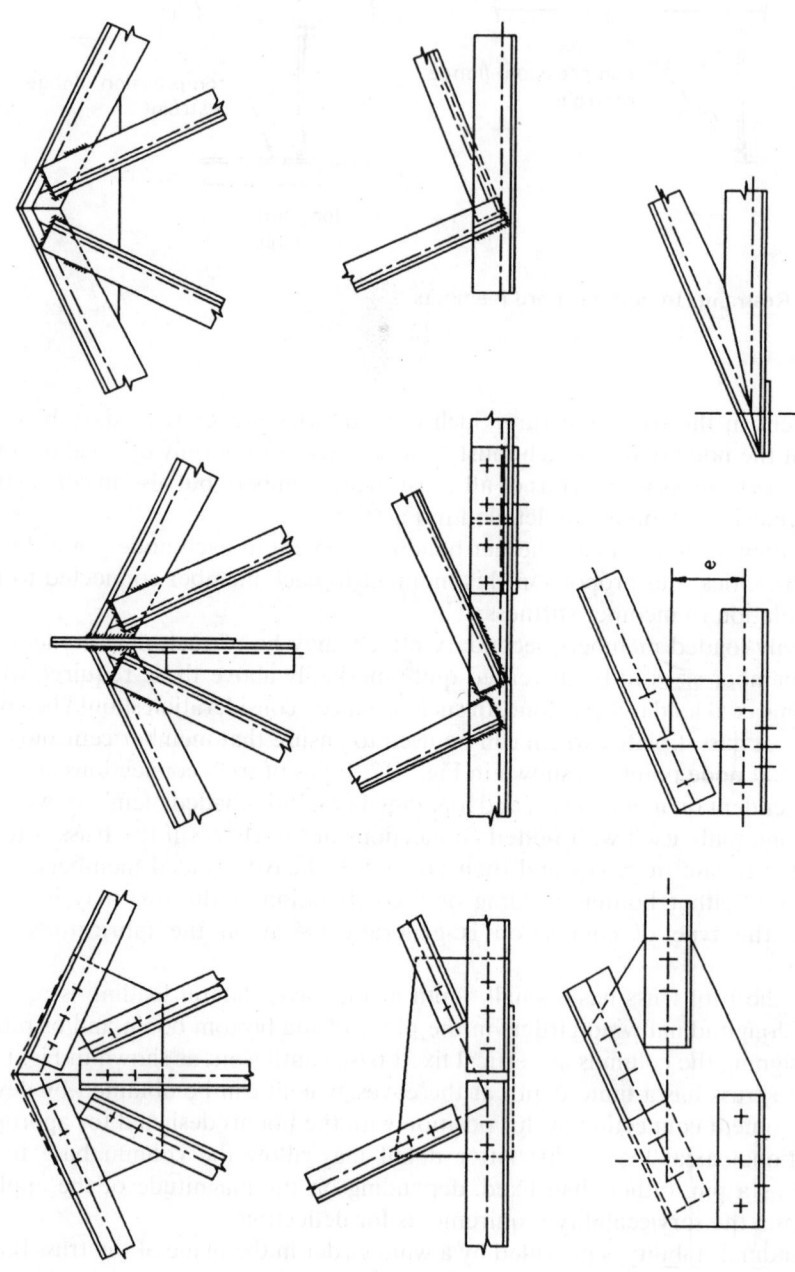

Fig. 1.16 Typical joints in trusses

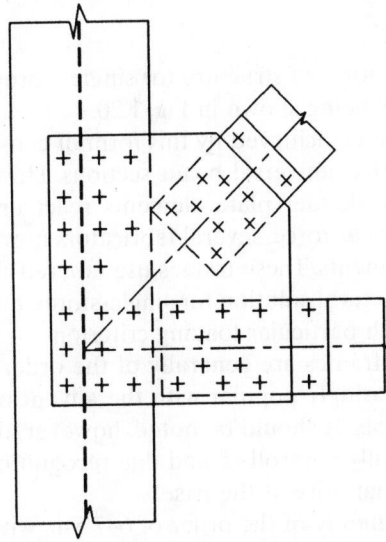

Fig. 1.17 Ideal joint with all member centroids coincident

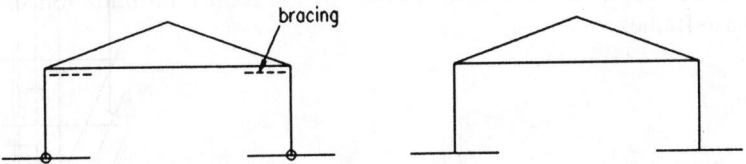

Fig. 1.18 Sway resistance for truss roofs

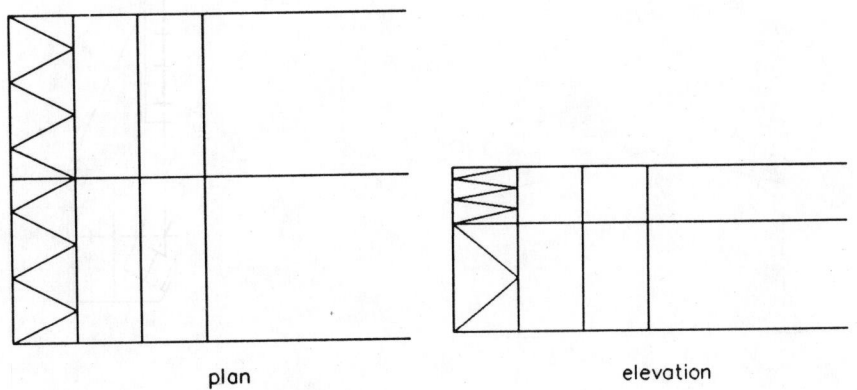

plan   elevation

Fig. 1.19 Gable-end bracing systems

1.4.4 Portal frames

By far the most common form of structure for single-storey structures is the portal frame, the principal types being shown in Fig. 1.20.

Spans of up to 60 m can be achieved by this form of construction, the frame generally comprising hot-rolled universal beam sections. However, with the increase in understanding of how slender plate elements react under combined bending moment, axial load and shear force, several fabricators now offer a structural frame fabricated from plate elements. These frames use tapered stanchions and rafters to provide an economic structural solution for single-storey buildings, the frame being 'custom designed' for each particular loading criterion.

Roof slopes for portal frames are generally of the order of 6° but slopes as low as 1° are becoming increasingly popular with the advent of new cladding systems such as standing seam roofs. It should be noted, however, that frame deflections at low slopes must be carefully controlled, and due recognition must be taken of the large horizontal thrusts that arise at the base.

Frame centres are commonly of the order of 6–7.5 m, with eaves heights ranging from 6 to 15 m in the case of aircraft hangars or similar structures.

Resistance to lateral loading is provided by moment-resisting connections at the eaves, stanchion bases being either pinned or fixed. Frames which are designed on the basis of having pinned bases are heavier than those having fixity at the bases, although the increase in frame cost is offset by the reduced foundation size for the pinned-base frames.

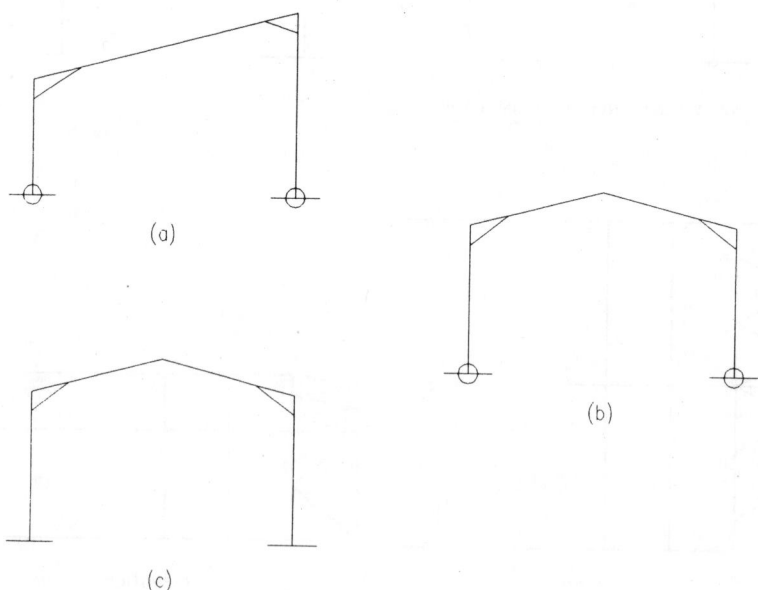

Fig. 1.20   Portal-frame structures

Parallel-flange universal sections, subject to meeting certain physical constraints regarding breadth-to-thickness ratios of both flanges and webs, lend themselves to rapid investigation by the plastic methods of structural analysis. The basis of the plastic method is the need to determine the load applied to the frame which will induce a number of 'plastic hinges' within the frame, thereby causing failure of the frame as a mechanism.

This requirement is best illustrated by the following simple example.

Considering the pinned-base frame shown in Fig. 1.21(a), subject to a uniform vertical load, w, per unit length: the reactions at the foundations are shown in Fig. 1.21(b). The frame has one degree of indeterminacy. In order that the frame fails as a mechanism, at least two plastic hinges must form (i.e. the degree of indeterminacy + 1) as shown in Fig. 1.21(c). (It should be noted that although four hinges are shown in Fig. 1.21(c), due to 'theoretical' symmetry only one pair either side of the apex will in fact form, due to the obvious imperfections in both loading and erection conditions.)

In many structures, other than the most simple, it is not clear where the plastic hinges will form. There are several methods available to the design engineer which greatly assist the location of these hinge positions, not least the abundance of proprietary software packages specifically relating to this form of analysis. Prior to the use of these packages, however, it is imperative that the engineer fully understands the fundamentals of plastic analysis by taking time to calculate, by hand, several design examples. The example which follows uses a graphical construction as a means of illustrating the applications of the method to a simple portal frame. Further information on plastic analysis is given in Chapter 11.

The frame shown in Fig. 1.22(a) has one degree of indeterminacy. It is made statically determinate by assuming a roller at the right-hand base as shown in Fig. 1.22(b), and the free-bending moment diagram drawn as shown in Fig. 1.22(c). The reactant line for the horizontal force 'removed' to achieve a statically determinate structure must now be drawn as follows:

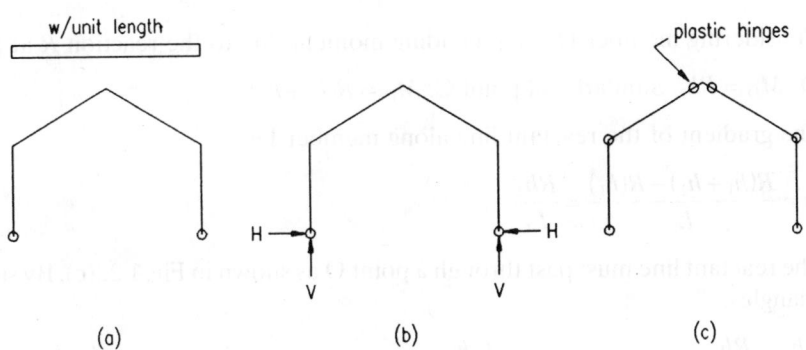

(a)   (b)   (c)

Fig. 1.21   Structural behaviour of pinned-base portal

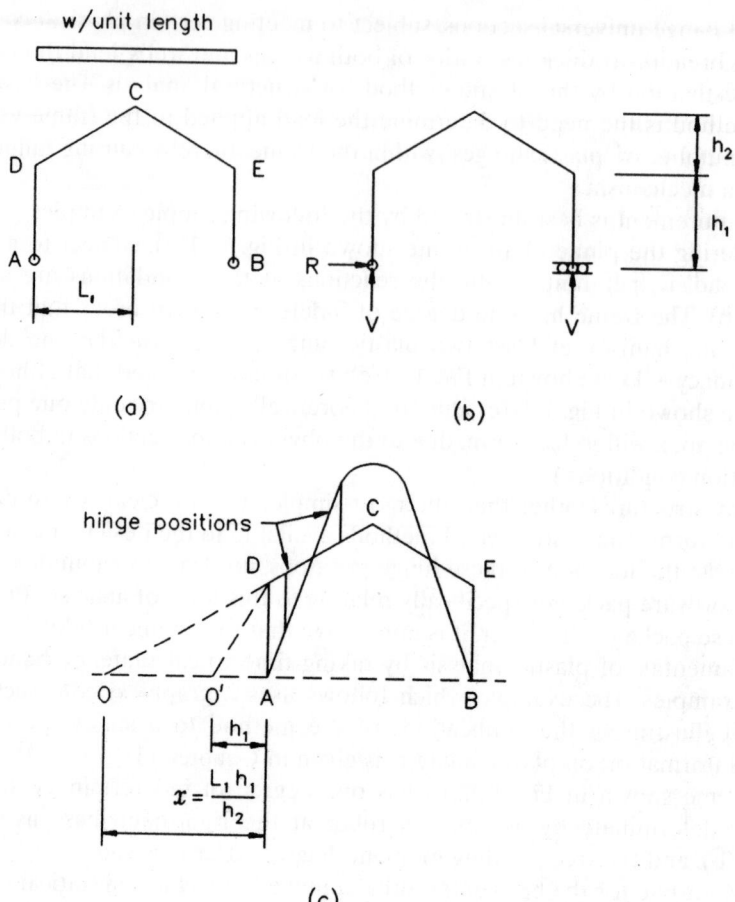

Fig. 1.22 Application of graphical method

(1) Considering member DC, the bending moment due to the reaction R at point D, $M_D = Rh_1$. Similarly, at point C, $M_C = R(h_1 + h_2)$.

(2) The gradient of the reactant line along member DC:

$$m = \frac{R(h_1 + h_2) - R(h_1)}{L_1} = \frac{Rh_2}{L_1}$$

(3) The reactant line must past through a point O as shown in Fig. 1.22(c). By similar triangles:

$$\frac{Rh_1}{x} = \frac{Rh_2}{L_1} \qquad \text{giving} \qquad x = \frac{L_1 h_1}{h_2}$$

Therefore, by rotating the reactant line through point O, the positions of the bending moment of equal magnitude in the positive and negative regions can be found and the member sized accordingly based on this value of bending moment.

Having found the positions of the reactant line which gives the number of hinges required for a mechanism to form, in this case two, the reactant line for the stanchions can be drawn through point O, a distance h_1 from the end of the free-bending moment diagram. The unknown reaction, H, is then calculated as $H = M_D/h_1$.

In a majority of instances, portal frames are constructed with a haunch at the eaves, as shown in Fig. 1.23(a).

Depending on the length/depth of the haunch, the plastic hinges required for a mechanism to form are shown in Figs. 1.23(b) and (c). The dimensional details for the haunch can be readily investigated by the graphical method by superimposing the dimensions of the haunch on the free-bending moment diagram. The reactant line can then be rotated accordingly until the required mechanism is achieved and members sized accordingly.

Haunches are generally fabricated from parallel beam sections (Fig. 1.24). In all cases, the haunch must remain in the elastic region. A detailed check is required along its length, from a stability point of view, in accordance with the requirements laid down in the relevant code of practice. In some instances, bracing of the bottom flange must be provided from a purlin position within its length, as shown in Fig. 1.25.

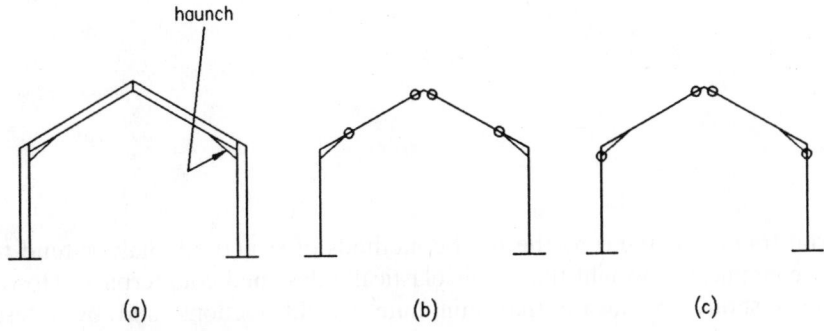

Fig. 1.23   Alternative hinge locations for haunched portal frames

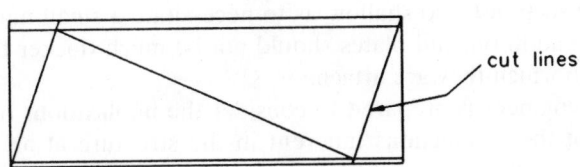

Fig. 1.24   Haunch fabrication

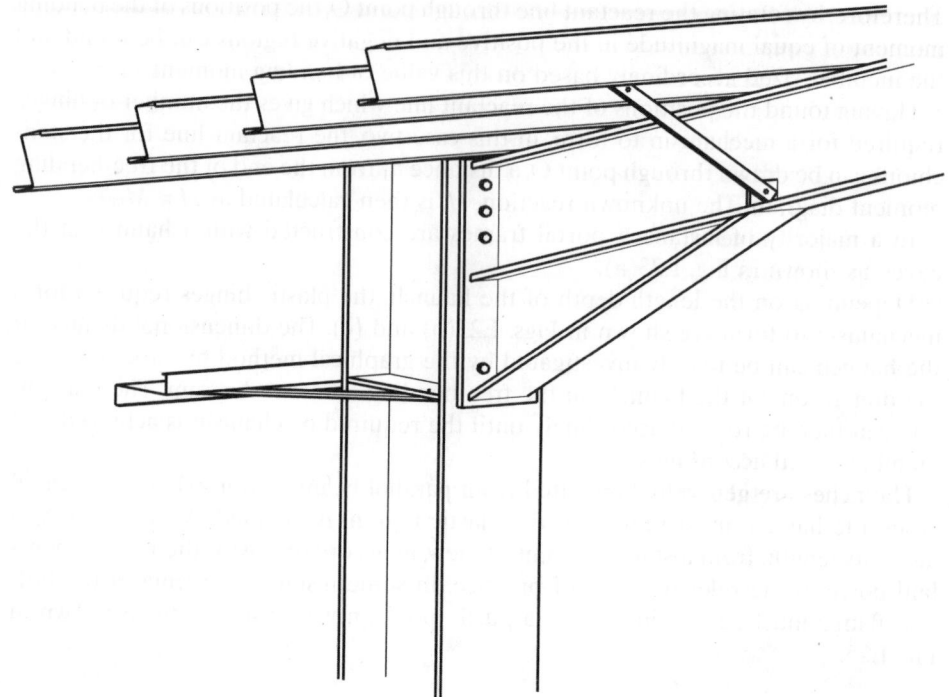

Fig. 1.25   Bottom flange restraint

Portal frames analysed by the plastic methods of structural analysis tend to be more economical in weight than their elastically designed counterparts. However, engineers should be aware that minimum weight sections, and by inference minimum depth sections, have to be connected together to withstand the moments and forces induced by the applied loading. Particular attention should be paid at an *early stage* in the design process to the economics of the connection. Cost penalties may be induced by having to provide, for example, gusset plates between pairs of bolts should the section be so shallow as to necessitate a small number of highly-stressed bolts. In addition, end plates should not be much thicker than the flanges of the sections to which they are attached.

Provided the engineer is prepared to consider the implications of his calculated member sizes on the connections inherent in the structure at an early stage, an economic solution will undoubtedly result. Leaving connection design solely to the fabricator, without any consideration as to the physical constraints of providing

a number of bolts in an extremely shallow depth, will undoubtedly result in a connection which is both difficult to design and fabricate, and costly.

Having sized the members based on the previous procedure, it is imperative that
an analysis at serviceability limit state is carried out (i.e. unfactored loads) to check
deflections at both eaves and apex. This check is required not only to ascertain
whether deflections are excessive, but also as a check to ensure that the deflections
and accompanying frame movement can be accommodated by the building envelope without undue cracking of any brickwork or tearing of metal cladding sheets
at fixing positions. Excessive lateral deflections can be reduced by increasing the
rafter size and/or by fixing the frame bases. It should be noted that the haunch has
a significant effect on frame stiffness due to its large section properties in regions
of high bending moment.

1.4.5 Tied portal

The tied portal, a variation of the portal frame, is illustrated in Fig. 1.26.

Economy of material can be achieved, albeit at the expense of reducing the allowable headroom of the building, by provision of a tie at eaves level. Under vertical
loading, the eaves spread is reduced due to the induced tensile force in the tie.
However, although the rafter size is reduced, because under vertical loading the only
possible mode of failure of the rafter is that of a fixed-end beam, deflections are
more critical. Lateral loading due to wind further complicates the problem since in
some cases the tie may not act in tension and becomes redundant. It is common,
therefore, for tied portals to have fixed bases, which provide greater stability and
resistance to horizontal loading.

1.4.6 Stressed-skin design

In addition to providing the weathertight membrane, the steel sheeting can also be
utilized as a structural element itself. Correct detailing of connections between individual cladding sheets (and their connections to the support steelwork) will induce
a stressed-skin effect which can offer both resistance to transverse wind loads, and
restraint to the compression flanges of the main frame elements.

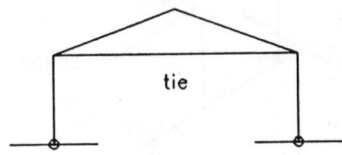

Fig. 1.26   Tied portal frame

It is clear that, when vertical load is applied to a pitched roof portal frame, the apex tends to deflect downwards, and the eaves spread horizontally. This displacement cannot occur without some deformation of the cladding. Since the cladding is fixed to the purlins, the behaviour of the cladding and purlins together is analogous to that of a deep plate girder, i.e. the purlins resist the bending moment, in the form of axial forces, and the steel sheeting resists the applied shear, as shown in Fig. 1.27.

As a consequence of this action, the load applied over the 'length' of the 'plate girder' has to be resisted at the ends of the span. In the case of stressed-skin action being used to resist transverse wind loads on the gable end of the structure, adequate connection must be present over the end bay to transmit the load from the assumed 'plate girder' into the braced bay, and into the foundations, as shown in Fig. 1.28.

Some degree of stressed-skin action is present in all portal frame structures where cladding is fixed to the supporting members by mechanical fasteners. Claddings which are either brittle (i.e. fibre cement) or are attached to the supporting structure by clips (i.e. standing seam systems) are not suitable for stressed-skin applications.

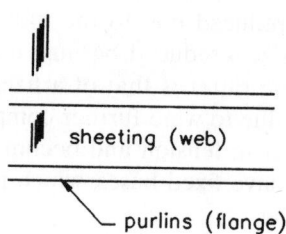

Fig. 1.27   Stressed-skin action

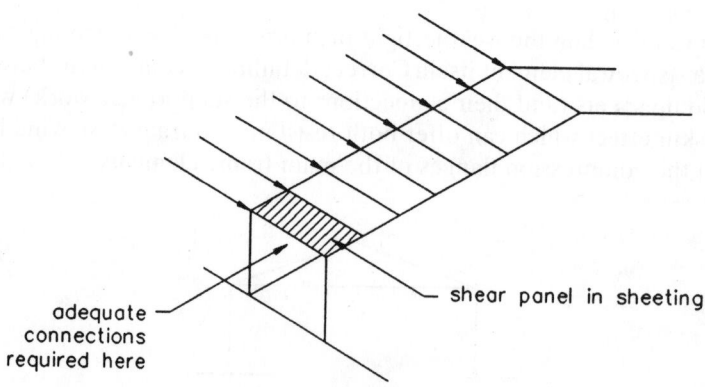

Fig. 1.28   Stressed-skin action for gable-end bracing

Correct detailing and correct fixing is essential to the integrity of stressed-skin applications:

(1) Suitable bracing members must be used to allow the applied loads from diaphragm action to be transmitted to the foundations.
(2) At the position of laps in the sheeting, suitable mechanical fasteners must be present to allow continuity of load between sheets. These fasteners can be screws or rivets, which must be capable of withstanding both shear and pull out due to the stressed-skin effect and applied loads respectively.
(3) The ends of the sheet must similarly be connected to the supporting members (i.e. purlins).
(4) All stressed-skin panels must be connected to adequately designed edge members, which must be capable of transmitting the axial forces induced by bending, in addition to the forces induced by imposed and wind load effects.

1.4.7 Purlins and siderails

Due to the increasing awareness of the load-carrying capabilities of sections formed from thin-gauge material, proprietary systems for both purlins and siderails have been developed by several manufacturers in the United Kingdom. Consequently, unless a purlin or siderail is to be used in a long-span or high-load application (when a hot-rolled angle or channel may be used), a cold-formed section is the most frequently used cladding support member for single-storey structures.

Cold-formed sections manufactured from thin-gauge material are particularly prone to twisting and buckling due to several factors which are directly related to the section's shape. The torsional constant of all thin-gauge sections is low (it is a function of the cube of the thickness); in the case of lipped channels the shear centre is eccentric to the point of application of load, thus inducing a twist on the section; in the case of Zeds the principal axes are inclined to the plane of the web, thus inducing bi-axial bending effects. These effects affect the load-carrying capacity of the section.

When used in service the support system is subject to downward loading due to dead and live loads such as cladding weight, snow, services, etc., and uplift if the design wind pressure is greater than the dead load of the system. Therefore, for a typical double-span system as shown in Fig. 1.29, the compression flange is restrained against rotation by the cladding for downward loading, but it is not so restrained in the case of load reversal.

In supporting the external fabric of the building, the purlins and siderails gain some degree of restraint against twisting and rotation from the type of cladding used and the method of its fastening to the supporting member. In addition, the connection of the support member to the main frame also has a significant effect on the load-carrying capacity of the section. Economical design, therefore, must take account of the above effects.

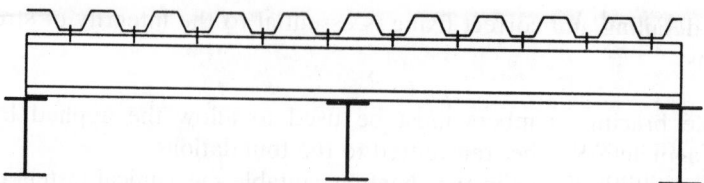

Fig. 1.29   Typical double-span purlin with cladding restraint

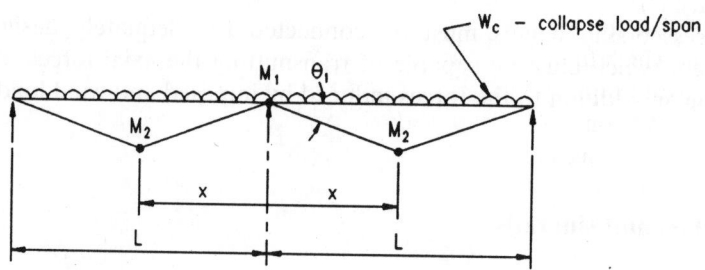

Fig. 1.30   Collapse mechanism for a two-span purlin system

There are four possible approaches to the design of a purlin system:

(1) Design by calculation based on an elastic analysis as detailed in the relevant code of practice BS 5950: Part 5.[6] This approach neglects any beneficial effect of cladding restraint for the wind uplift case.
(2) Empirical design based on approximate procedures for Zeds as given in the codes of practice. This approach leads to somewhat uneconomic design.
(3) Design by calculation based on a rational analysis which accounts for the stabilizing influence of the cladding, plasticity in the purlin as the ultimate load is approached, and the behaviour of the cleat at the internal support. The effects, however, are difficult to quantity.
(4) Design on the basis of full-scale testing.

Manufacturers differ in the methods: the example used here is from one manufacturer who has published the method used.[7]

For volume production, design by testing is the approach which is used. Although this approach is expensive, maximum economy of material can be achieved and the cost of the testing can be spread over several years of production.

Design by testing involves the 'fine-tuning' of theoretical expressions for the collapse load of the system. The method is based on the mechanism shown in Fig. 1.30 for a two-span system.

From the above mechanism it can be shown that:

the collapse load, $W_c = f(M_1, M_2, x, L)$
$$x = f(M_1, M_2, L)$$
$$\theta_1 = f(W_c, M_1, L)$$

The performance of, for example, a two-span system is considerably enhanced if some redistribution of bending moment from the internal support is taken into account. The moment–rotation characteristic at the support is very much dependent on the cleat detail and the section shape. The characteristics of the central support can be found by testing a simply-supported beam subject to a central point load, so as to simulate the behaviour of the central support of a double-span system.

From this test, the load–deflection characteristics can be plotted well beyond the deflection at which first yield occurs. A lower bound empirical expression can then be found for the support moment, M_1, based on an upper limit rotational capacity. A similar expression can be found for M_2, the internal span moment, again on the basis of a test on a simply-supported beam subject to a uniformly distributed load, applied by the use of a vacuum rig, or perhaps sandbags.

The design expressions can then be confirmed by the execution of numerous full-scale tests on double-span systems employing pairs of purlins supporting proprietary cladding.

As described earlier, load reversal under wind loading invariably occurs, thereby inducing compressive forces to the flange in the internal span, which is not restrained by the cladding as is the case in the downward load case. Anti-sag rods or the like are placed within the internal span, thereby reducing the overall buckling length of the member. The system is again tested in load reversal conditions and, as before, the design expressions can be further refined.

In some instances, sheeting other than conventional trapezoidal cladding (which is invariably through fixed to the purlin by self-drilling fastenings in alternate troughs) will not afford the full restraint to the compression flange: examples are standing seam roofs, brittle cladding, etc. The amount of restraint afforded by these latter types of cladding cannot easily be quantified. For the reasons outlined above, further full-scale testing and similar procedures of verification of design expressions are carried out.

The results of the full-scale tests are then condensed into easy to use load–span tables which are given in the purlin manufacturers' design and detail literature. The use of these tables is outlined in Table 1.1.

The tabular format is typical of that contained in all purlin manufacturers' technical literature. The table is generally prefaced by explanatory notes regarding fixing condition and lateral restraint requirements, the latter being particularly relevant to the load-reversal case. Conditions which arise in practice, and which are not covered in the technical literature, are best dealt with by the manufacturers' technical services department, which should be consulted for all non-standard cases.

Siderail design is essentially identical to that for purlins: load capacities are again arrived at after test procedures.

Self-weight deflections of siderails due to bending about the weak axis of the section are overcome by a tensioned wire system incorporating tube struts, typically

Table 1.1 Typical load table

Span (m)	Section	UDL (kN)[a]	Purlin centres (mm)				
			1000	1200	1400	1600	1800
6.0	CF170160	9.50	1.58	1.32	1.13	0.99	0.88
	CF170170	10.75	1.79	1.49	1.28	1.12	1.00
	CF170180	11.75	1.96	1.63	1.40	1.22	1.09
	CF200160	12.25	2.04	1.70	1.46	1.28	1.13
	CF200180	13.50	2.25	1.88	1.61	1.41	1.25

[a]UDL is the *total* uniformly distributed load on a single purlin of a 6 m double-span system which would produce a failure of the section.

Loads given beneath the columns under purlin centres give the allowable uniformly distributed load on the purlin at a given spacing. The figure is arrived at by dividing the UDL by the product of span and purlin centres, i.e. for CF170180, 6 m span, 1800 mm centre,

allowable load $= 11.75/(6 \times 1.8) = 1.09\,kN/m^2$

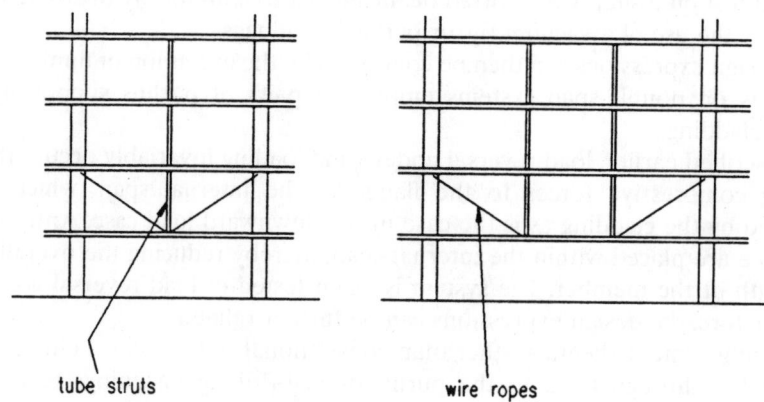

tube struts   wire ropes

Fig. 1.31   Anti-sag systems for side wall rails

at mid-span for spans of 6–7 m, and third-points of the span for spans of 7–8 m and above (Fig. 1.31).

1.4.8 Cladding

In addition to visual and aesthetic requirements, one of the fundamental roles of the cladding system is to provide not only the weatherproofing for the building but also the insulation requirements suited to both the building *and* the environment.

Cladding profiles are roll-formed and produced in volume, so giving excellent economy, and this means that, apart from an extremely rare occasion, the chosen

profile will be from a manufacturer's published range. The loads that the profile can withstand are determined by the manufacturer and published in his brochure. Until recently they would have been determined by test, but calculation methods have been developed which are satisfactory for trapezoidal shapes. These give slightly conservative answers but, in general, profiles that are walkable will have capacities in excess of those required. Sheeting can be erected satisfactorily without the need to walk indiscriminately over the whole area. The crowns at mid-span are the most susceptible to foot traffic but these positions can easily be avoided, even if crawl boards are not used.

In double-skin construction it is normal to assume all loads are taken on the outer sheet, and the inner liner merely has to be erectable and stiff enough to prevent noticeable sag once installed.

The specifier, therefore, has to select the profile appropriate to the design requirement and check the strength and deflection criteria against the manufacturer's published load tables. Care must also be taken to ensure the fasteners are adequate.

As mentioned briefly in 1.2.1 above, the use of higher specification cladding systems, by default, reduces the emissions of greenhouse gases; the pressing need to reduce such gases is an almost internationally accepted goal.

In recognition of this, the UK pledged, at the Kyoto summit, to reduce greenhouse gases by some 12% by 2010. In addition, the government's own manifesto included a commitment to reduce CO_2 emissions by 20%.

Consequently, in order to achieve this goal, significant changes have been proposed to the Building Regulations (England & Wales) – Approved Document Part L for Non-Domestic Buildings,[8] since it was recognized that a significant proportion of the UK's emission of greenhouse gases arises from energy used in the day-to-day use of building structures.

The significant changes are outlined below:

- Increase thermal insulation standards by use of *improved U* values
- Introduce an air leakage index
- Introduce *as-built* inspections
- Introduce *whole building design* by integrating the building fabric with heating, cooling and air-conditioning requirements
- Introduce the monitoring of *material alterations* to an existing structure
- Introduce operating log books and energy consumption meters.

1.4.8.1 Improved U values

Approved Document Part L embraces significant changes to the insulation requirements of the building fabric that will impact on the use of fuel and power.

The intended reduction in the *U* values of both roofs and walls is detailed below.

Roof systems   *U Value*
        Current   0.45 W/m²K
        Part L   0.35 W/m²K
        2007   0.16 W/m²K

Wall systems
        Current   0.45 W/m²K
        Part L   0.35 W/m²K
        2007   0.25 W/m²K

As a consequence of the above, the insulation thickness in either composite, glass fibre or rock fibre cladding systems will significantly increase. This will have an effect on the dead load that is applied to the supporting structure, and could result in some nominal increases in section sizes of the primary and secondary steelwork of the building.

1.4.8.2 *Air leakage index*

It will be mandatory for buildings to be designed and constructed so that an air leakage index will not be exceeded. This index will be set at a specific value of volume of air leakage per hour per square metre of external surface at a pressure difference of 50 pascals.

Present proposals for this air leakage index are as follows:

Maximum leakage rate: December 2002   10.0 m³/hr/m² @ 50 Pa
Maximum leakage rate: 2007   5.0 m³/hr/m² @ 50 Pa

To achieve the above, care must be taken to specify and effectively construct sealed junctions between elements, thereby minimizing air leakage.

The air leakage index will be quantified by *in situ* air pressurization tests, using fans, which must always be used on buildings with a floor area that exceeds 1000 m².

Should the building fail the test, remedial measures must be undertaken, and further tests carried out until the building is deemed to comply to the satisfaction of building control.

1.4.8.3 *'As-built' inspections*

In order that the revised insulation standards are adhered to, it will be necessary to ensure that the provision of insulation within the fabric itself does not exhibit zones that could compromise the specification. For example, there should be no signifi-

cant leakage paths between panels; thermal bridges should be minimized; insulation should be continuous, dry and as 'uncompressed' as possible.

One method of ascertaining compliance would be by use of infra-red thermography, where the external fabric of the building is 'photographed' using specialist equipment. By this method, areas of the external fabric that are 'hot' – implying poor insulation and heat loss *from* the building – will be shown as colours at the red end of the spectrum. Conversely, areas that are 'cold' will be shown at the blue end of the spectrum.

1.4.8.4 *Whole building design*

As the title of this subsection suggests, the building should be designed taking due account of how the constituent parts interact with each other. To this end, insulation, air tightness, windows, doors and rooflights, heating systems and the like should not be viewed as individual elements, but rather as elements that, in some way, have influence on the in-service performance of each other.

1.4.8.5 *Material alterations*

Material alteration is intended to cover substantial works to existing buildings that were designed and constructed prior to the changes in Approved Document Part L.

For example, should major works be required to roof or wall cladding, it will be necessary to provide insulation to achieve the U value of a new building. Following on from the latter eventuality will be the need to make provision for improving the building's air-tightness.

The replacement of doors and windows will also entail the use of products that meet the requirements of new buildings, as too will the installation of new heating systems, for example.

1.4.8.6 *Building log books and energy meters*

The building owner will initially be provided with details of all the products that constitute the building, including a forecast of annual energy consumption based on the building design specification. The owner in turn will be required to keep detailed maintenance records and the like to ensure that the products within the building are properly maintained, and, when replaced, fully comply with the new requirements.

Energy meters should be provided to ensure that comparisons of energy consumption can be made with those forecast at design stage. These meters in turn will provide any new owner/tenant with detailed information on which to base future energy forecasts.

The cladding has also to withstand the applied loads of snow, wind, and foot traffic during fixing and maintenance. It must also provide the necessary lateral stability to the supporting purlin and siderail systems. Occasionally it will form part of the lateral stability of the structure in the form of a stressed-skin diaphragm, mentioned above.

1.4.8.7 Cladding systems

A variety of systems is available to suit environmental and financial constraints. The most common are listed below.

Single-skin trapezoidal roofing

This was widely used in the past with plasterboard or similar material as the lining material, and fibreglass insulation in the sandwich. The construction is susceptible to the plasterboard becoming damp due to condensation. An alternative is the use of rigid insulation boards, which are impervious to damp, supported on tee bars between the purlins. Unless the joints are sealed, which is difficult to achieve, condensation is likely to form. Although inexpensive, this type is therefore limited in its applicability.

The minimum slope is governed by the need to provide watertight joints and fasteners. If manufacturers' instructions on the use of sealants and stitching to laps are rigorously followed, this type can be used down to slopes of approximately 4°.

Double-shell roof construction

In this form of construction the plasterboard has been replaced by a steel liner sheet of 0.4 mm thickness with some stiffening corrugations. The lining is first installed and fastened to the purlins, followed by the spacing Zeds, insulation and outer sheet. The liner tray is not designed to take full wind and erection loads, and therefore large areas should not be erected in advance of the outer skin. The liner tray is normally supplied in white polyester finish, providing a pleasing internal finish. The weatherproofing criteria are the same as for single-skin systems and generally the minimum slope is 4°. Differing thicknesses of insulation are accommodated by varying spacer depths. The norm is 80 mm of fibreglass giving a nominal U value of 0.44 W/m^2°C.

Standing seam systems

The traditional forms of construction described above suffer from the inherent disadvantage of having to be fixed by screw-type fasteners penetrating the sheet. Traditional fixing methods also limit the length of sheet that can be handled even if, in theory, long lengths can be rolled; thus laps are required.

The need for weathertightness at the lap constrains the minimum slope. A $5000\,m^2$ traditional roof has $20\,000$ through fasteners and has to resist around 1 million gallons of water a year. The difficulty in ensuring that this large number of fasteners is watertight demonstrates the desirability of minimizing the number of penetrations. This has led to the development of systems having concealed fastenings and the ability to roll and fix long lengths. In order to cater for the thermal expansion in sheets, which may be $30\,m$ long, the fastenings are in the form of clips which, while holding down the sheeting, allow it to move longitudinally. As discussed elsewhere, this may reduce the restraint available to the purlins and affect their design. When used in double-skin configuration the liner panel is normally conventionally fastened and provides sufficient restraint. The available permutations are too numerous to give general rules but purlin manufacturers will give advice.

It is necessary to fasten the sheets to the structure at one point to resist downslope forces and progressive movement during expansion and contraction. With the through fasteners reduced to the minimum and laps eliminated or specially detailed, roof slopes as low as 1° (after deflection) can be utilized. The roofs must be properly maintained since accumulation of debris is more likely and ponding leads to a reduced coating life.

Standing seam systems are used to replace the traditional trapezoidal outer sheets in single- and double-skin arrangements as described earlier.

Composite panels

This most recent development in cladding systems provides solutions for many of the potential problems with metal roofing. The insulating foam is integral with the sheets and so totally fills the cavity, and with good detailing at the joints condensation can be eliminated in most environments.

The strength of the panel is dependent on the composite action of the two metal skins in conjunction with the foam. Theoretical calculations are possible although there are no codified design procedures. Since both steel and foam properties can vary, and these are predetermined by the manufacturer, it is a question of selecting the panels from load tables provided rather than individual design. In addition to having to resist external loads, the effects of temperature differential must be taken into account. The critical combinations are wind suction with summer temperatures and snow acting with winter temperatures. The range of temperature considered is dependent on the colour and hence heat absorption of the outer skin; darker colours for roofs should only be considered in conjunction with the manufacturer, if at all.

Both standing seam and traditional trapezoidal forms are available with the same slope restrictions as non-composite forms.

A particular advantage is the erectability of the panels, which is a one-pass operation and, therefore, a rapid process. This is combined with inherent robustness and walkability.

Since the integrity of the panel is important, and it is difficult to inspect the foam and its adhesion once manufactured, quality control of the materials and manufacturing environment in terms of temperature and dust control is vital. Reputable manufacturers should, therefore, be specified and their manufacturing methods ascertained.

External firewall

Where buildings are close to the site boundary the Building Regulations require that the construction is such that reasonable steps are taken to prevent fire spreading to adjacent property. It has been demonstrated by tests that walls of double-skin steel construction with fibreglass or mineral wool insulation can achieve a four hour fire rating. The siderails and fixings require special details which were included in the test arrangements of the particular manufacturer and it is important that these are followed closely. They include such things as providing slotted holes to allow expansion of the rails rather than induce buckling, which may allow gaps to open in the sheeting at joints.

References to Chapter 1

1. Horridge, J.F. (1985) *Design of Industrial Buildings*. Civil Engineering Steel Supplement, November.
2. British Standards Institution (1998) Part 3: *Code of practice for imposed roof loads*. BS 6399. BSI, London.
3. British Standards Institution (1997) Part 2: *Code of practice for wind loads*. BS 6399. BSI, London.
4. British Standards Institution (2000) Part 1: *Code of practice for design – Rolled and welded sections*. BS 5950. BSI, London.
5. King C.M. (2001) *In-plane Stability of Portal Frames to BS 5950-1:2000*. (SCI-P292) The Steel Construction Institute, Ascot.
6. British Standards Institution (1998) Part 5: *Code of practice for design of cold formed thin gauge sections*. BS 5950. BSI, London.

7. Davies J.M. & Raven G.K. (1986) Design of cold formed purlins. *Thin Walled Metal Structures in Buildings*, pp. 151–60. IABSE Colloquium, Zurich, Switzerland.
8. The Building Regulations (2002) Part L: *Conservation of Fuel and Power*. HMSO, London.

Chapter 2
Multi-storey buildings

by ALAN HART and PHILIP PEACOCK

2.1 Introduction

The term *multi-storey building* encompasses a wide range of building forms. This chapter reviews some of the factors that should be considered when designing the type of multi-storey buildings commonly found in Europe, namely those less than 15 storeys in height. Advice on designing taller buildings may be found in the references to this chapter.[1-5]

2.1.1 The advantages of steel

In recent years the development of steel-framed buildings with composite metal deck floors has transformed the construction of multi-storey buildings in the UK. During this time, with the growth of increasingly sophisticated requirements for building services, the very efficiency of the design has led to the steady decline of the cost of the structure as a proportion of the overall cost of the building, yet the choice of the structural system remains a key factor in the design of successful buildings.

The principal reasons for the appeal of steel for multi-storey buildings are noted below.

- Steel frames are fast to erect.
- The construction is lightweight, particularly in comparison with traditional concrete construction.
- The elements of the framework are prefabricated and manufactured under controlled, factory conditions to established quality procedures.
- The accuracy implicit in the manufacturing process by which the elements are produced enables the designer to take a confident view of the geometric properties of the erected framework.
- The dryness of the form of construction results in less on-site activities, plant, materials and labour.
- The framework is not susceptible to drying-out movement or delays due to slow strength gain.
- Steel frames have potential for adaptability inherent in their construction. Later

42

modification to a building can be achieved relatively easily by unbolting a connection; with traditional concrete construction such modifications would be expensive, and more extensive and disruptive.
- The use of steel makes possible the creation of large, column-free internal spaces which can be divided by partitions and, by eliminating the external wall as a load-bearing element, allows the development of large window areas incorporated in prefabricated cladding systems.

2.1.2 Design aims

For the full potential of the advantages of steel-frame construction to be realized, the design of multi-storey buildings requires a considered and disciplined approach by the architects, engineers and contractors involved in the project. They must be aware of the constraints imposed on the design programme by the lead time between placing a contract for the supply of the steel frame and the erection of the first pieces on site. The programme should include such critical dates on information release as are necessary to ensure that material order and fabrication can progress smoothly.

The designer must recognize that the framework is the skeleton around which every other element of the building will be constructed. The design encompasses not only the structure but also the building envelope, services and internal finishes. All these elements must be co-ordinated by a firm dimensional discipline, which recognizes the modular nature of the components, to ensure maximum repetition and standardization. Consequently it is impossible to consider the design of the framework in isolation. It is vital to see the frame as part of an integrated building design from the outset: the most efficient solution for the structure may not be effective in achieving a satisfactory solution for the total building.

In principle, the design aims can be considered under three headings:

- Technical
- Architectural
- Financial.

Technical aims

The designer must ensure that the framework, its elements and connections are strong enough to withstand the applied loads to which the framework will be subjected throughout its design life. The system chosen on this basis must be sufficiently robust to prevent the progressive collapse of the building or a significant part of it under accidental loading. This is the primary technical aim. However, as issues related to strength have become better understood and techniques for the strength design of frameworks have been formalized, designers have progressively used

lighter and stronger materials. This has generated a greater need to consider serviceability, including dynamic floor response, as part of the development of the structural concept.

Other important considerations are to ensure adequate resistance to fire and corrosion. The design should aim to minimize the cost, requirements and intrusion of the protection systems on the efficiency of the overall building.

Architectural aims

For the vast majority of buildings the most effective structural steel frame is the one which is least obtrusive. In this way it imposes least constraint on internal planning, and produces maximum usable floor area, particularly for open-plan offices. It also provides minimal obstruction to the routeing of building services. This is an important consideration, particularly since building services are becoming more extensive and demanding on space and hence on the building framework.

Occasionally the structure is an essential feature of the architectural expression of the building. Under these circumstances the frame must achieve, among other aims, a balance between internal planning efficiency and an expressed structural form. However, these buildings are special, not appropriate to this manual, and will not be considered in more detail, except to give a number of references.

Financial aims

The design of a steel frame should aim to achieve minimum overall cost. This is a balance between the capital cost of the frame and the improved revenue from early occupation of the building through fast erection of the steel frame: a more expensive framework may be quicker to build and for certain uses would be more economic to a client in overall terms. Commercial office developments are a good example of this balance. Figure 2.1 shows a breakdown of construction costs for a typical development.

2.1.3 Influences on overall design concept

Client brief

Clients specify their requirements through a brief. It is essential for effective design to understand exactly the intentions of the client: the brief is the way in which the client expresses and communicates these intentions. As far as the frame designer is concerned, the factors which are most important are intended use, budget cost limits, time to completion and quality. Once these are understood a realistic basis for producing the design will be established. The designer should recognize however that in practice the brief is likely to evolve as the design develops.

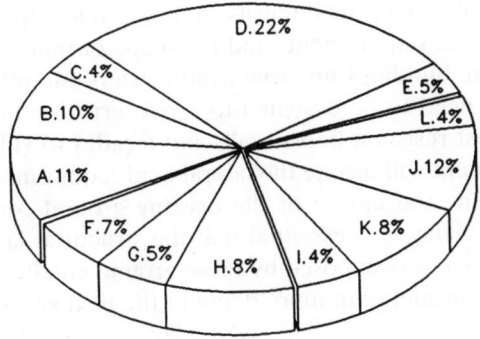

Fabric
A Demolition. Piling foundations and concrete work 11%
B Steel Frame. Deck and Fire Protection 10%
C Brickwork and Drywalling 4%
D External and internal Cladding and Sunscreens 22%
E Roofing and Rooflights 5%
                                                            52%

Finishes		Services	
F Ceiling and Floors	7%	I Plumbing and Sprinklers ...4%
G Stone	5%	J H.V.A.C. 12%
H Others	8%	K Electrics 8%
	20%	L Lifts 4%
		28%

Fig. 2.1   Typical cost breakdown

Statutory constraints

The design of all buildings is subject to some form of statutory constraint. Multi-storey buildings, particularly those in an urban environment, are subject to a high level of constraint, which will generally be included in the conditions attached to the granting of outline planning permission. The form and degree that this may take could have significant impact on the frame design. For example, street patterns and lighting restrictions may result in a non-rectilinear plan with a 'stepped-back' structure. If an appropriate layout is to be provided it is vital to understand these constraints from the outset.

Certain government buildings, financial headquarters and other strategically sensitive buildings may need to be designed to resist terrorist threats. Provision may be in the form of blast-resistant flows, walls and facades to vulnerable areas.

Physical factors

The building must be designed to suit the parameters determined by its intended use and its local environment.

Its intended use will dictate the intensity of imposed loadings, the fire protection and corrosion resistance requirements and the scope of building services.

Certain government buildings, financial headquarters and other strategically sensitive buildings may need to be designed to resist terrorist threats. Provision may be in the form of blast-resistant floors, walls and façades to vulnerable areas.

The local environment will dictate the lateral load requirements but more importantly it will determine the nature of the existing ground conditions. To achieve overall structural efficiency it is essential that the structural layout of the frame is responsive to the constraints imposed by these ground conditions.

These factors are considered in more detail in the next section.

2.2 Factors influencing choice of form

Environmental

There are a number of factors which influence the choice of structural form that are particular to the site location. These can have a dominant effect on the framing arrangement for the structure.

The most obvious site-dependent factors are related to the ground conditions. A steel-framed building is likely to be about 60% of the weight of a comparable reinforced concrete building. This difference will result in smaller foundations with a consequent reduction on costs. In some cases this difference in weight enables simple pad foundations to be used for the steel frame where the equivalent reinforced concrete building would require a more complex and expensive solution. For non-uniformly loaded structures it will also reduce the magnitude of differential settlements and for heavily loaded structures may make possible the use of a simple raft foundation in preference to a large capacity piled solution (Fig. 2.2).

Difficult ground conditions may dictate the column grid. Long spans may be required to bridge obstructions in the ground. Such obstructions could include, for example, buried services, underground railways or archaeological remains. Generally, a widely spaced column grid is desirable since it reduces the number of foundations and increases the simplicity of construction in the ground.

Other site-dependent constraints are more subtle. In urban areas they relate to the physical constraints offered by the surrounding street plan, and the rights of light of adjoining owners. They also relate to the planning and architectural objectives for specific sites. The rights of light issues or planning considerations may dictate that upper floors are set back from the perimeter resulting in stepped construction of the upper levels. Invariably the resulting framing plan is not rectilinear and may have skew grids, cantilevers and re-entrant corners.

These constraints need to be identified early in the design in order that they are accommodated efficiently into the framing. For example, wherever possible, stepped-back façades should be arranged so that steps take place on the column grid and hence avoid the need for heavy bridging structures. In other situations the designer should always investigate ways in which the impact of lack of uniformity in building form can be contained within a simple structural framing system which generates a minimum of element variations and produces simple detailing.

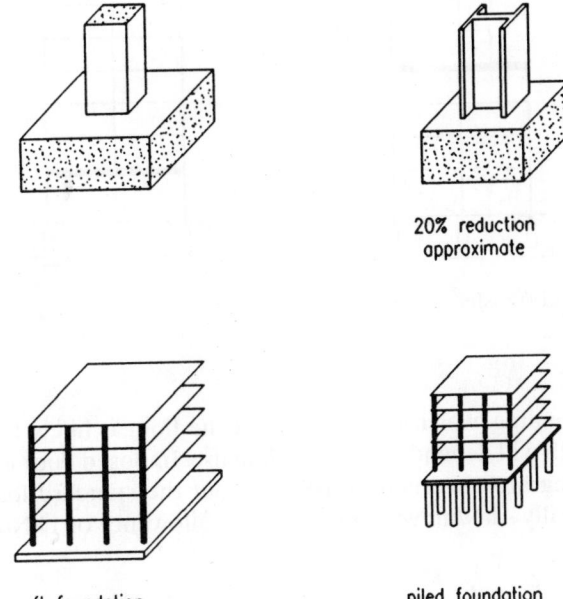

20% reduction
approximate

raft foundation   piled foundation

Fig. 2.2   Foundation savings

Building use

The building use will dictate the planning module of the building, which will in turn determine the span and column grids. Typical grids may be based on a planning module of 600/1200 mm or 500/1500 mm. However, the use has much wider impact, particularly on floor loadings and building services. The structural arrangement, and depth selected, must satisfy and accommodate these requirements.

For example, financial-dealing floors require clear open spaces located on the lower floors, which would dictate a different structural solution to the rest of the building. This may necessitate the use of a transfer structure to carry the upper floors on an economical column grid (Fig. 2.3).

Floor loadings

Because steel-framed buildings are relatively light in weight, excessive imposed loadings will have a greater effect on the sizing of structural components, particularly floor beams, than with reinforced concrete structures.

The floor loadings to be supported by the structure have two components:

- The permanent or dead loading comprising the self-weight of the flooring and the supporting structure together with the weight of finishes, raised flooring, ceiling, air-conditioning ducts and equipment.

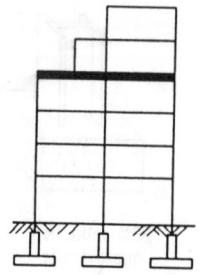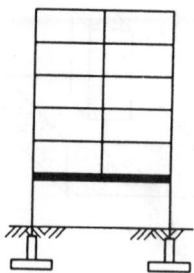

Fig. 2.3 Typical load transfer systems

- The imposed loading, which is the load that the floor is likely to sustain during its life and which will depend on the building use. Imposed floor loads for various types of building are governed by BS 6399 but the standard loading for office buildings is usually $4\,kN/m^2$ with an additional allowance of $1\,kN/m^2$ for movable partitioning.

For normal office loadings, dead and imposed loadings are roughly equal in proportion but higher imposed load allowances will be necessary in plantrooms or to accommodate special requirements such as storage or heavy equipment.

Floor beams will be designed to limit deflection under the imposed loadings. British Standard BS 5950 governing the design of structural steelwork sets a limit for deflection under imposed loading of (span/200) generally and (span/360) where there are brittle finishes. Edge beams supporting cladding will be subject to restriction on deflection of 10–15 mm. Deflections may be noticeable in the ceiling layout and should be taken into account when determining the available clearance for service routes. The designer should therefore check the cumulative effect of deflections in the individual members of a floor system although the actual maximum displacement is in practice almost always less than that predicted. In some instances, vibrations of floor components may cause discomfort or affect sensitive equipment, and the designer should check the fundamental response of the floor system. The threshold of perceptible vibrations in building is difficult to define, and present limits are rather arbitrary. There is some evidence that modern lightweight floors can be sensitive to dynamic loads, which may have an effect on delicate equipment. However, in most situations a simple check on the natural frequency of the floor system is all that is required.

Building services and finishes

In buildings requiring anything other than minimal electrical services distribution, the inter-relationship of the structure, the mechanical and electrical services and the building finishes will need to be considered together from the outset.

It is essential to co-ordinate the details of the building services, cladding and structure at an early stage of the project in order to produce a building which is simple to fabricate and quick to erect. Apparently minor variations to the steelwork, brought about by services and finishes requirements defined after a steel fabrication contract has been let, can have a disproportionate effect on the progress of fabrication and erection. Steel buildings impose a strict discipline on the designer in terms of the early production of final design information. If the designer fails to recognize this, the advantages of steel-framed building cannot be realized.

The integration of the building services with the structure is an important factor in the choice of an economic structural floor system. The overall depth of the floor construction will depend on the type and distribution of services in the ceiling void. The designer may choose to separate the structural and services zones or accommodate the services by integrating them with the structure, allowing for the structural system to occupy the full depth of the floor construction. (See Fig. 2.4.)

Separation of zones usually requires confining the ducts, pipes and cables to a horizontal plane below the structure, resulting in either a relatively deep overall floor construction or close column spacings. Integration of services with structure requires either deep perforated structural components or vertical zoning of the services and structure.

For the range of structural grids used in conventional building, traditional steel floor construction is generally deeper than the equivalent reinforced concrete flat slab: the difference is generally 100–200 mm for floor structures which utilize composite action and greater for non-composite floors (Fig. 2.5). The increased depth is only at the beam position; elsewhere, between beams, the depth is much less and the space between them may accommodate services, particularly if the beams may be penetrated (Fig. 2.6). The greater depth of steel construction does not therefore necessarily result in an increase in building height if the services are integrated within the zone occupied by the structure. A number of possible solutions exist for

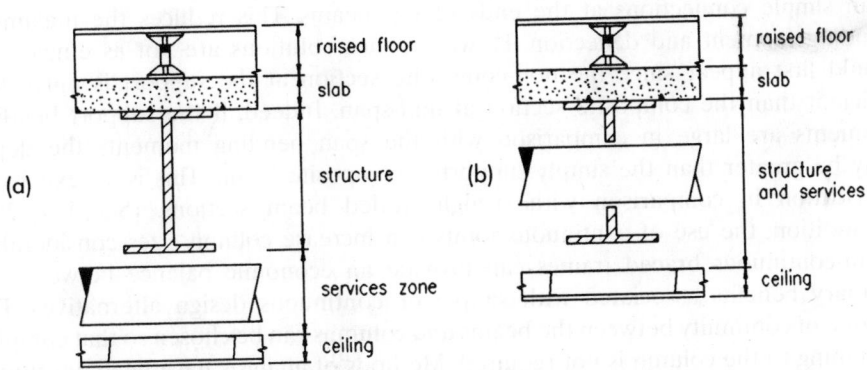

Fig. 2.4   Building services and floor structure: (a) separation of services and structure; (b) integration of services and structure

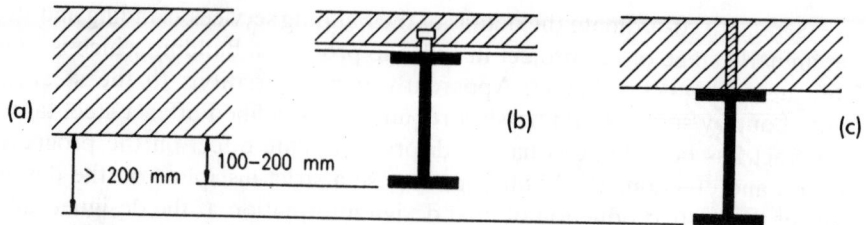

(a)

(b)

(c)

> 200 mm   100–200 mm

Fig. 2.5   Overall floor depths: (a) R.C. flat slab; (b) composite; (c) non-composite

(a)   (b)

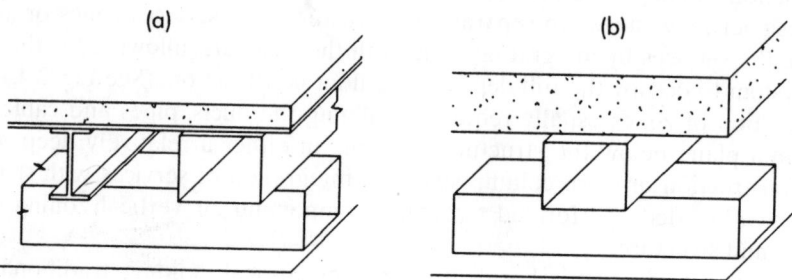

Fig. 2.6   Ceiling voids: (a) steel frame: variable void height; (b) concrete slab: constant void height

integrated systems, particularly in long-span structures utilizing castellated, cellular or stub-girder beams. (See Fig. 2.7.)

A number of solutions have been developed which allow long spans to co-exist with separation of the building services by profiling the steel beam to provide space for services, either at the support or in the span. Automated plate cutting and welding techniques are used to produce economical profiled plate girders, with or without web openings. (See Fig. 2.8.)

Overall depth may be reduced by utilizing continuous or semi-continuous rather than simple connections at the ends of the beams. This reduces the maximum bending moment and deflection. However, such solutions are not as efficient as would first appear since the non-composite section at the support is much less efficient than the composite section at mid-span. Indeed, if the support bending moments are large in comparison with the span bending moments the depth may be greater than the simply-supported composite beam. This is an expensive fabrication in comparison with straight rolled beam sections. (See Fig. 2.9.) In addition, the use of continuous joints can increase column sizes considerably. Semi-continuous braced frames can provide an economic balance between the primary benefits associated with simple or continuous design alternatives. The degree of continuity between the beams and columns can be chosen so that complex stiffening to the column is not required. Methods of analysis have been developed for non-composite construction to permit calculation by hand. It is possible to achieve reduced beam depths and reduced beam weights.

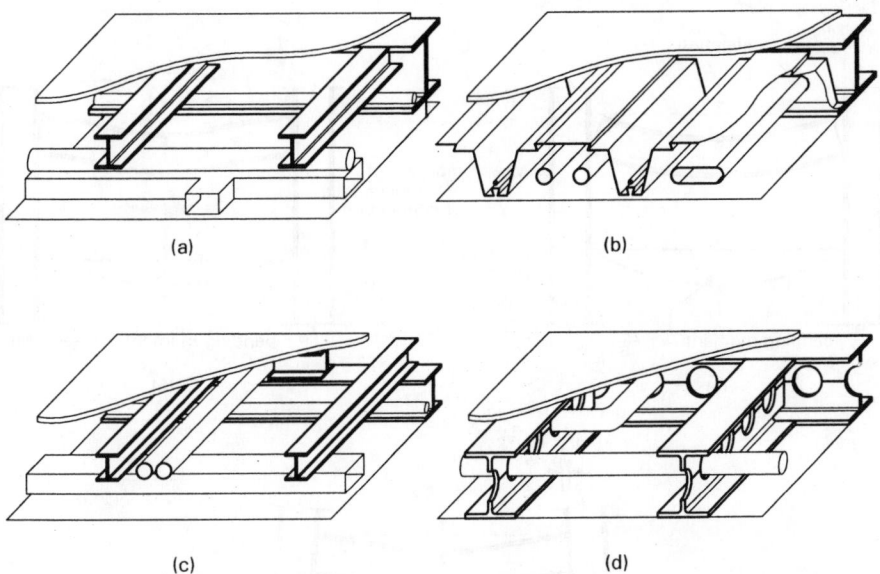

Fig. 2.7   Integration of services: (a) separated (traditional); (b) integrated (shallow floor 'Slimdek®' system); (c) integrated (long span 'primary' beam – stub girder); (d) integrated (long span 'secondary' beams)

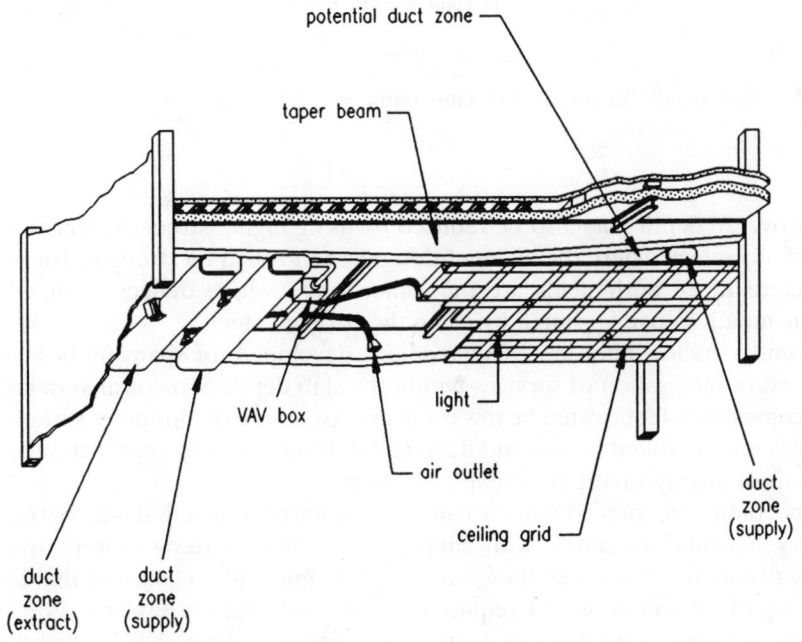

Fig. 2.8   Tapered beams and services

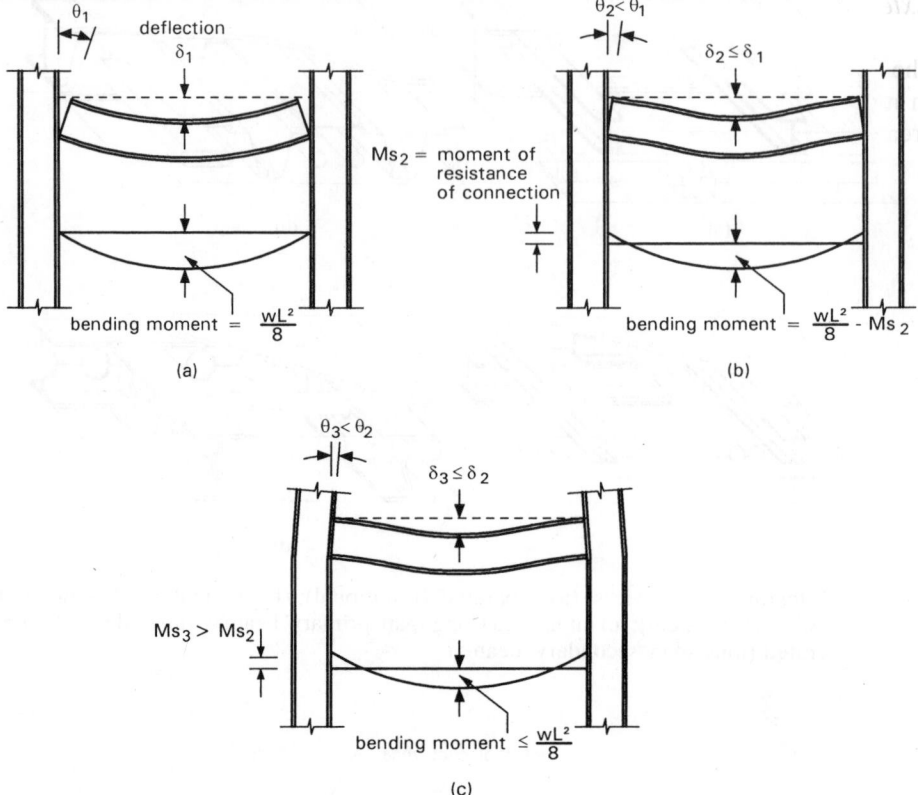

Fig. 2.9 Floor depth: (a) simple; (b) semi-continuous; (c) continuous

The overall depth may also be reduced by using higher-strength steel, but this is only of advantage where the element design is controlled by strength. The stiffness characteristics of both steels are the same: hence, where deflection or vibration govern, no advantage is gained by using the stronger steel.

Recently, shallow floor systems have been developed for spans up to about 9 m which allow integration of services within the slab depth. Structural systems range from conventional fabricated beams using precast units to proprietary systems using new asymmetric rolled beams and deep metal decking. These approaches can form the basis of energy-saving sustainable solutions.

Semi-continuous braced frames can provide an economic balance between the primary benefits associated with simple or continuous design alternatives. The degree of continuity between the beams and columns can be chosen so that complex stiffening to the column is not required. Methods of analysis have been developed for non-composite construction to permit calculation by hand. It is possible to achieve reduced beam depths and reduced beam weights.

External wall construction

The external skin of a multi-storey building is supported off the structural frame. In most high quality commercial buildings the cost of external cladding systems greatly exceeds the cost of the structure. This influences the design and construction of the structural system in a number of ways:

- The perimeter structure must provide a satisfactory platform to support the cladding system and be sufficiently rigid to limit deflections of the external wall.
- A reduction to the floor zone may significantly reduce the area and hence cost of cladding.
- Fixings to the structure should facilitate rapid erection of cladding panels.
- A reduction in the weight of cladding at the expense of cladding cost will not necessarily lead to a lower overall construction cost.

Lateral stiffness

Steel buildings must have sufficient lateral stiffness and strength to resist wind and other lateral loads. In tall buildings the means of providing sufficient lateral stiffness forms the dominant design consideration. This is not the case for low- to medium-rise buildings.

Most multi-storey buildings are designed on the basis that wind and/or notional horizontal forces acting on the external cladding are transmitted to the floors, which form horizontal diaphragms transferring the lateral load to rigid elements and then to the ground. These rigid elements are usually either braced-bay frames, rigid-jointed frames, reinforced concrete or steel–concrete–steel sandwich shear walls.

Low-rise unbraced frames up to about six storeys may be designed using the simplified wind-moment method. In this design procedure, the frame is made statically determinate by treating the connections as pinned under vertical load and rigid under horizontal loads. This approach can be used on both composite and non-composite frames, albeit with strict limitations on frame geometry, loading patterns and member classification.

British Standard BS 5950 sets a limit on lateral deflection of columns as height/300 but height/600 may be a more reasonable figure for buildings where the external envelope consists of sensitive or brittle materials such as stone facings.

Accidental loading

A series of incidents in the 1960s culminating in the partial collapse of a system-built tower block at Ronan Point in 1968 led to a fundamental reappraisal of the approach to structural stability in building.

Traditional load-bearing masonry buildings have many in-built elements providing inherent stability which are lacking in modern steel-framed buildings. Modern structures can be refined to a degree where they can resist the horizontal and

vertical design loadings with the required factor of safety but may lack the ability to cope with the unexpected.

It is this concern with the safety of the occupants and the need to limit the extent of any damage in the event of unforeseen or accidental loadings that has led to the concept of robustness in building design. Any element in the structure that supports a major part of the building either must be designed for blast loading or must be capable of being supported by an alternative load path. In addition, suitable ties should be incorporated in the horizontal direction in the floors and in the vertical direction through the columns. The designer should be aware of the consequences of the sudden removal of key elements of the structure and ensure that such an event does not lead to the progressive collapse of the building or a substantial part of it. In practice, most modern steel structures can be shown to be adequate without any modification.

Cost considerations

The time taken to realize a steel building from concept to completion is generally less than that for a reinforced concrete alternative. This reduces time-related building costs, enables the building to be used earlier, and produces an earlier return on the capital invested.

To gain full benefit from the 'factory' process and particularly the advantages of speed of construction, prefabrication, accuracy and lightness, the cladding and finishes of the building must have similar attributes. The use of heavy, slow and in situ finishing materials is not compatible with the lightweight, prefabricated and fast construction of a steel framework.

The cost of steel frameworks is governed to a great extent by the degree of simplicity and repetition embodied in the frame components and connections. This also applies to the other elements which complete the building.

The criterion for the choice of an economic structural system will not necessarily be to use the minimum weight of structural steel. Material costs represent only 30–40% of the total cost of structural steelwork. The remaining 60–70% is accounted for in the design, detailing, fabrication, erection and protection. Hence a choice which needs a larger steel section to avoid, say, plate stiffeners around holes or allows greater standardization will reduce fabrication costs and may result in the most economic overall system.

Because a steel framework is made up of prefabricated components produced in a factory, repetition of dimensions, shapes and details will streamline the manufacturing process and is a major factor in economic design (Fig. 2.10).

Fabrication

The choice of structural form and method of connection detailing have a significant impact on the cost and speed of fabrication and erection. Simple braced frameworks with bolted connections are considered the most economic and the fastest to build for low- to medium-rise buildings.

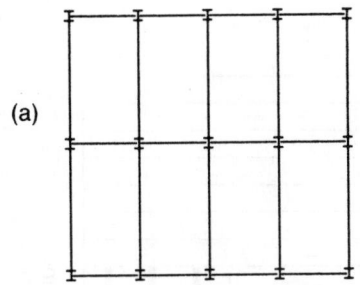   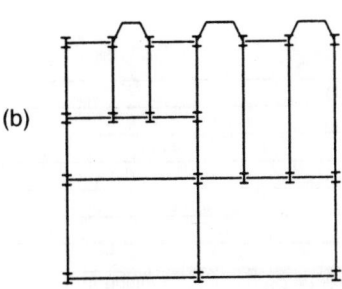

(a)   (b)

Fig. 2.10   Structural costs: (a) economic and (b) uneconomic layouts

Economy is generally linked to the use of standard rolled sections but, with the advent of automated cutting and welding equipment, special fabricated sections are becoming economic if there is sufficient repetition.

The development of efficient, automated, cold-sawing techniques and punching and drilling machines has led to the fabrication of building frameworks with bolted assemblies. Welded connections involve a greater amount of handling in the fabrication shop, with consequent increases in labour and cost.

Site-welded connections require special access, weather protection, inspection and temporary erection supports. By comparison, on-site bolted connections enable the components to be erected rapidly and simply into the frame and require no further handling.

The total weight of steel used in continuous frames is less than in semi-continuous or simple frames, but the connections for continuous frames are more complex and costly to fabricate and erect. On balance, the cost of a continuous frame structure is greater, but there may be other considerations which offset this cost differential. For example, in general the overall structural depth of continuous frames is less. This may reduce the height of the building or improve the distribution of building services, both of which could reduce the overall cost of the building.

Corrosion protection to internal building elements is an expensive and time-consuming activity. Experience has shown that it is unnecessary for most internal locations and consequently only steelwork in risk areas should require any protection. Factory-applied coatings of intumescent fire protection can be cost-effective and time-saving by removing the operation from the critical path.

Construction

A period of around 8–12 weeks is usual between placing a steel order and the arrival of the first steel components on site. Site preparation and foundation construction generally take a similar or longer period (see Fig. 2.11). Hence, by progressing fabrication in parallel with site preparation, significant on-site construction time may be saved, as commencement of shop fabrication is equivalent to start-on-site for an in situ concrete-framed building. By manufacturing the frame in a factory, the risks

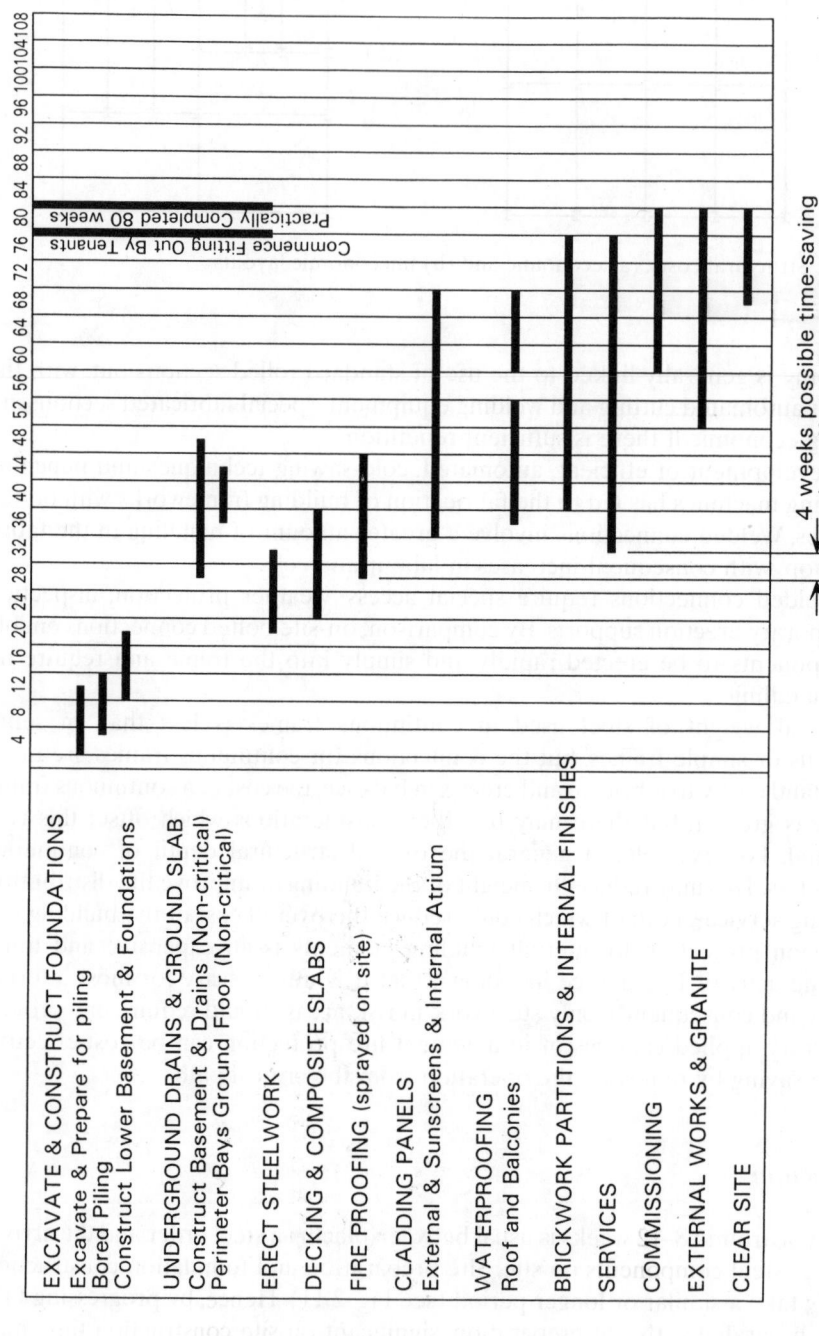

Fig. 2.11   Typical progress schedule (in weeks)

of delay caused by bad weather or insufficient or inadequate construction resources in the locality of the site are significantly reduced.

Structural steel frameworks should generally be capable of being erected without temporary propping or scaffolding, although temporary bracing will be required, especially for welded frames. This applies particularly to the construction of the concrete slab, which should be self-supporting at all stages of erection. Permanent metal or precast concrete shutters should be used to support the in situ concrete.

In order to allow a rapid start to construction, the structural steelwork frame should commence at foundation level, and preference should be given to single foundations for each column rather than raft or shared foundations (Fig. 2.12).

Speed of erection is directly linked to the number of crane hours available. To reduce the number of lifts required on site, the number of elements forming the framework should be minimized within the lifting capacity of the craneage provided on site for other building components. For similar-sized buildings, the one with the longer spans and fewer elements will be the fastest to erect. However, as has been mentioned earlier, longer spans require deeper, heavier elements, which will increase the cost of raw materials and pose a greater obstruction to the distribution of building services, thereby requiring the element to be perforated or shaped and hence increasing the cost of fabrication.

Columns are generally erected in multi-storey lengths: two is common and three is not unusual. The limitation on longer lengths is related more to erection than to restrictions on transportation, although for some urban locations length is a major consideration for accessibility.

To provide rapid access to the framework the staircases should follow the erection of the frame. This is generally achieved by using prefabricated stairs which are detailed as part of the steel frame.

The speed of installation of the following building elements is hastened if their connection and fixing details are considered at the same time as the structural steel frame design. In this way the details can be either incorporated in the framework or separated from it, whichever is the most effective overall: it is generally more efficient to separate the fixings and utilize the high inherent accuracy of the frame

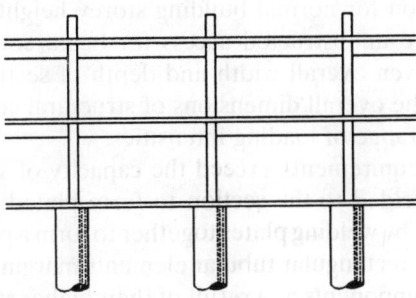

Fig. 2.12   Columns on large diameter bored piles

to use simple post-fixed details, provided these do not require staging or scaffolding to give access.

Finally, on-site painting extends the construction period and provides potential compatibility problems with following applied fire protection systems. Painting should therefore only be specified when absolutely necessary.

2.3 Anatomy of structure

In simple terms, the vertical load-carrying structure of a multi-storey building comprises a system of vertical column elements interconnected by horizontal beam elements which support floor-element assemblies. The resistance to lateral loads is provided by diagonal bracing elements, or wall elements, introduced into the vertical rectangular panels bounded by the columns and beams to form vertical trusses, or walls. Alternatively, lateral resistance may be provided by developing a continuous or semi-continuous frame action between the beams and columns. The flexibility of connections should be taken into account in the analysis. All structures should have sufficient sway stiffness, so that the vertical loads acting with lateral displacements of the structure do not induce excessive secondary forces or moments in the members or connections. A building frame may be classed as 'non-sway' if the sway deformation is sufficiently small for the resulting secondary forces and moments to be negligible. In all other cases the building frame should be classed as 'sway-sensitive'. The stiffening effect of cladding and infill wall panels may be taken into account by using the method of partial sway bracing. The floor-element assemblies provide the resistance to lateral loads in the horizontal plane.

In summary, the components of a building structure are columns, beams, floors and bracing systems (Fig. 2.13).

2.3.1   Columns

These are generally standard, universal column, hot-rolled sections. They provide a compact, efficient section for normal building storey heights. Also, because of the section shape, they give unobstructed access for beam connections to either the flange or web. For a given overall width and depth of section, there is a range of weights which enable the overall dimensions of structural components to be nominally maintained for a range of loading intensities.

Where the loading requirements exceed the capacity of standard sections, additional plates may be welded to the section to form plated columns, or fabricated columns may be formed by welding plates together to form a plate-column (Fig. 2.14).

The use of circular or rectangular tubular elements marginally improves the load-carrying efficiency of components as a result of their higher stiffness-to-weight ratio. Concrete filling significantly improves the axial load-carrying capacity and fire resistance.

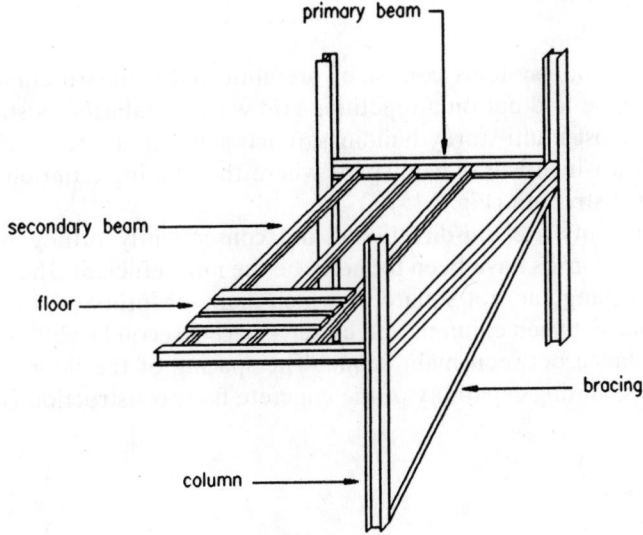

Fig. 2.13   Conventional steel frame components

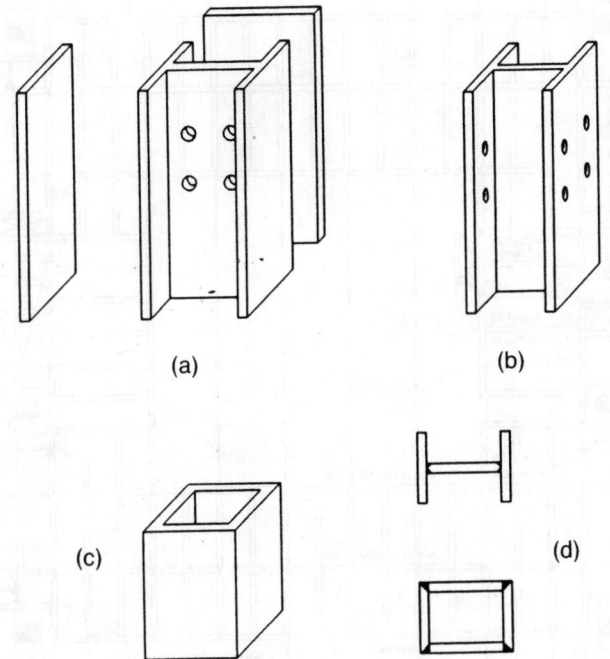

Fig. 2.14   Types of column: (a) plated (by addition of plates to U.C. section); (b) universal; (c) tubular; (d) fabricated plate

2.3.2 Beams

Structural steel floor systems consist of prefabricated standard components, and columns should be laid out on a repetitive grid which establishes a standard structural bay. For most multi-storey buildings, functional requirements will determine the column grid which will dictate spans where the limiting criterion will be stiffness rather than strength (Fig. 2.15).

Steel components are uni-directional and consequently orthogonal structural column and beam grids have been found to be the most efficient. The most efficient floor plan is rectangular, not square, in which main, or 'primary', beams span the shorter distance between columns and closely-spaced 'secondary' floor beams span the longer distance between main beams. The spacing of the floor beams is controlled by the spanning capability of the concrete floor construction (Fig. 2.16).

Steelwork Plan Level 3

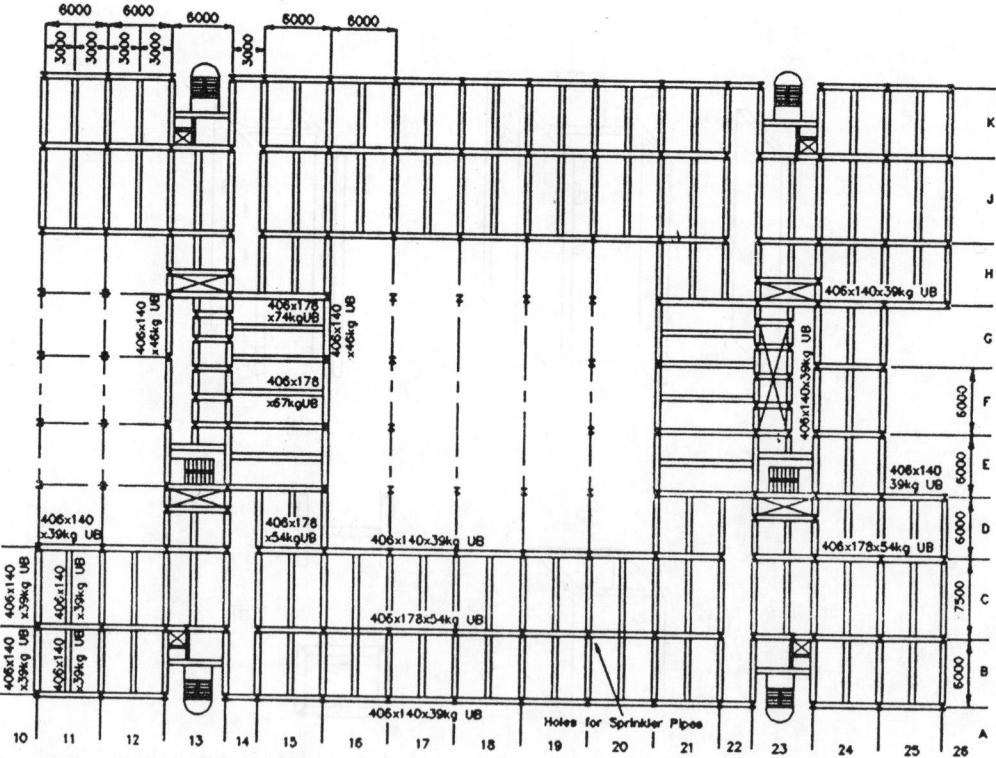

Fig. 2.15 Typical floor layout

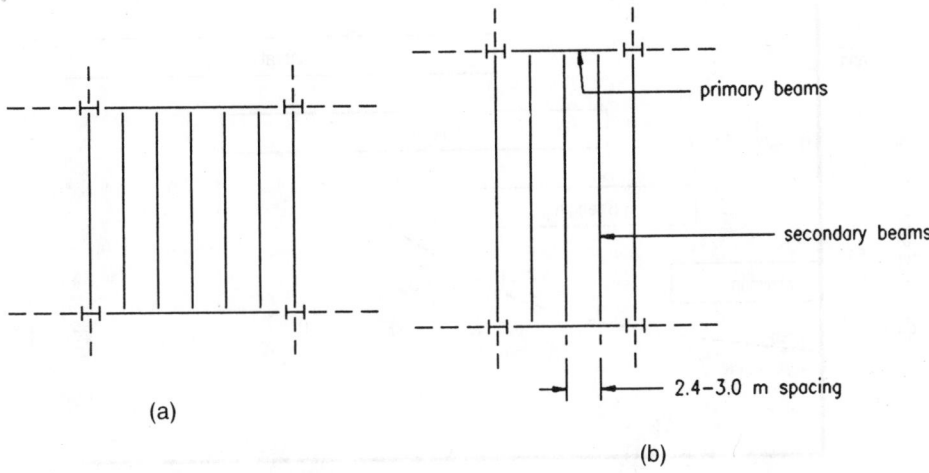

Fig. 2.16   Beam and shallow deck layout: (a) inefficient; (b) efficient

Having decided on the structural grid, the designer must choose an economic structural system to satisfy all the design constraints. The choice of system and its depth depends on the span of the floor (Fig. 2.17). The minimum depth is fixed by practical considerations such as fitting practical connections. As the span increases, the depth will be determined by the bending strength of the member and, for longer spans, by the stiffness necessary to prevent excessive deflection under imposed load or excessive sensitivity to induced vibrations (Fig. 2.18). For spans up to 9 m, shallow beams with precast floors or deep composite metal deck floors can be used to

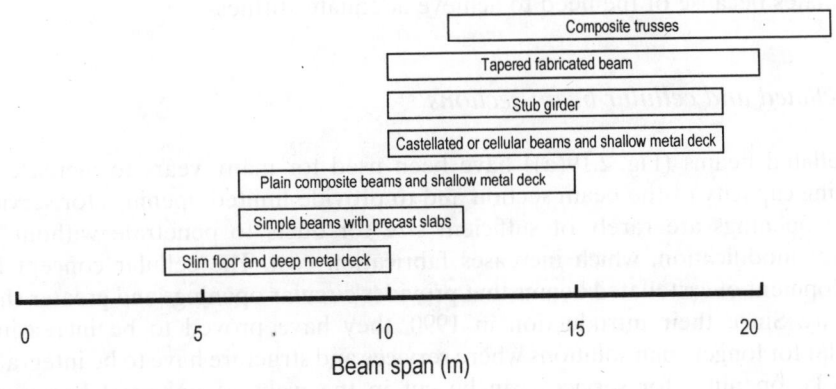

Fig. 2.17   Span ranges for different composite floor systems

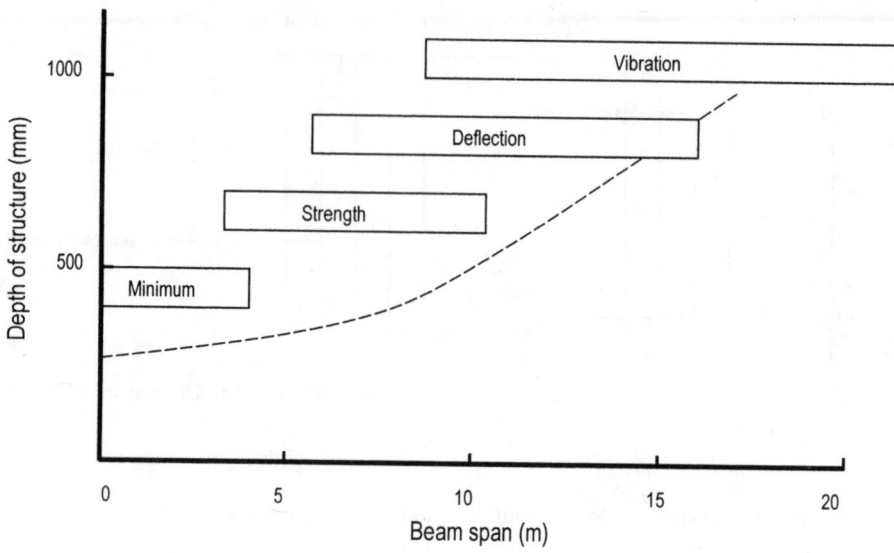

Fig. 2.18   Structural criteria governing choice

minimize the floor zone. Simple universal beams with precast floors or composite metal deck floors are likely to be most economic for spans up to 12 m. A range of section capacities for each depth enables a constant depth of construction to be maintained for a range of spans and loading. As with column components, plated beams and fabricated girders may be used for spans above 10–12 m. They are particularly appropriate where heavier loading is required and overall depth is limited. For medium to lightly loaded floors and long spans, beams may also take the form of castellated beams fabricated from standard sections, cellular sections or plates. Above 15 m, composite steel trusses may be economic. As the span increases, the depth and weight of the structural floor increase, and above 15 m spans depth predominates because of the need to achieve adequate stiffness.

Castellated and cellular beam sections

Castellated beams (Fig. 2.19(a)) have been used for many years to increase the bending capacity of the beam section and to provide limited openings for services. These openings are rarely of sufficient size for ducts to penetrate without significant modification, which increases fabrication cost. The cellular concept is a development of castellated beams that provides circular openings and greater shear capacity. Since their introduction in 1990, they have proved to be increasingly popular for longer span solutions where services and structure have to be integrated. Bespoke openings for services can be cut in the webs of universal beams and fabricated plate girders.

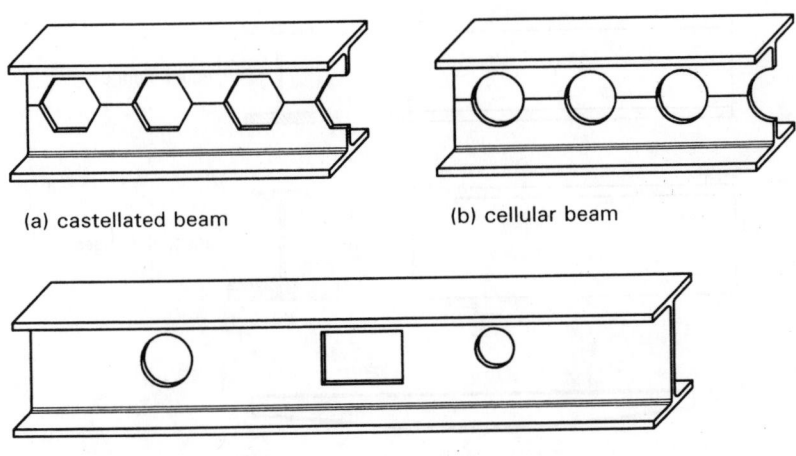

(a) castellated beam   (b) cellular beam

(c) beam with web openings

Fig. 2.19   Beams with web openings for service penetrations

Fabricated plate girders

Conventional universal beams span a maximum of about 15 m. Recent advances in automatic and semi-automatic fabrication techniques have allowed the economic production of plate girders for longer span floors. Particularly if a non-symmetric plate girder is used, it is possible to achieve economic construction well in excess of 15 m (Fig. 2.20). Such plate girders can readily accommodate large openings for major services. If these are in regions of high stress, single-sided web stiffening may be used. Away from regions of high stress, stiffening is usually not required.

The use of intumescent paints, applied offsite, is becoming increasingly popular. One major fabricator is now offering an integrated design and fabrication service for customized plate girders which can achieve a fire resistance of 2 hours when applied as a single layer in an off-site, factory-controlled process.

Taper beams

Taper beams (Fig. 2.20) are similar to fabricated plate girders except that their depth varies from a maximum in mid-span to a minimum at supports, thus achieving a highly efficient structural configuration. For simply-supported composite taper beams in buildings the integration of the services can be accommodated by locating the main ducts close to the columns. Alternative taper beam configurations can be used to optimize the integration of the building services.

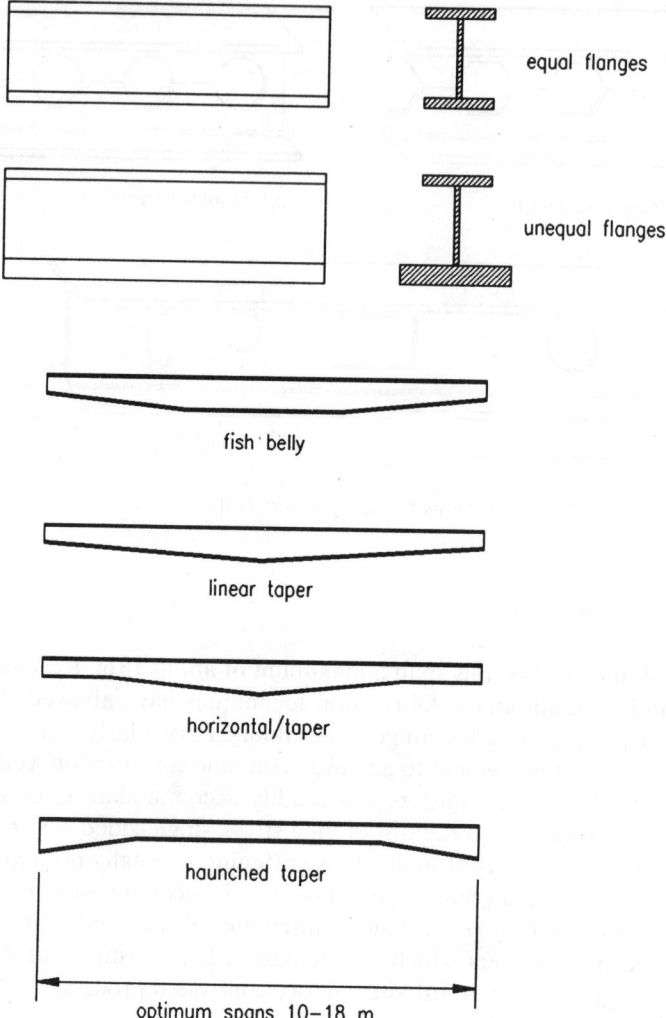

equal flanges

unequal flanges

fish belly

linear taper

horizontal/taper

haunched taper

optimum spans 10–18 m

Fig. 2.20 Fabricated plate girders and taper beams

Composite steel floor trusses

Use of composite steel floor trusses as primary beams in the structural floor system permits much longer spans than would be possible with conventional universal beams. The use of steel trusses for flooring systems is common for multi-storey buildings in North America but seldom is used in Britain. Although they are considerably lighter than the equivalent universal beam section the cost of fabrication is very

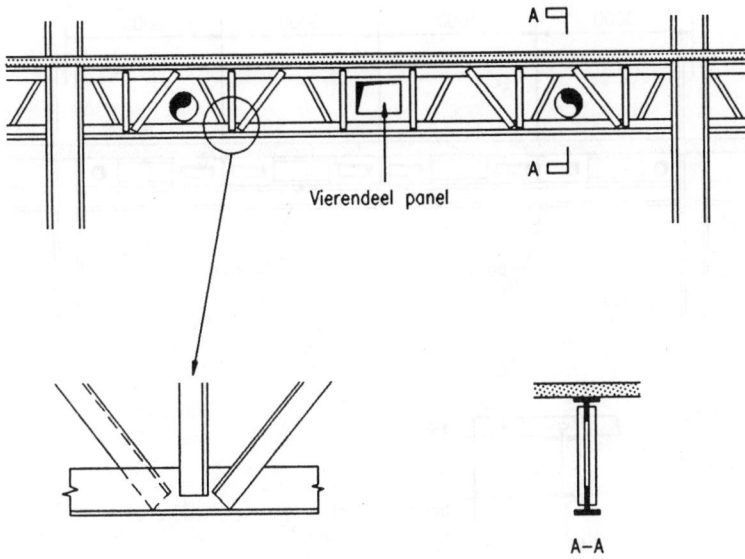

Fig. 2.21   Composite truss

much greater, as is the cost of fireproofing the truss members. For maximum economy, trusses should be fabricated from T-sections and angles using simple welded lap joints. The openings between the diagonal members should be designed to accept service ducts, and if a larger opening is required a Vierendeel panel can be incorporated at the centre of the span. Because a greater depth is required for floor trusses, the integration of the services is always within the structural zone (Fig. 2.21).

Stub girder construction

Stub girders were developed in North America in the 1970s as an alternative form of construction for intermediate range spans of between 10 and 14 m. They have not been used significantly in the UK. Figure 2.22 shows a typical stub girder with a bottom chord consisting of a compact universal column section which supports the secondary beams at approximately 3-metre centres. Between the secondary beams a steel stub is welded on to the bottom chord to provide additional continuity and to support the floor slab. The whole system acts as a composite Vierendeel truss. A disadvantage of stub girders is that the construction needs to be propped while the concrete is poured and develops strength. Arguably, a deep universal beam with large openings provides a more cost-effective alternative to the stub girder because of the latter's high fabrication content.

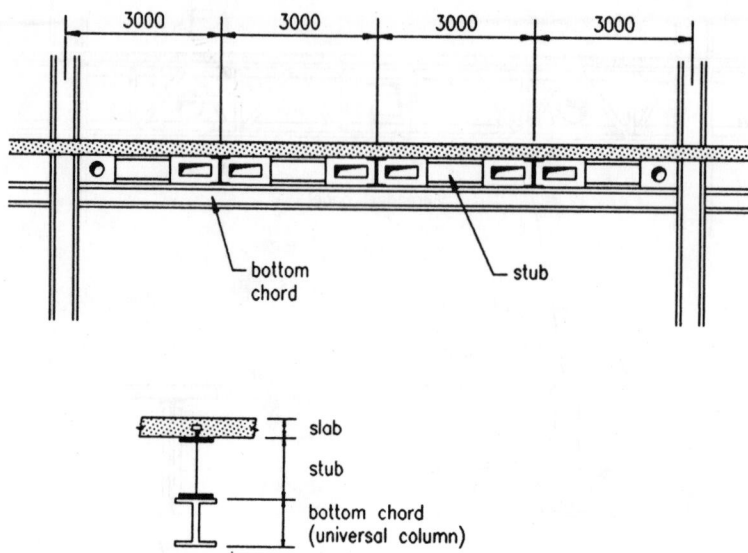

Fig. 2.22   Stub girder

2.3.3 Floors

These take the form of concrete slabs of various forms of construction spanning between steel floor beams (Fig. 2.23). The types generally found are:

- in situ concrete slab cast on to permanent profiled shallow or deep metal decking, acting compositely with the steel floor beams;
- precast concrete slabs acting non-compositely with the floor beams:
- in situ concrete slab, with conventional removable shuttering, acting compositely with the floor beams;
- in situ concrete slab cast on thin precast concrete slabs to form a composite slab, which in turn acts compositely with the floor beams.

The most widely used construction internationally is profiled shallow metal decking. Composite action with the steel beam is normally provided by shear connectors welded through the metal decking on to the beam flange. Shallow floor systems using deep metal decking are gaining popularity in the UK although precast concrete systems are still used extensively. Composite action enables the floor slab to work with the beam, enhancing its strength and reducing deflection (Fig. 2.24). Because composite action works by allowing the slab to act as the compression flange of the combined steel and concrete system, the advantage is greatest when the beam is sagging. Consequently composite floor systems are usually designed as simply supported.

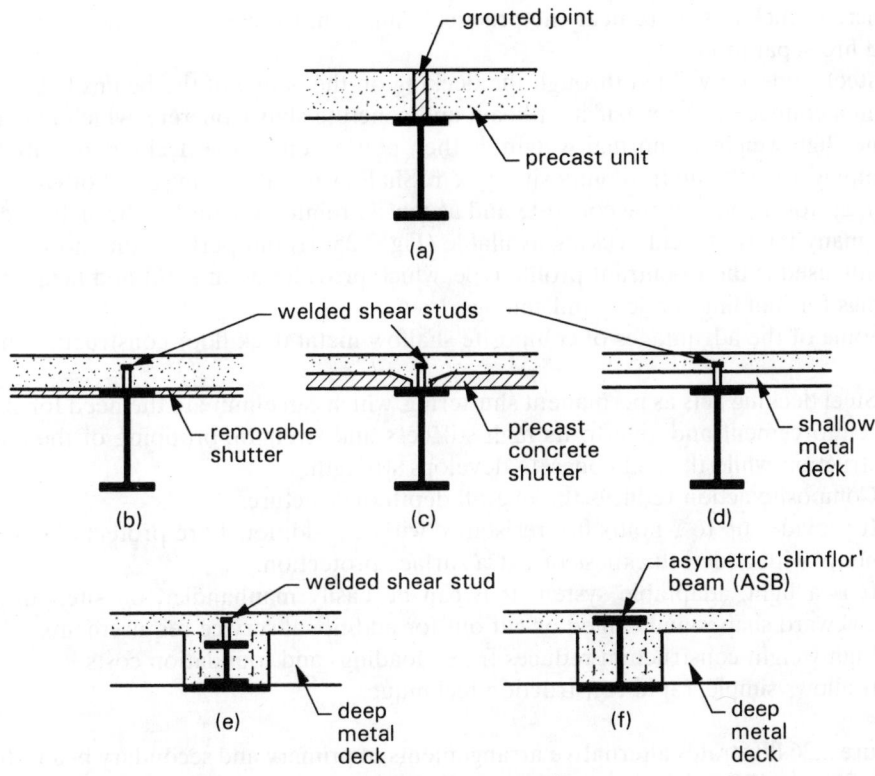

Fig. 2.23   Floor construction: (a) precast (non-composite); (b) in situ (composite); (c) in situ/precast (composite); (d) in situ/shallow metal decking (composite); (e) Slim-floor – in situ/deep metal decking (composite); (f) Slimdek® – in situ/deep metal decking (composite)

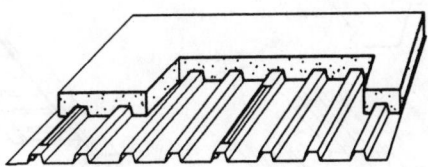

Fig. 2.24   Metal deck floor slabs

Shallow metal deck floor construction

Experience has shown that the most efficient floor arrangements are those using shallow metal decking spanning about 3–4.5 m between floor beams. For these spans the metal decking does not normally require propping during concreting and the

concrete thicknesses are near the practical minimum for consideration of strength and fire separation.

Steel studs are welded through the decking on the flange of the beams below to form a connection between steel beam and concrete slab. Concrete, which may be either lightweight or normal weight, is then poured on to the decking, usually by pumping, to make up the composite system. Shallow metal decking acts both as permanent formwork for the concrete and as tensile reinforcement for the slab. There are many types of steel decking available (Fig. 2.25(a)) but perhaps the most commonly used is the re-entrant profile type, which provides a flat soffit and facilitates fixings for building services and ceilings.

Some of the advantages of composite shallow metal deck floor construction are:

- Steel decking acts as permanent shuttering, which can eliminate the need for slab reinforcement and, due to its high stiffness and strength, propping of the construction while the wet concrete develops strength.
- Composite action reduces the overall depth of structure.
- It provides up to 2 hours fire resistance without additional fire protection·and 4 hours with added thickness or extra surface protection.
- It is a light, adaptable system that can be easily manhandled on site, cut to awkward shapes and drilled or cut out for additional service requirements.
- Lightweight construction reduces frame loadings and foundation costs.
- It allows simple, rapid construction techniques.

Figure 2.26 illustrates alternative arrangements of primary and secondary beams for a deck span of 3 m.

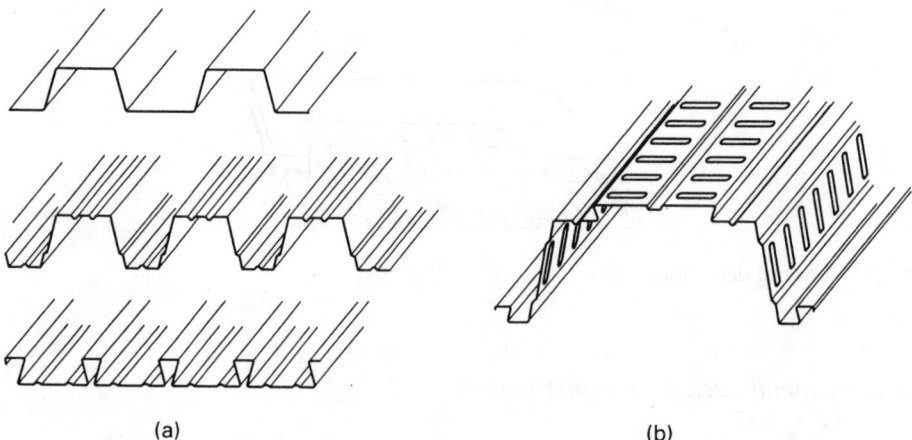

(a)   (b)

Fig. 2.25   Metal deck profiles: (a) shallow deck (50–100 mm); (b) deep deck (150–250 mm)

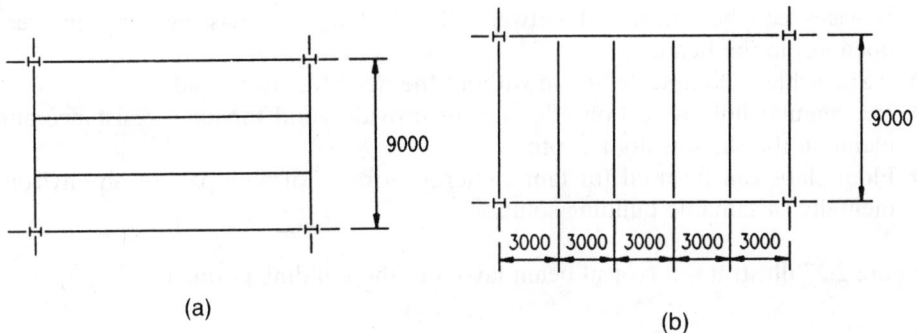

Fig. 2.26 Alternative framing systems for floors: (a) long span secondary beams; (b) long span primary beams

Deep metal deck, shallow floor construction

Deep metal decks are normally used to create shallow floors e.g. Slimflor® and Slimdek® construction. The deep metal deck extends the span capability up to 9 m; however, the deck and/or support beams may require propping during concreting. An additional tensile reinforcement bar is provided within the ribs of the deep decking to improve the load carrying capacity and fire resistance of the floor slab. Although there are several variants internationally, Slimflor® construction in the UK comprises a universal column with a plate welded to the underside supporting the deep metal decking. Shear studs are shop-welded to the top flange of the beam to form a connection between the steel beam and concrete slab. Slimdek® construction is a technologically advanced solution comprising a rolled asymmetric beam (ASB) which supports the deep metal decking directly. Shear connection is achieved through the bond developed between the steel beam and concrete encasure. These features reduce material and fabrication content. Partial concrete encasement of the steel beam provides up to 1 hour inherent fire resistance.

The range of deep metal deck profiles (Fig. 2.25(b)) is more limited than for shallow decks, and those available carry similar attributes and advantages. Some additional advantages of Slimflor® and Slimdek® construction are:

- The shallow composite slab achieves excellent load capacity, diaphragm action and robustness
- There are fewer frame components to erect, saving construction time
- The shallow floor construction allows more floors for a given building height or reduces the cost of cladding and vertical services, lift shafts, etc.
- It provides up to 1 hour inherent fire resistance and up to 2 hours using passive fire protection to the beam soffit only
- Large openings for vertical services can be formed in the floor slab without the need for secondary framing

- Services can be integrated between the decking ribs passing through web-openings in the beam
- ASBs achieve composite action without the need for shear studs
- Rectangular hollow section edge beams provide good torsional resistance and maintain the shallow floor depth
- Floor slabs can be used for fabric energy storage forming part of an environmentally sustainable building solution.

Figure 2.27 illustrates a typical beam layout at the building perimeter.

Precast floor systems

Universal beams supporting precast prestressed floor units (Fig. 2.28) have some advantages over other forms of construction. Although of heavier construction than comparable composite metal deck floors, this system offers the following advantages.

- Fewer floor beams since precast floor units can span up to 6–8 m without difficulty.
- No propping is required.
- Shallow floor construction can be obtained by supporting precast floor units on shelf angles or on wide plates attached to the bottom flanges of universal columns acting as beams (Slimfloor).
- Fast construction because no time is needed for curing and the development of concrete strength.

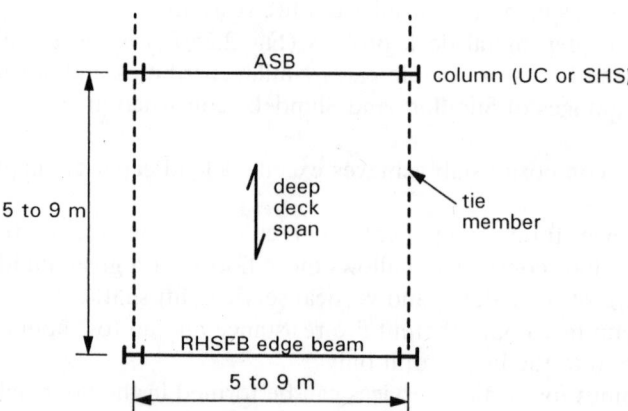

Fig. 2.27   Slimdek® floor arrangement

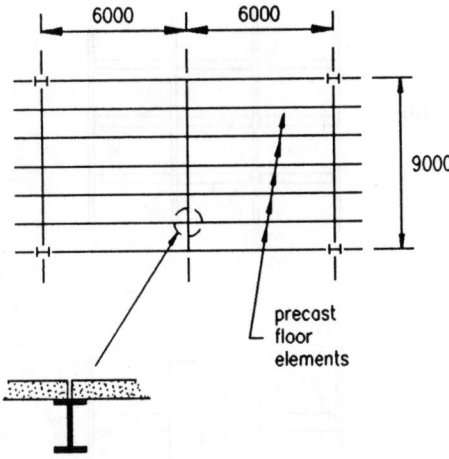

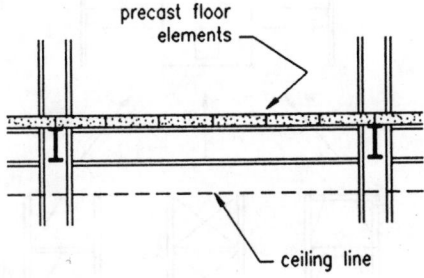

Fig. 2.28   Precast concrete floors

On the other hand the disadvantages are:

- Composite and diaphragm action is not readily achieved without a structural floor screed.
- Heavy floor units are difficult to erect in many locations and require the use of a tower crane, which may have implications for the construction programme.

2.3.4 Bracings

Three structural systems are used to resist lateral loads: continuous or wind-moment frames, reinforced concrete walls and braced-bay frames (Fig. 2.29). Combinations of these systems may also be used.

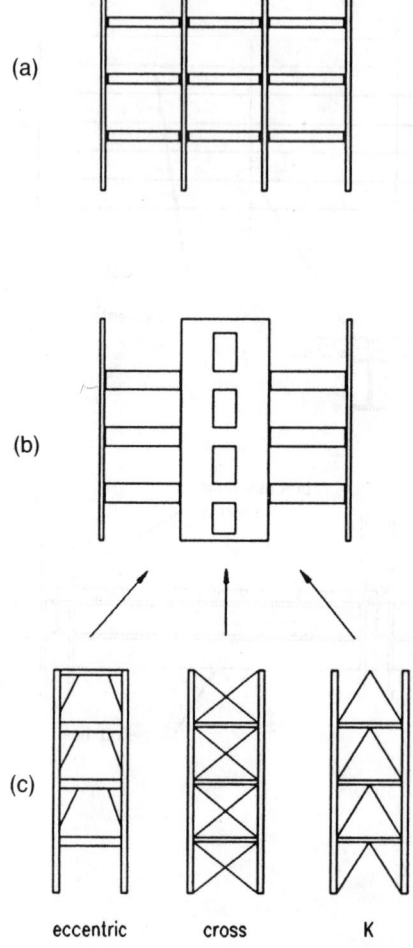

Fig. 2.29 Bracing structures: (a) continuous frame; (b) reinforced concrete wall; (c) braced bay frames

Continuous construction

Continuous frames are those with rigid moment-resisting connections between beam and columns. It is not necessary that all connections in a building are detailed in this way: only sufficient frames to satisfy the performance requirements of the building.

The advantage of a continuous frame is:

- Provides total internal adaptability with no bracings between columns or walls to obstruct circulation.

However, the disadvantages are:

- Increased fabrication for complex framing connections
- Increased site connection work, particularly if connections are welded
- Columns are larger to resist bending moments
- Generally, less stiff than other bracing systems.

Wind-moment frames are limited in application.

Shear walls

Reinforced concrete walls constructed to enclose lift, stair and service cores generally possess sufficient strength and stiffness to resist the lateral loading.

Cores should be located to avoid eccentricity between the line of action of the lateral load and the centre of stiffness of the core arrangement. However, the core locations are not always ideal because they may be irregularly shaped, located at one end of the building or are too small. In these circumstances, additional braced bays or continuous frames should be provided at other locations (Fig. 2.30).

Although shear walls have traditionally been constructed in in situ reinforced concrete they may also be constructed of either precast concrete or brickwork.

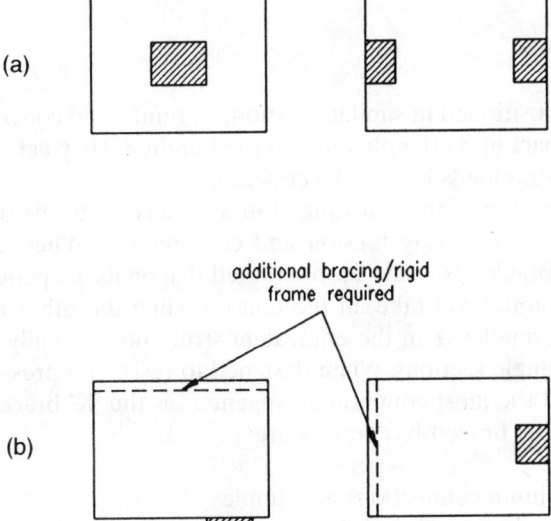

Fig. 2.30 Core locations: (a) efficient; (b) inefficient

The advantages of shear walls are:

- The beam-to-column connections throughout the frame are simple, easily fabricated and rapidly erected.
- Shear walls tend to be thinner than other bracing systems and hence save space in congested areas such as service and lift cores.
- They are very rigid and highly effective.
- They act as fire compartment walls.

The disadvantages are:

- The construction of walls, particularly in low- and medium-rise buildings, is slow and less accurate than steelwork.
- The walls are difficult to modify if alterations to the building are required in the future.
- They are a separate form of construction, which is likely to delay the contract programme.
- It is difficult to provide connections between steel and concrete to transfer the large forces generated.

Recent developments in steel–concrete–steel composite sandwich construction (Bi-steel®) largely eliminate these disadvantages and allow pre-fabricated and fully assembled lift shafts to be erected simultaneously with the main steel framing. Steel–concrete–steel construction can also be used for blast-resistant walls and floors.

Braced-bay frames

Braced-bays are positioned in similar locations to reinforced concrete walls, so they have minimal impact upon the planning of the building. They act as vertical trusses which resist the wind loads by cantilever action.

The bracing members can be arranged in a variety of forms designed to carry solely tension or alternatively tension and compression. When designed to take tension only, the bracing is made up of crossed diagonals. Depending on the wind direction, one diagonal will take all the tension while the other remains inactive. Tensile bracing is smaller than the equivalent strut and is usually made up of flat-plate, channel or angle sections. When designed to resist compression, the bracings become struts and the most common arrangement is the 'K' brace.

The advantages of braced-bay frames are:

- All beam-to-column connections are simple
- The braced bays are concentrated in location on plan
- The bracing configurations may be adjusted to suit planning requirements (eccentric bracing)

- The system is adjustable if building modifications are required in the future
- Bracing can be arranged to accommodate doors and openings for services
- Bracing members can be concealed in partition walls
- They provide an efficient bracing system.

A disadvantage is:

- Diagonal members with fire proofing can take up considerable space.

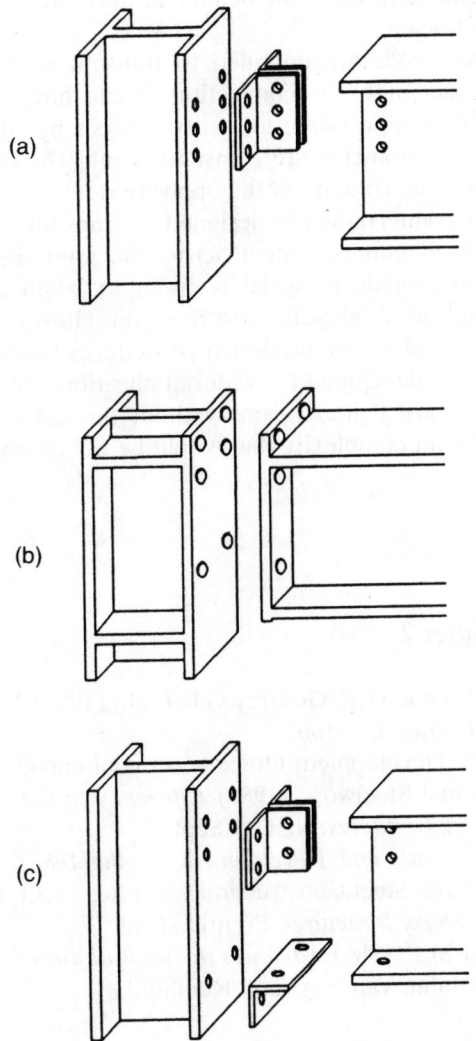

Fig. 2.31   Connections: (a) simple; (b) continuous; (c) semi-continuous

2.3.5 Connections

The most important aspect of structural steelwork for buildings is the design of the connections between individual frame components.

The selection of a component should be governed not only by its capability to support the applied load, but also by its ease of connection to other components.

Basically there are three types of connection, each defined by its structural behaviour: simple, continuous and semi-continuous (Fig. 2.31):

(1) Simple connections transmit negligible bending moment across the joint: the connection is detailed to allow the beam end to rotate. The beam behaves as a simply supported beam.
(2) Continuous connections are designed to transmit shear force and bending moment across the joint. The connection should have sufficient stiffness or moment capacity as appropriate to justify analysis by either elastic or plastic analysis. Beam end moments are transmitted into the column itself and any beam framing into the column on the opposite side.
(3) Semi-continuous connections are designed to transmit the shear force and a proportion of the bending moment across the joint. The principle of these connections is to provide a partial restraint to beam end-rotation without introducing complicated fabrication to the joint. However, the design of such joints is complex, and so simple design procedures based upon experimental evidence have been developed for wider application. The advantages of semi-continuous design are lighter beams without the corresponding increase in column size and joint complexity that would be the case with fully continuous connections.

References to Chapter 2

1. Hart F., Henn W., Sontag H. & Godfrey G.B. (Ed.) (1985) *Multi-Storey Buildings in Steel*, 2nd edn. Collins, London.
2. National Economic Development Office and Economic Development Committee for Constructional Steelwork (1985) *Efficiency in the Construction of Steel Framed Multi-Storey Buildings*. NEDO, Sept.
3. Owens G. (1987) *Trends and Developments in the Use of Structural Steel for Multi-Storey Buildings.* Steel Construction Institute, Ascot, Berks.
4. McGuire W. (1968) *Steel Structures.* Prentice-Hall.
5. Zunz G.J. & Glover M.J. (1986) *Advances in Tall Buildings.* Council on Tall Buildings and Urban Habitat. Van Nostrand Reinhold.

Further reading for Chapter 2

Brett P. & Rushton J. (1990) *Parallel beam approach – a design guide*. The Steel Construction Institute, Ascot, Berks.

Couchman G.H. (1997) *Design of semi-continuous braced frames*. The Steel Construction Institute, Ascot, Berks.

Hensman J.S. & Way A.G.J. (2000) *Wind-moment design of unbraced composite frames*. The Steel Construction Institute, Ascot, Berks.

Lawson R.M. & Rackham J.W. (1989) *Design of haunched composite beams in buildings*. The Steel Construction Institute, Ascot, Berks.

Lawson R.M. & McConnel R. (1993) *Design of stub girders*. The Steel Construction Institute, Ascot, Berks.

Lawson R.M., Mullett D.L. & Rackham J.W. (1997) *Design of asymmetric Slimflor beams using deep composite decking*. The Steel Construction Institute, Ascot, Berks.

Lawson *et al.* (2002) *Design of FABSEC Beams in Non-Composite Applications (Including Fire)*. The Steel Construction Institute, Ascot, Berks.

Mullett D.L. (1992) *Slim floor design and construction*. The Steel Construction Institute, Ascot, Berks.

Mullett D.L. (1998) *Composite Floor Systems*. Blackwell Science.

Mullett D.L. & Lawson R.M. (1999) *Design of Slimflor fabricated beams using deep composite decking*. The Steel Construction Institute, Ascot, Berks.

Narayanan R., Roberts T.M. & Naji F.J. (1994). *Design guide for steel–concrete–steel sandwich construction – Volume 1: General principles and rules for basic elements*. The Steel Construction Institute, Ascot, Berks.

Owens G.W. (1989) *Design of fabricated composite beams in buildings*. The Steel Construction Institute, Ascot, Berks.

Salter P.R., Couchman G.H. & Anderson D. (1999) *Wind-moment design of low rise frames*. The Steel Construction Institute, Ascot, Berks.

Skidmore, Owings & Merrill (1992) *Design of composite trusses*. The Steel Construction Institute, Ascot, Berks.

Slimdek® Manual (2001) Corus Construction Centre, Scunthorpe, Lincs.

Ward J.K. (1990) *Design of composite and non-composite cellular beams*. The Steel Construction Institute, Ascot, Berks.

Yandzio E. & Gough M. (1999) *Protection of buildings against explosions*. The Steel Construction Institute, Ascot, Berks.

A worked example follows which is relevant to Chapter 2.

Corus Construction Centre www.corusconstruction.com		Subject **MULTI-STOREY DESIGN EXAMPLE**	Chapter ref. **2**
		Design code	Made by **PEP**	Sheet no. **1**
		BS 5950: Part 1: 2000	Checked by **TRM**	

Building geometry

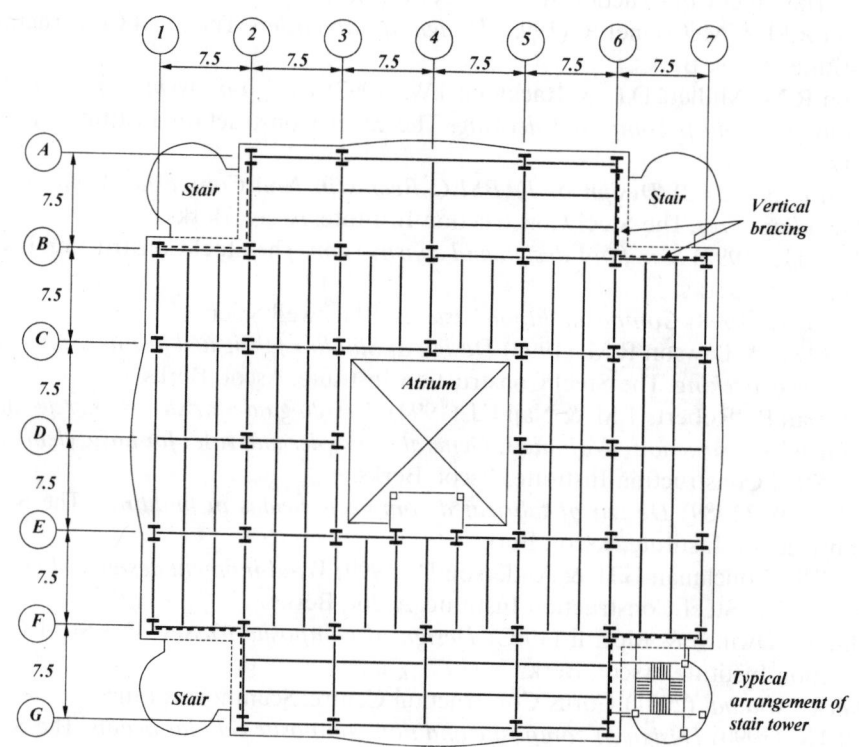

Typical Floor Plan

Building use: Office building with basement car parking and high-level plant room. Imposed loading for office floors exceeds minimum statutory loading given in BS 6399-1: 1996 at client's request. This design example illustrates the design of elements in the braced towers provided in four corners of the building to achieve lateral stability. The floor plate is generally 130 mm lightweight aggregate concrete on metal decking that acts compositely with the decking and floor beams and is assumed to provide diaphragm action. Fire protection is achieved with a sprayed intumescent coating. Alternatively, fire protection could be removed from a number of beams by adopting the fire safe design approach outlined in Chapter 34.

Corus Construction Centre www.corusconstruction.com	corus	Subject **MULTI-STOREY DESIGN EXAMPLE**		Chapter ref. **2**
		Design code **BS 5950: Part 1: 2000**	Made by **PEP** Checked by **TRM**	Sheet no. **2**

kN/m^2

Roof loading

Dead	Steelwork + metal deck	0.27	
	Concrete (130 mm lw)	1.80	
	Finishes	2.00	
	Services below	0.30	
		4.37	

Imposed	From BS 6399-3: 1988 (inc. Amd. 1–3 1997)	1.50

Plant room/B1 loading

Dead	Steelwork + metal deck	0.49	
	150 mm lw concrete slab	2.15	
	Suspended ceiling	0.20	*(not level B1)*
	Services	0.30	*(not level B1)*
		3.14	*(2.64 – level B1)*

Imposed	Floor	7.50	*(2.50 – B1 level)*

Office loading

Dead	Steelwork + metal deck	0.27	
	130 mm lw concrete slab	1.80	
	Suspended ceiling	0.20	*(not ground)*
	Raised floor	0.30	
	Services	0.30	*(not ground)*
		2.87	*(2.37 – ground)*

Imposed	Floor	4.00
	Partition allowance	1.00

Stair loading

Dead	As floor at level applied	
Imposed		4.00

Corus Construction Centre www.corusconstruction.com	corus	Subject **MULTI-STOREY DESIGN EXAMPLE**		Chapter ref. **2**
		Design code **BS 5950: Part 1: 2000**	Made by **PEP** Checked by **TRM**	Sheet no. **3**

Area supported

Total floor area = 1688 m²
Total stair area = 158 m²

	Area supported (m²)		
	Col. G6	*Col. F6*	*Col. F7*
Floor	*19.0*	*46.6*	*19.0*
Stairs	*0.8*	*1.6*	*0.8*

* The above areas apply to all levels.

Perimeter loads

Dead	*Roof*	*1.5 kN/m*
	Plant room	*18.8 kN/m*
	General	*9.6 kN/m (not ground, B1 or B2)*

Length	*External*	*120 + (3 × 4) = 132.0 m*
	Atrium	*14.4 × 4 = 57.6 m*
	Total	*189.6 m*

Length supported by column	*1*	*=*	*4.8 m*	*(G6)*
	2	*=*	*0*	*(F6)*
	3	*=*	*4.8 m*	*(F7)*

Notes:

The allocation of loads to columns 1, 2 and 3 based on the above data is shown on the following sheets.

Imposed load reductions in accordance with BS 6399-1: 1966 are taken for all floors but exclude the roof.

Corus Construction Centre	corus	Subject **MULTI-STOREY DESIGN EXAMPLE**		Chapter ref. **2**
www.corusconstruction.com		Design code **BS 5950: Part 1: 2000**	Made by **PEP** Checked by **TRM**	Sheet no. **4**

Column dead loads

Level	Unit loads		Column loads			Cumulative column loads		
	Floor	*Perim.*	*Col.G6*	*Col.F6*	*Col.F7*	*Col.G6*	*Col.F6*	*Col.F7*
	kN/m²	*kN/m*	*kN*	*kN*	*kN*	*kN*	*kN*	*kN*
R	4.37	1.5	93.7	210.6 *(2)*	93.7	93.7	210.6	93.7
P	3.14	18.8	*(1)* 152.4	151.3	152.4	246.1	*(3)* 361.9	246.1
"8"	2.87	9.6	102.9	138.3	102.9	349.0	500.2	349.0
"7"	2.87	9.6	102.9	138.3	102.9	451.9	638.5	451.9
"6"	2.87	9.6	102.9	138.3	102.9	554.8	776.8	554.8
"5"	2.87	9.6	102.9	138.3	102.9	657.7	915.1	657.7
"4"	2.87	9.6	102.9	138.3	102.9	760.6	1053.4	760.6
"3"	2.87	9.6	102.9	138.3	102.9	863.5	1191.7	863.5
"2"	2.87	9.6	102.9	138.3	102.9	966.4	1330.0	966.4
"1"	2.87	9.6	102.9	138.3	102.9	1069.3	1468.3	1069.3
G	2.37		46.9	114.2	46.9	1116.2	1582.5	1116.2
B1	2.64		52.3	127.2	52.3	1168.5	1709.7	1168.5
B2								

Examples:
(1) $(3.14 \times 19.0) + (3.14 \times 0.8) + (18.8 \times 4.8) = 152.4\,kN$
(2) $(4.37 \times 46.6) + (4.37 \times 1.6)$ $= 210.6\,kN$
(3) $210.6 + 151.3$ $= 362.9\,kN$

Corus Construction Centre www.corusconstruction.com	corus	Subject **MULTI-STOREY DESIGN EXAMPLE**	Chapter ref. **2**
		Design code **BS 5950: Part 1: 2000**	Made by **PEP** Checked by **TRM**	Sheet no. **5**

Column imposed loads

Level	Imposed load kN/m²	Imposed load reduction %	Column loads			*Cumulative column loads		
			Col.G6 kN	Col.F6 kN	Col.F7 kN	Col.G6 kN	Col.F6 kN	Col.F7 kN
R	1.5	0	31.7	76.3	31.7	31.7	76.3	31.7
8								
P	7.5	0	(1) 145.7	(2) 355.9	145.7	177.4	432.2	177.4
"8" 4	5	10	98.2	239.4	98.2	251.2	612.1	251.2
"7" 4	5	20	98.2	239.4	98.2	305.4	744.1	305.4
"6" 4	5	30	98.2	239.4	98.2	339.9	(3) 828.2	339.9
"5" 4	5	40	98.2	239.4	98.2	354.8	864.4	354.8
"4" 4	5	40	98.2	239.4	98.2	413.7	1008.0	413.7
"3" 4	5	40	98.2	239.4	98.2	472.6	1151.7	472.6
"2" 4	5	40	98.2	239.4	98.2	531.6	1295.3	531.6
"1" 4	5	40	98.2	239.4	98.2	590.5	1439.0	590.5
6								
G 4	5	40	98.2	239.4	98.2	649.4	1582.6	649.4
B1 4	2.5	50	50.7	122.9	50.7	571.8	1393.0	571.8
B2								

Values include % imposed load reductions

Examples:
(1) (7.5 × 19.0) + (4.0 × 0.8)   = 145.7 kN
(2) (7.5 × 46.6) + (4.0 × 1.6)   = 355.9 kN
(3) 76.3 + [(355.9 + 239.4 + 239.4 + 239.4) × (1 − 30/100)] = 828.2 kN

Corus Construction Centre www.corusconstruction.com	**corus**	Subject *MULTI-STOREY DESIGN EXAMPLE*		Chapter ref. *2*
		Design code *BS 5950: Part 1: 2000*	Made by *PEP* Checked by *TRM*	Sheet no. *6*

Wind load

BS 6399-2: 1977 (inc. Amd. 1 2002)

Using standard method (since H < B division by parts is not applicable and frictional drag is neglected as having little effect):

Basic wind speed V_b = 21 m/s (London)

Building length L = 48 m
Building width W = 48 m
Building wall height H = 46 m
Building reference height H_r = 46 m
Building type factor K_b = 1 (open plan office)
Dynamic augmentation factor C_r = 0.047
Altitude factor S_a = 1.02 (20 m above sea level)
Directional factor S_d = 1.00 (wind direction III – 210 deg)
Seasonal factor S_s = 1.00
Probability factor S_p = 1.00

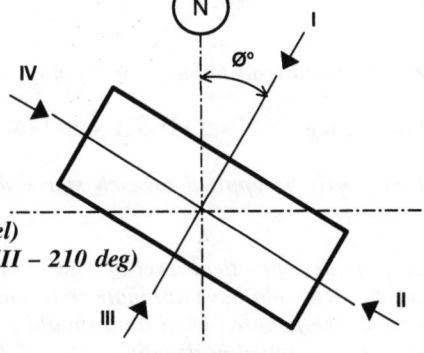

Wind directions

Site wind speed $V_s = V_b S_a S_d S_s S_p$ = 21.42 m/s

Average height of nearby buildings H_o = 20 m
Upwind spacing X_o = 20 m
Displacement height $H_d = 0.8 H_o$ ($X_o \le 2Ho$)
Effective height H_e = greater of $H_r - H_d$ or 0.4 H_r = 30.0 m
Distance to sea = 80 km (upwind @ 210 deg ± 45 deg)
Terrain = town
Terrain & building factor S_b = 1.874   *(Table 4 – by linear interpolation –*
Logarithmic interpolation is optional)

Effective wind speed $V_e = V_s S_b$ = 40.15 m/s

Note: normally, either all wind directions should be checked to establish the highest effective wind speed or a conservative approach may be taken by using a value of S_d = 1.0 together with the shortest distance to sea irrespective of direction. A lower value of S_b will be obtained for sites in town by using the hybrid approach.

Dynamic pressure $q_s = 0.613\ V_e^2$ = 988 N/m²

Breadth B = 48 m
Inwind depth D = 48 m
Ratio B/D = 1
Span ratio = D/H = 1.04 ≥ 1 but ≤ 4   *2.4.1.2*

Corus Construction Centre www.corusconstruction.com	corus	Subject ***MULTI-STOREY DESIGN EXAMPLE***	Chapter ref. *2*
		Design code *BS 5950: Part 1: 2000*	Made by *PEP* Checked by *TRM*	Sheet no. 7

Net pressure coefficient $C_p = 1.195$ *(Table 5a – Linear interpolation)*

Size effect factor for external pressure $C_a = 0.839$ Figure 4:

> *Effective height $H_e = 30.0\,m$*
> *Distance to sea $= 80\,km$*
> *Terrain $= town$*
> *Dimension $a = (48^2 + 46^2)^{\frac{1}{2}} = 66.48\,m$*
> *Site exposure type $= A$*

For net wind load to building $P = 0.85\ (\Sigma q_s C_p C_a A)(1 + C_r)$ Clause 2.1.3.6 NOTE 3

Simplifying $P = \Sigma 881.6\ A \times 10^{-3}\,kN$

Loads will be applied to each storey and transferred to each floor level (see table overleaf).

To provide a practical level of robustness against the effects of incidental loading, all structures should have adequate resistance to horizontal forces. The horizontal component of the factored wind load should not be taken as less than 1.0% of the factored dead load applied horizontally. (cl. 2.4.2.3)

Example to table on page 85

(1) $(4.37 \times 1688) + (4.37 \times 158) + (1.5 \times 189.6) = 8351\,kN$
(2) $8351 \times 1.4 \times 1\%$ $= 117\,kN$

Corus Construction Centre www.corusconstruction.com	corus	Subject MULTI-STOREY DESIGN EXAMPLE		Chapter ref. 2
		Design code BS 5950: Part 1: 2000	Made by *PEP* Checked by *TRM*	Sheet no. 8

Wind and minimum horizontal forces

	h (m)	Area (m^2)	Wind Force P (kN) (Unfactored)	Wind Force F (kN) (Unfactored)	Dead load (kN) (Unfactored)	Min. Force F' (kN) (Factored)
R	46			169.3	8351 (1)	117 (2)
F		384	338.5			
P	38			253.9	9361	131
		192	169.3			
"8"	34			169.3	7118	100
		192	169.3			
"7"	30			169.3	7118	100
		192	169.3			
"6"	26			169.3	7118	100
		192	169.3			
"5"	22			169.3	7118	100
		192	169.3			
"4"	18			169.3	7118	100
		192	169.3			
"3"	14			169.3	7118	100
		192	169.3			
"2"	10			169.3	7118	100
		192	169.3			
"1"	6			211.6	7118	100
		288	253.9			
G				127.0	4375	61
B1					4873	68
B2						
			$\Sigma 1947$	$\Sigma 1947$		

*F is the total force applied at that level. The factored minimum force ex dead load (F')
is clearly less than the value of factored wind load and will be ignored in further
calculations.*

*Loads are divided by 4 in the sub-model analysis to represent the force applied to one
braced tower. To justify this approach the behaviour of the whole building was analysed
using CSC Fastrak® Multi-storey to confirm that the sub-model for the bracing system
was sufficiently accurate. One of the vertical braced towers was then analysed using CSC
S-Frame™.*

N.B. Asymmetric frames may be subject to torsional and out-of-plane P delta effects.

Corus Construction Centre corus www.corusconstruction.com	Subject *MULTI-STOREY DESIGN EXAMPLE*	Chapter ref. *2*
	Design code *BS 5950: Part 1: 2000*	Made by *PEP* Checked by *TRM*	Sheet no. *9*

Notional horizontal forces (NHF)

To allow for the effects of practical imperfections such as lack of verticality, notional horizontal forces are considered.

At each level,

$F_{NHF} = 0.5\%$ *of the factored vertical dead and imposed loads at that level.*

For simplicity, calculations for the NHFs do not include any % imposed load reductions.

Roof   **kN**
D.L. $= 4.37 \times (1688 + 158) + (1.5 \times 189.6)$ $=$ **8351**
L.L. $= (1.5 \times 1688) + (4.0 \times 158)$ $=$ **3164**
NHF $= 0.5/(100 \times 4) \times [(1.4 \times 8351) + (1.6 \times 3164)]$ $=$ **20.9**

Plant room
D.L. $= 3.14 \times (1688 + 158) + (18.8 \times 189.6)$ $=$ **9361**
L.L. $= (7.5 \times 1688) + (4.0 \times 158)$ $=$ **13292**
NHF $= 0.5/(100 \times 4) \times [(1.4 \times 9361) + (1.6 \times 13292)] =$ **43.0**

Office floors *(typ.)*
D.L. $= 2.87 \times (1688 + 158) + (9.6 \times 189.6)$ $=$ **7118**
L.L. $= (5 \times 1688) + (4.0 \times 158)$ $=$ **9072**
NHF $= 0.5/(100 \times 4) \times [(1.4 \times 7118) + (1.6 \times 9072)]$ $=$ **30.6**

Ground
D.L. $= 2.37 \times (1688 + 158)$ $=$ **4375**
L.L. $= (5 \times 1688) + (4.0 \times 158)$ $=$ **9072**
NHF $= 0.5/(100 \times 4) \times [(1.4 \times 4375) + (1.6 \times 9072)]$ $=$ **25.8**

Basement 1
D.L. $= 2.64 \times (1688 + 158)$ $=$ **4873**
L.L. $= (2.5 \times 1688) + (4.0 \times 158)$ $=$ **4852**
NHF $= 0.5/(100 \times 4) \times [(1.4 \times 4873) + (1.6 \times 4852)]$ $=$ **18.2**

Corus Construction Centre	Subject		Chapter ref.
www.corusconstruction.com	corus	*MULTI-STOREY DESIGN EXAMPLE*	*2*
	Design code	Made by *PEP*	Sheet no. *10*
	BS 5950: Part 1: 2000	Checked by *TRM*	

Summary of NHF loads

Level	NHF kN
Roof	20.9
Plant	43.0
8	30.6
7	30.6
6	30.6
5	30.6
4	30.6
3	30.6
2	30.6
1	30.6
Ground	25.8
Base 1	18.2
Base 2	

The NHF load case is considered as acting simultaneously with dead and imposed loads when no wind forces are acting.

N.B. The NHFs should be applied in the two orthogonal directions (global X and Y) separately and may need to be applied in the positive and negative direction to give the worst effect

Sway stiffness P–δ effects

All structures (including portions between expansion joints) should have sufficient sway stiffness, so that the vertical loads acting with the lateral displacements of the structure do not induce excessive secondary forces or moments in the members or connections. Where such 'second order' (P–δ) effects are significant, they should be allowed for in the design of those parts of the structure that contribute to its resistance to horizontal forces.

2.4.2.5

Sway stiffness should be provided by the system of resisting horizontal forces. Whatever system is used, sufficient stiffness should be provided to limit sway deformation in any horizontal direction and also to limit twisting of the structure on plan.

Except for single-storey frames with moment-resisting joints, or other frames in which sloping members have moment-resisting connections, λ_{cr} should be taken as the smallest value, considering every storey.

2.4.2.6

In the following table, the elastic critical buckling load factor λ_{cr} has been determined from:

$$\lambda_{cr} = h/(200 \times \delta)$$

where h = *storey height and*
δ = *storey drift due to NHFs*

Corus Construction Centre www.corusconstruction.com	**corus**	Subject **MULTI-STOREY DESIGN EXAMPLE**		Chapter ref. **2**
		Design code BS 5950: Part 1: 2000	Made by **PEP** Checked by **TRM**	Sheet no. **11**

Because of symmetry one of the four braced towers only was modelled as a space frame analysed using CSC S-Frame™.

Basic load cases

1. **Dead load**
2. **Reduced imposed loads**
3. **Wind loading (Y direction)**
4. **NHF (Y direction)**

Results from NHF load case

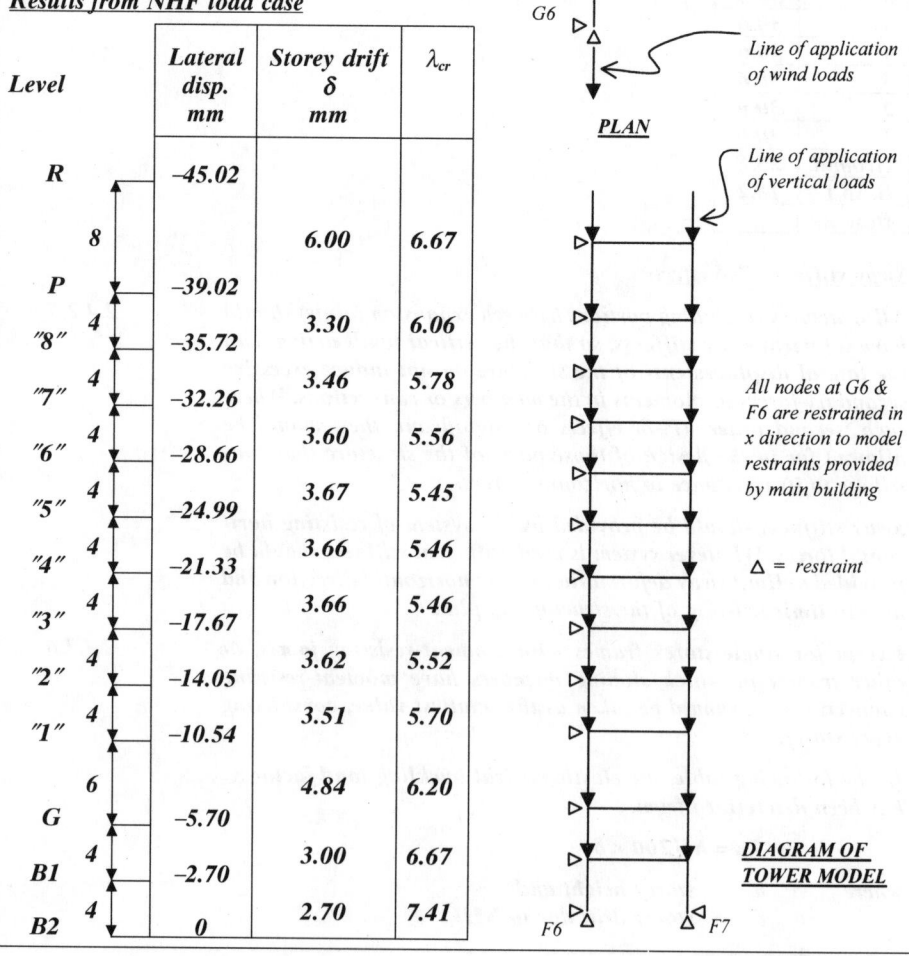

Level	Lateral disp. mm	Storey drift δ mm	λ_{cr}
R	–45.02		
8		6.00	6.67
P	–39.02		
"8"	–35.72	3.30	6.06
"7"	–32.26	3.46	5.78
"6"	–28.66	3.60	5.56
"5"	–24.99	3.67	5.45
"4"	–21.33	3.66	5.46
"3"	–17.67	3.66	5.46
"2"	–14.05	3.62	5.52
"1"	–10.54	3.51	5.70
		4.84	6.20
G	–5.70		
B1	–2.70	3.00	6.67
B2	0	2.70	7.41

Line of application of wind loads

PLAN

Line of application of vertical loads

All nodes at G6 & F6 are restrained in x direction to model restraints provided by main building

Δ = *restraint*

DIAGRAM OF TOWER MODEL

Corus Construction Centre www.corusconstruction.com	Subject **MULTI-STOREY DESIGN EXAMPLE**		Chapter ref. **2**
	Design code *BS 5950: Part 1: 2000*	Made by *PEP* Checked by *TRM*	Sheet no. **12**

The smallest value of λ_{cr} is 5.45, which is < 10. Therefore the frame   **2.4.2.7**
*is sway sensitive, and as λ_{cr} is not less than 4.0, the amplified sway
method may be used.*

This gives an amplification factor for NHFs in the Y direction of:

$$k_{amp} = \lambda_{cr} / (1.15\,\lambda_{cr} - 1.5) = 1.14$$

*(for clad structures, provided that the stiffening effect of masonry
infill wall panels or diaphragms of profiled steel sheeting is not
explicitly taken into account.)*

<u>Cautionary notes to analysis approach</u>

1. *Results are for sway only in the Y direction using NHF in the
   Y direction from the gravity combination. This analysis must
   normally be repeated for sway in the X direction using NHF in
   the X direction. The choice is then to either:*
   *(a) use the smallest value of λ_{cr} to determine one value of k_{amp}
       applied to loads in both directions, or*
   (b) calculate a λ_{cr} and hence k_{amp} for each direction.

   *Similarly a k_{amp} could be calculated for each load combination
   and applied to each appropriate combination.*

2. *The results from a simple tower constrained out-of-plane at all
   levels take no account of out-of-plane deflections and their
   influence on P–δ effects. An analysis of the whole building is
   required to spot this effect and determine whether it is significant. In this example the tower constrained out of plane is only
   a reasonable assumption owing to the very symmetric nature of
   the building as a whole. If the whole building is considered then
   there could be other columns (even simple ones not in braced
   bays) that lean over more. These will have a detrimental effect
   on the braced areas. That is, whilst not affected themselves
   (being simple columns), they will increase the P–δ effect on the
   braced towers. Taking the worst sway index for all columns also
   helps to compensate for any out-of-plane deflections.*

Corus Construction Centre corus www.corusconstruction.com	Subject **MULTI-STOREY DESIGN EXAMPLE**	Chapter ref. *2*
Design code **BS 5950: Part 1: 2000**	Made by **PEP** Checked by **TRM**	Sheet no. *13*

3. The horizontal members of the tower frame are constrained by the floor diaphragm in the whole building model. There are two effects.

   (a) The tower model has to be given horizontal members that are significantly stiffer than the beam stiffness alone. In the model analysed the 406 × 178 × 74 UB was retained but given an E value of 1000 times greater than that for steel.

   (b) In the whole building model the diaphragm transmits the horizontal loads to the tower and so there is no axial load present in the horizontal beams.

<u>Load case combinations</u>   *2.4.2.4*

The notional horizontal forces should be taken as acting simultaneously with the factored vertical dead and imposed loads but should not be combined with applied horizontal loads.

The load combinations are therefore:

$1.4 \times Dead + 1.6 \times Imposed$
$1.2 \times (Dead + Imposed \pm k_{amp} \times Wind)$
$1.4 \times Dead \pm 1.4 \times k_{amp} \times Wind$
$1.0 \times Dead \pm 1.4 \times k_{amp} \times Wind$
$1.4 \times Dead + 1.6 \times Imposed \pm k_{amp} \times NHF\ (Y\text{-}direction)$

<u>Proposed section sizes</u>

The columns have been spliced at several locations to ensure practical changes in section size and allow transport. The sections have been chosen to have sufficient stiffness to give reasonable sway behaviour. This means that the sections are over-designed for strength and buckling – see Sheet 15.

Stacks 1–3, B2 – Floor 1	*356 × 406 × 340 UC*
Stacks 4–6, Floor 1 – Floor 4	*356 × 406 × 235 UC*
Stacks 7–9, Floor 4 – Floor 7	*356 × 368 × 129 UC*
Stacks 10–12, Floor 7 – Roof	*305 × 305 × 97 UC*

Braces are HF 250 × 250 × 8 SHS and floor beams are 406 × 178 × 74 UB (see cautionary note 3(a) above).

Deflections under wind loading are given in the table below.

Corus Construction Centre www.corusconstruction.com	corus	Subject **MULTI-STOREY DESIGN EXAMPLE**		Chapter ref. **2**
		Design code *BS 5950: Part 1: 2000*	Made by **PEP** Checked by **TRM**	Sheet no. **14**

Lateral deflections due to wind load

Level		Lateral disp. (mm)	Storey drift δ (mm)	Height $\dfrac{\delta}{\delta}$
R		–67.5		
	8		9.4	851
P		–58.1		
	4		5.1	784
"8"		–53.0		
	4		5.3	755
"7"		–47.7		
	4		5.5	727
"6"		–42.2		
	4		5.5	727
"5"		–36.7		
	4		5.5	727
"4"		–31.2		
	4		5.4	741
"3"		–25.8		
	4		5.4	741
"2"		–20.4		
	4		5.1	784
"1"		–15.3		
	6		7.2	833
G		–8.1		
	4		4.3	930
B1		–3.8		
	4		3.8	1053
B2		0.0		

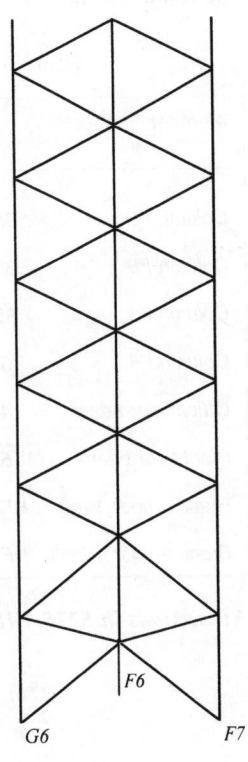

Maximum differential storey deflection due to wind loads is h/727 < h/300

Therefore, deflection is acceptable.

Corus Construction Centre	corus	Subject	Chapter ref.
www.corusconstruction.com		*MULTI-STOREY DESIGN EXAMPLE*	*2*

Design code	Made by *PEP*	Sheet no. *15*
BS 5950: Part 1: 2000	Checked by *TRM*	

Confirmation of member sizes

The columns can be treated as simple even though they are part of braced bays. The moments due to continuity of the columns are small and are ignored. The eccentricity moments due to the end reactions from the incoming beams are considered. As simple columns they are designed to Clause 4.7.7 of BS 5950: Part 1: 2000. The beams are designed as simply-supported with no axial load and the braces are designed for axial load only.

The results for the most critical members in the braced tower are given below.

Member	Section size	Critical position	Critical combination	Utilization ratio
Column F6	356 406 340 UC	Below Floor 1	$1.4 D + 1.6 I + NHF Y$	0.65
Column F6	356 406 235 UC	Above Floor 1	$1.4 D + 1.6 I + NHF Y$	0.67
Column F6	356 368 129 UC	Below Floor 5	$1.4 D + 1.6 I + NHF Y$	0.81
Column F6	305 305 97 UC	Below Floor 8	$1.4 D + 1.6 I + NHF Y$	0.67
Office floor beam	406 178 74 UB	Mid-span	$1.4 D + 1.6 I$	0.85
Plant floor beam	533 210 82 UB	Mid-span	$1.4 D + 1.6 I$	0.90
Brace – floor 1 to G	HF250 250 8 SHS	Compression	$1.4 k_{amp} Wind$	0.74
Brace – floor 1 to G	HF250 250 8 SHS	Tension	$1.4 k_{amp} Wind$	0.44

All sections in S275. SHS are hot-finished to BS EN 10210.

Corus Construction Centre www.corusconstruction.com corus	Subject ***MULTI-STOREY DESIGN EXAMPLE***		Chapter ref. ***2***
	Design code ***BS 5950: Part 1: 2000***	Made by ***PEP*** Checked by ***TRM***	Sheet no. ***16***

Robustness of frame

To provide a degree of robustness and to reduce the possibility of accidental damage, column ties should be capable of transmitting ***2.4.5.3***

(1) The general tying force given by 0.5 $w_f s_t L_a$ for internal ties or 0.25 $w_f s_t L_a$ for edge ties but not less than 75 kN, where:

w_f = ***factored dead and imposed load per unit area***

s_f = ***mean transverse spacing of ties***

L_a = ***maximum span within the overall length of the tie***

$$\textit{tying force (office)} = 0.5 \times [(2.87 \times 1.4) + (5.0 \times 1.6)] \times 7.5 \times 7.5$$
$$= 338\,kN > 75\,kN$$

or

(2) At the periphery, the general tying force or an anchor force of 1% of the factored vertical load in the column, whichever is the greater.

Max. factored column force (Cols. Cl, C7, E1 and C7) = 6360 kN

Therefore anchor force = 63.60 kN

Design connection for P = ± 338 kN

(NB these forces should not be considered as additive to other loads.)

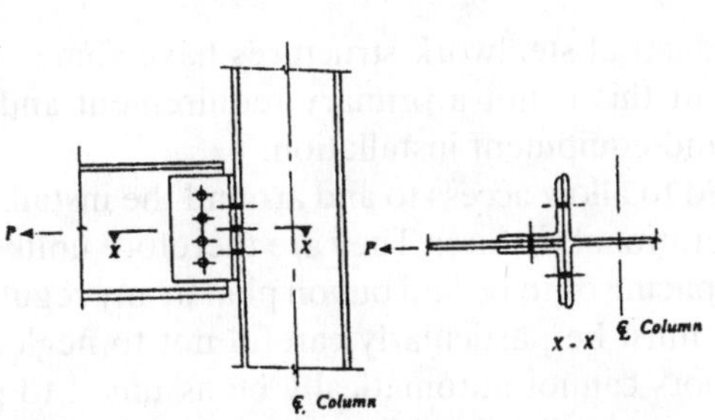

The author wishes to thank Mr Alan Rathbone of CSC (UK) Ltd. for his assistance in the preparation of this worked example.

Chapter 3
Industrial steelwork

by JOHN ROBERTS

3.1 Range of structures and scale of construction

3.1.1 Introduction

Structural steelwork for industrial use is characterized by its function, which is primarily concerned with the support, protection and operation of plant and equipment. In scale it ranges from simple support frameworks for single tanks, motors or similar equipment, to some of the largest integrated steel structures, for example complete electric power-generating facilities.

Whereas conventional single- and multi-storey structures provide environmental protection to space enclosed by walls and roof and, for multi-storey buildings, support of suspended floor areas, these features are never dominant in industrial steelwork. Naturally, in many industrial structures, the steel framework also provides support for wall and roof construction to give weather protection, but where this does occur the wall and roof profiles are designed to fit around and suit the industrial plant and equipment, frequently providing lower or different standards of protection in comparison with conventional structures. Many plant installations are provided only with rain shielding; high levels of insulation are unusual, and some plant and equipment is able to function and operate effectively without any weather protection at all. In such circumstances the requirements for operational and maintenance personnel dictate the provision of cladding, sheeting or decking.

Similarly, most industrial steelwork structures have some areas of conventional floor construction, but this is not a primary requirement and the flooring is incidental to the plant and equipment installation.

Floors are provided to allow access to and around the installation, being arranged to suit particular operational features. They are therefore unlikely to be constructed at constant vertical spacing or to be laid out on plan in any regular repetitive pattern. Steelwork designers must be particularly careful not to neglect the importance of two factors. First, floors cannot automatically be assumed to provide a horizontal wind girder or diaphragm to distribute lateral loadings to vertical-braced or framed bays: openings, missing sections or changes in levels can each destroy this action. Secondly, column design is similarly hampered by the lack of frequent and closely spaced two-directional lateral support commonly available in normal multi-storey structures.

Floors require further consideration regarding the choice of construction (see section 3.2.3) and loading requirements (see section 3.3.1).

3.1.2 Power station structures

Industrial steelwork for electrical generating plants varies considerably depending on the size of station and the fuel being used. These variations are most marked in boiler house structures; whereas coal-fired and oil-fired boilers are similar, nuclear power station boilers (reactors) are generally constructed in concrete for biological shielding purposes, steel being used normally in a secondary building envelope role. Turbine halls are, in principle, largely independent of fuel type, and many of the other plant structures (mechanical annexes, electrical switchgear buildings, coal hoppers, conveyors, pump-houses) are common in style to other industrial uses and so brief descriptions of the salient design features are of general interest.

Boiler houses (coal- or oil-fired) (Fig. 3.1) have to solve one overriding design criterion and as a result can be considered exercises in pure structural design.

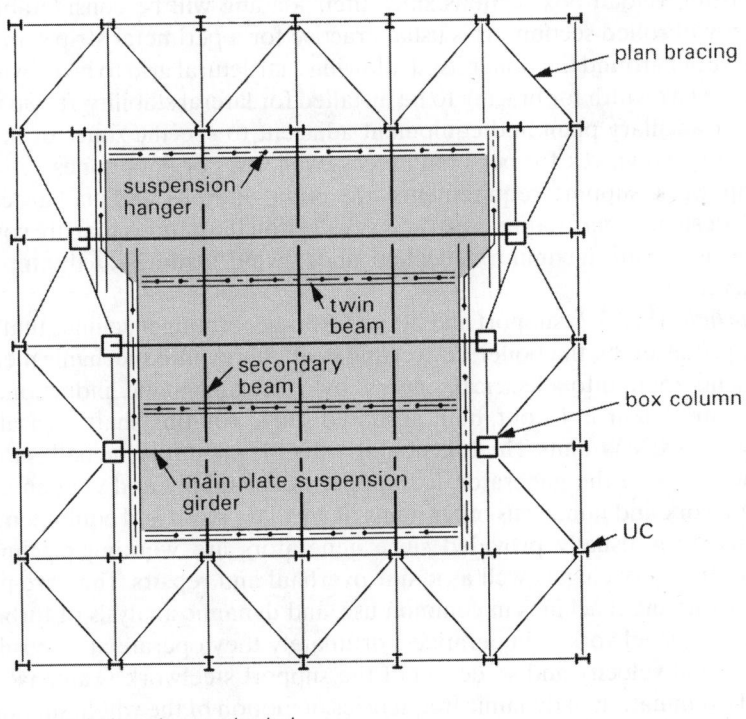

boiler shown shaded
vertical bracing not shown

Fig. 3.1 Plan of boiler house framing

Modern boilers are huge single pieces of plant with typical dimensions of 20 m × 20 m × 60 m high for a single 500 MW coal boiler. Where poor-quality coal is burnt or higher capacities are required, the dimensions can be even larger, up to about 25 m × 25 m × 80 m high for 900 MW size sets. As may be anticipated, the weights are equally massive, typically in the range of 7000–10 000 t for the plant sizes noted above.

Boilers are always *top-suspended* from the supporting structures and not built directly from foundation level upwards, nor carried by a combination of top and bottom support. This is because the thermal expansion of the boiler prevents dual support systems; unsurmountable buckling and stability problems on the thin-walled-tube structure of the casing would arise if the boiler was bottom-supported and hence in compression, rather than top-suspended and hence in tension. For obvious reasons no penetrations of the boiler can be acceptable and therefore no internal columns can be provided. The usual structural system is to provide an extremely deep and stiff system of suspension girders (plate or box) spanning across the boiler with an extensive framework of primary, secondary and (twin) tertiary beams terminating in individual suspension rods or hangers which support the perimeter walls and roof of the boiler itself.

Columns are massively loaded from the highest level and so are usually constructed from welded box-sections since their loading will be considerably above the capacity of rolled sections. It is usual practice for a perimeter strip some 5–10 m wide to be built around the boiler itself allowing a structural grid to be provided with an adequate bay width for bracing to be installed for lateral stability. It also provides support for ancillary plant and equipment adjacent to specific zones of the boiler, for pipework and valves, for personnel access walkways or floor zones.

The pipework support requirements are often onerous and in particular the pipework designers may require restrictive deflection limitations that are sometimes set as low as 50 mm maximum deflection under wind loadings at the top of 90 m high structures.

Turbine halls (Fig. 3.2) support and house turbo-generating machines that operate on steam produced by the boiler, converting heat energy into mechanical energy of rotation and then into electrical energy by electromagnetic induction. Turbo-generators are linear in layout, built around a single rotating shaft, typically some 25 m long for 500 MW units. The function of a steel-framed turbine hall is to protect and allow access to the generator, to support steam supply and condensed water return pipework and numerous other items of ancillary plant and equipment. Heavy crane capacity is usually provided since generators are working machines that require routine servicing as well as major overhaul and repairs. They are probably the largest rotating machines in common use, and dynamic analysis of turbine generator support steelwork is imperative. Fortunately they operate at a sensibly constant rotational velocity and so design of the support steelwork is amenable to an analytical examination of dynamic frequencies of motion of the whole support structure with plant loading in each possible mode, with similar consideration of any local frequency effect, such as vibration of individual elements of the structure or its framework.

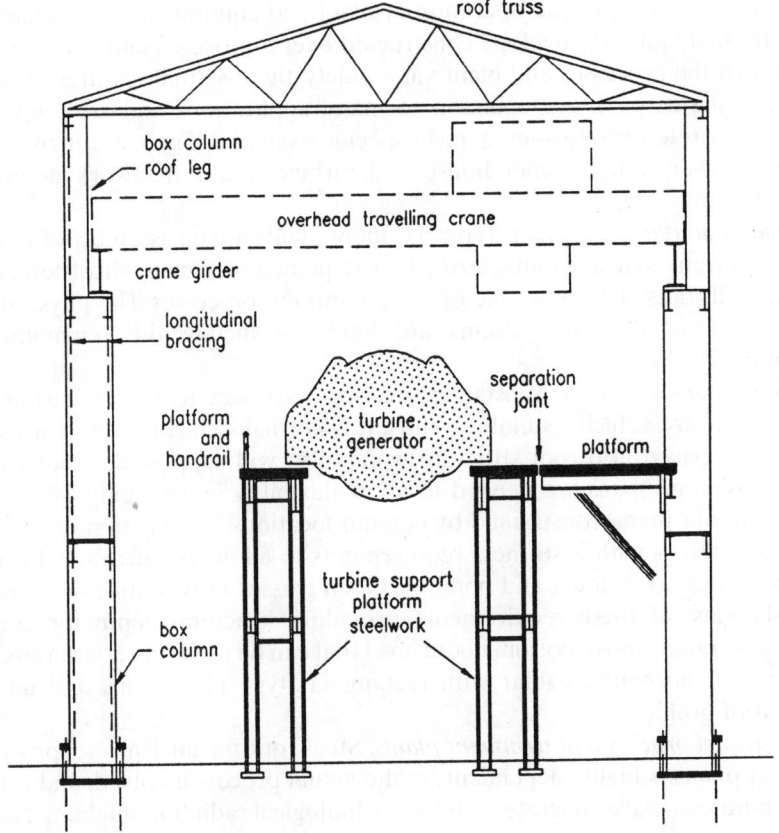

Fig. 3.2   Cross section through turbine hall

These frequencies of response are compared with the forcing frequency of motion of the machine, the design being adjusted so that no response frequencies exist in a band either side of the forcing frequency, to avoid resonance effects.

A complete physical separation is provided between the generator support steelwork and any adjoining main frame, secondary floor or support steel to isolate vibration effects. Zones on adjoining suspended floors are set aside for strip-down and servicing, and specific 'laydown' loading rates are allowed in the design of these areas to cater for heavy point and distributed loads.

3.1.3 Process plant steelwork

Steelwork for process and manufacturing plants varies across a wide spectrum of different industrial uses. Here it is considered to be steelwork that is intimately con-

nected with the support and operation of plant and equipment, rather than simply a steel-framed building envelope constructed over a process plant.

Although the processes and plant vary widely, the essential features of this type of industrial steelwork are common to many applications and are conveniently examined by reference to some typical specific examples. There are many similarities with power station boiler house and turbine house steelwork described in section 3.1.2.

Cement manufacturing plant. Typical cement plants are an assembly of functional structures arranged in a manufacturing flow sequence, with many short-term storage and material transfer facilities incorporated into the processes. The physical height and location of the main drums are likely to dictate the remaining plant orientation.

Vehicle assembly plants. Substantial overhead services to the various assembly lines characterize vehicle assembly plants. It is normal to incorporate a heavy-duty and closely-spaced grid roof structure which also will support the roof covering. Reasonably large spans are needed to allow flexibility in arranging assembly line layouts without being constrained by column locations. Automation of the assembly process brings with it stiffness requirements to allow use of robots for precise operations such as welding and bonding. Open trusses in two directions are likely to satisfy most of these requirements, providing structural depth for deflection control and a zone above bottom boom level that can be used for service runs. Building plans are normally regular with rectangular type plan forms and uniformly regular roof profiles.

Nuclear fuel process and treatment plants. Steelwork for nuclear fuel process and treatment plants is highly dependent on the actual process involved, and often has to incorporate massive concrete sections for biological radiation shielding purposes. Particular points to note are the importance of the paint or other finishes, both from the point of view of restricted access in certain locations, leading to maintenance problems, and also from the necessity for finishes in some areas to be capable of being decontaminated. Specialized advice is needed for the selection of suitable finishes or to give guidance on whether, for example, structural stainless steels would be appropriate. In addition, certain nuclear facilities must be designed for extreme events, the most relevant of which is seismic loading set by the statutory regulatory body. Frequently designs will have to be undertaken to comply with well-established codes of practice for seismically active zones in, for example, America. Seismic design requires the establishment of ductile structures to allow high levels of energy absorption prior to collapse, and local stability and ductile connection behaviour become of critical importance. It follows that joint and connection design has to be fully integrated with the structural steel design generally; the normal responsibilities of joint design assumed by the steelwork fabricator may have to be altered on these projects.

Petrochemical plants. Petrochemical plants tend to be open structures with little or no weather protection. The steelwork required is dedicated to providing support to plant and pipework, and support for access walkways, gangways, stairs and ladders. Plant layouts are relatively static over long periods of use, and the steel-

work is relatively economic in relation to the equipment costs. It is normal therefore to design the steelwork in a layout exactly suited to the plant and equipment without regard for a uniform structural grid. Benefits can still be gained from standardization of sizes or members and from maintaining an orthogonal grid to avoid connection problems. The access floors and interconnecting walkways and stairways must be carefully designed for all-weather access; use of grid flooring is almost universal. Protection systems for the steelwork should acknowledge both the threat from the potentially corrosive local environment due to liquid or gaseous emissions and the requirements to prevent closing down the facility for routine maintenance of the selected protection system.

Careful account needs to be taken of wind loading in terms of the loads that occur on an open structure and the lack of well-defined horizontal diaphragms from conventional floors. These two points are discussed further in sections 3.3.4 and 3.2.2.

3.1.4 Conveyors, handling and stacking plants

Many industrial processes need bulk or continuous handling of materials with a typical sequence as follows:

(1) loading from bulk delivery or direct from mining or quarrying work
(2) transportation from bulk loading area to short-term storage (stacking or holding areas)
(3) reclaim from short-term storage and transport to process plant.

Structural steelwork for the industrial plant which is utilized in these operations is effectively part of a piece of working machinery (Fig. 3.3). Design and construction standards must recognize the dynamic nature of the loadings and particularly must cater for out-of-balance running, overload conditions and plant fault or machinery failure conditions, any of which can cause stresses and deflections significantly higher than those resulting from normal operation. Most designers would adopt slightly lower factors of safety on loading for these conditions, but decisions need to be based on engineering judgement as to the relative frequency and duration of these types of loadings. The plant designer may well be unaware that such design decisions can be made for the steelwork and frequently will provide single maximum loading parameters that could incorporate a combination of all such possible events rather than a separate tabulation; the structural steelwork designer who takes the trouble to understand the operation of the plant can therefore ask for the appropriate information and use it to the best advantage.

A further feature of this type of steelwork is the requirement for frequent relocation of the loading area and hence the transportation equipment. Practical experience suggests that precise advance planning for specific future relocation is rarely feasible, so attention should be directed to both design and detailing so that future moves can cause the least disruption.

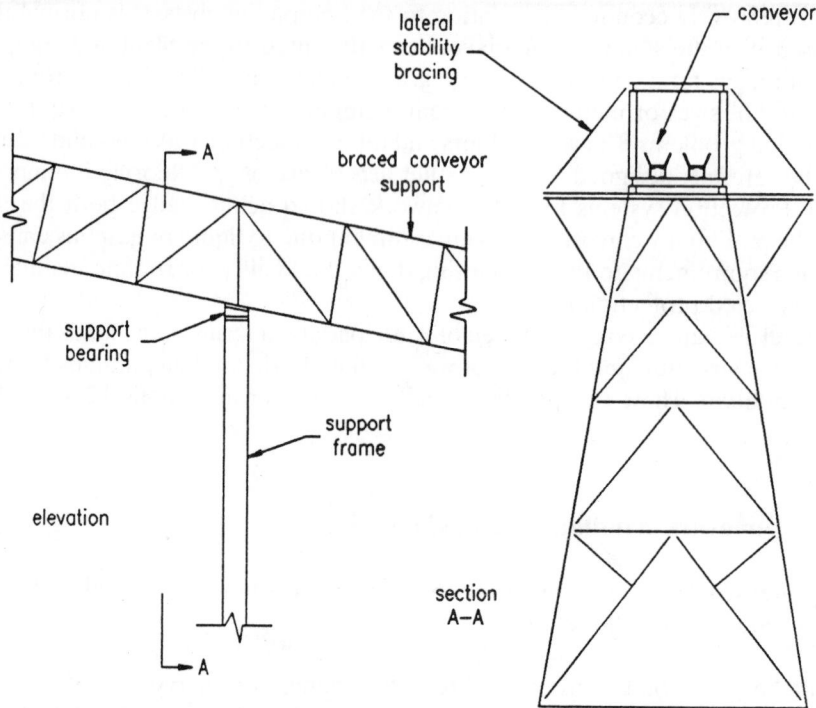

Fig. 3.3   Typical details of conveyor support

Foundation levels can be set at constant heights; or, if variations have to occur, then modular steps above or below a standard height should be adopted. The route should utilize standard plan angles between straight sections, and uniform vertical sloping sections between horizontal runs. Common base plate details and foundation bolt details can be utilized, even where this may be uneconomic for the initial layout installation. Some consideration should be given to allowing the supporting structure to be broken down into conveniently handled sections rather than into individual elements; the break-down joints can be permanently identified, for example, by painting in different colours.

3.1.5 General design requirements

Fatigue loadings must be considered, but even where fatigue turns out not to be a design criterion, it is vital to protect against vibration and consequent loosening of bolted assemblies. In this respect HSFG bolts are commonly used as they dispense with the necessity to provide a lock-nut; when correctly specified and installed they also display good fatigue-resistance.

However, just as plant engineers can display a lack of understanding about the necessity to design supporting steelwork afresh for each different structure even though the 'same' plant is being built, structural steelwork engineers must be aware that seemingly identical or repeat pieces of plant or equipment can in fact vary in significant details. Most plant installations are purpose-designed, and the layout and loadings are often provided initially to the structural engineer in terms of estimated or approximate values. It is vital to be aware of the accuracy of the information being used for design at any stage, and to avoid carrying out designs at an inappropriately advanced level. A considerable margin should be allowed, provided that it is established that the plant designers have not allowed a similar margin already in estimating the plant loadings. Experience shows that loadings are often overestimated at the preliminary design stage, but that this is compensated for by new loadings at new locations that were not originally envisaged.

3.2 Anatomy of structure

3.2.1 Gravity load paths

Vertical loadings on industrial steelwork can be extraordinarily heavy; some individual pieces of plant have a mass of 10000t or more. Furthermore, by their very nature these loadings generally act as discrete point loads or line loads rather than as uniformly distributed loadings. Load values are often ill-defined (see section 3.3.1) at the steelwork design stage and frequently additional vertical loadings are introduced at new locations late in the design process.

For these reasons, the gravity load paths must be established at an early stage to provide a simple, logical and well-defined system. The facility should exist to cater for a new load location within the general area of the equipment without the need to alter all existing main structural element locations. Typically this means that it is best to provide a layered system of beams or trusses with known primary span directions and spacings, and then with secondary (and sometimes in complex layouts, tertiary as well) beams, which actually provide vertical support to the plant.

While simplicity and a uniform layout of structure are always attractive to a structural designer, the non-uniform loadings and layout of plant mean that supporting columns may have to be positioned in other than a completely regular grid to provide the most direct and effective load path to the foundations. It is certainly preferable to compromise on a layout that gives short spans and a direct, simple route of gravity load to columns, than to proceed with designing on a regular grid of columns only to end up with a large range of member sizes and even types of beams, girders or trusses (Fig. 3.4).

Similarly, although it is clearly preferable for columns to run consistently down to foundation level, interference at low levels by further plant or equipment is quite common, which may make it preferable to transfer vertical loading to an offset column rather than use much larger spans at all higher levels in the structure. A

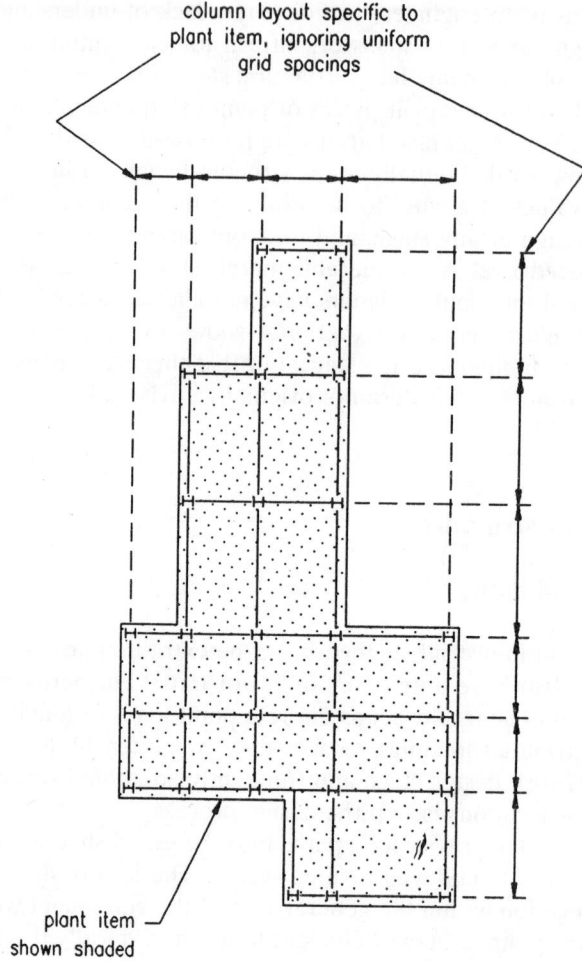

column layout specific to
plant item, ignoring uniform
grid spacings

plant item
shown shaded

Fig. 3.4 Support steelwork for an unsymmetrical plant item

conscious effort to be familiar with all aspects of the industrial process, and close liaison with the plant designers so that they are aware of the importance of an early and inviolate scheme for column locations, are both necessary to overcome this problem.

Different plant designers may be handling the equipment layout at differing levels in the structure, so it is important to ensure that loading information from all parties is received before the crucial decision on column spacing is made.

If a widely variable layout of potential loading attachment points exists, then consideration must be given to hanging or top-supporting certain types of plant. A typical example of this is a vehicle assembly line structure, where the overhead assembly line system of conveyors can sensibly be supported on a deep truss roof

with a regular column system. If this is contrasted with the multitude of columns and foundations that would be needed to support such plant from below and the prospect of having to reposition these supports during the design stage as more accurate information on the plant becomes available, then it can be an appropriate, if unorthodox, method of establishing a gravity load path.

3.2.2 Sway load paths

For two fundamental reasons sway load paths require particular consideration in industrial steelwork. First, the plant or equipment itself can produce lateral loading on the structure in addition to wind or seismic load where applicable. Therefore, higher total lateral loadings can exist whose points of application may differ from the usual cladding and floor intersection locations. The type and magnitudes of such additional loads are covered in section 3.3.3.

Second, many industrial steelwork structures lack a regular and complete (in plan) floor construction that provides a convenient and effective horizontal diaphragm. Therefore, the lateral load transfer mechanism must be considered carefully at a very early stage in the design (Fig. 3.5).

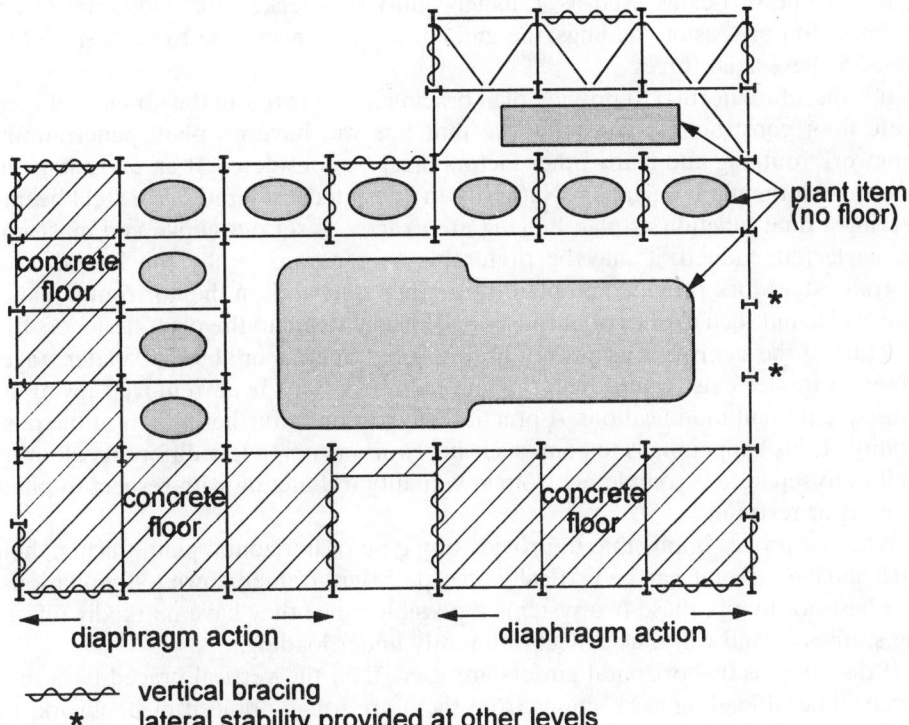

Fig. 3.5 Establishment of sway load paths

Naturally, there are many different methods of achieving lateral stability. Where floor construction is reasonably complete and regular through the height of the structure, then the design can be based on horizontal diaphragms transferring load to vertical stiff elements at intervals along the structure length or width. Where large openings or penetrations exist in otherwise conventional concrete floors, then it is important to design both the floor itself and the connections between the floor and steelwork for the actual forces acting, rather than simply relying automatically on the provision of an effective diaphragm as would often and justifiably occur in the absence of such openings.

It can be worthwhile deliberately to influence the layout to allow for at least a reasonable width of floor along each external face of building, say of the order of $\frac{1}{10}$ of the horizontal spacing between braced bays or other vertical stiff elements.

When concrete floors as described above are wholly or sensibly absent, then other types of horizontal girders or diaphragms can be developed. If solid plate or open mesh steel flooring is used, then it is possible, in theory at least, to design such flooring as a horizontal diaphragm, usually by incorporating steel beams as 'flange' members of an idealized girder, where the floor steel acts as the web. However, in practice, this is usually inadvisable as the flooring plate fixings are rarely found to be adequate for load transfer and indeed it may be a necessary criterion that some or all of the plates can be removable for operational purposes. A further factor is that any line of beams used as a 'flange' must be checked for additional direct compression or tension loadings; the end connections also have to be designed to transfer these axial forces.

It is therefore normal to provide plan bracing in steelwork in the absence of concrete floor construction. The influence that this will have on plant penetrations, pipework routeing and many other factors must be considered at an early stage in liaison with the plant engineers. Naturally, the design considerations of steel beams serving a dual function as plan bracing are exactly as set out above and must not be neglected. Indeed, it may be preferable to separate totally the lateral load restraint steelwork provided on plan from other steelwork in the horizontal plane. This will avoid such clashes of purposes and clearly signal to the plant designers the function of the steelwork, as a result preventing its misuse or abuse at a later stage. Many examples exist where plan bracing members have been removed owing to subsequent plant modifications. A practical suggestion to further minimize the possibility of this happening is to paint such steelwork a completely different colour as well as to separate it completely from any duality of function with respect to plant support or restraint.

Where it proves impossible to provide any type of horizontal plan bracing, then each and every frame can be vertically braced or rigid-framed down to foundations. It is best not to mix these two systems if possible, since they have markedly differing stiffnesses and will thus deflect differently under loading.

If diaphragms or horizontal girders are used, then the vertical braced bays that receive lateral loading as reactions from them are usually braced in steelwork. In conventional structures tension-only 'X' bracing is frequently used, and whereas this may be satisfactory for some straightforward industrial structures such as tank

support frames and conveyor support legs, it often proves necessary to design combined tension/compression bracing in 'N', 'K', 'M' or similar layouts depending on the relative geometry of the height to width of the bay and on what obstruction the bracing members cause to plant penetrations. Indeed, it is sound advice initially to provide significantly more bracing than may be considered necessary, for example by bracing in two, three or more bays on one line. When later developments in the plant and equipment layout mean that perhaps one or more panels must be altered or even removed, it is then still possible to provide lateral stability by rechecking or redesigning the bracing, without a major change in bracing location.

One particular aspect of vertical bracing design that requires care is in the evaluation of uplift forces in the tension legs of braced frames. Where plant and equipment provide a significant proportion of the total dead load, then it is important that *minimum* dead weights of the plant items are used in calculations. For virtually every other design requirement it is likely that rounding-up or contingency additions to loadings will have been made, especially at the early stages of the design. It is also important to establish whether part of the plant loading is a variable contents weight and, if so, to deduct this when examining stability of braced frames. In some cases, for example in hoppers, silos or tanks, this is obvious; but boilers or turbines which normally operate on steam may have weights expressed in a hydraulic test condition when flooded with water. The structural engineer has to be aware of these plant design features in order to seek out the correct data for design (see section 3.3.2).

3.2.3 Floors

As has already been discussed in section 3.2.2, regular and continuous concrete floors are not usual in industrial steelwork structures. Floors can consist of any of the following types:

(1) In situ concrete (cast on to removable formwork)
(2) In situ concrete (cast on to metal deck formwork)
(3) Fully precast flooring (no topping)
(4) Precast concrete units with in situ concrete topping
(5) Raised pattern 'Durbar' solid plate
(6) Flat solid steel plate
(7) Open grid steel flooring.

Typical sections of these types of floor construction are shown in Fig. 3.6.

The selection of floor type depends on the functional requirements and anticipated usage of the floor areas. Solid steel plates are used where transit or infrequent access is required or where the floor must be removable for future access to plant or equipment. They are normally used only internally (at least in the UK) to avoid problems with wet or waterlogged surfaces.

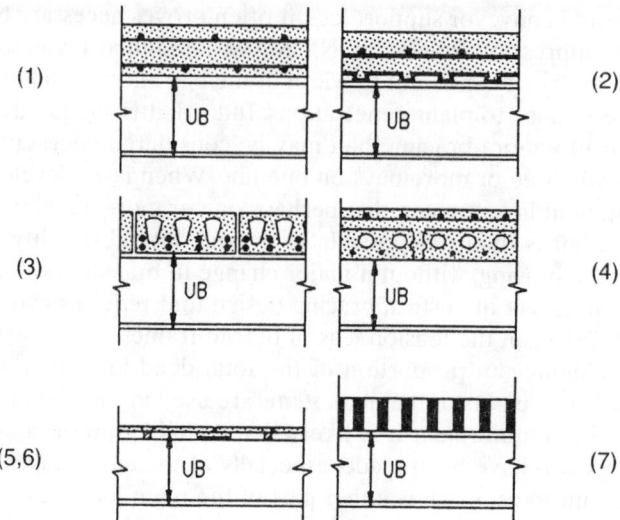

Fig. 3.6 Typical sections of floor construction

Open grid flooring is used inside for similar functions as solid steel plate, but with additional functional requirements where the flooring is subjected to spillage of liquids or where air flow through the floor is important. It is also used for stair treads and landings. Consideration should be given to making, say, landings and access strips from solid plate at intervals to assist in promoting a feeling of security among users. It is also in common use externally due to its excellent performance in wet weather.

Concrete floors are in use where heavy-duty non-removable floor areas are necessary. It is not normally advisable to use beam-and-pot-type floorings in any heavy industrial environment due to the damage that can be suffered by lightweight thin-walled blocks. For similar reasons, where precast concrete floors require a topping for finishes or to act as a structural diaphragm, it is advisable to use a fully-bonded small aggregate concrete topping with continuous mesh reinforcement, with a typical minimum thickness of 75–100 mm.

A particular advantage of both precast flooring and metal deck permanent formwork is that in many industrial structures the floor zones are irregular in plan and elevation and therefore cheap repetitive formwork is often not practicable.

Even where formwork can sensibly be used, the early installation of major plant items as steel frame erection proceeds can complicate in situ concrete formwork.

Floors must be able to accept holes, openings and plant penetrations on a random layout, and often must accept them very late in the design stage or as an alteration after construction. This provides significant problems for certain flooring, particularly precast concrete. In situ concrete can accept in a convenient manner most types of openings prior to construction, but it may be prudent deliberately to allow for

randomly positioned holes up to a certain size by oversizing reinforcement in both directions to act as trimming around holes within the specified units without extra reinforcement.

Metal deck formwork is not so adaptable as regards large openings and penetrations, since it is usually one-way spanning, especially where the formwork is of a type that can also act as reinforcement. If there is sufficient depth of concrete above the top of the metal deck profile, then conventional reinforcing bars can be used to trim openings. Many designers use bar reinforcement as a matter of course with metal deck formwork to overcome this problem and also to overcome fire protection problems that sometimes occur with unprotected metal decks used as reinforcement.

Steel-plate flooring should be designed to span two ways where possible, adding to its flexibility in coping with openings since it can be altered to span in one direction locally if required. Open-grid flooring is less adaptable in this respect as it only spans one way and therefore openings will usually need special trimming and support steelwork. Early agreement with plant and services engineers is vital to establish the likely maximum random opening required, and any structure-controlled restraint on location.

Other policy matters that should be agreed at an early stage are the treatment of edges of holes, edges of floor areas, transition treatments between different floor constructions and plant plinth or foundation requirements.

This last item is extremely important as many plant items have fixings or bearings directly on to steel and the exact interface details and limit of supply of structural steelwork must be agreed. Tolerances of erected structural steelwork are sometimes much larger than anticipated by plant and equipment designers and some method of local adjustment in both position and level must normally be provided. Where plant sits on to areas of concrete flooring then plinths are usually provided to raise equipment above floor level for access and pipe or cable connections. It is convenient to cast plinths later than the main floor, but adequate connection, for example by means of dowelled vertical bars and a scabbled or hacked surface, should always be provided. It is usually more practical to drill and fix subsequently all dowel bars, anchor bolts for small-scale steelwork and for holding-down plant items, and similar fixings, than to attempt to cast them into the concrete floor.

Except in particularly aggressive environments, floor areas are usually left unfinished in industrial structures. For concrete floors hard-trowelled finishes, floated finishes and ground surfaces are all used: selection depends on the use and wear that will occur. Steel plate (solid or open-grid) is normally supplied with either paint or hot-dip galvanized finishes depending on the corrosiveness of the environment; further guidance is given in BS 4592: Part 1[1] and by floor plate manufacturers.

Steel-plate floors are fixed to supporting steelwork by countersunk set screws, by countersunk bolts where access to nuts on the underside is practicable for removal of plates or by welding where plates are permanent features and unlikely to require replacement following damage.

Alternatively, proprietary clip fixings can be used for open-grid flooring plates to secure the plates to the underside of support beam or joist flanges.

3.2.4 Main and secondary beams

The plan arrangement of main and secondary beams in industrial steelwork structures normally follows both from the layout of the main items of plant and from the column locations. Thus a sequence of design decisions often occurs in which main beam locations dictate the column locations and not vice versa. If major plant occurs at more than one level then some compromise on column position and hence beam layout may be needed.

Since large plant items normally impose a line or point loading there are clear advantages in placing main or secondary beams directly below plant support positions. Brackets, plinths or bearings may be fitted directly to steelwork, and for major items of plant this is preferable to allowing the plant to sit on a concrete or steel floor. Where plant or machinery requires a local floor zone around its perimeter for access or servicing, it is common practice to leave out the flooring below the plant for access or because the plant protrudes below the support level.

Deflection requirements between support points should be ascertained. They may well control the beam design since stringent limits, for example relative deflections of 1 in 1000 of support spans, may apply. In addition when piped services are connected to the plant then total deflections of the support structure relative to the beam-to-column intersections may also need to be limited. Relative deflections can best be controlled by the use of deep beams in (lower-grade steel if necessary), and total deflections by placing columns as close as possible to the support positions.

It is preferable to avoid the necessity for load-hearing stiffeners at support points unless the plant dimensions are fixed before steelwork design and detailing take place. Where this is not possible then stiffened zones to prevent secondary bending of top (or bottom if supports are hung) flanges should be provided, even if the design requirements do not require load-bearing stiffeners. This then allows a measure of tolerance for aligning the support positions without causing local overstressing problems.

The stiffness of major plant items should be considered, at least qualitatively, in the steelwork design. Deep-walled tanks, bunkers or silos for example may well be an order of magnitude stiffer than the steel supporting structure.

The loading distribution given by the plant design engineers will automatically assume fully stiff (zero deflection) supports. When the stiffness of the supporting structure is not uniform in relation to the support point locations and loadings, then significant redistribution of loads can take place as the structure deflects. Where the plant support positions and loads and the structural steel layout are symmetrical then engineering judgement can be applied without quantitative evaluation. In extreme cases, however, a plant–structure interaction analysis may be required to establish the loadings accurately.

When hanger supports are required then pairs of beams or channels are a convenient solution which allows for random hanger positioning in the longitudinal direction (Fig. 3.7). Many hanger supports have springs bearings to minimize variations in support conditions due to plant temperature changes or to avoid the plant stiffness interactions described above.

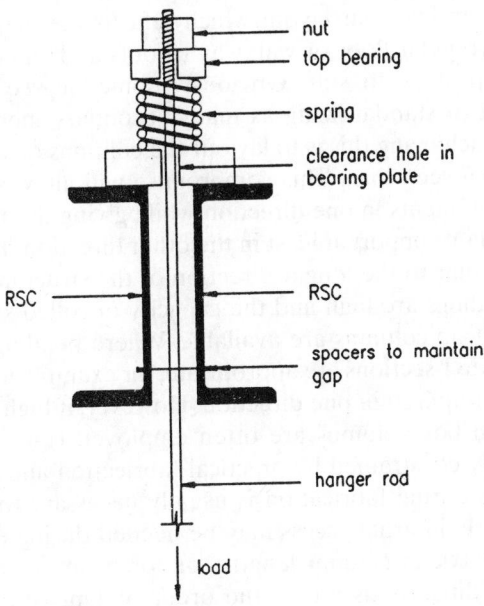

nut
top bearing
spring
clearance hole in
bearing plate
RSC
RSC
spacers to maintain
gap
hanger rod
load

Fig. 3.7 Detail at hanger support

The method of installation or removal of major items of plant frequently requires that beams above or, less commonly, below the plant must be designed to cater for hoisting or jacking-up the installed plant sections. Structural designers should query the exact method and route of plant installation to ensure that the temporary hoisting, jacking, rolling or set-down loads are catered for by the steel beam framework. Experience suggests that these data are not provided as a matter of course, and plant designers commonly believe that the steel structure can support these loads anywhere. Where, to ease the problems, plant installation occurs during steel erection then the method of removing plant during the building life-span may be the most significant temporary loading for the beam framework.

When main or secondary beams are specifically designed for infrequent but heavy lifting operations, it is good practice to fit a lifting connection to the beam to give positive location to the lifting position and to allow it to be marked with a safe working load. For design, smaller load safety factors are appropriate in these circumstances.

3.2.5 Columns

Column location in industrial buildings must be decided on practical considerations. Although regular grid layouts are desirable in normal structures, it is sometimes

impossible to avoid an irregular layout which reflects the major plant and equipment location, the irregular floor or walkway layouts and an envelope with irregular wall and roof profiles to suit. Obviously some degree of regularity is of considerable benefit in standardizing as many secondary members as possible; a common method of achieving this is to lay out the columns on a line-grid basis with a uniform spacing between lines. This compromise will allow standard lengths for beam or similar components in one direction while giving the facility to vary spans and provide direct plant support at least in the other direction. The line-grids should be set out perpendicular to the longer direction of the structure if possible.

When vertical loadings are high and the capacity of rolled sections is exceeded, several types of built-up columns are available. Where bending capacity is also of importance, large plate I-sections are appropriate, for example in frameworks where rigid frame action is required in one direction. However, if high vertical loads dominate then fabricated box columns are often employed (Fig. 3.8). Design of box columns is principally constrained by practical fabrication and erection considerations. Internal access during fabrication is usually necessary for fitting of internal stiffeners, and similarly internal access may be needed during erection for making splice connections between column lengths, or for beam-to-column connections. Preferred minimum dimensions are of the order of 1 m with absolute minimum dimensions of about 900 mm. Whenever possible, column plates should be sized to avoid the necessity of longitudinal and transverse stiffeners to control plate buckling. The simpler fabrication that results from the use of thick plates without stiffeners should lead to overall economies, and the increased weight of the member is not a serious penalty to pay in columns; the same argument does not apply to long-span box girders where increases in self-weight may well be of overriding importance. Under most conditions of internal exposure no paint protection to the box interior is necessary. If erection access is needed then simple fixed ladders should be provided. Transverse diaphragms are necessary at intervals (say 3–4 times the minimum column dimension) to assist in maintaining a straight, untwisted profile, and also at splices and at major beam-to-column intersections even if, as is usually the case, rigid connections are not being used. Diaphragms should be welded to all four box sides and be provided with manhole cut-outs if internal access is needed.

3.2.6 Connections

Connections between structural elements are similar to those in general structural practice. Specific requirements relating to industrial steel structures are considered here.

On occasions industrial plant and equipment may impose significant load variations on the structure and so consideration must be given to possible fatigue effects. Since basic steel members themselves are not susceptible to fatigue failure in normal

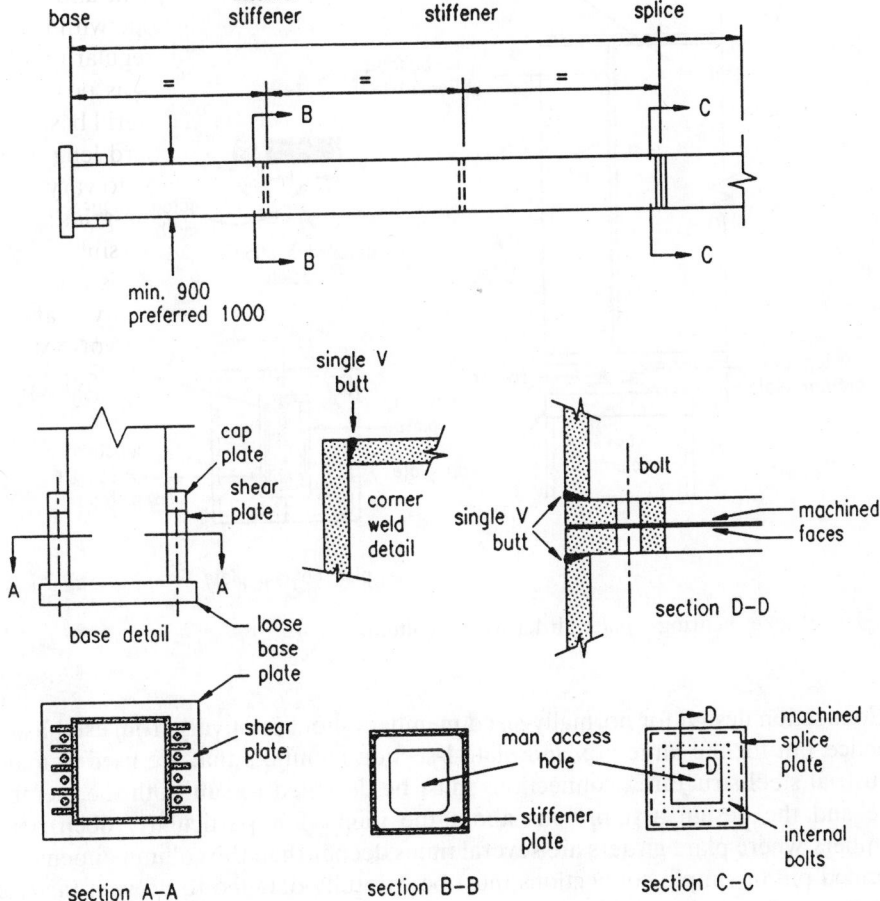

Fig. 3.8   Typical details of box columns

conditions, attention must be focused on fatigue-susceptible details, particularly those relating to welded and other connections. Specific guidance for certain types of structure is available, and where this is not relevant, general fatigue design guidance can be used.

Of general significance is the question of vibration and the possible damage to bolted connections that this can cause. It should be common practice for steelwork in close contact with any moving machinery to have vibration-resistant fixings. For main steelwork connections there is a choice between using HSFG bolts, which are inherently vibration-resistant, or using normal bolts with lock-nuts or lock-washer systems. A wide variety of locking systems is available which can be selected after consultation with the various manufacturers.

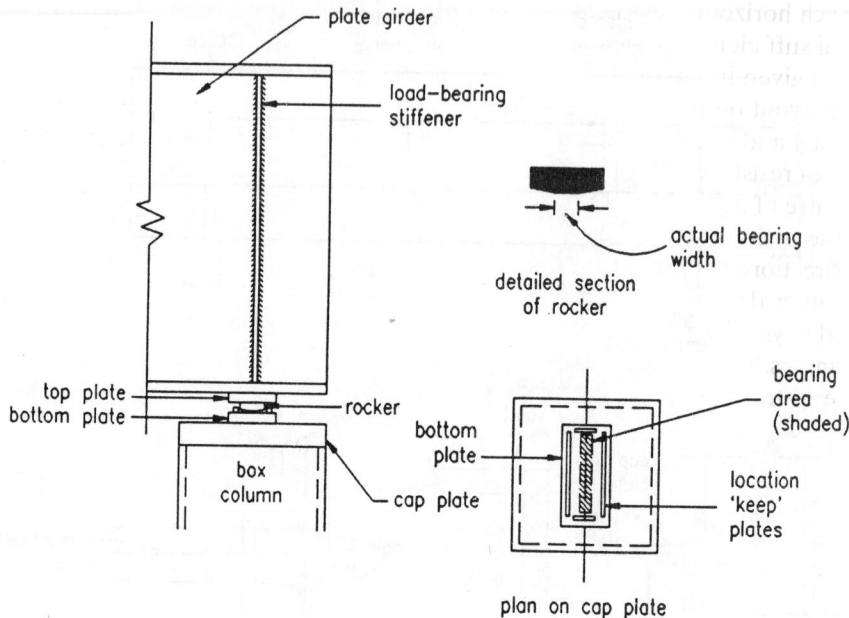

Fig. 3.9   Rocker bearing – plate girder to box column

Connection design for normally-sized members should not vary from established practice, but for the large box and plate I-section members that are used in major industrial steel structures, connections must be designed to suit both the member type and the design assumptions about the joints. For particularly deep beam members, where plate girders are several times deeper than the column dimensions, assumed pin or simple connections must be carefully detailed to prevent inadvertent moment capacity. If this care is not taken, significant moments can be introduced into column members even by notional simple connections due to the relative scale of the beam depth.

In certain cases it will be necessary to load a column centrally to restrict bending on it: a typical example is where deep suspension girders on power station boilers apply very high vertical loadings to their supporting columns. Here, a rocker cap plate detail is often used to assure centroidal load transfer into the column (Fig. 3.9). Conventional connections on smaller-scale members would not usually require such a precise connection as load eccentricities would be allowed for in the design.

3.2.7 Bracing, stiff walls or cores

Section 3.2.2 discusses the particular features of industrial steel structures in relation to achieving a horizontal or sway load path and describes the various methods

by which horizontal loads can be satisfactorily transferred to braced bays or other vertical stiff elements. General design requirements and some practical suggestions are also given in section 3.2.2 for braced steel bay design.

The layout on plan of vertically stiff elements is frequently difficult even in conventional and regularly framed structures. General guiding principles are that the centre of resistance of the bracing system in any direction should be coincident with the centre of action of the horizontal forces in that direction. In practice this means that the actions and resistances should be evaluated, initially qualitatively, in the two directions perpendicular to the structural frame layout.

Another desirable feature which is also common to many structures is that the braced bays or stiff cores should be located centrally on the plan rather than at the extremities. This is to allow for expansion and contraction of the structure without undue restraint from the stiff bays, and applies equally to a single structure or to an independent part of a structure separated by movement joints from other parts. It is difficult to achieve this ideal in a regular and uniform structure, and almost impossible in a typically highly irregular industrial steelwork structure. Fortunately, steelwork buildings, particularly those without extensive reinforced concrete floors and with lightweight cladding, are extremely tolerant of temperature movements and rarely suffer distress from what may be considered to be a less than ideal stiff bay layout.

The procedure for design purposes should be as follows. First a basic means of transferring horizontal loading to foundation level must be decided, and guidance on this is given in section 3.2.2. In either or both directions, where discrete braced bays or stiff walls or cores are being utilized, then initially a geometric apportionment of the total loading should be made by an imaginary division of the structure on plan into sections that terminate centrally between the vertically stiff structural element. The loadings thus obtained are used to design each stiff element.

When this process has been completed and if the means exist, by horizontal diaphragm or adequate plan bracing, to force equal horizontal deflections on to each element, a second-stage appraisal may be needed to investigate the relative stiffness of each stiff element. Then the horizontal loading can be distributed between vertical stiff elements on a more accurate basis and the step process repeated.

Considerable judgement can be applied to this procedure, since it is usually only of significance when fundamentally differing stiff elements are used together in one direction on the same structure. For example, where a horizontal diaphragm or plan bracing exists and where some of the stiff elements are braced steelwork and some are rigid frames, it will normally be found that the braced frames are relatively stiffer and will therefore carry proportionately more load than the rigid frames. Similarly, where a combination of concrete shear walls (or cores) and braced frames is used, then a relative stiffness distribution will be needed if an effective horizontal diaphragm or plan bracing ensures sensibly constant horizontal deflection or sway (Fig. 3.10).

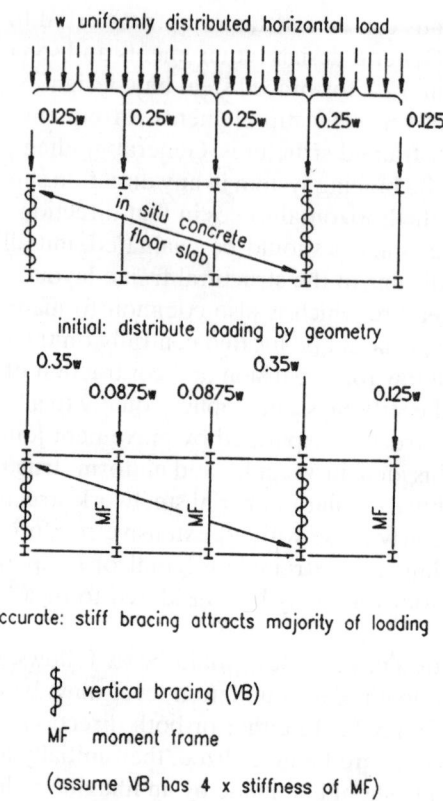

Fig. 3.10   Apportioning horizontal loads to vertical stiff elements

3.3 Loading

3.3.1 General

Two difficulties exist in defining the appropriate loadings on industrial buildings. First, the actual weight, and particularly the details of the position and method of load application, of items of plant or equipment must be established. Even for routine or replicated plant this information is difficult to obtain in a form that suits the structural engineer. When the plant or equipment is being custom-built then the problem becomes one of timing: information from the plant designer may not be available early enough for the structural design.

The second difficulty that frequently occurs is in the choice of a general, uniformly distributed imposed loading for any remaining floor areas not occupied by items of plant or equipment. Guidance from codes of practice must be used carefully when it is applied only to circulation spaces between all the fixed items of plant which are

known and whose loading has been evaluated and is considered separately as dead load. Further advice on both these problems is given in section 3.3.2.

The structural steelwork designer must take an open-minded approach to the loading information. It must be appreciated that early information from the plant designers will represent only estimated loadings, and an uncertainty allowance may well already be included in the values supplied, so that further allowances may be unnecessary. However, it frequently turns out that many secondary plant items are added at a later stage in the information process. The loadings from these should be satisfactorily absorbed into an initial uniformly distributed loading allowance provided that a logical scheme for dealing with loading is established at an early stage.

3.3.2 Process plant and equipment

The most important advice to the structural designer of steelwork for an industrial purpose is to ensure reasonable familiarity with the entire process or operation involved. Existing facilities can be visited, and the plant designers and operators will normally be more than willing to give a briefing which provides the opportunity to describe the form of structure envisaged to the plant designer at an early stage so avoiding later misunderstandings.

Some attempt should also be made to understand the jargon of the industry in question in order to gain the confidence of the plant designers and to allow effective intercommunication.

Particular examples of structural requirements which are not easily recognized as important by plant designers, and which should therefore be fully explained, are as follows:

(1) The physical space requirements of bracing members and the fact that they cannot be moved locally to give clearances.
(2) The actual size of finished steelwork taking account of splice plates, bolt heads and fittings projecting from the section sizes noted on drawings.
(3) The fact that a steel structure is not 100% stiff and that all loads cause deflections.
(4) The fact that a steel structure may interact with a dynamic loading and that dynamic overload multipliers calculated or allowed on the assumption of a fully rigid or infinite mass support are not always appropriate.

The basis of loading information for plant and equipment must be critically examined. Frequently loadings are given as a single all-up value, the components of which may not all act together or have only a very small probability of so doing. Alternatively, the maximum loadings may represent a peak testing condition or a fault, overload condition, whereas normal operating loadings may be considerably less in value. Worthwhile and justifiable savings in steelwork can be made if a statistically

based examination of the expected frequency and duration of such unusual conditions is made, leading to the adoption of reduced load factors without reducing the overall factor of safety. Comparisons with, for example, wind loading can be used to establish on a reasonably logical basis the appropriate load factors.

By definition, items of fixed plant can be treated as dead load in accordance with BS 6399: Part 1[2] when specific location and loads are known. At the early stages of design it is usual to adopt a relatively large imposed loading which will have to cater for fixed items of plant or equipment, the existence of which may not even be known at this stage and certainly not the location and loading data. The choice of what imposed loading to use at this stage can be assisted by the following guidance:

very light industrial processes	$7.5\,\mathrm{kN/m^2}$
medium/average industrial processes	$10–15\,\mathrm{kN/m^2}$
very heavy industrial processes	$20–30\,\mathrm{kN/m^2}$

One factor which will influence the choice within these values is the timing of the release of final plant and equipment design data in relation to the steelwork design and fabrication detailing process. If a second-stage design is possible then a lower imposed load can be allowed since local variations needed to account for specific items of fixed plant which exceed the imposed plan loading allowance can be accommodated.

Under these conditions it can also be worthwhile, particularly for designers with previous experience of the industrial process, to design columns and foundations for a lower imposed loading than beams. In certain layouts with long span main beams or girders at wide spacings this preliminary reduction can also be used for these members. It should be stressed that these proposals are not intended to contradict or override the particular reduced loading clauses in BS 6399,[2] but are a practical suggestion for the preliminary design stage.

When detailed plant layouts with location and loading data become available, fixed plant and equipment can be considered as dead load and subject therefore to the appropriate load factor, 1.4 instead of 1.6. Remaining zones of floor space without major items of plant should be allocated an imposed load which should reflect only the access and potential use of the floor and may therefore very well be reduced from the preliminary imposed loading, often in the range of $5–10\,\mathrm{kN/m^2}$.

Specific laydown areas for removal, replacement or maintenance of heavy plant are the only likely exceptions to this range of loading.

An alternative scheme for dealing with the second stage of loading information is to institute a checking procedure where the equivalent loading intensity of plant and equipment is calculated for each item as follows:

$$\frac{1.4}{1.6} \times \frac{\text{weight (kN)}}{\text{plan area (m}^2)}$$

the factor 1.4/1.6 being introduced to cater for the reclassification from imposed load to dead load.

Only in bays where this equivalent loading intensity exceeds the preliminary imposed loading will further evaluation be required, unless a scheme of reduced imposed loading for long-span beams and columns has been adopted, in which case a rigorous check on actual loading intensity must be carried out.

Many items of plant contain moving parts which are always liable to exert dynamic or vibratory loadings on to the structure. Assessing the structural response to these loadings is never easy, particularly when, as is usually the case, the plant and the structure are being designed out of sequence and by different designers.

Many rule-of-thumb approximations exist, in which dynamic effects are allowed for by percentage increases to dead load. Some instances of this are codified: for example, crane vertical loadings are factored by 1.25 (see BS 6399: Part 1[2]). Similar multipliers to static loading can be provided by plant suppliers but these must be used with great care and the limitations of such an approach should be appreciated. The dynamic load from one-off impact-type actions or from successive load applications at irregular time intervals depends both on the characteristics of the plant itself *and* on the mass and stiffness of the structure to which they are applied. Any dynamic load calculated independently from a knowledge of the structure will be based on assumed structure properties and often on limit values such as zero mass or infinite stiffness. Provided that the structure remains elastic and that resonance or similar frequency-related amplification does not occur the approximate dynamic factors normally represent upper limits. If more accurate evaluations are warranted for major items of plant, a plant–structure interaction dynamic analysis must be undertaken.

Vibrating plant or equipment will transmit vibrations to the structure, the effects of which can be dramatic. The frequency of vibration of the plant can induce resonance in the structure if the natural frequency of vibration of the whole or any part of the structure subjected to the vibrating force is equal or very close to the forcing frequency. Considerable judgement is needed to identify all possible modes of natural vibration that could be excited by the vibrating equipment or machinery. To safeguard against a resonant response there should be no natural frequency of vibration within the range of 0.5–1.5 times the forcing frequency. Useful guidance on calculating natural frequencies of structures is given by Bolton.[3]

A useful initial step is to calculate the lowest natural frequency of vertical motion of the floor construction and to ensure that it is higher than 3–4 Hz (cycles per second). This should avoid problems of response to human-induced vibration, the so-called 'springiness' of floors, caused by resonant amplification of footfalls, which lie in the range of 1–2 Hz. Then only plant-induced vibrations of lower than 1 Hz or higher than 2 Hz need to be investigated. Anti-vibration mountings for plant can be specified but these are not an automatic success as they only filter the induced forces and some form of variable loading will still be transmitted to the structure. Various types and grades of anti-vibration mountings are available; information on their suitability can be obtained from the manufacturers, based on details of both the item of plant and the structure.

Many of these devices are intended to reduce vibration of the plant or equipment itself. Care is required to ensure that the altered forcing vibrations which are

imposed on to the structure do not have a secondary adverse effect of inducing resonant vibrations into it.

3.3.3 Lateral loadings from plant

Lateral loadings imposed by plant on the structure derive from three sources. These are considered separately although there is a common theme throughout that the operating process undergoes a change in regime which is the cause of loading. Many of the actions which give rise to horizontal loads also cause vertical loads or at least vertical loading components which must be incorporated into the design. However, whereas vertical loadings are usually readily understood and allowed for in loadings provided by plant designers, one feature of the horizontal components of such loadings that causes confusion is the fundamental concept of equilibrium. Notwithstanding exotic situations where masses (projectiles) leave or impinge on a structure, or where motion energy is dissipated as heat of friction, equilibrium considerations dictate that lateral components of forces are in balance and consequentially they are often ignored. This is not satisfactory as balancing components may act a considerable distance apart (the load path must be examined in detail) or indeed the balancing components may act at different levels, leading to a more conventionally understood requirement to transfer lateral loading.

The three causes are as follows:

(1) temperature-induced restraints
(2) restraint against rotational or (more rarely) linear motion
(3) restraint against hydraulic or gaseous pressures.

Where plant undergoes a significant change in temperature, plant designers will typically assume that the structure is fully rigid and so can absorb the forces generated by application of restraints at the structure connection points. They will then design the plant itself for the additional stresses that are caused by preventing free thermal expansion or contraction. This is a safe upper bound procedure since the forces generated in both structure and plant represent maxima, with any deflection at the support reducing forces in both elements. Naturally the structural steelwork designer must be made aware both of the assumption of zero deformation so that the support can be made as stiff as possible in the required direction, and of the forces that are thus imposed.

In spite of the apparent complications of this approach, it is frequently adopted by plant designers for convenience on small items, and to avoid complexities in interconnection between plant items and with piped and ducted services on larger items.

The alternative approach, common on major plant subjected to significant thermal variation such as boilers and ovens, is to assume completely free supports with zero restraint against expansion or contraction. This is a lower bound solution

which needs some rational assessment of possible forces that could result from bearing or guide misalignment, malfunction or simple inefficiency. Where large plant items are involved the forces even at such guides or bearings can be significant if the balancing reaction is, for example, at ground level or even outside the structure itself.

The lateral forces from constant speed, rotating machinery, which are usually relatively low, are generally balanced by an essentially equal and opposite set of forces coming from the source of motive power. Nevertheless, their point of application must be considered and a load path established which transfers either them back to balance each other or alternatively down to foundation level in the conventional way. The structural steel designer must have a completely clear and unambiguous understanding of the source and effect of all the moving plant forces, to ensure that all of them are accounted for in the crucial interface between plant and equipment.

The start-up forces when inertia of the plant mass is being overcome, and the fault or 'jamming' loads which can apply when rotating or linear machinery is brought to a rapid halt, need consideration. They are obviously of short duration and so can justifiably be treated as special cases with a lower factor of safety. At the same time the load combinations that can actually co-exist should be established to avoid any loss of economy in design.

Pressure pipe loadings, significant in many plant installations, occur wherever there are changes in direction of pipework and associated pipework supports or restraints. Thermal change must also be considered. Pipework designers may well combine the effects to provide a schedule of the total forces acting at each support position. Since pipework forces are normally *not* reversible in direction, then on major installations the pipework designer may wish to make the installation by 'forcing' the pipework configuration, deliberately making sections too short or too long and then prestressing the pipework so that when installation is complete the operating conditions take the internal pipework stresses through a neutral stress zone and then reverse them.

Forcing is a complex procedure sensitive to lack of fit at the pipework supports and restraints and often to other factors such as the ambient temperature during installation and the exact sequence of connections. Where major pipe installations are planned and where the pipework designers are adopting these techniques, the structural steelwork designer should acknowledge that the loading values quoted may not be achieved in practice and make a further allowance.

3.3.4 Wind loadings

Wind loadings on fully enclosed industrial structures do not differ from wind loadings on conventional structures. The only special consideration that must be given applies to the assessment of pressure or force coefficients on irregular or unusual-shaped buildings. A number of sources give guidance on this topic, and specific

advice can be sought from the Building Research Establishment (BRE) Advisory Service or other specialist advisors.

However, on partly or wholly open structures with exposed plant or equipment great care must be exercised in dealing with wind loading.

It is frequently the case that the total wind loads are higher than on a fully clad building of the same size, due to two causes. First, small structural elements attract a higher force than equivalent exposed areas which form part of a large façade. Secondly, repetitive structural elements of plant items which are nominally shielded by any particular wind direction are not actually shielded and each element is subjected individually to a wind load. The procedure for carrying out this assessment is not covered in detail in BS 6399: Part 2[4] but guidance is available from the references listed in that document.

Loading on particularly large individual pieces of plant or equipment exposed to wind can be calculated by considering them to be small buildings and deriving overall force coefficients that relate to their size.

For smaller or more complex shapes, such as ductwork, conveyors and individual smaller plant items, it is sensible to take a conservative and easy to apply rule-of-thumb and use a net pressure coefficient $C_p = 2.0$ applied to the projected exposed area. The point of application of wind loadings from plant items on to the structure may be different from the vertical loading transfer points if sliding bearings or guides are being used.

3.3.5 Blast loadings

This section deals only with blast loadings from industrial processes and not with any generalized design requirements to survive blast loadings from unspecified sources or of unspecified values.

Varying requirements exist for blast loadings. Typical examples are:

(1) transformers, where the requirement is usually to deflect any blast away from other vulnerable pieces of plant but where frequently one or more walls and the roof are open, serving to dissipate much of the energy discharged
(2) dust or fine particle enclosures, which are often wholly inside enclosed buildings.

The decisions to be made are as follows.

(1) Can the potential source of the blast be relocated outside the building altogether, in a separate enclosure?
(2) If not, can it be placed against the external wall with arrangements to have a major permanent vented area or a specially designed blow-off panel, both of which will limit loadings on the remaining structure?
(3) Where the location cannot be controlled, it must be established which direction

or directions require full protection against damage and which can tolerate certain degrees of damage.

Loading data given by plant designers are usually stated in terms of peak pressures to be applied to projected areas in line with the potential source of the blast. The validity of the data must be treated as being highly suspect since blast loadings are classic examples of true dynamic loading where the time-dependent response of the structure actually determines the loading that is imposed. Quoted blast pressures, which are probably derived from theoretical considerations of high-rigidity high-mass targets fully enclosing the source, may be invalid for steel structures which have tremendous capacity to deform rapidly, absorbing energy and thereby reducing and smoothing out peak blast pressures. Whatever results are obtained from analysis or calculation, it is good practice to use a grade of steel which has good ductility, a lower yield stress from which to commence ductile behaviour and a long and reliable extensibility prior to fracture. While this will ensure reasonable material behaviour, overall ductility of the structure also depends on stability against premature buckling and the ductile behaviour of connections.

Steelwork designers should acknowledge the very imprecise nature of most blast loading data, even when the potential source of the blast is precisely located and specified. They should thus direct their attention to ensuring that collapse does not occur until major deflections and rotations have occurred, following normal guidelines for achieving plastic behaviour.

3.3.6 Thermal effects

This section deals with thermal effects from environmental factors; specific consideration of plant-induced thermal effects is given in section 3.3.3. Conventional guidance on the provision of structural expansion or contraction joints is often inappropriate and impractical to implement, and indeed joints frequently fail to perform as intended.

The key to avoiding damage or problems from thermal movements is to consider carefully the detailing of vulnerable finishes (for example, brickwork, blockwork, concrete floors, large glazing areas and similar rigid or brittle materials). Provided that conventional guidance is followed in the movement provisions for these materials, then structural joints in steel frames can usually be avoided unless there is particularly severe restraint between foundations and low-level steelwork.

Many industrial structures with horizontal dimensions of 100–200 m or more have been constructed without thermal movement joints, usually with lightweight, non-brittle cladding and roofing, and a lack of continuous suspended concrete floors.

When high restraint close to foundations or vulnerable plant or finishes are present, a thermal analysis can be carried out on the steel framework to examine the induced stresses and deflections and to evaluate options such as the introduction of joints or altering the structural restraints. Where steelwork is externally

exposed in the UK, the conditions vary locally but a minimum of –5°C to +35°C suffices for an initial sensitivity study. For many of the likely erection conditions a median temperature of 15°C can be assumed, and a range of ±20°C can thus be examined. Where the effects of initial investigations based on these values highlight a potential problem, more specific consideration can be given to the actual characteristic minimum and maximum temperatures (advice in the UK is available from the Meteorological Office), and steps may need to be taken to control erection, particularly foundation fixing, to take place at specified median temperatures.

3.4 Structure in its wider context

Industrial structural steelwork is inherently inflexible, being purpose-designed for a particular function or process, and indeed often being detailed to suit quite specific items of major plant and equipment. Nevertheless, it is important to try to cater for at least local flexibility to allow minor alterations in layout, upgrading or replacement of plant items. The most appropriate way to ensure this is to repeat the advice that has been given on numerous occasions already in this chapter. The designer must understand the industrial process involved and be aware of both structural and layout solutions that have been adopted elsewhere for similar processes. Previous structural solutions may not be right, but it is preferable to be aware of them and positively to reject them for a logical reason, than to reinvent the wheel at regular intervals.

General robustness in industrial buildings may be difficult to achieve by the normal route of adopting simple, logical shapes and structural forms, with well-defined load-paths and frequent effective bracing or other stability provisions. Instead of these provisions, then, it is sensible to ensure that a reasonable margin exists on element and connection design. Typically, planning for a 60–80% capacity utilization at the initial design stages will be appropriate, so that even when these allowances are reduced, as so frequently occurs, during the final design and checking stages, adequate spare capacity still exists to ensure that no individual element or joint can weaken disproportionately the overall structural strength of the building.

References to Chapter 3

1. British Standards Institution (1995) *Industrial type metal flooring, walkways and stair treads.* Part 1: *Specification for open bar gratings.* BS 4592, BSI, London.
2. British Standards Institution (1996) *Loading for buildings.* Part 1: *Code of practice for dead and imposed loads.* BS 6399, BSI, London.
3. Bolton A. (1978) Natural frequencies of structures for designers. *The Structural Engineer,* **56A**, No. 9 Sept., 245–53.

4. British Standards Institution (1995) *Loading for buildings.* Part 2: *Code of practice for wind loads* BS 6399, BSI, London.

Further reading for Chapter 3

Booth E.D., Pappin J.W. & Evans J.J.B. (1988) Computer aided analysis methods for the design of earthquake resistant structures – a review. *Proc. Instn Civ. Engrs*, **84**, Part 1, Aug., 671–91.

Fisher J.M. & Buckner D.R. (1979) *Light and Heavy Industrial Buildings.* American Institute of Steel Construction, Chicago, USA, Sept.

Forzey E.J. & Prescott N.J. (1989) Crane supporting girders in BS 15 – a general review. *The Structural Engineer*, **67**, No. 11, 6th June, 205–15.

Jordan G.W. & Mann A.P. (1990) THORP receipt and storage – design and construction. *The Structural Engineer*, **68**, No. 1, 9th Jan., 7–13.

Kuwamura H. & Hanzawa M. (1987) Inspection and repair of fatigue cracks in crane runway girders. *J. Struct. Engng, ASCE*, **113**, No. 11, Nov., 2181–95.

Mann A.P. & Brotton D.M. (1989) The design and construction of large steel framed buildings. *Proc. Second East Asia Pacific Conference on Structural Engineering and Construction*, Chaing Mai, Thailand, **2**, Jan., 1342–7.

Morris L.J. (Ed.) (1983) *Instability and plastic collapse of steel structures. Proc. M.R. Horne Conference.* Granada, St Albans.

Taggart R. (1986) Structural steelwork fabrication. *The Structural Engineer*, **64A**, No. 8, Aug., 207–11.

Chapter 4
Bridges

by ALAN HAYWARD

4.1 Introduction

Use of structural steel in bridges exploits its advantageous properties of economically carrying heavy loads over long spans with the minimum dead weight. Steel is however suitable for all span ranges, categorized in Table 4.1.

For long spans steel has been the natural solution since 1890 when the Firth of Forth cantilever railway bridge, the world's first major steel bridge, was completed. For short and medium spans concrete bridges held a monopoly from 1950 to 1980 because of the introduction of prestressing and precasting. Developments in steel during this period such as higher tensile strength and improved welding techniques were applied mainly to long spans. However, improvements in construction methods from 1980 have enabled steel to improve its market share within Europe and other continents to more than 50% for short and medium spans. Contributing factors to this trend are shown in Table 4.2.

Where traffic disruption during construction must be minimized then steel is always suitable. Most bridgework is now carried out under these conditions so that structural steel will continue as a primary choice. Steel has an advantage where speed of construction is vital; it is no coincidence that this usually results in cost economies. If rapidly erected steelwork is used as a skeleton from which the slab and finishes can be carried out without need for falsework then the advantages (summarized in Table 4.3) are fully realized.

For short and medium span highway bridges composite deck construction is economic because the slab contributes to the capacity of the primary members.[1] For continuous spans it uses the attributes of steel and concrete to best advantage. In cases where construction depth is restricted, for example in developed areas, then half-through or through construction is convenient; this is common for railway and pedestrian bridges.

For long spans, including suspension or cable-stayed bridges, all-steel orthotropic plate floors are used. Although the intrinsic costs of a steel orthotropic plate are higher (often up to about four times more) than an equivalent concrete slab, the advantage in dead weight reduction (approximately 1:3 ratio) may more than offset this when the overall economy is considered. Steel decks are also employed when erection must be completed in limited occupations, such as for railway bridges on existing routes. Movable bridges of swing, lift, rolling-lift ('bascule') or retractable type usually employ steel floors to rationalize the amount of counterweighting and

Table 4.1 Span ranges of bridges

Short	Up to 30 m
Medium	30 m to 80 m
Long	80 m

Table 4.2 Factors contributing to the improved market share for steel in short and medium spans from 1980

Factor	Reasons
(1) Stability in price of rolled steel products	Stability worldwide
(2) Automation of fabrication processes	Mechanized equipment for material preparation, welding and girder fabrication
(3) Faster erection with larger components	Availability of high-capacity cranes. Delivery in longer lengths
(4) Improved design codes	Easier yet more rigorous methods of design
(5) Evidence of durability problems with concrete	Life expiry of concrete bridges under 30 years old
(6) Better education in steel design	SCI and BCSA publications and training courses

Table 4.3 Advantages of steel bridges

Feature	. . . leading to . . .	Advantages
(1) Low weight of deck	Smaller foundations. Typical 30–50% reduction of weight compared with concrete decks.	Economy
(2) Light units for erection	Erection by mobile cranes	Rapid construction
(3) Bolted site connections	Minimal site inspection	Flexible site planning
(4) Prefabrication in factory	Effective quality control	More reliable product
(5) Modern methods of protective treatment	Application of treatment before erection. Long life to first maintenance.	Predictable whole life cost of structure
(6) Use of weathering steel		Minimum whole life cost
(7) Shallow construction depth	Minimum length of approach grades	Overall economy of highway. Slender appearance.
(8) Self-supporting steel	Elimination of falsework	Easier construction, especially for high structures
(9) Continuous spans	Fewer bearings and joints	Slender appearance. Reduced maintenance cost.

the mechanical equipment. Costs of this equipment usually exceed that of the bridge superstructure. For long spans of suspension or cabled-stayed form then special considerations affect the design including aerodynamic behaviour, the feasibility of deep foundations in estuarial conditions, cables with anchorages, non-linear struc-

tural behaviour and the absolute necessity to include the effects of the erection procedure in the design process.

This chapter on bridges gives emphasis to initial design, an important stage of the
process, because the basic decisions as to member proportions, spacings and splice
positions vitally affect economy of the structure. It is essential that the detailed
analysis is based upon optimized sizes which are as accurate as possible. If this is
not achieved then the detailed design will be inefficient because time consuming
repetitive work will have been expended, adversely affecting the economy of the
design and the construction costs. Guidance is given on the initial design of highway
bridges using composite deck construction, which is a significant proportion of the
number of steel bridges built, although since 1990 a significant number of railway
bridges have been constructed for new-build schemes such as the Channel Tunnel
Rail Link and in replacement of life expired structures.

Steel is particularly suitable for the strengthening and repair of existing bridges
arising from increases in highway traffic loadings and the incidence of accidental
impact damage from road vehicles. Steel is suitable for such work using welded or
bolted strengthening.

4.2 Selection of span

The majority of bridges fall within the category of short span because for many
crossings of rivers, railways or secondary highways a single span of less than 30 m is
sufficient. For multiple-span viaducts a decision on span length must be made, which
depends on factors shown in Table 4.4.

For long viaducts it is necessary to carry out comparative estimates for different
spans to determine the optimum choice, as shown in Fig. 4.1.

Table 4.4 Factors which decide choice of span for viaducts

Factor	Reasons
Location of obstacles	Pier positions are often dictated by rivers, railway tracks and buried services
Construction depth	Span length may be limited by the maximum available construction depth
Relative superstructure and substructure costs	Poor ground conditions require expensive foundations; economy favours longer spans
Feasibility of constructing intermediate piers in river crossings	(a) Tidal or fast-flowing rivers may preclude intermediate piers (b) For navigable waterways, accidental ship impact may preclude mid-river piers
Height of deck above ground	Where the height exceeds about 15 m, costs of piers are significant, encouraging longer spans
Loading	Heavier loadings such as railways encourage shorter spans

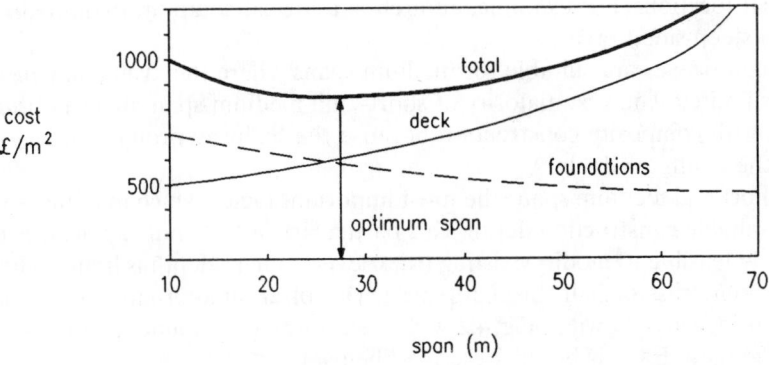

Fig. 4.1   Choice of optimum span for viaducts

Table 4.5   Typical optimum span ranges (m) for viaducts

Conditions	Highway	Railway
Simple foundations (spread footing or short piles)	25–45	20–30
Difficult foundations (piles >20 m long)	35–55	25–40
Piers >15 m high	45–65	30–45

Typical optimum spans are shown in Table 4.5.

Long spans are usually adopted only when a larger number of shorter spans are precluded by restrictions of the site. This is because the material content and cost per unit length of long spans is much greater. Therefore the bridging of a wide river or estuary should generally use a number of medium spans. Only if there is a high risk of shipping collision, or if the depth and speed of water flow is such as to make foundation construction very difficult, should a long-span bridge be chosen.

4.3 Selection of type

Suspension or cable-stayed bridges are suitable for the longest spans, but are less suitable to support heavy loading across short or medium spans. At the same time medium-span footbridges can appropriately be suspension, cable-stayed or arch types because concentrated loading is absent. Some of the considerations given to long spans such as aerodynamics need to be applied to footbridges. For footbridges the phenomenon of pedestrian-excited vibrations needs to be considered and tends to affect the design of spans exceeding about 25 m. The possibility of horizontal oscillation needs consideration if the structure is laterally flexible or is mounted on slender supports. Other bridge types such as arches or portals may be suitable

in special locations. For example, an arch is a logical solution for a medium span across a steep-sided ravine.

Through trusses are suitable for medium spans where the available construction depth is limited. The vast majority of short- and medium-span highway bridges are formed with composite construction because the highway profile can be arranged to suit the depth available.

For short and medium spans the most important factor which influences the type is the available construction depth. This particularly affects railway bridges because it is rarely feasible to modify existing track levels. Where depth is limited then types such as arch, truss or half-through plate girder offer an alternative solution. Types of steel bridge are shown in Fig. 4.2 with their normal economic span range and the world's longest. Each is briefly described below.

4.3.1 Suspension bridges

Suspension bridges (Fig. 4.3) are used for the longest spans across river estuaries where intermediate piers are not feasible. The cables form catenaries supporting both sides of the deck and are tied to the ground usually by gravity foundations sometimes combined with rock anchors. Thus ground conditions with firm strata at or close to the surface of the ground are essential. Towers are usually twin steel or concrete box members which are braced together above the roadway level. They are designed so as to be freestanding under wind loading during construction until the cables are installed. Cables are either a compacted bundle of parallel high tensile steel strands (commonly 5 mm diameter) installed progressively by 'spinning' or may be formed from a group of wire ropes. Deck hangers are wire ropes (or round steel rods for light loading as for a footbridge) clamped to the cable and connected to the deck at a spacing equal to the length of each deck unit erected, typically 18 m. The construction process for suspension bridges is more time consuming than for other types because the deck cannot be installed until the towers, anchorages, cable and hangers are constructed.

Depending upon ground conditions, the cables can be catenaries supporting side spans. Cables may alternatively be straight from tower top to the ground anchorages and merely support a main span, side spans being non-existent or formed as short-span viaducts. Decks are either trusses with a steel orthotropic plate floor spanning between or an aerofoil box girder. Footways are often cantilevered outside the two sets of cables.

Aerodynamic behaviour must be considered in design because of the tendency for the deck and cables to oscillate in flexure and torsion under 'vortex shedding' and other wind effects. This is due to the flexible nature and light weight of suspension bridges illustrated by the collapse of the USA Tacoma Narrows Bridge in 1940, which had a very flexible narrow deck consisting of twin plate girders forming a torsionally weak deck of 'bluff' shape prone to wind vortex shedding. Aerodynamic considerations usually justify wind tunnel testing of models. The advantage of an aerofoil box girder such as used on the Severn and Humber bridges is that

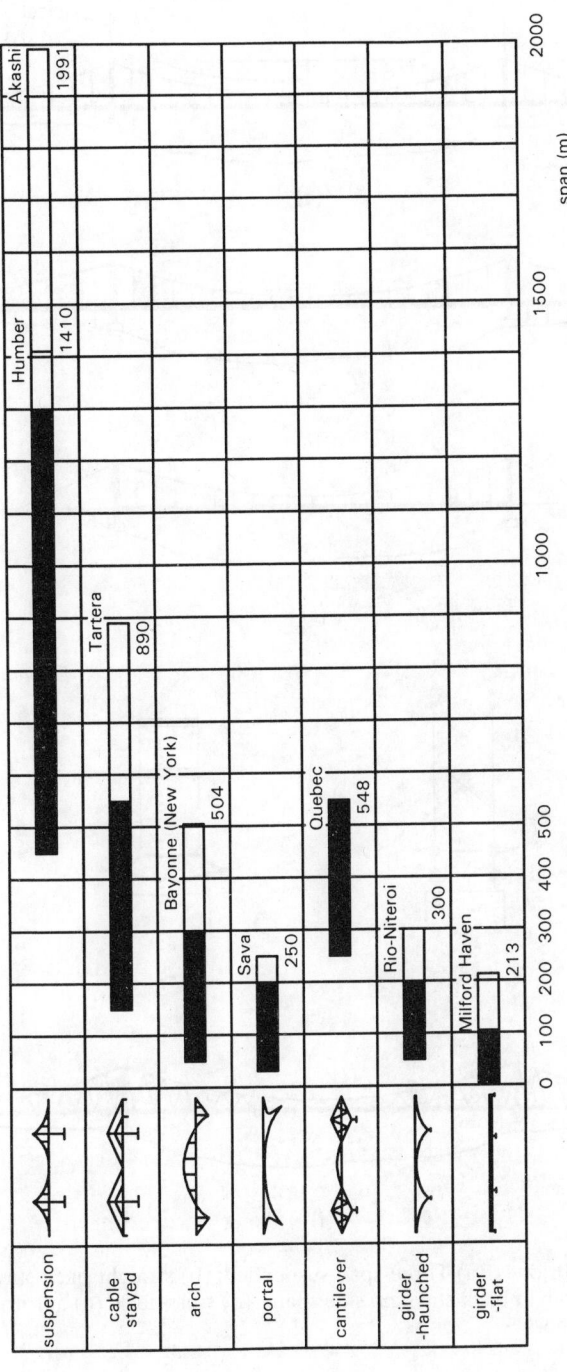

Fig. 4.2   Normal span range of bridge types

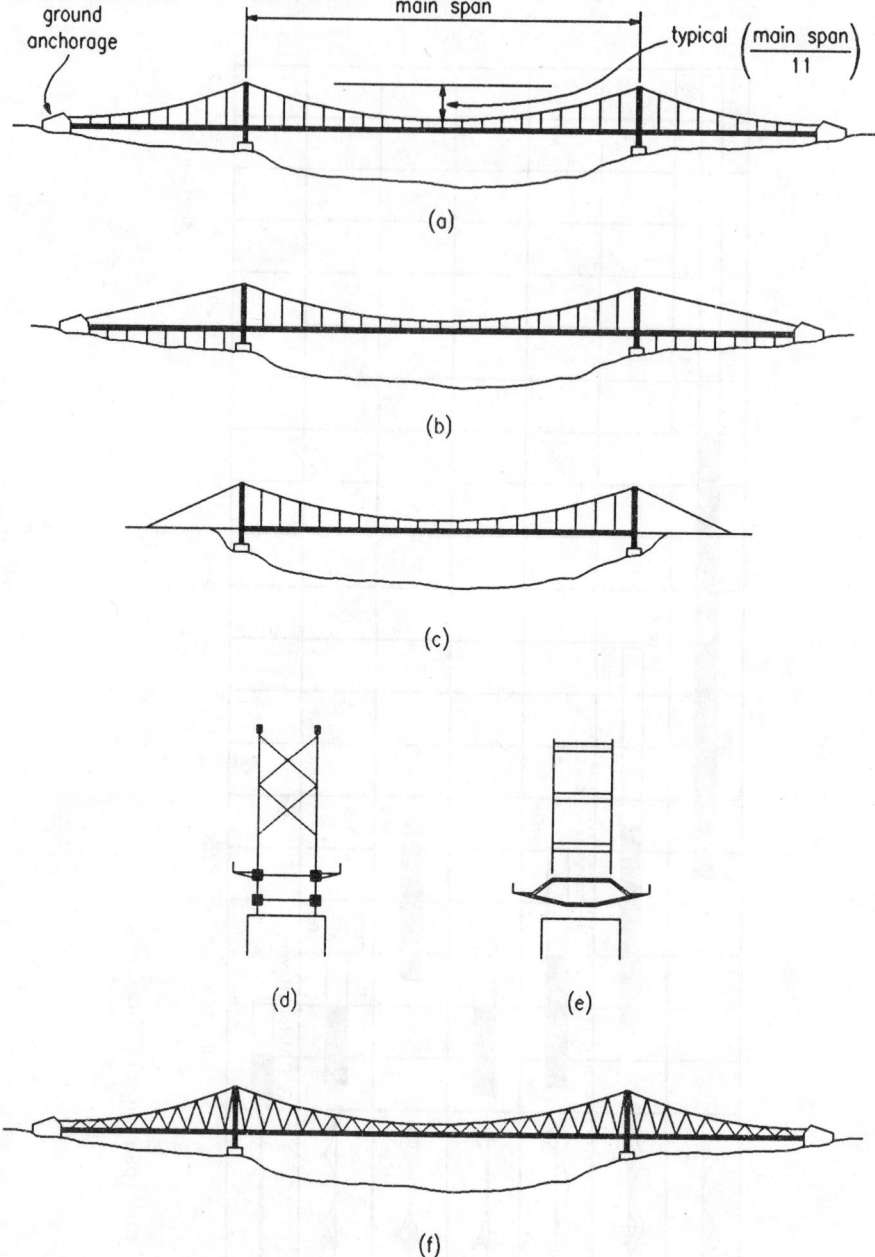

Fig. 4.3 Suspension bridges. (a) Three-span suspended; (b) straight back stays, viaduct side spans; (c) straight back stays, no side spans; (d) truss deck; (e) aerofoil box deck; (f) inclined hangers

it discourages vortex shedding and reduces the wind forces to be resisted by the towers and substructures. The Severn (1966) and Bosporus Bridge (1970) use inclined hangers that form a truss system which helps to reduce any tendency to oscillate. Suspension bridges behave as non-linear structures under asymmetric deck loading so that deflections may be significant. Behaviour under such loading depends upon the combined gravity stiffness and the flexural rigidity of the deck or stiffening girder. The type is less suitable for heavy loading such as railway traffic, especially for short spans. Suspension bridges are sometimes suitable for medium spans carrying pedestrian or light traffic.

4.3.2 Cable-stayed bridges

Cable-stayed bridges (Fig. 4.4) are of a suspension form using straight cables which are directly connected to the deck. The structure is self-anchoring and therefore less dependent upon good foundation conditions, but the deck must be designed for the significant axial stress from the horizontal component of the cable forces. The construction process is quicker than for a suspension bridge because the cables and deck are erected at the same time and the amount of temporary works is reduced. Either twin sets of cables are used or alternatively for dual carriageways a single plane of cables and tower can be located in the central reserve space. Two basic forms of cable configurations are used, either 'fan' or 'harp'. A fan layout minimizes bending effects in the structure due to its better triangulation but anchorages can be less easy to incorporate into the towers. The harp form is often preferred where there are more than, say, four cables. The number of cables depends on the span and cable size, which is often selected such that each fabricated length of deck (say, 20 m) contains an anchorage at one end to suit a cantilever erection method. Bridges either have two towers and are symmetrical in elevation or have a single tower as suited to the site.

Floors are generally an orthotropic steel plate but composite slabs can be used for spans up to about 250 m. A box girder is essential for bridges having a single plane of cables to achieve torsional stability, but otherwise either box girders or twin plate girders are suitable. Aerodynamic oscillation is a much less serious problem than with suspension bridges but must be considered. Some bridges with plate girders incorporate non-structural aerodynamic edge fairings.

It is essential to use cables of maximum strength and modulus at a high working stress so that sag due to self-weight, which produces non-linear effects, is negligible. Cables are normally of parallel wires or prestretched locked coil wire rope. During erection the cable lengths are adjusted or prestressed so as to counteract the dead load deflections of the deck arising from extension of the cables.

4.3.3 Arch bridges

Arch bridges (Fig. 4.5) are suitable in particular site conditions. An example is a medium single span over a ravine where an arch with spandrel columns will

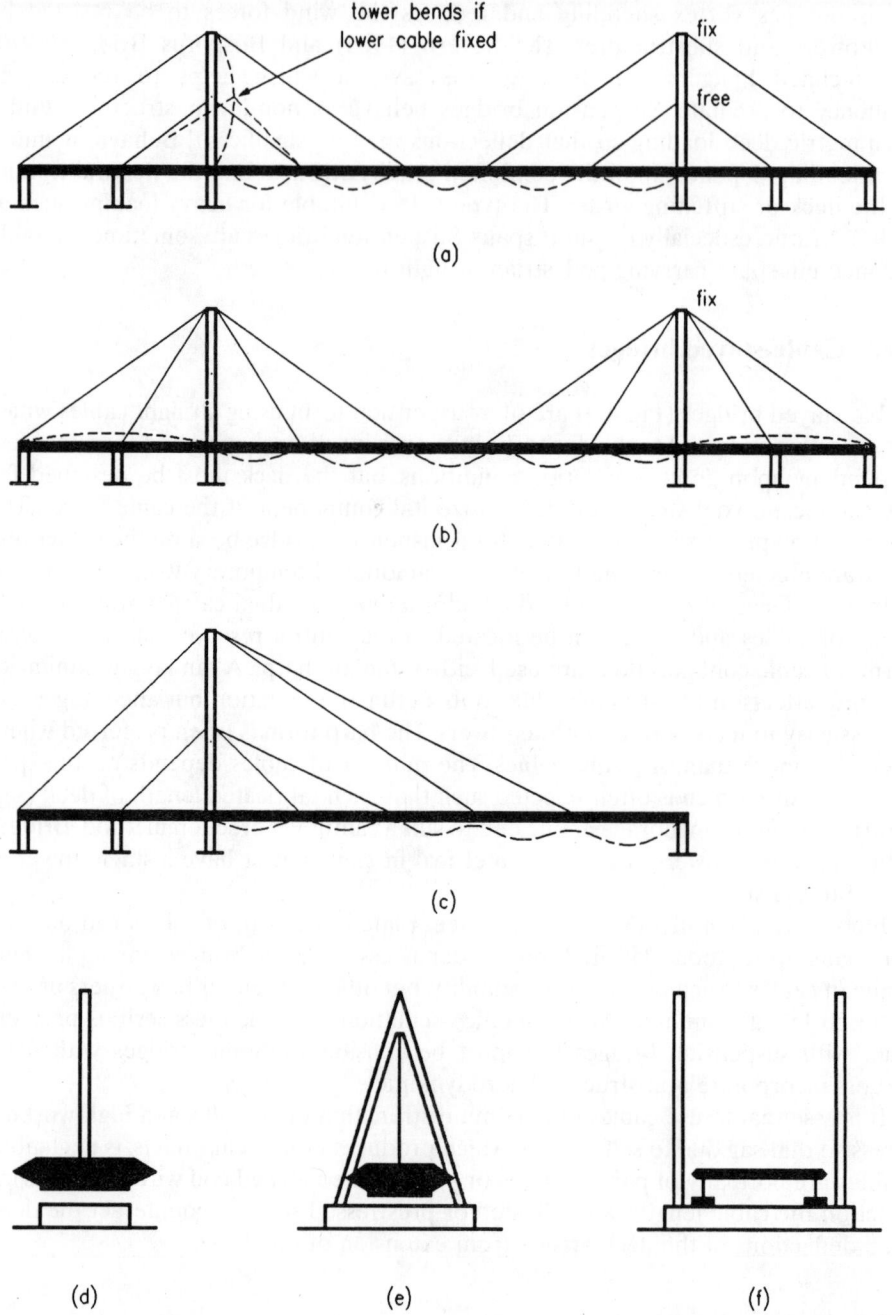

Fig. 4.4   Cable-stayed bridges. (a) Harp; (b) fan; (c) single tower; (d) single plane; (e) single plane; (f) twin plane

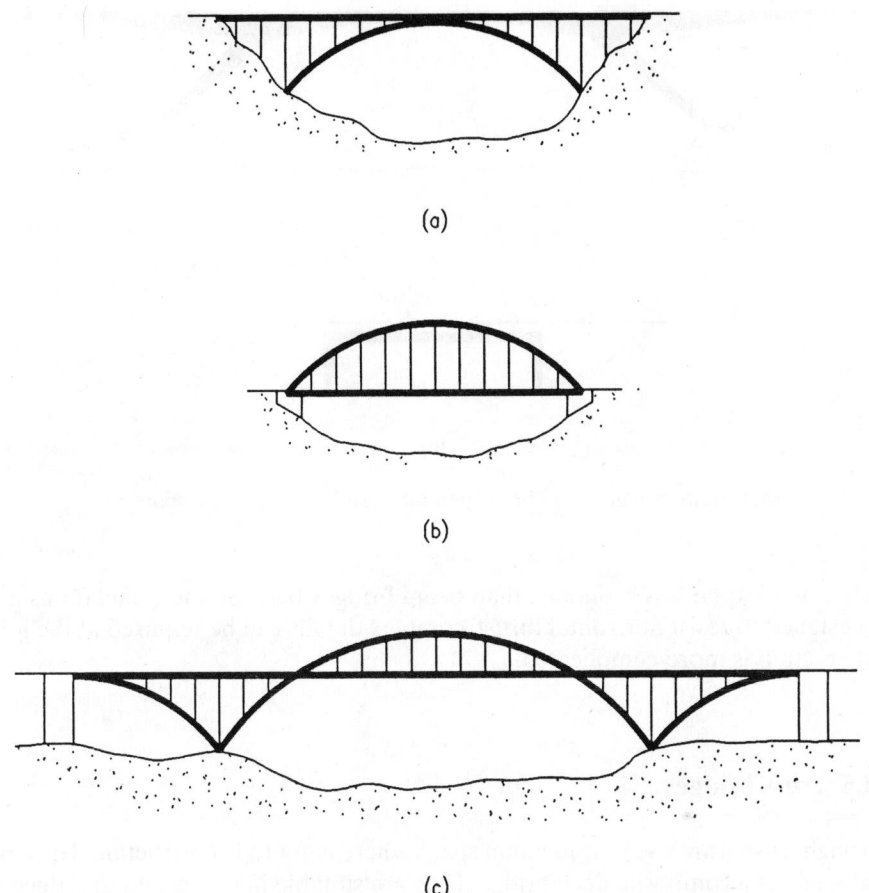

Fig. 4.5 Arch bridges. (a) Spandrel post arch; (b) tied arch; (c) part tied arch

efficiently carry a deck with the horizontal thrust taken directly to rock. A tied arch is suitable for a single span where construction depth is limited and presence of curved highway geometry or other obstruction to the approaches conflicts with the back stays of a cable-stayed bridge.

4.3.4 Portal frame bridges

Portal frame bridges (Fig. 4.6) are mainly suitable for short or medium spans. In a three-span form with sloping legs they can provide an economic solution by offering a reduction in span and have an attractive appearance. The risk of shipping collision with sloping legs must be considered for bridges over navigable rivers. Portal

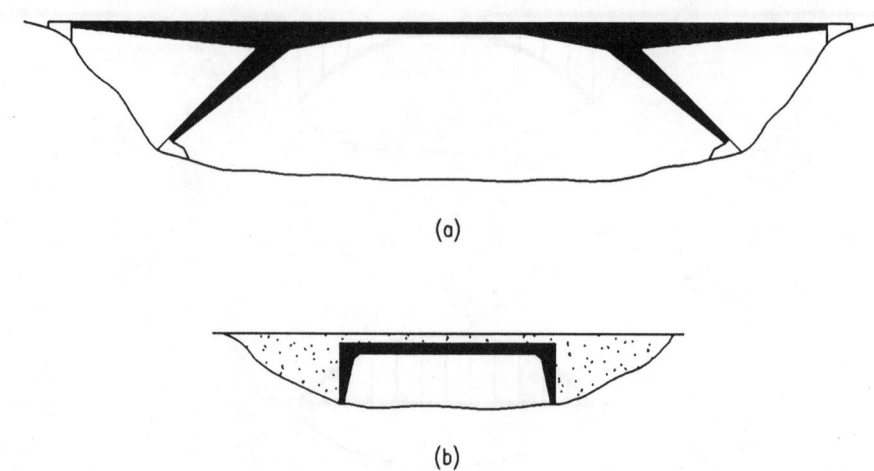

Fig. 4.6   Portal frame bridges. (a) Three-span inclined legs; (b) single span

bridges tend to be less economic than beam bridges because the foundations must be designed to resist horizontal thrust, complex details can be required at the joints and erection is more complicated.

4.3.5 Truss bridges

Through trusses are used for medium spans where a limited construction depth precludes use of a composite deck bridge. They are suitable in flat terrain to reduce the height and length of approach embankments and for railway bridges where existing gradients cannot be modified. A truss may be unacceptable visually; a bowstring truss is an alternative solution. For short spans and medium spans up to 50 m, trusses are generally less economic than plate girders because of higher fabrication cost. They are therefore adopted only where the available construction depth is not sufficient for composite beams. (See Fig. 4.7.)

4.3.6 Girder type

Girder bridges (Fig. 4.8) predominate over the previously described types for short and medium spans, and generally provide the most economic solution. For highway bridges composite deck construction is generally used unless the depth is very critical in which case half-through girders or through trusses may be necessary. Railway bridges frequently require to be of half-through or through form because of depth limitations; these are more suitable, being generally narrower. Pedestrian bridges

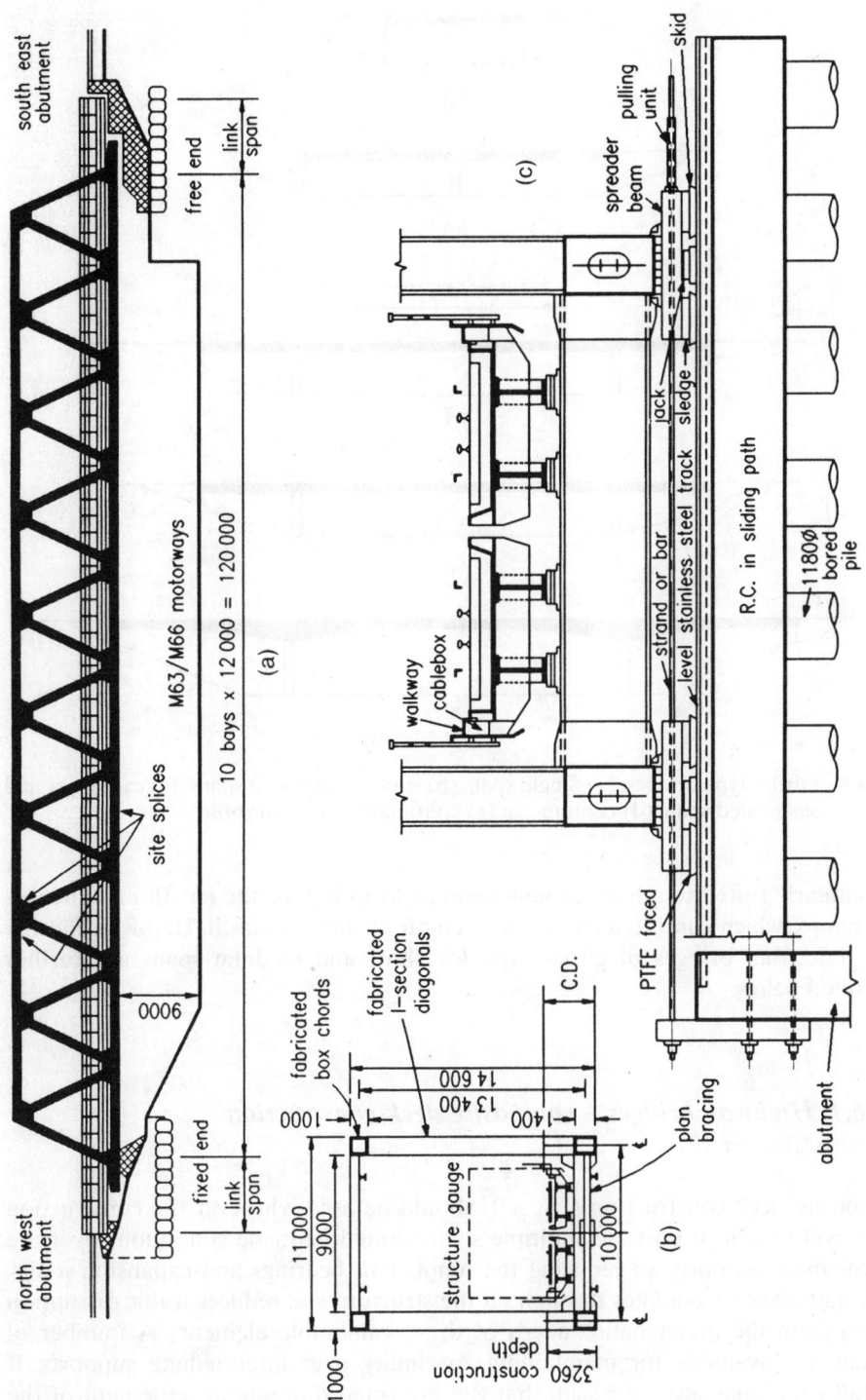

Fig. 4.7 Truss bridge: (a) elevation (b) cross section (c) arrangements for sliding into position

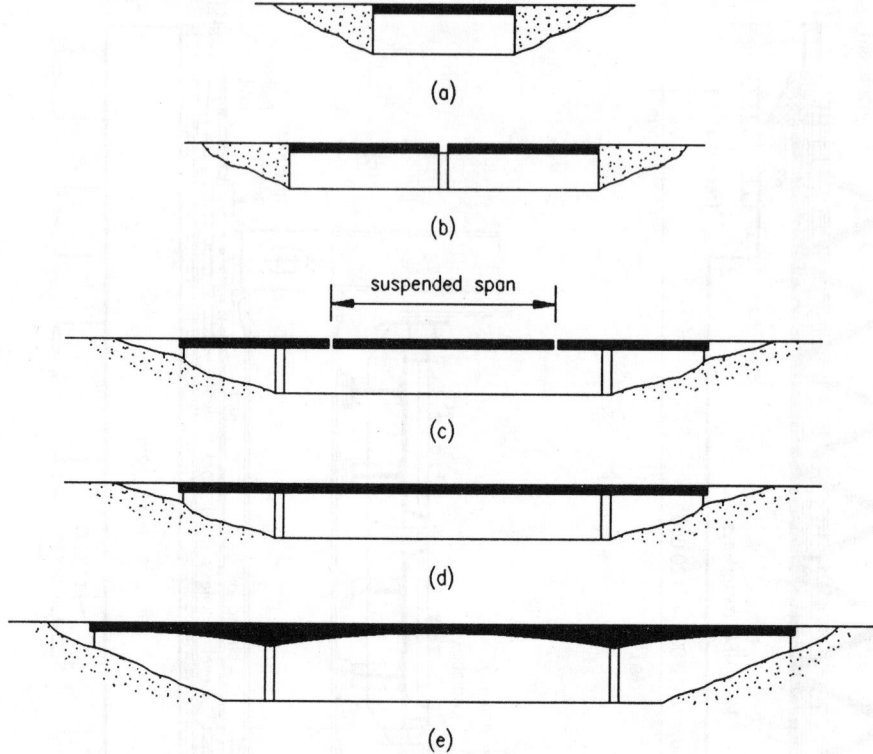

Fig. 4.8   Girder type bridges. (a) Single span; (b) simply-supported spans; (c) cantilever and suspended span; (d) continuous; (e) continuous – curved soffit

are similarly suited to a half-through form so as to reduce the length of staircases and ramps, which can often exceed the length of the span itself. Highway, railway and pedestrian bridges of girder type for short and medium spans are further described below.

4.3.6.1 Highway bridges – composite deck construction
(see also Chapter 17)

Composite deck construction (Fig. 4.9) should be used wherever the construction depth will permit. If possible, multiple spans should be made continuous over the intermediate supports, so reducing the number of bearings and expansion joints. Continuity gives economies throughout the structure and reduces traffic disruption arising from the maintenance needs of these vulnerable elements. A number of options are available for maintaining continuity over intermediate supports. If ground conditions are poor such that the predicted differential settlement of the

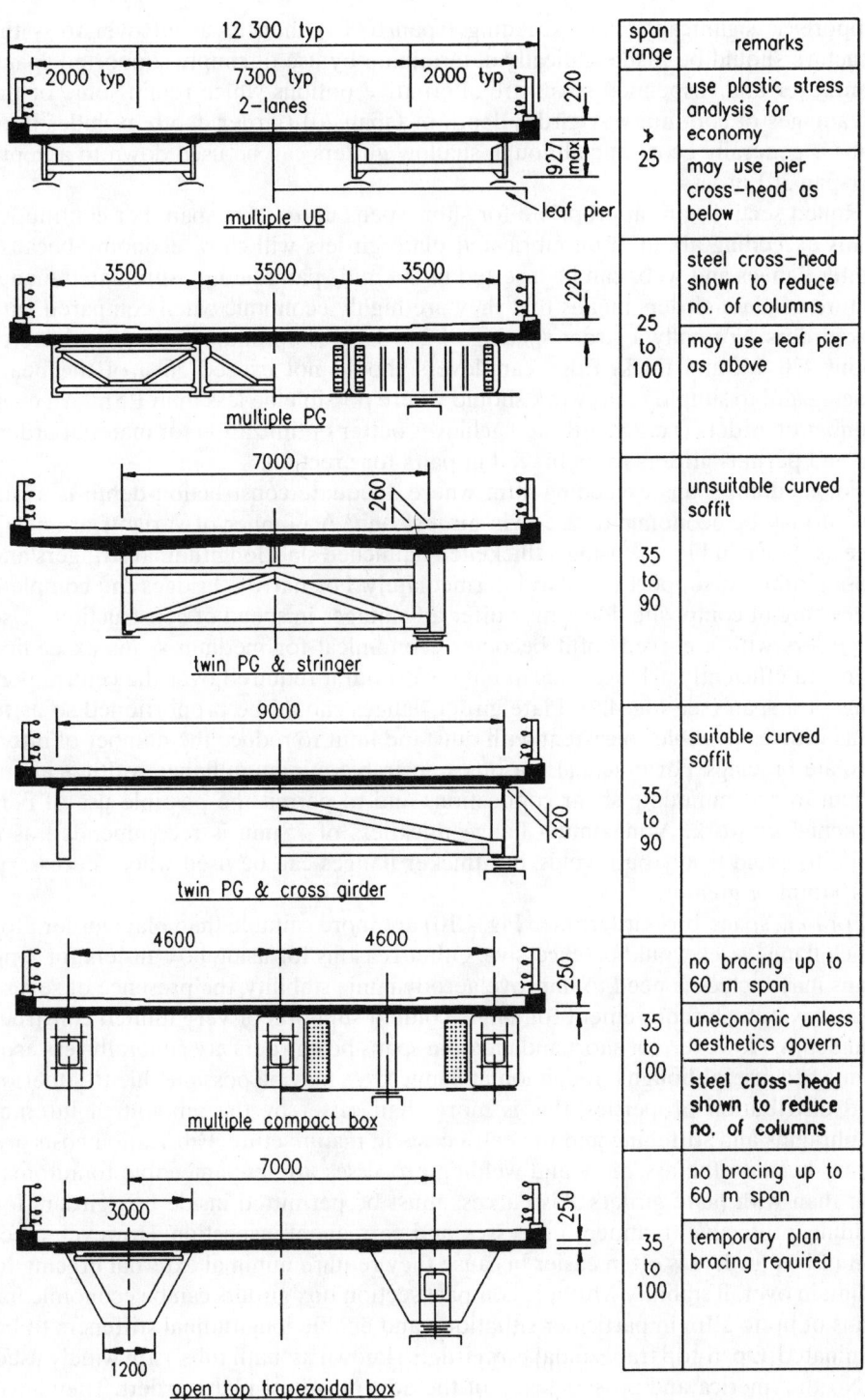

span range	remarks
> 25	use plastic stress analysis for economy
	may use pier cross-head as below
25 to 100	steel cross-head shown to reduce no. of columns
	may use leaf pier as above
35 to 90	unsuitable curved soffit
35 to 90	suitable curved soffit
35 to 100	no bracing up to 60 m span
	uneconomic unless aesthetics govern
	steel cross-head shown to reduce no. of columns
35 to 100	no bracing up to 60 m span
	temporary plan bracing required

Fig. 4.9   Highway bridges – composite deck construction

supports is significant (say, exceeding (span/1000)) then to avoid overstress the structure should be made statically determinate by use of simply-supported spans. Cantilever and suspended spans are alternative options which retain some of the advantages of continuity. A girder depth of (span/20) (girder depth excludes floor slab) is generally economic although shallow girders can be used down to a depth of (span/30) or less.

Rolled sections are appropriate for short spans up to 25 m span. For continuous spans exceeding about 22 m, fabricated plate girders will show economy because lighter flanges and webs can be inserted in the mid-span regions. Automated manufacture of plate girders means that they are highly economic when compared with box girders. Normally a girder spacing of 2.5–3.5 m is optimum with a floor slab of about 220–250 mm thick. Edge cantilevers should not exceed 50% of the beam spacing and to simplify falsework should where possible be less than 1.5 m. An even number of girders (i.e. 2,4, 6, 8, etc.) achieves better optimization for material ordering and permits girders to be braced in pairs for erection.

For medium spans exceeding 40 m where adequate construction depth is available it may be economic to use twin girders only. A number of variants are available as shown in Fig. 4.9, using a thickened haunched slab, longitudinal stringers and cross girders to support the slab intermediately. For narrow bridges the complete precasting of composite floors may offer advantages in speed of construction.[2] Use of girders with a curved soffit becomes economical for medium spans exceeding 45 m and efficiently achieves maximum headroom if required over the central portions of a span (see Fig. 4.9). Plate girder flanges should be proportioned so as to be as wide as possible consistent with outstand limit to reduce the number of intermediate bracings. For practical reasons a desirable minimum flange width is about 40 mm to accommodate shear connections and to permit the possible use of permanent formwork. A maximum flange thickness of 75 mm is recommended as a guide to avoid heavy butt welds, but thicker flanges can be used where necessary, to 100 mm or greater.

For long spans, box girders (see Fig. 4.10) are more suitable than plate girders, for which flange sizes would be excessive. Other reasons for using box girders for long spans may include a need to improve aerodynamic stability, the presence of severe plan curvature, a requirement for single column supports or very limited construction depth. However, for short and medium spans box girders are generally less economic because, although a reduction in flange sizes may be possible due to superior load distribution properties, this is more than offset by the amount of internal diaphragms and stiffening and the extra costs in manufacture. Fabrication costs are higher because the assembly and welding processes are less amenable to automation than with plate girders. Also access must be permitted inside box girders for welding, protective treatment processes, and permanent inspection. However, erection of box girders is often easier because they require minimal external bracing to maintain overall stability. Multiple compact section box girders can be economic for spans of up to 50 m in particular situations, and enable longitudinal stiffeners to be eliminated. Open-top trapezoidal box girders (known as 'bath tubs') are widely used in North America and possess some of the advantages of plate girders. They have

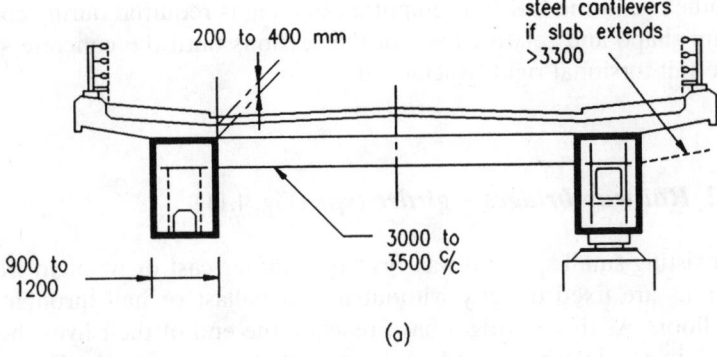

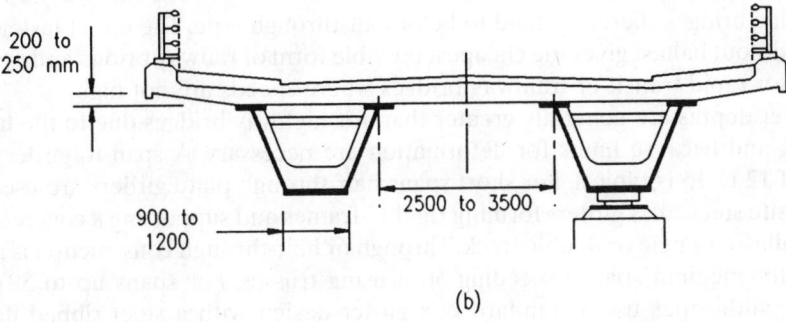

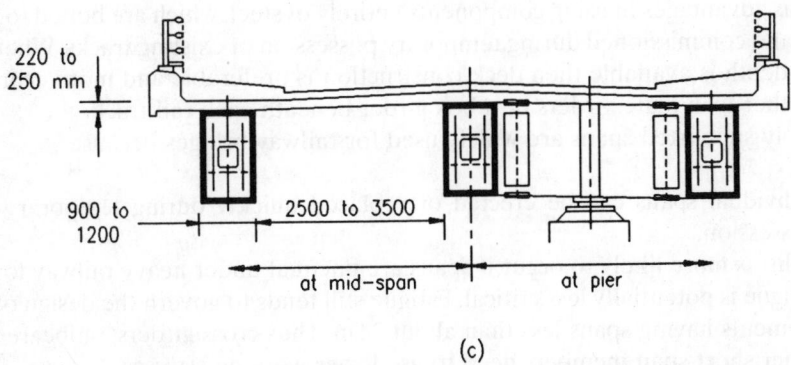

Fig. 4.10 Highway bridges – box girders. (a) Twin box and cross girders (spans 40–150 m); (b) open top box (spans 40–100 m); (c) multiple box (spans 30–60 m)

seen same use in the UK but temporary bracing is required during construction to maintain shape and relative twist of the sections until the concrete slab is placed and the full torsional rigidity achieved.

4.3.6.2 *Railway bridges – girder type* (Fig. 4.11)

Many existing small-span railway bridges weather cast or wrought iron girders to which rails are fixed directly without use of ballast or half-through girders with trough floors. As these bridges have reached the end of their lives they have been replaced in steelwork in modern form with ballasted track. The legacy of the original decks with their very shallow construction depth has influenced modern underline bridge practice in replacements and new bridges. The majority of railway underline bridges therefore tend to be of half-through type. The direct fastening of track without ballast gives the cheapest possible form of railway bridge and is appropriate for rapid transit or tramway bridges where speeds are not high.

Girder depths are generally greater than for highway bridges due to the heavier loading and because limits for deformation are necessary. A span-to-girder depth ratio of 12 to 15 is typical. For short spans half-through plate girders are used with composite steel cross girders forming rigid U-frames and supporting a concrete floor with ballasted single or double track. Through or half-through construction is appropriate for medium spans exceeding 50 m using trusses. For spans up to 39 m, the railway authorities use a standard box girder design with a steel ribbed floor of minimal depth which is achieved by spanning between the inner webs of trapezoidal box girders proportioned so as to fit closely within the station platform space. The type has advantages in using components entirely of steel, which are bolted together at site and commissioned during temporary possession of existing tracks. Where sufficient depth is available then deck construction is preferable and more economic, with either twin plate girders or a box girder beneath each rail track.

Simply-supported spans are widely used for railway bridges because:

(a) individual spans can be erected or replaced quickly during temporary track possession,
(b) uplift is more likely to occur if spans are unequal under heavy railway loading,
(c) fatigue is potentially less critical. Fatigue still tends to govern the design of steel elements having spans less than about 24 m. Thus cross girders, railbearers and other short-span members need to use lower working stresses.

4.3.6.3 *Pedestrian bridges* (Fig. 4.12)

A minimum clear deck width of 1.8 m is usual, increased to 3.0 m or 4.0 m in busy areas or if a cycleway is also present. Steel provides an efficient solution because

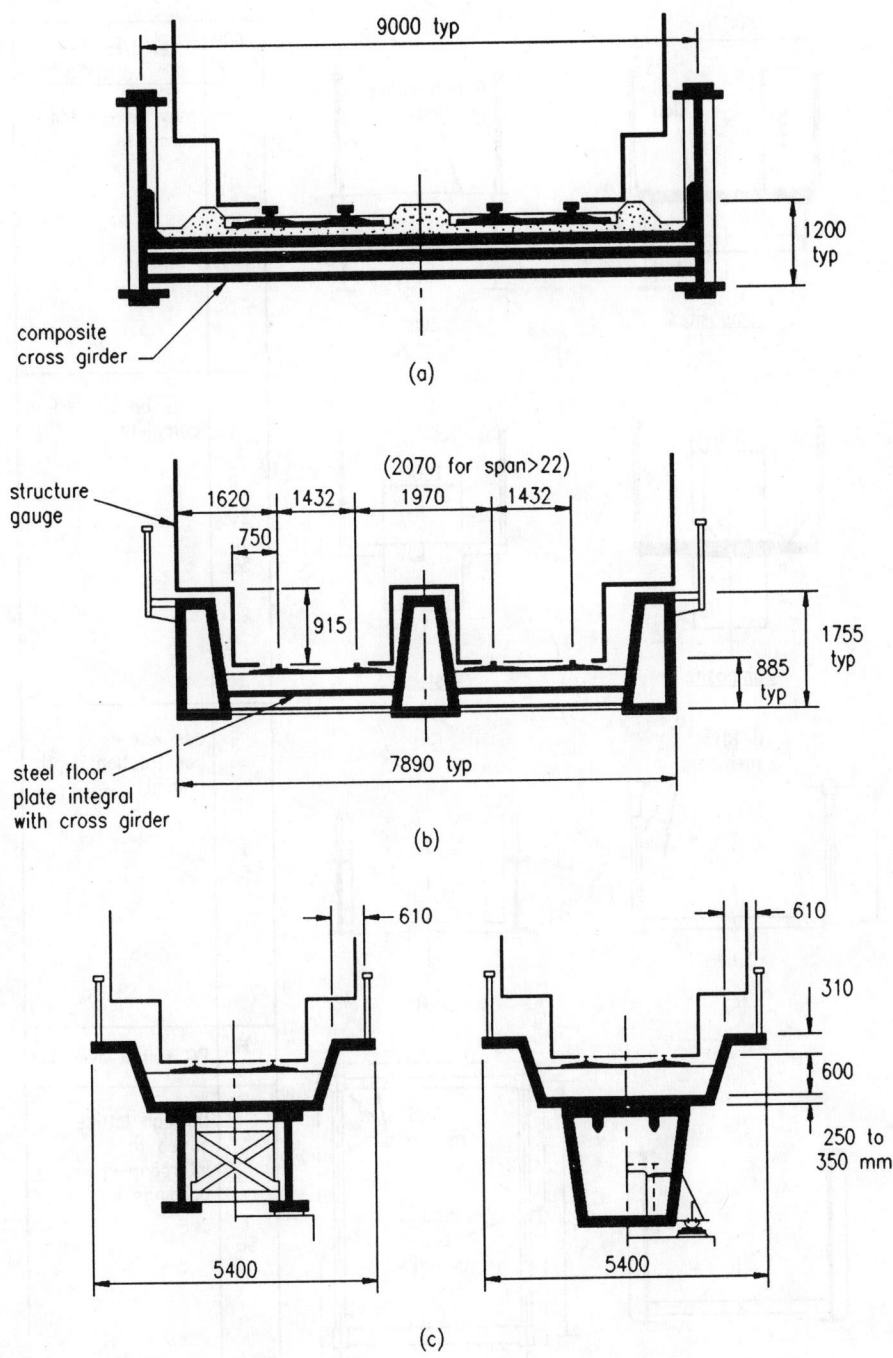

Fig. 4.11 Railway bridges. (a) Half-through plate girders; (b) half-through box girders; (c) composite deck type (spans > 30 m)

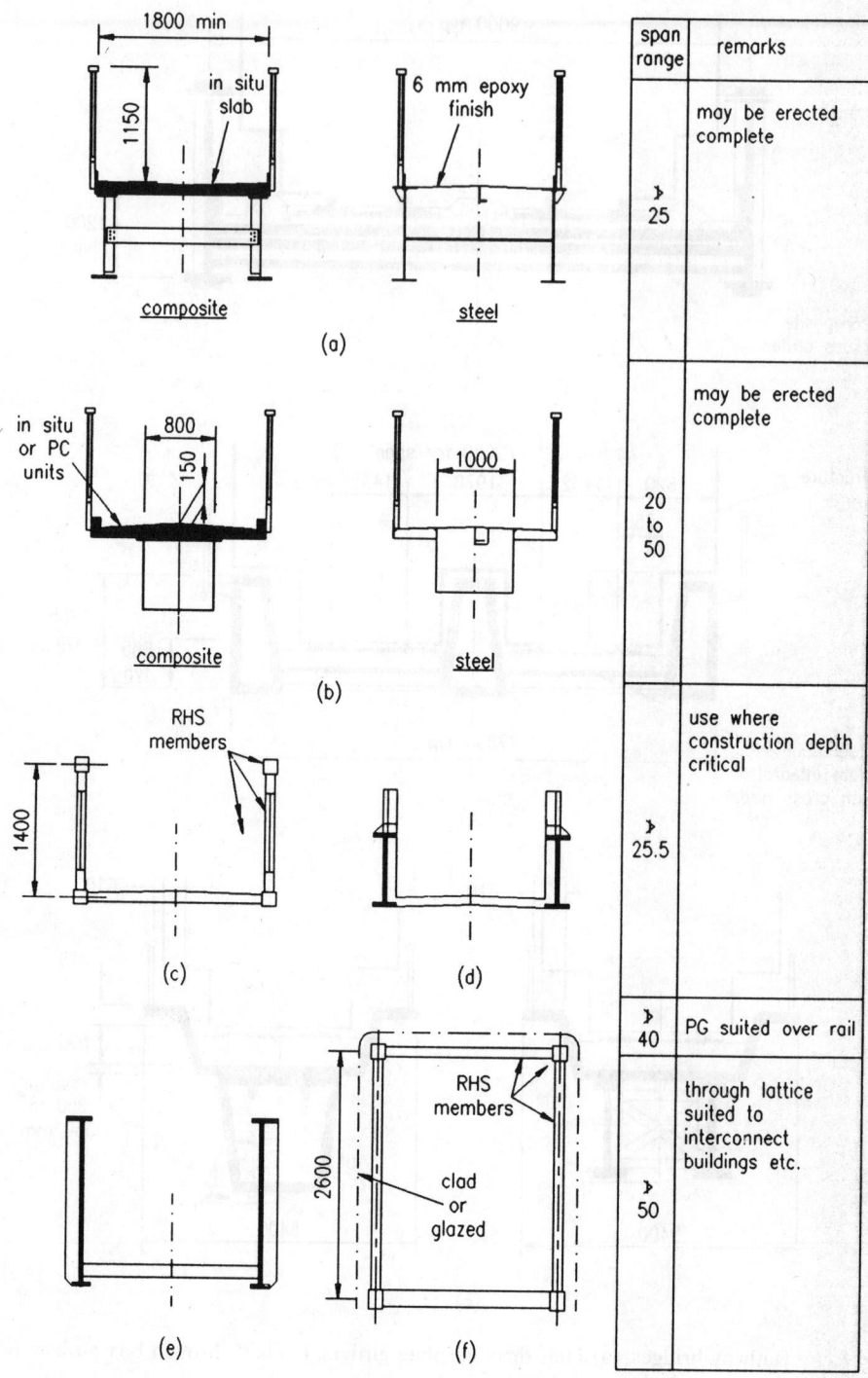

span range	remarks
⟩ 25	may be erected complete
20 to 50	may be erected complete
⟩ 25.5	use where construction depth critical
⟩ 40	PG suited over rail
⟩ 50	through lattice suited to interconnect buildings etc.

Fig. 4.12 Pedestrian bridges. (a) Twin universal beam; (b) box; (c) Vierendeel or Warren girder; (d) half-through universal beam; (e) half-through plate girder; (f) through lattice

the entire cross-section including parapets may be fabricated and erected in one unit. For this reason multiple spans tend to be simply supported and not continuous. Staircase and sloping or stepped ramp spans are also erected as complete units, and columns are often of steel.

Half-through cross-sections are often used because the shallow construction depth is able to provide the shortest lengths of staircase or ramp approaches in urban areas. Half-through rolled beams, Warren truss or Vierendeel girders are used with rolled hollow section (square or rectangular) members. Floors are steel stiffened plate, often surfaced with a factory-applied epoxy non-slip surfacing approximately 5 mm thick. Rigid connections between floor and girders provide for U-frame stability. Staircase approaches may be either steel stringers supporting steel plate treads or a central spine box with cantilever treads of steel or precast concrete. Ramps may be similarly formed, but where the span exceeds about 10 m then half-through construction tends to be used. Spiral approach ramps can be used, formed in steel using twin rolled sections curved in plan supporting a composite or steel plate floor. Where adequate construction depth is available, for example when a pedestrian bridge is required across a motorway or railway cutting, then deck cross-sections are appropriate with a composite or steel plate floor. Precast concrete floor units are also suitable. Primary members may be a single box girder, twin plate girders or twin rolled beams. A single box girder provides a structure of neat appearance.

4.4 Codes of practice

BS 5400[3] is used in the UK for the design of all bridges. Part 3 deals with the design of steel bridges and if composite construction is used then Part 5 must also be referred to. The Highways Agency and the railway authorities have their own particular requirements in the form of technical standards which implement and sometimes vary the clauses in BS 5400. It is expected that Eurocodes will eventually replace BS 5400.

Both BS 5400 and the Eurocodes are based on the limit state design concept, which means that the verification is carried out for both the serviceability and ultimate limit states.

The first steps for the preparation of the rules for the traffic loads for bridges, particularly for road bridges, have been undertaken. For the traffic loads on railway bridges the harmonized uniform load model UIC 71[4] has been previously adopted by the different national railway authorities.

4.5 Traffic loading

4.5.1 Highway bridges

Highway bridges in the UK are currently designed for HA loading (a uniformly distributed loading plus knife edge load applied to each traffic lane) together with HB
(abnormal vehicle) loading for structures carrying main highways. Details are given
in BD 37/01, which is a revision of the loading given in BS 5400: Part 2. HA and HB
loading are deemed to allow for dynamic and impact effects. For footways the
normal loading is $5\,kN/m^2$ reduced to $4\,kN/m^2$ where the highway is also loaded. It
is further reduced for longer loaded lengths, similarly to HA loads.

4.5.2 Railway bridges

The uniform load model UIC 71[4] used by the European national railway authorities is shown in Fig. 4.13: it is also used in BS 5400: Part 2 and reproduced in standard BD 37/01. Where specified, this loading is multiplied by a factor for bridges on
lines carrying heavier or lighter traffic.
   Dynamic factors must be applied as shown in Table 4.6. Dimension L is the length
(m) of the influence line for deflection of the element under consideration.
   Other loads arising from railway traffic are:

- centrifugal forces on curved track
- nosing – 100 kN force acting transversely at rail level

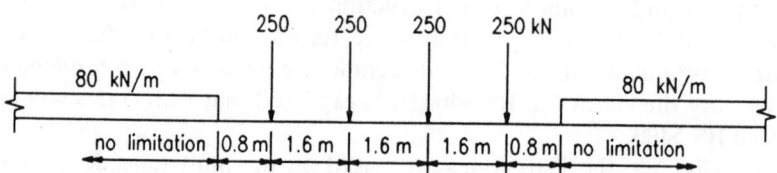

Fig. 4.13   UIC and BS 5400 load model (excluding impact)

Table 4.6 Dynamic factors for railway loading

Dimension L	Dynamic factor	
(m)	Bending moment	Shear
⩾3.6	2.00	1.67
>3.6 to 67	$0.73+\dfrac{2.16}{\sqrt{L}-0.2}$	$0.82+\dfrac{1.44}{\sqrt{L}-0.2}$
>67	1.00	1.00

- traction and braking
- derailed vehicles (for overturning consider 80 kN/m over 20 m lengths acting on the edge of the structure).

Fatigue is important for the design of railway bridges because a higher proportion of regular traffic attains the maximum loading compared with highway bridges, and the vehicles are constrained to run in the same lateral paths.

4.6 Other actions

Actions other than traffic loading which may need to be considered in one or more combinations are shown in Table 4.7. The term 'action' is used in Eurocodes to describe all loads or load effects.

4.7 Steel grades

Steels to BS EN 10025 of grade S355 are usual for bridges as they offer a better cost-to-strength ratio than grade S275. All parts subject to tensile stress are required

Table 4.7 Summary of actions other than those due to traffic loading

Action	Comments
Dead loads[a]	Weight of the structure
Superimposed dead loads[a]	Finishes and surfacings Services
Wind	Transverse, longitudinal and vertical Consider in presence of live load or otherwise
Temperature	Restraint and movement (e.g. flexure of columns) Frictional restraint of bearings Effect of temperature difference
Differential settlement	Foundation movements
Earth pressure	Vertical and horizontal pressures from retained material
Erection effects[a]	Strength during construction, e.g. stability of steelwork before composite floor slab cast
Snow load	May be relevant to moving bridges
Seismic	As may be specified by the national authority
Water flow	Flow against bridge supports

[a] These actions must be considered during initial design. For most bridge decks actions other than those from vertical traffic loading, dead and superimposed dead loads are unlikely to have a fundamental effect.

to achieve a specified notch toughness, depending upon design minimum temperature, stress level and material thickness.

Special consideration may be needed if tensile stresses occur only during erection, for example arising from lateral bending in girders due to wind loading, or during lifting of components. Judgement is necessary but it is often accepted that modest values of tensile stress can be permitted in such cases without the need for specific notch toughness provided that the work is not carried out during very low temperatures. In all other cases appropriate grades having specified notch toughness are necessary.

Weathering steel

To eliminate the need for painting, weathering grades in BS EN 10155 should be considered. Although it can be shown that the commuted cost of repainting steel bridges is not of great significance compared with the initial bridge cost, weathering steel is particularly useful in eliminating maintenance where access is difficult – over a railway and increasingly so over highways. Weathering steel is not suitable at or near the coast (i.e. within about 2 km of the sea). The Highways Agency requires sacrificial thickness to be added to all exposed surfaces for possible long-term corrosion of 1.5 mm per face in a severe industrial environment, 1 mm otherwise.

4.8 Overall stability and articulation

It is important to consider overall stability of the bridge and its articulation under temperature effects (see Fig. 4.14). For simply-supported bridges each span must transfer longitudinal and transverse loads to the foundations while being able to accommodate movement with suitable bearings and expansion joints. For continuous spans the deck must be pinned at one support with free bearings elsewhere. Normally only one bearing within the deck width should be pinned longitudinally so that each girder is free to articulate under traffic loading independently, unless the pier consists of a separate column beneath each bearing which gives flexibility. A number of choices are open to the designer but the system used will affect the design of bearings, bearing stiffeners, expansion joints and the foundations.

For temperature movements:
Movement range: $(12 \times 10^{-6}$ per °C$) \times$ temperature range $\times$ length from pinned bearing.
For typical conditions in Europe, for steel or composite decks, ultimate movement = ±4.5 mm per 10 m of length from mean temperature.

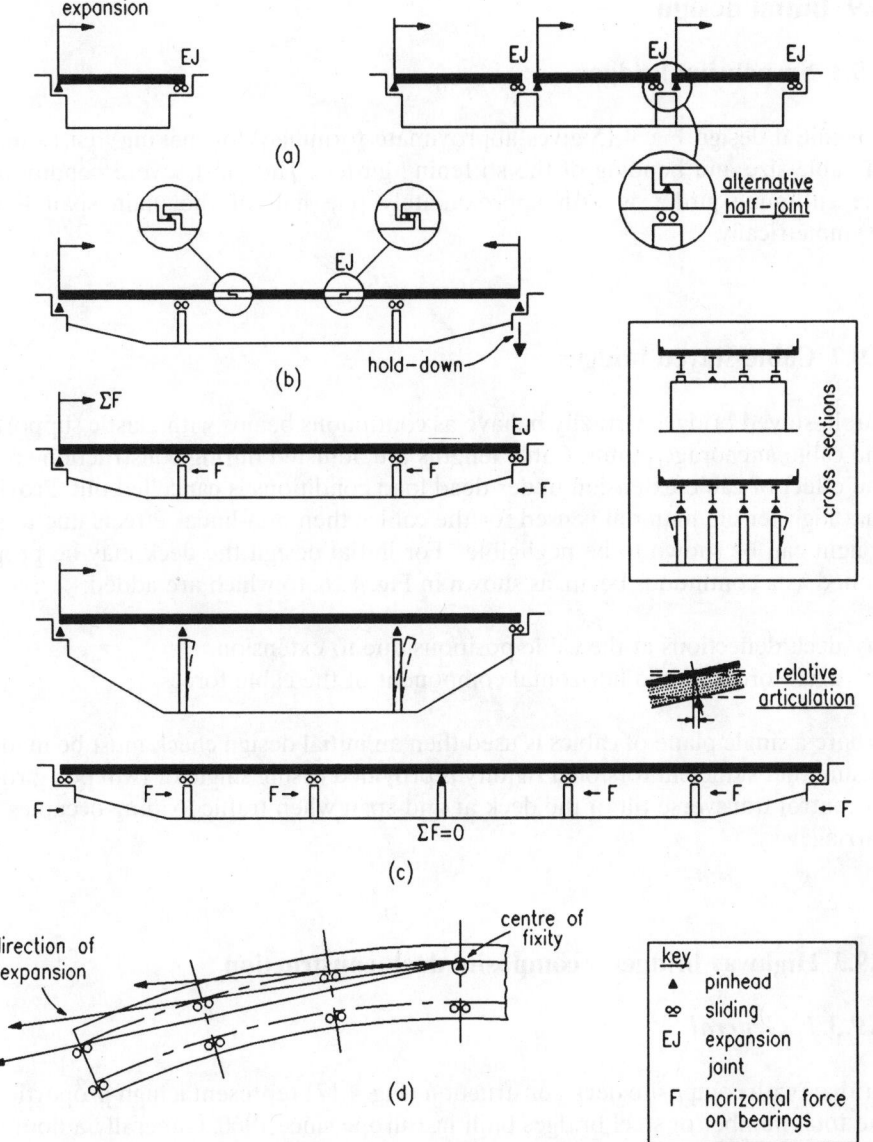

Fig. 4.14　Overall stability and articulation. (a) Simply supported, (b) cantilever and suspended span, (c) continuous, (d) curved viaducts

4.9 Initial design

4.9.1 Suspension bridges

For initial design Fig. 4.15 gives approximate formulae[5] for making first estimates of cable size and bending of the stiffening girders. The most severe condition for the stiffening girder is with approximately one half of the main span loaded asymmetrically.

4.9.2 Cable-stayed bridges

Cable-stayed bridges virtually behave as continuous beams with elastic supports at the cable anchorage points. Cable lengths are adjusted during construction so that the effect of cable extension under dead load conditions is cancelled out. Provided that high tensile material is used for the cables then non-linear effects due to self-weight can be shown to be negligible.[6] For initial design the deck may be proportioned as a continuous beam, as shown in Fig. 4.16, to which are added:

(a) deck deflections at the cable positions due to extension.
(b) axial forces due to horizontal component of the cable forces.

Where a single plane of cables is used then an initial design check must be made to ensure that sufficient torsional rigidity is provided (using single or twin box girders) to control transverse tilt of the deck at mid-span when traffic loading occupies one carriageway.

4.9.3 Highway bridges – composite deck construction

4.9.3.1 General

Bridges with composite deck construction (Fig. 4.17) represent a high proportion of the total number of steel bridges built in Europe since 1960. Generally a floor slab 220 mm to 250 mm thick is used composite with rolled sections or plate girders at spacings up to 3.5 m and depth between (span/20) and (span/30). For spans exceeding 40 m where adequate construction depth is available then twin girders can offer advantages with typical depth (span/18) to (span/25), and the floor slab is either haunched over the girders or supported by subsidiary stringers or cross girders. The slab thickness is determined by its requirement to resist local bending and punching shear effects from heavy wheel loads and needs to be reinforced in both directions: elastic design charts by Pucher are suitable.[7] In the hogging regions of continuous spans the slab will crack and be ineffective in overall flexure unless it is

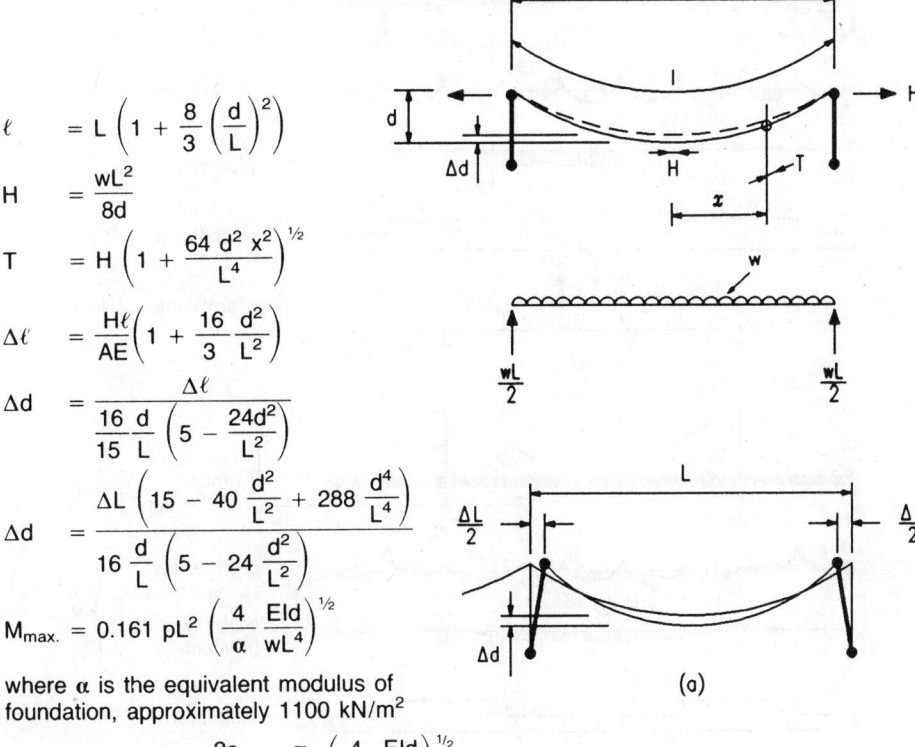

$$\ell = L\left(1 + \frac{8}{3}\left(\frac{d}{L}\right)^2\right)$$

$$H = \frac{wL^2}{8d}$$

$$T = H\left(1 + \frac{64\ d^2\ x^2}{L^4}\right)^{1/2}$$

$$\Delta\ell = \frac{H\ell}{AE}\left(1 + \frac{16}{3}\frac{d^2}{L^2}\right)$$

$$\Delta d = \frac{\Delta\ell}{\dfrac{16}{15}\dfrac{d}{L}\left(5 - \dfrac{24d^2}{L^2}\right)}$$

$$\Delta d = \frac{\Delta L\left(15 - 40\dfrac{d^2}{L^2} + 288\dfrac{d^4}{L^4}\right)}{16\dfrac{d}{L}\left(5 - 24\dfrac{d^2}{L^2}\right)}$$

$$M_{max.} = 0.161\ pL^2\left(\frac{4}{\alpha}\frac{EId}{wL^4}\right)^{1/2}$$

where α is the equivalent modulus of foundation, approximately 1100 kN/m^2

$M_{max.}$ occurs when $\dfrac{2a}{L} = \dfrac{\pi}{4}\left(\dfrac{4}{\alpha}\dfrac{EId}{wL^4}\right)^{1/2}$

(a)

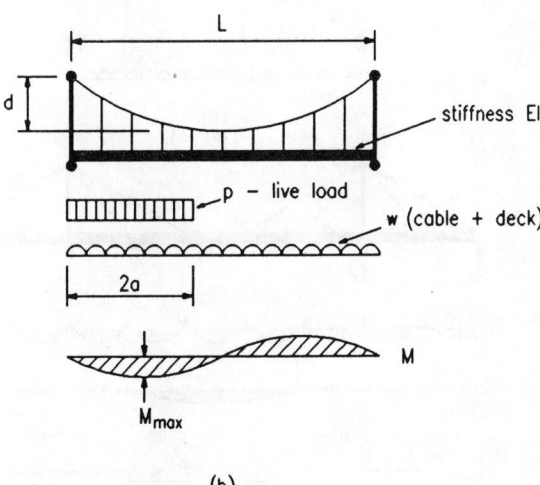

(b)

Fig. 4.15   Suspension bridges, approximate formulae (Pugsley). (a) Neglecting stiffness of deck; (b) including deck stiffness

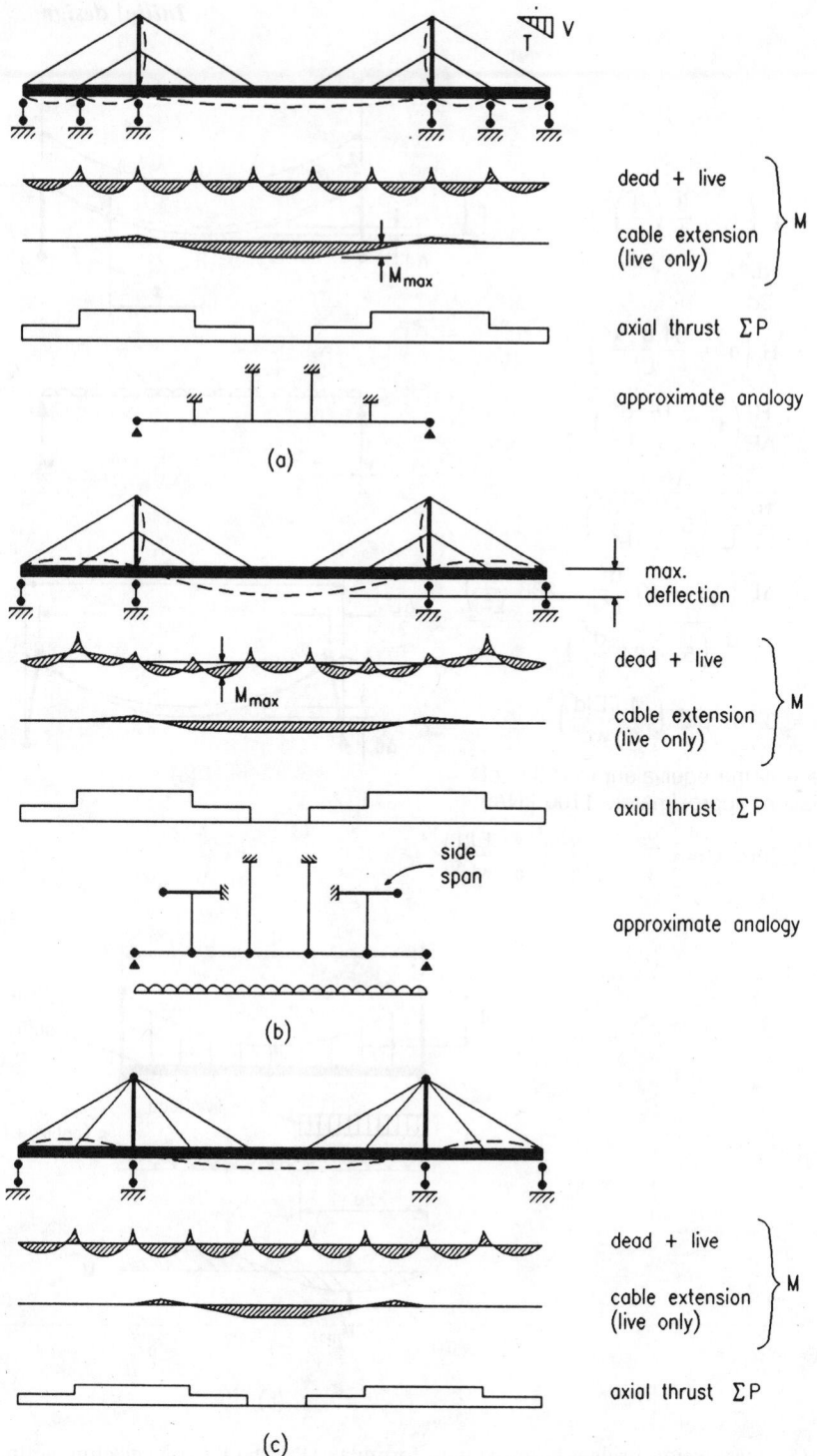

Fig. 4.16 Cable-stayed bridges, initial analysis. (a) Anchored side spans (harp shown); (b) harp; (c) fan

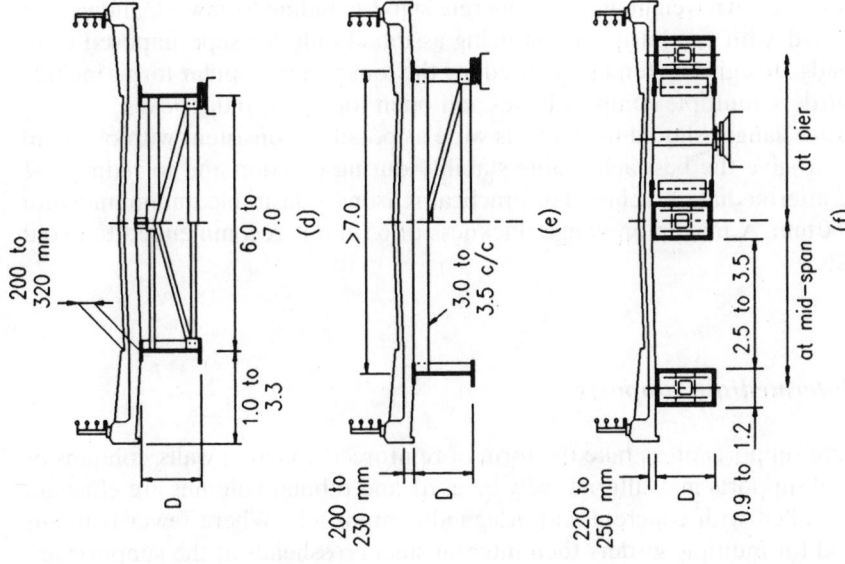

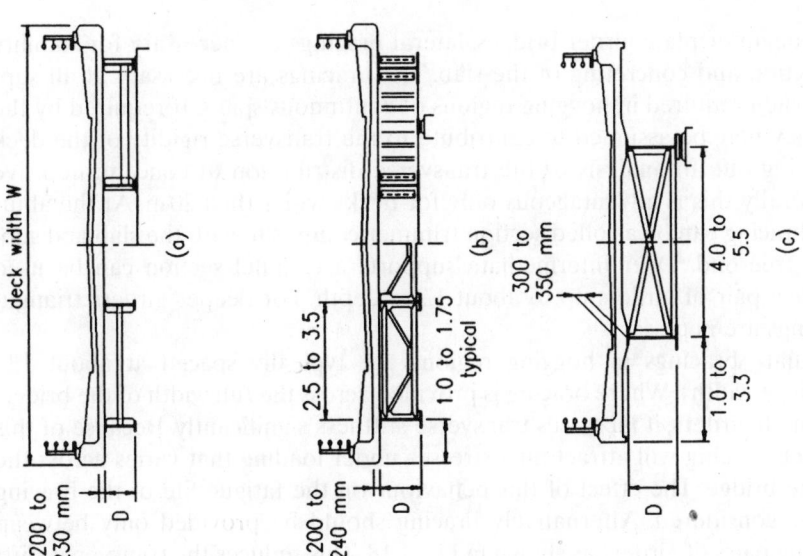

Fig. 4.17 Highway bridges – composite deck construction. (a) Multiple universal beam ($N = 4$); (b) multiple plate girder ($N = 4$); (c) twin plate girder, haunched slab ($N = 2$); (d) twin plate girder and stringer ($N = 2$); (e) twin plate girder and cross girder ($N = 2$); (f) multiple box ($N = 6$)

prestressed. The longitudinal reinforcement, however, acts as part of the composite section. Most composite bridges are designed as 'unpropped', i.e. the erected steel-work supports its own weight and the concrete slab (including formwork allowance) until hardened, with composite action being assumed only for superimposed dead and live loads. Box girders tend to be used for the long spans. Popular forms include twin box girders, multiple compact boxes and open top trapezoidal boxes.

Plate girder flanges should be made as wide as possible, consistent with outstand limitations, to give the best achievable stability during erection and to reduce the number of intermediate bracings. For practical reasons a desirable minimum width is about 400 mm. A maximum flange thickness of 65 mm is recommended to avoid heavy welds.

4.9.3.2 Intermediate supports

Intermediate supports often take the form of reinforced concrete walls, columns or portals. Steel supports may alternatively be used, and tubular columns are efficient, especially if filled with concrete and designed compositely. Where fewer columns are required for multiple girders then integral steel crossheads at the supports are sometimes used.

4.9.3.3 Bracings

For rolled beam or plate girder bridges, lateral bracings are necessary for stability during erection and concreting of the slab. The bracings are necessary at all supports and when required in hogging regions of continuous spans. If required by the designer they may be assumed to contribute to the transverse rigidity of the deck when carrying out an analysis of the transverse distribution of concentrated live loads. Generally this is advantageous only for decks wider than 20 m. At the abutments the bracing can be a rolled section trimmer composite with the slab and supporting its free end. Over intermediate supports a channel section can be used between each pair of girders up to about 1.2 m depth. For deeper girders triangulated bracings are necessary.

Intermediate bracings in hogging regions are typically spaced at about 12× (bottom flange width). Where bracing is provided across the full width of the bridge, i.e. between all girders, it increases transverse stiffness significantly. Because of this stiffness such bracing will attract high stresses under loading that varies across the width of the bridge. The effect of this behaviour on the fatigue life of the bracing needs to be considered. Alternatively bracing should be provided only between neighbouring pairs of girders as shown in Fig. 4.18. This reduces the transverse stiffness considerably and alleviates the problem of fatigue. Such a structure is also likely to be easier to erect. If the bridge is curved in plan with girders fabricated in straight

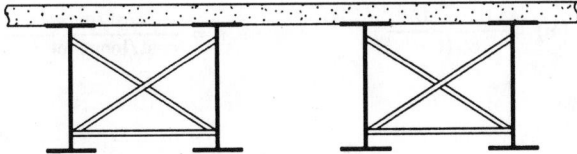

Fig. 4.18   Lateral bracing of bridge girders

chords they should be located adjacent to the site splices where torsion is induced. Bracings may be of a triangulated form or of single channel sections between each pair of girders where up to 1.2 m depth. Bracings are usually bolted to vertical web stiffeners in the main girders, which may need to be increased in size to accommodate the bolted connections. Angle sections are commonly used with lapped single shear connections, which permit tolerance in accommodating camber difference between adjacent girders.

Plan bracing systems may be required for spans exceeding 55 m for temporary stability under their own weight and that of the wet concrete; they may be removed after the floor slab is cast.

4.9.3.4 Locations of splices and change of section

Where rolled beams up to 1.0 m deep are used then for the maximum span range of about 33 m it is convenient to use a constant section with one splice within each span located at about 0.1–0.2 × span from the internal supports. The beam size will be determined by the maximum bending and shear effects at the supports where the slab is cracked.

For plate or box girders the component lengths for shop fabrication should be the maximum possible consistent with delivery and site restrictions to reduce the amount of site assembly. For spans up to about 55 m two splices per span are generally suitable, located at 0.15–0.25 × span from the internal supports, at which changes of flanges and web thickness should be made. A minimum number of other flange or web workshop joints should be made, consistent with plate length availability. The decision whether to introduce thickness changes within a fabricated length should take account of the cost of butt welds compared with the potential for material saving. Figure 4.19 indicates a basis for considering this optimization.

4.9.3.5 Curved bridge decks

Bridges which are curved in plan may be formed using straight fabricated girders, with direction changes introduced at each site splice. Alternatively the steel girders

$$L \ (m) = \frac{r \times 10^3}{7.85 \ (t_1 - t_2)} \qquad \text{where } r = \frac{cost/m \ weld}{cost/tonne \ of \ steel}$$

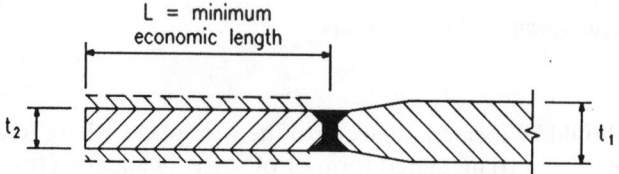

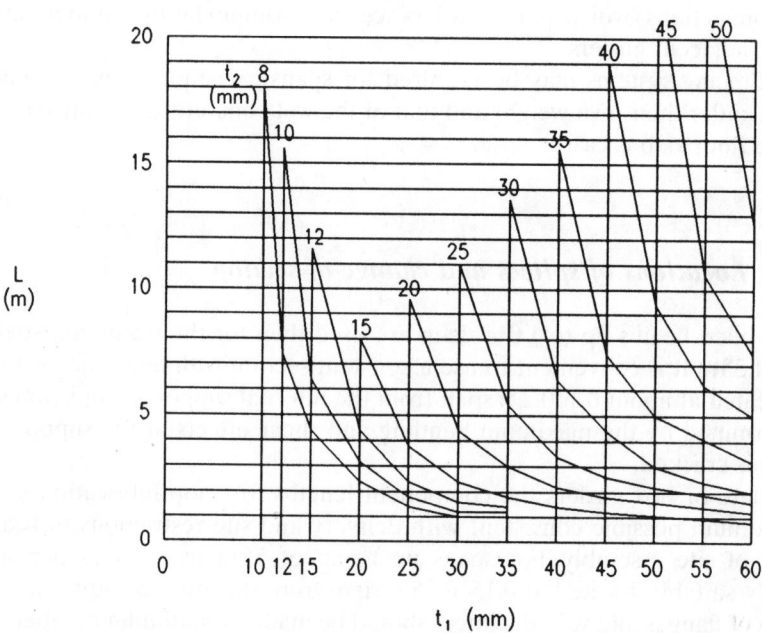

Fig. 4.19 Economy of flange and web thickness changes. The figure gives an indication of the minimum length (*L*) for which a selected thickness change will be economic for flanges and webs of girders. Below this length it will be more economic to continue the thicker plate (t_1)

can also be truly curved in plan, in which case the secondary stresses which arise from torsion must be evaluated. Truly curved girders are appropriate where the radius of plan curvature is less than about 500 m. It is likely that additional transverse bracings will be needed to reduce these stresses.

4.9.3.6 Initial sizes – composite plate girders

For economic design an analysis should be carried out of the transverse distribution of concentrated live loads between the main girders of the cross-section. The floor slab together with any continuous transverse bracing will significantly redistribute the maximum load applied to the most severely loaded girders. Prior to this the designer will wish to select initial sizes and make an estimate of the total weight of structural steel. Figures 4.20–4.24[8] provide initial estimates of flange area, web thickness and overall unit weight of steelwork (kg/m²) for typical composite bridge cross-sections as shown in Fig. 4.17.

The figures were derived from approximate designs using simplifying assumptions

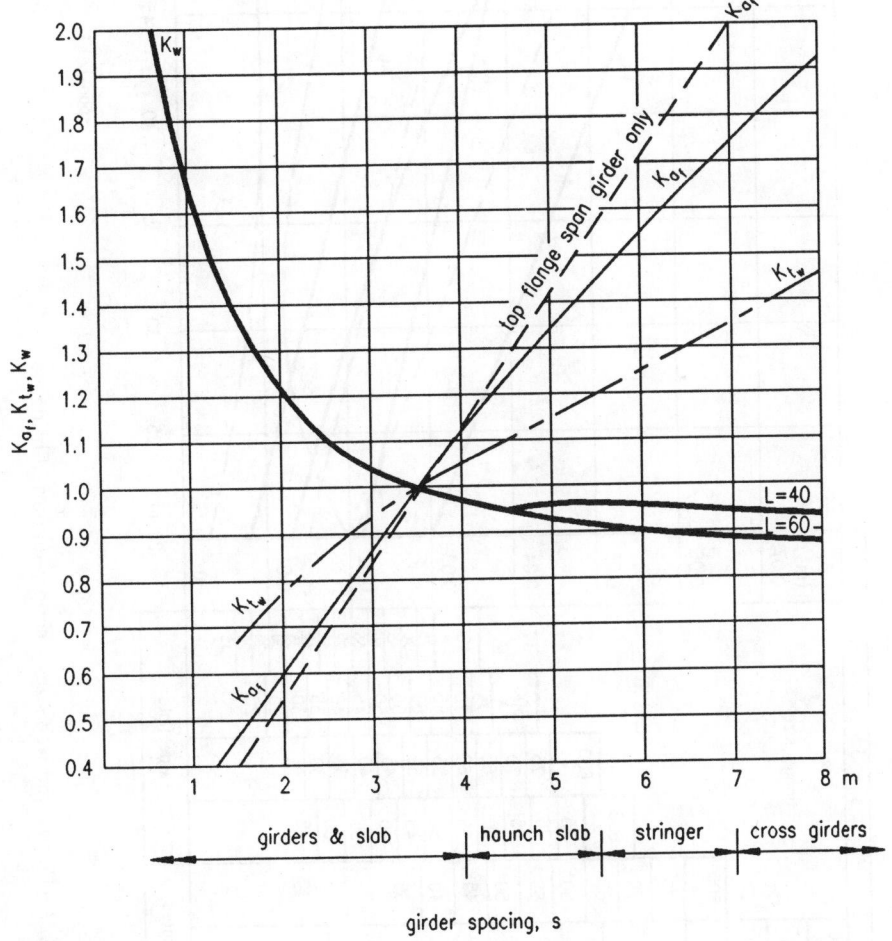

Fig. 4.20 Girder spacing factors

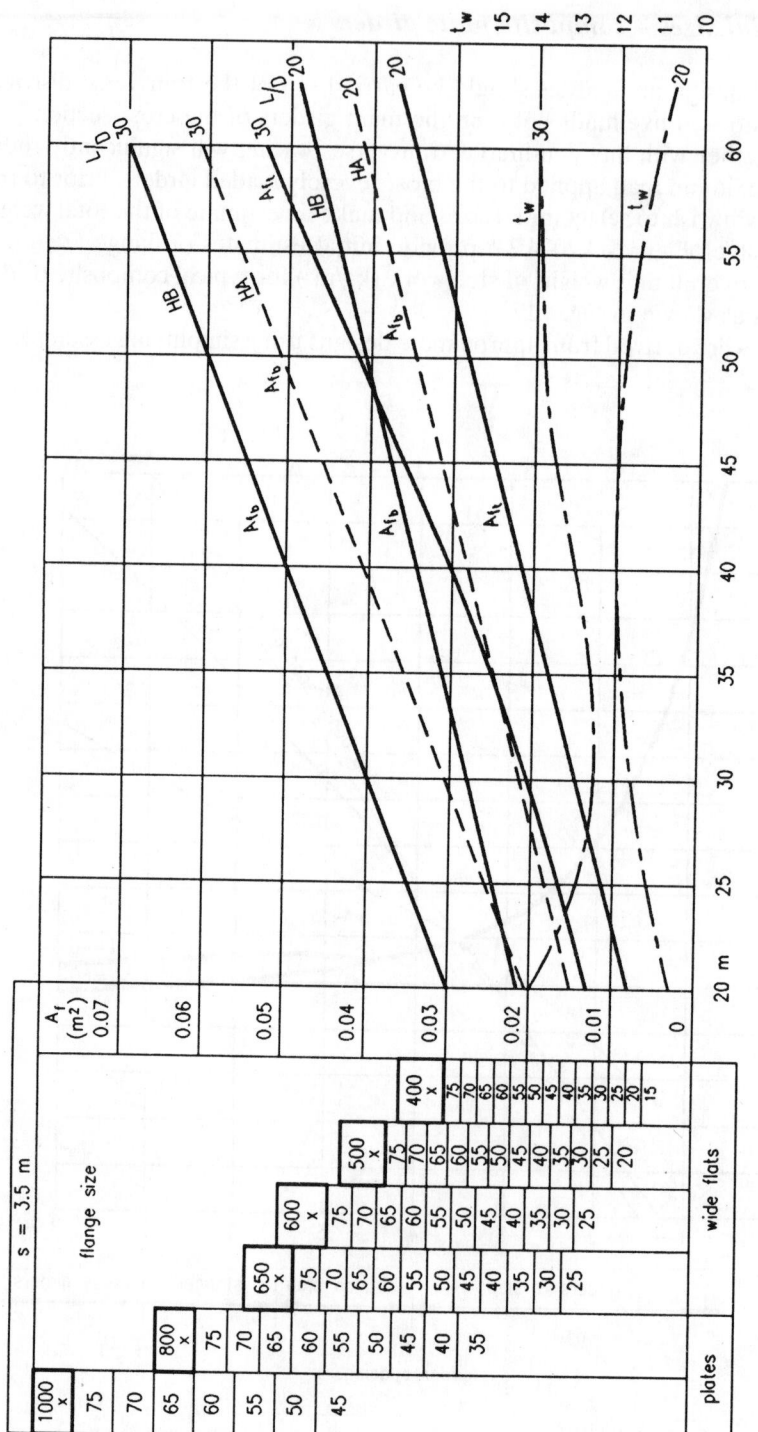

Fig. 4.21 Flange and web sizes – simply-supported bridges

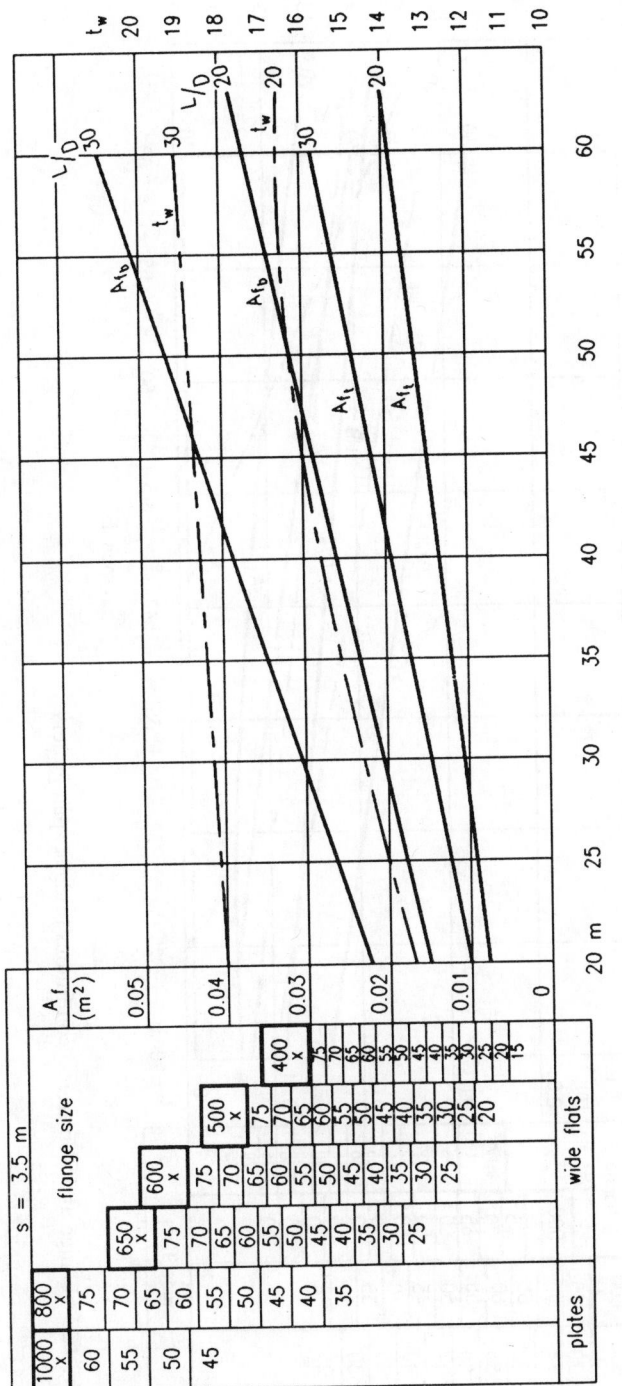

Fig. 4.22   Flange and web sizes – continuous bridges, pier girders

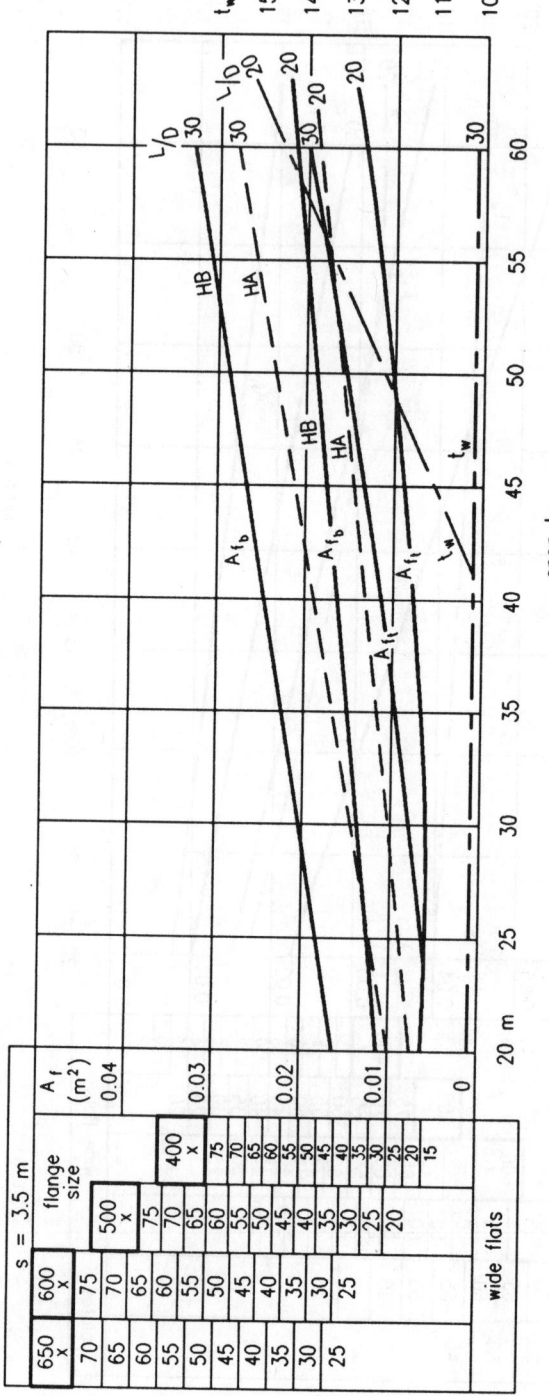

Fig. 4.23 Flange and web sizes – continuous bridges, span girders

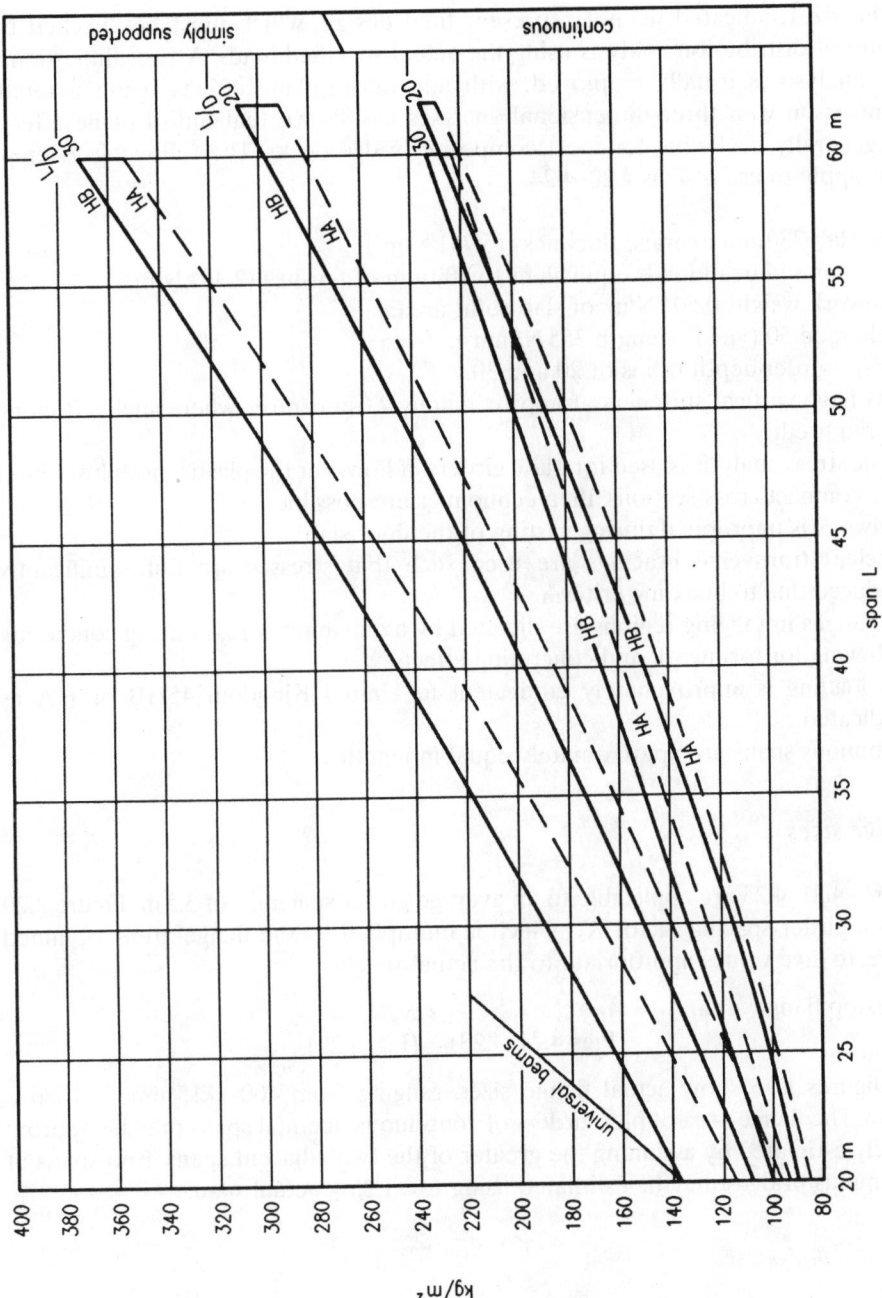

Fig. 4.24 Overall unit weights – plate girder bridges

for loads and transverse distribution and to achieve correlation with actual UK bridges.

The sizes indicated do not represent final design, which must be checked by a form of distribution analysis using the actual specified loads. A two-dimensional grid analysis is usually employed, with any out-of-plane effects being ignored. Comparison with three-dimensional analyses has shown that out-of-plane effects are generally negligible for most composite bridge decks. The following assumptions apply to use of Figs 4.20–4.24.

Deck slab 230 mm average thickness (5.75 kN/m²).
Superimposed dead loads equivalent to 100 mm of finishes (2.40 kN/m²).
Formwork weight 0.50 kN/m² of slab soffit area.
Steel grade 50 (yield strength 355 N/mm²).
Span-to-girder-depth ratios of 20 and 30.
Webs have vertical stiffeners at approximately 2.0 m centres where such stiffening is required.
Elastic stress analysis is used for plate girders. If however the plastic modulus is used for compact cross-sections, then economies are possible.
Steelwork is unpropped during casting of the floor slab.
Sufficient transverse bracings are used such that stresses are not significantly reduced due to buckling criteria.
Top flanges in sagging regions are dictated by a maximum stress during concreting allowing for formwork and concreting effects.
Live loading is approximately equivalent to United Kingdom 45HB or HA as indicated.
Continuous spans are approximately equal in length.

Flange sizes

Figures 4.21–4.23 are applicable to an average girder spacing s of 3.5 m. Figure 4.20 gives a girder spacing factor K_{a_f}, which is multiplied by the flange areas, obtained above, to give values appropriate to the actual spacing,

e.g. top flange area, $A_{f_t} = A_{f_t} \times K_{a_f}$
       (Figs 4.21–4.23) (Fig. 4.20)

The figures also show actual flange sizes, ranging from 400×15 mm to 1000×75 mm. The flange area of pier girders of continuous unequal spans may be approximately estimated by assuming the greater of the two adjacent spans. End spans of continuous bridges may be estimated using $L = 1.25 \times$ actual span.

Web thickness

Web thicknesses are obtained using Figs 4.21–4.23 applicable to $s = 3.5$ m. Adjustment for the actual girder spacing s is obtainable from Fig. 4.20,

i.e. web thickness, $t_w = t_w$ $\times K_{t_w}$
          (Figs 4.21–4.23)   (Fig. 4.20)

The thickness obtained may be regarded as typical. However, designers may prefer to opt for thicker webs to reduce the number of web stiffeners. Consideration should also be given to the use of unstiffened webs of appropriate thickness, i.e. $d/t \not< $ 60–100 depending on shear.

Overall unit weight

Overall unit weight (kg/m^2 of gross deck area) is read against the span L from Fig. 4.24 for simply-supported or continuous bridges with L/D ratios of 20 or 30, under HB or alternatively HA loading and applicable to $s = 3.5$ m.

     Adjustment for average girder spacing s other than 3.5 m is obtainable from Fig. 4.20.

i.e. kg/m^2 = kg/m^2 $\times K_w$
    (Fig. 4.24)   (Fig. 4.20)

The unit weight provides an approximate first estimate of steelwork weight allowing for all stiffeners, bracings, shear connectors, etc.

     For continuous bridges with variable depth, Figs 4.21 and 4.23 may be used to provide a rough guide, assuming a span-to-depth ratio (L/D) for each span based upon the average girder depth.

     For box-girder bridges a rough estimate may be obtained by replacing each box girder with an equivalent pair of plate girders.

     The mean span for use in Fig. 4.24 should be determined as follows:

$$\text{mean span, } L = \sqrt[+]{\left(\frac{L_1^4 + L_2^4 + \ldots + L_n^4}{n} \right)}$$

where n is the number of spans.

4.9.3.7 Initial sizes – rolled section beams

An estimate of size for simply-supported spans only may be obtained from Figs 4.25 and 4.26[10] for elastic or plastic stress analysis respectively, using universal beams up to $914 \times 419 \times 388$ kg/m size. For an estimate of the total weight of structural steel a factor of 1.1 applied to the main beams provides a reasonable allowance for bracings, bearing stiffeners and shear connectors. Figures 4.25 and 4.26 are based upon a concrete strength of 37.5 N/mm^2, and show the required mass per metre of universal beam. Table 4.8 gives reference to the relevant serial size.

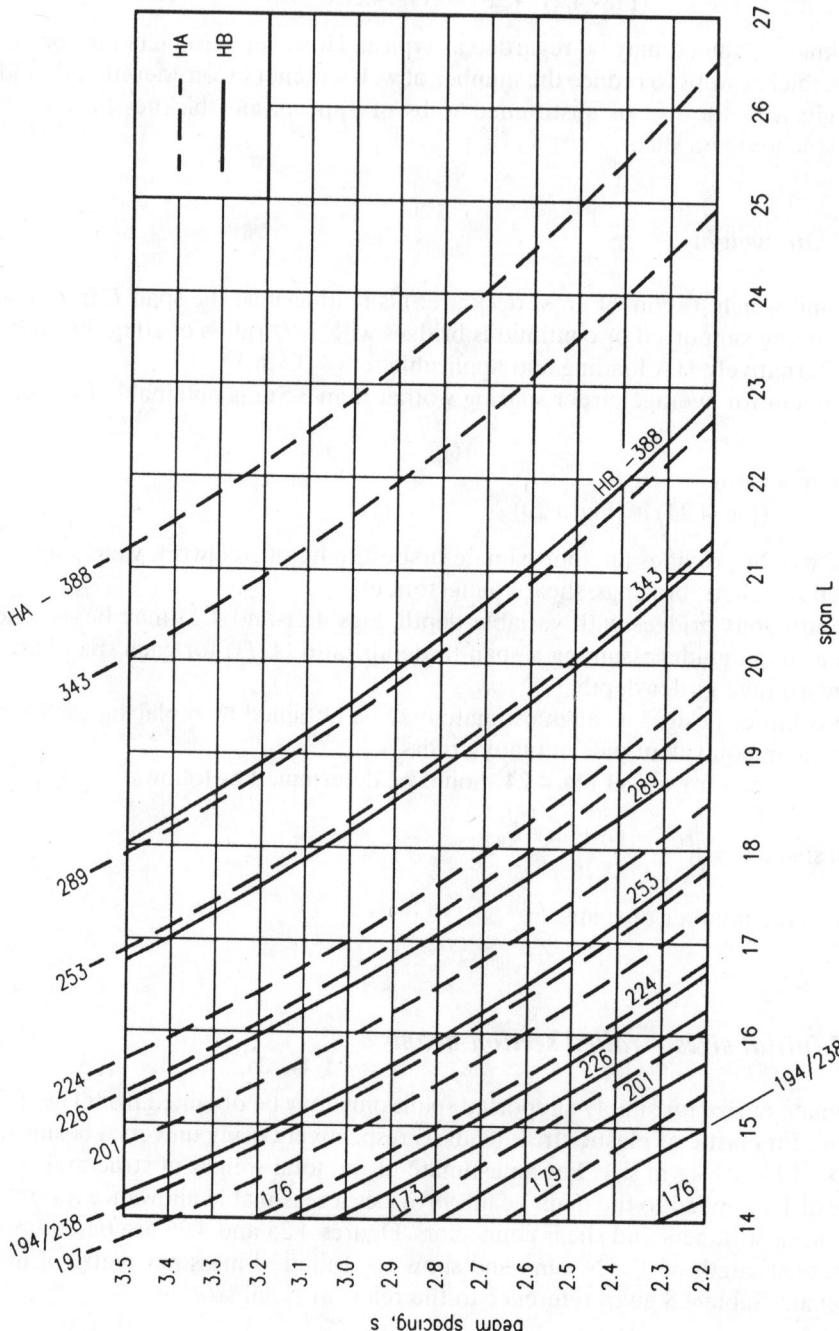

Fig. 4.25 Universal beam sizes – simply-supported bridges – elastic design

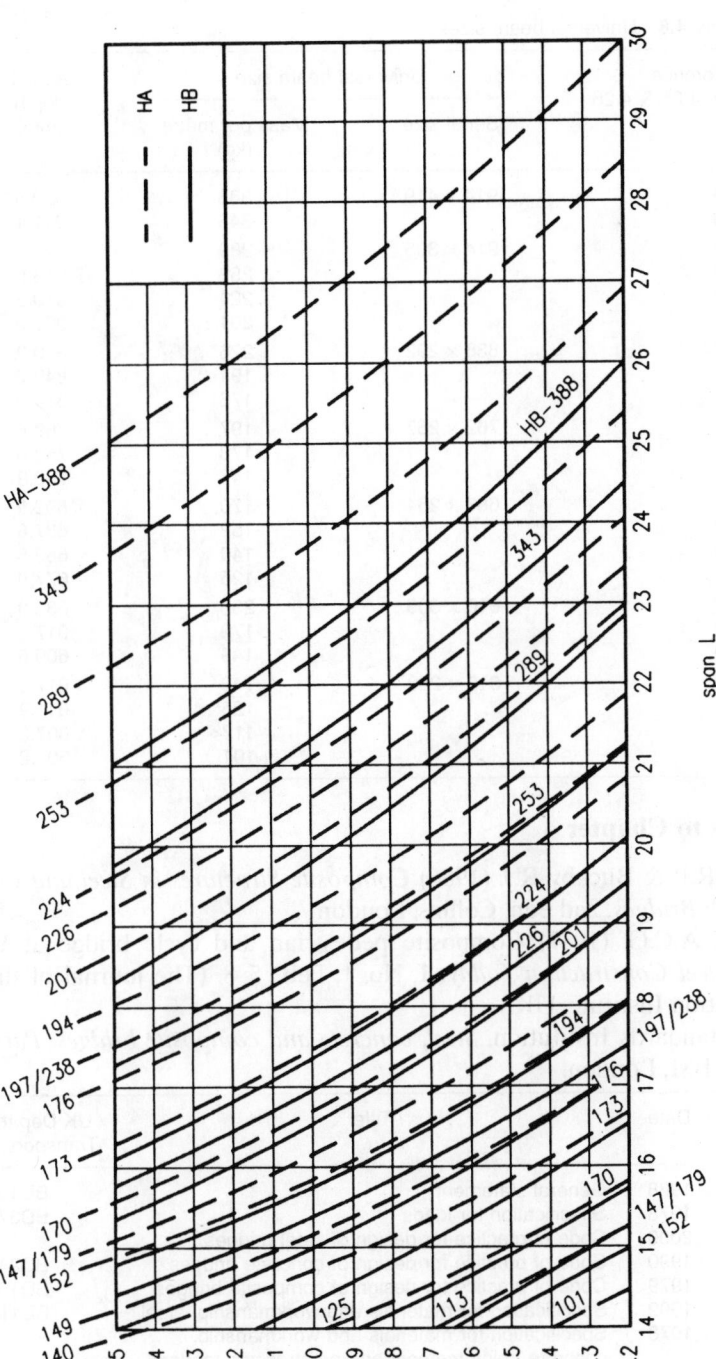

Fig. 4.26 Universal beam sizes – simply-supported bridges – plastic design

Table 4.8 Universal beam sizes

Reference Figs 4.25 & 4.26	Universal beam size		Actual depth (mm)
	Serial size	Mass per metre (kg)	
388	914 × 419	388	920.5
343		343	911.4
289	914 × 305	289	926.6
253		253	918.5
224		224	910.3
201		201	903.0
226	838 × 292	226	850.9
194		194	840.7
176		176	834.9
197	762 × 267	197	769.6
173		173	762.0
147		147	753.9
170	686 × 254	170	692.9
152		152	687.6
140		140	683.5
125		125	677.9
238	610 × 305	238	633.0
179		179	617.5
149		149	609.6
140	610 × 229	140	617.0
125		125	611.9
113		113	607.3
101		101	602.2

References to Chapter 4

1. Johnson R.P. & Buckby R.J. (1986) *Composite Structures in Steel and Concrete, Volume 2: Bridges*, 2nd edn. Collins, London.
2. Hayward A.C.G. (1987) Composite pedestrian and cycle bridge at Welham Green. *Steel Construction Today*, **1**, No. 1, Feb., 5–8. (The journal of the Steel Construction Institute, UK.)
3. British Standards Institution. *Steel, concrete and composite bridges: Parts 1–10.* BS 5400, BSI, London.

BS 5400: Part	Date	Title	UK Department of Transport Standard
1	1988	General statement	BD15/82
2	1978	Specification for loads	BD37/01
3	2000	Code of practice for design of steel bridges	
4	1990	Code of practice for design of concrete bridges	BD24/84
5	1979	Code of practice for design of composite bridges	BD16/82
6	1999	Specification for materials and workmanship, steel	BD11/82
7	1978	Specification for materials and workmanship, concrete, reinforcement and prestressing tendons	–
8	1978	Recommendations for materials and workmanship, concrete, reinforcement and prestressing tendons	–
9	1983	Bridge bearings	BD10/83
10	1980	Code of practice for fatigue	BD9/81

4. UIC (1971) Leaflet 702. UIC, 14 rue Jean-Ray F., 75015 Paris, France.
5. Pugsley A. (1968) *The Theory of Suspension Bridges*, 2nd edn. Edward Arnold, London.
6. Podolny W. (1976) *Construction and Design of Cable-Stayed Bridges*. John Wiley & Sons, New York.
7. Pucher A. (1973) *Influence Surfaces of Elastic Plates*, 4th edn. Springer-Verlag, New York.
8. Hayward A.C.G. (2002) *Composite Steel Highway Bridges*. Corus Construction Centre, Scunthorpe.

A worked example follows which is relevant to Chapter 4.

The **Steel Construction Institute** Silwood Park, Ascot, Berks SL5 7QN	Subject **INITIAL DESIGN OF HIGHWAY BRIDGE**	Chapter ref. **4**
Design code **BS 5400**	Made by **ACGH** Checked by **GWO**	Sheet no. **1**

Problem

A composite highway bridge has 3 continuous spans of 24, 40 and 32 m as shown. Overall deck width is 12 m and it carries 45 units of HB loading. There are 4 plate girders in the cross-section of 1.75 m depth. Estimate the main girder size and the weight of structural steel.

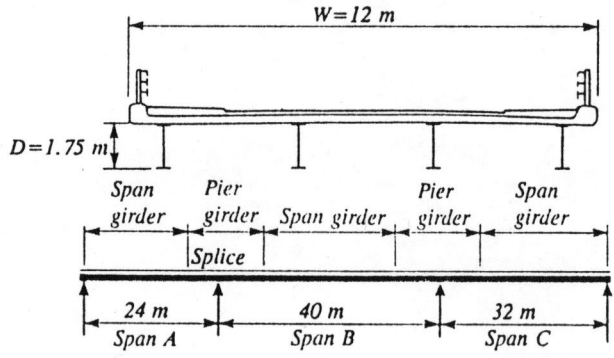

Average girder spacing 's' $=$ $\dfrac{12}{4}$ $=$ *3.0 m*

Flange and web sizes

From Figure 4.20 k_{af} $=$ *0.85 (top flange span girders)*
              k_{af} $=$ *0.87 (generally)*
              k_{tw} $=$ *0.95*

SPAN A

This is an end span so take

L $=$ $1.25 \times 24\,m$ $=$ *30 m*

therefore L/D $=$ $\dfrac{30\,m}{1.75\,m}$ $=$ *17*

so assume L/D $=$ *20*

The Steel Construction Institute Silwood Park, Ascot, Berks SL5 7QN 	Subject **INITIAL DESIGN OF HIGHWAY BRIDGE**	Chapter ref. **4**
	Design code **BS 5400**	Made by *ACGH* Checked by *GWO*	Sheet no. **2**

Top Flange, A_{ft} $=$ A_f *(from Figure 4.23)* $\times K_{af}$

$=$ 0.006×0.85 $=$ $0.0051\,m^2$ <u>400×15</u>

Bottom Flange, A_{fb} $=$ A_f *(from Figure 4.23)* $\times K_{af}$

$=$ 0.014×0.87 $=$ $0.012\,m^2$ <u>500×25</u>

Web, t_w $=$ t_w *(from Figure 4.23)* $\times K_{tw}$

$=$ 10×0.95 $=$ $9.5\,mm$ <u>$10\,mm\ web$</u>

<u>SPAN B</u>

Span girder

L/D $=$ $\dfrac{40\,m}{1.75\,m}$ $=$ 22.9

Top Flange, A_{ft} $=$ A_f *(from Figure 4.23)* $\times K_{af}$

$=$ 0.009×0.85 $=$ $0.0077\,m^2$ <u>400×20</u>

Bottom Flange, A_{fb} $=$ A_f *(from Figure 4.23)* $\times K_{af}$

$=$ 0.020×0.87 $=$ $0.0174\,m^2$ <u>500×35</u>

Web, t_w $=$ t_w *(see Figure 4.23)* $\times K_{tw}$

$=$ 10×0.95 $=$ $9.5\,mm$ <u>$10\,mm\ web$</u>

<u>SPAN C</u>

This is an end span so take

L $=$ $1.25 \times 32\,m$ $=$ $40\,m$

Therefore sizes as Span B.

The **Steel Construction Institute** Silwood Park, Ascot, Berks SL5 7QN	Subject ***INITIAL DESIGN OF HIGHWAY BRIDGE***	Chapter ref. **4**
	Design code **BS 5400**	Made by *ACGH*	Sheet no. **3**
		Checked by *GWO*	

Pier girders

Take L as the greater of the two adjacent spans i.e. assume L = 40 m at both supports.

Therefore L/D   =   *40/1.75*   =   *22.9*

Top Flange, A_{ft}   =   *A_f (see Figure 4.22) × K_{af}*

=   *0.015 × 0.87*   =   *0.0131 m^2*   <u>*400 × 35*</u>

Bottom Flange, A_{fb}   =   *A_f (see Figure 4.22) × K_{af}*

= *0.030 × 0.87*   =   *0.026 m^2*   <u>*500 × 55*</u>

Web, t_w   =   *t_w (see Figure 4.22) × K_{tw}*

= *16.5 × 0.95*   =   *15.7 mm*   <u>*18 mm web*</u>

Steel Weight

Span A: *L*   =   *1.25 × 24 m*   =   *30 m*

Span B: *L*   =   *40 m*

Span C: *L*   =   *1.25 × 32 m*   =   *40 m*

Mean Span   = $\left[\dfrac{L^4 + L_2^4 \ldots L_n^4}{n}\right]^{\frac{1}{4}} = \left[\dfrac{30^4 + 40^4 + 40^4}{3}\right]^{\frac{1}{4}}$

=   <u>*37.5*</u>

L/D   =   *37.5/1.75*   =   *21*

kg/m^2   =   *kg/m^2 (from Figure 4.24) × K_w (from Figure 4.20)*

=   *142 kg/m^2 × 1.03*   =   <u>*146 kg/m^2*</u>

Therefore steel weight

= $\dfrac{146\ kg/m^2}{1000}$ *× (24 m + 40 m + 32 m) × 12 m wide* =   <u>*168 tonnes*</u>

Chapter 5
Other structural applications of steel

Edited by IAN DUNCAN with contributions from MICHAEL GREEN, ERIC HINDHAUGH, IAN LIDDELL, GERARD PARKE, JOHN TYRRELL and MATTHEW LOVELL

5.1 Towers and masts

5.1.1 Introduction

Self-supporting and guyed towers have a wide variety of uses, from broadcasting of television and radio, telecommunications for telephone and data transmission to overhead power lines, industrial structures, such as chimneys and flares, and miscellaneous support towers for water supply, observation or lighting. These structures range from minor lighting structures, where collapse might have almost no further consequences, to major telecommunications links passing thousands of telephone calls or flare structures on which the safety of major chemical plant can depend. The term 'mast' describes a tower which depends for its stability on cable guys.

5.1.2 Structural types

Steel towers can be constructed in a number of ways but the most efficient use of material is achieved by using an open steel lattice. Typical arrangements for microwave radio and transmission towers are shown in Fig. 5.1. The use of an open lattice avoids presenting the full width of structure to the wind but enables the construction of extremely lightweight and stiff structures. Most power transmission, telecommunication and broadcasting structures fall into this class.

Lattice towers are typically square or triangular and have low redundancy. The legs are braced by the main bracings: both of these are often propped by additional secondary bracing to reduce the effective buckling lengths. The most common forms of main bracing are shown in Fig. 5.2.

Lattice towers for most purposes are made of bolted angles. Tubular legs and bracings can be economic, especially when the stresses are low enough to allow relatively simple connections. Towers with tubular members may be less than half the weight of angle towers because of the reduced wind load on circular sections.

169

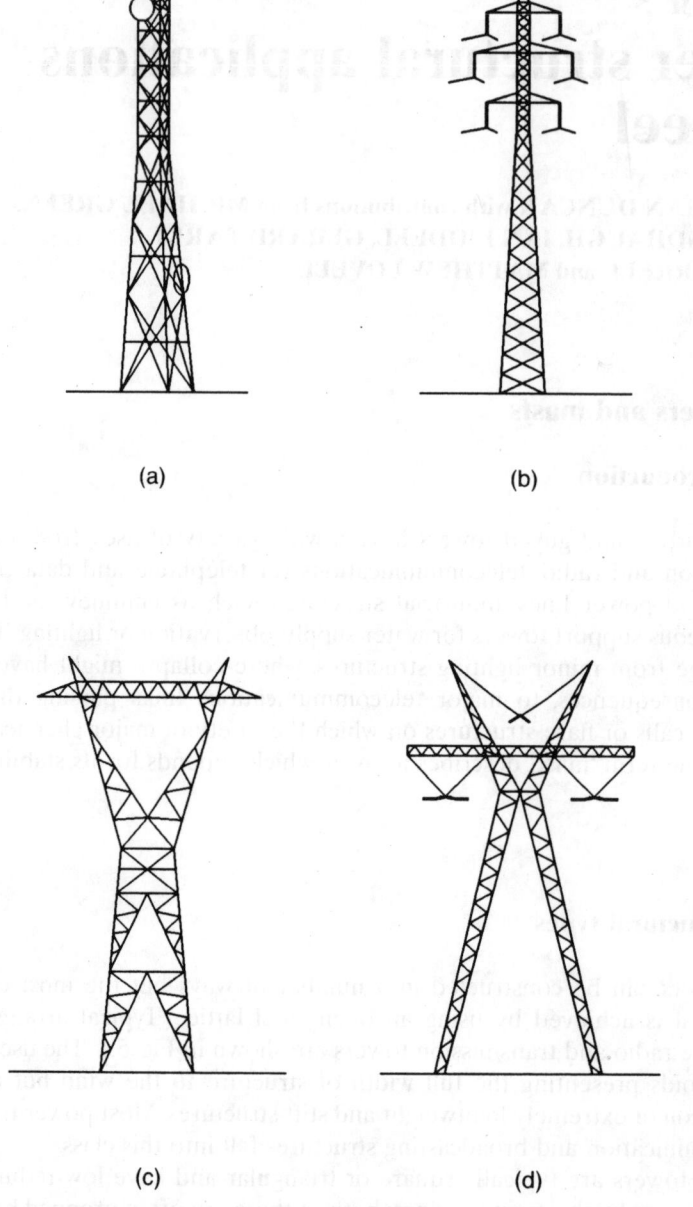

Fig. 5.1 Lattice towers: (a) microwave tower. (b), (c) and (d) transmission towers

However, the extra cost of the tube and the more complicated connection details can exceed the saving of steel weight and foundations.

Connections are usually arranged to allow site bolting and erection of relatively small components. Angles can be cut to length and bolt holes punched by machines

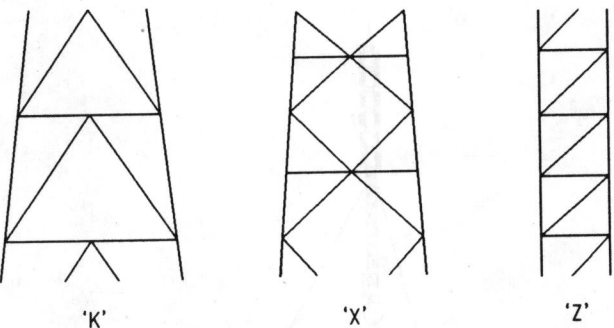

Fig. 5.2   Main bracing arrangements

as part of the same operation. Where heavy-lift cranes are available much larger segments of a tower can be erected but often even these are site bolted together.

Guyed towers provide height at a much lower material cost than self-supporting towers due to the efficient use of high-strength steel in the guys. Guyed towers are normally guyed in three directions over an anchor radius of typically $\frac{2}{3}$ of the tower height and have a triangular lattice section for the central mast. Tubular masts are also used, especially where icing is very heavy and lattice sections would ice up fully. A typical example of a guyed tower is shown in Fig. 5.3.

The range of structural forms is wide and varied. Other examples are illustrated in Figs 5.4 and 5.5. Figure 5.4 is a modular tower arrangement capable of extension for an increased number of antennas. The arrangement shown in Fig. 5.5 is adopted for supporting flare risers where maintenance of the flare tip is carried out at ground level.

A significant influence on the economics of tower construction is the method of erection, which should be carefully considered at the design stage.

5.1.3 Environmental loading

The primary environmental loads on tower structures are usually due to wind and ice, sometimes in combination. Earthquakes can be important in some parts of the world for structures of high mass, such as water towers. Loading from climatic temperature variations is not normally significant but solar radiation may induce local stresses or cause significant deflections, and temperatures can influence the choice of ancillary materials.

Most wind codes use a simple quasi-static method of assessing the wind loads, which has some limitations for calculating the along-wind responses but is adequate for the majority of structures. Tower structures with aerodynamically solid sections and some individual members can be subject to aeroelastic wind forces caused by vortex shedding, galloping, flutter and a variety of other mechanisms which are

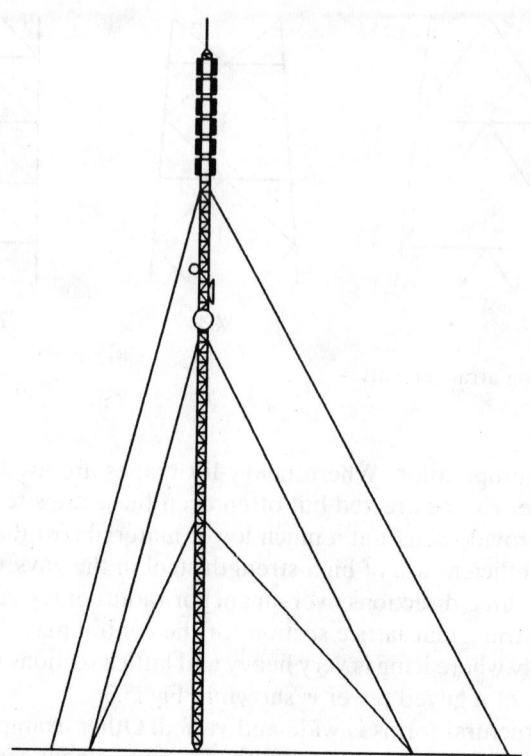

Fig. 5.3   Guyed tower

Fig. 5.4   Modular tower

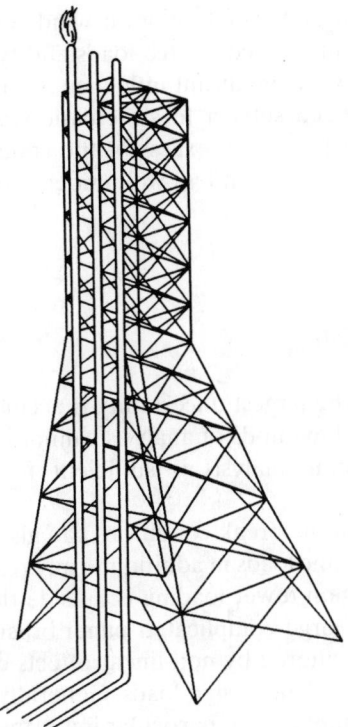

Fig. 5.5   Flare tower

either poorly covered or ignored by current codes of practice. Such factors have been responsible for more tower collapses and serviceability failures worldwide than any shortfall in resistance to along-wind loads.

Most national and international design codes now specify wind loads in terms of design wind speeds, either mean hourly or gust, that will recur on average once in a 50 year period (i.e. with an annual probability of 2%). Guidance is sometimes given on wind shape factors for typical sections and lattices. Consideration of dynamic response to the wind is not always covered in depth and there is still a mixture of limit state and working stress codes.

BS 8100[1] is a recent code in a limit state format specifically written for towers. Wind loads are specified in terms of a '50 year return' mean hourly wind pressure together with gust factors which convert the forces to an equivalent static gust. The overall wind forces calculated using BS 8100 are substantially similar to those that would be obtained using earlier codes such as CP3[2] but forces near the tops of towers are relatively higher due to an allowance for dynamic response. The code also gives guidance on means of allowing for the importance of particular structures by adjusting the partial factor on the design wind speed.

Guidance is limited for structures that have a significant dynamic response at their natural frequencies, and gust factors for guyed towers are specifically excluded from the scope of the code. Part 4 of BS 8100 is intended to address these aspects.

The influence of height and topography on wind speed can be significant; this is covered in some detail in both codes. Ice loads and types of ice are also covered but neither mentions the very significant influence of topography on the formation of ice. This has not yet been subject to systematic study but some hill sites are known to be subject to icing well in excess of the code requirements. The combination of wind and ice loads is even less well understood although some guidance is given.

5.1.4 Analysis

In the analysis of towers the largest uncertainty is accurate knowledge of the wind loads. Highly sophisticated methods of analysis cannot improve this. A static linear three-dimensional structural analysis is sufficient for almost all lattice tower structures.

For transmission towers, line break conditions can also be critical. Line breakage will in general induce dynamic loads in addition to any residual static loads. Detailed consideration of transmission tower loading is outside the scope of this section.

For lattice towers with large complicated panel bracing, the secondary bracing forces can be significantly altered by non-linear effects caused by curvature of the panels under the influence of the design loads. Generally the rules in the codes are sufficient, but where structures are of particular importance or where there is much repetition of a design, a non-linear analysis may be necessary.

Dynamic analyses of self-supporting lattice towers are rarely necessary unless there are special circumstances such as high masses at the top, use as a viewing platform, or circular or almost solid sections of mast which could be responsive to vortex shedding or galloping. Knowledge of the dynamic response is also necessary for assessment of fatigue of joints if this is significant.

For guyed towers the non-linear behaviour of the guys is a primary influence and cannot be ignored. The choice of initial tension, for example. can have a very great effect on the deflections (and dynamic behaviour). The effects of the axial loads in the mast on column stiffness can be significant. Methods of static analysis are given in the main international codes for the design of guyed towers. Guyed towers can also be particularly sensitive to dynamic wind effects especially those with cylindrical or solid sections.[2]

General guidance on the dynamic responses and aerodynamic instabilities of towers can be obtained from References 3, 4, 5 and 6.

5.1.5 Serviceability

Serviceability requirements vary greatly depending on the purpose of the structure and its location.

Steel towers and connections arc normally galvanized and are also painted with a durable paint system if the environment is likely to be polluted or otherwise corrosive. It is important that regular maintenance is carried out; climbing access is normally provided for inspections.

Deflections of towers are generally significant only if they would result in a loss of serviceability. This can be critical for the design of telecommunication structures using dish antennas. In the past signal losses due to deflection have often been assessed on the misunderstanding that the deflections under the design wind storm would occur sufficiently often to affect the signals. Studies have demonstrated that short periods of total loss of signal during storms smaller than the design wind storm have a negligible effect on the reliability of microwave links compared with losses due to regular atmospheric conditions.

5.1.6 Masts and towers in building structures

Consideration of a masted solution arises from the need to provide a greater flexibility in the plan or layout of the building coupled with its aesthetic value to the project as a whole. At the same time it offers the opportunity to utilize structural materials in their most economic and effective tensile condition. The towers or masts can also provide high-level access for maintenance and plant support for services. The plan form resulting from a mast structure eliminates the need for either internal support or a deeper structure to accommodate the clear span. By providing span assistance via suspension systems the overall structural depth is minimized giving a reduction in the clad area of the building perimeter. The concentration of structural loads to the mast or towers can also benefit substructure particularly in poor ground conditions where it is cost effective to limit the extent of substructures (Figs 5.6 and 5.7). However, differential settlement can have a significant effect on the structure by relaxing ties on suspension systems. The consequent load redistribution must be considered.

Most tension structure building forms consist of either central support or perimeter support, or a mixture of the two. Any other solutions are invariably a variation on a theme. Plan form tends to be either linear or a series of repetitive squares.

The forces and loads experienced by towers and masts are illustrated in Figs 5.8 and 5.9. In all cases it is advantageous but not essential that forces are balanced about the mast. Out-of-balance loads will obviously generate variable horizontal and vertical forces, which require resolution in the assessment of suitable structural sections.

Suspension ties must be designed to resist not only tension but also the effects of vibration, ice build-up and catenary sag. Ties induce additional compressive forces in the members they assist. These forces require careful consideration, often necessitating additional restraints in the roof plane in either the open sections or top chord of any truss.

Longitudinal stability is created either by twinning the masts and creating a vertical truss or by cross bracing preferably to ground (Fig. 5.10).

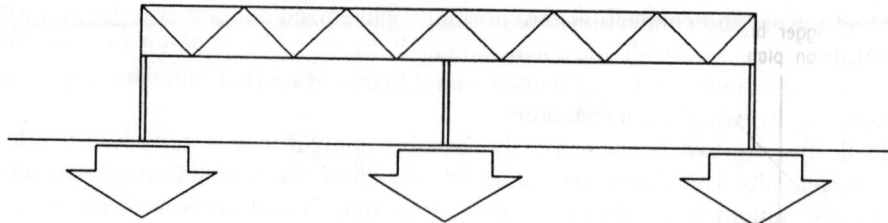

Fig. 5.6 Traditional long-span structure

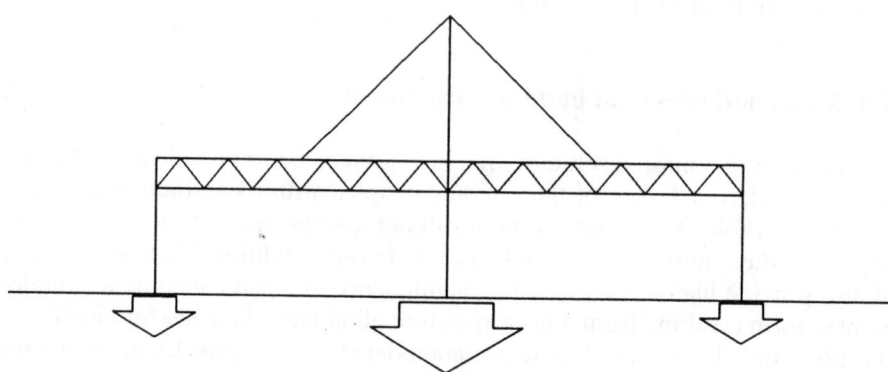

Fig. 5.7 Suspension structure

If outriggers are used, as is the case with the majority of masted structures, then the lateral stability of the outrigger can be resolved in a similar form to the masts by a stiff truss or Vierendeel section in plan or by plan or diagonal bracing at the extremity (Fig. 5.11).

5.2 Space frames

5.2.1 Introduction

Steel skeletal space frames are three-dimensional structures capable of very large column-free spans. These structures, constructed from either individual elements or prefabricated modules, process a high strength-to-weight ratio and stiffness. Steel space frames may be used efficiently to form roofs, walls and floors for projects such as shopping arcades, but their real supremacy is in providing roof cover for sports stadia, exhibition halls, aircraft hangars and similar major structures.

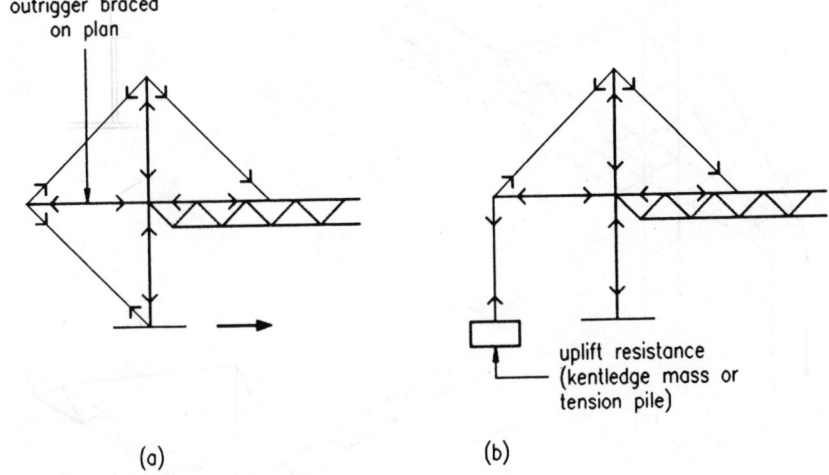

Fig. 5.8   Types of perimeter support

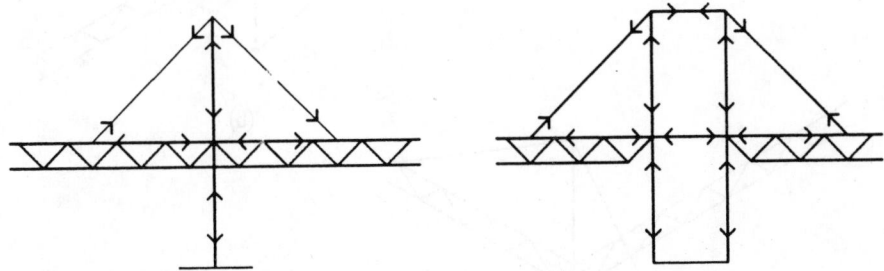

Fig. 5.9   Central support

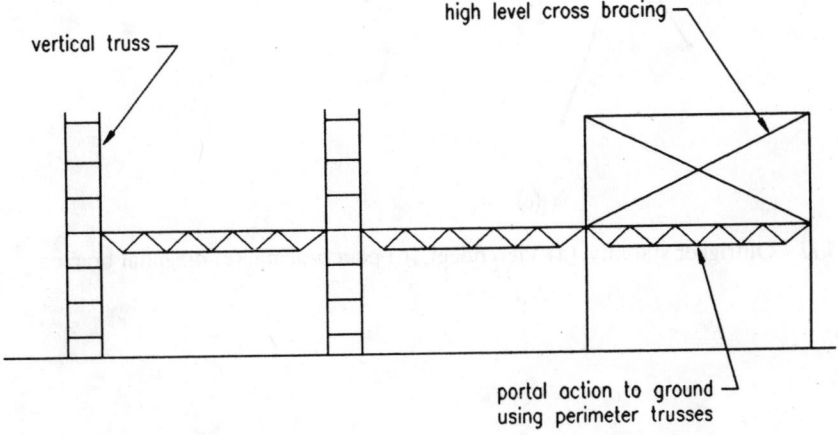

Fig. 5.10   Longitudinal stability

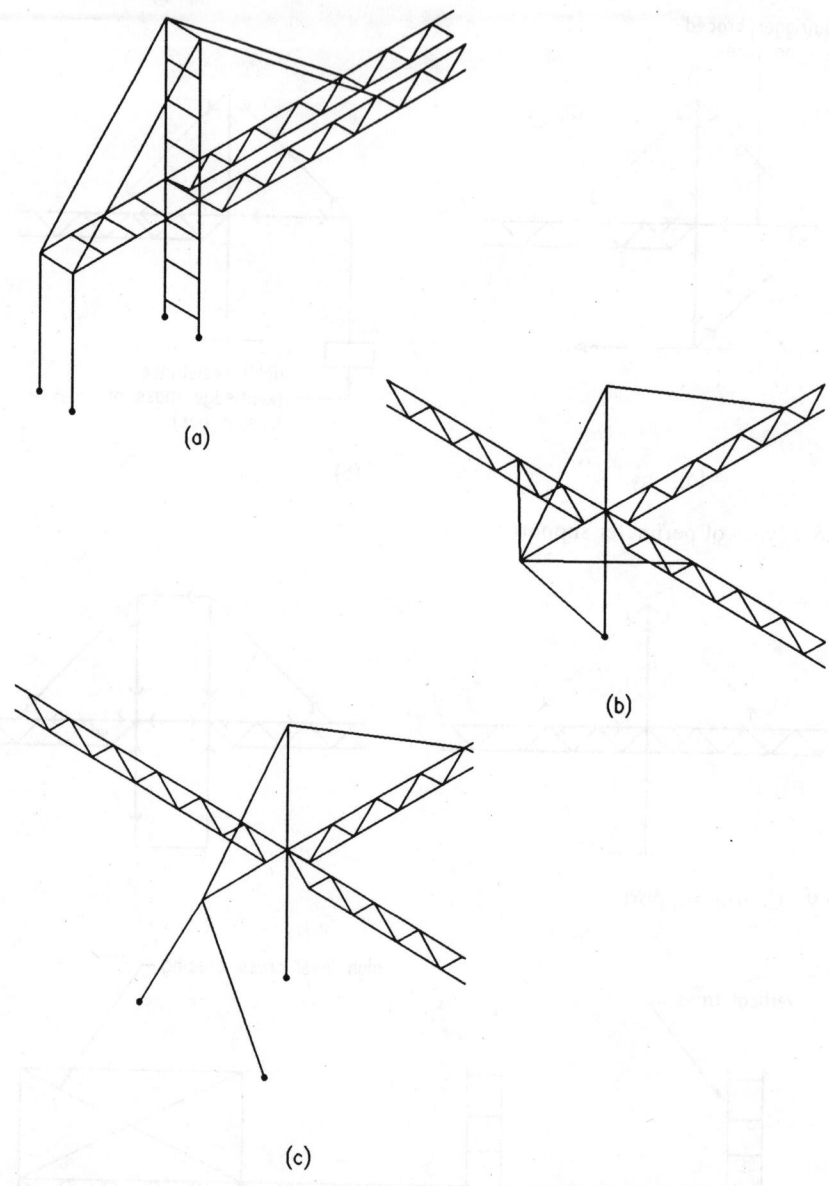

Fig. 5.11   Outrigger stability: (a) Vierendeel, (b) plan bracing, (c) diagonal bracing

5.2.2 Structural types

Space frames are classified as single-, double- or multi-layered structures, which may be flat, resulting in grid structures, or may be curved in one or two directions, forming barrel vaults and dome structures. Grid structures can be further categorized into lattice and space grids in which the members may run in two, three or four principal directions. In double-layer lattice grids the top and bottom grids are identical, with the top layer positioned directly over the bottom layer. Double-layer space grids are usually formed from pyramidal units with triangular or square bases resulting in either identical parallel top and bottom grids offset horizontally to each other, or parallel top and bottom grids each with a different configuration interconnected at the node points by inclined web members to form a regular stable structure.

Single-layer grids are primarily subject to flexural moments, whereas the members in double- and triple-layer grids are almost entirely subject to axial tensile or compressive forces. These characteristics of single-, double- and triple-layer grids determine to a very large extent their structural performance. Single-layer grids, developing high flexural stresses, are suitable for clear spans up to 15 m while double-layer grids have proved to be economical for clear spans in excess of 100 m. The main types of double-layer grids in common use are shown in Fig. 5.12.

Skeletal space frames curved in one direction forming single- or double-layer barrel vaults also provide elegant structures capable of covering large clear spans. Single-layer vaults are suitable for column-free spans of up to 40 m, which may be substantially increased by incorporating selected areas of double-layer structure forming stiffening rings. Double-layer barrel vaults are normally capable of clear spans in excess of 120 m. Figure 5.13 shows the main types of bracing used for single-layer barrel vaults.

Dome structures present a particularly efficient and graceful way of providing cover to large areas. Single-layer steel domes have been constructed from tubular members with spans in excess of 50 m while double-layer dome structures have been constructed with clear spans slightly greater than 200 m. Skeletal dome structures can be classified into several categories depending on the orientation and position of the principal members. The four most popular types usually constructed in steel are ribbed domes, Schwedler domes, three-way grid domes and parallel lamella domes.

Ribbed domes, as the name suggests, are formed from a number of identical rib members, which follow the meridian line of the dome and span from the foundations up to the top of the structure. The individual rib members may be of tubular lattice construction and are usually interconnected at the crown of the dome using a small diameter ring beam.

The Schwedler dome is also formed from a series of meridional ribs but, unlike the ribbed dome, these members are interconnected along their length by a series of horizontal rings. In order to resist unsymmetric loads the structure is braced by diagonal members positioned on the surface of the dome bisecting each trapezium formed by the meridional ribs and horizontal rings.

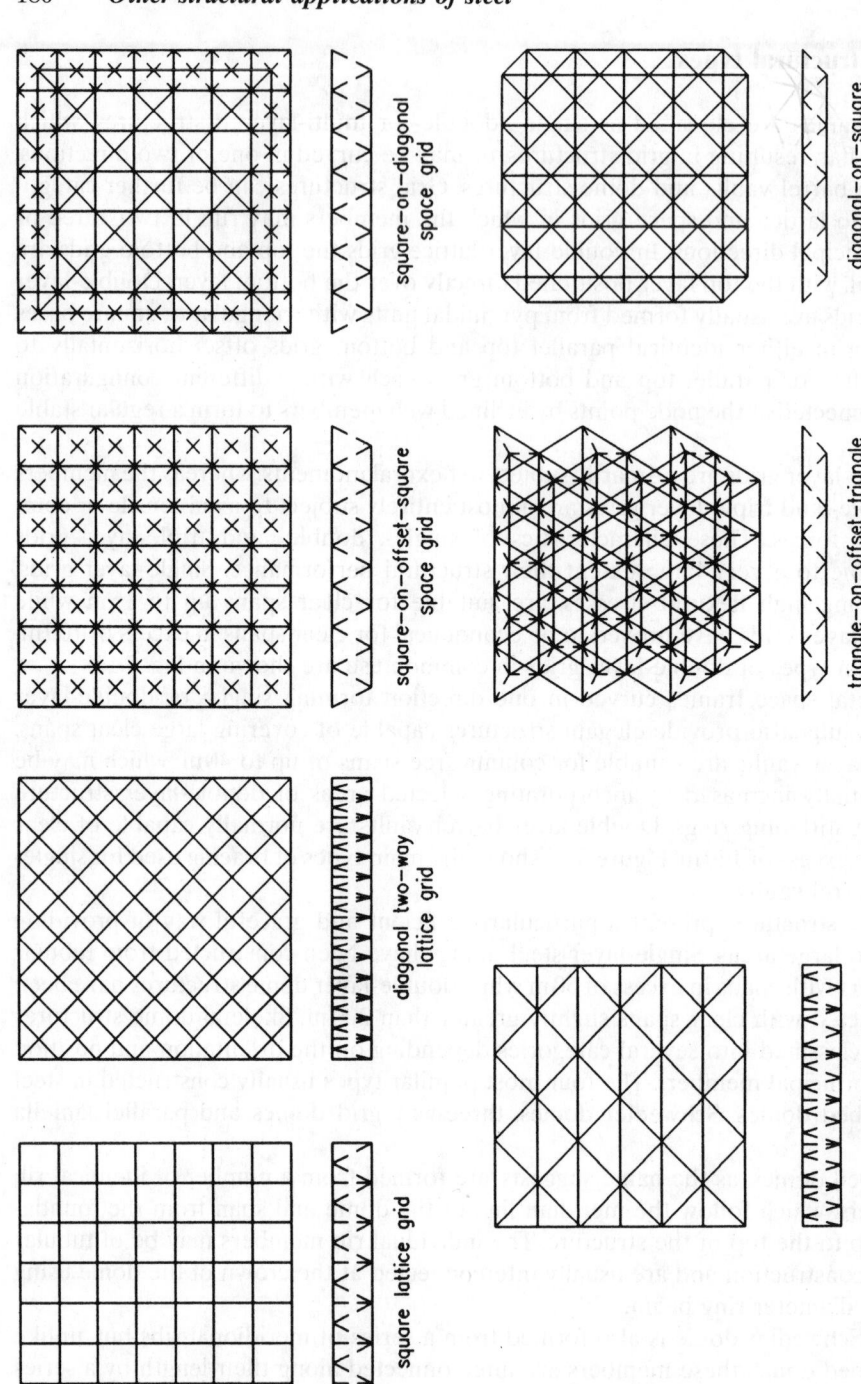

Fig. 5.12 Lattice and space grids

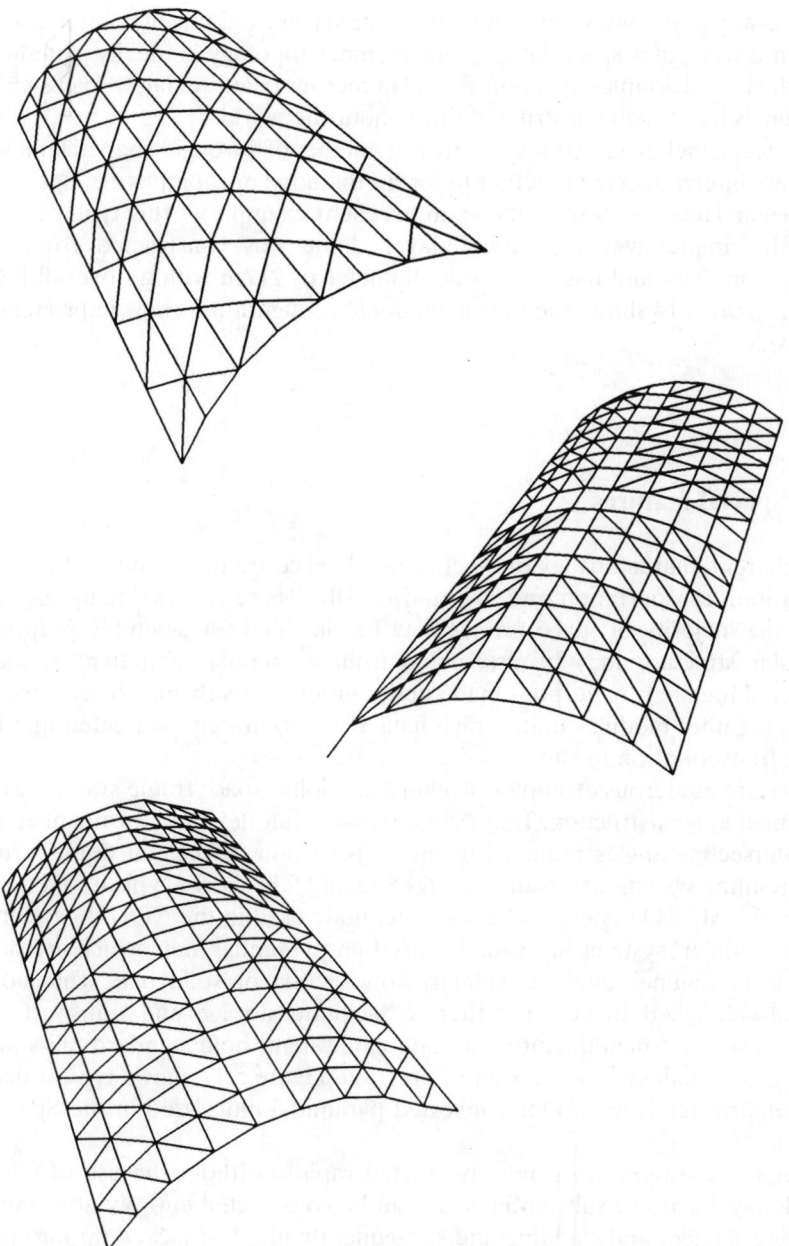

Fig. 5.13 Bracing of single-layer barrel vaults

Three-way grid domes are formed from three principal sets of members arranged to form a triangular space lattice. This member topology is ideally suited to both single-layer and double-layer domes, and numerous beautiful large-span steel three-way domes have been constructed throughout the world.

The steel lamella dome is formed from a number of 'lozenge'-shaped lamella units which are interconnected together to form a diamond or rhombus arrangement. The spectacular Houston Astrodome is an excellent example of this type of construction. This impressive steel double-layer dome was constructed from lamella units 1.52 m deep and has an outside diameter of 217 m with an overall height of 63.4 m. Figure 5.14 shows the four main dome configurations now in prominent use worldwide.

5.2.3 Special features ·

The inherent characteristics of steel skeletal space frames facilitate their ease of fabrication, transportation and erection on site. There are two main groups into which the majority of space frames may be classified for assembly purposes: the particular structure may be assembled from a number of individual members connected together by purpose-made nodes or alternatively may be constructed by joining together modular units which have been accurately fabricated in a factory before transportation to site.

There are numerous examples of 'chord and joint' space frame systems available for immediate construction. These systems offer full flexibility of member lengths and intersecting angles required in the construction of skeletal dome structures. Many jointing systems are available; Figs 5.15 and 5.16 show a typical spherical node used in the MERO system and a cast steel node used in the NODUS system.

The 'modular' systems are usually based on pyramidal units which are prefabricated from channel, angle, circular hollow section or solid bars. The individual units are designed to nest together to facilitate storage and transportation by road or sea. Most manufacturers of modular systems hold standard units in stock, which greatly enhances the speed of erection. Figure 5.17 shows typical details of the prefabricated steel modular inverted pyramidal units used in the Space Deck System.

· Steel space frames are generally erected rapidly without the use of falsework. Double-layer grids of substantial span can be constructed entirely at ground level including services and cladding and subsequently lifted or jacked up into the final position. Dome structures can be assembled from the top downwards using a central climbing column or tower. A novel approach adopted for the erection of a dome with a major axis of 110 m and a minor axis of 70 m involved fabrication of the dome on the ground in five sections, which were temporarily pinned to each other. The central section was then lifted and the remaining segments of the dome locked into position as shown in Fig. 5.18.

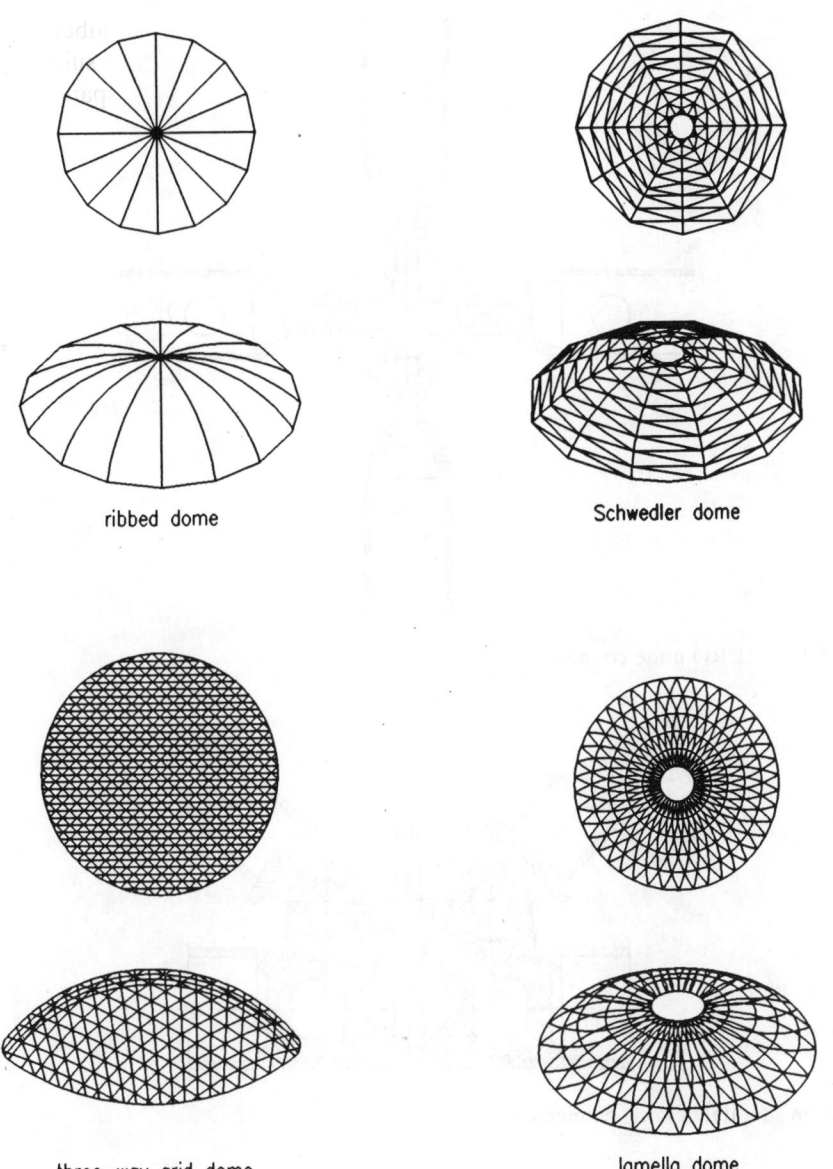

ribbed dome

Schwedler dome

three - way grid dome

lamella dome

Fig. 5.14 Dome configurations

5.2.4 Analysis

The analysis of space frames results in the production of large sets of linear simultaneous equations which must inevitably require the use of a computer for their solution. For these equations to be formulated it is necessary to input into the

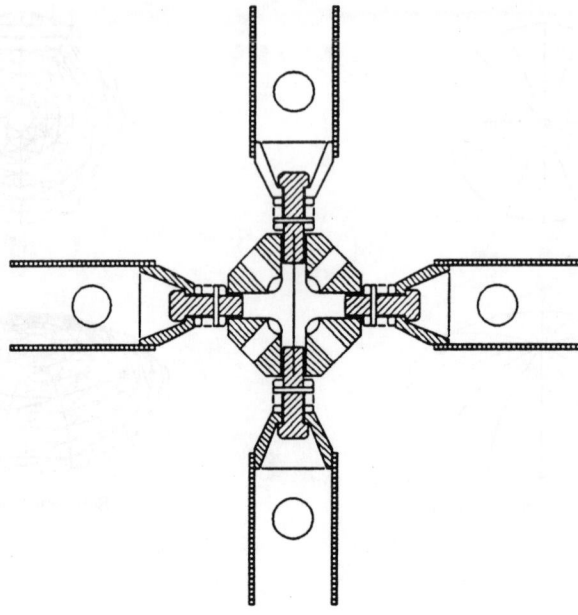

Fig. 5.15   MERO node connector

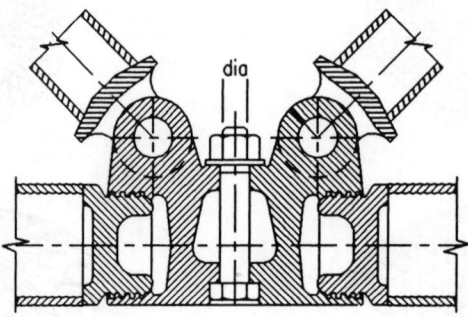

Fig. 5.16   NODUS node connector

computer significant amounts of information relating to the topology of the structure and properties of the individual members. This operation can be very time-consuming unless modern pre-processing techniques which allow rapid data generation are adopted.[7]

The members forming double-layer space frames are principally stressed in either axial tension or axial compression. Consequently it is usual in the analysis of double-layer space trusses to assume that the members in the structure are

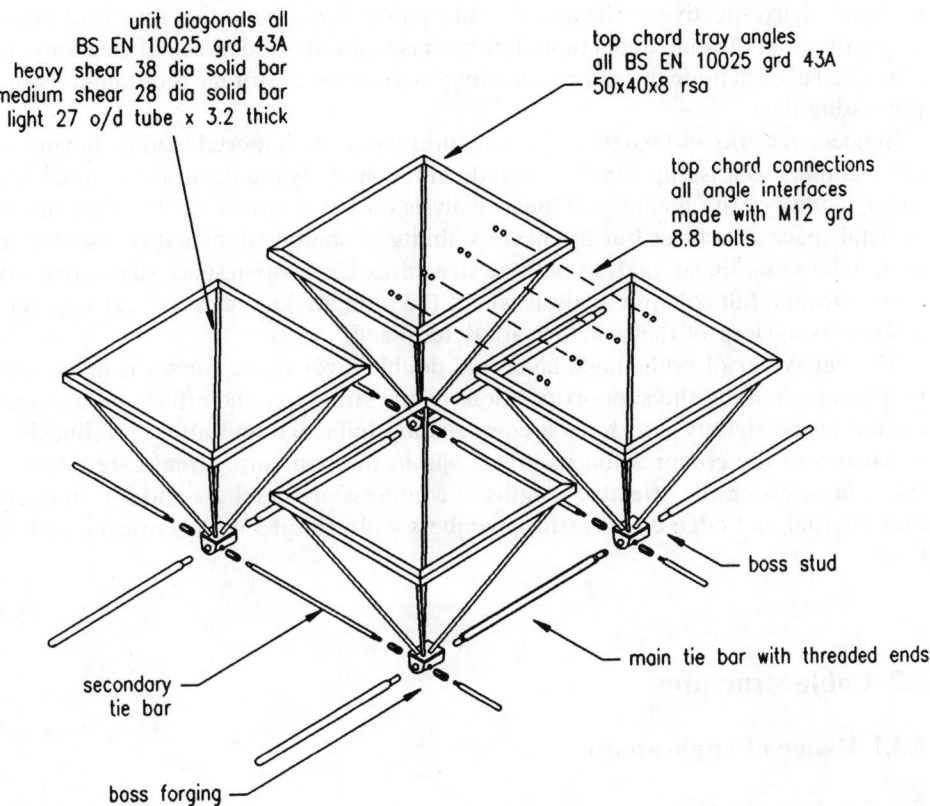

unit diagonals all
BS EN 10025 grd 43A
heavy shear 38 dia solid bar
medium shear 28 dia solid bar
light 27 o/d tube x 3.2 thick

top chord tray angles
all BS EN 10025 grd 43A
50x40x8 rsa

top chord connections
all angle interfaces
made with M12 grd
8.8 bolts

boss stud

main tie bar with threaded ends

secondary
tie bar

boss forging

Fig. 5.17   Exploded view of 2.4 m square section of Space Deck

assembly on the ground, with temporarily pinned joints

lift up

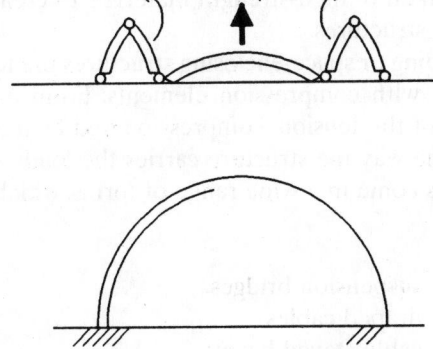

Fig. 5.18   Novel method of dome erection

pin-jointed, irrespective of the actual joint rigidity. This assumption may lead to the overestimation of node deflections, but because only three degrees of freedom are permitted at each node this approach minimizes computer storage requirements and processing time.

Single-layer grid and barrel vault structures carry the imposed load by flexure of the members so it is important to include in the analysis the flexural, torsional and shear rigidity of the members. Linear analysis only is required for the majority of skeletal space structures but to ensure stability of shallow domes it is essential to undertake a non-linear analysis of these structures. Large-span space structures may benefit from a full collapse analysis where the ultimate load-carrying capacity and collapse behaviour of the structure are determined.[8]

The behaviour of both single-layer and double-layer space trusses is influenced to a great extent by the support positions of the structure. The effects of joint and overall frame rigidity also have a commanding influence and affect the buckling behaviour of the compression members within the structure. Great care must be taken in assessing the effective lengths of compression members and it is unlikely that internal and edge compression members will exhibit similar critical buckling loads.

5.3 Cable structures

5.3.1 Range of applications

5.3.1.1 Introduction

Most structural elements are able to carry bending forces as well as tension and compression, and are hence able to withstand reversals in the direction of loading. Tension elements are unique in that they can carry only tension. In compressive or bending elements, the loading capacity is often reduced by buckling effects, while tension elements can work up to the full tensile stress of the material. Consequently full advantage can be taken of high-strength materials to create light, efficient and cost-effective long-span structures.

To create useful spanning or space-enclosing structures the tension elements have to work in conjunction with compression elements. From an architectural point of view, the separation of the tension, compression and bending elements leads to a visual expression of the way the structure carries the loads, or at least one set of loads. Tension structures come in a wide range of forms, which can be broadly categorized as follows:

(1) two-dimensional – suspension bridges
  – draped cables
  – cable-stayed beams
  – cable trusses

(2) three-dimensional – cable truss systems
(3) surface-stressed   – pneumatically-stressed
                         – prestressed.

5.3.1.2 Structural forms

Suspension bridges (draped cable)

A suspension bridge (Fig. 5.19) is essentially a catenary cable prestressed by dead weight only. Early suspension bridges with flexible decks suffered from large deflections and sometimes from unstable oscillation under wind. A system of inclined hangers proposed originally to reduce deflection under live load has been employed recently also to counteract wind effects. The suspension cable is taken over support towers to ground anchors. The stiffened deck is supported primarily by the vertical or inclined hangers. The system is ideally suited to resisting uniform downward loads. The principle has been used for buildings, mostly as draped cable structures.

Cable-stayed beams

Cables assist the deck beam by supporting its self-weight. Compression is taken in the deck beam so that ground anchors are not required (Fig. 5.20). The cable-stayed principle has recently been developed for single-storey buildings (see section 5.1 of this chapter). The cable system is designed for and primarily resists gravity loads; in buildings with lightweight roof construction the uplift forces, which are of similar magnitude, are resisted by bending of the stiffening girders. The system is suitable for spans of 30–90 m and has recently been widely used for industrial and sports buildings.

Cable truss

The hanging cable resists downloads and the hogging cable resists upload (Fig. 5.21). If diagonal bracing is used, non-uniform load can be resisted without large deflections but with larger fluctuation of force in the cables.

Three-dimensional cable truss

The classic form of this structure is the bicycle wheel roof, in which a circular ring beam is braced against buckling by a radial cable system. These radial cables are divided into an upper and lower set, providing support to the central hub (Fig. 5.22).

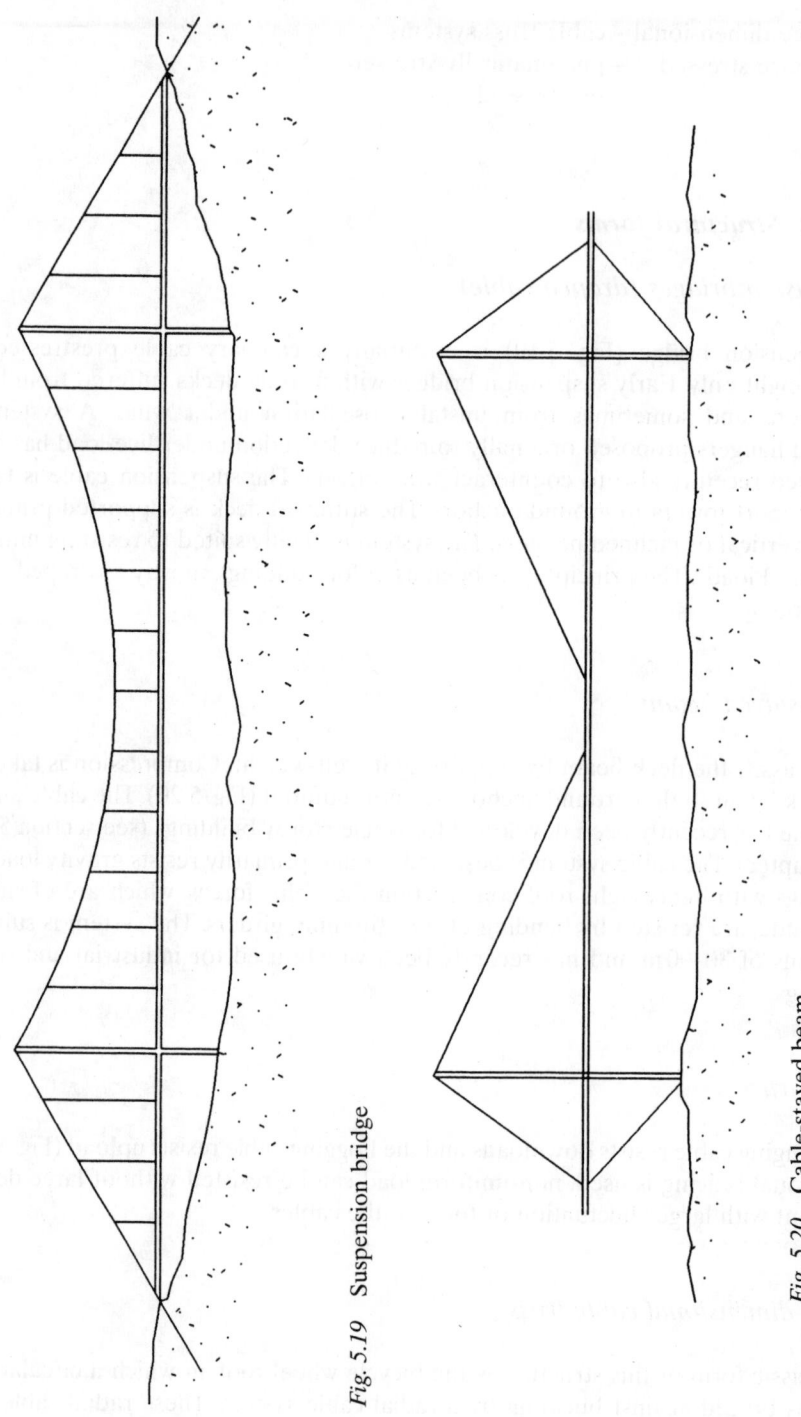

Fig. 5.19 Suspension bridge

Fig. 5.20 Cable-stayed beam

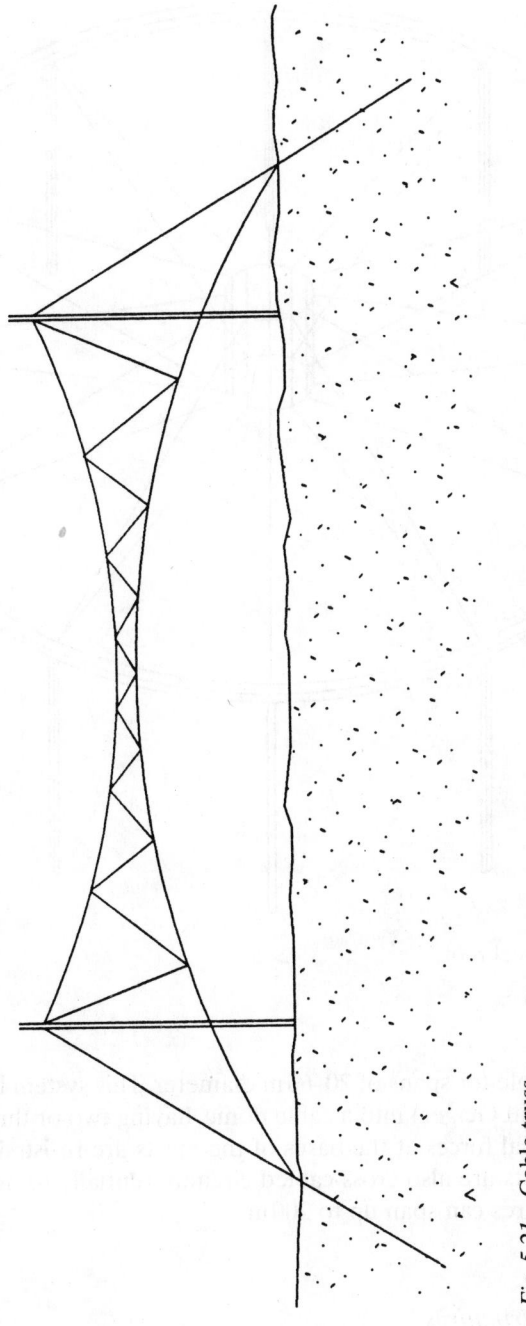

Fig. 5.21 Cable truss

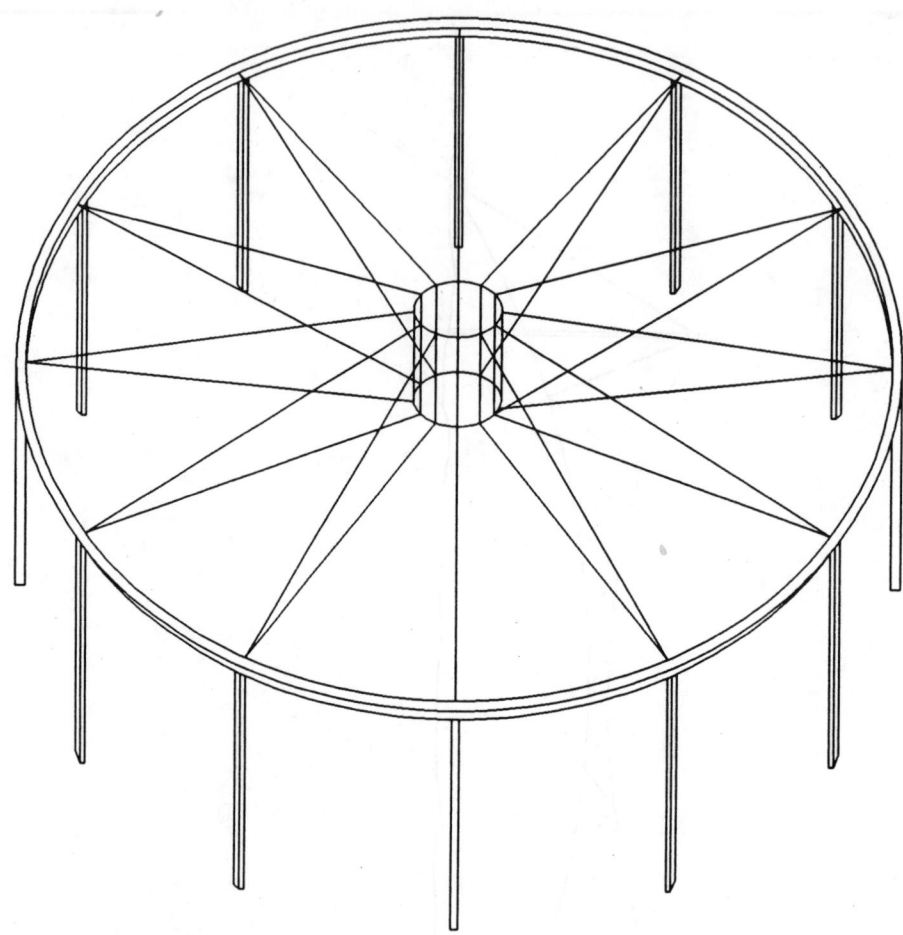

Fig. 5.22   Bicycle wheel roof

The system is suitable for spans of 20–60 m diameter. This system has recently been developed (by David Geiger) into a cable dome, having two or three rings of masts (Fig. 5.23). The radial forces at the bases of the masts are resisted by circumferential cables. The masts are also cross-cabled circumferentially to maintain their stability. These structures can span up to 200 m.

Surface-stressed structures

A cable network can be arranged to have a doubly-curved surface either by giving it a boundary geometry which is out-of-plane or by inflating it with air pressure. The

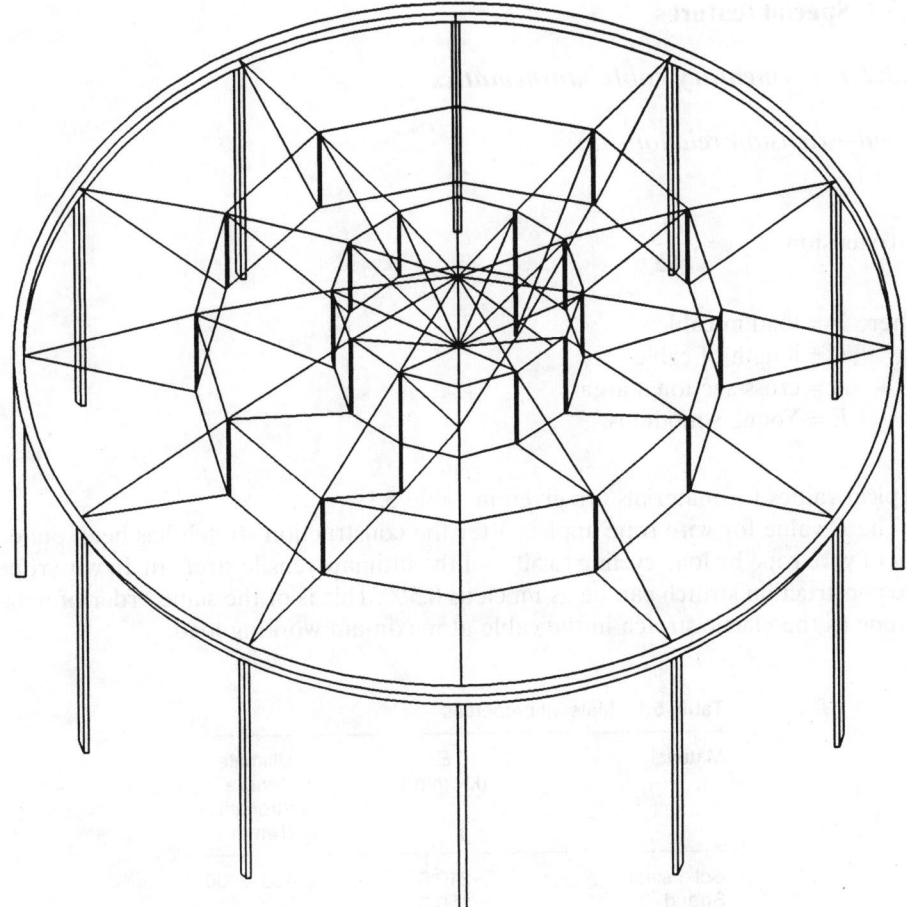

Fig. 5.23   Cable dome

cable net must be prestressed either by tensioning the cables to the boundary points or by the inflation pressure. The effect of the double curvature and the prestress is to stiffen the structure to prevent undue deflection and oscillation under loads. Cable net structures can create dramatic wide-span roofs very economically. They can be clad with fabric, transparent foil, metal decking or timber boarding, insulation and tiles.

Low-profile air-supported roofs can provide the most economical structure for covering very large areas (10–50 acres). In designing these structures the aerodynamic profile must be taken into consideration, as must snow loading and the methods of installation and maintenance.

5.3.2 Special features

5.3.2.1 Elementary cable mathematics

Load–extension relationship

$$\text{Extension} \quad e = \frac{TL}{AE}$$

where T = load in cable
  L = length of cable
  A = cross-sectional area
  E = Young's modulus.

Typical values for materials are given in Table 5.1.

The E value for wire rope applies after the construction stretch has been pulled out of wire rope by load cycling to 50% of the ultimate tensile strength. In wire rope the construction stretch can be as much as 0.5%. This is of the same order of magnitude as the elastic stretch in the cable at maximum working load.

Table 5.1 Material properties

Material	E (kN/mm²)	Ultimate tensile strength (N/mm²)
Solid steel	210.0	400–2000
Strand	150.0	2000
Wire rope	112.0	2000
Polyester fibres	7.5	910
Aramid fibres	112.0	2800

Circular arc loaded radially (Fig. 5.24)

$$\text{Tension} \quad T = PR$$

where P = load/unit length radial to cable
  R = radius of cable.

Radius of circular arc, R (Fig. 5.25)

$$\text{Radius} \quad R = \frac{S^2}{8d} + \frac{d}{2}$$

where S = span
d = dip.

Catenary loaded vertically (Fig. 5.26)

Horizontal force $H = \dfrac{WS^2}{8d}$

Vertical force $V = \dfrac{WS}{2}$

Maximum tension $T = (H^2 + V^2)^{\frac{1}{2}}$

where S = span
d = dip
W = vertical load/unit length.

These formulae permit initial estimates of forces in cables to be made. For full and accurate analysis it is necessary to use a non-linear computer analysis which takes into account the change of curvature caused by stretch. For well-curved cables the

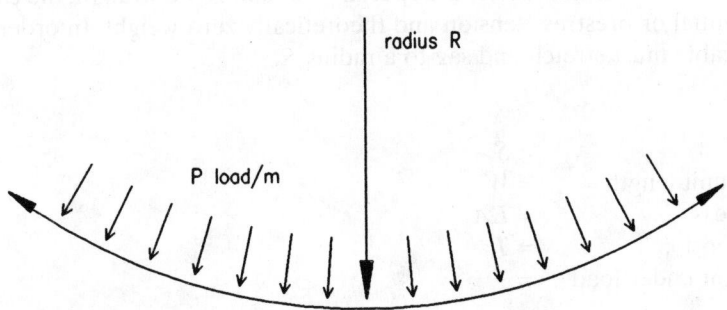

Fig. 5.24   Circular arc loaded radially

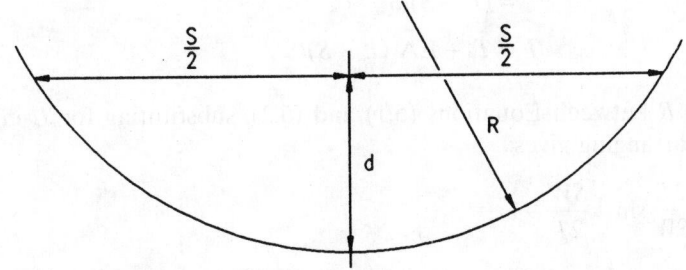

Fig. 5.25   Radius of circular arc

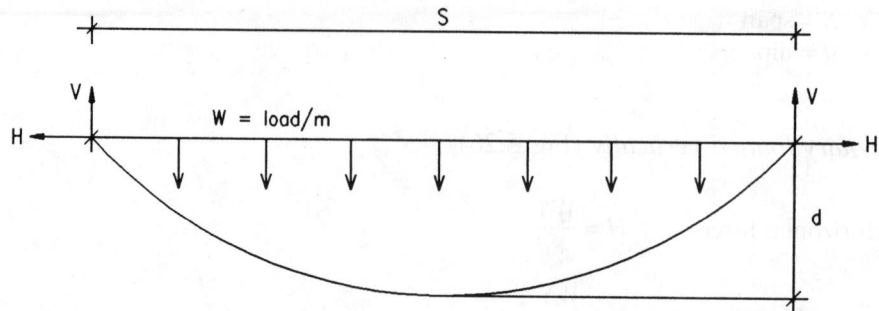

Fig. 5.26 Catenary loaded vertically

hand analysis is accurate enough and gives a useful guide to the forces involved and hence the sizes of cables and fittings.

Prestressed cable

The straight cable (or flat fabric) is a special problem. To be straight, the cable must have an initial or prestress tension and theoretically zero weight. In order to carry load the cable must stretch and sag to a radius R.

For

span	$= S$
load/unit length	$= W$
stiffness	$= EA$
pre-tension	$= T_0$
tension under load	$= T$

Equilibrium equation $T = RW$ (5.1)

New length $L = 2R\sin^{-1}(S/2R)$ (5.2)

Strain $= (L - S)/S$ (5.3)

Tension $T = T_0 + EA\,(L - S)/S$ (5.4)

Eliminating R between Equations (5.1) and (5.2), substituting for L in Equation (5.4) and rearranging gives:

$$\frac{T - T_0}{EA} = \frac{2T}{SW}\sin^{-1}\frac{SW}{2T} - 1 \qquad (5.5)$$

This equation can be solved iteratively for T. The deflection can be found from the earlier formulae for T and R for a circular arc. It should be noted that:

(1) the extension of a tie with low initial tension is considerably more than TL/EA.
(2) straight cables can be used for load carrying, but either the tension will be very high or there will be large deflections.

Two-way cable net (Fig. 5.27)

Approximate calculations can be carried out by hand in a similar way to the single cable calculations above. The basic equilibrium equation for two opposing cables is:

$$\frac{t_1}{R_1} + \frac{t_2}{R_2} = P$$

where P = load/surface area and t_1 and t_2 are the tensions/unit width, i.e.

$$t_1 = \frac{T_1}{a_1} \qquad t_2 = \frac{T_2}{a_2}$$

where a_1 and a_2 are the cable spacings and T_1 and T_2 are the cable tensions.

The loads on ridge cables and boundaries can be estimated from resolution of the cable forces if the geometry is known.

The full analysis of cable net structures is a specialized and complicated process which requires specially written computer programs. The procedure involves the following stages:

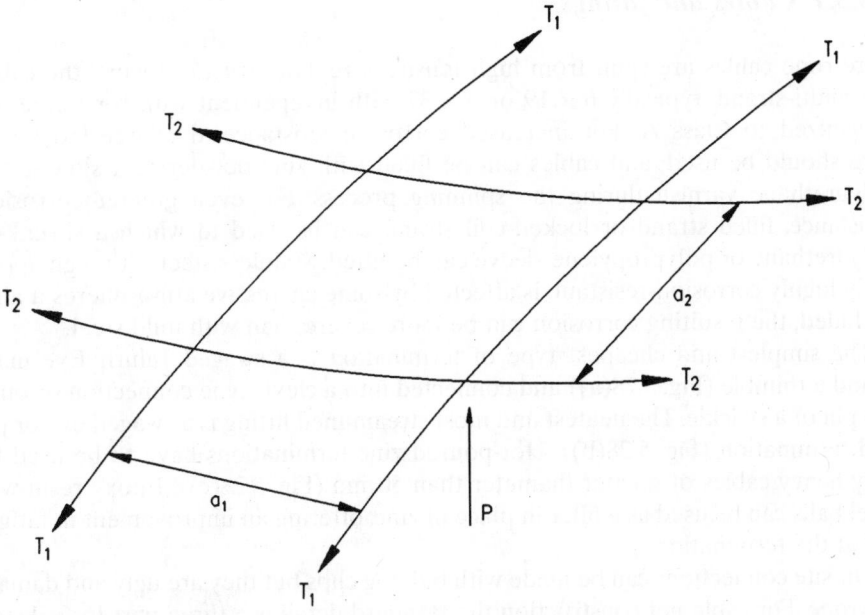

Fig. 5.27   Two-way cable net

(1) Formfinding: In this stage the cables are treated as constant tension elements and the geometry is allowed to move into its equilibrium position. On completion of the formfinding stage the cable net geometry under the prestress forces will be defined.
(2) Load analysis: The prestress model is converted into an elastic model. To do this the slack lengths, l, of all elements must be set so that

$$t_0 = l - \frac{T}{EA}$$

The model is then analysed for the defined dead, wind and snow loads.
(3) Cutting pattern definitions: Cable net structures are fully prefabricated with exact cable lengths so that the prestressed form can be realized. In this stage the form model will be refined so that the mesh lengths are exactly equal, etc. The offsets for the boundary clamps must be allowed for. The stressed cable lengths can then be defined.

During prefabrication of the cables they must be prestretched to eliminate construction struction stretch; they must then be marked at the specified tensions. Tolerances in fabrication are of the order of 0.02%.

5.3.3 Detailing and construction

5.3.3.1 Cables and fittings

Wire rope cables are spun from high tensile wire. For structural work the cables are multi-strand, typically 6 × 19 or 6 × 37 with independent wire rope core and galvanized to Class A. For increased corrosion-resistance, the largest diameter wire should be used, and cables can be filled with zinc powder in a slow-setting polyurethane varnish during the spinning process. For even greater corrosion-resistance, filled strand or locked-coil strand can be used to which a shrunk-on polyurethane or polypropylene sleeve can be fitted. Stainless steel, although apparently highly corrosion-resistant, is affected by some aggressive atmospheres if air is excluded; the resulting corrosion can be more severe than with mild steel.

The simplest and cheapest type of termination is a swaged Talurit Eye made round a thimble (Fig. 5.28(a)) and connected into a clevis type connection or on to the pin of a shackle. The neatest and most streamlined fitting is a swaged eye or jaw end termination (Fig. 5.28(b)). Hot-poured zinc terminations have to be used for very heavy cables of greater diameter than 50 mm (Fig. 5.28(c)). Epoxy resin with steel balls can be used as a filler in place of zinc, offering an improvement in fatigue life at the termination.

On-site connections can be made with bulldog clips but they are ugly and damage the rope. For cable net construction the standard detail is a three-part forged steel clamp, of which the two outer parts are identical (see Fig. 5.28(d)). Forging is expen-

sive for small numbers and so for smaller structures machined aluminium components may be preferred. Double cables can have a swaged aluminium extrusion prefixed to each pair of cables, which can then be connected with a single bolt. For the attachment of net cables to edge cables, forged steel clamps are generally used (Fig. 5.28(e)). Lower cost alternatives are bent plate or machined aluminium clamps.

Cable life is reduced by corrosion and fatigue. Galvanized cables under cover suffer very little corrosion; external cables properly protected should have a life of 50 years. Plastic sheathing has the great disadvantage of making inspection of the cable impossible.

Fatigue investigations have shown that it is wise to limit the maximum tension in a cable to 40% of its ultimate strength for long-life structures. For structures with a design life of up to ten years a limit of 50% is acceptable. Flexing of the cables at clamps or end termination will cause rapid fatigue damage.

5.3.3.2 Rods as tension members

Steel rods are often used as tension members in external situations since they are stiffer and can be given better protection against corrosion. The rods are usually threaded at the ends and screwed into special end fittings. In the case of long rods which can be vibrated by the wind, the end connection must be free to move in two directions, otherwise fatigue damage will occur.

Since tension members are usually critical components of the structural system, consideration should be given to using rods or cables in pairs to provide additional safety.

5.4 Steel in residential construction

5.4.1 Introduction to light steel construction

Light steel framing uses galvanized cold-formed steel sections as the main structural components. These sections are widely used in the building industry and are part of a proven technology. Light steel framing extends the range of steel framed options into residential construction, which has traditionally been in timber and masonry.

The Egan Task Force report called for improved quality, increased use of off-site manufacture, and reduced waste in construction, and the Egan principles have been adopted by The Housing Corporation and other major clients in the residential sector. Light steel framing satisfies these Egan principles and it combines the benefits of a reliable quality controlled product with speed of construction on site and the ability to create existing structural solutions.

Light steel framing is generally based on the use of standard C or Z shaped steel sections produced by cold rolling from strip steel. Cold formed sections are

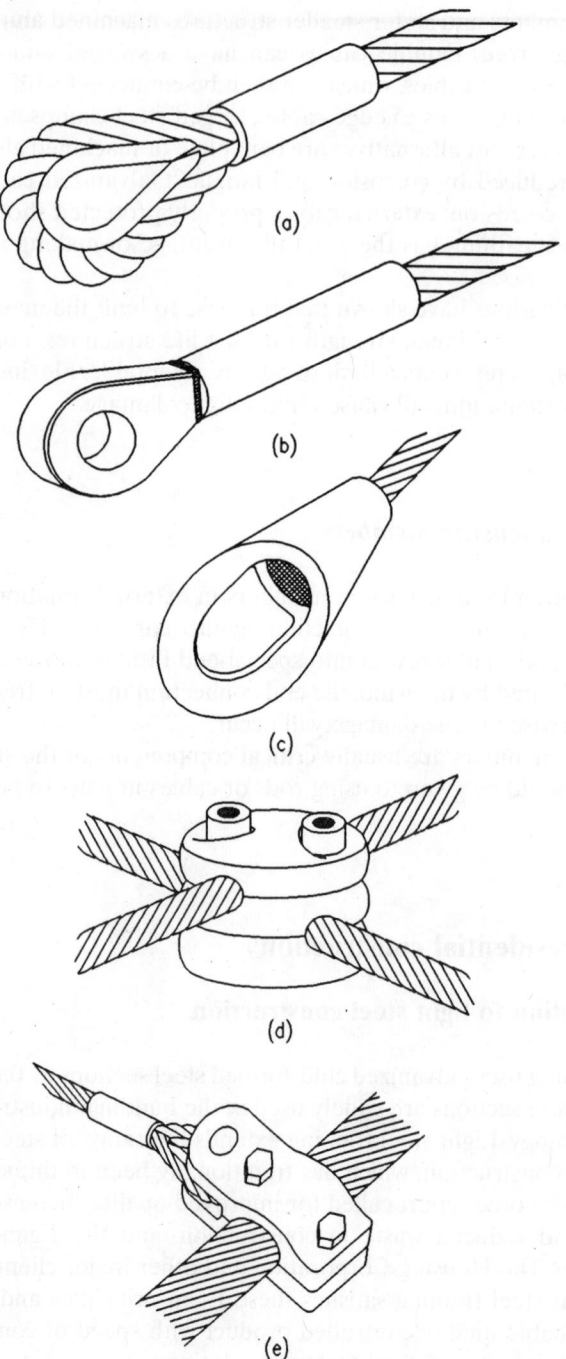

Fig. 5.28 Cable fittings: (a) swaged eye, (b) swaged terminator, (c) hot-poured white metal
eye, (d) cross clamp, (e) net and boundary connection

generically different from hot rolled steel sections, such as Universal Beams, which are used in fabricated steelwork. The steel used in cold formed sections is relatively thin, typically 0.9–3.2 mm, and is galvanized for corrosion protection.

Cold formed steel sections are widely used in many sectors of construction, including mezzanine floors, industrial buildings, commercial buildings and hotels, and are gaining greater acceptance in the residential sector. Light steel framing is already well established in residential construction in North America, Australia and Japan.

5.4.2 Methods of construction

The basic building elements of light steel framing are cold formed sections which can be prefabricated into panels or modules, or assembled on site using various methods of connection. The different forms of construction are reviewed in the following sections and are illustrated in Figs 5.29 to 5.31.

Fig. 5.29 Light steel framing using discrete members ('stick-build' construction)

Fig. 5.30 Light steel framing using prefabricated panels

5.4.3 'Stick-build' construction

In this method of construction (illustrated in Fig. 5.32), discrete members are assembled on site to form columns, walls, rafters, beams and bracing to which cladding, internal lining and other elements are attached. The elements are generally delivered cut to length, with pre-punched holes, but connections are made on site using self-drilling self-tapping screws, bolts, or other appropriate site techniques.

The main advantages of 'stick-build' construction are:

- construction tolerances and modifications can be accommodated on site
- connection techniques are relatively simple
- manufacturers do not require the workshop facilities associated with panel or modular construction
- large quantities of light steel members can be densely packed and transported in single loads
- components can be easily handled on site.

Fig. 5.31   Modular construction using light steel framing

'Stick-build' construction is generally labour intensive on site compared with the other methods, but can be useful in complex construction, where prefabrication is not feasible. This form of construction is widely used in North America and Australia, where there is an infrastructure of contractors skilled in the technique. This stems from a craft tradition of timber frame construction that now uses many power tools. In these countries, traditional timber contractors have changed to light steel framing with little difficulty.

5.4.4 Panel construction

Wall panels, floor cassettes and roof trusses may be prefabricated in a factory and later assembled on site, as in Fig. 5.33. For accuracy, panels are manufactured in purpose-made jigs. Some of the finishing materials may be applied in the factory, to speed on-site construction. Panels can comprise the steel elements alone or the facing materials and insulation can be applied in the factory. The panels are connected on site using conventional techniques (bolts or self drilling screws).

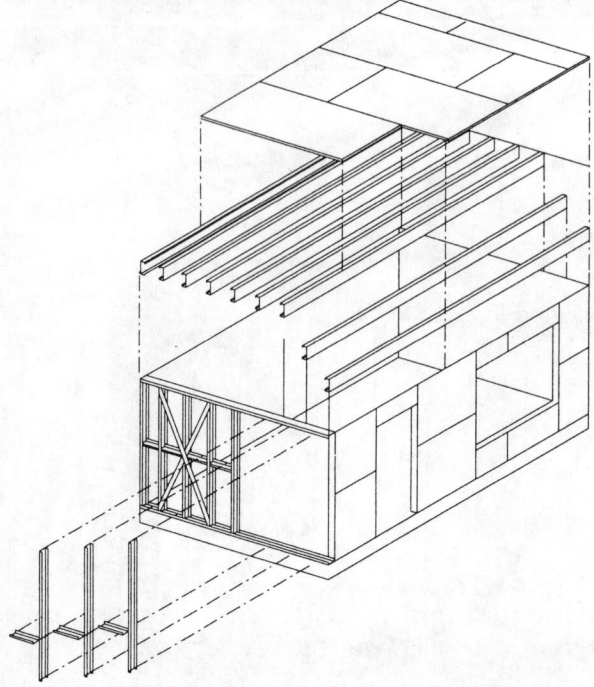

Fig. 5.32 'Stick-build' site construction using wall studs and floor joists

The main advantages of panel or sub-frame construction are:

- speed of erection of the panels or sub-frames
- quality control in production
- reduced site labour costs
- scope for automation in factory production.

The geometrical accuracy and reliability of the panels and other components is better than with stick-build construction because panels are prefabricated in a factory environment. The accurate setting out and installation of foundations is a key factor to achieve rapid assembly of the panels and to obtain the maximum efficiency of the construction process.

5.4.5 Modular construction

In modular construction, units are completely prefabricated in the factory and may be delivered to site with all internal finishes, fixtures and fittings in place, as illustrated in Fig. 5.34. Units may be stacked side by side, or one above the other, to form the stable finished structure.

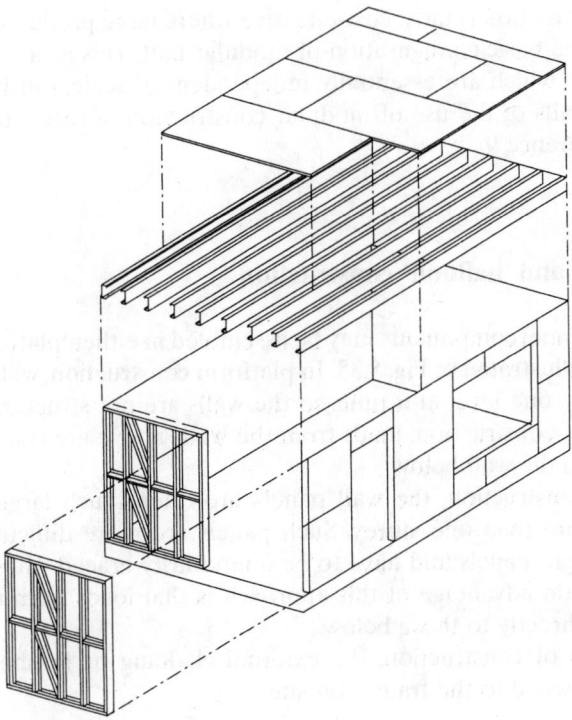

Fig. 5.33   Site assembly of light steel wall panels

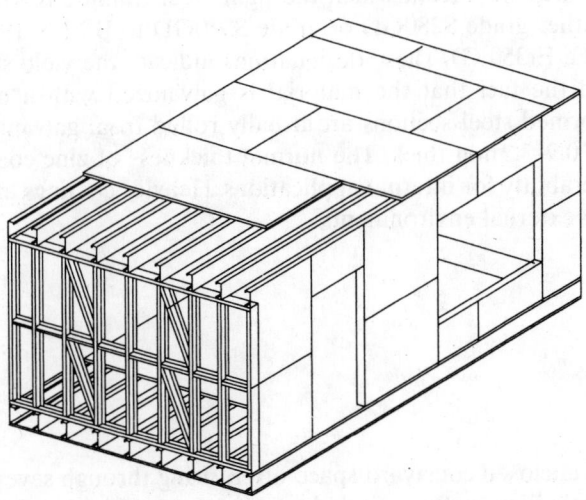

Fig. 5.34   Modular construction using light steel framing

Modular construction is most cost-effective where large production runs are possible for the same basic configuration of modular unit. This is because prototyping and set-up costs, which are essentially independent of scale, can be shared across many units. Details of the use of modular construction in residential construction are given in reference 9.

5.4.6 Platform and 'balloon' construction

'Stick-build' or panel components may be assembled in either 'platform' or 'balloon' construction, as illustrated in Fig. 5.35. In platform construction, walls and floors are built sequentially one level at a time, so the walls are not structurally continuous. In some forms of construction, loads from the walls above are transferred through the floor joists to the wall below.

In 'balloon' construction, the wall panels are often much larger and are continuous over more than one storey. Such panels are more difficult to erect than single storey height panels and have to be temporarily braced whilst the floors are installed. The main advantage of this approach is that loads from the walls above are transferred directly to those below.

In both forms of construction, the external cladding or finishes are generally installed and attached to the frames on-site.

5.4.7 Material properties

The galvanized strip steel from which the light steel framing is formed is usually designated as either grade S280GD or grade S350GD to BS EN 10147[10] (formerly Fe E 280 G or Fe E 350 G). These designations indicate the yield strength (280 or 350 N/mm^2) and the fact that the material is galvanized with a minimum G275 coating. Cold formed steel sections are usually rolled from galvanized sheet steel that is typically 0.9–3.2 mm thick. The normal thickness of zinc coating (275 g/m^2) has excellent durability for internal applications. Heavier coatings are available for more aggressive external environments.

5.5 Atria

5.5.1 General

An atrium is an enclosed courtyard space often rising through several storeys of a building. Its function is usually to provide daylighting within the plan of a building. Often it provides covered circulation space between floor levels which readily

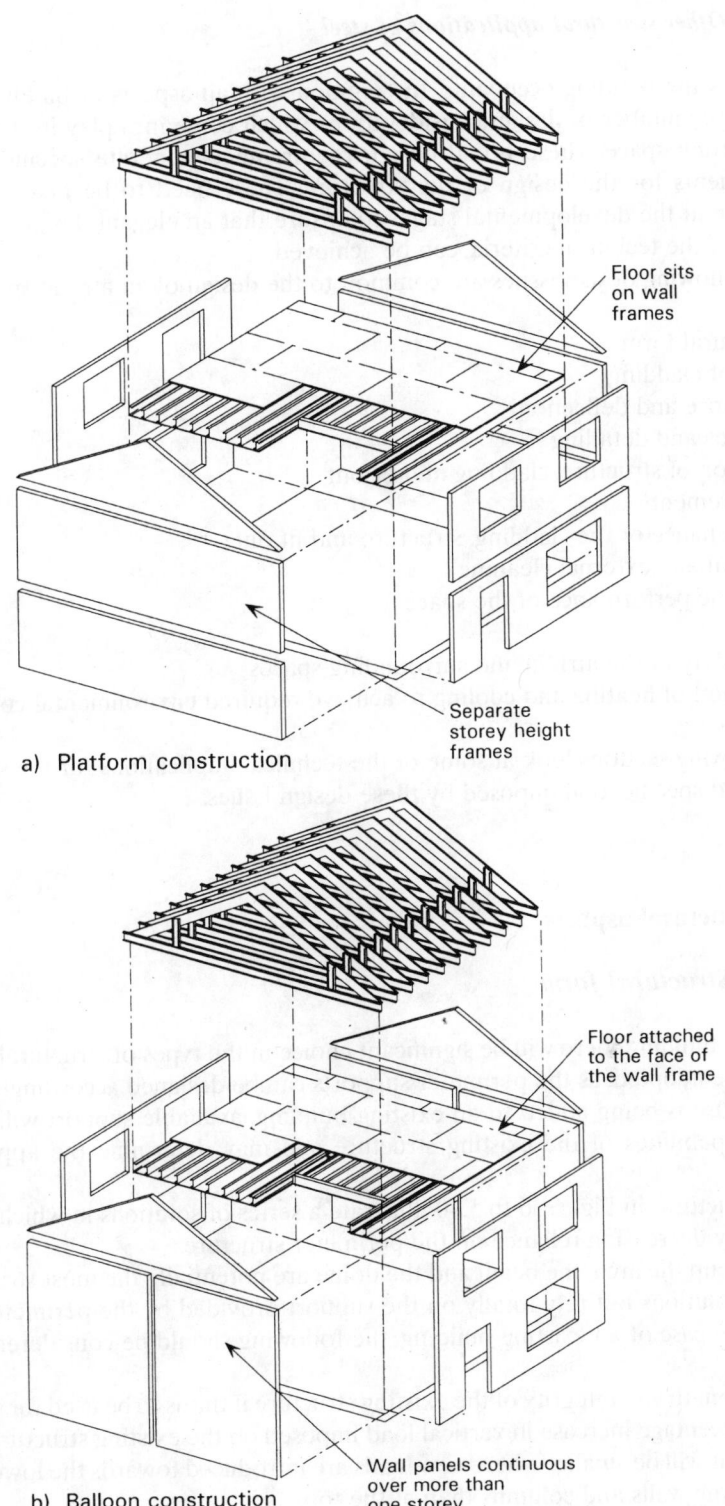

Floor sits
on wall
frames

Separate
storey height
frames

a) Platform construction

Floor attached
to the face of
the wall frame

Wall panels continuous
over more than
one storey

b) Balloon construction

Fig. 5.35 'Platform' and 'balloon' forms of panel construction

orientates the building occupants. In common with all aspects of building design, there are a number of design considerations which come into play in the creation of an atrium space. These broad design requirements generate specific technical requirements for the design of the structure. These need to be taken into consideration at the developmental stages to ensure that an elegant design, consistent with all of the technical criteria, can be achieved.

The following design issues are common to the design of an atrium space:

- Structural form
- Type of cladding
- Tolerance and deflection
- Finishes and detailing
- Erection of structure, cladding and fit out
- Procurement
- Maintenance of the cladding, structure and fit out
- Internal and external cleaning
- Acoustic performance of the space
- Cost
- Fire safety of the atrium and surrounding spaces
- Provision of heating and cooling to achieve required environmental conditions.

The following sections look at some of the technical implications for the structural design and specification imposed by these design issues.

5.5.2 Structural aspects

5.5.2.1 Structural form

For a new building there will be significant choice in the types of structural systems that can be adopted, as the perimeter supports can be designed accordingly. Where an enclosure is being added to an existing building, available support will depend on the capabilities of the existing structure and may determine the appropriate options.

The structures in Figs 5.36 to 5.40 illustrate a series of solutions in which there is a generally decreasing reliance on the perimeter structure.

As a group the arch, the beam and the dome are potentially the most structurally efficient solutions but rely totally on the support provided by the perimeter structure. In the case of an existing building, the following should be considered:

- The strength and integrity of the existing structure if this is to be used for support.
- The percentage increase in vertical load imposed on the existing structure by the new. This will be smaller where new loads are introduced towards the lower levels of existing walls and columns than at the top.

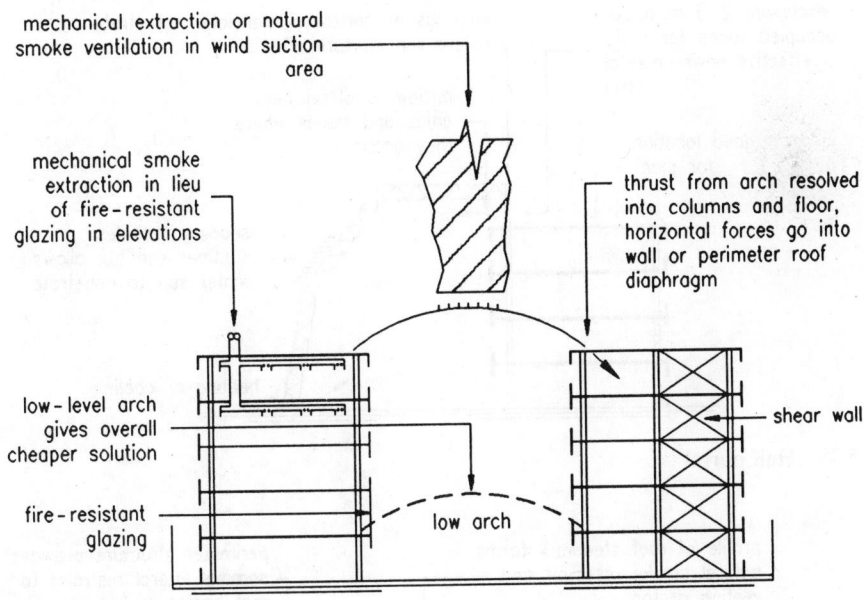

mechanical extraction or natural smoke ventilation in wind suction area

mechanical smoke extraction in lieu of fire-resistant glazing in elevations

thrust from arch resolved into columns and floor, horizontal forces go into wall or perimeter roof diaphragm

low-level arch gives overall cheaper solution

shear wall

fire-resistant glazing

low arch

Fig. 5.36 Arch

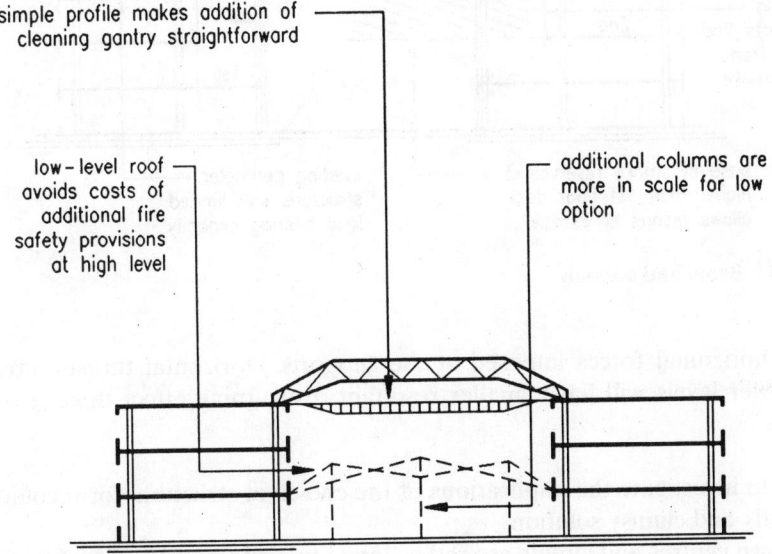

simple profile makes addition of cleaning gantry straightforward

low-level roof avoids costs of additional fire safety provisions at high level

additional columns are more in scale for low option

Fig. 5.37 Beam or truss

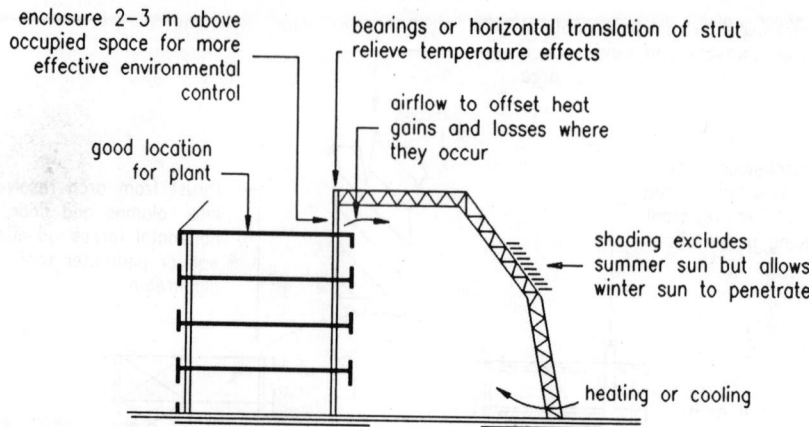

Fig. 5.38   Half portal

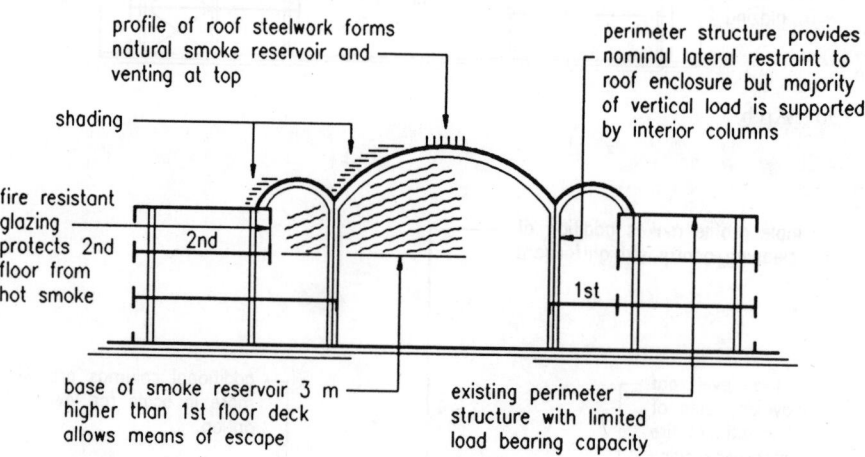

Fig. 5.39   Beam and column

- The horizontal forces imposed at the supports. Horizontal thrusts introduced at lower levels will have smaller resultant overturning effect than at a higher level.

Failure to investigate the implications of the choice of structural form could result in a costly and clumsy solution.

As town centres and streets are redeveloped to compete with out-of-town shopping, the need for enclosure is a common solution, often considered. In these cases, the adjacent buildings are typically of a wide variety of constructions of various heights and condition. The solutions defined in Fig. 5.39 may be appropriate to avoid the complexities of resolving the various boundary conditions. The perimeter struc-

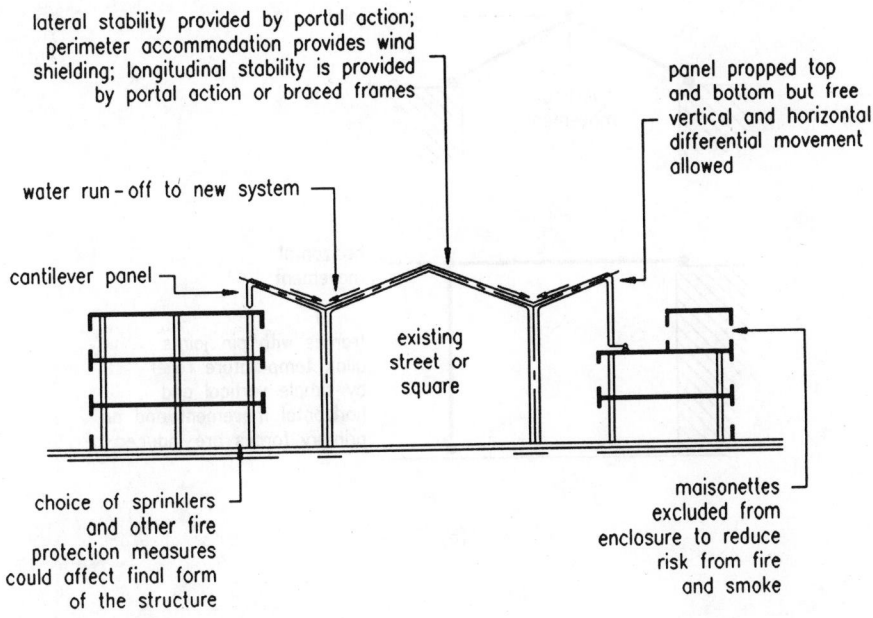

lateral stability provided by portal action; perimeter accommodation provides wind shielding; longitudinal stability is provided by portal action or braced frames

panel propped top and bottom but free vertical and horizontal differential movement allowed

water run – off to new system

cantilever panel

existing street or square

choice of sprinklers and other fire protection measures could affect final form of the structure

maisonettes excluded from enclosure to reduce risk from fire and smoke

Fig. 5.40   Portal frame

ture is required to support a small proportion of the vertical load and relatively small horizontal loads from wind and stabilizing forces for the new columns, which does not generally pose a problem since there is usually some redundant horizontal capacity due to the wind shielding effect of the new enclosure. The portal style solution (Fig. 5.40) is substantially self-supporting.

Temperature effects are often a governing design consideration for an atrium structure due to exposure to heating from sunlight. Temperatures in steelwork painted black can be 30° higher compared with white where directly exposed to the sun's radiation. It is important to allow for movement of the structure, to prevent locked-in stresses from building up. Releases of this sort are often incompatible with providing horizontal stability. The choice of structural form needs to reflect these opposing requirements. For example, a statically determinate structure such as a three pinned arch can accommodate temperature movements and differential settlements without generating secondary internal forces (see Fig. 5.41).

5.5.2.2 *Type of cladding*

Atria generally require transparent or translucent forms of cladding. Glazing is a traditional solution but is relatively heavy. If the space beneath is heated the thermal performance may dictate double-glazing, which is both heavy and costly. Other

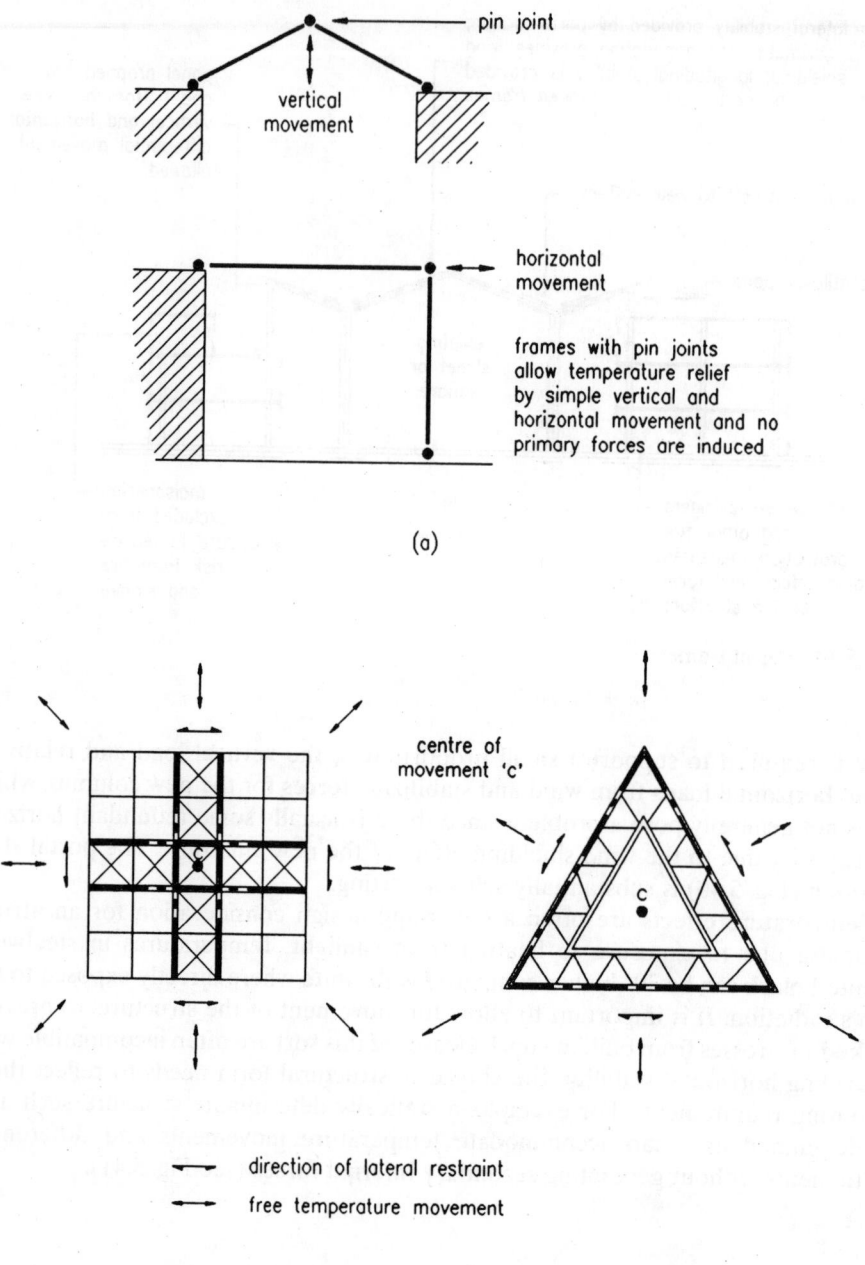

Fig. 5.41 Movements caused by temperature changes

forms of cladding such as ETFE foil cushions can offer a lightweight, thermally efficient and cost-effective alternative to glazing.

Glazing systems may be framed or unframed. The allowable spans of the systems vary. The relationship between the layout of the primary structure and the cladding system will determine whether there is a need for a secondary system of support. It is important to consider this relationship at concept stage to achieve the minimum of elegantly arranged members. The following dimensions provide guidance:

- Planar glazing – spans in either direction around 2 m
- Framed systems – spans 4–6 m by 2 m
- ETFE foil cushions – spans 6–8 m in either direction.

Manufacturers should be consulted for more precise information.

Framing systems are generally made from aluminium extrusions, and therefore have thermal expansion and contraction characteristics which differ from steelwork. This needs to be considered in the detailing of the connectivity between the two elements.

Tolerance between cladding and supporting steel framework also needs careful consideration. The accuracy with which steelwork can be fabricated and erected is generally lower than that of the cladding system. This incompatibility should also be accounted for in the detailing of connections.

Deflections of the atrium structure under loads can have significant bearing on the cladding system. Some glazing systems require a minimum slope to ensure their water tightness. The deflected shape of the supporting structure must satisfy these requirements under all relevant loading conditions. This may require the structure to incorporate a precamber from the theoretical final geometry. The cladding system must also be capable of accommodating any differential movements of the supporting structure.

5.5.2.3 Finishes and detailing

The detailing of individual members and their connections is crucial to achieving a satisfactory architectural treatment. It is commonly the fabricator's responsibility to detail connections, and often this expertise can inform the design process. However, it is important to ensure that aesthetic considerations are properly defined and controlled. The following should be considered:

- Nature of the connection between components of the primary structure
- Connection of the cladding system
- Connection of sundry components such as access equipment, lighting, etc.

Connections can be expressed or 'invisible'. The strategy for connecting components should take into account the practicality of achieving the desired outcome. Site

welding can offer 'invisible connections' within certain limitations. The following should be considered:

- The type of welding technique and position required to form a connection
- Butt welding can be ground flush whereas fillet welds cannot
- The type of weld testing required to verify connection adequacy
- The effects of heat distortion on the structural strength and geometry
- The effect of welding on shop-applied finishes.

Where connections are aesthetically critical, sample workmanship should always be called for as a basis for selecting a contractor. Samples should be kept as a record against which to judge finished workmanship.

Steelwork will require a paint finish for reasons of aesthetics and durability. Paint technology provides an ever increasing number of options. However, the choice can be guided by consideration of the following criteria:

- The steelwork fabricator who delivers and erects the steel will inevitably damage any paint applied at the works no matter how carefully the protection measures are prepared.
- The contractor who has a very tight programme may like to see primer, under-coat and finish coat on the steelwork applied at the works since it may be very difficult to apply the finish in high-level steelwork without extensive temporary works.
- Very hard paint finishes are more difficult to damage although when the inevitable occurs they are usually more difficult to repair to a good standard.
- If the finish-coat is applied on site, some parts of the steelwork may be difficult to reach and the standard of finish will not be as good as that achieved in the paint shop.
- Unless touching up to damaged areas is very minor, the result is unlikely to be satisfactory and therefore a full site applied decorative finish may prove necessary.
- The choice of system will also depend on the fabricator's paint shop and its relationship with the paint supplier.

5.5.2.4 Erection

The engineer should always consider the erection method for a structure. In the case of atria this may be of particular concern for the following reasons;

- The site is constrained by surrounding buildings
- The application of load to existing structures to be carefully controlled
- The need to precamber the geometry
- Restrictions on the location, reach and capacity of craneage

- Limits on the size of components
- The formation of neat site connections
- Safe access for operatives for the erection of the primary structure, cladding and other items.

The erection methodology may preclude the use of some structural forms. It is, therefore, essential to consider this aspect at concept stage.

5.5.2.5 *Procurement*

An atrium structure will have several complex interfaces:

- The cladding
- Cleaning equipment
- Supporting structure

The management of these interfaces is critical to the successful implementation of the design. If the structure, cladding, etc are procured through separate contracts it is even more critical to define the technical requirements and responsibilities at interfaces. The following issues should be addressed:

- Forces and movements to be accommodated
- Compatibility requirements of materials
- Design responsibility for components and fixings
- Definition of overall theoretical and as-built geometry.

It can be convenient to incorporate several packages of work into a single contract. This can reduce risk associated with the management of interfaces. This route may increase cost due to the increased responsibilities carried by the contractor.

5.5.2.6 *Maintenance*

Ongoing and long term cleaning and maintenance will be required and may impact on the design. Ongoing work will require a simple safe means of access. It will be necessary to account for the loading and connection of gantries, walkways, harness points, etc., in the design of the primary structure.

Long term maintenance could be facilitated by independent access systems such as scaffolding. The main impact on the design of the structure will be the quality of the specification. The life to first maintenance of the finishes and components will increase with initial cost. This needs to be balanced by the costs of the maintenance activity. The client should be consulted on the options available and choose the most

appropriate option. This will depend on his/her long term interest in the building fabric.

5.5.2.7 Acoustics

Large enclosed spaces may require acoustic treatment to limit reverberation times. Hard surface finishes can exacerbate the problem. These considerations may affect the choice of finishes within an atrium or necessitate the provision of acoustic baffles or other treatments. This is only of particular concern if the atrium structure is required to support the loads of such components. If so, these must be allowed for in the design.

ETFE foil cushions are substantially transparent from an acoustic point of view, which may make them unsuitable for city centre applications. They may also generate significant noise due to the impact of rain.

5.5.2.8 Cost

There are several factors that affect the cost of steel structures of this type:

- Type of section used
- Complexity of fabrication
- Complexity of installation
- Type of finish
- Quality of workmanship
- Procurement arrangements

It is common to budget steelwork on the basis of rates applied to each tonne of material. While this is a useful indication of cost, it is important to recognize that the appropriate rates for these types of structure may be two to four times greater than those for standard structural steelwork.

5.5.3 Fire engineering

5.5.3.1 Background

The fundamental principle of atrium fire safety design is to ensure that the design standard is not compromised by the presence of the atrium; the standard of the building with the atrium should not be less than the standard of an equivalent building without the atrium. This section describes the considerations and processes

involved in the development of the atrium fire strategy, concluded by summarizing the implications for the structural design.

The primary objective of the fire strategy is to achieve an acceptable standard of life safety. BS 5588: Part 7 is the relevant 'life-safety' guidance for the incorporation of atria into buildings, and is applicable in all but limited circumstances. A package of fire safety provisions is developed to address the following requirements:

- Means of escape
- Compartmentation
- Structural fire resistance
- External spread of fire
- Facilities for the fire service. Smoke clearance is relevant for this situation.

This package depends on the size and use of the building. As examples:

(1) In a small office building, the design is for simultaneous evacuation and there is no need for compartmentation between storeys. Thus in this situation the atrium can be open, as the consequences of a fire anywhere in the building with an atrium are no different (in terms of life safety) than those in the same building without an atrium.
(2) In tall office buildings, storey-to-storey compartmentation is defined to enable phased evacuation; this vertical compartmentation needs to be maintained in the building with an atrium.
(3) In a hospital, as well as vertical compartmentation there is a need for horizontal compartmentation. This enables progressive horizontal evacuation of the fire compartment, with all other compartments remaining in place.
(4) With an atrium it is necessary to consider thermal radiation across the atrium, to ensure that this does not create untenable conditions in a non-fire compartment on the same storey adjacent to the atrium.

In situations where occupants remain in the building, the psychological effects of smoke in the atrium must also be considered, even though there may be no other risk posed by the fire.

The property protection and business continuity implications must be carefully considered. Even in situations where there is an acceptable standard of life safety, the atrium provides the opportunity for substantial spread of smoke, especially with an unenclosed atrium. These additional requirements are usually specified by the client, operator, or insurer and not regulated under legislation. However, it should be noted that there are Local Government Acts (such as Section 20 in London) and national legislation outside England which have property protection implications.

The design procedure of the atrium can be generalized by consideration of:

- Use of the accommodation adjacent to the atrium
- Use of the atrium base

- Performance of the atrium façade
- Control of smoke within the atrium.

Effective smoke control is usually the key to the success of the atrium fire strategy. This can be achieved either by preventing smoke entering the atrium (e.g. an atrium façade with fire-resisting performance and a fire-sterile atrium base) or, alternatively, by managing any smoke within the atrium. The atrium can be divided into three zones:

(1) Atrium base
(2) The smoke reservoir at the top of the atrium, where any hot smoke will collect and can be extracted
(3) Remainder of the atrium.

Different design considerations will apply for these different zones.

The smoke management will normally take advantage of a smoke extract system. The design of this system depends on the quantity of smoke within the atrium and the extract capacity. The *quantity of smoke* is dependent on the fire size and width of spill, where the smoke spills into the atrium. Complex spill geometries create large quantities of smoke. *Smoke extract* can be by either mechanical extract or natural systems. In either case it is necessary to arrange for low-level make-up for replacement air. Wind-effects on natural systems must be addressed. The extract can serve either to limit the depth of the smoke reservoir, or to limit the smoke reservoir temperature, or both. In either case the fire size must be controlled.

The fire size is normally controlled either by use of sprinklers or by management control. Management control defines islands of fire load, the size of the island dependent on the amount of combustibles, separated to ensure that the fire does not spread between islands. Typical rules are $10\,m^2$ islands separated by $4\,m$. Sprinklers are usually assumed in smoke control calculations to limit the fire size to a maximum area equal to the sprinkler-grid area. In practice, sprinklers may extinguish the fire or control it to a substantially smaller area. As an alternative to sprinklers or management control, a fire-engineered approach can be used to account for the likelihood of a particular fire size occurring, given the use of the accommodation ('natural fire' based design).

It is important also to account for cool smoke, which cannot be assumed to rise (and may even drop). Where the smoke cannot be assumed to be hot enough to rise to the reservoir by buoyancy, it will be carried by the airflows that exist within the atrium. CFD analysis may be required to design the cold smoke system. Careful coordination with the building services engineer will enable the best value solution, particularly with natural ventilation systems, as the cold smoke is primarily carried by the environmental flows. An upward flow of air is beneficial for both the fire strategy and the environmental strategy.

Use of the atrium base can be enabled, provided that the rules for smoke and fire spread are respected. It is difficult to use sprinklers for controlling the fire size in the atrium base, as there is no ceiling above it (the atrium roof is usually too high

to enable effective operation of sprinklers), and thus fire load control is either by management control or by an engineered approach.

Where the atrium facade is required to have a fire performance, this is defined either to prevent the fire breaking into the atrium, or to prevent the smoke breaking back in. Where an enclosure is required to have a fire-resisting performance, this can either be achieved by a formal fire-resisting system, or alternatively a smoke-retarding construction may be used in combination with an extract system to limit smoke temperatures. There is currently no standard for smoke-retarding construction, and thus it is often preferable to specify fire-resisting construction to avoid complicated specifications and enhanced site control.

5.5.3.2 *Structural implications*

The structural implications of the above can be summarized as follows:

- **Structural fire resistance**: The required fire resistance for elements of structure is equivalent to that of the building without the atrium.
- **Facade**: The structure needs to ensure that the facade can achieve its required performance. This may equate to ensuring that the deflections in a fire condition are limited or, more simply, by specifying a fire-resisting system for the facade. Although BS 5588: Part 7 defines a performance criterion for 'smoke retarding' construction, consideration should be given to specifying fire-resisting construction instead; this is because there is no standard specification for smoke-retarding construction, and there will be necessity to develop ad-hoc material specification, construction methods and QA procedures to ensure that the system meets the required performance and to demonstrate this to the regulators.
- **Atrium roof**: Whilst structural elements solely supporting the roof neither formally nor usually require fire resistance, it is prudent to consider the effect of collapse in development of the fire strategy. Thus potentially there may be a requirement for fire resistance. Also, as normal, where the roof structural elements support elements requiring a fire-resistance (such as the atrium facade), then these structural elements must achieve the required fire performance.

The fire strategy for a building with an atrium is complex, particularly where there is a need to incorporate local government Acts or property protection requirements, or where there are non-standard client, architectural, or engineering objectives.

A fire-engineered approach will enable the optimum strategy and best value to be achieved. Structural fire engineering can be used with an atrium building in the same way as for a building without an atrium, taking account of the inherent performance of the structure to reduce or omit the need for fire protection. Fire engineering enables best integration of the fire strategy within the design, to ensure that all systems are tailored to meet the precise performance and reliability requirements (for fire and all other requirements).

5.5.4 Environmental engineering

5.5.4.1 General

There is an intimate relationship between thermal physics, smoke ventilation and structural form in atria, and the design of the larger structural shape is influenced by them. In addition, the materials used will have to meet the fire and physics criteria to perform as required.

Environmentally, an atrium can be used as a buffer zone between indoors and outdoors. This zone can be exposed to the external environment, providing protection from wind, rain, sun, etc., and/or a controlled environment similar to the indoor conditions of the building.

Important considerations in atrium design include:

- Characteristic height for stack ventilation
- Areas of glazing – optimization of natural light to the space beneath
- Need for temperature-controlled conditions
- Risk of condensation (winter) – heated or unheated
- Spatial separation (environmentally) from internal spaces
- Ventilation manually openable or BMS controlled
- BMS ventilation control – temperature or wind direction sensitive
- Shading (internal, external), material options
- Smoke reservoirs.

5.5.4.2 Shading

In the atrium, heat transfer by radiation is important, and shading systems may be required to reduce direct solar radiation and allow diffuse solar radiation for natural light. The type of shading – internal or external – will usually be related to its aesthetics, performance and maintenance costs. Shading can enhance stack effect by absorbing solar heat and contributing to buoyancy driven flows.

5.5.4.3 Ventilation

Stack ventilation is used when cross ventilation is not possible and single sided ventilation cannot provide sufficient air change rates. Strong temperature gradients can occur in atria, stack effect being enhanced by wind forces in some cases. Temperature differences can be used both during the day and during the night to enhance night cooling.

The atrium needs to extend above the height of the stack so that the neutral

pressure level is high enough to provide sufficient driving force to draw air in and out of the building.

It may be necessary to add an atrium to an existing building if the shape of the building is outside the limits for cross ventilation (15 m deep). This allows for natural ventilation to all areas, improved natural light penetration and beneficial effects on internal circulation.

5.5.4.4 Smoke

Smoke control can require a large space at the top of the atrium to act as a smoke reservoir. This reduces the amount of smoke moving into upper floors and allows for full evacuation of the building. Extraction of toxic gases can be by means of mechanical and/or natural vents located at the top of the atrium. Smoke venting via the atrium can be achieved by taking advantage of the strong buoyancy of smoke, but the use of channelling screens may be necessary. For instance, automatic window or curtain closing on the atrium perimeter can be used as a screen mechanism.

References to Chapter 5

1. British Standards Institution (1986) *Lattice towers and masts*. Part 1: *Code of practice for loading*. BS 8100, BSI, London.
2. British Standards Institution (1972) *Loading*. Chapter V: Part 2: *Wind loads*. CP3, BSI, London.
3. Construction Industry Research and Information Association (CIRIA) (1980) *Wind Engineering in the Eighties*. CIRIA Conference Report 12/13 Nov.
4. Engineering Sciences Data Unit. Wind engineering sub-series (4 volumes). ESDU International, London.
5. Vickery B.J. & Basu R.I. (1983) Across wind vibrations of structures of circular cross section. *Journal of Wind Engineering and Industrial Aerodynamics*, **12**, 49–97.
6. International Association for Shell and Spatial Structures (IASS) (1981) *IASS Recommendations for Guyed Masts*. IASS, Madrid.
7. Nooshin H. & Disney P. (1989) Elements of Formian. *Proceedings of 4th Intl Conf. Civ. and Struct. Engng Computing* (Ed. by H. Nooshin), pp. 528–32. University of Surrey, Guildford, UK.
8. Parke G.A.R. (1990) Collapse analysis and design of double-layer grids. In *Studies in Space Structures* (Ed. by H. Nooshin), pp. 153–79. Multi-Science Publishers.
9. Gorgolewski M.T., Grubb P.J. & Lawson R.M. (2001) *Modular Construction using light steel framing: Design of residential buildings*. SCI-P302. The Steel Construction Institute, Ascot, Berks.

10. British Standards Institution (1992) BS EN 10147: *Specification for continuously hot-dip coated structural steel sheet and strip*. BSI, London.

Further reading for Chapter 5

Section 5.2

Bell A.J. & Ho T.Y. (1984) NODUS spaceframe roof construction in Hong Kong. *Proceedings 3rd International Conference on Space Structures* (Ed. by H. Nooshin), pp. 1010–15. University of Surrey, Guildford, UK.

Bunni U.K., Disney P. & Makowski Z.S. (1980) *Multi-Layer Space Frames*. Constrado, London.

Makowski Z.S. (1984) *Analysis, Design and Construction of Braced Domes*. Granada, St Albans.

Makowski Z.S. (1985) *Analysis, Design and Construction of Braced Barrel Vaults*. Elsevier Applied Science Publishers, Barking, Essex.

Parke G.A.R. & Walker H.B. (1984) A limit state design of double-layer grids. *Proceedings 3rd International Conference on Space Structures* (Ed. by H. Nooshin), pp. 528–32. University of Surrey, Guildford, UK.

Supple W.J. & Collins I. (1981) Limit state analysis of double-layer grids. *Analysis, Design and Construction of Double-Layer Grids* (Ed. by Z.S. Makowski), pp. 93–117. Applied Science Publishers, Barking, Essex.

Section 5.3

Liddell W.I. (1988) Structural fabric and foils. *Kerensky Memorial Conference – Tension Structures*, June, Institution of Structural Engineers.

Troitsky M.S. (1988) *Cable Stayed Bridges: Theory and Design*, 2nd edn. BSP Professional, London.

For surface stressed structures refer to publications of the Institut für Leicht Flächentragwerke, Universität Stuttgart, Pfaffenwaldring 14 7000, Stuttgart 80
    IL 5 Convertible roofs
    IL 8 Nets in nature and technics
    IL 15 Air hall handbook

Section 5.4

Chung K.F. (1993) *Building design using cold formed steel sections: Worked examples*. The Steel Construction Institute, Ascot, Berks.

Clough R.H. & Ogden R.G. (1993) *Building design using cold formed steel sections: Acoustic insulation*. The Steel Construction Institute, Ascot, Berks.

Grubb P.J. & Lawson R.M. (1997) *Building design using cold formed steel sections: Construction detailing and practice.* The Steel Construction Institute, Ascot, Berks.

Grubb P.J., Gorgolewski M.T. & Lawson R.M. (2001) *Building design using cold formed steel sections: Light steel framing in residential construction.* The Steel Construction Institute, Ascot, Berks.

Lawson R.M. (1993) *Building design using cold formed steel sections: Fire protection.* The Steel Construction Institute, Ascot, Berks.

Lawson R.M., Grubb P.J., Prewer J. & Trebilcock P. (1999) *Building design using modular construction: An architect's guide.* The Steel Construction Institute, Ascot, Berks.

Rogan A.L. & Lawson R.M. (1998) *Value and benefit assessment of light steel framing in housing.* The Steel Construction Institute, Ascot, Berks.

Trebilcock P.J. (1994) *Building design using cold formed steel sections: An architect's guide.* The Steel Construction Institute, Ascot, Berks.

Section 5.5

Baker N. (1983) Atria and conservatories, Part 1. *Architect's Journal*, **177**, No. 20, 18 May, 67–9, and Part 2. *Architect's Journal*, **177**, No. 21, 25 May, 67–70.

Bednar M.J. (1986) *The New Atrium.* McGraw-Hill.

Chartered Institution of Building Services Engineers (CIBSE) (1999) *CIBSE Guide A Environmental Design*, 6th Edn. CIBSE.

DeCicco P.R. (1983) *Life Safety Considerations in Atrium Buildings. Fire Prevention 164*, November, 27–33. Polytechnic Institute NY.

Dickson M.G.T. & Green M.G. (1986) *Providing Intermediate Space.* National Structural Steel Conference. BCSA.

Hawkes D. (1983) Atria and conservatories, Part 1. *Architect's Journal*, **177**, No. 19, 11 May, 67–70.

Kendrick C., Martin A.J. & Booth W. (1998) *Refurbishment of air-conditioned buildings for natural ventilation*, Technical Note TN8/98, The Building Services Research and Information Association.

Land District Surveyors Association (1989) *Fire Safety in Atrium Buildings.*

Law M. & O'Brien T. (1989) *Fire Safety of Bare External Structural Steel.* Steel Construction Institute, Ascot, Berks.

Lloyds Chambers (1983) *Framed in Steel*, No. 11, Nov., Corus.

Martin A.J. (1995) *Control of natural ventilation*, Technical Note TN11/95, The Building Services Research and Information Association.

Chapter 6
Applied metallurgy of steel

by MICHAEL BURDEKIN

6.1 Introduction

The versatility of steel for structural applications rests on the fact that it can be readily supplied at a relatively cheap price in a wide range of different product forms, and with a useful range of material properties. The key to understanding the versatility of steel lies in its basic metallurgical behaviour. Steel is an efficient material for structural purposes because of its good strength-to-weight ratio. A diagram of strength-to-weight ratio against cost per unit weight for various structural materials is shown in Fig. 6.1. Steel can be supplied with strength levels from about $250\,N/mm^2$ up to about $2000\,N/mm^2$ for common structural applications, although the strength requirements may limit the product form. The material is normally ductile with good fracture toughness for most practical applications. Product forms range from thin sheet material, through optimized structural sections and plates, to heavy forgings and castings of intricate shape. Although steel can be made to a wide range of strengths it generally behaves as an elastic material with a high (and relatively constant) value of the elastic modulus up to the yield or proof strength. It also usually has a high capacity for accepting plastic deformation beyond the yield strength, which is valuable for drawing and forming of different products, as well as for general ductility in structural applications.

Steel derives its mechanical properties from a combination of *chemical composition*, *heat treatment* and *manufacturing processes*. While the major constituent of steel is always iron the addition of very small quantities of other elements can have a marked effect upon the type and properties of steel. These elements also produce a different response when the material is subjected to heat treatments involving cooling at a prescribed rate from a particular peak temperature. The manufacturing process may involve combinations of heat treatment and mechanical working which are of critical importance in understanding the subsequent performance of steels and what can be done satisfactorily with the material after the basic manufacturing process.

Although steel is such an attractive material for many different applications, two particular problems which must be given careful attention are those of corrosion behaviour and fire resistance, which are dealt with in detail in Chapters 35 and 34 respectively. Corrosion performance can be significantly changed by choice of a steel of appropriate chemical composition and heat treatment, as well as by corrosion protection measures. Although normal structural steels retain their strength at tem-

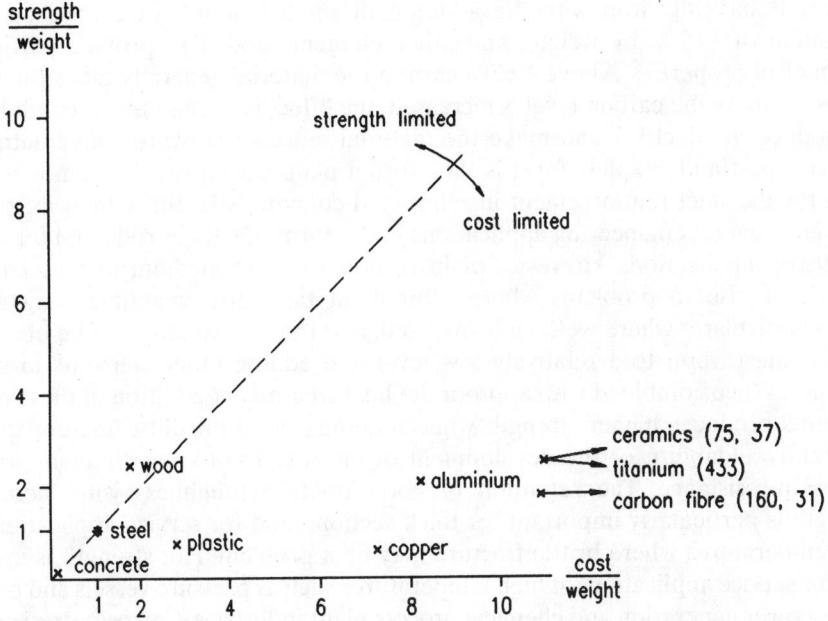

Fig. 6.1 Strength/weight and cost/weight ratios for different materials normalized to steel (1,1)

peratures up to about 300°C, there is a progressive loss of strength above this temperature so that in an intense fire, bare steel may lose the major part of its structural strength. Although the hot strength and creep strength of steels at high temperature can be improved by special chemical formulation, it is usually cheaper to provide fire protection for normal structural steels by protective cladding.

6.2 Chemical composition

6.2.1 General

The key to understanding the effects of chemical composition and heat treatment on the metallurgy and properties of steels is to recognize that the properties depend upon the following factors:

(1) microstructure
(2) grain size
(3) non-metallic inclusions
(4) precipitates within grains or at grain boundaries
(5) the presence of absorbed or dissolved gases.

Steel is basically iron with the addition of small amounts of carbon up to a maximum of 1.67% by weight, and other elements added to provide particular mechanical properties. Above 1.67% carbon the material generally takes the form of cast iron. As the carbon level is increased, the effect is to raise the strength level, but reduce the ductility and make the material more sensitive to heat treatment. The cheapest and simplest form is therefore a plain carbon steel commonly supplied for the steel reinforcement in reinforced concrete structures, for wire ropes, for some general engineering applications in the form of bars or rods, and for some sheet/strip applications. However, plain carbon steels at medium to high carbon levels give rise to problems where subsequent fabrication/manufacturing takes place, particularly where welding is involved, and more versatility can be obtained by keeping carbon to a relatively low level and adding other elements in small amounts. When combined with appropriate heat treatments, addition of these other elements produces higher strength while retaining good ductility, fracture toughness, and weldability, or the development of improved hot strength, or improved corrosion-resistance. The retention of good fracture toughness with increased strength is particularly important for thick sections, and for service applications at low temperatures where brittle fracture may be a problem. Hot strength is important for service applications at high temperatures such as pressure vessels and piping in the power generation and chemical process plant industries. Corrosion-resistance is important for any structures exposed to the environment, particularly for structures immersed in sea water. Weathering grades of steel are designed to develop a tight adherent oxide layer which slows down and stifles continuing corrosion under normal atmospheric exposure of alternate wet and dry conditions. Stainless steels are designed to have a protective oxide surface layer which re-forms if any damage takes place to the surface, and these steels are therefore designed not to corrode under oxidizing conditions. Stainless steels find particular application in the chemical industry.

6.2.2 Added elements

The addition of small amounts of carbon to iron increases the strength and the sensitivity to heat treatment (or hardenability, see later). Other elements which also affect strength and hardenability, although to a much lesser extent than carbon, are *manganese, chromium, molybdenum, nickel* and *copper*. Their effect is principally on the microstructure of the steel, enabling the required strength to be obtained for given heat treatment/manufacturing conditions, while keeping the carbon level very low. Refinement of the grain structure of steels leads to an increase in yield strength and improved fracture toughness and ductility at the same time, and this is therefore an important route for obtaining enhanced properties in steels. Although heat treatment and in particular cooling rate are key factors in obtaining grain refinement, the presence of one or more elements which promote grain refinement by aiding the nucleation of new grains during cooling is also extremely beneficial.

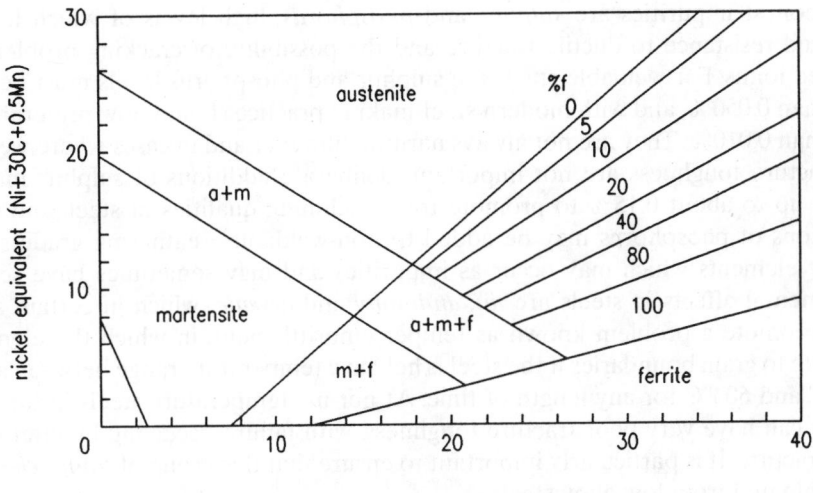

Fig. 6.2 Constitution (Schaeffler) diagram for stainless steels

Elements which promote grain refinement, and which may be added in small quantities up to about 0.050%, are *niobium, vanadium* and *aluminium.*

The major elements which may be added for hot strength and also for corrosion-resistance are chromium, nickel and molybdenum. Chromium is particularly beneficial in promoting corrosion-resistance as it forms a chromium oxide surface layer on the steel, which is the basis of stainless steel corrosion protection in oxidizing environments. When chromium and nickel are added in substantial quantities with chromium levels in the range 12% to 25%, and nickel content up to 20%, different types of stainless steel can be made. As with the basic effect of carbon in the iron matrix, certain other elements can have a similar effect to chromium or nickel but on a lesser scale. From the point of view of the effects of chemical composition, the type of stainless steel formed by different combinations of chromium and nickel can be shown on the Schaeffler diagram in Fig. 6.2. The three basic alternative types of stainless steel are ferritic, austenitic and martensitic stainless steels, which have different inherent lattice crystal structures and microstructures, and hence may show significantly different performance characteristics.

6.2.3 Non-metallic inclusions

The presence of non-metallic inclusions has to be carefully controlled for particular applications. Such inclusions arise as a residue from the ore in the steelmaking process, and special steps have to be taken to reduce them to the required level. The

commonest impurities are *sulphur* and *phosphorus*, high levels of which lead to reduced resistance to ductile fracture and the possibility of cracking problems in welded joints. For weldable steels the sulphur and phosphorus levels must be kept less than 0.050%, and with modern steel-making practice should now preferably be less than 0.010%. They are not always harmful however and in cases where welding or fracture toughness are not important, deliberate additions of sulphur may be made up to about 0.15% to promote free machining qualities of steel, and small additions of phosphorus may be added to non-weldable weathering grade steels. Other elements which may occur as impurities and may sometimes have serious detrimental effects in steels are *tin*, *antimony* and *arsenic*, which in certain steels may promote a problem known as temper embrittlement, in which the elements migrate to grain boundaries if the steel is held in a temperature range between about 500°C and 600°C for any length of time. At normal temperature steels in this condition can have very poor fracture toughness, with failure occurring by inter-granular fracture. It is particularly important to ensure that this group of *tramp elements* is eliminated from low alloy steels.

Steels with a high level of dissolved gases, particularly oxygen and nitrogen, can behave in a brittle manner. The level of dissolved gases can be controlled by addition of small amounts of elements with a particular affinity for them so that the element combines with the gas and either floats out in the liquid steel at high temperature or remains as a distribution of solid non-metallic inclusions. A steel with no such additions to control oxygen level is known as a *rimming steel*, but for most structural applications the elements *silicon* and/or *aluminium* are added as deoxidants. Aluminium also helps in controlling the free nitrogen level, which it is important to keep to low levels in cases where the phenomenon of strain ageing embrittlement may be important.

6.3 Heat treatment

6.3.1 Effect on microstructure and grain size

During the manufacture of steel the required chemical composition is achieved while it is in the liquid state at high temperature. As the steel cools, it solidifies at the melting temperature at about 1350°C, but substantial changes in structure take place during subsequent cooling and may also be affected by further heat treatments. If the steel is cooled slowly, it is able to take up the equilibrium type of lattice crystal structure and microstructure appropriate to the temperature and chemical composition.

These conditions can be summarized on a phase or equilibrium diagram for the particular composition; the equilibrium diagram for the iron–iron carbide system is shown in Fig. 6.3. Essentially this is a diagram of temperature against percentage of carbon by weight in the iron matrix. At 6.67% carbon, an inter-metallic compound called *cementite* is formed, which is an extremely hard and brittle material. At the

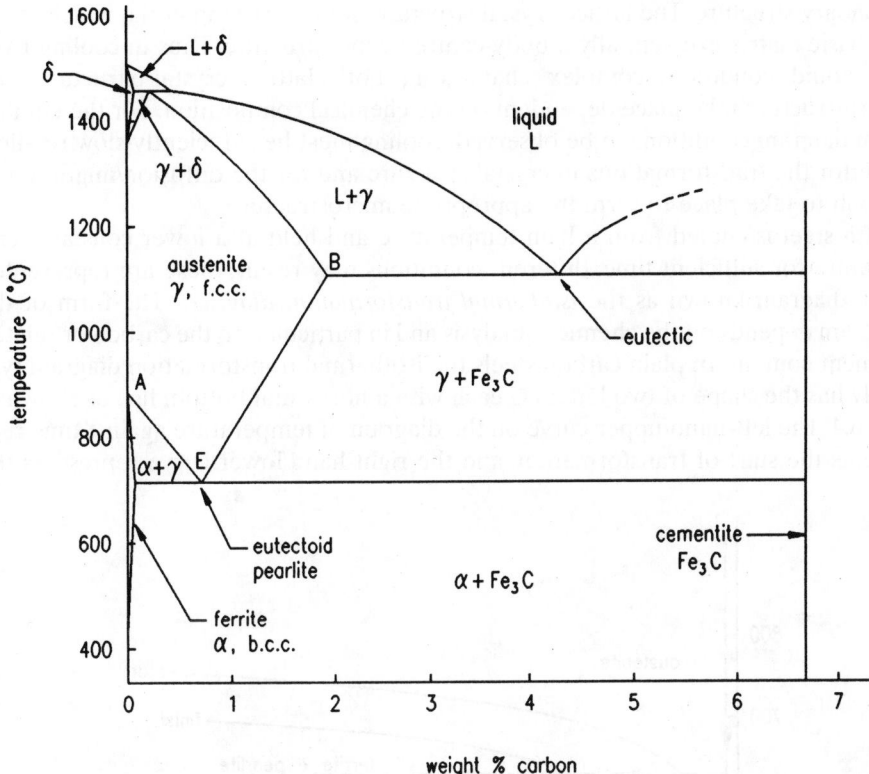

Fig. 6.3 Equilibrium phase diagram for iron – iron carbide system (f.c.c., face-centred cubic; b.c.c., body-centred cubic)

left-hand end of the diagram, with very low carbon contents, the equilibrium structure at room temperature is *ferrite*. At carbon contents between these limits the equilibrium structure is a mixture of ferrite and cementite in proportion depending on the carbon level. On cooling from the melting temperature, at low carbon levels a phase known as *delta ferrite* is formed first, which then transforms to a different phase called *austenite*. At higher carbon levels, the melting temperature drops with increasing carbon level and the initial transformation may be direct to austenite. The austenite phase has a face-centred cubic lattice crystal structure, which is maintained down to the lines AE and BE on Fig. 6.3. As cooling proceeds slowly the austenite then starts to transform to the mixture of ferrite and cementite which results at room temperature. However, point E on the diagram represents a eutectoid at a composition of 0.83% carbon at which ferrite and cementite precipitate alternately in thin laths to form a structure known as *pearlite*. At compositions less than 0.83% carbon, the type of microstructure formed on slow cooling transformation from austenite is a mixture of ferrite and pearlite. Each type of phase present at its appropriate temperature has its own grain size, and the ferrite/pearlite grains tend to precipitate in a network within and based on the previous austenite grain

boundary structure. The lattice crystal structure of the ferrite material which forms the basic matrix is essentially a body-centred cubic structure. Thus in cooling from the liquid condition, complex changes in both lattice crystal structure and microstructure take place dependent on the chemical composition. For the equilibrium diagram conditions to be observed, cooling must be sufficiently slow to allow time for the transformations in crystal structure and for the diffusion/migration of carbon to take place to form the appropriate microstructures.

If a steel is cooled from a high temperature and held at a lower constant temperature for sufficient time, different conditions may result; these are represented on a diagram known as the *isothermal transformation diagram*. The form of the diagram depends on the chemical analysis and in particular on the carbon or related element content. In plain carbon steels the isothermal transformation diagram typically has the shape of two letters C each with a horizontal bottom line as shown in Fig. 6.4. The left-hand/upper curve on the diagram of temperature against time represents the start of transformation, and the right-hand/lower curve represents the

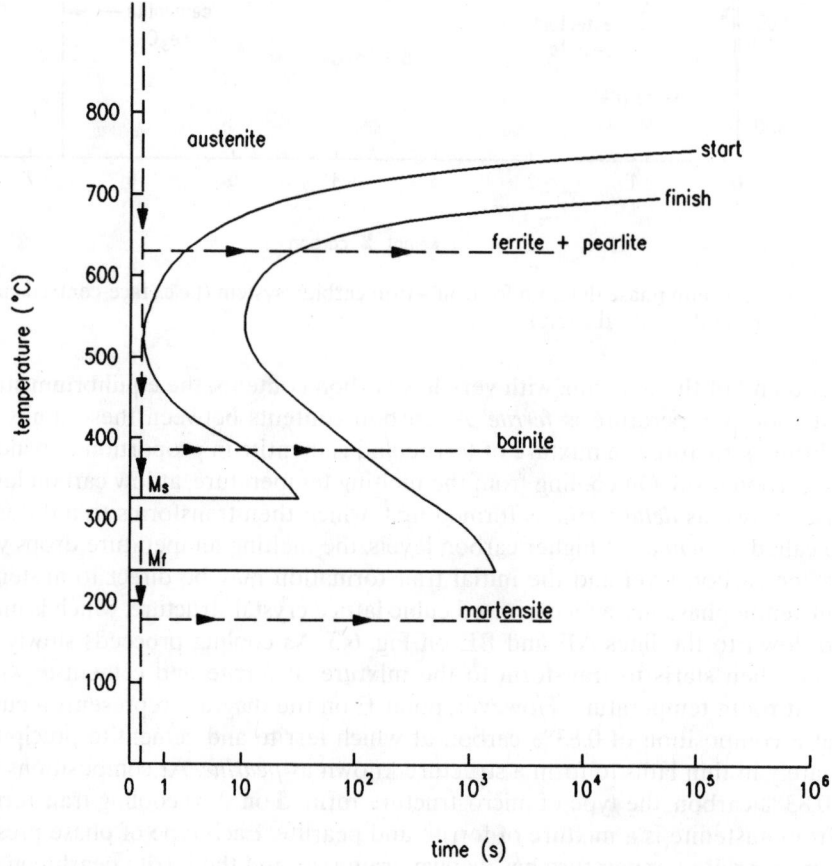

Fig. 6.4   Isothermal transformation diagram for 0.2% C, 0.9% Mn steel

completion of transformation with time. For steels with a carbon content below the eutectoid composition of 0.83% carbon, holding at a temperature to produce isothermal transformation through the top half of the letter C leads to the formation of a ferrite/pearlite microstructure. If the transformation temperature is lowered to pass through the lower part of the C curves, but above the bottom horizontal lines, a new type of microstructure is obtained, which is called *bainite*, which is somewhat harder and stronger than pearlite, but also tends to have poorer fracture toughness. If the transformation temperature is dropped further to lie below the two horizontal lines, transformation takes place to a very hard and brittle substance called *martensite*. In this case the face-centred cubic lattice crystal structure of the austenite is not able to transform to the body-centred cubic crystal structure of the ferrite, and the crystal structure becomes locked into a distorted form known as a body-centred tetragonal lattice. Bainite and martensite do not form on equilibrium cooling but result from quenching to give insufficient time for the equilibrium transformations to take place.

The position and shape of the C curves on the time axis depend on the chemical composition of the steel. Higher carbon contents move the C curve to the right on the time axis, making the formation of martensite possible at slower cooling rates. Alloying elements change the shape of the C curves, and an example for a low-alloy steel is shown in Fig. 6.5. Additional effects on microstructure, grain size and result-

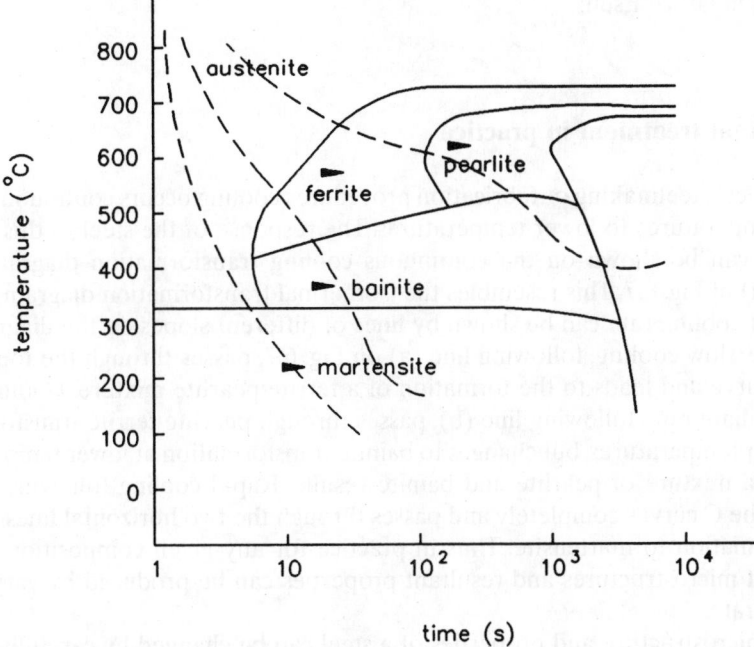

Fig. 6.5 Continuous cooling transformation diagram for 0.4% C, 0.8% Mn, 1% Cr, 0.2% Mo steel

ant properties can be obtained by combinations of mechanical work at appropriate temperatures during manufacture of the basic steel.

In addition to the effect of cooling rate on microstructure, the grain size is significantly affected by time at high temperatures and subsequent cooling rate. Long periods of time at higher temperatures within a particular phase lead to the merging of the grain boundaries and growth of larger grains. For ferritic crystal structures, grain growth starts at temperatures above about 600°C, and hence long periods in the temperature range 600°C to 850°C with slow cooling will tend to promote coarse grain size ferrite/pearlite microstructures. Faster cooling through the upper part of the C curves will give a finer grain structure but still of ferrite/pearlite microstructure.

The type of microstructure present in a steel can be shown and examined by the preparation of carefully polished and etched samples viewed through a microscope. Etching with particular types of reagent attacks different parts of the microstructure preferentially, and the etched parts are characteristic of the type of microstructure. Examples of some of the more common types of microstructure mentioned above are shown in Fig. 6.6. The basic microstructure of the steel is usually shown by examination in the microscope to magnifications of from 100 to about 500 times. Where it is necessary to examine the effects of very fine precipitates or grain boundary effects, it may be necessary to go to higher magnifications. With the electron microscope it is possible to reach magnifications of many thousands and, with specialized techniques, to reach the stage of seeing dislocations and imperfections in the crystal lattice itself.

6.3.2 Heat treatment in practice

In practical steelmaking or fabrication procedures cooling occurs continuously from high temperatures to lower temperatures. The response of the steel to this form of cooling can be shown on the continuous cooling transformation diagram (CCT diagram) of Fig. 6.7. This resembles the isothermal transformation diagram, but the effect of cooling rate can be shown by lines of different slopes on the diagram. For example, slow cooling, following line (a) on Fig. 6.7, passes through the top part of the C curve and leads to the formation of a ferrite/pearlite mixture. Cooling at an intermediate rate, following line (b), passes through pearlite/ferrite transformation at higher temperatures, but changes to bainite transformation at lower temperatures so that a mixture of pearlite and bainite results. Rapid cooling following line (c) misses the C curves completely and passes through the two horizontal lines to show transformation to martensite. Thus in practice for any given composition of steel different microstructures and resultant properties can be produced by varying the cooling rate.

The microstructure and properties of a steel can be changed by carefully chosen heat treatments after the original manufacture of the basic product form. A major group of heat treatments is effected by heating the steel to a temperature such that

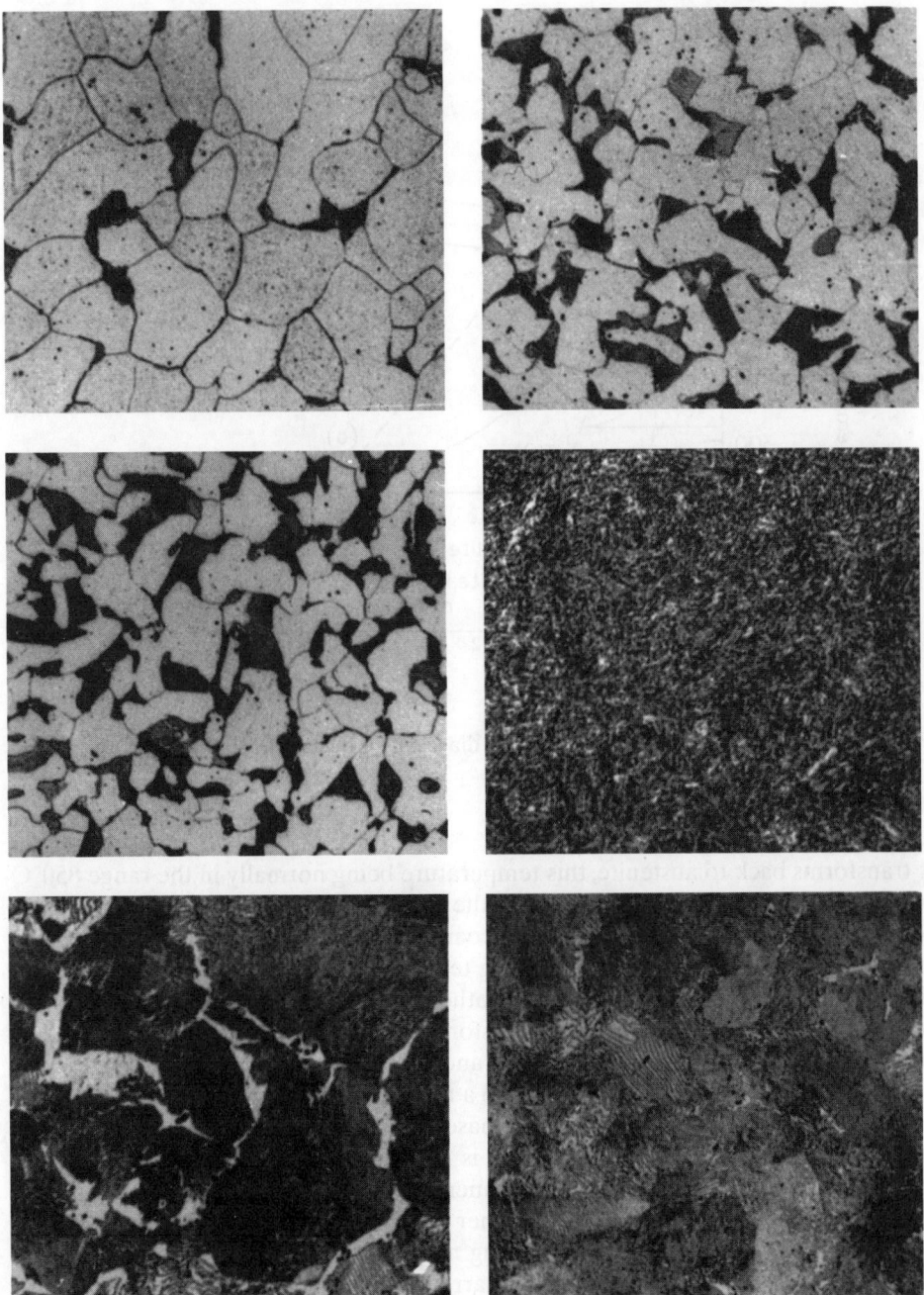

Fig. 6.6 Examples of common types of microstructure in steel (magnification × 500) (courtesy of Manchester Materials Science Centre, UMIST)

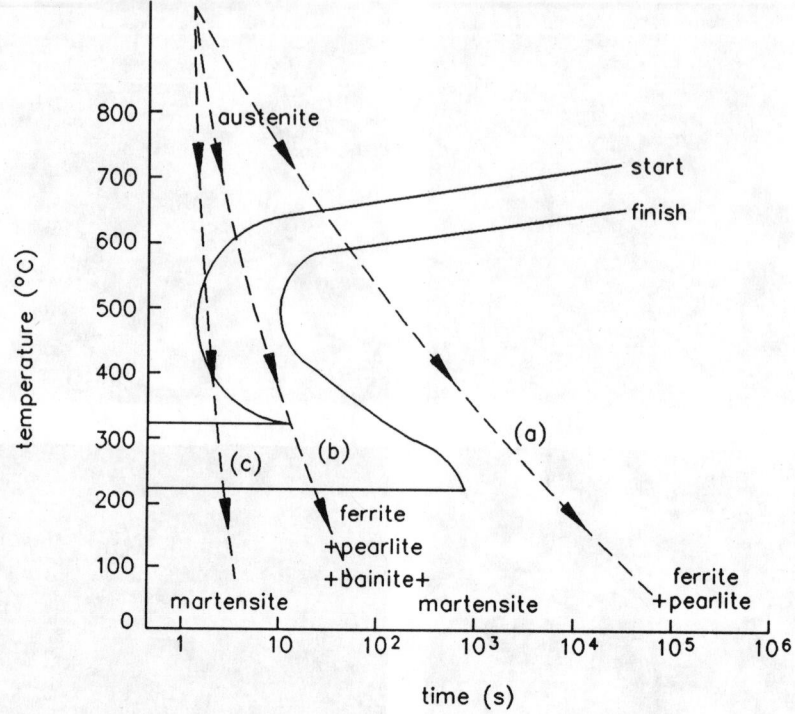

Fig. 6.7   Continuous cooling transformation diagram for 02% C, 0.9% Mn steel

it transforms back to austenite, this temperature being normally in the range 850°C to 950°C. It is important to ensure that the temperature is sufficient for full transformation to austenite, otherwise a very coarse-grained ferritic structure may result. It is also important that the austenitizing temperature is not too high, and that the time at this temperature is not too long, otherwise a coarse-grained austenite structure will form, making subsequent transformation to fine grains more difficult. A heat treatment in which cooling is slow and essentially carried out in a furnace is known as *annealing*. This tends to lead to a relatively coarse-grained final structure, as predicted by the basic equilibrium phase diagram, and is used to put materials into their softest condition. If the steel is allowed to cool freely in air from the austenitizing temperature, the heat treatment is known as *normalizing*, which gives a finer grain size and hence tends to higher yield strength and better toughness for a given composition of steel. Normalizing may be combined with rolling of a particular product form over a relatively narrow band of temperatures, followed by natural cooling in air, in which case it is known as *controlled rolling*. When the steel product form is cooled more rapidly by immersing it directly into oil or water, the heat treatment is known as *quenching*. Quenching into a water bath is generally more severe than quenching into an oil bath.

A second stage heat treatment to temperatures below the austenitizing range is frequently applied, known as *tempering*. This has the effect of giving more time for the transformation processes which were previously curtailed to develop further, and can permit changes in the precipitation of carbides, allowing them to merge together and develop into larger or spheroidal forms. These thermally activated events are highly dependent on temperature and time for particular compositions. The net effect of tempering is to soften previously hardened structures and make them tougher and more ductile.

Both plain carbon and low alloy steels can be supplied in the quenched and tempered condition for plates and engineering sections to particular specifications. The term 'hardenability' is used to describe the ability of steel to form martensite to greater depths from the surface, or greater section sizes. There are, therefore, practical limits of section thickness or size at which particular properties can be obtained.

In BS 970 (some sections of which have been replaced by BS EN standards as part of a phased transition), a range of compositions of engineering steels is given, together with the choice of heat treatments and limiting section sizes for which different properties can be supplied. The heat treatment condition is represented by a letter in the range P to Z. The more commonly supplied conditions are in the range P to T. It should be noted that the term 'hardenability' does not refer to the absolute hardness level which can be achieved, but to the ability to develop uniform hardening throughout the cross section. Cooling rates vary at different positions in the cross section as heat is conducted away in a quenching operation from the surface.

It is sometimes necessary to apply heat treatment to components or structures after fabrication, particularly when they have been welded. The aim is mainly to relieve residual stresses but heat treatment may also be required to produce controlled metallurgical changes in the regions where undesirable effects of welding have occurred. Applications at high temperatures may also lead to metallurgical changes taking place in service. It is vitally important that where any form of heat treatment is applied the possible metallurgical effects on the particular type of steel are taken into account.

Heat treatments are sometimes applied to produce controlled changes in shape or correction of distortion and again temperatures and times involved in these heat treatments must be carefully chosen and controlled for the particular type of steel being used.

6.4 Manufacture and effect on properties

6.4.1 Steelmaking

Manufacture of steel takes place mainly in massive integrated steelworks. The first stage starts with iron ore and coke, which are mixed and heated to produce a sinter. This mixture then has limestone added to form the burden or raw material fed into a blast furnace. Reactions which take place at high temperature in the blast furnace

lead to the formation of iron; the molten iron is tapped continuously from the bottom of the blast furnace. The molten metal at this stage is approximately 90% to 95% iron, the remainder being impurities, which have to he removed or reduced to acceptable levels at the next stage, that of steelmaking. This material is fed together with recovered scrap iron or steel into the steelmaking furnace, the common types of which are known as either a basic oxygen furnace or an electric arc furnace. In the basic oxygen furnace oxygen is blown on to the molten metal by a water-cooled lance. In the electric arc furnace heat is produced by an arc between electrodes over the metal surface and the molten metal itself conducting electricity. Chemical reactions take place following additions of selected materials to the molten metal, which lead to the reduction of the impurities and to the achievement of the required controlled chemical composition of the steel. The impurities are reduced by addition of elements which combine and float out to the surface of the molten metal in the slag or dross waste material on the surface. Deoxidation or killing of the steel takes place in the final stages before the furnace is tapped. Older steel manufacturing practice was to tap the steel from the furnace into ladies and then pour the molten steel into large moulds to produce ingots. These ingots would normally be allowed to solidify and cool before reprocessing at a later stage by rolling into the required product form. Modern steelmaking practice has now moved much more to a process known as *continuous casting*, in which molten steel is poured at a steady rate into a mould to form a continuous solid strand from which lengths of semi-finished product are cut for subsequent processing. Semi-finished products take the form of *slabs*, *billets* or *blooms*. Continuous casting has the advantage of eliminating the reheating and first stage rolling required in the ingot production route, and is generally more efficient, but ingot production is still required for some product forms.

6.4.2 Casting and forging

If the final product form is a casting the liquid steel is poured direct into a mould of the required geometry and shape. Steel castings provide a versatile way of achieving the required finished product, particularly where either many items of the same type are required and/or complex geometries are involved. Special skills are required in the design and manufacture of the moulds in order to ensure that good quality castings are obtained with the required mechanical properties and freedom from significant imperfections or defects. High-integrity castings for structural applications have been successfully supplied for critical components in bridges, such as the major cable saddles for suspension bridges, cast node and tubular sections for offshore structures, and the pump bowl casings for pressurized water reactor systems. The size of component which can be made in cast form is limited to a maximum of some 30–50t however, and only a small proportion of total steel production is completed as castings for direct application.

Another specialist route to the finished steel product is by forging, in which a bloom is heated to the austenitizing temperature range and formed by repeated mechanical pressing in different directions to achieve the required shape. The combination of temperature and mechanical work enables high-quality products with good mechanical properties to be obtained. An example of high-integrity forgings is the production of steel rings to form the shell/barrel of the reactor pressure vessel in a pressurized water reactor system. Again the proportion of steel production as forgings is a relatively small and specialized part of overall steel production.

6.4.3 Rolling

By far the largest amount of finished steel production is achieved by rolling. The semi-finished products cast from the steelmaking furnace are reheated to the austenitizing range and passed through a series of mills with rolls of the required profile to force the hot steel into the finished shape. An example of the distribution of steel products supplied to the UK construction industry is shown in the pie chart of Fig. 6.8. It can be seen that a major part is strip or sheet material which is produced by continuous rolling from slabs down to sheet of the required width and thickness which is first collected as a coil at the end of the rolling process, and subsequently cut into required lengths. The sheet material can be supplied either in bare steel form or with different types of coating. For example, it is possible to obtain steel sheet with a continuous galvanizing (zinc) coating for corrosion protection and it is also possible to obtain it with integral plastic coatings of different colours and patterns for decorative finish as well as corrosion protection.

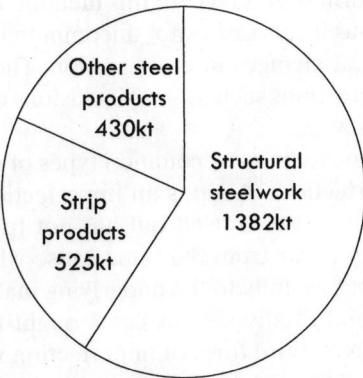

Fig. 6.8 Supply of steel products to the UK construction industry 2001 (UK Steel Association)

For the structural industry steel slabs can be rolled into plates of the required thickness, or into structural sections such as universal columns, universal beams, angle sections, rail sections, etc. Round blooms or ingots can be processed by a seamless tube rolling mill into seamless tubes of different diameters and thicknesses or solid bar subsequently drawn out into wire. Tubes can be used either for carrying fluids in small-diameter pipelines or as structural hollow sections of circular or rectangular shape. The shape of engineering structural sections is determined by the required properties of the cross section such as cross-sectional area and moments of inertia about different axes to give an effective distribution of the weight of the material for structural purposes. Rolled structural sections are supplied in a standard range of shapes detailed in BS 4: Part 1, and a selection of typical shapes and section properties is given in the Appendix *Geometrical properties of plane sections*.

6.4.4 Defects

In any bulk manufacturing process, such as the manufacture of steel, it is inevitable that a small proportion of the production will have imperfections which may or may not be harmful from the point of view of intended service performance of the product. In general the appropriate applications standards have clauses which limit any such imperfections to acceptable and harmless levels. In castings a particular family of imperfections can occur which are dependent on the material and the geometry being manufactured. The most serious types of imperfection are cracks caused by shrinkage stresses during cooling, particularly at sharp changes in cross section. A network of fine shrinkage cracks or tears can sometimes develop, again particularly at changes in cross section where the metal is subjected to a range of different cooling rates. The second type of imperfection in castings is solid inclusions, particularly in the form of sand where this medium has been used to form the moulds. Porosity, or gaseous inclusions, is not uncommon in castings to some degree and again tends to occur at changes in cross section. There is usually appreciable tolerance for minor imperfections such as sand inclusions or porosity provided these do not occur to extreme levels.

In rolled or drawn products, the most common types of defect are either cold laps or rolled-in surface imperfections. A lap is an imperfection which forms when the material has been rolled back on to itself but has not fully fused at the interface. Surface imperfections may occur from the same cause where a tongue of material is rolled down but does not fuse fully to the underlying material. Both of these faults are normally superficial and in any serious cases ought to be eliminated by final inspection at the steel mills. A third form of imperfection which can occur in plates, particularly when produced from the ingot route, is a *lamination*: the failure of the material to fuse together, usually at the mid-thickness of the plate. Laminations tend to arise from the rolling-out of pipes, or separation on the centreline of an ingot at either top or bottom which formed at the time of casting the ingot. Normal

practice is that sufficient of the top and bottom of an ingot is cut off before subsequent processing to prevent laminations being rolled into subsequent products, but nevertheless they do occur from time to time. Fortunately the development of cracks in rolled products is relatively rare although it may occasionally occur in drawn products or as a result of quenching treatments in heat-treated products.

Since much of the manufacture of steels involves processing at, and subsequent cooling from, high temperatures it will be appreciated that high thermal stresses can develop during differential cooling and these can lead to residual stresses in the finished product. In many cases these residual stresses are of no significance to the subsequent performance of the product but there are situations where their effect must be taken into account. The two in which residual stresses from the steel manufacture are most likely to be of importance are where close tolerance machining is required, or where compression loading is being applied to slender structural sections. For the machining case it may be necessary to apply a stress relief treatment, or alternatively to carry out the machining in a series of very fine cuts. The effect of the inherent manufacturing residual stresses on structural sections is taken account of in the design codes such as BS 5950 for steel buildings, by giving a varying factor on the limiting permissible stresses depending on the product shape. These factors have been determined by a series of research programmes on the buckling behaviour of different shaped sections coupled to measurements of the inherent residual stresses present.

6.5 Engineering properties and mechanical tests

As part of the normal quality control procedures of the steel manufacturer, and as laid down in the different specifications for manufacture of steel products, tests are carried out on samples representing each batch of steel and the results recorded on a test certificate. At the stage when the chemical analysis of the steel is being adjusted in the steelmaking furnace, samples are taken from the liquid steel melt at different stages to check the analysis results. Samples are also taken from the melt just before the furnace is tapped, and the analysis of these test results is taken to represent the chemical composition of the complete cast. The results of this analysis are given on test certificates for all products which are subsequently made from the same initial cast. The test certificate will normally give analysis results for C, Mn, Si, S and P for all steels, and where the specification requires particular elements to be present in a specific range the results for these elements will also be given. Even when additional elements are not specified, the steel manufacturer will often provide analysis results for residual elements which may have been derived from scrap used or which could affect subsequent fabrication of performance during fabrication particularly welding. Thus steel supplied to the *Specification for weldable structural steels*, BS 4360, will often have test certificates giving Cr, Ni, Cu, V, Mb and Al, as well as the main basic five elements.

In some specifications, the requirement is given for additional chemical analysis testing on each item of the final product form, and this is presented on the test certificates as product analysis in addition to the cast analysis. This does however incur additional costs. The *Specification for weldable structural steels* gives the opportunity for requiring the steelmaker to supply information on the carbon equivalent to assist the fabricator on deciding about precautions during welding (see later). In low-alloy and stainless steels, the test certificates will of course give the percentage of the alloying elements such as chromium, nickel, etc.

The test certificates should also give the results of mechanical tests on samples selected to represent each product range in accordance with the appropriate specification. The mechanical test results provided will normally include tensile tests giving the yield strength, ultimate strength and elongation to failure. In structural steels, where the fracture toughness is important, specifications include requirements for Charpy V-notch impact tests to BS 131: Part 2. The Charpy test is a standard notched bar impact test of 10mm square cross section with a 2mm deep V-notch in one face. A series of specimens is tested under impact loading either at one specification temperature or over a range of temperatures, and the energy required to break the sample is recorded. In the Euronorms, these notch ductility requirements are specified by letter grades JR, JO, J2 and K2. Essentially these requirements are that the steel should show a minimum of 27 J energy absorption at a specified testing temperature corresponding to the letter grade.

Charpy test requirements are also included in some of the general engineering steel specifications (BS 970) for pressure vessel steels and other important structural applications, particularly where welded structures are used.

The specifications normally require the steelmaker to extract specimens with their length parallel to the main rolling direction. In fact it is unlikely that the steel will be wholly isotropic, and significant differences in material properties may occur under different testing directions, which would not be evident from the normal test certificates unless special tests were carried out. It is possible in some specifications to have material tests carried out both transverse to the main rolling direction, and in the through-thickness direction of rolled products. Testing in the through-thickness direction is particularly important where the material may in fact be loaded in this direction in service by welded attachments. Since such tests are additional to the normal routine practice of the steel manufacturer, and cost extra both for the tests themselves and for the disruption to main production, it is not unexpected that steels required to be tested to demonstrate properties in other directions are more expensive than the basic quality of steel tested in one standard direction only. The quality-control system at the steel manufacturers normally puts markings in the form of stamped numbers or letters on each length or batch of products so that it can be traced back to its particular cast and manufacturing route. In critical structural applications it is important that this numbering system is transferred on through fabrication to the finished structure so that each piece can be identified and confirmed as being of the correct grade and quality. The test certificate for each batch of steel is therefore a most important document to the steel manufacturer, to the fabricator, and to the subsequent purchaser of the finished

component or structure. In addition to the chemical composition and mechanical properties, the test certificate should also record details of the steelmaking route and any heat treatments applied to the material by the steel manufacturer.

It is not uncommon for some semi-finished products to be sold by the steel manufacturer to other product finishers, or to stockholders. Unless these parties retain careful records of the supply of the material it may be difficult to trace specific details of the properties of steel bought from them subsequently, although some stockholders do maintain such records.

Where products are manufactured from semi-finished steel and subsequently given heat treatment for sale to the end user, the intermediate manufacturer should produce his own test certificates detailing both the chemical analysis of the steel and the mechanical properties of the finished product. For example, bolts used for structural connections are manufactured from bar material and are normally stamped with markings indicating the grade and type of bolt. Samples of bolts are taken from manufactured batches after heat treatment and subjected to mechanical tests to give reassurance that the correct strength of steel and heat treatment have been used.

6.6 Fabrication effects and service performance

Basic steel products supplied from the steel manufacturer are rarely used directly without some subsequent fabrication. The various processes involved in fabrication may influence the suitability for service of the steel, and over the years established procedures of good practice have been developed which are acceptable for particular industries and applications.

6.6.1 Cutting, drilling, forming and drawing

Basic requirements in the fabrication of any steel component are likely to be cutting and drilling. In thin sections, such as sheet material, steel can be cut satisfactorily by guillotine shearing, and although this may form a hardened edge it is usually of little or no consequence. Thicker material in structural sections up to about 15 mm thickness can also be cut by heavy-duty shears, useful for small part pieces such as gussets, brackets, etc. Heavier section thicknesses will usually have to be cut by cold saw or abrasive wheel or by flame cutting. Cold saw and abrasive wheel cutting produce virtually no detrimental effects and give good clean cuts to accurate dimensional tolerances. Flame cutting is carried out using an oxyacetylene torch to burn the steel away in a narrow slit, and this is widely used for cutting of thicker sections in machine-controlled cutting equipment. The intense heating in flame cutting does subject the edge of the metal to rapid heating and cooling cycles and so produces the possibility of a hardened edge in some steels. This can be controlled by either preheating just ahead of the cutting torch or using slower cutting speeds, or

alternatively, if necessary, any hardened edge can be removed by subsequent machining. In recent years laser cutting has become a valuable additional cutting method for thin material, in that intricate shapes and patterns can be cut out rapidly by steering a laser beam around the required shape.

Drilling of holes presents little problem and there are now available numerical/computer controlled systems which will drill multiple groups of holes to the required size and spacing. For thinner material hole punching is commonly used and although this, like shearing, can produce a hardened edge, provided the punch is sharp no serious detrimental effects occur in thinner material.

It is sometimes necessary to bend, form or draw steel into different shapes. Reinforcing steel for reinforced concrete structures commonly has to be bent into the form of hooks and stirrups. The curved sections of tubular members of offshore structures or cylindrical parts of pressure vessels are often rolled from flat plate to the required curvature. In these cases yielding and plastic strain take place as the material is deformed beyond its elastic limit. This straining moves the material condition along its basic stress–strain curve, and it is therefore important to limit the amount of plastic strain used up in the fabrication process so that that availability for subsequent service is not diminished to an unacceptable extent. The important variable in limiting the amount of plastic strain which occurs during cold forming is usually the ratio of the radius of any bend to the thickness or diameter of the material. Provided this ratio is kept high the amount of strain will be limited. Where the amount of cold work which has been introduced during fabrication is excessive, it may be necessary to carry out a reheat treatment in order to restore the condition of the material to give its required properties.

In the manufacture of wire, the steel is drawn through a series of dies gradually reducing its diameter and increasing the length from the initial rod sample. This cold drawing is equivalent to plastic straining and has the effect of both increasing the strength of the material and reducing its remaining ductility as the material moves along its stress/strain curve. In certain types of wire manufacture, intermediate heat treatments are necessary in order to remove damaging effects of cold work and enhance and improve the final mechanical properties.

6.6.2 Welding

One of the most important fabrication processes for use with steel is welding. There are many different types of welding and this subject is itself a fascinating multi-disciplinary world involving combined studies in physics, chemistry, electronics, metallurgy, and mechanical, electrical and structural engineering. Most welding processes involve fusion of the material being joined, by raising the temperature to the melting point of the material, either with or without the addition of separate filler metal. Although there is a huge variety of different welding processes, probably the most common and most important ones for general applications are the group of arc welding processes and the group of resistance welding processes. Among the newer

processes are the high energy density beam processes such as electron beam and laser welding.

Arc welding processes involve the supply of an intense heat source from an electric arc which melts the parent material locally, and may provide additional filler metal by the melting of a consumable electrode. These processes are extensively used in the construction industry, and for any welding of material thicknesses above the range of sheets. The resistance group of welding processes involve the generation of heat at the interface between two pieces of material by the passage of very heavy current directly between opposing electrodes on each face. The resistance processes do not involve additional filler metal, and can be used to produce local joints as spot welds, or a series of such welds to form a continuous seam. This group of processes is particularly suitable for sheet material and is widely used in the automotive and domestic equipment markets.

It will be appreciated that fusion welding processes involve rapid heating and cooling locally at the position where a joint is to be made. The temperature gradients associated with welding are intense, and high thermal stresses and subsequent residual stresses on cooling are produced. The residual stresses associated with welding are generally much more severe than those which result during the basic steel manufacturing process itself as the temperature gradients are more localized and intense. Examples of the residual stress distribution resulting from the manufacture of a butt weld between two plates and a T-butt weld with one member welded on to the surface of a second are shown in Fig. 6.9. As will be seen from other chapters, residual stresses can be important in the performance of steel structures because of their possible effects on brittle fracture, fatigue and distortion. If a steel material has low fracture toughness, and is operating below its transition temperature, residual stresses may be very important in contributing to failure by brittle fracture at low applied stresses. If on the other hand the material is tough and yields extensively before failure, residual stresses will be of little importance in the overall structural strength. These effects are summarized in Fig. 6.10.

In fatigue-loaded structures residual stresses from welding are important in altering the mean stress and stress ratio. Although these are secondary factors compared to the stress range in fatigue, the residual stress effect is sufficiently important that it is now commonly assumed that the actual stress range experienced at a weld operates with an upper limit of the yield strength due to locked-in residual stresses at this level. Thus although laboratory experiments demonstrate different fatigue performance for the same stress range at different applied mean stress in plain unwelded material, the trend in welded joints is for the applied stress ratio effect to be overridden by locked-in residual stresses.

The effect of welding residual stresses on distortion can be significant, both at the time of fabrication and in any subsequent machining which may be required. The forces associated with shrinkage of welds are enormous, and will produce overall shrinkage of components and bending/buckling deformations out of a flat plane. These effects have to be allowed for either by pre-setting in the opposite direction to compensate for any out-of-plane deformations or by making allowances with components initially over-length to allow for shrinkage.

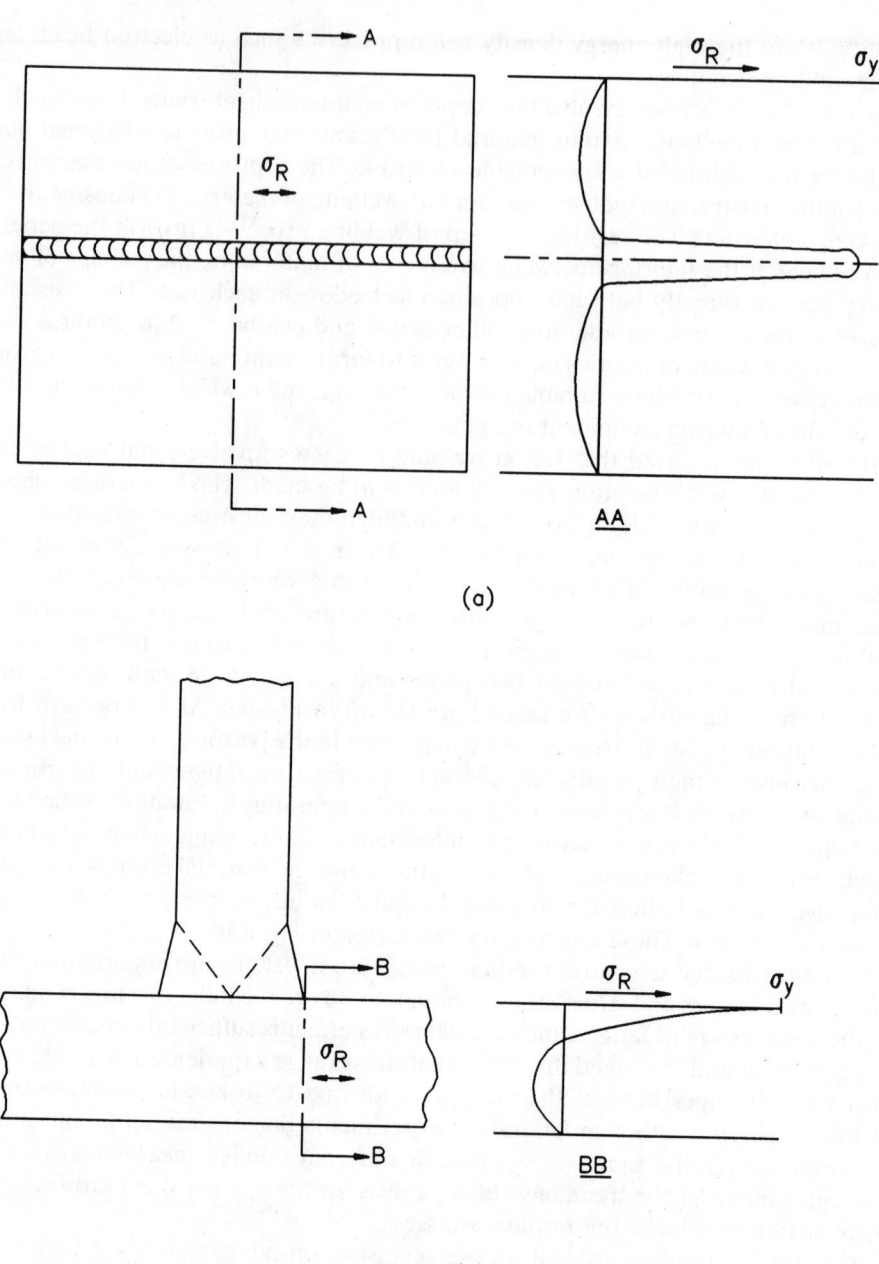

Fig. 6.9 Typical weld residual stress distributions: (a) butt weld, (b) T-butt weld (σ_R residual stress; σ_y yield stress)

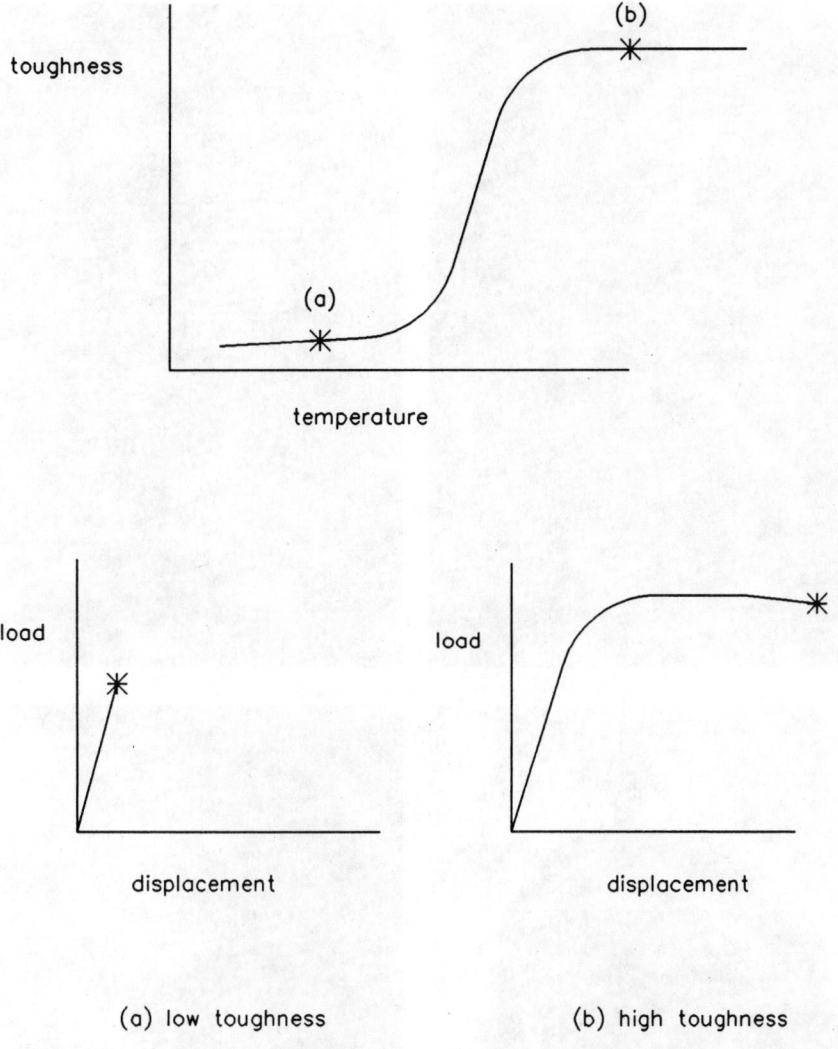

Fig. 6.10   Effects of toughness and residual stresses on strength in tension

Just as the basic steel manufacturing process can lead to the presence of imperfections, welding also can lead to imperfections which may be significant. The types
of imperfection can be grouped into three main areas: planar discontinuities, non-
planar (volumetric) discontinuities, and profile imperfections. By far the most
serious of these are planar discontinuities, as these are sharp and can be of a
significant size. There are four main types of weld cracking which can occur in
steels as planar discontinuities. These are *solidification (hot) cracking, hydrogen-
induced (cold) cracking, lamellar tearing*, and *reheat cracking*. Examples of these are
shown in Fig. 6.11. *Hot cracking* occurs during the solidification of a weld due to the

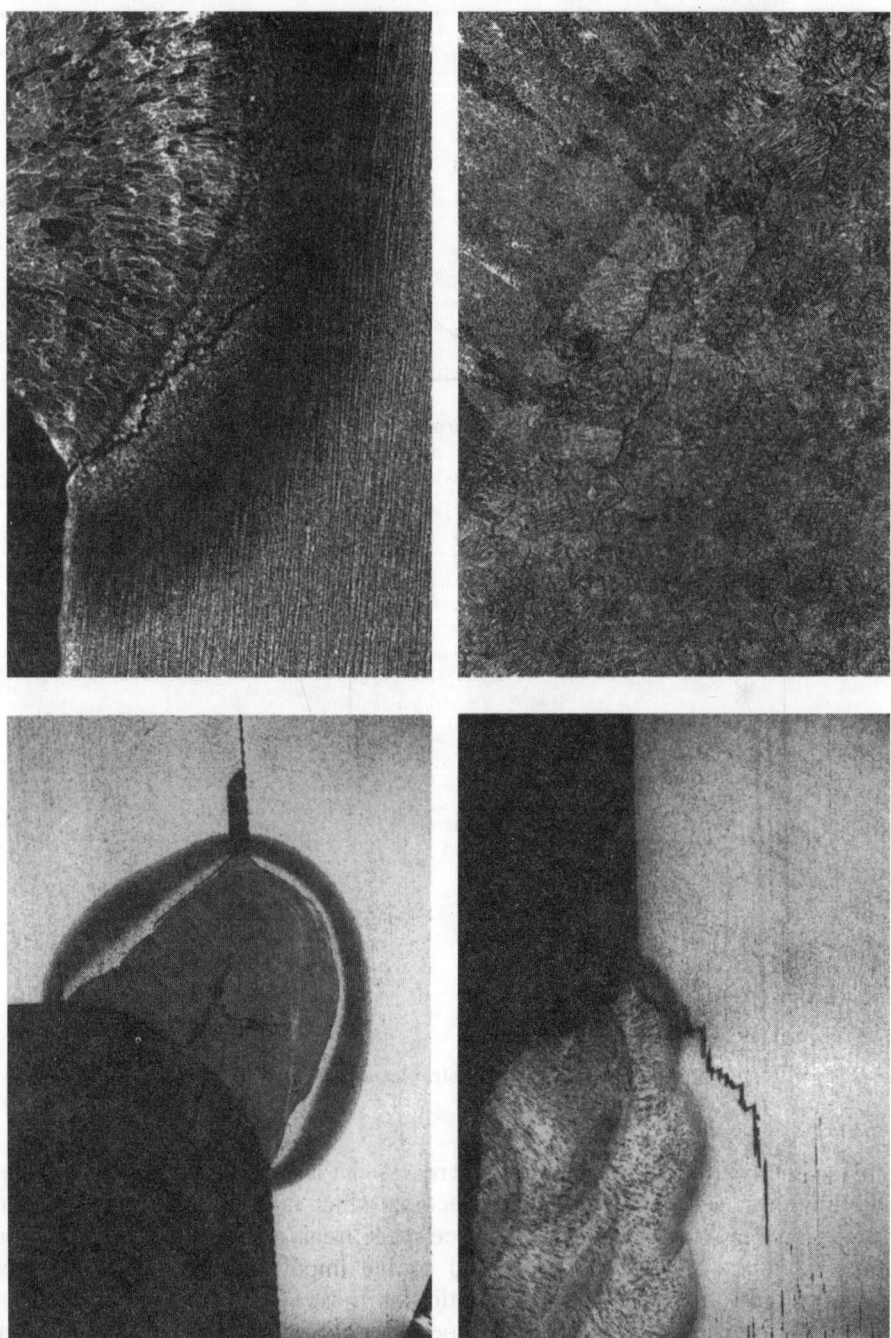

Fig. 6.11   Examples of different types of cracking which may occur in welded steel joints
(courtesy of The Welding Institute)

rejection of excessive impurities to the centreline. Impurities responsible are usually sulphur and phosphorus, and the problem is controlled by keeping them to a low level and avoiding deep narrow weld beads. *Cold cracking* is due to the combination of a susceptible hardened microstructure and the effects of hydrogen in the steel lattice. The problem is avoided by control of the steel chemistry, arc energy heat input, preheat level, quenching effect of the thickness of joints being welded, and by careful attention to electrode coatings to keep hydrogen potential to very low levels. Guidance on avoiding this type of cracking in the heat affected zones of weldable structural steels is given in BS 5135.

Lamellar tearing is principally due to the presence of excessive non-metallic inclusions in rolled steel products resulting in the splitting open of these inclusions under the shrinkage forces of welds made on the surface. The non-metallic inclusions usually responsible are either sulphides or silicates; manganese sulphides are probably the most common. The problem is avoided by keeping the impurity content low, particularly the sulphur level to below about 0.010%, and by specifying tensile tests in the through-thickness direction to show a minimum ductility by reduction of area dependent on the amount of weld shrinkage anticipated (i.e. size of welded attachment, values of R of A of 10% to 20% are usually adequate). *Reheat cracking* is a form of cracking which can develop during stress relief heat treatment or during high temperature service in particular types of steel (usually molybdenum or vanadium bearing) where secondary precipitation of carbides develops before relaxation of residual stresses has taken place.

Other forms of planar defect in welds are the operator or procedure defects of *lack of penetration* and *lock of* fusion. The volumetric/non-planar imperfections divide into the groups of *solid inclusions* and *gaseous inclusions*. The solid inclusions are usually slag from the electrode/flux coating and the gaseous inclusions result from porosity trapped during the solidification of the weld. In general the non-planar defects are much less critical than planar defects of the same size and are usually limited in their effect because their size is inherently limited by their nature.

6.7 Summary

6.7.1 Criteria influencing choice of steel

The basic requirement in the choice of a particular steel is that it must be fit for the product application and design conditions required. It must be available in the product form and shape required and it should be at the minimum cost for the required application. Clearly before the generic type of material is chosen as steel it must be shown to be advantageous to use steel over other contending materials, and therefore the strength-to-weight ratio and cost ratios must be satisfactory.

The steel must have the required strength, ductility and long-term service life in the required environmental service conditions. For structural applications the steel must also have adequate fracture toughness, this requirement being implemented by standard Charpy test quality control levels.

Where the steel is to be fabricated into components or structures, its ability to retain its required properties in the fabricated condition must be clearly established. One of the most important factors in this respect for a number of industries is the weldability of steel, and in this respect the chemical composition of the steel must be controlled within tight limits, and the welding processes and procedures adopted must be compatible with the material chosen.

The corrosion-resistance and potential fire-resistance/high temperature performance of the steel may be important factors in some applications. A clear decision has to be taken at the design stage as to whether resistance to these effects is to be achieved by external or additional protection measures, or inherently by the chemical composition of the steel itself. Stainless steels with high quantities of chromium and nickel are significantly more expensive than ferritic carbon or carbon manganese steels. Particular application standards generally specify the range of material types which are considered suitable for their particular application.

Increased strength of steels can be obtained by various routes, including increased alloying content, heat treatment, or cold working. In general as the strength increases so does the cost and there may be little advantage in using high-strength steels in situations where either fatigue or buckling are likely to be ruling modes of failure. It should not be overlooked that although there is some increase in cost of the basic raw material with increasing strength, there is likely to be a significant increase in fabrication costs, with additional precautions necessary for the more sophisticated types of higher-strength material.

Certain product forms are available only in certain grades of steel. It may not be possible to achieve high strength in some product shapes and retain dimensional requirements through the stage of heat treatment because of distortion problems.

Wherever possible, guidance should be sought on the basis of similar previous experience or prototype trials to ensure that the particular material chosen will be suitable for its required application.

6.7.2 Steel specifications and choice of grade

Structural steelwork, comprising rolled products of plate, sections and hollow sections, is normally of a weldable carbon or carbon-manganese structural steel to the new European based standard, BS EN 10025. Two strength grades are most commonly used, grade S275 and grade S355, having yield strengths typically of 275 N/mm^2 and 355 N/mm^2 respectively. To help designers adjust from the old BS 4360 deagnations, the Appendix *Properties of steel* contains a number of tables which compare the old standard with the new European standards for structural steels.

Further standards are cited in the Appendix *British Standards for steelwork*.

Further reading for Chapter 6

Baddoo N.R. & Burgon B.A. (2001) *Structural Design in Stainless Steel*. The Steel Construction Institute, Ascot, Berks.

Dieter G.E. (1988) *Mechanical Metallurgy*, 3rd edn (SI metric edition). McGraw-Hill.

Gaskell D. (1981) *Introduction to Metallurgical Thermodynamics*, 2nd edn. McGraw-Hill, New York.

Honeycombe R.W.K. (1995) *Steels: Microstructure and Properties*, 2nd edn. Edward Arnold, London.

Lancaster J.F. (1987) *Metallurgy of Welding*, 4th edn. Allen and Unwin.

Porter D.A. & Easterling K.F. (2001) *Phase Transformations in Metals and Alloys*, 2nd edn. Nelson Thomas, Cheltenhum.

Smallman R.E. (1999) *Modern Physical Metallurgy*, 6th edn. Butterworth-Heinemann, Oxford.

Szekely J. & Themelis N.J. (1971) *Rate Phenomena in Process Metallurgy*. Wiley.

Chapter 7
Fracture and fatigue

by JOHN YATES

Structural steelwork is susceptible to several failure processes. The principal amongst these are wet corrosion, plastic collapse, fatigue cracking and rapid fracture, often termed brittle fracture. In this chapter, failure by rapid fracture and fatigue cracking will be discussed.

7.1 Fracture

7.1.1 Introduction

The term brittle fracture is used to describe the fast, unstable fractures that occur with very little energy absorption. In contrast, ductile fracture is a relatively slow process that absorbs a considerable amount of energy, usually through plastic deformation. Some metals, such as copper and aluminium, have a crystalline structure that enables them to resist fast fracture under all loading conditions and at all temperatures. This is not the case for many ferrous alloys, particularly structural steels, which can exhibit brittle behaviour at low temperatures and ductile behaviour at higher temperatures. The consequence of a brittle fracture in a structure may be an unexpected, catastrophic failure. An understanding of the fundamentals of this subject is therefore important for all structural engineers.

The introduction and development of fracture mechanics technology allows the engineer to examine the susceptibility of steel structures, especially their welded joints, to failure assuming that a defect of a given size is present and knowing the operating conditions.

7.1.2 Ductile and brittle behaviour

Ductile fracture is normally preceded by extensive plastic deformation. Ductile fracture is slow, and generally results from the formation and coalescence of voids. These voids are often formed at inclusions due to the large tensile stresses set up at the inclusion/metal interface, as seen in Fig. 7.1(a). Ductile fracture usually goes through the grains but, if the density of inclusions or of pre-existing holes is higher

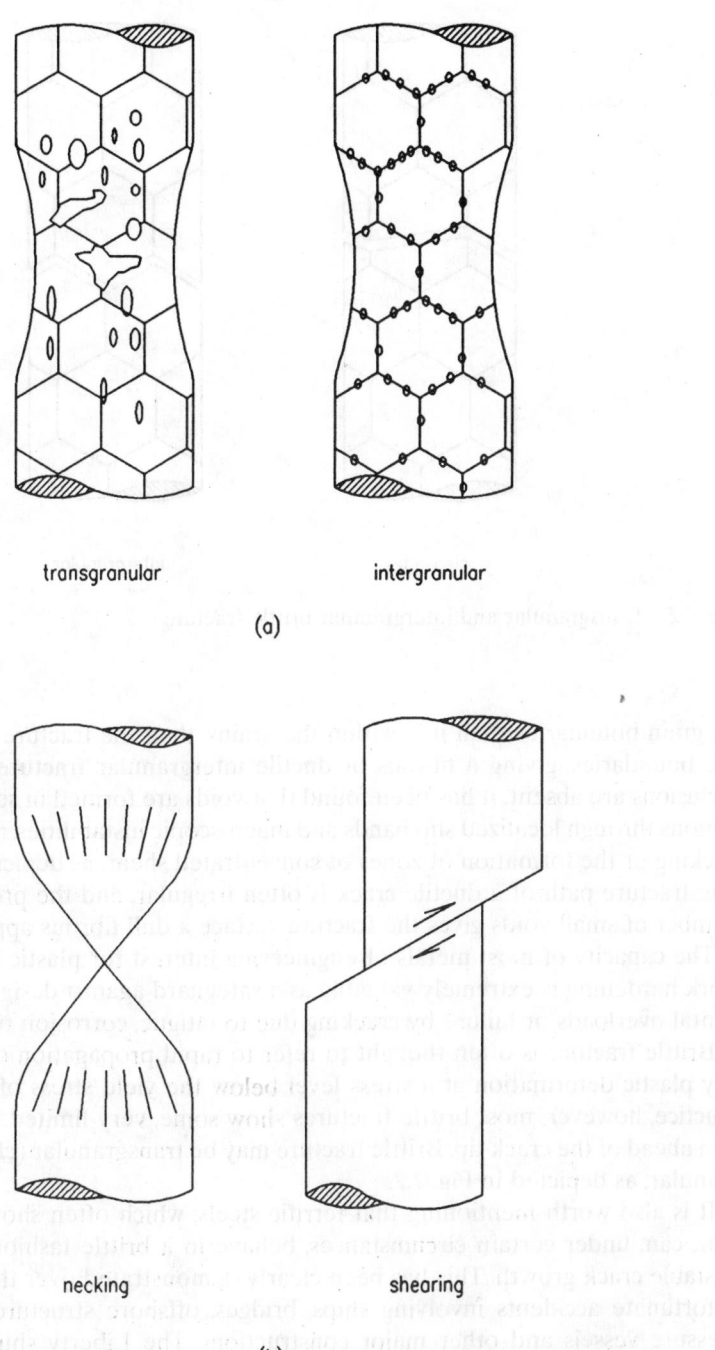

transgranular   intergranular

(a)

necking   shearing

(b)

Fig. 7.1 Plastic deformation (a) by voids growth and coalescence, (b) by necking or shearing

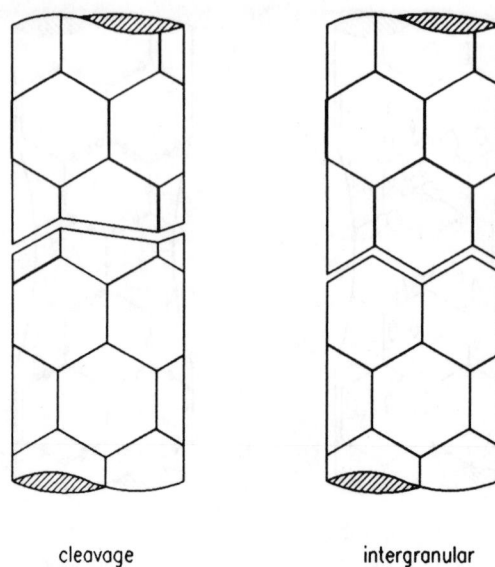

cleavage   intergranular

Fig. 7.2   Transgranular and intergranular brittle fracture

on grain boundaries than it is within the grains, then the fracture path may follow the boundaries, giving a fibrous or ductile intergranular fracture. In cases where inclusions are absent, it has been found that voids are formed in severely deformed regions through localized slip bands and macroscopic instabilities, resulting in either necking or the formation of zones of concentrated shear, as depicted in Fig. 7.1(b). The fracture path of a ductile crack is often irregular, and the presence of a large number of small voids gives the fracture surface a dull fibrous appearance.

The capacity of most metals of engineering interest for plastic deformation and work hardening is extremely valuable as a safeguard against design oversight, accidental overloads or failure by cracking due to fatigue, corrosion or creep.

Brittle fracture is often thought to refer to rapid propagation of cracks without any plastic deformation at a stress level below the yield stress of the material. In practice, however, most brittle fractures show some, very limited, plastic deformation ahead of the crack tip. Brittle fracture may be transgranular (cleavage) or intergranular, as depicted in Fig. 7.2.

It is also worth mentioning that ferritic steels, which often show ductile behaviour, can, under certain circumstances, behave in a brittle fashion leading to fast unstable crack growth. This has been clearly demonstrated over the years by some unfortunate accidents involving ships, bridges, offshore structures, gas pipelines, pressure vessels and other major constructions. The Liberty ships and the King Street bridge in Melbourne, Australia, the Sea Gem drilling rig for North Sea gas and the collapse of the Alexander Kielland oil rig are a few examples of the casualties of brittle fracture.

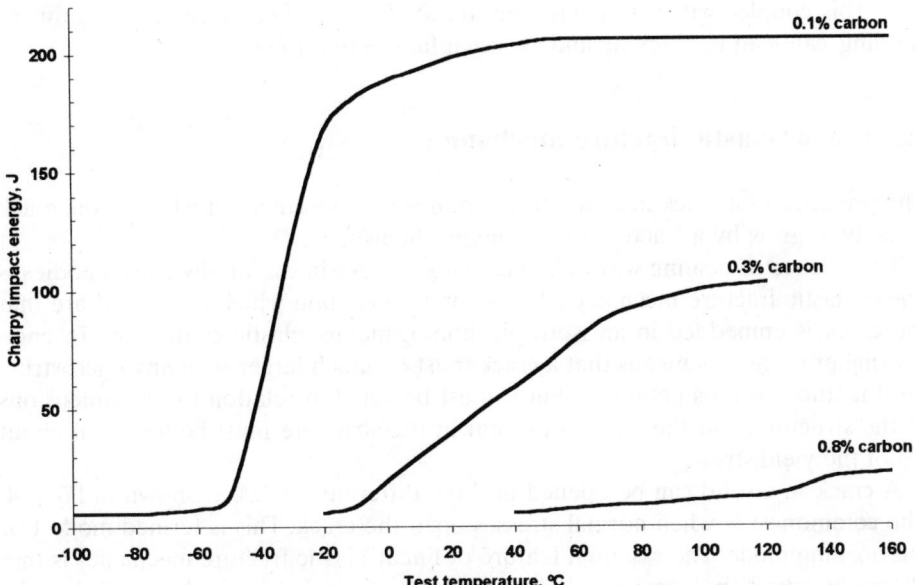

Fig. 7.3   The effect of temperature and carbon content on the impact energy of ferritic steels

An important feature of ferritic steels is the transition temperature between ductile and brittle fracture. Understanding the factors which influence the transition temperature allows designers to be able to select a material which will be ductile at the required operating temperatures for a given structure. The traditional procedure for assessing the ductile to brittle transition in steels is by impact testing small notched beams.[1] The energy absorbed during the fracture process is a measure of the toughness of the material and varies from a low value at low temperatures to a high value as the temperature is raised. The characteristic shape of the impact energy–temperature graph has led to the terminology of the *upper* and *lower shelf*. The low temperature, brittle behaviour is often referred to as the lower shelf and the high temperature, ductile behaviour the upper shelf.

The Charpy V-notch test[1] is the most popular impact testing technique and is described later in this chapter. The transition in impact toughness values obtained from Charpy tests on carbon steels at different temperatures is shown in Fig. 7.3. BS EN 10025[2] gives temperatures for minimum toughness values to be obtained for a range of structural steel grades. Impact transition curves are a simple way of defining the effect that variables such as heat treatment, alloying elements and effects of welding have on the fracture behaviour of a steel. Charpy values are useful for quality control but more sophisticated tests[3,4] are required if the full performance of a material is to be exploited.

An understanding of the fracture behaviour of steel is particularly important when considering welded structures. Welding can considerably reduce the toughness of plate in regions close to the fusion line and introduce defects in the weld

area. This, coupled with residual tensile stresses from the heating and cooling during welding, can lead to cracking and eventual failure of a joint.

7.2 Linear elastic fracture mechanics

The presence of a crack in a structure requires an assessment of whether the crack is likely to grow by a fracture or fatigue mechanism.

The science of dealing with relatively large cracks in essentially elastic bodies is linear elastic fracture mechanics. The assumptions upon which it is based are that the crack is embedded in an isotropic, homogeneous, elastic continuum. In engineering practice, this means that a crack must be much larger than any microstructural feature, such as grain size, but it must be small in relation to the dimensions of the structure, and the stresses present in the structure must be less than about 1/3 of the yield stress.

A crack in a solid can be opened in three different modes, as shown in Fig. 7.4. The commonest is when normal stresses open the crack. This is termed mode I or the opening mode. The essential feature of linear elastic fracture mechanics is that all cracks adopt the same parabolic profile when loaded in mode I and that the tensile stresses ahead of the crack tip decay as a function of $1/\sqrt{}$(distance) from the crack tip.

The absolute values of the opening displacements and the crack tip stresses depend on the load applied and the length of the crack. The scaling factor for the stress field and crack displacements is called the *stress intensity factor*, K_I, and is related to the crack length, a, and the remote stress in the body, σ, by

$$K_I = Y\sigma\sqrt{\pi a} \tag{7.1}$$

where Y is a geometric correction term to account for the proximity of the boundaries of the structure and the form of loading applied. The subscript I denotes the mode I, or opening mode, of loading.

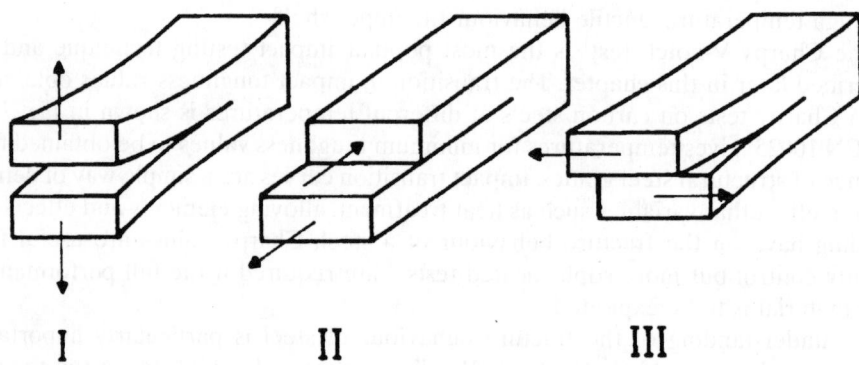

I   II   III

Fig. 7.4 Modes of crack opening

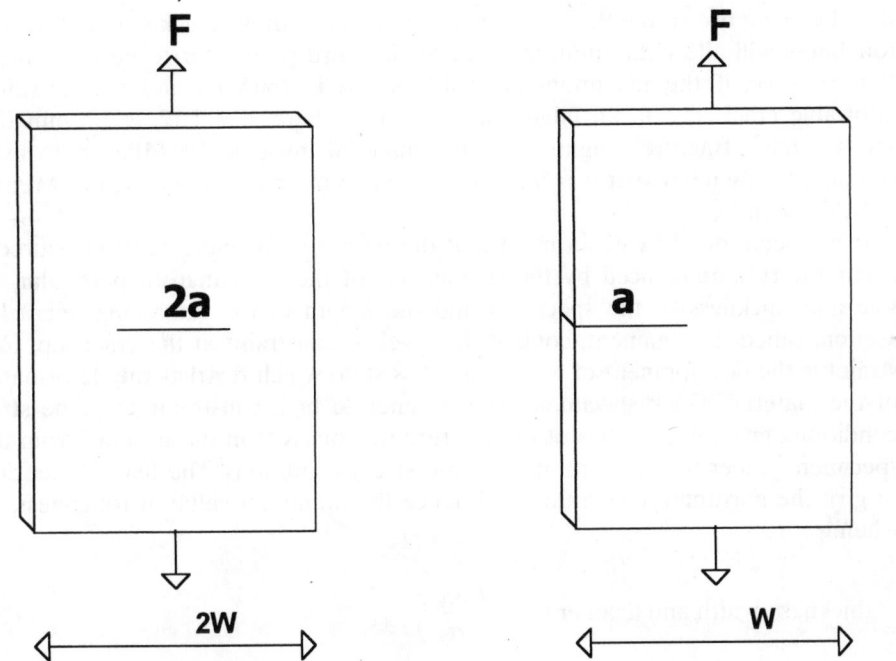

Fig. 7.5   Convention for describing the length of embedded and edge cracks

It is important to note that the convention is that an embedded crack that has two tips has a length of $2a$, and an edge crack which only has one tip has a length of a: see Fig. 7.5. Values of the correction term Y have been compiled for many geometries and load cases and are published in references 5 and 6. It is also important to take care over the units used in calculating stress intensity factors. Values may be quoted in MPa$\sqrt{m}$, MN.m$^{-1.5}$ or N.mm$^{-1.5}$. The conversion between them is

$$1\,\text{MPa}\sqrt{m} = 1\,\text{MN.m}^{-1.5} = 31.62\,\text{N.mm}^{-1.5}$$

The usefulness of the stress intensity factor lies in the concept of similitude. That is, if two cracks, one in a small laboratory specimen and one in a large structure, have the same value of K_I then they have identical opening displacements and identical crack tip stress fields. If the laboratory specimen fails in a brittle, catastrophic manner at a critical value of K_I then the structure will also fail when the stress intensity factor reaches that value. This critical value is a material property, termed the *fracture toughness*, and is represented by K_{Ic}.

The practical applications of linear elastic fracture mechanics are in assessing the likelihood that a particular combination of loading and crack size will cause a sudden fracture. Given that a structure is at risk of failing if

$$Y\sigma\sqrt{\pi a} \geq K_{Ic} \tag{7.2}$$

then knowing two of the three parameters maximum stress, crack size and fracture toughness will allow the limiting value of the third parameter to be determined. For example, if the maximum allowable stress is 160 MPa and the maximum allowable crack size in an edge cracked plate, where $Y = 1.12$, is 100 mm then the minimum fracture toughness of the material must be 100 MPa√m to avoid fracture. (Remember that it is important to keep track of the units and 1 MPa√m = 31.62 N.mm$^{-1.5}$.)

It has been found by experiment that the fracture toughness value measured in a laboratory is influenced by the dimensions of the specimen. In particular, the size and thickness of the specimen and the length of the remaining uncracked section, called the ligament, control the level of constraint at the crack tip. Constraint is the development of a triaxial stress state which restricts the deformation of the material. Thick specimens, which generate high constraint or plane strain conditions, give lower values of the fracture toughness than those found from thin specimens under low constraint or plane stress conditions. The least dimensions to give the maximum constraint, and hence the minimum value of toughness, are when:

$$\text{thickness, width and ligament} \geq 2.5 \left(\frac{K_{Ic}}{\sigma_y} \right)^2 \qquad (7.3)$$

This ensures that the localised plasticity at the crack tip is less than 2% of any of the dimensions of the body and therefore does not disturb the elastic crack tip stress field.

The size requirement of Equation (7.3) gives rise to practical problems in testing. For example, steel with a room temperature yield strength of 275 N.mm^{-2} and a minimum fracture toughness of 70 MPa√m needs to be tested using a specimen at least 160 mm thick to ensure plane strain conditions. In reality, much structural steelwork is made from sections substantially thinner than this and its fracture toughness will be significantly higher. It is good practice in many engineering disciplines to measure fracture toughness values using specimens of similar thickness to the proposed application.

Tables 3 to 7 in BS 5950: 2000[7] define, as a function of temperature, the maximum thickness of steel that should be used to avoid brittle fracture. This is the same as BS 5400-3: 2000[8] for steel bridge design. Both BS 5400 and BS 5950 are consistent with, but generally more conservative than, the minimum thickness to ensure plane strain conditions, as estimated from Equation (7.3). This means that the toughness of the steel in use should be better than the minimum value achievable in thick sections.

Linear elastic fracture mechanics is of much greater use on the lower shelf of the ductile-brittle behaviour of steels than on the upper shelf. At low temperatures, the yield strength is higher, the fracture toughness much lower and plane strain conditions can be achieved in relatively thin sections. On the upper shelf, linear elastic fracture mechanics tends to be inapplicable, and other techniques need to be used.

7.3 Elastic–plastic fracture mechanics

The need to consider fracture resistance of materials outside the limits of validity of plane strain linear elastic fracture mechanics (LEFM) is important for most engineering designs. To obtain valid K_{Ic} results for relatively tough materials it would be necessary to use a test piece of dimensions so large that they would not be representative of the sections actually in use.

Historically, the approaches to fracture mechanics when significant plasticity has occurred have been to consider either the crack tip opening displacement or the J-integral. The current British Standard on fracture mechanics toughness tests, BS 7448: Part 1: 1991, describes how a single approach to linear elastic and general yielding fracture mechanics tests can be adopted.[4]

Significant yielding at a crack tip leads to the physical separation of the surfaces of a crack, and the magnitude of this separation is termed the crack tip opening displacement (CTOD), and has been given the symbol δ. The CTOD approach enables critical toughness test measurements to be made in terms of δ_c, and then applied to determine allowable defect sizes for structural components.

The J-integral is a mathematical expression that may be used to characterize the local stress and strain fields around a crack front. Like the CTOD, the J-integral simplifies to be consistent with the stress intensity factor approach when the conditions for linear elastic fracture mechanics prevail. When non-linear conditions dominate either CTOD or J are useful parameters for characterizing the crack tip fields.

The relationships between K_I, δ and J_I under linear elastic conditions are

$$J_I = G_I = \frac{K_I^2}{E}(1 - v^2) \text{ in plane strain} \tag{7.4}$$

$$\delta = \frac{K_I^2}{\sigma_Y E} \tag{7.5}$$

where G_I is the strain energy release rate, which was the original, energy based approach to studying fracture.

In all cases, either linear elastic or general yielding, there is a parameter that describes the loading state of a cracked body. This might be K_I, δ or J_I as appropriate to the conditions. The limiting case for a particular material is the critical value at which failure occurs. Under elastic conditions this is the sudden fracture event, and the material property is K_{Ic}. Under elastic–plastic conditions, failure is not usually rapid, and the critical condition, δ_c or J_{Ic}, is usually associated with the onset of the ductile fracture process.

The assessment of flaws in fusion welded structures is covered by BS 7910: 1999,[9] which allows the use of material fracture toughness values in the form of K_{Ic}, δ_c or J_{Ic}. In the absence of genuine fracture mechanics derived toughness data, estimates of K_{Ic} may be made from empirical correlations with Charpy V-notch impact energies. Caution should be exercised when doing so as the degree of fit of the correlations tends to be poor.

The BS 7910: 1999 document includes design curves for assessing the acceptability of a known flaw at a given stress level. There are three levels of sophistication in the analysis requiring more precise information about the stresses and material properties as the assessment becomes more advanced. The procedure is very powerful as it considers the possibility of either fracture or plastic collapse as alternative failure processes. The first level uses a simple approach with built-in safety factors and conservative estimates of the material fracture toughness and the applied and residual stresses in the structure. The standard allows for toughness estimates from Charpy impact tests to be used at Level 1. Level 2 is the normal assessment route for steel structures and requires more accurate estimates of the stresses, material properties and defect sizes and shapes. Level 3 is much more sophisticated and can accommodate the tearing behaviour of ductile metals. The details of dealing with multiple flaws, residual stresses and combinations of bending and membrane stresses are all dealt with in the document. In many practical cases, a Level 1 assessment is sufficient.

Annex D in BS 7910: 1999 describes the manual procedure for determining the acceptability of a flaw in structure using the Level 1 procedure. An equivalent flaw parameter, $\bar{a}$, is defined as the half length of a through-thickness flaw in an infinite plate subjected to a remote tension loading. An equivalent tolerable flaw parameter, $\bar{a}_m$, can then be estimated and used to represent a variety of different defect shapes and sizes of equivalent severity:

$$\bar{a}_m = \frac{1}{2\pi}\left(\frac{K_{mat}}{\sigma_{max}}\right)^2 \tag{7.6}$$

Equivalent part-thickness flaw dimensions can then be estimated from graphical solutions presented in Annex D of the standard. The possibility of plastic collapse of the cracked sectioned must be checked by calculating the ratio, known as S_r, of a reference stress to the flow stress of the material. The flow stress is taken to be the average of the yield and tensile strengths. The reference stress is related to the applied and residual stresses in the structure and depends on the geometry of the structure and the defect. Details of its calculation are given in Annex C. Provided that the S_r parameter is less than 0.8, there is no risk of failure by collapse and any failure will be as a result of fracture.

The parameter K_{mat} in Equation (7.6) is a measure of the material fracture toughness, which may well not be valid plane strain K_{Ic} value but is appropriate to the conditions and section sizes under review. Furthermore, Level 1 allows for the possibility of estimating a fracture toughness value from Charpy impact data as described in Annex E of the standard.

Tough structural steels are extremely tolerant of the presence of cracks. If one considers a typical structural steel, with a yield stress of 275 N.mm^{-2} and a minimum fracture toughness of 70 MPa√m, subjected to a maximum allowable stress of 165 N.mm^{-2}, then the maximum allowable flaw from Equation (7.6) is about 60 mm long. (Remember that $\bar{a}_m$ is the half length of a through-thickness crack in an infinite plate.) The correction for a finite width plate makes less than 10% difference

to the crack size provided that the width of the plate is more than four times the length of the crack.

Reducing the maximum allowable stress in the structure has a big effect on the allowable flaw size. Halving the maximum stress increases the allowable flaw size to around 240 mm, a four-fold increase. The corollary, of course, is that the use of high strength metals makes a structure more sensitive to the presence of defects. High strength materials mean that higher structural stresses are allowed; high strength steels also tend to have lower fracture toughnesses than low strength steels. This combination means that the maximum allowable flaw size can become quite small.

7.4 Materials testing for fracture properties

There are, essentially, two approaches to fracture testing: the traditional impact test on a small notched bar, following the work of Izod or Charpy, or a fracture toughness test on a pre-cracked beam or compact tension specimen.

The principal advantage of an impact test is that it is relatively quick, simple and cheap to carry out. It provides qualitative information about the relative toughness of different grades of material and is well suited to quality control and material acceptance purposes.

Fracture toughness tests on pre-cracked specimens provide a direct measure of the fracture mechanics toughness parameters K_{Ic}, J_{Ic} or critical crack tip opening displacement. They are, however, more expensive to perform, use larger specimens and require more complex test facilities. The main advantage of fracture toughness tests is that they provide quantitative data for the design and assessment of structures.

7.4.1 Charpy test

Reference 1 specifies the procedure for the Charpy V-notch impact test. The test consists of measuring the energy absorbed in breaking a notched bar specimen by one blow from a pendulum as shown in Fig. 7.6. The test can be carried out at a range of temperatures to determine the transition between ductile and brittle behaviour for the material. The Charpy impact value is usually denoted by the symbol C_v and is measured in joules.

Many attempts have been made to correlate Charpy impact energies with fracture toughness values. The large scatter found in impact data makes such relationships difficult to describe with any great degree of confidence. No single method is currently able to describe the entire temperature–toughness response of structural ferritic steels. The correlation method that forms the basis of Annex E in BS 7910: 1999 and also the latest European guidelines on flaw assessment methods[10] is that

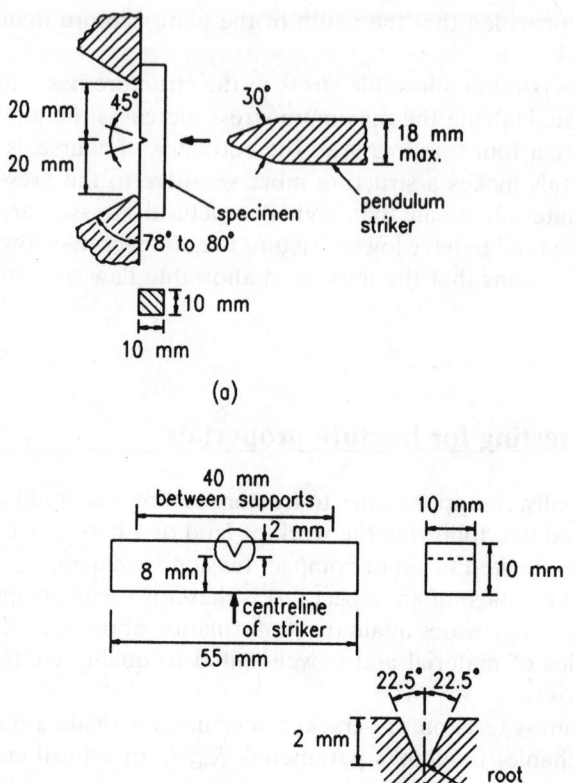

Fig. 7.6   Charpy test. (a) Test arrangement; (b) specimen

known as the *Master Curve*. The temperature at which a certain specified Charpy impact energy is achieved, usually 27J, is used as a fixed point, and the shape of the transition curve is generated from that temperature. This is also the principle behind the allowable section thickness guideline in BS 5400-3: 2000 and BS 5950: 2000.

7.4.2 Fracture mechanics testing

The recommended methods for determining fracture toughness values of metallic material are described in BS 7448: Part 1: 1991.[4] This covers both linear elastic and elastic–plastic conditions. The advantage of a single test procedure is that the results may be re-analysed to give a critical CTOD or critical *J* value if the test is found not to conform to the requirements for a valid plane strain K_{Ic} result.

The principle behind the tests is that a single edge-notched bend or compact

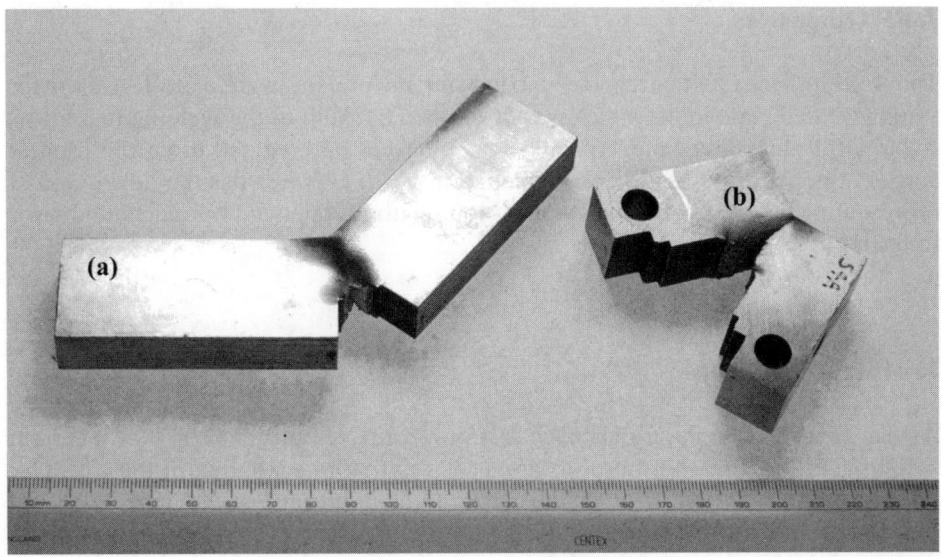

Fig. 7.7 Typical fracture toughness test specimens. (a) Three-point bend beam; (b) compact tension specimen

tension specimen (see Fig. 7.7) is cyclically loaded within prescribed limits until a sharp fatigue crack is formed. The specimen is then subjected to a displacement-controlled monotonic loading until either brittle fracture occurs or a prescribed maximum force is reached. The applied force is plotted against displacement and, provided specific validity criteria are met, a plane strain K_{Ic} may be found by analysis of the data. When the validity criteria are not met, the data may be re-analysed to evaluate a critical CTOD or critical J for that material. The determination of critical CTOD requires the relationship between applied load and the opening of the mouth of the crack, measured using a clip gauge. Critical J calculations need the load against load line displacement response, so the detailed arrangements of the two types of test are slightly different.

The specimen size requirements for a valid K_{Ic} are that:

$$\text{thickness, width and ligament} \geq 2.5\left(\frac{K_{Ic}}{\sigma_y}\right)^2 \tag{7.6}$$

The size requirements for the J value to be valid are that:

$$\text{thickness, width and ligament} \geq 25\left(\frac{J_{Ic}}{\sigma_y}\right) \tag{7.7}$$

This implies that significantly smaller specimens are required for critical J value tests than for K_{Ic} tests.

7.4.3 Other tests

There are other tests that can be carried out such as the wide plate test used for testing welded plate joints, which was developed by Wells at the Welding Institute.[11] A large full-thickness plate, typically 1 m square, is butt welded using the process and treatments to be used in the production weld. This test has the advantage of representing failure of an actual welded joint without the need for machining prior to testing.

7.4.4 Test specimens

As has already been described, each test procedure requires a sample of a certain size. In addition, the position of a sample in relation to a weld or, in the case of a thick plate, its position through the thickness is important. In modern structural steels toughness of the parent plate is rarely a problem. However, once a weld is deposited the toughness of the plate surrounding the weld, particularly in the heat-affected zone (HAZ), will be reduced. Although lower than for the parent plate, the C_v or fracture toughness values that can be obtained should still provide adequate toughness at all standard operating temperatures. In the case of thicker joints, appropriate post-weld heat treatments should be carried out.

7.5 Fracture-safe design

The design requirements for steel structures in which brittle fracture is a consideration are given in most structural codes. There are several key factors which need to be considered when determining the risk of brittle fracture in a structure. These are:

(1) minimum operating temperature
(2) loading – in particular, rate of loading
(3) metallurgical features such as parent plate, weld metal or HAZ
(4) thickness of material to be used.

Each of these factors influences the likelihood of brittle fracture occurring.

From experimental data and parametric studies examining the maximum tolerable defect sizes in steels under various operating conditions, codes such as BS 5400-3 and BS 5950-1 provide tables giving the maximum permissible thickness of steel at given operating temperatures. Where a design does not fit into these broad categories attempts have been made to use fracture mechanics to provide a criterion for material selection in terms of Charpy test energy absorption as this information is usually provided by the steel makers. When true fracture mechanics toughness

values are available, methods such as those described in References 8 and 9 should be employed to assess the acceptability of flaws in fusion welded structures.

As discussed earlier, at normal operating temperatures and slow rates of loading valid K_{Ic} values are not usually obtained for structural steels, and in the offshore and nuclear industries critical CTOD and J tests are widely used. Under these conditions, valid K_{Ic} values from low temperature tests will be conservative. If the structure is acceptable when assessed using such conservative data then there is often no great need to pursue the problem, particularly as all forms of fracture mechanics testing are expensive compared with routine quality control tests such as Charpy testing.

In general, the fracture toughness of structural steel increases with increasing temperature and decreasing loading rates. The effect of temperature on fracture toughness is well known. The effect of loading rate may be equally important, not only in designing new structures, but also in understanding the behaviour of existing ones which may have been built from material with low toughness at their service temperature. The shift in the ductile–brittle transition temperature for structural steels can be considerable when comparing loading rates used in slow bend tests with those in Charpy tests. Results from experimental work have shown that the transition temperature for a BS EN 10025 S355 J steel can change from around 0°C to −60°C with decreasing loading rate.

Materials standards set limits for the transition temperatures of various steel grades based on Charpy tests. When selecting an appropriate steel for a given structure it must be remembered that the Charpy values noted in the standards apply to parent plate. Material toughness varies in the weld and heat-affected zones of welded joints, and these should be checked for adequate toughness. Furthermore, since larger defects may be present in the weld area than the parent plate, appropriate procedures should be adopted to ensure that a welded structure will perform as designed. These could involve non-destructive testing including visual examination. In situations where defects are found, fracture mechanics procedures such as those in References 8 and 9 can be used to assess their significance.

7.6 Fatigue

7.6.1 Introduction

A component or structure which survives a single application of load may fracture if the application is repeated a large number of times. This would be classed as fatigue failure. Fatigue life can be defined as the number of cycles and hence the time taken to reach a pre-defined failure criterion. Fatigue failure is by no means a rigorous science and the idealizations and approximations inherent in it prevent the calculation of an absolute fatigue life for even the simplest structure.

In the analysis of a structure for fatigue there are three main areas of difficulty in prediction:

(1) The operational environment of a structure and the relationship between the environment and the actual forces on it;
(2) The internal stresses at a critical point in the structure induced by external forces acting on the structure;
(3) The time to failure due to the accumulated stress history at the critical point.

There are three approaches to the assessment of fatigue life of structural components. The traditional method, called the *S–N* approach, was first used in the mid 19th century. This relies on empirically derived relationships between applied elastic stress ranges and fatigue life. A development of the S–N approach is the *strain–life* method in which the plastic strains are considered important. Empirical relationships are derived between strain range and fatigue life. The third method, based on fracture mechanics, considers the growth rate of an existing defect. The concept of *defect tolerance* follows directly from fracture mechanics assessments.

7.6.2 Loadings for fatigue

Fluctuating loads arise from a wide range of sources. Some are intentional, such as road and rail traffic over bridges. Others are unavoidable, such as wave loading on offshore oil rigs, and some are accidental – a car striking a kerb, for example. Occasionally the loads are entirely unforeseen: resonance of a slender tower under gusting wind loads can induce large numbers of small amplitude loads. References 12–14 are useful sources of information on fatigue loading.

The designer's objective is to anticipate the sequence of service loading throughout the life of the structure. The magnitude of the peak load, which is vital for static design purposes, is generally of little concern as it represents only one cycle in millions. For example, highway bridge girders may experience 100 million significant cycles in their lifetime. The sequence is important because it affects the stress range, particularly if the structure is loaded by more than one independent load system.

For convenience, loadings are usually simplified into a load spectrum, which defines a series of bands of constant load levels and the number of times that each band is experienced, as shown in Fig. 7.8.

7.6.3 The nature of fatigue

Materials subject to a cyclically variable stress of a sufficient magnitude change their mechanical properties. In practice a very high percentage of all engineering failures are due to fatigue. Most of these failures can be attributed to poor design or manufacture.

Fatigue failure is a process of crack propagation due to the highly localized cyclic plasticity that occurs at the tip of a crack or metallurgical flaw. It is not a single

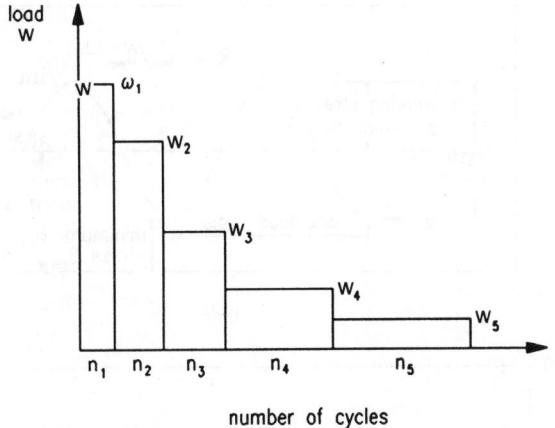

Fig. 7.8 Typical load spectrum for design

mechanism but the result of several mechanisms operating in sequence during the life of a structure: propagation of a defect within the microstructure of the material; slow incremental propagation of a long crack; and final unstable fracture.

Crack initiation is a convenient term to cover the early stages of crack growth that are difficult to detect. The reality is that a crack starts to grow from the first loading cycles and continues right through to failure. In welded structures or cast components the initiation phase is bypassed as substantial existing defects are already likely to be present.

7.6.4 S–N curves

The traditional form for presenting fatigue data is the S–N curve, where the total cyclic stress range (S) is plotted against the number of cycles to failure (N). A typical curve is shown in Fig. 7.9 with a description of the usual terminology used in fatigue. Logarithmic scales are conventionally used for both axes. However, this is not universal, and S–N data may also be presented on linear stress axes instead of logarithmic; stress amplitudes instead of ranges and reversals instead of cycles.

Fatigue endurance data are obtained experimentally. S–N curves can be obtained for a material, using smooth laboratory specimens, for components or for detailed sub-assemblies such as welded joints. In all cases, a series of specimens is subject to cycles of constant load amplitude to failure. A sufficient number of specimens are tested for statistical analysis to be carried out to determine both mean fatigue strength and its standard deviation. Depending on the design philosophy adopted, design strength is taken as mean minus an appropriate number of standard deviations.

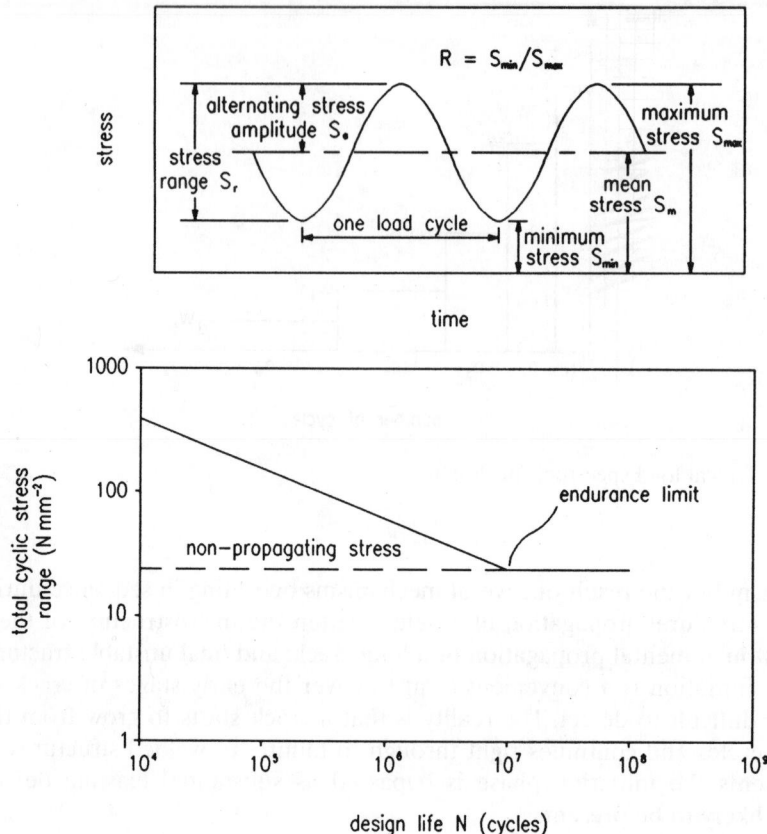

Fig. 7.9 Typical *S–N* curve and nomenclature used in fatigue

Under some circumstances, laboratory tests on steel specimens appear to have an infinite life below a certain stress range. This stress range is variously known as the *fatigue limit* or *endurance limit*. In practice, the tests are stopped after two, ten or one hundred million cycles, and if it has not broken then the stress range is assumed to be below the fatigue limit. The fatigue, or endurance, limit will tend to disappear under variable-amplitude loading or in the presence of a corrosive environment. This means that real components will eventually fail whatever the stress range of the loading cycles, but the life may be very long indeed.

Welded steel joints are usually regarded as containing small defects due to the welding process itself. It has been found after much experimental work that the relationship between fatigue life and applied stress range follows the form

$$N_f \Delta \sigma^a = b \tag{7.8}$$

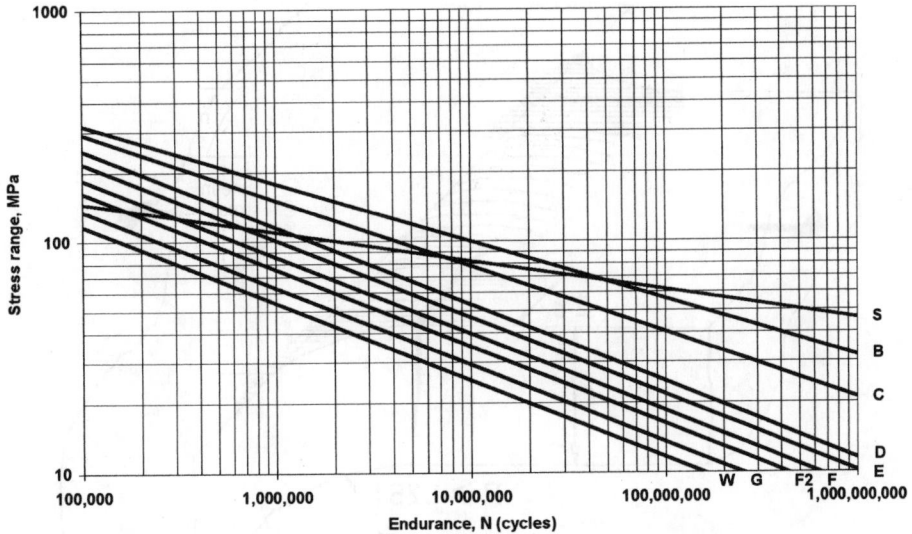

Fig. 7.10   Fatigue design curves for welded joints according to BS 5400-10: 1980 and BS 7608: 1993

where N_f is the number of cycles to failure, $\Delta\sigma$ is the applied stress range, and a and b are constants which depend on the geometry of the joint. The value of a is in the range of 3 to 4 for welded joints in ferritic steels.

The British Standard BS 7608: 1993 contains a code of practice for fatigue design of welded structures which is consistent with the assessments for fatigue in steel bridges in BS 5400-10: 1980[8] and other welded structures in BS 7910: 1999.[9] The basis is a series of design S–N curves for different welded joint configurations and crack locations. Each potential crack location in each type of joint is classified by a letter: S, B, C, D, E, F, F2, G or W. Each detail class has a specific design S–N curve (Fig. 7.10). This curve indicates the number of cycles at any given stress range that should be achieved with 97.7% probability of surviving. The design S–N curves for each class of joint are based on extensive experimental data and are suitable for ferritic steels. Details of dealing with other metals are given in BS 7910: 1999.

The procedure is to identify the worst weld detail in the design and the ranges of the stress cycles experienced by the structure. If there is only one stress range, then the corresponding lifetime can be read off from the appropriate S–N curve. This is the number of repeated cycles that the structure can endure and have a 97.7% probability of survival.

Consider the weld detail in Fig. 7.11. This is classified as type F2. If this is subjected to a cyclic stress with a range of $20\,\mathrm{N.mm^{-2}}$ every 7 seconds, then reading off the type F2 curve in Fig. 7.10 indicates that the structure should survive 66 million cycles. This corresponds to 4.8×10^8 seconds or about 14 years.

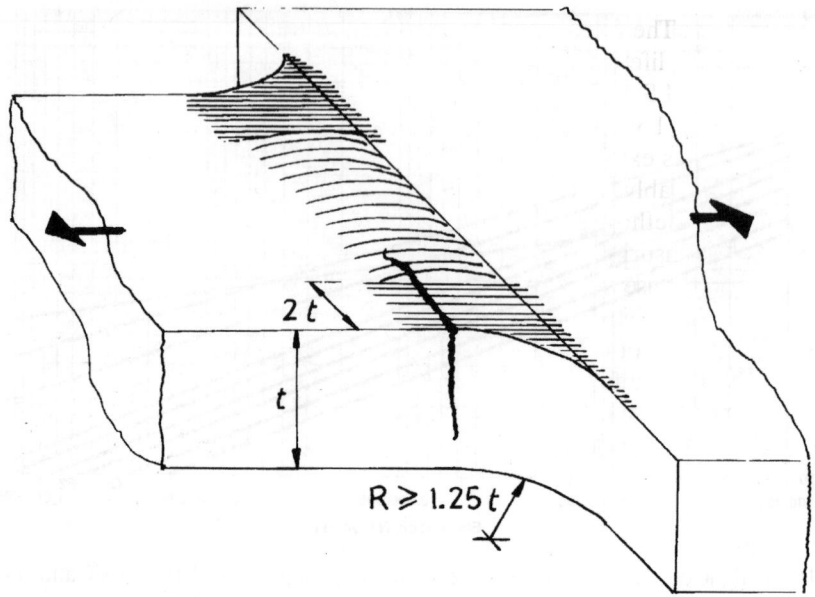

Fig. 7.11   An example of an F2 cracked weld

7.6.5 Variable-amplitude loading

For constant-amplitude loading, the permissible stress range can be obtained directly from Fig. 7.10 by considering the required design life. In practice it is more common for structures to be subjected to a loading spectrum of varying amplitudes or random vibrations. In such cases use is made of Miner's rule.[15]

Miner's rule is a linear summation of the fatigue damage accumulated during the life of the structure. For a joint subjected to a number of repetitions, n_i, each of several stress ranges $\Delta\sigma_i$, the value of n_i corresponding to each $\Delta\sigma_i$ should be determined from stress spectra measured on similar equipment or by making reasonable assumptions as to the expected service history. The permissible number of cycles, N_i, at each stress range, $\Delta\sigma_i$, should then be determined from Fig. 7.10 for the relevant joint class and the stress range adjusted so that the linear cumulative damage summation does not exceed unity:

$$\frac{n_1}{N_1} + \frac{n_2}{N_2} + \frac{n_3}{N_3} + \ldots + \frac{n_j}{N_j} = \sum_{i=1}^{i=j} \frac{n_i}{N_i} < 1.0 \tag{7.9}$$

The order in which the variable-amplitude stress ranges occur in a structure is not considered in this procedure.

An example would be a class F2 weld that is subjected to 3×10^7 cycles at a stress range of $20\,N.mm^{-2}$ and then the stress range increases to $30\,N.mm^{-2}$ and the remaining life needs to be estimated. The lifetime at $20\,N.mm^{-2}$ is read from Fig. 7.10 and

is 6.6×10^7 cycles. The fraction of life used is therefore $(3 \times 10^7)/(6.6 \times 10^7) = 0.455$ and the remaining life fraction is 0.545. At a stress range of $30\,\text{N.mm}^{-2}$ the lifetime from Fig. 7.10 is 1.9×10^7 cycles, so the remaining life for this welded joint is $0.545 \times 1.9 \times 10^7 = 1 \times 10^7$ cycles.

Various methods exist to sum the spectrum of stress cycles. The *rainflow* counting method is probably the most widely used for analysing long stress histories using a computer. This method separates out the small cycles that are often superimposed on larger cycles, ensuring both are counted. The procedure involves the simulation of a time history, or use of measure sequences, with appropriate counting algorithms. Once the spectrum of stress cycles has been determined, the load sequence is broken down into a number of constant load range segments. The *reservoir* method, which is easy to use by hand for short stress histories, is described in Reference 9.

7.6.6 Strain–life

The notion that fatigue is associated with plastic strains led to the strain–life approach to fatigue. The endurance of laboratory specimens is correlated to plastic strain range, or amplitude, in a strain-controlled fatigue test (Fig. 7.12). A common empirical curve fit to the fatigue lifetime data takes the form:

$$\varepsilon_a = \frac{\sigma'_\mathrm{f}}{E}(2N_\mathrm{f})^b + \varepsilon'_\mathrm{f}(2N_\mathrm{f})^c \qquad (7.10)$$

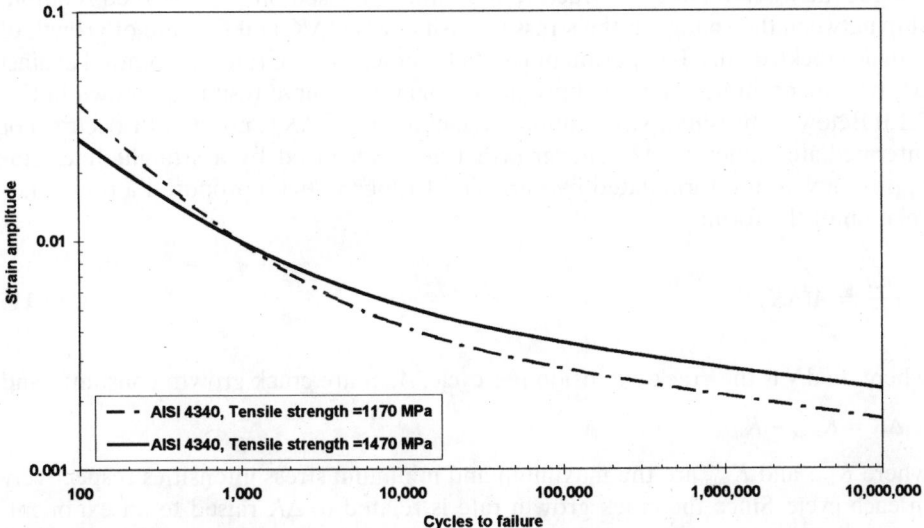

Fig. 7.12 Strain–life curves for a high strength steel in two conditions

where E is Young's modulus and σ_f', ε_f', b and c are considered to be material properties. It is worth noting that the first term of the right hand side is an elastic stress term which dominates at low loads and long lifetimes. The second term is a plastic strain term which dominates at high loads and short lifetimes.

The strain–life approach is preferred over the S–N method since it is almost identical to the S–N approach at long lives and elastic stresses, and is more general for problems of short lives, high strains, high temperatures or localized plasticity at notches.

The technique is:

(1) determine the strains in the structure, often by finite element analysis;
(2) identify the maximum local strain range, the 'hot spot' strain;
(3) read, from the strain–life curve, the lifetime to first appearance of a crack at that strain at that position.

Variable-amplitude loads are dealt with in the same way as in the S–N method, with the local hot spot strains and their associated lifetimes being determined for each block of loading.

The strain–life, or local strain, method has wider use in fatigue assessments in the engineering industry than the S–N method and is available in commercial computer software.

7.6.7 Fracture mechanics analysis

Fatigue life assessment using fracture mechanics is based on the observed relationship between the change in the stress intensity factor, ΔK, and the rate of growth of fatigue cracks, da/dN. If experimental data for crack growth rates are plotted against ΔK on a logarithmic scale, an approximate sigmoidal curve results, as shown in Fig. 7.13. Below a threshold stress intensity factor range, ΔK_{th}, no growth occurs. For intermediate values of ΔK, the growth rate is idealized by a straight line. This approach was first formulated by Paris and Erdogan,[16] who proposed a power law relation of the form:

$$\frac{da}{dN} = A(\Delta K)^m \tag{7.11}$$

where da/dN is the crack extension per cycle, A, m are crack growth constants, and

$$\Delta K = K_{max} - K_{min}$$

where K_{max} and K_{min} are the maximum and minimum stress intensities respectively in each cycle. Since the crack growth rate is related to ΔK raised to an exponent, it is important that ΔK should be known accurately if meaningful crack growth predictions are to be made.

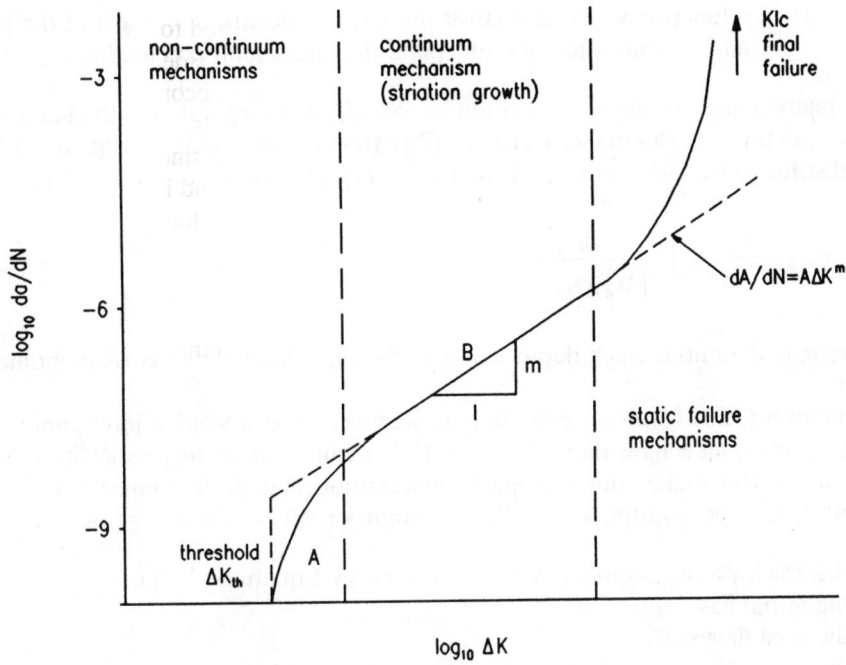

Fig. 7.13 Schematic presentation of crack growth

Values of A and m to describe the fatigue crack growth rate can be obtained from specific tests on the materials under consideration. From published data reasonable estimates of fatigue growth rates in ferritic steels are given by[9]

$m = 3$
$A = 5.21 \times 10^{-13}$ for non-aggressive environments at temperatures up to 100°C
$A = 2.3 \times 10^{-12}$ for marine environments at temperatures up to 20°C

for crack growth rates in mm/cycle and stress intensity factors in N.mm$^{-1.5}$.

Some care should be taken with fatigue lifetime estimates in corrosive environments to make sure that the fatigue crack growth data are appropriate to the particular combination of steel grade and environment. Small changes in steel composition or environmental conditions can result in very large changes in crack growth behaviour.

The fracture mechanics convention in the welded steel structures industry is slightly different from that of other practitioners. For a crack at the toe of a welded joint

$$\Delta K = M_K Y \Delta \sigma \sqrt{\pi a} \tag{7.12}$$

where $\Delta \sigma$ = applied stress range
   a = crack depth
   Y = a correction factor dependent on crack size, shape and loading

M_K = a function which allows for the stress concentration effect of the joint and depends on crack site, plate thickness, joint and loading.

In many industries the magnification factor, M_K, is incorporated in the geometric correction term, Y. This makes Equation (7.12) the fatigue version of Equation (7.1). Substituting Equation (7.12) in Equation (7.11), rearranging and integrating gives

$$N_f = \frac{1}{A\pi^{m/2}\Delta\sigma^m} \int_{a_i}^{a_f} \frac{da}{\left[M_K Y\sqrt{a}\right]^m} \tag{7.13}$$

where a_i is the initial crack depth and a_f is the final crack depth corresponding to failure.

Equation (7.13) forms an extremely powerful tool. If a welded joint contains a crack or crack-like flaw, then Equation (7.13) can be used to predict its fatigue endurance. This makes the reasonable assumption that the life consists of crack growth from a pre-existing crack. The techniques requires knowledge of:

(1) the crack propagation behaviour described by Equation (7. 11),
(2) the initial flaw size,
(3) the final flaw size,
(4) the geometry and loading correction terms Y and M_K.

The integration is straightforward if Y and M_K are independent of crack length, which is not usually the case. Otherwise some suitable numerical technique must be used. The life is fairly insensitive to the final crack length, but highly dependent on the initial flaw size. Engineering judgement is often required when selecting appropriate values of these crack sizes.

If Y and M_K are independent of crack length, and $m \neq 2$, then

$$N_f = \frac{1}{(m/2-1)A\pi^{m/2}(M_k Y\Delta\sigma)^m} \left[a_i^{\left(1-\frac{m}{2}\right)} - a_f^{\left(1-\frac{m}{2}\right)} \right] \tag{7.14}$$

The initial crack size can be taken as the largest flaw that escapes the detection technique used. This is likely to be several millimetres for visual or ultrasonic inspection of large welded structures.

The final allowable flaw size might be taken from Equation (7.5), or it may be a crack that penetrates the wall of a containment vessel and allows fluid to escape. Other failure conditions could be a crack that allows excessive displacements to occur or a crack that is observable to the naked eye and is therefore unacceptable to the customer.

The design S–N curves, described by Equation (7.8), and given in the British Standards BS 5400-10: 1980, BS 7608: 1993 and BS 7910: 1999, are effectively the same as Equation (7.14), the integration of the crack propagation equation.

7.6.8 Improvement techniques

7.6.8.1 Introduction

The fatigue performance of a joint can be enhanced by the use of weld improvement techniques. There is a large amount of data available on the influence of weld improvement techniques on fatigue life but as yet little progress has been made into developing practical design rules. Modern steelmaking has led to the production of structural steels with excellent weldability. The low fatigue strength of a welded connection is generally attributed to a short crack initiation period. An extended crack initiation life can be achieved by:

(1) reducing the stress concentration of the weld,
(2) removing crack-like defects at the weld toe,
(3) reducing tensile welding residual stresses or introducing compressive stresses.

The methods employed fall broadly into two categories:

weld geometry improvement: grinding: weld dressing; profile control
residual stress reduction: peening; thermal stress relief

Most of the current information relating to weld improvement has been obtained from small-scale specimens. When considering actual structures one important factor is size. In a large structure, long-range residual stresses due to the assembly of the members are present and will influence the fatigue life. In contrast to small joints, where peak stress is limited to the weld toe, the peak stress region in a large multi-pass joint may include several weld beads, and cracks may initiate anywhere in this highly stressed area.

However, it is good practice not to seek benefit from improvement techniques in the design office.

7.6.8.2 Grinding

The improvement of the weld toe profile and the removal of slag inclusions can be achieved by grinding either with a rotary burr or with a disc. To obtain the maximum benefit from this type of treatment it is important to extend the grinding to a sufficient depth to remove all small undercuts and inclusions. The degree of improvement achieved increases with the amount of machining carried out and the care taken by the operator to produce a smooth transition.

The performance of toe-ground cruciform specimens is fully investigated in Reference 11. Under freely corroding conditions the benefit from grinding is minimal. However, in air, the results appear to fall on the safe side of the mean curve: endurance is altered by a factor of 2.2. It is therefore recommended that an

increase in fatigue life by a factor of 2.2 can be taken if controlled local machining or grinding is carried out.

7.6.8.3 Weld toe remelting

Weld toe remelting by TIG and plasma arc dressing are performed by remelting the toe region with a torch held at an angle of 50° or 90° to the plate (without the addition of filler material). The difference between TIG and plasma dressing is that the latter requires a higher heat input.

Weld toe remelting can result in large increases in fatigue strength due to the effect of providing low contact angle in the transition area between the weld and the plate and by the removal of slag inclusions and undercuts at the toe.

7.6.8.4 Hammer peening

Improved fatigue properties of peened welds are obtained by extensive cold working of the toe region. These improved fatigue properties are due to:

(1) introduction of high compressive residual stresses,
(2) a flattening of crack-like defects at the toe,
(3) an improved toe profile.

It can be shown that weld improvement techniques greatly improve the fatigue life of weldments. For weldments subject to bending and axial loading, peening appears to offer the greatest improvement in fatigue life, followed by grinding and TIG dressing.

7.6.9 Fatigue-resistant design

The nature of fatigue is well understood and analytical tools are available to calculate the fatigue life of complex structures. The accuracy of any fatigue life calculation is highly dependent on a good understanding of the expected loading sequence during the whole life of a structure. Once a global pattern has been developed then a more detailed inspection of particular areas of a structure, where the effects of loading may be more important, due to the geometries of joints for example, should be carried out.

Data have been gathered for many years on the performance of bridges, towers cranes and offshore structures where fatigue is a major design consideration. Codes of practice, such as BS 5400-10 and BS 7608, give details for the estimation of fatigue

lives. Where a structure is subjected to fatigue it is important that welded joints are considered carefully. Fatigue and brittle fractures can be initiated at discontinuities of shape, at notches and cracks which give rise to high local stress. Avoidance of local structural and notch peak stresses by good design is the most effective means of increasing fatigue life. It is important that during the design process consideration is given to the manner in which the structure is to be fabricated. Acute-angled welds of less than 30° are difficult to fabricate, particularly in tubular structures. This could lead to defects in the weld. Furthermore it is also difficult to carry out non-destructive testing on such welds. Despite these problems, some recently designed structures have incorporated such features.

Repairs carried out to structures in service are expensive, and in the worst case may require that a facility is closed down temporarily. Care needs to be taken when specifying secondary attachments to main primary steelwork. These are often not considered in detail during the design process as they themselves are not complex. However, there have been a number of failures in offshore structures due to fatigue crack growth resulting from a welded attachment. The following general suggestions can assist in the development of an appropriate design of a welded structure with respect to fatigue strength:

(1) Adopt butt or single and double bevel butt welds in preference to fillet welds.
(2) Use double-sided in preference to single-sided fillet welds.
(3) Aim to place weld, particularly toe, root and weld end in area of low stress.
(4) Ensure good welding procedures are adopted and adequate non-destructive testing (NDT) undertaken.
(5) Consider the effects of localized stress concentration factors.
(6) Consider potential effects of residual stresses.

References to Chapter 7

1. British Standards Institution (1990) *Charpy impact test on metallic materials.* BS EN 10045-1, BSI, London.
2. British Standards Institution (1990) *Hot rolled products of non-alloy structural steels,* BS EN 10025, BSI, London.
3. American Society for Testing and Materials (1981) *The standard test for J_{Ic} a measure of fracture toughness.* ASTM E813-81.
4. British Standards Institution (1991) *Fracture mechanics toughness tests. Part 1. Method for determination of K_{Ic}, critical CTOD and critical J values of metallic materials.* BS 7448-1: 1991, BSI, London.
5. Murakami Y. (1987) *Stress Intensity Factors Handbook.* Pergamon Press, Oxford.
6. Tada H., Paris P.C. & Irwin G.R. (1985) *The Stress of Cracks Handbook.* Del Research Corporation, Hellertown, Pa., USA.

7. British Standards Institution (2000) *Structural use of steelwork in building. Code of practice for design.* BS 5950-1: 2000.
8. British Standards Institution (2000) *Steel, concrete and composite bridges. Code of practice for design of steel bridges.* BS 5400-3: 2000 and *Code of practice for fatigue.* BS 5400-10: 1980, BSI, London.
9. British Standards Institution (1999) *Guidance on methods for assessing the acceptability of flaws in fusion welded structures.* BS 7910: 1999, BSI, London.
10. SINTAP (1999) *Structural Integrity Assessment Procedures for European Industry.* Project BE 95-1426. Final Procedure, British Steel Report, Rotherham, UK.
11. American Society for Testing and Materials (1982) *Design of fatigue and fracture resistant structures.* ASTM STP 761.
12. British Standards Institution (1993) *Fatigue design and assessment of steel structures.* BS 7608: 1993, BSI, London.
13. Department of Energy (1990) *Offshore Installations: Guidance Design Construction and Certification,* 4th edn. HMSO.
14. British Standards Institution (1983 & 1980) *Rules for the design of cranes. Part 1: Specification for classification, stress calculations and design criteria for structures.* Part 2: *Specification for classification, stress calculations and design of mechanisms.* BS 2573, BSI, London.
15. Miner, M.A. (1945) Cumulative damage in fatigue. *Journal of Applied Mechanics,* **12**, A159–A164.
16. Paris P.C. & Erdogan F. (1963) A critical analysis of crack propagation laws. *Journal of Basic Engineering,* **85**, 528–534.

Further reading for Chapter 7

Dowling N.E. (1999) *Mechanical Behavior of Materials: Engineering Methods for Deformation, Fracture, and Fatigue,* 2nd edn. Prentice-Hall, Inc.
Gray T.F.G., Spence J. & North T.H. (1982) *Rational Welding Design,* 2nd edn. Butterworth, London.
Gurney T.R. (1979) *Fatigue of Welded Structures,* 2nd edn. Cambridge University Press.
Pellini W.S. (1983) *Guidelines for Fracture-Safe and Fatigue-Reliable Design of Steel Structures.* The Welding Institute.
Radaj D. (1990) *Design and Analysis of Fatigue Resistant Welded Structures.* Abington Publishing, Cambridge.

Chapter 8

Sustainability and steel construction

by GRAHAM RAVEN

8.1 Introduction

The UK Government, along with many others, has a very strong policy to encourage sustainable development. In May 1999, it published *A better quality of life – a strategy for sustainable development for the United Kingdom*. The principles in this document were further developed for the construction industry in *Building a better quality of life – a strategy for more sustainable construction*, which was published in April 2000. The underlying principles of sustainability lie in the appropriate balance of economic, social and environmental impacts, the so-called triple bottom line. The strong message is that for a viable long term future any enterprise must pay due concern to the following issues:

- Maintenance of high and stable levels of economic growth and employment
- Social progress which recognizes the needs of everyone
- Prudent use of natural resources
- Effective protection of the environment.

Clearly, to be successful all these issues need to be balanced, as over-commitment to any one of them at the expense of the others will lead to failure in the long term.

There are many examples of how the steel construction sector has progressed to be itself more sustainable and contribute to sustainable construction. Some of these are set out below.

Steel has a significant role in construction and has established a strong position. In the UK structural frames now have approximately 70% of the multi-storey frame and 95% of the single storey industrial buildings market.

8.2 Economic impacts

The UK steelwork sector has worked and continues to work hard to provide the most economic solutions and work methods. It has become world-class through its investments in IT and associated technologies, based on the use of 3-D modelling for detailing with direct links to CNC machines and suppliers. This is proving to be a strong foundation for the release of further benefits which stems from integration

with design activities through the adoption of data transfer protocols such as the CIMSteel Integration Standards.

Since long before the needs of sustainability were fully recognized, there has been a history of continuous improvement in structural design methods which have resulted in economies to clients over a period of years. This has been encouraged and has been complemented by the development of systems and components, many involving the use of light gauge steel in addition to the more traditional hot rolled sections.

8.3 Social impacts

For any enterprise to be successful over a reasonable period it needs skilled and conscientious people. The steel construction sector continues to invest heavily in the education and training of both its employees and its customers in the efficient and safe use of its products. Considerable importance in the sustainability agenda is attached to the effects of enterprises on the local community. A factory production environment which encourages a stable workforce is more conducive to both employers and employees in the promotion of skills development. It is obviously easier for a relationship with the local community to develop in such an environment, than where there is a predominantly casual or itinerant workforce.

On site, fast dry construction from off-site manufactured assemblies is less disruptive to neighbours and provides earlier weatherproofing and hence more reliable and acceptable working conditions for other trades. This contributes to safer working and a higher quality product. The risk of water pollution from wet trades is also minimized.

8.4 Environmental impacts

8.4.1 Effective protection of the environment

One of the key drivers for change is the need to reduce global warming gas emissions. This is largely achieved through conservation of both embodied and operational non-renewable energy. Operational energy is the most significant, and SCI is active in several areas. Research has shown that adaptable solutions can be designed using composite construction, which, even with its efficient use of material, still provides sufficient thermal capacity for fabric energy storage systems to be effective.[1] This work is being extended to take advantage of air and water cooling of floor systems. Of course precast concrete floors can also be most efficiently incorporated into steel framed buildings, as demonstrated by the Wessex Water headquarters building. This boasts one of the lowest energy consumptions in the UK for buildings of this type and utilizes a lightweight structural steel frame.[2]

For the occupancy patterns of sheds and residential buildings, fabric energy storage systems are less appropriate and high insulation values and air tightness are more relevant. These can be conveniently achieved in steel-framed construction and, in the case of sheds, through the use of well-detailed, insulated steel cladding systems.

Prefabrication has a major contribution in minimizing pollution. All forms of waste to landfill, as well as noise, dust, emissions and other potential contamination on building sites, can be more easily controlled when most of the work is carried out under factory conditions.

8.4.2 Prudent use of natural resources

Originally driven by economic reasons, the history of design development has led to methods which optimize the use of natural resources. UK steel design methods are among the most efficient internationally. Resources are conserved not only by minimizing them at first use but also by extending their lifetime. The life of steel structures is extended through both the ability to create flexible spaces from the use of long spans and the adaptability of the structure during its lifetime to accommodate change of usage. When decommissioning is necessary, the elements may be re-used or, if this is not viable, recycled. Recent studies[3] have shown that at present 10% of steel arising from construction demolition is re-used, with a further 85% recycled. Furthermore, it is practical to design structures in steel which are suited to deconstruction rather than demolition and so can be re-used. This further enhances the efficiency of the use of all the resources embodied in the components.

In general, off-site factory production is easier to control and manage and so leads to less waste of both materials and other resources. The steel construction sector is well advanced in the production of modules which include services and other trades as well as steel components.

This short review demonstrates that design for sustainability is an extremely broad and important area, and steel can help provide solutions in a wide variety of ways. The remainder of the chapter concentrates on more specific issues influencing the environmental impacts.

8.5 Embodied energy

Embodied energy is the energy consumed in the process of manufacturing, using and later disposing of, or recycling, materials. Whilst a substantial amount of energy is required to manufacture steel, it is a highly efficient structural material and relatively little energy is required for the rolling process to form structural sections. Overall therefore steel structures tend to have good embodied energy profiles. Furthermore, steel is a material that lends itself extremely well to the sustainability hierarchy of 'reduce, re-use and recycle'.

8.5.1 Reduction

Steel is a material which lends itself to prefabrication. Structural components such as beams and columns are delivered to site as finished items; wastage during the construction process is therefore minimal. Continuous improvement in structural design has led to British Standards that give design rules which result in optimum use of material while maintaining high standards of safety.

8.5.2 Re-use

Re-using materials requires less energy than recycling, although it is likely that some refurbishment will be necessary. Steel structures are conceptually a kit of parts; with appropriate design and detailing of connections both between steel elements and other components there is considerable potential to increase the 10% of structures currently re-used.[3]

The ability to dismantle steel structures on a piece-by-piece basis minimizes the environmental problems often associated with demolition such as noise, dust and safety hazards. Furthermore, steel foundation and substructure components can usually be extracted and recovered at the end of a building's life, eliminating ground contamination. The high monetary value of steel compared with other construction materials justifies the expense involved in recovery.

There are obviously barriers to overcome, for example assuring the next user of the history and quality of the components and the need to provide the necessary storage and accessibility to available components; however, these can be overcome.

To maximize the potential for re-use and minimize any fabrication needed for subsequent users a higher degree of standardization than is normal today is very desirable. Since solutions are technically feasible, designers and suppliers are reacting to the needs and pressures of society to put the necessary processes into place.

8.5.3 Recycling

Steel is 100% recyclable, and repeated melting, casting and rolling have no detrimental effect on quality. 50% of all steel production worldwide is from recycled sources. This considerably reduces embodied energy, pollution and resource depletion.

Scrap steel has a monetary value and there are established international markets for steel scrap. The waste products generated from the production of steel such as ferro-lime (steel slag) and blast-furnace slag are also recycled and used as substitutes for primary aggregates and cement. As mentioned previously, recent studies as part of a pan-European project into life cycle assessment of steel construction products

have shown that currently over 85% of steel from demolition of buildings is recycled and a further 10% re-used.[3]

8.5.4 Housing and embodied energy

In a typical new house complying with current Building Regulations, the embodied energy of construction may be equivalent to approximately 5 years of operational energy. However, with increasing insulation and air-tightness standards, which together are reducing operational energy requirements, the embodied energy is becoming more significant.

Residential light steel frame construction is usually based on close centre frameworks made of cold-formed studs and joists (generally 'C' or 'Sigma' sections). These sections, produced from galvanized steel by roll forming, are usually 1.2–3.2 mm thick. Sections and small panels can often be manhandled into position; large panels can be erected using small capacity cranes.

The inherent strength and flexibility of light steel framing allows the utilisation of roof spaces as additional useable area. This reduces the footprint of dwellings relative to their internal floor area, which is particularly advantageous for high density low rise urban developments.

Although steel has a relatively high embodied energy content (approximately 18 MJ/kg), it is used efficiently in light steel frame construction. Approximately 20 kg/m^2 of steel are used in a typical house frame, resulting in an overall embodied energy content of 360 MJ/m^2. Investigations suggest this is approximately equal to the figure for traditional construction, with any differences due in general to the particular circumstance rather than the choice of material.

Furthermore, the higher strength to weight ratio compared with alternative materials allows the design of buildings with greater spans, fewer internal load-bearing walls and fewer foundations.

Light steel house frames can easily be strengthened, extended, modified, dismantled, reassembled and repaired, and so have a long service life.

Light steel modular construction offers further advantages. It reduces transport requirements by around 50% compared with conventional construction. Building materials are delivered in bulk to the factory and prefabricated modules delivered to the site. The number of journeys to the site is therefore minimized, reducing noise and nuisance as well as energy in transportation. In the case of volumetric modular construction, whole rooms may be completed fitted out in the factory.

8.5.5 Non-residential buildings and embodied energy

The embodied energy burdens of commercial buildings vary considerably depending upon their specification, as illustrated in Fig. 8.1. For example, prestige office

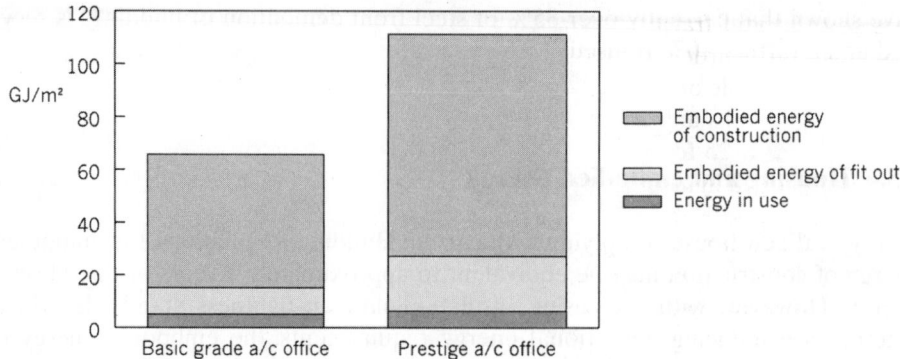

Fig. 8.1 Comparison of embedded energy and operating energy for a basic grade and prestige office over a 60-year life

buildings can have twice the embodied energy content of lower specification offices. This is due to the way they are clad and fitted out rather than the effects of the structural frame.

Embodied energy is typically small in comparison to operational energy over the life of a commercial building (generally an order of magnitude less). The balance is however changing as improved designs reduce operational energy and the rate of change of use increases so requiring more refurbishment.

The embodied energy of the structural components of a typical office building is only about 25–30% of the total embodied energy and tends to vary relatively little irrespective of the type of structure. A life cycle assessment (LCA)[4] study of a typical four storey office building compared five alternative steel and concrete structures. No significant difference was identified between the embodied energy burdens of the different options. Although steel has a high embodied energy content per kilogram, its high strength to weight ratio means that a much lower volume of steel is required to construct a steel frame than the volume of concrete needed to construct an equivalent frame. The cross-sectional area and mass of steel structural members is very much smaller than that of equivalent concrete members.

The comparative lightness of modern steel frame structures has been important in reducing their embodied energy burdens. In addition to savings associated with the frame materials, designers can often adopt lighter foundation systems than would be required for other forms of construction. Alternatively the number of foundations may be reduced, since the spanning capabilities of steel can allow larger structural bays, and consequently fewer column bases requiring support. The larger bays also allow for more flexible use of space; steel frames are easily adaptable, thereby prolonging the building life and reducing yet further the impact of embodied energy over the life cycle.

In composite construction, where beams act in tandem with the floor slab, the additional strength and stiffness that are achieved allow the use of relatively shallow beams. These have lower embodied energy and the volume of concrete in the floor is barely altered.

Modern portal frame design has played a significant role in reducing the embodied energy burden of single storey industrial and commercial buildings. Portal frames are capable of spanning considerable distances using relatively light sections. Roof beam depths are generally of the order of span divided by 60, as opposed to span divided by 25 for a conventional beam and column roof. The combination of steel frames with light gauge purlins and rails supporting insulated cladding systems provides a very efficient form of construction resulting in the very high (95%) major share. Systems are available to meet the requirements of the new Part L to the UK Building Regulations[5] in terms of both insulation and air tightness.

A key element in achieving low embodied energy is the pursuit of structural efficiency. By reducing the amount of structural steel required per unit area of commercial building, designers can not only achieve elegant and economic structural solutions, but also reduce embodied energy burdens.

8.6 Operational energy

Buildings require energy for heating and lighting, powering equipment and machines, and possibly mechanical ventilation and cooling services. This type of energy, known as *operational energy*, is greatly influenced by the design of the building, and the servicing strategy.

The threat of global climate change resulting from 'greenhouse gases', of which the most prevalent is carbon dioxide, is driving a concerted effort to reduce energy use in buildings. In the UK, buildings account for 40–50% of carbon dioxide emissions. Significant reductions in operational energy (and hence CO_2 emissions) are necessary to combat climate change. Through careful design, it is possible to reduce operational energy by at least one third, by reducing the need for air conditioning or even eliminating it altogether.[2]

8.6.1 Housing and operational energy

Residential buildings account for about 30% of the total energy used in the UK. This relatively high percentage of energy use, together with increasing demand for new housing, indicates the importance of achieving improved energy efficiency in this sector.

Light steel frame residential buildings have proved to have good standards of environmental performance, particularly in terms of energy conservation. Many have significantly exceeded the requirements of UK Building Regulations in terms of thermal performance, as shown in Table 8.1.

The following measures are recommended to reduce operational energy requirements:

Table 8.1 Comparison of thermal performance of light steel frame residential buildings with UK Building Regulations

U-values (W/m²K)	Walls	Floor	Roof	Windows
1995 Building Regulations	0.45	0.45	0.25	3.3
2002 Building Regulations	0.35	0.25	0.20	2.2
Good practice[1]	0.35	0.32	0.20	2.8
Ultra-low energy homes[2]	0.20	0.20	0.15	2.0
Low energy steel frame house[3]	0.20	0.35	0.20	1.8

Notes:
1. *Good Practice Guide 79 – Energy efficiency in housing, Department of Environment, 1993.*
2. *General Information Report 38 – Review of ultra-low energy homes, Best Practice Programme of the Department of Environment, 1996.*
3. *U-values of the Oxford Brookes University Demonstration Building.*

- Minimization of heat loss through the building envelope
- Provision of an airtight building envelope with controlled ventilation
- Provision of efficient and controlled heating
- Use of appropriate passive heating, cooling and ventilation methods.

These measures can often be achieved at little or no extra overall cost, particularly where savings can be made in the capital cost of heating systems as a result of additional insulation and airtight construction. To accompany the regulations, a document is being developed showing robust details of plans which will, if properly constructed, give compliance with the UK Building Regulations. To achieve compliance with the new regulations[5] will be easier with framed construction than with traditional methods.

Light steel frame housing has several generic advantages but as with all forms of construction attention should be paid to the detailing in order to maximize the benefits.

Placing the insulation on the outside of the frame is known as *warm frame construction*; this minimizes heat loss since there is minimal thermal bridging. Risk of condensation is eliminated since all the members are contained within the warm internal environment. High levels of insulation are possible however without significantly increasing wall thickness by the addition of insulation between the frame members. Heat loss resulting from air infiltration can also be conveniently minimized by the inclusion of an air tight membrane in the walls.

8.6.2 Commercial buildings and operational energy

Commercial buildings use energy primarily for space and water heating, lighting, cooling, ventilation and small power. In order to reduce operational energy, buildings should be designed to provide maximum occupant comfort throughout the year with minimum energy requirements. The charts in Fig. 8.2 show typical energy requirements and carbon dioxide emissions for two types of commercial building.

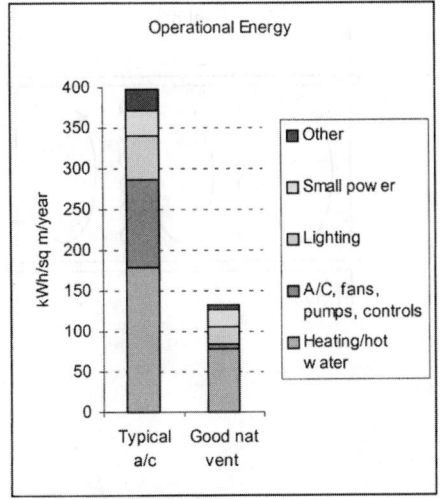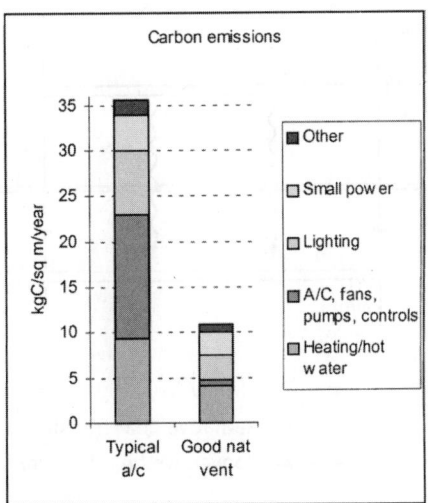

Fig. 8.2   Comparison of operational energy requirements and resulting carbon emissions for a typical air conditioned office and a good practice naturally ventilated open plan office (source: Econ 19, Energy use in offices, Energy Efficiency Best Practice Programme)

Although heating accounts for a significant part of the annual operational energy use, it produces a relatively small proportion of the overall carbon emissions since it is generally provided by burning natural gas. In the UK, gas produces only 41% of the carbon dioxide emissions per kilowatt hour that are produced by electricity. Therefore the greatest scope for reducing carbon emissions from commercial buildings is in reducing the energy required for cooling, mechanical ventilation and artificial lighting.

Energy use is heavily dependent upon building design, including such factors as layout, orientation, thermal capacity, glazing arrangements, solar shading, cooling and ventilation. The choice of structural system (steel or concrete frame[4]) has been found to have very little effect on operational energy requirements, so the following notes apply in general to all commercial buildings.

Structural steel gives the design team the flexibility it needs to exploit these factors to the full.

One of the major ways of providing the necessary comfort levels is to make use of fabric energy storage. This involves using the floor plate to absorb heat during the period when the building is occupied during the day and then purging it at night as shown in Fig. 8.3.

Initially there was a perception that thick concrete floors were needed for the system to be effective. However, it has been shown that, with a 24-hour cycle of heating and cooling, the necessary volume and disposition of concrete can be provided by composite construction.[2]

Perforated permeable ceilings can be used to permit the necessary contact between the warm air and the floor. Enhancements can be achieved by ducting air

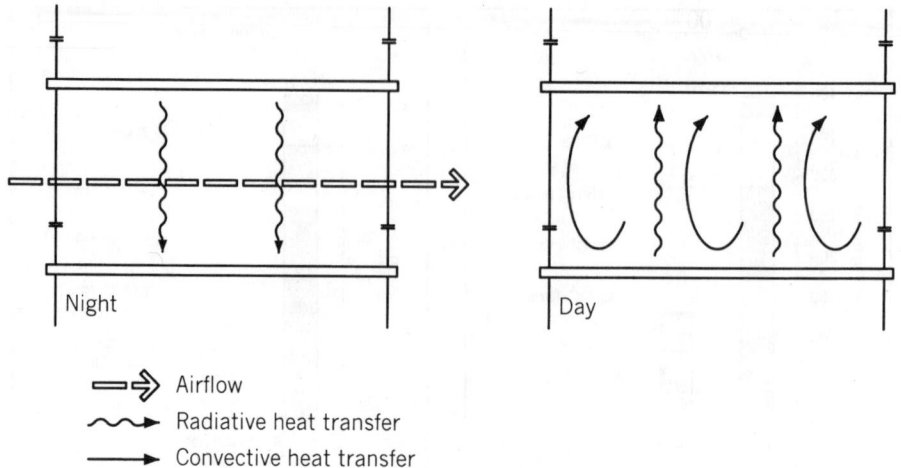

Night   Day

⊫⇒ Airflow
〜〜➤ Radiative heat transfer
———➤ Convective heat transfer

Fig. 8.3   Diurnal cycle for fabric energy storage

immediately above or below the slab. Examples are illustrated in the SCI publication *Environmental Floor Systems.*[7]

8.7 Summary

Steel in construction contributes in many ways to more sustainable solutions for the built environment. The major points are that the basic material is 100% recyclable without degradation, thereby contributing to conservation of resources. The building, in effect, becomes a warehouse of parts for the future.

Design and construction methods have been, and continue to be, refined to optimize the use of the material and the resources needed to process it. Steel structures allow the creation of flexible space, and if they require alteration they lend themselves to adaptation.

At the end of their viable life on a particular site, appropriate design and detailing mean that components can be deconstructed and re-used rather than recycled.

A broader exposition of the issues raised can be found in the SCI Publication *The Role of Steel in Environmentally Responsible Buildings.*[6]

References to Chapter 8

1. Eaton K.J. & Ogden R. (1995) Thermal and structural mass. *The Architects Journal,* **202**, No. 8, 24 Aug., 43.
2. Bennetts R. (2001) High watermark. *The Architects Journal,* **214**, No. 19, 22 Nov., 46–51.

3. Sansom M.R. (2001) Construction steel reuse and recycling rates approach 100%. *New Steel Construction*, Vol. 9 No. 4, July/Aug., pp. 12–13.
4. Eaton K.J. & Amato A. (1998) *A comparative environmental life cycle assessment of modern office buildings*. SCI P-182. The Steel Construction Institute, Ascot, Berks.
5. The Building Regulations. (2002) Part L Conservation of Fuel and Power, HMSO.
6. Gorgolewski M. (1999) *The Role of Steel in Environmentally Responsible Buildings*. SCI P-174. The Steel Construction Institute, Ascot, Berks.
7. Amato A., Ogden R. & Plank R. (1997) *Environmental floor systems*. SCI P-181. The Steel Construction Institute, Ascot, Berks.

Chapter 9
Introduction to manual and computer analysis

by RANGACHARI NARAYANAN

9.1 Introduction

The analysis of structures consists essentially of mathematical modelling of the response of a structure to the applied loading. Such models are based on idealizations of the structural behaviour of the material and of the components. They are, therefore, imperfect to a larger or smaller degree, depending upon the extent of inaccuracy built into the assumptions in modelling. This is not to imply that the calculations are meaningless, rather to emphasize the fact that the assessment of structural responses is the best estimate that can be obtained in the light of the assumptions implicit in the modelling of the system. Some of these assumptions are necessary in the light of inadequate data; others are introduced to simplify the calculation procedure to economic levels.

There are several idealizations introduced in the modelling process.

- Firstly, the physical dimensions of the structural components are idealized. For example, skeletal structures are represented by a series of line elements and joints are assumed to be of negligible size. The imperfections in the member straightness are ignored or at best idealized.
- Material behaviour is simplified. For example, the stress–strain characteristic is assumed to be linearly elastic, and then perfectly plastic. No account is taken of the variation of yield stress along or across the member. The influence of residual stresses due to thermal processes (such as hot rolling and flame cutting), as well as that due to cold working and roller straightening, is ignored.
- The implications of actions which are included in the analytical process itself are frequently ignored. For example, the development of local plasticity at connections or possible effects of change of geometry causing local instability are rarely, if ever, accounted for in the analysis.

However, it must be recognized that the design loads employed in assessing structural response are themselves approximate. The analysis chosen should therefore be adequate for the purpose and should be capable of providing the solutions at an economical cost.

The fundamental concepts employed in mathematical modelling are discussed next.

286

9.1.1 Equations of static equilibrium

From Newton's law of motion, the conditions under which a body remains in static equilibrium can be expressed as follows:

- The sum of the components of all forces acting on the body, resolved along any arbitrary direction, is equal to zero. This condition is completely satisfied if the components of all forces resolved along the x, y, z directions individually add up to zero. (This can be represented by $\Sigma P_x = 0$, $\Sigma P_y = 0$, $\Sigma P_z = 0$, where P_x, P_y and P_z represent forces resolved in the x, y, z directions.) These three equations represent the condition of zero translation.
- The sum of the moments of all forces resolved in any arbitrarily chosen plane about any point in that plane is zero. This condition is completely satisfied when all the moments resolved into xy, yz and zx planes all individually add up to zero. ($\Sigma M_{xy} = 0$, $\Sigma M_{yz} = 0$ and $\Sigma M_{zx} = 0$.) These three equations provide for zero rotation about the three axes.

If a structure is planar and is subjected to a system of coplanar forces, the conditions of equilibrium can be simplified to *three* equations as detailed below:

- The components of all forces resolved along the x and y directions will individually add up to zero ($\Sigma P_x = 0$ and $\Sigma P_y = 0$).
- The sum of the moments of all the forces about any arbitrarily chosen point in the plane is zero (i.e. $\Sigma M = 0$).

9.1.2 The principle of superposition

This principle is only applicable when the displacements are linear functions of applied loads. For structures subjected to multiple loading, the total effect of several loads can be computed as the sum of the individual effects calculated by applying the loads separately. This principle is a very useful tool in computing the combined effects of many load effects (e.g. moment, deflection, etc.). These can be calculated separately for each load and then summed.

9.2 Element analysis

Any complex structure can be looked upon as being built up of simpler units or components termed 'members' or 'elements'. Broadly speaking, these can be classified into *three* categories:

- *Skeletal structures* consisting of members whose one dimension (say, length) is much larger than the other two (viz. breadth and height). Such a *line element* is

variously termed as a bar, beam, column or tie. A variety of structures are obtained by connecting such members together using rigid or hinged joints. Should all the axes of the members be situated in one plane, the structures so produced are termed *plane structures*. Where all members are not in one plane, the structures are termed *space structures*.

- Structures consisting of members whose two dimensions (viz. length and breadth) are of the same order but much greater than the thickness fall into the second category. Such structural elements are called *plated structures*. Such structural elements are further classified as plates and shells depending upon whether they are plane or curved. In practice these units are used in combination with beams or bars. Slabs supported on beams, cellular structures, cylindrical or spherical shells are all examples of plated structures.

- The third category consists of structures composed of members having all the three dimensions (viz. length, breadth and depth) of the same order. The analysis of such structures is extremely complex, even when several simplifying assumptions are made. Dams, massive raft foundations, thick hollow spheres, caissons are all examples of three-dimensional structures.

For the most part the structural engineer is concerned with skeletal structures. Increasing sophistication in available techniques of analysis has enabled the economic design of plated structures in recent years. Three-dimensional analysis of structures is only rarely carried out. Under incremental loading, the initial deformation or displacement response of a steel member is elastic. Once the stresses caused by the application of load exceed the yield point, the cross section gradually yields. The gradual spread of plasticity results initially in an elasto-plastic response and then in plastic response, before ultimate collapse occurs.

9.3 Line elements

The deformation response of a line element is dependent on a number of cross-sectional properties such as area, A, second moment of area ($I_{xx} = \int y^2 \, dA$; $I_{yy} = \int x^2 \, dA$) and the product moment of area ($I_{xy} = \int xy \, dA$). The two axes xx and yy are orthogonal. For doubly symmetric sections, the axes of symmetry are those for which $\int xy \, dA = 0$. These are known as *principal axes*. For a plane area, the principal axes may be defined as a pair of rectangular axes in its plane and passing through its centroid, such that the product moment of area $\int xy \, dA = 0$, the co-ordinates referring to the principal axes. If the plane area has an axis of symmetry, it is obviously a principal axis (by symmetry $\int xy \, dA = 0$). The other axis is at right angles to it, through the centroid of the area.

Tables of properties of the section (including the centroid and shear centre of the section) are available as published data (e.g. SCI *Steelwork Design Guide*, Vol. 1).[1]

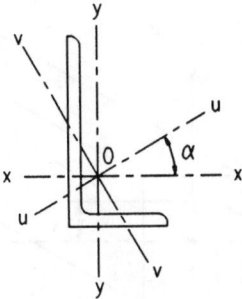

Fig. 9.1   Angle section (no axis of symmetry)

If the section has no axis of symmetry (e.g. an angle section) the principal axes will have to be determined. Referring to Fig. 9.1, if uOu and vOv are the principal axes, the angle α between the uu and xx axes is given by

$$\tan 2\alpha = \frac{-2I_{xy}}{I_{xx} - I_{yy}}$$

$$I_{uu} = \frac{I_{xx} + I_{yy}}{2} + \frac{I_{xx} - I_{yy}}{2}\cos 2\alpha - I_{xy}\sin 2\alpha \tag{9.1}$$

$$I_{vv} = \frac{I_{xx} + I_{yy}}{2} - \frac{I_{xx} - I_{yy}}{2}\cos 2\alpha + I_{xy}\sin 2\alpha \tag{9.2}$$

The values of α, I_{uu} and I_{vv} are available in published steel design guides (e.g. Reference 1).

9.3.1 Elastic analysis of line elements under axial loading

When a cross section is subjected to a compressive or tensile axial load, P, the resulting stress is given by the load/area of the section, i.e. P/A. Axial load is defined as one acting at the centroid of the section. When loads are introduced into a section in a uniform manner (e.g. through a heavy end-plate), this represents the state of stress throughout the section. On the other hand, when a tensile load is introduced via a bolted connection, there will be regions of the member where stress concentrations occur and plastic behaviour may be evident locally, even though the mean stress across the section is well below yield.

If the force P is not applied at the centroid, the longitudinal direct stress distribution will no longer be uniform. If the force is offset by eccentricities of e_x and e_y measured from the centroidal axes in the y and x directions, the equivalent set of actions are (1) an axial force P, (2) a bending moment $M_x = Pe_x$ in the yz plane and (3) a bending moment $M_y = Pe_y$ in the zx plane (see Fig. 9.2). The method of evaluating the stress distribution due to an applied moment is given in a later section.

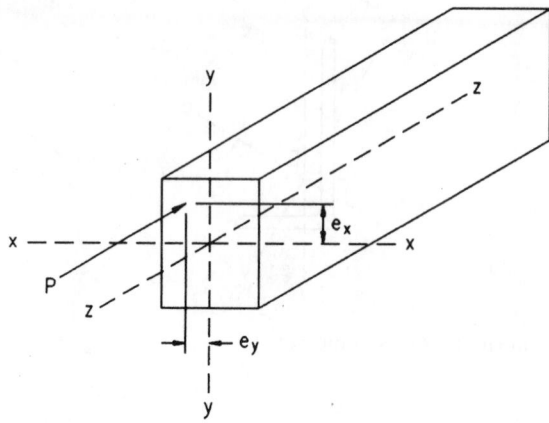

Fig. 9.2   Compressive force applied eccentrically with reference to the centroidal axis

The total stress at any section can be obtained as the algebraic sum of the stresses due to P, M_x and M_y.

9.3.2 Elastic analysis of line elements in pure bending

For a section having at least one axis of symmetry and acted upon by a bending moment in the plane of symmetry, the Bernoulli equation of bending may be used as the basis to determine both stresses and deflections within the elastic range. The assumptions which form the basis of the theory are:

- The beam is subjected to a pure moment (i.e. shear is absent). (Generally the deflections due to shear are small compared with those due to flexure; this is not true of deep beams.)
- Plane sections before bending remain plane after bending.
- The material has a constant value of modulus of elasticity (E) and is linearly elastic.

The following equation results (see Fig. 9.3).

$$\frac{M}{I} = \frac{f}{y} = \frac{E}{R} \tag{9.3}$$

where M is the applied moment; I is the second moment of area about the neutral axis; f is the longitudinal direct stress at any point within the cross section; y is the distance of the point from the neutral axis; E is the modulus of elasticity; R is the radius of curvature of the beam at the neutral axis.

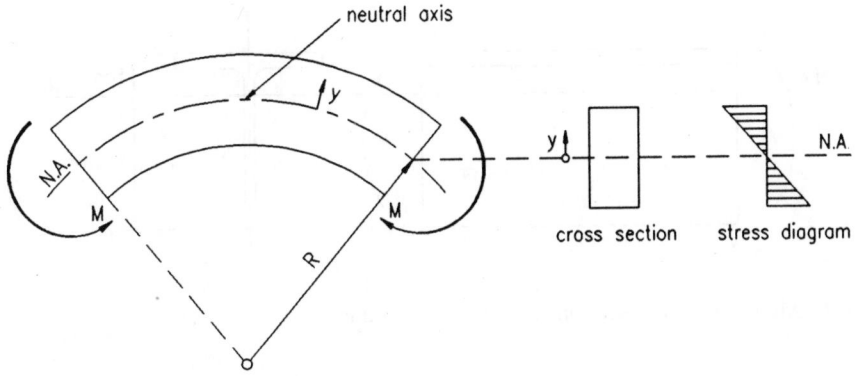

Fig. 9.3   Pure bending

From the above, the stress at any section can be obtained as

$$f = \frac{M\,y}{I}$$

For a given section (having a known value of I) the stress varies linearly from zero at the neutral axis to a maximum at extreme fibres on either side of the neutral axis:

$$f_{max} = \frac{M\,y_{max}}{I} = \frac{M}{Z} \qquad\qquad (9.4)$$

where $Z = \dfrac{I}{y_{max}}$.

The term Z is known as the *elastic section modulus* and is tabulated in section tables.[1] The elastic moment capacity of a given section may be found directly as the product of the elastic section modulus, Z, and the maximum allowable stress.

If the section is doubly symmetric, then the neutral axis is mid-way between the two extreme fibres. Hence, the maximum tensile and compressive stresses will be equal. For an unsymmetric section this will not be the case, as the value of y for the two extreme fibres will be different.

For a monosymmetric section, such as the T-section shown in Fig. 9.4, subjected to a moment acting in the plane of symmetry, the elastic neutral axis will be the centroidal axis. The above equations are still valid. The values of y_{max} for the two extreme fibres (one in compression and the other in tension) are different. For an applied sagging (positive) moment shown in Fig. 9.4, the extreme fibre stress in the flange will be compressive and that in the stalk will be tensile. The numerical values of the maximum tensile and compressive stresses will differ. In the case sketched in Fig. 9.4, the magnitude of the tensile stress will be greater, as y_{max} in tension is greater than that in compression.

Caution has to be exercised in extending the pure bending theory to asymmetric sections. There are two special cases where no twisting occurs:

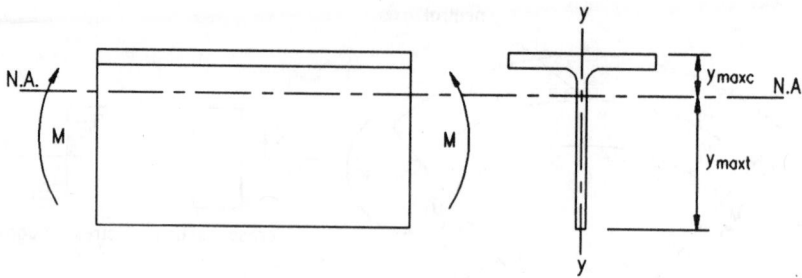

Fig. 9.4   Monosymmetric section subjected to bending

- Bending about a principal axis in which no displacement perpendicular to the plane of the applied moment results.
- The plane of the applied moment passes through the shear centre of the cross section.

When a cross section is subjected to an axial load and a moment such that no twisting occurs, the stresses may be determined by resolving the moment into components M_{uu} and M_{vv} about the principal axes uu and vv and combining the resulting longitudinal stresses with those resulting from axial loading:

$$f_{u,v} = \pm \frac{P}{A} \pm \frac{M_{uu}\, v}{I_{uu}} \pm \frac{M_{vv}\, u}{I_{vv}} \tag{9.5}$$

For a section having two axes of symmetry (see Fig. 9.2) this simplifies to

$$f_{x,y} = \pm \frac{P}{A} \pm \frac{M_{xx}\, y}{I_{xx}} \pm \frac{M_{yy}\, x}{I_{yy}}$$

Pure bending does *not* cause the section to twist. When the shear force is applied eccentrically in relation to the shear centre of the cross section, the section twists and initially plane sections no longer remain plane. The response is complex and consists of a twist and a deflection with components in and perpendicular to the plane of the applied moment. This is not discussed in this chapter. A simplified method of calculating the elastic response of cross sections subjected to twisting moments is given in an SCI publication.[2]

9.3.3 Elastic analysis of line elements subject to shear

Pure bending discussed in the preceding section implies that the shear force applied on the section is zero. Application of transverse loads on a line element will, in general, cause a bending moment which varies along its length, and hence a shear force which also varies along the length is generated.

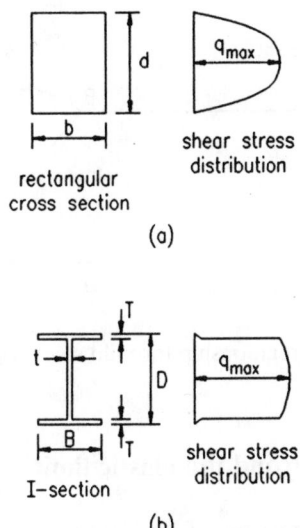

Fig. 9.5   Shear stress distribution: (a) in a rectangular cross section and (b) in an I-section

If the member remains elastic and is subjected to bending in a plane of symmetry (such as the vertical plane in a doubly symmetric or monosymmetric beam), then the shear stresses caused vary with the distance from the neutral axis.

For a narrow rectangular cross section of breadth b and depth d, subjected to a shear force V and bent in its strong direction (see Fig. 9.5(a)), the shear stress varies parabolically from zero at the lower and upper surfaces to a maximum value, q_{max}, at the neutral axis given by

$$q_{max} = \frac{3V}{2bd}$$

i.e. 50% higher than the average value.

For an I-section (Fig. 9.5(b)), the shear distribution can be evaluated from

$$q = \frac{V}{IB} \int_{y=h}^{y=h_{max}} by \, dy \qquad (9.6)$$

where B is the breadth of the section at which shear stress is evaluated. The integration is performed over that part of the section remote from the neutral axis, i.e. from $y = h$ to $y = h_{max}$ with a general variable width of b.

Clearly, for the I- (or T-) section, at the web/flange interface the value of the integral will remain constant. As the section just inside the web becomes the section just inside the flange, the value of the vertical shear abruptly changes as B changes from web thickness to flange width.

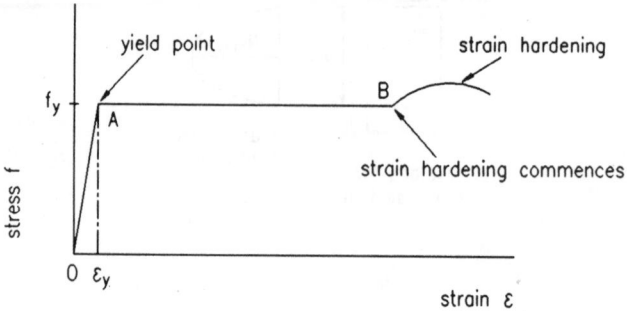

Fig. 9.6   Idealized stress–strain relationship for mild steel

9.3.4 Elements stressed beyond the elastic limit

The most important characteristic of structural steels (possessed by no other material to the same degree), is their capacity to withstand considerable deformation without fracture. A large part of this deformation occurs during the process of *yielding*, when the steel extends at a constant and uniform stress known as the *yield stress*.

Figure 9.6 shows, in its idealized form, the stress–strain curve for structural steels subjected to direct tension. The line 0A represents the elastic straining of the material in accordance with Hooke's law. From A to B, the material yields while the stress remains constant and is equal to the yield stress, f_y. The strain occurring in the material during yielding remains after the load has been removed and is called *plastic strain*. It is important to note that this plastic strain AB is at least ten times as large as the elastic strain, ε_y, at yield point.

When subjected to compression, various grades of structural steel behave in a similar manner and display the same property of yield. This characteristic is known as *ductility of steel*.

9.3.5 Bending of beams beyond the elastic limit

For simplicity, the case of a beam symmetrical about both axes is considered first. The fibres of the beam subjected to bending are stressed in tension or compression according to their position relative to the neutral axis and are strained as shown in Fig. 9.7.

While the beam remains entirely elastic, the stress in every fibre is proportional to its strain and to its distance from the neutral axis. The stress, f, in the extreme fibres cannot exceed the yield stress, f_y.

When the beam is subjected to a moment slightly greater than that which first produces yield in the extreme fibres, it does not fail. Instead, the outer fibres yield

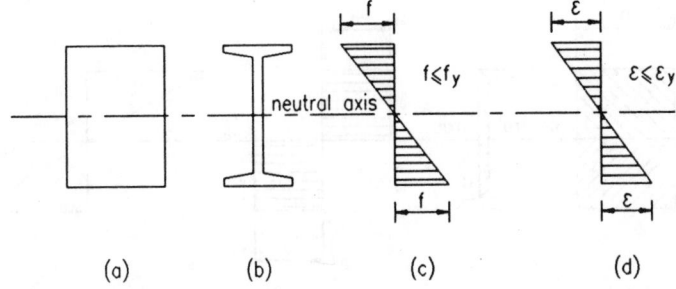

Fig. 9.7 Elastic distribution of stress and strain in a symmetric beam. (a) Rectangular section, (b) I-section, (c) stress distribution for (a) or (b), (d) strain distribution for (a) or (b)

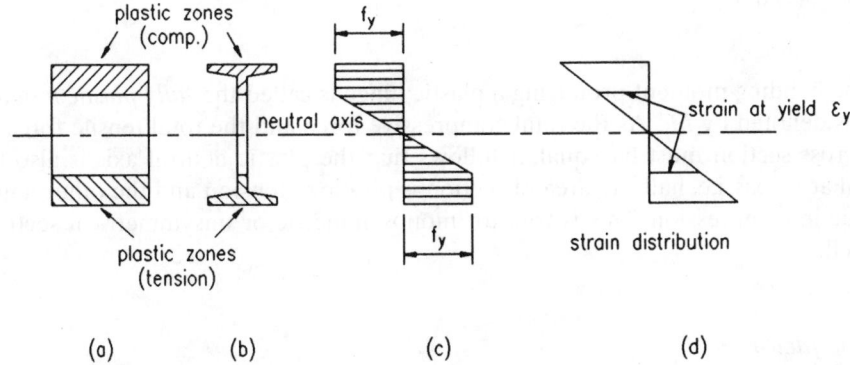

Fig. 9.8 Distribution of stress and strain beyond the elastic limit for a symmetric beam. (a) Rectangular section, (b) I-section, (c) stress distribution for (a) or (b), (d) strain distribution for (a) or (b)

at constant stress, f_y, while the fibres nearer to the neutral axis sustain increased elastic stresses. Figure 9.8 shows the stress distribution for beams subjected to such moments. Such beams are said to be *partially plastic* and those portions of their cross sections which have reached the yield stress are described as *plastic zones*.

The depths of the plastic zones depend upon the magnitude of the applied moment. As the moment is increased, the plastic zones increase in depth, and it is assumed that plastic yielding will continue to occur at yield stress, f_y, resulting in two stress blocks, one zone yielding in tension and one in compression. Figure 9.9 represents the stress distribution in beams stressed to this stage. The plastic zones occupy the whole area of the sections, which are then described as being *fully plastic*. When the cross section of a member is fully plastic under a bending moment, any attempt to increase this moment will cause the member to act as if hinged at that point. This point is then described as a *plastic hinge*.

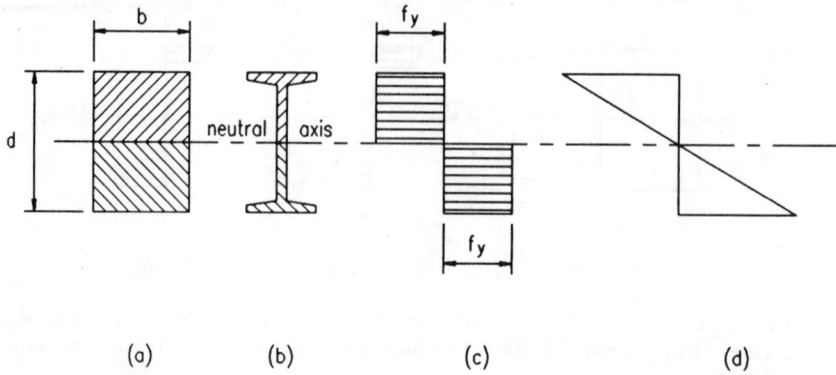

Fig. 9.9 Distribution of stress and strain in a fully plastic cross section. (a) Rectangular section, (b) I-section, (c) stress distribution for (a) or (b), (d) strain distribution for (a) or (b)

The bending moment producing a plastic hinge is called the *fully plastic moment* and is denoted by M_p. As the total compressive force and the total tensile force on the cross section must be equal, it follows that the plastic neutral axis is also the equal area axis, i.e. half the area of section is plastic in tension and the other half is plastic in compression. This is true for monosymmetric or unsymmetrical sections as well.

Shape factor

As described previously there will be two stress blocks, one in tension, the other in compression, each at yield stress. For equilibrium of the cross section, the areas in compression and tension must be equal. For a rectangular section the plastic moment can be calculated as

$$M_p = 2b\frac{d}{2}\frac{d}{4}f_y = \frac{bd^2}{4}f_y$$

which is 1.5 times the elastic moment capacity.

It will be noted that, in developing this increased moment, there is large straining in the external fibres of the section together with large rotations and deflections. The behaviour may be plotted as a moment–rotation curve. Curves for various sections are shown in Fig. 9.10.

The ratio of the plastic modulus, S, to the elastic modulus, Z, is known as the *shape factor*, v, and it will govern the point in the moment–rotation curve when non-linearity starts. For the ideal section in bending, i.e. two flange plates, this will have a value of unity. The value increases for more material at the centre of the section. For a universal beam, the value is about 1.15 increasing to 1.5 for a rectangle.

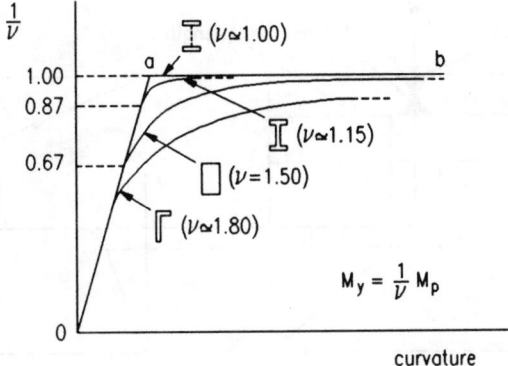

Fig. 9.10　Moment–rotation curves

Plastic hinges and rigid plastic analysis

In deciding the manner in which a beam may fail it is desirable to understand the concept of how plastic hinges form when the beam becomes fully plastic. The number of hinges necessary for failure does not vary for a particular structure subject to a given loading condition, although a part of a structure may fail independently by the formation of a smaller number of hinges. The member or structure behaves in the manner of a hinged mechanism, and, in doing so, adjacent hinges rotate in opposite directions.

As the plastic deformations at collapse are considerably larger than elastic ones, it is assumed that the line element remains rigid between supports and hinge positions i.e. all plastic rotation occurs at the plastic hinges.

Considering a simply-supported beam subjected to a point load at mid-span (Fig. 9.11), the maximum strain will take place at the centre of the span where a plastic hinge will be formed at yield of full section. The remainder of the beam will remain straight: thus the entire energy will be absorbed by the rotation of the plastic hinge.

Work done at the plastic hinge $\quad\quad\quad = M_p(2\theta)$

Work done by the displacement of the load $= W\left(\dfrac{L}{2}\theta\right)$

At collapse, these two must be equal:

$$2M_p\theta = \frac{WL}{2}\theta$$
$$W = 4M_p/L \quad \text{or} \quad M_p = WL/4$$

The moment at collapse of an encastré beam with a uniformly distributed load ($w = W/L$) is worked out in a manner similar to the above from Fig. 9.12.

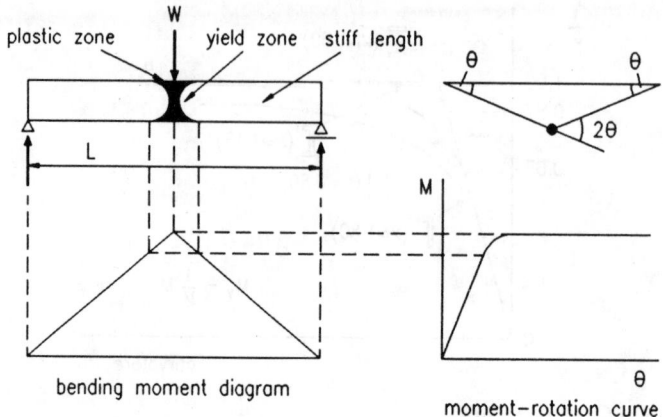

Fig. 9.11   Centrally-loaded simply-supported beam

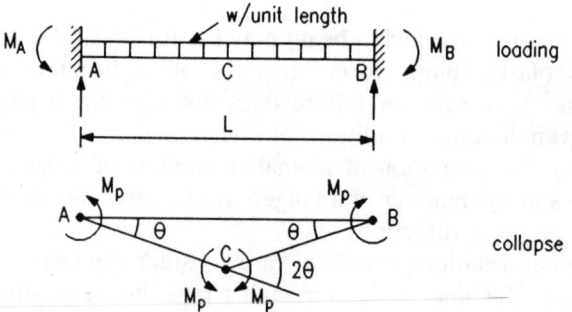

Fig. 9.12   Encastré beam with a uniformly distributed load

Work done at the three plastic hinges $= M_p(\theta + 2\theta + \theta) = 4M_p\theta$

Work done by the displacement of the load $(W) = \dfrac{W}{L}\dfrac{L}{2}\dfrac{L}{2}\theta = \dfrac{WL}{4}\theta$

Equating the two,

$$\frac{WL}{4}\theta = 4M_p\theta$$

$$W = 16M_p/L$$

or $M_p = WL/16$

The moments at collapse for other conditions of loading can be worked out by a similar procedure.

9.3.6 Load factor and theorems of plastic collapse

The load factor at rigid plastic collapse, λ_p, is defined as the lowest multiple of the design loads which will cause the whole structure, or any part of it, to become a mechanism.

In the limit-state approach, the designer seeks to ensure that at the appropriate factored loads the structure will *not* fail. Thus the rigid plastic load factor, λ_p, must not be less than unity, under factored loads.

The number of independent mechanisms, n, is related to the number of possible plastic hinge locations, h, and the degree of redundancy, r, of the skeletal structure, by the equation

$$n = h - r$$

The three theorems of plastic collapse are given below for reference:

(1) *Lower bound or static theorem*
A load factor, λ_s, computed on the basis of an *arbitrarily assumed* bending moment diagram which is *in equilibrium with the applied loads and where the fully plastic moment of resistance is nowhere exceeded*, will always be less than, or at best equal to, the load factor at rigid plastic collapse, λ_p.
λ_p is the highest value of λ_s which can be found.
(2) *Upper bound or kinematic theorem*
A load factor, λ_k, computed on the basis of an *arbitrarily assumed* mechanism will always be greater than, or at best equal to, the load factor at rigid plastic collapse, λ_p.
λ_p is the lowest value of λ_k which can be found.
(3) *Uniqueness theorem*
If *both* the above criteria ((1) and (2)) are satisfied, then $\lambda = \lambda_p$.

9.3.7 Effect of axial load and shear

If a member is subjected to the combined action of bending moment and axial force, the plastic moment capacity will be reduced.

The presence of an axial load implies that the sum of the tension and compression forces in the section is not zero (see Fig. 9.13). This means that the neutral axis moves away from the equal area axis, providing an additional area in tension or compression depending on the type of axial load. The presence of shear forces will also reduce the moment capacity. For the beam sketched in Fig. 9.13,

axial load resisted $= 2at f_y$

Defining $\quad n = \dfrac{\text{axial force resisted}}{\text{axial capacity of section}} = \dfrac{2at}{A}$,

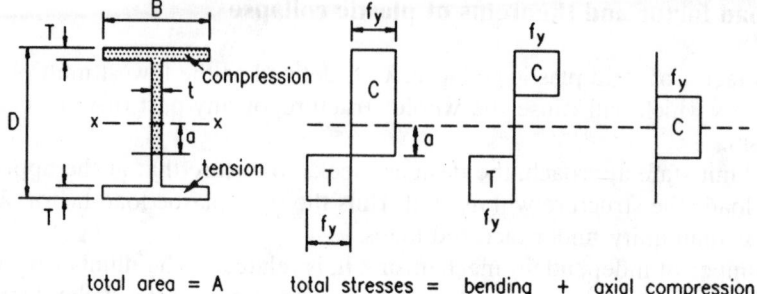

Fig. 9.13   The effect of combined bending and compression

$$a = \frac{nA}{2t}$$

For a given cross section, the plastic moment capacity, M_p, can be evaluated as explained previously. The reduced moment capacity, M'_p, in the presence of the axial load can be calculated as follows:

$$M'_p = M_p - t\,a^2 f_y$$

$$= M_p - t\frac{n^2 A^2}{4t^2} f_y = \left(S - \frac{n^2 A^2}{4t}\right) f_y \tag{9.7}$$

where S is the plastic modulus of the section.

Section tables provide the moment capacity for available steel sections using the approach given above.[1] Similar expressions will be obtained for minor axis bending.

9.3.8 Plastic analysis of beams subjected to shear

Once the material in a beam has started to yield in a longitudinal direction, it is unable to sustain applied shear. When a shear, V, and an applied moment, M, are applied simultaneously to an I-section, a simplifying assumption is employed to reduce the complexity of calculations; shear resistance is assumed to be provided by the web, hence the shear stress in the web is obtained as a constant value of V divided by the web area (see Fig. 9.14). The longitudinal direct stress to cause yield, f_1, in the presence of this shear stress, q, is obtained by using the von Mises yield criterion:[3]

$$f_y^2 = f_1^2 + 3q^2$$

The reduced plastic moment capacity is given by

$$M_r = M_p - \left(\frac{f_y - f_1}{f_y}\right) M_{pw} \tag{9.8}$$

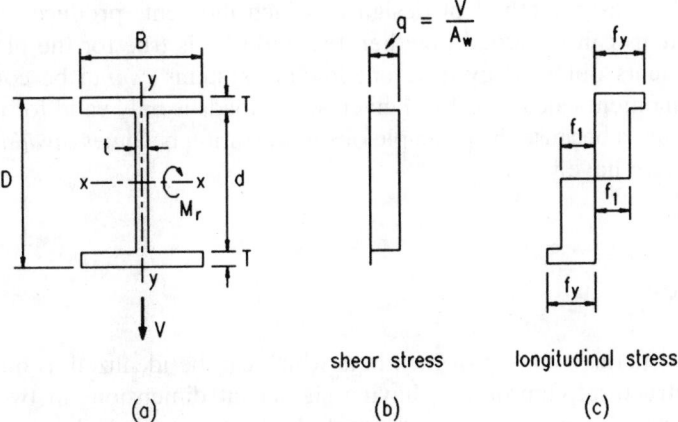

shear stress   longitudinal stress

(a)   (b)   (c)

Fig. 9.14   Combined bending and shear

where M_{pw} is the fully plastic moment of resistance of the web.

The addition of an axial load to the above condition can be dealt with by shifting the neutral axis, as was done in Fig. 9.13. The web area required to carry the axial load is now given by P/f_1 and the depth of the web, d_a, corresponding to this is given by

$$d_a = \frac{P}{f_1 t_w}$$

A further reduction in moment due to the introduction of the axial load is given by

$$\frac{t_w d_a^2}{4} f_1$$

Hence the reduced moment capacity of the section is given by

$$M_1 = M_p - \left(\frac{f_y - f_1}{f_y}\right) M_{pw} - \frac{t_w d_a^2}{4} f_1 \tag{9.9}$$

where $f_1 = \sqrt{(f_y^2 - 3q^2)}$.

9.3.9 Plastic analysis for more than one condition of loading

When more than one condition of loading is to be applied to a line element, it may not always be obvious which is critical. It is necessary then to perform separate calculations, one for each loading condition, the section being determined by the solution requiring the largest plastic moment.

Unlike the elastic method of design in which moments produced by different loading systems can be added together, the opposite is true for the plastic theory. Plastic moments obtained by different loading systems *cannot* be combined, i.e. the plastic moment calculated for a given set of loads is only valid for that loading condition. This is because the principle of superposition becomes *invalid* when parts of the structure have yielded.

9.4 Plates

Most steel structures consist of members which can be idealized as line elements. However, structural components having significant dimensions in two directions (viz. plates) are also encountered frequently. In steel structures, plates occur as components of I-, H-, T- or channel sections as well as in structural hollow sections. Sheets used to enclose lift shafts or walls or cladding in framed structures are also examples of plates.

With plane sheets, the stiffness and strength in all directions is identical and the plate is termed *isotropic*. This is no longer true when stiffeners or corrugations are introduced in one direction. The stiffnesses of the plate in the x and y directions are substantially different. Such a plate is termed *orthotropic*.

The x and y axes for the analysis of the plate are usually taken in the plane of the plate, as shown in Fig. 9.15, while the z axis is perpendicular to that plane. An element of the plate will be subjected to six stress components: *three* direct stresses (σ_x, σ_y and σ_z) and *three* shear stresses (τ_{xy}, τ_{yz} and τ_{zx}). There are six corresponding strains: three direct strains (ε_x, ε_y and ε_z) and three shear strains (γ_{xy}, γ_{yz} and γ_{zx}). These stresses and strains are related in the elastic region by the material properties Young's modulus *(E)* and Poisson's ratio *(v)*.

When considering the response of the plate, the approach customarily employed is termed *plane stress idealization*. As the thickness, t, of the plate is small compared with its other two dimensions in the x and y directions, the *stresses* having components in the z direction are negligible (i.e. σ_z, τ_{yz} and τ_{xz} are all zero). This implies that

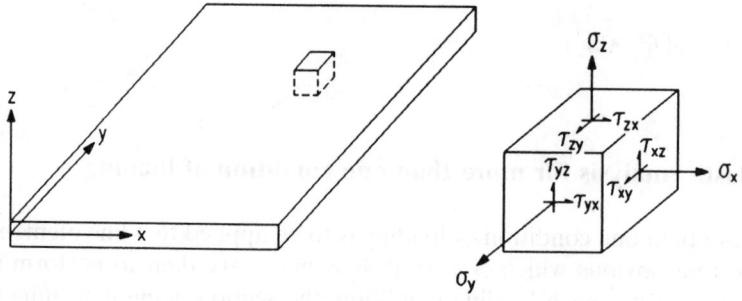

Fig. 9.15   Stress components on an element

the out-of-plane displacement is *not* zero, and this condition is referred to as plane stress idealization.

For an isotropic plate, the general equation relating the displacement, *w*, perpendicular to the plane of the plate element is given by

$$\frac{\partial^4 w}{\partial x^4} + 2\frac{\partial^4 w}{\partial x^2 \partial y^2} + \frac{\partial^4 w}{\partial y^4} = \frac{q}{D} \tag{9.10}$$

where *q* is the normal applied load per unit area in the *z* direction which will, in general, vary with *x* and *y*. The term *D* is the flexural rigidity of the plate, given by

$$D = \frac{Et^3}{12(1 - v^2)} \tag{9.11}$$

The main difficulty in using this approach lies in the choice of a suitable displacement function, *w*, which satisfies the boundary conditions. For loading conditions other than the simplest, an exact solution of this differential equation is virtually impossible. Hence approximate methods (e.g. multiple Fourier series) are utilized. Once a satisfactory displacement function, *w*, is obtained, the moments per unit width of the plate may be derived from

$$M_x = -D\left(\frac{\partial^2 w}{\partial x^2} + v\frac{\partial^2 w}{\partial y^2}\right)$$

$$M_y = -D\left(\frac{\partial^2 w}{\partial y^2} + v\frac{\partial^2 w}{\partial x^2}\right) \tag{9.12}$$

$$M_{xy} = -M_{yx} = D(1-v)\frac{\partial^2 w}{\partial x \partial y}$$

For orthotropic plates, the stiffness in *x* and *y* directions is different and the equations are suitably modified as given below:

$$D_x\frac{\partial^4 w}{\partial x^4} + 2D_{xy}\frac{\partial^4 w}{\partial x^2 \partial y^2} + D_y\frac{\partial^4 w}{\partial y^4} = q \tag{9.13}$$

where D_x and D_y are the flexural rigidities in the two directions.

In view of the difficulty of using classical methods for the solution of plate problems, finite element methods have been developed in recent years to provide satisfactory answers.

9.5 Analysis of skeletal structures

The evaluation of the stress resultants in members of skeletal frames involves the solution of a number of simultaneous equations. When a structure is in equilibrium, every element or constituent part of it is also in equilibrium. This property is made use of in developing the concept of the *free body diagram* for elements of a structure.

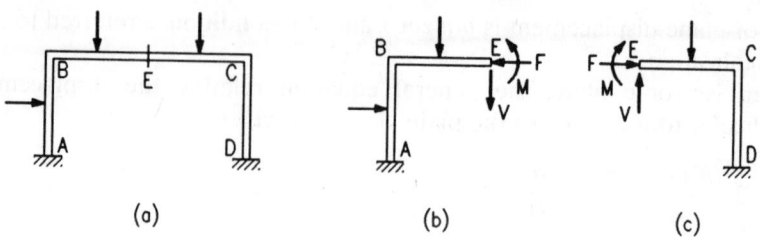

(a)   (b)   (c)

Fig. 9.16   Free body diagram

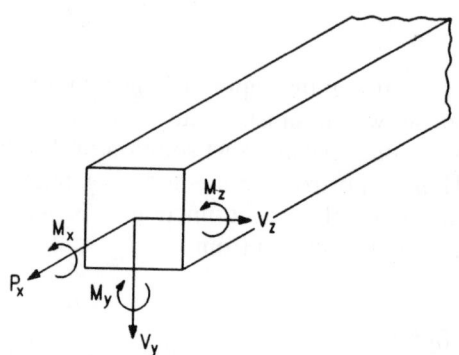

Fig. 9.17   Force and moments in x, y and z directions

The portal frame sketched in Fig. 9.16 will now be considered for illustrating the concept. Assuming that there is an imaginary cut at E on the beam BC, the part ABE continues to be in equilibrium if the two forces and moment which existed at section E of the uncut frame are applied externally. The internal forces which existed at E are given by (1) an axial force F, (2) a shear force V and (3) a bending moment M. These are known as *stress resultants*. The external forces on ABE, together with the forces F, V and M, keep the part ABE in equilibrium; Fig. 9.16(b) is called the free body diagram. On a rigid jointed plane frame there are three stress resultants at each imaginary cut. The part ECD must also remain in equilibrium. This consideration leads to a similar set of forces F, V and M shown in Fig. 9.16(c). It will be noted that the forces acting on the cut face E are equal and opposite. If the two free body diagrams are moved towards each other, it is obvious the internal forces F, V and M cancel out and the structure is restored to its original state of equilibrium. As previously stated, equilibrium implies $\Sigma P_x = 0$; $\Sigma P_y = 0$; $\Sigma M = 0$ for a planar structure. These equations can be validly applied by considering the structure as a whole, or by considering the free body diagram of a part of a structure.

In a similar manner, it can be seen that a three-dimensional rigid-jointed frame has six stress resultants across each section. These are the axial force, two shears in two mutually perpendicular directions and three moments, as shown in Fig. 9.17.

With pin-jointed frames, be they two- or three-dimensional, there is only one stress resultant per member, viz. its axial load. When forces act on an elastic structure, it undergoes deformations, causing displacements at every point within the structure.

The solution of forces in the frames is accomplished by relating the stress resultants to the displacements. The number of equations needed is governed by the *degrees of freedom*, i.e. the number of possible component displacements. At one end of the member of a pin-jointed plane frame, the member displacement has translational components in the x and y directions only, and no rotational displacement. The number of degrees of freedom is two. By similar reasoning it will be apparent that the number of degrees of freedom for a rigid-jointed plane frame member is three. For a member of a three-dimensional pin-jointed frame it is also three, and for a similar rigid-jointed frame it is six.

9.5.1 Stiffness and flexibility

Forces and displacements have a vital and interrelated role in the analysis of structures. Forces cause displacements and the occurrence of displacements implies the existence of forces. The relationship between forces and displacements is defined in one of two ways, viz. *flexibility* and *stiffness*.

Flexibility gives a measure of displacements associated with a given set of forces acting on the structure. This concept will be illustrated by considering the example of a spring loaded at one end by a static load P (see Fig. 9.18).

As the spring is linearly elastic, the extension, Δ, produced is directly proportional to the applied load, P. The deflection produced by a unit load (defined as the *flexibility* of the spring) is obviously Δ/P. Figure 9.18(b) illustrates the deflection response of a beam to an applied load P. Once again the flexibility of the beam is Δ/P.

In the simple cases considered above, flexibility simply gives the load–displacement response at a point. A more generalized definition applicable to the displace-

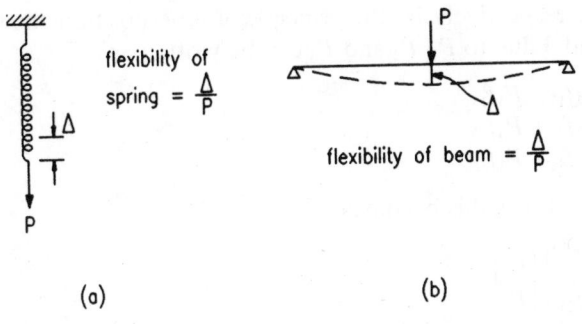

Fig. 9.18   Flexibility

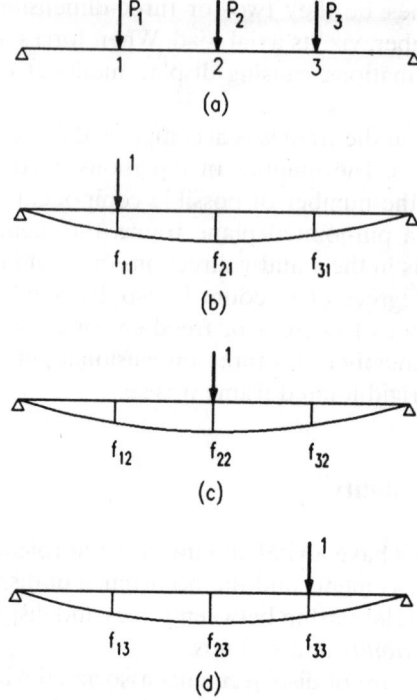

Fig. 9.19 Flexibility coefficients for a loaded beam

ment response at a number of locations will now be obtained by considering the beam sketched in Fig. 9.19.

Considering a unit load acting at point 1 (Fig. 9.19(b)), the corresponding deflections at points 1, 2 and 3 are denoted as f_{11}, f_{21} and f_{31} (the first subscript denotes the point at which the deflection is measured; the second subscript refers to the point at which the unit load is applied). The terms f_{11}, f_{21}, f_{31} are called *flexibility coefficients*. Figure 9.19(c) and (d) give the corresponding flexibility coefficients for load positions 2 and 3 respectively. By the principle of superposition, the total deflections at points 1, 2 and 3 due to P_1, P_2 and P_3 can be written as

$$\Delta_1 = P_1 f_{11} + P_2 f_{12} + P_3 f_{13}$$
$$\Delta_2 = P_1 f_{21} + P_2 f_{22} + P_3 f_{23}$$
$$\Delta_3 = P_1 f_{31} + P_2 f_{32} + P_3 f_{33}$$

Written in matrix form, this becomes

$$\begin{Bmatrix} \Delta_1 \\ \Delta_2 \\ \Delta_3 \end{Bmatrix} = \begin{bmatrix} f_{11} & f_{12} & f_{13} \\ f_{21} & f_{22} & f_{23} \\ f_{31} & f_{32} & f_{33} \end{bmatrix} \begin{Bmatrix} P_1 \\ P_2 \\ P_3 \end{Bmatrix}$$

(9.14)

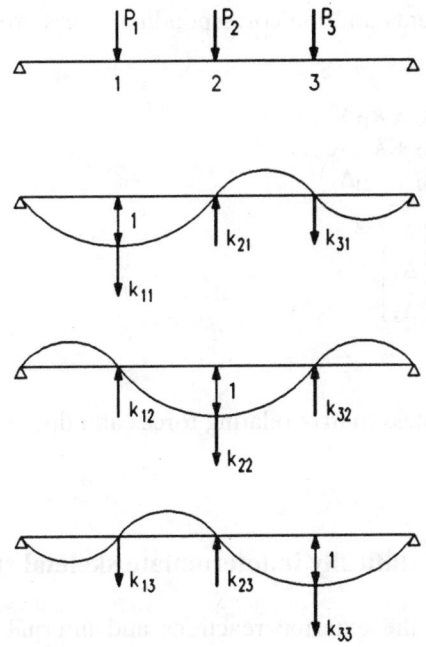

Fig. 9.20 Stiffness coefficients

or $\quad \{\Delta\} = [F]\{P\}$
where $\{\Delta\}$ = displacement matrix
$\quad [F]$ = flexibility matrix relating displacements to forces
$\quad \{P\}$ = force matrix
Hence $\{P\} = [F]^{-1}\{\Delta\}$

Stiffness is the inverse of flexibility and gives a measure of the forces corresponding to a given set of displacements. Considering the spring illustrated in Fig. 9.18(a), it is noted that the deflection response is directly proportional to the applied load, *P*. The force corresponding to unit displacement is obviously P/Δ. Likewise in Fig. 9.18(b) the load to be applied on the beam to cause a unit displacement at a point below the load is P/Δ. In its simplest form, stiffness coefficient refers to the load corresponding to a unit displacement at a given point and can be seen to be the reciprocal of flexibility. The concept is explained further using Fig. 9.20.

First the locations 2 and 3 are restrained from movement and a unit displacement is given at 1. This implies a downward force k_{11} at 1, an upward force k_{21} at 2 and a downward force k_{31} at 3. The forces at points 2 and 3 are necessary as otherwise there will be displacements at the locations 2 and 3.

The forces k_{11}, k_{21} and k_{31} are designated as stiffness coefficients. In a similar manner, the stiffness coefficients corresponding to unit displacements at points 2 and 3 are obtained.

The stiffness coefficients and the corresponding forces are linked by the following equations

$$
\begin{aligned}
P_1 &= k_{11}\Delta_1 + k_{12}\Delta_2 + k_{13}\Delta_3 \\
P_2 &= k_{21}\Delta_1 + k_{22}\Delta_2 + k_{23}\Delta_3 \\
P_3 &= k_{31}\Delta_1 + k_{32}\Delta_2 + k_{33}\Delta_3
\end{aligned}
$$

$$
\text{or} \quad \begin{Bmatrix} P_1 \\ P_2 \\ P_3 \end{Bmatrix} = \begin{bmatrix} k_{11} k_{12} k_{13} \\ k_{21} k_{22} k_{23} \\ k_{31} k_{32} k_{33} \end{bmatrix} \begin{Bmatrix} \Delta_1 \\ \Delta_2 \\ \Delta_3 \end{Bmatrix} \tag{9.15}
$$

$$
\text{or} \quad \{P\} = [K]\{\Delta\}
$$

where $[K]$ is the stiffness matrix relating forces and displacements.

9.5.2 Introduction to statically indeterminate skeletal structures

A structure for which the external reactions and internal forces and moments can be computed by using only the three equations of statics ($\Sigma P_x = 0$, $\Sigma P_y = 0$ and $\Sigma M = 0$) is known as *statically determinate*. A structure for which the forces and moments cannot be computed from the principles of statics alone is *statically indeterminate*. Examples of statically determinate skeletal structures are shown in Fig. 9.21.

In structures shown in Fig. 9.21(a), (b) and (c), the supporting forces and moments are just sufficient in number to withstand the external loading. For example, if one of the supports of (b) were to fail or if one of the members of (c) were to be removed, the structure would collapse.

However, when the beam or frame is provided with additional supports (see Fig. 9.21(d), (e)) or if the pin-jointed truss has more members than are required to make it 'perfect' (Fig. 9.21(b)), the structure becomes statically indeterminate.

The degree of indeterminacy (also termed the *degree of redundancy*) is obtained by the number of member forces or reaction components (viz. moments or forces) which should be 'released' to convert a statically indeterminate structure to a determinate one. If *n* forces or moments are required to be so released, the degree of indeterminacy is *n*. We need *n* independent equations (in addition to three equations of statics for a planar structure) to solve for forces and moments at all locations in the structure. The additional equations are usually written by considering the deformations or displacements of the structure. This means that the section properties (viz. area, second moment of area, etc.) have an important effect in evaluating the forces and moments of an indeterminate structure. Also, the settlement of a support or a slight lack of fit in a pin-jointed structure contributes materially to the internal forces and moments of an indeterminate structure.

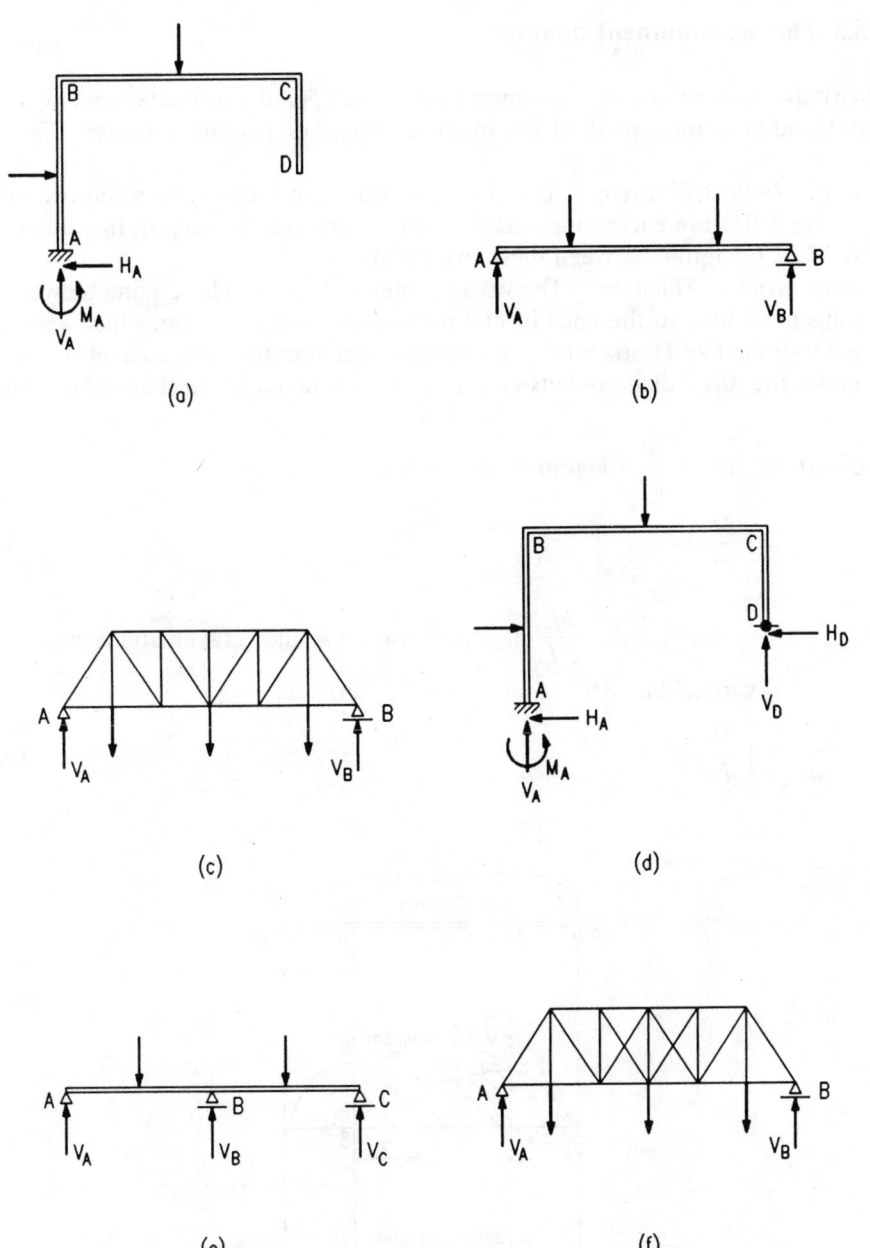

Fig 9.21   Statically determinate and indeterminate skeletal structures. (a), (b) and (c) are determinate; (d), (e) and (f) are indeterminate

9.5.3 The area moment method

The simplest technique of analysing a beam which is indeterminate to a low degree is by the area moment method. The method is based on two theorems (see Fig. 9.22):

- *Area Moment Theorem 1*: The change in slope (in radians) between two points of the deflection curve in a loaded beam is numerically equal to the area under the *M/EI* diagram between these two points.
- *Area Moment Theorem 2*: The vertical intercept on any chosen line between the tangents drawn to the ends of any portion of a loaded beam, which was originally straight and horizontal, is numerically equal to the first moment of the area under the *M/EI* diagram between the two ends taken about that vertical line.

$$\theta_B - \theta_A = \text{area of } \frac{M}{EI} \text{ diagram between A and B}$$

$$= \int_A^B \frac{M}{EI} \, dx \tag{9.16}$$

$$\Delta = \text{moment of the } \frac{M}{EI} \text{ diagram between A and B taken about the}$$

vertical line RS

$$= \int_A^B \frac{Mx}{EI} \, dx \tag{9.17}$$

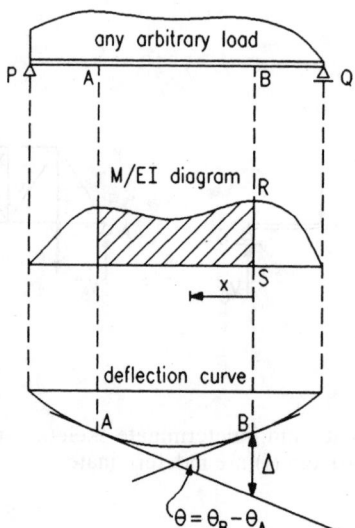

Fig. 9.22   Area moment theorems

(*Caution*: The vertical intercept is *not* the deflection of the beam from its original position.)

The area moment method can be used for solving problems like encastré beams, propped cantilevers, etc. The procedure is as follows:

(1) The redundant supports are removed, thereby releasing the redundant forces and moments. The statically determinate M/EI diagram for externally applied loads can then be drawn.
(2) The externally applied loads are removed, the redundant forces and moments are introduced one at a time, and the M/EI diagrams corresponding to each of these forces and moments are drawn.
(3) The slopes at supports and intercepts on a vertical axis passing through the supports are then calculated.
(4) A number of expressions are obtained. These are then equated to known values of slopes or displacements at supports. The equations so obtained can then be solved for the unknown redundant reactions. This enables the evaluation of the forces and moments in the structure.

9.5.4 The slope–deflection method

The slope–deflection method can be used to analyse all types of statically indeterminate beams and rigid frames. In this method all joints are considered rigid and the angles between members at the joints are considered not to change as the loads are applied. When beams or frames are deformed, the rigid joints are considered to rotate as a whole.

In the slope–deflection method, the rotations and translations of the joints are the unknowns. All end moments are expressed in terms of end rotations and translations. In order to satisfy the conditions of equilibrium, the sum of end moments acting on a rigid joint must total zero. Using this equation of equilibrium, the unknown rotation of each joint is evaluated, from which the end moments are computed.

For the span AB shown in Fig. 9.23, the object is to express the end moments M_{AB} arid M_{BA} in terms of end rotations θ_A and θ_B and translation Δ.

With the applied loading on the member, fixed end moments M_{FAB} and M_{FBA} are required to hold the tangents at the ends fixed in direction. (Counter clockwise end moments and rotations are taken as positive.) The slope–deflection equations for the case sketched in Fig. 9.23 are

$$M_{AB} = M_{FAB} + \frac{2EI}{L}\left(-2\theta_A - \theta_B + \frac{3\Delta}{L}\right)$$

$$M_{BA} = M_{FBA} + \frac{2EI}{L}\left(-2\theta_B - \theta_A + \frac{3\Delta}{L}\right)$$

(9.18)

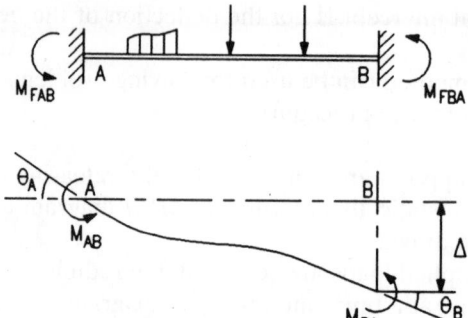

Fig. 9.23   Slope–deflection method

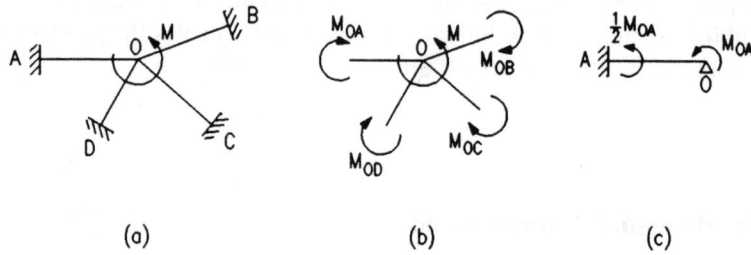

(a)   (b)   (c)

Fig. 9.24   Moment-distribution method

where M_{FAB}, M_{FBA} are the fixed end moments at A and B due to loading and settlement of supports (counter clockwise positive); θ_A and θ_B are end rotations; and Δ is the downward settlement of support B relative to support A.

9.5.5 The moment-distribution method

The moment-distribution method can be employed to analyse continuous beams or rigid frames. Essentially it consists of solving the simultaneous equations in the slope –deflection method by successive approximations. Since the solution is by successive iteration, it is not even necessary to determine the degree of redundancy.

Two facets of the method must be appreciated (see Fig. 9.24):

● When a stiff joint in a structural system absorbs an applied moment with rotational movement only (i.e. no translation), Fig. 9.24(a), the moment resisted by the various members meeting at the joint is in proportion to their respective stiffnesses (Fig. 9.24(b)).
● When a member is fixed at one end and a moment, M, is applied at the other freely supported end, the moment induced at the fixed end is half the applied

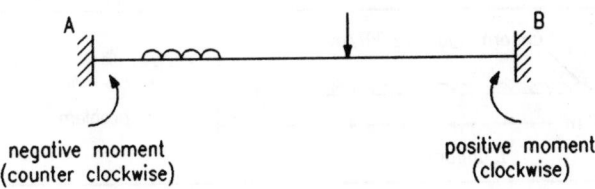

negative moment
(counter clockwise)

positive moment
(clockwise)

convention: clockwise moments are positive

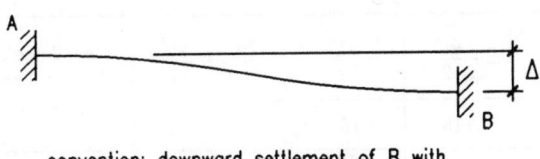

convention: downward settlement of B with
respect to A is positive

Fig. 9.25   Sign conventions used in moment-distribution method

moment and acts in the same direction as *M*. (This is frequently referred to as the carry over.) (Fig. 9.24(c).)

Figure 9.25 shows the sign conventions employed in the moment-distribution method and Fig. 9.26 illustrates the moment-distribution procedure.

The moment-distribution method consists of locking all joints first and then releasing them one at a time. To begin with, all joints are locked, which implies that the fixed end moments due to applied loading will be applied at each joint. By releasing one joint at a time, the unbalanced moment at each joint is distributed to the various members meeting at the joint. Half of these applied moments are then carried over to the other end of each member. This creates a further imbalance at each joint and the unbalanced moments are once again distributed to all members meeting at each joint in proportion to their respective stiffnesses.

This procedure is repeated until the totals of all moments at each joint are sufficiently close to zero. At this stage the moment-distribution process is stopped and the final moments are obtained by summing up all the numbers in the respective columns.

9.5.6 Unit load method

Energy methods provide powerful tools for the analysis of structures. The unit load method can be directly derived from the complementary energy theorem, which

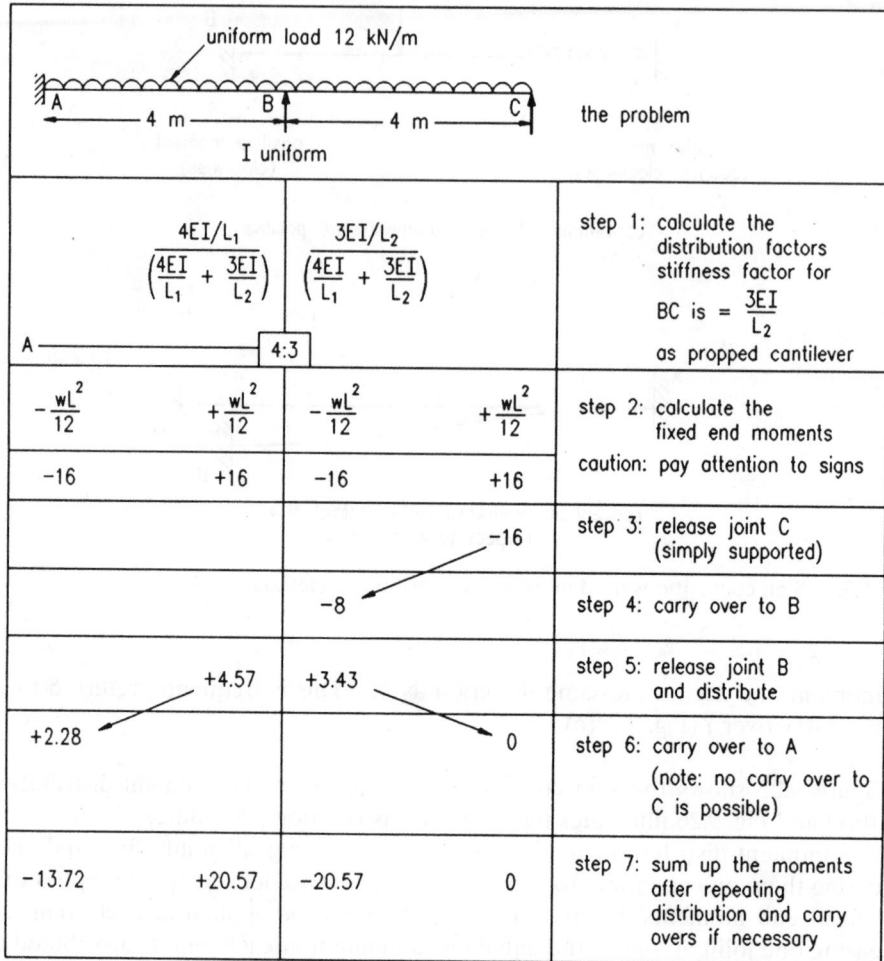

Fig. 9.26 An example of moment-distribution procedure

states that for any elastic structure in equilibrium under loads P_1, P_2, ..., the corresponding displacements x_1, x_2, ... are given by the partial derivatives of the complementary energy, C, with respect to the loads P_1, P_2, etc. In other words,

$$\frac{\partial C}{\partial P_1} = x_1 \qquad \frac{\partial C}{\partial P_2} = x_2 \tag{9.19}$$

For a linearly elastic system, the complementary energy is equal to the strain energy, U, hence

$$x_1 = \frac{\partial U}{\partial P_1} \qquad x_2 = \frac{\partial U}{\partial P_2} \tag{9.20}$$

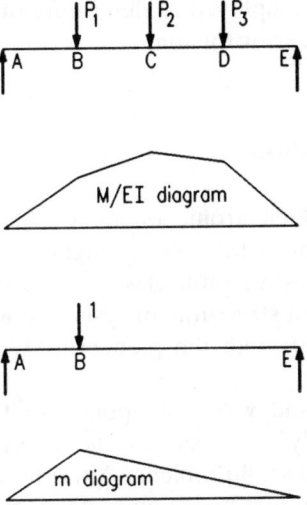

Fig. 9.27 Unit load method

The total strain energy of an elastic system is given by the sum of the strain energies stored in each member due to bending, shear, torsion and axial loading.

The use of this will be illustrated by considering a simply-supported beam of length l subject to an external loading (see Fig. 9.27). Strain energy stored in the beam is predominantly flexural and is given by

$$\int_0^l \frac{M^2 dx}{2EI}$$

$$x_1 = \text{deflection under } P_1 = \frac{\partial}{\partial P_1}\left(\int_0^l \frac{M^2 dx}{2EI}\right) = \int_0^l \frac{M}{EI}\frac{\partial M}{\partial P_1} dx \qquad (9.21)$$

$\dfrac{\partial M}{\partial P_1}$ is the bending moment due to a unit load and is denoted by m.

Hence the procedure of the unit load method can be outlined (see Fig. 9.27):

(1) $\dfrac{M}{EI}$ diagram due to the external loading is obtained.

(2) The external loads are now removed and the moment diagram (m) due to a unit load applied at the point of required deflection is drawn.

(3) These two diagrams should now be integrated; in other words, the ordinates of the two diagrams are multiplied to obtain the deflection, given by

$$x = \int_0^l \frac{M}{EI} m \, dx$$

The same principle can be employed to determine the displacement due to other causes, viz. axial load or shear or torsion.

9.6 Finite element method

The advent of high-speed electronic digital computers has given tremendous impetus to numerical methods for solving engineering problems. Finite element methods form one of the most versatile classes of such methods which rely strongly on the matrix formulation of structural analysis. The application of finite elements dates back to the mid-1950s with the pioneering work by Argyris,[4] Clough and others.

The finite element method was first applied to the solution of plane stress problems and subsequently extended to the analysis of axisymmetric solids, plate bending problems and shell problems. A useful listing of elements developed in the past is documented in text books on finite element analysis.[5]

Stiffness matrices of finite elements are generally obtained from an assumed displacement pattern. Alternative formulations are equilibrium elements and hybrid elements. A more recent development is the so-called strain based elements. The formulation is based on the selection of simple independent functions for the linear strains or change of curvature; the strain–displacement equations are integrated to obtain expressions for the displacements.

The basic assumption in the finite element method of analysis is that the response of a continuous body to a given set of applied forces is equivalent to that of a system of discrete elements into which the body may be imagined to be subdivided. From the energy point of view, the equivalence between the body and its finite element model is therefore exact if the strain energy of the deformed body is equal to that of its discrete model.

The energy due to straining of the element, U, written in two-dimensional form is

$$U = \frac{1}{2}\int\int(\varepsilon_x\sigma_x + \varepsilon_y\sigma_y + \gamma_{xy}\tau_{xy})\,\mathrm{d}(\text{vol})$$

or in matrix form

$$U = \frac{1}{2}\int\int\{\varepsilon\}^{\text{T}}\{\sigma\}\,\mathrm{d}(\text{vol}) \tag{9.22}$$

in which $\{\varepsilon\}^{\text{T}} = \{\varepsilon_x, \varepsilon_y, \gamma_{xy}\}$

$$\{\sigma\} \qquad = \{\sigma_x, \sigma_y, \tau_{xy}\}^{\text{T}} \tag{9.23}$$

and $\{\varepsilon\} \qquad = [f]\{\delta^c\}$

$$\{\delta^c\} \qquad = \{u_1, u_2, \ldots, u_n\}^{\text{T}} \tag{9.24}$$

where $\{\delta\}$ is the nodal displacement vector, $[f]$ is a function defining the strain distribution and e refers to a typical finite element. When the strain distribution within the model is exactly the same as that prevailing within the body, then the energy equation will be exactly satisfied.

The exact determination of the strain distribution function $[f]$ in Equation (9.24) presents considerable difficulties, since this can only be done by a rigorous solution of the equations of linear elasticity. It may not always be possible to obtain an *exact* shape function for the solution: however, a suitable function which is *adequate* to model strains can usually be selected. The derivation of simple membrane elements for plate problems is presented in the following pages.

9.6.1 Finite element procedure

As mentioned above, the basic concept of the finite element method is the idealization of the continuum as an assemblage of discrete structural elements. The stiffness properties of each element are then evaluated and the stiffness properties of the complete structure are obtained by superposition of the individual element stiffnesses. This gives a system of linear equations in terms of nodal point loads and displacements whose solution yields the unknown nodal point displacements.

The idealization governs the type of element which must be used in the solution. In many cases only one type of element is used for a given problem, but sometimes it is more convenient to adopt a 'mixed' subdivision in which more than one type of element is used.

The elements are assumed to be interconnected at a discrete number of nodal points or nodes. The nodal degrees of freedom normally refer to the displacement functions and their first partial derivatives at a node but very often may include other terms such as stresses, strains and second or even higher partial derivatives. For example, the triangular and rectangular membrane elements of Fig. 9.28(a) and (b) have 6 and 8 degrees of freedom respectively, representing the translations u and v at the corner nodes in the x and y directions.

A displacement function in terms of the co-ordinate variables x, y and the nodal displacement parameters (e.g. u_i, v_i, or δ_i) are chosen to represent the displacement variations within each element. By using the principle of virtual work or the principle of minimum total potential energy, a stiffness matrix relating the nodal forces to the nodal displacements can be derived. Hence the choice of suitable displacement functions is the most important part of the whole procedure. A good displacement function leads to an element of high accuracy with converging characteristics; conversely, a wrongly chosen displacement function yields poor or non-converging results.

A displacement function may conveniently be established from simple polynomials or interpolation functions. The displacement field in each element must be expressed as a function of nodal point displacements only, and this must be done in such a way as to maintain inter-element compatibility, since this condition is neces-

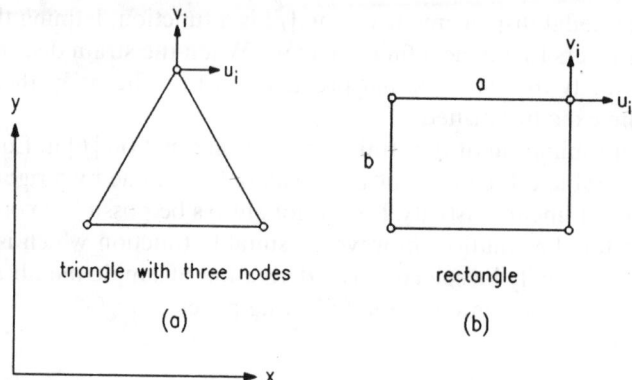

Fig. 9.28   Triangular and rectangular membrane elements

sary to establish a bound on the strain energy. Therefore, the displacement pattern and the nodal point degrees of freedom must be selected properly for each problem considered.

In general, it is not *always* necessary that the compatibility must be satisfied in order to achieve convergence to the true solution. If complete compatibility is not achieved, there exists an uncertainty as to the bound on the strain energy of the system. Therefore, in order to justify the performance and the ability of these elements to converge to the true solution, a critical test which indicates the performance in the limit must be carried out: that is, a convergence test with decreasing mesh size. If the performance is adequate, then convergence to the true solution is achieved within a reasonable computational effort. and the requirement for the complete compatibility can be relaxed.

Besides satisfying the compatibility requirement, the assumed displacement functions should include the following properties:

(1) Rigid body modes
(2) Constant strain and curvature states
(3) Invariance of the element stiffnesses.

9.6.2 Idealization of the structure

The finite element idealization should represent the real structure as closely as possible with regard to geometrical shape, loading and boundary conditions. The geometrical form of the structure is the major factor to be considered when deciding the shape of elements to be used. In two-dimensional analyses the most frequently used elements are triangular or rectangular shapes. The triangular element has the advantage of simplicity in use and the ability to fit into irregular boundaries. Figure

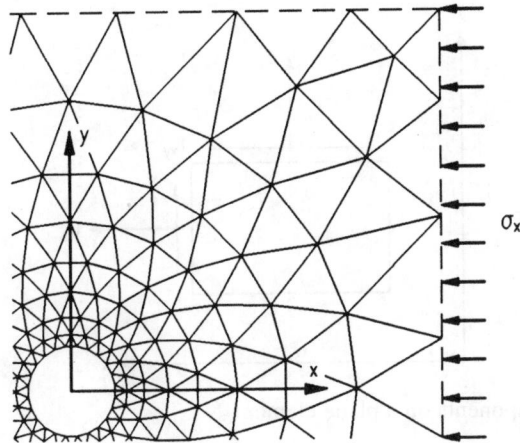

Fig. 9.29   Circular hole in uniform stress field

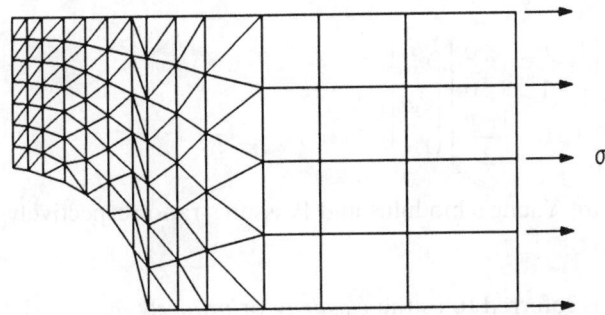

Fig. 9.30   Perforated tension strip (plane stress) – a quarter of a plate is analysed

9.29 shows an example using triangular elements and Fig. 9.30 shows a combination of triangular and rectangular elements. These figures also demonstrate the need to use relatively small elements in areas where high stress gradients occur. In many such cases the geometrical shape of the structure is such that a fine mesh of elements is required in order to match this shape.

9.6.3 Procedure for evaluating membrane element stiffness

The formulation of the stiffness matrix $[K^c]$ of the membrane element is briefly discussed below.

The energy due to straining of the element is given by Equation (9.22). The stress components are shown in Fig. 9.31 and ε_x, ε_y, γ_{xy} are the corresponding strain

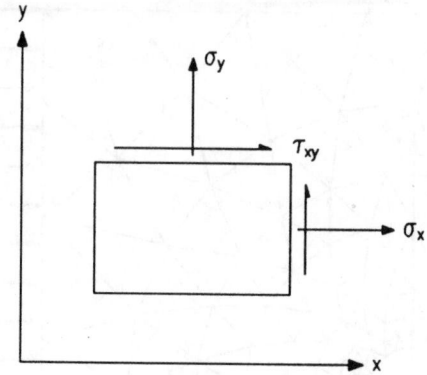

Fig. 9.31   Stress components on a plane element

components. For the state of plane stress, the stress components are related to the strain components as given below:

$$\begin{Bmatrix} \sigma_x \\ \sigma_y \\ \tau_{xy} \end{Bmatrix} = \frac{E}{1-\nu^2} \begin{bmatrix} 1 & \nu & 0 \\ \nu & 1 & 0 \\ 0 & 0 & \dfrac{1-\nu}{2} \end{bmatrix} \begin{Bmatrix} \varepsilon_x \\ \varepsilon_y \\ \gamma_{xy} \end{Bmatrix} \tag{9.25}$$

where E and ν are Young's modulus and Poisson's ratio respectively.

$$\therefore \{\sigma\} = [D]\{\varepsilon\}$$

The matrix $[D]$ is referred to as the *elasticity* or *property matrix*. The strain energy of the element is given by Equation (9.22) as

$$U = \frac{1}{2} \iint \{\varepsilon\}^T [D] \{\varepsilon\} \, d(\text{vol}) \tag{9.26}$$

Denoting the generalized nodal displacement by the vector $\{\delta^e\}$,

$$\{\delta^e\} = [C]\{A\} \tag{9.27}$$

where $\{A\} = [a_1, a_2, \ldots, a_n]^T$, which is a vector of polynomial constants, and $[C]$ is a transformation matrix.

The strains are obtained through making appropriate differentiations of the displacement function with respect to the relevant co-ordinate variable x or y. Thus,

$$\{\varepsilon\} = \begin{Bmatrix} \varepsilon_x \\ \varepsilon_y \\ \gamma_{xy} \end{Bmatrix} = \begin{Bmatrix} \partial u / \partial x \\ \partial v / \partial y \\ \dfrac{\partial u}{\partial y} + \dfrac{\partial v}{\partial x} \end{Bmatrix} \tag{9.28}$$

$$\therefore \{\varepsilon\} = [B]\{A\} = [B][C]^{-1}\{\delta^e\} \tag{9.29}$$

where $[B]$ is the transformation matrix.

Using Equations (9.26) and (9.29), the following expression for U in terms of the nodal point displacement vector $\{\delta^e\}$ is obtained:

$$U = \frac{1}{2}\{\delta^e\}^T [C]^{-1,T} \left[\int \int [B]^T [D][B]\mathrm{d}(\mathrm{vol}) \right][C]^{-1}\{\delta^e\} \tag{9.30}$$

Differentiation of U with respect to the nodal displacements yields the stiffness matrix $[K^e]$:

$$[K^e] = t[C]^{-1,T} \left[\int_A [B]^T [D][B]\mathrm{d}A \right][C]^{-1} \tag{9.31}$$

where t is the thickness of the element (assumed constant).

The calculation of the stiffness matrices is generally carried out in two stages. The first stage is to calculate the terms inside the square brackets of Equation (9.31) i.e. the integration part. The second stage is to multiply the resulting integrations by the inverse of the transformation matrix $[C]$ and its transpose.

Equation (9.31) can now be written as

$$[K^e] = t[C]^{-1,T}[Q][C]^{-1} \tag{9.32}$$

where $[Q] = \int_A [B]^T [D][B]\mathrm{d}A$

The simplest elements for plane stress analysis have nodal points at the corners only and have two degrees of kinematic freedom at each nodal point, i.e. u and v. This type of element proves simple to derive and has been widely used. The simplest elements of this type are rectangular and triangular in shape.

A triangular element with nodal points at the corners is shown in Fig. 9.28(a). The displacement function of this element has two degrees of freedom at each nodal point and the displacements are assumed to vary linearly between nodal points. This results in constant values of the three strain components over the entire element; the displacement functions are

$$u = a_1 + a_2 x + a_3 y \tag{9.33}$$
$$v = a_4 + a_5 x + a_6 y$$

The rectangular element with sides a and b, shown in Fig. 9.28(b), is used with the following displacement functions:

$$u = a_1 + a_2 x + a_3 y + a_4 xy$$
$$v = a_5 + a_6 x + a_7 y + a_8 xy \tag{9.34}$$

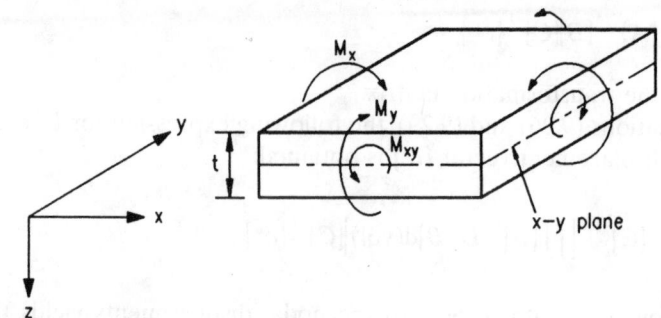

Fig. 9.32 Moments acting on a plane element

9.6.4 Procedure for evaluating plate bending element stiffness

The energy due to straining (bending) of the element is

$$U_b = \frac{1}{2}\int_A (\chi_x M_x + \chi_y M_y + 2\chi_{xy} M_{xy})\,\mathrm{d}A \tag{9.35}$$

or in matrix form

$$U_b = \frac{1}{2}\int_A (\chi_x, \chi_y, 2\chi_{xy}) \begin{Bmatrix} M_x \\ M_y \\ M_{xy} \end{Bmatrix}\mathrm{d}A$$

$$= \frac{1}{2}\int_A [\chi]^T \{M\}\,\mathrm{d}A$$

where the moments M_x, M_y, M_{xy} are given in Fig. 9.32 and χ_x, χ_y, and χ_{xy} are the corresponding curvatures, i.e.

$$\chi_x = \frac{\partial^2 w}{\partial x^2} \quad \chi_y = \frac{\partial^2 w}{\partial y^2} \quad \chi_{xy} = \frac{\partial^2 w}{\partial x \partial y} \tag{9.36}$$

where w is the transverse displacement of the plate element.

The conventional relationship between curvatures and moment is

$$\begin{Bmatrix} M_x \\ M_y \\ M_{xy} \end{Bmatrix} = [\overline{D}] \begin{Bmatrix} \chi_x \\ \chi_y \\ 2\chi_{xy} \end{Bmatrix} \tag{9.37}$$

or

$$\{M\} = [\overline{D}]\{\chi\}$$

For an isotropic plate, the rigidity matrix, $[\overline{D}]$, may be written as

$$[\overline{D}] = \begin{bmatrix} D_1 & \nu D_1 & 0 \\ \nu D_1 & D_1 & 0 \\ 0 & 0 & D_{xy} \end{bmatrix}$$

and

$$D_x = D_y = D_1 = \frac{Et^3}{12(1-\nu^2)}$$

$$D_{xy} = \frac{1}{2}(1-\nu)D_1$$

Finally the strain energy can be written as

$$U_b = \frac{1}{2}\int_A [\chi]^T[\overline{D}][\chi]dA \tag{9.38}$$

Denoting the generalized nodal displacement by the vector $\{\overline{\delta^e}\}$,

$$\{\overline{\delta^e}\} = [\overline{C}]\{\overline{A}\} \tag{9.39}$$

where $\{\overline{A}\}$ is a vector of polynomial constants. From Equations (9.36) and (9.39),

$$\{\chi\} = [\overline{B}]\{\overline{A}\} = [\overline{B}][\overline{C}]^{-1}\{\overline{\delta^e}\} \tag{9.40}$$

Using Equations (9.38) and (9.40), the following equation for U_b in terms of nodal point displacement parameters $\{\overline{\delta^e}\}$ is obtained:

$$U_b = \frac{1}{2}\{\overline{\delta^e}\}^T[\overline{C}]^{-1,T}\left[\int_A [\overline{B}]^T[\overline{D}][\overline{B}]dA\right][\overline{C}]^{-1}\{\overline{\delta^e}\} \tag{9.41}$$

Differentiation of U with respect to the nodal displacements yields the stiffness matrix $[\overline{K^e}]$:

$$[\overline{K^e}] = [\overline{C}]^{-1,T}\left[\int_A [\overline{B}]^T[\overline{D}][\overline{B}]dA\right][\overline{C}]^{-1} \tag{9.42}$$

Equation (9.42) can now be written as

$$[\overline{K^e}] = [\overline{C}]^{-1,T}[\overline{Q}][\overline{C}]^{-1} \tag{9.43}$$

where

$$[\overline{Q}] = \int_A [\overline{B}]^T[\overline{D}][\overline{B}]dA \tag{9.44}$$

As mentioned before, the accuracy is dependent on choosing a large number of elements. Many refined elements giving greater accuracy are described in standard books on finite element methods, which also provide details of assembling the

elements and analysing the structure.[5-7] These methods are used for solving a wide range of problems.

References to Chapter 9

1. The Steel Construction Institute (SCI) (2001) *Steelwork Design Guide to BS 5950: Part 1: 2000, Vol. 1: Section Properties Member Capacities* (6th edn) SCI, Ascot, Berks.
2. Nethercot D.A., Salter P.R. & Malik A.S. (1989) *Design of Members Subject to Combined Bending and Torsion.* The Steel Construction Institute, Ascot, Berks.
3. Timoshenko S. (1976) *Strength of Materials – Part 2*, 3rd edn. Van Nostrand & Co., New York.
4. Argyris J.H. (1960) *Energy Theorems and Structural Analysis.* Butterworths, London.
5. Zienkiewicz O.C. & Cheung Y.K. (2000) *The Finite Element Method in Structural and Continuum Mechanics*, 5th edn. Butterworth-Heinemann, Oxford.
6. Coates R.C., Coutie M.G. & Kong F.K. (1988) *Structural Analysis*, 3rd edn. Chapman & Hall, London.
7. Nath B. (1974) *Fundamentals of Finite Elements for Engineers.* Athlone Press, London.

Further reading for Chapter 9

Brown D.G. (1995) *Modelling of Steel Structures for Computer Analysis.* The Steel Construction Institute, Ascot, Berks.

Chapter 10
Beam analysis

by JOHN RIGHINIOTIS

10.1 Simply-supported beams

The calculations required to obtain the shear forces (SF) and bending moments (BM) in simply-supported beams form the basis of many other calculations required for the analysis of built-in beams, continuous beams and other indeterminate structures.

Appropriate formulae for simple beams and cantilevers under various types of loads are presented in the Appendix *Bending moment, shear and deflection tables for cantilevers* and *simply-supported beams*.

In the case of simple beams it is necessary to calculate the support reactions before the bending moments can be evaluated; the procedure is reversed for built-in or continuous beams. The following rules relate to the SF and BM diagrams for beams:

(1) the shear force at any section is the algebraic sum of normal forces acting to one side of the section
(2) shear is considered positive when the shear force calculated as above is upwards to the left of the section
(3) the BM at any section is the algebraic sum of the moments about that section of all forces to one side of the section
(4) moments are considered positive when the middle of a beam sags with respect to its ends or when tension occurs in the lower fibres of the beam
(5) for point loads only the SF diagram will consist of a series of horizontal and vertical lines, while the BM diagram will consist of sloping straight lines, changes of slope occurring only at the loads
(6) for uniformly distributed loads (UDL) the SF diagram will consist of sloping straight lines, while the BM diagram will consist of second-degree parabolas
(7) the maximum BM occurs at the point of zero shear, where such exists, or at the point where the shear force curve crosses the base line.

10.2 Propped cantilevers

Beams which are built-in at one end and simply-supported at the other are known as *propped cantilevers*. Normally, the ends of the beams are on the same level, in

which case bending moments and reactions may be derived in two ways: by employing the Theorem of Three Moments or by deflection formulae. Appropriate formulae for propped cantilevers under various types of loading are presented in the Appendix *Bending moment, shear and deflection tables for propped cantilevers.*

10.2.1 Solution by the Theorem of Three Moments

Consider the propped cantilever AB in Fig. 10.1.

The bending moment at B may be found by using the Theorem of Three Moments, and assuming that AB is one span of a two-span continuous beam ABC which is symmetrical in every way about B.

Then the loads on AB and BC will produce free BM diagrams whose areas are A_1 and A_2 respectively, the centres of gravity (CG) of the areas being distances x_1 and x_2 from A and C respectively.

Now $$M_A L_1 + 2M_B(L_1 + L_2) + M_C L_2 = 6\left(\frac{A_1 x_1}{L_1} + \frac{A_2 x_2}{L_2}\right)$$

where M_A, M_B and M_C are the numerical values of the hogging moments at the supports A, B and C.

But
$$L_1 = L_2 = L$$
$$M_A = M_C = 0$$
$$A_1 = A_2 = A$$

and
$$x_1 = x_2 = x$$

Hence
$$2M_B(2L) = 6 \times 2\left(\frac{Ax}{L}\right)$$

and
$$M_B = \frac{3Ax}{L^2}$$

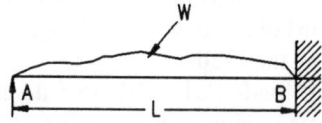

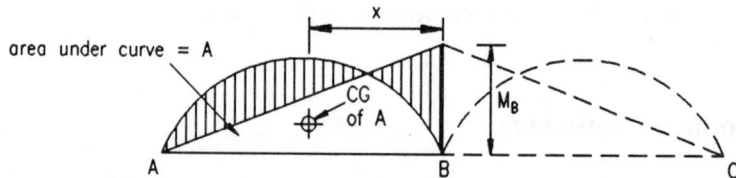

Fig. 10.1   Bending moment diagram for propped cantilever

Therefore the moment at the fixed end of a propped cantilever

$$= 3Ax/L^2$$

where A = the area of the free BM diagram, AB being considered as a simply-supported beam

x = the distance from the prop to the CG of the free BM diagram

and L = the span.

The reactions at each support may be found by employing a modified form of the formula used for beams built-in at both ends:

$$\text{SF}_A = \text{the simple support reaction at A} = -\frac{M_B}{L}$$

$$\text{SF}_B = \text{the simple support reaction at B} = +\frac{M_B}{L}$$

where A is the propped end and B is built-in.

10.2.2 Sinking of supports

When the supports for a loaded propped cantilever do not maintain the same relative levels as in the unloaded condition, the BM and SF may be obtained by using the deflection method (Fig. 10.2). When the prop, B, sinks the load which it takes is reduced, while the fixing moment at the other end is increased. Two special cases arise: the first when the prop sinks so much that no load is taken by the prop, and the second when the built-in end sinks so much that the fixing moment is reduced to zero, i.e. the cantilever resembles a simple support beam. The two special cases are shown in Fig. 10.2.

10.3 Fixed, built-in or encastré beams

When the ends of a beam are firmly held so that they cannot rotate under the action of the superimposed loads, the beam is known as a fixed, built-in or encastré beam. The BM diagram for such a beam is in two parts: the free or positive BM diagram, which would have resulted had the ends been simply-supported, i.e. free to rotate, and the fixing or negative BM diagram which results from the restraints imposed upon the ends of the beam.

Normally, the supports for built-in beams are at the same level and the ends of the beams are horizontal. This type will be considered first.

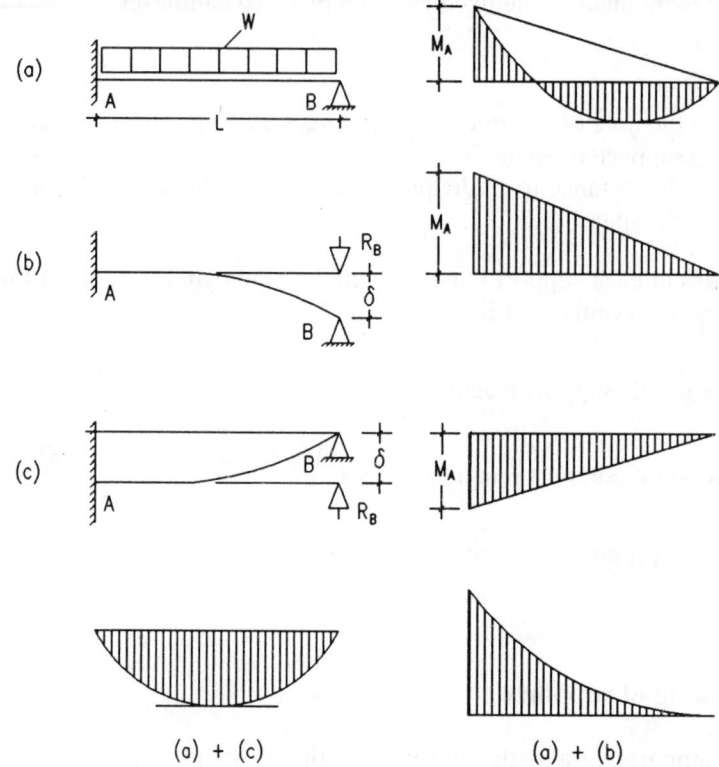

Fig. 10.2　Bending moment diagram for propped cantilever with sinking support

10.3.1 Beams with supports at the same level

The two conditions for solution, derived by Mohr, are:

(1) the area of the fixing or negative BM diagram is equal to that of the free or positive BM diagram
(2) the centres of gravity of the two diagrams lie in the same vertical line, i.e. are equidistant from a given end of the beam.

Figure 10.3 shows a typical BM diagram for a built-in beam.

ACDB is the diagram of the free moment M_s and the trapezium AEFB is the diagram of the fixing moment M_i, the portions shaded representing the final diagram.

Let A_s = the area of the free BM diagram
and A_i = the area of the fixing moment diagram.

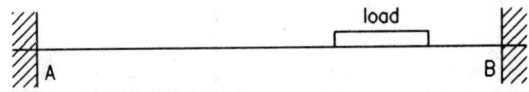

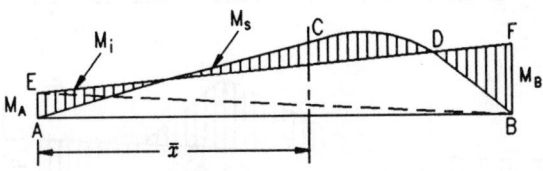

Fig. 10.3   Bending moment diagram for fixed-end beam

Then from condition (1) above, $A_s = A_i$, while from condition (2) their centres of gravity lie in the same vertical line, say, distance $\bar{x}$ from the left-hand support A.

Now　　　　　AE = the fixing moment M_A

and　　　　　BE = the fixing moment M_B.

Therefore　　$\dfrac{M_A + M_B}{2} \times L = A_i$

and　　　　　$M_A + M_B = \dfrac{2A_i}{L}$ 　　　　　　　　　**(10.1)**

Divide the trapezium AEFB by drawing the diagonal EB and take area moments about the support A.

Then　　$A_i\bar{x}$　　$= \left(\dfrac{M_A \times L}{2} \times \dfrac{L}{3} \right) + \left(\dfrac{M_B \times L}{2} \times \dfrac{2L}{3} \right)$

$$= \frac{L^2}{6}(M_A + 2M_B)$$

$$M_A + 2M_B = \frac{6A_i\bar{x}}{L^2} \qquad\qquad \textbf{(10.2)}$$

But　　$M_A + M_B = \dfrac{2A_i}{L}$ 　　　　　　　　　**(10.3)**

Also　　　　　$A_s = A_i$

Subtracting Equation (10.1) from Equation (10.2) and substituting A_s for A_i gives

$$M_B = \frac{6A_s\bar{x}}{L^2} - \frac{2A_s}{L}$$

Similarly　　$M_A = \dfrac{4A_s}{L} - \dfrac{6A_s\bar{x}}{L^2}$

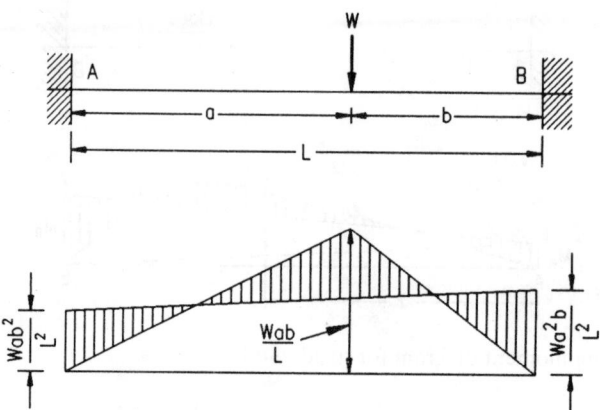

Fig. 10.4 Bending moment diagram for point load on fixed-end beam

It will be seen, therefore, that the fixing moments for any built-in beam on level supports can be calculated provided that the area of the free BM diagram and the position of its centre of gravity are known.

For point loads, however, the principle of reciprocal moments provides the simplest solution.

With reference to Fig. 10.4,

$$M_A = \frac{Wab}{L} \times \frac{b}{L} = \frac{Wab^2}{L^2}$$

$$M_B = \frac{Wab}{L} \times \frac{a}{L} = \frac{Wa^2b}{L^2}$$

i.e. the fixing moments are in reciprocal proportion to the distances of the ends of the beam from the point load.

In the case of several isolated loads, this principle is applied to each load in turn and the results summed.

It should be noted that appropriate formulae for built-in beams are given in the Appendix *Bending moment, shear and deflection tables for built-in beams.*

10.3.2 Beams with supports at different levels

The ends are assumed, as before, to be horizontal.

The bent form of the unloaded beam as shown in Fig. 10.5 is similar to the bent form of two simple cantilevers, which can be achieved by cutting the beam at the centre C, and placing downward and upward loads at the free ends of the cantilevers such that the deflection at the end of each cantilever is $d/2$.

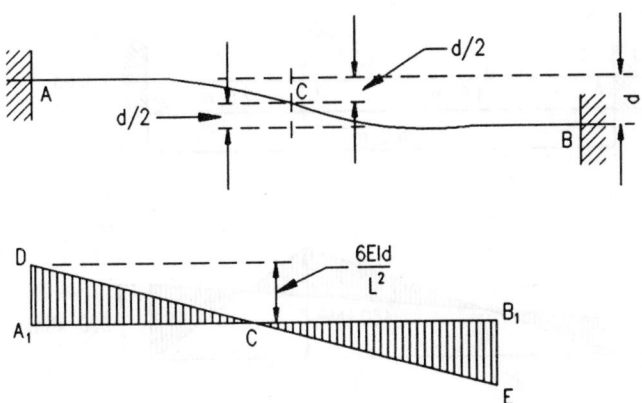

Fig. 10.5   Bending moment diagram for fixed-end beam with supports at different levels

Therefore $\dfrac{d}{2} = \dfrac{P(L/2)^3}{3EI}$ (being the standard deflection formula)

or $P = \dfrac{12EId}{L^3}$

This load would cause a BM at A or B equal to

$$P \times \frac{L}{2} = \frac{12EId}{L^3} \times \frac{L}{2} = \frac{6EId}{L^2}$$

The solution in any given case consists of adding to the ordinary diagram of BM, the BM diagram A_1DCEB_1.

Shear forces in fixed beams

In the case of fixed beams it is necessary to evaluate the BM before the SF can be determined. This is the converse of the procedure for the case of simply-supported beams.

The SF at the ends of a beam is found in the following manner:

$$SF_A = \text{the simple support reaction at A} = +\frac{M_A - M_B}{L}$$

$$SF_B = \text{the simple support reaction at B} = +\frac{M_B - M_A}{L}$$

where M_A and M_B are the numerical values of the moments at the ends of the beam.

These formulae must be followed exactly with respect to the signs shown since if M_A is smaller than M_B the signs will adjust themselves.

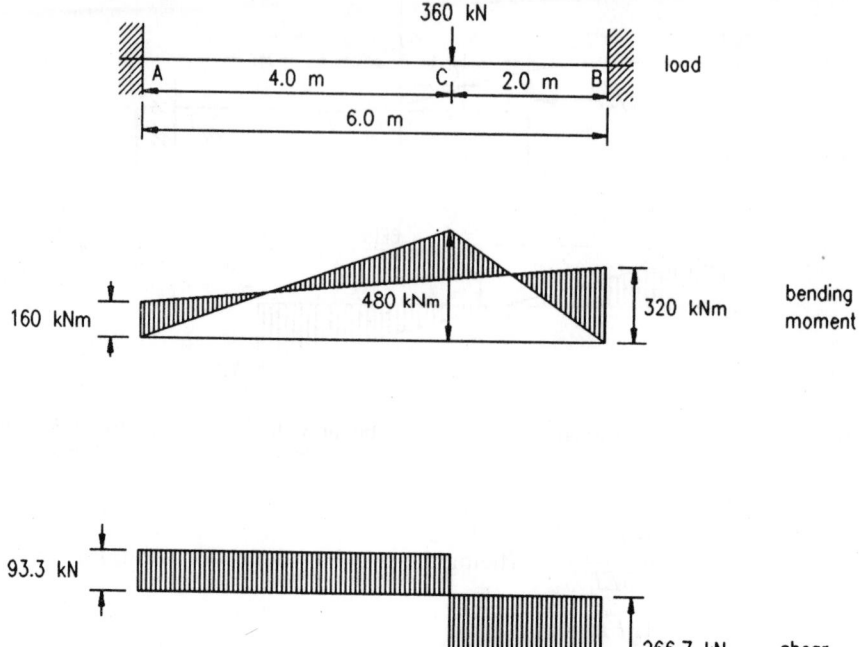

Fig. 10.6   Bending moment and shear force diagrams for fixed-end beam

It will be seen that for symmetrical loads, where $M_A = M_B$, the reactions will be the same as for simply-supported beams.

An example of bending moment and shear force diagrams for a built-in beam carrying a point load is given in Fig. 10.6.

10.4 Continuous beams

The solution of this type of beam consists, in the first instance, of the evaluation of the fixing or negative moments at the supports.

The most general method is the use of Clapeyron's Theorem of Three Moments.

The theorem applies only to any two adjacent spans in a continuous beam and in its simplest form deals with a beam which has all the supports at the same level, and has a constant section throughout its length.

The proof of the theorem results in the following expression:

$$M_A \times L_1 + 2M_B(L_1 + L_2) + M_C \times L_2 = 6\left(\frac{A_1 \times x_1}{L_1} + \frac{A_2 \times x_2}{L_2}\right)$$

where M_A, M_B and M_C are the numerical values of the hogging moments at the supports A, B and C respectively, and the remaining terms are illustrated in Fig. 10.7.

In a continuous beam the conditions at the end supports are usually known, and these conditions provide starting points for the solution.

The types of end conditions are three in number:

(1) simply supported
(2) partially fixed, e.g. a cantilever
(3) completely fixed, i.e. the end of the beam is horizontal as in the case of a fixed beam.

The SF at the end of any span is calculated after the support moments have been evaluated, in the same manner as for a fixed beam, each span being treated separately.

It is essential to note the difference between SF and reaction at any support, e.g. with reference to Fig. 10.7 the SF at support B due to span AB added to the SF at B due to span BC is equal to the total reaction at the support.

If the section of the beam is not constant over its whole length, but remains constant for each span, the expression for the moments is rewritten as follows:

$$M_A \times \frac{L_1}{I_1} + 2M_B\left(\frac{L_1}{I_1} + \frac{L_2}{I_2}\right) + M_C \times \frac{L_2}{I_2} = 6\left(\frac{A_1 \times x_1}{L_1 \times I_1} + \frac{A_2 \times x_2}{L_2 \times I_2}\right)$$

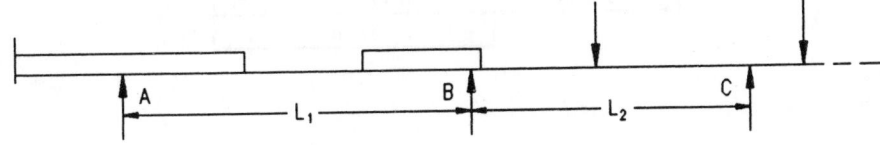

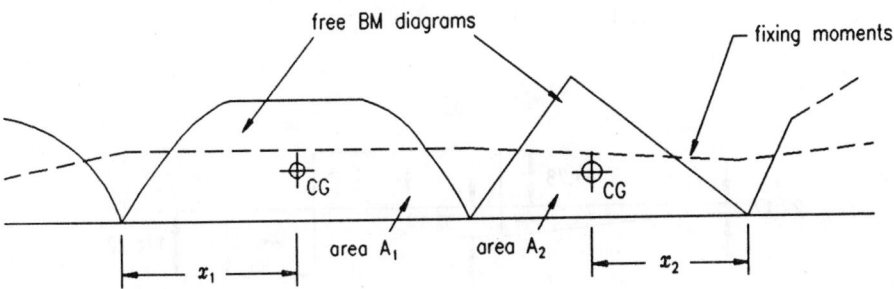

Fig. 10.7   Clapeyron's Theorem of Three Moments

in which I_1 is the second moment of area for span L_1 and I_2 is the second moment of area for span L_2.

Example

A two-span continuous beam ABC, of constant cross section, is simply supported at A and C and loaded as shown in Fig. 10.8.
 Applying Clapeyron's theorem,

$L_1 = 2.0\,\text{m}$

$L_2 = 3.0\,\text{m}$

$A_1 = \dfrac{10 \times 2}{8} \times \dfrac{2 \times 2}{3} = \dfrac{10}{3}\,\text{kN m}^2$

$A_2 = \dfrac{200 \times 1 \times 2}{3} \times \dfrac{3}{2} = 200\,\text{kN m}^2$

$x_1 = 1.0\,\text{m}$

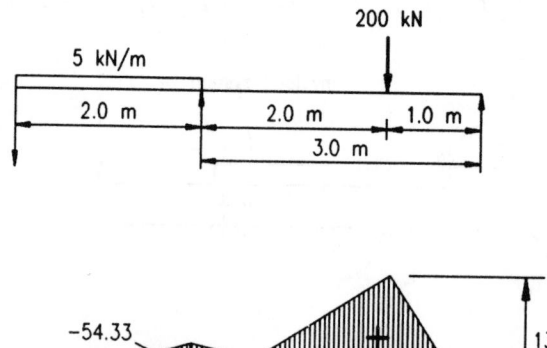

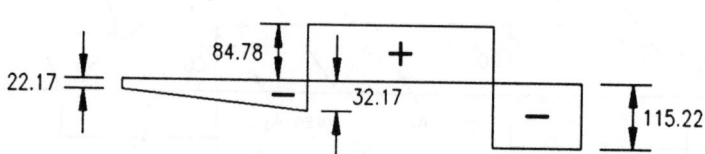

Fig. 10.8 Bending moment and shear force diagrams for two-span beam

$$x_2 = \frac{4.0}{3}\,\text{m}$$

Therefore $M_A \times 2 + 2M_B(2 + 3) + M_C \times 3 = 6\left(\dfrac{10 \times 1}{3 \times 2} + \dfrac{200 \times 4}{3 \times 3}\right)$

Since A and C are simple supports

$$M_A = M_C = 0$$

Therefore $M_B = \dfrac{6}{10}\left(\dfrac{10}{6} + \dfrac{800}{9}\right) = \dfrac{6}{10}(90.56) = 54.33\ \text{kN m}$

$\text{SF}_A = 5 + \dfrac{0 - 54.33}{2} = 5 - 27.17 = -22.17\ \text{kN}$

SF_B for span $AB = 5 + 27.17 = 32.17\ \text{kN}$

$\text{SF}_C = \dfrac{200 \times 2}{3} + \dfrac{0 - 54.33}{3} = 133.33 - 18.11 = 115.22\ \text{kN}$

SF_B for span $BC = \dfrac{200}{3} + 18.11 = 66.67 + 18.11 = 84.78\ \text{kN}$

Note that the negative reaction at A means that the end A will tend to lift off its support and will have to be held down.

10.5 Plastic failure of single members

The concept of the plastic hinge, capable of undergoing large rotation once the applied moment has reached the limiting value M_p, constitutes the basis of plastic design. This concept may be illustrated by examining the development of the collapse mode of a fixed-end beam subjected to a uniformly distributed load of increasing intensity w (Fig. 10.9(a)). Such a member is statically indeterminate, having three redundancies which however reduce to two unknowns if the axial thrust in the member is assumed to be zero. It will be assumed that the two unknown quantities are the fixing moments M_A and M_B. As the load increases, the beam initially behaves in an elastic manner and the value of the redundant moments can be derived by applying the three general conditions used in elastic structural analysis, namely those of

(1) equilibrium (application of statics)
(2) moment–curvature ($EI\ \mathrm{d}^2y/\mathrm{d}x^2 = M$)
(3) compatibility condition (continuity, including geometric conditions at the supports).

For the beam in Fig. 10.9(a), the first condition (equilibrium) is satisfied by drawing the bending moment diagram (shaded in Fig. 10.9(b)) as a superposition of

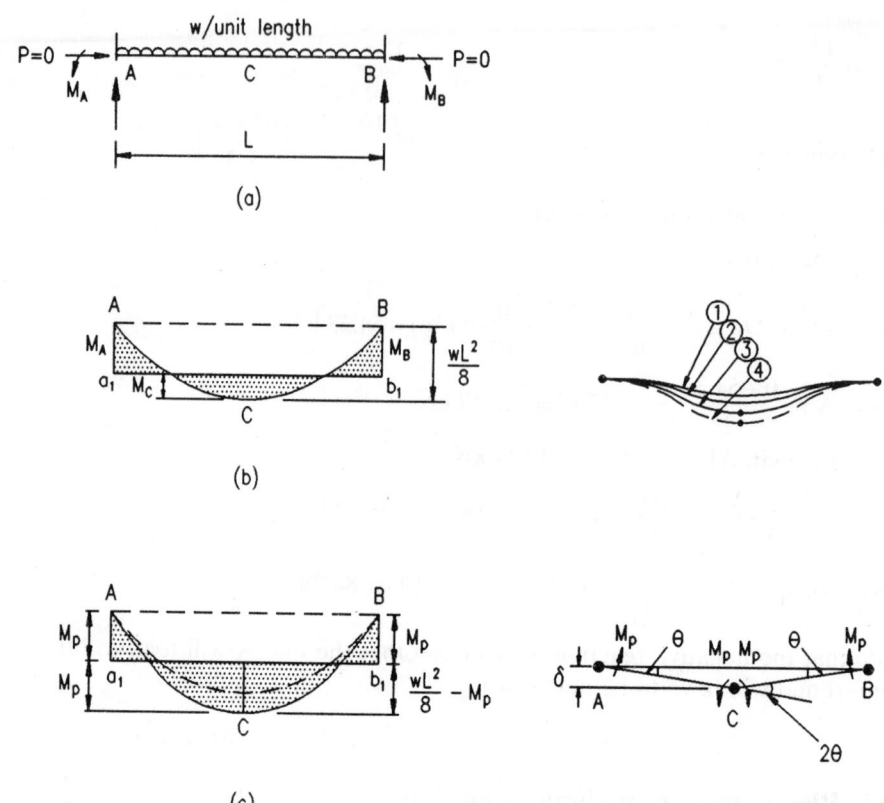

Fig. 10.9 Plastic failure of a fixed-end beam

the simply-supported sagging parabolic moment diagram ACB (peak value of $wL^2/8$) on the uniform moment diagram Aa_1b_1B due to the end hogging moments M_A and M_B. From the moment–curvature relation applied over the full length of the beam, it is readily deduced, by any one of a number of standard methods, that the rotations of the end sections at A (θ_A clockwise) and at B (θ_B anti-clockwise) and the central deflection Δ_C are given by

$$\theta_A = \frac{wL^3}{24EI} - \frac{M_A L}{3EI} - \frac{M_B L}{6EI} \tag{10.4}$$

$$\theta_B = \frac{wL^3}{24EI} - \frac{M_A L}{6EI} - \frac{M_B L}{3EI} \tag{10.5}$$

$$\Delta_C = \frac{5}{384}\frac{wL^4}{EI} - \frac{M_A L^2}{16EI} - \frac{M_B L^2}{16EI} \tag{10.6}$$

The compatibility conditions that now have to be satisfied are the directional restraint of the end sections of the beam, i.e. $\theta_A = \theta_B = 0$, giving the well-known

result $M_A = M_B = wL^2/12$. Equilibrium considerations now lead to the derivation of the central sagging moment, M_C. It follows from Fig. 10.9(b) that

$$M_C = \frac{wL^2}{8} - \left(\frac{M_A}{2} + \frac{M_B}{2}\right) \qquad (10.7)$$

whence $M_C = wL^2/24$, i.e. the end moments are twice the central moment. Finally, application of Equation (10.3) gives the central deflection as $wL^4/384EI$.

Suppose the load intensity w is increased until the fibres yield at some point in the beam. It is assumed that the cross section has an idealized moment–curvature relationship, i.e. the shape factor is unity (see Section 9.3.5). Up to this stage the ratio of the end to central moments remains at 2, as represented by the dashed line in Fig. 10.9(c). Due to the symmetry of the structure and loading, plastic hinges will develop simultaneously at the fixed ends, i.e. at the points of maximum moment. At the moment when the hinges form at the ends, the fixing moments have become equal to the M_p of the beam and the loading intensity w has reached a value of $12M_p/L^2$. A slight load increase then causes the plastic hinges to rotate while sustaining this constant moment M_p. This means that thereafter the beam behaves as a simply-supported beam with constant end moments of M_p. The structure is now statically determinate, and therefore the two degrees of redundancy in the original problem no longer exist. In other words, two plastic hinges have formed, eliminating a corresponding number of redundancies. At the same time two compatibility requirements have been eliminated; the condition that end slopes are zero is no longer correct because the ends of the member are now rotating as plastic hinges.

The central deflection at this stage, derived from Equation (10.6), has the value

$$\Delta_C = \frac{5}{384}\left(\frac{12M_p}{L^2}\right)\frac{L^4}{EI} - \frac{M_pL^2}{8EI} - \frac{M_pL^2}{32EI}$$

and, from Equation (10.7), the central sagging moment M_C becomes $[(wL^2/8) - M_p]$. Further increases in the load intensity cause a third plastic hinge to form at midspan: see the full line in Fig. 10.9(c). This means that the moment M_C attains a value of M_p and therefore at C

$$M_p = \frac{wL^2}{8} - M_p$$

hence

$$M_p = wL^2/16 \qquad \text{or} \qquad w = 16M_p/L^2$$

The ratio of the end and central moments is now unity; as a result of the formation of the plastic hinges there has been a redistribution of moments.

Substituting $w = 16M_p/L^2$ and $M_A = M_B = M_p$ in Equation (10.6) gives a central deflection of $\Delta_C = M_pL^2/12EI$. When the final hinge has formed the deflection increases rapidly without any further increase in the load. The beam is said to have failed as a hinged mechanism.

Now consider the same fixed-end beam with an initial settlement of $wL^4/144EI$ at end A (see Fig. 10.10(a)). The results derived from moment–curvature considerations [Equations (10.1)–(10.3)] can still be applied provided θ_A, θ_B and Δ_C are taken with reference to the chord AB between the ends of the member, the chord having rotated through an angle $wL^3/144EI$ relative to its initial position A_0B. Thus the compatibility conditions at the ends of the member are now

$$\theta_A = -\theta_B = wL^3/144EI$$

whence, by substitution in Equations (10.4), (10.5) and (10.7),

$$M_A = wL^2/24 \qquad M_B = wL^2/8 \qquad M_C = wL^3/24$$

The largest elastic moment occurs at B and therefore the first hinge starts there at a load intensity given by $M_p = wL^2/8$, i.e. $w = 8M_p/L^2$. Equation (10.6) shows that the central deflection is $M_pL^2/144EI$.

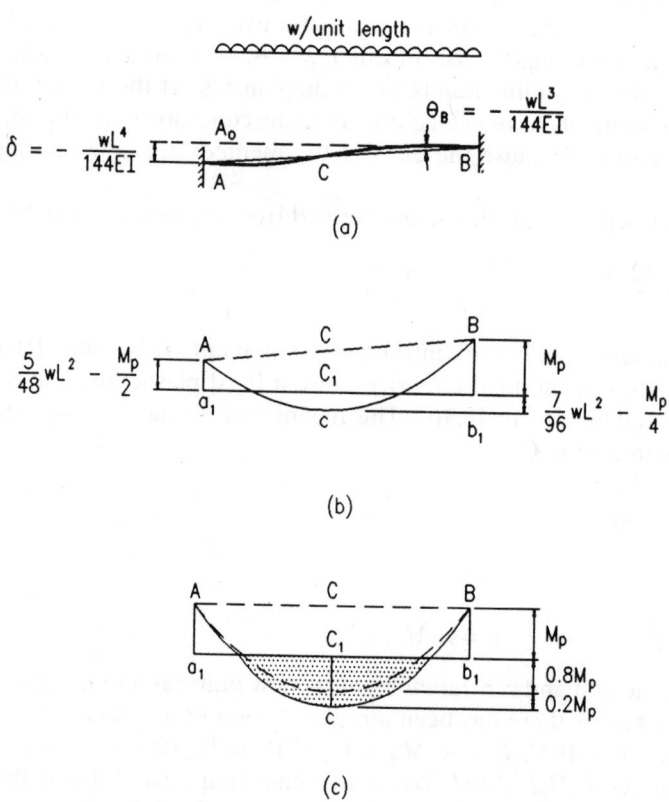

Fig. 10.10   Plastic failure of a fixed-end beam with initial settlement

The number of redundancies has now been reduced by one (since $M_B = M_p$) while the compatibility condition represented by Equation (10.7) no longer applies. Substitution of $\theta_A = wL^3/144EI$ and $M_B = M_p$ in Equations (10.4) and (10.7) gives

$$M_A = \frac{5wL^2}{48} - \frac{M_p}{2}$$

$$M_C = \frac{7wL^2}{96} - \frac{M_p}{4}$$

Using these new values for M_A and M_C the bending-moment diagram (Fig. 10.10(b)) is obtained, and inspection of the diagram shows that the second plastic hinge can be expected to occur at end A. Putting $M_A = M_p$ gives

$$w = \frac{3M_p}{2} \times \frac{48}{5L^2} = \frac{14.4M_p}{L^2} \qquad M_C = 0.8M_p$$

while Equation (10.3) gives $\Delta_C = M_p L^2/16EI$. The beam is now statically determinate with $M_A = M_B = M_p$, and eventually the third hinge forms at C when w reaches $16M_p/L^2$ (see Fig. 10.10(c)) and $\Delta_C = M_p L^2/12EI$. The important point to note is that, despite the difference in initial conditions, the failure pattern, the failure load and the deflection at the point of failure (relative to the ends) are the same for the two cases. The uniqueness of the plastic limit load, i.e. its independence of initial conditions of internal stress or settlement of supports, is a general feature of plastic analysis. Deflections at the point of collapse can however be affected, as indeed they are in the second case just described, when considered relative to the original support position A_0B (Fig. 10.10(a)).

10.6 Plastic failure of propped cantilevers

The case of the propped cantilever under a uniformly distributed load, Fig. 10.11(a), cannot be solved quite so simply, and both upper and lower bound methods described in Chapter 9 have to be used. A possible equilibrium condition is shown in Fig. 10.11(b) where the reactant line a_1B has been arranged so that the co-ordinate at the left-hand support is equal to the co-ordinate of the resultant moment diagram at mid-span. At the right-hand support the condition of zero resultant moment has to be satisfied. If the equal moments at A and C are regarded as plastic hinge (M_p) values, the mechanism condition is satisfied. Hence M_p is an upper bound (unsafe) value and should be denoted by M_u. Considering the geometry of the moment diagram at mid-span,

$$\frac{\lambda wL^2}{8} = Cc_1 + c_1c = \frac{M_u}{2} + M_u \qquad (M_u = M_p)$$

where λ is the load factor at rigid plastic collapse (see Section 9.3.6)

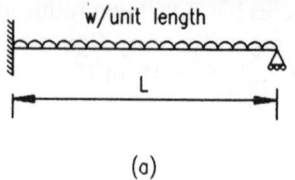

(a)

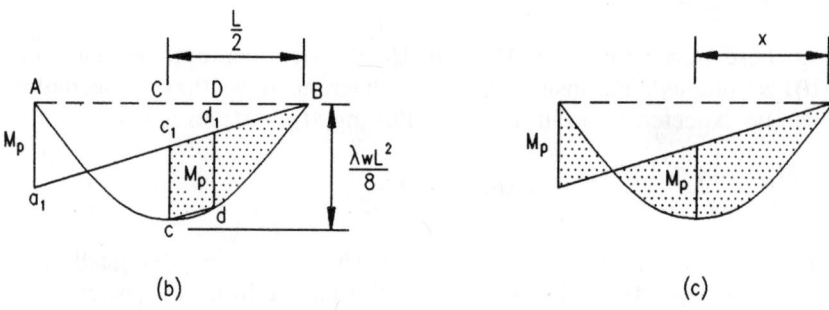

(b)   (c)

Fig. 10.11   Plastic failure of a propped cantilever

whence

$$M_c = \frac{\lambda w L^2}{12} = 0.0833\lambda w L^2$$

By a closer inspection of the resultant moment diagram, it is readily shown that the maximum resultant moment does not occur at the mid-span, but between C and D, at $5L/12$ from the propped support. The value of this moment is $\lambda w L^2/11.52$ (= $0.0868\lambda w L^2$), which is in excess of M_u. Note that this value is based on the moment at the fixed end being $\lambda w L^2/12$. By making the plastic moment of the beam equal to this higher value, a bending moment diagram satisfying the equilibrium and plastic moment condition is derived, i.e. a static solution $M_1 = 0.0868\lambda w L^2$. The required M_p lies between these limits, M_1 and M_u, i.e.

$$0.0833\lambda w L^2 \leq 0.0868\lambda w L^2$$

To obtain the exact value of M_p, the sagging hinge is positioned at an unknown distance x from the right-hand support (Fig. 10.11(c)). Considering the total ordinate of the free-moment diagram at this hinge position, then

$$\frac{x}{L}M_p + M_p = \frac{\lambda w L}{2}x - \frac{\lambda w}{2}x^2$$

whence

$$M_p = \frac{\lambda w L}{2}x\left(\frac{L-x}{L+x}\right) \tag{10.8}$$

As the mechanism condition is satisfied, then any value of x inserted into Equation (10.8) will give an upper bound solution, but the safest design will be achieved when M_p is a maximum. By differentiating with respect to x, the solution $x = (\sqrt{2} - 1)L = 0.414L$ is obtained; substituting for x in Equation (10.5) gives $M_p = \lambda w L^2/11.66 \ (= 0.0858\lambda w L^2)$. Note that this exact value of M_p lies within the range indicated by the upper and lower bounds previously calculated. This particular result is useful and can be applied directly to continuous beam problems where an end span carries a uniformly distributed load.

Further reading for Chapter 10

Horne M.R. & Morris L.J. (1981) *Plastic Design of Low-Rise Frames*. Constrado Monograph, Collins, London.
Kleinlogel A. (1948) *Mehrstielige Rahmen*, 6th edn. Ungar, New York.
Neal B.G. (1977) *The Plastic Methods of Structural Analysis*, 3rd edn. Chapman & Hall, London.

Chapter 11
Plane frame analysis

by JOHN RIGHINIOTIS

11.1 Formulae for rigid frames

11.1.1 General

The formulae given in this section are based on Professor Kleinlogel's *Rahmen-formeln* and *Mehrstielige Rahmen*.[1] The formulae are applicable to frames which are symmetrical about a central vertical axis, and in which each member has constant second moment of area.

Formulae are given for the following types of frame:

Frame I   Hingeless rectangular portal frame.
Frame II   Two-hinged rectangular portal frame.
Frame III   Hingeless gable frame with vertical legs.
Frame IV   Two-hinged gable frame with vertical legs.

The loadings are so arranged that dead, snow and wind loads may be reproduced on all the frames. For example, wind suction acting normal to the sloping rafters of a building may be divided into horizontal and vertical components, for which appropriate formulae are given, although all the signs must be reversed because the loadings shown in the tables act inwards, not outwards as in the case of suction.

It should be noted that, with few exceptions, the loads between node or panel points are uniformly distributed over the whole member. It is appreciated that it is normal practice to impose loads on frames through purlins, siderails or beams. By using the coefficients in Fig. 11.1, however, allowance can be made for many other symmetrically placed loads on the cross-beams of frames I and II shown, where the difference in effect is sufficient to warrant the corrections being made. The indeterminate BMs in the whole frame are calculated as though the loads were uniformly distributed over the beam being considered, and then all are adjusted by multiplying by the appropriate coefficient in Fig. 11.1. It may be of interest to state why these adjustments are made. In any statically indeterminate structure the indeterminate moments vary directly with the value of the following quantity:

$$\frac{\text{area of the free BM diagram}}{EI}$$

Where the loaded member is of constant cross section, EI may be ignored.

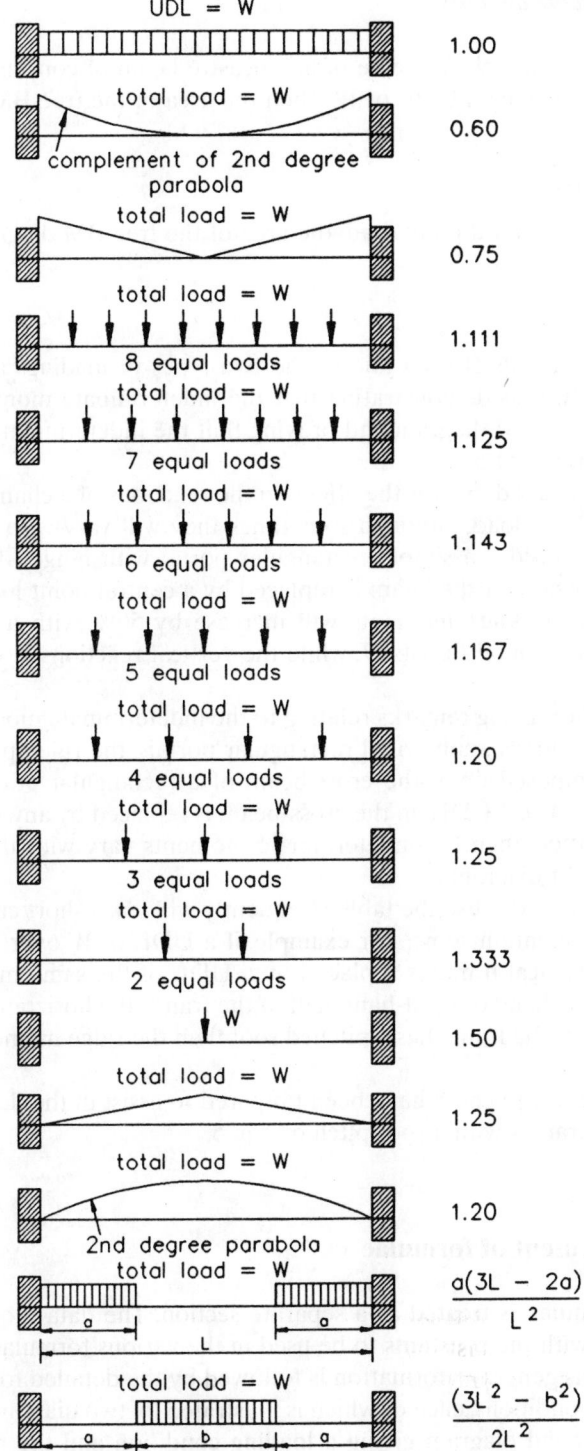

Fig. 11.1 Conversion coefficients for symmetrical loads

Consider, as an example, the case of an encastré beam of constant cross section and of length L carrying a UDL of W. Then the area of the free BM diagram is

$$\frac{WL}{8} \times \frac{2L}{3} = \frac{WL^2}{12}$$

If, however, W is a central point load, the area of the free BM diagram is

$$\frac{WL}{4} \times \frac{L}{2} = \frac{WL^2}{8}$$

The fixed end moments (FEM) due to the two types of loadings are $WL/12$ and $WL/8$ respectively, thus demonstrating that the indeterminate moments vary with the area of the free BM diagram and proving that the indeterminate moments are in the proportion of $1:1.5$.

No rules can be laid down for the effect on the reactions of a change in the mode of application of the load, although sometimes they will vary with the indeterminate moments. Consider a simple rectangular portal with hinged feet. If a UDL placed over the whole of the beam is replaced by a central point load of the same magnitude, then the knee moments will increase by 50% with a corresponding increase in the horizontal thrusts H, while the vertical reactions V will remain the same.

Although the foregoing remarks relating to the indeterminate moments resulting from symmetrical loads apply to all rectangular portals, the rule applies for asymmetrical loads imposed upon the cross-beam of a rectangular portal frame with hinged feet. If a vertical UDL on the cross-beam is replaced by any vertical load of the same magnitude, then the indeterminate moments vary with the areas of the respective free BM diagram.

No doubt readers who use the tables frequently will learn short cuts, but it is not inappropriate to mention some. For example, if a UDL of W over the whole of a single-bay symmetrical frame is replaced by a UDL of the same magnitude of W over either the left-hand or right-hand half of the frame, the horizontal thrust at the feet is unaltered. If the frame has a pitched roof then the ridge moment will also be unaltered.

The charts in the Appendix have been prepared to assist in the design of rectangular frames or frames with a roof pitch of 1 in 5.

11.1.2 Arrangement of formulae

Each set of formulae is treated as a separate section. The data required for each frame, together with the constants to be used in the various formulae, are given on the first page. This general information is followed by the detailed formulae for the various loading conditions, each of which is illustrated by two diagrams placed side-by-side, the left-hand diagram giving a loading condition and the right-hand one giving the appropriate BM and reaction diagram. It should be noted, however, that

some BMs change their signs as the frames change their proportions. This will be appreciated by examining the charts.

For simple frames, i.e. for single-storey frames, the formulae for reactions immediately follow the formulae for BMs for each load.

Considering the simple frames only, the type of formula depends on the degree of indeterminacy and the shape of the frame. Auxiliary coefficients X are introduced whenever the direct expressions become complicated or for other reasons of expediency.

No hard and fast rules can be laid down for the notation and it must be noted that each set of symbols and constants applies only to the particular frame under consideration, although, of course, an attempt has been made to produce similarity in the types of symbols.

11.1.3 Sign conventions

All computations must be carried out algebraically, hence every quantity must be given its correct sign. The results will then be automatically correct in sign and magnitude.

The direction of the load or applied moment shown in the left-hand diagram for each load condition is considered to be positive. If the direction of the load or moment is reversed, the signs of all the results obtained from the formulae as printed must be reversed.

For simple frames, the moments causing tension on the inside faces of the frame are considered to be positive. Upward vertical reactions and inward horizontal reactions are also positive.

For multi-storey or multi-bay frames the same general rules apply to moments and vertical reactions.

11.1.4 Checking calculations for indeterminate frames

Calculations for indeterminate frames may be checked by using some other method of analysis, but it is also possible to check any frame or portion of a frame, such as that above the line AB in Fig. 11.2, by ensuring that the following rules are obeyed:

(1) the three fundamental statical equations, i.e. $\Sigma H = 0$, $\Sigma V = 0$ and $\Sigma M = 0$, have been satisfied, and, in addition, either that
(2) the sum of the areas of the M/EI diagram above any line, such as AB, is zero if A and B are fully fixed; or
(3) the sum of the moments, with respect to the base AB, of the areas of the M/EI diagram above the line AB is zero if A and B are partially restrained (as shown in Fig. 11.2) or are hinged.

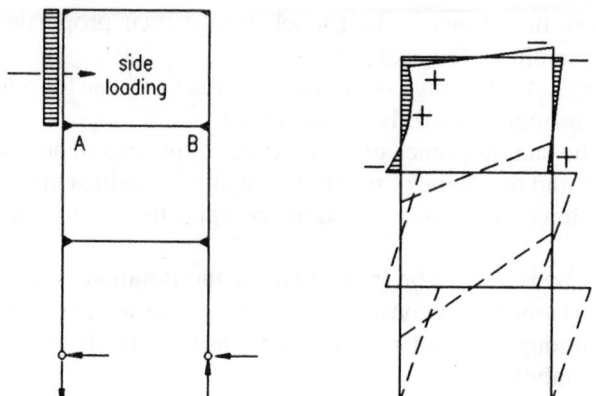

Fig. 11.2 Checking an indeterminate frame

The underlying principles in rules (2) and (3) above are those used in the application of the 'column analogy' method of analysis.

As an example of rule (2), consider the frame in Fig. 11.3 where EI is constant. Then the sum of the areas of the M/EI diagram, considering the legs first, is

$$\frac{2}{EI}\left[\frac{(+0.0736-0.0826)\times 6.0}{2}\right]+\frac{2}{EI}\left[\frac{(-0.0826+0.0893)\times 8.078}{2}\right]$$

$$=\frac{-0.054+0.054}{EI}=0$$

thus demonstrating that the moments calculated are correct.

Now consider the frame in Fig. 11.4 as an example for rule (3).

Then the sum of the moments of the areas of the M/EI diagram, working from A round to D, is

$$\frac{1}{EI}\left\{\left[\frac{11.25\times 4.8}{2}\times\frac{2\times 4.8}{3}\right]+\left[\frac{6\times 4.8\times 2}{3}\times\frac{4.8}{2}\right]\right.$$

$$+\left[\left(\frac{11.25-12.75}{2}\right)9.6\times 4.8\right]+\left[\frac{-12.75\times 4.8}{2}\times\frac{2\times 4.8}{3}\right]\right\}=0$$

$$\frac{1}{EI}\left[\frac{4.8^2\times 2}{2\times 3}(11.25+6-4.5-12.75)\right]=0$$

demonstrating again that the calculations are correct.

11.2 Portal frame analysis

The design process explained here assumes that the reader is familiar with the basic concepts of limit state steel design, and the basic analysis and design of continuous

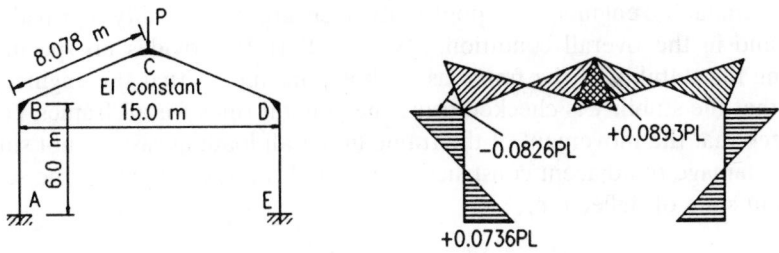

Fig. 11.3   Checking a single-bay portal – rule (2)

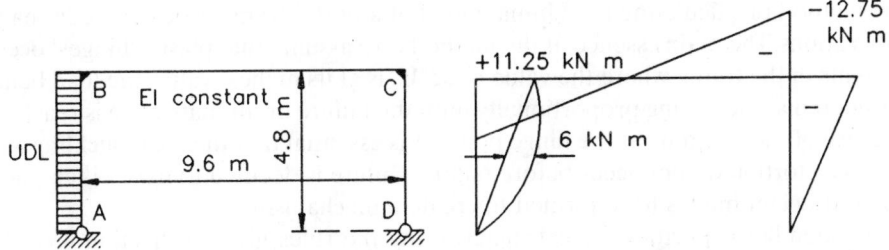

Fig. 11.4   Checking a rectangular portal – rule (3)

beams. These notes take the form of a worked example. A method of determining preliminary member sizes has been included.

11.2.1 Methods of analysis

BS 5950: Part 1[2] allows two main methods of analysis of a structure:

(1) *Linear elastic.* The frame is analysed either by hand or by computer assuming linear elastic behaviour. Once the forces, moments and shears have been derived by elastic analysis the ultimate capacity of each section is checked using the rules given in Section 4 of the Code.
(2) *Simple plastic theory.* The frame is analysed using the basic principles of simple plastic theory. Once the forces, moments and shears have been derived by analysis the member capacities are checked. Those containing plastic hinges are checked in accordance with Section 4.

Elastic analysis

Using elastic analysis it is to be expected that the structure will be heavier than that designed by plastic methods, but less stability bracing will be needed. It may well be that the final details will also be more simple.

It will remain the engineer's responsibility to ensure that stability is provided both locally and in the overall condition. BS 5950: Part 1[2] provides no specific rules regarding the stability of the frame as a whole; this means that the engineer must ensure that the stability is checked using the general rules for all frames. He must also check that the movement of the frame under all loading cases is not sufficient to cause damage to adjacent construction, i.e. brick walls or cladding, the serviceability limit state of deflection.

Plastic analysis

The method of calculating the ultimate load of a portal frame is described in many publications. The main essence of the method is to assume that plastic 'hinges' occur at points in the frame where the value of M/M_p is at its highest value, the load being considered as increasing proportionally until the failure or ultimate state is reached. Because of the straining at the hinge points it is essential that the local buckling and lateral distortion do not occur before failure. Failure is deemed to have taken place when sufficient hinges have formed to create a mechanism.

The member capacities are calculated using the rules given in Section 4 of the Code but with the additional restrictions applied to hinge positions. In addition, positive requirements are put on checking frame stability for both single-bay and multi-bay frames. Plastic designed frames are lighter than elastic designed frames, providing deflection is not a governing point; however, additional bracing may well be required.

11.2.2 Stability

With the use of lighter frames, various aspects of stability take a more prominent part in the design procedures. As far as portal frames are concerned the following areas are important:

(1) overall frame stability, in that the strength of the frame should not be affected by changes in geometry during loading ($P\Delta$ effect)
(2) snap-through stability, in multi-bay frames (three or more), where the effects of continuity can result in slender rafters
(3) plastic hinge stability, where the member must be prevented from moving out of plane or rotating at plastic hinges
(4) rafter stability, ensuring that the rafter is stable in bending as an unrestrained beam
(5) leg stability, where the leg below the plastic hinge must be stable
(6) haunch stability, where the tapered member is checked to ensure that the inner (compression flange) is stable.

11.2.3 Selecting suitable members for a trial design

The design of a portal frame structure is in reality a process of selecting suitable members and then proving their ability to perform in a satisfactory manner. Inexperienced engineers can be given some guidance to estimate initial member sizes. In order to speed the initial selection of members, three graphs have been produced to enable simple pin-based frames to be sized quickly.

These graphs have been prepared making the following assumptions:

(1) plastic hinges are formed at the bottom of the haunch in the leg and near the apex in the rafter, the exact position being determined by the frame geometry
(2) the depth of the rafter is approximately span/55 and the depth of the haunch below the eaves intersection is 1.5 times rafter depth
(3) the haunch length is 10% of the span of the frame, a limit generally regarded as providing a balance between economy and stability
(4) the moment in the rafter at the top of the haunch is $0.87M_p$, i.e. it is assumed that the haunch area remains elastic
(5) the calculations assume that the calculated values of M_p are provided exactly by the sections and that there are no stability problems. Clearly these conditions will not be met, and it is the engineer's responsibility to ensure that the chosen sections are fully checked for all aspects of behaviour.

The graphs cover the range of span/eaves height between 2 and 5 and rise/span of 0 to 0.2 (where 0 is a flat roof). Interpolation is permissible but extrapolation is not. The three graphs give:

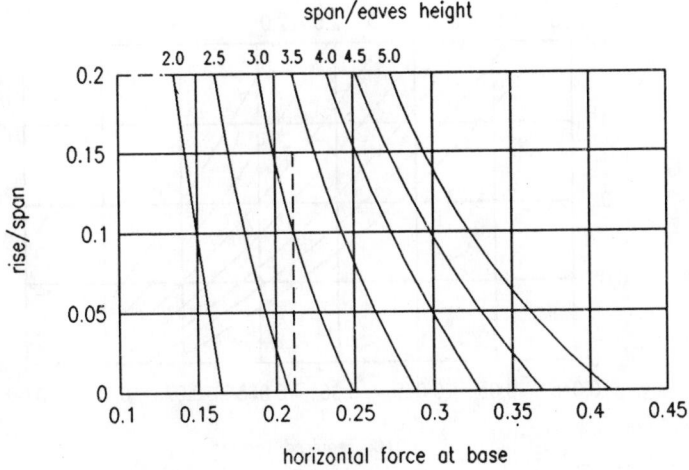

Fig. 11.5   Rise/span against horizontal force at base for various span/eaves heights

Figure 11.5: the horizontal force at the feet of the frame as a proportion of the total factored load wL, where w is the load/unit length of rafter and L is the span of the frame

Figure 11.6: the value of the moment capacity required in the rafters as a proportion of the load times span wL^2

Figure 11.7: the value of the moment capacity required in the legs as a proportion of the load times span wL^2.

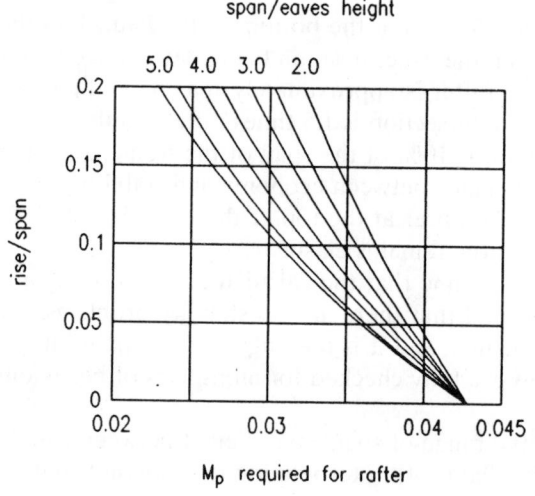

Fig. 11.6 Rise/span against required M_p of rafter for various span/eaves heights

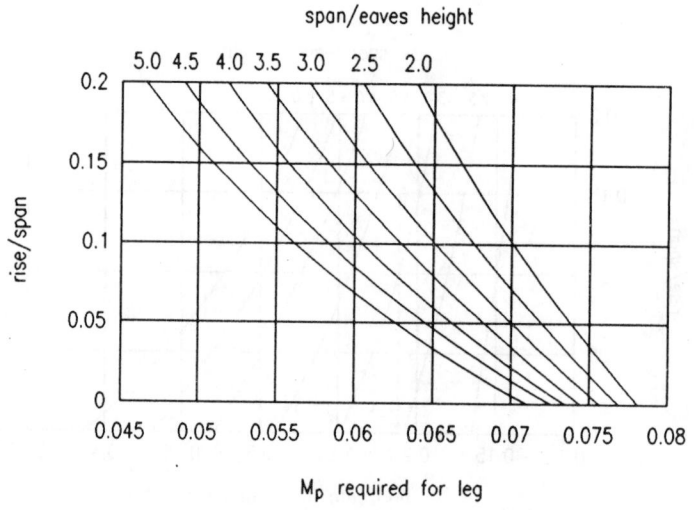

Fig. 11.7 Rise/span against required M_p of leg for various span/eaves heights

The graphs are non-dimensional and may be used with any consistent set of units. In the worked example kilonewtons and metres are used.

Method of use of the graphs

(1) Determine the ratio span/height to eaves.
(2) Determine the ratio rise/span.
(3) Calculate wL (total load) and wL^2.
(4) Look up the values from the graphs.
(5) Horizontal force at root of frame = value from Fig. 11.5 $\times wL$.
(6) M_p required in rafter = value from Fig. 11.6 $\times wL^2$.
(7) M_p required in leg = value from Fig. 11.7 $\times wL^2$.

11.2.4 Worked example of plastic design (see Fig. 11.8)

Determination of member sizes

Although the engineer may use his experience or other methods to determine pre-liminary member sizes, the graphs in Figs 11.5, 11.6 and 11.7 will be used for this example. In order to use these graphs, four parameters are required:

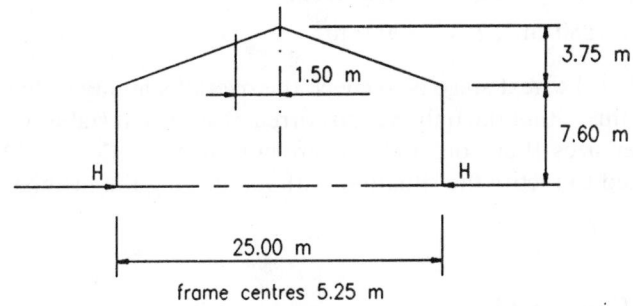

loading	unfactored design	load factor	factored load	total factored load
imposed	0.75 kN/m²	1.6	1.2 kN/m²	158 kN
dead	0.43 kN/m²	1.4	0.6 kN/m²	79 kN
total				237 kN

factored load/m (w) = 237/25 = 9.48 kN/m

Fig. 11.8 Portal frame design example

span/height to eaves	$= 25/7.6$	$= 3.29$
rise/span	$= 3.75/25$	$= 0.15$
wL (total load on frame)	$= 9.48 \times 25$	$= 237$ kN
wL^2	$= 9.48 \times 25^2$	$= 5925$ kN m

Horizontal thrust at feet of frame (Fig. 11.5)

$$= 0.21 \times 237 \qquad = 49.8 \text{ kN m}$$

Moment capacity of rafter (Fig. 11.6)

$$= 0.0305 \times 5925 \qquad = 181 \text{ kN m}$$

Moment capacity of leg (Fig. 11.7)

$$= 0.059 \times 5925 \qquad = 350 \text{ kN m}$$

Assuming a design strengh of 275 N/mm^2 (S275 steel):

S_x required for rafter $= 181 \times 1000/275 = 658$ cm^3

S_x required for leg $\quad = 350 \times 1000/275 = 1270$ cm^3

Trial sections (*NB* these are first trials and may not be adequate):

Rafter $\quad 406 \times 140 \times 39$ kg UB (S275 steel);
$\qquad\qquad S_x = 721$ cm^3, $I_x = 12\,500$ cm^4

Leg $\qquad 457 \times 152 \times 60$ kg UB (S275 steel);
$\qquad\qquad S_x = 1280$ cm^3, $I_x = 25\,500$ cm^4

It is suggested that the next stage is to check the overall stability of the frame. The main reason for this is that the only way to correct insufficient stability is to change the main member sizes. If any other checks are not satisfied, additional bracing can frequently be used to rectify the situation, without altering the member sizes.

References to Chapter 11

1. Kleinlogel A. (1931) *Mehrstielige Rahmen*. Ungar. New York.
2. British Standards Institution (2000) *Structural use of steelwork in building*. Part 1: *Code of practice for design in simple and continuous construction: hot rolled sections*. BS 5950, BSI, London.

Further reading for Chapter 11

Baker J.F. (1954) *The Steel Skeleton, Vol. 1*, 1st edn. Cambridge University Press.

Baker J.F., Horne M.R. & Heyman J. (1956) *The Steel Skeleton, Vol. 2.* Cambridge University Press.

Horne M.R. & Morris L.S. (1981) *Plastic Design of Low-Rise Frames.* Constrado Monograph, Collins, London.

Neal B.G. (1977) *The Plastic Methods of Structural Analysis*, 3rd edn. Chapman & Hall, London.

Chapter 12
Applicable dynamics

by MICHAEL WILLFORD

12.1 Introduction

The dynamic performance of steel structures has traditionally only been an area of interest for special classes of structure, particularly slender wind-sensitive structures (masts, towers and stacks), structures supporting mechanical equipment, offshore structures and earthquake-resistant structures. However, a wider interest in dynamic behaviour has recently come about as a result of the widespread adoption of long span composite floors in buildings, for which vibration criteria must be satisfied.

Dynamic loads in structures may arise from a number of sources including the following:

Forces generated inside a structure – Machinery
– Impacts
– Human activity (walking, dancing, etc.)

External forces – Wind buffeting and other aerodynamic effects
– Waves (offshore structures)
– Impacts from vehicles, etc.

Ground motions – Earthquakes
– Ground-borne vibration due to railways, roads, pile driving, etc.

The principal effects of concern are:

Strength – The structure must be strong enough to resist the peak dynamic forces that arise.

Fatigue – Fatigue cracks can initiate and propagate when large numbers of cycles of vibration inducing significant stress are experienced, leading to reduction in strength and failure.

Perception – Human occupants of a building can perceive very low amplitudes of vibration, and, depending on the circumstances, may find vibration objectionable. Certain items of precision equipment are also extremely sensitive to vibration. Perception will generally be the most onerous dynamic criterion in occupied buildings.

It is beyond the scope of this book to address all these issues, and references are suggested for more detailed guidance. The following sections are intended to give an overview of the fundamental features of dynamic behaviour and to introduce some of the terminology employed and analysis procedures available. It must be borne in mind that dynamic behaviour is influenced by a larger number of parameters than static behaviour, and that some of the parameters cannot be predicted precisely at the design stage. It is often advisable to investigate the effects of varying initial assumptions to ensure that the most critical situations that may occur have been examined.

12.2 Fundamentals of dynamic behaviour

The principal features of dynamic behaviour may be illustrated by the examination of a very simple dynamic system, a concentrated mass M supported on a light cantilever of flexural rigidity EI and length L as illustrated in Fig. 12.1. The mass is subjected to forces P which vary with time t.

12.2.1 Dynamic equilibrium

One of the basic methods of dynamic analysis is the examination of dynamic equilibrium to formulate an equation of motion. Consider the dynamic equilibrium of the mass illustrated in Fig. 12.1. If at some time t the mass is displaced upwards from its static equilibrium position by y, and has velocity $\dot{y}$ and acceleration $\ddot{y}$ (positive upwards) the mass is in general subjected to the following forces:

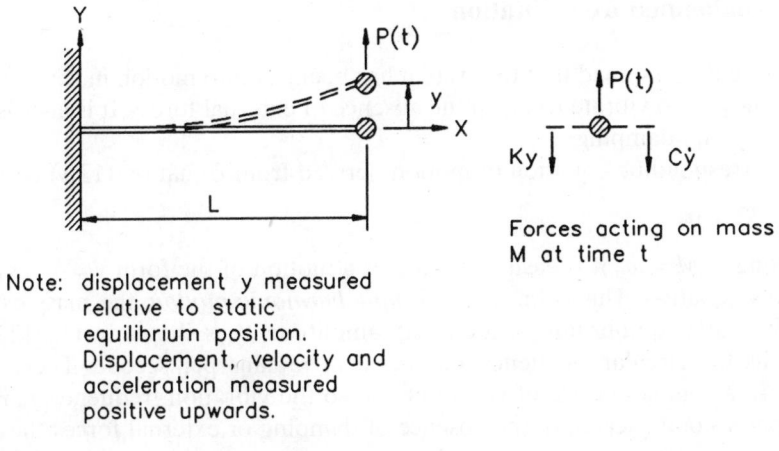

Note: displacement y measured
relative to static
equilibrium position.
Displacement, velocity and
acceleration measured
positive upwards.

Fig. 12.1   Example of a dynamic system

External force $P(t)$

Stiffness force Ky ($K = 3EI/L$ for uniform cantilever)

Damping force $C\dot{y}$ (a dissipative force assumed to act in the opposite direction to the velocity).

The resultant of these forces will cause the mass to accelerate according to Newton's 2nd law of motion. Hence the dynamic equilibrium equation may be written as

$$M\ddot{y} = P(t) - Ky - C\dot{y}$$

or

$$M\ddot{y} + C\dot{y} + Ky = P(t) \tag{12.1}$$

Equation (12.1) is the general equation of motion for a single degree of freedom dynamic system (a system whose behaviour can be defined by a single quantity, in this case the deflection of the mass, y). It is clearly most important that the quantities in this equation are defined in dynamically consistent units. Examples of consistent units are given in Table 12.1.

Table 12.1

Mass	Force	Displacement	Time
kg	N	m	s
tonnes	kN	m	s

Solutions to Equation (12.1) for different assumptions regarding the force $P(t)$ can give an insight into the principal features of dynamic behaviour.

12.2.2 Undamped free vibration

In this case it is assumed that the system has been set into motion in some way and is then allowed to vibrate freely in the absence of external forces. It is also assumed that there is no damping.

The corresponding equation of motion derived from Equation (12.1) is:

$$M\ddot{y} + Ky = 0 \tag{12.2}$$

By putting $K/M = \omega_n^2$ it is easily shown that a motion of the form $y = Y\cos\omega_n t$ satisfies this equation. This is known as *simple harmonic motion*; the mass oscillates about its static equilibrium position with amplitude Y as shown in Fig. 12.2. ω_n is known as the circular frequency, measured in radians per second. There are 2π radians in a complete cycle of vibration and so the vibration frequency, f_n, is $\omega_n/2\pi$ cycles per second (hertz). In the absence of damping or external forces the system will vibrate in this manner indefinitely.

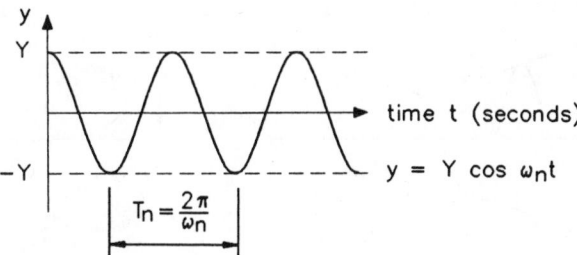

Fig. 12.2 Simple harmonic motion

The amplitude of vibration Y is the peak displacement of the mass relative to its static equilibrium position. Vibration amplitudes are sometimes referred to as root mean square (rms) quantities, where $y_{rms} = (1/T)\sqrt{(\int_0^T y^2 \, dt)}$, and T is the total time over which the vibration is considered. For continuous simple harmonic motion $y_{rms} = Y/\sqrt{2}$.

12.2.3 Damped free vibration

Energy is always dissipated to some extent during vibration of real structures. Inclusion of the damping force in the free vibration equation of motion leads to

$$M\ddot{y} + C\dot{y} + Ky = 0 \qquad (12.3)$$

The solution to this equation for a lightly damped system when the mass is initially displaced by Y and then released is

$$y = Ye^{-\xi\omega_n t} \cos(\omega_d t - \phi) \qquad (12.4)$$

The motion takes the form shown in Fig. 12.3(a), and it can be seen that the vibration amplitude decays exponentially with time. The rate of decay is governed by the amount of damping present.

If the damping constant C is sufficiently large then oscillation will be prevented and the motion will be as in Fig. 12.3(b). The minimum damping required to prevent overshoot and oscillation is known as critical damping, and the damping constant for critical damping is given by $C_o = 2\sqrt{(KM)}$, where K and M are the stiffness and mass of the system.

Practical structures are lightly damped, and the damping present is often expressed as a proportion of critical. In Equation (12.4)

$$\xi = \text{critical damping ratio} = \frac{C}{C_o}$$

The phase shift angle ϕ is small when the damping is small.

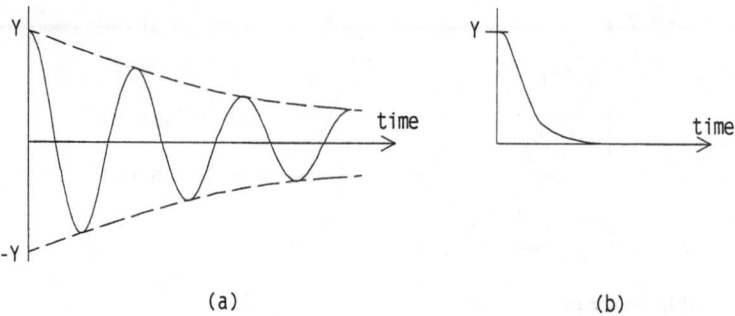

Fig. 12.3   Damped free vibration: (a) lightly damped system, (b) critically damped system

The frequency of damped oscillation is

$$\omega_d = \omega_n \left(1 - \xi^2\right)^{\frac{1}{2}} \text{ radians/s}$$

where ω_n is the undamped natural frequency. With low damping ($\xi \ll 1$) the reduction of natural frequency (increase in natural period) resulting from damping is negligible.

12.2.4 Response to harmonic loads

A load which varies sinusoidally with time at a constant frequency is known as a harmonic load. This form of dynamic load is characteristic of machinery operating at constant speed, and many other types of continuous vibration can be approximated to this form.

When such a force of amplitude P and frequency f Hz (or $\omega = 2\pi f$ radians/s) is applied to the simple structure described in previous sections the structure will be caused to vibrate. After some time the motions of the structure will reach a steady state: that is, vibration of a constant amplitude and frequency will be achieved. The dynamic equation of motion is:

$$M\ddot{y} + C\dot{y} + Ky = P\cos\omega t \qquad\qquad (12.5)$$

and it can be shown that the steady-state response motion is described by:

$$y = Y\cos(\omega t - \phi) \qquad\qquad (12.6)$$

Note that the steady-state vibration occurs at the frequency of the harmonic force exciting the motion, not at the natural frequency of the structure.

The displacement amplitude Y can be shown to be:

$$Y = \frac{P}{K} \cdot \frac{1}{\left\{ \left[1 - \left(\frac{\omega}{\omega_n} \right)^2 \right]^2 + \left(2\xi \frac{\omega}{\omega_n} \right)^2 \right\}^{\frac{1}{2}}} \tag{12.7}$$

P/K is the deflection of the structure under a static force P, and the multiplying term can be regarded as a magnification factor. The variation of magnification factor with frequency ratio and damping ratio is shown in Fig. 12.4.

When the dynamic force is applied at a frequency much lower than the natural frequency of the system ($\omega/\omega_n \ll 1$), the response is quasi-static; the response is governed by the stiffness of the structure and the amplitude is close to the static deflection P/K.

When the dynamic force is applied at a frequency much higher than the natural frequency ($\omega/\omega_n \gg 1$), the response is governed by the mass (inertia) of the structure; the amplitude is less than the static deflection.

When the dynamic force is applied at a frequency close to the natural frequency, the stiffness and inertia forces in the vibrating system are almost equal and opposite at any instant, and the external force is resisted by the damping force. This is the condition known as *resonance*, when very large dynamic magnification factors are possible. When damping is low the maximum steady-state magnification occurs when $\omega \approx \omega_n$, and then

$$Y = \frac{P}{K} \frac{1}{2\xi} \tag{12.8}$$

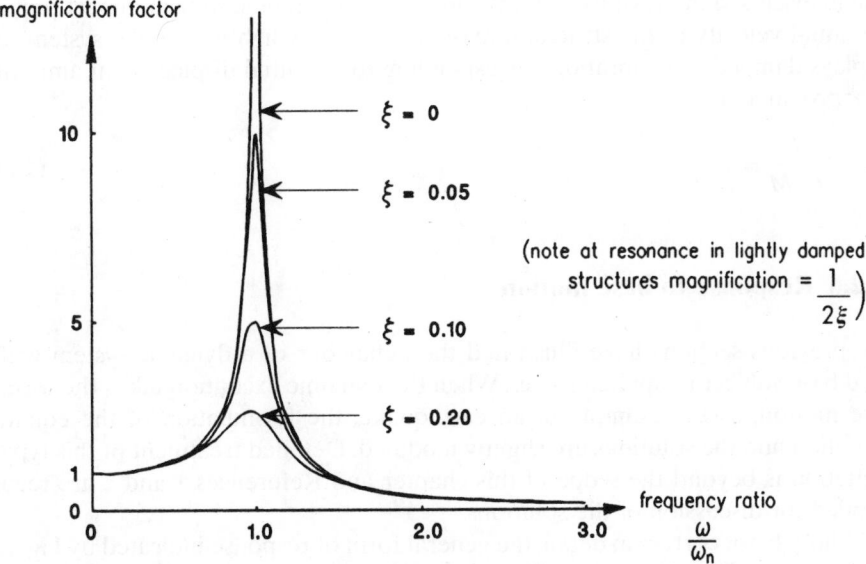

Fig. 12.4   Steady-state response to harmonic loads

Since in many structural systems ξ is of the order of 0.01, magnification factors of the order of 50 may result. The force in the structure is proportional to the displacement so the same magnification factor applies to structural forces.

Human perception of motion is usually related to acceleration levels rather than displacement. The peak acceleration amplitude at steady state is given by

$$\hat{a} = \omega^2 Y$$
$$= (2\pi f)^2 Y \tag{12.9}$$

In practice, there is usually advantage in avoiding the possibility of resonance whenever possible by ensuring that structural frequencies are well away from the frequencies of any known sources of substantial dynamic force.

12.2.5 Response to an impact

Another dynamic loading case of interest is the response of a structure to an impact, say from an object falling on to the structure. A full discussion of impact loading is given in Reference 1, but a simple approximate method is useful for many practical situations when the mass of the impacting object is small compared with the mass of the structure, and the impact duration is short compared with the natural period of the structure. In these cases the effect of the impact can be assessed as an impulse I acting on the structure. The magnitude of I may be calculated as $m \, \Delta v$, where m is the mass of the falling object and Δv its change in velocity at impact. If there is no rebound Δv can be taken as the approach velocity. For the simple system discussed in previous sections conservation of momentum at impact requires the initial velocity of the structural mass to be I/M. A lightly damped system then displays damped free vibration corresponding to an initial displacement amplitude of approximately

$$Y = \frac{I}{\omega_n M} = \frac{I}{2\pi f_n M} \tag{12.10}$$

12.2.6 Response to base motion

The previous sections have illustrated the behaviour of a dynamic system with a fixed base subject to applied forces. When the dynamic excitation takes the form of base motion, as for example in an earthquake, the formulation of the equation of motion and the solutions are slightly modified. Detailed treatment of this type of excitation is beyond the scope of this chapter and References 1 and 2 are recommended for discussion of the solutions.

Although not correct in detail, the general form of response indicated by Fig. 12.4 for harmonic applied forces is still relevant. Resonance occurs for those components of the base motion close in frequency to the natural frequency of the structure.

12.2.7 Response to general time-varying loads

Although some dynamic loads (e.g. impacts and harmonic loads) are simple and can be dealt with analytically, many forces and ground motions that occur in practice are complex.

In general, numerical analysis procedures are required for the evaluation of responses to these effects, and these are often available in finite element analysis programs.

The techniques employed include:

* direct step-by-step integration of the equations of motion using small time increments,
* Fourier analysis of the forcing function followed by solution for Fourier components in the frequency domain,
* for random forces, random vibration theory and spectral analysis.

The background and application of these techniques are discussed in References 1 and 3.

12.3 Distributed parameter systems

The dynamic system considered in section 12.2 was simple in that its entire mass was concentrated at one point. This enabled Newton's second law (and therefore the equation of motion) to be written in terms of a single variable.

Although some practical structures can be represented adequately in this way, in other cases it is not realistic to assume that the mass is concentrated at one point. The classical treatment of distributed systems is illustrated below by analysis of a uniform beam; similar techniques can be used for two-dimensional (plate) structures. The analysis calculates the natural frequency of the beam and, by reference to section 12.2.2, this can be done by ignoring external forces and damping forces.

12.3.1 Dynamic equilibrium

The beam under consideration is shown in Fig. 12.5. For a small element of the beam of length dx the variation of shear force V along the beam results in a net force on the element of $\dfrac{\partial V}{\partial x}\,dx$ which must cause the mass $m\,dx$ to accelerate.

Dynamic equilibrium therefore requires:

$$\frac{\partial V}{\partial x}\,dx = m\,dx\,\frac{\partial^2 y}{\partial t^2} \tag{12.11}$$

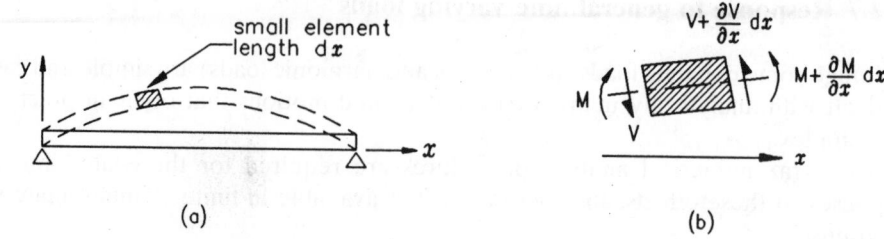

Fig. 12.5   Dynamic equilibrium of vibrating beams. (a) Uniform beam mass/unit length = m, (b) small element mass = $m\,dx$

For a flexural beam the deflection y is a function of the bending moment distribution, with:

$$\frac{\partial^2 y}{\partial x^2} = \frac{M}{EI}$$

noting also that for moment equilibrium on the element:

$$V = -\frac{\partial M}{\partial x}$$

$$\frac{\partial V}{\partial x} = -\frac{\partial^2 M}{\partial x^2}$$

Substituting these relationships into Equation (12.11) leads to the equation of motion for a flexural beam:

$$\frac{\partial^4 y}{\partial x^4} + \frac{m}{EI}\frac{\partial^2 y}{\partial t^2} = 0 \tag{12.12}$$

It is sometimes convenient to represent an entire framed structure as a shear beam. In this case the deflection y is a function of the shear force, with

$$\frac{\partial y}{\partial x} = \frac{V}{GA_s}$$

The equation of motion for a shear beam is therefore:

$$\frac{\partial^2 y}{\partial x^2} - \frac{m}{GA_s}\frac{\partial^2 y}{\partial t^2} = 0 \tag{12.13}$$

12.3.2 Modes of vibration

The partial differential equations derived above can be solved for given sets of boundary conditions. The solution in each case identifies a family of modes of vibra-

tion (eigenvectors) with corresponding natural frequencies (eigenvalues). The mode shape describes the relative displacements of the various parts of the beam at any instant in time.

A number of standard results are given in Tables 12.2 and 12.3, and more comprehensive lists are contained in References 4 and 5.

Table 12.2 Natural frequencies of uniform flexural beams

Length = L (m) (typical units)
Flexural rigidity = EI (Nm^2)
Mass/unit length = m (kg/m)

$$f_n = \frac{k_n}{2\pi}\sqrt{\left(\frac{EI}{mL^4}\right)}\ \text{Hz: values of } k_n \text{ given below}$$

	Mode	Shape	k_n	Nodal points at $x/L =$			
Simply-supported	1		9.87	0	1		
	2		39.5	0	0.5	1	
	3		88.8	0	0.333	0.667	1
	n	$x = \sin\left(n\pi\frac{x}{L}\right)$	$n^2\pi^2$	0	$\frac{1}{n}$ ... $\frac{n-1}{n}$	1	
Encastré	1		22.4	0	1		
	2		61.7	0	0.5	1	
	3		121	0	0.359	0.641	1
Cantilever	1		3.52	0			
	2		22.0	0	0.774		
	3		61.7	0	0.5	0.868	

Table 12.3 Natural frequencies for uniform shear cantilevers

$$\text{Length} = L \quad \text{(m) (typical units)}$$
$$\text{Shear rigidity} = GA_s \quad \text{(N)}$$
$$\text{Mass/unit length} = m \quad \text{(kg/m)}$$

$$f_n = \frac{k_n}{2\pi}\sqrt{\left(\frac{GA_s}{mL^2}\right)} \text{ Hz: values of } k_n \text{ given below}$$

Mode	Shape	k_n	Nodal points at $x/L =$
1		1.57	0
2		4.71	0 0.667
3		7.85	0 0.4 0.8
n	$y = \sin\left((2n-1)\dfrac{\pi x}{2L}\right)$	$(2n-1)\dfrac{\pi}{2}$	$0 \quad \dfrac{2}{(2n-1)} \ \dfrac{4}{(2n-1)} \ \dfrac{6}{(2n-1)}$ etc.

12.3.3 Calculation of responses

For linear elastic behaviour, once the mode shapes and frequencies have been established, dynamic responses can be calculated treating each mode as a single degree of freedom system such as the one described in section 12.2.2. In theory, distributed systems have an infinite number of modes of vibration, but in practice only a few modes, usually those of lowest frequency, will contribute significantly to the overall response.

It is convenient to describe a mode shape by a displacement parameter ϕ defined (or normalized) such that the maximum value at any point is 1.0. If the mode is excited by dynamic forces resulting in a modal response amplitude of Y_m, then the displacement amplitude at a point i on the structure is

$$y_i = \phi_i Y_m$$

where ϕ_i is the mode shape value at the point.

The magnitude of Y_m can be calculated for simple dynamic loads treating the mode as a single degree of freedom system having the following properties. These are often referred to as modal or generalized properties:

modal mass $M^* = \Sigma M_i \phi_i^2$ sum over whole structure
modal stiffness $K^* = \omega_n^2 M^*$
modal force $P^* = \Sigma P_i \phi_i$ sum for all loaded points

Note that for mode n of a uniform simply-supported beam of span L and mass/unit length m with $\phi = \sin(n\pi x/L)$, the modal mass M^* is:

$$M^* = m \int_0^L \left[\sin\left(n\pi x/L\right)\right]^2 dx$$

which is $0.5\,mL$ (half the total mass of the beam).

The application of these concepts to the problems of floor vibration and wind-induced vibration is described in References 6 and 7.

12.3.4 Approximate methods to determine natural frequency

Approximate methods are useful for estimating the natural frequencies of structures not conforming with one of the special cases for which standard solutions exist, and for checking the predictions of computer analyses when these are used.

One of the most useful approximate methods relates the natural frequency of a system to its static deflection under gravity load, δ. With reference to section 12.2.2 the natural frequency of a single lumped mass system is:

$$f_n = \frac{1}{2\pi}\sqrt{\left(\frac{K}{M}\right)}$$

This may be rewritten, replacing the mass term M by the corresponding *weight Mg*, as:

$$f_n = \frac{\sqrt{g}}{2\pi}\sqrt{\left(\frac{K}{Mg}\right)}$$

Since Mg/K is the static deflection under gravity load (δ):

$$f_n = \frac{\sqrt{g}}{2\pi\sqrt{\delta}}$$

$$= 15.76/\sqrt{\delta} \quad \text{when } \delta \text{ is measured in mm} \tag{12.14}$$

This formula is exact for any single lumped mass system.

For distributed parameter systems a similar correspondence is found, although the numerical factor in Equation (12.14) varies from case to case, generally between 16 and 20. For practical purposes a value of 18 will give results of sufficient accuracy.

When applying these formulae the following points should be noted.

(1) The static deflection should be calculated assuming a weight corresponding to the loading for which the frequency is required. This is usually a dead load with an allowance for expected imposed load.
(2) For horizontal modes of vibration (e.g. lateral vibration of an entire structure) the gravity force must be applied laterally to obtain the appropriate lateral deflection, as shown in Fig. 12.6(a).

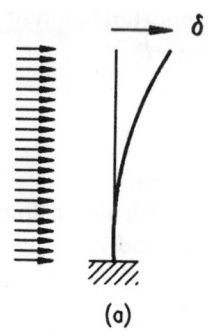

(a)

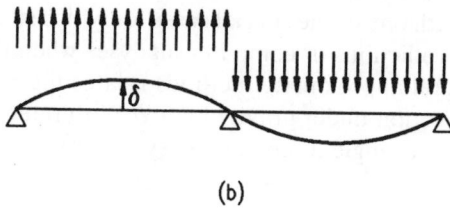

(b)

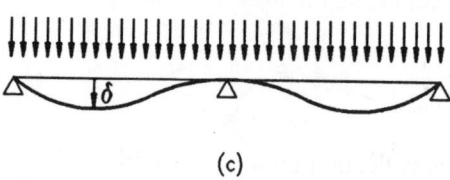

(c)

Fig. 12.6 Use of gravity deflections to estimate natural frequencies. In each case $f_n \simeq 18/\sqrt{\delta}$. (a) For a vertical structure the gravity load is applied horizontally. (b) Continuous structure gravity load applied in opposite directions on alternate spans. (c) Continuous structure – symmetric mode has higher frequency.

(3) The mode shape required must be carefully considered in multi-span structures. In the two-span beam of Fig. 12.6(b) the lowest frequency will correspond to an asymmetrical mode as illustrated; the corresponding δ must be obtained by applying gravity in opposite directions on the two spans. The normal gravity deflection will correspond to the symmetrical mode with a higher natural frequency, Fig. 12.6(c).

These concepts can be extended to estimating the natural frequencies of primary beam – secondary beam systems. If the static deflection of the primary beam is δ_p and the static deflection of a secondary beam is δ_s (relative to the primary) the combined natural frequency is approximately

$$f_n = \frac{18}{\sqrt{(\delta_p + \delta_s)}}$$

from which it can be shown that

$$\frac{1}{f_n^2} = \frac{1}{f_p^2} + \frac{1}{f_s^2} \tag{12.15}$$

where f_p and f_s are the natural frequencies of the primary and secondary beams alone.

Care must always be taken in using these formulae so that a realistic mode shape is implied. There will generally be continuity between adjacent spans at small amplitudes even in simply-supported designs, and in many situations a combined mode where both the primaries and secondaries are vibrating together in a 'simply-supported' fashion is not possible.

12.4 Damping

Damping arises from the dissipation of energy during vibration. A number of mechanisms contribute to the dissipation, including material damping, friction at interfaces between components and radiation of energy from the structure's foundations.

Material damping in steel provides a very small amount of dissipation and in most steel structures the majority of the damping arises from friction at bolted connections and frictional interaction with non-structural items, particularly partitions and cladding. Damping is found to increase with increasing amplitude of vibration.

The amount of damping that will occur in any particular structure cannot be calculated or predicted with a high degree of precision, and design values for damping are generally derived from dynamic measurements on structures of a corresponding type.

Damping can be measured by a number of methods, including:

- rate of decay of free vibration following an impact (Fig. 12.3(a))
- forced excitation by mechanical vibrator at varying frequency to establish the shape of the steady-state resonance curve (Fig. 12.4)
- spectral methods relying on analysis of response to ambient random vibration such as wind loading.

All these methods can run into difficulty when several modes close in frequency are present. One result of this is that on floor structures (where there are often several closely spaced modes) the apparent damping seen in the initial rate of decay after impact can be substantially higher than the true modal damping.

Table 12.4 Typical modal damping values by structure type

Structure type	Structure damping (% critical)
Unclad welded steel structures (e.g. steel stacks)	0.3%
Unclad bolted steel structures	0.5%
Composite footbridges	1%
Floor (fitted out), composite and non-composite	1.5%–3% (may be higher when many partitions on floor)
Clad buildings (lateral sway)	1%

Damping is usually expressed as a fraction or percentage of critical (ξ), but the logarithmic decrement (δ) is also used. The relationship between the two expressions is $\xi = \delta/2\pi$.

Table 12.4 gives typical values of modal damping that are suggested for use in calculations when amplitudes are low (e.g. for occupant comfort). Somewhat higher values are appropriate at large amplitudes where local yielding may develop, e.g. in seismic analysis.

12.5 Finite element analysis

Many simple dynamic problems can be solved quickly and adequately by the methods outlined in previous sections. However, there are situations where more detailed numerical analysis may be required and finite element analysis is a versatile technique widely available for this purpose. Numerical analysis is often necessary for problems such as:

(1) determination of natural frequencies of complex structures
(2) calculation of responses due to general time-varying loads or ground motions
(3) non-linear dynamic analysis to determine seismic performance.

12.5.1 Basis of the method

As explained in Chapter 9 the finite element method describes the state of a structure by means of deflections at a finite number of node points. Nodes are connected by elements which represent the stiffness of the structural components.

In static problems the equilibrium of every degree of freedom at the nodes of the idealization is described by the stiffness equation:

$$F = KY$$

where F is the vector of applied forces, Y is the vector of displacements for every degree of freedom, and K is the stiffness matrix. Solution of unknown displacements for a known force vector involves inversion of the stiffness matrix.

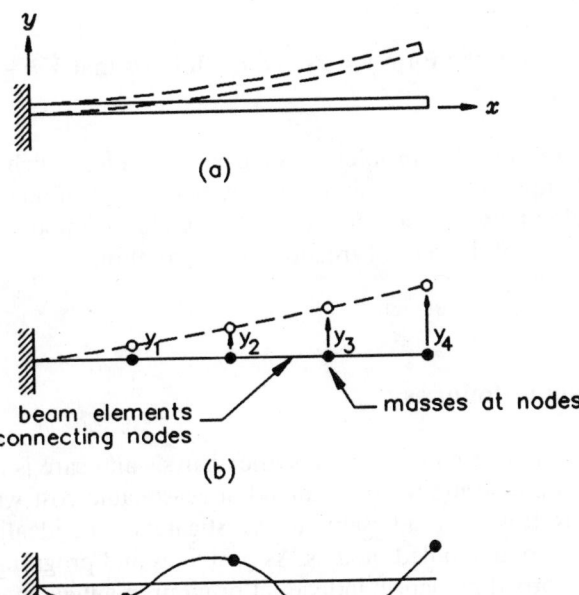

Fig. 12.7   Finite element idealization. (a) Uniform beam in free vibration, (b) finite element representation, (c) fourth mode not accurately represented

The extension of the method to dynamic problems can be visualized in simple terms by considering the dynamic equilibrium of a vibrating structure.

Figure 12.7(a) shows the instantaneous deflected shape of a vibrating uniform cantilever, and Fig. 12.7(b) shows a finite element idealization of this condition. The shape is described by the deflections of the nodes, Y, and a mass is associated with each degree of freedom of the idealization.

At the instant considered when the deflection vector is Y the forces at the nodes provided by the stiffness elements must be KY.

These forces may in part be resisting external instantaneous nodal forces P and may in part be causing the mass associated with each node to accelerate. The equations of motions of all the nodes may therefore be written as

$$P - M\ddot{Y} = KY$$

or

$$KY + M\ddot{Y} = P$$

where M is the mass matrix, $\ddot{Y}$ is the acceleration vector, and P is the external force vector.

The natural frequencies and mode shapes are obtained by solving the undamped free vibration equations:

$$KY + M\ddot{Y} = 0$$

Assuming a solution of the form $y = Y\cos\omega_n t$, it follows that $\ddot{Y} = -\omega_n^2 Y$ and hence

$$[K - \omega_n^2 M]Y = 0$$

This is a standard eigenvalue problem of matrix algebra for which various numerical solution techniques exist. The solution provides a set of mode shape vectors Y with corresponding natural frequencies ω_n. The number of modes possible will be equal to the number of degrees of freedom in the solution.

12.5.2 Modelling techniques

Dynamic analysis is more complex than static analysis and care is required so that results of appropriate accuracy are obtained at reasonable cost when using finite element programs. It is often advisable to investigate simple idealizations initially before embarking upon detailed models. As problems and programs vary it is possible to give only broad guidance; individual program manuals must be consulted and experience with the program being used is invaluable. More detailed background is given in Reference 8.

The first stage in any dynamic analysis will invariably be to obtain the natural frequencies and mode shapes of the structure. As can be seen from Fig. 12.7(c) a given finite element model will represent higher modes with decreasing accuracy. If it is only necessary to obtain a first mode frequency accurately then a relatively coarse model, such as that illustrated in Fig. 12.7(b), will be perfectly adequate. In order to obtain an accurate estimate of the fourth mode a greater subdivision of the structure would be necessary, since the distribution of inertia load along the uniform beam in this mode is not well represented by just four masses.

Probably the most widely used approach for eigenvalue problems is subspace iteration. This is a robust solution method which maps problems with a large number of degrees of freedom on to a 'subspace', with a much smaller number of degrees of freedom, to reduce the problem size. The structure's eigenvalues (frequencies) are the same as those of the subspace, and the eigenvectors (modeshapes) of the structure are calculated from the eigenvectors of the subspace. This method resolves the smallest eigenvalues with the highest accuracy. Most eigensolvers will also include Sturm sequence checks to ensure that the eigenvalues found are the ones required, and that there are none missing from the sequence.

Two approaches exist for calculating the element mass matrix. The consistent mass matrix is the most accurate way of representing the mass and inertia of the element, but when the aim is to calculate the dynamic response of the structure the simpler lumped mass approach can be more effective as it avoids single element modes of vibration.

It is important to ensure that all the relevant mass is accounted for in a modal dynamic analysis. In many cases the engineer starts from a model, built for static

analysis, where 'mass' is applied as loading. It is important to ensure that all the mass present is correctly included as mass (not loading) in the dynamic analysis.

Clearly, dynamically consistent units must be used throughout, and these units may not be the same as those used for a static analysis using the same model. A hand check of the first mode frequency using an approximate or empirical method is strongly advisable to ensure that the results are realistic. In addition, there is even more need than with static analysis to view computer analysis results as approximate. It is very difficult to predict natural frequencies of real structures with a high degree of precision unless the real boundary conditions and structural stiffness can be defined with confidence. This is rarely the case and these are uncertainties that finite element analysis cannot resolve.

12.6 Dynamic testing

Calculation of the dynamic properties and dynamic responses of structures still presents some difficulties, and testing and monitoring of structures has a significant role in structural dynamics. Testing is the only way by which the damping of structures can be obtained, by which analytical methods can be calibrated and many forms of dynamic loading can be estimated. It is often an essential part of the assessment and improvement of structures where dynamic response is found to be excessive in practice. Further details are contained in references 9 and 10.

References to Chapter 12

1. Clough R.W. & Penzien J. (1993) *Dynamics of Structures*, 2nd edn. McGraw-Hill.
2. Dowrick D.J. (1987) *Earthquake Resistant Design for Engineers and Architects*, 2nd edn. John Wiley & Sons.
3. Warburton G.B. (1976) *The Dynamical Behaviour of Structures*, 2nd edn. Pergamon Press, Oxford.
4. Harris C.M. & Crede C.E. (1976) *Shock and Vibration Handbook*. McGraw-Hill.
5. Roark R.J. & Young W.C. (1989) *Formulas for Stress and Strain*, 6th edn. McGraw-Hill.
6. Construction Industry Research & Information Association (CIRIA)/The Steel Construction Institute (SCI) (1989) *Design Guide on the Vibration of Floors*. SCI Publication 076, SCI, Ascot, Berks.
7. Bathe K.J. (1996) *Finite Element Procedures*. Prentice Hall, Englewood Cliffs, NJ.
8. Bachmann H. (1995) *Vibration Problems in Structures*. Birkhauser Verlag AG.
9. Ewins D.J. & Inman D.J. (2001) *Structural dynamics @2000: Current status and future directions*. Research Studies Press, Baldock.

10. Ewins D.J. (2001) *Modal Testing: Theory, practice and application*, 2nd edn. Research Studies Press, Baldock.

Further reading for Chapter 12

Blevins R.D. (1995) *Formulas for natural frequency and mode shape* (corrected edition). Krieger, Malubar, FL.

Chopra A.K. (2001) *Dynamics of Structures – Theory and Applications to Earthquake Engineering*, 2nd edn. Prentice Hall, Upper Saddle River, NJ.

National Building Code of Canada (1995) Commentary A – Serviceability Criteria for Deflections and Vibrations.

Chapter 13
Local buckling and cross-section classification

by DAVID NETHERCOT

13.1 Introduction

The efficient use of material within a steel member requires those structural properties which most influence its load-carrying capacity to be maximized. This, coupled with the need to make connections between members, has led to the majority of structural sections being thin-walled as illustrated in Fig. 13.1. Moreover, apart from circular tubes, structural steel sections (such as universal beams and columns, cold-formed purlins, built-up box columns and plate girders) normally comprise a series of flat plate elements. Simple considerations of minimum material consumption frequently suggest that some plate elements be made extremely thin but limits must be imposed if certain potentially undesirable structural phenomena are to be avoided. The most important of these in everyday steelwork design is local buckling.

Figure 13.2 shows a short UC section after it has been tested as a column. Considerable distortion of the cross-section is evident with the flanges being deformed out of their original flat shape. The web, on the other hand, appears to be comparatively undeformed. The buckling has therefore been confined to certain plate elements, has not resulted in any overall deformation of the member, and its centroidal axis has not deflected. In the particular example of Fig. 13.2, local buckling did not develop significantly until well after the column had sustained its 'squash load' equal to the product of its cross-sectional area times its material strength. Local buckling did not affect the load-carrying capacity because the proportions of the web and flange plates are sufficiently compact. The fact that the local buckling appeared in the flanges before the web is due to these elements being the more slender.

Terms such as *compact* and *slender* are used to describe the proportions of the individual plate elements of structural sections based on their susceptibility to local buckling. The most important governing property is the ratio of plate width to plate thickness, β, often referred to as the *b/t ratio*. Other factors that have some influence are material strength, the type of stress system to which the plate is subjected, the support conditions provided, and whether the section is produced by hot-rolling or welding.

Although the rigorous treatment of plate buckling is a mathematically complex topic,[1] it is possible to design safely and in most cases economically with no direct consideration of the subject. For example, the properties of the majority of standard hot-rolled sections have been selected to be such that local buckling effects are

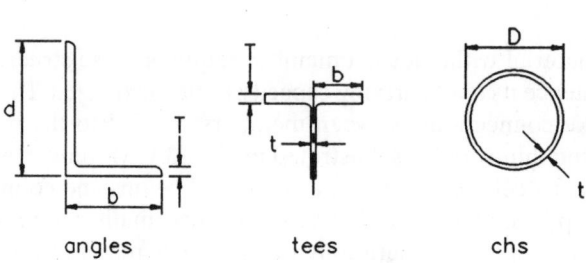

rolled beams and columns

rolled channels

rhs   b=B−3t
       d=D−3t

angles

tees

chs

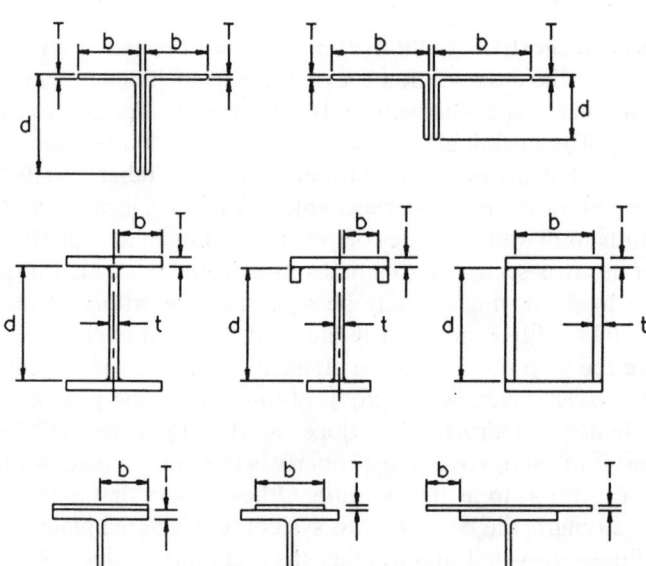

Fig. 13.1   Structural cross-sections

Fig. 13.2 Local buckling of column flange

unlikely to affect significantly their load-carrying capacity when used as beams or columns. Greater care is, however, necessary when using fabricated sections for which the proportions are under the direct control of the designer. Also, cold-formed sections are often proportioned such that local buckling effects must be accounted for.

13.2 Cross-sectional dimensions and moment–rotation behaviour

Figure 13.3 illustrates a rectangular box section used as a beam. The plate slenderness ratios for the flanges and webs are b/T and d/t, and elastic stress diagrams for both components are also shown. If the beam is subject to equal and opposite end moments M, Fig. 13.4 shows in a qualitative manner different forms of relationship between M and the corresponding rotation θ.

Assuming d/t to be such that local buckling of the webs does not occur, which of the four different forms of response given in Fig. 13.4 applies depends on the compression flange slenderness b/T. The four cases are defined as:

(a) $b/T \leq \beta_1$, full plastic moment capacity M_p is attained and maintained for large rotations and the member is suitable for plastic design – *plastic* cross-section (Class 1).

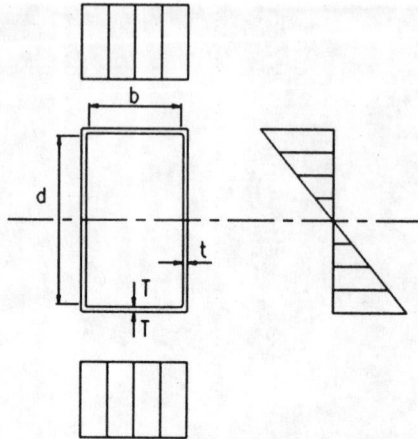

Fig. 13.3   Rectangular hollow section used as a beam

(b) $\beta_1 < b/T < \beta_2$, full plastic moment capacity M_p is attained but is only maintained for small rotations and the member is suitable for elastic design using its full capacity – *compact* cross-section (Class 2).

(c) $\beta_2 < b/T \le \beta_3$, full elastic moment capacity M_y (but not M_p) is attained and the member is suitable for elastic design using this limited capacity – *semi-compact* cross-section (Class 3).

(d) $\beta_3 < b/T$, local buckling limits moment capacity to less than M_y – *slender* cross-section (Class 4).

The relationship between moment capacity M_u and compression flange slenderness b/T indicating the various β limits is illustrated diagrammatically in Fig. 13.5. In the figure the value of M_u for a semi-compact section is conservatively taken as the moment corresponding to extreme fibre yield M_y for all values of b/T between β_2 and β_3. This is more convenient for practical calculation than the more correct representation shown in Fig. 13.4 in which a moment between M_y and M_p is indicated. Since the classification of the section as plastic, compact, etc., is based on considerations of the compression flange alone, the assumption concerning the web slenderness d/t is that its classification is the same as or better than that of the flange. For example, if the section is semi-compact, governed by the flange proportions, then the web must be plastic, compact or semi-compact; it cannot be slender.

If the situation is reversed so that the webs are the controlling elements, then the same four categories, based on the same definitions of moment–rotation behaviour, are now determined by the value of web slenderness d/t. However, the governing values of β_1, β_2 and β_3 change since the web stress distribution differs from the pure compression in the top flange. Since the rectangular fully plastic condition, the triangular elastic condition and any intermediate condition contain less compression, the values of β are larger. Thus section classification also depends upon the type of

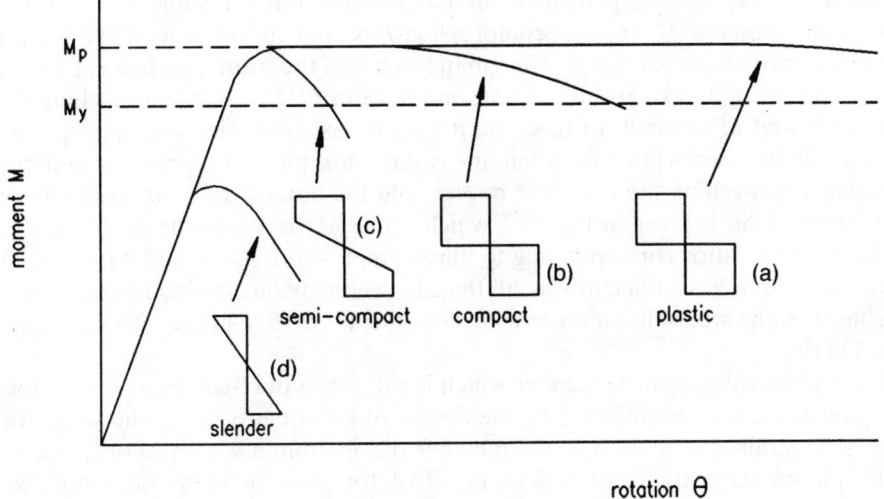

Fig. 13.4   Behaviour in bending of different classes of section

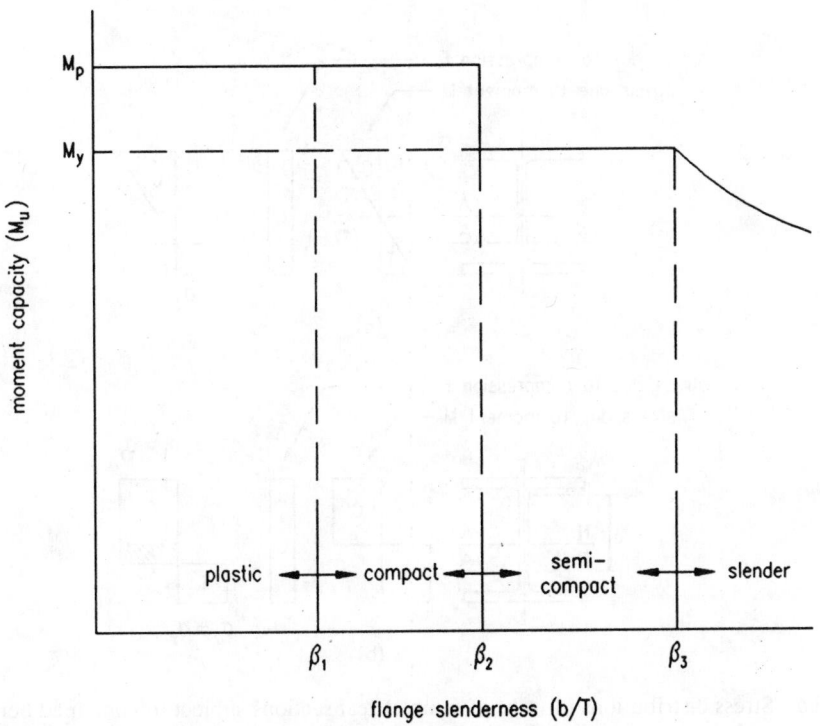

Fig. 13.5   Moment capacity as a function of flange slenderness

stress system to which the plate element under consideration is subjected. If, in addition to the moment M, an axial compression F is applied to the member, then for elastic behaviour the pattern of stress in the web is of the form shown in Fig. 13.6(a). The values of σ_1 and σ_2 are dependent on the ratio F/M with σ_2 approaching σ_y, if F is large and M is small. In this case it may be expected that the appropriate β limits will be somewhere between the values for pure compression and pure bending, approaching the former if $\sigma_2 \approx \sigma_y$, and the latter if $\sigma_2 \approx -\sigma_2$. A qualitative indication of this is given in Fig. 13.7, which shows M_u as a function of d/t for three different σ_2/σ_1 ratios corresponding to pure compression, $\sigma_2 = 0$ and pure bending. If the value of d/t is sufficiently small that the web may be classified as compact or plastic, then the stress distribution will adopt the alternative plastic arrangement of Fig. 13.6(b).

For a plate element in a member which is subject to pure compression the load-carrying capacity is not affected by the degree of deformation since the scope for a change in strain distribution as the member passes from a wholly elastic to a partially plastic state, as illustrated in Fig. 13.4 for pure bending, does not exist. The plastic and compact classifications do not therefore have any meaning; the only decision required is whether or not the member is slender, and specific values are only required for β_3.

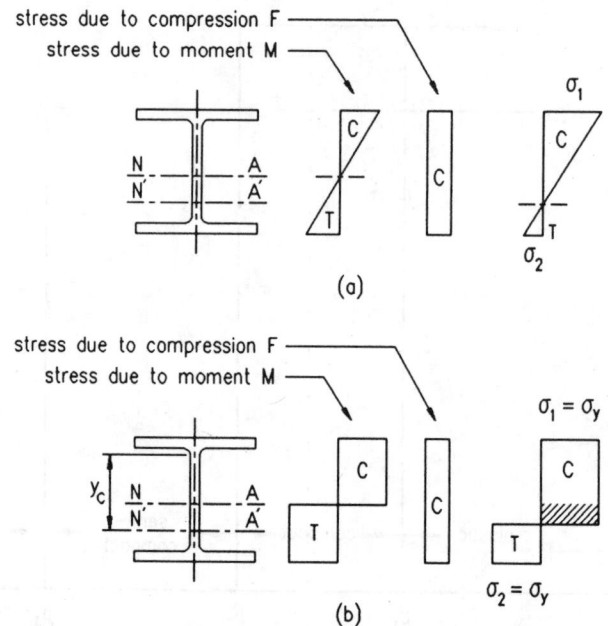

Fig. 13.6   Stress distributions in webs of symmetrical sections subject to combined bending and compression. (a) Semi-compact, elastic stress distribution. (b) Plastic or compact, plastic stress distribution

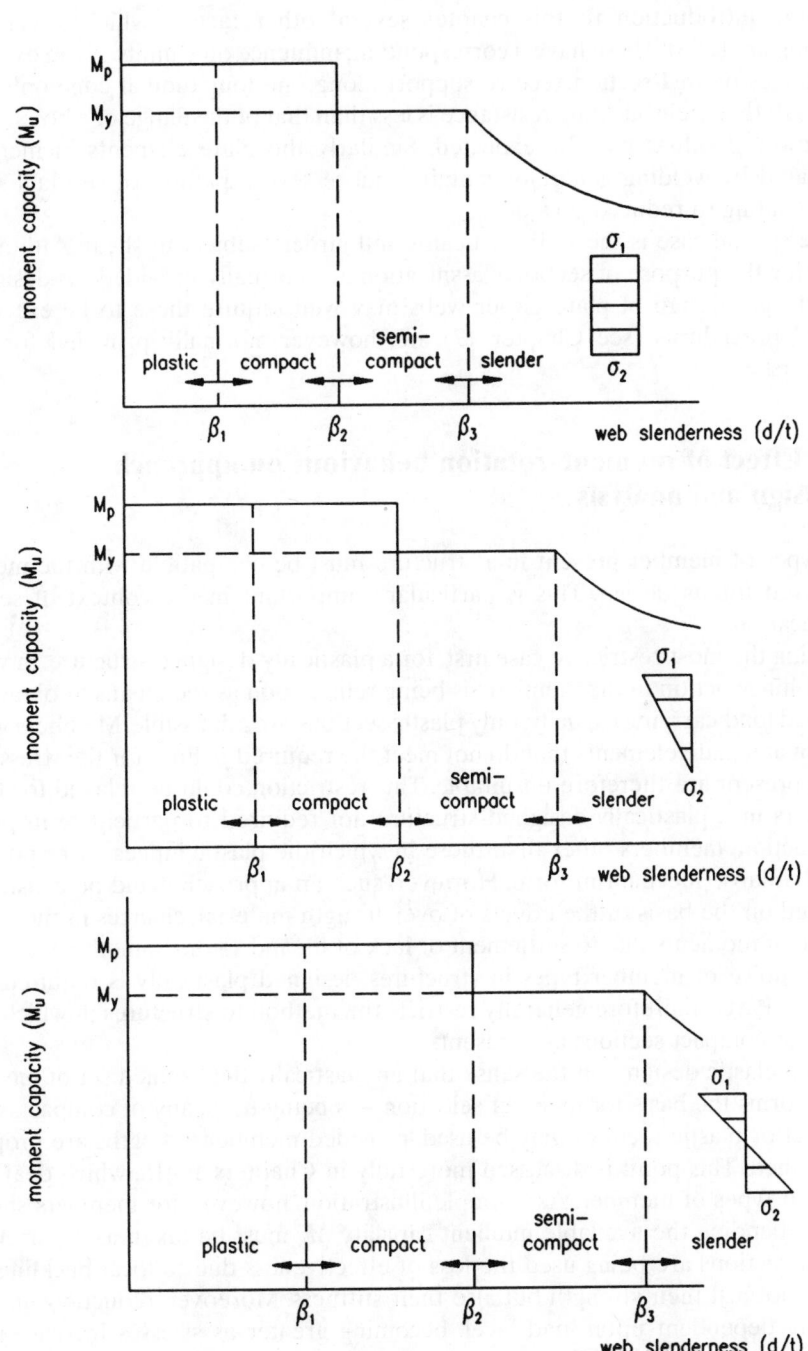

Fig. 13.7 Moment capacity as a function of *d/t* for different web stress patterns

In the introduction to this chapter several other factors which affect local buckling are listed. These have a corresponding influence on β limits. As an example, the flanges of an I-section receive support along one longitudinal edge only, with the result that their buckling resistance is less than that of the flange of a box section, and lower β values may be expected. Similarly the plate elements in members fabricated by welding generally contain a more severe pattern of residual stress, again leading to reduced β values.

One special case is the webs of beams and girders subject to shear. Although β limits for the purpose of section classification are normally provided for designers, the efficient design of plate girder webs may well require these to be exceeded. Special procedures (see Chapter 17) are, however, normally provided for such members.

13.3 Effect of moment–rotation behaviour on approach to design and analysis

The types of member present in a structure must be compatible with the method employed for its design. This is particularly important in the context of section classification.

Taking the most restrictive case first, for a plastically designed structure, in which plastic hinge action in the members is being relied upon as the means to obtain the required load-carrying capacity, only plastic sections are admissible. Members which contain any plate elements that do not meet the required β_1 limit for the stress condition present are therefore unsuitable. This restriction could be relaxed for those members in a plastically-designed structure not required to participate in plastic hinge action: members other than those in which the plastic hinges corresponding to the collapse mechanism form. However, such an approach could be considered unsound on the basis of the effects of overstrength material, changes in the elastic pattern of moments due to settlement or lack of fit, and so on. Something less than a free choice of member types in structures designed plastically is required, and BS 5950: Part 1 therefore generally restricts the method to structures in which only plastic or compact sections are present.

When elastic design – in the sense that an elastically determined set of member forces forms the basis for member selection – is being used, any of compact, semi-compact or plastic sections may be used, provided member strengths are properly determined. This point is discussed more fully in Chapters 14–18, which deal with different types of member. As a simple illustration, however, for members subject to pure bending the available moment capacity M_u must be taken as M_p or M_y. If slender sections are being used the loss of effectiveness due to local buckling will reduce not just their strength but also their stiffness. Moreover, reductions in stiffness are dependent upon load level, becoming greater as stresses increase sufficiently to cause local buckling effects to become more significant. Strictly speaking, such changes should be included when determining member forces. However, any attempt to do this would render design calculations prohibitively difficult and it is

therefore usual to make only a very approximate allowance for the effect. It is of most importance for cold-formed sections.

In practice the designer, having decided upon the design approach (essentially either elastic or plastic), should check section classification first using whatever design aids are available. Since most hot-rolled sections are at least compact in both S275 and S355 steel, this will normally be a relatively trivial task. When using cold-formed sections, which will often be slender, sensible use of manufacturer's literature will often eliminate much of the actual calculation. Greater care is required when using sections fabricated from plate, for which the freedom to select dimensions and thus b/T and d/t ratios means that any class is possible.

13.4 Classification table

Part of a typical classification table, extracted from BS 5950: Part 1, is given in Table 13.1. Values of β_1, β_2, and β_3 for flanges, defined as plates supported along one longitudinal edge, and webs, defined as plates supported along both longitudinal edges, under pure compression, pure bending and combined compression and bending are listed. While the third case reduces to the second as the compression component reduces to zero, it does not accord with the first case when the web is wholly subject to uniform compression. The reason for this is that the neutral axis of a member subject to bending and compression in which the web is wholly in compression must lie in the flange, or at least at the web/flange junction, with the result that the tensile strains in the flange provide some degree of stabilizing influence. A slightly higher set of limits than those provided for a plate supported by other elements which are themselves in compression, such as the compression flange of a box beam, is therefore appropriate.

13.5 Economic factors

When design is restricted to a choice of suitable standard hot-rolled sections, local buckling is not normally a major consideration. For plastically designed structures

Table 13.1 Extract from table of section classification limits (BS 5950: Part 1)

Type of element	Class of section		
	Plastic (β_1)	Compact (β_2)	Semi-compact (β_3)
Outstand element of compression flange	$b/T \leq 9\varepsilon$	$b/T \leq 10\varepsilon$	$b/T \leq 15\varepsilon$
Internal element of compression flange	$b/T \leq 28\varepsilon$	$b/T \leq 32\varepsilon$	$b/T \leq 40\varepsilon$
Web with neutral axis at mid-depth	$d/t \leq 80\varepsilon$	$d/t \leq 100\varepsilon$	$d/t \leq 120\varepsilon$
Web, generally	$d/t \leq 80\varepsilon/(1 + r_1)$	$d/t \leq 100\varepsilon/(1 + r_1)$	$d/t = 120\varepsilon/(1 + r_1)$

in which $r_1 = F_c/dtp_{yw}$ lies between −1 and +1 and is a measure of the stress ratio within the web

only plastic sections are suitable: thus the designer's choice is slightly restricted, although no UBs and only 4 UCs in S275 steel and 7 UBs and 9 UCs in S355 steel are outside the limits of BS 5950: Part 1 when used in pure bending. Although considerably more sections are unsatisfactory if their webs are subject to high compression, the number of sections barred from use in plastically designed portal frames is, in practice, extremely small. Similarly for elastic design no UB is other than semi-compact or better, provided it is not required to carry high compression in the web, while all UCs are at least semi-compact even when carrying their full squash load.

The designer should check the class of any trial section at an early stage. This can be done most efficiently using information of the type given in Reference 2. For webs under combined compression and bending the first check should be for pure compression as this is the more severe. Provided the section is satisfactory no additional checks are required; if it does not meet the required limit a decision on whether it is likely to do so under the less severe combined load case must be made.

The economic use of cold-formed sections, including profiled sheeting of the type used as decking and cladding, often requires that members are non-compact. Quite often they contain plate elements that are slender, with the forming process being exploited to provide carefully proportioned shapes. Since cold-formed sections are proprietary products, manufacturers normally provide design literature in which member capacities which allow for the presence of slender plate elements are listed. If rigorous calculations are, however, required, then Parts 5 and 6 of BS 5950 contain the necessary procedures.

When using fabricated sections the opportunity exists for the designer to optimize on the use of material. This leads to a choice between three courses of action:

(1) eliminate all considerations of local buckling by ensuring that the width-to-thickness ratios of every plate element are sufficiently small;
(2) if employing higher width-to-thickness ratios, use stiffeners to reduce plate proportions sufficiently so that the desired strength is achieved;
(3) determine member capacities allowing for reductions due to exceeding the relevant compact or semi-compact limits.

Effectively only the first of these is available if plastic design is being used. For elastic design when the third approach is being employed and the sections are slender, then calculations inevitably are more involved as even the determination of basic cross-sectional capacities requires allowances for local buckling effects through the use of concepts such as the effective width technique.[1]

References to Chapter 13

1. Bulson P.S. (1970) *The Stability of Flat Plates*. Chatto and Windus, London.
2. The Steel Construction Institute (SCI) (2001) *Steelwork Design Guide to BS 5950: Part 1: 2000, Vol. 1: Section Properties. Member Capacities*, 6th edn. SCI, Ascot, Berks.

Chapter 14
Tension members

by JOHN RIGHINIOTIS and ALAN KWAN

14.1 Introduction

Theoretically, the tension member transmitting a direct tension between two points in a structure is the simplest and most efficient structural element. In many cases this efficiency is seriously impaired by the end connections required to join tension members to other members in the structure. In some situations (for example, in cross-braced panels) the load in the member reverses, usually by the action of wind, and then the member must also act as a strut. Where the load can reverse, the designer often permits the member to buckle, with the load then being taken up by another member.

14.2 Types of tension member

The main types of tension member, their applications and behaviour are:

(a) open and closed single rolled sections such as angles, tees, channels and the structural hollow sections. These are the main sections used for tension members in light trusses and lattice girders for bracing.
(b) compound sections consisting of double angles or channels. At least one axis of symmetry is present and so the eccentricity in the end connection can be minimized. When angles or other shapes are used in this fashion, they should be interconnected at intervals to prevent vibration, especially when moving loads are present.
(c) heavy rolled sections and heavy compound sections of built-up H- and box sections. The built-up sections are tied together either at intervals (batten plates) or continuously (lacing or perforated cover plates). Batten plates or lacing do not add any load-carrying capacity to the member but they do serve to provide rigidity and to distribute the load among the main elements. Perforated plates can be considered as part of the tension member.
(d) bars and flats. In the sizes generally used, the stiffness of these members is very low; they may sag under their own weight or that of workmen. Their small cross-sectional dimensions also mean high slenderness values and, as a consequence, they may tend to flutter under wind loads or vibrate under moving loads.

(e) ropes and cables. Further discussion on these types of tension members is included in section 14.7 and Chapter 5, section 5.3.

The main types of tension members are shown in Fig. 14.1.
   Typical uses of tension members are:

(a) tension chords and internal ties in trusses and lattice girders in buildings and bridges.
(b) bracing members in buildings.
(c) main cables and deck suspension cables in cable-stayed and suspension bridges.
(d) hangers in suspended structures.

Typical uses of tension members in buildings and bridges are shown in Fig. 14.2.

14.3 Design for axial tension

Rolled sections behave similarly to tensile test specimens under direct tension (Fig. 14.1).
   For a straight member subject to direct tension, F:

tensile stress, $f_t = \dfrac{F}{A}$

elongation, $\delta_L = \dfrac{FL}{AE}$ (in the linear elastic range)

load at yield, $P_y = P_y A$ = load at failure (neglecting strain hardening)

For typical stress–strain curves for structural steel and wire rope see Fig. 14.3.

14.3.1 BS 5950: Part 1

The design of axially loaded tension members is given in Clause 4.6.1. The tension capacity is

$P_t = p_y A_e$

where A_e is the sum of the net effective areas (defined in Clause 3.4.3). Here a steel grade dependent factor K_e is used to determine the effective net area from the actual net area of a member with holes, i.e. the gross area less deductions for fastener holes. Reference should be made to clause 3.4.4.3 for members with staggered holes.
   The values for coefficient K_e, given below for steels complying with BS5950-2, come from results which show that the presence of holes does not reduce the effective capacity of a member in tension provided that the ratio of the net area to the

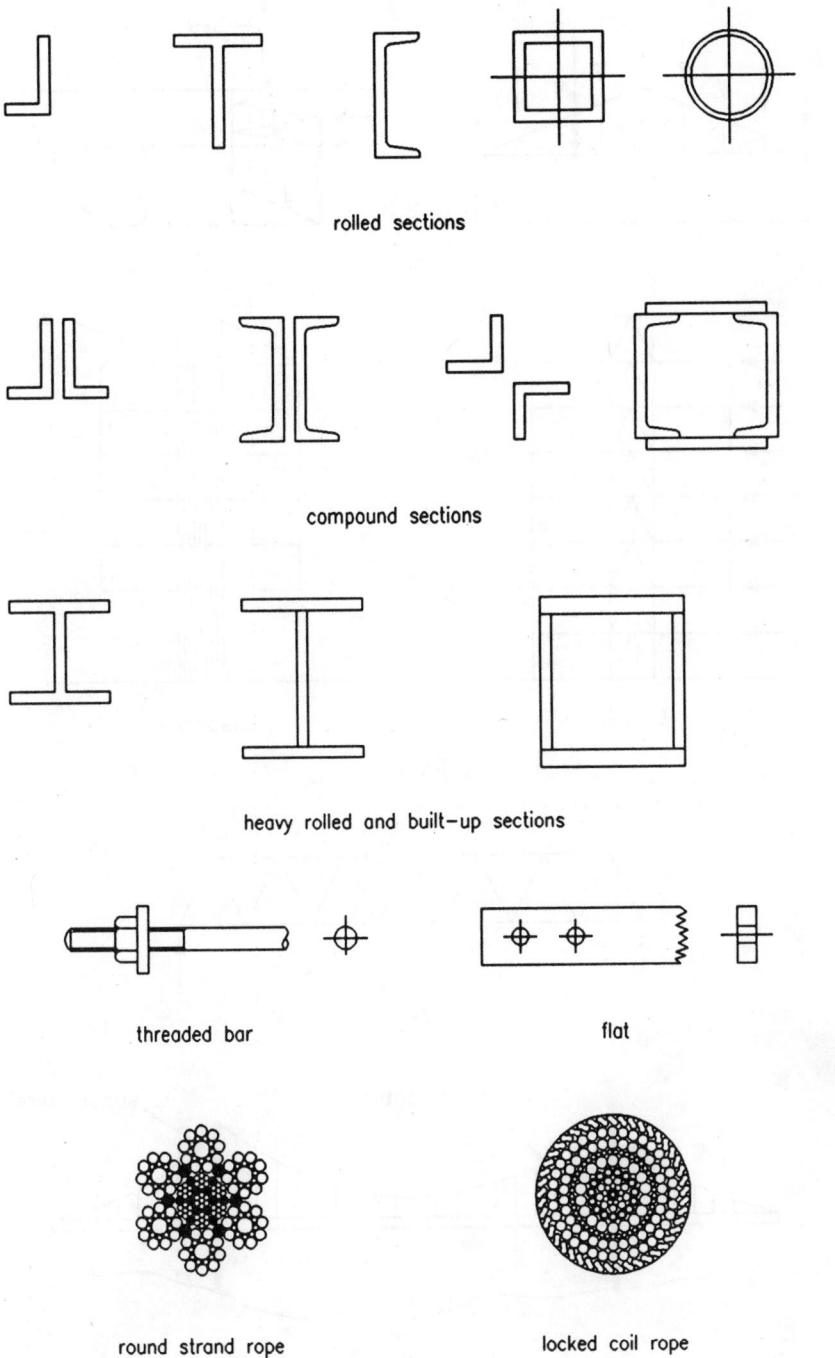

rolled sections

compound sections

heavy rolled and built—up sections

threaded bar   flat

round strand rope   locked coil rope

Fig. 14.1 Tension members

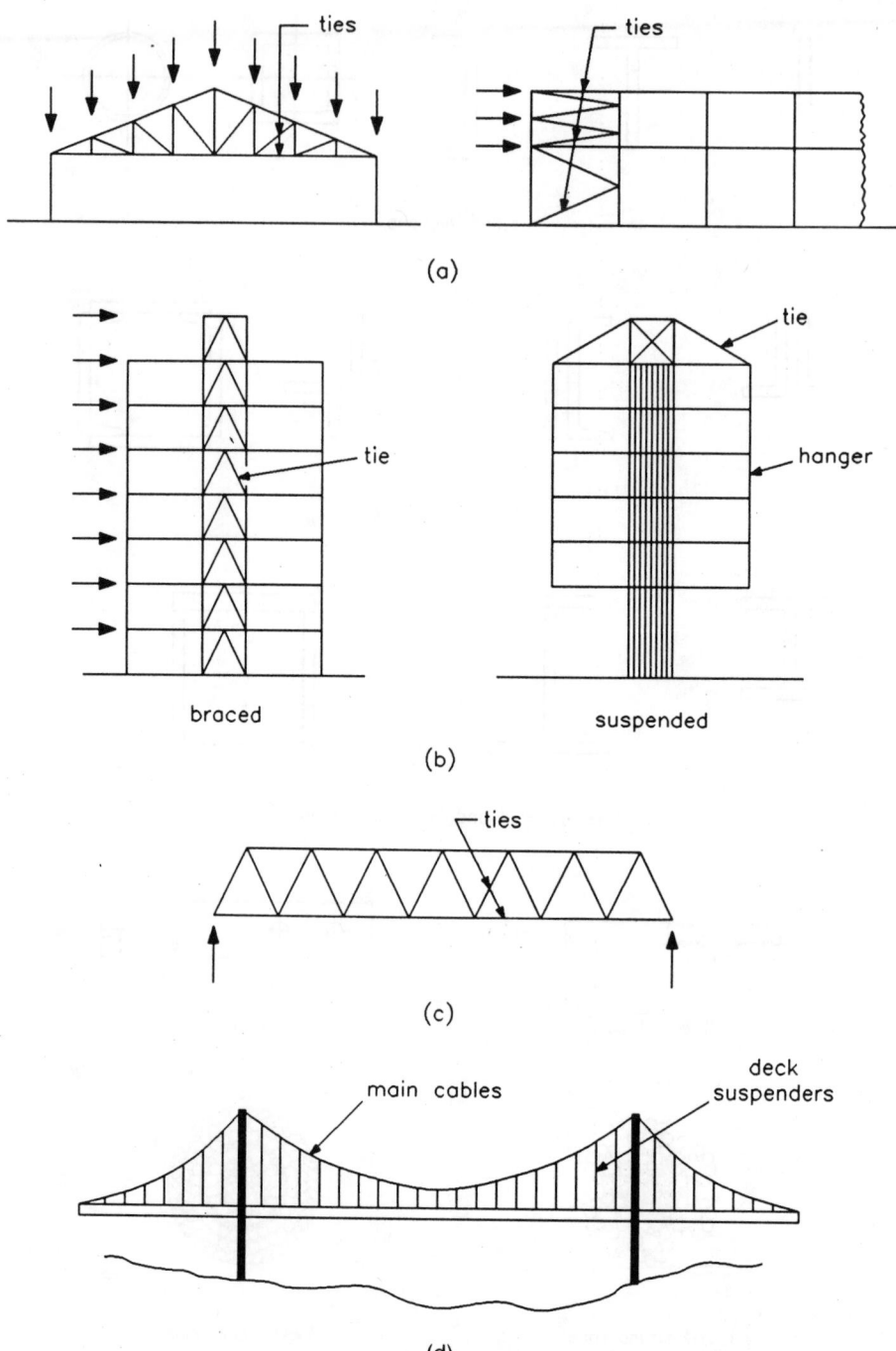

Fig. 14.2 Tension members in buildings and bridges. (a) Single-storey building – roof and truss bracing. (b) Multi-storey building. (c) Bridge truss. (d) Suspension bridge

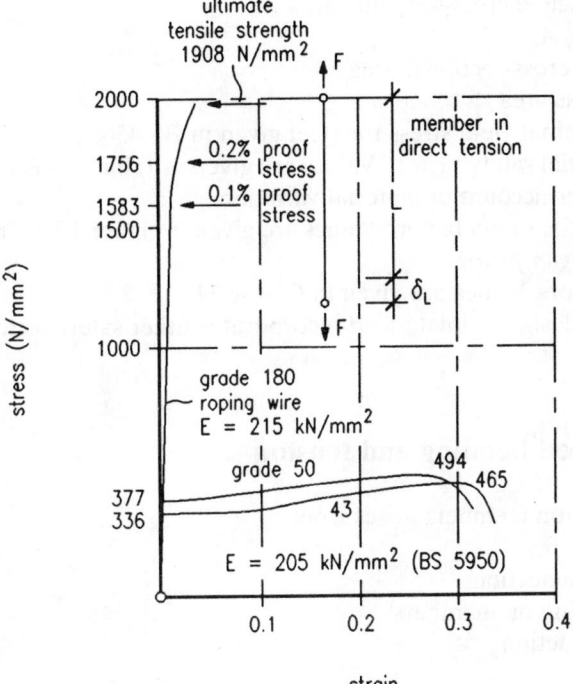

Fig. 14.3 Stress–strain curves for structural steels and wire rope

gross area is suitably greater than the ratio of the yield strength to the ultimate strength.

K_e = 1.2 for grade S275
K_e = 1.1 for grade S355
K_e = 1.0 for grade S460

For other steels $K_e = (U_s/1.2)/p_y$, where U_s is the specified minimum ultimate tensile strength and p_y is the design strength.

14.3.2 BS 5400: Part 3

The member should be such that the design ultimate axial load does not exceed the tensile resistance P_D, given by (using BS 5400 notation):

$$P_D = \frac{\sigma_y A_e}{\gamma_m \gamma_{f3}}$$

where A_e = effective cross-sectional area
           = $k_1 k_2 A_t$
   A_t = net cross-sectional area
   A = gross area
   σ_y = nominal yield stress for steel given in BS 4360
   γ_m = partial safety factor.[a] Values are given in Table 2 of BS 5400: Part 3. This takes account of material variation.
   γ_{f_3} = partial safety factor.[a] Values are given in clause 4.3.3. This is often called the *gap factor*.
   k_1, k_2 = factors. Values are given in Clause 11.
[a] Note that the design ultimate load incorporates other safety factors (γ_{fL}).

14.4 Combined bending and tension

Bending in tension members arises from

(a) eccentric connections
(b) lateral loading on members
(c) rigid frame action

When a structural member is subjected to axial tension combined with bending about axes xx and yy, then the total stress in the section ($f_{max} = f_r + f_{bx} + f_{by}$) comes from the sum of the stress from each of the separate actions: tensile ($f_t = F/A$); bending stress about xx axis ($f_{bx} = M_x/Z_x$); and bending stress about yy axis ($f_{by} = M_y/Z_y$). The separate stress diagrams are shown in Fig. 14.4(a). The load values causing yield when each of the three actions act alone are:

tensile load at yield, $P_y = p_y A$
moment at yield (xx axis), $M_{xx} = p_y Z_x$
moment at yield (yy axis), $M_{yy} = p_y Z_y$

The values P_y, M_{xx} and M_{yy} form part of a three-dimensional interaction surface, and any point on this surface gives a combination of F, M_x and M_y for which the maximum stress equals the yield stress.

An elastic interaction surface for a rectangular hollow section is shown in Fig. 14.4(b).

The maximum values for axial tension and plastic moment for the same member are:

axial tension, $P_y = p_y A$
plastic moment (xx axis), $M_{px} = p_y S_x$
plastic moment (yy axis), $M_{py} = p_y S_y$

The difference between the elastic and plastic interaction surfaces represents the additional design strength available if plasticity is taken into account (Fig. 14.5).

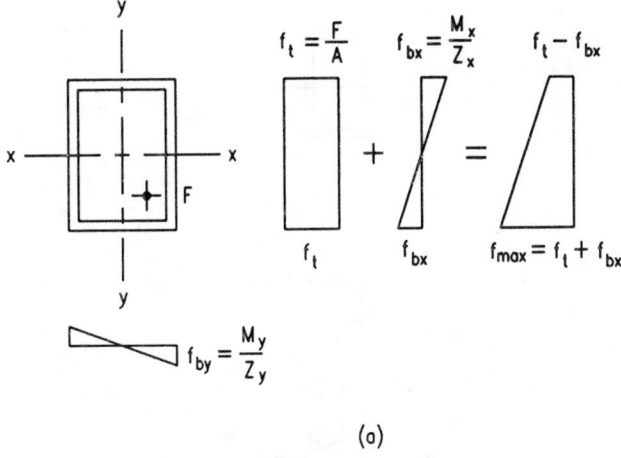

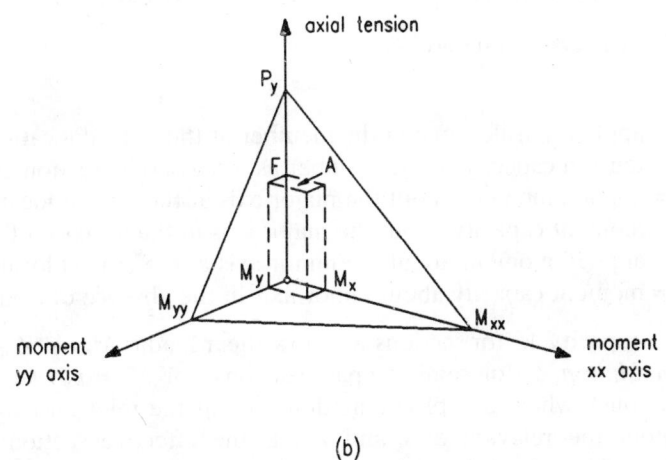

Fig. 14.4   Combined bending and tension – elastic analysis. (a) Stress diagram. (b) Elastic interaction surface

14.4.1 BS 5950: Part 1

The design of tension members with moments is given in Clause 4.8.2. This states that tension members should be checked for capacity at the points of greatest bending moments and axial loads, usually at the ends. The check can be carried out using the 'simplified method' (Clause 4.8.2.2) or the 'more exact method'. In the simplified method, the following relationship should be satisfied:

$$\frac{F_t}{P_t} + \frac{M_x}{M_{cx}} + \frac{M_y}{M_{cy}} \le 1$$

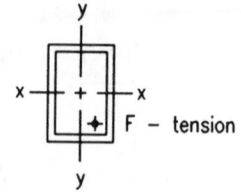

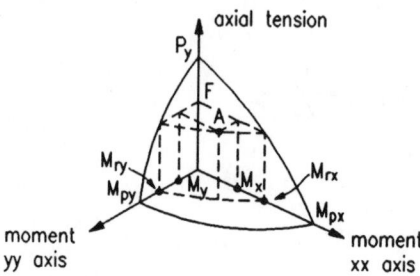

Fig. 14.5   Plastic interaction surface

where F_t = applied tensile force in the member at the critical location
$\quad\quad\ P_t$ = tension capacity of the member as discussed in Section 14.3.1 above
$\quad\quad\ M_x$ = applied moment about the major axis at the critical location
$\quad\quad\ M_{cx}$ = moment capacity about the major axis in the absence of axial load
$\quad\quad\ M_y$ = applied moment about the minor axis at the critical location
$\quad\quad\ M_{cy}$ = moment capacity about minor axis in the absence of axial load.

The moment capacity M_c for sections with low shear load is $M_c = p_y Z_{eff}$ (for slender sections), or $M_c = p_y Z$ (for semi-compact sections), or $M_c = p_y S$ (for plastic and compact sections), where S = plastic modulus about the relevant axis, Z = elastic modulus about the relevant axis, and Z_{eff} is the effective section modulus as determined in Clause 3.6.2. In all cases, the moment capacity should be limited to $1.2p_y Z$.

Greater economy may be achieved for plastic or compact cross sections subject to single plane bending through use of the 'more exact method' (Clause 4.8.2.3). The bending moment is not to exceed the 'reduced plastic moment capacity' M_{rx} or M_{ry} for bending about the major and minor axes respectively. The bending capacities are reduced due to the presence of axial load, and the reduction comes through a decrease in the plastic section modulus. Formulae are provided in Appendix J.2 of the Code for symmetric I- and H-sections.

The 'more exact method' may be used for bending in two planes only in doubly-symmetric members, in which case

$$\left(\frac{M_x}{M_{rx}}\right)^{z1} + \left(\frac{M_y}{M_{ry}}\right)^{z2} \le 1$$

where the constants z_1 and z_2 have values of 2.0 for solid and hollow circular sections, 1.0 for solid and hollow rectangular sections, $z_1 = 2.0$ and $z_2 = 1.0$ for symmetric I- and H-sections, and 1.0 for all other sections.

Additionally, the Code requires a member under tension and bending to be checked for resistance to lateral-torsional buckling. In this case, the axial force is removed, and the member is treated as a beam under bending alone (see 16.3.6).

14.4.2 BS 5400: Part 3

A member subjected to co-existent tension and bending should be such that at all cross sections the following relationship is satisfied:

$$\frac{P}{P_D} + \frac{M_x}{M_{Dxt}} + \frac{M_y}{M_{Dyt}} \leq 1.0$$

where P = axial tensile force
        P_D = tensile resistance, see section 14.3.2
        M_x, M_y = bending moments at the section about xx and yy axes, respectively
        M_{Dxt}, M_{Dyt} = corresponding bending resistances with respect to the extreme tensile fibres.

Additionally, if at any section within the middle third of the length of the member the maximum compressive stress due to bending exceeds the tensile stress due to axial load, the design should be such that:

$$\frac{M_{x\max}}{M_{Dxc}} + \frac{M_{y\max}}{M_{Dyc}} < 1 + \frac{P}{P_D}$$

where $M_{x\max}, M_{y\max}$ = maximum moments anywhere within the middle third
        M_{Dxc}, M_{Dyc} = corresponding bending resistances with respect to the extreme compression fibres.

For compact sections the bending resistance is given in clause 9.9.1.2. This is:

$$M_D = \frac{Z_{pe}\sigma_{Ic}}{\gamma_m \gamma_{f3}}$$

where σ_{Ic} = limiting compressive stress (see clause 9.8)
        Z_{pc} = plastic modulus of the effective section (see clause 9.4.2).

Clause 9.9.1.3 should be referred to for the resistance of non-compact sections.

14.5 Eccentricity of end connections

Simplified design rules are given in both BS 5950 and BS 5400 for the effects of combined tension and bending caused solely by the eccentric load introduced into the member by the end connection.

14.5.1 BS 5950: Part 1

BS 5950 provides simplified design rules for the common single- and double-angles, channels and T-sections. Eccentrically connected tension members may still be designed as axially loaded members but with a reduction in the cross sectional area.

Capacities of single angles connected through one leg, single channels connected only through the web and T-sections connected only through the flange, are given by

for bolted connections, $P_t = p_y(A_e - 0.5a_2)$

for welded connections, $P_t = p_y(A_g - 0.3a_2)$

where a_2 is the gross cross-sectional area (A_g) less the gross area of the connected element (= thickness × overall leg width for an angle, the overall depth for a channel, or the flange width for a T-section).

Capacities of back-to-back double angles connected through one leg only, double channels connected through the web, or double T-sections connected through the flange only, are given by

for bolted connections, $P_t = p_y(A_e - 0.25a_2)$

for welded connections, $P_t = p_y(A_g - 0.15a_2)$

provided that the two component sections are sufficiently connected along their lengths so as to enable them to act compositely (i.e. connected to both sides of a gusset or section, and the components are interconnected by bolts or welds and held apart, by battens or solid packing pieces, in at least two locations within their length). Where the two component sections are both connected to the same side of a gusset or section, or insufficiently interconnected along their lengths, then they should be treated as single sections and their individual capacity determined as in the previous paragraph.

14.5.2 BS 5400: Part 3

The bending moment resulting from an eccentricity of the end connections should be taken into account. For single angles connected by one leg these provisions may be considered to be met if the effective area of the unconnected leg is taken as:

$$\left(\frac{3A_1}{3A_1 + A_2}\right)A_2$$

where A_1 = net area of the connected leg
$\quad\quad A_2$ = net area of the unconnected leg.

14.6 Other considerations

14.6.1 Serviceability, fatigue and corrosion

Ropes and bars are not normally used in building construction because they lack stiffness, but they have been used in some cases as hangers in suspended buildings. Very light, thin tension members are susceptible to excessive elongation under direct load as well as lateral deflection under self-weight and lateral loads. Special problems may arise where the members are subjected to vibration or conditions leading to failure by fatigue, such as can occur in bridge deck hangers. Damage through corrosion is undesirable; adequate protective measures must be adopted. All these factors can make the design of tension members a complicated process in some cases.

The light-rolled sections used for tension members in trusses and for bracing are easily damaged during transport. It is customary to specify a minimum size for such members to prevent this happening. For angle ties, a general rule is to make the leg length not less than one-sixtieth of the member length.

Ties subject to load reversal, e.g. under the action of wind, could buckle. Where such buckling is dangerous (e.g. due to a lack of alternative load paths) or merely unsightly, tension members should also be checked as a compressive member (see 15.3).

In a fatigue assessment of a member, the stress range considered is the greatest algebraic difference between principal stresses occurring during any one stress cycle. The above applies when the stresses occur on principal planes of more than 45° apart.

The stresses should be calculated using elastic theory and taking account of all axial, bending and shear stresses occurring under the design loadings. Alternatively the design loadings in BS 5400: Part 10 can be used.

In assessing the fatigue behaviour of a member, the following effects have to be included:

- shear lag, restrained torsion and distortion, transverse stresses and flange curvatures,
- effective width of steel plates,
- stresses in triangulated skeletal structures due to load applications away from the joints, member eccentricities to joints and rigidity of joints,
- cracking of concrete in composite elements.

The following effects however, can be ignored:

- residual stresses,
- eccentricities unnecessarily arising in a standard detail,
- plate buckling.

14.6.2 Stress concentration factors

In cases of geometrical discontinuity, such as a change of cross section or an aperture, the resulting stress concentrations may be determined either by special analysis or by the use of stress concentration factors. Stress concentrations are not usually important in ductile materials but can be the cause of failure due to fatigue or brittle fracture in certain considerations. A hole in a flat member can increase the stresses locally on the net section by a factor which depends on the ratio of hole diameter to net plate width.

When designing against fatigue it is convenient to consider three levels of stress concentration:

(a) stress concentrations from structural action due to the difference between the actual structural behaviour and the static model chosen,
(b) macroscopic stress concentrations due to large scale geometric interruptions to stress flow,
(c) microscopic and local geometry stress concentrations due to imperfections within the weld or the heat affected zone.

For the design of welded details and connections further reference should be made to BS 5400: Part 10. However, for non-standard situations it may be necessary to determine the stress concentration factor directly from a numerical analysis study or from an experimental model.

In many cases the detail under consideration is very likely not to fit neatly into one of the classes. On site, the actual stress range for a particular loading occurrence is likely to be strongly influenced by detailed fit of the joint and overall fit of the structure. Therefore the overall form should be such that load paths are as smooth as possible, and unintended load paths should be avoided, particularly where fit could significantly influence behaviour. Discontinuities must be avoided by tapering and appropriate choice of radii.

14.6.3 Fabrication and erection

The behaviour of tension members in service depends on the fabrication tolerances and the erection sequence and procedure. Care must be taken to ensure that no

tension member is slack after erection, so that they are all immediately active in resisting service loads.

Screwed ends and turnbuckles can be used to adjust lengths of bars and cables after they are in place. Bracing members fabricated from rolled sections should be installed and properly tightened before other connections and column base plates are bolted up, to bring the structure into line and square. Bracing members are usually specified slightly shorter than the exact length to avoid sagging and allow them to be immediately effective.

Complete or partial trial shop assembly is often specified in heavy industrial trusses and bridge members to ensure that the fabrication is accurate and that erection is free from problems.

14.7 Cables

This section is directed mainly towards bridge cables; see Chapter 5, section 5.3 for cables in building structures.

14.7.1 Composition

A cable may be composed of one or more structural ropes, structural strands, locked coil strands or parallel wire strands. A strand, with the exception of a parallel wire strand, is an assembly of wires formed vertically around a central wire in one or more symmetrical layers. A strand may be used either as an individual load-carrying member, where radius of curvature is not a major requirement, or as a component in the manufacture of structural rope.

A rope is composed of a plurality of strands vertically laid around a core. In contrast to the strand, a rope provides increased curvature capability and is used where curvature of the cable becomes an important consideration. The significant differences between strand and rope are as follows:

- at equal sizes, a rope has lower breaking strength than a strand,
- the modulus of elasticity of a rope is lower than that of a strand,
- a rope has more curvature capability than a strand,
- the wires in a rope are smaller than those in a strand of the same diameter: consequently, a rope for a given size coating is less corrosion-resistant because of the thinner coating on the smaller diameter wires.

14.7.2 Application

Cables used in structural applications, namely for suspension systems in bridges, fall into the following categories:

(a) parallel-bar cables;
(b) parallel-wire cables;
(c) stranded cables (see Fig. 14.1);
(d) locked-coil cables (see Fig. 14.1).

The final choice depends on the properties required by the designer, i.e. modulus of elasticity, ultimate tensile strength, durability. Other criteria include economic and structural detailing, i.e. anchorages, erection, etc.

14.7.3 Parallel-bar cables

Parallel-bar cables are formed of steel rods or bars, parallel to each other in metal ducts, kept in position by polyethylene spacers. The process of tensioning the bar or rods individually is simplified by the capability of the bars to slide longitudinally. Cement grout, injected after erection, makes sure that the duct plays its part in resisting the stresses due to live loads.

Transportation in reels is only possible for the smaller diameters while for the larger sizes delivery is made in straight bars 15.0–20.0 m in length. Continuity of the bars has to be provided by the use of couplers, which considerably reduces the fatigue strength of the stay.

The use of mild steel necessitates larger sections than when using high-strength wires or strands. This leads to a reduction in the stress variation and thus lessens the risk of fatigue failure.

14.7.4 Parallel-wire cables

Parallel wires are used for cable-stayed bridges and pre-stressed concrete. Their fatigue strength is satisfactory, mainly because of their good mechanical properties.

14.7.5 Corrosion protection

Wires in the cables should be protected from corrosion. The most effective protection is obtained by hot galvanizing by steeping or immersing the wires in a bath of melted zinc, automatically controlled to avoid overheating. A wire is described as *terminally galvanized* or *galvanized re-drawn* depending on whether the operation has taken place after drawing or in between two wire drawings prior to the wire being brought to the required diameter. For reinforcing bars and cables, the first method is generally adopted. A quantity of zinc in the range of 250–330 g/m^2 is deposited, providing a protective coating 25–45 μm thick.

14.7.6 Coating

The coating process, used currently for locked-coil cables, consists of coating the bare wires with an anti-corrosion product with a good bond and long service life. The various substances used generally have a high dropping point so as not to run back towards the lower anchorages. They are usually high viscosity resins or oil-based grease, paraffins or chemical compounds.

14.7.7 Protection of anchorages

The details of the connections between the ducts and the anchorages must prevent any inflow or accumulation of water. The actual details depend on the type of anchorages used, on the protective systems for the cables, and on their slope. There are different arrangements intended to ensure water tightness of vital zones.

14.7.8 Protection against accidents

Cables should be protected against various risks of accident, such as vehicle impact, fire, explosion and vandalism. Measures to be taken may be based on the following:

(a) protection of the lower part of the stay, over a height of about 2.0 m, by a steel tube fixed into the deck and fixed into the duct; the tube dimensions (thickness and diameter) must be adequate.
(b) strength of the lower anchorage against vehicle impact.
(c) replacement of protective elements possible without affecting the cables them-selves and, as far as possible, without interrupting traffic.

Further reading for Chapter 14

Adams P.F., Krentz H.A. & Kulak G.L. (1973) *Canadian Structural Steel Design.* Canadian Institute of Steel Construction, Ontario.
Bresler B., Lin T.Y. & Scalzi J.B. (1968) *Design of Steel Structures.* Wiley, Chichester.
Dowling P.J., Knowles P. & Owens G.W. (1988) *Structural Steel Design.* Butterworths, London.
Horne M.R. (1971) *Plastic Theory of Structures*, 1st edn. Nelson, Walton-on-Thames.
Owens G.W. & Cheal B.D. (1989) *Structural Steelwork Connections.* Butterworths, London.
Timoshenko S.P. & Goodier J.N. (1970) *Theory of Elasticity.* McGraw-Hill, London.

Toy M. (1995) *Tensile Structures*. Academy Editions, London.

Trahair N.S. (2001) *The Behaviour and Design of Steel Structures to BS 5950*, 3rd edn. Spon Press, London.

Troitsky M.S. (1988) *Cable-Stayed Bridges*, 2nd edn. BSP Professional Books, London.

Vandenberg M. (1988) *Cable Nets*. Academy Editions, London.

A series of worked examples follows which are relevant to Chapter 14.

The **Steel Construction Institute** Silwood Park, Ascot, Berks SL5 7QN	Subject **TENSION MEMBERS**		Chapter ref. **14**
	Design code **BS 5950**	Made by **ASKK**	Sheet no. **1**
		Checked by **BD**	

BS 5950: Part 1

Problem 1

Design a single angle tie for the member AB shown.

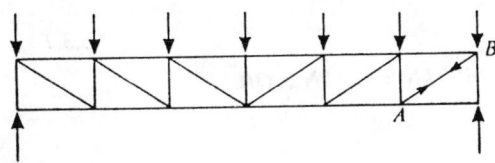

Tensile force in member AB

Dead load	= 122 kN
Imposed load	= 220 kN
Total	= 342 kN

Material: *Use Steel Grade S355*

Connections: *a) Welded*
                                   b) Bolted

Factored load

$F = (1.4 \times 122) + (1.6 \times 220) = 523\,kN$ *Table 2*

Material Steel Grade S355, thickness $\leq 16\,mm$ *Table 9*
Design strength $p_y = 355\,N/mm^2$

a) Welded connections

Try $125 \times 75 \times 10$ connected by long leg. Code allows angle section with eccentric loading to be treated with a reduced tension capacity as given in Clause 4.6.3.   *4.6.2*

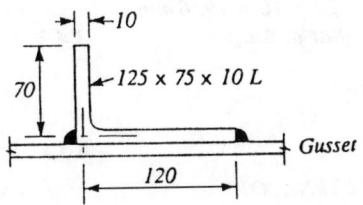

The **Steel Construction Institute** Silwood Park, Ascot, Berks SL5 7QN	Subject **TENSION MEMBERS**	Chapter ref. **14**
	Design code **BS 5950**	Made by **ASKK**	Sheet no. **2**
		Checked by **BD**	

Area for reduced tension capacity   4.6.3.1

$$a_1 = 120 \times 10 = 1200 \, mm^2$$
$$\text{Gross area } A_g = 1910 \, mm^2$$
$$a_2 = A_g - a_1 = 1910 - 1200 = 710 \, mm^2$$

Tension capacity $P_t = p_y(A_g - 0.3a_2)$ 4.6.3.1

$P_t = 355(1900 - 0.3 \times 710) = 598885 \, N \approx 599 \, kN > 523 \, kN \therefore OK$

$\therefore$ *Use 125 × 75 × 10 Steel Grade S355*

b) *Bolted connections*

Try 150 × 75 × 10 connected by long leg. Code allows angle section with eccentric loading to be treated with a reduced tension capacity as given in Clause 4.6.3.   4.6.2

Area for reduced tension capacity   4.6.3.1

$$a_1 = 120 \times 10 = 1200 \, mm^2$$

Effective net area A_e = *sum of effective net areas* a_e 4.6.1
Here, there is only one element, with net effective area

$$a_e = K_e a_n$$ 3.4.3

For steel grade S355, $K_e = 1.1$.

Hence
$$a_n = \text{Gross area} - \text{bolt hole} = 2160 - 24 \times 10 = 1920 \, mm^2$$
$$a_e = 1.1 \times 1920 = 2112 \, mm^2 \text{ (Check that } a_e \le a_g)$$ 3.4.3
$$A_e = 2112 \, mm^2$$
$$a_2 = A_g - a_1 = 2160 - 1200 = 960 \, mm^2$$

Tension capacity $P_t = p_y(A_e - 0.5a_2)$ 4.6.3.1

$P_t = 355(2112 - 0.5 \times 960) = 579360 \, N \approx 579 \, kN > 523 \, kN \therefore OK$

$\therefore$ *Use 150 × 75 × 10 Steel Grade S355*

The **Steel Construction Institute** Silwood Park, Ascot, Berks SL5 7QN	Subject *TENSION MEMBERS*		Chapter ref. *14*
	Design code *BS 5950*	Made by *ASKK*	Sheet no. *3*
		Checked by *BD*	

Problem 2

Design a double angle tie for the member AB in Problem 1. Assume the use of double angles, connected back to back through a 8 mm gusset. Assume welded connections.

Code allows angle section with eccentric loading to be treated with a reduced tension capacity as given in Clause 4.6.3.　　　　**4.6.2**

Try 2 × (75 × 50 × 8), connected through long leg.

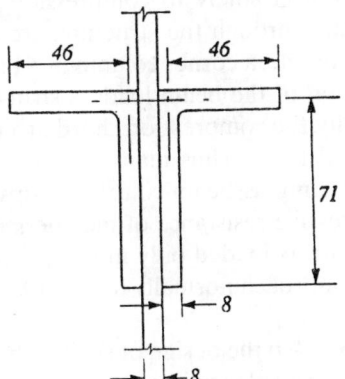

Effective Area, A_e

$a_1 = 71 \times 8 = 568 \ mm^2$
$a_2 = 46 \times 8 = 368 \ mm^2$

$$A_e = 2 \times \left[a_1 + \left[\frac{5 \times a_1}{5 \times a_1 + a_2} \right] \times a_2 \right]$$

$$\therefore A_e = 2 \times \left[568 + \left[\frac{5 \times 568}{5 \times 568 + 368} \right] \times 368 \right]$$

$$= 1788 \ mm^2$$

Area for reduced tension capacity　　　　**4.6.3.2**

$$a_1 = 2 \times (75 \times 8) = 1200 \ mm^2$$
$$Gross \ area \ A_g = 2 \times 941 = 1882 \ mm^2$$
$$a_2 = A_g - a_1 = 1882 - 1200 = 682 \ mm^2$$

Tension capacity $P_t = p_y(A_g - 0.15a_2)$　　　　**4.6.3.2**

$P_t = 355(1882 - 0.15 \times 682) = 631793 \ N \approx 632 \ kN > 523 \ kN \therefore OK$

$\therefore$ *Use 2 × (75 × 50 × 8) Steel Grade S355*
Note: The two angles must be connected at regular intervals.　　　　**4.6.3.2**

Chapter 15
Columns and struts

by DAVID NETHERCOT

15.1 Introduction

Members subject to compression, referred to as either 'columns' or 'struts', form one of the basic types of load-carrying component. They may be found, for example, as vertical columns in building frames, in the compression chords of a bridge truss or in any position in a space frame.

In many practical situations struts are not subject solely to compression but, depending upon the exact nature of the load path through the structure, are also required to resist some degree of bending. For example, a corner column in a building is normally bent about both axes by the action of the beam loads, a strut in a space frame is not necessarily loaded concentrically, the compression chord of a roof truss may also be required to carry some lateral loads. Thus many compression members are actually designed for combined loading as beam-columns. Notwithstanding this, the ability to determine the compressive resistance of members is of fundamental importance in design, both for the struts loaded only in compression and as one component in the interaction type of approach normally used for beam-column design.

The most significant factor that must be considered in the design of struts is buckling. Depending on the type of member and the particular application under consideration, this may take several forms. One of these, local buckling of individual plate elements in compression, has already been considered in Chapter 13. Much of this chapter is devoted to the consideration of the way in which buckling is handled in strut design.

15.2 Common types of member

Various types of steel section may be used as struts to resist compressive loads; Fig. 15.1 illustrates a number of them. Practical considerations such as the methods to be employed for making connections often influence the choice, especially for light members. Although closed sections such as tubes are theoretically the most efficient, it is normally much easier to make simple site connections, using the minimum of skilled labour or special equipment, to open sections. Typical arrangements include:

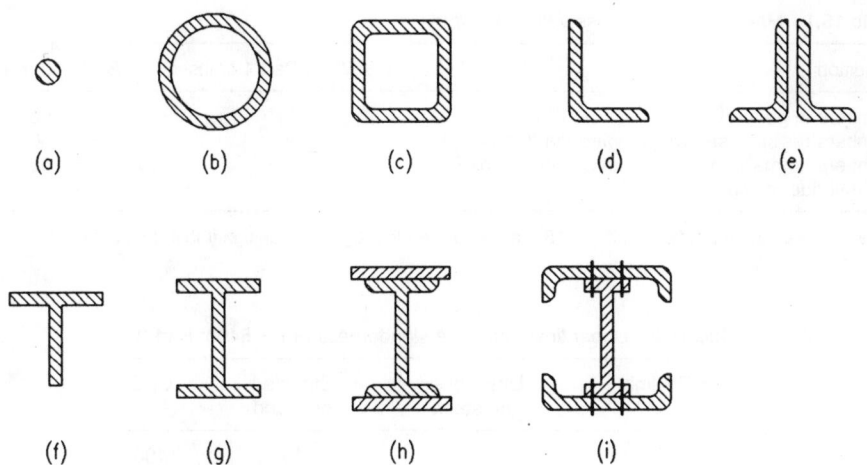

(a)   (b)   (c)   (d)   (e)

(f)   (g)   (h)   (i)

Fig. 15.1   Typical column cross sections

(1) light trusses and bracing – angles (including compound angles back to back) and tees
(2) larger trusses – circular hollow sections, rectangular hollow sections, compound sections and universal columns
(3) frames – universal columns, fabricated sections e.g. reinforced UCs
(4) bridges – box columns
(5) power stations – stiffened box columns.

15.3 Design considerations

The most important property of a strut as far as the determination of its load-carrying capacity is concerned is its slenderness, λ, defined as the ratio of its effective length, L_E, divided by the appropriate radius of gyration, r. Codes of practice such as BS 5950 used to place upper limits on λ, as indicated in Table 15.1, so as to avoid the use of flimsy construction, i.e. to ensure that a member which will ordinarily be subject only to axial load does have some limited resistance to an accidental lateral load, does not rattle, etc. Although not now explicitly stated in the 2000 version of the code, it is still good practice to aim for robust construction. Similarly, for BS 5400: Part 3, no actual limits are specified but the user is provided with a general advisory note to the effect that construction should be suitably robust. By contrast, BS 5400 does place upper bounds on local plate slenderness to avoid consideration of local buckling (see Table 15.2).

   Strut design will normally require that, once a trial member has been selected and its loading and support conditions determined, attention be given to whichever of the following checks are relevant for the particular application:

Table 15.1 Maximum slenderness values for struts

Condition	BS 5950: Part 1 limits	BS 5950: Part 5
Members in general	180[a]	180
Members resisting self-weight wind loads only	250[a]	250
Members normally in tension but subject to load reversal due to wind	–	350

[a] Check for self-weight deflection if $\lambda > 180$; allow for bending effects in design if this deflection $> L/1000$

Table 15.2 Upper limits on plate slenderness of BS 5400: Part 3

b/t or D/t limit[a]	Unstiffened outstand	Stiffened outstand	CHS
$p_y = 355\,N/mm^2$	12	14	100
$p_y = 275\,N/mm^2$	14	16	114

[a] See Fig. 13.1 for definitions of b; D = outside diameter

(1) Overall flexural buckling – largely controlled by the slenderness ratio, λ, which is a function of member length, cross-sectional shape and the support conditions provided; also influenced by the type of member
(2) Local buckling – controlled by the width-to-thickness ratios of the component plate elements (see Chapter 13); with some care in the original choice of member this need not involve any actual calculation
(3) Buckling of component parts – only relevant for built-up sections such as laced and battened columns; the strength of individual parts must be checked, often by simply limiting distances between points of interconnection
(4) Torsional or torsional-flexural buckling – for cold-formed sections and in extreme cases of unusually shaped heavier open sections, the inherent low torsional stiffness of the member may make this form of buckling more critical than simple flexural buckling.

In principle, local buckling and overall buckling (flexural or torsional) should always be checked. In practice, provided cross sections that at least meet the semi-compact limits for pure compression are used, then no local buckling check is necessary since the cross section will be fully effective.

15.4 Cross-sectional considerations

Since the maximum attainable load-carrying capacity for any structural member is controlled by its local cross-sectional capacity (factors such as buckling may prevent this being achieved in practice), the first step in strut design must involve considera-

tion of local buckling as it influences axial capacity. Only two classes of section are relevant for purely axially compressed members: either the section is not slender, in which case its full capacity, $p_y A$, is available, or it is slender and some allowance in terms of a reduced capacity is required. The distinctions between plastic, compact and semi-compact as described in Chapter 13 therefore have no relevance when the type of member under consideration is a strut.

The general approach for a member containing slender plate elements designed in accordance with BS 5950: Part 1 utilizes an effective cross-section as illustrated in Fig. 15.2. Essentially the effective portion of any slender plate element is taken as the maximum width of that class of element that corresponds to the semi-compact limit. A slightly simpler, but generally much more conservative option requires the use in all calculations relating to that member, apart from those concerned with components of connections to the member, of a reduced design strength, with the magnitude of the reduction being dependent on the extent to which the semi-compact limits (the boundary between not slender and slender) are exceeded.

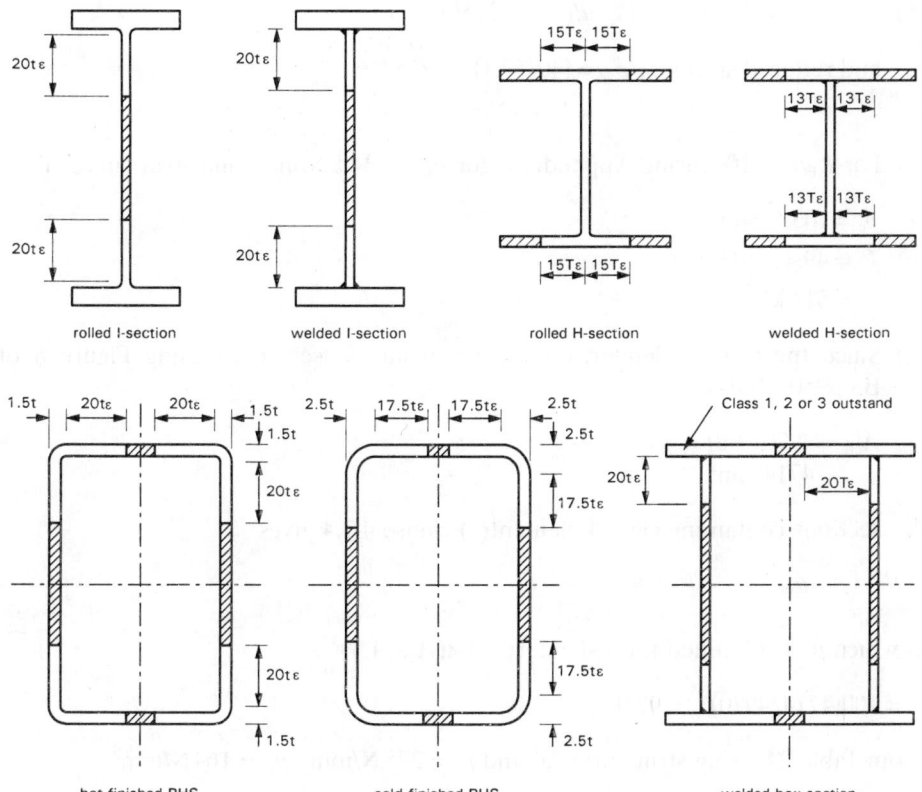

Fig. 15.2   Effective cross-sections under pure compression. (Based on Figure 8(a) in BS 5950: 2000-1)

Effectively this means that the member is assumed to be made of a steel having a lower material strength than is actually the case. As noted above, BS 5400: Part 3 does not permit the use of columns with slender outstands.

15.4.1 Columns with slender webs

Reference 1 indicates that most UB and a few RHS, even in S275 steel, will be slender according to BS 5950: Part 1 if used as struts. A suitable allowance for the weakening effect of local buckling should therefore be made. Two different procedures are given, outlined in the following example.

Consider a $406 \times 140 \times 39$ UB (the UB with the highest d/t) in S275 material. This has a d/t of 57.1 compared with the limit in Table 11 of 40 assuming the section to be fully stressed: i.e. r_2 as defined in clause 3.5.5 is equal to unity (for $r_2 < 1$ a higher limit is applicable).

(1) Use clause 3.6.5, limiting $d/t = 40(275/p_{yr})^{0.5}$

and reduced strength, $p_{yr} = (40/57.1)^2 \times 275$
$$= 134.8 \text{ N/mm}^2$$

For $L_E/r_y = 100$, using Appendix C for $p_{yr} = 134.8 \text{ N/mm}^2$ and strut curve 'a'

$p_c = 103.5 \text{ N/mm}^2$
$P_c = 49.4 \times 103.5/10$
$\quad = 511 \text{ kN}$

(2) Since the web is slender, obtain A_{eff} from clause 3.6.2.2 using Figure 8 of BS 5950: 2001-1.

$A_{eff} = 4970 - 40 \times 6.4$
$\quad = 4714 \text{ mm}^2$

For sections containing class 4 element(s) clause 4.7.4 gives

$P_c = A_{eff} p_{cs}$

in which p_{cs} is obtained for a slenderness $\lambda(A_{eff}/A_g)^{0.5}$

$= 100(4714/4970)^{0.5} = 97.0$

From Table 24, using strut curve 'a' and $p_y = 275 \text{ N/mm}^2$, $p_{cs} = 164 \text{ N/mm}^2$

$\therefore \quad P_c = 4714 \times 164/1000$
$\quad = 773 \text{ kN}$

This compares with the value for a 'fully effective equivalent' of 780 kN.

The second method leads to a significantly higher result, being only some 1% below the fully effective figure. The main reason for this is that the first method applies the same reduction in strength, based on the proportions of the most slender plate element, to the whole of the cross section, whereas the second method makes a reduction only for the plate elements which are slender.

Both BS 5400: Part 3 and BS 5950: Part 5 employ a method for slender sections in compression that uses the concept of effective area. Its implementation has been illustrated in general terms by the second method of the above example. The bridge code prohibits the use of slender outstands by specifying the absolute upper limits on *b*/*t* given in Table 15.2 with no procedure to cover arrangements outside these, and determining the effective area of the cross section by summing the full areas of any outstands and compact webs and the effective areas, $K_c A_c$, of the slender webs. Values of K_c are obtained as a function of *b*/*t* from a graph.

It is worth emphasizing that for design using hot-rolled sections the majority of situations may be treated simply by ensuring that the proportions of the cross section lie within the semi-compact limits. Reference 1, in addition to listing flange *b*/*T* and web *d*/*t* values for all standard sections, also identifies those sections that are slender according to BS 5950: Part 1 when used in either S275 or S355 steel. Table 15.3 lists these.

Table 15.3   Sections classified as 'slender' when used as struts

Section	S275	S355
	All *except*	All *except*
Universal	1016 × 305 × 487	1016 × 305 × 487
beam[1]	1016 × 305 × 437	1016 × 305 × 437
	1016 × 305 × 393	1016 × 305 × 393
	914 × 419 × 388	
	610 × 305 × 238	610 × 305 × 238
	610 × 305 × 179	
	533 × 210 × 122	
	457 × 191 × 98	
	457 × 191 × 89	
	457 × 152 × 82	
	406 × 178 × 74	
	356 × 171 × 67	356 × 171 × 67
	356 × 171 × 57	
	305 × 165 × 54	305 × 165 × 54
	305 × 165 × 46	
	305 × 127 × 48	305 × 127 × 48
	305 × 127 × 42	305 × 127 × 42
	305 × 127 × 37	
	254 × 146 × 43	254 × 146 × 43
	254 × 146 × 37	254 × 146 × 37
	254 × 146 × 31	
	254 × 102 × 28	
	254 × 102 × 25	
	254 × 102 × 22	
	203 × 133 × 30	203 × 133 × 30
	203 × 133 × 25	203 × 133 × 25
	203 × 102 × 23	203 × 102 × 23
	178 × 102 × 19	178 × 102 × 19

Table 15.3   Continued

Section	S275	S355
	152 × 89 × 16 127 × 76 × 13	152 × 89 × 16 127 × 76 × 13
Universal column	none	none
Circular hollow section[1,2]		508 × 8.0 406.4 × 6.3
Rectangular hollow section[1,2]	• 250 × 150 × 5 300 × 200 × 6.3 400 × 200 × 8 450 × 250 × 8 and 10 500 × 300 × 8 and 10	160 × 80 × 4 200 × 100 × 5 200 × 120 × 5 250 × 150 × 5 and 6.3 300 × 200 × 6.3 400 × 200 × 8 and 10 450 × 250 × 8 and 10 500 × 300 × 8 to 12.5
Square hollow sections[1,2]	300 × 300 × 6.3 350 × 350 × 8	200 × 200 × 5 250 × 250 × 6.3 300 × 300 × 6.3 350 × 350 × 8 400 × 400 × 10
Rolled steel angles	200 × 150 × 12 200 × 100 × 12 200 × 100 × 10 125 × 75 × 8 100 × 50 × 6	200 × 150 × 15 200 × 150 × 12 200 × 100 × 15 200 × 100 × 12 200 × 100 × 10 150 × 90 × 10 150 × 75 × 10 125 × 75 × 8 100 × 75 × 8 100 × 65 × 7 80 × 40 × 6 65 × 50 × 5
	200 × 200 × 16 150 × 150 × 12 150 × 150 × 10 120 × 120 × 8 100 × 100 × 8 90 × 90 × 7 75 × 75 × 6 50 × 50 × 4	200 × 200 × 18 200 × 200 × 16 150 × 150 × 12 150 × 150 × 10 120 × 120 × 10 120 × 120 × 8 100 × 100 × 8 90 × 90 × 8 90 × 90 × 7 75 × 75 × 6 70 × 70 × 6 60 × 60 × 5 50 × 50 × 4

[1] BS 5950-1: 2000 clause 3.5.5 presents equations for calculating stress ratios r_1 and r_2, which are used for the classification of the web of I-, H- and box sections in Table 11 and RHS in Table 12. These ratios allow the influence of the applied axial force on the local buckling resistance of the section to be taken into account during the section classification process. The cross-sections noted as slender in this table are assumed to be fully loaded in compression.
[2] Hot finished tubular steel to BS EN 10210-2: 1997

15.5 Compressive resistance

The axial load-carrying capacity for a single compression member is a function of its slenderness, its material strength, cross-sectional shape and method of manufacture. Using BS 5950: Part 1, the compression resistance, P_c, is given by clause 4.7.4 as

$$P_c = A_g p_c$$

in which A_g is the gross area and p_c is the compressive strength.

Values of p_c in terms of slenderness λ and material design strength p_y are given in Table 24. Slenderness is defined as the ratio of the effective length, L_E (taken as the geometrical length L for the present but see section 15.7), to the least radius of gyration, r.

The basis for Table 24 is the set of four column curves shown in Fig. 15.3. These have resulted from a comprehensive series of full-scale tests, supported by detailed numerical studies, on a representative range of cross sections.[2] They are often referred to as the *European Column Curves*. Four curves are used in recognition of the fact that for the same slenderness certain types of cross section consistently perform better than others as struts. This is largely due to the arrangement of the material but is also influenced by the residual stresses that form as a result of differential cooling after hot rolling. It is catered for in design by using the strut curve selection table given as Table 23 in BS 5950: Part 1.

The first step in column design is therefore to consult Table 23 to see which strut curve of Table 24 is appropriate. For example, if the case being checked is a UC liable to buckle about its minor axis, Table 24 for strut curve c should be used. Selection of a trial section fixes r and A_g; the geometrical length will be defined by the application required, so λ and thus p_c and P_c may be obtained.

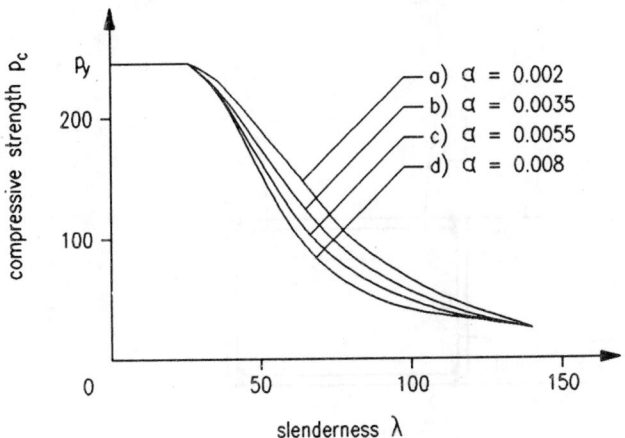

Fig. 15.3 Strut curves of BS 5950

The above process should be used for all types of rolled section, including those reinforced by the addition of welded cover plates for which Fig. 14 should be used to supplement Table 23 when deciding which curve of Table 24 to use for determining p_c. When sections are fabricated by welding plates together, however, the pattern of residual stresses produced by the heating and cooling of the welding process will be rather different from that typically found in a hot-rolled section[3] as illustrated in Fig. 15.4. This tends to produce lower buckling strengths. It is allowed for in BS 5950: Part 1 by the modification to the column curves shown in Fig. 15.5 in which the basic curve of Fig. 15.3 is replotted for a reduced material design strength. This has the effect of reducing the basic curve down to the level of that for the welded member in the important medium slenderness region. Clause 4.7.5 thus requires that when checking the axial resistance of a member that has been fabricated from plate by welding, the value of p_c be obtained using a p_y value of $20\,\text{N/mm}^2$ less than the actual figure.

BS 5400: Part 3 is intended to cover a wider range of construction than BS 5950: Part 1. Specifically in this context it recognizes the greater likelihood that struts may contain slender elements (other than outstands) and so it defines compression resistance, P_c, as

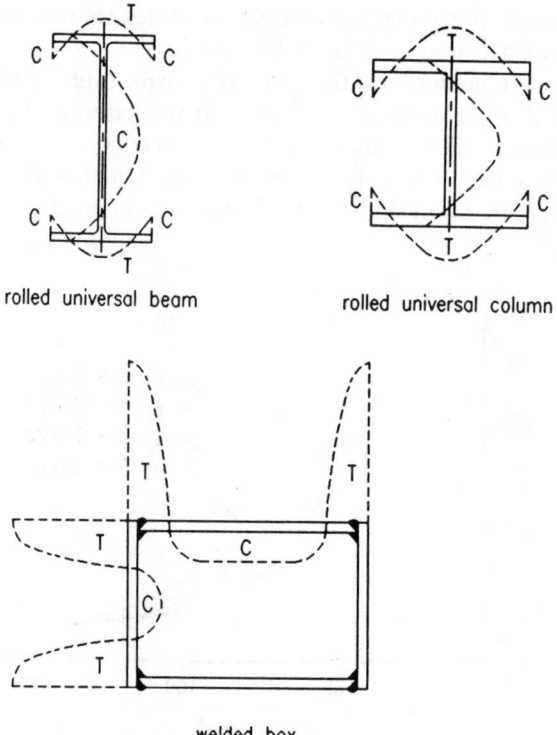

rolled universal beam   rolled universal column

welded box

Fig. 15.4   Distribution of residual stresses

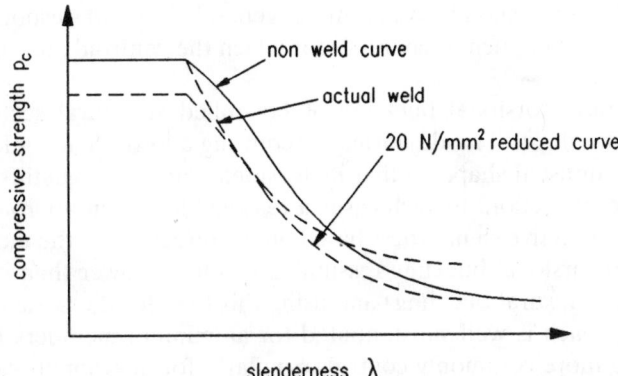

Fig. 15.5   Modified strut curve for welded sections used in BS 5950: Part 1

$P_c = A_e p_c$ (clause 10.6.1.1)

in which A_e is the effective area defined as

$A_e = \Sigma K_c (k_h A_c)$ for members other than CHSs

where K_c allows for loss of effectiveness in slender plate elements, determined from Figure 36, k_h allows for the presence of holes, and A_c is the net area; and

$A_e = A_c$ for CHSs for which $\dfrac{D}{t}\sqrt{\left(\dfrac{p_y}{335}\right)} \leq 50$

$A_e = A_c\left[1.15 - 0.003\dfrac{D}{t}\sqrt{\left(\dfrac{p_y}{355}\right)}\right]$ for CHSs for which $\dfrac{D}{t}\sqrt{\left(\dfrac{p_y}{355}\right)} > 50$

p_c is obtained from a set of curves (Figure 37) similar to Fig. 15.3.

Selection of the appropriate column curve is made using a simpler selection table than that of Part 1 of BS 5950. Essentially it distinguishes between curves on the basis of the ratio of radius of gyration to distance from neutral axis to extreme fibre, apart from heavy sections and hot-finished SHS, which are universally allocated to the lowest and highest curves respectively.

As noted in the example of 15.4.1, for class 4 sections the slenderness λ may be reduced in the ratio of the square root of the effective to the gross area.

15.6 Torsional and flexural-torsional buckling

In addition to the simple flexural buckling described in the previous section, struts may buckle due to either pure twisting about their longitudinal axis or a combination of bending and twisting. The first type of behaviour is only possible for centrally-loaded doubly-symmetrical cross sections for which the centroid and shear

centre coincide. The second, rather more general, form of response occurs for centrally-loaded struts such as channels for which the centroid and shear centre do not coincide.

In practice pure torsional buckling of hot-rolled structural sections is highly unlikely, the pure flexural mode normally requiring a lower load, unless the strut is of a somewhat unusual shape so that its torsional and warping stiffnesses are low as for a cruciform section. In such cases a reasonable design approach consists of determining an effective slenderness based on the direct use of the member's elastic critical load for torsional buckling (assuming this to be lower than its elastic critical load for pure flexural buckling) and using this to enter the basic column design curve. This approach is well substantiated for aluminium members for which torsional buckling more commonly controls. Similarly, for unsymmetrical sections use of an effective slenderness based on the member's lowest elastic critical load, corresponding to flexural-torsional buckling in this case, permits the basic column design curve to be retained and used in a more general way. For certain types of section, such as hot-rolled angles, special empirically-based design approaches are provided in BS 5950: Part 1 which recognize the possibility of some torsional influence. They are discussed in detail in section 15.8.

Torsional-flexural buckling is of greater practical significance in the design and use of cold-formed sections. This arises for two reasons:

(1) Because the torsion constant, J, depends on t^3, the use of thin material results in the ratio of torsional to flexural stiffness being much reduced as compared with hot-rolled sections
(2) The forming process leads naturally to a preponderance of singly-symmetrical or unsymmetrical open sections as these can be produced from a single sheet.

Procedures are given in BS 5950: Part 5 for determining the axial strength of singly-symmetrical sections using a factored effective length αL_E, in which α is obtained from

$$\alpha = 1 \qquad \text{for } P_{EY} \leq P_{TF}$$
$$\alpha = P_{EY}/P_{TF} \qquad \text{for } P_{EY} > P_{TF}$$

Formulae for P_{TF} in terms of basic cross-sectional properties are also provided. The use of $\alpha \neq 1$ is only required for those situations illustrated in Fig. 15.6 for which P_{TF} is the lowest elastic critical load.

15.7 Effective lengths

Basic design information relating column strength to slenderness is normally founded on the concept of a pin-ended member, e.g. Fig. 15.3. Stated more precisely, this means a member whose ends are supported such that they cannot translate

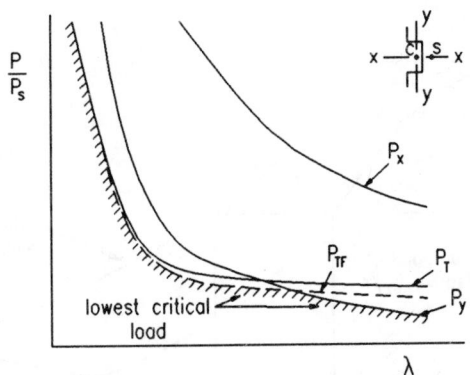

Fig. 15.6   Elastic critical load curves for a member subject to flexural–torsional buckling

relative to one another but are able to rotate freely. Compression members in actual structures are provided with a variety of different support conditions which are likely to be less restrictive in terms of translational restraint, giving fixity in position, with or without more restriction in terms of rotational restraint, giving fixity in direction.

The usual way of treating this topic in design is to use the concept of an effective column length, which may be defined as 'the length of an equivalent pin-ended column having the same load-carrying capacity as the member under consideration provided with its actual conditions of support'. This engineering definition of effective length is illustrated in Fig. 15.7, which compares a column strength curve for a member with some degree of rotational end restraint with the basic curve for the same member when pin-ended.

In determining the column slenderness ratio the geometrical length, L, is replaced by the effective length, L_E. Values of effective length factors $k = L_E/L$ for a series of standard cases are provided in BS 5950: Part 1, BS 5400: Part 3, etc.; Fig. 15.8 illustrates typical values. When compared with values given by elastic stability theory,[2] these appear to be high for those cases in which reliance is being placed on externally provided rotational fixity; this is in recognition of the practical difficulties of providing sufficient rotational restraint to approach the condition of full fixity. On the other hand, translational restraints of comparatively modest stiffness are quite capable of preventing lateral displacements. A certain degree of judgement is required of the designer in deciding which of these standard cases most nearly matches his arrangements. In cases of doubt the safe approach is to use a high approximation, leading to an overestimate of column slenderness and thus an underestimate of strength. The idea of an effective column length may also be used as a device to deal with special types of column, such as compound or tapered members, the idea then being to convert the complex problem into one of an equivalent simple column for which the basic design approach of the relationship between compressive strength and slenderness may be employed.

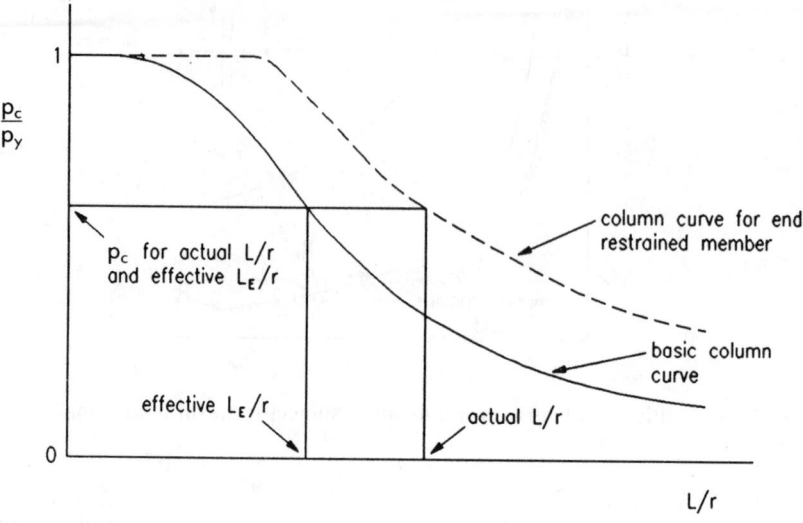

Fig. 15.7 Use of effective length with column curve to allow for end restraint

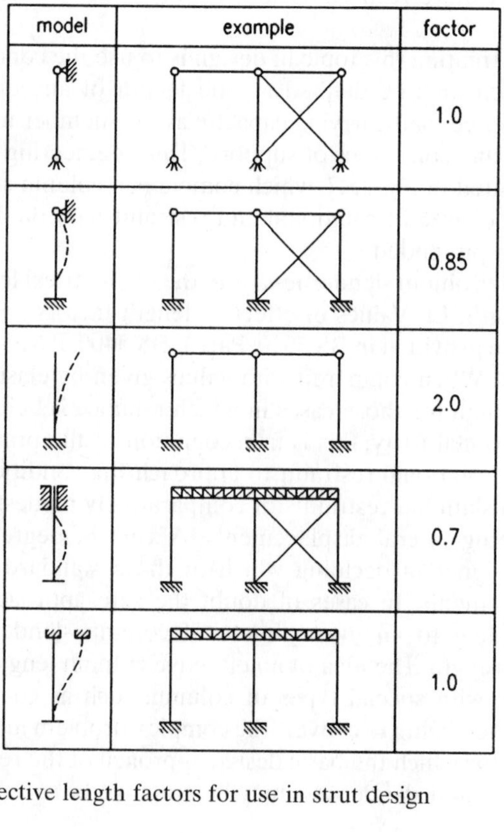

model	example	factor
		1.0
		0.85
		2.0
		0.7
		1.0

Fig. 15.8 Typical effective length factors for use in strut design

Of fundamental importance when determining suitable effective lengths is the classification of a column as either a sway case for which translation of one end relative to the other is possible or a non-sway case for which end translation is prevented. For the first case, effective lengths will be at least equal to the geometrical length, tending in theory to infinity for a pin-base column with no restraint at its top, while for the non-sway case, effective lengths will not exceed the geometrical length, decreasing as the degree of rotational fixity increases.

For non-standard cases, it is customary to make reference to published results obtained from elastic stability theory. Provided these relate to cases for which buckling involves the interaction of a group of members with the less critical restraining the more critical, as illustrated in Fig. 15.9, such evidence as is available suggests that the use of effective lengths derived directly from elastic stability theory in conjunction with a column design curve of the type shown in Fig. 15.3 will lead to good approximations of the true load-carrying capacity.

For compression members in rigid-jointed frames the effective length may, in both cases, be directly related to the restraint provided by the surrounding members by using charts presented in terms of the stiffness of these members provided in Appendix E of BS 5950: Part 1. Useful guidance on effective column lengths for a variety of more complex situations is available from several sources.[2-4]

When designing compression members in frames configured on the basis of simple construction, the use of effective column lengths provides a simple means of recognizing that real connections between members will normally provide some degree of rotational end restraint, leading to compressive strengths somewhat in excess of those that would be obtained if columns were treated as pin-ended. If axial load levels and unsupported lengths change within the length of a member that is continuous over several segments, such as a building frame column spliced so as to act as a continuous member but carrying decreasing compression with height or a compression chord in a truss, then the less heavily loaded segments will effectively restrain the more critical segments.

Even though the distribution of internal member forces has been made on the assumption of pin joints, some allowance for rotational end restraint when designing the compression members is therefore appropriate. Thus the apparent contradiction of regarding a structure as pin-jointed but using compression member effective lengths that are less than their actual lengths does have a basis founded upon an approximate version of reality. Figure 15.10 resents results obtained from elastic stability theory for columns continuous over a number of storeys which show

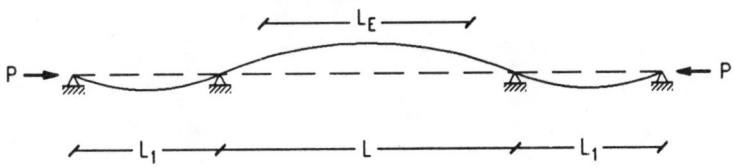

Fig. 15.9   Restraint to critical column segment from adjacent segments

a/L	0	0.1	0.2	0.3	0.4	0.5	0.6	0.7	0.8	0.9	1.0	frame	column
L_E/L	1.0	1.11	1.24	1.40	1.56	1.74	1.93	2.16	2.31	2.50	2.70		
L_E/L	2.0	2.07	2.13	2.20	2.27	2.34	2.41	2.48	2.55	2.62	2.70		
L_E/L	0.70	0.72	0.74	0.77	0.79	0.81	0.84	0.87	0.91	0.95	1.0		

column	frame	a/L	0	0.1	0.2	0.3	0.4	0.5	0.6	0.7	0.8	0.9	1.0
		L_E/L	0.70	0.73	0.76	0.79	0.82	0.85	0.88	0.91	0.94	0.97	1.0
		L_E/L	0.50	0.53	0.57	0.61	0.65	0.70	0.75	0.81	0.87	0.93	1.0

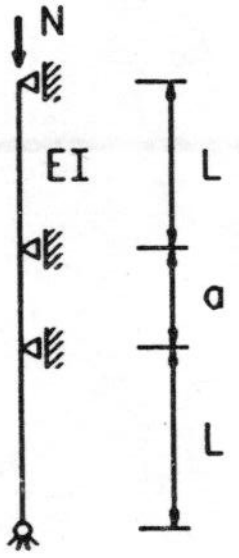

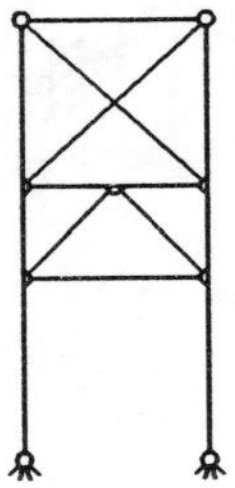

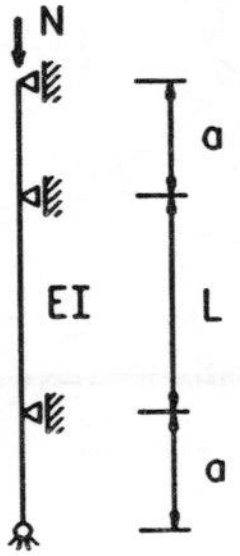

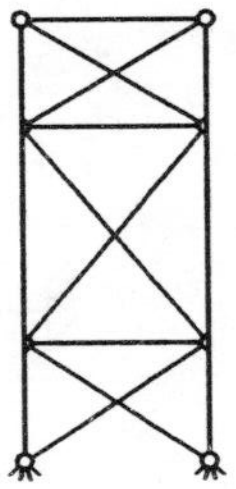

Fig. 15.10 Effective length factors for continuous columns based on elastic stability theory

how the effective length of the critical segment will be reduced if more stable segments (shorter unbraced lengths in this case) are present. A practical equivalent for each case in terms of simple braced frames with pinned beam-to-column connections is also shown. It is also necessary to recognize that practical equivalents of pin joints may also be capable of transferring limited moments. This point is considered explicitly in BS 5950: Part 1 for both building frames and trusses; the effect on the design of compression members is considered in detail in Chapter 18 for the former and in Chapter 19 for the latter.

For compression members in rigid-jointed frames the effective length is directly related to the restraint provided by all the surrounding members. Strictly speaking an interaction of all the members in the frame occurs because the real behaviour is one of frame buckling rather than column buckling, but for design purposes it is often sufficient to consider the behaviour of a limited region of the frame. Variants of the 'limited frame' concept are to be found in several codes of practice and design guides. That used in BS 5950: Part 1 is illustrated in Fig. 15.11. The limited frame comprises the column under consideration and each immediately adjacent member treated as if its far end were fixed. The effective length of the critical column is then obtained from a chart which is entered with two coefficients k_1 and k_2, the values of which depend on the stiffnesses of the surrounding members K_U, K_{TL}, etc., relative to the stiffness of the column K_C, a concept similar to the well-known moment distribution method. Two distinct cases are considered: columns in non-sway frames and columns in frames that are free to sway. Figures 15.12 (a) and (b) and 15.12 (c) and (d) illustrate both cases as well as giving the associated effective

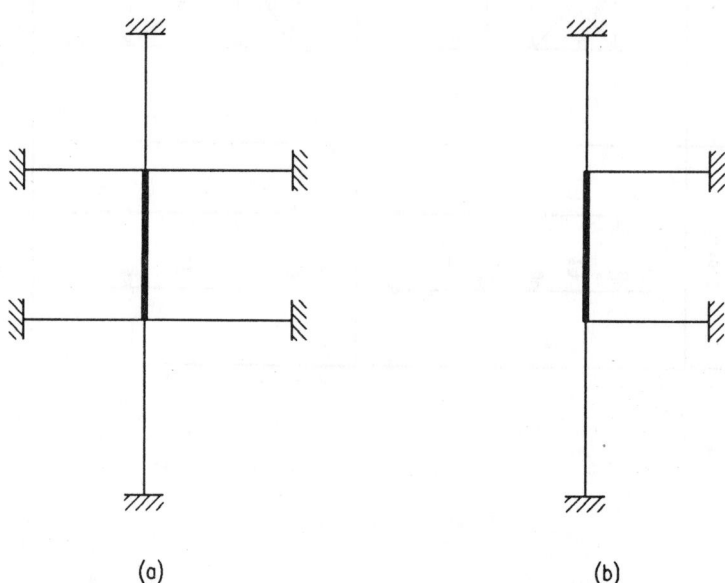

(a)   (b)

Fig. 15.11   Limited frames. (a) Internal column. (b) External column

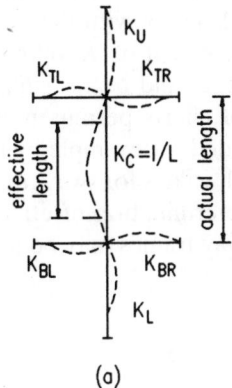

(a)

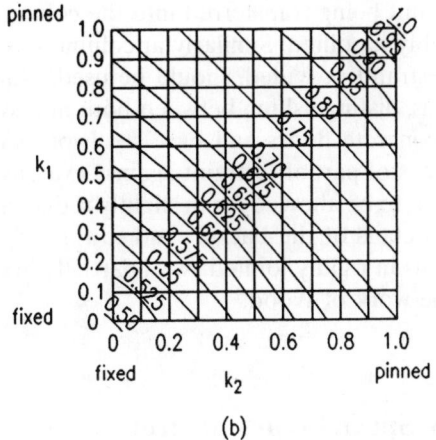

(b)

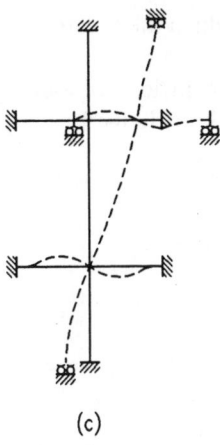

(c)

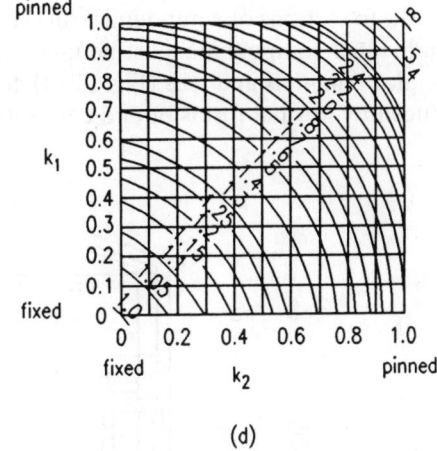

(d)

Fig. 15.12 Limited frames and corresponding effective length charts of BS 5950: Part 1. (a) Limited frame and (b) effective length ratios ($k_3 = \infty$), for non-sway frames. (c) Limited frame and (d) effective length ratios (without partial bracing, $k_3 = 0$), for sway frames

length charts. For the former, the factors will vary between 0.5 and 1.0 depending on the values of k_1 and k_2, while for the latter, the variation will be between 1.0 and ∞. These end points correspond to cases of: rotationally fixed ends with no sway and rotationally free ends with no sway; rotationally fixed ends with free sway and rotationally free ends with free sway.

For beams not rigidly connected to the column or for situations in which significant plasticity either at a beam end or at either column end would prevent the

restraint being transferred into the column, the K (and thus the k) values must be suitably modified. Similarly at column bases, k_2 values, in keeping with the degree of restraint provided, should be used. Guidance is also provided on K values for beams, distinguishing between both non-sway and sway cases and beams supporting concrete floors and bare steelwork. A further pair of charts permits modest degrees of partial restraint against sway, as might be provided for example by infill panels, to be allowed for in slightly reducing effective length values for sway frames. Full details of the background to this approach to the determination of effective lengths in rigidly jointed sway, partially braced and non-sway frames may be found in the work of Wood.[4]

15.8 Special types of strut

The design of two types of strut requires that certain additional points be considered:

(1) built-up sections or compound struts (Fig. 15.13), for which the behaviour of the individual components must be taken into account
(2) angles, channels and tees (Fig. 15.14), for which the eccentricity of loading produced by normal forms of end connection must be acknowledged.

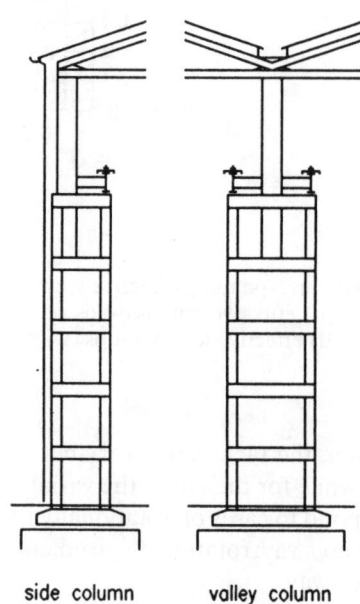

side column   valley column

Fig. 15.13   Built-up struts

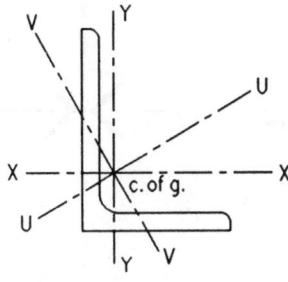

<pre>
axis X–X }   rectangular axes
axis Y–Y }

axis U–U   maximum principal axis
axis V–V   minimum principal axis
</pre>

Fig. 15.14 Geometrical properties of an angle section

In both cases, however, it will often be possible to design this more complex type of member as an equivalent single axially-loaded strut.

15.8.1 Design of compound struts

Individual members may be combined in a variety of ways to produce a more efficient compound section. Figure 15.15 illustrates the most common arrangements. In each case the concept is one of providing a compound member whose overall slenderness will be such that its load-carrying capacity will significantly exceed the sum of the axial resistances of the component members, i.e. for the case of Fig. 15.15(b) the laced strut will be stronger than the four corner angles treated separately.

Thus the design approach of BS 5950: Part 1 is to set conditions which when met permit the compound member to be designed as a single integral member. The following cases are considered explicitly:

(1) Laced struts conforming with the provisions of clause 4.7.8
(2) Battened struts conforming with the provisions of clause 4.7.9
(3) Batten-starred angles conforming with the provisions of clause 4.7.11, which uses much of clause 4.7.9
(4) Batten parallel angle struts conforming with the provisions of clause 4.7.12
(5) Back-to-back struts conforming with the provisions of clause 4.7.13.1 if the components are separated and of clause 4.7.13.2 if the components are in contact.

The detailed rules contained in these clauses are essentially of two types:

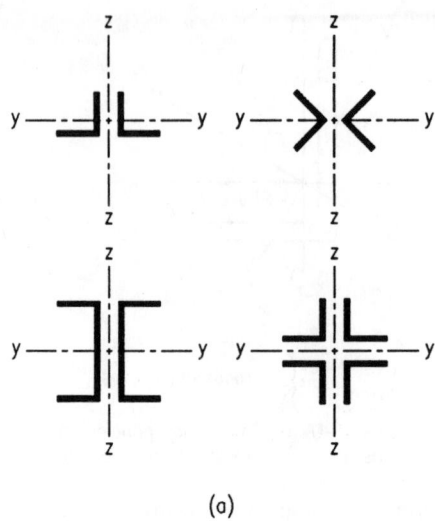

(a)

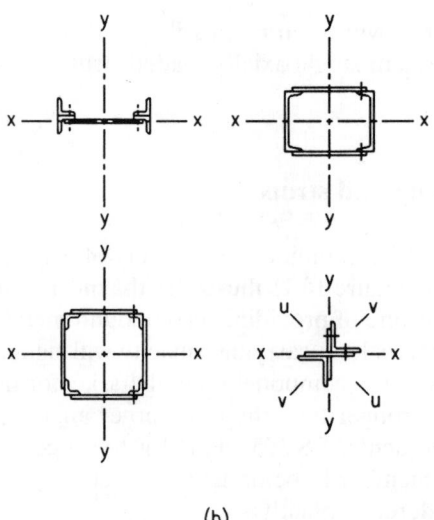

(b)

Fig. 15.15 Typical arrangements for compound struts: (a) closely spaced, (b) laced or battened

(1) Covering construction details such as the arrangements for interconnection in a general 'good practice' manner
(2) Quantitative rules for the determination of the overall slenderness, limits necessary for component slenderness, forces for which the interconnections should be designed, etc.

BS 5400: Part 3 also contains specific rules for the design of:

(1) Batten struts (clause 10.8)
(2) Laced struts (clause 10.9)
(3) Struts connected by perforated cover plates (clause 10.10)
(4) Struts consisting of back-to-back components (clause 10.11).

These are somewhat more detailed than those of BS 5950: Part 1, particularly in the matter of determining suitable design forces for the interconnections, i.e. battens, lacings, etc.

15.8.2 Design of angles, channels and tees

Four specific cases are covered in detail by BS 5950: Part 1:

(1) Single angles in clause 4.7.10.2
(2) Double angles in clause 4.7.10.3
(3) Single channels in clause 4.7.10.4
(4) Single tees in clause 4.7.10.5.

In all cases, guidance is provided on the determination of the slenderness to be used when obtaining p_c for each of the more common forms of fastening arrangement. In cases where only a single fastener is used at each end the resulting value of p_c should then be reduced to 80% so as to allow for the combined effects of load eccentricity and lack of rotational end restraint. For ease of use the whole set of slenderness relationships is grouped together in Table 25.

   BS 5400: Part 3 only gives specific consideration to single angles. This distinguishes between single bolt and 'other' forms of end connection. In both cases load eccentricity may be ignored but for the former only 80% of the calculated resistance may be used.

15.9 Economic points

Strut design is a relatively straightforward design task involving choice of cross-sectional type, assessment of end restraint and thus effective length, calculation of slenderness, determination of compressive strength and hence checking that the trial section can withstand the design load. Certain subsidiary checks may also be required part way through this process to ensure that the chosen cross section is not slender (or make suitable allowances if it is) or to guard against local failure in compound members. Thus only limited opportunities occur for the designer to use judgement and to make choices on the grounds of economy. Essentially these are

restricted to control of the effective length, by introducing intermediate restraints where appropriate, and the original choice of cross section.

However, certain other points relating to columns may well have a bearing on the overall economy of the steel frame or truss. Of particular concern is the need to be able to make connections simply. In a multi-storey frame, the use of heavier UC sections thus may be advantageous, permitting beam-to-column connections to be made without the need for stiffening the flanges or web. Similarly in order to accommodate beams framing into the column web an increase in the size of UC may eliminate the need for special detailing.

While compound angle members were a common feature of early trusses, maintenance costs due both to the surface area requiring painting and to the incidence of corrosion caused by the inherent dirt and moisture traps have caused a change to the much greater use of tubular members. If site joints are kept to the minimum tubular trusses can be transported and handled on site in long lengths and a more economic as well as a visually more pleasing structure is likely to result.

References to Chapter 15

1. The Steel Construction Institute (SCI) (2001) *Steelwork Design Guide to BS 5950: Part 1: 2000 Vol. 1: Section Properties, Member Capacities*, 6th edn. SCI, Ascot, Berks.
2. Ballio G. & Mazzolani F.M. (1983) *Theory and Design of Steel Structures*. Chapman and Hall.
3. Allen H.G. & Bulson P.S. (1980) *Background to Buckling*. McGraw-Hill.
4. Wood R.H. (1974) Effective lengths of columns in multi-storey buildings. *The Structural Engineer*, **52**, Part 1, July, 235–44, Part 2, Aug., 295–302, Part 3, Sept., 341–6.

Further reading for Chapter 15

ECCS (1986) *Behaviour and Design of Steel Plated Structures* (Ed. by P. Dubas & E. Gehri). ECCS Publication No. 44.

Galambos T.V. (Ed.) (1998) *Guide to Stability Design Criteria for Metal Structures*, 5th edn. Wiley, New York.

Hancock G.J. (1988) *Design of Cold-Formed Steel Structures*. Australian Institute of Steel Construction, Sydney.

Kirby P.A. & Nethercot D.A. (1985) *Design for Structural Stability*. Collins, London.

Trahair N.S. & Nethercot D.A. (1984) *Bracing Requirements in Thin-Walled Structures, Developments in Thin-Walled Structures – 2* (Ed. by J. Rhodes & A.C. Walker), pp. 92–130. Elsevier Applied Science Publishers, Barking, Essex.

A series of worked examples follows which are relevant to Chapter 15.

The **Steel Construction** **Institute** Silwood Park, Ascot, Berks SL5 7QN 	Subject **COLUMN EXAMPLE 1** **ROLLED UNIVERSAL** **COLUMN**	Chapter ref. **15**
	Design code **BS 5950: Part 1**	Made by **DAN**	Sheet no. **1**
		Checked by **GWO**	

Problem

Check the ability of a 203 × 203 × 52 UC in S275 steel to withstand an axial compressive load of 1250 kN over an unsupported height of 3.6 m assuming that both ends are held in position but are provided with no restraint in direction. Design to BS 5950: Part 1.

The problem is as shown in the sketch.

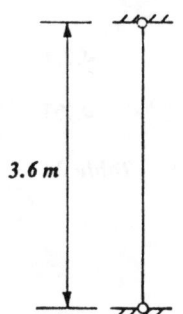

Take effective length	L_E = *1.0L*	*Table 22*
∴	L_E = *1.0 × 3.6 × 10³*	
	= *3600 mm*	

On the assumption that weak axis flexural buckling will govern determine compressive strength p_c from Table 24 curve c — *Table 23*

A = *66.4 cm²*	*b/T* = *8.16*	***Steelwork***
		Design Guide
r_y = *5.16 cm*	*d/t* = *20.1*	***Vol 1***

Since T < 16 mm   *take* p_y = *275 N/mm²*   *Table 9*

Check section classification for pure compression.   *3.5*

Need only check section is not slender;   *Table 11*

for outstand   *b/T* ≤ *15ε*

The Steel Construction Institute Silwood Park, Ascot, Berks SL5 7QN	Subject **COLUMN EXAMPLE 1** **ROLLED UNIVERSAL** **COLUMN**	Chapter ref. **15**
	Design code **BS 5950: Part 1**	Made by **DAN**	Sheet no. **2**
		Checked by **GWO**	

for web $d/t \leq 40\varepsilon$

$\varepsilon = \sqrt{(275/p_y)} = 1.0$

actual b/T   =   8.16, within limit

actual d/t   =   20.1, within limit

∴ *Section is not slender.*

$P_c = A_g p_c$ 4.7.4

$\lambda = L_E/r_y = 3600/51.6 = 70$ 4.7.3

For $\lambda = 70$ and $p_y = 275\,N/mm^2$ *Table 24*

value of $p_c = 202\,N/mm^2$

∴ $P_c = 6640 \times 202 = 1341 \times 10^3\,N$

$= 1341\,kN$

This exceeds required resistance of 1250 kN and section is therefore OK.

∴ *Use 203 × 203 × 52 UC*

	Subject	Chapter ref.
The Steel Construction Institute Silwood Park, Ascot, Berks SL5 7QN	*COLUMN EXAMPLE 2* *WELDED BOX*	**15**
	Design code *BS 5950: Part 1*	Made by *DAN* Checked by *GWO*	Sheet no. **1**

Problem

Check the ability of a 960 mm square box column fabricated from 30 mm thick S355 plate to withstand an axial compressive load of 22000 kN over an unsupported height of 15 m assuming that both ends are held in position but are provided with no restraint in direction. Design to BS 5950: Part 1.

The problem is as shown in the sketches.

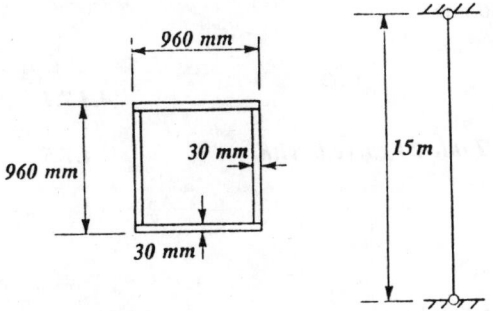

Take effective length $L_E = 1.0L$ Table 22

$$\therefore \quad L_E = 1.0 \times 15 \times 10^3$$

$$= 15000 \, mm$$

Determine p_c from Table 24, curve b   Table 23

Since $T > 16\,mm$, take $p_y = 345\,N/mm^2$ Table 11

Check section classification for pure compression.   3.5

Need only check section is not slender;   Table 11

for flange $b/T \le 40\varepsilon$

$$\varepsilon = \sqrt{(275/345)} = 0.9$$

The Steel Construction Institute	Subject	Chapter ref.
Silwood Park, Ascot, Berks SL5 7QN	*COLUMN EXAMPLE 2 WELDED BOX*	*15*
	Design code	Made by *DAN*	Sheet no. *2*
	BS 5950: Part 1	Checked by *GWO*	

actual b/T = (960 – 2 × 30)/30 = 30

limit   = 0.9 × 40   = 36.0

∴ Section is not slender.

$I_y = (960 \times 960^3 - 900 \times 900^3)/12 = 1.61 \times 10^{10} \, mm^4$

$A_g = 2 \times 30(960 + 900) = 111600 \, mm^2$

$r_y = \sqrt{(1.61 \times 10^{10}/111600)} = 380 \, mm$

$\lambda = L_E/r_y = 15000/380$
$= 39.5$ *4.3.7.1*

Since section fabricated by welding, use Table 24 curve b with   *4.7.5*

$p_y = 345 - 20$

$= 325 \, N/mm^2$

For λ = 39.5   and p_y = 325 N/mm²   *Table 27b*

value of p_c = 293 N/mm²

$P_c = A_g p_c = 111600 \times 293$ *4.7.4*

$= 32.7 \times 10^6 \, N$

$= 32700 \, kN$

This exceeds required resistance of 22000 kN and section is OK.

∴ Use section shown in figure.

The Steel Construction Institute Silwood Park, Ascot, Berks SL5 7QN	Subject *COLUMN EXAMPLE 3 ROLLED UNIVERSAL BEAM*	Chapter ref. *15*

Design code *BS 5950: Part 1*	Made by *DAN*	Sheet no. *1*
	Checked by *GWO*	

Problem

A 457 × 191 × 89 UB in S275 steel is to be used as an axially loaded column over a free height of 7.0 m. Both ends will be held in position but not direction for both planes. The possibility exists to provide discrete bracing members capable of preventing deflection in the plane of the flanges only. Investigate the advisability of using such bracing.

Clearly strength cannot exceed that for major axis failure. Check no. of intermediate (minor-axis) braces needed to achieve this.

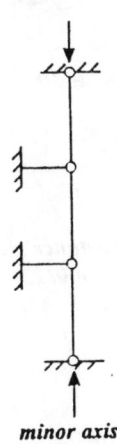

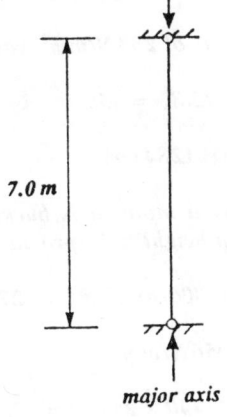

7.0 m

minor axis   major axis

Take L_{Ex} = *1.0L*

∴ L_{Ex} = $1.0 \times 7.0 \times 10^3$ = *7000 mm*

A = *114 cm²* b/T = *5.42*   *Steelwork Design Guide Vol 1*

r_y = *4.28 cm* d/t = *38.5*

r_x = *19.0 cm* p_y = *265 N/mm² (for T = 17.7 mm)*

Section not slender   *Table 11*

P_c = $A_g p_c$ *4.7.4*

λ_x = L_{Ex}/r_x = *7000/190*   *4.7.3*
       = *37*

Use Table 24, curve a for p_{cx}   *Table 23*

The Steel Construction Institute Silwood Park, Ascot, Berks SL5 7QN	Subject **COLUMN EXAMPLE 3** **ROLLED UNIVERSAL BEAM**		Chapter ref. **15**
	Design code *BS 5950: Part 1*	Made by *DAN*	Sheet no. **2**
		Checked by *GWO*	

For $\lambda = 37$ and $p_y = 265\,N/mm^2$ *Table 24*

value of $p_{cx} = 253\,N/mm^2$

$P_{cx} = 11400 \times 253 = 2.884 \times 10^6\,N$

$\qquad\qquad\qquad\quad = \underline{2884\,kN}$

Use Table 24, curve b for p_{cy} *& Table 23*

For a p_{cy} value of $253\,N/mm^2$, value of $\lambda_y \not> 30$ *Table 24*

$\therefore \qquad L_{Ey}/42.8 = 30$

and $L_{Ey} \not> 1284\,mm$

$\therefore$ to achieve a minor axis buckling resistance $\not< 2987\,kN$ would require 7.0 m height to be provided with $7000/1284 \rightarrow 5$ restraints

$L_{Ey}/r_y = (7000/6)/42.8 = 27$

$p_{cy} = 256\,N/mm^2$ *Table 24*

$P_{cy} = 11400 \times 256 \quad = 2.918 \times 10^6$

$\qquad\qquad\qquad\qquad\quad = \underline{2918\,kN}$

Not however that substantial improvements on the basic minor axis resistance for $L_{Ey} = 7.0\,m$ of $741\,kN$ may be achieved for rather less restraints as shown below.

no. of restraints	P_{cy} (kN)	P_{cy}/P_{cx}
0	730	0.25
1	1972	0.68
2	2531	0.88
3	2645	0.92

Chapter 16
Beams

by DAVID NETHERCOT

16.1 Common types of beam

Beams are possibly the most fundamental type of member present in a civil engineering structure. Their principal function is the transmission of vertical load by means of flexural (bending) action into, for example, the columns in a rectangular building frame or the abutments in a bridge which support them.

Table 16.1 provides some idea of the different structural forms suitable for use as beams in a steel structure; several of these are illustrated in Fig. 16.1. For modest spans, including the majority of those found in buildings, the use of standard hot-rolled sections (normally UBs but possibly UCs if minimizing floor depth is a prime consideration or channels if only light loads need to be supported) will be sufficient. Lightly loaded members such as the purlins supporting the roof of a portal-frame building are frequently selected from the range of proprietary cold-formed sections produced from steel sheet only a few millimetres thick, normally already protected against corrosion by galvanizing, in a variety of highly efficient shapes, advantage being taken of the roll-forming process to produce sections with properties carefully selected for the task they are required to perform. For spans in excess of those that can be achieved sensibly using ready-made sections some form of built-up member is required. Castellated beams, formed by profile cutting of the web and welding to produce a deeper section, typically 50% deeper using the standard UK geometry, are visually attractive but cannot withstand high shear loads unless certain of the castellations are filled in with plate. The range of spans for which UBs may be used can be extended if cover plates are welded to both flanges.

Alternatively a beam fabricated entirely by welding plates together may be employed allowing variations in properties by changes in depth, for example, flange thickness, or, in certain cases where the use of very thin webs is required, stiffening to prevent premature buckling failure is necessary. A full treatment of the specialist aspects of plate-girder design is provided in Chapter 17. If spans are so large that a single member cannot economically be provided, then a truss may be a suitable alternative. In addition to the deep truss fabricated from open hot-rolled sections, SHS or both, used to provide long clear spans in sports halls and supermarkets, smaller prefabricated arrangements using RHS or CHS provide an attractive alternative to the use of standard sections for more modest spans. Truss design is discussed in Chapter 19.

431

Table 16.1　Typical usage of different forms of beam

Beam type	Span range (m)	Notes
(0) Angles	3–6	Used for roof purlins, sheeting rails, etc., where only light loads have to be carried
(1) Cold-formed sections	4–8	Used for roof purlins, sheeting rails, etc., where only light loads have to be carried
(2) Rolled sections: UBs, UCs, RSJs, RSCs	1–30	Most frequently used type of section; proportions selected to eliminate several possible types of failure
(3) Open web joists	4–40	Prefabricated using angles or tubes as chords and round bar for web diagonals, used in place of rolled sections
(4) Castellated beams	6–60	Used for long spans and/or light loads; depth of UB increased by 50%; web openings may be used for services, etc.
(5) Compound sections e.g. UB + RSC	5–15	Used when a single rolled section would not provide sufficient capacity; often arranged to provide enhanced horizontal bending strength as well
(6) Plate girders	10–100	Made by welding together 3 plates sometimes automatically; web depths up to 3–4 m sometimes need stiffening
(7) Trusses	10–100	Heavier version of (3); may be made from tubes, angles or, if spanning large distances, rolled sections
(8) Box girders	15–200	Fabricated from plate, usually stiffened; used for OHT cranes and bridges due to good torsional and transverse stiffness properties

Since the principal requirement of a beam is adequate resistance to vertical bending, a very useful indication of the size of section likely to prove suitable may be obtained through the concept of the span to depth ratio. This is simply the value of the clear span divided by the overall depth. An average figure for a properly designed steel beam is between 15 and 20, perhaps more if a particularly slender form of construction is employed or possibly less if very heavy loadings are present.

When designing beams, attention must be given to a series of issues, in addition to simple vertical bending, that may have some bearing upon the problem. Torsional loading may often be eliminated by careful detailing or its effects reasonably regarded as of negligible importance by a correct appreciation of how the structure actually behaves; in certain instances it should, however, be considered. Section classification (allowance for possible local buckling effects) is more involved than is the case for struts since different elements in the cross-section are subject to different patterns of stress; the flanges in an I-beam in the elastic range will be in approximately uniform tension or compression while the web will contain a stress gradient. The possibility of members being designed for an elastic or a plastic state, including the use of a full plastic design for the complete structure, also affects section classification. Various forms of instability of the beam as a whole or of parts subject to locally high stresses, such as the web over a support, also require attention. Finally certain forms of construction may lend themselves to the appearance of unacceptable vibrations; although this is likely to be affected by the choice of beams, its coverage is left to Chapter 20 on floors.

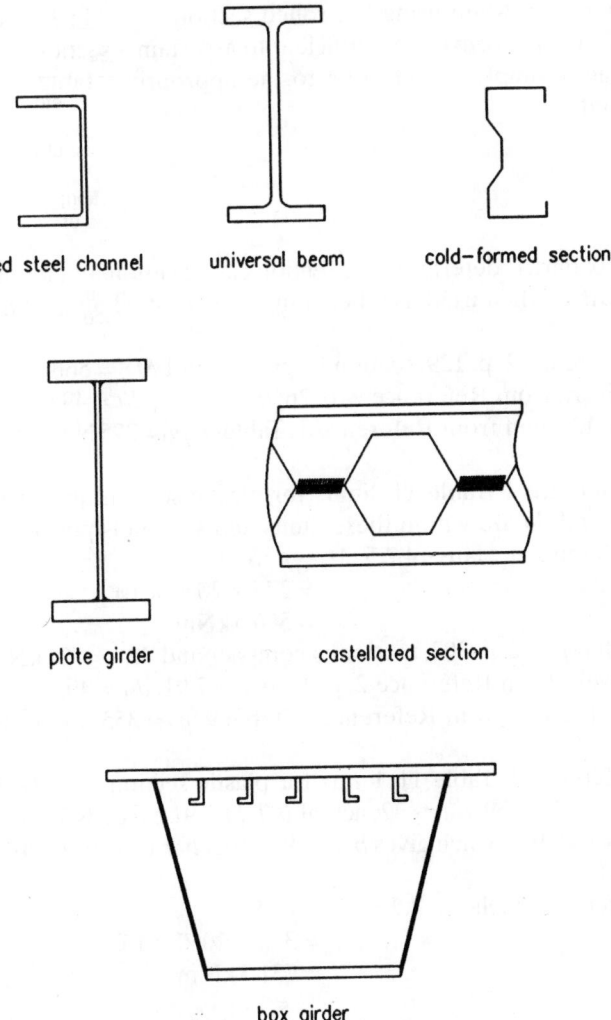

Fig. 16.1 Types of beam cross-section

16.2 Cross-section classification and moment capacity, M_c

The possible influences of local buckling on the ability of a particular cross-section to attain, and where appropriate also to maintain, a certain level of moment are discussed in general terms in Chapter 13. In particular, section 13.2 covers the influence of flange (b/T) and web (d/t) slenderness on moment–rotation behaviour (Fig. 13.4) and moment capacity (Fig. 13.5). When designing beams it is usually sufficient to consider the web and the compression flange separately using the appropriate sets of limits.[1]

In building design when using hot-rolled sections, for which relevant properties are tabulated,[2] it will normally be sufficient to ascertain a section's classification and moment capacity simply by referring to the appropriate table. An example illustrates the point.

Example

Using BS 5950: Part 1[1] determine the section classification and moment capacity for a 533 × 210 UB82 when used as a beam in (1) S275 steel and (2) S355 steel.

(1) From Reference 2, p. 129, section is 'plastic' and $M_{cx} = 566$ kN m
  Alternatively, from Reference 2, p. 26, $b/T = 7.91$, $d/t = 49.6$
  Since $T = 13.2$ mm from Reference 1, Table 9, $p_y = 275$ N/mm^2 and $\varepsilon = (275/p_y)^{\frac{1}{2}}$
  $= 1.0$
  From Reference 1, Table 11, limits for plastic section are $b/t \leqslant 9\varepsilon$, $d/t \leqslant 80\varepsilon$; actual b/t and d/t are within these limits and section is plastic.
  From Reference 1, clause 4.2.5, $M_c = p_y S$
  $$= 275 \times 2060 \times 10^{-3}$$
  $$= 566.5 \text{ kNm}$$

(2) From Reference 2, p. 303, section is compact and $M_{cx} = 731$ kN m
  Alternatively, from Reference 2, p. 26, $b/T = 7.91$, $d/t = 49.6$
  Since $T = 13.2$ mm from Reference 1, Table 9, $p_y = 355$ N/mm^2 and $\varepsilon = (275/p_y)^{\frac{1}{2}}$
  $= 0.88$
  From Reference 1, Table 11, limits for plastic section are $b/T \leqslant 9\varepsilon$, $d/t = 80\varepsilon$, which give $b/T \leqslant 7.9$, $d/t \leqslant 72$; actual b/T of 7.91 exceeds 7.9 so check compact limits of $b/T \leqslant 10\varepsilon$, which gives $b/T \leqslant 9$; actual b/T of 7.91 meets this and section is compact
  From Reference 1, clause 4.2.5, $M_c = p_y S$
  $$= 355 \times 2060 \times 10^{-3}$$
  $$= 731.3 \text{ kNm}$$

Thus whichever grade of steel is being used, the moment capacity, M_c, will be the maximum attainable value corresponding to the section's full plastic moment capacity, M_p. Provided that S275 material is used, the beam is capable of redistributing moments, since the plastic cross section behaves as illustrated in Fig. 13.4(a) and so could be used in a plastically designed frame. Use of the higher strength S355 material, while not affecting the beam's ability to attain M_p, precludes the use of plastic design since redistribution of moments cannot occur due to the lack of rotation capacity implied by the behaviour illustrated in Fig. 13.4(b). For a continuous structure designed on the basis of elastic analysis to determine the distribution of internal forces and moments or for simple construction, the question of rotation capacity is not relevant and in both cases moment capacity should be taken as M_p. The fact that the section is plastic for S275 material is then of no particular relevance; a more appropriate classification would be to regard the section as compact or better.

Table 16.2 Non-plastic UBs and UCs in bending

	Compact	Semi-compact
	S275	
UC	305 × 305 × 97 203 × 203 × 46	356 × 368 × 129 152 × 152 × 23
	S355	
UB	762 × 267 × 134 406 × 178 × 54 406 × 140 × 39 305 × 165 × 40 254 × 146 × 31 203 × 133 × 25	356 × 171 × 45
UC	305 × 305 × 118 203 × 203 × 52 152 × 152 × 30	356 × 368 × 153 356 × 368 × 129 305 × 305 × 97 254 × 254 × 73 203 × 203 × 46 152 × 152 × 23

Inspection of the relevant tables in Reference 2 shows that when used as beams:

S275 steel
all UBs are plastic
all but 4 UCs are plastic
(2 are semi-compact)
S355 steel
all but 7 UBs are plastic
(1 is semi-compact)
all but 9 UCs are plastic
(6 are semi-compact)

In those cases of semi-compact sections, moment capacities based on M_y are listed. The full list of non-plastic beam sections is given as Table 16.2.

When using fabricated sections individual checks on the web and compression flange using the actual dimensions of the trial section must be made. It normally proves much simpler if proportions are selected so as to ensure that the section is compact or better since the resulting calculations need not then involve the various complications associated with the use of non-compact sections. This point is particularly noticeable when designing bridge beams to BS 5400: Part 3. When slender sections are used the amount of calculation increases considerably due to the need to consider loss of effectiveness of some parts of the cross section due to local buckling when determining M_c. Probably the most frequent use of slender sections involves cold-formed shapes used for example as roof purlins. Because of their

proprietary nature, the manufacturers normally provide design information, much of it based on physical testing, listing such properties as moment capacity. In the absence of design information, reference should be made to Part 5 of BS 5950 for suitable calculation methods.

Although the part of the web between the flange and the horizontal edge of the castellation in a castellated beam frequently exceeds the compact limit for an outstand, sufficient test data exist to show that this does not appear to influence the moment capacity of such sections. Section classification should therefore be made in the same way as for solid web beams, the value of M_c being obtained using the net modulus value for the section at the centre of a castellation.

16.3 Basic design

One (or more) of a number of distinct limiting conditions may, in theory, control the design of a particular beam as indicated in Table 16.3, but in any particular practical case only a few of them are likely to require full checks. It is therefore convenient to consider the various possibilities in turn, noting the conditions under which each is likely to be important. For convenience the various phenomena are first considered principally within the context of using standard hot-rolled sections, i.e. UBs, UCs, RSJs and channels; other types of cross-section are covered in the later parts of this chapter.

Table 16.3 Limiting conditions for beam design

Ultimate	Serviceability
Moment capacity, M_e (including influence of local buckling)	Deflections due to bending (and shear if appropriate)
Shear capacity, P_v	Twist due to torsion
Lateral–torsional buckling, M_b	Vibration
Web buckling, P_w	
Web bearing, P_{yw}	
Moment–shear interaction	
Torsional capacity, M_T	
Bending–torsion interaction	

16.3.1 Moment capacity, M_c

The most basic design requirement for a beam is the provision of adequate in-plane bending strength. This is provided by ensuring that M_c for the selected section exceeds the maximum moment produced by the factored loading. Determination of M_c, which is linked to section classification, is fully covered in section 16.2.

For a statically determinate structural arrangement, simple considerations of statics provide the moment levels produced by the applied loads against which M_c must be checked. For indeterminate arrangements, a suitable method of elastic analysis such as moment distribution or slope deflection is required. The justification for using an elastically obtained distribution of moments with, in the case of compact or plastic cross-sections, a plastic cross-sectional resistance has been fully discussed by Johnson and Buckby.[3]

16.3.2 Effect of shear

Only in cases of high coincident shear and moment, found for example at the internal supports of continuous beams, is the effect of shear likely to have a significant influence on the design of beams.

Shear capacity P_v is normally calculated as the product of a shear strength p_v, often taken for convenience as $0.6p_y$ which is close to the yield stress of steel in shear of $1/\sqrt{3}$ of the uniaxial tensile yield stress, and an appropriate shear area A_v. The process approximates the actual distribution of shear stress in a beam web as well as assuming some degree of plasticity. While suitable for rolled sections, it may not therefore be applicable to plate girders. An alternative design approach, more suited to webs containing large holes or having variations in thickness, is to work from first principles and to limit the maximum shear stress to a suitable value; BS 5950: Part 1 uses $0.7p_y$. In cases where $d/t > 63$, shear buckling limits the effectiveness of the web and reference to Chapter 17 should be made for methods of determining the reduced capacity.

In principle the presence of shear in a section reduces its moment capacity. In practice the reduction may be regarded as negligibly small up to quite large fractions of the shear capacity P_v. For example, BS 5950: Part 1 requires a reduction in M_c for plastic or compact sections only when the applied shear exceeds $0.6P_v$, and permits the full value of M_c to be used for all cases of semi-compact or slender sections.

Figure 16.2 illustrates the application of the BS 5950: Part 1 rule for plastic or compact sections to a typical UB and UC having approximately equal values of plastic section modulus S_x. Evaluation of the formula in cases where $F_v/P_v > 0.6$ first requires that the value of S_v, the plastic modulus of the shear area A_v (equal to tD in this case), be determined. This is readily obtained from the tabulated values of S for rectangles given in Reference 2 corresponding to a linear reduction from S_x to $(S_x - S_v)$.

16.3.3 Deflection

When designing according to limit state principles it is customary to check that deflections at working load levels will not be such as to impair the proper function

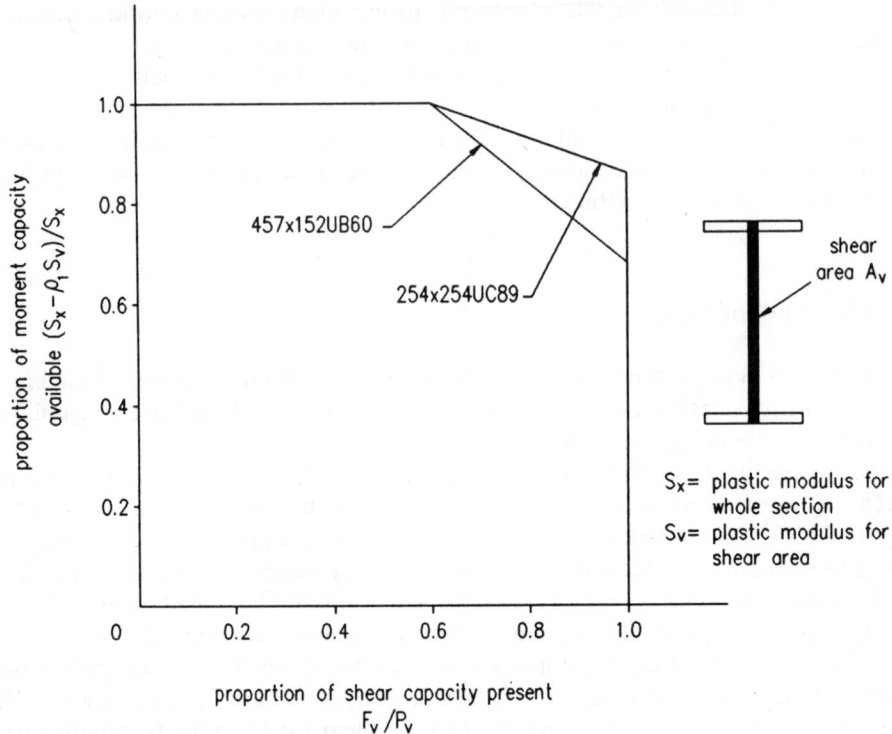

Fig. 16.2   Moment–shear interaction for plastic or compact sections to BS 5950: Part I

of the structure. For beams, examples of potentially undesirable consequences of excessive serviceability deflections include:

- cracking of plaster ceilings
- allowing crane rails to become misaligned
- causing difficulty in opening large doors.

Although earlier codes of practice specified limits for working load deflections, the tendency with more recent documents[1] is to draw attention to the need for deflection checks and to provide advisory limits to be used only when more specific guidance is not available. Table 8 of BS 5950: Part 1 gives 'recommended limitations' for certain types of beams and crane gantry girders and states that 'Circumstances may arise where greater or lesser values would be more appropriate.' Not surprisingly surveys of current practice[4] reveal large variations in what is considered appropriate for different circumstances.

When checking deflections of steel structures under serviceability loading, the central deflection Δ_{max} of a uniformly-loaded simply-supported beam, assuming linear–elastic behaviour, is given by

Table 16.4   Minimum *I* values for uniformly-loaded simply-supported beams for various deflection limits

α	200	240	250	325	360	400	500	600	750	1000
K	1.24	1.49	1.55	2.02	2.23	2.48	3.10	3.72	4.65	6.20

$$\Delta_{max} = \frac{5}{384} \frac{WL^3}{EI} 10^{12} \, (\text{mm}) \tag{16.1}$$

in which W = total load (kN)

$\quad\quad\quad\quad E$ = Young's modulus (N/mm^2)

$\quad\quad\quad\quad I$ = second moment of area (mm^4)

$\quad\quad\quad\quad L$ = span (m).

If Δ_{max} is to be limited to a fraction of L, Equation (16.1) may be rearranged to give

$$I_{rqd} = 0.62 \times 10^{-2} \, \alpha WL^2 \tag{16.2}$$

in which α defines the deflection limit as

$$\Delta_{max} = L/\alpha$$

and I_{rqd} is now in cm^4.

Writing $I_{rqd} = KWL^2$, Table 16.4 gives values of K for a range of values of α.

Since deflection checks are essentially of the 'not greater than' type, some degree of approximation normally is acceptable, particularly if the calculations are reduced as a result. Converting complex load arrangements to a roughly equivalent UDL permits Table 16.4 to be used for a wide range of practical situations. Table 16.5 gives values of the coefficient K by which the actual load arrangement shown should be multiplied in order to obtain an approximately equal maximum deflection.

Tables of deflections for a number of standard cases are provided in the Appendix.

16.3.4 Torsion

Beams subjected to loads which do not act through the point on the cross-section known as the shear centre normally suffer some twisting. Methods for locating the shear centre for a variety of sectional shapes are given in Reference 5. For doubly symmetrical sections such as UBs and UCs it coincides with the centroid while for channels it is situated on the opposite side of the web from the centroid; for rolled channels its location is included in the tables of Reference 2. Figure 16.3 illustrates its position for a number of standard cases.

The effects of torsional loading may often be minimized by careful detailing, particularly when considering how loads are transferred between members. Proper attention to detail can frequently lead to arrangements in which the load transfer

Table 16.5 Equivalent UDL coefficients

a/L	K_Δ	No. of equal loads	b/L	c/L	K
0.5	1.0	2	0.2	0.6	0.91
0.4	0.86		0.25	0.5	1.10
0.375	0.82		0.333	0.333	1.3
0.333	0.74	3	0.167	0.333	1.05
0.3	0.68		0.2	0.3	1.14
0.25	0.58		0.25	0.25	1.27
0.2	0.47	4	0.125	0.25	1.03
0.1	0.24		0.2	0.2	1.21
		5	0.1	0.2	1.02
			0.167	0.167	1.17
		6	0.083	0.167	1.01
			0.143	0.143	1.15
		7	0.071	0.143	1.01
			0.125	0.125	1.12
		8	0.063	0.125	1.01
			0.111	0.111	1.11

$$K_\Delta = 1.6\frac{a}{L}\left[3 - 4\left(\frac{a}{L}\right)^2\right]$$

a/L	0.01	0.05	0.1	0.15	0.2	0.25	0.3	0.35	0.4	0.45	0.5
K_Δ	0.05	0.24	0.47	0.70	0.91	1.10	1.27	1.41	1.51	1.58	1.60

is organized in such a way that twisting should not occur.[5] Whenever possible this approach should be followed as the open sections normally used as beams are inherently weak in resisting torsion. In circumstances where beams are required to withstand significant torsional loading, consideration should be given to the use of a torsionally more efficient shape such as a structural hollow section.

16.3.5 Local effects on webs

At points within the length of a beam where vertical loads act, the web is subject to concentrations of stresses, additional to those produced by overall bending. Failure by buckling, rather in the manner of a vertical strut, or by the development of unacceptably high bearing stresses in the relatively thin web material

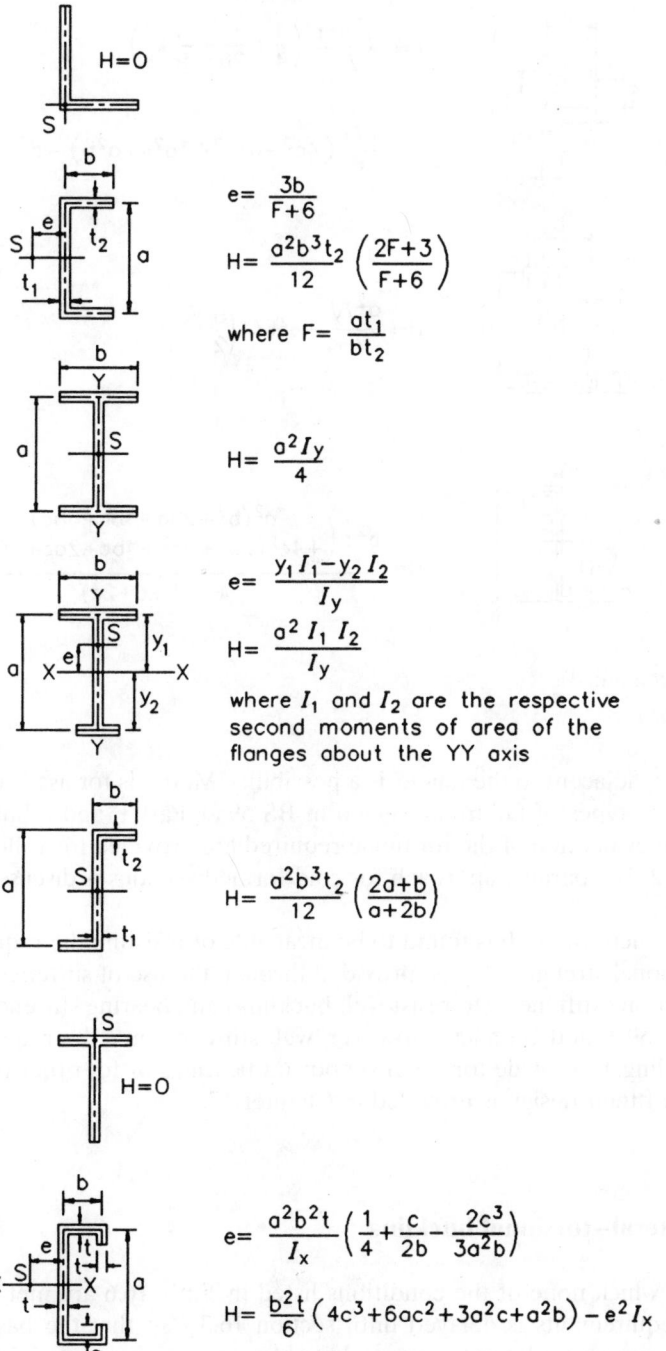

Fig. 16.3   Location of shear centre for standard sections (*H* is warping constant)

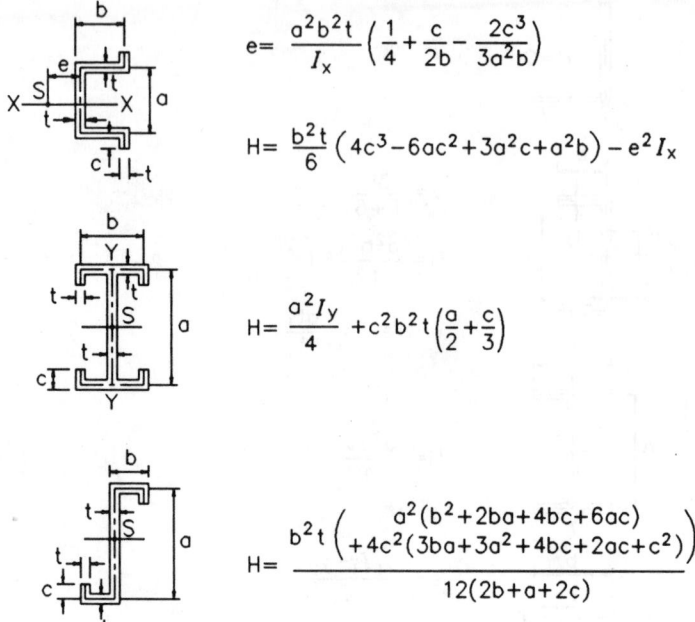

$$e = \frac{a^2b^2t}{I_x}\left(\frac{1}{4} + \frac{c}{2b} - \frac{2c^3}{3a^2b}\right)$$

$$H = \frac{b^2t}{6}\left(4c^3 - 6ac^2 + 3a^2c + a^2b\right) - e^2 I_x$$

$$H = \frac{a^2 I_y}{4} + c^2 b^2 t\left(\frac{a}{2} + \frac{c}{3}\right)$$

$$H = \frac{b^2t\left(\begin{array}{c} a^2(b^2 + 2ba + 4bc + 6ac) \\ + 4c^2(3ba + 3a^2 + 4bc + 2ac + c^2) \end{array}\right)}{12(2b + a + 2c)}$$

Fig. 16.3   (continued)

immediately adjacent to the flange, is a possibility. Methods for assessing the likelihood of both types of failure are given in BS 5950: Part 1, and tabulated data to assist in the evaluation of the formulae required are provided for rolled sections in Reference 2. The parallel approach for cold-formed sections is discussed in section 16.7.

In cases where the web is found to be incapable of resisting the required level of load, additional strength may be provided through the use of stiffeners. The design of load-carrying stiffeners (to resist web buckling) and bearing stiffeners is covered in both BS 5950 and BS 5400. However, web stiffeners may be required to resist shear buckling, to provide torsional support at bearings or for other reasons; a full treatment of their design is provided in Chapter 17.

16.3.6 Lateral–torsional buckling

Beams for which none of the conditions listed in Table 16.6 are met (explanation of these requirements is delayed until section 16.3.7 so that the basic ideas and parameters governing lateral–torsional buckling may be presented first) are liable to have their load-carrying capacity governed by the type of failure illustrated in Fig. 16.4. Lateral–torsional instability is normally associated with beams subject to

Table 16.6 Types of beam not susceptible to lateral–torsional buckling

loading produces bending about the minor axis

beam provided with closely spaced or continuous lateral restraint

closed section

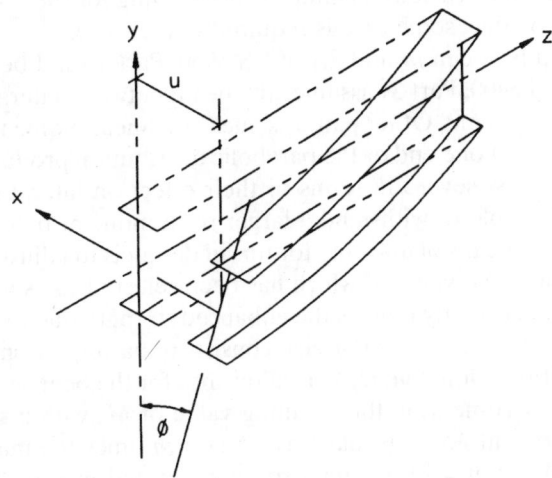

Fig. 16.4 Lateral–torsional buckling

vertical loading buckling out of the plane of the applied loads by deflecting sideways and twisting; behaviour analogous to the flexural buckling of struts. The presence of both lateral and torsional deformations does cause both the governing mathematics and the resulting design treatment to be rather more complex.

The design of a beam taking into account lateral–torsional buckling consists essentially of assessing the maximum moment that can safely be carried from a knowledge of the section's material and geometrical properties, the support conditions provided and the arrangement of the applied loading. Codes of practice, such as BS 5400: Part 3, BS 5950: Parts 1 and 5, include detailed guidance on the subject. Essentially the basic steps required to check a trial section (using BS 5950: Part I for a UB as an example) are:

(1) assess the beam's effective length L_E from a knowledge of the support conditions provided (clause 4.3.5)
(2) determine beam slenderness λ_{LT} using the geometrical parameters u (tabulated in Reference 2), L_E/r_y, v (Table 19 of BS 5950: Part 1) using values of x (tabulated in Reference 2).
(3) obtain corresponding bending strength p_b (Table 16)
(4) calculate buckling resistance moment $M_b = p_b \times$ the appropriate section modulus, S_x (class 1 or 2), Z_x (class 3), $Z_{x,eff}$ (class 4).

The central feature in the above process is the determination of a measure of the beam's lateral–torsional buckling strength (p_b) in terms of a parameter (λ_{LT}) which represents those factors which control this strength. Modifications to the basic process permit the method to be used for unequal flanged sections including tees, fabricated Is for which the section properties must be calculated, sections containing slender plate elements, members with properties that vary along their length, closed sections and flats. Various techniques for allowing for the form of the applied loading are also possible; some care is required in their use.

The relationship between p_b and λ_{LT} of BS 5950: Part 1 (and between σ_{li}/σ_{yc} and $\lambda_{LT} \sqrt{(\sigma_{yc}/355)}$ in BS 5400: Part 3) assumes the beam between lateral restraints to be subject to uniform moment. Other patterns, such as a linear moment gradient reducing from a maximum at one end or the parabolic distribution produced by a uniform load, are generally less severe in terms of their effect on lateral stability; a given beam is likely to be able to withstand a larger peak moment before becoming laterally unstable. One means of allowing for this in design is to adjust the beam's slenderness by a factor n, the value of which has been selected so as to ensure that the resulting value of p_b correctly reflects the enhanced strength due to the non-uniform moment loading. An alternative approach consists of basing λ_{LT} on the geometrical and support conditions alone but making allowance for the beneficial effects of non-uniform moment by comparing the resulting value of M_b with a suitably adjusted value of design moment $\overline{M}$. $\overline{M}$ is taken as a factor m times the maximum moment within the beam M_{max}; $m = 1.0$ for uniform moment and $m < 1.0$ for non-uniform moment. Provided that suitably chosen values of m and n are used, both methods can be made to yield identical results; the difference arises simply in the way in which the correction is made, whether on the slenderness axis of the p_b versus λ_{LT} relationship for the n-factor method or on the strength axis for the m-factor method. Figure 16.5 illustrates both concepts, although for the purpose of the figure the m-factor method has been shown as an enhancement of p_b by $1/m$ rather than a reduction in the requirement of checking M_b against $\overline{M} = mM_{max}$. BS 5950: Part 1 uses the m-factor method for all cases, while BS 5400: Part 3 includes only the n-factor method.

When the m-factor method is used the buckling check is conducted in terms of a moment $\overline{M}$ less than the maximum moment in the beam segment M_{max}; then a separate check that the capacity of the beam cross-section M_c is at least equal to M_{max} must also be made. In cases where $\overline{M}$ is taken as M_{max}, then the buckling check will be more severe than (or in the ease of a stocky beam for which $M_b = M_c$, identical to) the cross-section capacity check.

Allowance for non-uniform moment loading on cantilevers is normally treated somewhat differently. For example, the set of effective length factors given in Table 14 of Reference 1 includes allowances for the variation from the arrangement used as the basis for the strength–slenderness relationship due to both the lateral support conditions and the form of the applied loading. When a cantilever is subdivided by one or more intermediate lateral restraints positioned between its root and tip, then segments other than the tip segment should be treated as ordinary beam segments when assessing lateral–torsional buckling strength. Similarly a cantilever subject to

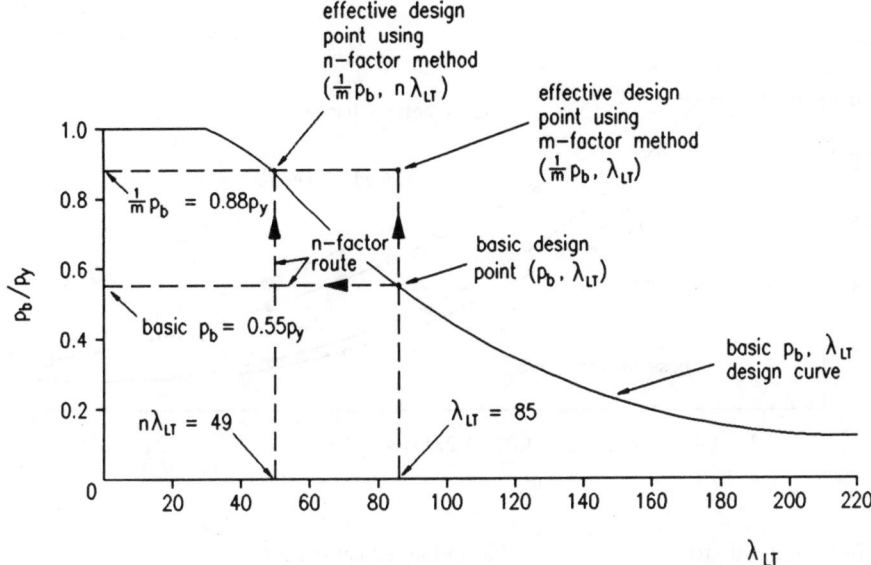

Fig. 16.5   Design modifications using *m*-factor or *n*-factor methods

an end moment such as horizontal wind load acting on a façade, should be regarded as an ordinary beam since it does not have the benefit of non-uniform moment loading.

For more complex arrangements that cannot reasonably be approximated by one of the standard cases covered by correction factors, codes normally permit the direct use of the elastic critical moment M_E. Values of M_E may conveniently be obtained from summaries of research data.[6] For example, BS 5950: Part 1 permits λ_{LT} to be calculated from

$$\lambda_{LT} = \sqrt{(\pi^2 E/p_y)} \sqrt{(M_p/M_E)} \tag{16.3}$$

As an example of the use of this approach Fig. 16.6 shows how significantly higher load-carrying capacities may be obtained for a cantilever with a tip load applied to its bottom flange, a case not specifically covered by BS 5950: Part 1.

16.3.7 Fully restrained beams

The design of beams is considerably simplified if lateral–torsional buckling effects do not have to be considered explicitly – a situation which will occur if one or more of the conditions of Table 16.6 are met.

In these cases the beam's buckling resistance moment M_b may be taken as its moment capacity M_c and, in the absence of any reductions in M_c due to local buck-

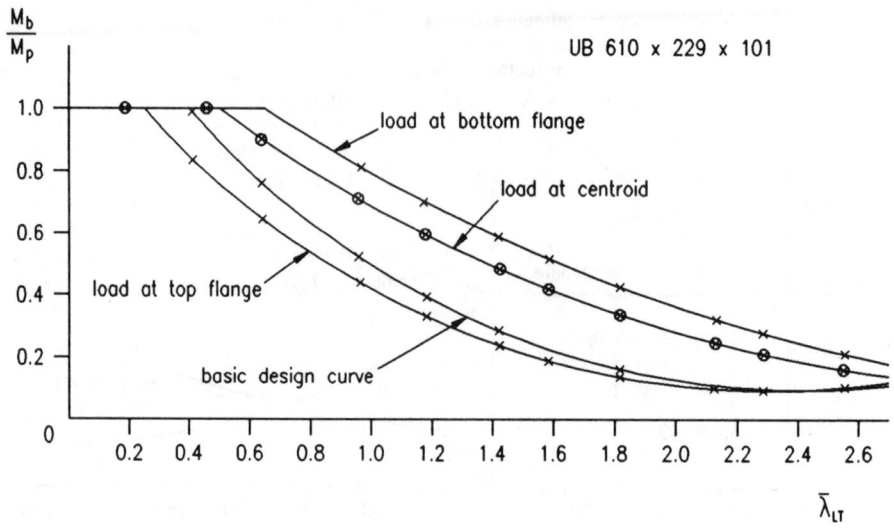

Fig. 16.6   Lateral–torsional buckling of a tip-loaded cantilever

ling, high shear or torsion, it should be designed for its full in-plane bending strength. Certain of the conditions corresponding to the case where a beam may be regarded as 'fully restrained' are virtually self-evident but others require either judgement or calculation.

Lateral–torsional buckling cannot occur in beams loaded in their weaker principal plane; under the action of increasing load they will collapse simply by plastic action and excessive in-plane deformation. Much the same is true for rectangular box sections even when bent about their strong axis. Figure 16.7, which is based on elastic critical load theory analogous to the Euler buckling of struts, shows that typical RHS beams will be of the order of ten times more stable than UB or UC sections of the same area. The limits on λ below which buckling will not affect M_b of Table 38 of BS 5950: Part 1, are sufficiently high ($\lambda = 340, 225$ and 170 for D/B ratios of 2, 3 and 4, and $p_y = 275\,\text{N/mm}^2$) that only in very rare cases will lateral–torsional buckling be a design consideration.

Situations in which the form of construction employed automatically provides some degree of lateral restraint or for which a bracing system is to be used to enhance a beam's strength require careful consideration. The fundamental requirement of any form of restraint if it is to be capable of increasing the strength of the main member is that it limits the buckling type deformations. An appreciation of exactly how the main member would buckle if unbraced is a prerequisite for the provision of an effective system. Since lateral–torsional buckling involves both lateral deflection and twist, as shown in Fig. 16.4, either or both deformations may be addressed. Clauses 4.3.2 and 4.3.3 of BS 5950: Part 1 set out the principles governing the action of bracing designed to provide either lateral restraint or torsional restraint. In common with most approaches to bracing design these clauses assume

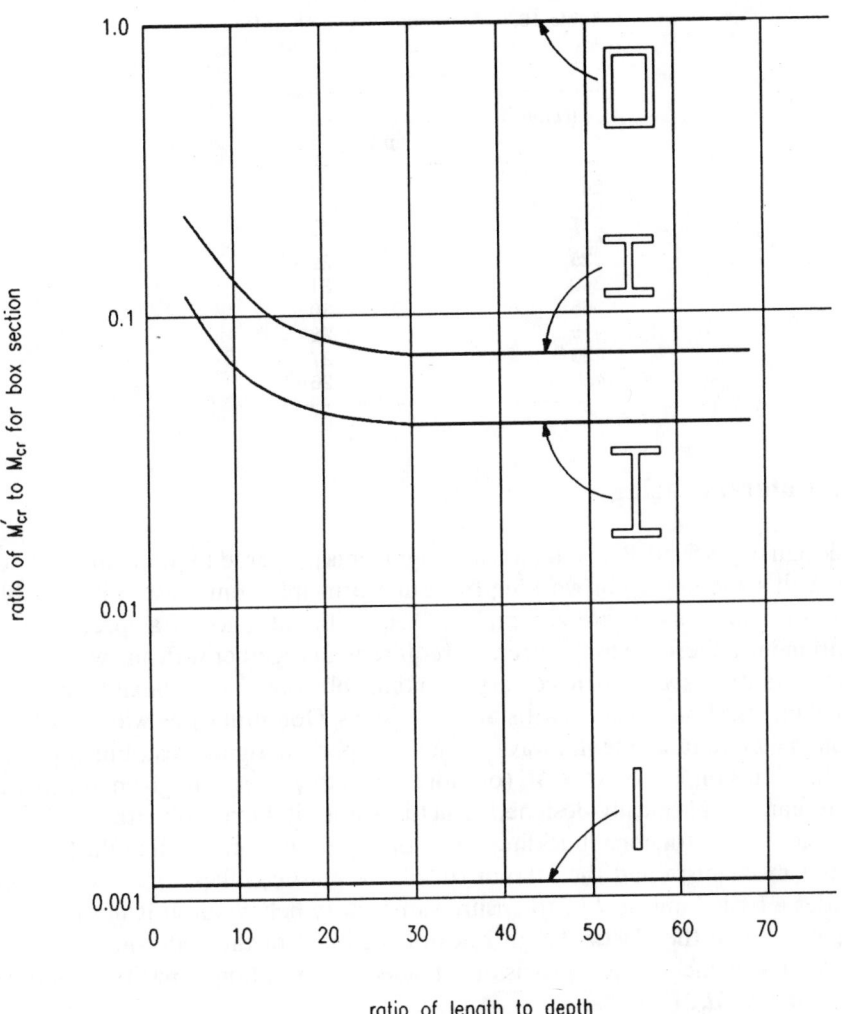

Fig. 16.7 Effect of type of cross-section on theoretical elastic critical moment

that the restraints will effectively prevent movement at the braced cross-sections, thereby acting as if they were rigid supports. In practice, bracing will possess a finite stiffness. A more fundamental discussion of the topic, which explains the exact nature of bracing stiffness and bracing strength, may be found in References 7 and 8. Noticeably absent from the code clauses is a quantitative definition of 'adequate stiffness', although it has subsequently been suggested that a bracing system that is 25 times stiffer than the braced beam would meet this requirement. Examination of Reference 7 shows that while such a check does cover the majority of cases, it is still possible to provide arrangements in which even much stiffer bracing cannot supply full restraint.

Table 16.7   Maximum values of λ_{LT} for which $p_b = p_y$ for rolled sections

p_y (N/mm^2)	Value of λ_{LT} up to which $p_b = p_y$
245	37
265	35
275	34
325	32
340	31
365	30
415	28
430	27
450	26

16.4 Lateral bracing

For design to BS 5950: Part 1, unless the engineer is prepared to supplement the code rules with some degree of working from first principles, only restraints capable of acting as rigid supports are acceptable. Despite the absence of a specific stiffness requirement, adherence to the strength requirement together with an awareness that adequate stiffness is also necessary, avoiding obviously very flexible yet strong arrangements, should lead to satisfactory designs. Doubtful cases will merit examination in a more fundamental way.[7,8] Where properly designed restraint systems are used the limits on λ_{LT} for $M_b = M_c$ (or more correctly $p_b = p_y$) are given in Table 16.7.

For beams in plastically-designed structures it is vital that premature failure due to plastic lateral–torsional buckling does not impair the formation of the full plastic collapse mechanism and the attainment of the plastic collapse load. Clause 5.3.3 provides a basic limit on L/r_y to ensure satisfactory behaviour; it is not necessarily compatible with the elastic design rules of section 4 of the code since acceptable behaviour can include the provision of adequate rotation capacity at moments slightly below M_p.

The expression of clause 5.3.3 of BS 5950: Part 1,

$$L_m \leq \frac{38 r_y}{\left[f_c/130 + (p_y/275)^2 (x/36)^2 \right]^{\frac{1}{2}}} \tag{16.4}$$

makes no allowance for either of two potentially beneficial effects:

(1) moment gradient
(2) restraint against lateral deflection provided by secondary structural members attached to one flange as by the purlins on the top flange of a portal frame rafter.

The first effect may be included in Equation (16.4) by adding the correction term

of Brown,[9] the basis of which is the original work on plastic instability of Horne.[10] This is covered explicitly in clause 5.3.3. A method of allowing for both effects when the beam segment being checked is either elastic or partially plastic is given in Appendix G of BS 5950: Part 1; alternatively the effect of intermittent tension flange restraint alone may be allowed for by replacing L_m with an enhanced value L_s obtained from clause 5.3.4 of BS 5950: Part 1.

In both cases the presence of a change in cross-section, for example, as produced by the type of haunch usually used in portal frame construction, may be allowed for. When the restraint is such that lateral deflection of the beam's compression flange is prevented at intervals, then Equation (16.4) applies between the points of effective lateral restraint. A discussion of the application of this and other approaches for checking the stability of both rafters and columns in portal frames designed according to the principles of either elastic or plastic theory is given in section 18.7.

16.5 Bracing action in bridges – U-frame design

The main longitudinal beams in several forms of bridge construction will, by virtue of the structural arrangement employed, receive a significant measure of restraint against lateral–torsional buckling by a device commonly referred to as *U-frame* action. Figure 16.8 illustrates the original concept based on the half-through girder form of construction. (See Chapter 4 for a discussion of different bridge types.) In a simply-supported span, the top (compression) flanges of the main girders, although laterally unbraced in the sense that no bracing may be attached directly to them, cannot buckle freely in the manner of Fig. 16.4 since their lower flanges are restrained by the deck. Buckling must therefore involve some distortion of the girder web into the mode given in Fig. 16.8 (assuming that the end frames prevent lateral movement of the top flange). An approximate way of dealing with this is to regard each longitudinal girder as a truss in which the tension chord is fully

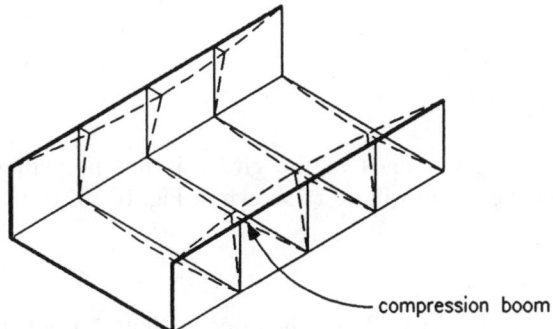

Fig. 16.8   Buckling of main beams of half-through girder

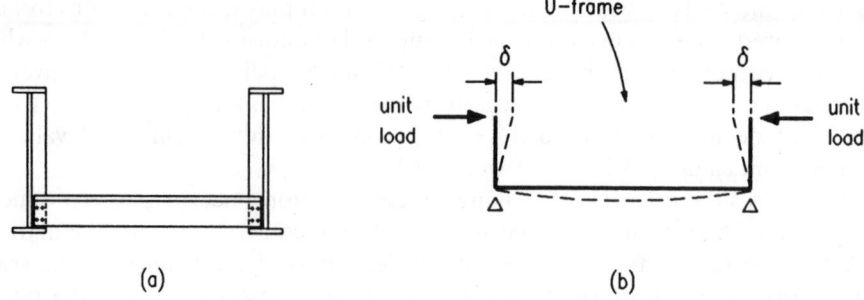

Fig. 16.9   U-frame restraint action. (a) Components of U-frame. (b) U-frame elastic support stiffness

laterally restrained and the web members, by virtue of their lateral bending stiffness, inhibit lateral movement of the top chord. It is then only a small step to regard this top chord as a strut provided with a series of intermediate elastic spring restraints against buckling in the horizontal plane. The stiffness of each support corresponds to the stiffness of the U-frame comprising the two vertical web stiffeners and the cross-girder and deck shown in Fig. 16.9.

The elastic critical load for the top chord is

$$P_{cr} = \pi^2 EI / L_E^2 \tag{16.5}$$

in which L_E is the effective length of the strut.

If the strut receives continuous support of stiffness $(1/\delta\, L_R)$ per unit length, in which L_R is the distance between U-frame restraints, and buckles in a single half-wave, this load will be given by

$$P_{cr} = \left(\pi^2 EI / L^2\right) + \left(L^2 / \pi^2 \delta L_R\right) \tag{16.6}$$

which gives a minimum value when

$$L = \pi (EI\delta L_R)^{0.25} \tag{16.7}$$

giving

$$P_{cr} = 2(EI / \delta L_R)^{0.5} \tag{16.8}$$

or

$$L_E = (\pi / \sqrt{2})(EI\delta L_R)^{0.25} \tag{16.9}$$

If lateral movement at the ends of the girder is not prevented by sufficiently stiff end U-frames, the mode will be as shown in Fig. 16.10. The effective length is then:

$$L_E = \pi (EI\delta L_R)^{0.25} \tag{16.10}$$

In clause 9.6.4.1.1.2 of BS 5400: Part 3 the effective length is given by

$$L_E = k(EI_c L_R \delta)^{0.25}$$

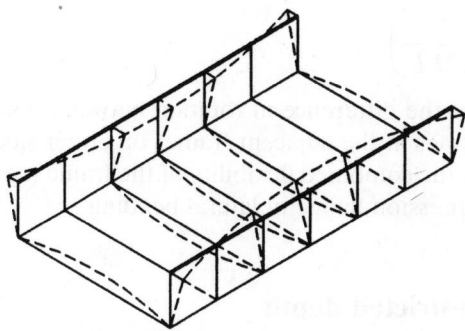

Fig. 16.10   Buckling mode for half-through construction with flexible end frames

where K is a parameter that takes account of the stiffness of the end U-frames. For effectively rigid frames, $K = 2.22$, which is the same as $\pi/\sqrt{2}$.

For unstiffened girders a similar approach is possible with the effective U-frame now comprising a unit length of girder web plus the cross-member. In all cases the assessment of U-frame stiffness via the δ parameter is based on summing the deflections due to bending of the horizontal and vertical components, including any flexibility of the upright to cross-frame connections. Clauses 9.6.4.1.3 and 9.6.4.2.2 deal respectively with the cases where actual vertical members are either present or absent.

Because the U-frames are required to resist the buckling deformations, they will attract forces which may be estimated as the product of the additional deformation, as a proportion of the initial lateral deformation of the top chord, and the U-frame stiffness as

$$F_{R} = \left(\frac{1}{1 - \sigma_{fc}/\sigma_{ci}} - 1 \right) \frac{L_{E}}{667\delta} \quad \text{or} \quad F_{R} = \left(\frac{\sigma_{fc}}{\sigma_{ci} - \sigma_{fc}} \right) \frac{L_{E}}{667\delta} \tag{16.11}$$

in which the assumed initial bow over an effective length of flange (L_{E}) has been taken as $L_{E}/667$, and $1/(1 - \sigma_{fc}/\sigma_{ci})$ is the amplification, which depends in a non-linear fashion on the level of stress σ_{fc} in the flange.

For a frame spacing L_{R} and a flange critical stress corresponding to a force level of $\pi^{2}EI_{c}/L_{R}^{2}$ the maximum possible value of F_{R} given in clause 9.12.2 of BS 5400: Part 3 is

$$F_{R} = \left(\frac{\sigma_{fc}}{\sigma_{ci} - \sigma_{fc}} \right) EI_{c}/16.7L_{R}^{2} \tag{16.12}$$

Additional forces in the web stiffeners are produced by rotation θ of the ends of the cross beam due to vertical loading on the cross beam. Clause 9.12.2.3 of BS 5400: Part 3 evaluates the additional force as:

$$F_{c} = 3EI_{1}\theta/d_{2}^{2} \tag{16.13}$$

$$F_c = \theta d \left(\frac{1}{1.5\delta + L_R^3 / 12EI_c} \right) \tag{16.14}$$

In this expression, θ is the difference in rotation between that at the U-frame and the mean of the rotations at the adjacent frames on either side. The division in the expression represents the combined flexibility of the frame (conservatively taken as 1.5δ) and of the compression flange in lateral bending.

16.6 Design for restricted depth

Frequently beam design will be constrained by a need to keep the beam depth to a minimum. This restriction is easy to understand in the context of floor beams in a multi-storey building for which savings in overall floor depth will be multiplied several times over, thereby permitting the inclusion of extra floors within the same overall building height or effecting savings on expensive cladding materials by reducing building height for the same number of floors. Within the floor zone of buildings with large volumes of cabling, ducting and other heavy services, only a fraction of the depth is available for structural purposes.

Such restrictions lead to a number of possible solutions which appear to run contrary to the basic principles of beam design. However, structural designers should remember that the main framing of a typical multi-storey commercial building typically represents less than 10% of the building cost and that factors such as the efficient incorporation of the services and enabling site work to proceed rapidly and easily are likely to be of greater overall economic significance than trimming steel weight.

An obvious solution is the use of universal columns as beams. While not as structurally efficient for carrying loads in simple vertical bending as UB sections, as illustrated by the example of Table 16.8, their design is straightforward. Problems of web bearing and buckling at supports are less likely due to the reduced web d/t ratios. Lateral–torsional buckling considerations are less likely to control the design of laterally unbraced lengths because the wider flanges will provide greater lateral stiffness (L/r_y values are likely to be low). Wider flanges are also advantageous for supporting floor units, particularly the metal decking used frequently as part of a composite floor system.

Difficulties can occur, because of the reduced depth, with deflections, although dead load deflections may be taken out by precambering the beams. This will not assist in limiting deflections in service due to imposed loading, although composite action will provide a much stiffer composite section. Excessive deflection of the floor beams under the weight of wet concrete can significantly increase slab depths at mid-span, leading to a substantially higher dead load. None of these problems need cause undue difficulty provided they are recognized and the proper checks made at a sufficiently early stage in the design.

Another possible source of difficulty arises in making connections between shallow beams and columns or between primary and secondary beams. The reduced

Table 16.8 Comparison of use of UB and UC for simple beam design

71 kN/m

6 m

M_{max} = 320 kN m

F_v = 213 kN

beams at 3 m spacing

457 × 152 × UB 60	254 × 254 × UC 89
M_c = 352 kN m	M_c = 326 kN m
P_v = 600 kN	P_v = 435 kN
$F_v < 0.6\ P_v$ – no interaction	$F_v < 0.6\ P_v$ – no interaction

From Equation (16.2) and Table 16.5, assuming deflection limit is $L/360$ and service load is 47 kN m

I_{rqd} = 2.23 (47 × 3) 6^2 = 11 319 cm⁴

I_x = 25 500 cm⁴	I_x = 14 300 cm⁴

web depth can lead to problems in physically accommodating sufficient bolts to carry the necessary end shears. Welding cleats to beams removes some of the dimensional tolerances that assist with erection on site as well as interfering with the smooth flow of work in a fabricator's shop that is equipped with a dedicated saw and drill line for beams. Extending the connection beyond the beam depth by using seating cleats is one solution, although a requirement to contain the connection within the beam depth may prevent their use.

Beam depths may also be reduced by using moment-resisting beam-to-column connections which provide end fixity to the beams; a fixed end beam carrying a central point load will develop 50% of the peak moment and only 20% of the central deflection of a similar simply-supported beam. Full end fixity is unlikely to be a realistic proposition in normal frames but the replacement of the notionally pinned beam-to-column connection provided by an arrangement such as web cleats, with a substantial end plate that functions more or less as a rigid connection, permits the development of some degree of continuity between beams and columns. These arrangements will need more careful treatment when analysing the pattern of internal moments and forces in the frame since the principles of simple construction will no longer apply.

An effect similar to the use of UC sections may be achieved if the flanges of a UB of a size that is incapable of carrying the required moment are reinforced by welding plates over part of its length. Additional moment capacity can be provided where it is needed as illustrated in Fig. 16.11; the resulting non-uniform section is stiffer and deflects less. Plating of the flanges will not improve the beam's shear capacity since this is essentially provided by the web and the possibility of shear or indeed local web capacity governing the design must be considered. A further

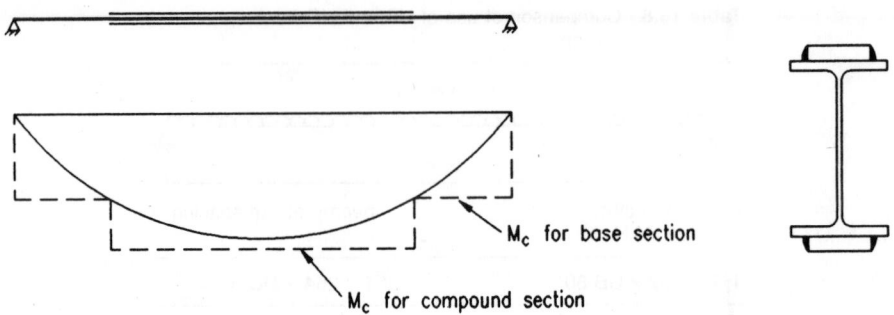

M_c for base section

M_c for compound section

Fig. 16.11 Selective increase of moment capacity by use of a plated UB

development of this idea is the use of tapered sections fabricated from plate.[11] To be economic, tapered sections are likely to contain plate elements that lie outside the limits for compact sections.

Because of the interest in developing longer spans for floors and the need to improve the performance of floor beams, a number of ingenious arrangements have developed in recent years.[12] Since these all utilize the benefits of composite action with the floor slab, they are considered in Chapter 21.

16.7 Cold-formed sections as beams

In situations where a relatively lightly loaded beam is required such as a purlin or sheeting rail spanning between main frames supporting the cladding in a portal frame, it is common practice to use a cold-formed section produced cold from flat steel sheet, typically between about 1 mm and 6 mm in thickness, in a wide range of shapes of the type shown in Fig. 16.12. A particular feature is that normally each section is formed from a single flat bent into the required shape; thus most available sections are not doubly symmetric but channels, zeds and other singly symmetric shapes. The forming process does, however, readily permit the use of quite complex cross-sections, incorporating longitudinal stiffening ribs and lips at the edges of flanges. Since the original coils are usually galvanized, the members do not normally require further protective treatment.

The structural design of cold-formed sections is covered by BS 5950: Part 5, which permits three approaches:

(1) design by calculation using the procedures of the code, section 5, for members in bending
(2) design on the basis of testing using the procedures of section 10 to control the testing and section 10.3 for members in bending
(3) for three commonly used types of member (zed purlins, sheeting rails and lattice joists), design using the simplified set of rules given in section 9.

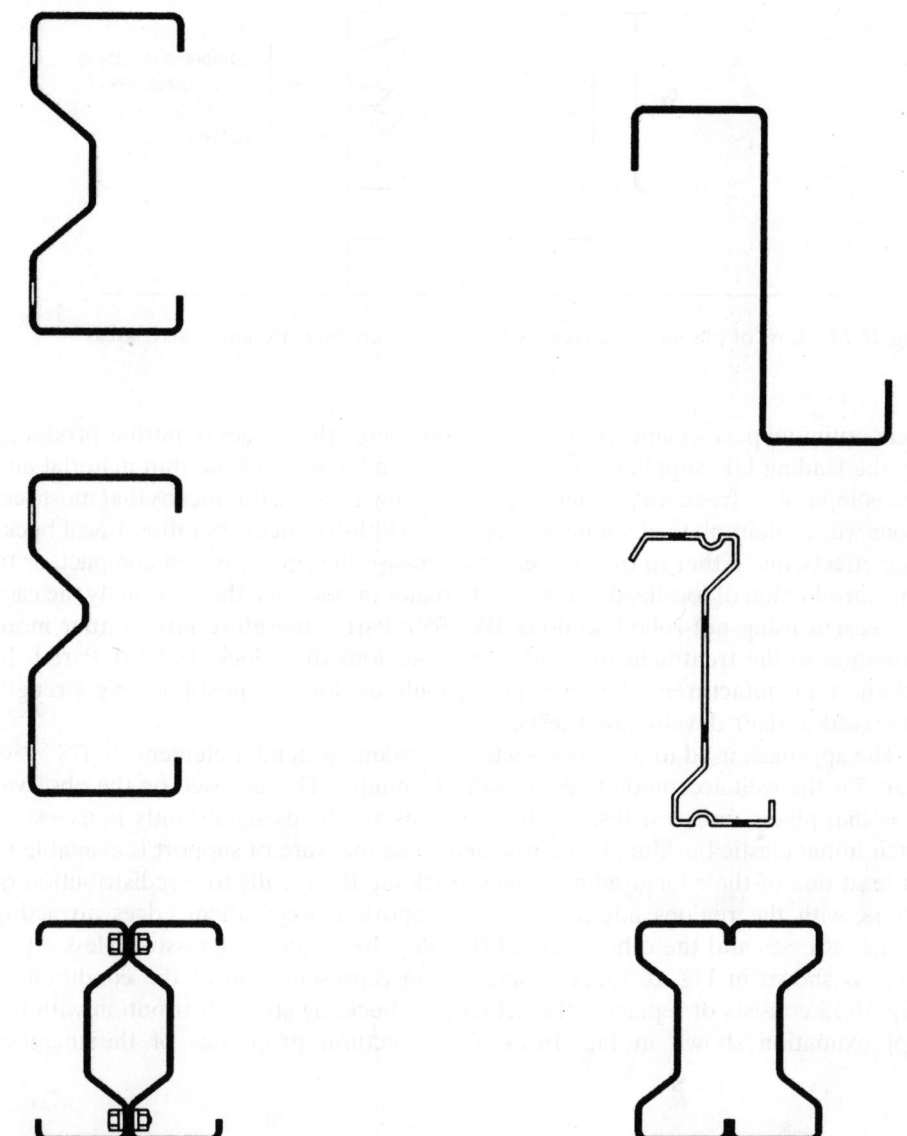

Fig. 16.12 Typical cold-formed section beam shapes

In practice option (2) is the most frequently used, with all the major suppliers providing design literature, the basis of which is usually extensive testing of their product range, design being often reduced to the selection of a suitable section for a given span, loading and support arrangement using the tables provided.

Most cold-formed section types are the result of considerable development work by their producers. The profiles are therefore highly engineered so as to produce a

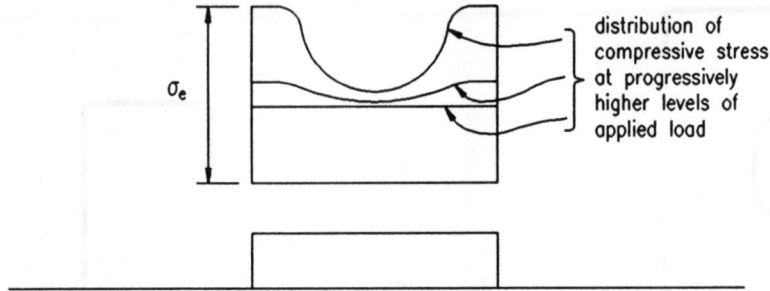

Fig. 16.13   Loss of plating effectiveness at progressively higher compressive stress

near optimum performance, a typical example being the ranges of purlins produced by the leading UK suppliers. Because of the combination of the thin material and the comparative freedom provided by the forming process, this means that most sections will contain plate elements having high width-to-thickness ratios. Local buckling effects, due either to overall bending because the profile is non-compact, or to the introduction of localized loads, are of greater importance than is usually the case for design using hot-rolled sections. BS 5950: Part 5 therefore gives rather more attention to the treatment of slender cross-sections than does BS 5950: Part 1. In addition, manufacturers' design data normally exploit the post-buckling strength observed in their development tests.

The approach used to deal with sections containing slender elements in BS 5950: Part 5 is the well accepted effective width technique. This is based on the observation that plates, unlike struts, are able to withstand loads significantly in excess of their initial elastic buckling load, provided some measure of support is available to at least one of their longitudinal edges. Buckling then leads to a redistribution of stress, with the regions adjacent to the supported longitudinal edges attracting higher stresses and the other parts of the plate becoming progressively less effective, as shown in Fig. 16.13. A simple design representation of the condition of Fig. 16.13 consists of replacing the actual post-buckling stress distribution with the approximation shown in Fig. 16.14. The structural properties of the member

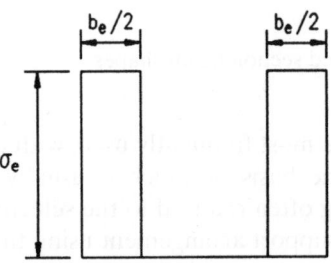

Fig. 16.14   Effective width design approximation

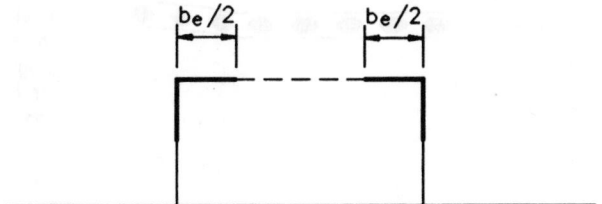

Fig. 16.15   Effective cross-section

(strength and stiffness) are then calculated for this effective cross-section as illustrated in Fig. 16.15. Tabulated information in BS 5950: Part 5 for steel of yield strength 280N/mm^2 makes the application of this approach simpler in the sense that effective widths may readily be determined, although cross-sectional properties have still to be calculated. The use of manufacturer's literature removes this requirement. For beams, Part 5 also covers the design of reinforcing lips on the usual basis of ensuring that the free edge of a flange supported by a single web behaves as if both edges were supported; web crushing under local loads, lateral–torsional buckling and the approximate determination of deflections take into account any loss of plating effectiveness.

For zed purlins or sheeting rails section 9 of BS 5950: Part 5 provides a set of simple empirically based design rules. Although easy to use, these are likely to lead to heavier members for a given loading, span and support arrangement than either of the other permitted procedures. A particular difference of this material is its use of unfactored loads, with the design conditions being expressed directly in member property requirements.

16.8 Beams with web openings

One solution to the problem of accommodating services within a restricted floor depth is to run the services through openings in the floor beams. Since the size of hole necessary in the beam web will then typically represent a significant proportion of the clear web depth, it may be expected that it will have an effect on structural performance. The easiest way of visualizing this is to draw an analogy between a beam with large rectangular web cut-outs and a Vierendeel girder. Figure 16.16 shows how the presence of the web hole enables the beam to deform locally in a similar manner to the shear type deformation of a Vierendeel panel. These deformations, superimposed on the overall bending effects, lead to increased deflection and additional web stresses.

A particular type of web hole is the castellation formed when a UB is cut, turned and rewelded as illustrated in Fig. 16.17. For the normal UK module geometry this leads to a 50% increase in section depth with a regular series of hexagonal holes. Other geometries are possible, including a further increase in depth through the use

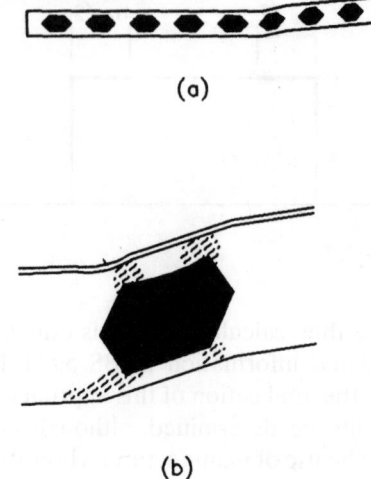

(a)

(b)

Fig. 16.16   Vierendeel-type action in beam with web openings: (a) overall view, (b) detail of deformed region

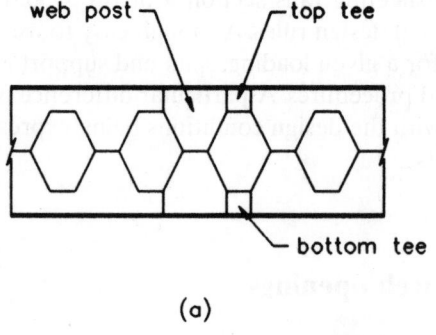

(a)

Fig. 16.17   Castellated beam: (a) basic concept, (b) details of normal UK module geometry

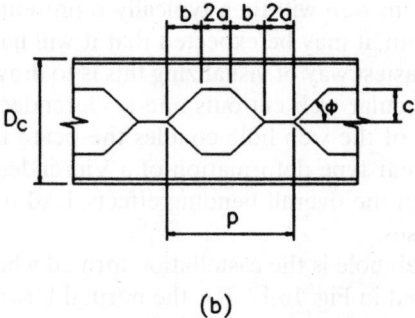

(b)

of plates welded between the two halves of the original beam. Some aspects of the design of castellated beams are covered by the provisions of BS 5950: Part 1, while more detailed guidance is available in a Constrado publication.[13]

Based on research conducted in the USA, a comparatively simple elastic method for the design of beams with web holes, including a fully worked example, is available.[14] This uses the concept of an analysis for girder stresses and deflections that neglects the effects of the holes, coupled with checking against suitably modified limiting values. The full list of design checks considered in Reference 14 is:

(1) web shear due to overall bending acting on the reduced web area
(2) web shear due to local Vierendeel bending at the hole
(3) primary bending stresses (little effect since overall bending is resisted principally by the girder flanges)
(4) local bending due to Vierendeel action
(5) local buckling of the tee formed by the compression flange and the web adjoining the web hole
(6) local buckling of the stem of the compression tee due to secondary bending
(7) web crippling under concentrated loads or reactions near a web hole; as a simple guide, Reference 14 suggests that for loads which act at least $(d/2)$ from the edge of a hole this effect may be neglected
(8) shear buckling of the web between holes; as a simple guide, Reference 14 suggests that for a clear distance between holes that exceeds the hole length this effect may be neglected
(9) vertical deflections; as a rough guide, secondary effects in castellated beams may be expected to add about 30% to the deflections calculated for a plain web beam of the same depth $(1.5D)$. Beams with circular holes of diameter $(D/2)$ may be expected to behave similarly, while beams with comparable rectangular holes may be expected to deflect rather more.

As an alternative to the use of elastic methods, significant progress has been made in recent years in devising limit state approaches based on ultimate strength conditions. A CIRIA/SCI design guide[15] dealing with the topic principally from the point of composite beams is now available. If some of the steps in the 24-point design check of Reference 15 are omitted, the method may be applied to non-composite beams, including composite beams under construction. Much of the basis for Reference 15 may be traced back to the work of Redwood and Choo,[16] and the following treatment of bare steel beams is taken from Reference 16.

The governing condition for a stocky web in the vicinity of a hole is taken as excessive plastic deformation near the opening corners and in the web above and below the opening as illustrated in Fig. 16.18. A conservative estimate of web strength may then be obtained from a moment–shear interaction diagram of the type shown as Fig. 16.19. Values of M_0 and V_1 in terms of the plastic moment capacity and plastic shear capacity of the unperforated web are given in Reference 14 for both plain and reinforced holes; M_1 may also be determined in this way. Solution of these equations is tedious, but some rearrangement and simplification are possible

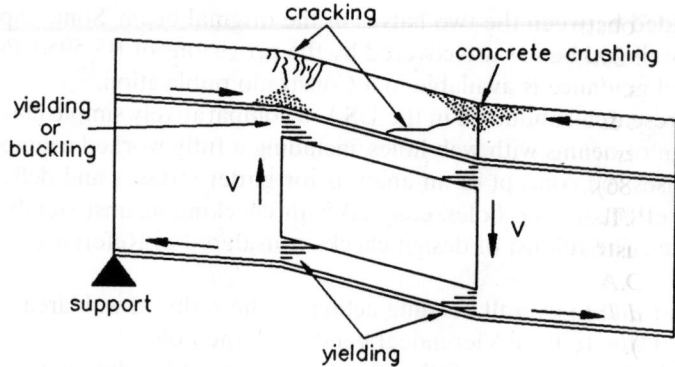

Fig. 16.18   Hole-induced failure

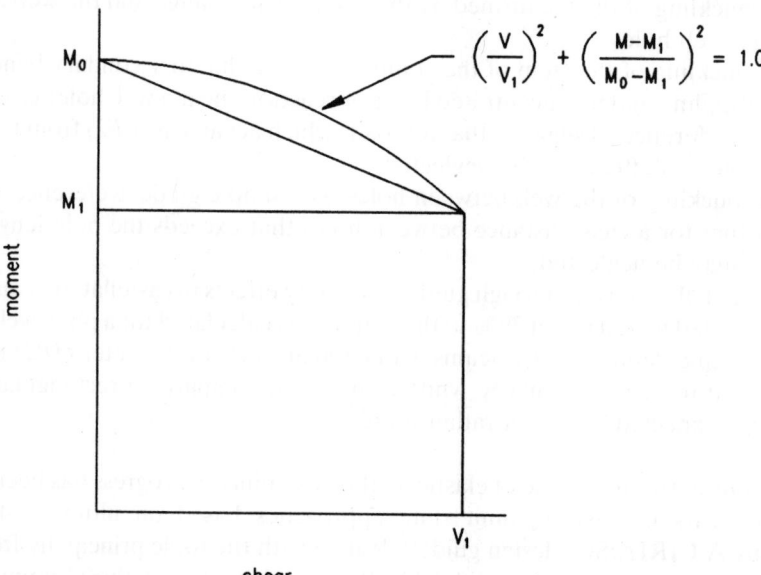

$$\left(\frac{V}{V_1}\right)^2 + \left(\frac{M-M_1}{M_0-M_1}\right)^2 = 1.0$$

Fig. 16.19   Moment–shear interaction for a stocky web in the vicinity of a hole

so that an explicit solution for the required area of reinforcement may be obtained. However, the whole approach is best programmed for a microcomputer, and a program based on the full method of Reference 15 is available from the SCI.

References to Chapter 16

1. British Standards Institution (2000) Part 1: *Code of practice for design in simple and continuous construction: hot rolled sections*. BS 5950, BSI, London.

2. The Steel Construction Institute (SCI) (2001) *Steelwork Design Guide to BS 5950: Part 1: 2000, Vol. 1, Section Properties, Member Capacities*, 6th edn. SCI, Ascot, Berks.
3. Johnson R.P. & Buckby R.J. (1979) *Composite Structures of Steel and Concrete, Vol. 2: Bridges with a Commentary on BS 5400: Part 5*, 1st edn. Granada, London. (2nd edn, 1986).
4. Woodcock S.T. & Kitipornchai S. (1987) Survey of deflection limits for portal frames in Australia. *J. Construct. Steel Research*, **7**, No. 6, 399–418.
5. Nethercot D.A., Salter P. & Malik A. (1989) *Design of Members Subject to Bending and Torsion*. The Steel Construction Institute, Ascot, Berks (SCI Publication 057).
6. Dux P.F. & Kitipornchai S. (1986) Elastic buckling strength of braced beams. *Steel Construction*, (AISC), **20**, No. 1, May.
7. Trahair N.S. & Nethercot D.A. (1984) Bracing requirements in thin-walled structures. In *Developments in Thin-Walled Structures – 2* (Ed. by J. Rhodes & A.C. Walker), pp. 93–130. Elsevier Applied Science Publishers, Barking, Essex.
8. Nethercot D.A. & Lawson R.M. (1992) *Lateral stability of steel beams and columns – common cases of restraint*. SCI Publication 093. The Steel Construction Institute, Ascot, Berks.
9. Brown B.A. (1988) The requirements for restraints in plastic design to BS 5950. *Steel Construction Today*, **2**, No. 6, Dec., 184–6.
10. Horne M.R. (1964) Safe loads on I-section columns in structures designed by plastic theory. *Proc. Instn. Civ. Engrs*, **29**, Sept., 137–50.
11. Raven G.K. (1987) The benefits of tapered beams in the design development of modern commercial buildings. *Steel Construction Today*, **1**, No. 1, Feb., 17–25.
12. Owens G.W. (1987) Structural forms for long span commercial building and associated research needs. In *Steel Structures, Advances, Design and Construction* (Ed. by R. Narayanan), pp. 306–319. Elsevier Applied Science Publishers, Barking, Essex.
13. Knowles P.R. (1985) *Design of Castellated Beams for use with BS 5950 and BS 449*. Constrado.
14. Constrado (1977) *Holes in Beam Webs: Allowable Stress Design*. Constrado.
15. Lawson R.M. (1987) *Design for Openings in the Webs of Composite Beams*. CIRIA Special Publication S1 and SCI Publication 068. CIRIA/Steel Construction Institute.
16. Redwood R.G. & Choo S.H. (1987) Design tools for steel beams with web openings. In: *Composite Steel Structures, Advances, Design and Construction* (Ed. by R. Narayanan), pp. 75–83. Elsevier Applied Science Publishers, Barking, Essex.

A series of worked examples follows which are relevant to Chapter 16.

The Steel Construction Institute Silwood Park, Ascot, Berks SL5 7QN	Subject **BEAM EXAMPLE 1 LATERALLY RESTRAINED UNIVERSAL BEAM**	Chapter ref. **16**
Design code **BS 5950: Part 1**	Made by **DAN** Checked by **GWO**	Sheet no. **1**

Problem

Select a suitable UB section to function as a simply supported beam carrying a 140 mm thick solid concrete slab together with an imposed load of 7.0 kN/m². Beam span is 7.2 m and beams are spaced at 3.6 m intervals. The slab may be assumed capable of providing continuous lateral restraint to the beam's top flange.

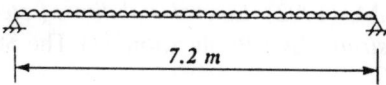

7.2 m

Due to restraint from slab there is no possibility of lateral-torsional buckling, so design beam for:

i)   **Moment capacity**
ii)   **Shear capacity**
iii) **Deflection limit**

Loading

D.L.	= (2.4 × 9.81 × 0.14)	= 3.3 kN/m²	
I.L.		= 7.0 kN/m²	
Total serviceability loading		= 10.3 kN/m²	*Table 2*

Total load for ultimate limit state

$$= 1.4 \times 3.3 + 1.6 \times 7.0 \qquad = 15.8\,kN/m^2$$

Design ultimate moment $\qquad = (15.8 \times 3.6) \times 7.2^2/8$

$$= 369\,kNm$$

Design ultimate shear $\qquad = (15.8 \times 3.6) \times 7.2/2$

$$= 205\,kN$$

The **Steel Construction Institute** Silwood Park, Ascot, Berks SL5 7QN	Subject **BEAM EXAMPLE 1** **LATERALLY RESTRAINED** **UNIVERSAL BEAM**	Chapter ref. **16**
	Design code **BS 5950: Part 1**	Made by **DAN** Checked by **GWO**	Sheet no. **2**

Assuming use of S275 steel and no material greater than 16 mm thick,

take p_y $\quad = 275\,N/mm^2$

Required $S_x = 369 \times 10^6/275$
$\qquad\qquad = 1.34 \times 10^6\,mm^3 = 1340\,cm^3$

A $457 \times 152 \times 67\,UB$ has a value of S_x of $1440\,cm^3$

$T = 15.0 < 16.0\,mm$

$\therefore \quad p_y = 275\,N/mm^2$

Check section classification

Actual $b/T = 5.06 \quad d/t = 44.7$

$\varepsilon = (275/p_y)^{1/2} = 1$

Limit on b/T for plastic section $= 9 > 5.06$

Limit on d/t for shear $= 63 > 44.7$

$\therefore$ _Section is plastic_

Actual $\quad M_c = 275 \times 1440 \times 10^3$
$\qquad\qquad = 396 \times 10^6\,Nmm$
$\qquad\qquad = \underline{396\,kNm} > \underline{369\,kNm}\,\,OK$

Vertical shear capacity

$P_v = 0.6\,p_y A_v$

where $\quad A_v = tD$

$\therefore \qquad P_v = 0.6 \times 275 \times 9.1 \times 457.2 = 686 \times 10^3\,N$

$\qquad\qquad = \underline{686\,kN} > \underline{205\,kN}\,\,OK$

Table 9

Steelwork Design Guide Vol 1

3.5.2

Table 11

4.2.5

4.2.3

The **Steel Construction Institute** Silwood Park, Ascot, Berks SL5 7QN	Subject **BEAM EXAMPLE 1 LATERALLY RESTRAINED UNIVERSAL BEAM**	Chapter ref. **16**
	Design code **BS 5950: Part 1**	Made by **DAN**	Sheet no. **3**
		Checked by **GWO**	

Check serviceability deflections under imposed load   2.5.1

$$\delta = \frac{5 \times (7.0 \times 3.6) \times 7200^4}{384 \times 205000 \times 32400 \times 10^4}$$

$$= 13.3mm \quad = \quad span/541$$

From Table 5, limit is span/360   ∴   *δ OK*

∴   Use 457 × 152 × 67UB Grade 43

The **Steel Construction Institute** Silwood Park, Ascot, Berks SL5 7QN		Subject **BEAM EXAMPLE 2 LATERALLY UNRESTRAINED UNIVERSAL BEAM**	Chapter ref. **16**
Design code **BS 5950: Part 1**		Made by **DAN**	Sheet no. **1**
		Checked by **GWO**	

Problem

For the same loading and support conditions of example 1 select a suitable UB assuming that the member must be designed as laterally unrestrained.

It is not now possible to arrange the calculations in such a way that a direct choice is possible; a guess and check approach must be adopted.

Try $610 \times 229 \times 125\,UB$

$u = 0.873$ $\qquad\qquad r_y = 4.98\,cm$

$x = 34.0$ $\qquad\qquad S_x = 3680\,cm^3$

$\lambda = L_E / r_y = 7200 / 49.8$
$\qquad = 145$

$\lambda / x = 145 / 34 = 4.26$

$v = 0.85$

$\lambda_{LT} = u\,v\,\lambda$

$\therefore\ \lambda_{LT} = 0.873 \times 0.85 \times 145$

$\qquad = 108$

$P_b = 116\,N/mm^2$

$M_b = S_x p_b = 3680 \times 10^3 \times 116$

$\qquad = 427 \times 10^6\,N\,mm$

$\qquad = \underline{427\,kN\,m} > \underline{369\,kN\,m}\ OK$

Steelwork Design Guide Vol 1

4.3.7.5

Table 19

Table 20

4.3.7.3

The **Steel Construction Institute** Silwood Park, Ascot, Berks SL5 7QN	Subject **BEAM EXAMPLE 2 LATERALLY UNRESTRAINED UNIVERSAL BEAM**	Chapter ref. **16**
	Design code **BS 5950: Part 1**	Made by **DAN** Checked by **GWO**	Sheet no. **2**

Since section is larger than before, P_v and δ will also be satisfactory

∴   *Adopt $610 \times 229 \times 125\,UB$*

However, for a UDL, $M_{LT} = 0.925$ Table 18

Refer to member capacities section for values of M_b directly, noting S275 material.   *Steelwork Design Guide Vol 1*

Value of M_b required is $369\,kNm$; interpolating for $M_{LT} = 0.925$ & $L_E = 7.2\,m$ suggests as possible sections:

$533 \times 210 \times 122\,UB$ M_b OK

$610 \times 229 \times 113\,UB$ M_b OK

Since both are larger than that checked for shear capacity and serviceability deflection in the previous example, either may be adopted.

∴   *Adopt $533 \times 210 \times 122\,UB$ or $610 \times 229 \times 113\,UB$ Grade S275*

The Steel Construction Institute	Subject	Chapter ref.
Silwood Park, Ascot, Berks SL5 7QN	**BEAM EXAMPLE 3 UNIVERSAL BEAM SUPPORTING POINT LOADS**	**16**

Design code	Made by *DAN*	Sheet no. *1*
BS 5950: Part 1	Checked by *GWO*	

Problem

Select a suitable UB section in S275 steel to carry the pair of point loads at the third points transferred by crossbeams as shown in the accompanying sketch. Design to BS 5950: Part 1.

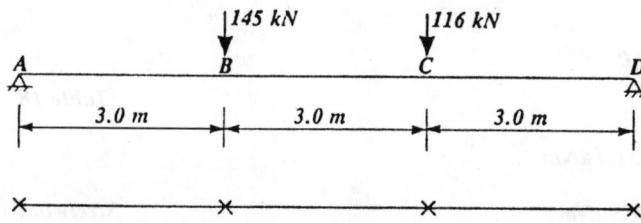

The crossbeams may reasonably be assumed to provide full lateral and torsional restraint at B and C; assume further that ends A and D are similarly restrained. Thus the actual level of transfer of load at B and C (relative to the main beam's centroid) will have no effect, the lateral-torsional buckling aspects of the design being one of considering the 3 segments AB, BC and CD separately.

From statics the BMD and SFD are

BMD

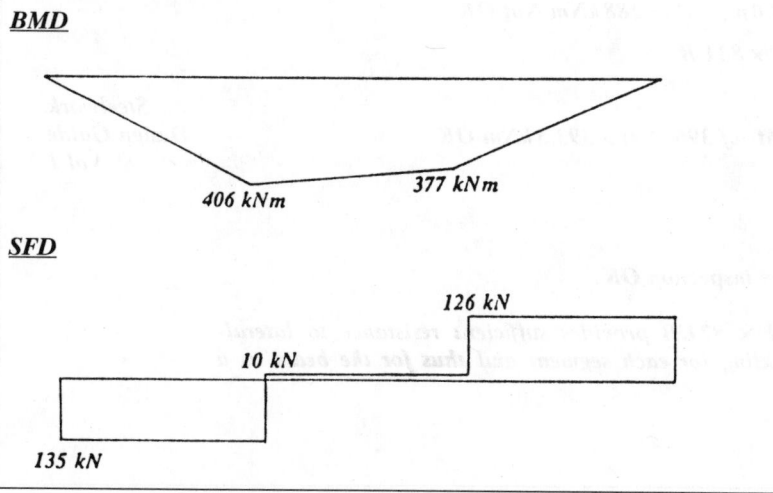

SFD

The Steel Construction Institute  Silwood Park, Ascot, Berks SL5 7QN	Subject **BEAM EXAMPLE 3 UNIVERSAL BEAM SUPPORTING POINT LOADS**	Chapter ref. **16**
	Design code **BS 5950: Part 1**	Made by **DAN**	Sheet no. **2**
		Checked by **GWO**	

For initial trial section select a UB with $M_{cx} > 406$ kNm.

A $457 \times 152 \times 74$ UB provides M_{cx} of 429 kNm. Now check lateral-torsional buckling strength for segments AB, BC & CD.

Steelwork Design Guide Vol 1

AB

β = 0/406   = 0.0

m_{LT} = 0.57

Table 18

$\overline{M}$ = 0.57 × 406 = 231.4 kNm

For $L_E = 3.0$ m, $M_b = 288$ kNm
∴   *288 > 231.4 kNm OK*

Steelwork Design Guide Vol 1

BC

β = 377/406   = 0.93

m_{LT} = 0.97

Table 18

$\overline{M}$ = 0.97 × 406 = 393.8 kNm

But for $L_E = 3.0$ m, $M_b = 288$ kNm <u>Not</u> OK

Try $457 \times 191 \times 82$ UB

this provides M_b of 396 kNm > 393.8 kNm OK

Steelwork Design Guide Vol 1

CD

Satisfactory by inspection OK

A $457 \times 191 \times 82$ UB provides sufficient resistance to lateral-torsional buckling for each segment and thus for the beam as a whole.

The **Steel Construction Institute** Silwood Park, Ascot, Berks SL5 7QN 	Subject **BEAM EXAMPLE 3 UNIVERSAL BEAM SUPPORTING POINT LOADS**		Chapter ref. **16**
	Design code **BS 5950: Part 1**	Made by **DAN**	Sheet no. **3**
		Checked by **GWO**	

Check shear capacity; maximum shear at Λ is 135 kN

$P_v = 752 \, kN > 135 \, kN$ *OK*

Check bearing and buckling capacity of web at the supports –

required capacity is 135 kN.

Since C_1 values exceed 135 kN in both cases, section is clearly adequate.

Steelwork Design Guide Vol 1

For initial check on serviceability deflections assume as equivalent UDL and factor down all loads by 1.5 to obtain

$$w = \frac{145 + 116}{1.5 \times 9} = 19.3 \, kN/m$$

$$\delta = \frac{5 \times 19.3 \times 9000^4}{384 \times 205000 \times 37100 \times 10^4}$$

$$= 21.68 \, mm = span/415$$

Limiting deflection is span/360 ∴ *OK*

Table 5

Beam is clearly satisfactory for deflection since these (approximate) calculations have used the full load and not just the imposed load.

∴ *Adopt 457 × 191 × 82 UB*

Chapter 17
Plate girders

by TERENCE M. ROBERTS and RANGACHARI NARAYANAN

17.1 Introduction

Plate girders are employed to support heavy vertical loads over long spans for which the resulting bending moments are larger than the moment resistance of available rolled sections. In its simplest form the plate girder is a built-up beam consisting of two flange plates, fillet welded to a web plate to form an I-section (see Fig. 17.1). The primary function of the top and bottom flange plates is to resist the axial compressive and tensile forces caused by the applied bending moments; the main function of the web is to resist the shear. Indeed this partition of structural action is used as the basis for design in some codes of practice.

For a given bending moment the required flange areas can be reduced by increasing the distance between them. Thus for an economical design it is advantageous to increase the distance between flanges. To keep the self-weight of the girder to a minimum the web thickness should be reduced as the depth increases, but this leads to web buckling considerations being more significant in plate girders than in rolled beams.

Plate girders are sometimes used in buildings and are often used in small to medium span bridges. They are designed in accordance with the provisions contained in BS 5950: Part 1: 2000[1] and BS 5400: Part 3[2] respectively. This chapter explains current practice in designing plate girders for buildings and bridges; references to the relevant clauses in the codes are made.

17.2 Advantages and disadvantages

The development of highly automated workshops in recent years has reduced the fabrication costs of plate girders very considerably; box girders and trusses still have to be fabricated manually, with consequently high fabrication costs. Optimum use of material is made, compared with rolled sections, as the girder is fabricated from plates and the designer has greater freedom to vary the section to correspond with changes in the applied forces. Thus variable depth plate girders have been increasingly designed in recent years. Plate girders are aesthetically more pleasing than trusses and are easier to transport and erect than box girders.

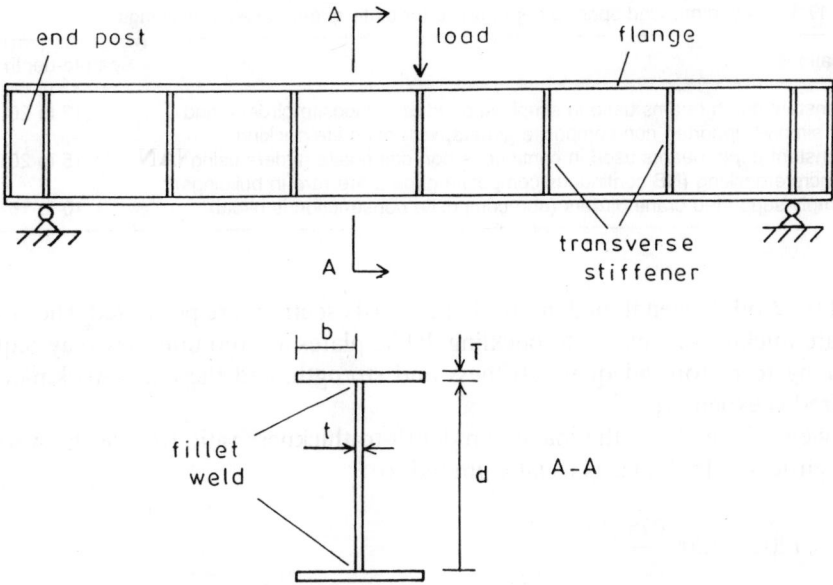

Fig. 17.1 Elevation and cross-section of a typical plate girder

There are only a very few limitations in the use of plate girders. Compared with trusses they are heavier, more difficult to transport and have larger wind resistance. The provision of openings for services is also more difficult. The low torsional stiffness of plate girders makes them difficult to use in bridges having small plan radius. Plate girders can sometimes pose problems during erection because of concern for the stability of compression flanges.

17.3 Initial choice of cross-section for plate girders in buildings

17.3.1 Span-to-depth ratios

Advances in fabrication methods allow the economic manufacture of plate girders of constant or variable depth. Traditionally, constant-depth girders were more common in buildings; however, this may change as designers become more inclined to modify the steel structure to accommodate services.[3] Recommended span-to-depth ratios are given in Table 17.1.

17.3.2 Recommended plate thickness and proportions

In general the slenderness of the cross-sections of plate girders used in buildings should not exceed the limits specified for class 3 semi-compact cross-sections (clause

Table 17.1 Recommended span-to-depth ratios for plate girders used in buildings

Applications	Span-to-depth ratio
(1) Constant-depth beams used in simply-supported composite girders, and for simply-supported non-composite girders, with concrete decking	12 to 20
(2) Constant-depth beams used in continuous non-composite girders using concrete decking (NB continuous composite girders are rare in buildings)	15 to 20
(3) Simply-supported crane girders (non-composite construction is usual)	10 to 15

3.5 of the Code), even though more slender cross-sections are permitted. The choice of plate thickness is related to buckling. If the plates are too thin they may require stiffening to restore adequate stiffness and strength, and the extra workmanship required is expensive.

In view of the above the maximum depth-to-thickness ratio (d/t) of the webs of plate girders in buildings is usually limited to

$$d/t < 120\, \varepsilon = 120 \left(\frac{275}{p_{yw}} \right)^{1/2}$$

where p_{yw} is the design strength of the web plate. The outstand width-to-thickness ratio of the compression flange (b/T) is usually limited to

$$b/T \le 13\, \varepsilon = 13 \left(\frac{275}{p_{yf}} \right)^{1/2}$$

where p_{yf} is the design strength of the compression flange.

Changes in flange size along the girder are not usually worthwhile in buildings. For non-composite girders the flange width is usually within the range 0.3–0.5 times the depth of the section (0.4 is most common). For simply-supported composite girders these guidelines can still be employed for preliminary sizing of compression flanges. The width of tension flanges can be increased by 30%.

17.3.3 Stiffeners

Horizontal web stiffeners are not usually required for plate girders used in buildings. Vertical web stiffeners may be provided to enhance the resistance to shear near the supports. Intermediate stiffening at locations far away from supports will, in general, be unnecessary due to reduced shear.

The provision of vertical or transverse web stiffeners increases both the critical shear strength q_{cr} (initial buckling strength) and the shear buckling strength q_w (post-buckling strength) of web panels. The critical shear strength is increased by a reduction in the web panel aspect ratio a/d (width/depth). Shear buckling strength is increased by enhanced tension field action, whereby diagonal tensile membrane

stresses, which develop during the post-buckling phase, are resisted by the boundary members (vertical stiffeners and flanges).

Transverse stiffeners are usually spaced such that the web panel aspect ratio is between 1.0 and 2.0, since there is little increase in strength for larger panel aspect ratios. For end panels designed without utilizing tension field action the aspect ratio is reduced to 0.6–1.0. Sometimes double stiffeners are employed as bearing stiffeners at the end supports, to form what is known as an *end post*. The overhang of the girder beyond the support is generally limited to a maximum of one eighth of the depth of the girder.

17.4 Design of plate girders used in buildings to BS 5950: Part 1: 2000

17.4.1 General

Any cross-section of a plate girder will normally be subjected to a combination of shear force and bending moment, present in varying proportions. BS 5950: Part 1: 2000[1] specifies that the design of plate girders should satisfy the relevant provisions given in clauses 4.2 (members subject to bending) and 4.3 (lateral-torsional buckling) together with the additional provisions in 4.4 (plate girders). The additional provisions in clause 4.4 of the Code are related primarily to the susceptibility of slender web panels to local buckling.

17.4.2 Dimensions of webs and flanges

Minimum web thickness requirements (clause 4.4.3 of the Code) are based on serviceability considerations, such as adequate stiffness to prevent unsightly buckles developing during erection and in service, and also to avoid the compression flange buckling into the web.

The buckling resistance of slender webs can be increased by the provision of web stiffeners. In general the webs of plate girders used in buildings are either unstiffened or have transverse stiffeners only (see Fig. 17.1).

The following minimum web thickness values are prescribed to avoid serviceability problems.

(1) Unstiffened webs, $t \geq \dfrac{d}{250}$

(2) Transversely stiffened webs,

For $a > d$ $t \geq \dfrac{d}{250}$

For $a \leq d$ $t \geq \left(\dfrac{d}{250}\right)\left(\dfrac{a}{d}\right)^{1/2}$

The following minimum web thickness values are prescribed to avoid the compression flange buckling into the web.

(3) Unstiffened webs $t \geq \left(\dfrac{d}{250} \right) \left(\dfrac{p_{yf}}{345} \right)$

(4) Transversely stiffened webs

For $a > 1.5d$ $t \geq \left(\dfrac{d}{250} \right) \left(\dfrac{p_{yf}}{345} \right)$

For $a \leq 1.5d$ $t \geq \left(\dfrac{d}{250} \right) \left(\dfrac{p_{yf}}{455} \right)^{1/2}$

Local buckling of the compression flange may also occur if the flange plate is of slender proportions. In general there is seldom good reason for the b/T ratio of the compression flanges of plate girders used in buildings to exceed the class 3 semi-compact limit (clause 3.5 of the Code) $b/T \leq 13\,\varepsilon$.

17.4.3 Moment resistance

17.4.3.1 Web not susceptible to shear buckling

Determination of the moment resistance M_c of laterally restrained plate girders depends upon whether or not the web is susceptible to shear buckling. If the web depth to thickness ratio $d/t \leq 62\,\varepsilon$ the web should be assumed not to be susceptible to shear buckling, and the moment resistance of the section should be determined in accordance with clause 4.2.5 of the Code.

17.4.3.2 Web susceptible to shear buckling

If the web depth to thickness ratio $d/t > 62\,\varepsilon$ it should be assumed susceptible to shear buckling. The moment resistance of the section M_c should be determined taking account of the interaction of shear and moment, using the following methods.

(1) Low shear

Provided that the applied shear force $F_v \leq 0.6\,V_w$, where V_w is the simple shear buckling resistance from clause 4.4.5.2 of the Code (see section 17.4.4.2 (1)), the moment resistance should be determined from clause 4.2.5 of the Code. For class 1 plastic and class 2 compact cross-sections:

$M_c = p_y S$

where p_y is the design strength of the steel and S is the plastic modulus. For class 3 semi-compact cross-sections:

$$M_c = p_y Z$$

where Z is the elastic section modulus. For class 4 slender cross-sections:

$$M_c = p_y Z_{eff}$$

where Z_{eff} is the effective section modulus determined in accordance with clause 3.6.2 of the Code.

(2) High shear flanges-only method

If $F_v > 0.6 V_w$ but the web is designed for shear only, provided that the flanges are not class 4, a conservative value M_f for the moment resistance may be obtained by assuming that the moment is resisted by the flanges only. Hence:

$$M_f = p_{yf} S_f$$

where S_f is the plastic modulus of the flanges only.

(3) High shear general method

If $F_v > 0.6 V_w$ and the applied moment does not exceed the low shear value given by (1), the web should be designed using Annex H.3 of the Code for the applied shear combined with any additional moment beyond the flanges-only moment resistance M_f given by (2).

17.4.4 Shear resistance

17.4.4.1 Web not susceptible to shear buckling

If the web depth to thickness ratio $d/t \leq 62\varepsilon$ it is not susceptible to shear buckling and the shear resistance P_v should be determined in accordance with clause 4.2.3 of the Code, i.e.

$$P_v = 0.6 p_{yw} A_v = 0.6 p_{yw} td$$

where $A_v = td$ is the shear area.

17.4.4.2 Web susceptible to shear buckling

If $d/t > 62\varepsilon$ the shear buckling resistance of the web should be determined in accordance with clause 4.4.5 of the Code.

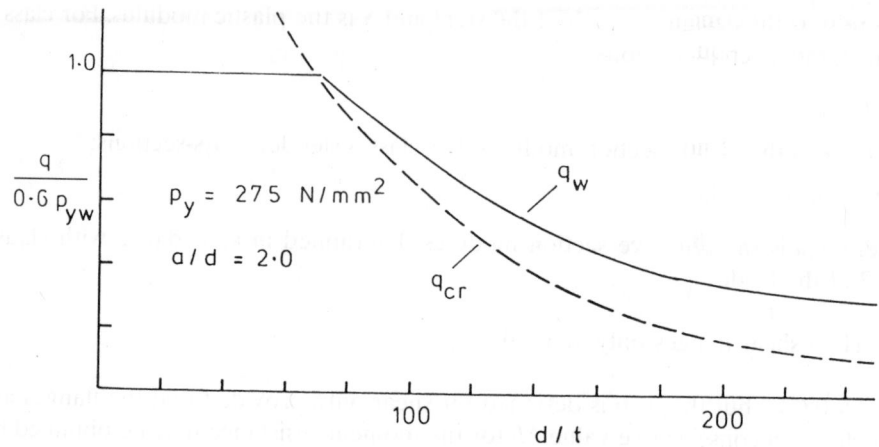

Fig. 17.2　Relationship between shear buckling strength q_w and critical shear strength q_{cr}

The procedures for determining the shear buckling resistance of slender plate girder webs, given in BS 5950: Part 1: 2000,[1] have been updated and are now consistent with the procedures given in Eurocode 3: Design of steel structures.[4] Two methods are specified, namely the simplified method and the more exact method. Both methods are based on the post-critical shear buckling strength of the web q_w, and result in significantly greater values of shear resistance than the elastic critical method of BS 5950: Part 1: 1985. The relationship between q_w and q_{cr}, the elastic critical shear strength, is illustrated in Fig. 17.2. The more exact method incorporates a flange related component of shear resistance and is comparable with the method based on tension field theory in BS 5950: Part 1: 1985. The critical shear buckling resistance V_{cr}, based on q_{cr}, is retained only as a reference value and for the design of particular types of end panel.

The two methods for determining the shear buckling resistance are as follows.

(1)　Simplified method

The shear buckling resistance V_b of a web, with or without intermediate transverse stiffeners, may be taken as the simple shear buckling resistance V_w given by:

$$V_w = dtq_w$$

where q_w is the shear buckling strength of the web. Values of q_w are tabulated in Table 21 of the Code for web panel aspect ratios from 0.4 to infinity. The equations on which q_w is based are specified in Annex H.1 of the Code.

(2)　More exact method

Alternatively the shear buckling resistance V_b of a web panel between two transverse stiffeners may be determined as follows. If the flanges of the panel are fully

stressed, i.e. the mean longitudinal stress in the smaller flange due to moment or axial force f_f is equal to the flange strength p_{yf}, then:

$$V_b = V_w = dtq_w$$

If the flanges are not fully stressed ($f_f < p_{yf}$):

$$V_b = V_w + V_f \quad \text{but} \quad V_b \le P_v = 0.6 p_{yw} td$$

in which V_f is the flange-dependent shear buckling resistance given by

$$V_f = \frac{P_v(d/a)\left[1 - (f_f/p_{yf})^2\right]}{1 + 0.15(M_{pw}/M_{pf})}$$

M_{pf} is the plastic moment resistance of the smaller flange about its own equal area axis, perpendicular to the plane of the web. M_{pw} is the plastic moment resistance of the web about its equal area axis perpendicular to the plane of the web, determined using p_{yw}. For a rectangular flange and a web of uniform thickness:

$$M_{pf} = 0.25 p_{yf}(2b + t)T^2$$
$$M_{pw} = 0.25 p_{yw} td^2$$

(3) Critical shear buckling resistance

The critical shear buckling resistance V_{cr} of the web of an I-section is given by

$$V_{cr} = dtq_{cr}$$

where q_{cr} is the critical shear buckling strength determined in accordance with clause 4.4.5.4 of the Code as follows (see also Annex H.2 of the Code).

If $V_w = P_v$ then $V_{cr} = P_v$

If $P_v > V_w > 0.72 P_v$ then $V_{cr} = (9 V_w - 2 P_v)/7$

If $V_w \le 0.72 P_v$ then $V_{cr} = \left(\dfrac{V_w}{0.9}\right)^2 \Big/ P_v$

17.4.4.3 Panels with openings

Web openings frequently have to be provided in girders used in building construction for service ducts etc. When any dimension of such an opening exceeds 10% of the minimum dimensions of the panel in which it is located, reference should be made to clause 4.15 of the Code. Panels with openings should not be used as anchor panels, and the adjacent panels should be designed as end panels.

Guidance on the design of plate girders with openings is provided in Reference 5.

17.4.5 Resistance of a web to combined effects

If the moment resistance of a plate girder is determined using the high shear general method (see section 17.4.3.2 (3)) the resistance of the web to combined effects should be checked in accordance with Annex H.3 of the Code.

The interaction between bending and shear in plate girders is illustrated in Fig. 17.3. The broken line represents the high shear flanges only method (see section 17.4.3.2 (2)) while the full line represents the low shear and high shear general methods.

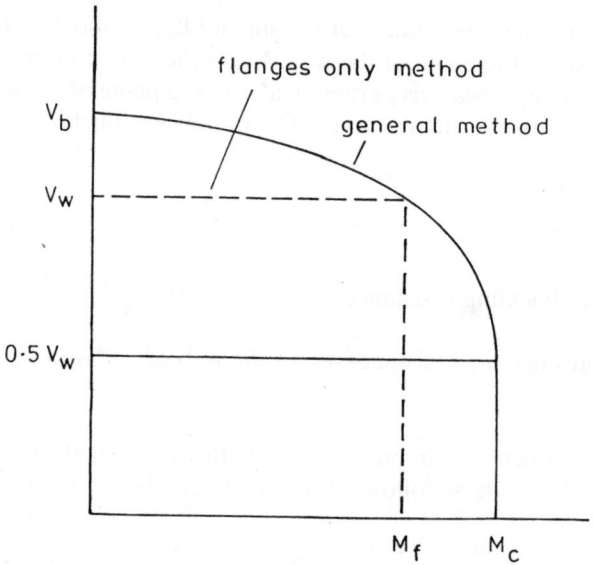

Fig. 17.3   Interaction between shear and moment resistance of plate girder webs

17.4.6 End panels and end anchorage

17.4.6.1 General

End anchorage need not be provided if either the shear resistance P_v (see section 17.4.4.1) not the shear buckling resistance V_w (see section 17.4.4.2 (1)) is the governing design criterion, indicated by $P_v \leq V_w$, or the applied shear force F_v is less than the critical shear buckling resistance V_{cr} (see section 17.4.4.2 (3)). For all other situations some form of end anchorage is required to resist the post-critical membrane stresses (tension field) which develop in a buckled web.

Three alternatives for providing end anchorage are recommended in Annex H.4 of the Code, namely single stiffener end posts, twin stiffener end posts and anchor

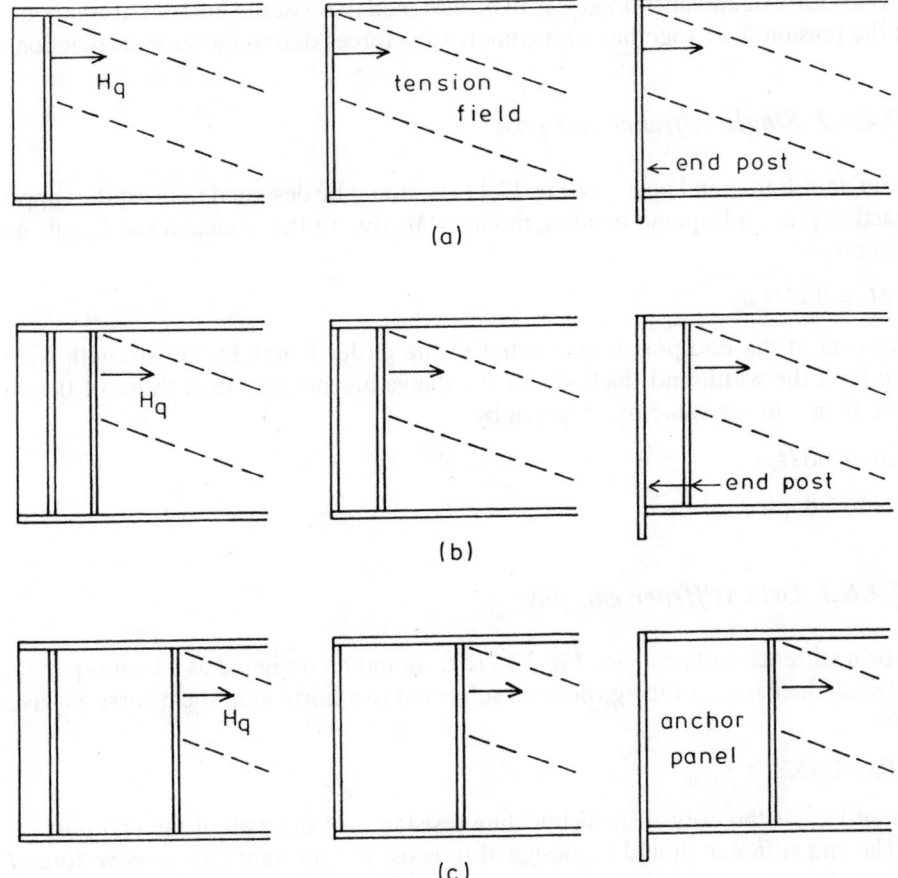

Fig. 17.4   End anchorage (a) single stiffener end post (b) twin stiffener end post (c) anchor panel

panels (see Fig. 17.4). End anchorage should be provided for a longitudinal anchor force H_q representing the longitudinal component of the tension field, at the ends of webs without intermediate stiffeners and at the end panels of webs with intermediate transverse stiffeners.

If the web is fully loaded in shear ($F_v \geq V_w$):

$$H_q = 0.5 dt p_{yw} \left(1 - \frac{V_{cr}}{P_v}\right)^{1/2}$$

If the web is not fully loaded in shear:

$$H_q = 0.5 dt p_{yw} \left(\frac{F_v - V_{cr}}{V_w - V_{cr}}\right)\left(1 - \frac{V_{cr}}{P_v}\right)^{1/2}$$

Each form of end anchorage has to be designed to resist the horizontal component of the tension field, together with compressive forces due to the support reactions.

17.4.6.2 Single stiffener end post

A single stiffener end post (see Fig. 17.4 (a)) should be designed to resist the support reaction plus an in-plane bending moment M_{tf} due to the anchor force H_q given in general by:

$$M_{tf} = 0.15H_q d$$

However, if the end post is connected to the girder flange by full strength welds, and both the width and thickness of the flange are not less than those of the end post, then a lower value of M_{tf} given by:

$$M_{tf} = 0.1H_q d$$

may be adopted.

17.4.6.3 Twin stiffener end post

A twin stiffener end post (see Fig. 17.4 (b)) should be designed as a beam spanning between the flanges of the girder and subjected to a horizontal shear force R_{tf} given by:

$$R_{tf} = 0.75H_q \le V_{cr.ep}$$

where $V_{cr.ep}$ is the critical shear buckling resistance of the web of the end post.

The end stiffener should be designed to resist the relevant compressive force F_e due to the support reaction plus a compressive force F_{tf} due to bending of the end post given by:

$$F_{tf} = \frac{0.15H_q d}{a_e}$$

where a_e is the spacing of the two end stiffeners.

The other stiffener forming part of the end post should be designed to resist the relevant compressive force F_s due to the support-reaction. However, if the tensile force due to bending of the end post $F_{tf} > F_s$ then it should be checked for a tensile force equal to $(F_{tf} - F_s)$.

17.4.6.4 Anchor panel

An anchor panel (see Fig. 17.4 (c)) should satisfy the condition that the applied shear force F_v is less than the critical shear buckling resistance V_{cr}, i.e.

$$F_v \le V_{cr}$$

In addition the anchor panel should be considered as a beam spanning between the flanges of the girder and satisfy the condition (see section 17.4.6.3)

$$R_{tf} = 0.75H_q \leq V_{cr.ep}$$

The two transverse stiffeners bounding the anchor panel should be designed in a similar manner to the stiffeners of a twin end post.

17.4.7 Web stiffeners

17.4.7.1 Types of stiffeners

Transverse stiffeners are generally required to ensure the satisfactory performance of the web panels of slender plate girders. The three most important types of stiffeners are as follows.

(a) Intermediate transverse
(b) Load-carrying
(c) Bearing

Stiffeners of each of these types are subjected to compression and should be checked for bearing and buckling. A particular stiffener may also serve more than one function and should therefore be designed for combined effects: e.g. an intermediate transverse stiffener may also be load carrying.

The design of each of the above types of stiffener is detailed in clauses 4.4.6 and 4.5 of the Code.

17.4.7.2 Intermediate transverse web stiffeners

Intermediate transverse web stiffeners are used to increase both the critical shear buckling strength q_{cr} and post-critical shear buckling strength q_w of slender web panels. They are designed for minimum stiffness and buckling resistance, as specified in clause 4.4.6 of the Code.

Intermediate transverse web stiffeners not subject to external loads should have a minimum second moment of area I_s about the centreline of the web given by:

$$\text{For } \frac{a}{d} \geq \sqrt{2} \quad I_s = 0.75 dt^3_{min}$$

$$\text{For } \frac{a}{d} < \sqrt{2} \quad I_s = 1.5(d/a)^2 dt^3_{min}$$

where t_{min} is the minimum required web thickness for the actual stiffener spacing a.

The buckling resistance of intermediate transverse web stiffeners not subject to external forces and moments should satisfy the condition

$$F_q = V_{max} - V_{cr} \le P_q$$

where F_q is the compressive axial force, assumed equal to the larger of the shears in the two web panels adjacent to the stiffener minus the critical shear buckling resistance of the same web panel, and P_q is the buckling resistance of the stiffener from clause 4.5.5 of the Code.

Additional requirements are specified for the minimum stiffness and buckling resistance of intermediate transverse web stiffeners subject to external forces and moments.

17.4.7.3 *Load-carrying stiffeners*

Load-carrying stiffeners are provided to prevent local buckling of the web due to concentrated loads or reactions applied through the flange. They should be positioned wherever such actions occur, if the resistance of the unstiffened web would otherwise be exceeded.

Load-carrying stiffeners must be checked for both bearing and buckling, as specified in clauses 4.5.2 and 4.5.3 of the Code. When checking for buckling an effective web width of $15t$ on either side of the centreline of the stiffener is considered to act with it to form a cruciform section. The resulting stiffener strut is then assumed to have an effective length of 0.7 times its actual length if the loaded flange is restrained against rotation in the plane of the stiffener. If the loaded flange is not restrained the effective length is taken equal to the actual length. The buckling resistance of the stiffener strut is then determined as for a normal compression member.

When checking the stiffener for bearing, only that area of the stiffener in contact with the flange should be taken into account.

Sometimes it is not possible to provide stiffeners immediately under the externally applied loading, e.g. a travelling load on a gantry girder or loads applied during launching. In such cases of patch loading (loads applied between the stiffeners) an additional check is required to ensure that the specified compressive strength of the web for edge loading is not exceeded.

17.4.7.4 *Bearing stiffeners*

Bearing stiffeners are provided to prevent local crushing of the web due to concentrated loads or reactions applied through the flange. If such actions do not exceed the buckling resistance of the web, then the load-carrying stiffeners of section 17.4.7.3 are not needed. However, the local bearing resistance may still be exceeded.

If so then bearing stiffeners should be provided and designed in accordance with clause 4.5.2 of the Code.

17.4.8 Gantry girders

Plate girders used to support cranes should be designed in accordance with clauses 4.4 and 4.11 of the Code.

17.5 Initial choice of cross-section for plate girders used in bridges

17.5.1 Choice of span

Plate girders are frequently employed to support railway and highway loadings on account of their economic advantages and ease of fabrication, and are considered suitable for spans in the region of 25–100 m. Plate girders are considered when rolled sections are not big enough to carry the loads over the chosen span.

Spans are usually fixed by site restrictions and clearances. If there is freedom for the designer, simply-supported spans within the range of 25–45 m will be found to be appropriate; the optimum for continuous spans is about 45 m, as 27 m long girders can be spliced with pier girders 18 m long. However, plate girders can be employed for spans of up to about 100 m in continuous construction.

17.5.2 Span-to-depth ratios

For supporting highways a composite concrete decking, having a concrete thickness in the region of 250 mm, is commonly employed. A span-to-depth ratio of 20 for simply-supported spans serves as an initial choice for such girders; where continuous spans are employed the depth of the girder can be reduced at least by one-third compared with a simply-supported span.

Through girders of constant depth are rarely employed in continuous construction. For simply supported plate girders designed for highway loading, a span-to-depth ratio of 20 can serve as an initial choice; for through girders supporting railway loading larger depths are required, and a span-to-depth ratio of 15 is more appropriate. Variable-depth plate girders are more appropriate where spans in excess of 30 m are required.

For continuous girders in composite construction, a span-to-depth ratio of 25 at the pier and 40 at mid-span is suitable. For highway bridges provided with an orthotropic deck, the corresponding values are increased to 30 and 60 respectively.

Table 17.2   Guide for selection of deck thickness

Transverse girder spacing (m)	Deck	Thickness (mm)
2.5–3.8	Reinforced concrete slab on permanent formwork	Constant 225–275
3.3–5.5	Haunched slab	Min. 250 up to 350 at haunch
5.0–7.0	Stringer	Constant 225–275
6.5+	Cross girders	Typically 250

17.5.3 Initial sizing of the flanges

It is important to keep the flanges as wide as possible consistent with outstand limitations. The oustand should not ordinarily be wider than 12 times its thickness if the flange is fully stressed in compression, or 16 times its thickness if the flange is not fully stressed or is in tension. This ensures that the flange is stable during erection with the minimum number of bracings. For practical reasons, e.g. to accommodate detailing for certain types of permanent formwork, it is desirable to use a minimum flange width of 400 mm. A maximum flange thickness of 65 mm is recommended to avoid heavy welds and the consequent distortion.

Changes in the width of the top flange can be incorporated easily in composite decks to suit design requirements, and these do not invite criticism on appearance grounds. On the other hand, changes in widths of bottom flanges are less acceptable visually. In any case it is desirable not to change flange sizes frequently lest the economy achieved by saving in material should be offset by expensive butt welding.

For composite girders having concrete decks it is usually necessary to allow for at least two rows of shear connectors on top flanges of beams; for longer spans three rows may be required at piers where high shear transfer takes place.

The cross-sections chosen affect deck thickness and overall structural form. Table 17.2 provides useful guidance for the initial selection of deck thickness.

17.5.4 Initial sizing of the web

The initial choice of web thickness is related to stiffening. There is no special advantage in using too slender a web, as the material saved will have to be replaced by stiffening; moreover, the workmanship with a thicker web plate is often superior. Probably the biggest single problem when determining the girder layout for a concrete decked plate girder road bridge is to achieve a solution with optimum transverse deck spans and minimum cost of web steel, as this involves the balancing of weight against workmanship.

The following advice is intended as general guidance.

Table 17.3   Initial values of web thickness

Beam depth (mm)	Web thickness (mm)
Up to 1200	10
1200–1800	12
1800–2250	15
2250–3000	20

Consideration should also be given to the economy of providing thicker webs, without any intermediate stiffening.

(a) Two or three vertical stiffeners should be provided, at a spacing of approximately one times the girder depth, close to the bearings; thereafter their spacing should be increased to 1.5 times the girder depth.

(b) In general horizontal stiffeners should be avoided. In long span continuous plate girders they may be necessary in locations close to the piers.

(c) It is desirable to provide vertical stiffeners on one side and horizontal stiffeners on the other side, where possible.

(d) Initial values for web thickness are suggested in Table 17.3, if it is intended to provide web stiffening. Consideration should also be given to the economy of providing thicker webs, without any intermediate stiffening.

(e) Compact girders are more economical for most simply supported spans and for shorter continuous spans; economical plate girders for longer continuous spans will be non-compact.

17.6 Design of steel bridges to BS 5400: Part 3

17.6.1 Global analysis

The Code requires that the global analysis of the structure should be carried out elastically to determine the load effects, i.e. bending moments, shear forces etc. The section properties to be used will generally be those of the gross section (see clauses 7.1 and 7.2 of the Code[2]).

Plastic analysis of the structure, i.e. redistribution of moments due to plastic hinge formation, is not allowed under BS 5400: Part 3.

Analysis should be carried out for individual and unfactored load cases. Summation of the load effects in different combinations and with different load factors can then be carried out in tabular form, as required by the design process.

17.6.2 Design of beams at the ultimate limit state

17.6.2.1 Basis of design

In the design of beam cross sections at the ultimate limit state the following need to be considered.

(a) Material strength
(b) Limitations on shape on account of local buckling of individual elements (webs and flanges)
(c) Effective section (reductions for compression buckling and holes)
(d) Lateral torsional buckling
(e) Web buckling (governed by depth-to-thickness ratio of web and panel aspect ratio)
(f) Combined effects of bending and shear.

17.6.2.2 Material strength (clause 4.3 of the Code)

The nominal material strength is the yield stress of steel. The partial factor on strength γ_m depends on the structural component and behaviour; gernerally it is 1.05 at the ultimate limit state, but higher values are specified in the Code for compressive stress in bending or buckling.

17.6.2.3 Shape limitations (clause 9.3 of the Code)

The resistance of a section can be limited by local buckling if the flange outstand to thickness ratio is large. The Code limits this ratio so that local buckling will not govern. Similarly, limits are given for outstands in tension to limit local shear lag effects.

Where the compression region of a web without stiffeners has a very large depth to thickness ratio, i.e. > 68, the bending resistance of the section is modified by the requirement that webs are effectively reduced to take account of local buckling. Webs with longitudinal stiffeners are not so reduced.

When calculating the bending resistance of beams, it is important to recognize the contribution made by the webs and make appropriate provision for it in computing the modulus of the 'effective section'. Slender webs without horizontal stiffeners buckle in the compression zone when subjected to high bending stresses. Based on parametric studies, the bending resistance contributed by the web (associated with a flange) is assessed by reducing the thickness of the web using the formulae

$$\frac{t_{we}}{t_w} = 1 \qquad \text{if } \frac{y_c}{t_w}\left(\frac{\sigma_{yw}}{355}\right)^{1/2} \leq 68$$

$$\frac{t_{we}}{t_w} = \left[1.425 - 0.00625\frac{y_c}{t_w}\left(\frac{\sigma_{yw}}{355}\right)^{1/2}\right] \qquad \text{if } 68 < \frac{y_c}{t_w}\left(\frac{\sigma_{yw}}{355}\right)^{1/2} < 228$$

$$\frac{t_{we}}{t_w} = 0 \qquad \text{if } \frac{y_c}{t_w}\left(\frac{\sigma_{yw}}{355}\right)^{1/2} \geq 228$$

where y_c is the depth of the web measured in its plane from the elastic neutral axis of the gross section of the beam to the compressive edge of the web.

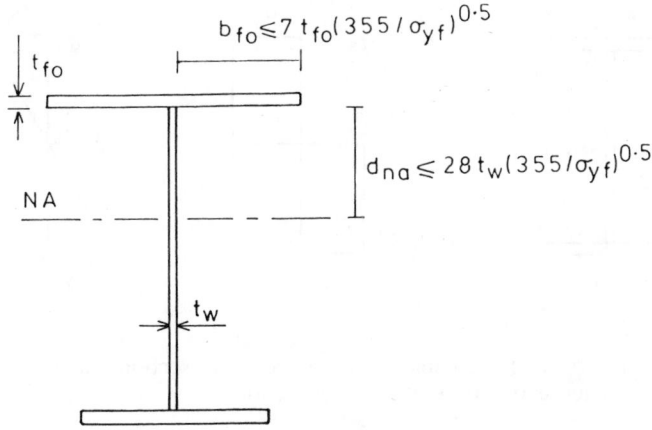

Fig. 17.5   Section geometry and classification – compact sections

The Code further defines cross-sections as compact or non-compact, and a different design approach is required for each classificatiod.[7]

The classification of a section depends on the width to thickness ratio of the elements of the cross-section considered, as shown in Fig. 17.5 for a compact plate girder.

A compact cross-section can develop the full plastic moment resistance of the section, i.e. a rectangular stress block, and local buckling of the individual elements of the cross section will not occur before this stage is reached (see Fig. 17.6 (b)).

However, in a non-compact section local buckling of elements of the cross section may occur before the full moment resistance is reached, and hence the design of such sections is limited to first yield in the extreme fibre i.e. a triangular stress block (see Fig. 17.6 (c)).

These classifications can best be illustrated by considering a beam in fully restrained bending. This gives the idealized moment resistance of the cross sections as follows.

Compact sections $\qquad M_1 = \dfrac{Z_{pe}\sigma_y}{\gamma_m\gamma_{f3}}$

Non-compact sections $\quad M_1 = \dfrac{Z\sigma_y}{\gamma_m\gamma_{f3}}$

where σ_y is the yield stress, Z_{pe} is the plastic section modulus, Z is the elastic section modulus, γ_m is the partial factor on strength, and γ_{f3} is the partial factor on loads, reflecting the uncertainty of loads.

The use of the plastic section modulus for compact sections does not imply that plastic analysis can be employed; in fact it is specifically excluded by BS 5400: Part 3. The achievement of a rectangular stress block does not necessarily mean that there has been redistribution of moments along the member.

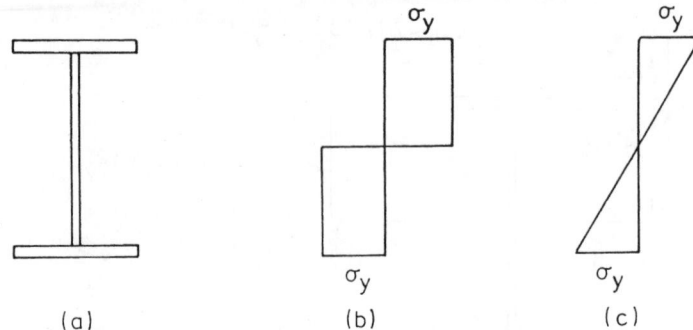

Fig. 17.6 Design stresses for compact and non-compact sections (a) beam cross-section
(b) compact section (c) non-compact section

17.6.2.4 Lateral torsional buckling (clause 9.6 of the Code)

Where a member has portions of its length with unrestrained elements in compression, lateral torsional buckling may occur (see Fig. 17.7). The Code deals with these effects by the use of a slenderness parameter λ_{LT} defined by

$$\lambda_{LT} = \frac{L_e}{r_y} K_4 \eta v$$

where L_e is the effective length for lateral torsional buckling, r_y is the radius of gyration of the whole beam about its minor y-y axis, K_4 is a torsion factor (taken as 0.9 for rolled I or channel sections and 1.0 for all other sections), η is a factor that takes account of moment gradient, i.e. the shape of the moment diagram and the fact that a uniform moment over the unrestrained length will cause buckling more readily than a non-uniform moment, and v is a torsion factor dependent on the shape of the beam.

The process of calculating the value of the limiting compressive stress σ_{lc} corresponding to the value of λ_{LT} is described in the Code.

Portions of beams between restraints can deflect downwards, sideways and rotate. Failure may then occur before the full moment resistance of the section is reached. The possibility of this type of failure is dictated by the unrestrained length of the compression flange, the cross-section geometry of the beam and the moment gradient.

17.6.2.5 Shear resistance (clause 9.9.2 of the Code)

The ultimate shear resistance V_D of a web panel under pure shear should be taken as

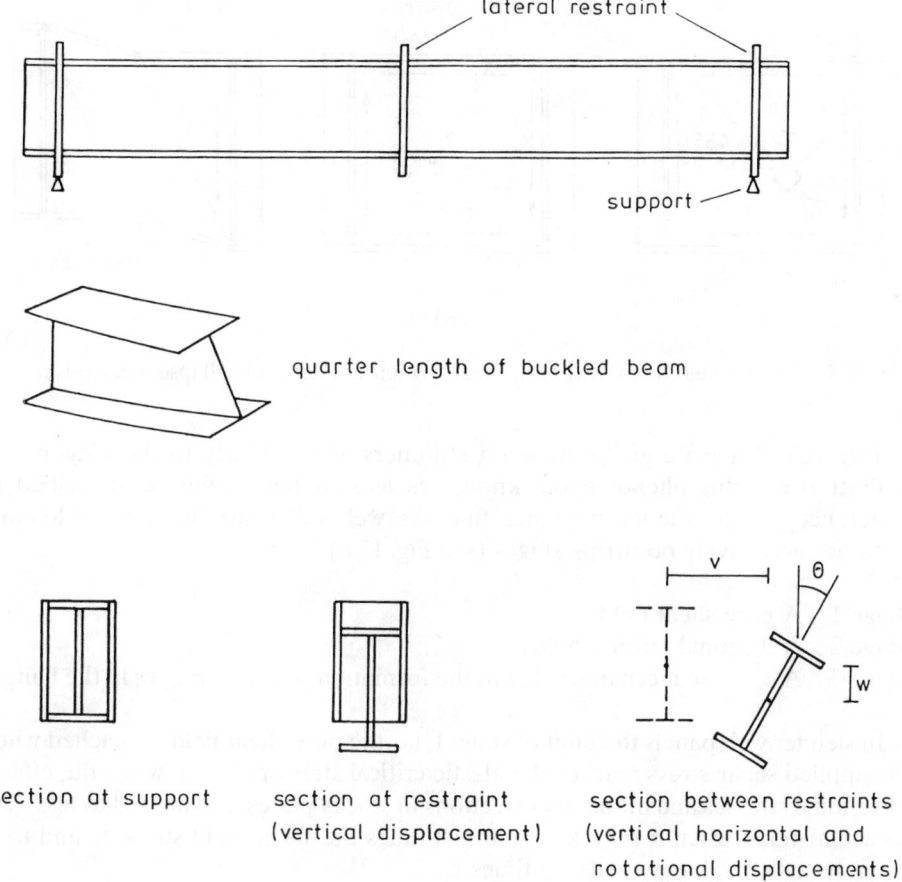

Fig. 17.7   Lateral torsional buckling

$$V_{\mathrm{D}} = \left[\frac{t_{\mathrm{w}}(d_{\mathrm{w}} - h_{\mathrm{h}})}{\gamma_{\mathrm{m}} \gamma_{\mathrm{f3}}} \right] \tau_1$$

where t_{w} and d_{w} are the thickness and depth of the web respectively, h_{h} is the height of the largest cut-out within the panel being considered, γ_{m} and γ_{f3} are partial material and load factors respectively, and τ_1 is the limiting shear stress. For slender webs τ_1 is governed by shear buckling, which becomes significant in webs with depth to thickness ratios greater than about 80.

The actual shear strength of a web panel is dependent on the following:

(a) Yield stress (suitably factored)
(b) Depth-to-thickness ratio of web
(c) Spacing of stiffeners
(d) Conditions of restraint provided by flanges

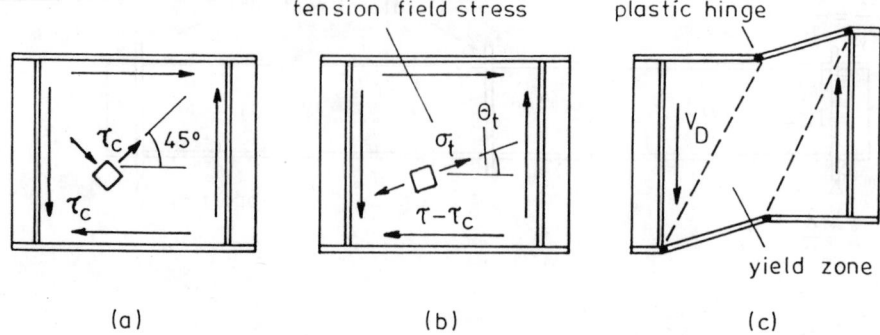

Fig. 17.8   Tension field theory (a) pure shear (b) tension field (c) collapse mechanism

The web of a plate girder between stiffeners acts similarly to the diagonal of a Pratt truss (this phenomenon, known as *tension field action*, is described in Reference 6). The theory stipulates that the web will resist the applied loading in three successively occurring stages (see Fig. 17.8):

Stage 1   A pure shear field
Stage 2   A diagonal tension field
Stage 3   A collapse mechanism due to the formation of plastic hinges in the flanges

In slender web panels the limit of stage 1, i.e. the pure shear field, is reached when the applied shear stress reaches the elastic critical stress τ_c. To allow for the effects of residual stresses and initial imperfections in stocky plates, τ_c is limited to less than its actual value when it is greater than 0.8 times the shear yield stress τ_y, and to τ_y when its actual value exceeds 1.5 times τ_y.

The elastic critical stress τ_c for a plate loaded in shear is given by

$$\tau_c = K \frac{\pi^2 E}{12(1-v^2)} \left(\frac{t_w}{d_{we}} \right)^2$$

where E and v are the elastic modulus and Poisson's ratio of the steel respectively, and

$$K = 5.34 + \frac{4}{\phi^2} \quad \text{when } \phi \geq 1$$

$$K = 4 + \frac{5.34}{\phi^2} \quad \text{when } \phi < 1$$

$$\phi = \text{panel aspect ratio} = \frac{a}{d_{we}}$$

The criteria outlined previously have been incorporated in computing the value of τ_c as follows:

$$\frac{\tau_c}{\tau_y} = \frac{904}{\beta^2} \qquad \text{when } \beta \geq 33.62$$

$$= 1 \qquad \text{when } \beta \leq 24.55$$

$$= 1.54 - 0.022\beta \qquad \text{when } 24.55 < \beta < 33.62$$

where

$$\beta = \frac{\lambda}{K^{1/2}}$$

$$\lambda = \frac{d_{we}}{t_w} \left(\frac{\sigma_{yw}}{355} \right)^{1/2}$$

In stage 2 a tensile membrane stress field develops in the panel, the direction of which does not necessarily coincide with the diagonal of the panel. The maximum shear resistance is reached in stage 3, when the pure shear stress of stage 1 and the membrane stress of stage 2 cause yielding of the panel according to the von Mises yield criterion, and plastic hinges are formed in the flanges.

The magnitude of the membrane tensile stress σ_t in terms of its assumed direction θ and the first stage limiting stress τ_c is given by

$$\frac{\sigma_t}{\tau_y} = \left[3 + (2.25\sin^2 2\theta - 3)\left(\frac{\tau_c}{\tau_y} \right)^2 \right]^{1/2} - 1.5 \frac{\tau_c}{\tau_y} \sin 2\theta$$

Adding the resistance in the three stages, the ultimate shear strength τ_u is obtained as

$$\frac{\tau_u}{\tau_y} = \left[\frac{\tau_c}{\tau_y} + 5.264 \sin\theta \left(m_{fw} \frac{\sigma_t}{\tau_y} \right)^{1/2} + \frac{\sigma_t}{\tau_y}(\cot\theta - \phi)\sin^2\theta \right]$$

when

$$m_{fw} \leq \frac{\phi^2}{4\sqrt{3}} \frac{\sigma_t}{\tau_y} \sin^2\theta$$

and

$$\frac{\tau_u}{\tau_y} = \left[\frac{4\sqrt{3} m_{fw}}{\phi} + \frac{\sigma_t}{2\tau_y} \sin^2\theta + \frac{\tau_c}{\tau_y} \right]$$

when

$$m_{fw} > \frac{\phi^2}{4\sqrt{3}} \frac{\sigma_t}{\tau_y} \sin^2\theta$$

To provide an added measure of safety in respect of slender webs, the above values are multiplied by a varying correction factor f to obtain the limiting shear stress.

$$f = 1 \qquad\qquad\qquad \text{when} \quad \lambda \le 56$$

$$f = \frac{1.15}{1.35} \qquad\qquad\quad \text{when} \quad \lambda \ge 156$$

$$f = \frac{1.15}{1.15 + 0.002(\lambda - 56)} \quad \text{when} \quad 56 < \lambda < 156$$

The value of τ_l/τ_y is taken as the lower of τ_u/τ_y computed as above and 1.0.

The term m_{fw} in the above equations is a non-dimensional representation of the plastic moment resistance of the flange (taking the smaller value of the top and bottom flanges, ignoring any concrete).

$$m_{fw} = \frac{M_p}{d_{we}^2 t_w \sigma_{yw}}$$

$$= \frac{\sigma_{yf} b_{fe} f_f^2}{2 d_{we}^2 t_w \sigma_{yw}}$$

Values of τ_l/τ_y versus λ are plotted in Figures 11–17 of the Code corresponding to various values of m_{fw} and ϕ. These can be used directly by designers.

The term b_{fe} is the width of flange associated with the web and is limited to

$$b_{fe} \le 10 t_f \left(\frac{355}{\sigma_{yf}} \right)^{1/2}$$

Where two flanges are unequal, the value of m_{fw} is taken conservatively as the value corresponding to the weaker flange. Moreover, only a section symmetric about the mid-plane is taken as effective; any portion of the flange plate outside this plane of symmetry is ignored so that complexities due to the torsion of the flange are eliminated.

The above procedure has to be repeated for several values of θ, and the highest value of τ_u/τ_y is to be used. From parametric studies it has been established that θ is never less than $(1/3)\cot^{-1}\phi$ or more than $(4/3)\cot^{-1}\phi$.

When tension field action is used, consideration must be given to the anchorage of the tension field forces in the end panels, and special procedures must be adopted for designing the end stiffener.

17.6.2.6 Combined bending and shear (clause 9.9.3 of the Code)

The Code has simplified the procedure for girders without longitudinal stiffeners by allowing shear and bending resistance to be calculated independently and then combined by employing an interaction relationship on the basis given below.

(a) The bending resistance of the whole section is determined with and without contribution from the web (M_l and M_R respectively).

(b) The shear resistance using the tension field theory discussed above is determined with and without contribution from the flanges (V_D and V_R respectively).

(c) The bending and shear resistance without any contribution from the web and flanges respectively (M_R and V_R) can be mobilized simultaneously.

(d) The pure bending resistance of the whole section M_D can be obtained, even when there is a coincident shear on the section, provided the latter is less than $\frac{1}{2}V_R$. Similarly the theoretical design shear resistance V_D of the whole beam section is attained provided the coexisting bending moment is not greater than $\frac{1}{2}M_R$.

The interaction relationship is linear between this set of values of shear force and bending moment and is shown graphically in Fig. 17.9. The important feature from the designer's point of view is that full values of calculated bending resistance can be utilized in the presence of a moderate magnitude of shear.

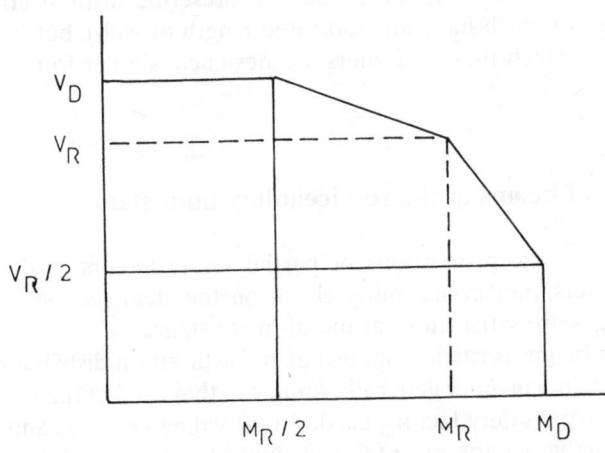

Fig. 17.9   Interaction between shear and bending resistance

For girders with longitudinal stiffeners, account is taken of the combined effects of bending and shear by comparison of the stresses in the web panels with the relevant buckling strength of the panel. The buckling coefficients are based upon large-deflection elastic–plastic finite element studies. An interaction expression is then used which is based on an equivalent stress check. This makes suitable allowance for webs with longitudinal stiffeners and having large depth-to-thickness ratios; no reduction need be made to account for local buckling.

17.6.2.7 *Bearing stiffeners (clause 9.14 of the Code)*

BS 5400: Part 3 requires such stiffeners to be provided at supports. The design forces to be applied to the stiffener include the direct reaction from the bearing (less the

force component in the flange if the beam soffit is haunched), a force to account for the destabilizing effect of the web, forces from any local cross bracing (including forces from restraint systems), and a force due to the tension field. The derivation of these is far more precise and lengthy than BS 5950: Part 1: 2000 requires, largely because plate girders in bridges are usually part of a much more complex structural system than in buildings.

Bearing stiffeners are almost always placed symmetrically about the web centreline. Stiffeners that are symmetrical about the axis perpendicular to the web, i.e. flats or tees rather than angles, are preferred.

17.6.2.8 *Intermediate stiffeners (clause 9.13 of the Code)*

Intermediate stiffeners are usually placed on one side of the web only. Standard flat sizes should be used. BS 5400: Part 3 does not prescribe stiffness criteria for intermediate stiffeners (which have an associated length of web), but it does prescribe design forces for which these stiffeners are designed, similar to those outlined for bearing stiffeners.

17.6.3 Design of beams at the serviceability limit state

In practice, due to the proportions of partial safety factors at the ultimate and serviceability limits, the serviceability check on the steel section is automatically satisfied if designs are satisfactory at the ultimate stage.

The design of beams is carried out using an elastic strain distribution throughout the cross-section, in a manner generally similar to that used at the ultimate limit but with different partial safety factors, i.e. different values of γ_{fL}, γ_m and γ_{f3}. The determination of effective section must take account of shear lag, where appropriate.

Stresses should be checked at critical points in the steel member to ensure that no permanent deformation due to yielding takes place.[8-10] Plastic moment resistance is not considered at the serviceability limit state.

In composite construction, crack widths in the concrete deck may often govern the design at the serviceability limit state. For non-compact sections the analysis at the ultimate limit state is carried out using a triangular (elastic) stress distribution. Consequently the serviceability limit check for bending is always satisfied, since the factors are less than those at the ultimate limit state. The Code lists the clauses which require checking at both limit states.

17.6.4 Fatigue

Fatigue should be checked with reference to BS 5400: Part 10. Generally it is only bracing connections, stiffeners and shear connectors, and their welding to the

girders, that have to be checked for fatigue. The details of the connection may have a significant effect on the fatigue life, and by careful detailing fatigue may be 'designed out' of the bridge.

17.6.5 Design format

One of the significant features of BS 5400: Part 3 is that plastic redistribution across the section is now permitted, in certain circumstances, prior to the attainment of the ultimate limit state. Such redistribution is much less than permitted in simple plastic design of buildings, but can give a considerable enhancement in strength. For example, the design bending strength of compact sections is based on the plastic section modulus, and the basic shear strength is based on rules which take account of tension field action, which only develops in the presence of considerable plasticity. The interaction between shear strength and bending strength of a cross-section is empirically based and implicitly assumes that plastic strains may occur prior to the attainment of the design strength. Finally, the design strength of compact sections that are built in stages assumes that a redistribution of stresses may take place within the cross-section. The ultimate limit state check for the completed structure is simply to ensure that the bending resistance (given by the limiting stress times the plastic section modulus divided by partial factors) is greater than the maximum design moment (obtained by summing all the moments due to the various design loads).

This recognition of the reserve of strength beyond first yield is limited to situations where plasticity may be permitted safely. The principal limitations are as follows.

(a) Plastic section moduli may only be used for compact sections, i.e. those that can sustain local compressive yielding without any local buckling.
(b) Where the structure is built in stages and is non-compact, plasticity is permitted implicitly only in considering the interaction between bending and shear strength.
(c) Where longitudinal stiffeners are present, neither the plastic section modulus nor the plastic bending / shear interaction may be used.
(d) If the flanges are curved in elevation, similar restrictions to (c) apply.

References to Chapter 17

1. British Standards Institution (2000) BS 5950: *Structural use of steelwork in building*: Part 1: *Code of practice for design: rolled and welded sections*. BSI, London.
2. British Standards Institution (1982) BS 5400: *Steel concrete and composite bridges*: Part 3: *Code of practice for the design of steel bridges*. BSI, London.

3. Owens G.W. (1989) *Design of Fabricated Composite Beams in Buildings.* The Steel Construction Institute, Ascot, Berks.
4. Narayanan R., Lawless V., Naji F.J. & Taylor J.C. (1993) *Introduction to Concise Eurocode 3 (C-EC3) with worked examples.* SCI Publication 115, The Steel Construction Institute, Ascot, Berks.
5. Lawson R.M. & Rackham J.W. (1989) *Design for Openings in Webs of Composite Beams.* The Steel Construction Institute, Ascot, Berks.
6. Evans H.R. (1988) Design of plate girders. In *Introduction to Steelwork Design to BS 5950: Part 1*, pp. 12.1–12.10. The Steel Construction Institute, Ascot, Berks.
7. Chaterjee S. (1981) Design of webs and stiffeners in plate and box girders. In *The Design of Steel Bridges* (Ed. by K.C. Rockey & H.R. Evans), Chapter 11. Granada.
8. Iles D.C. (1989) *Design Guide for Continuous Composite Bridges 1: Compact Sections.* SCI Publication 065, The Steel Construction Institute, Ascot, Berks.
9. Iles D.C. (1989) *Design Guide for Continuous Composite Bridges 2: Non-Compact Sections.* SCI Publication 066, The Steel Construction Institute, Ascot, Berks.
10. Iles D.C. (1991) *Design Guide for Simply Supported Composite Bridges.* SCI Publication 084, The Steel Construction Institute, Ascot, Berks.

Further reading for Chapter 17

Iles D.C. (2001) *Design of Composite Bridges: General Guidance.* SCI Publication 289, The Steel Construction Institute, Ascot, Berks.
Iles D.C. (2001) *Design of Composite Bridges: Worked Examples.* SCI Publication 290, The Steel Construction Institute, Ascot, Berks.

A worked example follows which is relevant to Chapter 17.

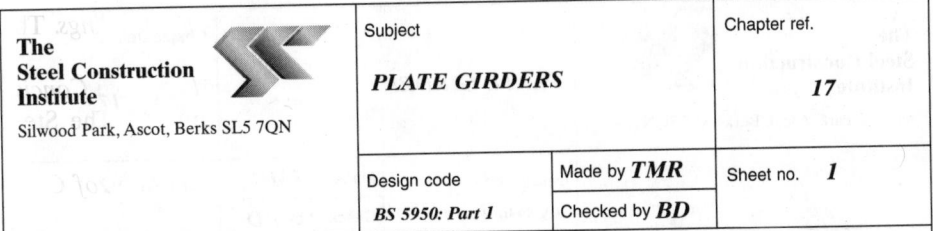

	Subject	Chapter ref.
The Steel Construction Institute Silwood Park, Ascot, Berks SL5 7QN	*PLATE GIRDERS*	**17**

Design code	Made by *TMR*	Sheet no. **1**
BS 5950: Part 1	Checked by *BD*	

WE 17.1 Design brief

The girder shown in Fig.WE 17.1 is fully restrained throughout its length. For the loading shown and specified in WE 17.2 design a transversely stiffened plate girder in grade S275 steel. Girder depth is unrestricted.

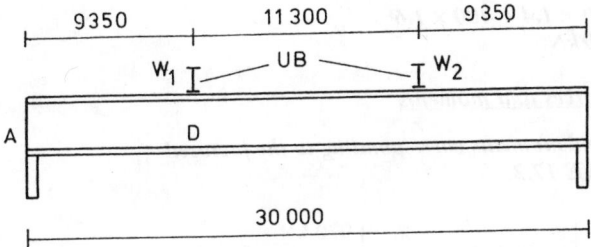

Fig.WE 17.1   *Plate girder span and loading*

$w = w_d + w_i$

$W_1 = W_{1d} + W_{1i}$

$W_2 = W_{2d} + W_{2i}$

WE 17.2 Loading

Dead loads	UDL	$w_d = 20\,kN/m$
	Point loads	$W_{1d} = 200\,kN$ $W_{2d} = 200\,kN$
Imposed loads	UDL	$w_i = 40\,kN/m$
	Point loads	$W_{1i} = 400\,kN$ $W_{2i} = 400\,kN$

WE 17.3 Load factors

2.4.1.1
Table 2

Dead load factor $\gamma_{fd} = 1.4$

Imposed load factor $\gamma_{fi} = 1.6$

The **Steel Construction Institute** Silwood Park, Ascot, Berks SL5 7QN	Subject **PLATE GIRDERS**	Chapter ref. **17**
Design code **BS 5950: Part 1**	Made by *TMR* / Checked by *BD*	Sheet no. **2**

WE 17.4 Factored loads

2.4.1.1
Table 2

$$w = w_d \gamma_{fd} + w_i \gamma_{fi} = 20 \times 1.4 + 40 \times 1.6$$
$$= 92\,kN/m$$

$$W_1' = W_{1d}\gamma_{fd} + W_{1i}\gamma_{fi} = 200 \times 1.4 + 400 \times 1.6$$
$$= 920\,kN$$

$$W_2' = W_{2d}\gamma_{fd} + W_{2i}\gamma_{fi} = 200 \times 1.4 + 400 \times 1.6$$
$$= 920\,kN$$

WE 17.5 Design shear forces and moments

The design shear forces and moments corresponding to the factored loads are shown in Fig.WE 17.2.

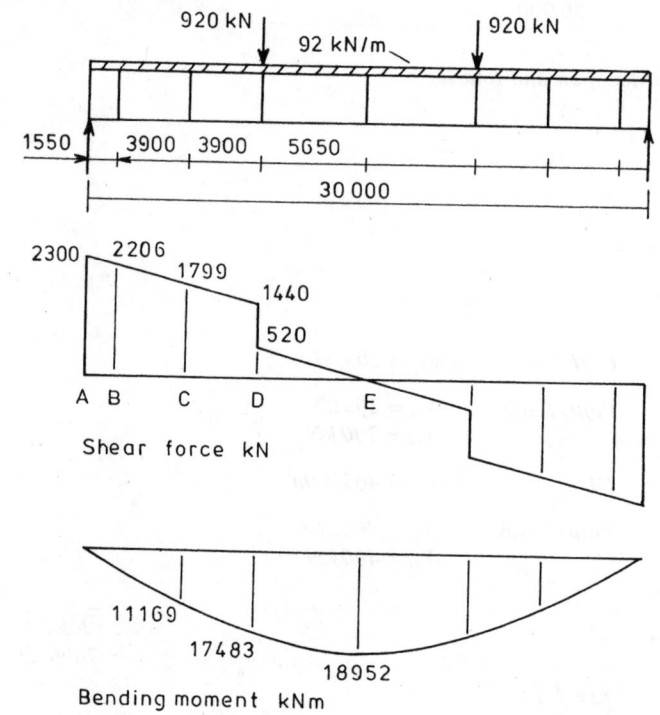

Fig.WE 17.2   Design shear forces, moments and stiffener spacing

The **Steel Construction Institute** Silwood Park, Ascot, Berks SL5 7QN 	Subject **PLATE GIRDERS**	Chapter ref. **17**
	Design code **BS 5950: Part 1**	Made by **TMR** Checked by **BD**	Sheet no. **3**

WE 17.6 Initial sizing of plate girder

The recommended span/depth ratio for simply supported non-composite girders varies between 12 for short span girders and 20 for long span girders. Herein the depth is assumed to be span/15.

$$d = \frac{span}{15} = \frac{30\,000}{15} = 2000\,mm$$

Estimate flange area assuming $p_{yf} = 255\,N/mm^2$ 3.1.1   Table 9

$$A_f = \frac{M_{max}}{dp_{yf}} = \frac{18952 \times 10^6}{2000 \times 255} = 37160\,mm^2$$ 4.4.4.2(b)

For non-composite girders the flange width is usually within the range 0.3 to 0.5 of the depth. Assume a flange $750 \times 50 = 37500\,mm^2$.

The minimum web thickness for plate girders in buildings usually satisfies $t \geq d/120$. Assume the web thickness $t = 15\,mm$, slightly less than d/120 due to the proposed transverse stiffening.

WE 17.7 Section classification

Assume girder with flanges $750 \times 50\,mm$ and web $2000 \times 15\,mm$

Flange

For $T = 50\,mm$ $p_{yf} = 255\,N/mm^2$ Table 9

$$\varepsilon = \left(\frac{275}{p_{yf}}\right)^{1/2} = \left(\frac{275}{255}\right)^{1/2} = 1.04$$ Table 11 Note 2

$$b = \frac{750 - 15}{2} = 367.5\,mm$$

$$\frac{b}{T} = \frac{367.5}{50} = 7.35$$

*Compact limiting value of b/T for welded sections is $9\varepsilon = 9 \times 1.04$ Table 11
= 9.36.*

The Steel Construction Institute	Subject	Chapter ref.
Silwood Park, Ascot, Berks SL5 7QN	*PLATE GIRDERS*	17

	Design code	Made by *TMR*	Sheet no. 4
	BS 5950: Part 1	Checked by *BD*	

$\dfrac{b}{T} = 7.35 < 9.36$

∴ *Flange is compact*

Web

For t = 15 mm $p_{yw} = 275 \, N/mm^2$ $\varepsilon = 1$ *Table 9*

$\dfrac{d}{t} = \dfrac{2000}{15} = 133$

$\dfrac{d}{t} = 133 > 120 \, \varepsilon = 120$

∴ *Web is slender*   *Table 11*

Since $\dfrac{d}{t} > 63\varepsilon$ the web must be checked for shear buckling   *4.4.5.1*

<u>*WE 17.8 Dimensions of web and flanges*</u>   *4.4.3*

Assume a stiffener spacing a > 1.5d

Minimum web thickness to avoid serviceability problems   *4.4.3.2(b)*

$t = 15 \geq \dfrac{d}{250} = \dfrac{2000}{250} = 8.0 \, mm$ *OK*

To avoid the flanges buckling into the web   *4.4.3.3(b)*

$t = 15 \geq \left(\dfrac{d}{250}\right)\left(\dfrac{p_{yf}}{345}\right)$

$= \left(\dfrac{2000}{250}\right)\left(\dfrac{255}{345}\right) = 5.91 \, mm$

<u>*WE 17.9 Moment resistance*</u>   *4.4.4*

For sections with slender webs (and flanges which are not slender)   *4.4.4.2(b)*
three methods of design are specified. Method (b) is used herein.
The moment is assumed to be resisted by the flanges alone while
the web is designed for shear only.

$M_f = p_{yf} A_f h_s$

where $h_s = 2000 + 50 = 2050$ is the distance between the centroids
of flanges

$M_f = 255 \times 750 \times 50 \times 2050 \times 10^{-6} = 19603 \, kNm$

$M_f = 19603 > M_{max} = 18952 \, kNm$ *OK*

The **Steel Construction Institute** Silwood Park, Ascot, Berks SL5 7QN	Subject **PLATE GIRDERS**		Chapter ref. **17**
	Design code **BS 5950: Part 1**	Made by **TMR**	Sheet no. **5**
		Checked by **BD**	

WE 17.10 Shear buckling resistance of web

4.4.5

Webs with intermediate transverse stiffeners may be designed according to 4.4.5.2 or 4.4.5.3. The simplified method of 4.4.5.2 is used herein. The assumed stiffener spacing is shown in Fig. WE 17.2.

WE 17.11 Shear buckling resistance of end panel AB

The Code allows three alternative methods for providing end anchorage. The most commonly employed method is to design end panels without accounting for tension field action. For end panel AB:

$$d = 2000\,mm \qquad t = 15\,mm \qquad F_v = 2300\,kN$$

$$\frac{a}{d} = \frac{1550}{2000} = 0.775 \quad \frac{d}{t} = \frac{2000}{15} = 133$$

$$p_{yw} = 275\,N/mm^2$$

$$P_v = 0.6p_{yw}d\,t = 0.6 \times 275 \times 2000 \times 15 \times 10^{-3} = 4950\,kN \qquad 4.2.3$$

$$q_w = 132\,N/mm^2 \qquad Table\ 21$$

$$V_w = q_w d\,t = 132 \times 2000 \times 15 \times 10^{-3} = 3960\,kN \qquad 4.4.5.2$$

$$\frac{V_w}{P_v} = \frac{3960}{4950} = 0.8$$

$$V_{cr} = \frac{9V_w - 2P_v}{7} = \frac{9 \times 3960 - 2 \times 4950}{7} = 3677\,kN \qquad 4.4.5.4$$

$$F_v = 2300 < V_{cr} = 3677\,kN \qquad OK$$

$$F_v = 2300 < P_v = 4950\,kN \qquad OK$$

End panel AB should also be checked as a beam spanning between the flanges of the girder capable of resisting a shear force R_{tf} due to the anchor forces in panel BC.

H.4.4
Fig. 17.4

	Subject		Chapter ref.
The **Steel Construction** **Institute** Silwood Park, Ascot, Berks SL5 7QN	*PLATE GIRDERS*		*17*
	Design code	Made by *TMR*	Sheet no. *6*
	BS 5950: Part 1	Checked by *BD*	

$R_{tf} = 0.75\ Hq \geq V_{cr.ep}$

For panel BC H_q is given conservatively by   *H.4.1*

$H_q = 0.5d\ t\ p_{yw}(1 - V_{cr}/P_v)^{1/2}$

Calculations presented in WE 17.12 give

$V_{cr} = 1983\,kN$

$H_q = 0.5 \times 2000 \times 15 \times 275 \times 10^{-3}\left(1 - \dfrac{1983}{4950}\right)^{1/2} = 3193\,kN$

For panel AB spanning between the flanges

$a = 2000\,mm \qquad d = 1550\,mm \qquad t = 15\,mm$

$P_v = 0.6 \times 275 \times 1550 \times 15 \times 10^{-3} = 3836\,kN$ *4.2.3*

$\dfrac{a}{d} = \dfrac{2000}{1550} = 1.29 \quad \dfrac{d}{t} = \dfrac{1550}{15} = 103$

$q_w = 132\,N/mm^2$ *Table 21*

$V_w = 1550 \times 15 \times 132 \times 10^{-3} = 3069\,kN$ *4.4.5.2*

$\dfrac{V_w}{P_v} = \dfrac{3069}{3836} = 0.8$

$V_{cr.ep} = \dfrac{9 \times 3069 - 2 \times 3836}{7} = 2850\,kN$ *4.4.5.4*

Check

$R_{tf} = 0.75 \times 3193 = 2395 < V_{cr.ep} = 2850\,kN \qquad OK$

The Steel Construction Institute  Silwood Park, Ascot, Berks SL5 7QN	Subject **PLATE GIRDERS**	Chapter ref. **17**
	Design code **BS 5950: Part 1**	Made by **TMR** / Checked by **BD**	Sheet no. **7**

WE 17.12 Shear buckling resistance of panels BC and DE

The shear buckling resistance of panel BC is determined using the simplified method.

$d = 2000\,mm \qquad t = 15\,mm \qquad a = 3900\,mm$

$\dfrac{a}{d} = \dfrac{3900}{2000} = 1.95 \quad \dfrac{d}{t} = \dfrac{2000}{15} = 133$

$q_w = 94\,N/mm^2$

$V_w = d\,t\,q_w = 2000 \times 15 \times 94 \times 10^{-3} = 2820\,kN$

$\dfrac{V_w}{P_v} = \dfrac{2820}{4950} = 0.57$

$V_{cr} = \dfrac{\left(\dfrac{V_w}{0.9}\right)^2}{P_v} = \dfrac{\left(\dfrac{2820}{0.9}\right)^2}{4950} = 1983\,kN$ 4.4.5.4

$F_v = 2206 < V_w = 2820\,kN \qquad OK$

For panel DE

$\dfrac{a}{d} = \dfrac{5650}{2000} = 2.825 \quad \dfrac{d}{t} = \dfrac{2000}{15} = 133$

$q_w = 89\,N/mm^2$

$V_w = 2000 \times 15 \times 89 \times 10^{-3} = 2670\,kN$

$V_{cr} = \dfrac{\left(\dfrac{2670}{0.9}\right)^2}{4950} = 1778\,kN$

$F_v = 520 < V_w = 2670\,kN \qquad OK$

The **Steel Construction Institute** Silwood Park, Ascot, Berks SL5 7QN	Subject **PLATE GIRDERS**	Chapter ref. **17**
	Design code **BS 5950: Part 1**	Made by **TMR**	Sheet no. **8**
		Checked by **BD**	

WE 17.13 Load-carrying stiffener at A

The stiffener should be designed for a compressive force F_e due to the support reaction plus a compressive force F_{tf} due to the anchor force   H.4.4

$$F_x = F_e + F_{tf} = F_e + 0.15\ H_q\ \frac{d}{a_e}$$

$F_e = 2300\,kN \quad H_q = 3193\,kN \quad (see\ WE\ 17.11)$

$$F_x = 2300 + 0.15 \times 3193 \times \frac{2000}{1550} = 2918\,kN$$

Try stiffener consisting of two flats 280×22 mm.

Check oustands   4.5.1.2

For $t_s = 22 < 40$ mm $p_{ys} = 265\,N/mm^2$ Table 9

$$\varepsilon = \left(\frac{275}{p_{ys}}\right)^{1/2} = \left(\frac{275}{265}\right)^{1/2} = 1.02$$

$b_s = 280 \le 13\ \varepsilon\ t_s = 13 \times 1.02 \times 22 = 292\,mm$

Since $b_s < 13\ \varepsilon\ t_s$ design should be based on the net cross-section area $A_{s,net}$. Allowing 15 mm to cope for the web flange weld:

$A_{s,net} = (280 - 15)\ 22 \times 2 = 11660\,mm^2$

Check bearing resistance of stiffener, neglecting resistance of web   4.5.2.2

$P_s = p_{ys}A_{s,net} = 265 \times 11660 \times 10^{-3} = 3090\,kN$

The **Steel Construction Institute** Silwood Park, Ascot, Berks SL5 7QN	Subject *PLATE GIRDERS*	Chapter ref. *17*
	Design code **BS 5950: Part 1**	Made by *TMR* Checked by *BD*	Sheet no. **9**

$F_x = 2918 < P_s = 3090\,kN$ **OK**

Check buckling resistance of stiffener   4.5.3.3

$P_x = A_s p_c$

The effective area A_s includes the area of the stiffeners plus an effective width of web equal to $15t = 15 \times 15 = 225\,mm$ (see Fig. WE 17.3).

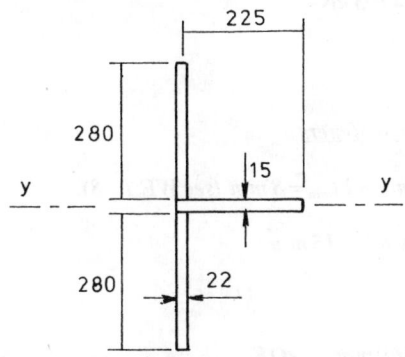

225

280

15

y ————————— y

280   22

Fig. WE 17.3   Load-carrying stiffener at A

$A_s = 2 \times 280 \times 22 + (225 + 11) \times 15 = 15860\,mm^2$

Assume the flange of the girder to be restrained against rotation in the plane of the stiffener.

$L_E = 0.7d = 0.7 \times 2000 = 1400\,mm$

Determine I_y for stiffener neglecting small contribution from the web   4.7.5

$$I_y = \frac{22 \times 575^3}{12} = 348 \times 10^6\,mm^4$$

Determine compressive strength p_c.

The Steel Construction Institute Silwood Park, Ascot, Berks SL5 7QN	Subject **PLATE GIRDERS**	Chapter ref. **17**
	Design code **BS 5950: Part 1**	Made by **TMR** Checked by **BD**	Sheet no. **10**

Since the stiffeners themselves are not welded sections, the reduction of p_{ys} by 20 N/mm² (see 4.7.5) should not be applied

4.5.3.3

$$\lambda = \frac{L_E}{\left(\dfrac{I_y}{A_s}\right)^{1/2}} = \frac{1400}{\left(\dfrac{348 \times 10^6}{15860}\right)^{1/2}} = 9.45$$

$p_c = 265 \, N/mm^2$

Table 24

$P_x = A_s p_c = 15860 \times 265 \times 10^{-3} = 4203 \, kN$

$F_x = 2918 < P_x = 4203 \, kN \qquad OK$

WE 17.14 Intermediate transverse stiffeners

4.4.6

For panels BC and CD a = 3900 mm and $t_{min} = 8$ mm (see WE 17.8).

Try stiffener consisting of two flats 80 × 15 mm

$p_s = 275 \, N/mm^2 \qquad \varepsilon = 1$

$b_s = 80 < 13 \, \varepsilon \, t_s = 13 \times 1 \times 13 = 169 \, mm \qquad OK$

4.5.1.2

Check minimum stiffness

4.4.6.4

$$\frac{a}{d} = \frac{3900}{2000} = 1.95 > \sqrt{2}$$

$I_s = 0.75 d \, t_{min}^3 = 0.75 \times 2000 \times 8^3$

$\quad = 0.768 \times 10^6 \, mm^4$

Actual I_y of stiffener neglecting web

$$I_y = \frac{15 \times 175^3}{12} = 6.70 \times 10^6 \ mm^4$$

$I_y = 6.70 \times 10^6 > I_s = 0.768 \times 10^6 \, mm^4 \qquad OK$

Check buckling resistance of stiffeners

4.4.6.6

The Steel Construction Institute Silwood Park, Ascot, Berks SL5 7QN	Subject **PLATE GIRDERS**		Chapter ref. **17**
	Design code **BS 5950: Part 1**	Made by **TMR**	Sheet no. **11**
		Checked by **BD**	

Intermediate transverse stiffeners should be designed for a compressive axial force F_q given by

$$F_q = V_{max} - V_{cr}$$

For panels BC and CD (see WE 17.12)

$V_{max} = 2206\,kN \quad V_{cr} = 1983\,kN$

$F_q = 2206 - 1983 = 223\,kN$

Since F_q is less than the external load applied to the stiffener at D see WE 17.15 for design.

WE 17.15 Intermediate load-carrying stiffener at D

This stiffener is subjected to an external compressive force $F_x = 920\,kN$ applied in line with the web.

Try stiffener consisting of two flats $80 \times 15\,mm$.

Minimum stiffness requirement as for other transverse stiffeners is satisfied (see WE 17.14).

Check buckling resistance assuming a width of web equal to 15t on each side of the centreline of the stiffener (see Fig. WE 17.4).   4.5.3.3

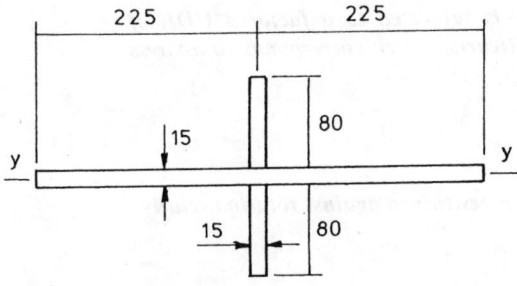

Fig.WE 17.4   Load-carrying stiffener at D

The **Steel Construction Institute** Silwood Park, Ascot, Berks SL5 7QN	Subject **PLATE GIRDERS**	Chapter ref. **17**
	Design code **BS 5950: Part 1**	Made by *TMR*	Sheet no. **12**
		Checked by *BD*	

For panels CD and DE

$(V_{cr})_{CD} = 1983\,kN$ $(V_{cr})_{DE} = 1778\,kN$ H.2

For both panels $F_q = V_{max} - V_{cr}$ is negative and therefore $< F_x =$ 4.4.6.6
920 kN. Stiffener should therefore be designed for   4.5.3.3

$F_x < P_x = A_s p_c$ 4.7.5

$A_s = 160 \times 15 + 30 \times 15 \times 15 = 9150\,mm^2$

Neglecting the small contribution of the web:

$I_y = \dfrac{15 \times 175^3}{12} = 6.70 \times 10^6\ mm^4$

$r_y = \left(\dfrac{I_y}{A_s}\right)^{1/2} = \left(\dfrac{6.70 \times 10^6}{9150}\right)^{1/2} = 27.06\ mm$

$L_E = 0.7 \times 2000 = 1400\,mm$

$\lambda = \dfrac{L_E}{r_y} = \dfrac{1400}{27.06} = 51.73$ $p_{ys} = 275\ N/mm^2$

$p_c = 249\ N/mm^2$

$P_x = A_s p_c = 9150 \times 249 \times 10^{-3} = 2278\,kN$ Table 24

$F_x = 920 < P_x = 2278\,kN$ *OK*

WE 17.16 Loads applied between stiffeners   4.5.3.2

The web of the plate girder is subjected to a factored UDL of 92 kN/m applied between stiffeners, which corresponds to a stress f_{ed} given by

$f_{ed} = \dfrac{92}{15} = 6.13\ N/mm^2$

When the compression flange is restrained against rotation relative to the web

$p_{ed} = \left[2.75 + \dfrac{2}{(a/d)^2}\right] \dfrac{E}{(d/t)^2}$

The Steel Construction Institute Silwood Park, Ascot, Berks SL5 7QN	Subject **PLATE GIRDERS**		Chapter ref. **17**
	Design code **BS 5950: Part 1**	Made by **TMR** Checked by **BD**	Sheet no. **13**

Check for largest panel only

$$p_{ed} = \left[2.75 + \frac{2}{(5650/2000)^2}\right]\frac{205000}{(2000/15)^2} = 34.6 \, N/mm^2$$

$f_{ed} = 6.13 < p_{ed} = 34.6 \, kN/mm^2$ **OK**

WE 17.17 Stiffener bounding anchor panel at B   H.4.4

Since the compressive force F_s due to the support reaction is zero, the stiffener should be checked for a tensile force equal to F_{tf}

$F_{tf} = 0.15H_q$

$H_q = 3193 \, kN$ *(see WE 17.11)*

$d = 2000 \, mm$ $a_e = 1550 \, mm$

$$F_{tf} = 0.15 \times 3193 \times \frac{2000}{1550} = 618 \, kN$$

Since this force is significantly less than the compressive resistance of the transverse stiffeners (see WE 17.15) stiffener is adequate.

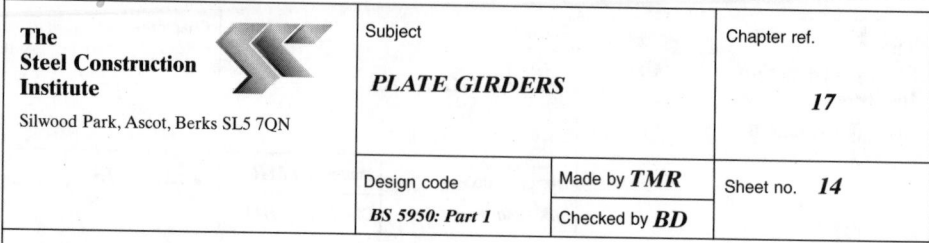

The Steel Construction Institute Silwood Park, Ascot, Berks SL5 7QN	Subject **PLATE GIRDERS**		Chapter ref. **17**
	Design code **BS 5950: Part 1**	Made by **TMR**	Sheet no. **14**
		Checked by **BD**	

WE 17.18 Final girder

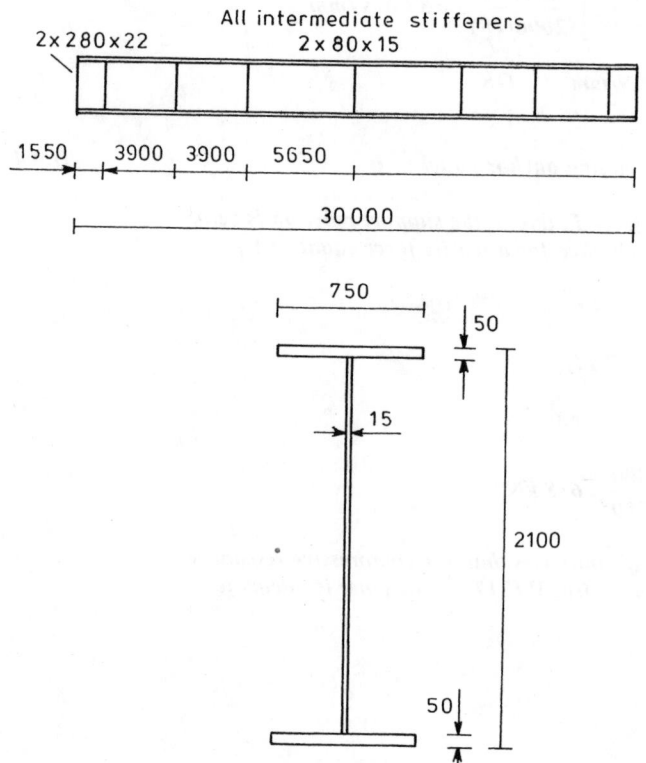

Fig. WE 17.5   *Final plate girder details*

Chapter 18
Members with compression and moments

by **DAVID NETHERCOT**

18.1 Occurrence of combined loading

Chapters 15 and 16 deal respectively with the design of members subject to compression and bending when these loadings act in isolation. However, in practice a combination of the two effects is frequently present. Figure 18.1 illustrates a number of common examples.

The balance between compression and bending, which may be induced about one or both principal axes, depends on a number of factors, the most important ones being the type of structure, the form of the applied loading, the member's location in the structure and the way in which the connections between the members function.

For building frames designed according to the principles of simple construction, it is customary to regard column moments as being produced only by beam reactions acting through notional eccentricities as illustrated in Fig. 18.2. Thus, column axial load is accumulated down the building but column moments are only ever generated by the floor levels under consideration with the result that they typically contain increasing ratios of compression to moment. Many columns are therefore designed for high axial loads but rather low moments. Corner columns suffer bending about both axes, but may well carry less axial load; edge columns are subject to bending about at least one axis; and internal columns may, if both the beam framing arrangements and the loading are balanced, be designed for axial load only.

Conversely the columns of portal frames are required to carry high moments in the plane of the frame but relatively low axial loads, unless directly supporting cranes. Rafters also attract some small axial load, particularly when wind loading is being considered. Portals employ rigid connections between members permitting the transfer of moments around the frame. Similarly, multi-storey frames designed on the basis of rigid beam-to-column connections are likely to contain columns with large moments.

Members required to carry combined compression and moments are not restricted to rectangular building frames. Although trusses are often designed on the basis that member centrelines intersect at the joints, this is not always possible and the resulting eccentricities induce some moments in the predominantly axially

511

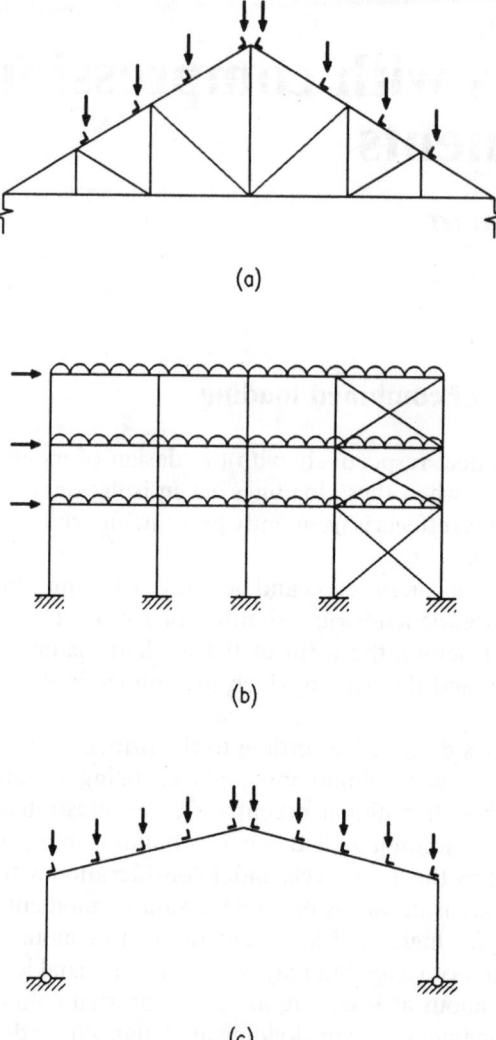

(a)

(b)

(c)

Fig. 18.1   Occurrence of beam-columns in different types of steel frames. (a) Roof truss –
top chord members subject to bending from purlin loads and compression due to
overall bending. (b) Simple framing – columns subject to bending from eccentric
beam reactions and compression due to gravity loading. (c) Portal frame – rafters
and columns subject to bending and compression due to frame action

loaded members. Sometimes the transfer of loads from secondary members into the
main booms of trusses is arranged so that they do not coincide with joints, leading
to beam action between joints being superimposed on the compression produced
by overall bending.

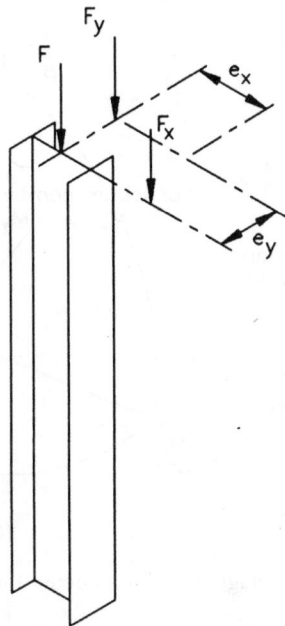

Fig. 18.2   Loading on a beam-column in a 'simply-designed' frame

18.2 Types of response – interaction

It is customary in many parts of the world, but not in the UK, to designate members subject to combined compression and bending as *beam-columns*. The term is helpful in appreciating member response because the combination of loads produces a combination of effects incorporating aspects of the behaviour of the two extreme examples: a beam carrying only bending loads and a column carrying only compressive load. This then leads naturally to the concept of interaction as the basis for design, an approach in which the proportions of the member's resistance to each component load type are combined using diagrams or formulae. Figure 18.3 illustrates the concept in general terms for cases of two and three separate load components. Any combination is represented by a point on the diagram, and an increasing set of loads with fixed ratios between the components corresponds to a straight line starting from the origin. Points that fall inside the boundary given by the design condition are safe, those that fall on the boundary just meet the design condition and those that lie outside the boundary represent an unsafe load combination. For the two load component case if one load type is fixed and the maximum safe value of the other is required, the vertical co-ordinate corresponding to the specified load is first located; projecting horizontally to meet the design boundary the horizontal co-ordinate can be read off.

The exact version of Fig. 18.3 appropriate for a design basis depends upon several factors, which include the form of the applied loading, the type of response that

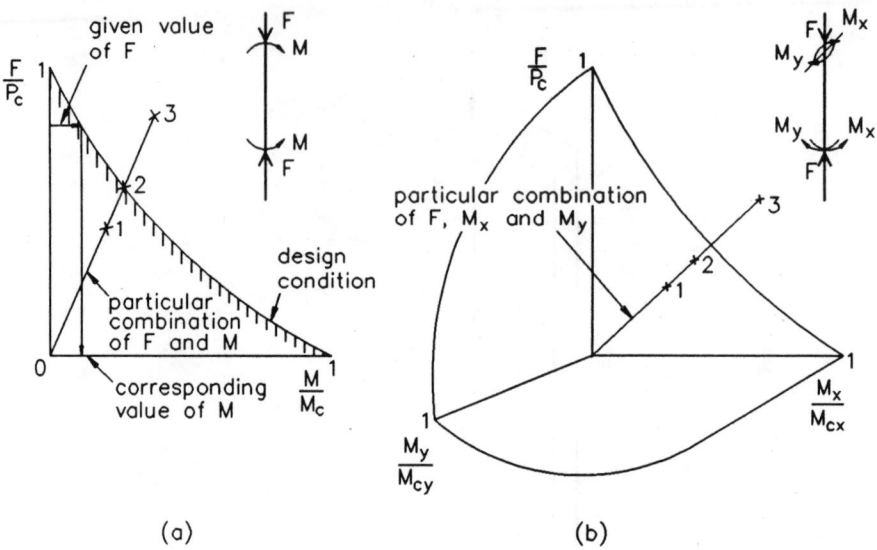

Fig. 18.3   Concept of interaction diagrams for combined loading: (a) two-dimensional, (b) three-dimensional

is possible, the member slenderness and the cross-sectional shape. Design methods for beam-columns must therefore seek to balance the conflicting requirements of rigour, which would try to adjust the form of the design boundary of Fig. 18.3 to reflect the influence of each of these factors, and simplicity. However, some appreciation of the role of each factor is necessary if even the simplest design approach is to be properly appreciated and applied.

The importance of member slenderness may be appreciated readily with reference to the two-dimensional example illustrated in Fig. 18.4. The member is loaded by compression plus equal and opposite end moments and is assumed to respond simply by deflecting in the plane of the applied loading. Under the action of the applied moments bending occurs, leading to a lateral deflection v. The moment at any point within the length comprises two components: a constant primary moment M due to the applied end moments plus a secondary moment Fv due to the axial load F acting through the lateral deflection v. Summing the effects of compression and bending gives

$$F/P_c + M_{max}/M_c = 1.0 \tag{18.1}$$

in which P_c and M_c are the resistance as a strut and a beam respectively and M_{max} is the total moment.

Analysis of beam-column problems shows that M_{max} may be closely approximated by

$$M_{max} = M/(1 - F/P_{cr}) \tag{18.2}$$

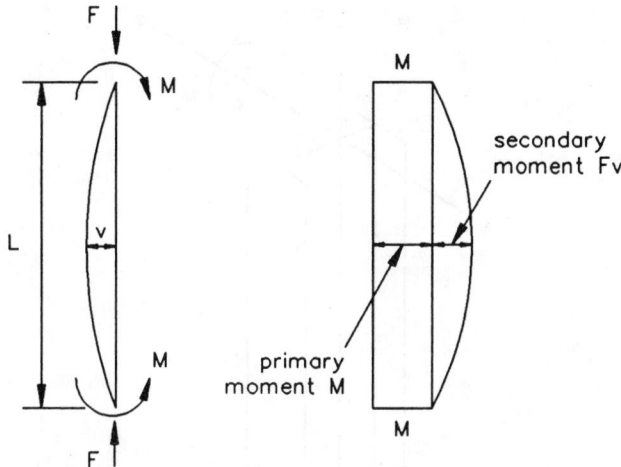

Fig. 18.4   In-plane behaviour of beam-columns

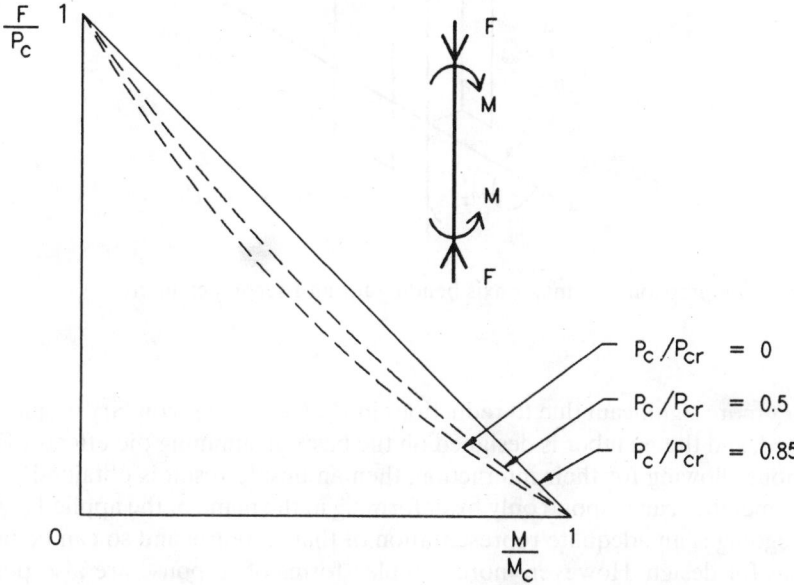

Fig. 18.5   Effect of slenderness on form of interaction according to Equation (18.3)

in which $P_{cr} = \pi^2 EI/L^2$ is the elastic critical load.

Combining these two expressions gives

$$F/P_c + M/(1 - F/P_{cr})M_c = 1.0 \qquad \qquad \textbf{(18.3)}$$

Figure 18.5 shows how this expression plots in an increasingly concave fashion as member slenderness increases and the amplification of the primary moments M

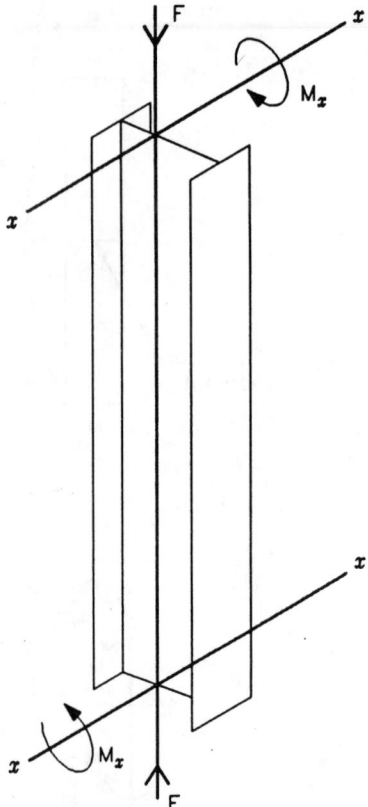

Fig. 18.6   Compression plus major axis bending for an I-section column

becomes more significant due to reductions in P_{cr}. Clearly if secondary moments are neglected, and the member is designed on the basis of summing the effects of F and M without allowing for their interaction, then an unsafe result is obtained.

If the member can respond only by deforming in the plane of the applied bending, the foregoing is an adequate representation of that response and so can be used as the basis for design. However, more complex forms of response are also possible. Figure 18.6 illustrates an I-section column subject to compression and bending about its major axis. Reference to Chapters 15 and 16 on columns and beams indicates that either the compression or the bending if acting alone would induce failure about the minor axis, in the first case by simple flexural buckling at a load P_{cy} and in the second by lateral–torsional buckling at a load M_b. Both tests and rigorous analysis confirm that the combination of loads will also produce a minor axis failure. Noting the presence of the amplification effect of the axial load acting through the bending deflections in the plane of the web, and not the out-of-plane deformations associated with the eventual failure mode, leads therefore to a modified form of

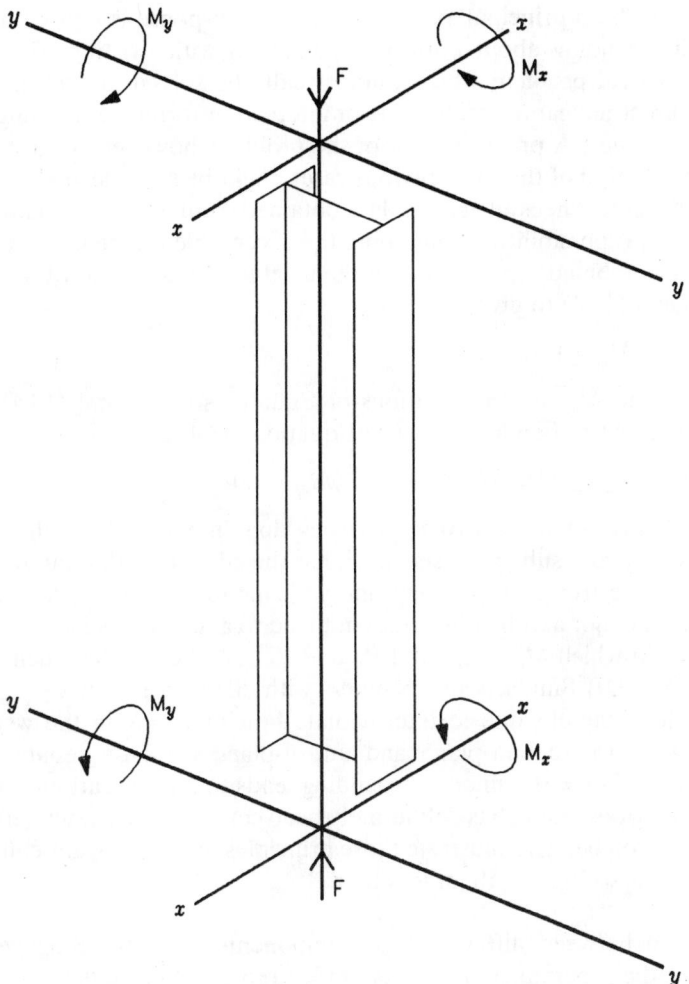

Fig. 18.7 Compression plus major and minor axis bending for an I-section column

Equation (18.3) as an interaction equation that might form a suitable basis for design:

$$F/P_{cy} + M/(1 - F/P_{crx})M_b = 1.0 \tag{18.4}$$

Note that both resistances P_{cy} and M_b relate to out-of-plane failure but that the amplification depends upon the in-plane Euler load. Consideration of the term F/P_{crx}, given that F is limited to P_{cy} which is less than P_{cx} and which is, in turn, less than P_{crx}, suggests that amplification effects are less significant than for the purely in-plane case.

Applied moments are not necessarily restricted to a single plane. For the most general case illustrated in Fig. 18.7, in which compression is accompanied by

moments about both principal axes, the member's response is a three-dimensional one involving bending about both axes combined with twisting. This leads to a complex analytical problem which cannot really be solved in such a way that it provides a direct indication of the type of interaction formulae that might be used as a basis for design. A practical view of the problem, however, suggests that some form of combination of the two previous cases might be suitable, provided any proposal were properly checked against data obtained from tests and reliable analyses. This leads to two possibilities: combining the acceptable moments M_x and M_y that can safely be combined separately with the axial load F obtained by solving Equations (18.3) and (18.4) to give

$$M_x/M_{ax} + M_y/M_{cy} = 1.0 \tag{18.5}$$

in which M_{ax} and M_{ay} are the solutions of Equations (18.3) and (18.4); or simply adding the minor axis bending effect to Equation (18.4) as

$$F/P_{cy} + M_x/(1 - F/P_{crx})M_b + M_y/(1 - F/P_{cry})M_{cy} = 1.0 \tag{18.6}$$

Although the first of these two approaches does not lead to such a seemingly straightforward end result as the second, it has the advantage that interaction about both axes may be treated separately, and so leads to a more logical treatment of cases for which major axis bending does not lead to a minor axis failure as for a rectangular tube in which $M_b = M_{cx}$, and P_{cx} and P_{cy} are likely to be much closer than is the case for a UB. Similarly for members with different effective lengths for the two planes, for example, due to intermediate bracing acting in the weaker plane only, the ability to treat in-plane and out-of-plane response separately and to combine the weaker with minor axis bending leads to a more rational result.

The foregoing discussion has deliberately been conducted in rather general terms, the main intention being to illustrate those principles on which beam-column design should be based. Collecting them together:

(1) interaction between different load components must be recognized; merely summing the separate components can lead to unsafe results
(2) interaction tends to be more pronounced as member slenderness increases
(3) different forms of response are possible depending on the form of the applied loading.

Having identified these three principles it is comparatively easy to recognize their inclusion in the design procedures of BS 5950 and BS 5400.

18.3 Effect of moment gradient loading

Returning to the comparatively simple in-plane case, Fig. 18.8 illustrates the patterns of primary and secondary moments in a pair of members subject to unequal end moments that produce either single- or double-curvature bending. For the first

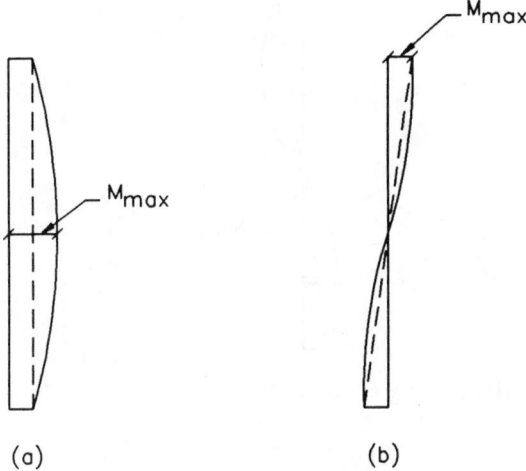

Fig. 18.8   Primary and secondary moments 1: (a) single curvature, (b) double curvature

case, the point of maximum combined moment occurs near mid-height where secondary bending effects are greatest. On the other hand, for double-curvature bending the two individual maxima occur at quite different locations, and for the case illustrated, in which the secondary moments have deliberately been shown as small, the point of absolute maximum moment is at the top. Had larger secondary moments been shown, as is the case in Fig. 18.9, then the point of maximum moment moves down slightly but is still far from that of the single-curvature case.

Theoretical and experimental studies of steel beam-columns constrained to respond in-plane and subject to different moment gradients, as represented by the factor of the ratio of the numerically smaller end moment to the numerically larger end moment (β), show clearly that, when all other parameters are held constant, failure loads tend to increase as β is varied from +1 (uniform single-curvature bending) to −1 (uniform double-curvature bending). Figure 18.10 illustrates the point in the form of a set of interaction curves. Clearly, if all beam-column designs were to be based upon the β = +1 case, safe but rather conservative designs would result.

Figure 18.10 also shows how for high moments the curves for $\beta \neq$ +1 tend to merge into a single line corresponding to the condition in which the more heavily stressed end of the member controls design. Reference to Figs 18.8 and 18.9 illustrates the point. The left-hand and lower parts of Fig. 18.10 correspond to situations in which Fig. 18.9 controls, while the right-hand and upper parts represent failure at one end. The two cases are sometimes referred to as 'stability' and 'strength' failure respectively. While this may offend the purists, for in both cases the limiting condition is one of exhausting the cross-sectional capacity, but at different locations within the member length, it does, nonetheless, serve to draw attention to the principal difference. Also shown in Fig. 18.10 is a line corresponding to the cross-sectional inter-

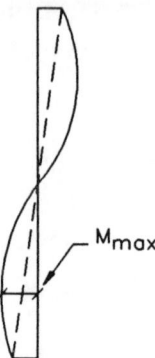

Fig. 18.9   Primary and secondary moments 2

action: the combinations of F and M corresponding to the full strength of the cross-section. This is the 'strength' limit, representing the case where the primary moment acting in conjunction with the axial load accounts for all the cross-section's capacity.

The substance of Fig. 18.10 can be incorporated within the type of interaction formula approach of section 18.2 through the concept of *equivalent uniform moment* presented in Chapter 16 in the context of the lateral–torsional buckling of beams; its meaning and use for beam-columns are virtually identical. For moment gradient loading member stability is checked using an equivalent moment $\overline{M} = mM$, as shown in Fig. 18.11. Coincidentally, suitable values of m, based on both test data and rigorous ultimate strength analyses, for the in-plane beam-column case are almost the same as those for laterally unrestrained beams (see section 16.3.6); m may conveniently be represented simply in terms of the moment gradient parameter β.

The situation corresponding to the upper boundary or strength failure of Fig. 18.10 must be checked separately using appropriate means of determining cross-sectional capacity under F and M. The strength check is superfluous for $\beta = +1$ as it can never control, while as $\beta \to -1$ and $M \to M_c$ it becomes increasingly likely that the strength check will govern. The procedure is:

(1) check stability using an interaction formula in terms of buckling resistance P_c and moment capacity M_c with axial load F and equivalent moment $\overline{M}$,
(2) check strength using an interaction procedure in terms of axial capacity P_s and moment capacity M_c with coincident values of axial load F and maximum applied end moment M_1 (this check is unnecessary if $m = 1.0$; $\overline{M} = M_1$ is used in the stability check).

Consideration of other cases involving out-of-plane failure or moments about both axes shows that the equivalent uniform moment concept may also be applied. For simplicity the same m values are normally used in design, although minor

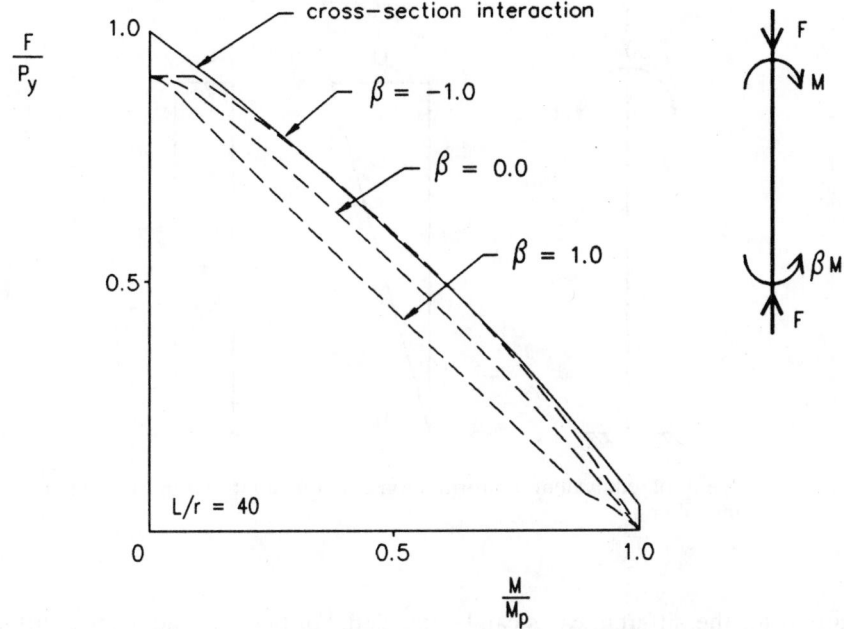

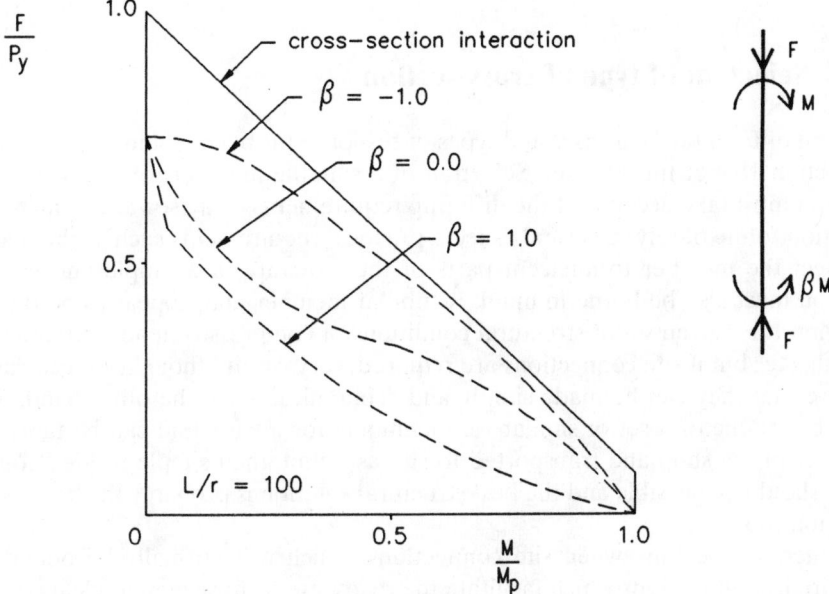

Fig. 18.10 Effect of moment gradient on interaction

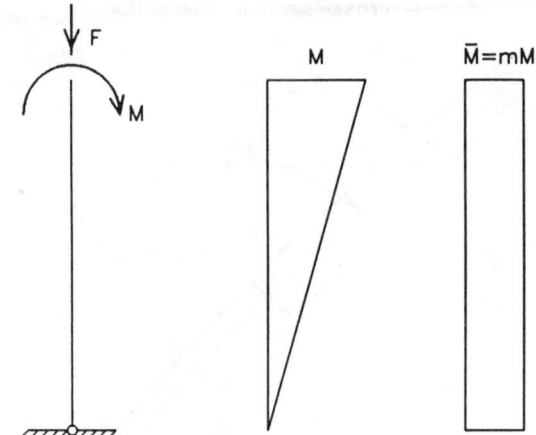

Fig. 18.11   Concept of equivalent uniform moment applied to primary moments on a beam-column

variations for the different cases can be justified. For biaxial bending, two different values, m_x and m_y, for bending about the two principal axes may be appropriate. An exception occurs for a column considered pinned at one end about both axes for which $\beta_x = \beta_y = 0.0$, whatever the sizes of the moments at the top.

18.4 Selection of type of cross-section

Several different design cases and types of response for beam-columns are outlined in section 18.2 of this chapter. Selection of a suitable member for use as a beam-column must take account of the differing requirements of these various factors. In addition to the purely structural aspects, practical requirements such as the need to connect the member to adjacent parts of the structure in a simple and efficient fashion must also be borne in mind. A tubular member may appear to be the best solution for a given set of structural conditions of compressive load, end moments, length, etc., but if site connections are required, very careful thought is necessary to ensure that they can be made simply and economically. On the other hand, if the member is one of a set of similar web members for a truss that can be fabricated entirely in the shop and transported to site as a unit, then simple welded connections should be possible and the best structural solution is probably the best overall solution too.

Generally speaking when site connections, which will normally be bolted, are required, open sections which facilitate the ready use of, for example, cleats or end-plates are to be preferred. UCs are designed principally to resist axial load but are also capable of carrying significant moments about both axes. Although buckling in

the plane of the flanges, rather than the plane of the web, always controls the pure axial load case, the comparatively wide flanges ensure that the strong-axis moment capacity M_{cx} is not reduced very much by lateral–torsional buckling effects for most practical arrangements. Indeed the condition $M_b = M_{cx}$ will often be satisfied.

In building frames designed according to the principles of simple construction, the columns are unlikely to be required to carry large moments. This arises from the design process by which compressive loads are accumulated down the building but the moments affecting the design of a particular column lift are only those from the floors at the top and bottom of the storey height under consideration. In such cases preliminary member selection may conveniently be made by adding a small percentage to the actual axial load to allow for the presence of the relatively small moments and then choosing an appropriate trial size from the tables of compressive resistance given in Reference 1. For moments about both axes, as in corner columns, a larger percentage to allow for biaxial bending is normally appropriate, while for internal columns in a regular grid with no consideration of pattern loading, the design condition may actually be one of pure axial load.

The natural and most economic way to resist moments in columns is to frame the major beams into the column flanges since, even for UCs, M_{cx} will always be comfortably larger than M_{cy}. For structures designed as a series of two-dimensional frames in which the columns are required to carry quite high moments about one axis but relatively low compressive loads, UBs may well be an appropriate choice of member. The example of this arrangement usually quoted is the single-storey portal building, although here the presence of cranes, producing much higher axial loads, the height, leading to large column slenderness, or a combination of the two, may result in UCs being a more suitable choice. UBs used as columns also suffer from the fact that the d/t values for the webs of many sections are non-compact when the applied loading leads to a set of web stresses that have a mean compressive component of more than about 70–100 N/mm^2.

18.5 Basic design procedure

When the distribution of moments and forces throughout the structure has been determined, for example, from a frame analysis in the case of continuous construction or by statics for simple construction, the design of a member subject to compression and bending consists of checking that a trial member satisfies the design conditions being used by ensuring that it falls within the design boundary defined by the type of diagram shown as Fig. 18.3. BS 5950 and BS 5400 therefore contain sets of interaction formulae which approximate such boundaries, use of which will automatically involve the equivalent procedures for the component load cases of strut design and beam design, to define the end points. Where these procedures permit the use of equivalent uniform moments for the stability check, they also require a separate strength check.

BS 5950: Part 1 requires that stability be checked using

$$\frac{F_c}{P_c} + \frac{m_x M_x}{p_y Z_x} + \frac{m_y M_y}{p_y Z_y} \leq 1 \tag{18.7a}$$

$$\frac{F_c}{P_{cy}} + \frac{m_{LT} M_{LT}}{M_b} + \frac{m_y M_y}{p_y Z_y} \leq 1 \tag{18.7b}$$

The first equation applies when major-axis behaviour is governed by in-plane effects and the second when lateral-torsional buckling controls. Both should normally be checked.

In Equation (18.7) the use of $p_y Z$, rather than M_c, makes some allowance in the case of plastic and compact sections for the effects of secondary moments as described in section 18.2. For non-compact sections, for which $M_c = p_y Z$, no such allowance is made and an unconservative effect is therefore present. Evaluation of Equation (18.7) may be effected quite rapidly if the tabulated values of P_{cy}, P_{cx}, M_b and $p_y Z_y$ given in Reference 1 for all UB, UC, RSJ and SHS are used. In the cases where m values of less than unity are being used it is essential to check that the most highly stressed cross section is capable of sustaining the coincident compression and moment(s). BS 5950: Part 1 covers this with the expression

$$\frac{F}{A_g p_y} + \frac{M_x}{M_{cx}} + \frac{M_y}{M_{cy}} \not> 1 \tag{18.8}$$

Clearly when both M_{cx} and M_b values are the same Equation (18.7) is always a more severe check, or in the limit is identical, and only Equation (18.7) need be used. Values of $A_g p_y$, M_{cx} and M_{cy} are also tabulated in Reference 1.

An an alternative to the use of Equations (18.7) and (18.8), BS 5950: Part 1 permits the use of more exact interaction formulae. For I- or H-sections with equal flanges these are presented in the form:

$$\frac{F_c}{P_{cx}} + \frac{m_x M_x}{M_{cx}}\left[1 + 0.5\frac{F_c}{P_{cx}}\right] + 0.5\frac{m_{yx} M_y}{M_{cy}} \leq 1 \tag{18.9a}$$

$$\frac{F_c}{P_{cy}} + \frac{m_{LT} M_{LT}}{M_b} + \frac{m_y M_y}{M_{cy}}\left[1 + \frac{F_c}{P_{cy}}\right] \leq 1 \tag{18.9b}$$

$$\frac{m_x M_x(1 + 0.5 F_c/P_{cx})}{M_{cx}(1 - F_c/P_{cx})} + \frac{m_y M_y(1 + F_c/P_{cy})}{M_{cy}(1 - F_c/P_{cy})} \leq 1 \tag{18.9c}$$

in which the three expressions cover respectively:

(a) Major axis buckling
(b) Lateral-torsional buckling
(c) Interactive buckling

All three should normally be checked.

The local capacity of the cross-section should also be checked. Class 1 and class 2 doubly-symmetric sections may be checked using:

$$\left(\frac{M_x}{M_{rx}}\right)^{z1}+\left(\frac{M_y}{M_{ry}}\right)^{z2} \not> 1 \tag{18.10}$$

In Equation (18.10) the denominators in the two terms are a measure of the moment that can be carried in the presence of the axial load F.

For fabricated sections, the principles of plastic theory may be applied first to locate the plastic neutral axis for a given combination of F, M_x and M_y and then to calculate M_{rx} and M_{ry}. This is manageable for uniaxial bending – F and M_x or F and M_y – but it is tedious for the full three-dimensional case and some use of approximate results[1] may well be preferable.

18.6 Cross-section classification under compression and bending

It is assumed in the discussion of the use of the BS 5950: Part 1 procedure that the designer has conducted the necessary section classification checks so as to ensure that the appropriate values of M_{cx}, M_{cy}, etc. are used. When the tabulated data of Reference 1 are being employed, any allowances for non-compactness are included in the listed values of M_{cx} and M_{cy}, but only if P_{cx} and P_{cy} have been taken from the strut tables rather than the beam-column tables will these contain any reduction. The reason is that for pure compression the stress pattern is known, whereas under combined loading the requirement may be to sustain only a very small axial load; to reduce P_{cx} and P_{cy} on the basis of uniform compression in each plate element of the section is much too severe. For simplicity, section classification may initially be conducted under the most severe conditions of pure axial load; if the result is either plastic or compact nothing is to be gained by conducting additional calculations with the actual pattern of stresses. However, if the result is a non-compact section, possibly when checking the web of a UB, then it is normally advisable for economy of both design time and actual material use to repeat the classification calculations more precisely.

18.7 Special design methods for members in portal frames

18.7.1 Design requirements

Both the columns and the rafters in the typical pitched roof portal frame represent particular examples of members subject to combined bending and compression. Provided such frames are designed elastically, the methods already described for assessing local cross-sectional capacity and overall buckling resistance may be employed. However, these general approaches fail to take account of some of the special features present in normal portal frame construction, some of which can, when properly allowed for, be shown to enhance buckling resistance significantly.

When plastic design is being employed, the requirements for member stability change somewhat. It is no longer sufficient simply to ensure that members can safely

resist the applied moments and thrust; rather for members required to participate in plastic hinge action, the ability to sustain the required moment in the presence of compression during the large rotations necessary for the development of the frame's collapse mechanism is essential. This requirement is essentially the same as that for a 'plastic' cross-section discussed in Chapter 13. The performance requirement for those members in a plastically designed frame actually required to take part in plastic hinge action is therefore equivalent to the most onerous type of response shown in Fig. 13.4. If they cannot achieve this level of performance, for example because of premature unloading caused by local buckling, then they will prevent the formation of the plastic collapse mechanism assumed as the basis for the design, with the result that the desired load factor will not be attained. Put simply, the requirement for member stability in plastically-designed structures is to impose limits on slenderness and axial load level, for example, that ensure stable behaviour while the member is carrying a moment equal to its plastic moment capacity suitably reduced so as to allow for the presence of axial load. For portal frames, advantage may be taken of the special forms of restraint inherent in that form of construction by, for example, purlins and sheeting rails attached to the outside flanges of the rafters and columns respectively.

Figure 18.12 illustrates a typical collapse moment diagram for a single-bay pin-base portal subject to gravity load only (dead load + imposed load), this being the usual governing load case in the UK. The frame is assumed to be typical of UK practice with columns of somewhat heavier section than the rafters and haunches of approximately 10% of the clear span and twice the rafter depth at the eaves. It is further assumed that the purlins and siderails which support the cladding and are attached to the outer flanges of the columns and rafters provide positional restraint to the frame, i.e. prevent lateral movement of the flange, at these points. Four regions in which member stability must be ensured may be identified:

(1) full column height AB
(2) haunch, which should remain elastic throughout its length

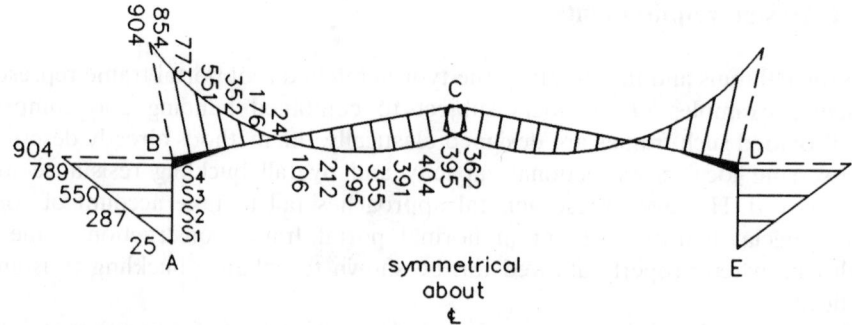

Fig. 18.12 Moment distribution for dead plus imposed load condition

(3) eaves region of rafter for which the lower unbraced flange is in compression due to the moments, from end of haunch
(4) apex region of the rafter between top compression flange restraints.

18.7.2 Column stability

Figure 18.13 provides a more detailed view of the column AB, including both the bracing provided by the siderails and the distribution of moment over the column

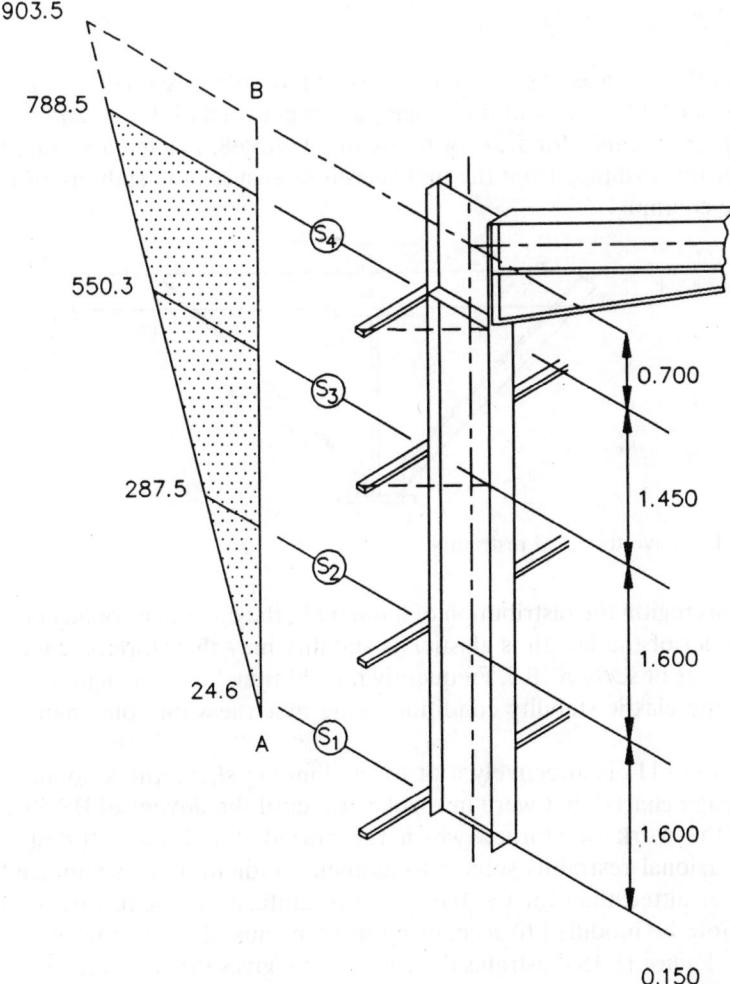

Fig. 18.13   Member stability – column

height. Assuming the presence of a plastic hinge immediately below the haunch, the design requirement is to ensure stability up to the formation of the collapse mechanism.

According to clause 5.3.2 of BS 5950: Part 1, torsional restraint must be provided no more than $D/2$, where D is the overall column depth, measured along the column axis, from the underside of the haunch. This may conveniently be achieved by means of the knee brace arrangement of Fig. 18.14. The simplest means of ensuring adequate stability for the region adjacent to this braced point is to provide another torsional restraint within a distance of not more than L_m, where L_m is taken as equal to L_u obtained from clause 5.3.3 as

$$L_u \le \frac{38r_y}{\left[f_c/130+(p_y/275)^2(x/36)^2\right]^{\frac{1}{2}}} \tag{18.11}$$

Noting that the mean axial stress in the column f_c is normally small, that p_y is around $275\,\text{N/mm}^2$ for S275 steel and that x has values between about 20 and 45 for UBs, gives a range of values for L_u/r_y of between 30 and 68. Placing a second torsional restraint at this distance from the first therefore ensures the stability of the upper part of the column.

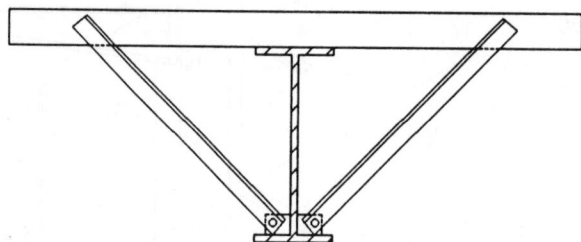

Fig. 18.14 Effective torsional restraints

Below this region the distribution of moment in the column normally ensures that the remainder of the length is elastic. Its stability may therefore be checked using the procedures of section 18.5. Frequently no additional intermediate restraints are necessary, the elastic stability condition being much less onerous than the plastic one.

Equation (18.11) is effectively a fit to the limiting slenderness boundary of the column design charts[3] that were in regular use until the advent of BS 5950: Part 1, based on the work of Horne,[4] which recognized that for lengths of members between torsional restraints subject to moment gradient, longer unbraced lengths could be permitted than for the basic case of uniform moment. Equation (18.11) may therefore be modified to recognize this by means of the coefficients proposed by Brown.[5] Figure 18.15 illustrates the concept and gives the relevant additional for-mulae. For a 533 × 210 UB82 of S275 steel for which $x = 41.6$ and assuming $f_c = 15\,\text{N/mm}^2$, the key values become:

$$L_u \quad = 31.55 r_y$$
$$KL_u \quad = 122.7 r_y$$
$$K_0 KL_u = 90.6 r_y$$
$$\beta_m \quad = 0.519$$

When checking a length for which the appropriate value of β is significantly less than +1.0, use of this modification permits a more relaxed approach to the provision of bracing. Some element of trial and error is involved since the exact value of β to be used is itself dependent upon the location of the restraints.

Neither the elastic nor the plastic stability checks described above take account of the potentially beneficial effect of the tension flange restraint provided by the

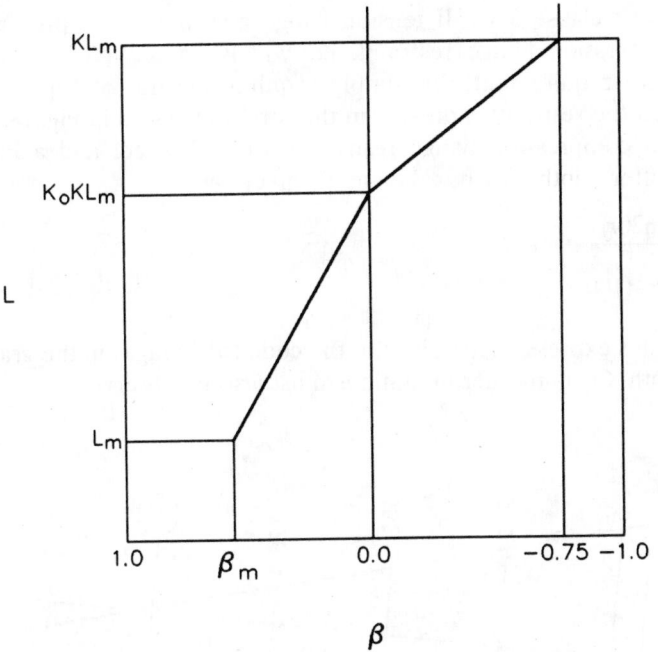

$$20 < x \leqslant 30 \quad K = 2.3 + 0.03\, x - x\, f_c / 3000$$

$$30 < x < 50 \quad K = 0.8 + 0.08\, x - (x - 10) f_c / 2000$$

$$K_0 = (180 + x)/300$$

$$\text{S275 steel } \beta_m = 0.44 + x/270 - f_c/200$$

$$\text{S355 steel } \beta_m = 0.47 + x/270 - f_c/250$$

Fig. 18.15 Modification to Equation (18.11) to allow for moment gradient

sheeting rails. This topic has been extensively researched,[6] with many of the findings being distilled into the design procedures of Appendix G of BS 5950: Part 1. Separate procedures are given for both elastic and plastic stability checks. Although significantly more complex than the use of Equation (18.11) or the methods of section 18.5, their use is likely to lead to significantly increased allowable unbraced lengths, particularly for the plastic region.

18.7.3 Rafter stability

Stability of the eaves region of the rafter may most easily be ensured by satisfying the conditions of clause 5.3.4. If tension flange restraint is not present between points of compression flange restraint, i.e. widely spaced purlins and a short unbraced length requirement, this simply requires the use of Equation (18.11). However, when the restraint is present in the form illustrated in Fig. 18.16, the distance between compression flange restraints for S275 steel and a haunch that doubles the rafter depth may be taken as L_s, given by

$$L_s = \frac{620r_y}{1.25\left[72 - (100/x)^2\right]^{\frac{1}{2}}}$$

(18.12)

Variants of this expression are given in the code for changes in the grade of steel or haunch depth. Certain other limitations must also be observed:

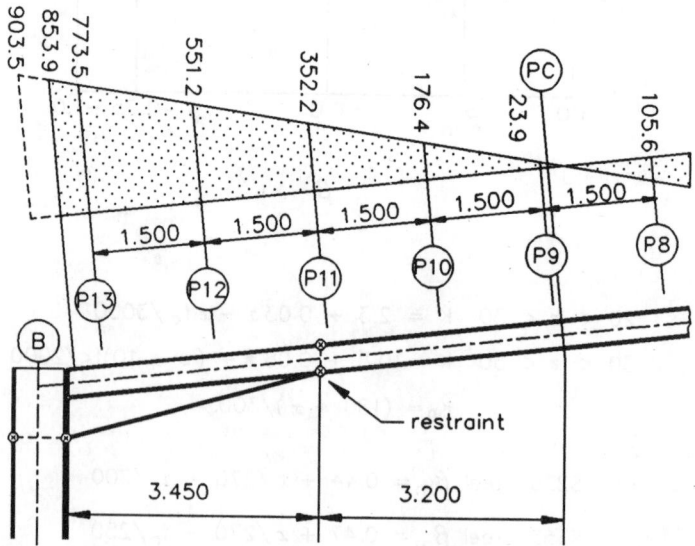

Fig. 18.16   Member stability in haunched rafter region

(1) the rafter must be a UB
(2) the haunch flange must not be smaller than the rafter flange
(3) the distance between tension flange restraints must be stable when checked as a beam using the procedure of section 16.3.6.

Equation (18.12) is less sensitive than Equation (18.11) to changes in x, with the result that it gives an average value for L_s/r_y of about 65. It is often regarded as good practice to provide bracing at the toe of the haunch since this region corresponds to major changes in the pattern of force transfer due to the change in the line of action of the compression in the bottom flange. In cases where the use of clause 5.3.4 does not give a stable haunch because the length from eaves to toe exceeds L_s, Appendix G may be used to obtain a larger value of L_s. If this is still less than the haunch length, then additional compression flange restraints are required.[6]

18.7.4 Bracing

The general requirements of lateral bracing systems have already been referred to in Chapter 16 – sections 16.3, 16.4 and 16.5 in particular. When purlins or siderails are attached directly to a rafter or column compression flange it is usual to assume that adequate bracing stiffness and strength are available without conducting specific calculations. In cases of doubt the ability of the purlin to act as a strut carrying the design bracing force may readily be checked. Definitive guidance on the appropriate magnitude to take for such a force is noticeably lacking in codes of practice. A recent suggestion for members in plastically-designed frames[7] is 2% of the squash load of the compression flange of the column or rafter: $0.02p_yBT$ at every restraint. In order that bracing members possess sufficient stiffness a second requirement that their slenderness be not more than 100 has also been proposed.[7] Both suggestions are largely based on test data. For elastic design the provisions of BS 5950: Part 1 may be followed.

When purlins or siderails are attached to the main member's tension flange, any positional restraint to the compression flange must be transferred through both the bracing to main member interconnection and the webs of the main member. Both effects are allowed for in the work on which the special provisions in BS 5950: Part 1 for tension flange restraint are based.[8] When full torsional restraint is required so that interbrace buckling may be assumed, the arrangement of Fig. 18.17 is often used. The stays may be angles, tubes (provided simple end connections can be arranged) or flats (which are much less effective in compression than in tension). In theory a single member of sufficient size would be adequate, but practical considerations such as hole clearance[6] normally dictate the use of pairs of stays. It should also be noted that for angles to the horizontal of more than 45° the effectiveness of the stay is significantly reduced.

Reference 9 discusses several practical means of bracing or otherwise restraining beam-columns.

References to Chapter 18

1. The Steel Construction Institute (SCI) (2001) *Steelwork Design Guide to BS 5950: Part 1: 2000, Vol. 1, Section Properties, Member Capacities*, 6th edn. SCI, Ascot, Berks.
2. Advisory Desk (1988) *Steel Construction Today*, **2**, Apr., 61–2.
3. Morris L.J. & Randall A.L. (1979) *Plastic Design*. Constrado. (See also *Plastic Design (Supplement)*, Constrado, 1979.)
4. Horne M.R. (1964) Safe loads on I-section columns in structures designed by plastic theory. *Proc. Instn Civ. Engrs*, **29**, Sept., 137–50 and *Discussion*, **32**, Sept. 1965, 125–34.
5. Brown B.A. (1988) The requirements for restraint in plastic design to BS 5950. *Steel Construction Today*, **2**, 184–96.
6. Morris L.J. (1981 & 1983) A commentary on portal frame design. *The Structural Engineer*, **59A**, No. 12, 394–404 and **61A**, No. 6. 181–9.
7. Morris L.J. & Plum D.R. (1988) *Structural Steelwork Design to BS 5950*. Longman, Harlow, Essex.
8. Horne M.R. & Ajmani J.L. (1972) Failure of columns laterally supported on one flange. *The Structural Engineer*, **50**, No. 9, Sept., 355–66.
9. Nethercot D.A. & Lawson R.M. (1992) *Lateral stability of steel beams and columns – common cases of restraint*. SCI publication 093, The Steel Construction Institute.

Further reading for Chapter 18

Chen W.F. & Atsuta T. (1977) *Theory of Beam-Columns, Vols 1 and 2*. McGraw-Hill, New York.
Davies J.M. & Brown B.A. (1996) *Plastic Design to BS 5950*. Blackwell Science, Oxford.
Galambos T.V. (1998) *Guide to Stability Design Criteria for Metal Structures*, 5th edn. Wiley, New York.
Horne M.R. (1979) *Plastic Theory of Structures*, 2nd edn. Pergamon, Oxford.
Horne M.R., Shakir-Khalil H. & Akhtar S. (1967) The stability of tapered and haunched beams. *Proc. Instn Civ. Engrs*, **67**, No. 9, 677–94.
Morris L.J. & Nakane K. (1983) Experimental behaviour of haunched members. In *Instability and Plastic Collapse of Steel Structures* (Ed. by L.J. Morris), pp. 547–59. Granada.

A series of worked examples follows which are relevant to Chapter 18.

The Steel Construction Institute  Silwood Park, Ascot, Berks SL5 7QN	Subject **BEAM-COLUMN EXAMPLE 1 ROLLED UNIVERSAL COLUMN**	Chapter ref. **18**
	Design code **BS 5950: Part 1**	Made by **DAN** / Checked by **GWO**	Sheet no. **1**

Problem

Select a suitable UC in S275 steel to carry safely a combination of 940 kN in direct compression and a moment about the minor axis of 16 kNm over an unsupported height of 3.6 m.

Problem is one of uniaxial bending producing failure by buckling about the minor axis. Since no information is given on distribution of applied moments make conservative (& simple) assumption of uniform moment (b = 1.0).

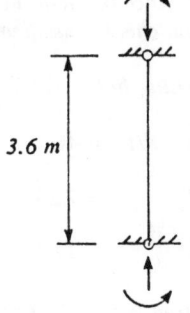

3.6 m

Try 203 × 203 × 60 UC – member capacities suggest P_{cy} of approximately 1400 kN will provide correct sort of margin to carry the moment

Steelwork Design Guide Vol 1

$r_y = 5.19\,cm$ $Z_y = 199\,cm^3$

$A = 75.8\,cm^2$ $S_y = 303\,cm^3$

$\lambda_y = 3600/51.9 = 69.4$ 4.7.2

Use Table 24 curve c for p_c Table 23

For $p_y = 275\,N/mm^2$ and $\lambda = 69.4$

value of $p_c = 183\,N/mm^2$

$P_{cy} = 183 \times 7580 = 1387 \times 10^3\,N$ 4.7.4
$= 1387\,kN$

$$\frac{940}{1387} + \frac{16}{275 \times 199000 \times 10^{-6}} = 0.68 + 0.29$$ 4.8.3.3.1

$$= 0.97$$

∴ *Adopt 203 × 203 × 60 UC*

The Steel Construction Institute	Subject **BEAM-COLUMN EXAMPLE 1 ROLLED UNIVERSAL COLUMN**	Chapter ref. **18**
Silwood Park, Ascot, Berks SL5 7QN		

Design code **BS 5950: Part 1**	Made by **DAN**	Sheet no. **2**
	Checked by **GWO**	

The determination of P_{cy} assumed that the section is not slender; similarly the use of Clause 4.8.3.3.1 in the present form presumes that the section is not slender. The actual stress distribution in the flanges will vary linearly due to the minor axis moment component of the load. Since the actual case cannot be more severe than uniform compression, check classification for pure compression.   3.5

Flange limiting b/T = 15   *Table 11*

Web limiting d/t   = 40

Actual b/T   = 7.23

Actual d/t   = 17.3

∴   section is not slender

		Subject	Chapter ref.
The Steel Construction Institute Silwood Park, Ascot, Berks SL5 7QN		***BEAM-COLUMN EXAMPLE 2 ROLLED UNIVERSAL BEAM***	**18**
		Design code **BS 5950: Part 1**	Made by *DAN*	Sheet no. **1**
			Checked by *GWO*	

Problem

Check the suitability of a 533 × 210 × 82 UB in S355 steel for use as the column in a portal frame of clear height 5.6 m if the axial compression is 160 kN, the moment at the top of the column is 530 kNm and the base is pinned. The ends of the column are adequately restrained against lateral displacement (i.e. out of the plane) and rotation.

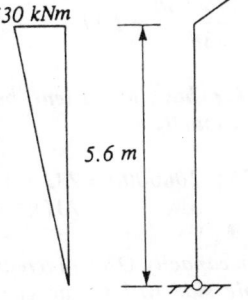

530 kNm

5.6 m

Loading corresponds to compression and major axis moment distributed as shown. Check initially over full height.

r_y = 4.38 cm	u = 0.865	*Steelwork Design Guide vol 1*
S_x = 2060 cm³	x = 41.6	
A = 104 cm²		
p_y = 355 N/mm²		*Table 9*
λ_y = 5600/43.8 = 128		*4.7.2*
λ/x = 128/41.6 = 3.08		
v = 0.91		*Table 19*
λ_{LT} = 0.865 × 0.91 × 128 = 101		*4.3.6.7*
p_b = 139 N/mm²		*Table 16*
M_b = 139 × 2060000 = 286 × 10⁶ Nmm = 286 kNm		*4.3.6.4*

The Steel Construction Institute	Subject	Chapter ref.
Silwood Park, Ascot, Berks SL5 7QN	*BEAM-COLUMN EXAMPLE 2 ROLLED UNIVERSAL BEAM*	*18*

Design code	Made by *DAN*	Sheet no. *2*
BS 5950: Part 1	Checked by *GWO*	

Use Table 24 curve b for p_c   **Table 23**

for $\lambda_y = 128$ $p_c = 103\,N/mm^2$ **Table 24**

P_{cy} *$= 103 \times 10400 = 1071 \times 10^3\,N = 1071\,kN$*   **4.7.4**

For $\beta = 0/530 = 0$ take $m_{LT} = 0.60$ **Table 18**

$$\frac{\overline{M}}{M_b} = \frac{0.60 \times 530}{286} = 1.11$$

∴ member has insufficient buckling resistance moment. Check moment capacity

$M_{cx} = 355 \times 2060000 = 731 \times 10^6\,Nmm$
$= 731\,kNm$

∴ section capacity OK so increase stability by inserting a brace from a suitable side rail to the compression flange. Estimate suitable location as 1.6m below top.

For uppr part of column

λ_y *$= 1600/43.8 = 37$*   **4.7.2**

$\lambda/x = 37/41.6 = 0.9$

v *$= 0.99$*   **Table 19**

$\lambda_{LT} = 0.865 \times 0.99 \times 37 = 32$ **4.3.6.7**

p_b *$= 350\,N/mm^2$*   **Table 18**

$M_b = 350 \times 2060000 = 721 \times 10^6\,Nmm$ **4.3.6.4**
$= 721\,kNm$

p_c *$= 320\,N/mm^2$*   **Table 24**

$P_{cy} = 320 \times 10400 = 3328 \times 10^3\,N$ **4.7.4**
$= 3328\,kN$

	Subject	Chapter ref.
The **Steel Construction** **Institute** Silwood Park, Ascot, Berks SL5 7QN 	**BEAM-COLUMN EXAMPLE 2** **ROLLED UNIVERSAL BEAM**	**18**
	Design code **BS 5950: Part 1**	Made by **DAN** Checked by **GWO**	Sheet no. **3**

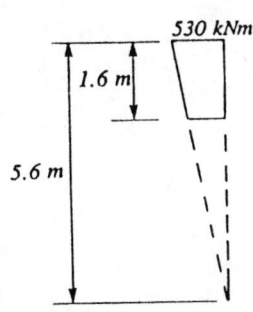

Table 18

$$\beta = \frac{5.6 - 1.6}{5.6} = 0.72$$

$$m_{LT} = 0.86$$

$$\frac{\overline{M}}{M_b} = \frac{0.86 \times 530}{721} = 0.63$$

$$\frac{P}{P_c} = \frac{160}{3328} = 0.05$$

$$0.05 + 0.63 = 0.68 \, OK$$

4.8.3.3.1

Check lower part of column for moment of

$$0.72 \times 530 = 382 \, kNm$$

$$\lambda_y = 4000/43.8 = 91$$

4.7.2

$$\lambda/x = 91/41.6 = 2.2$$

$$v = 0.96$$

Table 19

$$\lambda_{LT} = 0.865 \times 0.96 \times 91 = 76$$

4.3.6.7

$$p_b = 202 \, N/mm^2$$

$$M_b = 416 \, kNm$$

4.3.6.4

The **Steel Construction Institute** Silwood Park, Ascot, Berks SL5 7QN	Subject **BEAM-COLUMN EXAMPLE 2 ROLLED UNIVERSAL BEAM**	Chapter ref. **18**
Design code **BS 5950: Part 1**	Made by **DAN**	Sheet no. **4**
	Checked by **GWO**	

$p_c = 178 \, N/mm^2$ *Table 24*

$P_{cy} = 1851 \, kN$ *4.7.4*

$$\frac{\overline{M}}{M_b} = \frac{0.60 \times 382}{416} = 0.55$$

$$\frac{P}{P_{cy}} = \frac{160}{1851} = 0.09$$

$0.09 + 0.55 = 0.64 \, OK$ *Use 1 brace*   *4.8.3.3.1*

Capacity of cross-section under compression and bending should also be checked at point of maximum coincident values. However, since $M_{cx} = 355 \times 2060000 \times 10^{-6} = 731 \, kNm$ and compression is small by inspection, capacity is OK.

As before, use of M_b presumes section is at least compact.

b/T limit = 10ε   *Table 7*

d/t limit (pure compression) = 40ε

Since $\varepsilon = (275/355)^{1/2} = 0.88$ *these are:*

8.4 and 34.3

Actual b/T = 5.98

Actual d/t = 41.2

∴ d/t greater than limit for pure compression. However, actual loading is principally bending for which limit is $100\varepsilon = 88$

∴ without performing a rigorous check (by locating plastic neutral axis position etc.) it is clear that section will meet the limit for principally bending.

Section compact

∴ Adopt $533 \times 210 \times 82 \, UC$

		Subject		Chapter ref.
The **Steel Construction** **Institute** Silwood Park, Ascot, Berks SL5 7QN		***BEAM-COLUMN EXAMPLE 3*** ***RHS IN BIAXIAL BENDING***		**18**
		Design code **BS 5950: Part 1**	Made by **DAN** Checked by **GWO**	Sheet no. **1**

Problem

Select a suitable RHS in S355 material for the top chord of the 26.2 m span truss shown below.

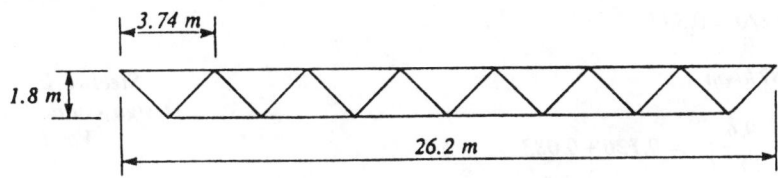

Trusses are spaced at 6 m intervals with purlins at 1.87 m intervals; these may be assumed to prevent lateral deflection of the top chord at these points. Under the action of the applied loading the chord loads in the most severely loaded bay are:

compression   *664 kN*

vertical moment   *24.4 kNm*

horizontal moment   19.6 kNm

It is necessary to consider a length between nodes, allowing for the lateral restraint at mid-length under the action of compression plus biaxial bending.

Take L_{Ex} at distance between nodes and L_{Ey} as distance between purlins

$L_{Ex} = 3.74\,m$

$L_{Ey} = 1.87\,m$

Try 150 × 150 × 10 RHS

For $L_{Ex} = 3.74\,m$ $P_{cx} = 1560\,kN$ *Steelwork*

For $L_{Ey} = 1.87\,m$ $P_{cy} = 1900\,kN$ *Design Guide*

$P_z = 1970\,kN$ $M_{cx} = M_{cy} = 102\,kN$ *Vol 1*

The **Steel Construction Institute** Silwood Park, Ascot, Berks SL5 7QN	Subject **BEAM-COLUMN EXAMPLE 3 RHS IN BIAXIAL BENDING**	Chapter ref. **18**
	Design code **BS 5950: Part 1**	Made by **DAN**	Sheet no. **2**
		Checked by **GWO**	

Check local capacity using "more exact" method for plastic section   4.8.2.3

$$\left(\frac{M_x}{M_{rx}}\right)^{5/3} + \left(\frac{M_y}{M_{ry}}\right)^{5/3} \leq 1$$

$F/P_z = 664/1970 = 0.337$

$M_{rx} = M_{ry} = 87\,kNm$

$$\left(\frac{24.4}{87}\right)^{5/3} + \left(\frac{19.6}{87}\right)^{5/3} = 0.120 + 0.083$$

$$= 0.203 < 1 \text{ local capacity OK}$$

Steelwork Design Guide Vol I

Check overall buckling using "more exact" method   4.8.3.3.2

Major axis

$$\frac{F_c}{P_{cx}} + \frac{m_x M_x}{M_{cx}}\left[1 + 0.5\frac{F_c}{P_{cx}}\right] + 0.5\frac{m_{yx}M_y}{M_{cy}} \leq 1$$

$$\frac{664}{1560} + \frac{24.4}{102}\left[1 + 0.5\frac{664}{1560}\right] + 0.5\frac{19.6}{102}$$

$$= 0.426 + 0.290 + 0.096 = 0.812 \text{ OK}$$

Lateral-torsional buckling check is not required for a closed section.

Interactive buckling

$$\frac{m_x M_x(1 + 0.5F_c/P_{cx})}{M_{cx}(1 - F_c/P_{cx})} + \frac{m_y M_y(1 + 0.5F_c/P_{cy})}{M_{cy}(1 - F_c/P_{cy})} \leq 1$$

$$\frac{24.4(1 + 0.5 \times 664/1560)}{102(1 - 664/1560)} + \frac{19.6(1 + 0.5 \times 664/1900)}{102(1 - 664/1900)}$$

$$= 0.505 + 0.350 = 0.855 \text{ OK}$$

For this example since m = 1.0 has been used throughout overall buckling will always control.

Chapter 19
Trusses

by PAUL TASOU

19.1 Common types of trusses

19.1.1 Buildings

The most common use of trusses in buildings is to provide support to roofs, floors and such internal loading as services and suspended ceilings. There are many types and forms of trusses; some of the most widely used are shown in Fig. 19.1. The type of truss adopted in design is governed by architectural and client requirements, varied in detail by dimensional and economic factors.

The Pratt truss, Fig. 19.1(a) and (e), has diagonals in tension under normal vertical loading so that the shorter vertical web members are in compression and the longer diagonal web members are in tension. This advantage is partially offset by the fact that the compression chord is more heavily loaded than the tension chord at mid-span under normal vertical loading. It should be noted, however, that for a light-pitched Pratt roof truss wind loads may cause a reversal of load thus putting the longer web members into compression.

The converse of the Pratt truss is the Howe truss (or English truss), Fig. 19.1(b). The Howe truss can be advantageous for very lightly loaded roofs in which reversal of load due to wind will occur. In addition the tension chord is more heavily loaded than the compression chord at mid-span under normal vertical loading. The Fink truss, Fig. 19.1(c), offers greater economy in terms of steel weight for long-span high-pitched roofs as the members are subdivided into shorter elements. There are many ways of arranging and subdividing the chords and web members under the control of the designer.

The mansard truss, Fig. 19.1(d), is a variation of the Fink truss which has the advantage of reducing unusable roof space and so reducing the running costs of the building. The main disadvantage of the mansard truss is that the forces in the top and bottom chords are increased due to the smaller span-to-depth ratio.

However, it must not escape the designer's mind that any savings achieved in steel weight by introducing a greater number of smaller members may, as is often the case, substantially increase fabrication and maintenance costs.

The Warren truss, Fig. 19.1(f), has equal length compression and tension web members, resulting in a net saving in steel weight for smaller spans. The added advantage of the Warren truss is that it avoids the use of web members of differing length and thus reduces fabrication costs. For larger spans the modified Warren truss,

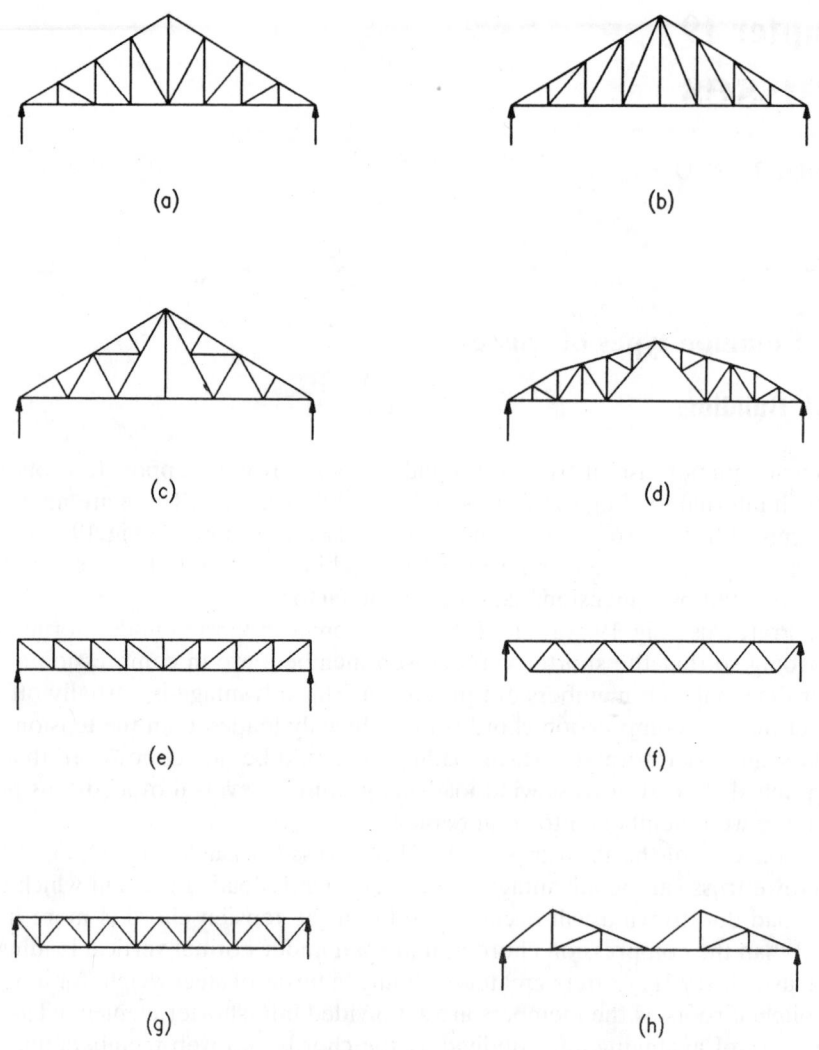

Fig. 19.1   Common types of roof trusses: (a) Pratt – pitched, (b) Howe, (c) Fink, (d) mansard, (e) Pratt – flat, (f) Warren, (g) modified Warren, (h) saw-tooth

Fig. 19.1(g), may be adopted where additional restraint to the chords is required (this also reduces secondary stresses). The modified Warren truss requires more material than the parallel-chord Pratt truss, but this is offset by its symmetry and pleasing appearance. The saw-tooth or butterfly truss, Fig. 19.1(h), is just one of many examples of trusses used in multi-bay buildings, although the other types described above are equally suitable.

19.1.2 Bridges

Trusses are now infrequently used for road bridges in the UK because of high fabrication and maintenance costs. However, the recent award-winning Brinnington railway bridge (Fig. 19.2) demonstrates that they can still be used to create efficient and attractive railway structures. In many parts of the world, particularly in developing countries where labour costs are low and material costs are high, trusses are often adopted for their economy in steel. Their structural form also lends itself to transportation in small components and piece-small erection, which may be suitable for remote locations.

Some of the most commonly used trusses suitable for both road and rail bridges are illustrated in Fig. 19.3. Pratt, Howe and Warren trusses, Fig. 19.3(a), (b) and (c), which are discussed in section 19.2.1, are more suitable for short to medium spans.

Fig. 19.2 Brinnington Railway Bridge

The economic span range of the Pratt and Howe trusses may be extended by subdividing the diagonals and the deck support chord as shown for the Petit truss, Fig. 19.3(h), although this often gives rise to high secondary stresses for short to medium spans. In the case of the Warren truss the unsupported length of the chords may be too great for the economic span and depth range; in such a case the modified Warren truss, Fig. 19.3(d), is more appropriate.

The variable depth type truss such as the Parker, Fig. 19.3(e), offers an aesthetically pleasing structure. With this type of truss its structural function is emphasized and the material is economically distributed at the cost of having expensive fabrication due to the variable length and variable inclination of the web and top chord members.

For economy the truss depth is ideally set at a fixed proportion of the span. As the span increases the truss depth and bay width increase accordingly. The bay width is usually fixed by providing truss nodes on the centrelines of the deck cross-

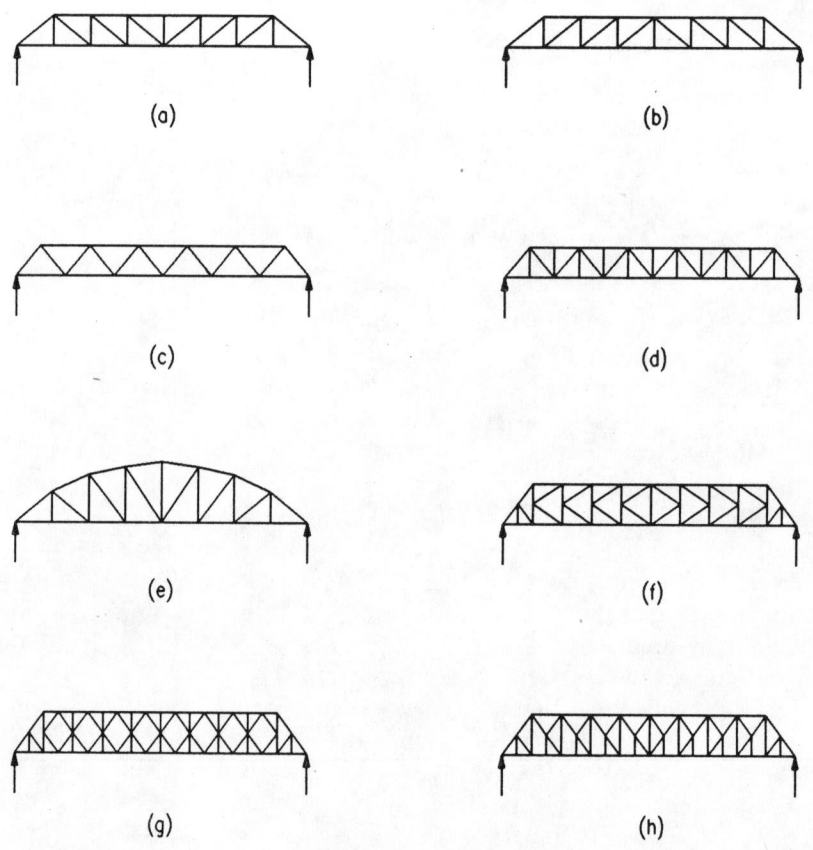

Fig. 19.3 Common types of bridge trusses: (a) Pratt, (b) Howe, (c) Warren, (d) modified Warren, (e) Parker, (f) K, (g) diamond, (h) Petit

beams, thus avoiding high local bending stress in the deck chord. For large-span bridges with an economical spacing for the deck cross-beams, the height of the truss may be as much as four times the bay width. In such a case a subdivided form of truss will be required to avoid very long uneconomical compression web members, or tensile members subjected to load reversal due to moving live loads. The diamond, Petit and K-trusses, Fig. 19.3(g), (h) and (f), are just three types which can be used.

The diamond and Petit trusses have the advantage of having shorter diagonals than the K-truss. The main disadvantage of trusses such as the diamond or Petit, which have intermediate bracing members connected to the chords away from the main joints, is that they give rise to high secondary stresses for short to medium spans due to differential joint deformation caused by moving live loads. The K-truss is far superior in this respect.

19.2 Guidance on overall concept

19.2.1 Buildings

For pitched-roof trusses such as the Pratt, Howe or Fink, Fig. 19.1(a), (b) and (c), the most economical span-to-depth ratio (at apex) is between 4 and 5, with a span range of 6 m to 12 m, the Fink truss being the most economical at the higher end of the span range. Spans of up to 15 m are possible but the unusable roof space becomes excessive and increases the running costs of the building. In such circumstances the span-to-depth ratio may be increased to about 6 to 7, the additional steel weight (increase in initial capital expenditure) being offset by the long-term savings in the running costs. For spans of between 15 m and 30 m, the mansard truss, Fig. 19.1(d), reduces the unusable roof space but retains the pitched appearance and offers an economic structure at span-to-depth ratios of about 7 to 8.

The parallel (or near parallel) chord trusses (also known as lattice girders) such as the Pratt or Warren, Fig. 19.1(e) and (f), have an economic span range of between 6 m and 50 m, with a span-to-depth ratio of between 15 and 25 depending on the intensity of the applied loads. For the top end of the span range the bay width should be such that the web members are inclined at approximately 50° or slightly steeper. For long, deep trusses the bay widths become too large and are often subdivided with secondary web members.

For roof trusses the web member intersection points with the chords should ideally coincide with the secondary transverse roof members (purlins). In practice this is not often the case for economic truss member arrangements, thus resulting in the supporting chord being subject to local bending stresses.

The most economical spacing for roof trusses is a function of the span and load intensity and to a lesser extent the span and spacing of the purlins, but as a general rule the spacing should be between $\frac{1}{4}$ and $\frac{1}{5}$ of the span, which results in a spacing of between 4 m and 10 m for the economic range of truss spans.

For short-span roof trusses between 6 m and 15 m the minimum spacing should be limited to 3–4 m.

19.2.2 Bridges

Road and railway truss bridges can either be underslung (deck at top chord level), through (deck at bottom chord level) or semi-through.

Limits on headroom, navigation height and construction depth will determine whether an underslung or through truss will be the most appropriate. For large-span bridges the through type is often adopted as ample headroom will be available to permit direct lateral restraint to the top compression chord using cross bracing. For short-span bridges, however, the underslung type is most appropriate provided the navigation height or construction depth limits are satisfied. Underslung trusses are usually more economical than either through or semi-through trusses as the deck structure performs the dual function of directly supporting the traffic loads and providing lateral restraint to the compression chords. In the case of short-span through trusses the span-to-depth ratio may be uneconomically low if the top chord is restrained by cross-beams, as sufficient traffic headroom must be provided. In such a situation it is more economical to brace the top compression chords by U-frame action.

A span-to-depth ratio of between 6 and 8 for railway bridges and between 10 and 12 for road bridges offers the most economical design. In general terms the proportions should be such that the chords and web members have approximately an equal weight.

The bay widths should be proportioned so that the diagonal members are inclined at approximately 50° or slightly steeper. For large-span trusses subdivision of the bays is necessary to avoid having excessively long web members.

The Pratt, Howe and Warren trusses, Fig. 19.3(a)–(c), have an economic span range of between 40 m and 100 m for railway bridges and up to 150 m for road bridges. For the shorter spans of the range the Warren truss requires less material than either the Pratt or the Howe trusses. For medium spans of the range the Pratt or Howe trusses are both more favourable and by far the most common types. For large spans the modified Warren truss and subdivided Parker (inclined chord) truss are the most economical.

For spans of between 100 m and 250 m the depth of the truss may be up to four times the economic bay width, and in such a case the K-, diamond or Petit (subdivided Pratt or Howe) trusses are more appropriate. For the shorter spans of the range the diamond or Petit trusses, by their nature, are subject to very high secondary stresses. In such a case the K-truss, with primary truss members at all nodes, is more appropriate as joint deflections are uniform, greatly reducing the secondary stresses.

For spans greater than 150 m variable depth trusses are normally adopted for economy.

The spacing of bridge trusses depends on the width of the carriageway for road bridges and the required number of tracks for railway bridges, in addition to considerations regarding lateral strength and rigidity. However, in general the spacing should be limited to between $\frac{1}{18}$ and $\frac{1}{20}$ of the span, with a minimum of 4 m to 5 m for through trusses and approximately $\frac{1}{15}$ of the span, with a minimum of 3 m to 4 m, for underslung trusses.

19.3 Effects of load reversal

For buildings with light pitched roofs, load reversal is often caused by wind suction and internal pressure. Load reversal caused by wind load is of particular importance as light sections normally acting as ties under dead and imposed loads may be severely overstressed or even fail by buckling when required to act as struts. For heavy pitched or flat roofs load reversal is rarely a problem because the dead load usually exceeds the wind uplift forces.

For bridge trusses, load reversal in the component elements may be caused by the erection technique adopted or by moving live loads, particularly in continuous bridges. During the detailed design stage, consideration should be given to the method of erection to ensure stability and adequacy of any member likely to experience load reversal. For short-span simply-supported trusses erected whole, load reversal in the chords and web members is attained if the crane pick-up points during erection are at or near mid-span, Fig. 19.4(a) and (b). For large-span bridges, erection by the cantilever method causes load reversal in the chords and web members. Load reversal caused by moving loads is usually more significant in continuous trusses. A convenient way of overcoming the problem of load reversal in web elements which are likely to buckle is to provide either temporary or permanent counter bracing, Fig. 19.4(c). This will ensure that the web elements are always in tension under all load conditions and avoids the use of heavy compression elements.

19.4 Selection of elements and connections

19.4.1 Elements

For light roof trusses in buildings the individual members are normally chosen from rolled sections for economy; these are illustrated in Fig. 19.5(a). Structural hollow sections are becoming more popular due to their efficiency in compression and their neat and pleasing appearance in the case of exposed trusses. Structural hollow sections, however, have higher fabrication costs and are only suited to welded construction. For larger-span heavily-loaded roof trusses and small-span bridge trusses it often becomes necessary to use heavier sections such as rolled universal beams

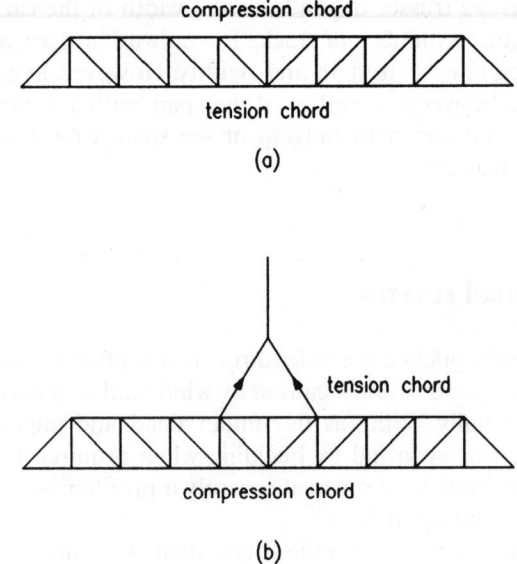

compression chord

tension chord

(a)

tension chord

compression chord

(b)

counter bracing

(c)

Fig. 19.4 Effects of load reversal. (a) Normal loading; (b) reversal during erection; (c) counter bracing

and columns and multiples of the smaller rolled sections such as back-to-back angles and channels, Fig. 19.5(b).

For large-span bridge trusses, compound or fabricated sections are normally necessary, particularly for the chords and the compression web members. Figure 19.5(c) illustrates some typical arrangements often used, although the choice available to the designer is very wide. For heavily-loaded bridge trusses, fabricated box sections offer economy in material due to their high efficiency in compression. The open sections shown assist the connections to the web members of the truss but require lacing or battening and are prone to distortion during fabrication.

Providing suitable access to all members and surfaces for inspection, cleaning and painting should be a primary consideration in deciding on the sections and details

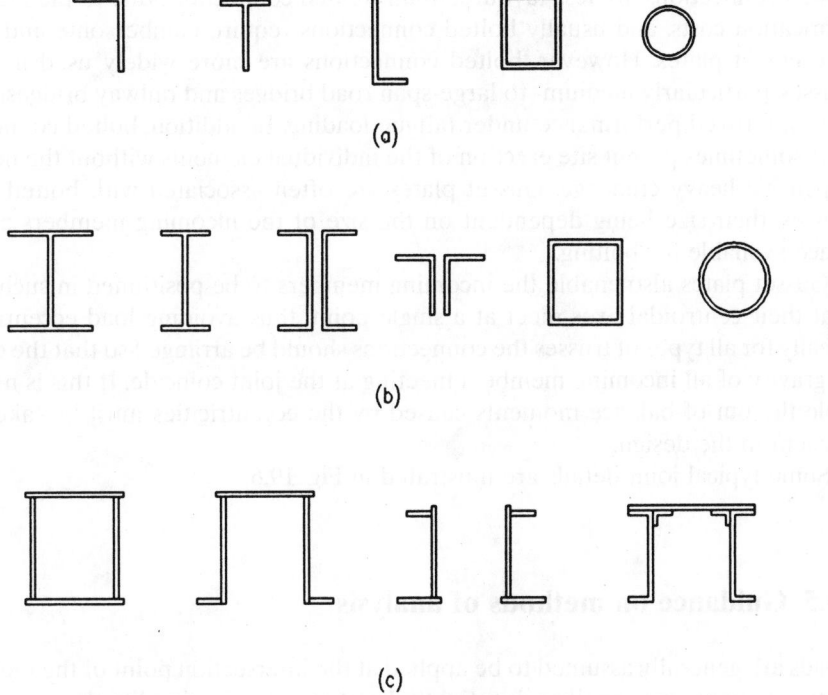

Fig. 19.5   Typical element cross sections. (a) Light building trusses; (b) heavy building trusses and light, small-span bridges; (c) road and railway bridges

to be incorporated in the design. Laced sections are disadvantageous in this respect as access can be severely restricted. In highly corrosive environments welded closed box or circular hollow sections with welded connections are usually used in order to reduce maintenance costs as all exposed surfaces are readily accessible.

19.4.2 Connections

There are basically three types of connections used for connecting truss elements to each other, that is, welding, bolting and riveting. Riveting is rarely used in the UK due to the very high labour costs involved, although it is still widely used in developing countries where labour costs are low. Small-span trusses which can be transported whole from the fabrication shop to the site can be entirely welded. In the case of large-span roof trusses which cannot be transported whole, welded subcomponents are delivered to site and are either bolted or welded together on site. Generally in steelwork construction bolted site splices are much preferred to welded splices for economy and speed of erection. In light-building roof trusses entirely

bolted connections are less favoured than welded connections due to the increased fabrication costs, and usually bolted connections require cumbersome and obtrusive gusset plates. However, bolted connections are more widely used in bridge trusses, particularly medium- to large-span road bridges and railway bridges, due to their improved performance under fatigue loading. In addition, bolted connections may sometimes permit site erection of the individual elements without the need for expensive heavy craneage. Gusset plates are often associated with bolted bridge trusses, their size being dependent on the size of the incoming members and the space available for bolting.

Gusset plates also enable the incoming members to be positioned in such a way that their centroidal axes meet at a single point, thus avoiding load eccentricities. Ideally for all types of trusses the connections should be arranged so that the centres of gravity of all incoming members meeting at the joint coincide. If this is not possible the out-of-balance moments caused by the eccentricities must be taken into account in the design.

Some typical joint details are illustrated in Fig. 19.6.

19.5 Guidance on methods of analysis

Loads are generally assumed to be applied at the intersection point of the members, so that they are principally subjected to direct stresses. To simplify the analysis the weights of the truss members are assumed to be apportioned to the top and bottom chord panel points and the truss members are assumed to be pinned at their ends, even though this is usually not the case. Normally chords are continuous and the connections are either welded or contain multiple bolts; such joints tend to restrict relative rotations of the members at the nodes and end moments develop.

Generally, in light building trusses secondary stresses are negligible and are often ignored. Secondary stresses in light building trusses may be neglected provided that:

- the slenderness of the chord members in the plane of the truss is greater than 50, and
- the slenderness of most of the web members, about the same axis, is greater than 100.

However, in bridge trusses the secondary stresses can be a significant proportion of the primary stresses and must be taken into account. The British Standard for the design of steel bridges, BS 5400: Part 3: 2000, requires the fixity of the joints to be taken into account although axial deformation of the members may be ignored for the ultimate limit state.

The magnitude of the secondary stresses depends on a number of factors including member layout, joint rigidity, the relative stiffness of the incoming members at the joints and lack of fit.

Manual methods of analysis may be used to analyse the stresses, particularly in

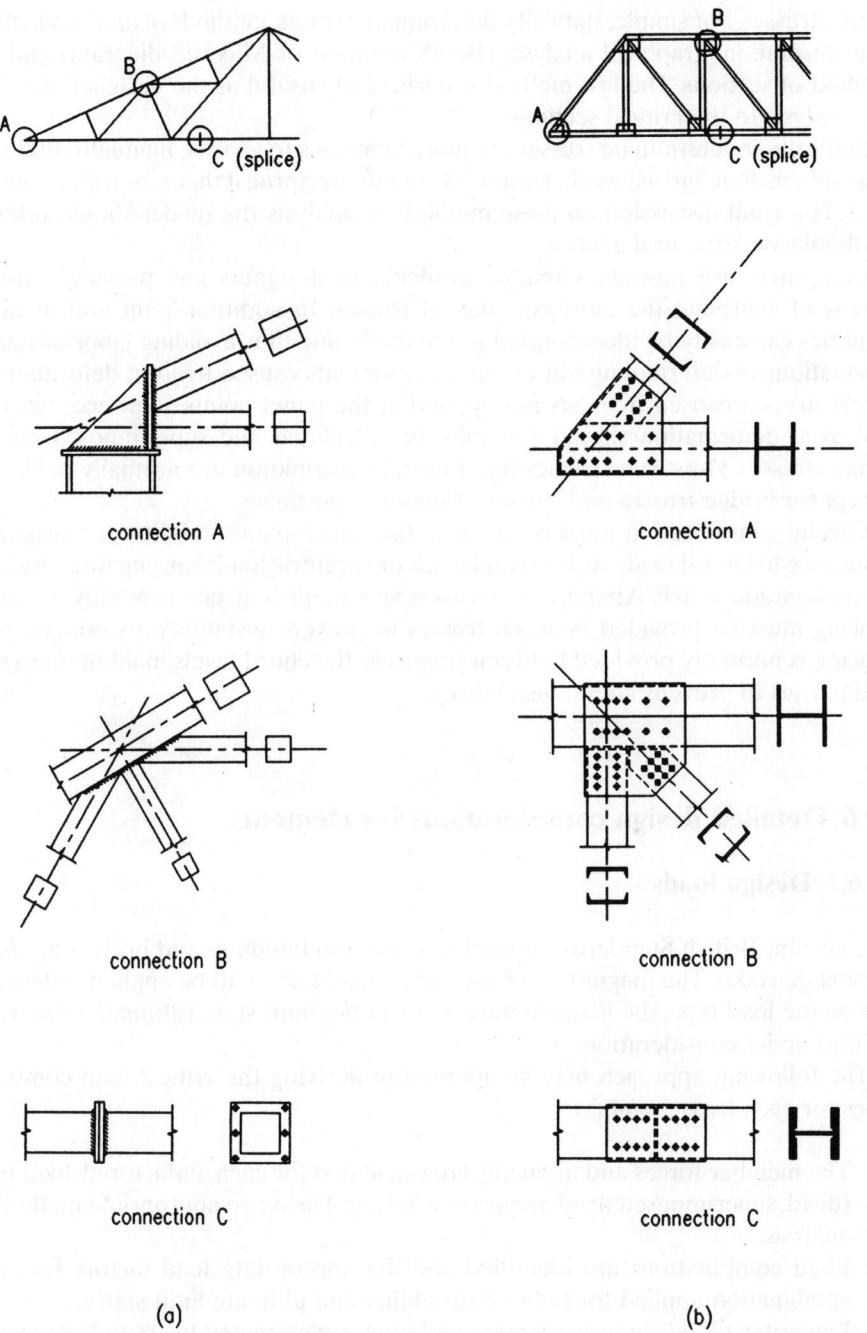

Fig. 19.6   Typical joints in trusses. (a) Welded RHS building roof truss; (b) bolted bridge truss

simple trusses. For simple, statically-determinate trusses, methods of analysis include joint resolution, graphical analysis (Bow's notation or Maxwell diagram) and the method of sections. The last method is particularly useful as the designer can limit the analysis to the critical sections.

Statically-indeterminate trusses are more laborious to analyse manually; methods available include virtual work, least work and the reciprocal theorem with influence lines. For a full discussion on these methods of analysis the reader should refer to textbooks on structural analysis.

Computers are nowadays readily available to designers and provide a useful means of analysing the most complex of trusses. In addition, joint and member rigidities can easily be incorporated in the modelling thus avoiding laborious hand calculations in determining out-of-balance moments caused by joint deformations. Local stresses caused by loads not applied at the panel points, joint eccentricities and axial deformation should generally be calculated and superimposed on the direct stresses. However, stresses due to axial deformation are normally neglected except for bridge trusses and trusses of major importance.

Careful consideration must be given to the out-of-plane stability of a truss and resistance to lateral loads such as wind loads or eccentric loads causing torsion about their longitudinal axis. An individual truss is very inefficient, and generally sufficient bracing must be provided between trusses to prevent instability. In bridges, plan bracing is normally provided between trusses at the chord levels in addition to stiff end portals to prevent lateral instability.

19.6 Detailed design considerations for elements

19.6.1 Design loads

The current British Standards for steel structures in buildings and bridges are both limit-state codes. The magnitude of the partial load factors to be applied is dependent on the load type, the load combination and the limit state (ultimate or serviceability) under consideration.

The following approach may be adopted in deriving the critical load combinations for each truss member:

(1) The member forces and moments are calculated for each, unfactored, load type (dead, superimposed dead, imposed, wind, etc.) using an appropriate method of analysis.
(2) Load combinations are identified and the appropriate load factors for each combination applied for both serviceability and ultimate limit states.
(3) The critical loads in each element and joint are extracted for both limit states.

The above process is long-winded but with experience the designer can often take short cuts in determining the critical load combinations for each element.

In the analysis the member forces and moments due to joint fixity should be calculated and superimposed on the global member forces. For trusses in buildings the secondary effects due to joint fixity may normally be ignored provided the slenderness, in the plane of the truss, is greater than 50 for the chord elements and 100 for most of the web members. If this condition is satisfied the members are assumed to be pin jointed in the analysis. Secondary effects due to axial deformations are usually ignored in building trusses. Local effects due to joint eccentricities and where loads are not applied at nodes should be taken into account.

For bridge trusses to BS 5400: Part 3: 2000, the effects of joint rigidity are required to be taken into account. Secondary stresses due to axial deformations may be ignored at the ultimate limit state but should be considered at the serviceability limit state and for fatigue checks. As for building trusses, the local effects due to joint eccentricities and cases where loads are not applied at nodes must be considered in bridge trusses.

19.6.2 Effective length of compression members

For building trusses the fixity of the joints and the rigidity of adjacent members may be taken into account for the purpose of calculating the effective length of compression members. The designer should be careful to ensure that the critical slenderness is identified. For chords, out-of-plane unrestrained lengths do not necessarily relate to the truss nodes, and effective length factors are usually unity; in-plane effective length factors may be demonstrated to be less than unity if the restraining actions of tension members and non-critical compression members are mobilized at the ends of the member. Single angle elements, for both the webs and chord, have minimum radii of gyration that do not lie either in, or normal to, the plane of the truss.

For compression members in bridge trusses the effective lengths may either be obtained from Table 11 of BS 5400: Part 3: 1982 or be determined by an elastic critical buckling analysis of the truss.

In the case of simply supported underslung trusses the top compression chord will be effectively restrained laterally throughout its length provided the connection between the chord and the deck is capable of resisting a uniformly distributed lateral force of 2.5% of the maximum force in the chord. The effective length in such a case is taken as zero where friction provides the restraint, or as equal to the spacing of discrete connections where these are provided.

The economic advantages of underslung trusses over through or semi-through trusses is obvious in this respect, due to the dual function of the deck structure.

In the case of unbraced compression chords, that is, chords with no lateral restraints, the provision of U-frames is necessary. The effective length of the compression chord is a function of the stiffness of the chord and the spacing and

stiffness of the U-frame members. Clause 12.5 of BS 5400: Part 3: 1982 gives guidance on the calculation of the effective length of compression chords restrained by U-frames.

19.6.3 Detailed design

For building trusses to BS 5950 the members need only be designed at the ultimate limit state for strength, stability and fatigue where applicable, and at the serviceability limit state for deflection and durability.

Compression members in bridge trusses to BS 5400 are designed at both the ultimate and serviceability limit states. Certain compression members, however, are exempt from the serviceability limit state check as defined in clause 12.2.3 of BS 5400: Part 3. Tension members need only be designed at the ultimate limit state.

For guidance on the detailed design of axially loaded members the reader should refer to Chapters 14 and 15, and to Chapter 18 for members subject to combined axial load and bending.

19.7 Factors dictating the economy of trusses

Some of the general factors dictating the economy of trusses relating to truss type, spacing, span-to-depth ratios, pitch, etc. have already been discussed earlier. However, factors such as the location of the structure, contractors' experience and material availability may have a significant effect on the choice of truss type and details adopted. When designing trusses for overseas locations, particularly the developing countries or for remote areas with difficult access, the designer should consider the following:

- Material available locally, i.e. weldable or unweldable steel
- Preferred connections, i.e. welded, bolted or riveted
- Maximum size of elements that can be transported to site
- Method of erection, capacity and type of plant available
- Use of rolled, compound or fabricated sections
- Redundancy of structure in case of overloading and lack of maintenance
- Experience of local contractors
- Simple design with maximum repetition.

The designer should always try to maximize the use of local materials, labour and expertise so as to avoid expensive importation of materials and trained manpower. In addition, the relative costs between materials and labour should be reflected in the design.

19.8 Other applications of trusses

Trusses are often used as secondary structures in buildings and bridges in the form of triangulated bracing. Bracing is generally required to resist horizontal loading in buildings or bridges or to prevent deformations and provide torsional rigidity to stiffening girders or box girders.

In buildings, bracing is often required for stability and to transmit horizontal wind loads or crane surges down to foundation level. To avoid the use of heavy compression bracing members, the members are usually arranged so that they always act in tension. Although this requires a high degree of redundancy it is normally more economical than providing compression members. Some examples of wind bracing to single-storey building are illustrated in Fig. 19.7(a). For multi-storey buildings with 'simple' connections, vertical bracing is required on all elevations to stabilize the building. Normally the floor slabs act as horizontal bracing which transmits the lateral wind loads to the vertical bracing. If the steel frame to the building is erected before the floors are constructed, temporary horizontal bracing must be supplied which can be removed once the floors are in place. Bracing at floor levels may be required if the slabs are discontinuous. Although horizontal bracing in buildings can often be hidden within the depth of the floor, vertical bracing can be obtrusive and undesirable, particularly if the building is clad in glass. In such an instance rigidly-jointed frames are adopted in which the wind loading is resisted by bending in the beams and stanchions. However, if the joints to such frames are made with friction-grip bolts, then temporary bracing may be required for stability prior to the joints being completed. Figure 19.7(b) illustrates some typical bracing systems to multi-storey buildings.

In bridges, secondary truss or bracing frames are often required for stability and to resist lateral loads due to wind in addition to loads due to road and railway loading such as centrifugal or braking forces. Depending on the type of structure the bracing may be temporary or permanent, usually placed in both horizontal and vertical planes. For trusses, permanent plan bracing is provided at both chord levels for underslung bridges and at deck level for through or semi-through bridges. Where headroom permits, plan bracing is also provided at the top chord level for through trusses, thus conveniently providing restraint to the top compression chord and avoiding the need for stiff U-frames. In addition, vertical bracing is also provided between trusses to reduce differential loading and therefore distortion between trusses and to provide added restraint to the compression chord.

In composite steel plate girder and concrete slab decks temporary vertical bracing may be required when the concrete is poured to provide lateral restraint to the plate girder compression flanges. It may be removed once the concrete has gained sufficient strength to act compositely with the steelwork. In stiffening girders or box girders bracing is often provided in place of plated diaphragms to avoid torsional distortion and to maintain the shape of the cross section under service loads. Figure 19.7(c) illustrates some uses of trusses in bridgeworks. Other uses of trusses in bridgeworks include trestling, i.e. triangulated temporary support frames normally

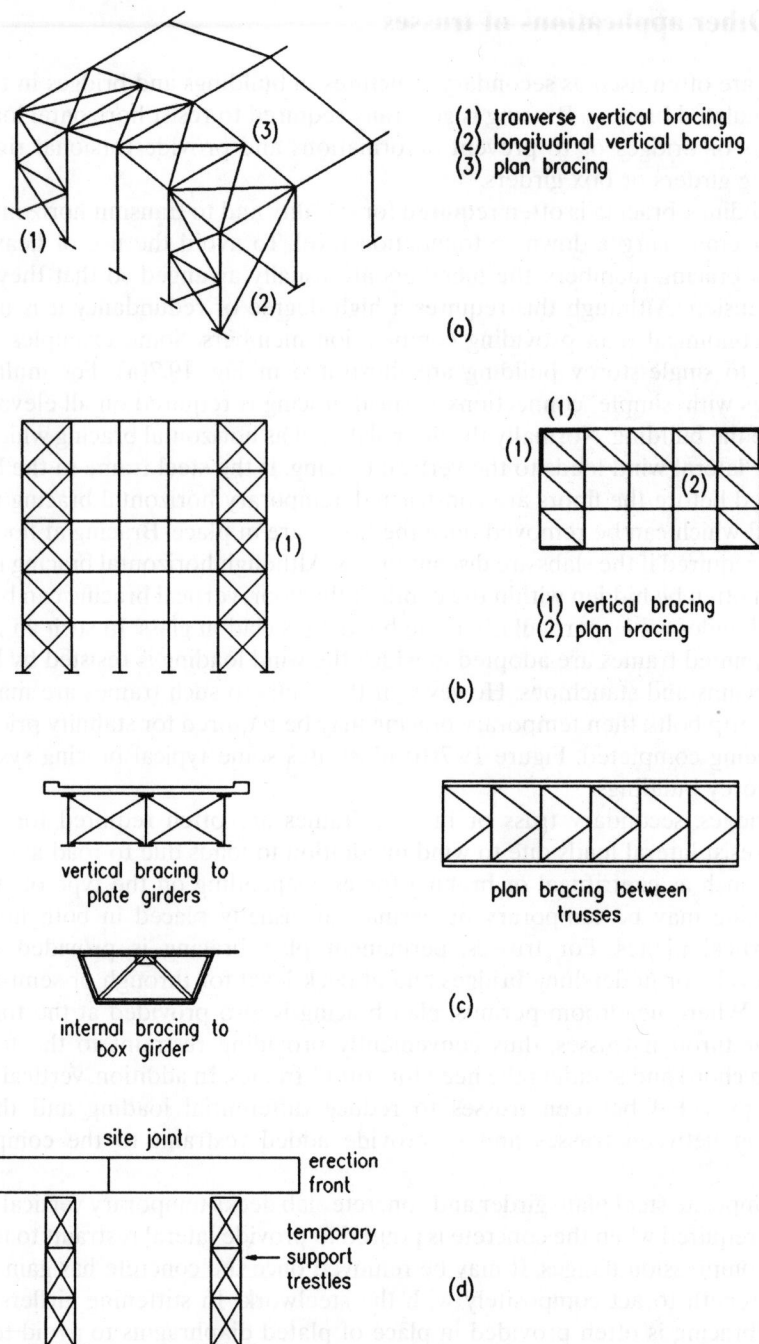

(1) tranverse vertical bracing
(2) longitudinal vertical bracing
(3) plan bracing

(a)

(1) vertical bracing
(2) plan bracing

(b)

vertical bracing to
plate girders

plan bracing between
trusses

internal bracing to
box girder

(c)

site joint

erection
front

temporary
support
trestles

(d)

Fig. 19.7 Other applications of trusses. (a) Bracing to single-storey building; (b) bracing to multi-storey building; (c) typical bracing to bridges; (d) other uses

used to support medium- to large-span bridges over land during erection: see Fig. 19.7(d).

19.9 Rigid-jointed Vierendeel girders

19.9.1 Use of Vierendeel girders

Vierendeel girders, unlike trusses or lattice girders, are rigidly-jointed open-web girders having only vertical members between the top and bottom chords. The chords are normally parallel or near parallel; some typical forms are shown in Fig. 19.8(a).

The elements in Vierendeel girders are subjected to bending, axial and shear stress, unlike conventional trusses with diagonal web members where the members are primarily designed for axial loads. Vierendeel girders are usually more expensive than conventional trusses and their use is limited to instances where diagonal web members are either obtrusive or undesirable. Vierendeel girders in bridges are rare; they are more commonly used in buildings where access for circulation or a large number of services is required within the depth of the girder.

The economic proportions and span lengths are similar to those of the parallel chord trusses already discussed in section 19.2.

19.9.2 Analysis

Vierendeel girders are statically indeterminate structures but various manual methods of analysis have been developed. The statically determinate method assumes pin joints at the mid-points of the verticals and chords of each panel. The method, however, is only suitable for girders with parallel chords of constant stiffness and when the loads are applied at the node points. Various modified moment distribution methods have been developed for the analysis of Vierendeel girders which allow for inclined chords, chords of different stiffness in the panels and member widening at the node positions.

The use of computers offers the most accurate and efficient way of analysing Vierendeel girders, particularly those with inclined chords, chords of varying stiffness and when the loading is not applied at the node positions. A further advantage of computer analysis is that joint rotations and deflections are easily calculated.

Plastic theory may be applied to the design of Vierendeel girders in a similar way to its application to other rigid frames such as portal frames. Failure of the structure, as a whole, generally results from local failure of a small number of its members to form a mechanism. Once the failure mode is established the chords and vertical are designed against failure. Computer programs are available for the plastic analysis of plane frameworks including Vierendeel arrangements.

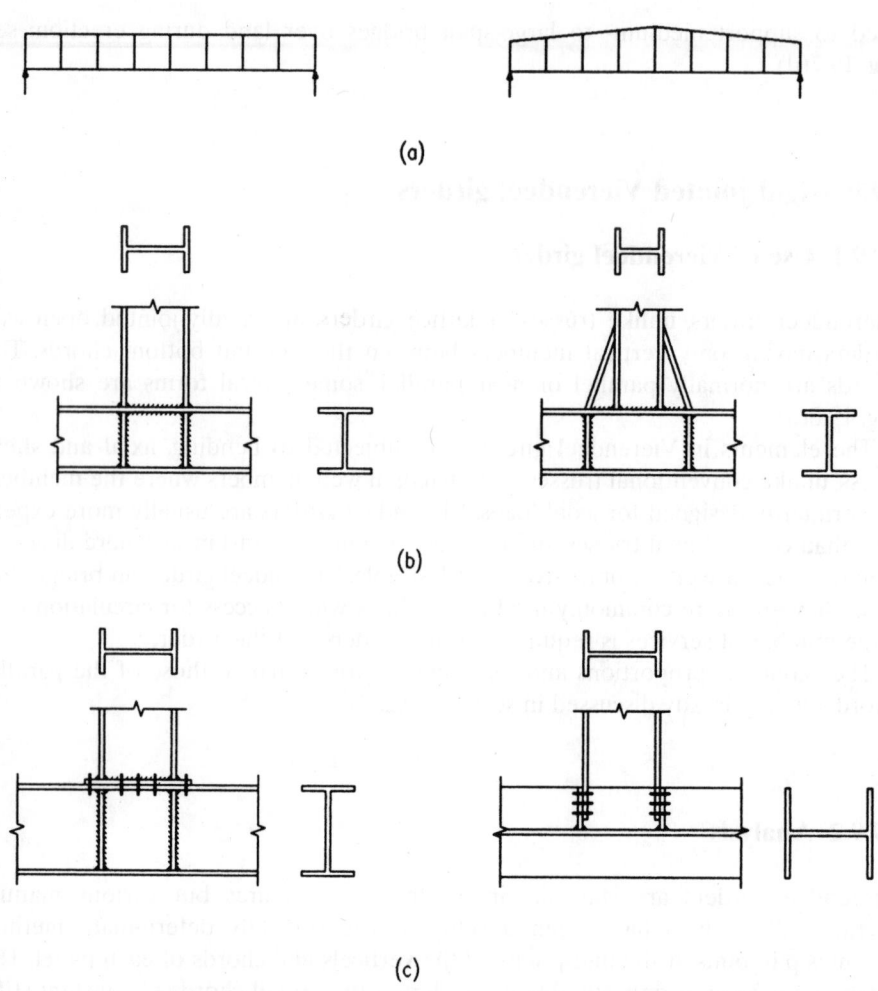

Fig. 19.8 Typical details of Vierendeel girders: (a) typical forms, (b) welded connections, (c) bolted connections

19.9.3 Connections

Vierendeel girders have rigid joints with full fixity and so the connections must be of the type which prevents rotation or slip of the incoming members, such as welded or friction-grip bolted connections. Welded connections are usually the most efficient and compact although undesirable if the connections are required to be made on site. Normally site splices are bolted for economy. For very large Vierendeel girders delivered and erected piecemeal, fully bolted connections are normally used. For member and joint efficiency the ends of the verticals are often splayed. This is

of advantage in heavily-loaded girders as the high concentrated local stresses are reduced thus avoiding the need for heavy stiffening.

Some typical joint examples are illustrated in Fig. 19.8(b) and (c).

A series of worked examples follows which are relevant to Chapter 19.

The **Steel Construction** **Institute** Silwood Park, Ascot, Berks SL5 7QN		Subject ***ROOF TRUSS***	Chapter ref. ***19***
		Design code ***BS 5950: Part 1***	Made by ***RT*** Checked by ***GWO***	Sheet no. ***1***

Problem

Design the roof trusses for an industrial building with two 25 m spans, 120 m long. The roofing is insulated metal sheeting with purlins at node positions. The building is 9 m to the eaves.

Structural form:

For 25 m span a pitched roof will not be economical as the height at the apex would be in the order of 5.5 m.

Ideal solutions would either be a Mansard truss or a parallel chord Pratt, Howe or Warren truss.

For the prupose of this design example a Mansard truss will be adopted.

Economical span to depth ratio between 7 and 8

For 3.5 m depth $\dfrac{span}{depth}$ = $\dfrac{25}{3.5}$

  = *7.14 acceptable*

Truss spacing should be in the region of 1/4 to 1/5th of the span

For 6 m truss centres

spacing/span = *6/25* = *1/4.17 acceptable*

A truss spacing of 6 m conveniently suits the length of the building.

		Subject		Chapter ref.
The **Steel Construction** **Institute** Silwood Park, Ascot, Berks SL5 7QN		**ROOF TRUSS**		**19**
		Design code **BS 5950: Part 1**	Made by **RT** Checked by **GWO**	Sheet no. **2**

Number of bays   =   120/6   =   20 No.

Truss dimensioning:

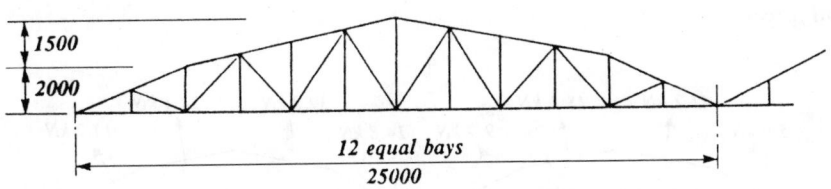

		kN/m^2	**BS 648**
Dead load – Steel sheeting	=	**0.075**	
Insulation	=	**0.020**	
Fixings	=	**0.025**	
Services etc	=	**0.100**	
Total	=	**0.22 kN/m²**	

Loading:

for 6 m bays

Roof dead load	**=**	**0.22 × 25 × 6**	**=**	**33.0 kN**
Allow for purlins			**=**	**11.8 kN**
Allow for own weight			**=**	**30.0 kN**
		Total load	**=**	**74.8 kN**
nodal load	**=**	$\dfrac{74.8}{12}$	**=**	**6.2 kN**

The **Steel Construction Institute** Silwood Park, Ascot, Berks SL5 7QN	Subject ***ROOF TRUSS***	Chapter ref. ***19***
	Design code ***BS 5950: Part 1***	Made by ***RT***	Sheet no. ***3***
		Checked by ***GWO***	

Imposed load: ***BS 6399***

No access to roof, ∴ snow load $= 0.75\,kN/m^2$

nodal load $= 0.75 \times 6 \times 25/12 = 9.4\,kN$

Wind forces:

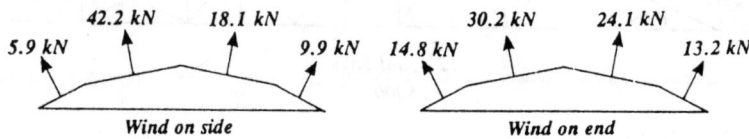

42.2 kN   18.1 kN   30.2 kN   24.1 kN

5.9 kN   9.9 kN   14.8 kN   13.2 kN

Wind on side   *Wind on end*

The **Steel Construction Institute** Silwood Park, Ascot, Berks SL5 7QN 	Subject ***ROOF TRUSS***	Chapter ref. *19*
	Design code *BS 5950: Part 1*	Made by *RT*	Sheet no. *4*
		Checked by *GWO*	

Member forces (unfactored)

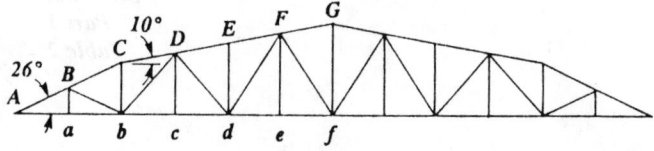

Member	Dead Load kN	Imposed Load kN	Wind on Side kN	Wind on End kN
A-B	−78.8	−118.2	104.0	91.0
B-C	−72.7	−109.1	101.7	85.2
C-D	−67.0	−100.5	96.2	81.5
D-E	−76.8	−115.2	110.6	87.4
E-F	−76.8	−115.2	113.2	89.2
F-G	−66.6	−99.9	102.0	79.1
A-a	72.5	108.8	−106.9	−97.2
a-b	72.5	108.8	−106.9	−97.2
b-c	74.3	111.5	−115.3	−94.4
c-d	74.3	111.5	−115.3	−94.4
d-e	71.9	107.9	−105.9	−82.8
e-f	71.9	107.9	−105.9	−82.8
B-a	0	0	0	0
C-b	12.7	19.1	−15.1	−11.8
D-c	0	0	0	0
E-d	−6.2	−9.3	11.9	8.5
F-e	0	0	0	0
G-f	18.6	27.9	−37.4	−29.4
B-b	−7.4	−11.1	6.3	9.5
b-D	−12.8	−19.2	17.8	8.8
D-d	1.7	2.6	−1.9	−2.8
d-F	6.2	9.3	−14.9	−12.8
F-f	−11.5	−17.3	22.4	17.8

The above forces have been calculated assuming pin joints

The **Steel Construction Institute** Silwood Park, Ascot, Berks SL5 7QN	Subject **ROOF TRUSS**	Chapter ref. **19**

Design code **BS 5950: Part 1**	Made by **RT**	Sheet no. **5**
	Checked by **GWO**	

Load factors & combinations:

For Dead + Imposed
$1.4 \times DL + 1.6 \times IL$

For Dead + Wind
$1.0 \times DL + 1.4 \times WL$

For Dead + Imposed + Wind

not critical as wind loads act in opposite direction to Dead and Imposed loads

BS 5950:
Part 1
Table 2

Member forces (factored)

Member	Dead + Imposed kN	Dead + Wind kN
A-B	−299.4	66.8
B-C	−276.3	69.7
C-D	−254.6	67.7
D-E	−291.8	78.0
E-F	−291.8	81.7
F-G	−253.1	76.2
A-a	275.6	−77.2
a-b	275.6	−77.2
b-c	282.4	−87.1
c-d	282.4	−87.1
d-e	273.3	−76.4
e-f	273.3	−76.4
B-a	0	0
C-b	48.3	−8.4
D-c	0	0
E-d	−23.6	10.5
F-e	0	0
G-f	70.7	−33.8
B-b	−28.1	5.9
b-D	−48.6	12.1
D-d	6.5	−2.2
d-F	23.6	−14.7
F-f	−43.8	19.9

The **Steel Construction Institute** Silwood Park, Ascot, Berks SL5 7QN		Subject *ROOF TRUSS*	Chapter ref. *19*
		Design code *BS 5950: Part 1*	Made by *RT*	Sheet no. **6**
			Checked by *GWO*	

Member Design:

(A) <u>*Top Chord*</u> *– member A-B*

Maximum compressive force $= 299.4\,kN$

Effective length L_E $= 1.0\,L$ $= 2311\,mm$ *Table 22*

Try $90 \times 90 \times 5\,RHS\ S275$

$$\lambda \quad = \quad \frac{L_E}{r_y} \quad = \quad \frac{2311}{34.6}$$

$$= \quad 66.8$$

$p_y \quad = \quad 275\,N/mm^2$

$p_c \quad = \quad 228\,N/mm^2$ *Table 24(a)*

Allowable compressive force P_c $=$ $A_g P_c$ *4.7.4*

$A_g \quad = \quad 16.9\,cm^2$

$P_c \quad = \quad 16.9 \times 10^2 \times 228 \times 10^{-3} \quad = \quad 385\,kN > 299.4\,kN\ OK$

$\therefore$ <u>*Use* $90 \times 90 \times 5$ *RHS Grade 43*</u>

(B) <u>*Bottom Chord*</u> *– Member b-c*

Maximum tensile force $=$ $282.4\,kN$
Maximum compressive force $=$ $87.1\,kN$

For welded connections allowable tensile force, P_t *4.6.1*

$P_t \quad = \quad A_e P_y \qquad A_e \quad = \quad$ *gross area in this case*

Assume the same size section as the top chord

$\therefore$ *Try* $90 \times 90 \times 5$ *RHS Grade 43*

Allowable tensile force $=$ $16.9 \times 10^2 \times 275 \times 10^{-3}$
                    $=$ $464.8\,kN > 282.4\,kN\ OK$

The Steel Construction Institute Silwood Park, Ascot, Berks SL5 7QN	Subject *ROOF TRUSS*	Chapter ref. *19*
	Design code *BS 5950: Part 1*	Made by *RT* Checked by *GWO*	Sheet no. 7

Under Wind + Dead load the bottom chord goes into compression.

By inspection, the bottom chord will require longitudinal ties. Assume that these ties are at quarter positions i.e. 6.25 m spacing.

$\therefore L_E \quad = \quad 6.25 \times 1.0$

$\therefore p_c \quad = \quad 56\,N/mm^2$ *Table 24(a)*

$P_c \quad = \quad 16.9 \times 10^2 \times 56 \times 10^{-3}$

$\quad = \quad 94.6\,kN > 87.1\,kN\ OK$

$\therefore$ *Use 90 × 90 × 5 RHS S275* *with longitudinal ties at 6.25 m intervals.*

(C)   Web members – member G-f

Maximum tensile force $\quad = \quad 70.7\,kN$

Maximum compressive force $\quad = \quad 33.8\,kN$

Try using 40 × 40 × 3.2 RHS

Allowable tensile force $P_t \quad = \quad 466 \times 275 \times 10^{-3}$ *4.6.1*

$\quad = \quad 128.2\,kN > 70.7\,kN \therefore OK$

Effective length $\quad L_E \quad = \quad 1.0L$ *Table 22*

Slenderness $\lambda \quad = \quad \dfrac{L_E}{r_y} \quad = \quad \dfrac{3500}{15.0} \quad = \quad 233.3$

$p_c \quad = \quad 35\,N/mm^2$ *Table 24(a)*

Allowable compressive force $P_c \quad = \quad 466 \times 35 \times 10^{-3}$

$P_c \quad = \quad 16.3\,kN < 33.8\,kN$

section is overstressed due to wind load reversal

The Steel Construction Institute Silwood Park, Ascot, Berks SL5 7QN 	Subject **ROOF TRUSS**	Chapter ref. **19**
	Design code **BS 5950: Part 1**	Made by **RT**	Sheet no. **8**
		Checked by **GWO**	

Try using 80 × 40 × 4 RHS

$$\lambda = \frac{3500}{15.9} = 220$$

$$\therefore p_c = 39\,N/mm^2 \qquad\qquad\qquad\qquad\qquad Table\ 27(a)$$

Allowable compressive force $P_c = 39 \times 888 \times 10^{-3}$

$$= 34.6\,kN > 33.8\,kN\ OK$$

<u>*Use 80 × 40 × 4 RIHS*</u>

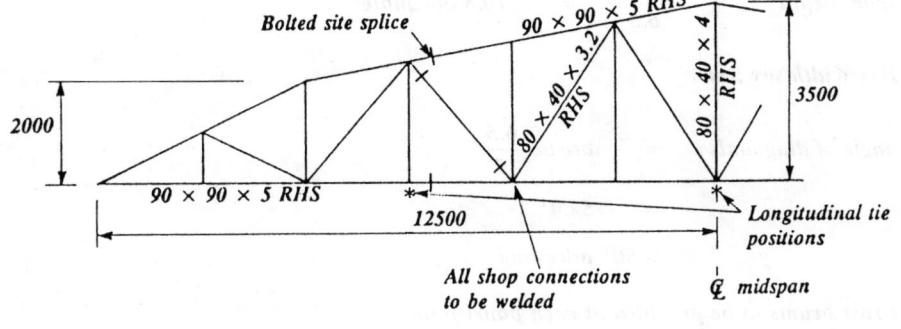

	Subject		Chapter ref.
The **Steel Construction** **Institute** Silwood Park, Ascot, Berks SL5 7QN	*ROAD BRIDGE*		**19**
	Design code **BS 5450**	Made by *RT*	Sheet no. *1*
		Checked by *GWO*	

Problem

Design the principal structure for a bridge span of 70 m, carrying a 2 lane carriageway 7.5 m wide with 2 footways 1.5 m wide. The loading is to be HA plus 37.5 units of HB. Construction depth below the road level is limited to 2.5 m maximum.

Structural form:

For road bridge span to depth ratio ~ 10 to 12

Try using 6.5 m depth

$$span/depth \quad = \quad \frac{70}{6.5} \quad = \quad 10.8 \; adequate$$

Bay width say 5 m

$$angle \; of \; diagonals \quad = \quad arc \; tan \frac{6.5}{5}$$

$$= \quad 52.4°$$

$$> 50° \; adequate$$

Cross beams to be provided at each panel point.

From the limitations imposed on the construction depth below road level an underslung truss will not be appropriate. A through truss will be adopted with an unbraced compression chord to meet the unlimited headroom requirement. Restraint to the compression chord to be provided by U-frame action.

A Pratt truss will be appropriate for a span of 70 m although a Warren or Howe truss could also be adopted.

The **Steel Construction Institute** Silwood Park, Ascot, Berks SL5 7QN	Subject ***ROAD BRIDGE***	Chapter ref. **19**
	Design code **BS 5450**	Made by ***RT*** Checked by ***GWO***	Sheet no. **2**

Proposed arrangement:

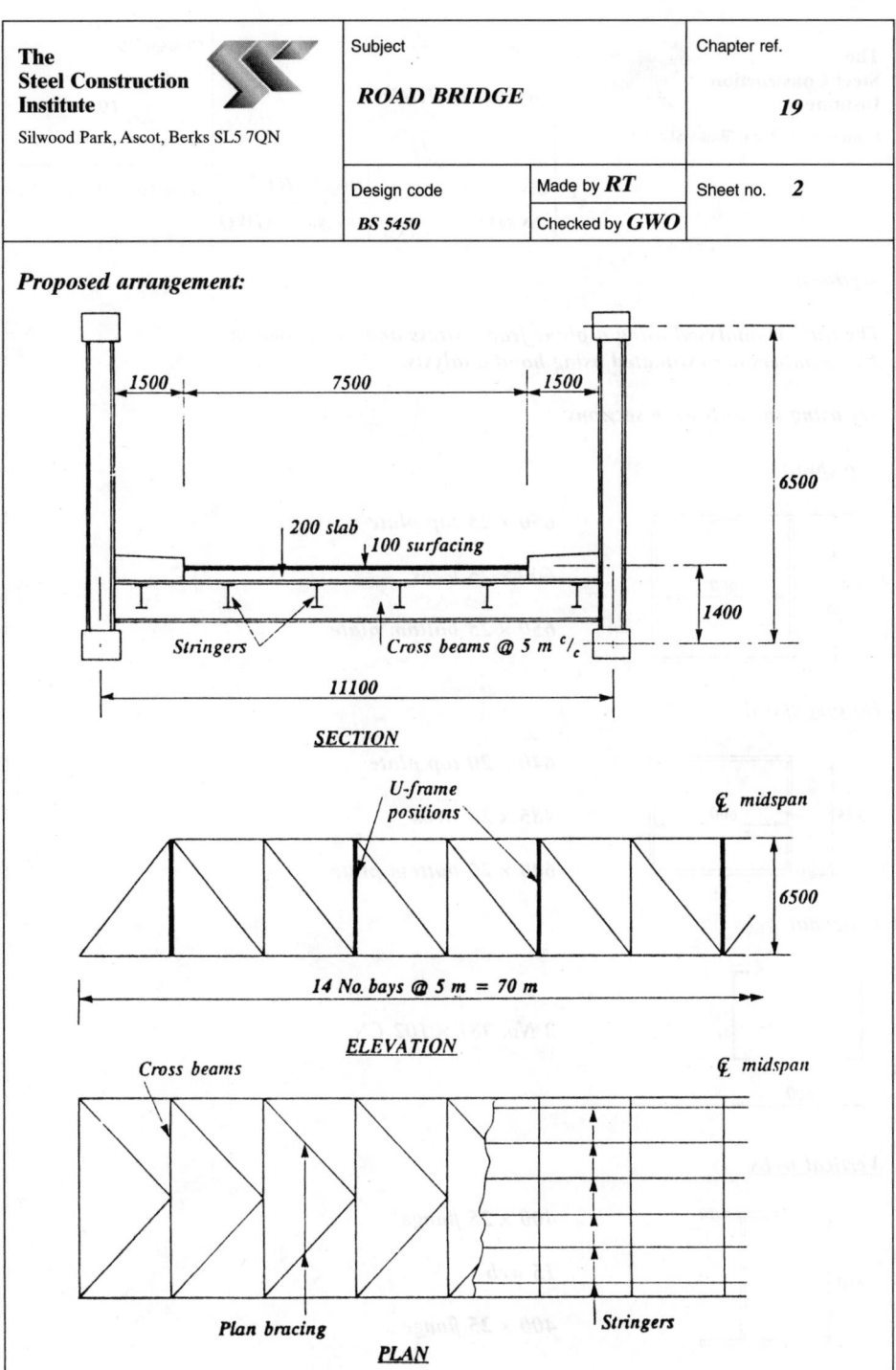

SECTION

ELEVATION

PLAN

The **Steel Construction Institute** Silwood Park, Ascot, Berks SL5 7QN	Subject **ROAD BRIDGE**	Chapter ref. **19**
	Design code **BS 5450**	Made by **RT**	Sheet no. **3**
		Checked by **GWO**	

Sections:

The truss is analysed using a plane frame stress analysis program. The members are estimated using hand analysis.

Try using the following sections:

Top chord

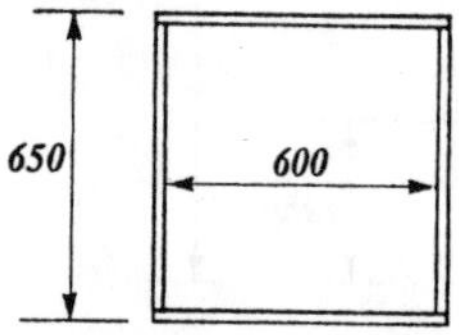

650 × 25 top plate

600 × 25 webs

650 × 25 bottom plate

Bottom chord

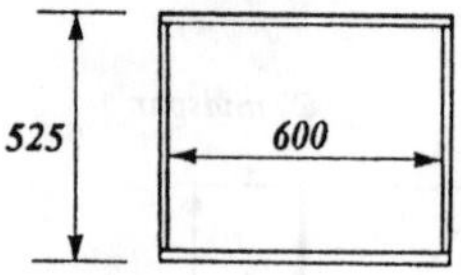

640 × 20 top plate

485 × 20 webs

640 × 20 bottom plate

Diagonal webs

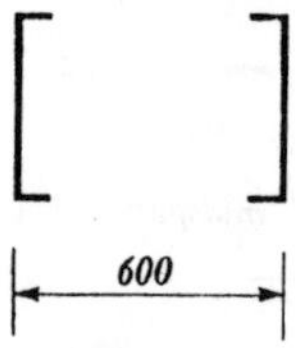

2 No. 381 × 102 C's

Vertical webs

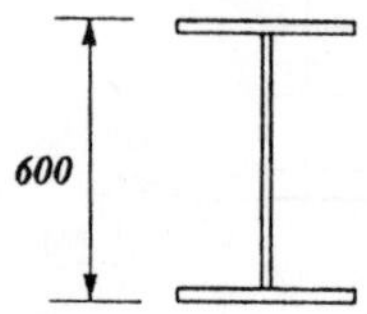

400 × 25 flange

15 web

400 × 25 flange

The **Steel Construction Institute**  Silwood Park, Ascot, Berks SL5 7QN	Subject **ROAD BRIDGE**	Chapter ref. **19**
	Design code **BS 5450**	Made by **RT** Sheet no. **4** Checked by **GWO**

Section properties:

Member	Area mm²	$\bar{y}$ mm	I In plane mm⁴	I Out of plane mm⁴	wt/m kN/m
Top chord	62500	325	4.074 E9	4.074 E9	4.95
Bottom chord	45000	263	2.012 E9	2.738 E9	3.53
Diagonals	14038		0.298 E9	1.072 E9	1.10
Verticals	28250		0.267 E9	1.861 E9	2.16

For the purpose of this example the members will be kept constant throughout the span.

Design for load combination   *Part 2, 4.4*

Load cases:

(a)   Steel Dead Load – top nodal points   = 72/2   = **36 kN**

       Steel Dead load – Bottom nodal points   = 36 + 24   = **60 kN**

(b)   Concrete dead load – bottom nodal points   = 126 + 36   = **162.0 kN**

(c)   Superimposed dead – bottom nodal points   = 41.3 + 5   = **46.3 kN**

(d)   HA UDL + KEL – bottom nodal points   = **112.5 kN**

       Travelling load   = **120 kN**

(e)   HB + Associated HA UDL = bottom nodal points   = **36.2 kN**

       & 4 No. Travelling loads @ 257 kN

(f)   Footway loading – bottom nodal points   = **28.2 kN**

The Steel Construction Institute	Subject	Chapter ref.
Silwood Park, Ascot, Berks SL5 7QN	*ROAD BRIDGE*	**19**

	Design code	Made by *RT*	Sheet no. 5
	BS 5450	Checked by *GWO*	

Nodal points notation:

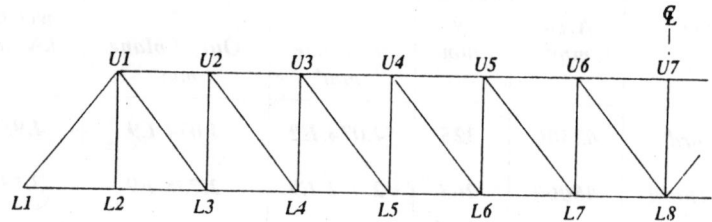

The design will be limited to the following members

U6–U7	**Top chord**
L7–L8	**Bottom chord**
U1–L3	**Diagonal**
U2–L3	**Vertical**

From the computer output,

Member	Dead Steel kN	Dead Concrete kN	Super Dead kN	HA kN	HB kN	FTWY kN	
U6–U7	−1695	−3041	−869	−2428	−3113	−530	*Unfactored axial loads*
L7–L8	1658	2974	850	2342	3244	518	
U1–L3	580	1039	297	836	1157	181	
U2–L3	−419	−812	−232	−648	−843	−141	
(U6–U7) U6	32.5	58.3	16.7	60.1	145.8	10.1	
U7	34.1	60.4	17.5	62.1	87.0	10.7	
(L7–L8) L7	15.2	27.2	7.8	33.3	95.8	4.7	*Unfactored bending moments kNm*
L8	16.2	28.9	8.3	32.7	34.0	5.1	
(U1–L3) UI	25.3	45.4	13.0	36.7	48.5	7.9	
L3	−24.4	−43.8	−12.5	−35.3	−50.0	−7.6	
(U2–L3) U2	22.1	39.7	11.4	32.2	44.4	6.9	
L3	−20.5	−36.6	−10.5	−29.9	−41.5	−6.4	

The **Steel Construction Institute** Silwood Park, Ascot, Berks SL5 7QN	Subject *ROAD BRIDGE*		Chapter ref. **19**
	Design code **BS 5450**	Made by **RT** Checked by **GWO**	Sheet no. **6**

For load combination 1   **Part 2, 4.4**

load combination 1a   **Part 2, T1**

at SLS   **1.0DS + 1.0DC + 1.25DL + 1.2HA + 1.0FTWY**

at ULS   **1.05DS + 1.2DC + 1.75SDL + 1.5HA + 1.0FTWY**

load combination 1b

at SLS   **1.0DS + 1.0DC + 1.25DL + 1.1HB + 1.0FTWY**

at ULS   **1.0DS + 1.2DC + 1.75SDL + 1.3HB + 1.5FTWY**

Factored Axial loads & Bending Moments

Member	Load Combination 1a				Load Combination 1b			
	ULS		SLS		ULS		SLS	
	AXIAL	BM	AXIAL	BM	AXIAL	BM	AXIAL	BM
	kN	kNm	kN	kNm	kN	kNm	kN	kNm
(U6–U7) U6 U7	−11387	239 248	−9222	193 201	−11792	338 268	−9733	281 222
(L7–L8) L7 L8	11087	119 123	8980	96 99	11791	194 118	9738	162 98
(U1–U3) U1 L3	3901	171 −164	3160	138 −133	4151	179 −177	3429	148 −146
(U2–L3) U2 L3	−3004	149 −138	−2428	121 −112	−3128	159 −147	−2578	131 −122

Load combination (1b) critical for members selected.

The tensile members are designed at the ULS only.   **Part 3, 12.2.3**

The Steel Construction Institute Silwood Park, Ascot, Berks SL5 7QN	Subject **ROAD BRIDGE**	Chapter ref. **19**
	Design code **BS 5450**	Made by **RT**	Sheet no. **7**
		Checked by **GWO**	

The compression members are designed at the ULS and the SLS to meet the requirements of Clause 12.2.3(a) and (b).

Compression chord effective length:

U-frame restraints are 10 m centres

Effective length $l_e = k_2 k_3 k_5 = (EI_c l_R \delta_R)$

Part 3
12.5.1
9.6.4.1.1.2

Ignoring the contribution of stiffness from the concrete slab

I_c = $4.074 \times 10^9 \, mm^4$

l_R = $10000 \, mm$

E = $205000 \, N/mm^2$

k_2 = 1.0 k_3 = 1.0

Take $k_5 = 2.5$ (assumes a moderately flexible end frame)

$$\delta_R = \frac{d_1^3}{3\,EI_1} + \frac{UB\,d_2^2}{EI_2} + f\,d_2^2 \qquad \textit{(Taking account only of verticals)}$$

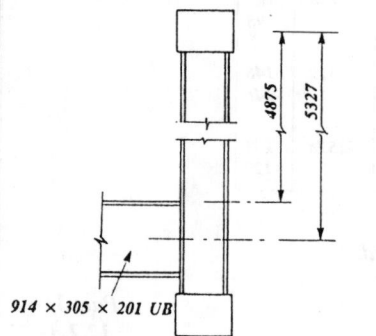

914 × 305 × 201 UB

d_1 = $4875 \, mm$

d_2 = $5327 \, mm$

I_1 = $1.861 \times 10^9 \, mm^4$

I_2 = $3.255 \times 10^9 \, mm^4$

U = 0.5

B = $11100 \, mm$

f = 0.1×10^{-10}

The **Steel Construction Institute** Silwood Park, Ascot, Berks SL5 7QN		Subject ***ROAD BRIDGE***		Chapter ref. ***19***
		Design code ***BS 5450***	Made by ***RT***	Sheet no. ***8***
			Checked by ***GWO***	

$$\delta_R = \frac{4875^3}{3 \times 2.05 \times 10^5 \times 1.861 \times 10^9} + \frac{0.5 \times 11100 \times 5327^2}{2.05 \times 10^5 \times 3.255 \times 10^9} + 0.1 \times 10^{-10} \times 5327^2$$

$$\delta_R = 6.21 \times 10^{-4}$$

$\therefore$ *effective length* $l_e = 2.5 \times 1.0 \times (2.05 \times 10^5 \times 4.074 \times 10^9 \times 10000 \times 6.21 \times 10^{-4})^{0.25}$

$\underline{l_e = 21216\,mm}$

Effective length of vertical web members:

For buckling in plane of truss l_e = *0.7l*   *Part 3, T11*
For buckling out of plane of truss l_e = *1.0l*

Member design:

The individual members may now be designed. For guidance on the design of axially loaded members with bending, reference should be made to Chapters 14, 15 and 18. Further analysis may be required if the chosen members are found to be over or understressed.

The Steel Construction Institute	Subject	Chapter ref.
Silwood Park, Ascot, Berks SL5 7QN	**ROAD BRIDGE**	**19**
Design code	Made by **RT**	Sheet no. **9**
BS 5450	Checked by **GWO**	

Typical connection detail:

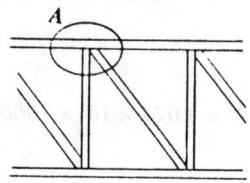

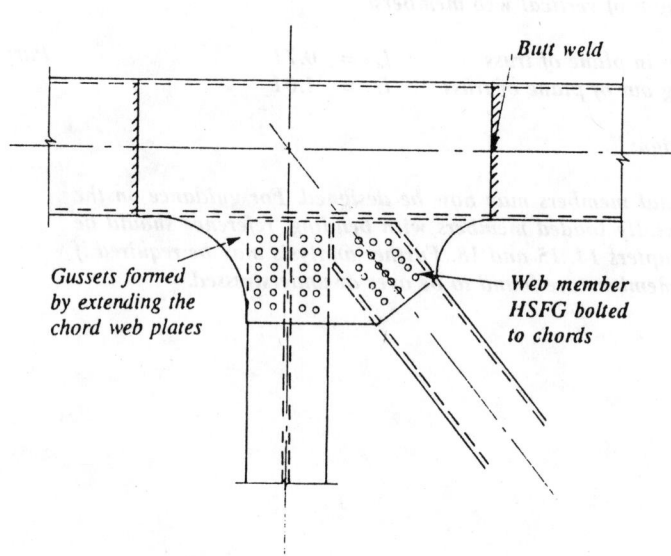

Enlarged Detail A

For guidance on the detailed design of connections the reader should refer to Chapter 26.

Chapter 20
Composite deck slabs

by MARK LAWSON and PETER WICKENS

20.1 Introduction

20.1.1 Form of construction

Composite floor is the general term used to denote the composite action of steel beams and concrete or composite slabs that form a structural floor. *Composite deck slabs*, in this context, comprise profiled steel decking (or sheeting) as the permanent formwork to the underside of a concrete slab spanning between support beams. The decking acts compositely with the concrete under service loading. It also supports the loads applied to it before the concrete has gained adequate strength. A light mesh reinforcement is placed in the concrete. A cross-section through a typical composite slab is shown in Fig. 20.1. Shear-connectors are used to develop composite action between the concrete slab and steel beams (see also Chapter 21).

The decking has a number of roles. It

(1) supports the loads during construction
(2) acts as a working platform
(3) develops adequate composite action with the concrete
(4) transfers in-plane loads by diaphragm action to vertical bracing or walls (in conjunction with the concrete topping)
(5) stabilizes the beams against lateral buckling provided that the ribs run perpendicular to the beam or are at an angle of at least 45° to the beam
(6) acts as transverse reinforcement to the composite beams
(7) distributes shrinkage strains preventing serious cracking.

In addition it has a number of advantages over precast or in situ concrete alternatives:

(1) construction periods are reduced
(2) decking is easily handled
(3) attachments (e.g. ceiling hangers) can be made easily
(4) openings can be formed

577

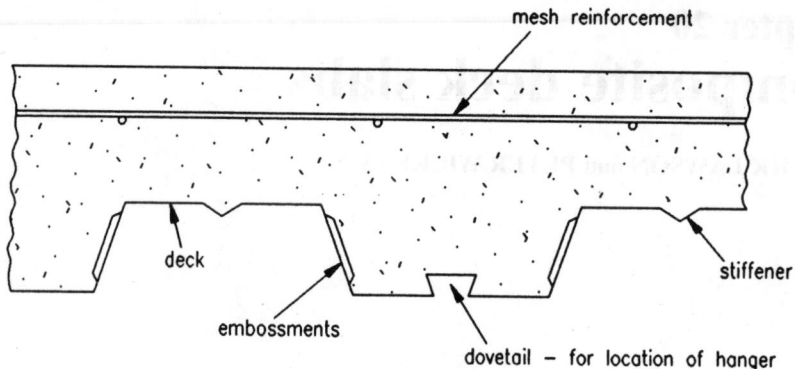

Fig. 20.1   Section through a typical composite deck slab

(5) shear-connectors can be welded through the decking
(6) decking can be cut to length and is less prone to tolerance problems.

The main economy sought in buildings is speed of construction and for this reason slabs and beams are generally designed to be unpropped during the construction stage. Spans of the order of 3 m to 3.6 m between support beams are most common, but can be increased to over 4 m if propping is used. Design of composite slabs is covered by BS 5950: Part 4.[1]

20.2 Deck types

Deck profiles are usually in the range of 38–80 mm height and 150–300 mm trough spacing with sheet thicknesses between 0.8 mm and 1.5 mm, use of the lower thickness being limited by local buckling and the upper thickness by difficulties in rolling. A summary of the different decking profiles marketed for use in composite slabs is shown in Fig. 20.2. There are two well-known generic types: the dovetail profile and the trapezoidal profile with web indentations.
    The shape of the profile is controlled by a number of criteria:

(1) the need to maximize the efficiency of the cross-section in bending (both positive and negative moments)
(2) the need to develop adequate composite action with the concrete by use of embossments or indentations or by the shape of the profile itself
(3) the efficient transfer of shear from the beam into the concrete slab (a similar problem to the haunch design in reinforced concrete slabs).

Practice in North America has been to use profiles with a trough pitch typically 300 mm and 50 mm height to achieve a 3 m span without temporary propping.[2]

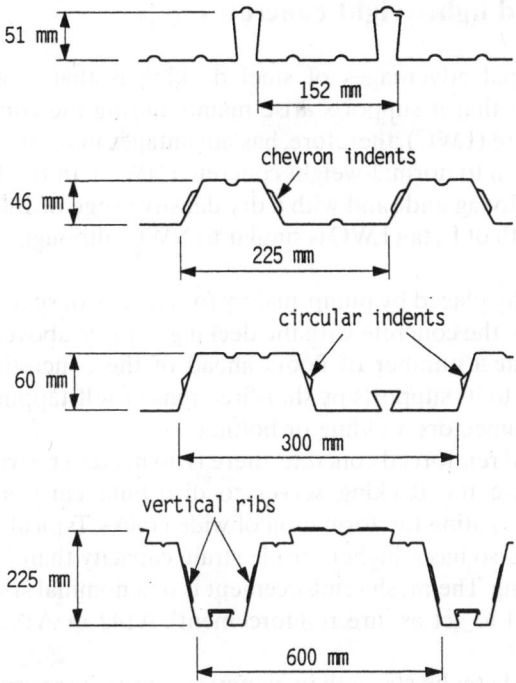

Fig. 20.2 Examples of modern composite deck profiles

In the UK, spans of 3.5–3.75 m are achieved with profiles of typically 60 mm depth.

Slimflor® construction[2] has also led to the development of deep deck profiles of 210–225 mm depth and 1.0–1.25 mm thickness. The total slab depth of 280–310 mm is such that the steel beam is fully encased in the slab except for its bottom flange. The deep deck slab is designed for spans of 6 m when unpropped during construction, or up to 7.5 m when propped. The beam may take the form of a universal column section with a welded plate or three welded plates. Recent developments by Corus have resulted in the rolling of asymmetric beams.[3] These are rolled with a patterned top flange which enhances the composite action between the beam and the overlying in situ concrete thus eliminating the need for welded shear connectors. The rolled sections have webs of greater thickness than that of the flanges to achieve superior fire resistance properties, and this extra web thickness also enhances the torsional properties of the section.

The grades of steel used for decking are specified in BS EN 10147.[4] The common grades in the UK are S280 and S350. Steel decking is usually galvanized to a standard of 275 g/m² (roughly 0.04 mm per face), which gives adequate protection for internal use.

20.3 Normal and lightweight concretes

One of the principal advantages of steel decking is that it acts as a working platform. The loads that it supports arise mainly during the concreting operation. Lightweight concrete (LWC), therefore, has advantages in terms of its reduced self-weight in comparison to normal-weight concrete (NWC). In the UK the main form of LWC comprises Lytag and sand with a dry density range of 1800–1900 kg/m³. The compressive strength of Lytag LWC is similar to NWC, although the elastic modulus is lower.

Concrete is usually placed by pump, mainly for reasons of speed, but also because it is difficult to 'skip' the concrete with the decking in place above. Indeed, it is usual practice to deck-out a number of floors ahead of the concreting operation. The decking is attached to its supports by shot-fired pins or self-tapping screws and later by welded shear-connectors, welding or bolting.

Unlike traditional reinforced concrete, there is no need to restrict bay sizes during construction because the decking serves to distribute early age and shrinkage strains, thereby eliminating the formation of wide cracks. Typical pumped pours are 500–1000 m². LWC also has a higher tensile strain capacity than NWC, reducing the tendency for cracking. The mesh reinforcement is of a nominal size to control cracking at supports, and to act as 'fire reinforcement'. A142 or A193 are the common sizes specified.

The concrete grade (cube strength in N/mm²) is normally specified as C30 to C40. Pumped concrete also contains additives to aid lubrication. For this reason the slump test is not a good measure of workability and so the *flow-meter* method is often used. The concrete is usually tamped level by fixing the tamping rails to the support beams. As these beams deflect, the slab level adopts that of the support beams. In propped construction, the slab level deflects further on removal of the props.

20.4 Selection of floor system

Slab depths are normally in the range of 110–150 mm for shallow profiles. Clearly, the minimum depth for structural adequacy is usually selected, but the slab depth is affected by insulation requirements in a fire (see Section 20.6.3) and serviceability criteria as influenced by the ratio of slab span to depth. As a general rule, span:depth ratios (considering the overall, slab depth) of continuous composite slabs should be less than 30 for LWC and 35 for NWC. Longer spans may suffer from excessive deflections and vibration of the supporting beams.

In designing the composite slab it is difficult to develop the full flexural resistance of the slab determined by the area of the deck acting as effective tensile reinforcement. Failure is normally one of shear-bond rupture resulting in slip between the deck and the concrete. Nevertheless, load capacities of composite slabs with the above proportions are normally adequate for most imposed loads up to

$10 \, kN/m^2$. The shear-bond resistance of composite slabs is determined from tests (see section 20.5.3); this resistance may be enhanced by end-anchorage provided by shear-connectors attached to the support beams.

The concrete slab also acts as the principal compressive element in the design of the composite beams (see Chapter 21). This interaction between the flexural behaviour of the slab and the beams is only important where both the slab and beam span in the same direction.

20.5 Basic design

20.5.1 Construction condition

The decking supports the weight of concrete in the finished slab, excess weight from concrete placement and ponding, the weight of operatives and any impact loads during construction. There is a variety of recommendations for this temporary construction load to be considered in the design of the decking: these are expressed in terms of either a uniformly-distributed load or a single line load transverse to the span being checked. In BS 5950: Part 4,[1] the standard used in the design of the decking and composite slab, these construction loads are specified as $1.5 \, kN/m^2$ or a transverse line load of $2 \, kN/m^2$ (for spans up to 3 m) respectively. Self-weight loads are multiplied by a load factor of 1.4[1] and construction loads by 1.6.

However, significant loads can be developed before concreting due to storage of equipment and materials.[5] Similarly, loads applied to the composite slab before it has gained adequate strength for fully composite action should not exceed $1.5 \, kN/m^2$. The definition of 'adequate strength' is considered to be a cube strength of 75% of the specified value, which is often achieved in 5–7 days after concreting.

The design of single-span decking is usually controlled by deflection and ponding of concrete. Continuous decking is generally designed on strength rather than deflection criteria, and longer unpropped spans can be achieved. The construction load is applied in design as a pattern load even though only one span is likely to be loaded to this extent, in view of the progressive nature of concreting. The 1994 version of BS 5950: Part 4 reduced the construction load to $0.5 \, kN/m^2$ on deck spans adjacent to the fully loaded one.

In BS 5950[1] a limit on the residual deflection of the soffit of the slab after concreting of the smaller of (span/180) or (slab depth/10) is specified, increased to (span/130) if the effects of ponding are included. Greater deflections are likely to be experienced during concreting.

The design of the decking in bending is dependent on the properties of the profile, and particularly the thin plate elements in compression. Where profiles are stiffened by one or two folds in the compression plate, the section is more efficient in bending.

The design of continuous decking is based, according to code requirements, on an elastic distribution of moment, as a safe lower bound to the collapse strength. Elastic moments are normally greatest at internal supports. For the two equal span

case this negative (hogging) moment is equal to that of the positive (sagging) moment of single-span decking. Many profiles with wide troughs are weaker under negative moment than positive, and the effect of the localized reaction at the internal support is to reduce further the negative moment resistance. In design to code requirements, therefore, continuous decking may appear to be weaker than simply-supported decking even though in reality it must be considerably stronger.

In general the load capacity of continuous decking is obtained from tests. The reduced construction load on adjacent spans has the effect of reducing the design negative moment so that elastic design more closely approximates the results of tests under the full design load (see following section).

20.5.2 Bending resistance of stiffened profiles

The compressive strength of a thin plate is presented in terms of an effective width of plate acting at its full yield strength. This effective width b_e is based on a semi-empirical formula[1] considering the post-buckling strength of the plate, and is given by

$$b_c = \frac{857t}{\sqrt{p_y}}\left(1 - \frac{187t}{b\sqrt{p_y}}\right) \quad \text{but} \quad \not> b \tag{20.1}$$

where t is the plate thickness, b is the plate width, both in mm, and p_y is the design strength of the steel in N/mm².

Where a stiffener is introduced, then the dimension b is based on the flat plate width between the stiffener and the web (see Fig. 20.3). However, there is a limiting size of stiffener below which it does not contribute to the increased capacity of the unstiffened plate. In most practical cases the web of the profile is fully effective.

The moment resistance of the section is evaluated using this effective width. In the calculations to BS 5950: Part 4 the design strength of the steel is taken as 0.93 times the specified yield strength. Under positive moment the top plate is in com-

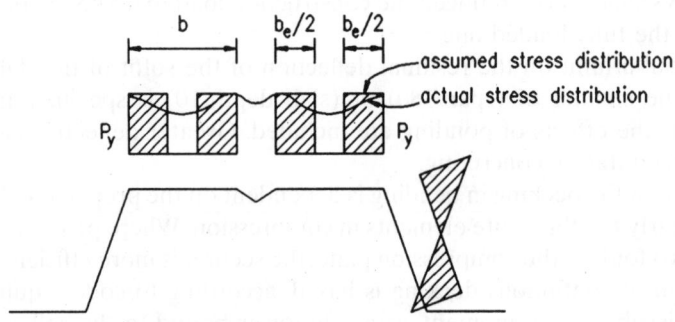

Fig. 20.3 Behaviour of stiffened plate in compression

pression, whereas under negative moment the bottom plate is in compression. This suggests that the most efficient profiles are symmetric in shape. For serviceability calculations, the actual stress replaces the design stress in Equation (20.1). For most serviceability calculations, this stress may be taken as 65% of the design stress.

The construction condition generally determines the permissible spans of the slabs; this is the case which has the greatest influence on the economy of the method of construction. Design based on elastic moments in continuous decking is very conservative because at failure there is a significant redistribution of moment from the most highly stressed areas at the supports to the mid-span area, initially as a result of elastic effects because the stiffness of the section changes with applied moment. Beyond the point at which yield takes place, plate collapse and some 'plastic' deformation occurs with increasing deformation.

If the decking were able to deform in an ideal plastic manner, as in Fig. 20.4(a), then the 'plastic' failure load of an end-span would be approximately

$$w = \frac{8}{L^2}(M_p + 0.46KM_n)$$

where M_p, and M_n are the elastic moment resistances of the profile under positive and negative moment respectively. The factor K (less than unity) is introduced because only a proportion of M_n can be developed at the large support rotations corresponding to failure. The key to this is the post-elastic behaviour of the section illustrated in Fig. 20.4(b).

From tests on different profile shapes it can be shown that the value of K has a lower bound over the normal range of sheet thicknesses and spans. Typically, the reduction in M_n (as reflected in the K value) corresponding to a support rotation of 2–3° to develop M_p is 50%. This method of design is covered in more detail in CIRIA Technical Note 116.[6] Design to BS 5950: Part 4: 1994 more closely approximates to this failure load, although it is still conservative by 10 to 15%.

20.5.3 Composite condition

The cross-sectional area of the decking A_p acts as conventional reinforcement at a lever arm determined by the centroid of the profile, as shown in Fig. 20.5. The tensile forces in the decking are developed as a result of the bond between the decking and the concrete, enhanced by some form of mechanical connection, such as by indenting the profile.

If the shear-connectors used to develop composite action between the slab and the steel beam are also included in the design, then the full flexural capacity of the slab can usually be mobilized because of the end anchorage provided. If not, failure is usually by slip between the decking and the concrete, known as *incomplete shear connection* as indicated in Fig. 20.6. This means that the full bending resistance of the slab is not developed. Nevertheless, the degree of composite action is sufficiently

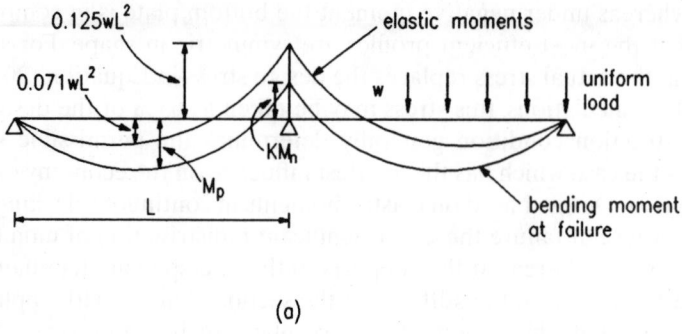

(a)

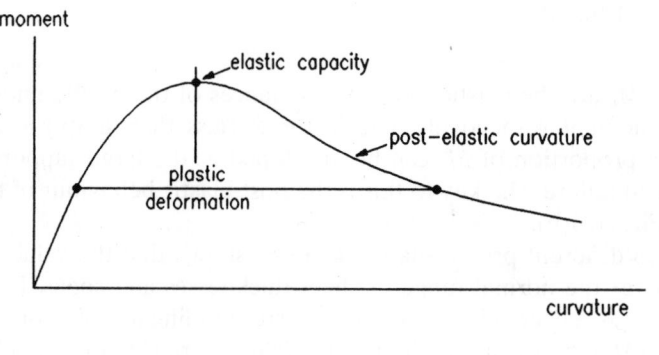

(b)

Fig. 20.4 Ultimate behaviour of continuous decking: (a) moment on two-span decking, (b) typical moment – curvature relationship of decking

good that, for all but very heavy imposed loads, permissible spans are controlled by the construction condition. Vertical shear rarely influences the design.

Composite slabs are usually designed as simply-supported elements with no account taken of continuity provided by the reinforcement in the concrete at internal beams, reflecting the behaviour of an end-bay slab. However, internal bays will not fail by incomplete shear connection provided the decking is laid continuously over the supports, an observation which also applies to end anchorage, which need only be developed at the edge beams or at sheet discontinuities in a continuous composite slab.

The ultimate resistance of composite slabs, in the absence of end anchorage, is controlled by a combination of friction and chemical bond, followed by mechanical interlock after initial slippage. As a result, the performance of a particular decking

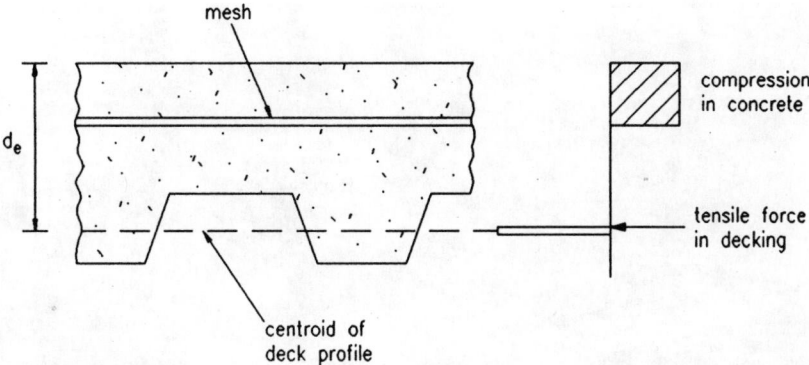

Fig. 20.5   Behaviour of composite deck slab as an equivalent reinforced concrete section

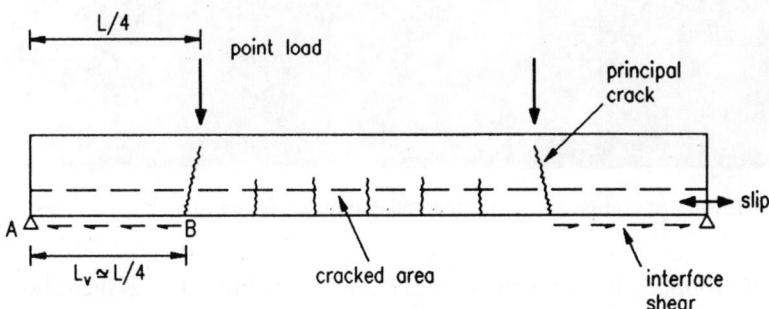

Fig. 20.6   Failure of composite slab by incomplete shear connection

system can only be readily assessed by testing. A typical composite slab under test is shown in Fig. 20.7.

Mechanical interlock is partly dependent on the local plate stiffness and so indentations are best situated close to the more rigid sections of the profile, such as corners or narrow plate elements. Slippage is associated with separation; some profiles, such as the dovetail section, achieve good shear-bond capacity by preventing separation. Different trapezoidal profiles incorporate a wide variety of indent and embossment shapes, as illustrated in Fig. 20.2, which have varying degrees of effectiveness.

If design were to be carried out on elastic principles, permissible bond strengths between the deck profile and the concrete would be of the order of $0.05\,\text{N/mm}^2$ for plain profiles rising to about $0.2\,\text{N/mm}^2$ for some indented profiles. First slip and flexural failure of plain trapezoidal profiles are coincident, whereas there is often a considerable reserve following initial slip in properly designed composite slabs.

If failure of a composite slab that is propped during construction occurs by incomplete shear connection, then the applied load to be considered in the analysis of its shear bond strength is the imposed load plus self-weight loads on depropping. If the

Fig. 20.7   Testing of composite slab to establish composite action

slab is not propped, then only the imposed load contributes to the shear bond mode of failure. If flexure governs strength, as in an equivalent reinforced concrete section, then the ultimate strength of the section is independent of the sequence of loading, and therefore the total load is to be considered in both propped and unpropped construction.

20.5.4 Requirements of BS 5950

The method of determining the ultimate shear bood capacity of composite slabs with a particular deck profile is based on load tests. A simply-supported slab of the appropriate proportions is subject to 2 or 4 point loads to simulate a uniformly distributed load (see Fig. 20.7). Crack inducers are cast into the slab so that a predetermined length of the outer portion of the slab can slip under load relative to the deck, without developing the tensile reserve of the concrete. These crack inducers are placed at the desired position for the form of loading (normally quarter span points).

Testing to BS 5950: Part 4[1] requires that a minimum of six composite slabs are tested covering a range of design parameters. The slabs are first subject to a dynamic loading between 50% and 150% of the desired working load, and then the load is increased statically to failure. The objective of the dynamic part of the test is to iden-

tify those cases where there is inherent fragility in the concrete–deck connection. The ultimate load is then recorded for each test.

The problem remains of how to use the test information in design, where the parameters may be different from those tested. This is achieved by using an empirical design formula in terms of experimentally derived constants.[7]

The vertical shear resistance (per unit width) of a composite slab is given by the formula

$$V_u = 0.8 A_c (m_d p d_c / L_v + k_d \sqrt{f_{cu}}) \tag{20.2}$$

where p is the ratio of the cross-sectional area of the profile to that of the concrete A_c per unit width of slab; f_{cu} is the cube strength of the concrete; d_c is the effective slab depth to the centroid of the profile; and L_y is the shear span length, taken as one quarter of the slab span L.

The empirical constants m_d and k_d are calculated from the slope and intercept, respectively, of the reduced regression line of Fig. 20.8. Tests are carried out at the extremes of the regression line (such as low and high L). The regression line is to be reduced by 15% if fewer than eight slab tests are performed over the range of spans. Extrapolation outside the limits of the tests is not permitted. The constant of 0.8 represents 1/(material factor) and takes account of the potential variability of this mode of failure. The slab shear resistance should exceed the applied shear forces using the load factors in Part 1 of BS 5950.

Physically, m_d is a broad measure of the effect of mechanical interlock and k_d broadly represents the friction bond. Despite this, k_d can be negative. In reality, it is unlikely that interlock is linearly dependent on profile area but Equation (20.2) may

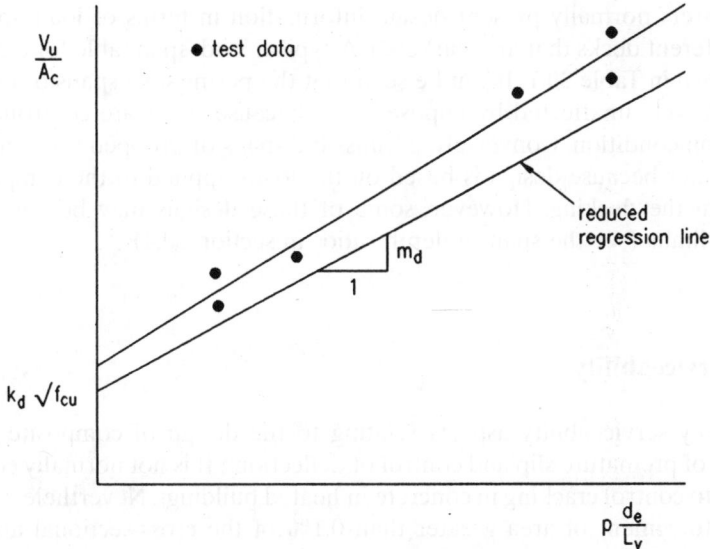

Fig. 20.8 Regression line for design of composite slabs

Table 20.1 Permissible spans (m) for typical 50 mm deep deck of thickness *t*

Support condition	Slab depth (mm)	*t* = 0.9 mm			*t* = 1.2 mm		
		Imposed loading (kN/m²)					
		2.5	5.0	7.5	2.5	5.0	7.5
Single span – no props	100	2.6	2.6	2.6	3.1	3.1	3.1
	150	2.2	2.2	2.2	2.7	2.7	2.7
Multiple span	100	2.7	2.7	2.7	3.2	3.2	3.1
	150	2.3	2.3	2.3	2.8	2.8	2.8
Single span – one prop	100	3.5	3.3	2.8	3.5	3.5	3.1
	150	4.6	4.1	3.5	5.2	4.5	3.8

Maximum span-to-depth ratio limited to 35 for normal weight concrete slabs

be considered to be reasonably accurate for small variations from the test parameters p, d_c, L_y and f_{cu}.

20.5.5 Design tables

Manufacturers normally present design information in terms of load–span tables for the different decks that are marketed. A typical load–span table for a composite slab is shown in Table 20.1. It can be seen that the permissible spans of unpropped slabs are largely unaffected by imposed load because spans are controlled by the construction condition. Conversely, permissible spans of propped slabs are considerably greater because design is based on the loads applied to the composite slab rather than the decking. However, some of these designs may be controlled by deflection limits (see the span-to-depth ratios in section 20.4).

20.5.6 Serviceability

The two key serviceability aspects relating to the design of composite slabs are avoidance of premature slip and control of deflections. It is not normally considered necessary to control cracking in concrete in heated buildings. Nevertheless, standard mesh reinforcement of area greater than 0.1% of the cross-sectional area of the concrete is placed at about 25 mm from the slab surface. This reinforcement also acts as 'fire-reinforcement' (see section 20.6). However, this minimum amount of

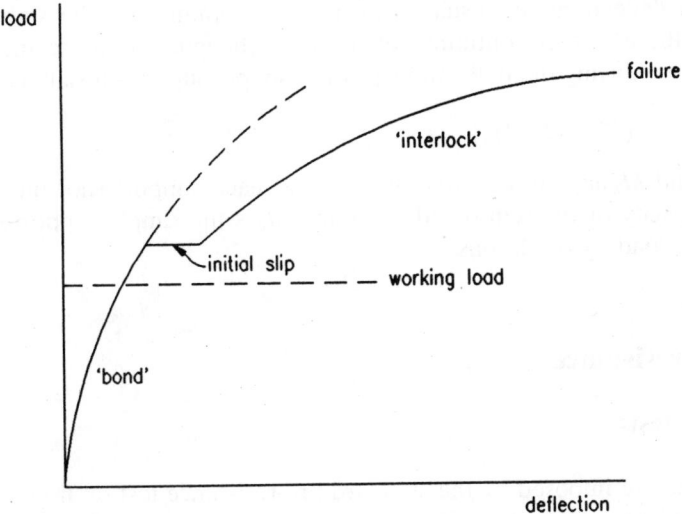

Fig. 20.9   Typical load–deflection behaviour of composite deck slabs

reinforcement would in theory be insufficient to control cracking at the supports of continuous slabs.

A typical load–deflection relationship of a composite slab is shown in Fig. 20.9. Initial slip occurs well before the ultimate load is reached in a well-designed slab. Nevertheless, it would be inadvisable if first slip occurred at below half of the ultimate load, because this might suggest poor serviceability performance.

The deflection of a composite slab is usually calculated on the assumption that it behaves as a reinforced concrete element with the deck area acting as an equivalent reinforcing bar. The section is assumed to be cracked in mid-span. The neutral axis depth below the upper slab surface is determined from

$$x_e = d_c \left\{ \sqrt{\left[(\alpha_c p)^2 + 2\alpha_c p\right]} - \alpha_c p \right\}$$
(20.3)

where α_c is the ratio of the elastic moduli of steel to concrete. The selection of an appropriate modular ratio is discussed in Chapter 21 (section 21.7.2). The values p and d_c are defined in section 20.5.3 and Fig. 20.5 respectively.

The cracked second moment of area of the composite slab (in steel units) is therefore given by:

$$l_c = x_c^3/(3\alpha_c) + p d_c (d_c - x_c)^2 + I_s$$
(20.4)

where I_s is the second moment of area of the deck profile per unit width.

In determining deflections it is common practice to use the average value of the cracked and uncracked second moments of area of the section, reflecting the fact that only part of the slab is cracked. Shrinkage-induced deflection of composite slabs is usually ignored.

Although deflections are usually calculated assuming that the slab is simply-supported, the effect of continuity of negative (hogging) reinforcement may be estimated by reducing the deflection δ_0 of the simply-supported slab according to:

$$\delta_c = \delta_0[1 - 0.6(M_1 + M_2)/M_0]$$

(20.5)

where M_1 and M_2 are the negative moments at each support (not more than the moment capacity of the reinforced slab) and M_0 is the simply-supported moment for the same loading conditions.

20.6 Fire resistance

20.6.1 Fire tests

Three criteria are imposed by the standard fire-resistance test on floors to BS 476: Part 20.[8] These are:

(1) strength or stability under load
(2) integrity to the passage of smoke or flame
(3) insulation, so that the rise of temperature on the upper surface of the slab is not excessive.

Failure of a floor or beam is deemed to occur when the maximum deflection exceeds (span/20) or when the rate of deflection exceeds a specified amount.[8] To satisfy the insulation criterion, the temperature on the upper surface of the slab should not exceed a maximum of 1.85°C or an average of 140°C above ambient.

In principle, the minimum depth of composite slabs is based on an insulation criterion, and the amount of reinforcement required in a fire is based on the strength criterion. It would be reasonable to assume that most of the tensile resistance of the deck would be lost in a severe fire and therefore the load-bearing capacity derives mainly from the reinforcement.

There are a number of beneficial factors which make fires in real buildings generally less onerous than in a standard fire test. From a structural point of view, the most important factor to be modelled in a fire test is the effect of structural continuity. Membrane action (or in-plane restraint) is generally ignored because of its indeterminate magnitude. Until recently, few fire tests had included the effect of continuity of the slab and its reinforcement.

Simply-supported composite slabs with nominal reinforcement rarely exceed a fire-resistance period of 30 minutes whereas tests on continuous slabs with the same reinforcement can achieve over 60 minutes fire resistance. Indeed, a series of ten fire tests carried out in the UK showed that 90 minutes fire resistance can be achieved for slabs with standard mesh reinforcement and subject to imposed loads of up to 6.7 kN/m². These imposed load and fire resistance requirements are typical of those specified currently for most commercial buildings.

20.6.2 Fire engineering method

The phrase 'fire engineering method' (FEM) is the term given to the means of calculating the amount of emergency reinforcement needed for a certain fire resistance period. This is described fully in a Steel Construction Institute publication.[9] In principle, the moment of resistance of the section is calculated taking into account the temperatures in the section and the reduced strength of the various elements at elevated temperatures. Temperatures are determined from tables or indicative tests. The plastic capacity of continuous members can be evaluated and compared to the applied moment with load factors of unity. In some areas, BS 5950: Part 8[10] permits a further 20% reduction in imposed load in a fire. Typically, therefore, the applied moment in a fire would be 50–60% of the ultimate design moment.

The critical element determining the moment of resistance of the section is the reinforcement. Mesh reinforcement is normally well-insulated from the effects of the fire and contributes significantly to the fire resistance of the slab. Where fire resistance periods greater than 90 minutes are required (see below) then additional reinforcing bars or heavier mesh can be introduced. In such cases it may be economic to design the slab as a reinforced concrete ribbed slab to BS 8110,[11] and treat the decking as permanent formwork. Bars of diameter 10 mm or 12 mm are normally placed singly in the deck troughs at the appropriate cover.

20.6.3 Design recommendations

Simple design tables covering common design cases are given in Reference 12. A minimum slab depth and mesh size are given for different profile types. Data for only two spans (i.e. 3 m and 3.6 m) and one imposed load (6.7 kN/m^2) are presented, but these may be converted to other cases by using the same equivalent moment as the tabulated cases. The data are reproduced in Table 20.2.

20.7 Diaphragm action

The decking serves to transfer lateral loads to vertical bracing of concrete walls. It is normally attached on all four sides to support beams at spacings of not less than 600 mm. However, deck–deck seam fasteners are not normally installed. The shear stiffness of the 'diaphragm' is very high and the strength is normally determined by the capacity of the fixings. Typical ultimate strengths of the standard screws or shot-fired pins are 6 kN/mm sheet thickness. Where through-deck welding of shear-connectors is used the ultimate shear strength of the decking is considerably enhanced.

The steel beams are laterally supported in simple bending by the decking, provided the decking crosses the beams and is attached to them by shear-connectors or shot-fired pins at spacings not exceeding 600 mm. However, beams running

Table 20.2 Simplified rules for fire resistance of composite deck slabs[12]

(a) Trapezoidal profiles (depth not exceeding 60 mm)

Max. span (m)	Fire rating (h)	Sheet thickness (mm)	Minimum dimensions		Mesh size
			Slab depth (mm)		
			NWC	LWC	
2.7	1	0.8	130	120	A142
3.0	1	0.9	130	120	A142
	1½	0.9	140	130	A142
3.6	1	1.0	130	120	A193
	1½	1.0	140	130	A193

(b) Dovetail profiles (depth not exceeding 50 mm)

Max. span (m)	Fire rating (h)	Sheet thickness (mm)	Minimum dimensions		Mesh size
			Slab depth (mm)		
			NWC	LWC	
2.5	1	0.8	100	100	A142
	1½	0.8	110	105	A142
3.0	1	0.9	120	110	A142
	1½	0.9	130	120	A142
3.6	1	1.0	125	120	A193
	1½	1.2	135	125	A193

parallel to the decking are laterally supported only at transverse beam connections. All beams are laterally supported under positive moment in their composite state.

20.8 Other constructional features

In the role of the decking as formwork, shuttering at the edge of the slab and at large openings is usually in the form of light cold-formed channels which are attached by shot-fired pins or screws to the decking. Small openings can be formed by leaving 'boxed out' voids in the slab and the decking cut away once the concrete has gained adequate strength. Guidance should be sought from the decking manufacturer regarding the maximum dimensions of unreinforced holes. Additional reinforcement should be placed to redistribute the loads that would otherwise have been resisted by the composite slabs in this zone.[5]

Propping of composite slabs is rarely used, but if so, it is important to ensure that the slab beneath can resist the self-weight and construction load of the slab being cast. This may be critical if the support slab has not gained its full strength for composite action. Timber spreaders should be used to avoid damage to the deck.

References to Chapter 20

1. British Standards Institution (1994) *Structural use of steelwork in building.* Part 4: *Code of practice for design of floors with profiled steel sheeting.* BS 5950, BSI, London.
2. Lawson R.M. & Mullet D.L. (1993) *Slim Floor Construction Using Deep Decking.* SCI, Ascot, Berks.
3. Lawson R.M., Mullet D.L. & Rackham J.W. (1997) *Design of Asymmetric Slimflor® Beams Using Deep Composite Decking.* SCI, Ascot, Berks.
4. British Standards Institution (1992) *Specification for continuously hot-dip zinc coated steel sheet and strip.* Technical delivery conditions. BS EN 10147, BSI, London.
5. Couchman G.H., Mullet D.L. & Rackham J.W. (2000) *Composite slabs and beams using steel decking: Best practice for design and construction.* Metal Cladding & Roofing Manufacturers Association and the Steel Construction Institute, Ascot, Berks.
6. Bryan E.R. & Leach P. (1984) *Design of Profiled Sheeting as Permanent Formwork.* Construction Industry Research and Information Association (CIRIA) Technical Note 116.
7. Shuster R.M. (1976) Composite steel-deck concrete floor systems. *Proc. Am. Soc. Civ. Engrs, J. Struct. Div.*, **102**, No. ST5, May, 899–917.
8. British Standards Institution (1987) *Fire tests on building materials and structures.* Part 20: *Method for determination of the fire resistance of elements of construction (general principles).* BS 476, BSI, London.
9. Steel Construction Institute (SCI) (1988) *Fire resistance of composite floors with steel decking.* SCI, Ascot, Berks.
10. British Standards Institution (1990) *Structural use of steelwork in building.* Part 8: *Code of practice for fire resistant design.* BS 5950. BSI, London.
11. British Standards Institution (1985) *Structural use of concrete.* Part 2: *Code of practice for special circumstances.* BS 8110, BSI, London.
12. Construction Industry Research and Information Association (CIRIA) (1986) *Data Sheet: Fire Resistance of Composite Slabs with Steel Decking.* CIRIA Special Publication 42.

A worked example follows which is relevant to Chapter 20.

The **Steel Construction Institute** Silwood Park, Ascot, Berks SL5 7QN		Subject **DESIGN OF COMPOSITE SLAB**	Chapter ref. **20**
		Design code **BS 5950: Part 4**	Made by **RML**	Sheet no. **1**
			Checked by **GWO**	

Problem

Check the design of a composite slab 125 mm deep, with beams at 3 m centres, and using the deck shown below. (Note: this design is carried out to BS 5950: Part 4: 1994.)

Deck shape:

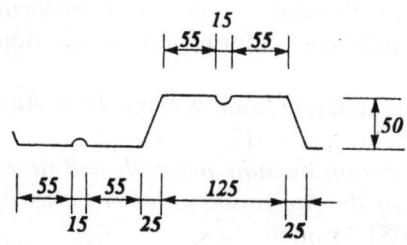

Steel thickness t = 1.2 mm (bare thickness of steel)

Steel grade $\quad p_y = 280\,N/mm^2$

Trough spacing = 300 mm

Loads

Imposed load	$= 5.0\,kN/m^2$
Partitions (imposed load)	$= 1.0\,kN/m^2$
Ceiling	$= 0.5\,kN/m^2$
Self weight (slab)	$= 2.0\,kN/m^2$ – see later

Construction load (temporary) = 1.5 kN/m² considered as imposed load

Concrete

Cube strength $f_{cu} = 30\,N/mm^2$

Density (dry) = 1800 kg/m³ (lightweight)

Density (wet) = 1900 kg/m³ (lightweight)

The Steel Construction Institute	Subject	Chapter ref.
Silwood Park, Ascot, Berks SL5 7QN	*DESIGN OF COMPOSITE SLAB*	*20*

Design code	Made by *RML*	Sheet no. 2
BS 5950: Part 4	Checked by *GWO*	

Construction condition

Self weight of slab

$$w_d = 1900 \times 9.81 \times 10^{-6}(75 + 50/2) = 1.86\,kN/m^2$$

Self weight of deck = 0.12 kN/m²

Total self weight ≃ 2.0 kN/m²

Clear span = 3.0 − 0.1 = 2.9 m

Design moment in construction condition

$$M_u = (2.0 \times 1.4 + 1.5 \times 1.6) \times \frac{2.9^2}{8}$$
$$= 5.47\,kN.m$$

Moment resistance of deck in sagging (positive bending)

Assume in the worst case that the deck is laid simply supported. The moment resistance derives from the positive (sagging) elastic moment resistance of the section.

Effective breadth of flat compression plate

Equation 20.1

$$b_e = \frac{857t}{\sqrt{p_y}}\left(1 - \frac{187t}{b\sqrt{p_y}}\right) \not> b$$

where b = 55 mm

$$b_e = \frac{857 \times 1.2}{\sqrt{280}}\left(1 - \frac{187 \times 1.2}{55\sqrt{280}}\right)$$
$$= 46.5\,mm < b$$

According to BS 5950: Part 4 the stiffener area is ignored when calculating the compressive resistance of the plate.

The Steel Construction Institute Silwood Park, Ascot, Berks SL5 7QN	Subject **DESIGN OF COMPOSITE SLAB**	Chapter ref. **20**
	Design code **BS 5950: Part 4**	Made by **RML**	Sheet no. **3**
		Checked by **GWO**	

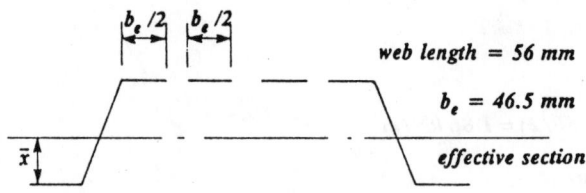

web length = 56 mm

b_e = 46.5 mm

effective section

Neutral axis position

$$\bar{x} = \frac{(46.5 \times 2 + 56) \times 50}{(125 + 2 \times 56 + 2 \times 46.5)} = 22.5 \, mm$$

Second moment of area

$$I = 1.2 \left[46.5 \times 2 \times (48.8 - 22.5)^2 + 125 \times 22.5^2 \right.$$
$$\left. + 2 \times 56 \times \frac{50^2}{12} + 2 \times 56 \times (24.4 - 22.5)^2 \right]$$
$$= 1.2(64.3 + 63.3 + 23.3 + 0.4) \times 10^3$$
$$= 181.6 \times 10^3 \, mm^4/trough$$
$$= 605 \times 10^3 \, mm^4/m \, width$$

Depth of web in compression = 26.3 mm

$d/t = 21.9$ – it is not necessary to check for local buckling of web

Elastic section modulus

$$Z_e = \frac{605 \times 10^3}{(49.4 - 22.5)} = 22.5 \times 10^3 \, mm^3/m$$

Elastic moment resistance of section (sagging moment)

$$M_c = Z_e \times 0.93 \, p_y = 22.4 \times 10^3 \times 0.93 \times 280 \times 10^{-6}$$
$$= 5.8 \, kNm > M_u \quad (see \ sheet \ 2)$$

(using a design strength of 0.93 p_y as in BS 5950: Part 4)

The Steel Construction Institute Silwood Park, Ascot, Berks SL5 7QN 	Subject ***DESIGN OF COMPOSITE SLAB***	Chapter ref. **20**
	Design code ***BS 5950: Part 4***	Made by ***RML***	Sheet no. **4**
		Checked by ***GWO***	

Check deflection of slab after construction

Bending moment on deck after construction

$= 2.0 \times 2.9^2/8 = 2.10\,kNm$

Equivalent stress in compression plate

$$\sigma \simeq \frac{2.10}{6.3} \times 280 = 93\,N/mm^2$$

Repeating the calculation of second moment of area using this stress (rather than p_y) in equation 20.1 gives:

$I = 649 - 10^3\,mm^4/m\ width$

Deflection of soffit of deck (simply supported)

$$\delta_2 = \frac{5 \times 2.0 \times 2900^4}{384 \times 205 \times 10^3 \times 649 \times 10^3}$$
$$= 13.9\,mm \quad (span/209)$$

This is less than the limit of span/180 in BS 5950: Part 4. If this limit had been exceeded it would be necessary to include for the effects of ponding of concrete at greater deflections.

Composite condition

Check moment resistance of the composite section as a reinforced concrete slab, assuming full shear connection.

Area of deck $\quad = (125 + 56) \times 2 \times 1.2$

$\qquad\qquad\qquad = 434\,mm^2/trough$

$\quad A_p \quad = 1448\,mm^2/m$

Tensile resistance $= 0.93 \times 280 \times 1448 \times 10^{-3}$

$\qquad\qquad\qquad = 377\,kN$

The **Steel Construction Institute** Silwood Park, Ascot, Berks SL5 7QN	Subject **DESIGN OF COMPOSITE SLAB**	Chapter ref. **20**
	Design code **BS 5950: Part 4**	Made by **RML**	Sheet no. **5**
		Checked by **GWO**	

Neutral axis depth into concrete

$$x_c = \frac{377 \times 10^3}{0.4 \times f_{cu} \times 10^3} = 31.4 mm$$

Plastic moment resistance of composite section

$$M_p = 377 \times (125 - 25 - 31.4/2) \times 10^{-3}$$
$$= 31.8 \, kNm$$

Applied moment in composite condition

$$M_u = [1.6(5.0 + 1.0) + 1.4(2.0 + 0.5)] \times 2.9^2/8$$
$$= 13.8 \, kNm$$

Therefore, the plastic moment resistance of the section is more than adequate.

Check for incomplete shear connection   *Equation 20.2*

$$V_u = 0.8 A_c \left(m_d \times p \times \frac{d_e}{L_v} + k_d \sqrt{f_{cu}} \right)$$

where $p = \dfrac{A_p}{A_c} = \dfrac{1448}{100 \times 1000} = 0.0145$

$d_e = 100 \, mm$

$L_v = 3000/4 = 750 \, mm$

Typical values of empirical constants are:

$M_d = 130$ *and* $k_d = 0.004$

The Steel Construction Institute Silwood Park, Ascot, Berks SL5 7QN 	Subject **DESIGN OF COMPOSITE SLAB**		Chapter ref. **20**
	Design code **BS 5950: Part 4**	Made by **RML**	Sheet no. **6**
		Checked by **GWO**	

$$V_u = 0.8 \times 100 \times \left(130 \times 0.0145 \times \frac{100}{750} + 0.004\sqrt{30} \right)$$

$$= 80(0.251 + 0.022)$$

$$= 21.8 \, kN$$

Applied ultimate shear force

$$V = 4 M_u / L = 4 \times 13.8 / 2.9 = 19.0 \, kN$$

The slab can resist the applied loads adequately without failure by incomplete shear connection.

Check shear stress on concrete

$$v_c = \frac{19 \times 10^3}{100 \times 10^3} = 0.19 \, N/mm^2$$

This is relatively small according to BS 8110.

<u>*Check deflection of composite slab*</u>

Properties of cracked section:

Neutral axis depth below upper surface of slab

$$x_e = d_e \left(\sqrt{(\alpha_e p)^2 + 2\alpha_e p} - \alpha_e p \right)$$

Equation 20.3

$$\alpha_e = modular \ ratio = 15 \ for \ LWC$$

$$\alpha_e p = 0.0145 \times 15 = 0.218$$

$$x_e = 100 \left(\sqrt{0.218^2 + 2 \times 0.218} - 0.218 \right)$$

$$= 51.5 \, mm$$

The **Steel Construction Institute** Silwood Park, Ascot, Berks SL5 7QN		Subject ***DESIGN OF COMPOSITE SLAB***	Chapter ref. *20*
		Design code **BS 5950: Part 4**	Made by *RML*	Sheet no. *7*
			Checked by *GWO*	

Second moment of area of cracked section   *Equation*
                                                              20.4

$$I_c = \frac{51.3^3}{3 \times 15} + 0.0145 \times 100 \times (100 - 51.5)^2 + 649$$

$$= 3035 + 3411 + 649$$

$$= 7095 \ mm^4 / mm \ width \ (in \ steel \ units)$$

Service load on composite slab

$$w_s = 5.0 + 1.0 + 0.5 = 6.5 \ kN/m^2$$

Deflection under service load

$$\delta = \frac{5 \times 6.5 \times 2900^4 \times 10^{-3}}{384 \times 205 \times 10^3 \times 7095}$$

$$= 4.1 \ mm \ (or \ span/707)$$

This is satisfactory.

Note:

*Mesh reinforcement is required for crack control and fire resistance
requirements.*

Chapter 21
Composite beams

by MARK LAWSON and PETER WICKENS

21.1 Applications of composite beams

In buildings and bridges, steel beams often support concrete slabs. Under load each component acts independently with relative movement or slip occurring at the interface. If the components are connected so that slip is eliminated, or considerably reduced, then the slab and steel beam act together as a composite unit (Fig. 21.1). There is a consequent increase in the strength and stiffness of the composite beam relative to the sum of the components.

The slab may be solid in situ concrete or the composite deck slab considered in Chapter 20. It may also comprise precast concrete units with an in situ concrete topping. In buildings, steel beams are usually of standard UB scction, but UC and asymmetric beam sections are sometimes used where there is need to minimize the beam depth. A typical building under construction is shown in Fig. 21.2. Welded fabricated sections are often used for long-span beams in buildings and bridges.

Design of composite beams in buildings is now covered by BS 5950: Part 3,[1] although guidance was formerly available in an SCI publication.[2] The design of composite beams incorporating composite slabs is affected by the shape and orientation of the decking, as indicated in Fig. 21.3.

One of the advantages of composite construction is smaller construction depths. Services can usually be passed beneath, but there are circumstances where the beam depth is such that services can be passed through the structure, either by forming large openings, or by special design of the structural system. A good example of this is the stub-girder.[3] The bottom chord is a steel section and the upper chord is the concrete slab. Short steel sections or 'stubs' are introduced to transfer the forces between the chords.

Openings through the beam webs can be provided for services. Typically, these can be up to 70% of the beam depth and can be rectangular or circular in shape. Guidance on the design of composite beams with web openings is given in Reference 4. Examples of the above methods of introducing services within the structure are shown in Fig. 21.4.

21.2 Economy

Composite beam construction has a number of advantages over non-composite construction:

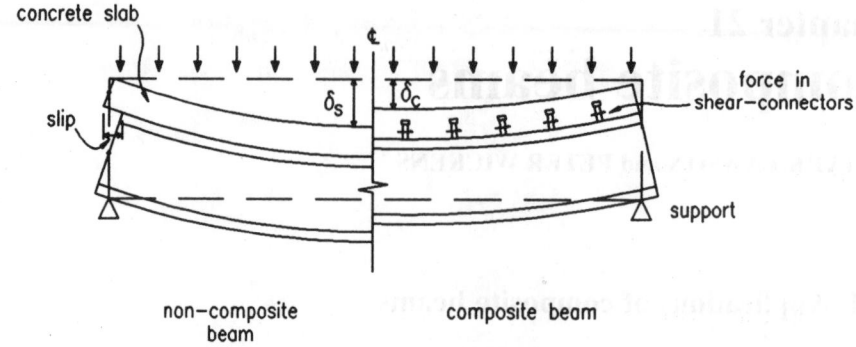

Fig. 21.1   Behaviour of composite and non-composite beams

Fig. 21.2   Composite building under construction showing decking and shear-connectors

(1) savings in steel weight are typically 30% to 50% over non-composite beams.
(2) the greater stiffness of the system means that beams can be shallower for the same span, leading to lower storey heights and savings in cladding, etc.

It also shares the advantage of rapid construction.

The main disadvantage is the need to provide shear-connectors at the interface between the steel and concrete. There may also be an apparent increase in complexity of design. However, design tables have been presented to aid selection of member sizes.[2]

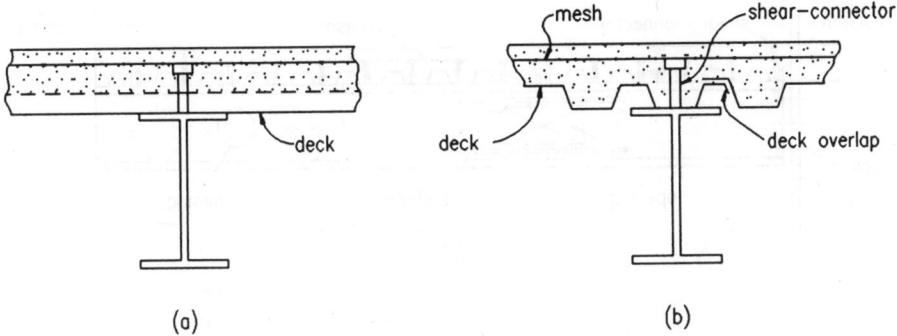

Fig. 21.3   Composite beams incorporating composite deck slabs: (a) deck perpendicular to beam, (b) deck parallel to beam

The normal method of designing simply-supported beams for strength is by plastic analysis of the cross-section. Full shear connection means that sufficient shear-connectors are provided to develop the full plastic capacity of the section. Beams designed for full shear connection result in the lightest beam size. Where fewer shear-connectors are provided (known as *partial shear connection*) the beam size is heavier, but the overall design may be more economic.

Partial shear connection is most attractive where the number of shear-connectors is placed in a standard pattern, such as one per deck trough or one per alternate trough where profiled decking is used. In such cases, the resistance of the shear-connectors is a fixed quantity irrespective of the size of the beam or slab.

Conventional elastic design of the section results in heavier beams than with plastic design because it is not possible to develop the full tensile resistance of the steel section. Designs based on elastic principles are to be used where the compressive elements of the section are non-compact or slender, as defined in BS 5950 Part 1. This mainly affects the design of continuous beams (see section 21.6.3).

21.3 Guidance on span-to-depth ratios

Beams are usually designed to be unpropped during construction. Therefore, the steel beam is sized first to support the self-weight of the slab before the concrete has gained adequate strength for composite action. Beams are assumed to be laterally restrained by the decking in cases where the decking crosses the beams (at an angle of at least 45° to the beam) and is directly attached to them. These beams can develop their full flexural capacity.

Where simply-supported unpropped composite beams are sized on the basis of their plastic capacity it is normally found that span-to-depth ratios can be in the

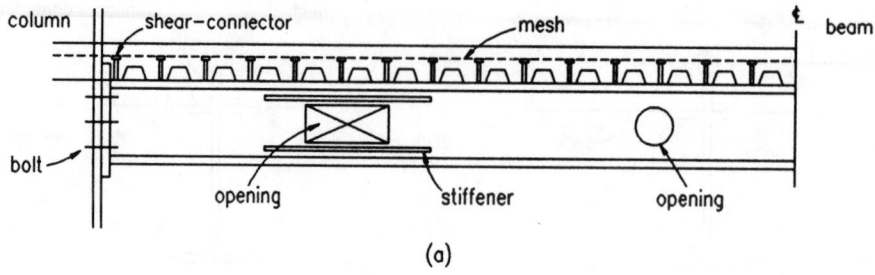

(a)

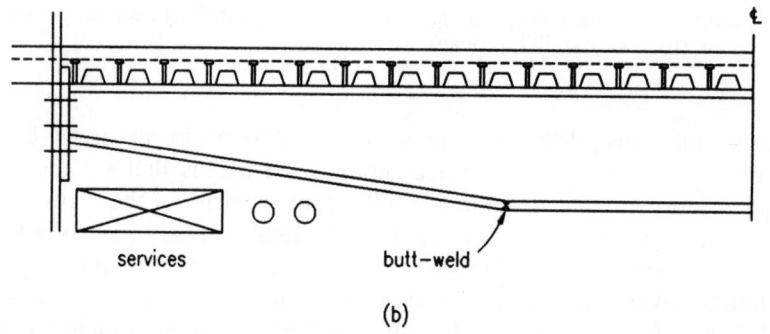

(b)

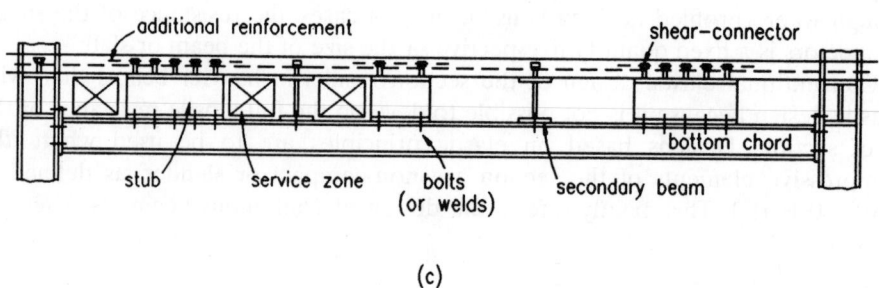

(c)

Fig. 21.4   Different methods of incorporating services within the structural depth

range of 18 to 22 before serviceability criteria influence the design. The 'depth' in these cases is defined as the overall depth of the beam and slab. S355 steel is often specified in preference to S275 steel in composite beam design because the stiffness of a composite beam is often three to four times that of the non-composite beam, justifying the use of higher working stresses.

The span-to-depth ratios of continuous composite beams are usually in the range of 22 to 25 for end spans and 25 to 30 for internal spans before serviceability criteria influence the design. Many continuous bridges are designed principally to satisfy the serviceability limit state.

21.4 Types of shear connection

The modern form of shear-connector is the welded headed stud ranging in diameter from 13 to 25 mm and from 65 to 125 mm in height. The most popular size is 19 mm diameter and 100 mm height before welding. When used with steel decking, studs are often welded through the decking using a hand tool connected via a control unit to a power generator. Each stud takes only a few seconds to weld in place. Alternatively, the studs can be welded directly to the steel beams in the factory and the decking butted up to or slotted over the studs.

There are, however, some limitations to through-deck welding: the top flange of the beam must not be painted, the galvanized steel should be less than around 1.25 mm thick, the deck should be clean and free of moisture, and there should be no gap between the underside of the decking and the top of the beam. The minimum flange thickness must not be less than the diameter of the stud divided by 2.5 (typically, 19/2.5 = 7.6 mm). The power generator needs 415 V electrical supply, and the maximum cable length between the weld gun and the power control units should be limited to around 70 m to avoid loss of power. Currently, only 13, 16 or 19 mm diameter studs can be through-deck welded on site.

Where precast concrete planks are used, the positions of the shear-connectors are usually such that they project through holes in the slab which are later filled with concrete. Alternatively, a gap is left between the ends of the units sitting on the top flange of the beam on to which the shear connectors are fixed. Reinforcement (usually in the form of looped bars) is provided around the shear-connectors.

There is a range of other forms of welded shear-connector, but most lack practical applications. The 'bar and hoop' and 'channel' welded shear-connectors have been use in bridge construction.[5] Shot-fired shear-connectors may be used in smaller building projects where site power might be a problem. All shear-connectors should be capable of resisting uplift forces; hence the use of headed rather than plain studs.

The number of shear-connectors placed along the beam is usually sufficient to develop the full flexural resistance of the member. However it is possible to reduce the number of shear-connectors in cases where the moment resistance exceeds the applied moment and the shear-connectors have adequate ductility (or deformation capacity). This is known as *partial shear connection* and is covered in section 21.7.4.

21.5 Span conditions

In buildings, composite beams are usually designed to be simply-supported, mainly to simplify the design process, to reduce the complexity of the beam-to-column connections, and to minimize the amount of slab reinforcement and shear-connectors that are needed to develop continuity at the ultimate limit state.

However, there are ways in which continuity can be readily introduced, in order to improve the stiffness of composite beams. Figure 21.5 shows how a typical connection detail at an internal column can be modified to develop continuity. The stub

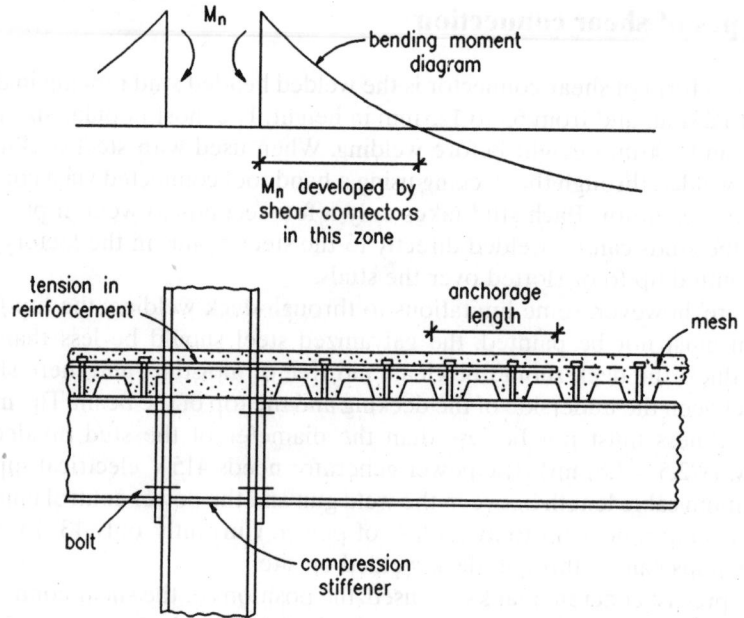

Fig. 21.5 Representation of conditions at internal column of continuous beam

girder system also utilizes continuity of the secondary members (see Fig. 21.4(c)). Other methods of continuous design are presented in References 3 and 6.

Continuous composite beams may be more economic than simply-supported beams where plastic hinge analysis of the continuous member is carried out, provided the section is plastic according to BS 5950: Part 1. However, where the lower flange or web of the beam is non-compact or slender in the negative (hogging) moment region, then elastic design must be used, both in terms of the distribution of moment along the beam, and also for analysis of the section. Lateral instability of the lower flange is an important design condition, although torsional restraint is developed by the web of the section and the concrete slab.[6]

In bridges, continuity is often desirable for serviceability reasons, both to reduce deflections, and to minimize cracking of the concrete slab, finishes and wearing surface in road bridges. Special features of composite construction appropriate to bridge design are covered in the publication by Johnson and Buckby[7] based on BS 5400: Part 5.[5]

21.6 Analysis of composite section

21.6.1 Elastic analysis

Elastic analysis is employed in establishing the serviceability performance of composite beams, or the resistance of beams subject to the effect of instability, for

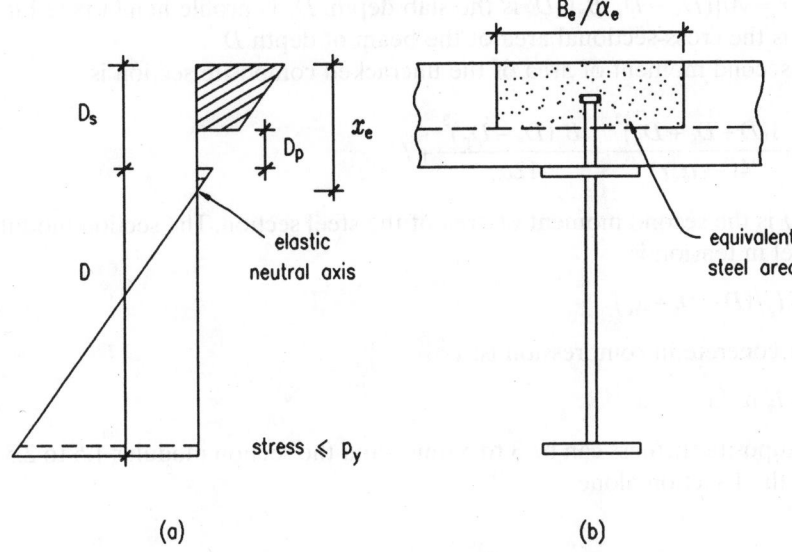

Fig. 21.6 Elastic behaviour of composite beam. (a) Elastic stress distribution. (b) Transformed section

example, in continuous construction, or in beams where the ductility of the shear connection is not adequate.

The important properties of the section are the section modulus and the second moment of area. First it is necessary to determine the centroid (elastic neutral axis) of the transformed section by expressing the area of concrete in steel units by dividing the concrete area within the effective breadth of the slab, B_e, by an appropriate modular ratio (ratio of the elastic modulus of steel to concrete).

In unpropped construction, account is taken of the stresses induced in the non-composite section as well as the stresses in the composite section. In elastic analysis, therefore, the order of loading is important. For elastic conditions to hold, extreme fibre stresses are kept below their design values, and slip at the interface between the concrete and steel should be negligible.

The elastic section properties are evaluated from the transformed section as in Fig. 21.6. The term α_e is the modular ratio. The area of concrete within the profile depth is ignored (this is conservative where the decking troughs lie parallel to the beam). The concrete can usually be assumed to be uncracked under positive moment.

The elastic neutral axis depth, x_e, below the upper surface of the slab is determined from the formula:

$$x_e = \frac{\dfrac{D_s - D_p}{2} + \alpha_e r\left(\dfrac{D}{2} + D_s\right)}{(1 + \alpha_e r)} \qquad (21.1)$$

where $r = A/[(D_s - D_p)B_e]$, D_s is the slab depth, D_p is profile height (see Fig. 21.6) and A is the cross-sectional area of the beam of depth D.

The second moment of area of the uncracked composite section is:

$$I_c = \frac{A(D+D_s+D_p)^2}{4(1+\alpha_e r)} + \frac{B_e(D_s-D_p)^3}{12\alpha_e} + I \tag{21.2}$$

where I is the second moment of area of the steel section. The section modulus for the steel in tension is:

$$Z_t = I_c/(D+D_s-x_e) \tag{21.3}$$

and for concrete in compression is:

$$Z_e = I_c\,\alpha_e/x_e \tag{21.4}$$

The composite stiffness can be 3 to 5 times, and the section modulus 1.5 to 2.5 times that of the I-section alone.

21.6.2 Plastic analysis

The ultimate bending resistance of a composite section is determined from its plastic resistance. It is assumed that the strains across the section are sufficiently high that the steel stresses are at yield throughout the section and that the concrete stresses are at their design strength. The plastic stress blocks are therefore rectangular, as opposed to linear in elastic design.

The plastic moment resistance of the section is independent of the order of loading (i.e. propped or unpropped construction) and is compared to the moment resulting from the total factored loading using the load factors in BS 5950: Part 1.

The plastic neutral axis of the composite section is evaluated assuming stresses of p_y in the steel and $0.45f_{cu}$ in the concrete. The tensile resistance of the steel is therefore $R_s = p_y A$, where A is the cross-sectional area of the beam. The compressive resistance of the concrete slab depends on the orientation of the decking. Where the decking crosses the beams the depth of concrete contributing to the compressive resistance is $D_s - D_p$ (Fig. 21.6(a)). Clearly, D_p is zero in a solid slab. Where the decking runs parallel to the beams (Fig. 21.6(b)), then the total cross-sectional area of the concrete is used. Taking the first case:

$$R_c = 0.45f_{cu}(D_s - D_p)B_e \tag{21.5}$$

where B_e is the effective breadth of the slab considered in section 21.7.1.

Three cases of plastic neutral axis depth x_p (measured from the upper surface of the slab) exist. These are presented in Fig. 21.7. It is not necessary to calculate x_p explicitly if the following formulae for the plastic moment resistance of I-section beams subject to positive (sagging) moment are used. The value R_w is the axial resistance of the web and R_f is the axial resistance of one steel flange (the section is

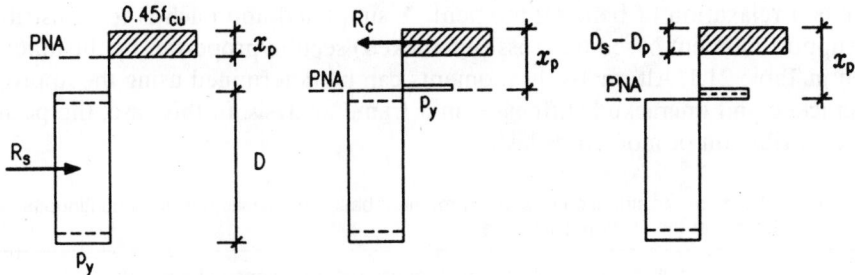

Fig. 21.7 Plastic analysis of composite section under positive (sagging) moment (PNA: plastic neutral axis)

assumed to be symmetrical). The top flange is considered to be fully restrained by the concrete slab.

The moment resistance, M_{pc}, of the composite beam is given by:

Case 1: $R_c > R_s$ (plastic neutral axis lies in concrete slab):

$$M_{pc} = R_s\left[\frac{D}{2} + D_s - \frac{R_s}{R_c}\left(\frac{D_s - D_p}{2}\right)\right] \tag{21.6}$$

Case 2: $R_s > R_c > R_w$ (plastic neutral axis lies in steel flange):

$$M_{pc} = R_s\frac{D}{2} + R_c\left(\frac{D_s + D_p}{2}\right) - \frac{(R_s - R_c)^2}{R_f}\frac{T}{4} \tag{21.7}$$

NB the last term in this expression is generally small *(T* is the flange thickness).

Case 3: $R_c < R_w$ (plastic neutral axis lies in web):

$$M_{pc} = M_s + R_c\left(\frac{D_s + D_p + D}{2}\right) - \frac{R_c^2}{R_w}\frac{D}{4} \tag{21.8}$$

where M_s is the plastic moment resistance of the steel section alone. This formula assumes that the web is compact i.e. not subject to the effects of local buckling. For this to be true the depth of the web in compression should not exceed $78t\varepsilon$, where *t* is the web thickness (ε is defined in BS 5950: Part 1). If the web is non-compact, a method of determining the moment resistance of the section is given in BS 5950: Part 3, Appendix B.[1]

21.6.3 Continuous beams

Bending moments in continuous composite beams can be evaluated from elastic global analysis. However, these result in an overestimate of moments at the supports because cracking of the concrete reduces the stiffness of the section and

permits a relaxation of bending moment. A simplified approach is to redistribute the support moment based on gross (uncracked) section properties by the amounts given in Table 21.1. Alternatively, moments can be determined using the appropriate cracked and uncracked stiffnesses in a frame analysis. In this case, the permitted redistribution of moment is less.

Table 21.1 Maximum redistribution of support moment based on elastic design of continuous beams at the ultimate limit state

	Classification of compression flange at supports				
				Plastic	
Assumed section properties at supports	Slender	Semi-compact	Compact	Generally	Beams (with nominal slab reinforcement)
Gross uncracked	10%	20%	30%	40%	50%
Cracked	0%	10%	20%	30%	30%

The section classification is expressed in terms of the proportions of the compression (lower) flange at internal supports. This determines the permitted redistribution of moment. A special category of plastic section is introduced where the section is of uniform shape throughout and nominal reinforcement is placed in the slab which does not contribute to the bending resistance of the beam. In this case the maximum redistribution of moment under uniform loading is increased to 50%. A simplified elastic approach is to use the design moments in Table 21.2 assuming that:

Table 21.2 Moment coefficients (multiplied by free moment of $WL/8$) for elastic design of continuous beams

		Classification of compression flange at supports				
					Plastic	
Location		Slender	Semi-compact	Compact	Generally	Beams (with nominal slab reinforcement)
Middle of end span	2 spans	0.71	0.71	0.71	0.75	0.79
	3 or more spans	0.80	0.80	0.80	0.80	0.82
First internal support	2 spans	0.91	0.81	0.71	0.61	0.50
	3 or more spans	0.86	0.76	0.67	0.57	0.48
Middle of internal spans		0.65	0.65	0.65	0.65	0.67
Internal supports (except first)		0.75	0.67	0.58	0.50	0.42
Redistribution		10%	20%	30%	40%	50%

(1) the unfactored imposed load does not exceed twice the unfactored dead load;
(2) the load is uniformly distributed;
(3) end spans do not exceed 115% of the length of the adjacent span;
(4) adjacent spans do not differ in length by more than 25% of the longer span.

An alternative to the elastic approach is plastic hinge analysis of plastic sections. Conditions on the use of plastic hinge analysis are presented in BS 5950: Part 3[1] and Eurocode 4 (draft).[8] However, large redistributions of moment may adversely affect serviceability behaviour (see section 21.7.8).

The ultimate load resistance of a continuous beam under positive (sagging) moment is determined as for a simply-supported beam. The effective breadth of the slab is based on the effective span of the beam under positive moment (see section 21.7.1). The number of shear-connectors contributing to the positive moment capacity is ascertained knowing the point of contraflexure.

The negative (hogging) moment resistance of a continuous beam or cantilever should be based on the steel section together with any properly anchored tension reinforcement within the effective breadth of the slab. This poses problems at edge columns, where it may be prudent to neglect the effect of the reinforcement unless particular measures are taken to provide this anchorage. The behaviour of a continuous beam is represented in Fig. 21.5.

The negative moment resistance is evaluated from plastic analysis of the section:

Case 1: $R_r < R_w$ (plastic neutral axis lies in web):

$$M_{nc} = M_s + R_s \left(\frac{D}{2} + D_r \right) - \frac{R_q^2}{R_w} \frac{D}{4} \qquad (21.9)$$

where R_r is the tensile resistance of the reinforcement over width B_e, R_q is the capacity of the shear-connectors between the point of contraflexure and the point of maximum negative moment (see section 21.7.3), and D_r is the height of the reinforcement above the top of the beam.

Case 2: $R_r > R_w$ (plastic neutral axis lies in flange):

$$M_{nc} = R_s \frac{D}{2} + R_r D_r - \frac{(R_s - R_r)^2}{R_f} \frac{T}{4} \qquad (21.10)$$

NB the last term in this expression is generally small.

The formulae assume that the web and lower flange are compact i.e. not subject to the effects of local buckling. The limiting depth of the web in compression is $78t\varepsilon$ (where ε is defined in Chapter 2) and the limiting breath: thickness ratio of the flange is defined in Table 11 of BS 5950: Part 1.

If these limiting slendernesses are exceeded then the section is designed elastically – often the situation in bridge design. The appropriate effective breadth of slab is used because of the sensitivity of the position of the elastic neutral axis and hence the zone of the web in compression to the tensile force transferred by the reinforcement. The elastic section properties are determined on the assumption that the concrete is cracked and does not contribute to the resistance of the section.

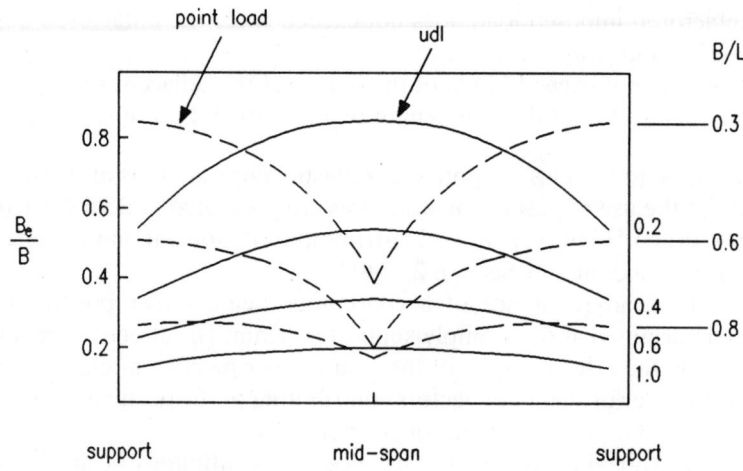

Fig. 21.8   Variation of effective breadth along beam and with loading

21.7 Basic design

21.7.1 Effective breadths

The structural system of a composite floor or bridge deck is essentially a series of parallel T beams with wide thin flanges. In such a system the contribution of the concrete flange in compression is limited because of the influence of 'shear lag'. The change in longitudinal stress is associated with in-plane shear strains in the flanges.

The ratio of the effective breadth of the slab to the actual breadth (B_e/B) is a function of the type of loading, the support conditions and the cross-section under consideration as illustrated in Fig. 21.8. The effective breadth of slab is therefore not a precise figure but approximations are justified. A common approach in plastic design is to consider the effective breadth as a proportion (typically 20%–33%) of the beam span. This is because the conditions at failure are different from the elastic conditions used in determining the data in Fig. 21.8, and the plastic bending capacity of a composite section is relatively insensitive to the precise value of effective breadth used.

Eurocode 4 (in its ENV or pre-norm version)[8] and BS 5950: Part 3[1] define the effective breadth as (span/4) (half on each side of the beam) but not exceeding the actual slab breadth considered to act with each beam. Where profiled decking spans in the same direction as the beams, as in Fig. 21.3(b), allowance is made for the combined flexural action of the composite slab and the composite beam by limiting the effective breadth to 80% of the actual breadth.

In building design, the same effective breadth is used for section analysis at both the ultimate and serviceability limit states. In bridge design to BS 5400: Part 5,[5] tabular data of effective breadths are given for elastic design at the serviceability

Table 21.3 Modular ratios (α_e) of steel of concrete

Type of concrete	Duration of loading		
	Short-term	Long-term	'Office' loading
Normal weight	6	18	10
Lightweight (density > 1750 kg/m³)	10	25	15

limit state. If plastic design is appropriate, the effective breadth is modified to (span/3).[5]

In the design of continuous beams, the effective breadth of the slab may be based conservatively on the effective span of the beam subject to positive or negative moment. For positive moment, the effective breadth is 0.25 times 0.7 span, and for negative moment, the effective breadth is 0.25 times 0.25 times the sum of the adjacent spans. These effective breadths reduce to 0.175 span and 0.125 span respectively for positive and negative moment regions of equal-span beams.

21.7.2 Modular ratio

The modular ratio is the ratio of the elastic modulus of steel to the creep-modified modulus of concrete, which depends on the duration of the load. The short- and long-term modular ratios given in Table 21.3 may be used for all grades of concrete. The effective modular ratio used in design should be related to the proportions of the loading that are considered to be of short- and long-term duration. Typical values used for office buildings are 10 for normal weight and 15 for lightweight concrete.

21.7.3 Shear connection

The shear resistance of shear-connectors is established by the push-out test, a standard test using a solid slab. A typical load–slip curve for a welded stud is shown in Fig. 21.9. The loading portion can be assumed to follow an empirical curve.[9] The strength plateau is reached at a slip of 2–3 mm.

The shear resistance of shear-connectors is a function of the concrete strength, connector type and the weld, related to the diameter of the connector. The purpose of the head of the stud is to prevent uplift. The common diameter of stud which can be welded easily on site is 19 mm, supplied in 75 mm, 100 mm or 125 mm heights. The material properties, before forming, are typically:

Ultimate tensile strength   450 N/mm²
Elongation at failure   15%

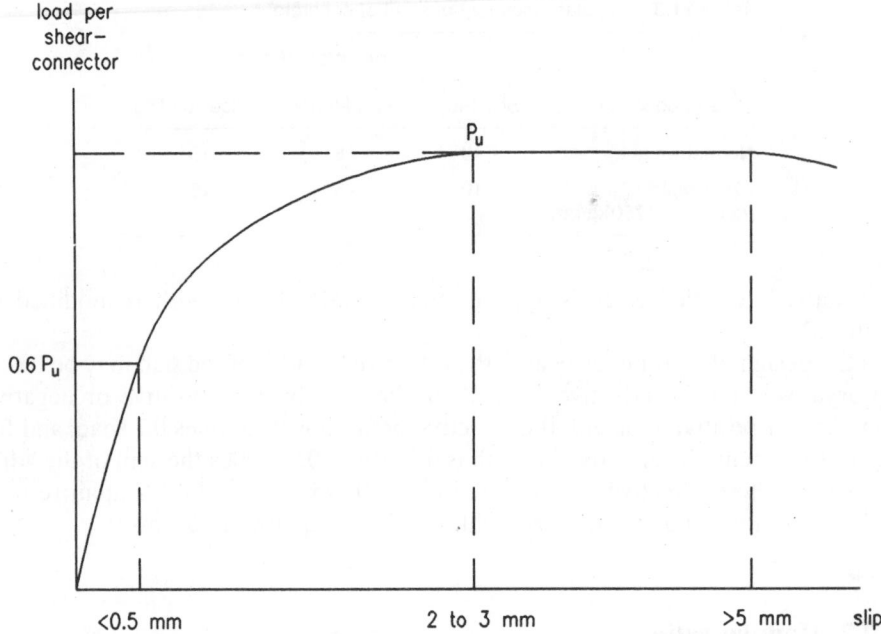

Fig. 21.9 Load–slip relationship for ductile welded shear-connector

Higher tensile strengths (495 N/mm^2) are required in bridge design.[5] Nevertheless, the 'push-out' strength of the shear-connectors is relatively insensitive to the strength of the steel because failure is usually one of the concrete crushing for concrete grades less than 40. Also the weld collar around the base of the shear-connector contributes to increased shear resistance.

The modern method of attaching studs in composite buildings is by through-deck welding. An example of this is shown in Fig. 21.10. The common in situ method of checking the adequacy of the weld is the bend-test, a reasonably easy method of quality control which should be carried out on a proportion of studs (say 1 in 50) and the first 2 to 3 after start up.

Other forms of shear-connector such as the shot-fired connector have been developed (Fig. 21.11). The strength of these types is controlled by the size of the pins used. Typical strengths are 30%–40% of the strengths of welded shear-connectors, but they demonstrate greater ductility.

The static resistances of stud shear-connectors are given in Table 21.4, taken from BS 5400: Part 5 and also incorporated in BS 5950: Part 3. As-welded heights are some 5 mm less than the nominal heights for through-deck welding; for studs welded directly to beams the length after welding (LAW) is taken as the nominal height. The strengths of shear-connectors in structural lightweight concrete (density >1750 kg/m^2) are taken as 90% of these values.

In Eurocode 4,[8] the approach is slightly different. Empirical formulae are given

Fig. 21.10   Welding of shear-connector through steel decking to a beam

based on two failure modes: failure of the concrete and failure of the steel. The
upper bound strength is given by shear failure of the shank and therefore there is
apparently little advantage in using high-strength concrete.

In BS 5950: Part 3 and Eurocode 4, the design resistance of the shear-connectors
is taken as 80% of the nominal static strength. Although this may broadly be considered to be a material factor applied to the material strength, it is, more correctly,
a factor to ensure that the criteria for plastic design are met (see below). The design
resistance of shear-connectors in negative (hogging) moment regions is conservatively taken as 60% of the nominal resistance. In BS 5400: Part 5, an additional
material factor of 1.1 is introduced to further reduce the design resistance of the
shear connectors.

Fig. 21.11   Shot-fired shear-connector

Table 21.4   Characteristic resistances of headed stud shear-connectors in normal weight concrete

Dimensions of stud shear-connectors (mm)			Characteristic strength of concrete (N/mm²)			
Diameter	Nominal height	As-welded height	25	30	35	40
25	100	95	146	154	161	168
22	100	95	119	126	132	139
19	100	95	95	100	104	109
19	75	70	82	87	91	96
16	75	70	70	74	78	82
13	65	60	44	47	49	52

For concrete of characteristic strength greater than 40 N/mm² use the values for 40 N/mm²
For connectors of heights greater than tabulated use the values for the greatest height tabulated

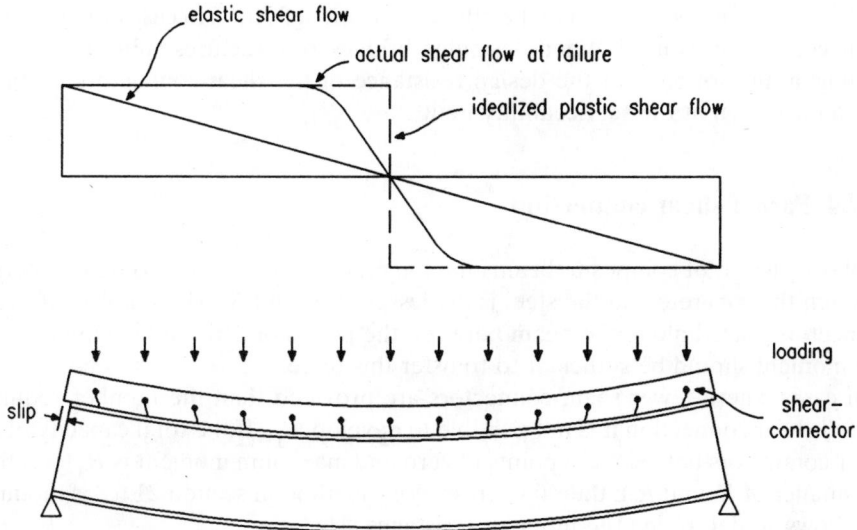

Fig. 21.12   Idealization of forces transferred between concrete and steel

In plastic design, it is important to ensure that the shear-connectors display adequate ductility. It may be expected that shear-connectors maintain their design resistances at displacements of up to 5 mm. A possible exception is where concrete strengths exceed C 40, as the form of failure may be more brittle.

For beams subject to uniform load, the degree of shear connection that is provided by uniformly-spaced shear-connectors (defined in section 21.7.4) reduces more rapidly than the applied moment away from the point of maximum moment. To ensure that the shear connection is adequate at all points along the beam, the design resistance of the shear-connectors is taken as 80% of their static resistance. This also partly ensures that flexural failure will occur before shear failure. For beams subject to point loads, it is necessary to design for the appropriate shear connection at each major load point.

In a simple composite beam subject to uniform load the elastic shear flow defining the shear transfer between the slab and the beam is linear, increasing to a maximum at the ends of the beam. Beyond the elastic limit of the connectors there is a transfer of force among the shear-connectors, such that, at failure, each of the shear-connectors is assumed to be subject to equal force, as shown in Fig. 21.12. This is consistent with a relatively high slip between the concrete and the steel. The slip increases as the beam span increases and the degree of shear connection reduces. For this reason BS 5400: Part 5[5] requires that shear-connectors in bridges are spaced in accordance with elastic theory. In building design, shear-connectors are usually spaced uniformly along the beam when the beam is subject to uniform load.

No serviceability limit is put on the force in the shear-connectors, despite the fact that consideration of the elastic shear flow suggests that such forces can be high at

working load. This is reflected in the effect of slip on deflection in cases where partial shear connection is used. When designing bridges[5] or structures subject to fatigue loading, a limit of 55% of the design resistance of the shear-connectors is appropriate for design at the serviceability limit state.

21.7.4 Partial shear connection

In plastic design of composite beams the longitudinal shear force to be transferred between the concrete and the steel is the lesser of R_c and R_s. The number of shear-connectors placed along the beam between the points of zero and maximum positive moment should be sufficient to transfer this force.

In cases where fewer shear-connectors are provided than the number required for full shear connection it is not possible to develop M_{pc}. If the total capacity of the shear-connectors between the points of zero and maximum moment is R_q (less than the smaller of R_s and R_c), then the stress block method in section 21.6.2 is modified as follows, to determine the moment resistance, M_c:

Case 4: $R_q > R_w$ (plastic neutral axis lies in flange):

$$M_c = R_s \frac{D}{2} + R_q\left[D_s - \frac{R_q}{R_c}\left(\frac{D_s - D_p}{2}\right)\right] - \frac{(R_s - R_q)^2}{R_f}\frac{T}{4} \tag{21.11}$$

NB the last term in this expression is generally small.

Case 5: $R_q < R_w$ (plastic neutral axis lies in web):

$$M_c = M_s + R_q\left[\frac{D}{2} + D_s - \frac{R_q}{R_c}\left(\frac{D_s - D_p}{2}\right)\right] - \frac{R_q^2}{R_w}\frac{D}{4} \tag{21.12}$$

The formulae are obtained by replacing R_c by R_q and re-evaluating the neutral axis position. The method is similar to that used in the American method of plastic design,[10] which predicts a non-linear increase of moment capacity with degree of shear connection K defined as:

$$K = R_q/R_s \quad \text{for } R_s < R_c$$
$$\text{or } K = R_q/R_c \quad \text{for } R_c < R_s$$

An alternative approach, which has proved attractive, is to define the moment resistance in terms of a simple linear interaction of the form:

$$M_c = M_s + K(M_{pc} - M_s) \tag{21.13}$$

The *stress block* and *linear interaction* methods are presented in Fig. 21.13 for a typical beam. It can be seen that there is a significant benefit in the stress block method in the important range of $K = 0.5$ to 0.7.

In using methods based on partial shear connection a lower limit for K of 0.4 is specified. This is to overcome any adverse effects arising from the limited deformation capacity of the shear-connectors.

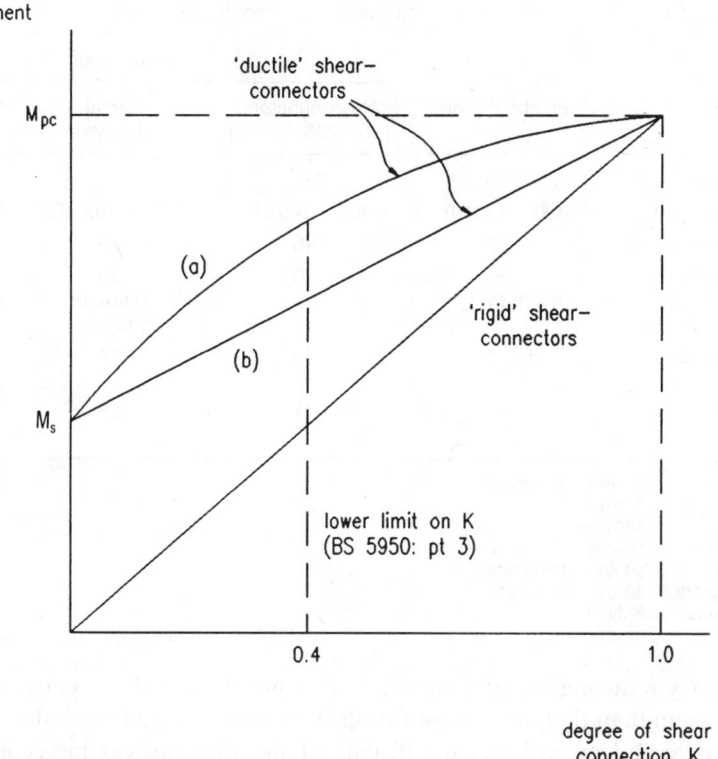

Fig. 21.13 Interaction between moment capacity and degree of shear connection. (a) Stress block method. (b) Linear interaction method

In BS 5950: Part 3, the limiting degree of shear connection increases with beam span (L in metres) such that:

$$K \leq (L-6)/10 \geq 0.4 \tag{21.14}$$

This formula means that beams longer than 16 m span should be designed for full shear connection, and beams of up to 10 m span designed for not less than 40% shear connection, with a linear transition between the two cases. Partial shear connection is also not permitted for beams subject to heavy off-centre point loads, except where checks are made, as below.

A further requirement is that the degree of shear connection should be adequate at all points along the beam length. For a beam subject to point loads, it follows that the shear-connectors should be distributed in accordance with the shear force diagram.

Comparison of the method of partial shear connection with other methods of design is presented in Table 21.5. Partial shear connection can result in overall

Table 21.5 Comparison of designs of simply-supported composite beams

Beam data	Elastic design	Plastic design		
		No connectors (BS 5950: Part 1)	Partial connection	Full shear connection
Full depth (mm)	536	536	435	435
Beam size (mm)	406 × 140 UB	406 × 140 UB	305 × 102 UB	305 × 102 UB
Beam weight (kg/m)	39	46	33	28
Number of 19 mm diameter shear-connectors	50 (every trough)	0	25 (alternate troughs)	50 (every trough)
Imposed load deflection (mm)	7	19	14	13
Self-weight deflection (mm)	14	11	25	30

Beam span:	7.5 m (unpropped)
Beam spacing:	3.0 m
Slab depth:	130 mm
Deck height:	50 mm
Steel grade:	50 ($p_y = 355\,\text{N/mm}^2$)
Concrete grade:	30 (normal weight)
Imposed load:	5 kN/m²

economy by reducing the number of shear-connectors at the expense of a slightly heavier beam than that needed for full shear connection. Elastic design is relatively conservative and necessitates the placing of shear-connectors in accordance with the elastic shear flow. Deflections are calculated using the guidance in section 21.7.8.

21.7.5 Influence of deck shape on shear connection

The efficiency of the shear connection between the composite slab and the composite beam may be reduced because of the shape of the deck profile. This is analogous to the design of haunched slabs where the strength of the shear-connectors is strongly dependent on the area of concrete around them. Typically, there should be a 45° projection from the base of the connector to the core of the solid slab to transfer shear smoothly in the concrete without local cracking.

The model for the action of a shear-connector placed in the trough of a deck profile is shown in Fig. 21.14. For comparison, also shown is the behaviour of a connector in a solid slab. Effectively, the centre of resistance in the former case moves towards the head of the stud and the couple created is partly resisted by bending of the stud but also by tensile and compressive forces in the concrete, encouraging concrete cracking, and consequently the strength of the stud is reduced.

A number of tests on different stud heights and profile shapes have been performed. The following formula has been adopted by most standards worldwide. The

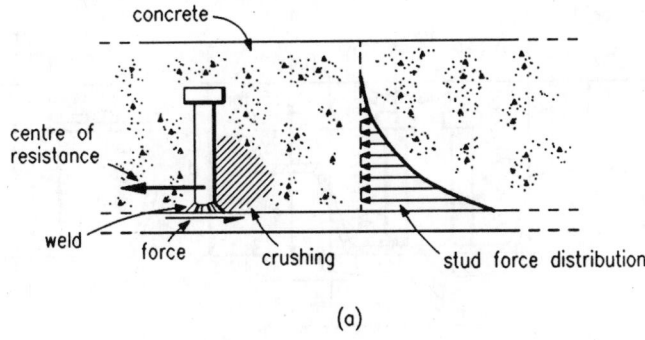

concrete

centre of resistance

weld

force   crushing

stud force distribution

(a)

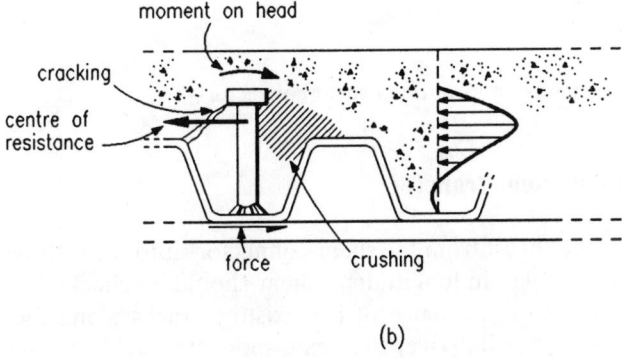

moment on head

cracking

centre of resistance

force   crushing

(b)

Fig. 21.14   Model of behaviour of shear-connector. (a) Shear-connector in plain slab. (b) Shear-connector fixed through profile sheeting

strength reduction factor (relative to a solid slab) for the case where the decking crosses the beams is:

$$r_\mathrm{p} = \frac{0.85}{\sqrt{n}} \frac{b_\mathrm{a}}{D_\mathrm{p}} \left(\frac{h - D_\mathrm{p}}{D_\mathrm{p}} \right) \begin{array}{l} \leq 1.0 \quad \text{for } n = 1 \\ \leq 0.8 \quad \text{for } n = 2 \end{array} \tag{21.15}$$

where b_a is the average width of the trough, h is the stud height, and n is the number of studs per trough ($n < 3$). The limit for pairs of studs is given in BS 5950 Part 3 and takes account of less ductile behaviour when $n > 1$. This formula does not apply in cases where the shear-connector does not project at least 35 mm above the top of the deck. A further limit is that $h \ngtr 2D_\mathrm{p}$ in evaluating r_p.

Where the decking is placed parallel to the beams no reduction is made for the number of connectors but the constant in Equation (21.15) is reduced to 0.6 (instead of 0.85). No reduction is made in the second case when $b_\mathrm{a}/D_\mathrm{p} > 1.5$.

Further geometric limits on the placing of the shear-connectors are presented in Fig. 21.15. The longitudinal spacing of the shear-connectors is limited to a maximum of 600 mm and a minimum of 100 mm.

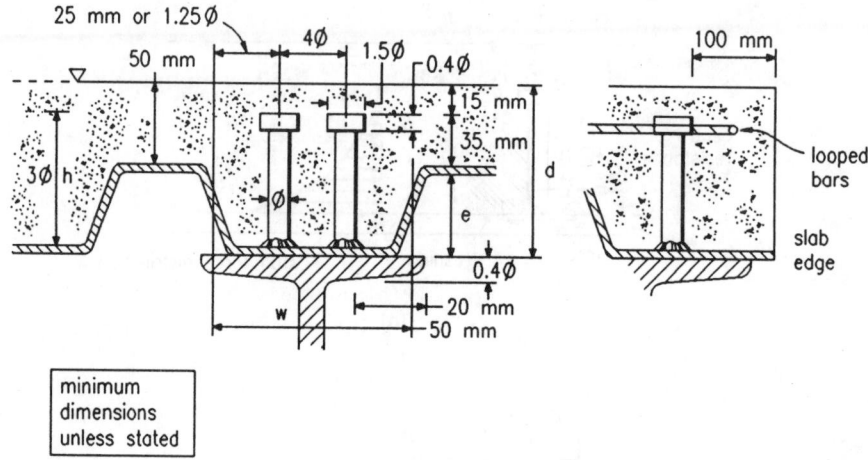

Fig. 21.15   Geometric limits on location of shear-connectors

21.7.6 Longitudinal shear transfer

In order to transfer the thrust from the shear-connectors into the slab, without split-ting, the strength of the slab in longitudinal shear should be checked. The strength is further influenced by the presence of pre-existing cracks along the beam as a result of the bending of the slab over the beam support.

The design recommendations used to check the resistance of the slab to longi-tudinal shear are based on research into the behaviour of reinforced concrete slabs. The design longitudinal shear stress which can be transferred is taken as $0.9\,\text{N/mm}^2$ for normal weight and $0.7\,\text{N/mm}^2$ for lightweight concrete; this strength is relatively insensitive to the grade of concrete.[11]

It is first necessary to establish potential planes of longitudinal shear failure around the shear-connectors. Typical cases are shown in Fig. 21.16. The top rein-forcement is assumed to develop its full tensile resistance, and is resisted by an equal and opposite compressive force close to the base of the shear-connector. Both top and bottom reinforcement play an important role in preventing splitting of the concrete.

The shear resistance per unit length of the beam which is equated to the shear force transferred through each shear plane (in the case of normal weight concrete) is:

$$V = 0.9L_s + 0.7A_r f_y \le 0.15\,L_s f_{cu} \tag{21.16}$$

where L_s is the length of each shear plane considered on a typical cross-section, which may be taken as the mean slab depth in Fig. 21.16(a) or the minimum depth in Fig. 21.16(b). The total area of reinforcement (per unit length) crossing the shear plane is A_r. For an internal beam, the slab shear resistance is therefore $2V$.

The effect of the decking in resisting longitudinal shear is considerable. Where

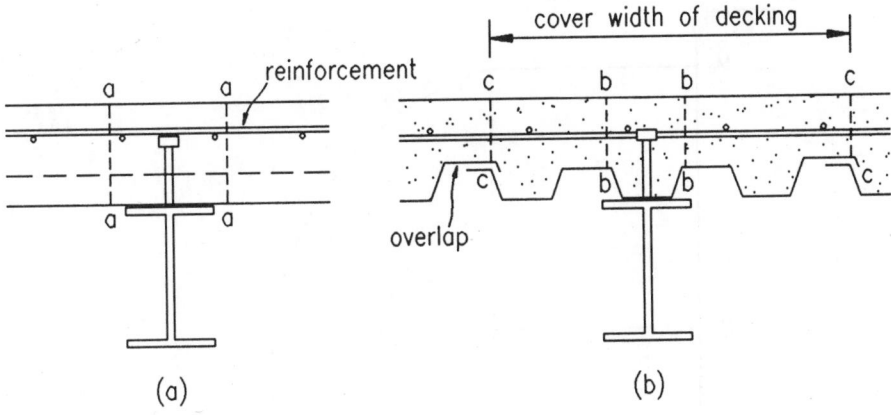

Fig. 21.16   Potential failure planes through slab in longitudinal shear

the decking is continuous over the beams, as in Fig. 21.3(a), or rigidly attached by shear-connectors, there is no test evidence of the splitting mode of failure. It is assumed, therefore, that the deck is able to provide an important role as transverse reinforcement. The term $A_t f_y$ may be enhanced to include the contribution of the decking, although it is necessary to ensure that the ends of the deck (butt joints) are properly welded by shear-connectors. In this case, the anchorage provided by each weld may be taken as $4\phi t_s p_y$, where ϕ is the stud diameter, t_s is the sheet thickness, and p_y is the strength of the sheet steel. Dividing by the stud spacing gives the equivalent shear resistance per unit length provided by the decking.

Where the decking is laid parallel to the beams, longitudinal shear failure can occur through the sheet-to-sheet overlap close to the line of studs. Failure is assessed assuming the shear force transferred diminishes linearly across the slab to zero at a distance of $B_e/2$. Therefore, the effective shear force that is transferred across the longitudinal overlaps is (1 – cover width of decking/effective breadth) × shear force. The shear resistance at overlaps excludes the effect of the decking.

To prevent splitting of the slab at edge beams, the distance between the line of shear-connectors and the edge of the slab should not be less than 100 mm. When this distance is between 100 mm and 300 mm, additional reinforcement in the form of U bars located below the head of the shear-connectors is to be provided.[12] No additional reinforcement (other than that for transverse reinforcement) need be provided where the edge of the slab is more than 300 mm from the shear-connectors.

21.7.7 Interaction of shear and moment in composite beams

Vertical shear can cause a reduction in the plastic moment resistance of a composite beam where high moment and shear co-exist at the same position along the beam

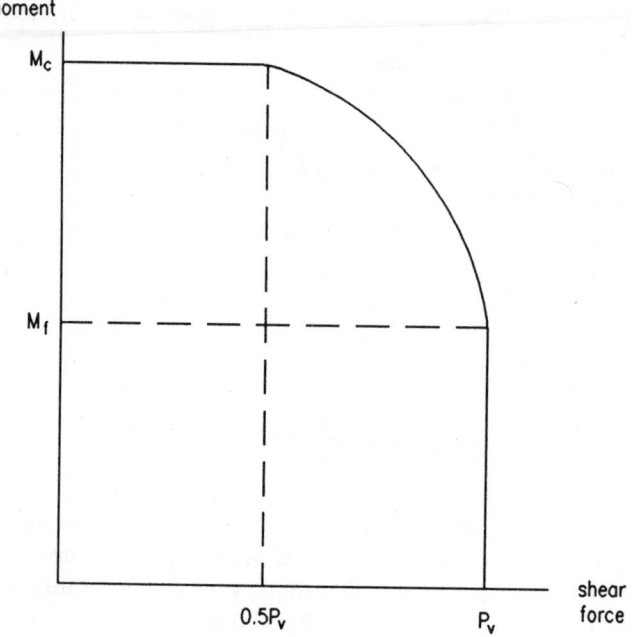

Fig. 21.17   Interaction between moment and shear

(i.e. the beam is subject to one or two point loads). Where the shear force F_v exceeds $0.5P_v$ (where P_v is the lesser of the shear resistance and the shear buckling resistance, determined from Part 1 of BS 5950), the reduced moment resistance is determined from:

$$M_{cv} = M_c - (M_c - M_f)(2F_v/P_v - 1)^2$$

(21.17)

where M_c is the plastic moment resistance of the composite section, and M_f is the plastic moment resistance of the composite section having deducted the shear area (i.e. the web of the section).

The interaction is presented diagrammatically in Fig. 21.17. A quadratic relationship has been used, as opposed to the linear relationship in BS 5950: Part 1, because of its better agreement with test data, and because of the need for greater economy in composite sections which are often more highly stressed in shear than non-composite beams.

21.7.8 Deflections

Deflection limits for beams are specified in BS 5950: Part 1. Composite beams are, by their nature, shallower than non-composite beams and often are used in structures where long spans would otherwise be uneconomic. As spans increase, so tra-

ditional deflection limits based on a proportion of the beam span may not be appropriate. The absolute deflection may also be important and pre-cambering may need to be considered for beams longer than 10 m.

Elastic section properties, as described in section 21.6.1, are used in establishing the deflection of composite beams. Uncracked section properties are considered to be appropriate for deflection calculations. The appropriate modular ratio is used, but it is usually found that the section properties are relatively insensitive to the precise value of modular ratio. The effective breadth of the slab is the same as that used in evaluating the bending resistance of the beam.

The deflection of a simple composite beam at working load, where partial shear connection is used, can be calculated from:[13]

$$\left.\begin{array}{ll} \delta'_c = \delta_c + 0.5(1-K)(\delta_s - \delta_c) & \text{for propped beams} \\ \delta'_c = \delta_c + 0.3(1-K)(\delta_s - \delta_c) & \text{for unpropped beams} \end{array}\right\} \qquad \textbf{(21.18)}$$

where δ_c and δ_s are the deflections of the composite and steel beam respectively at the appropriate serviceability load; K is the degree of shear connection used in the determining of the plastic strength of the beam (section 21.7.4). The difference between the coefficients in these two formulae arises from the different shear-connector forces and hence slip at serviceability loads in the two cases. These formulae are conservative with respect to other guidance.[10]

The effect of continuity in composite beams may be considered as follows. The imposed load deflection at mid-span of a continuous beam under uniform load or symmetric point loads may be determined from the approximate formula:

$$\delta_{cc} = \delta'_c(1 - 0.6(M_1 + M_2)/M_0) \qquad \textbf{(1.19)}$$

where δ'_c is the deflection of the simply-supported composite beam for the same loading conditions; M_0 is the maximum moment in a simply-supported beam subject to the same loads; M_1 and M_2 are the end moments at the adjacent supports of the span of the continuous beam under consideration.

To determine appropriate values of M_1 and M_2, an elastic global analysis is carried out using the flexural stiffness of the uncracked section.

For buildings of normal usage, these support moments are reduced to take into account the effect of pattern loading and concrete cracking. The redistribution of support moment under imposed load should be taken as the same as that used at the ultimate limit state (see Table 21.1), but not less than 30%.

For buildings subject to semi-permanent or variable loads (e.g. warehouses), there is a possibility of alternating plasticity under repeated loading leading to greater imposed load deflections. This also affects the design of continuous beams designed by plastic hinge analysis, where the effective redistribution of support moment exceeds 50%. In such cases a more detailed analysis should be carried out considering these effects (commonly referred to as 'shakedown') as follows:

(1) evaluate the support moments based on elastic analysis of the continuous beam under a first loading cycle of dead load and 80% imposed load (or 100% for semi-permanent load);

(2) evaluate the excess moment where the above support moment exceeds the plastic resistance of the section under negative moment;
(3) the net support moments based on elastic analysis of the continuous beam under imposed load are to be reduced by 30% (or 50% for semi-permanent loads) and further reduced by the above excess moment;
(4) these support moments are input into Equation (21.19) to determine the imposed load deflection.

21.7.9 Vibration

Shallower beams imply greater flexibility, and although the in-service performance of composite beams and floors in existing buildings is good, the designer may be concerned about the susceptibility of the structure to vibration-induced oscillations. The parameter commonly associated with this effect is the natural frequency of the floor or beams. The damping of the vibration by a bare steel–composite structure is often low. However, when the building is occupied, damping increases considerably.

The lower the natural frequency, the more the structure may respond dynamically to occupant-induced vibration. A limit of 4 Hz (cycles per second) is a commonly accepted lower bound to the natural frequency of each element of the structure. Clearly, vibrating machinery or external vibration effects pose particular problems and in such cases it is often necessary to isolate the source of the vibration.

In practice, the mass of the structure is normally such that the exciting force is very small in comparison, leading to the conclusion that long-span structures may respond less than light short-span structures. Guidance is given in Reference 14 and Chapter 12.

21.7.10 Shrinkage, cracking and temperature

It is not normally necessary to check crack widths in composite floors in heated buildings, even where the beams are designed as simply-supported, provided that the slab is reinforced as recommended in BS 5950: Part 4 or BS 8110 as appropriate. In such cases crack widths may be outside the limits given in BS 8110, but experience shows that no durability problems arise.

In other cases additional reinforcement over the beam supports may be required to control cracking, and the relevant clauses in BS 8110: Part 2[15] and BS 5400: Part 5[5] should be followed. This is particularly important where hard finishes are used.

Questions of long-term shrinkage and temperature-induced effects often arise in long-span continuous composite beams, as they cause additional negative (hogging) moments and deflections. In buildings these effects are generally neglected, but in bridges they can be important.[5,7]

The curvature of a composite section resulting from a free shrinkage (or temperature induced) strain ε_s in the slab is:

$$K_s = \frac{\varepsilon_s(D + D_s + D_p)A}{2(1 + \alpha_e r)I_c} \tag{21.20}$$

where I_c and r are defined in section 21.6.1 and α_e is the appropriate modular ratio for the duration of the action considered. The free shrinkage strain may be taken to vary between 100×10^{-6} in external applications and 300×10^{-6} in dry heated buildings. A creep reduction factor is used in BS 5400: Part 5[5] when considering shrinkage strains. This can reduce the effective strain by up to 50%. The central deflection of a simply-supported beam resulting from shrinkage strain is then $0.125K_sL^2$.

References to Chapter 21

1. British Standards Institution (1990) *Structural use of steelwork in building.* Part 3, Section 3.1: *Code of practice for design of composite beams.* BS 5950, BSI, London.
2. The Steel Construction Institute (SCI) (1989) *Design of Composite Slabs and Beams with Steel Decking.* SCI, Ascot, Berks.
3. Chien E.Y.L. & Ritchie J.K. (1984) *Design and Construction of Composite Floor Systems.* Canadian Institute of Steel Construction.
4. Lawson R.M. (1988) *Design for Openings in the Webs of Composite Beams.* The Steel Construction Institute, Ascot, Berks.
5. British Standards Institution (1979) *Steel, concrete and composite bridges.* Part 5: *Code of practice for design of composite bridges.* BS 5400, BSI, London.
6. Brett P.R., Nethercot D.A. & Owens G.W. (1987) Continuous construction in steel for roofs and composite floors. *The Structural Engineer*, **65A**, No. 10, Oct.
7. Johnson R.P. & Buckby R.J. (1986) *Composite Structures of Steel and Concrete, Vol. 2: Bridges*, 2nd edn. Collins.
8. British Standards Institution (1994) *Eurocode 4: Design of Composite steel and concrete structures. General rules and rules for buildings.* DD ENV 1994-1-1, BSI, London.
9. Yam L.C.P. & Chapman J.C. (1968) The inelastic behaviour of simply supported composite beams of steel and concrete. *Proc. Instn Civ. Engrs*, **41**, Dec., 651–83.
10. American Institute of Steel Construction (1986) *Manual of Steel Construction: Load and Resistance Factor Design.* AISC, Chicago.
11. Johnson R.P. (1975 & 1986) *Composite Structures of Steel and Concrete. Vol. 1: Beams. Vol. 2: Bridges*, 2nd edn. Granada.
12. Johnson R.P. & Oehlers D.J. (1982) Design for longitudinal shear in composite L beams. *Proc. Instn Civ. Engrs*, **73**, Part 2, March, 147–70.
13. Johnson R.P. & May I.M. (1975) Partial interaction design of composite beams. *The Structural Engineer*, **53**, No. 8, Aug., 305–11.

14. Wyatt T.A. (1989) *Design Guide on the Vibration of Floors.* The Steel Construction Institute (SCI)/CIRIA.
15. British Standards Institution (1985) *Structural use of concrete.* Part 2: *Code of practice for special circumstances.* BS 8110, BSI, London.
16. Lawson R.M., Mullet D.L. & Rackham J.W. (1997) *Design of Asymmetric Slimflor® Beams Using Deep Composite Decking.* The Steel Construction Institute, Ascot, Berks.

A series of worked examples follows which are relevant to Chapter 21.

The **Steel Construction** **Institute** Silwood Park, Ascot, Berks SL5 7QN 	Subject ***DESIGN OF SIMPLY*** ***SUPPORTED COMPOSITE*** ***BEAM***		Chapter ref. ***21***
	Design code ***BS 5950: Part 3***	Made by ***RML*** Checked by ***GWO***	Sheet no. ***1***

Problem

Design a composite floor with beams at 3 m centres spanning 12 m. The composite slab is 130 mm deep. The floor is to resist an imposed load of 5.0 kN/m², partition loading of 1.0 kN/m² and a ceiling load of 0.5 kN/m². The floor is to be unpropped during construction.

Deck:

Profile height	D_p	=	*50 mm*
Trough spacing		=	*300 mm*
Trough width (average)		=	*150 mm*

Shear connectors:

19 mm diameter

95 mm as-welded length

Concrete:

Compressive strength f_{cu} =   30 N/mm²

Density (lightweight concrete) is 1800 kg/m³ (dry)
(no extra allowance is made for wet weight – assumed to be included in construction load)

Loads:

Imposed load	=	*5.0 kN/m²*	
Partitions (imposed load)	=	*1.0 kN/m²*	
Ceiling	=	*0.5 kN/m²*	
Self weight (slab)	=	*2.00 kN/m²*	*see later*
Self weight (beam)	=	*0.67 kN/m²*	
Construction load *(temporary)*	=	*0.5 kN/m²*	

Finishes and other loads are neglected.

The Steel Construction Institute  Silwood Park, Ascot, Berks SL5 7QN	Subject **DESIGN OF SIMPLY SUPPORTED COMPOSITE BEAM**	Chapter ref. **21**
	Design code **BS 5950: Part 3**	Made by **RML**	Sheet no. **2**
		Checked by **GWO**	

Initial selection of beam size

Choose beam depth so that ratio of span/overall depth is about 20
Choose 457 × 191 × 67 kg/m beam steel grade S355
$p_y = 355 \, N/mm^2$ *as flange thickness is less than 16 mm*
Alternatively, use 'Design of Composite Slabs and Beams with   **SCI**
Steel Decking'.   **Publication**

Construction condition

Self weight of slab (symmetric deck)
Self weight = 105 × 1800 × 9.81 × 10⁻⁶ = 1.85 kN/m²
Self weight of deck is taken as 0.14 kN/m²
Total slab weight is 2.0 kN/m²

Design load in construction condition

$$= (1.6 \times 0.5 + 1.4 \times 2.0) \times 3.0 + 0.67 \times 1.4$$
$$= 11.74 \, kN/m$$

Design moment in construction condition

$$= 11.74 \times 12.0^2/8 = 211.0 \, kNm$$

Moment resistance of steel section

$$M_s = 1470 \times 10^3 \times 355 \times 10^{-6}$$
$$= 521.9 \, kNm > 211.0 \, kNm$$

(the beam is assumed to be laterally restrained by the decking)

Check stress due to self weight

$$Moment = (2.0 \times 3.0 + 0.67) \times 12^2/8$$
$$= 120.1 \, kNm$$

$$Bending \; stress = \frac{120.1 \times 10^6}{1300 \times 10^3} = 92 \, N/mm^2$$

Deflection of soffit of beam after construction

$$\delta = \frac{5}{384} \times \frac{(2.0 \times 3.0 + 0.67) \times 12000^4}{205 \times 10^3 \times 29400 \times 10^4}$$

		Subject	Chapter ref.
The Steel Construction Institute Silwood Park, Ascot, Berks SL5 7QN		*DESIGN OF SIMPLY SUPPORTED COMPOSITE BEAM*	*21*
		Design code **BS 5950: Part 3**	Made by **RML** Checked by **GWO**	Sheet no. **3**

Composite condition – ultimate load

Factored load on composite beam

$$w_u = (5.0 \times 1.6 + 1.0 \times 1.6 + 0.5 \times 1.4 + 2.0 \times 1.4) \times 3.0 + 0.67 \times 1.4$$
$$= 40.2 \, kN/m$$

Design moment on composite beam

$$M_u = \frac{40.2 \times 12^2}{8} = 723.6 \, kNm$$

Effective breadth of slab

$$B_e = 12000/4 = 3000 \, mm \not> b = 3000 \, mm$$

Resistance of slab in compression

$$R_c = 0.45 f_{cu} B_e (D_s - D_p)$$
$$= 0.45 \times 30 \times 3000 \times (130 - 50) \times 10^{-3}$$
$$= 3240 \, kN$$

Resistance of steel section in tension

$$R_s = A p_y$$
$$= 8540 \times 355 \times 10^{-3}$$
$$= 3032 \, kN$$

As $R_c > R_s$ *plastic neutral axis lies in concrete.*

Moment resistance of composite beam (equation 21.6) for full shear connection is:

$$M_{pc} = R_s \left\{ \frac{D}{2} + D_s - \frac{R_s}{R_c} \frac{(D_s - D_p)}{2} \right\} \quad \text{for } R_s \leq R_c$$
$$= 3032 \left\{ \frac{453.6}{2} + 130 - \frac{3032}{3240} \times \frac{80}{2} \right\} \times 10^{-3}$$
$$= 968.3 \, kNm > M_u$$

The Steel Construction Institute Silwood Park, Ascot, Berks SL5 7QN	Subject **DESIGN OF SIMPLY SUPPORTED COMPOSITE BEAM**	Chapter ref. *21*
	Design code *BS 5950: Part 3*	Made by *RML*	Sheet no. *4*
		Checked by *GWO*	

Design code *BS 5950: Part 3*	Made by *RML*	Sheet no. *4*
	Checked by *GWO*	

Degree of shear connection

Use two shear connectors per trough
(i.e. 2 per 300 mm spacing)

Number of shear connectors in half span $= \dfrac{6000}{150} = 40$

Resistance of shear connector:

  From Table 21.4, characteristic strength = 100 kN

  Design strength = 0.8 × 100 = 80 kN

  For lightweight concrete multiply by 0.9

  Design strength (LWC) = 0.9 × 80 = 72 kN

Reduction factor for profile shape (equation 21.15)

$$r = \frac{0.85}{\sqrt{2}} \times \frac{150}{50} \times \frac{(95-50)}{50} = 1.62 > 0.8$$

Therefore use r = 0.8 for pairs of shear connectors as required in BS 5950: Part 3 Clause 5.4.7.2.

Total resistance of shear connectors

$$R_q = 0.8 \times 72 \times 40 = 2304 \, kN$$

As $R_q < R_s$ and $R_q < R_c$

redesign beam for partial shear connection.

The Steel Construction Institute Silwood Park, Ascot, Berks SL5 7QN	Subject ***DESIGN OF SIMPLY SUPPORTED COMPOSITE BEAM***	Chapter ref. ***21***
	Design code *BS 5950: Part 3*	Made by ***RML***	Sheet no. ***5***
		Checked by ***GWO***	

Moment resistance of composite beam with partial shear connection (equation 21.11)

Tensile resistance of web

$$R_w = 8.5 \times (453.6 - 2 \times 12.7) \times 355 \times 10^{-3}$$
$$= 1292\,kN$$

As $R_q > R_w$, plastic neutral axis lies in steel flange

$$M_e = R_s\frac{D}{2} + R_q\left[D_s - \frac{R_q}{R_c}\frac{(D_s - D_p)}{2}\right] - \frac{(R_s - R_q)^2}{R_f}\frac{T}{4}$$

(normally the final term can be ignored)

$$M_c = 3032 \times \frac{453.6}{2} \times 10^{-3} + 2304 \times \left(130 - \frac{2304}{3240} \times \frac{80}{2}\right) \times 10^{-3}$$
$$- \frac{(3032 - 2304)^2}{(3032 - 1292) \times 0.5} \times \frac{12.7}{4} \times 10^{-3}$$
$$= 687.6 + 234.0 - 1.9 = 919.7\,kNm > 723.6\,kNm$$

Degree of shear connection provided

$$K = R_q/R_s = 2304/3032 = 0.76$$

Check possibility of providing fewer shear connections. Limit on degree of shear connection is:

$$K \geq \frac{L - 6}{10}; \quad but \quad K \geq 0.4$$
For L =   *12 m* $K \geq 0.6$

One shear connector every trough (i.e. 300 mm spacing) corresponds to K = 0.47 < 0.6

which does not provide adequate shear connection.

The Steel Construction Institute  Silwood Park, Ascot, Berks SL5 7QN	Subject **DESIGN OF SIMPLY SUPPORTED COMPOSITE BEAM**	Chapter ref. **21**
	Design code **BS 5950: Part 3**	Made by **RML**	Sheet no. **6**
		Checked by **GWO**	

<u>*Composite condition – service load*</u>

Determine elastic section properties

Modular ratio $\alpha_e = 15$ for LWC (see Table 21.3)

Elastic neutral axis depth (equation 21.1):

$$x_e = \frac{\dfrac{D_s - D_p}{2} + \alpha_e r\left(\dfrac{D}{2} + D_s\right)}{(1 + \alpha_e r)}$$

where $r = \dfrac{A}{(D_s - D_p)B_e} = \dfrac{8540}{80 \times 3000} = 0.0355$

$$x_e = \frac{\dfrac{80}{2} + 0.0355 \times 15 \times \left(\dfrac{453.6}{2} + 130\right)}{(1 + 0.0355 \times 15)}$$

$$= \ 150.1\,mm \ below \ top \ of \ slab$$

Second moment of area (equation 21.2):

$$I_c = \frac{A(D + D_s + D_p)^2}{4(1 + \alpha_e r)} + \frac{B_e(D_s - D_p)^3}{12\alpha_e} + I_s$$

$$= \frac{8540(453.6 + 130 + 50)^2}{4(1 + 0.0355 \times 15)} + \frac{3000 \times 80^3}{12 \times 15} + 294 \times 10^6$$

$$= \ 861.8 \times 10^6 \ mm^4$$

Elastic section modulus – steel flange (equation 21.3)

$$Z_t = \frac{861.8 \times 10^6}{(453.6 + 130 - 150.1)} = 1.98 \times 10^6 \ mm^3$$

Elastic section modulus – concrete (equation 21.4)

$$Z_c = \frac{861.8 \times 10^6}{150.1} \times 15 = 86.1 \times 10^6 \ mm^3$$

		Subject	Chapter ref.
The **Steel Construction** **Institute** Silwood Park, Ascot, Berks SL5 7QN 		*DESIGN OF SIMPLY SUPPORTED COMPOSITE BEAM*	*21*
		Design code BS 5950: Part 3	Made by *RML* Checked by *GWO*	Sheet no. 7

Service load on composite section

$$= (5.0 + 1.0 + 0.5) \times 3.0 = 19.5\,kN/m$$

Service moment

$$= 19.5 \times 12^2/8 = 351\,kNm$$

$$\textit{Stress in steel} = \frac{351 \times 10^6}{1.98 \times 10^6} = 177\,N/mm^2$$

Additional stress on non-composite section $= 92\,N/mm^2$

$$\textit{Total serviceability stress} = 177 + 92$$
$$= 269\,N/mm^2 < p_y$$

$$\textit{Stress in concrete} = \frac{351 \times 10^6}{86.1 \times 10^6}$$
$$= 4.1\,N/mm^2 < 0.5\,f_{cu}$$

Deflection – imposed load

Imposed load $= (5.0 + 1.0) \times 3.0 = 18.0\,kN/m$

$$\textit{Deflection} \quad \delta_c = \frac{5}{384} \times \frac{18.0 \times 12000^4 \times 10^{-3}}{205 \times 10^3 \times 861.8 \times 10^6}$$
$$\delta_c = 27.5\,mm$$

Additional deflection from partial shear-connection (equation 21.18)

$$\delta_s = \frac{861.8}{294.0} \times 27.5 = 80.6\,mm$$

$$\delta_c^I = \delta_c + 0.3(1 - K)(\delta_s - \delta_c)$$
$$= 27.5 + 0.3(1 - 0.76)(80.6 - 27.5)$$
$$= 31.3\,mm \quad (span/383)$$

Deflection is the limiting criterion.
It would be possible to show that the imposed load could be increased to 6 kN/m² for the same beam size.

The Steel Construction Institute		Subject	Chapter ref.
Silwood Park, Ascot, Berks SL5 7QN		**DESIGN OF SIMPLY SUPPORTED COMPOSITE BEAM**	*21*
Design code	Made by *RML*	Sheet no. **8**	
BS 5950: Part 3	Checked by *GWO*		

Check natural frequency of beam

Consider weight of floor in dynamic calculations to include: self weight of slab and beam + 10% imposed load + ceiling load but excluding partitions

$$= \ (2.0 + 0.1 \times 5.0 + 0.5) \times 3.0 + 0.67 \ = \ 9.67\,kN/m$$

Accurate calculations show that the ratio of the dynamic to static stiffness of a composite beam used in deflection calculations is 1.1 to 1.15. Consider deflection reduction of 1.1 for the same section properties as calculated under imposed load.

Deflection of composite beam, when subject to instantaneously applied self weight (as above)

$$\delta_{sw} \ = \ \frac{5 \times 9.67 \times 12^4 \times 10^3}{384 \times 205 \times 10^6 \times 1.1 \times 861.8 \times 10^{-6}}$$

$$= \ 13.4\,mm$$

Natural frequency of beam

$$f \ \simeq \ 18/\sqrt{\delta_{sw}} \ = \ 18/\sqrt{13.4} \ = \ 4.91\,Hz \ \underline{Satisfactory}$$

(A detailed check to Reference 21.14 demonstrates that this floor is satisfactory for general office use).

The Steel Construction Institute  Silwood Park, Ascot, Berks SL5 7QN	Subject **DESIGN OF CONTINUOUS COMPOSITE BEAM**	Chapter ref. **21**
	Design code **BS 5950: Parts 1 & 3**	Made by **RML**	Sheet no. **1**
		Checked by **TCC**	

Use same design data as for simply-supported composite beams.

Choose 406 × 178 × 60 kg/m UB grade S355 N.B. Section is plastic. (span/depth ratio = 22, and 12% weight saving).

<u>*Construction condition*</u>

Design load in construction condition

$$w_u = (1.6 \times 0.5 + 1.4 \times 2.0) \times 3.0 + 0.60 \times 1.4 = 11.6 \, kN/m$$

Check stability of bottom flange under worst combination of loading

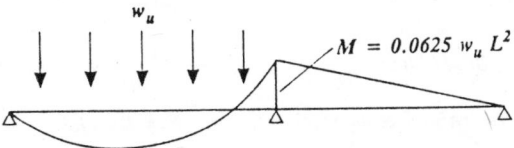

$$M = 0.0625 \times 11.6 \times 12^2 = 104.4 \, kNm$$

Slenderness $\lambda = 12000/39.7 = 302$

From SCI publication 'Steelwork Design: Guide to BS 5950: Part 1: Volume 1' (page B5)

$$u = 0.880 \quad x = 33.8$$

Assume top flange is laterally restrained by the decking. Use design formula for equivalent slenderness in BS 5950: Part 1 (Annex G.2.4.1).

$$\lambda_{TB} = n_t u v_t \lambda \quad \text{where} \quad v_t = \left[\frac{\frac{4a}{h_s}}{1 + \left(\frac{2a}{h_s}\right)^2 + \frac{1}{20}\left(\frac{\lambda}{x}\right)^2} \right]^{0.5}$$

and a = *distance of restraint above shear centre of beam* (= $h_s/2$ *in this case*)

h_s = *distance between shear centre of flanges*

The Steel Construction Institute Silwood Park, Ascot, Berks SL5 7QN	Subject ***DESIGN OF CONTINUOUS COMPOSITE BEAM***	Chapter ref. **21**
Design code *BS 5950: Parts 1 & 3*	Made by *RML* Checked by *TCC*	Sheet no. **2**

Hence:

$$v_t = \left[\cfrac{1}{1+\cfrac{1}{40}\left(\cfrac{\lambda}{x}\right)^2}\right]^{0.5} = \left[\cfrac{1}{1+\cfrac{1}{40}\left(\cfrac{302}{33.8}\right)^2}\right]^{0.5} = 0.58 \; and \; n_t = 1.0$$

$\lambda_{TB} = 1.0 \times 0.882 \times 0.58 \times 302 = 155$

From Table 16 of BS 5950: Part 1, bending strength
$p_b = 67\,N/mm^2$

Buckling resistance moment
$M_b = 67 \times 1200 \times 10^3 \times 10^{-6} = 80.4\,kNm$

For uniform member with unrestrained compression flange ($F_c = 0$), G.2.1

$$\frac{m_t M_x}{M_b} \le 1$$

$m_t =$ *equivalent uniform moment factor (G.4.2)*

With $\beta_t = 0$, y = 0.58 (see G.2.3)

Table G.1 gives $m_t = 0.56$

$$\frac{m_t M_x}{M_b} = \frac{0.56 \times 104.4}{80.4} = 0.73 \le 1$$

i.e. stable during construction.

Maximum negative (hogging) moment (conservative)
$\quad = 0.125 w_u L^2 \quad = 208.8\,kNm$

Plastic moment resistance of steel section
$\quad = 355 \times 1200 \times 10^3 \times 10^{-6}$
$\quad = 426\,kNm > 208.8\,kNm$

The Steel Construction Institute Silwood Park, Ascot, Berks SL5 7QN 	Subject **DESIGN OF CONTINUOUS COMPOSITE BEAM**	Chapter ref. **21**
	Design code	Made by *RML*	Sheet no. **3**
	BS 5950: Parts 1 & 3	Checked by *TCC*	

Hence, the steel beam can support the loads during construction.

Moment in mid-span of two-span beam after construction

$$= 0.0703 \times (2.0 \times 3.0 + 0.6) \times 12^2 = 66.8\,kNm$$

Bending stress (elastic)

$$= \frac{66.8 \times 10^6}{1058 \times 10^3} = 63\,N/mm^2$$

Deflection of underside of beam after construction

$$\delta_o = \frac{2.1 \times (2.0 \times 3.0 + 0.6) \times 12000^4}{384 \times 205 \times 10^3 \times 21600 \times 10^4}$$

$$= 16.9\,mm \quad (span/710)$$

<u>*Composite condition – ultimate loading*</u>

Factored loading on composite beam

$$w_u = [(5.0 + 1.0) \times 1.6 + (0.5 + 2.0) \times 1.4] \times 3.0 + 0.6 \times 1.4$$
$$= 40.1\,kN/m$$

Design moment on composite beam

Ratio of imposed load: dead load

$$= \frac{5.0 + 1.0}{2.0 + 0.5 + 0.6/3.0} = 2.22 > 2$$

Therefore, the use of Table 21.2 is not permitted. However, calculate moments from pattern load cases as follows:

The Steel Construction Institute Silwood Park, Ascot, Berks SL5 7QN	Subject **DESIGN OF CONTINUOUS COMPOSITE BEAM**	Chapter ref. **21**
Design code *BS 5950: Parts 1 & 3*	Made by **RML** Checked by **TCC**	Sheet no. **4**

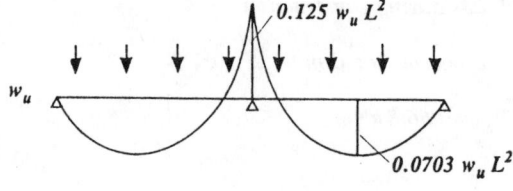

No redistribution

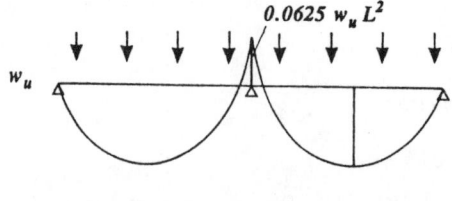

50% redistribution

$(0.0703 + 0.032)\ w_uL^2$
$= 0.103\ w_uL^2$
(accurate value = 0.096 w_uL^2)

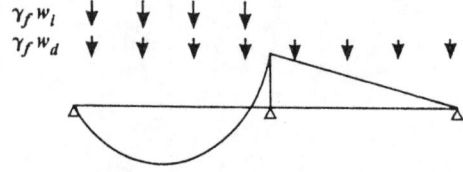

Imposed load on one span; dead load on both spans

The section is 'plastic' when grade S355 steel is used. According to Table 21.1, 50% redistribution of the negative (hogging) moment can be made provided the slab has nominal reinforcement.

Negative moment $=\ 0.0625 \times 40.1 \times 12^2$

$=\ 360.9\,kNm$

Positive moment $=\ 0.096 \times 40.1 \times 12^2$

$=\ 554.3\,kNm\ (50\%\ redistribution)$

Alternatively, single-span loading gives:

Positive (sagging) moment

$= \{0.096 \times 1.6 \times (5.0 + 1.0) \times 3.0 + 0.070 \times 1.4 \times [(2.0 + 0.5) \times 3.0 + 0.6]\} \times 12^2$

$=\ 512.4\,kNm$

The **Steel Construction Institute** Silwood Park, Ascot, Berks SL5 7QN 	Subject **DESIGN OF CONTINUOUS COMPOSITE BEAM**	Chapter ref. *21*
Design code **BS 5950: Parts 1 & 3**	Made by *RML* Checked by *TCC*	Sheet no. *5*

Plastic moment resistance of steel section

$= 355 \times 1200 \times 10^3 \times 10^{-6}$

$= 426 \, kNm > 360.9 \, kNm$

This exceeds the design moment. It may be shown that 42% redistribution of moment is adequate. Check moment resistance of composite section.

<u>*Moment resistance of composite section*</u>

Effective breadth of slab (for internal span)

$B_e = 0.7 \times 12000/4 = 2100 \, mm \not> beam \ spacing$

(note 0.7 has been used for internal span, rather than 0.8 for external span)

Resistance of slab in compression

$R_c = 0.45 f_{cu} B_e (D_s - D_p)$

$= 0.45 \times 30 \times 2100 \times (130 - 50) \times 10^{-3}$

$= 2268 \, kN$

Resistance of steel section in tension

$R_s = A p_y = 7650 \times 355 \times 10^{-3}$

$= 2715 \, kN$

As $R_s > R_c$, plastic neutral axis lies in steel flange

Plastic moment resistance of composite section

$$M_{pc} = R_s \frac{D}{2} + R_c \frac{(D_s + D_p)}{2} - \left[\frac{(R_s - R_c)^2}{R_f} \times \frac{T}{4} \right]$$

where $R_f = 12.8 \times 177.9 \times 355 \times 10^{-3}$

$= 808 \, kN$

The **Steel Construction Institute** Silwood Park, Ascot, Berks SL5 7QN	Subject ***DESIGN OF CONTINUOUS COMPOSITE BEAM***	Chapter ref. **21**
	Design code **BS 5950: Parts 1 & 3**	Made by **RML** Checked by **TCC**	Sheet no. **6**

The final term is small and may be neglected

$$M_{pc} = \left[2715 \times \frac{406.4}{2} + 2268 \times \frac{(130+50)}{2} - \frac{(2715-2268)^2}{808} \times \frac{12.8}{4} \right] \times 10^{-3}$$
$$= 551.7 + 204.1 - 0.8$$
$$= 755\,kNm > 554.3\,kNm$$

As there is sufficient reserve in the composite section, consider reducing the amount of shear connectors.

Partial shear connection

Zone of beam subject to positive (sagging) moment

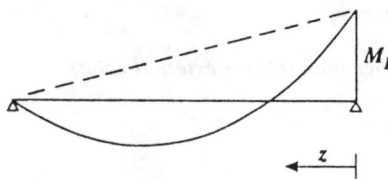

$$M = 0;\ when\ z = L\left(\frac{2M_1}{w_u L^2}\right)$$

If M_1 = $0.0625 w_u L^2$ *(after redistribution)*

     z = $L/8$

Zone of beam from point of zero to max. positive moment $\simeq 0.43L$

Use two shear connectors every 300 mm

$$Number\ of\ shear\ connectors = \frac{0.43 \times 12000 \times 2}{300} = 34$$

Resistance of one shear connector in LWC

(see simple beam calculations)   =   *72 kN*

R_q = $0.8 \times 72 \times 34$ =   *1958 kN*

	Subject	Chapter ref.
The Steel Construction Institute Silwood Park, Ascot, Berks SL5 7QN	*DESIGN OF CONTINUOUS COMPOSITE BEAM*	**21**
	Design code **BS 5950: Parts 1 & 3**	Made by **RML** Checked by **TCC**	Sheet no. **7**

Degree of shear connection $(R_c < R_s)$

K = $1958/2268$ = 0.86

Minimum shear connection

K > $\dfrac{L-6}{10}$ = $0.6 < 0.86$ **OK**

As $R_q > R_w$; *plastic neutral axis lies in steel flange*

Plastic moment resistance of section

$$M_c = R_s \frac{D}{2} + R_q \left[D_s - \frac{R_q}{R_c} \frac{(D_s - D_p)}{2} \right]$$

$$M_c = \left\{ 2715 \times \frac{406.4}{2} + 1958 \left[130 - \frac{1958}{2268} \frac{(130 - 50)}{2} \right] \right\} \times 10^{-3}$$

$\quad = 551.7 + 186.9$

$\quad = 738.6 \, kNm > 554.3 \, kNm$

<u>*Vertical shear resistance*</u>

Shear resistance (to BS 5950: Part 1, Clause 4.2.6)

P_y = $0.6 \times 355 \times 406.4 \times 7.9 \times 10^{-3}$

$\quad$ = $684 \, kN$

Applied shear force at internal support

F_v = $w_u \dfrac{L}{2} + \dfrac{M_1}{L}$

$\quad$ = $1.25 \left(\dfrac{w_u L}{2} \right)$

$\quad$ = $1.25 \times 40.1 \times 12/2$ = $300.7 \, kN$

F_y = $0.44 \, P_y < 0.5 \, P_y$

So no reduction in moment resistance is necessary.

The **Steel Construction Institute** Silwood Park, Ascot, Berks SL5 7QN	Subject **DESIGN OF CONTINUOUS COMPOSITE BEAM**	Chapter ref. *21*
	Design code **BS 5950: Parts 1 & 3**	Made by **RML** Checked by **TCC**	Sheet no. **8**

Composite condition – service loading

Determine elastic section properties under positive moment

Elastic neutral axis depth

$$x_e = \frac{\dfrac{D_s - D_p}{2} + \alpha_e r\left(\dfrac{D}{2} + D_s\right)}{1 + \alpha_e r}$$

Take $\alpha_e = 15$ *for LWC*

$$r = \frac{A}{(D_s - D_p)B_e} + \frac{7650}{(130 - 50)2100} = 0.045$$

$$x_e = \frac{\left[\dfrac{80}{2} + 0.045 \times 15 \times \left(\dfrac{406.4}{2} + 130\right)\right]}{(1 + 0.045 \times 15)}$$

$$= 158.1\,mm$$

Second moment of area of section (equation 21.2)

$$I_c = \frac{A(D + D_s + D_p)^2}{4(1 + \alpha_e r)} + \frac{B_e(D_s - D_p)^3}{12\alpha_e} + I_x$$

$$= \left[\frac{7650(406.4 + 130 + 50)^2}{4(1 + 0.045 \times 15)} + \frac{2100 \times (130 - 50)^3}{12 \times 15} + 216\right] \times 10^6$$

$$= (393.0 + 6.0 + 216) \times 10^6$$

$$= 615 \times 10^6\,mm^4$$

Elastic section modulus – steel flange (equation 21.3)

$$Z_t = \frac{615 \times 10^6}{(406.4 + 130 - 158.1)}$$

$$= 1.62 \times 10^6\,mm^3$$

The **Steel Construction Institute** Silwood Park, Ascot, Berks SL5 7QN	Subject *DESIGN OF CONTINUOUS COMPOSITE BEAM*	Chapter ref. **21**
	Design code **BS 5950: Parts 1 & 3**	Made by *RML* Checked by *TCC*	Sheet no. **9**

Service load on composite section

$$= (5.0 + 1.0 + 0.5) \times 3 = 19.5\,kN/m$$

Mid-span moment for serviceability conditions should take into account 'shakedown effects' as degree of moment redistribution at ultimate loads exceeds 30%.

Shakedown load = Dead load + 80% Imposed load

$$w_s = [2.0 + 0.5 + 0.8 \times (5.0 + 1.0)] \times 3 + 0.6$$
$$= 22.5\,kN/m$$

Shakedown moment (elastic negative (hogging) moment)

$$= 0.125 w_s L^2 = 0.125 \times 22.5 \times 12^2$$
$$= 405\,kNm$$

This is less than the plastic moment resistance of the beam (= 426 kNm), so no plastification occurs at working load.

Service moment under imposed loading on all spans (allowing for 30% redistribution of negative moment)

$$= (0.0703 + 0.15 \times 0.125) \times 19.5 \times 12^2$$
$$= 250.0\,kNm$$

Steel stress (lower flange)

$$= \frac{250.0 \times 10^6}{1.62 \times 10^6} = 154\,N/mm^2$$

Additional stress in steel beam arising from self weight of floor

$$= 63\,N/mm^2$$

Total stress $= 63 + 154 = 217\,N/mm^2 < p_y$

Concrete stress

$$= \frac{250 \times 10^6}{615 \times 10^6} \times \frac{158.1}{15} = 4.3\,N/mm^2 < 0.5 f_{cu}$$

The Steel Construction Institute Silwood Park, Ascot, Berks SL5 7QN 	Subject **DESIGN OF CONTINUOUS COMPOSITE BEAM**	Chapter ref. **21**
	Design code **BS 5950: Parts 1 & 3**	Made by **RML** Checked by **TCC**	Sheet no. **10**

Deflection – imposed loading

Imposed loading w_i $=$ $(5.0 + 1.0) \times 3$ $=$ $18.0 \, kN/m$

Free moment M_o $=$ $w_i L^2 / 8$ $=$ $18.0 \times 12^2 / 8$

  $=$ $324 \, kNm$

Negative (hogging) moment $=$ $324 \, kNm$ *for a two-span beam*

According to section 21.7.8, reduce negative moment by 30% to represent the effects of pattern loading

M_1 $=$ 0.7×324 $=$ $226.8 \, kNm$

M_2 $=$ 0

Deflection of simply supported beam

$$\delta_o = \frac{5 \times 18.0 \times 12^4 \times 10^3}{384 \times 205 \times 10^6 \times 615 \times 10^{-6}}$$

  $=$ $38.8 \, mm$

Allowing for partial shear connection K = 0.86

Deflection of steel beam $\delta_s = \dfrac{615}{215.1} \times 38.8 = 110.9 \, mm$

Modified deflection $\delta_0' = \delta_0 + 0.3(1 - K)\delta_s$

  $=$ $38.8 + 0.3 \times (1 - 0.86) \times 110.9$

  $=$ $43.4 \, mm$

Deflection of continuous beam

$$\delta_c = \delta_o\left[1 - 0.6\frac{(M_1 + M_2)}{M_0}\right] = 43.4\left(1 - 0.6 \times \frac{226.8}{324}\right)$$

  $=$ $25.2 \, mm$ *(span/476) __Satisfactory__*

The **Steel Construction** **Institute** Silwood Park, Ascot, Berks SL5 7QN 	Subject ***DESIGN OF CONTINUOUS*** ***COMPOSITE BEAM***	Chapter ref. ***21***
	Design code ***BS 5950: Parts 1 & 3***	Made by ***RML*** Checked by ***TCC***	Sheet no. ***11***

Stability of bottom flange

Check the stability of the bottom flange in the negative (hogging) moment region. The worst case for instability is when only one span is loaded:

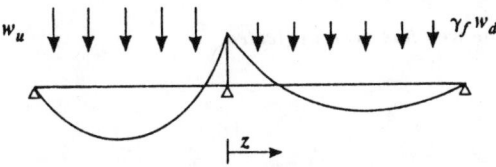

Negative moment $= 0.0625(w_u + \gamma_f w_d)L^2$

$\gamma_f w_d = 1.4[(2.0 + 0.5)3 + 0.6] = 11.3 \, kN/m$

$w_u = 40.1 \, kN/m$

Negative moment $= 0.0625(40.1 + 11.3)12^2$

$= 462.6 \, kNm$

This exceeds the plastic resistance moment of the beam,
$M_p = 426 \, kNm$

Use this moment M_p at the support

$M = 0$ *when* $z = L\left(\dfrac{2 \times M_p}{\gamma_f w_d L^2}\right)$

$z = 12\left(\dfrac{2 \times 426}{11.3 \times 12^2}\right) = 6.3 \, m$

It is necessary to introduce an additional lateral restraint close to the support such that $\lambda_{LT} < 30.2$ for S355 steel – Table 16

This is a safe approximation for a falling bending moment. However the beam should be checked in accordance with clauses

The **Steel Construction Institute** Silwood Park, Ascot, Berks SL5 7QN	Subject **DESIGN OF CONTINUOUS COMPOSITE BEAM**	Chapter ref. **21**
Design code **BS 5950: Parts 1 & 3**	Made by **RML** Checked by **TCC**	Sheet no. **12**

5.3.3 and 5.3.4 as appropriate for a segment adjacent to a plastic hinge where $\lambda_{LT} < \lambda_{LO}$.

Further, the restraint detail can only be considered effective where sufficient shear studs are welded around the web stiffeners on the top flange of the beam to allow a load path for the restraining forces to develop.

The distortional restraint provided by the web has been neglected. However, it is reasonable to take:

$n = 0.95; \quad u = 0.88; \quad v_t = 0.9$

such that $\lambda_{max} \quad = \quad \dfrac{30}{0.95 \times 0.88 \times 0.9} \quad = \quad 39.9$

and $L_{max} \quad = \quad 39.9 \times 39.7 \quad = \quad 1582\,mm$

(say 1500 mm)

Check the stability of the remaining portion of the beam

Moment at the above position

$= \quad \dfrac{6.3 - 1.5}{6.3} \times 426 \quad = \quad 325\,kNm$

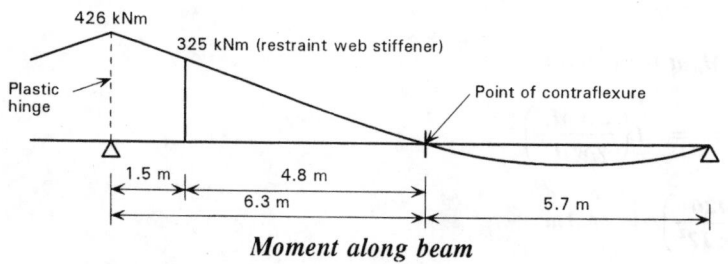

426 kNm

325 kNm (restraint web stiffener)

Plastic hinge

Point of contraflexure

1.5 m   4.8 m

6.3 m

5.7 m

Moment along beam

In accordance with Clause 4.3.5.3(d)-BS 5950-1:2000, the point of contraflexure for a composite beam ($D_b \leq 550^{mm}$) may be assumed to be a virtual lateral restraint to the bottom flange when determining the segment length L_{LT}

The Steel Construction Institute Silwood Park, Ascot, Berks SL5 7QN	Subject *DESIGN OF CONTINUOUS COMPOSITE BEAM*	Chapter ref. **21**
	Design code *BS 5950: Parts 1 & 3*	Made by **RML** Checked by **TCC**	Sheet no. **13**

$$L_{LT} = 4.8m \Rightarrow \lambda = \frac{4800}{39.7} = 121$$

G.2.4.1 $\lambda_{TB} = n_t u v_t \lambda$ $n_t = 1.0, \quad u = 0.88$

$$v_t = \left[\frac{1}{1 + \frac{1}{40}\left(\frac{\lambda}{x}\right)^2} \right]^{1/2} = 0.87$$

$\Rightarrow \quad \lambda_{TB} = 1.0 \times 0.88 \times 0.87 \times 121 = 93 \Rightarrow p_b = 155\,N/mm^2$ *(Table 16)*

$\Rightarrow \quad M_b = p_b S_x \Rightarrow M_b = 186\,kNm$

$\& \quad y = \left[\frac{1}{1 + \frac{1}{40}\left(\frac{\lambda}{x}\right)^2} \right] = 0.87, \quad since \quad a = \frac{h_s}{2}$

$\& \quad \beta_t = 0, from\ G.4.2 \Rightarrow m_t = 0.54$

For uniform member from G.2.1 ($F_c = 0$)   *(Table G.1)*

$$\frac{m_t M_x}{M_b} = \frac{0.54 \times 325}{186} = 0.94 < 1.0$$

The beam segment is stable.

So adequate restraint is provided by a single lateral restraint 1.5 m from the internal support. This is done by introducing a web stiffener at this point.

<u>*Check natural frequency of beam*</u>

To calculate the natural frequency of a continuous beam it is necessary to consider the mode shape of vibration. Because of the influence of the asymmetric inertial forces, the natural frequency is taken as the same as for a simply supported beam.

The Steel Construction Institute	Subject	Chapter ref.
Silwood Park, Ascot, Berks SL5 7QN	**DESIGN OF CONTINUOUS COMPOSITE BEAM**	**21**

	Design code	Made by **RML**	Sheet no. **14**
	BS 5950: Parts 1 & 3	Checked by **TCC**	

Considering the same loads as previously:

$$\delta_{sw} = \frac{5 \times 9.67 \times 12^4 \times 10^3}{384 \times 205 \times 10^6 \times 1.1 \times 615 \times 10^{-6}}$$

$$= 17.7\,mm$$

Natural frequency $= 18/\sqrt{\delta_{sw}} = 18/\sqrt{17.7} \simeq 4.3\,Hz$

However, when carrying out the full response factor analysis of Reference 21.14, account can be taken of the effective mass of both spans, thus reducing the amplitude of the vibration from a given excitation. This reduces the sensitivity of the floor to vibration.

Conclusion:

The design using a 406 × 178 × 60 kg/m UB is adequate. However, reducing the section to 406 × 178 × 54 kg/m would not result in an acceptable design because this section is 'compact' in S355 steel. Consequently, only 40% redistribution of support moment is permitted and design is limited by the moment resistance of the steel section at the supports.

Chapter 22
Composite columns

by MARK LAWSON and PETER WICKENS

22.1 Introduction

22.1.1 Form of construction

Composite columns comprise steel sections with a concrete encasement or core. Encased columns usually consist of standard I-beam or H-column sections with a rectangular or square concrete section encasement to form a solid composite section. Additional reinforcement is placed in the concrete cover around the steel section in order to prevent spalling under axial stress and in fire.

Concrete-filled columns consist of circular, square or rectangular hollow sections filled with concrete. Additional reinforcement is not normally required, except in columns of large section. Examples of these different types of composite columns are shown in Fig. 22.1.

The steel structure is normally constructed in advance of the formation of the composite section. The minimum size of the steel element is therefore controlled by the construction condition. For practical reasons it is normally necessary to fill the tubular sections shortly after their erection but I-sections can be encased much later. However, there are practical problems in concreting around a column with the floors in place. Alternatively, columns can be pre-encased in the factory and in situ concrete used to fill the zone around the beam–column connections.

22.1.2 Advantages of composite columns

In many cases of design, concrete or some other protective material is required around steel columns for reasons of fire-resistance and durability. It would seem appropriate, therefore, to develop composite action between the steel and concrete, thereby taking advantage of the inherent compressive strength of the concrete, increasing the compressive resistance of the section and leading to considerable savings in the cost of the steelwork. Even ignoring this composite action, the slenderness of the steel column in lateral buckling is reduced, thereby increasing the compressive stress that can be resisted by the steel section.

Much research has gone into the behaviour of concrete-filled tubular sections. Architecturally, tubular columns have many attractive features; concrete filling has

651

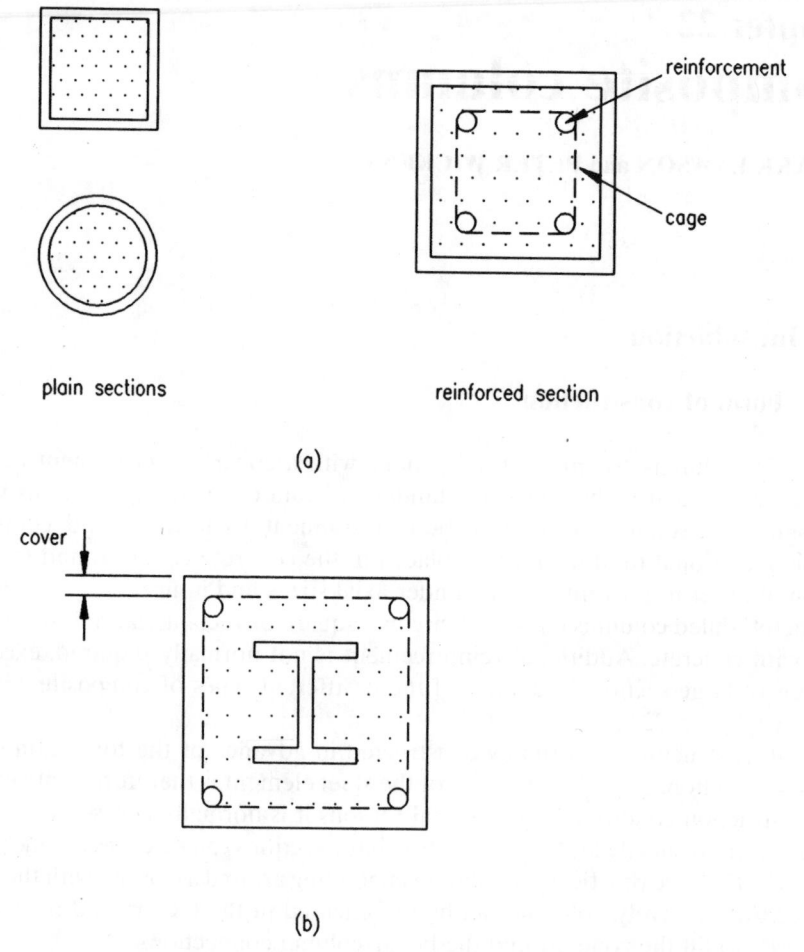

Fig. 22.1 Examples of composite columns: (a) RHS and CHS concrete-filled section, (b) concrete-encased section

no visual effect on their appeal. The advantages from a structural point of view are, first, the triaxial confinement of the concrete within the section, and second, the fire resistance of the column, which largely depends on the residual capacity of the concrete core.

22.1.3 Principles of design

The axial resistance of a stocky column, defined as a column that is not subject to the effects of instability, is determined by adding the ultimate compressive strengths

of the steel and concrete components. In traditional reinforced concrete design, a 'short' column is defined as one whose effective height to least cross-sectional dimension is less than 12, and a nominal allowance is made for eccentricity of axial load.

The axial resistance of concrete-filled sections is greater because the concrete is not able to expand laterally (Poisson's ratio effect) under load, and triaxial stresses are developed in the concrete. This causes an increase in the compressive strength of the concrete by an amount dependent on the proportions of the cross-section. The hoop tensions created in the steel have a small adverse affect on its strength.

The effect of eccentricity of axial load is to develop a bending moment in the section. The moment resistance of the section (in the absence of axial load) can be calculated considering plastic stress blocks (see Fig. 22.2). Formulae are given in BS 5400: Part 5, Appendix C.[1] The interaction between axial load and bending moment can be considered in terms of a simplified interaction formula (section 22.2.2).

Slender columns require a more refined treatment. The effective slenderness of a column is determined from the proportions of the composite section. The second moment of area is obtained by adding the second moments of area of the steel and concrete (divided by an appropriate modular ratio). This represents a considerable increase over the properties of steel alone. The axial stress that the section can resist is then determined from the column buckling curves for the steel section under consideration. The resulting axial stress, relative to the yield stress of the steel, is effectively a resistance reduction factor to be applied to the 'stocky column' compressive resistance.

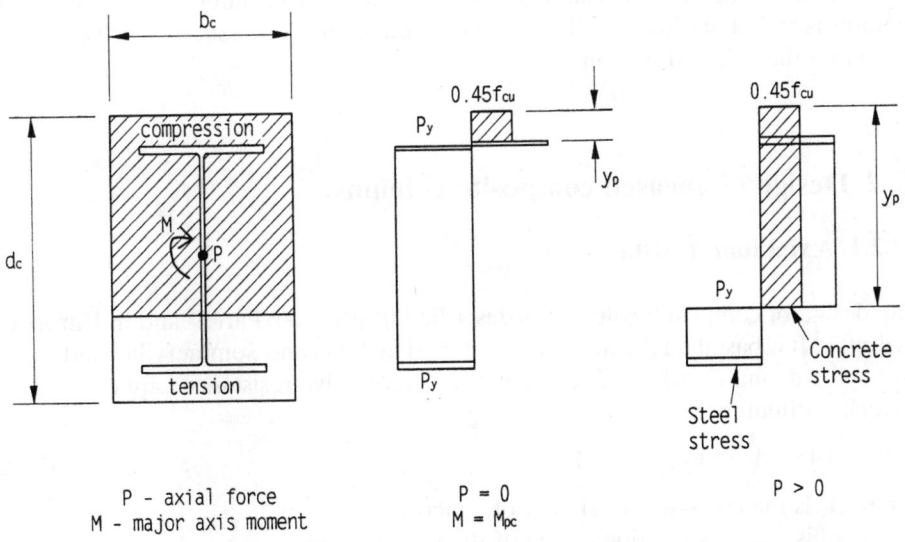

Fig. 22.2 Plastic stress blocks acting on cross-section of composite column subject to moment and axial force

It should be noted that the above approach assumes that loads are not applied laterally over the column length. Concentric and eccentric axial loads cause relatively low interface shear stresses between the steel and the concrete. Lateral loads cause greater shear stresses and may necessitate the introduction of shear-connectors (as for composite beams – see Chapter 21, section 21.4).

22.1.4 Cased strut method

The traditional method of designing cased columns, presented in BS 449: 1969,[2] can be very conservative but is readily accepted as a method of simple design.

The minimum width of the concrete casing, according to BS 449, should be the flange width b of the steel section plus 100 mm. At least four reinforcing bars are to be located in the concrete at a cover not exceeding 50 mm, to which 5 mm diameter steel stirrups are attached at spacings not exceeding 200 mm. The normal aggregate size is 10 mm and the minimum concrete grade is 21 N/mm^2.

To establish the axial capacity of the cased column (or strut), the radius of gyration of the solid section is taken as 0.2 (b + 100) mm, or alternatively that of the major axis of the steel section. The cross section excludes any concrete cover in excess of 75 mm from the overall dimension of the steel section. The net area of concrete is replaced by an equivalent area of steel by dividing by a modular ratio of 30.

Therefore, knowing the permissible axial stress, as a function of the effective slenderness of the cased column, and the equivalent area of steel, the axial capacity of the section can be easily calculated. However, an onerous limit on the use of this method is that axial load on the cased column should not exceed twice that permitted on the uncased section.

22.2 Design of encased composite columns

22.2.1 Axial load resistance

The design of composite columns is described in BS 5400: Part 5[1] and in Eurocode 4 (draft).[3] It is based on the method developed by Basu and Sommerville[4] and modified by Virdi and Dowling.[5] The maximum compressive resistance (squash load) of a stocky column is:

$$P_u = 0.45 f_{cu} A_c + A_s p_y + 0.87 A_r f_y \qquad (22.1)$$

where A_c is the cross-sectional area of concrete

A_s is the cross-sectional area of the steel section

A_r is the cross-sectional area of the reinforcement

f_{cu} is the cube strength of concrete
p_y is the design strength of the steel
f_y is the yield strength of the reinforcement.

In BS 5400: Part 5 a factor of 0.91 (corresponding to a material factor of 1.1) is used to modify the term $A_s p_y$. This formula is restricted to concrete contribution factors, defined as $0.45 f_{cu} A_c / P_u$, between 0.15 and 0.8. In order for any reinforcement to contribute to the axial load resistance of the column, shear links of not less than 5 mm diameter should be provided at not more than 150 mm spacing. The value of A_r should not exceed $0.03 A_c$.

Limits on the proportion of the cross section[3] are that the concrete casing should provide a minimum cover to the steel section of 40 mm, and that the dimensions of the concrete section used in determining P_u should not exceed $1.8 \times b$ or $1.6 \times D$ (where b is the flange width, and D is the depth of the steel section).

The relative slenderness of a composite column of length L is defined by the slenderness factor:

$$\bar{\lambda} = \frac{L}{\pi} \sqrt{\left(\frac{P_u}{E_s \Sigma I} \right)} \tag{22.2}$$

where ΣI is the combined second moments of area of the concrete, steel section and reinforcement expressed in steel units. To do this, the second moment of area of the concrete is divided by the modular ratio $E_s/(450 f_{cu})$, where E_s is the elastic modulus of steel in N/mm².

The slenderness factor $\bar{\lambda}$, converted to an effective slenderness λ by multiplying $\bar{\lambda}$ by $\pi \sqrt{(E_s / P_y)}$, can then be used to determine the design stress, p_c, for the steel section. The selection of the appropriate column design curve in BS 5950: Part 1 depends on the steel section used (i.e. Tables 24(a)–(d) in BS 5950: Part 1).

The design stress p_c divided by p_y is then taken as a resistance reduction factor K_1 to be applied to all the components of P_u in Equation (22.1). Hence, this method can be used to determine the axial resistance of a composite column, which is $K_1 P_u$ in the absence of applied moment. This method can be used for slenderness factors $\bar{\lambda}$ less than 2.0, as proposed in Eurocode 4;[3] for greater slendernesses second-order effects are underestimated. Columns of slenderness factor less than 0.2 may be taken as 'stocky' (i.e. $K_1 = 1.0$).

In BS 5400: Part 5,[1] short columns are defined as those where the ratio of the length L to the least lateral dimension b_c of the composite column does not exceed 12. Where short columns are not designed to be subject to significant applied moment, the axial resistance may be taken as $0.85 K_1 P_u$, in which the factor 0.85 allows for slight eccentricity of axial load.

Slender columns are those where L/b_c exceeds 12. In design to BS 5400: Part 5,[1] the analysis of the member should consider a minimum additional moment given by the axial load times an eccentricity of $0.03 b_c$. This, together with any further applied moments, may be treated as in the following section.

22.2.2 Combined axial load and bending moments

The moment resistance of a composite column may be determined by establishing plastic stress blocks defining the resistances of the portions of the section under tension and compression. The plastic neutral axis depth y_p is defined as below the extreme edge of the concrete in compression. Three cases of neutral axis position exist: in the concrete, through the steel flange, and in the web of the section. The position depends on the relative proportions of steel and concrete (see Fig. 22.2). Commonly, y_p lies within the steel flange (i.e. $y_p \approx (d_c - D)/2$) for major axis bending of a composite section. In this case:

$$A_s p_y \geq \left(\frac{d_c - D}{2} \right) \times (0.45 f_{cu} b_c) \geq A_w p_y + A_f (0.45 f_{cu}) \tag{22.3}$$

The moment resistance of the composite section is then given by:

$$M_{pc} = 0.5 A_s p_y (d_c - y_p) + 0.5 A_r (0.87 f_y)(d_c - 2d_r) \tag{22.4}$$

where d_c is the depth and b_c is the breadth of the concrete section, d_r is the cover to the reinforcement, D is the depth of the steel section, A_w is the cross-sectional area of the web, A_f is the cross-sectional area of the flange, and A_r is the cross-sectional area of any additional reinforcement.

Other cases are defined in BS 5400: Part 5, Appendix C.[1]

Because the above plastic stress block method slightly overestimates the bending resistance of the composite section, M_{pc} is multiplied by 0.9 for design purposes.

In the presence of axial load, the plastic neutral axis depth increases. For small to medium axial loads, the plastic neutral axis remains within the steel web, but for higher axial loads most of the section is in compression. A typical interaction diagram representing the variation of moment resistance with axial load is shown in Fig. 22.3. An interesting phenomenon is that there is a slight increase in moment resistance with increasing axial load (compression), and there is a certain axial load where the moment resistance of the section equals that in the absence of axial load (i.e. M_{pc} from Equation (22.4)).

For simple design, the possibility exists of defining the interaction between moment and axial force in terms of three intercepts A, B and C on the moment and axial load axes, and also point D, which corresponds to the axial load at which the moment capacity remains unchanged. Therefore, a trilinear relationship AD, DC, CB closely models the real interaction diagram. It is normal practice to ignore the beneficial effect of axial load as it cannot always be assumed to be coincident with the applied moment. The curvature of the interaction diagram at higher axial loads can also be ignored without much loss of economy. The method developed by Basu and Sommerville[4] empirically follows the shape of the interaction diagram using coefficients K_1, K_2 and K_3 (see Fig. 22.3).

The value of axial load P at which the moment resistance remains unchanged (point D) may be evaluated as follows.

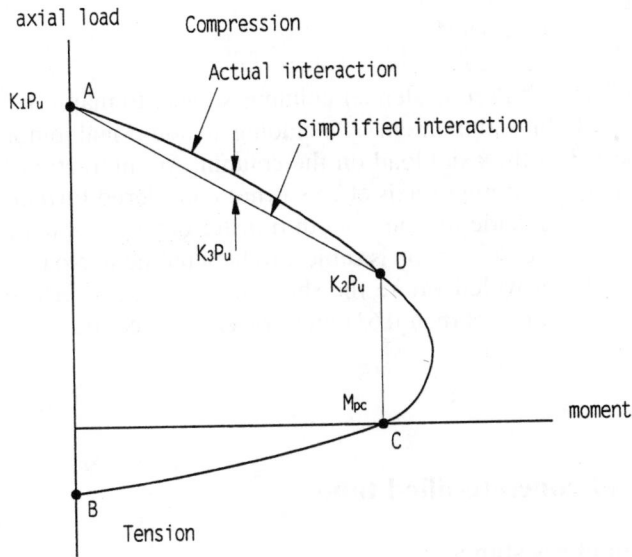

Fig. 22.3   Interaction between moment and axial force

Determine the neutral axis depth y_p for P of zero, corresponding to equal tension and compression forces across the composite section. Redefine a neutral axis depth, $y'_p = d_c - y_p$, which corresponds to an axis symmetrically placed with respect to y_p around the centre of the section. The net compressive resistance of the section with neutral axis y'_p corresponds to no change in moment resistance because the net moment effect of the section contained between depths y_p and y'_p is zero, as illustrated in Fig. 22.2.

It may be shown that the axial resistance of the section (termed P_0), corresponding to depth y'_p, is, in fact, the axial resistance of the concrete section ignoring the contribution of the steel member and the reinforcement. Hence,

$$P_0 = 0.45 f_{cu} d_c b_c \tag{22.5}$$

Dividing by P_u from Equation (22.1) gives a non-dimensional ordinate on the moment/axial-force diagram, corresponding to the moment resistance of the composite section. This ordinate is also equivalent to the concrete contribution factor for major axis bending, and can also be used for minor axis bending. For stocky columns it also corresponds to the appropriate value of K_2 in the Basu and Sommerville approach.[4]

For slender columns, account is to be taken of the moments arising from eccentricity of axial load in addition to the applied moments. The Basu and Sommerville method[4] has been codified in BS 5400: Part 5.[1] Because the method uses empirical formulae for K_1, K_2 and K_3 as a function of the slenderness of the column, it appears to be relatively complicated, but it may be simplified by taking K_3 as zero so that the moment resistance at any value of axial load is given by:

$$M = \left(\frac{K_1 - P/P_u}{K_1 - K_2} \right) M_{pc} \leq 0.9 M_{pc} \tag{22.6}$$

In design to BS 5400: Part 5,[1] slender columns subject to major axis moment are treated as subject to biaxial moment by including an additional minor axis moment of $0.03Pb_c$, where P is the axial load on the column. An interaction formula given for combining major and minor axis effects is not considered further here.

Provision should be made for the smooth transfer of force between the concrete and steel in cases where the section is subject to high moment. No mechanical shear connection need be provided where the shear stress at the interface between the concrete and the steel is less than $0.6 \, \text{N/mm}^2$ for encased columns.

22.3 Design of concrete-filled tubes

22.3.1 Axial load resistance

The compressive resistance of a concrete-filled rectangular or circular section is enhanced by the confining effect of the steel section on the concrete, which depends in magnitude on the shape of the section and the length of the column. Buckling tends to reduce the benefit of confinement on the squash load as the column slenderness increases. To account for this, modification factors are introduced. In circular sections it is possible to develop the cylinder strength ($0.83f_{cu}$) of the concrete.

The 'squash' load resistance of a circular concrete-filled column is:

$$P_u = C_1 p_y A_s + 0.87 f_y A_r + (0.83 f_{cu}/\gamma_{mc}) A_c \{1 + C_2 (t/\phi)[p_y/(0.83 f_{cu})]\} \tag{22.7}$$

where t is the thickness and ϕ is the diameter of the tubular section, and γ_{mc} is the material factor for concrete (= 1.5). The terms C_1 and C_2 are coefficients which are a function of the slenderness factor $\bar{\lambda}$ of the column, defined in Equation (22.2); C_1 is less than unity because of the effect of hoop tensions created in the steel. The values of C_1 and C_2 are presented in Table 22.1.

The method derives from research[6,7] carried out by CIDECT and CIRIA and has been incorporated into Eurocode 4.[3] Limits to the use of the method are that the term $A_s p_y$ should represent between 20% and 90% of P_u. To avoid local buckling, $\phi \leq 85t\varepsilon$ (where ε is $\sqrt{(275/P_y)}$).

The effect of slenderness on the axial resistance of a concrete-filled column may be treated as in section 22.2.1. The slenderness factor $\bar{\lambda}$ may be determined from Equation (22.2) as a function of P_u. This involves iteration as P_u is partly dependent on $\bar{\lambda}$. As a reasonable approximation, $\bar{\lambda}$ may be determined assuming that C_1 = 1.0 and C_2 = 0.0. Having evaluated $\bar{\lambda}$, the resistance reduction factor K_1 can be determined from the column buckling curve given in Table 27(a) of BS 5950: Part 1.

Table 22.1 Values of C_1 and C_2 defining the behaviour of concrete-filled circular hollow sections

λ	0	0.1	0.2	0.3	0.4	≥ 0.5
C_1	0.75	0.80	0.85	0.90	0.95	1.00
C_2	4.90	3.22	1.88	0.88	0.22	0.00

No design formulae are readily available for the axial resistance of rectangular filled sections. A limit on the proportions of rectangular sections is that $b \leq 52t\varepsilon$ and $d \leq 52t\varepsilon$.

22.3.2 Combined axial load and bending moment

The effect of eccentric axial load or applied moment may be treated conservatively by evaluating the moment resistance of the section taking the concrete strength as $0.45f_{cu}$, and ignoring the benefits of confinement of the concrete (i.e. $C_1 = 1.0$ and $C_2 = 0.0$). Formulae for M_{pc} are given in Reference 7. The interface shear stress between the steel and concrete infill is limited to $0.4 \, \text{N/mm}^2$.[3]

The reduced moment resistance of a concrete-filled section subject to axial load may be determined from the method in section 22.2.2. Design tables for both circular and rectangular hollow sections are given in Reference 7 and are based on the Basu and Sommerville approach.[4]

22.3.3 Fire-resistant design

The fire resistance of encased composite columns may be treated in the same way as reinforced concrete columns. The steel is insulated by an appropriate concrete cover, as given in BS 8110: Part 2.[8] To preserve the cover in the event of a fire, light reinforcement to the steel section is required. In such cases, two-hour fire resistance can usually be achieved.

The fire resistance of concrete-filled columns has been the subject of extensive research.[9] During a fire, sufficient redistribution of stress occurs between the hot steel section and the relatively cool concrete cover, such that a minimum of 30 minutes' fire resistance can be achieved.

The fire resistance of a concrete-filled hollow section may be established by defining the load ratio in fire conditions as:[10]

$$\eta = \frac{P_f}{K_1 P_u} \tag{22.8}$$

where $P_u = 0.83f_{cu}A_c + f_y A_r$ $\tag{22.9}$

Table 22.2 Fire-resistance of concrete-filled hollow sections as a function of load ratio[10]

Fire-resistance (min)	Load ratio, η	
	Plain concrete	5% fibre concrete
30	–	–
60	0.51	0.67
90	0.40	0.53
120	0.36	0.49

The axial load P_f in fire conditions is determined for load factors of unity. The concrete strength of $0.83f_{cu}$ is the cylinder strength of concrete, which can be developed in both circular and rectangular sections in fire. The material factors for concrete and for reinforcement are taken as 1.0 in this formula.

The term K_1 is determined as in section 22.2.1 or as in Table 10 of BS 5950 Part 8.[10] It excludes the contribution of the steel hollow section and includes the enhanced strength of the concrete. The reinforcing bars are assumed to be fully effective, provided they are located at the appropriate cover.[8] A method of including the effect of applied moment is given in BS 5950: Part 8,[10] which is described more fully in Reference 11.

Fire resistance periods that can be achieved for plain (or bar-reinforced) or fibre-reinforced concrete-filled hollow sections are presented in Table 22.2.

Where enhanced periods of fire resistance or greater load resistance are required, external fire protection may be provided. Normally, the effect of concrete filling offers a considerable reduction in the thickness of fire protection material that would be required for an unfilled hollow section.[9]

An important practical requirement for the use of concrete-filled sections is the provision of vent holes at the top and bottom of each column. These are to prevent dangerous build-up of steam pressure inside the columns in the event of a fire. They also permit seepage of any excess moisture in the concrete after construction. Two 12 mm diameter holes placed diametrically opposite each other at the top and bottom of each storey height have been used in testing and proved to be adequate.

Further guidance on fire resistance of concrete-encased and concrete-filled columns is given in ENV 1994-1-2 (Eurocode 4 Part 1.2) and in Reference 13.

References to Chapter 22

1. British Standards Institution (1979) *Steel, concrete and composite bridges.* Part 5: *Code of practice for design of composite bridges.* BS 5400, BSI, London.
2. British Standards Institution (1969) *Specification for the use of structural steelwork in building.* BS 449, BSI, London.

3. ENV 1994-1-1 Eurocode 4: Part 1.1 *Design of composite steel and concrete structures*. BSI, London.
4. Basu A.K. & Sommerville W. (1969) Derivation of formulae for the design of rectangular composite columns. *Proc. Instn Civ. Engrs*, supplementary volume, Paper 7206S, 233–80.
5. Virdi K.S. & Dowling P.J. (1973) The ultimate strength of composite columns in biaxial bending. *Proc. Instn Civ. Engrs*, **55**, Part 2, March, 251–72.
6. Sen H.K. & Chapman J.C. (1970) *Ultimate Load Tables for Concrete-Filled Tubular Steel Columns*. Construction Industry Research and Information Association (CIRIA), Technical Note 13.
7. Corus Tubes, Structural & Conveyance Business (2002) *Design of SHS concrete filled columns*. CT23: 2002. Corus Tubes.
8. British Standards Institution (1985) *Structural use of concrete*. Part 2: *Code of practice for special circumstances*. BS 8110, BSI, London.
9. British Steel, Tubes & Pipes (1998) *Fire resistant design – a guide to evaluation of Structural Hollow Sections using BS 5950 Part 8*. TD409: 1998. Corus Tubes.
10. British Standards Institution (1990) *Structural use of steelwork in building*. Part 8: *Code of practice for fire-resistant design*. BS 5950, BSI, London.
11. Lawson R.M. & Newman G.M. (1990) *Fire resistance design of steel structures – a handbook to BS 5950: Part 8*. The Steel Construction Institute, Ascot, Berks.
12. ENV 1994-1-2: Eurocode 4 Part 1.2 *Design of composite steel and concrete structures. General rules – structural fire design*. British Standards Institution, London.
13. Lawson R.M. & Newman G.M. (1996) *Structural fire design to EC3 and EC4 and comparison with BS 5950*. The Steel Construction Institute, Ascot, Berks.

A worked example follows which is relevant to Chapter 22. It is based on the 1990 version of EC4. There is no national code for composite columns in buildings and the 1990 version of EC4 was the latest available when the 6th edition was prepared.

The **Steel Construction Institute** Silwood Park, Ascot, Berks SL5 7QN	Subject **DESIGN OF COMPOSITE COLUMN**		Chapter ref. **22**
	Design code **EUROCODE 4 (1990 DRAFT)**	Made by **RML**	Sheet no. **1**
		Checked by **GWO**	

Problem

Determine the envelope of design resistance to axial load, major axis moment and minor axis moment of the composite column shown below. Its effective length is 4 m.

Column size   *400 mm square*

Steel section   *254 × 254 × 89 kg/m UC*

Grade S355 steel p_y =   *355 N/mm²*

Concrete grade f_{cu} =   *30 N/mm² (normal weight)*

Reinforcement   *4 No. T12 bars*

f_y =   *460 N/mm²*

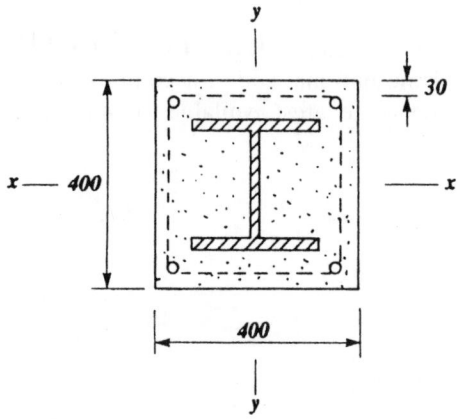

Cross-section through column

Note: EC3 does not limit the design strength for steel for t ≤ 40 mm.

The Steel Construction Institute Silwood Park, Ascot, Berks SL5 7QN		Subject *DESIGN OF COMPOSITE COLUMN*		Chapter ref. *22*
		Design code *EUROCODE 4 (1990 DRAFT)*	Made by **RML** Checked by **GWO**	Sheet no. **2**

Column section properties

Area of section $\qquad$ A_s $\quad = \quad$ $11400\,mm^2$

Area of reinforcement $\quad A_r$ $\quad = \quad 4 \times \pi \times \dfrac{12^2}{4} \quad = \quad 452\,mm^2$

Area of concrete $\qquad A_c$ $\quad = \quad 400^2 - 11400 - 452$

$\qquad\qquad\qquad\qquad\qquad = \quad 148.1 \times 10^3\,mm^2$

Squash load (compressive resistance) of stocky column – equation 22.1

$P_u \quad = \quad A_s p_y + A_r \times 0.87 f_y + A_c \times 0.45 f_{cu}$

$\qquad = \quad (11.4 \times 10^3 \times 355 + 452 \times 0.87 \times 460$
$\qquad\qquad + 148.1 \times 10^3 \times 0.45 \times 30) \times 10^{-3}$

$\qquad = \quad 6228\,kN$

Tensile resistance of column

$P_t \quad = \quad A_s p_y + A_r \times 0.87 f_y$

$\qquad = \quad (11.4 \times 10^3 \times 355 + 452 \times 0.87 \times 460) \times 10^{-3}$

$\qquad = \quad 4228\,kN$

Moment resistance of column – major axis

Determine position of plastic neutral axis (PNA) parallel to x-x axis of column. Assume that PNA lies at a distance Y into the web of the steel section. Concrete tensile strength is neglected.

Equating tension and compression, it follows that the forces in flanges and reinforcement cancel out.

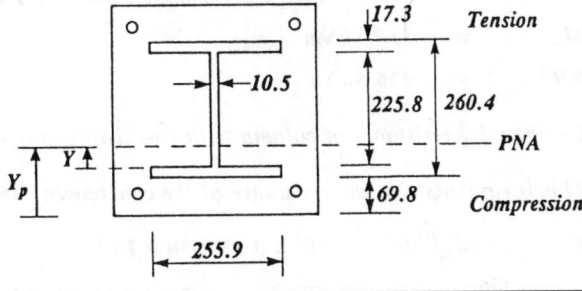

The Steel Construction Institute Silwood Park, Ascot, Berks SL5 7QN	Subject ***DESIGN OF COMPOSITE COLUMN***	Chapter ref. *22*
Design code ***EUROCODE 4 (1990 DRAFT)***	Made by *RML* Checked by *GWO*	Sheet no. *3*

Therefore

tensile force in web depth (225.8 – 2Y) = compressive force in concrete

$$(225.8 - 2Y) \times 10.5 \times 355 \ = \ [(Y + 69.8 + 17.3) \times 400 - \frac{452}{2} - 17.3$$

$$\times 225.9 - Y \times 10.5] \times 0.45 \times 30$$

$$841.7 \times 10^3 - 7.46 \times 10^3 Y \ = \ 5.4 \times 10^3 Y + 470.3 \times 10^3 - 3.05 \times 10^3$$
$$- 52.7 \times 10^3 - 0.14 \times 10^3 Y$$

or $(5.4 + 7.46 - 0.14) Y \ = \ 841.7 - 470.3 + 3.05 + 52.7$

$\therefore$ $Y = 33.6\,mm$

Hence plastic neutral axis depth from edge of slab is

Y_p $=$ $33.6 + 69.8 + 17.3 = 120.7\,mm$

Plastic moment of resistance about x-x axis:

Take moments about PNA at depth Y_p

M_{px} $=$ $17.3 \times 255.9 \times 355 \times 10^{-6} \times (260.4 - 17.3)$

$$+ \frac{452}{2} \times 0.87 \times 460 \times 10^{-6} \times (400 - 30 \times 2 - 12)$$

$$+ 10.5 \times \frac{Y^2}{2} \times 355 \times 10^{-6} + 10.5 \times \frac{(225.8 - Y)^2}{2} \times 355 \times 10^{-6}$$
$$+ (5.4 Y + 470.3 - 3.05 - 52.7 - 0.14 Y) \times 10^{-3} \times Y_p/2$$

M_{px} $=$ $382.0 + 29.7 + 2.1 + 68.8 + 35.7$ *for Y = 33.6 mm*

M_{px} $=$ $518.3\,kNm$

$0.9\,M_{px}$ $=$ $466.5\,kNm$

Compressive resistance of column with co-existing moment (equation 22.5)

This is equivalent to the resistance of the concrete in compression (see text)

$P_{o,x}$ $=$ $148.1 \times 10^3 \times 0.45 \times 30 \times 10^{-3}$ $=$ $1999\,kN$

The **Steel Construction Institute**  Silwood Park, Ascot, Berks SL5 7QN	Subject ***DESIGN OF COMPOSITE COLUMN***	Chapter ref. **22**
Design code ***EUROCODE 4*** ***(1990 DRAFT)***	Made by **RML** Checked by**GWO**	Sheet no. **4**

Moment resistance of column – minor axis

**Assume PNA at distance X
from edge of web.**

**Equating tensile and
compressive forces.**

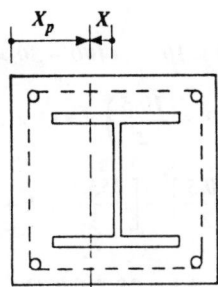

If $A_w p_y > 0.45 f_{cu} \left(b^2 - A_f - \dfrac{A_r}{2} \right)$

then PNA lies in web

$10.5 \times 225.8 \times 355 \times 10^{-3} > 0.45 \times 30 \times (400^2/2 - 17.3 \times 255.9 - 452/2)$

841.7 < 1017.2

So PNA lies just outside web

Therefore distance X is obtained from:

$(2X + 10.5) \times 2 \times 17.3 \times 355 \times 10^{-3} + 10.5 \times 225.8 \times 355 \times 10^{-3}$

$= 0.45 \times 30 \times \left[\left(200 - \dfrac{10.5}{2} - X \right) \times 400 - \dfrac{452}{2} - \left[\dfrac{255.9}{2} - \dfrac{10.5}{2} - X \right] \times 17.3 \right]$

$24.6X + 129.0 + 841.7 \quad = \quad 1051.6 - 5.4X - 3.1 - 28.7 + 0.2X$

$X (24.6 + 5.4 - 0.2) \quad = \quad 1051.6 - 841.7 - 129.0 - 3.1 - 28.7$

$X = 1.7\,mm$

Therefore X_p $\quad = \quad 200 - 10.5/2 - 1.7 \quad = \quad 193.0\,mm$

The **Steel Construction Institute** Silwood Park, Ascot, Berks SL5 7QN 	Subject **DESIGN OF COMPOSITE COLUMN**	Chapter ref. **22**
	Design code **EUROCODE 4 (1990 DRAFT)**	Made by **RML**	Sheet no. **5**
		Checked by **GWO**	

Plastic moment resistance about y-y axis

Take moments about X_p

$$M_{py} = \frac{452}{2} \times 0.87 \times 460 \times 10^{-6} \times (400 - 30 \times 2 - 12)$$

$$+ 10.5 \times 225.8 \times \left(X + \frac{10.5}{2}\right) \times 355 \times 10^{-6}$$

$$+ \left[\left[\frac{255.9}{2} - X - \frac{10.5}{2}\right]^2 + \left[\frac{255.9}{2} + X + \frac{10.5}{2}\right]^2\right] \times 17.3 \times 355 \times 10^{-6}$$

$$+ (1051.6 - 5.4X - 3.1 - 28.7 + 0.2X)\frac{X_p}{2} \times 10^{-3}$$

$$M_{py} = 29.7 + 5.8 + 201.6 + 97.6$$

$$= 334.7\,kNm$$

$$0.9\,M_{py} = 301.2\,kNm$$

Compressive resistance of column subject to co-existing moment of M_{py}

$$P_{o,y} = P_{o,x} = 1999\,kN$$

The **Steel Construction Institute** Silwood Park, Ascot, Berks SL5 7QN	Subject **DESIGN OF COMPOSITE COLUMN**		Chapter ref. **22**
	Design code **EUROCODE 4 (1990 DRAFT)**	Made by **RML**	Sheet no. **6**
		Checked by **GWO**	

Effect of column slenderness

Modular ratio α_e $= \dfrac{E_s}{450 f_{cu}} = \dfrac{205 \times 10^3}{450 \times 30} = 15.2$

Second moment of area of column – major axis bending

$I_{xx} \approx \left[I_{xs} + \dfrac{452}{4} \times (400 - 60 - 12)^2 \right]\left(1 - \dfrac{1}{\alpha_e} \right) + \dfrac{400^4}{12} \times \dfrac{1}{\alpha_e}$

$I_{xs} = 143.0 \times 10^6 \, mm^4$

$I_{xx} = 285.3 \times 10^6 \, mm^4$

Second moment of area of column – minor axis bending

$I_{yy} = \left[I_{ys} + \dfrac{452}{4} \times (400 - 60 - 12)^2 \right]\left(1 - \dfrac{1}{\alpha_e} \right) + \dfrac{400^4}{12} \times \dfrac{1}{\alpha_e}$

$I_{ys} = 48.5 \times 10^6 \, mm^4$

$I_{yy} = 197.0 \times 10^6 \, mm^4$

$P_u = 6228 \, kN$ squash load – see calculation sheet no. 2

Slenderness factor

$\bar{\lambda} = \dfrac{L}{\pi}\left(\dfrac{P_u}{E_s \Sigma I_{yy}} \right)^{0.5}$

$= \dfrac{4000}{\pi}\left(\dfrac{6228 \times 10^3}{205 \times 10^3 \times 197.0 \times 10^6} \right)^{0.5} = 0.5$

$\lambda_{eff} = \bar{\lambda}\pi(E_s/p_y)^{0.5} = 0.5 \, \pi (205 \times 10^3/355)^{0.5} \simeq 38$

From BS 5950: Part 1 Table 27(c)

Axial strength $p_c = 308 \, N/mm^2$

The **Steel Construction Institute** Silwood Park, Ascot, Berks SL5 7QN	Subject ***DESIGN OF COMPOSITE COLUMN***	Chapter ref. ***22***
Design code ***EUROCODE 4 (1990 DRAFT)***	Made by　***RML*** Checked by ***GWO***	Sheet no.　**7**

Factor K_1 representing reduced axial resistance of the composite column is 308/355 = 0.87

Hence $K_1 P_u = 0.87 \times 6228 = 5418\,kN$

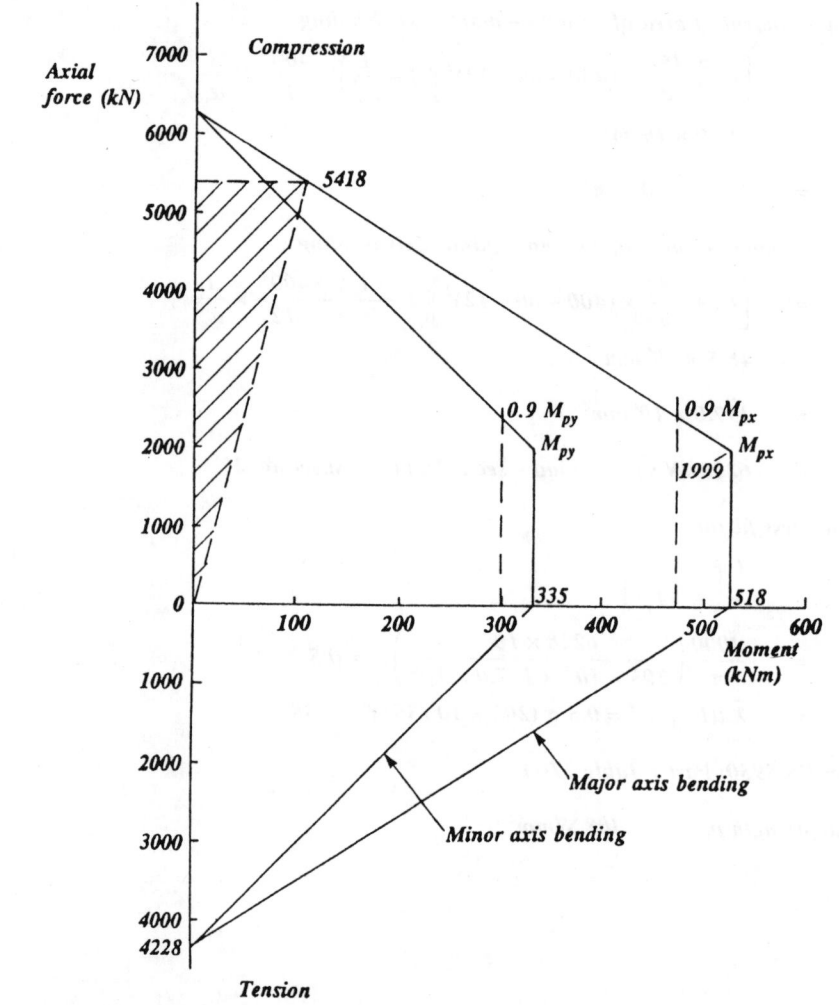

Interaction between axial and bending resistances of cross-section

	Subject	Chapter ref.
The Steel Construction Institute  Silwood Park, Ascot, Berks SL5 7QN	*DESIGN OF COMPOSITE COLUMN*	**22**

Design code *EUROCODE 4 (1990 DRAFT)*	Made by ***RML***	Sheet no. **8**
	Checked by ***GWO***	

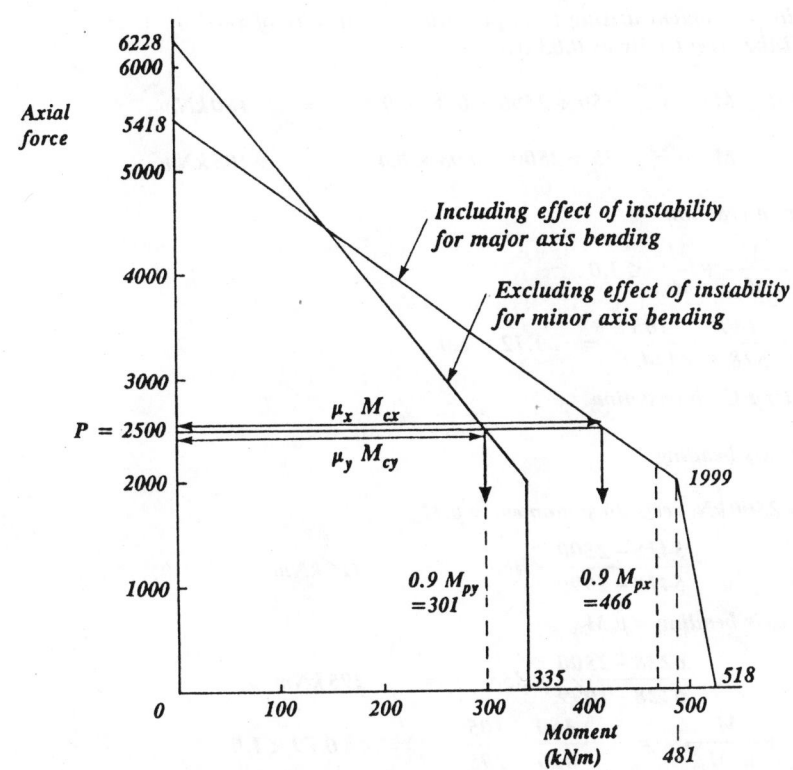

Reduced interaction of axial force and moment in slender column

Example of use of interaction diagram

Assume the following design loads

Axial force P   =   *2500 kN*

Bending moment about major axis   =   *150 kNm*

Bending moment about minor axis   =   *75 kNm*

The Steel Construction Institute Silwood Park, Ascot, Berks SL5 7QN	Subject ***DESIGN OF COMPOSITE COLUMN***	Chapter ref. **22**
Design code *EUROCODE 4 (1990 DRAFT)*	Made by **RML** Checked by **GWO**	Sheet no. **9**

Additional moment arising from potential eccentricity of vertical load. Take eccentricity as 0.03 b_c.

Moment M_x = $150 + 2500 \times 0.03 \times 0.4$ = $180\,kNm$

         M_y = $75 + 2500 \times 0.03 \times 0.4$ = $105\,kNm$

Simple interaction

$$\frac{P}{K_1 P_u} + \frac{M_x}{M_{px}} + \frac{M_y}{M_{py}} \leq 1.0$$

$$\frac{2500}{5418} + \frac{180}{518.3} + \frac{105}{334.7} \;=\; 1.12 > 1.0$$

Consider EC4 interaction

Major axis bending

At P = 2500 kN, resistance moment = $\mu_y M_{px}$

$$\mu_y M_{px} \;=\; \frac{5418 - 2500}{5418 - 1999} \times 488 \;=\; 416\,kNm$$

Minor axis bending = $\mu_y M_{py}$

$$\mu_y M_{py} \;=\; \frac{6228 - 2500}{6228 - 1999} \times 335 \;=\; 295\,kNm$$

$$\frac{M_x}{\mu_y M_{px}} + \frac{M_y}{\mu_y M_{py}} \;=\; \frac{180}{416} + \frac{105}{295} \;=\; 0.79 < 1.0$$

Therefore, the design is satisfactory for biaxial bending.

Chapter 23
Bolts

by HUBERT BARBER

23.1 Types of bolt

23.1.1 Non-preloaded bolts

The most frequently used bolts in structural connections are non-preloaded bolts of strength grades 4.6 and 8.8 used in 2 mm clearance holes. These bolts are termed *ordinary bolts* and are specified in BS 4190[1] and other standards as shown in Table 23.1, which is reproduced from BS 5950-2.[2] Precision bolts, manufactured to BS 3692,[3] for use in close tolerance holes are not widely used.

The strength grade designation system is in accordance with BS EN ISO 898-1.[4] It consists of two figures: the first is one-hundredth of the minimum tensile strength in N/mm^2, and the second is one-hundredth of the ratio between the minimum yield stress and the minimum tensile strength, expressed as a percentage. Multiplication of these two figures will give the yield stress in N/mm^2. A grade 4.6 bolt has a minimum tensile strength of 400 N/mm^2 and a yield stress of 240 N/mm^2 (0.6 × 400).

There is a similar designation system for nuts. This is a single figure, being one-hundreth of the specified proof load stress in N/mm^2. Hence a grade 8 nut has a proof load stress of 800 N/mm^2 and would normally be used with a grade 8.8 bolt. It is permissible however to use a nut of higher grade than the bolt to minimize the risk of thread stripping. Where bolts are galvanized or sheradized nuts of a higher grade must be used (see footnotes e–h in Table 23.1). Grade 8.8 bolts to BS 4190 are commonly available and are recommended for all main structural connections, with the standard bolt being 20 mm diameter. Grade 4.6 bolts are generally used only for fixing lighter components such as purlins or sheeting rails, when 12 mm or 16 mm bolts may be adopted. There may be situations, for example a column splice subjected to large load reversals in a braced bay, where the engineer feels that joint slip is unacceptable. In these cases preloaded high-strength friction-grip bolts may be used. HSFG bolts may also be used as non-preloaded in the same way as ordinary bolts.

23.1.2 Preloaded high-strength friction-grip bolts

High-strength friction-grip bolts are manufactured to the requirements of BS 4395.[5] This British Standard covers two distinctly different types of fastener.

Table 23.1 Matching bolts, nuts and washers (taken from BS 5950-2:2001)

Type of bolts		Nuts		Washers	
Grade	Standard	Class or grade[c]	Standard	Class	Standard
Non-preloaded bolts					
4.6	BS EN ISO 4016 BS EN ISO 4018	Class 4[d]	BS EN ISO 4034	100HV	BS EN ISO 7091
	BS 4190 BS 4933[a]	Grade 4	BS 4190	–	BS 4320[i]
8.8	BS EN ISO 4014[b] BS EN ISO 4017[b]	Class 8[e]	BS EN ISO 4032[b]	100HV	BS EN ISO 7091
	BS 4190	Grade 8[f]	BS 4190	–	BS 4320[i]
10.9	BS EN ISO 4014[b] BS EN ISO 4017[b]	Class 10[g]	BS EN ISO 4032[b]	100HV	BS EN ISO 7091
	BS 4190	Grade 10[h]	BS 4190	–	BS 4320[i]
Non-preloaded HSFG bolts					
General grade	BS 4395-1	General grade	BS 4395-1	–	BS 4320[j]
Higher grade	BS 4395-2	Higher grade	BS 4395-2		BS 4320[j]
Preloaded HSFG bolts					
General grade	BS 4395-1	General grade	BS 4395-1	–	BS 4395-1[k]
Higher grade	BS 4395-2	Higher grade	BS 4395-2		BS 4395-2[k]
Holding-down bolts					
4.6	BS 7419	Grade 4	BS 4190	–	BS 4320[j]
8.8	BS 7419	Grade 8[f]	BS 4190		BS 4320[j]
8.8 preloaded	BS 7419	General grade	BS 4395-1		BS 4395-1[k]

[a] BS 4933 has been declared obsolescent, but should still be used for 90° countersunk head bolts and cup head bolts until corresponding BS EN standards are available.
[b] Grade 8.8 and 10.9 bolts to the strength grades of BS EN ISO 4014 or BS EN ISO 4017 but with the dimensions and tolerances specified in BS EN ISO 4016 or BS EN ISO 4018 may also be used, with matching nuts to the strength classes of BS EN ISO 4032 but the dimensions and tolerances of BS EN ISO 4034.
[c] Nuts of a higher class or grade may also be used.
[d] Class 5 nuts for size M 16 and smaller.
[e] Nuts for galvanized or sherardized 8.8 bolts shall be class 10.
[f] Nuts for galvanized or sherardized 8.8 bolts shall be grade 10 to BS 4190.
[g] Nuts for galvanized or sheradrdized 10.9 bolts shall be class 12 to BS EN ISO 4033.
[h] Nuts for galvanized or sherardized 10.9 bolts shall be grade 12 to BS 4190.
[i] Black steel washers to section 2 of BS 4320, normal diameter series.
[j] Black steel washers to BS 4320:1968, Section 2, large diameter series.
[k] Direct tension indicators to BS 7644 may also be used with preloaded HSFG bolts.

Part 1 – General grade. The most commonly used type in general structural steelwork. The strength grade is about that of 8.8 bolts up to 24 mm diameter. For larger sizes, the minimum ultimate strength is reduced to approximately 74 kgf/mm² (725 N/mm²) and the yield to approximately 77% of that figure. Their use is governed by BS 4604: Part 1.[6]

Part 2 – Higher grade. These are made from 10.9 grade and although a higher tensile load can be applied to them there is a much reduced margin between the yield load and the ultimate strength. It is therefore not permissible to use them for connec-

tions in which there is applied tension. The 'part turn' method of tightening is not permitted by BS 4604: Part 2, which requires in Table 2 of that Standard that the prestressing is within strictly controlled limits, given as from 0.85 to 1.15 of the proof load.

23.1.3 Fully threaded bolts

Common practice in the past has been to use bolts with a short thread length, i.e. $1.5d$, and to specify them in 5 mm length increments. This can result in an enormous number of different bolts, which is both costly to administer and can prevent rapid erection.

It is recommended that fully threaded bolts (technically known as screws) be used as the industry standard. They can be provided longer than necessary for a particular connection and can therefore dramatically reduce the range of bolt lengths specified.

Research has demonstrated that the very marginal increase in deformation with fully threaded bolts in bearing has no significant effect on the performance of a typical joint. In the specific instances where this additional deformation might be of concern, it is normal and recommended practice to use preloaded (HSFG) bolts. These can be used for example in tension and compression splices where the bolts are in shear/bearing or in column splices where the column ends are not in bearing. A paper by Owens[7] gives the background to the use of fully threaded bolts in both tension and shear conditions.

23.2 Methods of tightening and their application

The usual methods of tightening of friction-grip bolts – part-turn of the nut, torque control and load-indicating washers – are outlined in section 33.4.5. Further details are given in Reference 8.

23.3 Geometric considerations

23.3.1 Hole sizes

Ordinary bolts should be used in holes having a suitable clearance in order to facilitate insertion. For bolts up to and including diameters of 24 mm the clearance should be 2 mm, and above 24 mm should be 3 mm. Table 33 in BS 5950-1 gives standard dimensions of holes for use with non-preloaded bolts. When using oversize or slotted holes care should be taken that the washers used are sufficiently

large and thick to span the hole. Large diameter washers to BS 4320[10] may be required.

Normal clearance holes as given for ordinary bolts are usually used for HSFG fasteners (see BS 5950-1 Table 36), but it is permissible to use oversize, short or long slotted holes provided standard hardened washers are used over the holes in the outer plies and not just under the turned part (see clause 6.4.6.2).

Oversize and short slotted holes may be used in all plies but long slotted holes may only be used in one single ply in any connection. If a long slotted hole occurs in an outer ply it should be covered with a washer plate longer than the slot and at least 8 mm thick.

The assessment of the slip resistance is affected when oversize or slotted holes are used. The constant K_s (clause 6.4.2 of BS 5950), which is 1.0 for bolts in clearance holes, is reduced to 0.85 and 0.7 as given in paragraph 23.5.2.

23.3.2 Spacing of fasteners, end and edge distances

Spacing is covered fully in Section 6.2 of BS 5950-1:2000, summarized as follows:

Minimum requirements (see Fig. 23.1)

Centres of fasteners	2.5d
Edge or end distances	
rolled, sawn, planed or machine flame cut	1.25D
sheared or hand flame cut	1.4D

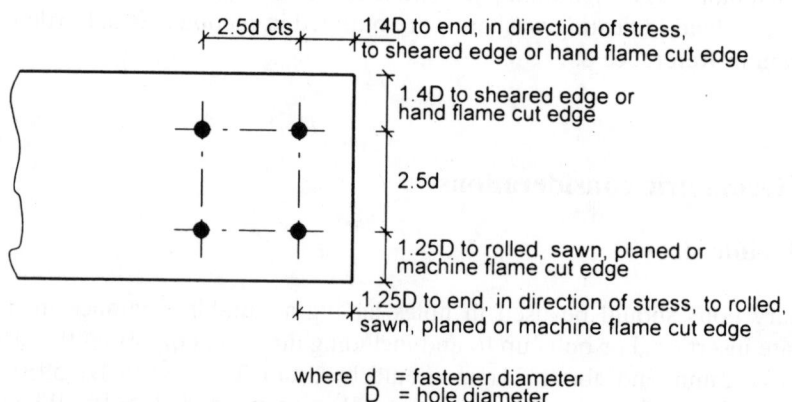

where d = fastener diameter
       D = hole diameter

Fig. 23.1   Minimum dimensions (see Table in Appendix)

The full bearing value of a bolt through a connected part cannot be developed if the end distance is less than $2D$. For end distances between $1.4D$ and $2D$ the bearing value is proportionally reduced. The minimum end distance for the full bearing value in the case of parallel shank HSFG bolts is $3d$ (see below).

Maximum requirements (see Fig. 23.2)

Centres of fasteners in the direction of stress $14t$

where t is the thickness of the thinner element.

Edge distances

Distance to the nearest line of fasteners from the edge of an unstiffened part $11t\varepsilon$

In corrosive environments

Centres of fasteners in any direction $16t$ or $200\,\text{mm}$
Edge distances $4t$ + $40\,\text{mm}$

d = fastener diameter
D = hole diameter
t = thickness of the thinner outside ply

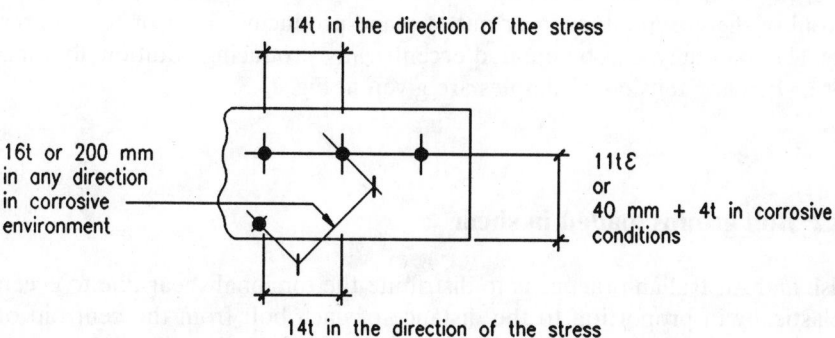

Fig. 23.2   Maximum dimensions (see Table in Appendix)

$$\varepsilon = \left[\frac{275}{p_y}\right]^{\frac{1}{2}}$$ where p_y is the design strength of the element concerned.

In the case of high-strength friction-grip bolts the slip resistance is a serviceability criterion and the connection could slip into bearing between working and failure load. It is necessary therefore to check the bearing strength of the joint. The end distance to attain the full bearing strength is increased to $3d$.

23.3.3 Back marks and cross centres

The back mark is the distance from the back of an angle or channel web to the centre of a hole through the leg or flange. This dimension is determined so as to allow the tightening of a bolt with a standard podger spanner, to be as near as possible to the centroidal axis and to allow the required edge distance. Recommended back marks and diameters are given in the tables for channels and for angles (Appendix *Connection design: Back marks in channel flanges* and *Back marks in angles*).

The distances between centres of holes (cross centres) in the flanges of joists, universal beams and universal columns are similarly determined after consideration of accessibility and edge distances.

Recommended cross centres and diameters are given in the Appendix *Connection design: Cross centres through flanges*.

23.4 Methods of analysis of bolt groups

23.4.1 Introduction

Any group of bolts may be required to resist an applied load acting through the centroid of the group either in or out of plane producing shear or tension respectively. The load may also be applied eccentrically producing additionally torsional shear or bending tension. Examples are given in Fig. 23.3.

23.4.2 Bolt groups loaded in shear

British and Australian practice is to distribute the torsional shear due to eccentricity elastically in proportion to the distance of each bolt from the centroid of the group. This is referred to as the *polar inertia method*.

(In some countries, notably Canada and in some cases the USA, the *instantaneous*

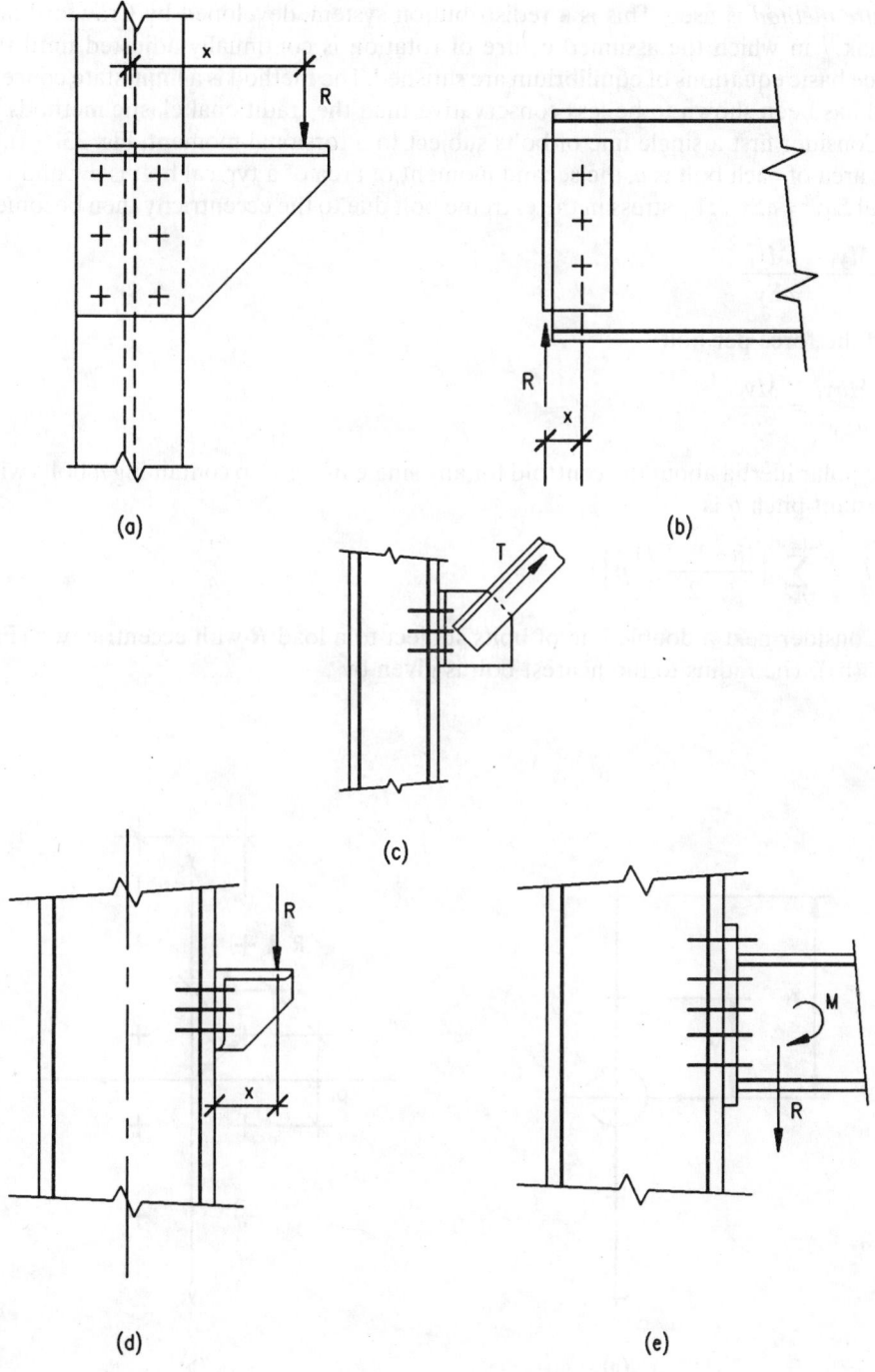

Fig. 23.3 Bolt groups

centre method is used. This is a redistribution system, developed by Crawford and Kulak,[11] in which the assumed centre of rotation is continually adjusted until the three basic equations of equilibrium are satisfied. The method is a limit-state concept and has been shown to be less conservative than the traditional elastic methods.)

Consider first a single line of bolts subject to a torsional moment, Fig. 23.4(a). If the area of each bolt is a, the second moment of area of a typical bolt is ay^2 and the total $\Sigma ay^2 = a\Sigma y^2$. The stress in the extreme bolt due to the eccentricity then becomes:

$$\frac{My_1}{I} = \frac{My_1}{a\Sigma y^2}$$

and the force per bolt

$$\frac{May_1}{a\Sigma y^2} = \frac{My_1}{\Sigma y^2}$$

The polar inertia about the centroid for any single line group containing n bolts with constant pitch p is

$$I_0 = \sum_{J=0}^{J=(n-1)} \left[\frac{(n-1-2J)}{2}p \right]^2$$

Consider next a double line of bolts subject to a load R with eccentricity x (Fig. 23.4(b)). The radius to the nearest bolt is given by

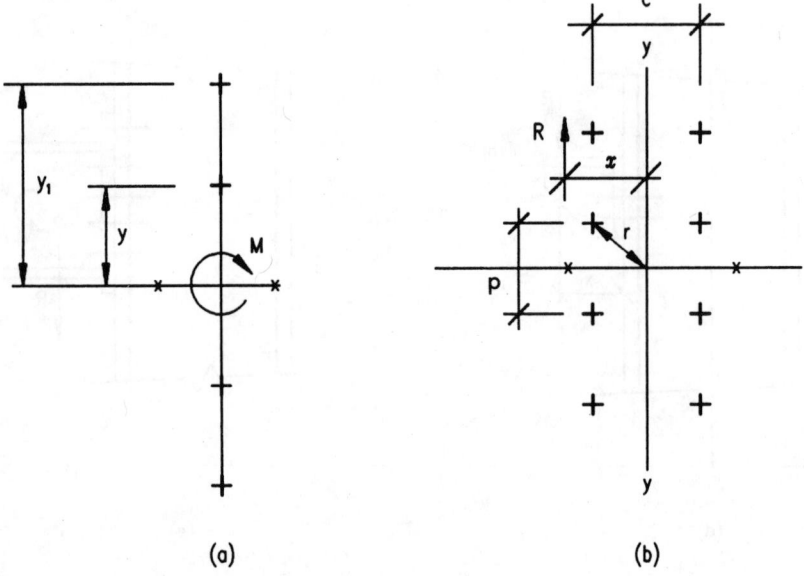

(a)   (b)

Fig. 23.4 Bolt group analysis

$$r = \sqrt{\left[\left(\frac{p}{2}\right)^2 + \left(\frac{c}{2}\right)^2\right]}$$

$$r^2 = \left(\frac{p}{2}\right)^2 + \left(\frac{c}{2}\right)^2$$

I_0 for a typical bolt is then ar^2 and it follows that I_0 for the whole group becomes $I_{xx} + I_{yy}$ which, if there are m vertical rows and n horizontal rows, is

$$I_{00} = m \sum_{J=0}^{J=(n-1)} \left[\frac{(n-1-2J)}{2}p\right]^2 + n \sum_{J=0}^{J=(m-1)} \left[\frac{(m-1-2J)}{2}c\right]^2$$

where c is the cross centres between the vertical lines. The distance to the extreme bolt is

$$r = \sqrt{\left\{\left[\frac{(n-1)}{2}p\right]^2 + \left[\frac{(m-1)}{2}c\right]^2\right\}}$$

The force in the extreme bolt due to the moment is

$$f_m = \frac{Rxr}{I_{00}}$$

The force in each bolt due to the shear (assumed equally divided between all bolts) is

$$f_v = \frac{R}{mn}$$

The combined force per bolt is the resultant of these two: see Fig. 23.5.

The resultant bolt force is then checked against the bolt strength in single shear, double shear or bearing as is appropriate. In the case of bearing, however, it should be remembered that the full strength cannot be achieved if the end distance meas-

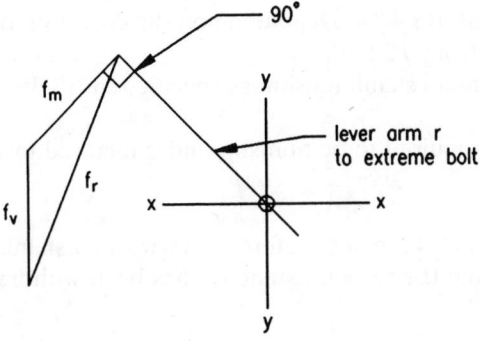

Fig. 23.5   Resultant force

ured along the line of the resultant is less than twice the diameter of the bolt. In this case the bearing strength is reduced in proportion.

23.5 Design strengths

23.5.1 General

Non-preloaded bolts have to resist forces in shear and bearing or tension or combinations of these. Preloaded HSFG bolts resist shear by developing friction between the plies; they may also resist external tension. The load capacity of a joint may be affected if it is excessively long, has a large grip length or thickness of packing, or is subjected to prying action.

23.5.2 Shear

When partially threaded bolts are loaded in shear some of the threads of the bolt may lie within the shear plane. The shear strength is then assessed on the tensile area of the bolt. If however it is known with certainty that no part of the threaded section of the bolt will fall within the shear plane the full shank area may be used in determining the strength.

Until a preloaded HSFG bolted shear connection slips, shear force is resisted by friction. The slip resistance, P_{sL}, is given by

$$P_{sL} = 1.1 \, K_s \, \mu \, P_0$$

where $K_s = 1.0$ for clearance holes;

$\qquad = 0.85$ for oversized holes, short slotted holes or long slotted holes loaded perpendicular to the slot direction;

$\qquad = 0.7$ for long slotted holes loaded parallel to the slot direction

$\quad \mu =$ slip factor from BS 5950-1 Table 35 or determined from tests in accordance with BS 4604. Depending on the condition of the faying surfaces μ varies from 0.2 to 0.5.

$\quad P_0 =$ the minimum shank tension as specified in BS 4604.

If the connection is required to be non-slip under factored loads,

$$P_{sL} = 0.9 \, K_s \, \mu \, P_0$$

BS 5950-1:2000 clause 6.4.2 makes reference to waisted-shank HSFG bolts. These are no longer used and the relevant standard has been withdrawn.

23.5.3 Bearing

Non-preloaded bolts

In cases where the strength is assessed from the bearing of the bolt on one of the connected parts it is necessary to limit the bearing strength in accordance with the end distance requirements.

For end distances between the minimum ($1.4D$) and the minimum value for full bearing ($2.0D$) the bearing value is proportional to the end distance.

Preloaded HSFG bolts

After slip has occurred an HSFG bolt acts in shear and bearing.

When the end distance is at least three times the nominal diameter of the bolt, the bearing strengths, p_{bg}, given in Table 32 of BS 5950-1:2000 are 460, 550 and 670 N/mm² for S275, S355 and S460 steel, respectively.

This is a condition which can only apply after joint slip has occurred, and the joint is beyond the serviceability limit.

23.5.4 Tension

When bolts are loaded in tension additional axial forces are sometimes induced due to prying action. In some circumstances these may be allowed for by reducing the nominal tension capacity of a bolt by 20% (see BS 5950-1 clause 6.3.4.2). In other cases, the prying force should be calculated and the total applied tension in the bolt should be compared with its capacity (clause 6.3.4.3).

It is permissible to subject HSFG bolts complying with Part 1 of BS 4395 to externally applied tension. Fasteners to Part 2 of that standard however should not be required to carry external tension.

23.5.5 Combined shear and tension

Non-preloaded bolts which are subject to both tension and shear should satisfy one of the following relationships depending on how prying forces have been accounted for:

– for the simple method

$$\frac{F_s}{P_s} + \frac{F_t}{P_{nom}} \leq 1.4$$

–   for the more exact method

$$\frac{F_s}{P_s} + \frac{F_{tot}}{P_t} \leq 1.4$$

where F_s is the applied shear, F_t is the applied tension, P_s is the shear capacity, and P_t is the tensile capacity. The shear capacity, P_s, may be based on the tension stress area or on the full shank area as appropriate (see section 23.5.2).

This expression allows that when either the shear or tensile capacity is fully taken up, the other may be stressed to 40% of its capacity.

Preloaded bolts in friction grip connections that are also subject to externally applied tension should satisfy:

–   for connections designed to be non-slip in service ($P_{sL} = 1.1\ K_s\ \mu\ P_0$):

$$\frac{F_s}{P_{sL}} + \frac{F_{tot}}{1.1 P_0} \leq 1 \qquad \text{but} \qquad F_{tot} \leq A_t p_t$$

–   for connections designed to be non-slip under factored loads ($P_{sL} = 0.9\ K_s\ \mu\ P_0$):

$$\frac{F_s}{P_{sL}} + \frac{F_{tot}}{0.9 P_0} \leq 1$$

where

A_t is the tensile stress area;
F_s is the applied shear;
F_{tot} is the total applied tension in the bolt, including the calculated prying force;
P_0 is the specified minimum preload, see BS 4604;
P_t is the tension strength of the bolt given in Table 34 of BS 5950-1:2000.

23.5.6 Long joints, large grips and packing

When the joint length in a splice or end connection, L_j, defined as the distance between the first and last bolt on either side of the joint is greater than 500 mm, the strength of the joint is reduced by the factor $(5500 - L_j)/5000$, where L_j is the joint length as defined in Fig. 23.6. Similarly when the grip length, T_g, the total thickness of the connected plies, is greater than five times the nominal diameter of the bolts, the strength of the joint is reduced by the factor $8d/(3d + T_g)$, where d is the nominal bolt diameter and T_g is as defined above (see Fig. 23.7).

The total thickness of steel packing, t_{pa}, at a shear plane should not exceed $4d/3$, where d is the nominal bolt diameter. Where t_{pa} exceeds $d/3$ the shear capacity, P_s, should be taken as:

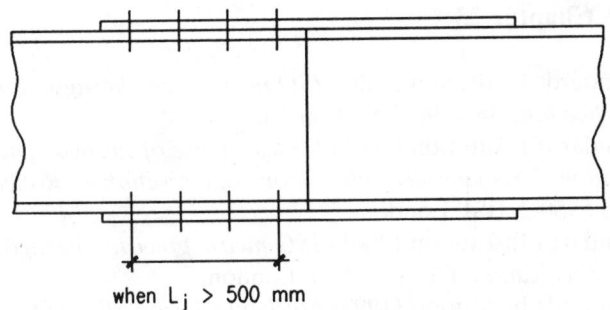

when L$_j$ > 500 mm

Fig. 23.6   Long joint

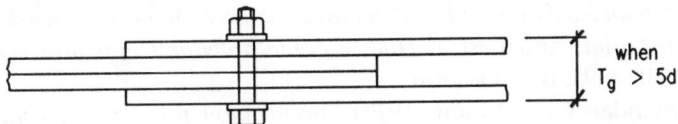

when
T$_g$ > 5d

Fig. 23.7   Large grip

$$P_s = p_s A_s \left(\frac{gd}{8d + 3t_{pa}} \right)$$

In cases when more than one of the above conditions apply it is only necessary to apply the factor producing the greater reduction.

The large grip reduction does not apply to HSFG bolts except when bolts are being used in bearing for considerations beyond serviceability. Similarly, 'long joint' effects in HSFG bolted joints apply only after slipping, and the same factor, $(5500 - L_j)/5000$, is used (see 6.4.4.).

23.6 Tables of strengths

23.6.1 Bolt strengths

Tables 31 and 32 of BS 5950 give respectively the strength of bolts in clearance holes and the bearing strength of connected parts for ordinary bolts in clearance holes. The values are given in the Appendix *Connection design: Bolt strengths*.

23.6.2 Bolt capacities

Tables of bolt capacities for various grades and applications are given in the Appendix *Connection design*.

References to Chapter 23

1. British Standards Institution (2001) *ISO metric black hexagon bolts, screws and nuts – specification.* BS 4190, BSI, London.
2. British Standards Institution (2001) *Structural use of steelwork in building.* Part 2: *Specification for materials, fabrication and erection – Rolled and welded sections.* BS 5950-2, BSI, London.
3. British Standards Institution (2001) *ISO metric precision hexagon bolts, screws and nuts – specification,* BS 3692, BSI, London.
4. British Standards Institution (1999) *Mechanical properties of fasteners made of carbon steel and alloy steel. Bolts, screws and studs.* BS EN ISO 898-1, BSI, London.
5. British Standards Institution (1969) *Specification for high-strength friction-grip bolts and associated nuts and washers for structural engineering metric series.* Part 1: *General grade* and Part 2: *Higher grade bolts and nuts and general grade washers.* BS 4395, BSI, London.
6. British Standards Institution (1970) *Specification for the use of high-strength friction-grip bolts in structural steelwork metric series.* Part 1: *General grade* and Part 2: *Higher grade.* BS 4604, BSI, London.
7. Owens G.W. (1992) Use of fully threaded bolts for connections in structural steelwork for buildings. *The Structural Engineer,* 1 September, pp. 297–300.
8. Owens G.W. & Cheal B.D. (1989) *Structural Steelwork Connections.* Butterworths, London.
9. British Standards Institution (2000) *Structural use of steelwork in buildings.* Part 1: *Code of practice for design – Rolled and welded sections.* BS 5950-1, BSI, London.
10. British Standards Institution (1968) *Specification for metal washers for general engineering purposes metric series.* BS 4320, BSI, London.
11. Crawsford S.F. and Kulak G.L. (1971) Eccentrically loaded bolted connections. *J. Struct. Div., ASCE,* **97**, No. ST3, March, 765–83.

Further reading for Chapter 23

BCSA (2002) *Joints in steel construction. Simple Connections.* British Constructional Steelwork Association/The Steel Construction Institute, Publication No. P212.
BCSA (2002) *National Structural Steelwork Specification for Building Construction,* 4th Edn. British Constructional Steelwork Association, Publication No. 203/02.
Kulak G.L., Fisher J.W. & Struik J.H.A. (1987) *Design Criteria for Bolted and Riveted Joints,* 2nd edn. John Wiley & Sons.

Chapter 24
Welds and design for welding

by RALPH B. G. YEO and HUBERT BARBER

Without welding, even under site conditions, modern buildings and bridges and the most critical of structures, such as offshore platforms and nuclear power stations, could not be constructed. Good designs lead to cost-effective fabrications that can be made to required standards by the use of coordinated specifications, which provide means for quantitative control of weld quality. This chapter briefly discusses the advantages of welding, the means to control weld quality, and some design recommendations.

24.1 Advantages of welding

Welding offers many advantages over other joining methods:

- Freedom of design, and the opportunity to develop innovative structures
- Easy introduction of stiffening elements
- Less weight than in bolted joints because fewer plates are required
- Welded joints allow increased usable space in a structure
- Protection against the effects of fire and corrosion are easier and more effective.

24.1.1 Aesthetics and freedom of design

Probably the main benefit of welded construction is the freedom of design, compared with bolted joints. Some important types of structure, such as orthotropic bridge decks, Vierendeel trusses, tubular frames, tapered beams, and even most T-joints, could not be made as easily by any other method. Welded joints allow more freedom in the use of rolled shapes, high strength steels, and corrosion-resistant steels. Provided that the joints are made using appropriate materials and practices, designers should feel free to develop structures that are aesthetic as well as functional.

Hollow sections, both rectangular and circular, can sensibly be joined only by welding, even if the final connection is by bolting.

24.1.2 Stiffness

Welded construction allows a designer to introduce stiffness and strength where required, in the most discreet and/or the most structurally efficient manner. The transfer of load and the stiffness can, with welding, be introduced in a gradual continuous manner, instead of in step changes through bolted pieces.

24.1.3 Weight, volume and size

Bolting requires the use of lap joints, which add weight to a structure. For example, the use of site welds, instead of bolted joints, to join plate girders to crossheads saved about 50 tonnes of connection plates, bolts, and nuts in the construction of the new Thelwall Viaduct.

In addition to savings in weight, the use of welded joints provides more usable volume in a typical building. Bolted joints reduce usable volume significantly because current legislation and fire engineering can demand that the fireproofing coatings that must be applied give a minimum thickness over every protruding nut and bolt head.

24.1.4 Durability and corrosion resistance

Total life cycle costing, including the costs of maintenance, is becoming increasingly important in all structures. For one aspect of this, the durability, the corrosion of structural steelwork can be retarded by three means – controlling the surrounding environment, effective coating systems, and the use of steels (and weld metals) with improved inherent corrosion resistance. Coating systems, however effective they are on plate surfaces, can only with great difficulty prevent corrosion in all the crevices of bolted joints, but the clean lines of welded structures allow the full effectiveness of modern coatings to be demonstrated. The use of corrosion resistant (weathering) steels (from BS EN 10155: 1993[1]) imparts long life, and although they can be used in bolted connections, they perform poorly in crevices and thus they work far better in welded solutions.

24.2 Ensuring weld quality and properties by the use of standards

The flow of information required for the successful design, fabrication, and inspection of welded buildings is shown schematically in Fig. 24.1.

Designers are responsible for the selection of joint type, weld size, weld properties, and required quality. The choice of welding process, the details of the edge preparation suitable for the process, and the achievement of the desired quality are the responsibility of the steelwork contractor. A comprehensive series of standards has been developed and harmonized across Europe to ensure that the performance

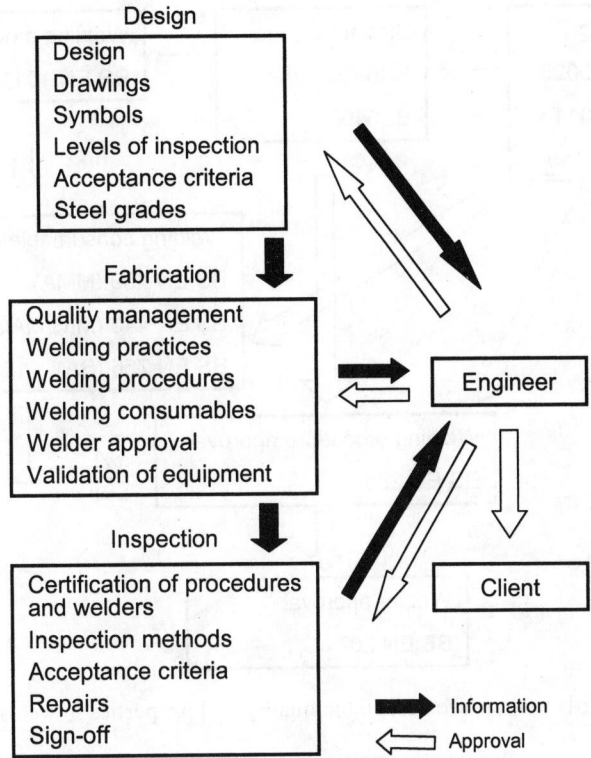

Fig. 24.1   Information required to ensure weld quality and performance

of welded joints will satisfy design requirements. Many of the notes added to design drawings, in attempts to define or clarify welding details, cause more problems than they solve.

The use of BS 5950[2] and BS 5400[3] automatically initiates a chain of standards that do not require further input from the designer, other than the approval of proposals.

As shown in Fig. 24.2, information on the welding process (from BS EN 1011–2: 2001[4]) and the consumables is included in a Welding Procedure Specification (WPS), which, when used by a suitably approved welder, will produce a weld with the required quality and properties.

24.2.1 Standards – joint type, weld type, welding symbols, and edge preparation

Designers should identify joint line, weld types, and throat dimensions using symbols shown in BS EN 22553: 1995,[5] *Welded, brazed and soldered joints – Symbolic representation on drawings*, which has replaced BS 499: Part 2a: 1980.

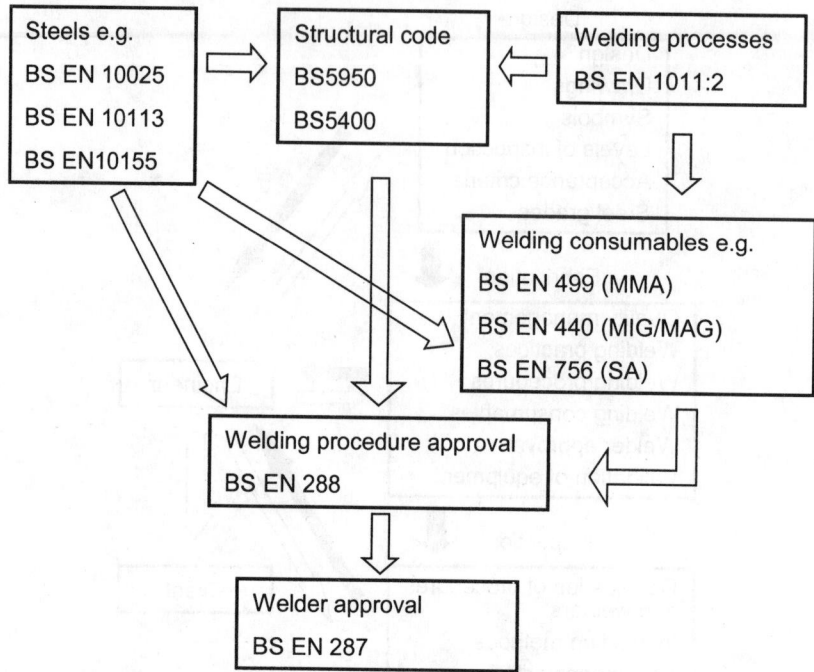

Fig. 24.2 Standards that ensure acceptable quality and properties in welds

Steelwork contractors use recommended edge preparations that are included in BS EN 1011-2: 2001,[4] *Welding – Recommendations for welding of metallic materials – Part 2: Arc welding of ferritic steels*, and BS EN 29692: 1994,[6] *Specification for metal-arc welding with covered electrode, gas-shielded metal-arc welding and gas welding. Joint preparation for steel.* Part 2 of this standard, which is being developed, will include submerged arc welding joint preparations. Joint preparations should be appropriate for the welding process to be used, and the designer should not impose a preparation that may not be suitable for all fabricators.

24.2.2 Standards – steel grade, steel selection

Designers should specify weldable structural steels, with grades and grade suffixes for yield strengths in the range 185–460 N/mm², shown in European standards such as:

BS EN 10025: 1993,[7] *Hot rolled products of non-alloy structural steels*
BS EN 10113: 1993,[8] *Hot rolled products in weldable fine grained structural steels*

BS EN 10137: 1996,[9] *Plates and wide flats made of high strength structural steels in the quenched and tempered or precipitation hardened conditions*

BS EN 10155: 1993,[1] *Structural steels with improved atmospheric corrosion resistance*

BS EN 10210: 1994,[10] *Hot finished structural hollow sections of non-alloy and fine grain structural steels*

BS EN 10219: 1997,[11] *Cold formed structural hollow sections of non-alloy and fine grain structural steels*

BS 7668: 1994,[12] *Specification for weldable structural steels – Hot finished structural steels in weather resistant steels.*

The designations of strength, toughness, type of steel, and supply condition, e.g. BS EN 10025: 1993,[7] S355K2G4, are more informative than previous systems (such as BS 4360: 1990, Grade 50D), and they provide better guidance to the steelwork contractor in the choice of welding consumables.

Steels have their Charpy V-notch impact toughness (at least 27J) tested at either room temperature, 0°C, or –20°C, and are adequate for most structural work in the UK. For applications at lower temperatures, some grades covered by BS EN 10113-2[8] have 40J impact toughness at –20°C and 27J toughness at –50°C. (For further details refer to Chapter 7.)

In order to prevent cold cracking and lamellar tearing see the recommendations of BS EN 1011–2: 2001.[4]

The use of the various product forms, standard sizes, and grades available in structural steels can result in considerable savings in cutting and welding costs, but not all shapes can be produced in all grades. Designers should check availability and cost before specifying.

24.2.3 Substitutions – thickness, yield strength, impact toughness, weldability, quality

Fabricators may request, for reasons of availability and cost, the substitution of the specified grade and thickness by another grade and thickness. Steelwork contractors and engineers should be familiar with the importance of several factors: thickness, yield strength, ratio of yield to ultimate strengths, impact toughness, weldability, and quality. Although it may appear that a substitute steel with higher yield strength and/or thicker sections may be beneficial because it will provide more strength, its impact energy may not comply with the code requirements and weldability might be adversely affected.

Thicker steel imposes several penalties. Both BS 5950[2] (clause 2.4.4) and BS 5400[3] (clause 6.5) show that moving to a thicker substitute and/or higher yield strength may require higher impact toughness and higher weld preheat temperatures to prevent cracking.

Steels with inferior impact toughness should not be accepted. For instance, a steel with a designation for 27J at 0°C should not be used to replace a steel designated

to have 27J at −20°C. Moving in the direction of lower test temperatures however is acceptable.

Steelwork contractors should always be seeking cost reduction, but cheaper steels with inferior quality generally lead to more expensive welds, especially if lamellar tearing is a possibility. Many connections require good through-thickness ductility to avoid tearing, and non-destructive testing before welding does not reveal these potential problems. Furthermore, impurities at levels well below the specified maxima will reduce impact toughness. Inferior rolling and levelling practices will cause dimensional problems when the steel is cut and welded. Moving to cheaper steels rarely saves overall costs.

24.2.4 Standards – welding processes and practices

BS EN 1011-2: 2001[4] provides guidance for welding practices used in the various codes for buildings and bridges, especially in the avoidance of cold cracks, hot cracks, and other unacceptable discontinuities. Significantly, it provides the fabricator with the means to estimate preheat temperatures required to avoid cold cracking caused by hydrogen, and it contains guidance on acceptance levels for weld discontinuities. When making recommendations to avoid cold cracking BS EN 1011 incorporates the principles of diffusible hydrogen content of the weld metal, the carbon equivalent value (CEV) of the parent metal, the combined thickness of the members of the joint being welded, the heat input (determined from the energy input to the weld), and the preheat temperature. The recommendations of BS EN 1011 are incorporated into a Welding Procedure Specification (WPS) to be followed during welding. This ensures that weld properties and soundness (for example the absence of cold cracking, controlled by the combination of carbon equivalent value, combined thickness, diffusible hydrogen content and preheat temperature) are achieved.

24.2.5 Welding standards – welding consumables

A harmonized set of European standards provides common designations for the yield strength and impact energy of weld metal deposited by the consumables used by the various processes, and additional information specific to the processes (shielding gas, flux type, etc).

BS EN 440: 1995,[13] *Wire electrodes and deposits for gas shielded metal arc welding of non alloy and fine grain steels. Classification*
BS EN 439: 1994,[14] *Shielding gases for arc welding and cutting*
BS EN 499: 1994,[15] *Covered electrodes for manual arc welding of non-alloy and fine grain steels. Classification*

BS EN 758: 1997,[16] *Tubular cored electrodes for metal arc welding with and without a gas shield of non-alloy and fine grain steels. Classification*
BS EN 756: 1995,[17] *Wire electrodes and wire-flux combinations for submerged arc welding of non-alloy and fine grain steels. Classification*
BS EN 760: 1996,[18] *Fluxes for submerged arc welding. Classification.*

The standards classify the properties of undiluted weld metal deposited from consumables in test welds under standardized welding conditions. The yield strength and impact toughness of the weld metal, and a series of operating characteristics, such as capability to weld in flat, horizontal, vertical and overhead positions, are used to give classification designations to the consumable. The strength and toughness of a production weld may differ from the test values, depending on dilution and heat input, but the designations guide the selection of consumables to produce weld metal properties that will match those of the parent plate.

Details of the consumables and the welding conditions to be used are entered into the WPS.

24.2.6 Standards – welding procedures

To comply with the requirements of BS 5950[2] and BS 5400,[3] all welding procedures must be approved, using BS EN 288,[19] *Specification and approval of welding procedures for metallic materials*, Part 3: 1992, *Welding procedure tests for the arc welding of steels*. This standard replaces BS 4870: Part 1, and it shows the requirements and methods for the approval of welding procedure specifications. All new welding procedure approvals are to be in accordance with this standard. An approved Welding Procedure Specification (WPS) provides all the information required by an approved welder to make a joint with the desired properties.

24.2.7 Standards – welder approval

BS 5950[2] and BS 5400[3] require the use of approved welders. The use of an approved WPS is only part of the quality assurance required in the production of satisfactory welds. The other essential ingredient is welder skill, which is guaranteed by ensuring that welders conform with the requirements of BS EN 287,[20] *Approval testing of welders for fusion welding*, Part 1: 1992, *Steels*. All the various criteria (process, welding position, steel thickness, etc.) specified in this standard are necessary to identify the ability of the welder to make specific welds. The welder must be capable of following written instructions in the production of sound welds.

24.2.8 Standards – inspection and weld quality

All structural materials contain imperfections, and a series of standards defines the methods of inspection and the extent to which the imperfections can be accepted. The significance of an imperfection depends mainly on its size, shape and position, and the local applied stresses and temperature. Imperfections that cause failure should be rejected and repaired, but many common imperfections, such as minor porosity, are acceptable. A useful guide to the scope of inspection and the weld quality acceptance criteria and corrective action is shown in Tables 1 and 2 of the National Structural Steelwork Specification for Building Construction.[21]

24.3 Recommendations for cost reduction

Small changes in design details can have significant effects on welding productivity and costs, without adversely affecting analysed stresses. This section is directed at design improvements that are qualitative in nature, but which reduce costs by paying attention to a series of simple principles which are summarized below in 24.3.5.

24.3.1 Overall principles

Modern computer aided design is very effective in the analysis of structures to gain 'maximum efficiency' in the use of materials, but the associated fabrication costs are not generally included in the software. Improved efficiency in the use of steel, especially for stiffening, usually leads to increased complication, the need for many short welds with difficult access, and a general increase in fabrication costs that will exceed the cost savings associated with the use of less steel. The use of standard rolled sections can save significant cutting and welding costs. Design priorities vary from one structure to another, but unless weight saving is crucial and is given highest priority, designers should aim for the minimum total cost which will, among other things, be a function of the number of welds to be made. When comparing design alternatives, the fabrication costs will be reduced if the number of individual pieces is minimized. For many structures the use of 'standard' 6 mm leg length fillet welds is convenient, but the associated heat input may be insufficient to prevent cold cracking of thick sections. BS EN 1011-2: 2001[4] indicates that welds are liable to crack adjacent to 6 mm fillet welds when the combined thickness of the elements being joined exceeds about 40 mm. In such cases the higher heat input from larger fillet sizes will avoid the need for preheating. Welding engineers should be consulted for advice when in doubt.

The sizes of welds should be no larger than required to transmit design stresses. The effective throat size a of a fillet weld should be taken as the perpendicular distance from the root of the weld to a straight line joining the fusion faces that lies

just within the cross-section of the weld; it should not be taken as greater than 0.7 times the effective leg length, which is what is measured during inspection. Unnecessary weld metal raises costs, and it increases distortion to no good effect. Designers should therefore indicate the throat size required for fillet and butt welds, and give steelwork contractors the responsibility to produce and confirm consistent production of that size.

Care is required in the design of welds for regions where members are closely spaced, and where joints have to be made inside assemblies. Any weld joint should be designed for easy access, and the welder or machine operator must be able to see where the weld is to be made, and be able to apply an MMA electrode or MIG/MAG gun so that the arc is directed down to the bottom of the joint and at the correct angle to ensure root penetration. A MIG/MAG gun is essentially a large pistol with a 20 mm barrel, 100 mm long, at 60° to the handle, attached to a heavy cable. MMA welds need space for the manipulation of electrodes that are either 350 mm or 450 mm long, with a flux coating diameter that often exceeds 6 mm, and held in tongs or a holder attached to the welding cable. Where access might be restricted it is recommended that designers should consult a welding engineer to confirm feasibility.

Welding position, shown in Fig. 24.3, is one of the important variables that influence the ease of fabrication, the costs of fabrication, and the mechanical properties of the weld.

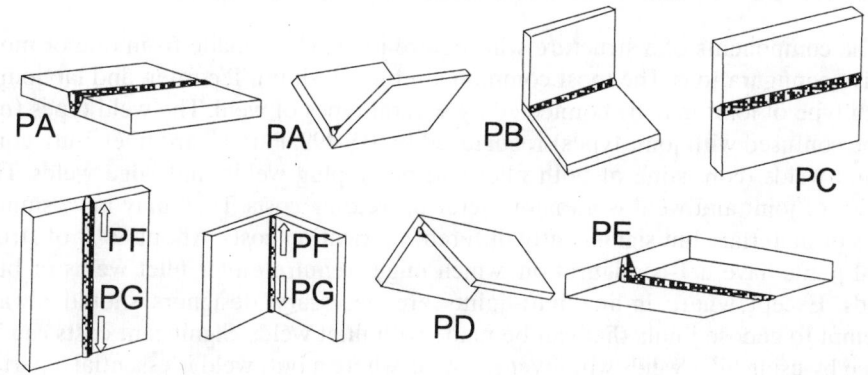

Fig. 24.3   Designations of welding positions

Welding position influences many important factors:

Deposition rate

The PA and PB welding positions allow the highest deposition rates. High deposition rates cannot be used when making welds in the vertical (PF and PG) and especially the overhead (PD and PE) positions.

Welder approval and availability of approved welders

Welds are more difficult in the overhead (PD and PE) positions than in the downhand (PA and PB) positions. Consequently it is more difficult for a welder to gain approval, and fewer approved welders are available for the more difficult positions.

Weld quality

Welds made in the more difficult PD, PE, and PF positions are likely to contain more defects than similar welds made in easier positions. Aim for the maximum number of welds to be made in the PA or PB positions with minimum re-positioning of the components.

Weld metal properties

Weld metal toughness, strength, and hardness are influenced not only by the choice of consumables but also by the size of the weld pool, which controls the heat input into the weld. When impact properties and hardness limits are specified, they must be tested at the low and high ends of the heat input range. If all welds can be made in the PA and PB positions only one set of tests need be made.

The components of a structure will require joints to be made from one or more of five configurations. The most common are butt (in-line), T, corner, and lap joints. Each type of joint may be connected by several types of weld. The weld types (not to be confused with joint types) recognized by BS 499: Part 1[22] are fillet, butt, compound welds (consisting of both fillet and butt), plug welds, and edge welds. The choice of joint and weld is a major factor in welding costs. They may have similar costs of materials, but significantly different fabrication costs. About 80% of structural joints have a T-configuration, which might require either fillet welds or butt welds. Except where in-line butt joints are necessary, designers should always attempt to choose joints that can be made with fillet welds. Significant costs can be saved by using fillet welds wherever possible; where a butt weld is essential a partial penetration weld should be selected if possible, bearing in mind strength, fatigue and corrosion limitations.

24.3.2 Fillet welds

Fillet welds have triangular cross section and they are commonly used to make T-joints, corner joints with several variations, and lap joints, shown in Fig. 24.4.

In this welding position, 8 mm is the maximum leg length that can be made in a single-pass weld. Single fillet welds should not be used where tension would

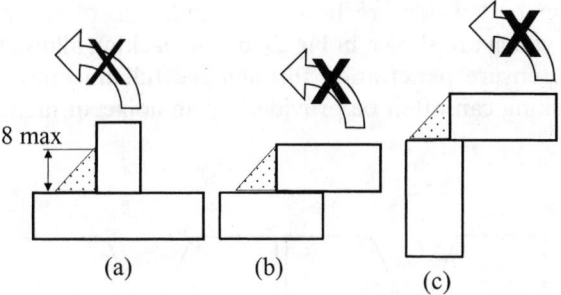

8 max

(a)   (b)   (c)

Fig. 24.4   Fillet weld configurations for (a) T-joints, (b) lap joints, (c) corner joints

introduce a bending stress that would open the root. Bending to close the root is admissible, and fillet welds on both sides of a member will prevent root opening.

24.3.3 Butt welds

Whereas fillet welds join the surfaces of adjacent members, butt welds join all or part of their cross section, and are consequently called *full* or *partial penetration welds*. Even where the joint configuration requires the use of a butt weld, partial penetration is often sufficient, instead of full penetration. Designers should state the weld throat dimension required, instead of routinely demanding full penetration with its extra difficulties and costs. Current practice is to use suitable weld preparations, procedures, and approved welders to ensure the required depth of fusion for the throat size. When partial penetration is adequate the designer should not state, 'All butt welds must have full penetration'. To a welding engineer, this means full penetration through the sections, when all that is intended is for the steelwork contractor to show proof of achieving the required throat depth. Figure 24.5 shows different approaches to butt welds. The partial penetration weld (Fig. 24.5(a)) is the most economical to prepare and weld. Throat size can be ensured by use of high-current MIG/MAG, flux-cored arc, or submerged arc welding, and the fabricator can be asked to provide proof of consistent penetration. Where full penetration is justified, welds should be made with backing (as Fig. 24.5(b)). Where backing is not permissible, welds are made by welding one side, grinding or gouging the root to sound metal for completion from the other side (Fig. 24.5(c)).

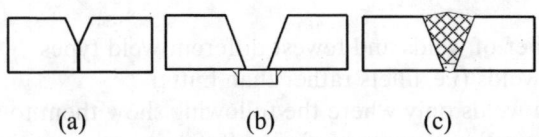

(a)   (b)   (c)

Fig. 24.5   Partial penetration butt welds

Full-penetration butt welds are best made with one of several types of weld backing, some of which are shown in Fig. 24.6. The backing allows the welder to use sufficient heat to ensure penetration through the full thickness of the members being joined. Backing can often be provided by an adjacent member, especially in corner welds.

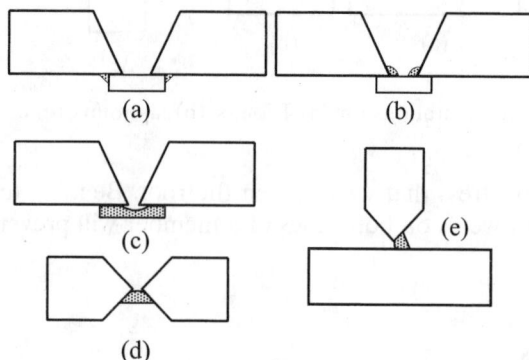

Fig. 24.6   Full penetration in butt welds is best achieved with root backing such as with closely fitting steel strips attached by fillet welds at the back (a), or within the groove (b), by a copper bar (c), or by ceramic tiles (d) and (e).

24.3.4 Consultation with fabricators

Even though steelwork contractors hold designers to be responsible for many welding problems, welding staff are reluctant (and often not qualified profession-ally) to recommend improvements. On the other hand, designers cannot be expected to be familiar with all welding innovations, nor the capabilities of individual welding shops and their personnel. The best way to ensure efficient welding details is for designers to initiate genuine dialogue with fabricators and to consider suggestions for alternative approaches to welding.

24.3.5 Summary of recommendations

The items discussed above can be summarized as follows:

- minimum number of welds and fewest different weld types
- minimum cost welds (i.e. fillets rather than butts)
- full penetration welds only where the following show them to be essential:
    (a) partial penetration throat area is insufficient to support the stresses,
    (b) for a one-sided weld the loading might open the root of the weld,

(c) the loading is cyclic and fatigue cracks could be initiated in the unfused part of the joint,

(d) corrosive attack might occur in the non-welded crevice.

- smallest size of welds (while recognizing that certain sizes are 'preferred')
- easy welds in terms of access and welding position
- root backing for easy production of full-penetration butt welds
- balanced welding (on both sides of neutral axes) for minimum distortion and least sensitivity to dimensional discrepancies
- easy inspection.

24.4 Welding processes

24.4.1 Introduction

All welding processes have their inherent advantages and disadvantages. The choice of welding process is generally the responsibility of the steelwork contractor, and it depends on the availability of equipment and welders. The objective of this section is to outline some of the features and characteristics of the popular arc welding processes.

National statistics provide an idea of their relative popularity. In the UK market, MIG/MAG welding currently accounts for about 50%, cored wire welding (with and without gas shielding) has a further 10%, and MMA welding now has about 30% of welding consumables consumption. Submerged arc welding has accounted for 7–8% for many years. Popularity depends mainly on productivity, which has the largest influence on welding costs.

The important processes used in steel fabrication are differentiated technically by making reference to the type of consumable electrode and the method of protecting the arc. They are defined and numbered in BS EN 24063 as follows:

111   metal-arc welding with covered electrode
114   flux-cored metal-arc welding without gas shield
121   submerged arc welding with electrode
135   metal-arc active gas welding, MAG welding
136   flux-cored wire metal-arc welding with active gas shield

24.4.2 Manual metal arc (MMA) welding

MMA welding dominated the structural steelwork industry for many years because of its versatility and suitability for both shop and site application, and the widespread availability of welders approved for its use. MMA welding is colloquially

Fig. 24.7   Manual metal-arc welding of structural components (Photograph courtesy of The Lincoln Electric Company)

known as 'stick' welding. MMA equipment is widely available and inexpensive in comparison with other processes.

A typical MMA electrode, which usually has a length in the range of 230–460 mm, is held firmly in tongs or holders that are connected by flexible cables to the power source, which may be a transformer, rectifier, inverter, or engine-driven generator, as shown in Fig. 24.7. The electrode coating melts in the arc and (1) protects the weld metal from the surrounding air, (2) forms a slag that protects and supports the weld metal, and (3) usually adds alloying elements to the weld metal.

24.4.3 MIG/MAG welding

MIG/MAG welding wire is generally copper-coated to provide good electrical contact with tip in the welding gun. The majority of MIG/MAG wires contain sufficiently high levels of silicon and manganese to prevent porosity that would be caused by oxygen that enters the weld metal from the active shielding gases used. A limited range of low-alloy wires is available for high strength and toughness applications.

Fig. 24.8   MIG/MAG welding – note the rectifier power source and feeder spool of welding
wire behind the welder (Photograph courtesy of The Lincoln Electric Company)

The essential components of equipment for MIG/MAG welding are the hand-held gun (or mechanized welding head) through which the continuous wire is fed into the arc, the wire feeder, and a constant-voltage welding power source, as shown in Fig. 24.8. The continuous nature of the process allows it to be used for semi-automatic, mechanized, automatic, and robotic welding. The arc conditions (voltage and current) are controlled mainly by the equipment, giving rise to the use of the term *semi-automatic* welding.

The MIG/MAG welding gun feeds the wire, conducts the current and delivers a gas shield. The wire leaves the gun from a copper contact tube, which is surrounded by a concentric gas nozzle. All components of the gun should have adequate current-carrying capacity and be tightly connected to provide good electrical and thermal conductivity.

The wire feeder, fitted with V-grooved rolls for solid MIG/MAG wire or knurled rolls for cored and submerged arc wires, pulls the wire from a spool or drum and pushes it through the cable (or welding head on a mechanized machine) to the gun.

24.4.4 Cored wire welding

Cored wire welding is a modification of MIG/MAG welding in which the electrode wires are hollow steel tubes filled with powders. Cored wires have some specific

benefits, especially productivity and weld quality, compared with solid wire MIG/MAG or MMA welding. Most cored wires require the use of a gas shield, but some are self-shielding, thereby making them ideal for site welding where winds would blow the gas shield away and cause weld metal embrittlement.

24.4.5 Submerged arc welding

In the submerged arc process shown in Fig. 24.9 the weld is protected by a flux instead of a gas. The flux powder is continually fed around the electrode, and distributed over the area to be welded. Consequently this process is virtually confined to welds that can be made in the flat position. Submerged arc welding is the preferred process for fabricating girders because it gives a combination of productivity and high quality that cannot be matched by other processes. The process is generally used for mechanized welding with machines for making accurately placed passes in straight lines or circles, such as girders, drums, and pipes, but fillet welds can be made with manual equipment in which flux is fed to the weld instead of a shielding gas.

Fig. 24.9   Automatic submerged arc welding machine fitted with two welding heads feeding flux from hoppers to cover the wire and arc (Photograph courtesy of The Lincoln Electric Company)

24.4.6 Welding productivity

Major improvements in productivity have been made possible by the use of processes that use continuous welding wires to provide the filler metal. The MMA process, which was the first to be used for widespread fabrication, deposits individual electrodes made from cut lengths of steel rod. Each rod runs for only about one minute before the stub must be discarded and a new electrode fitted into the holder. This intermittent process has been replaced by the use of the continuous processes (MIG/MAG and cored wire) where the welder can continue until lack of access or reach require the welder to stop welding and move to a new position. If feasible, the weld could continue until the spool (generally about 15 kg) is empty. Stops and starts are the main sites of defects in MMA welding. The continuous wire processes have generally better productivity and fewer stops and starts. Consequently, reduced welding times and improved quality are not incompatible.

24.4.7 Weld quality

The cost of inspection, repair, and reinspection of a defective weld is about ten times the cost of getting it right first time. Efforts should always be made to achieve acceptable quality. Quality is measured in terms of weld shape, soundness, and mechanical properties.

Weld quality is assessed by a variety of non-destructive tests with the objective of finding, identifying, and measuring the size of imperfections.

Mechanical properties are measured in pre-production procedure tests, and may be confirmed by test welds made under production conditions.

It is impossible to avoid imperfections in commercially produced materials, but not all imperfections will adversely affect the performance of a structure. The acceptability of a weld, therefore, relies on a 'fitness for purpose' assessment. Acceptance criteria are specified in national standards, including the National Structural Steelwork Specification (NSSS).

Imperfections are classified as:

- lack of fusion
- cracks
- porosity
- inclusions
- spatter
- weld geometry and profile

Of these, two types of defect are generally unacceptable – lack of fusion, and cracks. Both defects reduce the cross sectional area of a weld, lowering their load carrying capacity, and they significantly reduce resistance to fatigue failure in cyclic loading.

24.4.8 Distortion

Even if a weld is sound and acceptable mechanically, it can be rejected if the result-ing structure has been distorted to an unacceptable shape. Distortion is caused by the shrinkage of the weld metal and HAZ as it cools. Shrinkage cannot be elimi-nated, but the resultant distortion can be minimized. The most common means of reducing distortion are:

(1) selecting the joint type that requires the least amount of weld filler metal,
(2) using edge preparations that reduce the amount of filler metal required
(3) balancing the welding of individual members on either side of the neutral axis,
(4) balancing the welding on either side of butt welds in thick plate by use of double-V instead of single-V preparations,
(5) making welds at fast travel speeds (by mechanization) to minimize the heat input into the adjacent metal,
(6) using tack welds and jigs to restrain movement,
(7) using a back-step sequence to restrain movement, as shown in Fig. 24.10.

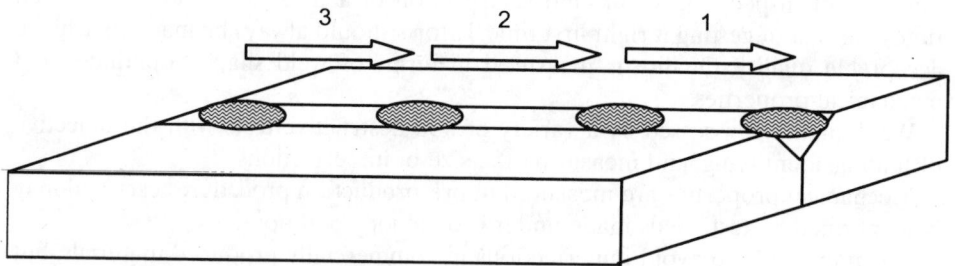

Fig. 24.10 Back-step welding sequence to restrain movement

24.5 Geometric considerations

24.5.1 Effective throats

The throat thickness of fillet welds is given in Table 24.1. The factors given in the table are approximately equal to the cosine of the half angle between the fusion faces for welds of equal leg length, and when multiplied by the leg length will give the perpendicular distance between the root and a line joining the intersections of the weld with the fusion faces. This by definition is the throat thickness. For welds with an angle less than 90° between the fusion faces, for which the defined distance would be greater than that for a right angle weld, there is an upper limit to the factor of 0.7. Similarly in the case of welds of unequal leg lengths the assumed throat thick-ness is not to exceed 0.7 multiplied by the shorter of the two legs (Fig. 24.11). Where

Table 24.1 Throat thickness of fillet welds

Angle between fusion faces (degrees)	Factor (to be applied to the leg lengths)
60 to 90	0.7
91 to 100	0.65
101 to 106	0.6
107 to 113	0.55
114 to 120	0.5

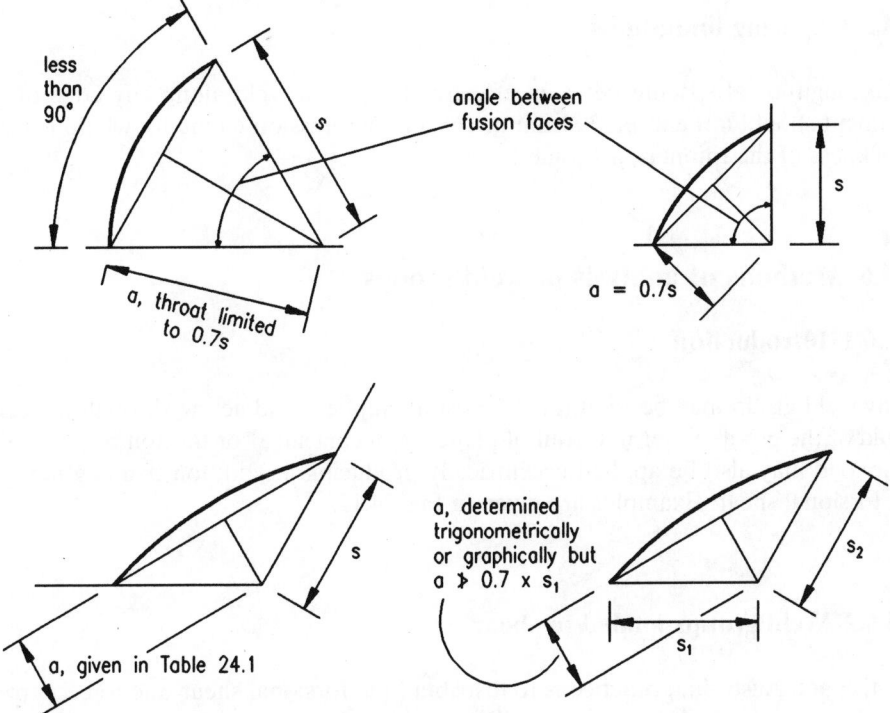

Fig. 24.11 Throat thickness of fillet welds

deep penetration welds are produced by submerged arc welding, the effective throat thickness may be measured to the minimum depth of fusion (see BS5950-1: 2000 clause 6.8.4).

24.5.2 Effective lengths

The effective length of a fillet weld is the actual length less twice the leg length to allow for the starting and stopping of the weld. It should not be less than four

times the leg length. When a fillet weld terminates at the end or edge of a plate it should be returned continuously round the corner for a distance of twice the leg length.

Intermittent fillet welds are laid in short lengths with gaps between. They should not be used in fatigue situations or where capillary action could lead to the formation of rust pockets. The effective length of each run within a length is calculated in accordance with the general requirements for fillet welds.

24.5.3 Spacing limitations

The longitudinal spacing between effective lengths of weld along any edge of an element should not exceed 300 mm or $16t$ for compression elements, where t is the thickness of the thinner part joined.

24.6 Methods of analysis of weld groups

24.6.1 Introduction

Any weld group may be required to resist an applied load acting through the centroid of the group either in or out of plane producing shear or tension respectively. The load may also be applied eccentrically producing in addition bending tension or torsional shear. Examples are given in Fig. 24.12.

24.6.2 Weld groups loaded in shear

British and Australian practice is to distribute the torsional shear due to eccentricity elastically in proportion to the distance of each element of the weld from the centroid of the group. This is referred to as the *polar inertia method*.

In some countries, notably Canada and in some cases the USA, the instantaneous centre method, referred to in Chapter 23 for bolt groups, is also used for weld groups.

The polar inertia method

Consider the four-sided weld group shown in Fig. 24.13(a). Assume the throat thickness is unity.

$$I_{xx} = \frac{2b^3}{12} + 2a\left(\frac{b}{2}\right)^2$$

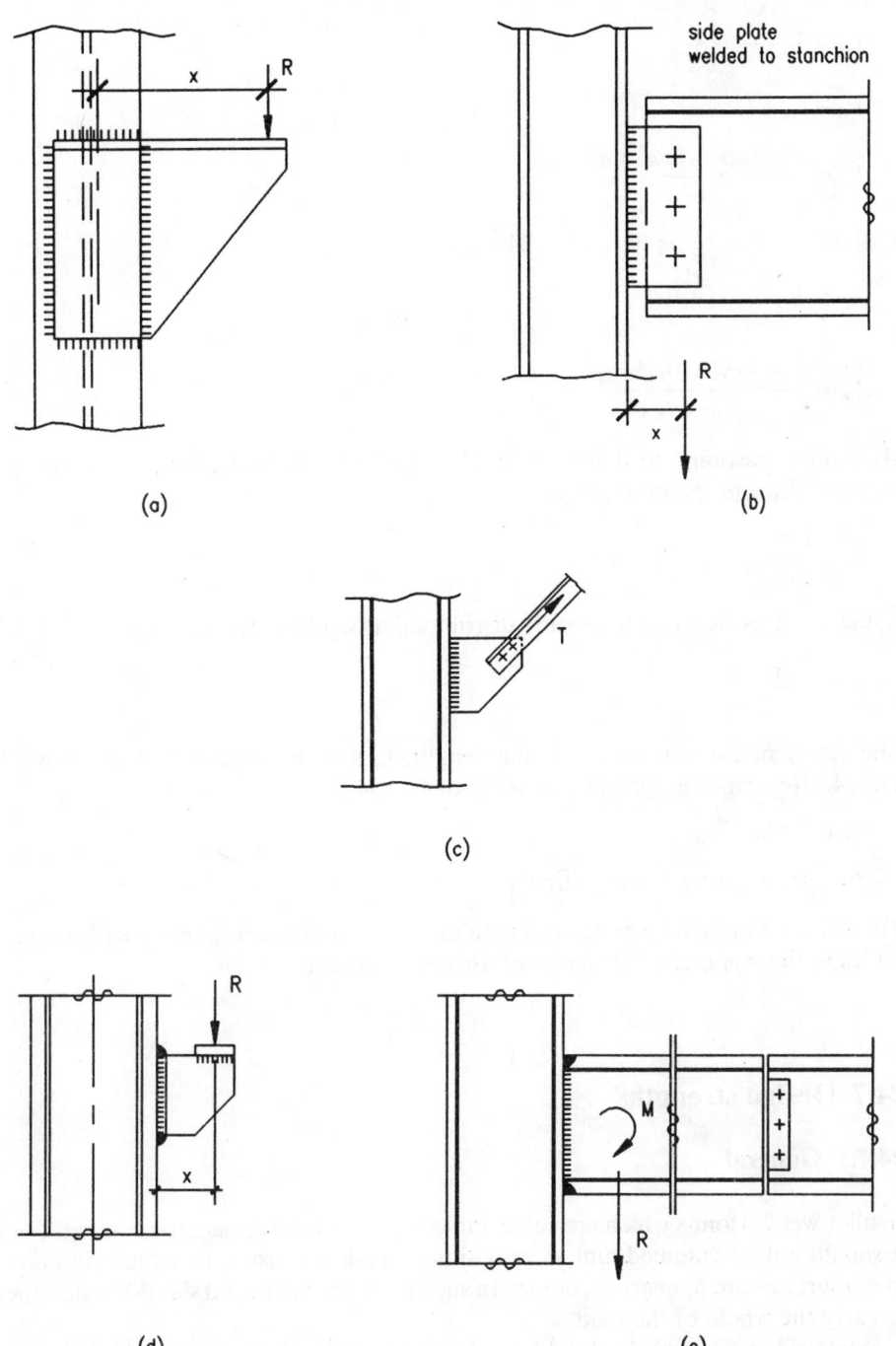

Fig. 24.12   Weld groups loaded eccentrically

$$I_{yy} = \frac{2a^3}{12} + 2b\left(\frac{a}{2}\right)^2$$

$$I_{00} = I_{xx} + I_{yy}$$

$$I_{00} = \frac{b^3 + 3ab^2 + 3ba^2 + a^3}{6}$$

Distance r to extreme fibre (Fig. 24.13(b)):

$$r = \frac{1}{2}\sqrt{(a^2 + b^2)}$$

$$Z_{00} = \frac{b^3 + 3ab^2 + 3ba^2 + a^3}{3\sqrt{(a^2 + b^2)}}$$

By similar reasoning to that given in Chapter 23 for bolts, f_m, the force vector per unit length from the moment, is

$$f_m = \frac{Rxr}{Z_{00}}$$

f_v, the shear, is assumed uniformly distributed around the weld group:

$$f_v = \frac{R}{2a + 2b}$$

The resultant force vector per unit length, f_r, may be determined as shown in Fig. 24.13(c) either graphically or trigonometrically:

$$\tan\alpha = b/a$$

$$f_r = \sqrt{\left[(f_m \cos\alpha + f_v)^2 + (f_m \sin\alpha)^2\right]}$$

The value f_r can then be compared with the strength of the weld proposed from the tables in the Appendix *Capacities of fillet welds* as appropriate.

24.7 Design strengths

24.7.1 General

In fillet-welded joints which are subject to compression forces as shown in Fig. 24.14, it should not be assumed, unless provision is made to ensure this, that the parent metal surfaces are in bearing contact. In such cases the fillet weld should be designed to carry the whole of the load.

Single-sided fillet welds should not be used in cases where there is a moment about the longitudinal axis: see Fig. 24.15. Ideally they should not be used to transmit tension.

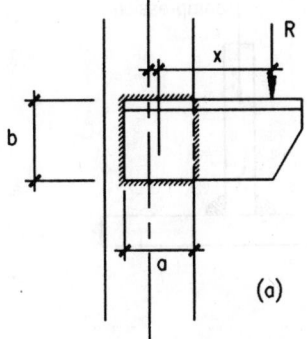

(a)

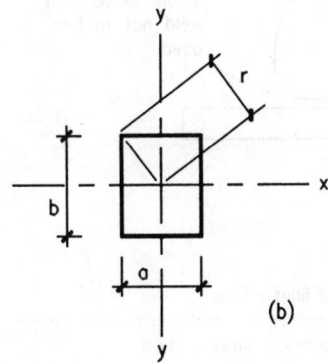

(b)

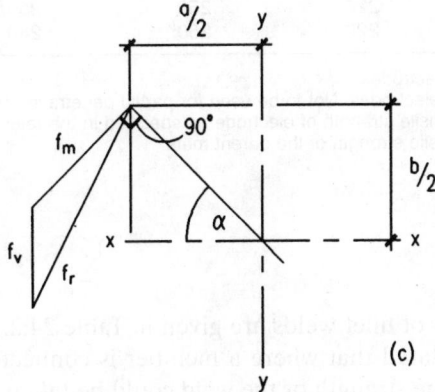

(c)

Fig. 24.13   Weld groups loaded in shear

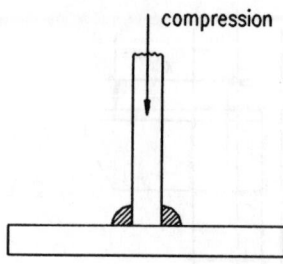

Fig. 24.14   Weld in compression

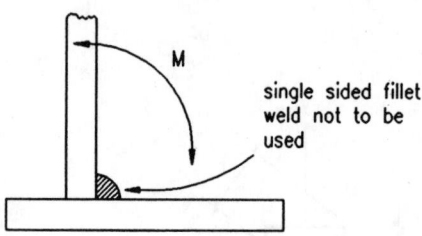

Fig. 24.15   Weld with moment

Table 24.2   Design strength, P_w, of fillet welds

Steel grade	Electrode classification			Other types[c]
	35 (N/mm²)	42 (N/mm²)	50 (N/mm²)	(N/mm²)
S275	220	220[a]	220[a]	$0.5U_e$ but
S355	220[b]	250	250[a]	$\leq 0.55U_s$
S460	220[b]	250[b]	280	

[a] Over-matching electrodes
[b] Under-matching electrodes. Not to be used for partial penetration butt welds.
[c] U_e = minimum tensile strength of electrode as specified in the relevant product standard
U_s = minimum tensile strength of the parent metal

24.7.2 Strength

The design strengths of fillet welds are given in Table 24.2.

BS 5950-1: 1990 stated that where a member is connected to a plate by a symmetrical fillet weld the strength of the weld could be taken as equal to the strength of the parent metal, if the weld was made with a suitable, the sum of the throat sizes was not less than the connected plate thickness, and the weld was principally subject

to direct compression or tension. In BS 5950-1: 2000 this rule has been deleted, and designers must use the design strengths given in Table 24.2. These values are significantly lower than the design strengths of the parent material (p_w = 220 N/mm² for S275 and p_w = 250 N/mm² for S355). This reduction in capacity has been partly offset by the allowance in BS 5950-1: 2000 for the higher transverse strength of fillet welds, which may be taken into account using the directional method when determining the weld capacity (see clause 6.8.7.3). For the case of two plates connected at right angles by symmetrical welds, transverse strengths of 1.25 × 220 = 275 N/mm² and 1.25 × 250 = 312.5 N/mm² are achieved for grade S275 and S355 respectively. Clearly, where S275 steel is used, the transverse strength is still equal to that of the parent metal, but for S355 the strength is significantly lower.

The Appendix *Capacities of fillet welds* shows fillet weld capacities using appropriate strengths for S275 and S355 steel.

References to Chapter 24

1. British Standards Institution (1993) *Structural steels with improved atmospheric corrosion resistance.* BS EN 10155, BSI, London.
2. British Standards Institution (2000) *Structural use of steelwork in building. Code of practice for design – rolled and welded sections.* BS5950-1, BSI, London.
3. British Standards Institution (2000) *Steel, concrete and composite bridges. Code of practice for design of steel bridges.* BS5400-3, BSI, London.
4. British Standards Institution (2001) *Welding. Recommendations for welding of metallic materials. Arc welding of ferritic steels.* BS EN 1011-2, BSI, London.
5. British Standards Institution (1995) *Welded, brazed and soldered joints. Symbolic representation on drawings.* BS EN 22553, BSI, London.
6. British Standards Institution (1994) *Specification for metal-arc welding with covered electrode, gas-shielded metal-arc welding and gas welding. Joint preparation for steel.* BS EN 29692, BSI, London.
7. British Standards Institution (1993) *Hot rolled products of non-alloy structural steels.* BS EN 10025, BSI, London.
8. British Standards Institution (1993) *Hot rolled products in weldable fine grained structural steels.* BS EN 10113, BSI, London.
9. British Standards Institution (1996) *Plates and wide flats made of high strength structural steels in the quenched and tempered or precipitation hardened conditions.* BS EN 10137, BSI, London.
10. British Standards Institution (1994) *Hot finished structural hollow sections of non-alloy and fine grain structural steels.* BS EN 10210, BSI, London.
11. British Standards Institution (1997) *Cold formed structural hollow sections of non-alloy and fine grain structural steels.* BS EN 10219, BSI, London.
12. British Standards Institution (1994) *Specification for weldable structural steels – Hot finished structural steels in weather resistant steels.* BS 7668, BSI, London.
13. British Standards Institution (1995) *Welding consumables. Wire electrodes and*

deposits for gas shielded metal arc welding of non-alloy and fine grain steel. *Classification*. BS EN 440, BSI, London.

14. British Standards Institution (1994) *Welding consumables. Shielding gases for arc welding and cutting.* BS EN 439, BSI, London,

15. British Standards Institution (1995) *Welding consumables: Covered electrodes for manual metal arc welding of non alloy and fine grain steels: Classification.* BS EN 499, BSI, London.

16. British Standards Institution (1997) *Welding consumables: Tubular cored electrodes for metal arc welding with and without a gas shield of non-alloy and fine grain steels: Classification.* BS EN 758, BSI, London.

17. British Standards Institution (1996) *Welding consumables: Wire electrodes and wire-flux combinations for submerged arc welding of non-alloy and fine grain steels: Classification.* BS EN 756, BSI, London.

18. British Standards Institution (1996) *Welding consumables. Fluxes for submerged arc welding. Classification.* BS EN 760, BSI, London.

19. British Standards Institution (1992) *Specification and approval of welding procedures for metallic materials.* BS EN 288, BSI, London.

20. British Standards Institution (1992) *Approval testing of welders for fusion welding. Steels.* BS EN 287-1, BSI, London.

21. The British Constructional Steelwork Association (2002) *National structural steelwork specification*, 4th edition, BCSA/SCI Ascot.

22. British standards Institution (1991) *Welding terms and symbols. Glossary for welding, brazing and thermal cutting.* BS 499–1, BSI, London

Further reading for Chapter 24

Blodgett, O.W. (1982) *Design of Welded Structures*, 12th edn. The James F. Lincoln Arc Welding Foundation, Mansfield Road, Aston, Sheffield S31 0BS

The Lincoln Electric Company (2000) *The Procedure Handbook of Arc Welding*, 14th edn. The James F. Lincoln Arc Welding Foundation, Mansfield Road, Aston, Sheffield S31 0BS

Chapter 25
Plate and stiffener elements in connections

by BRIAN CHEAL

25.1 Dispersion of load through plates and flanges

Where loads (or reactions) are applied to the flanges of beams, columns or girders the web adjacent to the flange must be checked for its local bearing capacity. The effective length of web to be used for checking the bearing capacity is obtained by assuming a dispersion of the load through the plates and flanges. Generally the dispersion to find the *stiff bearing length* is taken at an angle of 45° through solid material as shown in Fig. 25.1(a) and (b).[1] The dispersion depends upon the local bending resistance of the plate and so the dispersion can only occur when there is some restraint to balance the bending moment. For example, in Fig. 25.1(c) where the loose pack is not symmetrical about the point of application of the load, the 45° dispersion should not be taken through the pack.

In the case of a flange which is integral with or is connected to the web a greater angle of dispersion at a slope of 1:2.5 to the plane of the flange is allowed.[1] The dispersion of 1:2.5, taken to the web-to-flange connection, has been verified by tests with loads applied to columns, remote from the column ends. It can also be established by calculation, assuming that at failure the web crushes and four plastic hinges form in the flange (Fig. 25.2). The theoretical formula that is obtained is given in the draft Eurocode 3:[2]

$$P = (b_1 + n)t \, p_{yw} / \gamma_{M1}$$

in which the length due to dispersion *(n)* is given as (Fig. 25.3):

$$n = 2T(B/t)^{\frac{1}{2}} (p_{yf} / p_{yw})^{\frac{1}{2}} \left[1 - (f_a / p_{yf})^2 \right]^{\frac{1}{2}}$$

where P = crushing resistance of the web,
b_1 = length of stiff bearing,
t = thickness of web,
B = width of flange, but not greater than $25T$,
T = thickness of flange,
p_{yf} = yield stress of flange,
p_{yw} = yield stress of web,
f_a = longitudinal stress in the flange,
γ_{M1} = partial safety factor.

711

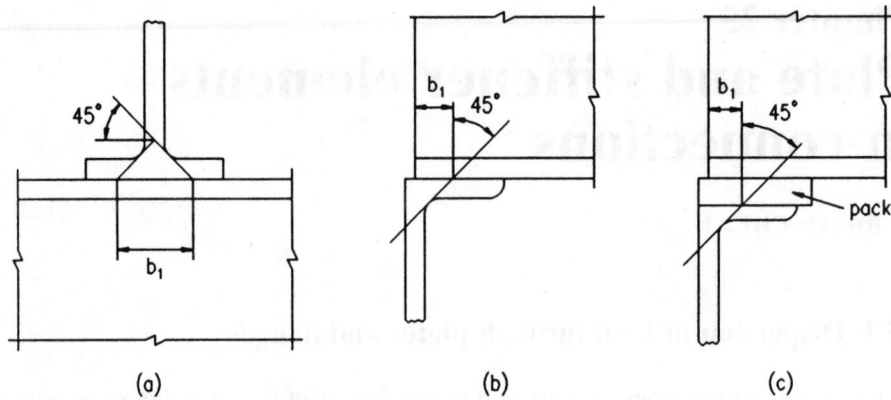

(a)   (b)   (c)

Fig. 25.1   Dispersion of load, b_1 = stiff bearing length

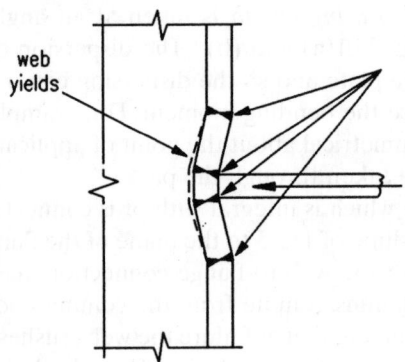

Fig. 25.2   Failure mechanism of flange

Conventionally for rolled sections dispersion has been taken to the K-line, i.e. through a distance equal to the flange thickness (T) plus the flange to web root radius (r). To adjust the formula to this practice the minimum value of $T(B/t)^{\frac{1}{2}}$ in terms of ($T + r$) for rolled sections is substituted in the equation, i.e. $T(B/t)^{\frac{1}{2}} = 2.5(T + r)$, hence the 1:2.5 dispersion.

BS5950-1: 2000 defines the bearing capacity of an unstiffened web, P_{bw}, as ($b_1 + nk)tp_{yw}$ where n is taken as 5 if the load is applied remote from the end of a member of $2 + 0.6b_e/k$ (but less than 5) where the load is applied near the end of a member and b_e is the distance to the end of the member from the edge of a stiff bearing; k is as shown in Fig. 25.3 for a rolled section but taken as only the flange thickness for welded sections. Where the applied load or reaction exceeds the bearing capacity of the unstiffened web, bearing stiffeners should be provided. These should be designed to carry the applied force minus the bearing capacity of the unstiffened web.

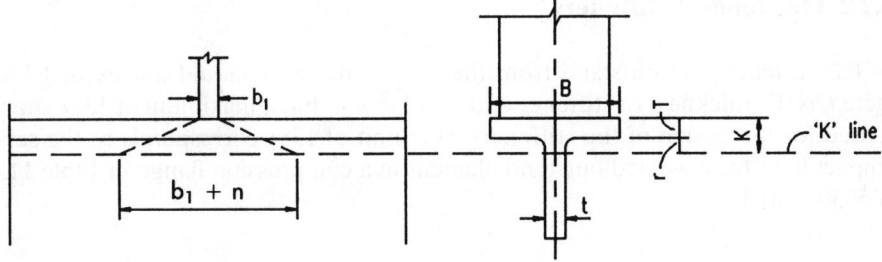

Fig. 25.3 Dispersion of load through flange

Compared with British and American sections European sections have a relatively larger root radius and the simplification of a $1:2.5$ dispersion does not apply.

Where there are axial stresses in a flat plate the plastic bending moment capacity is reduced. As the dispersion depends upon the bending capacity of the flange, when there are relatively high longitudinal stresses in the flange caused by axial load and bending moments the angle of dispersion will be significantly reduced. To allow for this effect the $1:2.5$ spread should be modified to $1:2.5\sqrt{\mu}$, where $\mu = 1 - (f_a/p_y)^2$ and f_a = average longitudinal stress in the flange. This modification is included in the formula in the draft Eurocode 3.[2]

25.2 Stiffeners

25.2.1 General

At connections to beams and columns web stiffeners are provided where compressive forces applied through a flange by loads or reactions exceed the buckling resistance of the unstiffened web, or where the compressive or tensile forces applied through the flange exceed the local capacity of the web at its connection with the flange. They may also be provided to stiffen the flange where it is inadequate in bending, e.g. in a bolted tension connection.

The rules in BS 5950-1 relating to the buckling resistance of an unstiffened web have been revised. In the 1990 version, the buckling resistance of an unstiffened web was given by $P_w = (b_1 + n_1)tp_c$, where b_1 and t are as defined earlier, n_1 is the length obtained by 45° dispersion through half the depth of the section and p_c is the compressive strength obtained from Table 27(c). In BS 5950-1: 2000 clause 4.5.3.1, the buckling resistance of the web, P_x, is obtained directly from the bearing capacity P_{bw} and the geometry of the section. There is no longer a need to refer to the strut curve (i.e. Table 24(c) in BS 5950-1: 2000). Three equations are presented for P_x, depending on the restraint of the flange and the location of the applied load relative to the end of the member.

25.2.2 Outstand of stiffeners

For flat stiffeners the outstand from the face of the web should not exceed $19t_s\varepsilon$, where t_s is the thickness of stiffener and $\varepsilon = (275/p_y)^{\frac{1}{2}}$; but a maximum of $13t_s\varepsilon$ should be used for the design of the stiffeners. The limit of $13t_s\varepsilon$ corresponds to the semi-compact limit for a welded outstand element in a compression flange in Table 11 of BS 5950: Part 1.

25.2.3 Buckling resistance

The buckling resistance of a stiffener should be based on the compressive strength of a strut using Table 24(c) in BS 5950: Part 1, the radius of gyration being taken about the axis parallel to the web. The effective section is the area of the stiffener together with an effective length of web on each side of the centreline of the stiffener limited to 15 times the web thickness. The effective length of the strut is taken as $0.7L$ (where L is the length of the stiffener) provided that the flange is restrained against rotation in the plane of the stiffener and that the stiffener is fitted or welded to the flange so that the restraint is in fact applied to the stiffener. Where the stiffener is not so restrained the effective length is taken as L. These effective lengths assume that the flange through which the load or reaction is applied is effectively restrained against lateral movement relative to the other flange. If this is not so the stiffener should be designed as part of the compression member applying the load, and the design of the connection should include the moments due to strut action (see BS 5950: Part 1, clause 4.5.3.3 and Annex C.3).

25.2.4 Local bearing

Web stiffeners are often *sniped* (the internal corners are cut) to clear the root radius or the web-to-flange welds. Since this reduces the effective bearing area of the stiffeners, it is necessary to check the local bearing capacity.

For the design of stiffeners where the web alone would be inadequate in bearing, and for the design of tension stiffeners, BS 5950: Part 1 allows the stiffeners to be designed to carry the applied load or reaction less the load capacity of the web without stiffeners.

In a compression detail, when stiffeners are used, a greater proportion of the load is initially carried by the stiffeners and it may well be advisable to limit the dispersion to 45° when calculating the portion of load carried by the web. This reservation follows from the stiffener outstand limitation being for a semi-compact section, which by definition has limited ductility. Alternatively, the full $1:2.5$ dispersion can be used with the compact criteria limitation $b \not> 8.5t_s\varepsilon$ applied to the stiffener outstand. In the case of a tensile load the potential problem is that the fillet welds

attaching the stiffeners to the flange have limited ductility. To overcome this problem, for a pair of symmetrical fillet welds, the leg length of the fillet welds should be not less than $0.85t_s$ for grade S275 steel and $1.0t_s$ for grade S355 steel, or the welds can be designed for a net load based on the 45° dispersion.

BS 5950-1: 1990 included a bearing check for load-carrying stiffeners (sub-clause 4.5.4.2). This stated that load-carrying stiffeners should be designed to resist 80% of the total applied force, irrespective of the capacity of the unstiffened web, i.e. $A > 0.8F_x/p_{ys}$, where A is the area of the stiffener in contact with the flange, F_x is the applied load, p_{ys} is the design strength of the stiffener. As a result of the reduction in the effective width of the web from $20t$ to $15t$ (see 25.2.3 above), this bearing check was removed. This was a key change since, in most practical cases, the size of the stiffeners was governed by this rule. There is still a requirement to check the bearing capacity, as BS 5950-1: 2000 states that load-carrying stiffeners should also be checked as bearing stiffeners. However, this requirement is not as onerous as the previous 80% rule, since bearing stiffeners are only designed to carry the external load minus the bearing capacity of the unstiffened web and not the full external load.

25.2.5 Bracket stiffeners

The outside edge of the welded column bracket, shown in Fig. 25.4, is in compression, and provided that a reasonably conservative approach is used in calculating the extreme fibre stress, semi-compact criteria for the outstand are appropriate (i.e. $b \ngtr 13t_s\varepsilon$). With relatively light loads, this may require an unnecessarily thick stiffener. To overcome the problem, the simple design approach of substituting f_a for p_y

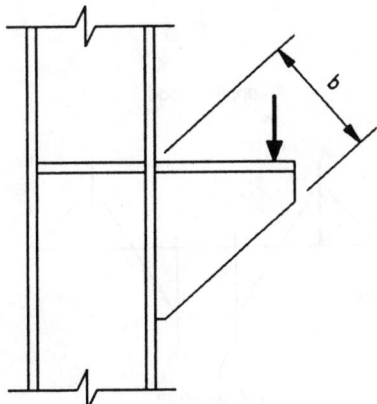

Fig. 25.4 Column bracket

in the formula for ε, so that $b \not> 13t_s \, (275/f_a)^{\frac{1}{2}}$, where $f_a = 1.5 \times$ the extreme fibre compressive stress (at factored loading), can be used.

25.3 Prying forces

Where bolts are used to carry tensile forces they almost invariably connect together plates or flanges which flex in bending and are as a result subject to prying action. This prying action is best illustrated by considering a tee-connection (Fig. 25.5). When the flange of the tee is relatively flexible it bends under the applied load and the flanges outside the bolts are pressed against the supporting plate. The reactions generated at the points of contact are referred to as prying forces. For equilibrium, the total force in the bolts must equal the applied force plus the prying forces. Only

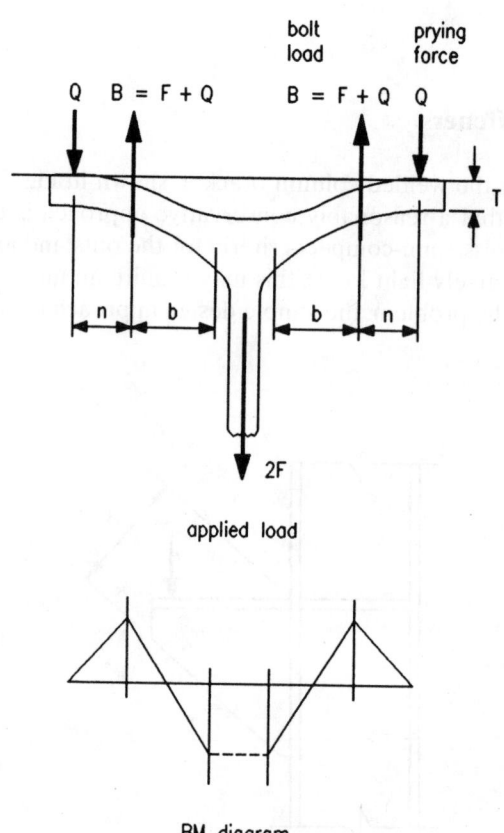

Fig. 25.5   Tee-connection prying forces

when the flange is very stiff, with no plastic deformation so that its flexural deformation is small relative to the elongation of the bolts, will the prying forces be
insignificant.

In BS5950-1: 2000 the tension capacity of a connection using bolts (including 90°
countersunk head bolts) may be checked using either a simple method (clause
6.3.4.2), which includes an allowance for prying forces, or a more exact method
(clause 6.3.4.3) which explicitly accounts for prying action. The simple method may
only be used in a connection where the cross-centre spacing of the bolt lines does
not exceed 55% of the flange width or end-plate width. When using the simple
method, if a connected part is designed assuming double curvature bending, its
moment capacity per unit width should be restricted to the elastic resistance, $p_y t_p^2/6$,
where t_p is the thickness of the connected part. In the simple method the prying force
need not be calculated but the tensile force per bolt, F_t, transmitted by the connection should not exceed the nominal tension capacity P_{nom} of the bolt, obtained from
$P_{nom} = 0.8 p_t A_t$, where A_t is the tensile stress area as specified in the appropriate bolt
standard. For bolts where the tensile stress area is not defined, A_t should be taken
as the area at the bottom of the threads and p_t is the tension strength of the bolt
obtained from Table 34.

The more exact method may be used for a connection in which both of the connected parts satisfy one or more of the following:

(a) the connected part spans between two or more supporting parts;
(b) the outstand of the connected part is designed assuming single curvature
    bending;
(c) the outstand of the connected part is designed assuming double curvature
    bending and the resulting prying force Q is calculated and included in the total
    applied tension F_{tot} in the bolt;
(d) the connected part spans between two or more supporting parts in one direc-
    tion, but acts as an outstand in the other direction, and the resulting prying force
    Q is calculated and included in the total applied tension F_{tot} in the bolt.

In cases (a) and (b) no prying force is necessary for equilibrium. In the more exact
method the moment capacity per unit width of the connected part should be taken
as the plastic resistance, $p_y t_p^2/4$, and the total applied tension F_{tot} in the bolt, including the calculated prying force, should not exceed the tension capacity P_t, taken as
$p_t A_t$.

25.4 Plates loaded in-plane

25.4.1 Deductions for holes

Where holes are staggered the effective net section through the line of the holes is
taken as the area of the gross cross section less a deduction of the area of the holes
in any (zig-zag) section less $s^2 t/4g$ for each gauge space in the line of holes (Fig.

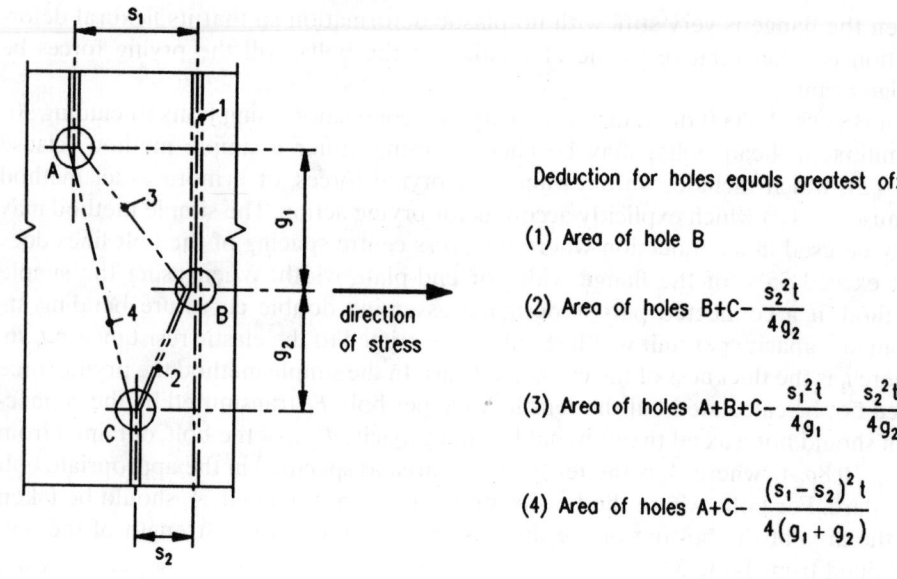

Deduction for holes equals greatest of:

(1) Area of hole B

(2) Area of holes $B+C- \dfrac{s_2^2 t}{4g_2}$

(3) Area of holes $A+B+C- \dfrac{s_1^2 t}{4g_1} - \dfrac{s_2^2 t}{4g_2}$

(4) Area of holes $A+C- \dfrac{(s_1-s_2)^2 t}{4(g_1+g_2)}$

Fig. 25.6   Deduction factors

25.6),[1] where s is the bolt pitch in the direction of stress, t is the thickness of the material and g is the bolt gauge transverse to the pitch.

Where plates are loaded in shear, the beneficial effects of strain hardening permit the presence of bolt holes to be ignored provided that $A_{v.net} \geq 0.85 A_v / K_e$, where $A_{v.net}$ is the net area of the plate and K_e is the effective net area coefficient, taken as 1.2 for S275 steel and 1.0 for S355 steel. If $A_{v.net}$ is less than $0.85 A_v \sqrt{K_e}$ then the shear capacity is given by $0.7 p_y K_e A_{v.net}$.

Where holes are present in a compression flange no allowance need be made for them. The presence of holes in a tension flange may be ignored if the net area of the tension element (after deducting the holes), a_{net}, is greater than the area of the tension element divided by K_e, the effective net area coefficient. Where this is not the case, the effective net area may be taken as $K_e a_{net}$.

25.4.2 Gusset plates

The design of gusset plates can be carried out by developing separate rules for each of the possible modes of failure (e.g. in Fig. 25.7, failure on line A–F or failure by tearing out of the section G–C–D–H, etc.). A simpler procedure is normally adopted in which dispersion of the load is assumed and only transverse sections are checked in tension or compression (with the addition of bending where appropriate). It is considered that 30° is a satisfactory maximum angle of dispersion.

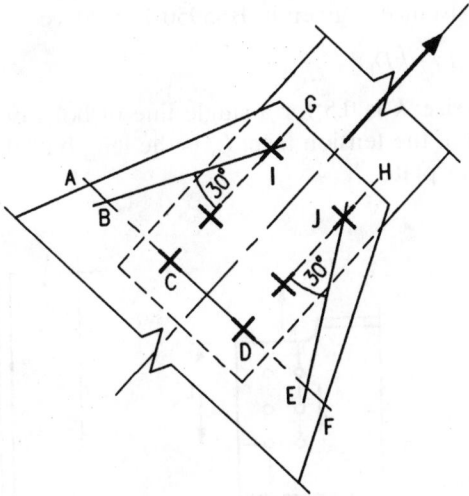

Fig. 25.7   Gusset plate load dispersion

The ratio of D–E to D–J is 0.577 (i.e. tan 30°); if this is compared with the relationship between shear strength and tensile strength the reason why the check on B–E also covers the other modes of failure is apparent.

25.4.3 Notched beams

When checking the shear capacity of beams and channels it is often possible to ignore the presence of holes in the web (see 25.4.1). However, if the beam or channel has been notched or the flanges stripped it is possible that failure may occur by failure in shear at the row of bolt holes along the shear face of the hole group, accompanied by tensile rupture along the line of bolt holes on the tension face of the hole group. This form of failure, known as block shear, is illustrated in Fig. 25.8.

The block shear resistance is given in BS5950-1: 2000 as:

$$P_r = 0.6p_y t[L_v + K_e(L_t - kD_t)]$$

where D_t is the hole size; k is 0.5 for a single line of bolts or 2.5 for two lines of bolts; L_t is the length of the tension face; L_v is the length of the shear face; t is the thickness of the web or plate.

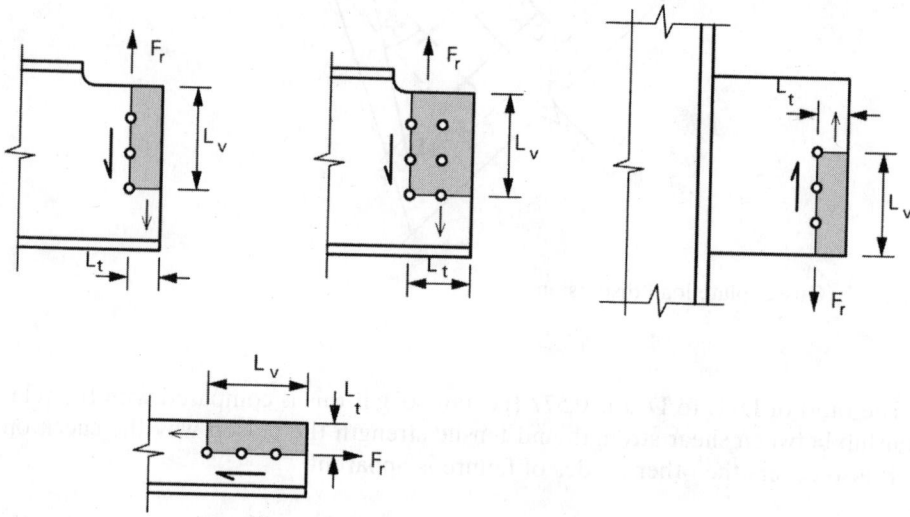

Fig. 25.8 Block shear (Based on Figure 22 BS 5950: 2000)

References to Chapter 25

1. British Standards Institution (2000) *Structural use of steelwork in building.* Part 1: *Code of practice for design – Rolled and welded sections.* BS 5950, BSI, London.
2. Commission of the European Communities (1988) *Design of steel structures.* Part 1: *General rules and rules for buildings.* Eurocode 3 (final draft), Dec.

Chapter 26
Design of connections

by DAVID MOORE

26.1 Introduction

In general the cost of the design, fabrication and erection of the structural frame in a steel framed building is approximately 30% of the total cost of construction. Of these three items, fabrication and erection account for approximately 67%. Any savings in the fabrication and erection costs can significantly reduce the overall cost of construction. The majority of the fabrication costs are absorbed by the connections, and the choice of connection also has a significant influence on the speed, ease, and, therefore, the cost of erection. It is evident that the potential for reducing the cost of steel construction lies in the suitable choice of the beam-to-column and beam-to-beam connections. Indeed, because of the repetitive nature of connections, even small material and labour savings in one connection can have an important effect on the overall economy of the building.

In view of the significance of design and detailing it is remarkable that it is often regarded as being of secondary importance in the design process. Current codes of practice do little to redress this undesirable situation as many of them give little guidance on connection design. Consideration of the local effects at connections is usually left to the designer and this has led to a diversity of both connection types and design methods. The traditional split of responsibilities, where the consultant designs the members of the frame and the connections are designed and detailed by the fabricator, has further compounded this problem.

Some of the many types of beam-to-column connections used in multi-storey steel frame construction are shown in Fig. 26.1. Choice of connection type is usually based on simplicity, duplication and ease of erection – all for economic reasons. Welded joints provide full moment continuity but are expensive due to the on-site welding involved. In recent years bolted connections have increased in popularity. They have the advantages of requiring less supervision than welded joints, having a shorter assembly time and support the load as soon as the bolts are in position. They also have a geometry that is easy to comprehend and can accommodate minor discrepancies in the dimensions of the beams and columns. However, when large forces are involved, bolted connections can be criticized for requiring extensive space, which may conflict with the architectural need for a 'clean line'.

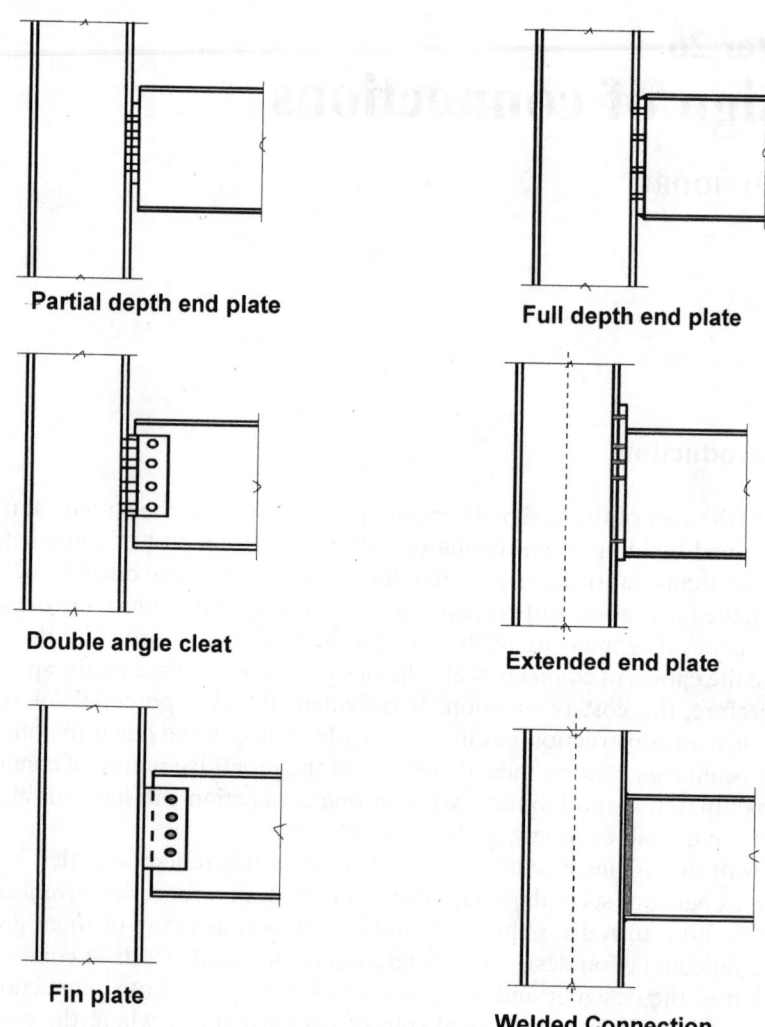

Partial depth end plate

Full depth end plate

Double angle cleat

Extended end plate

Fin plate

Welded Connection

Fig. 26.1   Typical beam-to-column connections

With such a large number of connection types and the variety within each type (each having different characteristics) it is not surprising that design engineers have difficulty in choosing a suitable connection and the most appropriate design method. Furthermore, the detailer is confronted with a bewildering choice of components to use in the connection and this has done little to help the move towards standardization.

Over the last decade there has been a growing awareness that the steel construction industry could improve its effectiveness and efficiency by increasing the repetition of elements within a structure and promoting the use of standard connections. To define a range of standard connections and standard design methods,

and to develop a framework which would lead to the widespread adoption of rationalized connections using standardised components, the Steel Construction Institute (SCI) and the British Constructional Steelwork Association (BCSA) in 1987 formed the SCI/BCSA connection group. This group has produced the following two publications:

> *Joints in Steel Construction: Simple Connections*[1]
> *Joints in Steel Construction: Moment Connections*[2]

Both of the above publications present standard design methods for the most commonly used connections. These design methods are based on a combination of the latest design approaches and practical aspects associated with current fabrication and erection techniques and produce realistic estimates of a connection's strength. The design checks presented in sections 26.2 and 26.3 for simple and moment connections respectively are based on the methods detailed in these publications. Some of the design methods are based on the approach and recommendations given in Eurocode 3: *Design of Steel Structures*. Part 1.1 *General Rules and Rules for buildings*[3], which may not yet be familiar to many practising engineers. This chapter therefore presents the design principles and the classification systems for connections used in Eurocode 3.

26.1.1 Design principles

There is a need for the design of connections to be consistent with the engineer's assumptions regarding the structural behaviour of the steel frame. Therefore, when choosing and proportioning connections the engineer should always bear in mind the basic requirements such as the stiffness or flexibility of the connection, strength and the required rotational capacity.

Care should also be taken to ensure that the assumptions made for the design of the various elements of the connection are compatible. For example, in a cover plate splice, sharing the load between ordinary bolts (in shear and tension) and a fillet weld is not acceptable because the deformation characteristics of ordinary bolts and fillet welds are incompatible. The fillet welds are stiff compared with the bolts in clearance holes, and as the load is applied the welds will initially carry most of the load. Then due to their limited ductility, the fillet welds will break before the bolts can take up their share of the load.

The engineer should also consider economy, which is at least as important as the structural considerations. As a general rule, if fabrication and erection costs are kept to a minimum then the overall cost will also tend to be a minimum. However, it should be realized that the costs are fabricator-dependent and rather than impose a particular connection type on a fabricator it is more cost effective to state a range of standard connection types that satisfy the design assumptions. The fabricator will then choose that connection which can be fabricated economically.

Finally, connections should be simple and good access should be provided for any welding operations and/or the placing and tightening of bolts.

26.1.2 Classification of connections

Connection design depends very much on the designer's decision regarding the method by which the structure is analysed. The UK steel design code, BS5950[4] and Eurocode 3 give four approaches for the design of a structure in which the behaviour of the connection is fundamental. These design methods are defined as *simple design, semi-continuous design, continuous design* and *experimental verification.* Elastic, plastic and elastic–plastic methods of global analysis can be used with any of the first three approaches, and Table 26.1 shows how the joint classification, the type of framing and the method of global analysis are related.

Table 26.1   Relationship between frame types, analysis and joint classification

Method of global analysis	Classification of joint		
Elastic	Nominally-pinned	Rigid	Semi-rigid
Plastic	Nominally-pinned	Full-strength	Partial-strength and ductile
Elastic–plastic	Nominally-pinned	Rigid and full-strength	Semi-rigid and/or partial strenght
Type of framing	Simple	Continuous	Semi-continuous

The simple method is based on the assumption that the beams are simply supported and implies that beam-to-column connections must be sufficiently flexible to restrict the development of significant end moment. Any horizontal forces must be resisted by bracing or shear walls, etc. When using this approach the connections are classified as nominally pinned no matter what method of global analysis is used. However, research has shown that most connections – even so-called simple connections – are capable of developing some moment capacity.

If the continuous approach is adopted the type of connection used will depend on the method of global analysis. When elastic analysis is used the joints are classified according to their stiffness, and rigid connection must be used. When plastic analysis is used the connections are classified according to their strength (moment capacity) and full-strength connections must be used. The term 'full-strength' relates the strength of the connection to that of the connected beam. If the moment capacity of the connection is higher than that of the connected beam then the connection is termed full-strength. The purpose of this comparison is to determine whether the joint or the connected member will limit the resistance of the structure. If the elastic–plastic method of global analysis is used then the connections are classified according to both their stiffness and strength, and rigid, full-strength connections must be used. These connections must be capable of carrying the design bending moment, shear force and axial load while maintaining the original angle between the connected members. While continuous design can produce economies in beam

size with respect to simple design most of these savings are offset by the need to supply joints with adequate rigidity.

While the simple method ignores stiffness and the continuous method only allows full-strength connections, the semi-continuous method accepts the fact that most practical connections are capable of providing some degree of stiffness and that their moment capacity may be limited. Once again the type of connection used will depend on the method of global analysis. When elastic analysis is used the connections are classified according to their stiffness, and semi-rigid connections can be used. It should be noted that the term 'semi-rigid' is a general classification and can be used to encompass all connections including pinned and rigid connections. If plastic global analysis is used the connections are classified according to their strength. Connections that have a lower moment capacity than the connected member are termed *partial-strength*. In this case the connection will fail before the connected member and must therefore possess sufficient ductility to allow plastic hinges to form in other parts of the structures. Where the elastic–plastic method of global analysis is used the connections are classified according to both their stiffness and strength, and semi-rigid, partial strength connections are used.

From the above discussion it is clear that a connection has three fundamental properties:

(1) Moment resistance
   The connection may be either full strength, partial strength or nominally pinned (i.e. not moment resisting)
(2) Rotational stiffness
   The connection may be rigid, semi-rigid or nominally pinned (i.e. no rotational stiffness)
(3) Rotational capacity
   Connections may need to be ductile. This criterion is less familiar to most designers and introduces the concept that a connection may need to rotate plastically at some stage of the loading cycle without failure. Pinned connections have to perform in this way, and the principle also applies to some moment connections.

These three properties are also used to classify connections, and Fig. 26.2 illustrates the different ways in which a connection may be classified.

Some guidance on the properties that are needed for connections in frames designed using one of the methods described above are given in Table 26.2.

26.1.3 Definitions

Definitions for some of the terms used in this chapter are given below:

(1) *Full strength connections* have a moment of resistance at least equal to that of the member.

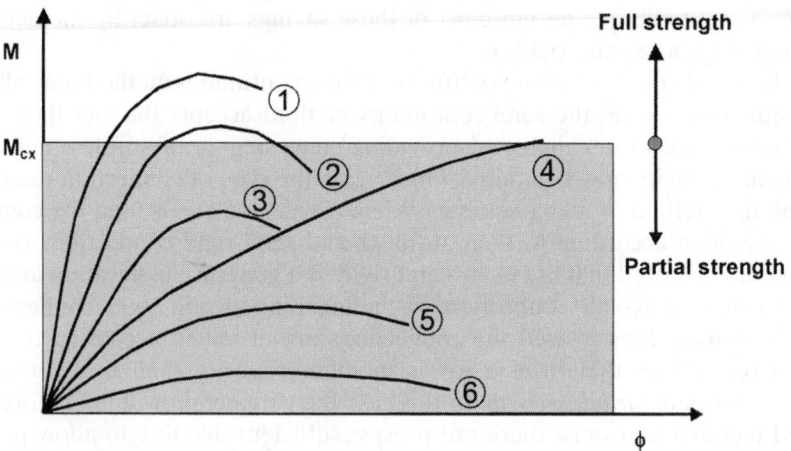

a. Classification by strength

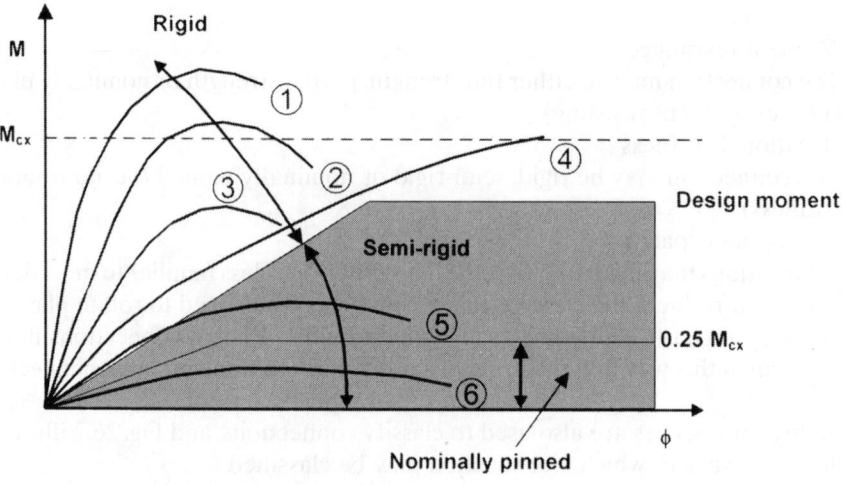

b. Classification by rigidity

Fig. 26.2   Classification of connections

(2) *Partial strength connections* have a moment of resistance which is less than that of the member.
(3) A *rigid connection* is stiff enough for the effect of its flexibility on the distribution of the bending moments in the frame to be neglected.
(4) A *semi-rigid connection* is too flexible to qualify as rigid, but is not a pin.

Table 26.2   Methods of frame design

Design		Connections			Notes
Type of framing	Global analysis	Properties	Fig. 26.2 Example	Section	
Simple	Pin joints	Nominally pinned	⑥	Section 26.2	Economic method of braced multi-storey frames. Connection design is made for shear strength only.
Continuous	Elastic	Rigid	①②③④	Section 26.3	Conventional elastic analysis
	Plastic	Full strength	①②④	Section 26.3	Plastic hinges form in the adjacent member, not in the connections. Popular for portal frame designs.
	Elastic–plastic	Full strength and rigid	①②④	Section 26.3	
Semi-continuous	Elastic	Semi-rigid	⑤⑥	Not covered	Connections are modelled as rotational springs. Prediction of connection stiffness presents difficulties.
	Plastic	Partial strength and ductile	⑤⑥	Not covered	Wind-moment design is a variant of this method.
	Elastic–plastic	Partial strength and/or semi-rigid	Any	Not covered	Full connection properties are modelled in the analysis. A research tool rather than a practical design method at the present time.

(5) *Nominally pinned connections* are sufficiently flexible to be regarded as pins for analysis. These connections are by definition flexible, not moment connections, although partial strength connections able to resist less than 25% of the plastic moment capacity of the beam may be regarded as nominally pinned.

(6) A *ductile connection* has sufficient rotation capacity to act as a plastic hinge. Connection ductility should not be confused with material ductility.

(7) In the *simple design* method of frame design the connections are assumed not to develop moments that adversely affect either the members of the structure or the structure as a whole.

(8) The *continuous design* method of frame design does not model the connection properties in the frame analysis. This covers either elastic analysis, where the connections are rigid, or plastic analysis, where the connections are full strength.

(9) For *semi-continuous design* of frames the connection properties have to be modelled in the analysis. This covers elastic analysis where semi-rigid connections are modelled as rotational springs, or plastic analysis where partial strength connections are modelled as plastic hinges.

26.2 Simple connections

26.2.1 Design philosophy

Simple connections are defined as those connections that transmit end shear only and have negligible resistance to rotation and therefore do not transfer significant moments at the ultimate limit state. This definition underlies the design of the overall structure in which the beams are designed as simply-supported and the columns are designed for axial load and the small moments induced by the end reactions from the beams. In practice, however, the connections do have a degree of fixity, which although not taken into account in the design is often sufficient to allow erection to take place without the need for temporary bracing.

The following three principal forms of simple connection are considered in this section:

(1) Double angle web cleats
(2) Flexible end-plates
(3) Fin plates

To comply with the design assumptions, simple connections must allow adequate end rotation of the beam as it takes up its simply-supported deflected profile and practical lack of fit. At the same time this rotation must not impair the shear and tying (for structural integrity – see below) capacities of the connection. In theory a 457 mm deep, simply supported beam spanning 6.0 m will develop an end rotation of 0.022 radians (1.26°) when carrying its maximum factored load. In practice this rotation will be considerably smaller because of the restraining action of the connection. When the beam rotates it is desirable to avoid the bottom flange of the beam bearing against the column as this can induce large forces in the connection. The usual way of achieving this is to ensure that the connection extends at least 10 mm beyond the end of the beam.

26.2.2 Structural integrity

The partial collapse of Ronan Point in 1968 alerted the construction industry to the problem of progressive collapse arising from a lack of positive attachment between principal elements in a structure. This resulted in amendments to both the Building Regulations and the UK's steelwork design code. Essentially, these changes take cognizance of this failure and require structures to have a minimum robustness to resist accidental loading. One method of achieving this is by tying all the principal elements of a structure together. This means that the beam-to-column connections of a steel frame must be capable of transferring a horizontal tying force in order to preserve the integrity of the structure and prevent progressive collapse in the event of accidental damage.

The current steel design code states that disproportionate collapse can be avoided provided the member and its connections are capable of transferring a tying force equal to the member's end reaction under factored loads but not less than 75 kN. There are more onerous requirements for certain multi-storey buildings.

26.2.3 Design procedures

The design of these simple connections is based on BS5950: Part 1. The capacities and design strengths of the fasteners and fittings are based on the rules given in clause 6.3 of BS 5950: Part 1. The spacing of the fasteners and their distances comply with the recommendations given in clause 6.2 of BS 5950: Part 1.

When fabricating and erecting these connections it is general practice to use the following components:

- Untorqued bolts in clearance holes, usually M20 grade 8.8 bolts.
- Cleats, end-plates, fin-plates and other fittings made from grade S275 steel.
- 6 mm fillet welds
- Punched holes on fittings using semi-automatic equipment.

Wherever possible, general industrial practice should be followed, and guidance is given in the National Structural Steelwork Specification.[6]

26.2.4 Beam-to-column connections

26.2.4.1 Double angle web cleats

Typical bolted double angle cleat connections about both the major and minor axis of a column are shown in Figs 26.3 and 26.4 respectively. These types of connection are popular because they have the facility to provide for minor site adjustments

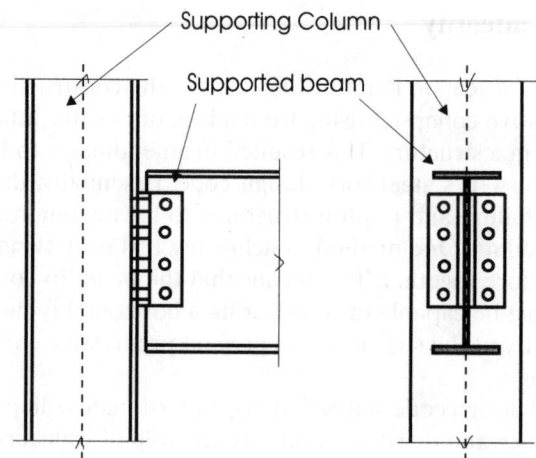

Fig. 26.3 Typical major axis double angle cleat connections

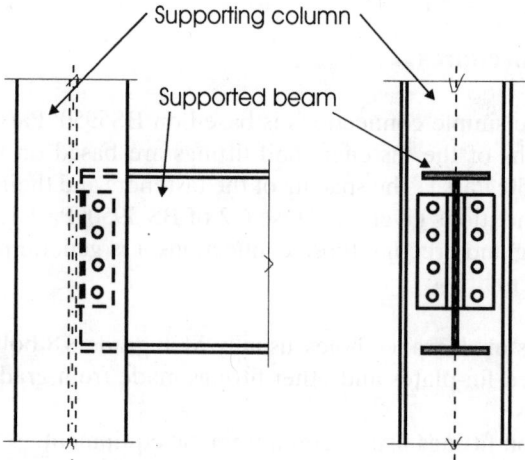

Fig. 26.4 Typical minor axis double angle cleat connections

when using untorqued bolts in 2 mm clearance holes. Normally the cleats are used in pairs. Any simple equilibrium analysis is suitable for the design of this type of connection. The one recommended in this publication assumes that the line of action of shear transfer between the beam and the column is at the face of the column. Using this model the bolt group connecting the cleats to the beam web must be designed for the shear force and the moment produced by the product of the end shear and the eccentricity of the bolt group from the face of the column. The bolts connecting the cleats to the face of the column should be designed for the applied shear only. In practice the cleats to the column are rarely critical and

the design is almost always governed by the bolts bearing on to the web of the beam. The rotational capacity of this connection is governed largely by the deformation capacity of the angles and the slip between the connected parts. Most of the rotation of the connections comes from the deformation of the angles while fastener deformation is very small. To minimize rotational resistance (and increase rotational capacity) the thickness of the angle should be kept to a minimum and the bolt cross-centres should be as large as is practically possible.

When connecting to the minor axis of a column it may be necessary to trim the flanges of the beam but this does not change the shear capacity of the beam. During erection the beam with the cleats attached is lowered down the column between the column flanges.

Single angle web cleats

Single angle web cleats are normally only used for small connections or where access precludes the use of double angle or end-plate connections.

This type of connection is not desirable from an erector's point of view because of the tendency of the beam to twist during erection. Care should be taken when using this type of connection in areas where axial tension is high (e.g. where the axial tension due to structural integrity requirements of BS 5950: Part 1 is high). The design checks for this type of connection are similar to those shown in Table 26.3. In addition to these checks the bolts connecting the cleat to the column must also be checked for the moment produced by the product of the end shear force and the distance between the bolts and the centreline of the beam.

The essential detailing requirements for this connection are shown in Fig. 26.5, and Table 26.3 shows the detailed design checks. The design procedure given in Table 26.3 applies to beams connected to either the column flange or the column web.

26.2.4.2 Flexible end-plates

Typical flexible end-plate connections about the major and minor axis of a column are shown in Figs 26.6 and 26.7 respectively. These connections consist of a single plate fillet welded to the end of the beam and site bolted to a supporting column. This connection is relatively inexpensive but has the disadvantage that there is no room for site adjustment. Overall beam lengths need to be fabricated within tight limits although packs can be used to compensate for fabrication and erection tolerances. The end-plate is often detailed to extend to the full depth of the beam but there is no need to weld the end-plate to the flanges of the beam.

Sometimes the end-plate is welded to the beam flanges to improve the stability of the frame during erection and avoid the need for temporary bracing. This type of connection derives its flexibility from the use of relatively thin end-plates combined with large bolt cross-centres. An 8 mm thick end-plate combined with 90 mm cross-centres

Table 26.3 Design checks for double angle cleat connections

Check no.	Description	Design rule	
1	Essential detailing requirements	See Fig. 26.5	
2	Shear capacity of bolt group connecting cleats to beam web	The shear capacity of a single bolt (P_s) must be greater than half the resultant force on the outermost bolt due to direct shear and moment ($F_s/2$).	
3	Shear and bearing capacity of cleat connected to supported beam	*For shear* The shear capacity of the leg of the angle cleat (P_v) must be greater than half the reaction at the end of the beam $\left(\dfrac{F_v}{2}\right)$.	*For bearing* The bearing capacity of the leg of the angle cleat per bolt (P_{bs}) must be greater than half the resultant force on the outermost bolt $\left(\dfrac{F_s}{2}\right)$.
4	Shear and bearing capacity of supporting beam	*For shear* The shear capacity of the beam (P_v) must be greater than the reaction at the end of the beam (F_v)	*For bearing* The bearing capacity of the beam web per bolt (P_{bs}) must be greater than the resultant force on the outermost bolt (F_s)
5	Shear capacity of bolt group connecting cleats to column	The shear capacity of the bolt group connecting the cleats to the column (ΣP_s) must be greater than the reaction at the end of the beam (F_v)	
6	Shear and bearing capacity of cleats connected to column	*For shear* The shear capacity of the leg of the angle cleat (P_v) must be greater than half the reaction at the end of the beam $\left(\dfrac{F_v}{2}\right)$.	*For bearing* The total bearing capacity of the leg of a single angle cleat (ΣP_{bs}) must be greater than half the reaction at the end of the beam $\left(\dfrac{F_v}{2}\right)$.
7	Local shear and bearing capacity of column web	*For shear* The local shear capacity of the column web (P_v) must be greater than half the sum of the beam end reactions either side of the column web $\dfrac{(F_{v1}+F_{v2})}{2}$.	*For bearing* The bearing capacity of the column web must be greater than half the sum of the beam end reactions either side of the column web $\dfrac{\left(\dfrac{F_{v1}}{n_1}\right)+\left(\dfrac{F_{v2}}{n_2}\right)}{2}$.
8	Structural integrity – tension capacity of pair of web angle cleats	The tying capacity of double angle web cleats must be greater than the tie force	
9	Structural integrity – tension and bearing capacity of beam web	*For tension* The net tension capacity of the beam web must be greater than the tie force.	*For bearing* The bearing capacity of the beam web must be greater than the tie force
10	Structural integrity – tension capacity of bolts	The tension capacity of the bolt group must be greater than the tie force	

Note: The design check numbers may not match the check numbers given in Ref 1.

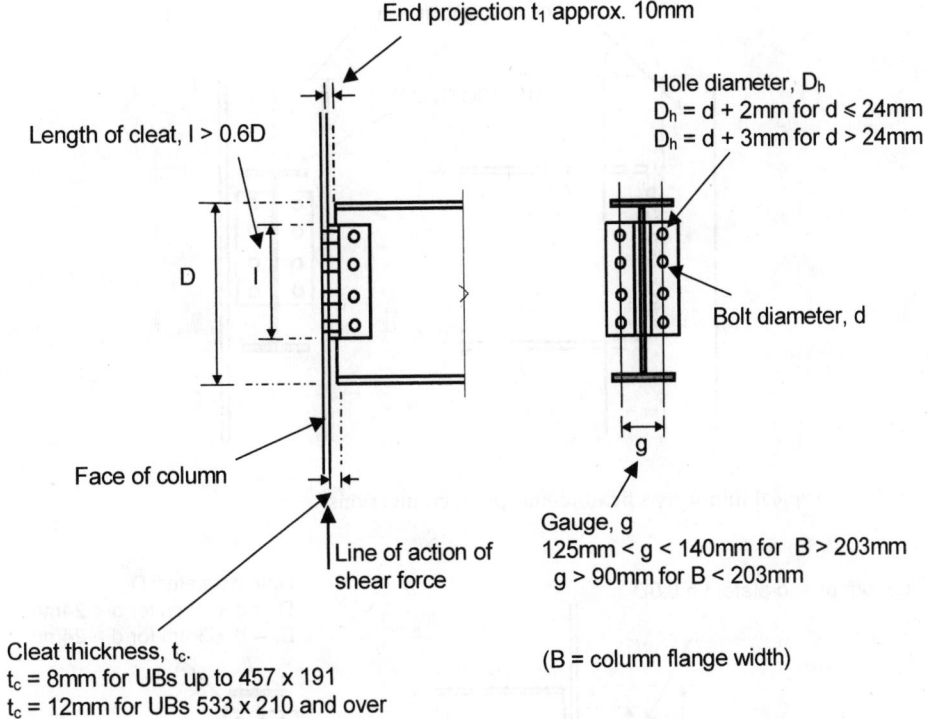

End projection t₁ approx. 10mm

Hole diameter, D_h
$D_h = d + 2mm$ for $d \leqslant 24mm$
$D_h = d + 3mm$ for $d > 24mm$

Length of cleat, $l > 0.6D$

D   l

Bolt diameter, d

Face of column

Gauge, g
$125mm < g < 140mm$ for $B > 203mm$
$g > 90mm$ for $B < 203mm$

Line of action of
shear force

Cleat thickness, t_c.
$t_c = 8mm$ for UBs up to 457 x 191
$t_c = 12mm$ for UBs 533 x 210 and over

(B = column flange width)

Fig. 26.5   Essential detailing requirements for double angle cleat connections

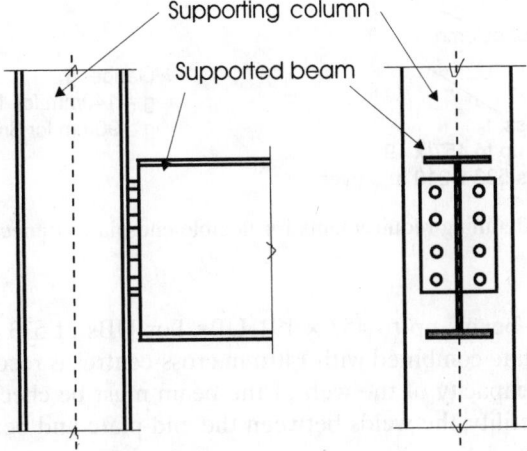

Supporting column
Supported beam

Fig. 26.6   Typical major axis flexible end-plate connections

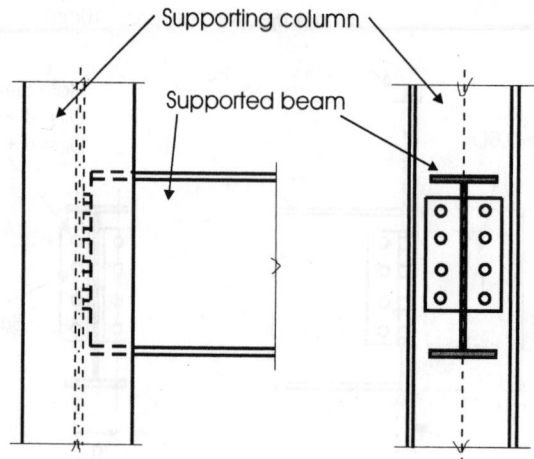

Fig. 26.7   Typical minor axis flexible end-plate connections

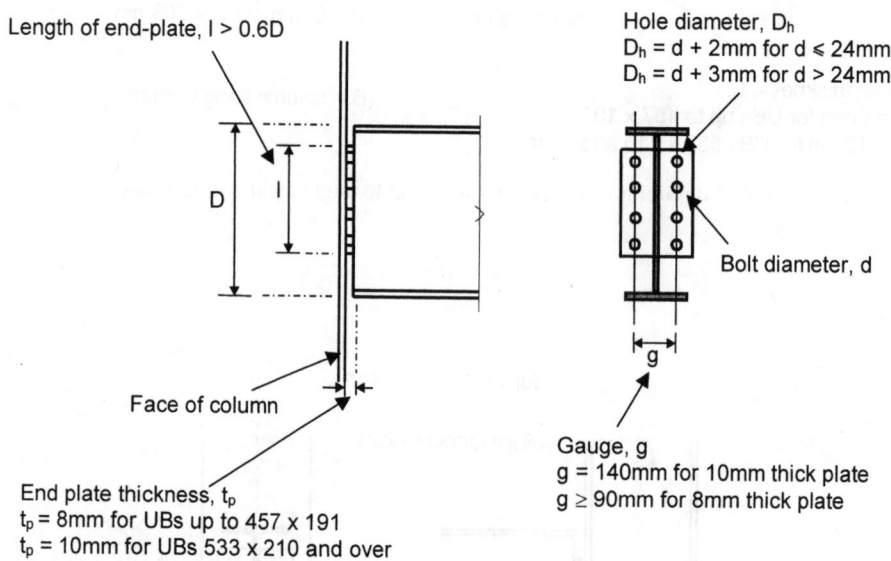

Fig. 26.8   Essential detailing requirements for flexible end-plate connections

is usually used for beams up to 457×191 UBs. For UBs of 533×210 and over a 10 mm thick end-plate combined with 140 mm cross-centres is recommended.

The local shear capacity of the web of the beam must be checked and, because of their lack of ductility, the welds between the end-plate and beam web must not be the weakest link.

The essential detailing requirements for this connection are shown in Fig. 26.8

Table 26.4 Design check for flexible end-plate connections

Check no.	Description	Design rule	
1	Essential detailing requirements	See Fig. 26.8	
2	Shear capacity of bolt group	The shear capacity of the bolt group (ΣP_s) must be greater than the reaction at the end of the beam (F_v).	
3	Shear and bearing capacity of end-plate	*For shear* The shear capacity of the end-plate (P_v) must be greater than half the reaction at the end of the beam $\left(\dfrac{F_v}{2}\right)$.	*For bearing* The bearing capacity of the end-plate (ΣP_{bs}) must be greater than half the reaction at the end of the beam $\left(\dfrac{F_v}{2}\right)$.
4	Shear capacity of the beam web at the end-plate	The shear capacity of the beam web connected to the end-plate (P_v) must be greater than the reaction at the end of the beam (F_v).	
5	Capacity of fillet welds connecting end-plate to beam web	The capacity of the fillet weld (P_{weld}) must be greater than the reaction at the end of the beam (F_v).	
6	Local shear and bearing capacity of column web	*For shear* The local shear capacity of the column web (P_v) must be greater than half the sum of the beam end reactions either side of the column web $\dfrac{(F_{v1}+F_{v2})}{2}$.	*For bearing* The bearing capacity of the column web (P_{bs}) must be greater than half the sum of the reactions either side of the column web divided by the number of bolt rows $\dfrac{\left(\dfrac{F_{v1}}{n_1}\right)+\left(\dfrac{F_{v2}}{n_2}\right)}{2}$.
7	Structural integrity – tension capacity of end-plate	The tying capacity of the end-plate must be greater than the tie force.	
8	Structural integrity – tension capacity of beam web	The tension capacity of the beam web must be greater than the tie force.	
9	Structural integrity – weld tension capacity.	The tension capacity of the beam web to end-plate weld must be greater than the tie force.	
10	Structural integrity – tension capacity of the bolts	The tension capacity of the tension bolt group must be greater than the tie force.	

Note: The design check numbers may not match the check numbers given in Ref 1.

and the design checks are detailed in Table 26.4. The design procedure in Table 26.4 applies to beams connected to either a column flange or a column web and the procedure is equally applicable for partial and full depth end-plates.

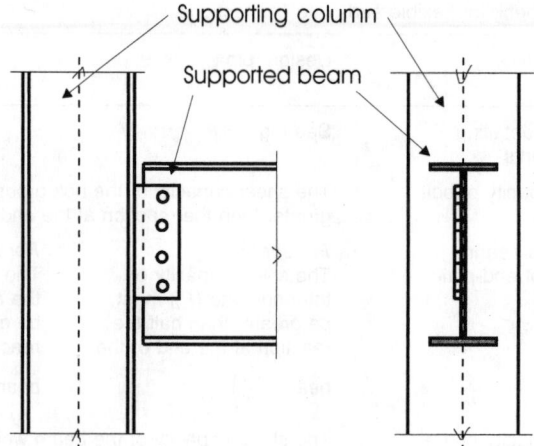

Fig. 26.9   Typical major axis fin-plate connection

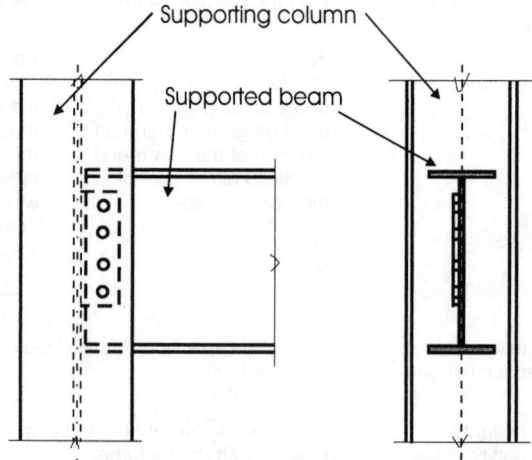

Fig. 26.10   Typical minor axis fin-plate connection

26.2.4.3 Fin plates

A more recent development which follows both Australian and American practice has been the introduction of the fin plate connection. This type of connection is primarily used to transfer beam end reactions and is economical to fabricate and simple to erect. There is clearance between the ends of the supported beam and the supporting column, thus ensuring an easy fit. Figures 26.9 and 26.10 show a typical bolted fin plate connection to the major and minor axes of a column respectively. These connections comprise a single plate with either pre-punched or pre-drilled holes that is shop welded to the supporting column flange or web.

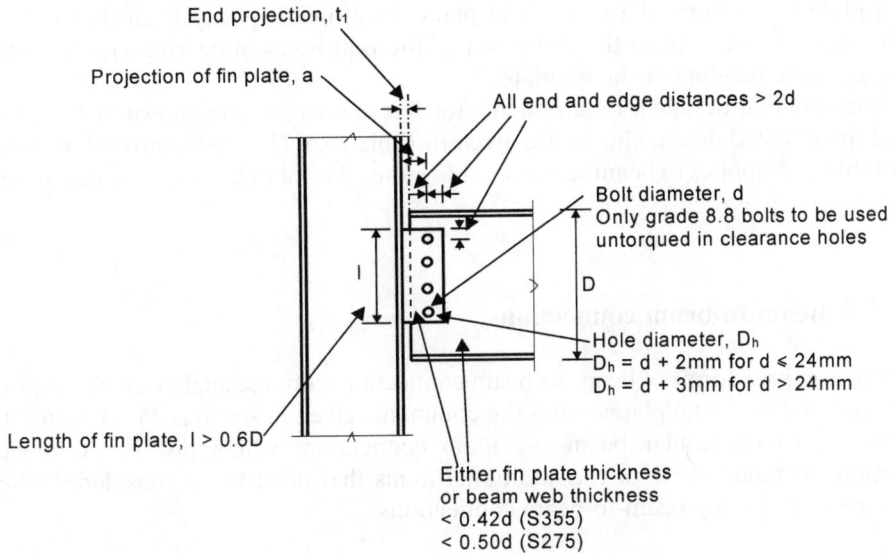

Fig. 26.11   Essential detailing requirements for fin-plate connections

Before developing a UK design method for fin plates, an extensive literature survey was carried out to identify practice in other countries. The most appropriate and best-researched design method was found to be that developed by the Australian Institute of Steel Construction (AISC).[6] This was adopted as the starting point for the UK approach. The principal difference between the AISC method and that adopted in the UK concerns the line of action for the shear at the end of the beam. Considerable effort has been invested in trying to identify the appropriate line of action for the shear. There are two possibilities: either the shear acts at the face of the column or it acts along the centre of the bolt group connecting the fin plate to the beam web. For this reason all critical sections should be checked for a minimum moment taken as the product of the vertical shear and the distance between the face of the column and the centre of the bolt group. The critical sections are then checked for the resulting moment combined with the vertical shear. The validation of this and other design assumptions were checked against a series of tests on fin plate connections. The results of these tests concluded that the design approach was conservative and gave adequate predictions of strength. The tests also showed that fin plates with long projections (dimension 'a' in Fig. 26.11) had a tendency to twist and fail by lateral torsional buckling. Therefore the following simple check was introduced to limit the projected length of the fin plate and obviate the need to carry out a lateral torsional bucking check:

$$t \geq 0.15a$$

where t is the thickness of the fin plate
        a is the distance between the weld line and the centre of the bolt group

Fin plate connections derive their in plane rotational capacity from the bolt deformation in shear, from the distortion of the bolt holes in bearing and from the out-of-plane bending of the fin plate.

The essential detailing requirements for this connection are shown in Fig. 26.11 and the detailed design checks are given in Table 26.5. The design procedure given in Table 26.5 applies to beams connected to either the column flange or column web.

26.2.5 Beam-to-beam connections

There are three forms of beam-to-beam connection – double angle web cleats, flexible end-plates and fin plates – and the comments given in sections 26.2.4.1, 26.2.4.2 and 26.2.4.3 on similar beam-to-column connections will apply. The following sections highlight some of the additional items that need to be considered when designing and using beam-to-beam connections.

26.2.5.1 Double angle web cleats

Figures 26.12 and 26.13 show typical beam-to-beam double angle web cleat connections with single notched and double notched beams respectively. Where the top flanges of the connected beams are at the same level, as in the case of the connection shown in Fig. 26.12, the flange of the supported beam is notched and the web must be checked, allowing for the effect of the notch. The top of the web of the notch, which is in compression, must be checked for local buckling of the unrestrained web. Provided that the supported beam is laterally restrained by a floor slab to ensure that the web at the top of the notch does not buckle, it is recommended that the length of the notch should not exceed the following limits.

For a beam with one flange notched:

$$d_{c1} \leq \frac{D}{5} \quad \text{and}$$

$$c \leq D \qquad \text{for} \quad \frac{D}{t_w} \leq 54.3 \quad \text{(S275 steel)}$$

$$c \leq \frac{160\,000D}{\left(D/t_w\right)^3} \qquad \text{for} \quad \frac{D}{t_w} > 54.3 \quad \text{(S275 steel)}$$

$$c \leq D \qquad \text{for} \quad \frac{D}{t_w} \leq 48.0 \quad \text{(S355 steel)}$$

$$c \leq \frac{110\,000D}{\left(D/t_w\right)^3} \qquad \text{for} \quad \frac{D}{t_w} > 48.0 \quad \text{(S355 steel)}$$

where c is the length of the notch

D is the depth of the beam

t_w is the thickness of the supported beam web, and

d_{c1} and d_{c2} are shown in Figs 26.12 and 26.13.

Table 26.5 Design checks for fin plate connections

Check no.	Description	Design rule	
1	Essential detailing requirements	See Fig. 26.11	
2	Capacity of bolt group connecting fin plate to the web of the supported beam	The bearing capacity per bolt (P_{bs}) must be greater than the resultant force on the outermost bolt due to direct shear and moment (F_s)	
3	Strength of the supported beam at the net section	*For shear* The shear capacity of the supported beam (P_v) must be greater than the reaction at the end of the beam (F_v)	*For bending* For long fin plates the resistance of the net section must be greater than the applied moment ($F_v a$)
4	Strength of the fin plate at the net section under bending and shear	*For shear* The shear capacity of the fin plate (P_v) must be greater than the reaction at the end of the beam (F_v)	*For bending* The elastic modulus of the net section of the fin plate ($p_y Z$) must be greater than the moment ($F_v a$) due to the end reaction and the distance a
5	Lateral torsional buckling resistance of long fin plates	The lateral torsional buckling resistance moment of the fin plate (M_b) must be greater than the moment due to the end reaction and the distance a ($F_v a$)	
6	Strength of weld connecting fin plate to supporting column	The leg length of the fillet weld(s) must be greater than 0.8 times the thickness of the fin plate ($0.8t$)	
7	Local shear check of column web	The local shear capacity of the column web (P_v) must be greater than half the sum of the beam end reactions either side of the column web $\dfrac{(F_{v1}+F_{v2})}{2}$.	
8	Structural integrity – tension capacity of fin plate	The tension capacity of the fin plate must be greater than the tie force	
9	Structural integrity – tension capacity of beam web	The net tension capacity of the beam web must be greater than the tie force	
10	Structural integrity – bearing capacity of beam web or fin plate	The bearing capacity of the beam web or fin plate must be greater than the tie force	
11	Structural integrity – capacity of column web	The tying capacity of the column web must be greater than the tie force	

Note: The design check numbers may not match the check numbers given in Ref 1.

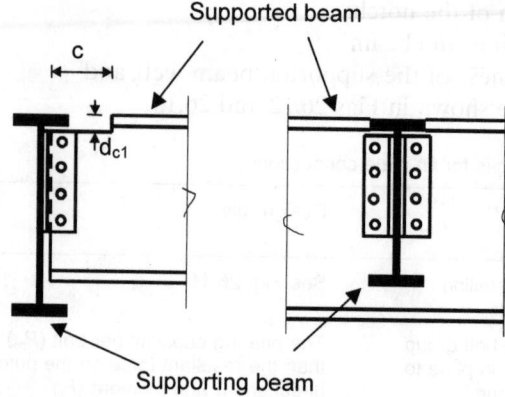

Fig. 26.12   Single notched beam to beam connection for double angle cleat connections

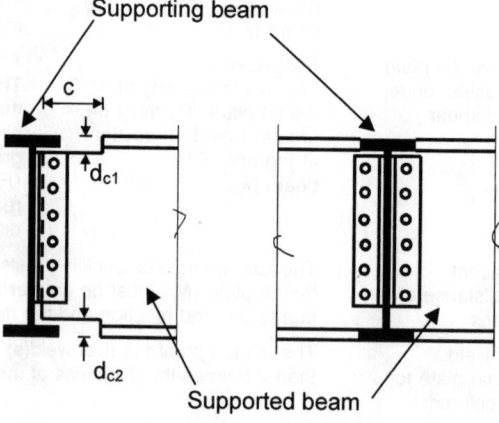

Fig. 26.13   Double notched beam to beam connection for double angle cleat connections

Where both flanges of the beam are notched, as in the case of the connection shown in Fig. 26.13, it is recommended that each notch should not exceed the limits set for a single notch as stated above.

For supported beams which are not laterally restrained, a more detailed investigation is required on the overall stability of the beam with notched ends against lateral torsional buckling.

Beams with only one flange notched should be checked for lateral torsional buckling using the method given in clause 4.3 of BS 5950 with a modified effective length (L_e) which takes account of the reduced stiffness of the beam at the notches.

The equivalent effective length L_e given below is based on the information given in References 7, 8 and 9 and is only valid for $c/L_b < 0.15$ and $d_c/D < 0.2$, where L_b is the length of the beam and d_c is the depth of the notch. All the other symbols are defined above. Beams with notches outside these limits should be checked as tee sections or stiffened.

$$L_e = L_b\left(1+\frac{2c(K^2+2K)}{L_b}\right)^{\frac{1}{2}}$$

$$K = \frac{K_0}{\lambda_b} \quad \text{(for values of } K_0 \text{ refer to [1])}$$

$$\lambda_b = \frac{uvL_b}{r_y}$$

u, v and r_y are for I-beam sections and are defined in BS 5950: Part 1.

The web angle cleat can become cumbersome when used to connect unequal sized beams. In this case it is necessary to notch the bottom flange of the smaller beam to prevent fouling of the bolts. Alternatively the cleat of the larger beam could be extended and the bolts placed below the bottom of the smaller beam.

The essential detailing requirements for double angle web cleat connections are given in Fig. 26.14 and Table 26.6 gives the recommended design procedure.

26.2.5.2 Flexible end-plates

This type of connection is shown in Fig. 26.15. Like the double angle cleat connection, the top flange of the supported beam is notched to allow it to fit to the web of the supporting beam.

If both beams are of a similar depth both flanges are notched. In either case, if the length of the notches exceed the limits given in section 26.2.5.1, the unrestrained web and beam must be checked for lateral torsional bucking. In practice the end-plate is often detailed to extend to the full depth of the notched beam and welded to the bottom flange. This makes the connection relatively stiffer than a partial depth end-plate but provided the end-plate is relatively thin and the bolt cross centres are large, the end-plate retains sufficient flexibility to be classified as a simple connection.

If the supporting beam is free to twist there will be adequate rotational capacity even with a thick end-plate. In the cases where the supporting beam is not free to twist, for example in a double sided connection, the rotational capacity must be provided by the connection itself. In such cases thick, full depth end-plates may lead to overstressing of the bolts and welds. Both partial and full depth end-plates derive their flexibility from the use of relatively thin end-plates combined with large bolt

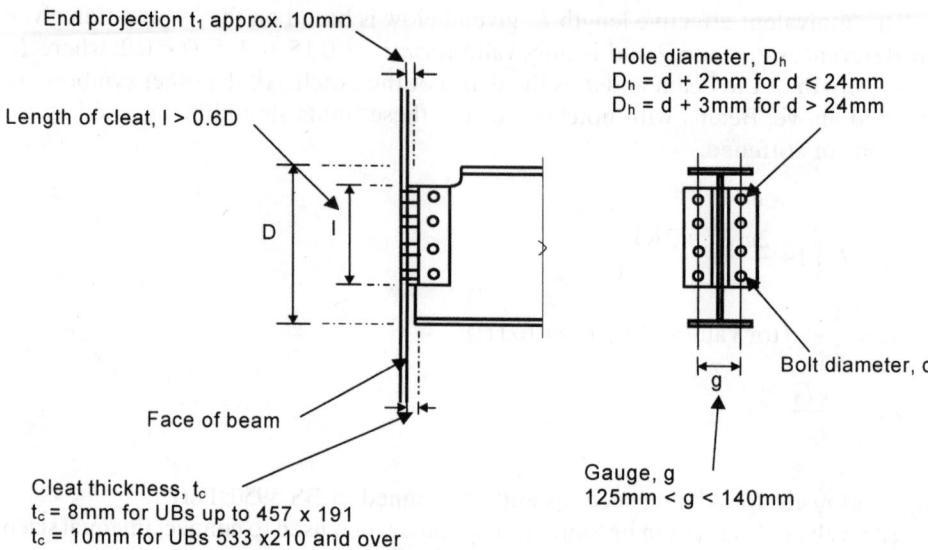

End projection t₁ approx. 10mm

Length of cleat, l > 0.6D

Hole diameter, D$_h$
D$_h$ = d + 2mm for d ⩽ 24mm
D$_h$ = d + 3mm for d > 24mm

D

l

Face of beam

Bolt diameter, d

Gauge, g

Cleat thickness, t$_c$
t$_c$ = 8mm for UBs up to 457 x 191
t$_c$ = 10mm for UBs 533 x210 and over

Gauge, g
125mm < g < 140mm

Fig. 26.14   Essential detailing requirements for beam-to-beam double angle web cleat connections

cross centres. Normally end-plates no more than 8mm or 10mm thick should be used.

The essential detailing requirements for these connections are shown in Fig. 26.16 and Table 26.7 gives the design checks. The procedure in Table 26.7 can used for both partial depth and full depth end-plates.

26.2.5.3 Fin plate connections

Typical bolted fin plate connections are shown in Figs 26.17 and 26.18. The comments made in section 26.2.4.3 on beam-to-column fin plates apply to beam-to-beam fin plates. In addition, a beam-to-beam fin plate connection requires either a long fin plate as shown in Fig. 26.18 or a notched beam as shown in Fig. 26.17. The designer must therefore choose between the reduced capacity of a long fin plate and the reduced capacity of a notched beam. Another minor consideration is the torsion induced when fin plates are attached to one side of the supported beam web. However, tests have shown that in these cases the torsional moments are small and can be neglected.

The essential detailing requirements for beam-to-beam fin plate connections are shown in Fig. 26.19 and the detailed design checks are given in Table 26.8. The design procedures given in Table 26.8 can be applied to supporting beams loaded from one or two sides.

Table 26.6 Design checks for beam-to-beam double angle cleat connections

Check no.	Description	Design rule	
1	Essential detailing requirements	See Fig. 26.14.	
2	Shear capacity of bolt group connecting cleats to web of supported beam	The shear capacity of a single bolt in double shear $(2P_s)$ must be greater than the resultant force on the outermost bolt due to direct shear and bending (F_s)	
3	Shear and bearing capacity of cleats connected to supported beam	*For shear* The shear capacity of the leg of the angle cleat (P_v) must be greater than half the reaction at the end of the beam $\left(\dfrac{F_v}{2}\right)$.	*For bearing* The bearing capacity of the leg of the angle cleat per bolt (P_{bs}) must be greater than half the resultant force on the outermost bolt $\left(\dfrac{F_s}{2}\right)$.
4a	Plain end beams – shear and bearing capacity of the supported beam	*For shear* The shear capacity of the beam (P_v) must be greater than the reaction at the end of the beam (F_v)	*For bearing* The bearing capacity of the supported beam web per bolt (P_{bs}) must be greater than the resultant force on the outermost bolt (F_s)
4b	Notched beams – shear and bearing capacity of the supported beam	*For shear* The shear capacity of the beam (P_v) must be greater than the reaction at the end of the beam (F_v).	*For bearing* The bearing capacity of the supported beam web per bolt (P_{bs}) must be greater than the resultant force on the outermost bolt (F_s)
4c	Notched beams – bending capacity of reduced beam section at the notch	The elastic moment capacity of the notched section (p_yZ) must be greater than the applied moment $F_v(t_1 + c)a$	
4d	Notched beams – local stability of notched beams restrained against lateral torsional buckling	Basic requirements for the notch – see section 26.2.5.1.	
4e	Notched beams – overall stability of notched beams unrestrained against lateral torsional buckling	See section 26.2.5.1 for checking the overall stability of a notched beam	
5	Shear capacity of bolt group connecting cleats to supporting beam	The shear capacity of the bolt group (ΣP_s) must be greater than the reaction at the end of the beam (F_v)	

Table 26.6 (*continued*)

Check no.	Description	Design rule	
6	Shear and bearing capacity of cleats connected to supporting beam	*For shear* The shear capacity of the leg of the angle cleat (P_v) must be greater than half the reaction at the end of the beam (F_v)	*For bearing* The total bearing capacity of the leg of the single angle cleat (ΣP_{bs}) must be greater than half the reaction at the end of the beam $\left(\dfrac{F_v}{2}\right)$.
7	Local shear and bearing capacity of supporting beam	*For shear* The local shear capacity of the beam web (P_v) must be greater than half the sum of the reactions at the end of the beams either side of the web $\dfrac{(F_{v1}+F_{v2})}{2}$.	*For bearing* The bearing capacity of the beam web (P_{bs}) must be greater than the sum of reactions at the ends of the beams either side of the web divided by the relevant number of bolt rows $\dfrac{\left(\dfrac{F_{v1}}{n_1}\right)+\left(\dfrac{F_{v2}}{n_2}\right)}{2}$.

Note: The design check numbers may not match the check numbers given in Ref 1.

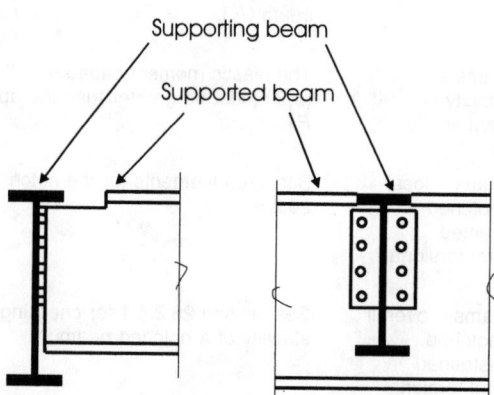

Supporting beam

Supported beam

Fig. 26.15   Typical beam to beam flexible end-plate connection

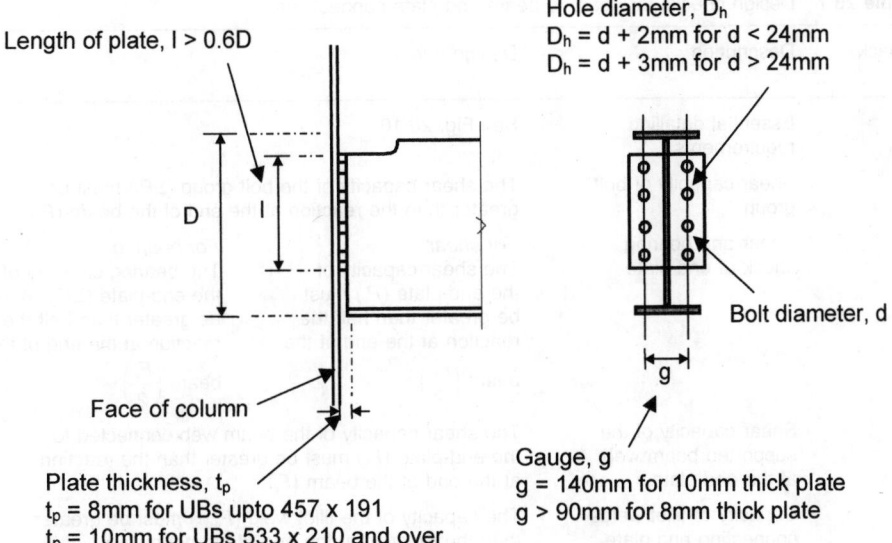

Length of plate, l > 0.6D

Hole diameter, D_h
$D_h = d + 2mm$ for $d < 24mm$
$D_h = d + 3mm$ for $d > 24mm$

D

l

Bolt diameter, d

Face of column

Gauge, g

Plate thickness, t_p
$t_p = 8mm$ for UBs upto 457 x 191
$t_p = 10mm$ for UBs 533 x 210 and over

Gauge, g
$g = 140mm$ for 10mm thick plate
$g > 90mm$ for 8mm thick plate

Fig. 26.16   Essential detailing requirements for beam to beam end-plate connections

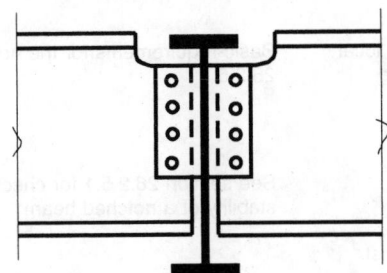

Fig. 26.17   Short fin-plate with single notched beams

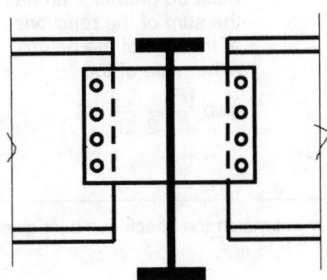

Fig. 26.18   Long fin-plate

Table 26.7   Design checks for beam-to-beam end-plate connections

Check no.	Description	Design rule	
1	Essential detailing requirements	See Fig. 26.16.	
2	Shear capacity of bolt group	The shear capacity of the bolt group (ΣP_s) must be greater than the reaction at the end of the beam (F_v)	
3	Shear and bearing check of end-plate	*For shear* The shear capacity of the end-plate (P_v) must be greater than half the reaction at the end of the beam $\left(\dfrac{F_v}{2}\right)$.	*For bearing* The bearing capacity of the end-plate (ΣP_{bs}) must be greater than half the reaction at the end of the beam $\left(\dfrac{F_v}{2}\right)$.
4	Shear capacity of the supported beam web at the end-plate	The shear capacity of the beam web connected to the end-plate (P_v) must be greater than the reaction at the end of the beam (F_v).	
5	Capacity of fillet welds connecting end-plate to supported beam web	The capacity of the fillet weld (P_{weld}) must be greater than the reaction at the end of the beam (F_v)	
6a	Notched beams – bending capacity of reduced beam section at notch	The elastic moment capacity at the notched section ($p_y Z$) must be greater than the applied moment ($F_v(t + c)$)	
6b	Notched beams – local stability of notched beams restrained against lateral torsional buckling	Basic requirements for the notch – see section 26.2.5.1	
6c	Notched beams – overall stability of notched beams unrestrained against lateral torsional buckling	See section 26.2.5.1 for checking the overall stability of a notched beam	
7	Local shear and bearing capacity of supporting beam	*For shear* The local shear capacity of the beam web (P_v) must be greater than half the sum of the reactions at the end of the beams either side of the web $\dfrac{(F_{v1} + F_{v2})}{2}$.	*For bearing* The bearing capacity of the beam web (P_{bs}) must be greater than the sum of reactions at the ends of the beams either side of the web divided by the relevant number of bolt rows $\dfrac{\left(\dfrac{F_{v1}}{n_1}\right) + \left(\dfrac{F_{v2}}{n_2}\right)}{2}$:

Note: The design check numbers may not match the check numbers given in Ref 1.

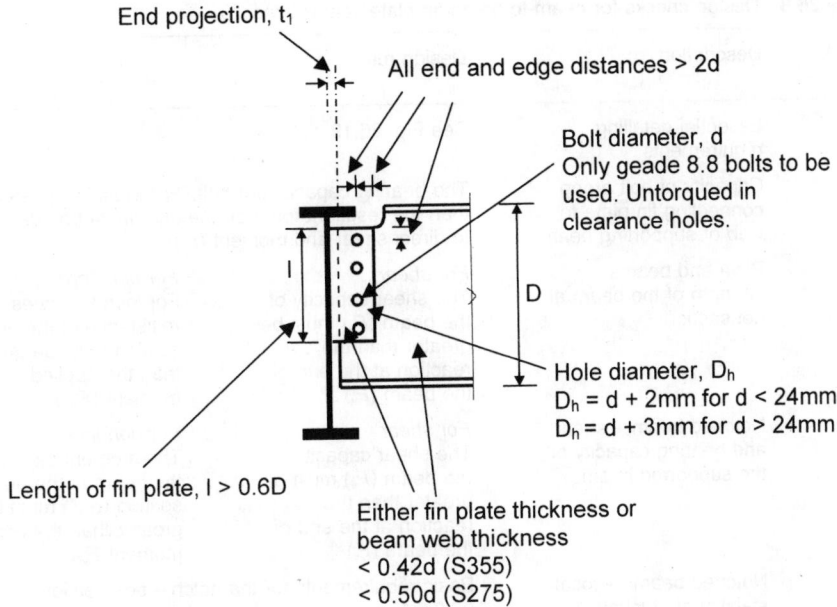

End projection, t₁

All end and edge distances > 2d

Bolt diameter, d
Only geade 8.8 bolts to be
used. Untorqued in
clearance holes.

D

Hole diameter, D_h
$D_h = d + 2mm$ for $d < 24mm$
$D_h = d + 3mm$ for $d > 24mm$

Length of fin plate, l > 0.6D

Either fin plate thickness or
beam web thickness
< 0.42d (S355)
< 0.50d (S275)

Fig. 26.19   Essential detailing requirements for beam to beam fin plate connections

26.2.6 Column splices

This section presents design requirements for column splices in braced multi-storey buildings. In this type of building column splices are required to provide continuity of both strength and stiffness about both axes of the columns. In general they are subject to both axial compression and moments resulting from the end reactions of the beams. If a splice is positioned near to a point of lateral restraint (i.e. within say 500 mm above the floor level), and the column is designed as pinned at that point, the splice may simply be designed for the axial load and any applied moments. If, however, the splice is positioned away from a point of lateral restraint (i.e. more than 500 mm above the level of the floor), or end fixity or continuity has been assumed when calculating the effective length of the column, the additional moment that can be induced by strut action must be taken in to account. The procedure for calculating these additional moments is given in Annex C.3 of BS 5950: Part 1.

Two types of splices are considered in this section, those where the ends of the members are prepared for contact in bearing, and those where ends of the members are not prepared for contact in bearing.

In both cases the column splices should hold the connected members in line and wherever practicable the members should be arranged so that the centroidal axis of the splice material coincides with the centroidal axes of the column sections above and below the splice.

Table 26.8 Design checks for beam-to-beam fin plate connections

Check no.	Description	Design rule	
1	Essential detailing requirements	See Fig. 26.19.	
2	Capacity of bolt group connecting fin plate to web of supporting beam	The bearing capacity per bolt (P_{bs}) must be greater than the resultant force on the outermost bolt due to direct shear and moment (F_s)	
3a	Plain end beams – strength of the beam at net section	*For shear* The shear capacity of the beam (P_v) must be greater than the reaction at the end of the beam (F_v)	*For bending* For long fin plates the resistance of the net section must be greater than the applied moment ($F_v a$)
3b	Notched beam – shear and bearing capacity of the supported beam	*For shear* The shear capacity of the beam (P_v) must be greater than the reaction at the end of the beam (F_v)	*For bending* The moment capacity of the beam at the notched section ($p_y Z$) must be greater than the applied moment $F_v a$
3c	Notched beams – local stability of notched beams restrained against lateral torsional buckling	Basic requirements for the notch – see section 26.2.5.1.	
3d	Notched beams – overall stability of notched beams unrestrained against lateral torsional buckling	See section 26.2.5.1 for checking the overall stability of a notched beam	
4	Strength of fin plate at net section under bending and shear	*For shear* The shear capacity of the fin plate (P_v) must be greater than the reaction at the end of the beam (F_v)	*For bending* The moment capacity of the fin plate ($p_y Z$) must be greater than the applied moment ($F_v a$)
5	Lateral torsional buckling resistance of long fin plates	The lateral torsional buckling resistance moment of the fin plate (M_b) must be greater than the moment due to the end reaction and the distance a ($F_v a$)	
6	Strength of weld connecting fin plate to supporting beam under bending and shear	The leg length of the fillet weld must be greater than 0.8 times the thickness of the fin plate	
7	Local shear capacity of supporting beam web	The local shear capacity of the supporting beam (P_v) must be greater than half the sum of the reactions at the ends of the beams either side of the supporting beam web $\dfrac{(F_{v1} + F_{v2})}{2}$.	

Note: The design check numbers may not match the check numbers given in Ref 1.

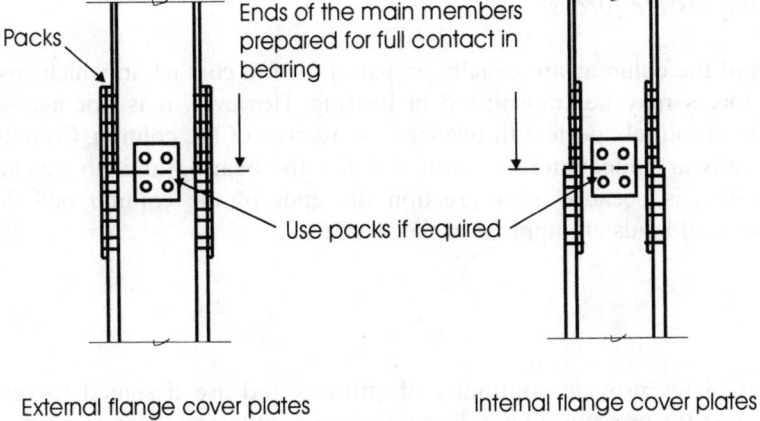

External flange cover plates   Internal flange cover plates

Fig. 26.20   Column splice with ends prepared for bearing – same serial size

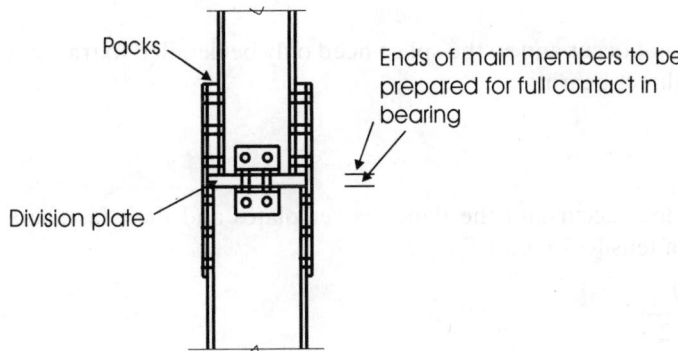

Fig. 26.21   Column splice with ends prepared for bearing – different serial size

26.2.6.1 Ends prepared for contact in bearing

Typical details for this type of column splice are shown in Figs 26.20 and 26.21. In both cases the splice is constructed using web and flange cover plates, and packs are used to make up any differences in the thicknesses of the web and the flanges. The flange cover plates may be placed on either the outside or the inside of the column. Placing the cover plates on the inside has the advantage of reducing the overall depth of the column.

Each column splice must be designed to carry axial compressive forces, the tension (if any) resulting from the presence of bending moments and any horizontal shear forces.

Axial compressive forces

The ends of the columns are usually prepared for full contact, in which case compressive forces may be transmitted in bearing. However, it is not necessary to achieve an absolutely perfect fit over the entire area of the column. Columns with saw cut ends are adequately smooth and flat for bearing and no machining is required. This is because after erection the ends of the column bed down as successive dead loads are applied to the structure.

Tension

The cover plates provide continuity of stiffness and are designed to resist any tension where the presence of bending moments is sufficiently high to overcome the compressive forces in the column. The presence of tension due to axial load and moment can be calculated as follows:

If $M < \dfrac{F_{cd} D}{2}$

tension does not occur and so the splice need only be detailed to transmit axial compression by direct bearing.

If $M > \dfrac{F_{cd} D}{2}$

net tension does occur and the flange cover plates and their fasteners should be checked for a tensile force of F_t:

$$F_t = \frac{M}{D} - \frac{F_{cd}}{2}$$

where M is the nominal moment due to factored dead and imposed load at the floor
immediately below the splice
F_{cd} is the axial compressive force due to factored dead load only
D is taken conservatively as the overall depth of the smaller column.

Where it is necessary to comply with the recommendations for structural integrity given in clause 2.4.5 of BS 5950: Part 1 then checks 3, 4 and 5 given in Table 26.9 should be checked, replacing F_t with the tensile force obtained from clause 2.4.5.3(c) of BS 5950; Part 1.

Shear forces

The horizontal shear forces that arise from the moment gradient in the column are normally resisted by the friction across the bearing surfaces of the two columns and by the web cover plates. Wind forces on the external elevations of buildings are

Table 26.9 Design checks for column splices – ends prepared for contact in bearing

Check no.	Description	Design rule
1a	Essential detailing requirements – flange cover plate on the outside	See Fig. 26.22(a)
1b	Essential detailing requirements – flange cover plate on the inside	See Fig. 26.22(b)
2	The presence of tension due to axial load and moment	See section on tension
3	Tension capacity of the cover plate	The net tension capacity of the cover plate ($p_y A_{tp}$) must be greater than the applied tensile force (F_t)
4	Shear capacity of bolt group connecting flange cover plate to column flange	The reduced shear capacity of the bolt group (reduction factor $\times \Sigma P_s$) must be greater than the applied tensile force (F_t)
5	Bearing capacity of flange cover plate connected to column flange	The bearing capacity of the flange cover plates (ΣP_{bs}) must be greater than the applied tensile force (F_t)
6	Structural integrity of splice	The tension capacity of the splice (checks 3, 4 and 5) must be greater than the tying force (F_{tie}) obtained from clause 2.4.5.3(c) of BS 5950: Part 1

Note: The design check numbers may not match the check numbers given in Ref 1.

normally taken directly in the floor slab. It is rare for column splices in simple construction to transmit wind shears.

The design checks given in Table 26.9 are based on the above assumptions. These design checks and the essential detailing requirements given in Fig. 26.22 give empirical rules which preserve the continuity of the finished structure, give a splice with a minimum of robustness and ensure erection stiffness. These design recommendations can be used for splices with internal or external flange cover plates.

26.2.6.2 *Ends not prepared for contact in bearing*

Typical details for this type of column splice are shown in Fig. 26.23. Both parts of the figure show that the column above and below the splice is the same serial size. In this case the splice is constructed using web and flange cover plates and, where required, packs are used to make up the differences in the web and flange thicknesses. Where columns of different serial size are to be connected multiple packs are necessary to take up the dimensional variations.

For this type of splice all the forces and moments are carried by the cover plates and no load is transferred through direct bearing. The axial load in the column is normally shared between the flange and web cover plates in proportion to their

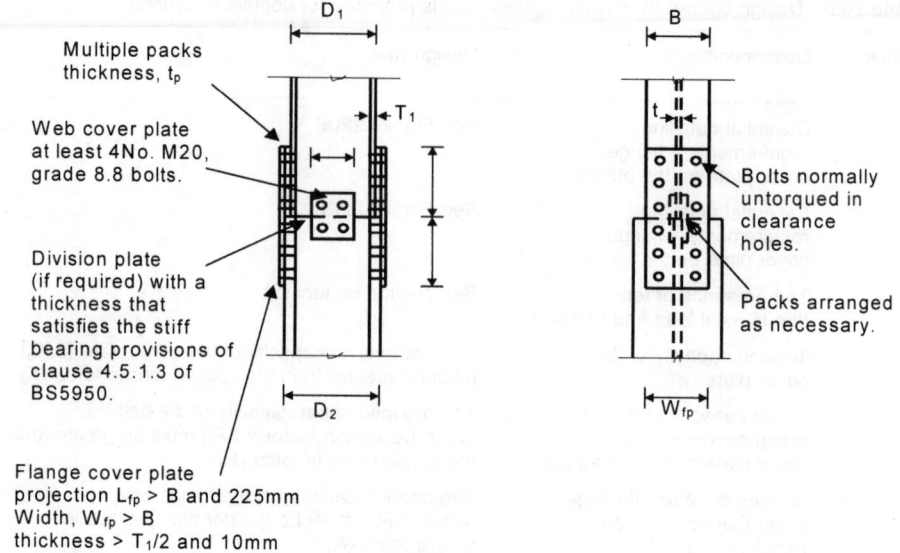

Fig. 26.22(a)   Essential detailing requirements for column splices with ends prepared for bearing (external flange cover plates)

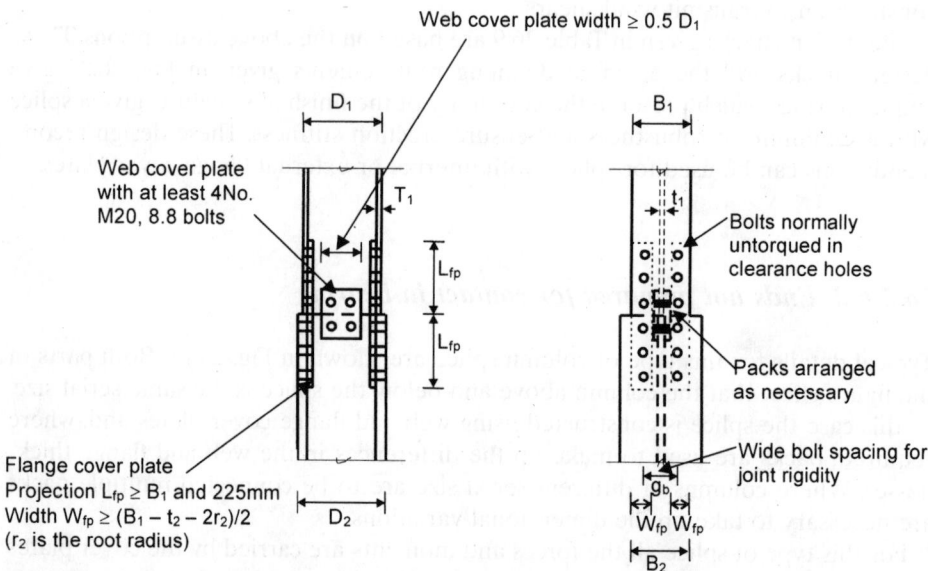

Fig. 26.22(b)   Essential detailing requirements for column splices with ends prepared for bearing (internal flange cover plates)

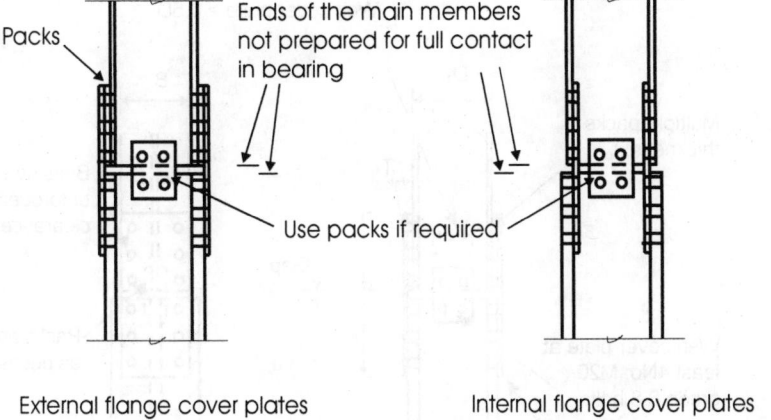

External flange cover plates   Internal flange cover plates

Fig. 26.23 Column splice with ends not prepared for bearing

areas while any bending moments are normally carried by the flange cover plates alone.

The maximum compressive and tensile forces in the flange cover plate is given by the following two expressions:

$$F_{1max} = \frac{M}{D} + F_c\left(\frac{A_f}{A}\right) \quad \text{for maximum compressive force}$$

$$F_{2max} = \frac{M}{D} - F_{cd}\left(\frac{A_f}{A}\right) \quad \text{for maximum tensile force}$$

where M is the nominal moment due to factored dead and live imposed loads at the floor level immediately below the splice,

F_c is the axial compressive force due to factored dead and imposed loads,

F_{cd} is the axial compressive force due to factored dead loads,

D is the overall depth of the smaller column (for external flange cover plates) or the centreline to centreline distance between the flange cover plates (for internal flange cover plates),

A_f is the area of one flange of the smaller column,

A is the total area of the smaller column.

26.3 Moment connections

26.3.1 Introduction

The following three principal forms of beam to column moment connection are considered in this section:

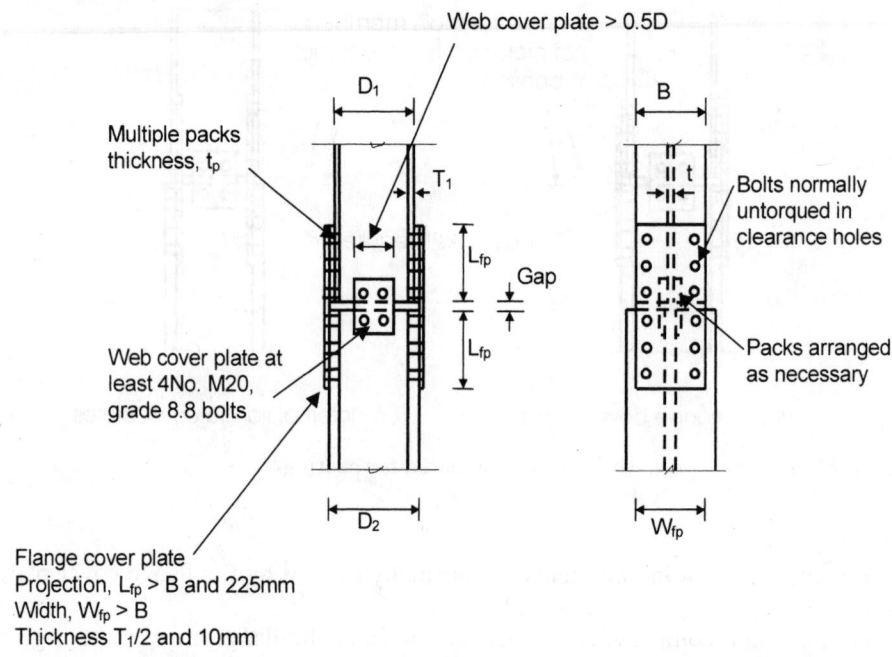

Fig. 26.24(a)   Essential detailing requirements for a column splice with ends not prepared for bearing (external flange cover plates)

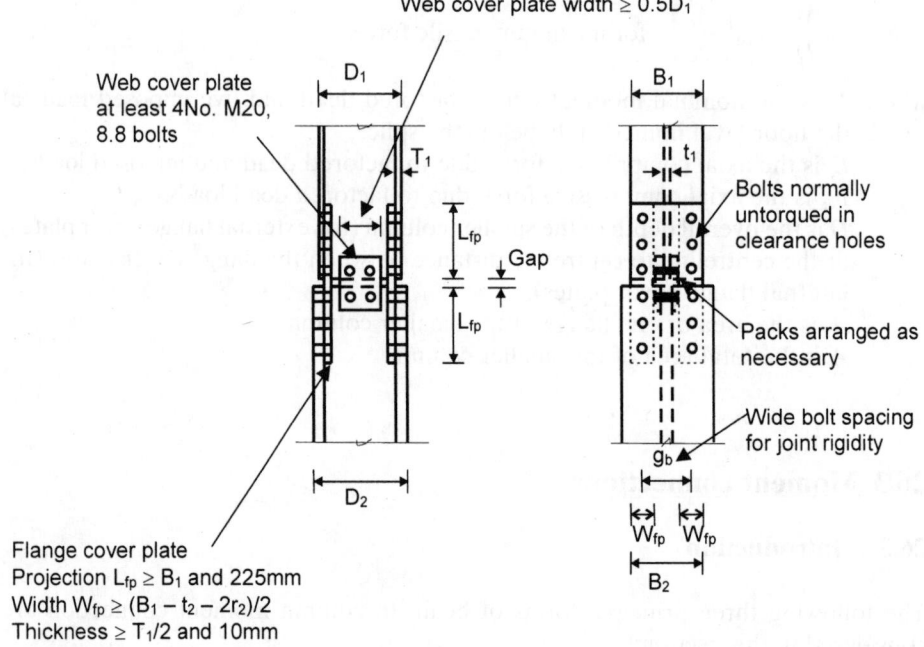

Fig. 26.24(b)   Essential detailing requirements for a column splice with ends not prepared for bearing (internal flange cover plates)

Table 26.10 Design checks for column splices – ends not prepared for contact in bearing

Check no.	Description	Design rule	
1a	Essential detailing requirements – flange cover plate on the outside	See Fig. 26.24	
1b	Essential detailing requirements – flange cover plates on the inside	See Fig. 26.24	
2	Axial capacity of the flange cover plates	*For compression* The axial compressive capacity of the cover plate ($p_y A_{fp}$) must be greater than the maximum applied compressive force (F_{1max})	*For tension* The effective axial tensile capacity of the cover plate ($p_y A_{fp}$) must be greater than the maximum applied tensile force (F_{2max})
3	Shear capacity of bolt group connecting the flange cover plate to column flange	The shear capacity of the bolt group (reduction factor $\times \Sigma P_s$ of the bolt shear capacities) must be greater than the maximum applied compressive force (F_{1max})	
4	Bearing capacity of flange cover plate connected to column flange	The bearing capacity of the flange cover plates (ΣP_{bs}) must be greater than the maximum applied compressive force ($F_{1,max}$)	
5	Shear capacity of bolt group connecting web cover plate to column web	The shear capacity of the connecting bolt group (ΣP_s) must be greater than half the axial compressive load (due to factored loads) in the web of the smaller column $\left(F_3 = \dfrac{F_c A_w}{2A}\right)$	
6	Axial capacity of web cover plate	The axial capacity of the web cover plate ($p_y A_{wp}$) must be greater than half the axial load in the web of the smaller column (F_3)	
7	Bearing capacity of web cover plate connected to column web	The bearing capacity of the web cover plate (ΣP_{bs}) must be greater than half the axial compressive load in the web of the smaller column (F_3)	
8	Structural integrity of splice	As Table 26.9	

Note: The design check numbers may not match the check numbers given in Ref 1.

Flush end-plate connections
Extended end-plate connections
Welded connections

Each of these connection types is shown in Fig. 26.1. Choice of connection type is usually based on simplicity, duplication and ease of fabrication, all for economic reasons. Site welded joints are used extensively in the USA and Japan for the con-

struction of buildings in seismic areas. Such connections can provide full moment continuity but can be expensive to produce. This type of connection is rarely used in the UK but with careful planning there is no reason why they cannot be used for a number of framing systems. A procedure for the design of both site welded and shop welded beam-to-column connections is given in the SCI/BCSA design guide *Joints in Steel Construction – Moment Connections*.[2]

The wind-moment method[10] for the design of unbraced frames has been extensively used to design many multi-storey steel framed buildings. In this design method the connections are assumed to behave as pins under gravity loads and the beams they support are designed assuming simple supports. However, under lateral wind loads the connections are assumed to behave rigidly. Connections which behave in this manner are called *wind-moment connections*. These connections generally consist of flush or extended end-plates and have little or no stiffening in the columns. They are easy to fabricate and provide a cost-effective solution for low-rise unbraced buildings. The main requirement for wind-moment connections is that they should be ductile. That is, they must be able to rotate as plastic hinges under gravity loading and still have sufficient strength to resist the moments from wind loads. To ensure this type of behaviour it is important to design the connection in such a way that the end-plate is thin enough to be the weak link and that it will fail before either the bolts or the welds. The end-plate must also have sufficient strength to resist the moments from the wind loads. The size of the end-plate in relation to the size and strength of the bolts must, therefore, be selected with great care. If it is too thick the bolts will fail first, making the connection brittle. If it is too thin, the strength of the connection will be too small to support the wind moments. When using grade 8.8 bolts the appropriate end-plate thickness is about 60% of the bolt diameter.

The design philosophy and design methods described in the next section can be used to design either a flush, extended or wind-moment connection.

26.3.2 Design philosophy

BS 5950: Part 1 gives little detailed information on the design of moment connections and therefore the method used in this publication is based on a combination of the capacity checks given in BS 5950: Part 1[4] and the design model given in Annex J of Amendment No. 1 to ENV Eurocode 3: Part 1.1.[11] The capacity checks for the bolts, welds and sections are all based on BS 5950: Part 1. However, the checks for calculating the capacity of either the end-plate or column flange in the tension region of a bolted end-plate connection are based on the design model given in Annex J of ENV Eurocode 3: Part 1.1/A1.

An end-plate connection transmits moment by coupling tension in the bolts with compression in the bottom flange. The bolt row furthest from the compression (bottom) flange may therefore attract the most tension, and traditional practice in the UK has been to assume a triangular distribution of bolt forces. However, the

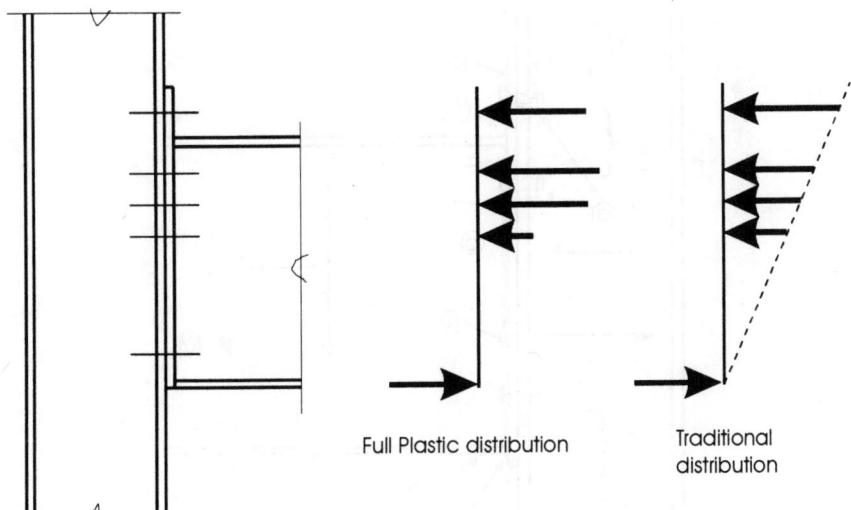

Full Plastic distribution

Traditional distribution

Fig. 26.25 Distribution of bolt forces

method adopted here is to use a plastic distribution of bolt forces. These two approaches are illustrated in Fig. 26.25. In the traditional UK approach the centre of compression is assumed to be in line with the compression flange of the beam, with the bolt row furthest from the centre of compression attracting the most tension.

In the Eurocode approach no assumption is made about the distribution of bolt forces. Instead, each bolt row is allowed to attain its full design strength (on the basis of the strength of the column flange or end-plate, whichever is the lowest). This model relies on adequate ductility of the connecting part in the uppermost bolt rows to develop the design strength of the lower bolt rows. To ensure adequate ductility a limit is set on the thickness of the column flange or end-plate relative to the strength of the bolts. Where S275 steel is used with grade 8.8 bolts the thickness of either end-plate or column flange should be less than 18.3 mm, 21.9 mm or 27.5 mm for M20, M24 and M30 bolts respectively. If this criterion is not satisfied then the force in the lower bolt rows is limited to a value resulting from the linear distribution.

The design method in Eurocode 3 uses what is called the *component approach*. In this approach the potential resistance of each component is calculated. These potential forces are then converted to the actual forces by considering equilibrium. If there is a surplus capacity in the bolt forces then these forces should be reduced, starting with the lowest row of bolts and working upwards progressively until equilibrium is achieved. The moment capacity of the connection is then calculated by summing the product of all the bolt row forces and their distance from the centre of compression.

In all, there are 15 principal checks to be made on the beam, the column, the bolts and the welds. These checks are shown in Fig. 26.26 and can conveniently be split

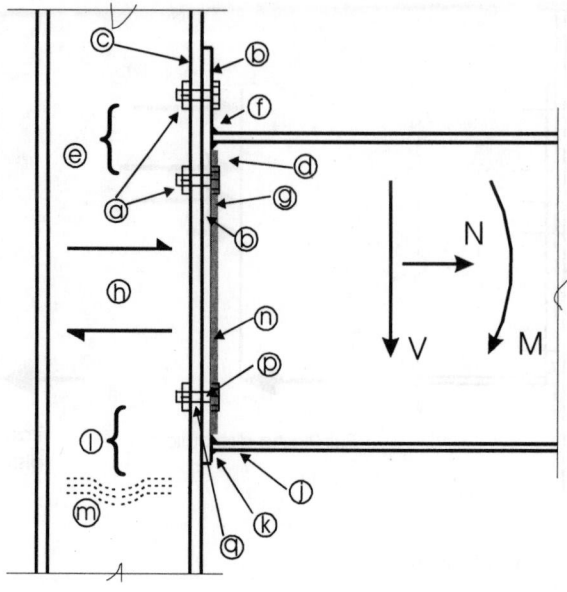

Zone	Ref	Checklist item
Tension	a	*Bolt tension*
	b	*End plate bending*
	c	*Column flange bending*
	d	*Beam web tension*
	e	*Column web tension*
	f	*Flange to end plate weld*
	g	*Web to end plate weld*
Horizontal Shear	h	*Column web panel shear*
Compression	j	*Beam flange compression*
	k	*Beam flange weld*
	l	*Column web crushing*
	m	*Column web buckling*
Vertical Shear	n	*Web to end plate weld*
	p	*Bolt shear*
	q	*Bolt bearing (plate or flange)*

Fig. 26.26   Design checks for a bolted moment connection

into four zones – the tension zone, the compression zone, horizontal shear and vertical shear. The checks associated with these zones are outlined in the sections given below.

26.3.2.1 Tension zone

The resistance of each bolt row in the tension zone is determined from a consideration of the following modes of failure:

- Bolt tension
- End-plate in bending
- Column flange in bending
- Beam web in tension
- Column web in tension.

In addition to the above the designer should also check the adequacy of the flange to end-plate welds and the web to end-plate welds in the tension region.

For determining the capacity of the bolts, the end-plate in bending and the column flange in bending the method outlined in Annex J of ENV Eurocode 3: Part 1 is used. This approach is based on many years of experimental and theoretical research on end-plate connections. The results of this work suggest that an equivalent tee-stub with an effective length of L_{eff} can be used to model the tension region of either the column flange or end plate. Figure 26.27 shows how an equivalent tee-stub can be used to model the failure of an end-plate. A full description of the effective lengths used for different bolt configurations is beyond the scope of this chapter but for further information the designer is referred to either Eurocode 3[12] or the SCI/BCSA design guide on *Joints in Steel construction – Moment Connections*.[2]

In a tee-stub, failure can occur by one of three mechanisms depending on the relative stiffness of the flange or end-plate and the bolts. These mechanisms are usually referred to as *complete yielding of the flange* (Mode 1), *simultaneous bolt failure and*

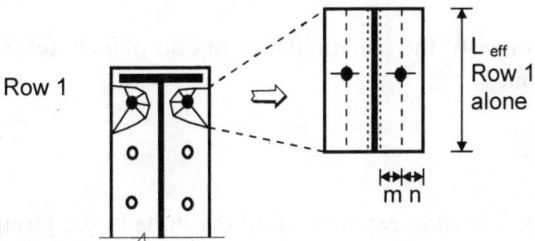

Fig. 26.27 Equivalent tee-stub

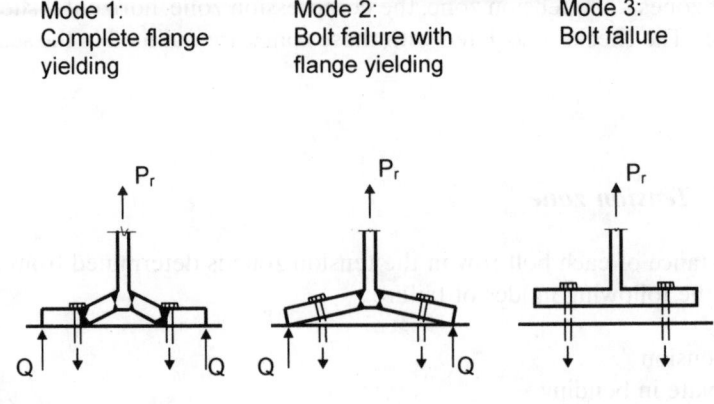

Mode 1:
Complete flange
yielding

Mode 2:
Bolt failure with
flange yielding

Mode 3:
Bolt failure

Q is the prying force

Fig. 26.28   Tee-stub modes of failure

yielding of the flange (Mode 2) and *bolt failure* (Mode 3). These modes are shown diagrammatically in Fig. 26.28. The equations for calculating the potential resistance for each of these modes of failure are given below.

Mode 1 Complete flange yielding

The potential resistance of either the column flange or end-plate, P_r, can be determined from the following expression:

$$P_r = \frac{4M_p}{m}$$

where M_p is the plastic moment capacity of the equivalent tee-stub representing the column flange or end-plate
m is the distance from the bolt centre to a line located 20% into either the column root or end-plate weld.

Mode 2 Bolt failure and yielding of the flange

The potential resistance of the column flange or end-plate in tension is given by the following expression:

$$P_r = \frac{2M_p + n(\Sigma P_t)}{M + n}$$

where ΣP_t is the total tension capacity of all the bolts in the group
n is the effective edge distance.

Mode 3 Bolt failure

The potential resistance of the bolts in the tension zone is give by the following expression:

$$P_r = \Sigma P_t$$

In each of the above modes no specific mention is made of prying action nor are any equations given to calculate its value. This is because prying action is implicit in the expressions for the calculation of the effective length L_{eff}. The principal author of this method, Zoetemijer, addresses the problem of prying action in a background publication.[12] In this publication Zoetemijer develops the following three expressions for the equivalent effective length of an unstiffened column flange taking into account different levels of prying action:

For prying force = 0.0	$L_{eff} = (p + 5.5m + 4n)$
For maximum prying force	$L_{eff} = (p + 4m)$
For an intermediate value	$L_{eff} = (p + 4m + 1.25n)$

where p is the bolt pitch

Zoetemijer explains that the first expression has an inadequate margin of safety against bolt failure while the margin of safety in the second is too high. He therefore suggests using the third equation, which allows for approximately 33% prying action. This approach simplifies the calculations by omitting complicated expressions for determining prying action.

BS 5950: Part 1 allows two approaches for calculating the tension capacity of a bolt in the presence of prying forces. The simple method given in clause 6.3.4.2 places certain restrictions on the centre-to-centre bolt spacing and on the capacity of the connected part to reduce prying action. A reduced bolt capacity is also used by including a 0.8 factor in the expression for calculating the nominal tension capacity (p_{nom}) of the bolt. One advantage of this approach is that it obviates the need to calculate prying forces directly. In the more exact approach given in clause 6.3.4.3 bolt tension capacities (p_t) are allowed provided the connection is not subject to prying action or the prying forces are included in the design method. As explained above the design method described here allows for prying action without the need to calculate prying forces directly and therefore the enhanced bolt capacities of the more exact method can be used.

Beam web/column web in tension

The resistance of either an unstiffened beam or column web in tension, P_t, at each row or group of bolt rows is given by the following expression:

$$P_t = L_t \times t_w \times p_y$$

where L_t is the effective length of web assuming a maximum spread at 60° from the bolts to the centre of the web

t_w is the thickness of the column or beam web

p_y is the design strength of the steel in the column or beam

26.3.2.2 Compression zone

The checks in the compression zone are similar to those traditionally adopted for web bearing and buckling and include the following:

- Column web in bearing
- Column web buckling
- Beam flange in compression

Column web in compression

In many designs it is common for the column web to be loaded to such an extent that it governs the design of the connection. However, this can be avoided either by choosing a heavier column or by strengthening the web with one of the compression stiffeners shown in section 26.3.2.4.

The resistance of an unstiffened column web subject to compressive forces, P_c, is given by the smaller of the expressions for column web bearing and column web buckling.

Column web bearing

The resistance of the column web to bearing is based on an area of web calculated by assuming the compression force from the beam's flange is dispersed over a length shown in Fig. 26.29. From this the resistance of the column web to crushing is given by the following expression:

$$P_{bw} = (b_1 + nk) \times t_c \times p_{yw}$$

where b_1 is the stiff bearing length based on a 45° dispersion through the end-plate from the edge of the welds

n except at the end of a member $n = 5$

k is obtained as follows:

– for a rolled I- or H-section $k = T_c + r$

– for a welded I- or H-section $k = T_c$

T_c is the column flange thickness

r is the root radius

t_c is the thickness of the column web

p_{yw} is the design strength of the column web

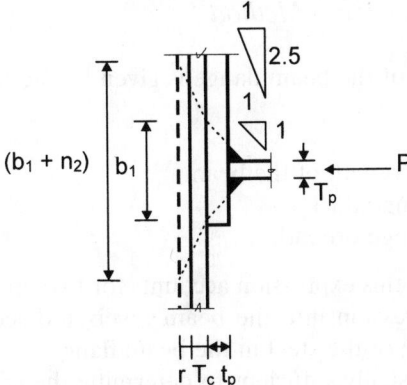

Fig. 26.29   Distribution of compressive force

Column web buckling

The resistance of the column web to buckling is based on bearing capacity of the column web and is given by the following expression in clause 4.5.3.1 of BS 5950: Part 1:

$$P_x = \frac{25\varepsilon t_c}{[(b_1 + nk) \times d]^{0.5}} P_{bw}$$

where b_1, n, k and t are defined above

d is the depth of the web

P_{bw} is the bearing capacity of the unstiffened web at the web-to-flange connection given in the section above.

$$P_c = \min(P_{bw}, P_x)$$

Beam flange in compression

Two approaches are presented for checking the resistance of the beam flange in compression. In the first method it is assumed that only the compression flange carries the compression in the beam and that the centre of compression acts at the centre of this flange. In the second approach the compression zone is allowed to spread up the beam and into the beam web. In this method the centre of compression will move towards the web. This latter approach is usually used in cases with either high moments or combined high moment and axial load.

Beam flange in compression – Method 1

The potential resistance of the beam flange is given by the following expression:

$$P_c = 1.4 \times p_{yb} \times T_b \times B_b$$

where p_{yb} is the design strength of the beam
T_b is the beam flange thickness
B_b is the beam flange breadth

The factor 1.4 in front of this expression accounts for two effects. Firstly, it accounts for the spread of compression into the beam's web and secondly it accounts for possible strain-hardening of the steel in the beam flange.

This simple check is usually sufficient to determine the crushing capacity of the beam's flange. However, where high moments are present or moment is combined with axial load, method 2 is more appropriate.

Beam flange in compression – Method 2

In this approach the potential resistance of the beam flange is given by the following expression:

$$P_c = 1.2 \times p_{yb} \times A_c$$

where A_c is the area in compression shown in Fig. 26.30.

It should be noted that in this approach the factor 1.4 is reduced to 1.2 since the contribution of the web is now taken into account directly. It should also be noted that the centre of compression is now at the centroid of the area A_c, and the lever-arm of the bolts is reduced accordingly. Changing the position of the centre of compression will also affect the moment, and an iterative calculation procedure becomes necessary.

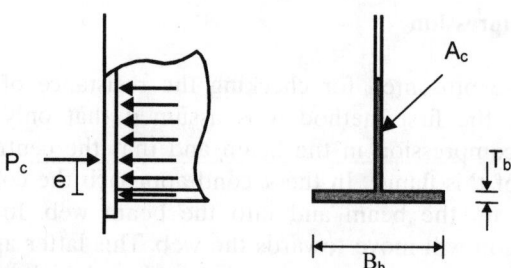

Fig. 26.30   Area of beam flange and web in compression

26.3.2.3 Shear zone

The column web panel must be designed to resist the resulting horizontal shear forces. To calculate these resultant forces the designer must take account of any connection to the opposite column flange. In a single sided connection with no axial force the resultant shear force will be equal to the compressive force at the beam flange level (i.e. the sum of the bolt forces due to the moment). For a symmetrical two-sided column connection with balanced moments the resultant shear force will be zero. However, in the case of a two sided connection subject to moments acting in the same sense the resultant shears will be additive. For any connection the resulting shear force can be obtained from the following expression:

$$F_{vp} = \frac{M_{b1}}{Z_1} - \frac{M_{b2}}{Z_2}$$

where M_{b1} and M_{b2} are the moments in connections 1 and 2 (hogging positive)

Z_1 and Z_2 are the lever-arms for connections 1 and 2. It is usually assumed that the moment can be represented by equal and opposite forces in the beam's tension and compression flanges. In this case Z_1 and Z_2 are equal to the distances between the centroids of the beam flanges for connections 1 and 2 respectively.

The resistance of an unstiffened column web panel in shear is given by the following expression:

$$P_v = 0.6 \times p_{yc} \times t_c \times D_c$$

where p_{yc} is the design strength of the column

t_c is the thickness of the column web

D_c is the depth of the column section

Webs of most UC sections will fail in panel shear before they fail in either bearing or buckling and therefore most single sided connections are likely to fail in shear. The strength of a column web can be increased either by choosing a heavier column section or by using one of the shear stiffeners shown in Fig. 26.13c.

26.3.2.4 Stiffeners

Most stiffening can be avoided through careful selection of the members during the design process. This will usually lead to a more cost-effective solution. However, where stiffening is unavoidable, one or more of the stiffener types shown in Fig. 26.31 may be used.

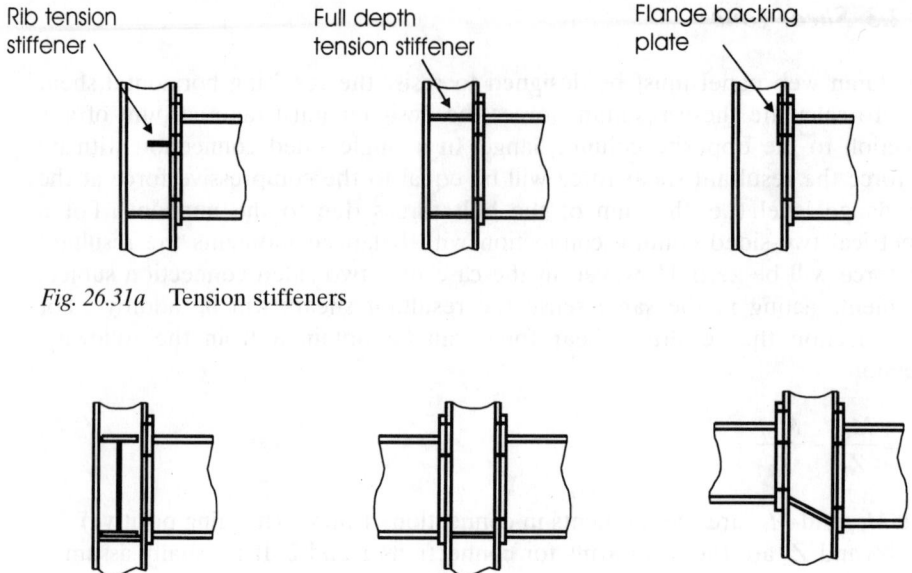

Rib tension stiffener

Full depth tension stiffener

Flange backing plate

Fig. 26.31a   Tension stiffeners

Fig. 26.31b   Compression stiffeners

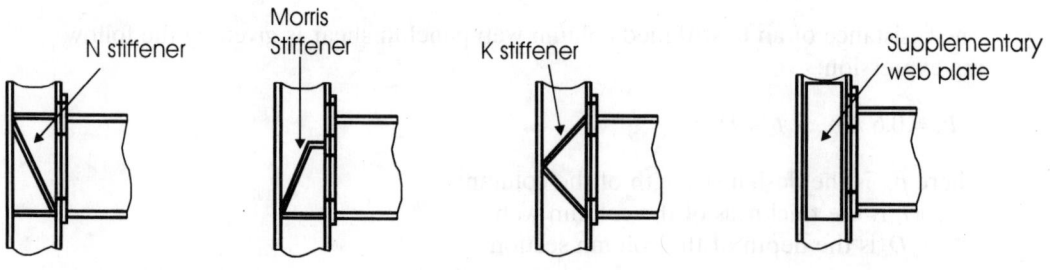

N stiffener

Morris Stiffener

K stiffener

Supplementary web plate

Fig. 26.31c   Commonly used method of shear stiffening

Horizontal stiffeners such as those shown in Fig. 26.31(a) and (b) are used where the concentrated loads from the beam flanges overstress the column web.

There is often a high shear stress in the column web, particularly in single sided connections, and stiffening is required. Diagonal or supplementary web plates can be used (see Fig. 26.31c). Wherever possible the angle of diagonal stiffeners should be between 30° and 60°. However, if the depth of the column is considerably less than the depth of the beam 'K' stiffening may be used.

In general the type of strengthening must be chosen so that it does not clash with other components at the connections.

26.4 Summary

The successful performance of every structural steel frame is dependent as much on its connections as on the size of its structural members. Bolted connections and in particular moment connections are complex in their behaviour. The distribution of the stresses and forces within the connection depends on both the capacity of the welds, bolts etc and on the relative ductility of the connected parts. It is therefore necessary for the design of connections to be consistent with the designer's assumptions regarding the structural behaviour of the steel frame. When choosing and proportioning connections the engineer should always consider the basic requirements such as the stiffness/flexibility of the connection, strength and the required rotational capacity. The design philosophy presented in this chapter together with the detailed design checks provide the engineer with a basic set of tools that can be used to design connections which are better able to meet the design assumptions. To aid the designer further this chapter concludes with a set of worked examples for simple connections.

References to Chapter 26

1. The Steel Construction Institute/The British Constructional Steelwork Association LTD (2002) *Joints in Steel Construction: Simple Connections*, Publication No. 212. SCI, BCSA.
2. The Steel Construction Institute/The British Constructional Steelwork Association LTD (1995) *Joints in Steel Construction: Moment Connections*, Publication No. 207. SCI, BCSA.
3. British Standards Institution (1993) DD ENV1993-1-1:1992 Eurocode 3: *Design of steel structures* Part 1.1 *General rules and rules for buildings*, BSI, London.
4. British Standards Institution (2000) BS 5950: *Structural use of steelwork in building* Part 1: *Code of practice for design – Rolled and welded sections*. BSI, London.
5. The British Constructional Steelwork Association & the Steel Construction Institute (1994) *National Structural Steelwork Specification for building construction*, Publication No. 203/94, BCSA, SCI, London.
6. Hogan T. J. & Thomas I. R. (1994) *Design of structural connections*, 4th edn, Australian Institute of Steel Construction.
7. Cheng J. J. R. & Yura J. A. (1988) Lateral buckling tests on coped steel beams, *Journal of Structural Engineering, ASCE*, **114**, No. 1, 1–15 January.
8. Gupta A. K. (1984) Buckling of coped beams, *Journal of Structural Engineering, ASCE*, **110**, No. 9, 1977–87.
9. Cheng J. J. R., Yura J. A. & Johnson C. P. (1988) Lateral buckling of coped steel beams, *Journal of Structural Engineering, ASCE*, **114**, No. 1, 16–30.
10. Salter P. R., Couchman G. H. & Anderson A. (1999) *Wind-moment design of Low Rise Frames*, Publication No. 263, The Steel Construction Institute, Ascot, Berks.

11. British Standards Institution (1992) ENV1993-1-1/A1: 1992 Eurocode 3: *Design of steel structures* Part 1.1 *General rules and rules for buildings*, BSI, London.
12. Zoetemijer P. (1974) *A design method for the tension side of statically loaded bolted beam-to-column connections*, Heron 20, No. 1, Delft University, Delft, The Netherlands.

A series of worked examples follows which is relevant to Chapter 26.

	Subject		Chapter ref.
BRE	**No. 1** **COLUMN SPLICE**		**26**
	Design code **BS 5950**	Made by **DBM** Checked by **BD**	Sheet no. **1**

The splice is between a 305 × 305 × 97 UC upper column and a 356 × 356 × 153 UC lower column. Both columns are grade S275 steel.

The factored forces and moments acting on the splice are:

Axial compression – 600 kN
Bending moment – 100 kNm
Shear force – 60 kN

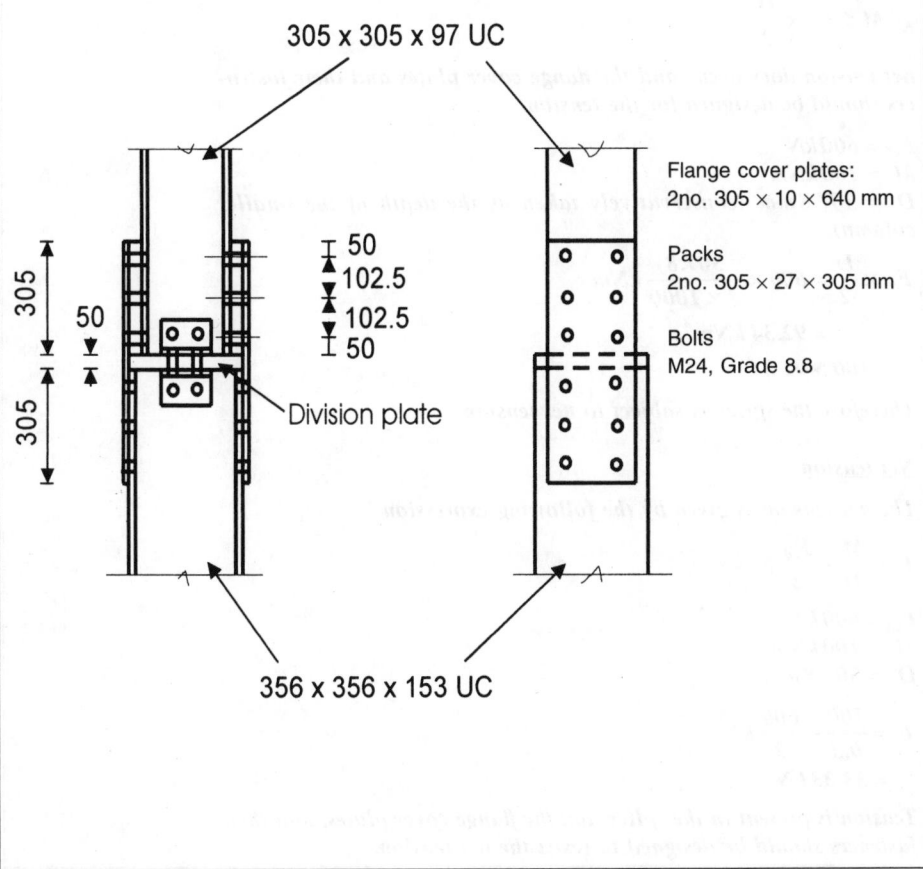

305 x 305 x 97 UC

50
102.5
102.5
50

50

Division plate

Flange cover plates:
2no. 305 × 10 × 640 mm

Packs
2no. 305 × 27 × 305 mm

Bolts
M24, Grade 8.8

305

305

356 x 356 x 153 UC

	Subject		Chapter ref.
BRE	*No. 1* *COLUMN SPLICE*		**26**
	Design code **BS 5950**	Made by **DBM** Checked by **BD**	Sheet no. **2**

Check 2 – The presence of tension due to axial load and moment

The basic requirement is to check for the presence of tension in the splice due to axial load and moment.

If $M < F_{cd} \times \dfrac{D}{2}$

tension does not occur and the splice need only be detailed to transmit axial compression in direct bearing.

If $M > F_{cd} \times \dfrac{D}{2}$

net tension does occur and the flange cover plates and their fasteners should be designed for the tension.

$F_{cd} = 600\,kN$
$M = 100\,kNm$
$D = 307.8\,mm$ *(conservatively taken as the depth of the smaller column)*

$$F_{cd} \times \frac{D}{2} = 600 \times \frac{307.8}{2 \times 1000}\,kNm$$

$$= 92.34\,kNm$$

$$100 > 92.34$$

Therefore the splice is subject to net tension.

<u>*Net tension*</u>

The net tension is given by the following expression:

$$F_t = \frac{M}{D} - \frac{F_{cd}}{2}$$

$F_{cd} = 600\,kN$
$M = 100\,kNm$
$D = 307.8\,mm$

$$F_t = \frac{100}{0.3} - \frac{600}{2}\,kN$$

$$= 33.33\,kN$$

Tension is present in the splice and the flange cover plates, and their fasteners should be designed to resist the net tension.

	Subject	Chapter ref.
BRE	No. 1 *COLUMN SPLICE*	*26*
	Design code **BS 5950**	Made by **DBM** Checked by **BD**	Sheet no. **3**

Check 3 – Tension capacity of the cover plate

The basic requirement is that the net tension capacity of the cover plate must be greater than or equal to the applied net tension.

The tension capacity of the cover plate is given by the following expression:

$$F_t = p_y A_{fp}$$

where p_y is the tensile strength of the plate

   A_{fp} *is the effective area of the cover plate*

   $A_{fp} = K_e A_{net} \le A_{gross}$

where K_e is the effective net area coefficient (for S275 steel $K_e = 1.2$)   **3.4.3**

$$A_{net} = 305 \times 10 - 2 \times 22 \, mm^2$$
$$= 3006 \, mm^2$$

$$A_{gross} = 305 \times 10 \, mm^2$$
$$= 3050 \, mm^2$$

$$A_{fp} = 1.2 \times 3006$$
$$= 3607.2 \, mm^2$$

Net area of the flange plate is greater than the gross area of the section therefore take the net area equal to the gross area.

$$A_{fp} = 3050 \, mm^2$$

$$F_t = 275 \times \frac{3050}{1000} \, kN$$
$$= 838.8 \, kN$$

   $$838.8 \, kN > 33.33 \, kN$$

The tension capacity of the flange cover plates is satisfactory.

	Subject		Chapter ref.
BRE	*No. 1* *COLUMN SPLICE*		**26**
	Design code	Made by *DBM*	Sheet no. *4*
	BS 5950	Checked by *BD*	

Check 4 – Shear capacity of bolt group connecting flange cover plate
to column flange

The basic requirement is that the shear capacity of the bolt group
must be greater than or equal to the applied tensile force.

The shear capacity of the bolt group is given by the following
expression:

Shear capacity of bolt group = Reduction factor × ΣPs

where P_s is the shear capacity of a single bolt

The reduction factor is a factor to allow for the effects of packing.

Shear strength of bolts

a.   Top bolts

The shear capacity of the top bolts is limited by the bearing capac-
ity of the flange plate. The bearing capacity is given by the smaller
of the following two expressions:

$$P_{bs} = k_{bs} \, d \, t_p \, p_{bs} \qquad\qquad 6.3.3.3$$

but

$$P_{bs} \leq 0.5 \, k_{bs} \, e \, t_p \, p_{bs} \qquad\qquad 6.3.3.3$$

$k_{bs} = 1.0$
$d = 24 \, mm$
$t_p = 10 \, mm$ *Table 32*
$p_{bs} = 460 \, N/mm^2$

$$P_{bs} = 1.0 \times 24 \times 10 \times \frac{460}{1000} \, kN$$

$$= 110.4 \, kN$$

but

$$P_{bs} \leq 0.5 \times 1.0 \times 50 \times \frac{460}{1000} \, kN$$

$$= 115 \, kN$$

$$P_{bs} = 110.4 \, kN$$

	Subject		Chapter ref.
BRE	**No. 1** **COLUMN SPLICE**		**26**
	Design code **BS 5950**	Made by **DBM** Checked by **BD**	Sheet no. **5**

Intermediate bolts

The shear capacity is given by the following expression:

$$P_s = p_s A_s$$ *6.3.2.1*

where A_s is the shear area = 353 mm²   *Table 30*
$p_s = 375 N/mm^2$

$$P_s = 375 \times \frac{353}{1000} kN$$

$$= 132.4 \, kN$$

Packing   *6.3.2.2*

The reduction factor is an empirical factor which allows for the effects of bending in the bolt due to thick packing

$$\text{Reduction factor} = \frac{9d}{8d + 3t_{pa}}$$

$$= \frac{9 \times 24}{8 \times 24 + 3 \times 27} = 0.79$$

Reduced shear capacity of bolt = 0.79 × 132.4
$$= 104.6 \, kN < P_{bs} \; (110.4 \, kN)$$

Shear capacity of joint

Since the reduced shear capacity is less than the bearing capacity, shear bearing capacity will be critical for all the bolts.

Shear capacity = 0.79 × 6 × 132.4 kN
$$= 627.6 \, kN$$

$$627.6 \, kN > 33.33 \, kN$$

Therefore the shear capacity is adequate.

	Subject		Chapter ref.
BRE	**No. 1** **COLUMN SPLICE**		**26**
	Design code	Made by **DBM**	Sheet no.　**6**
	BS 5950	Checked by **BD**	

Check 5 – Bearing capacity of flange cover plate connected to column flange

The basic check is that the bearing capacity of the flange cover plates (ΣP_{bs}) must be greater than the applied tensile force (F_t).

$$\Sigma P_{bs} \geq F_t$$

Top bolts

The bearing capacity of the top bolts is given by the following expressions:

$$P_{bs} = k_{bs}\, d\, t_p\, p_{bs}$$　　　　　　6.3.3.5

but

$$P_{bs} \leq 0.5\, k_{bs}\, e\, t_p\, p_{bs}$$

where $k_{bs} = 1.0$　　　　　　6.3.3.5
$\quad\quad d = 24\,mm$
$\quad\quad t_p = 10\,mm$
$\quad\quad p_{bs} = 460\,N/mm^2$　　　　　　*Table 32*

$$P_{bs} = 1.0 \times 24 \times 10 \times \frac{460}{1000}\,kN$$

$$= 110.4\,kN$$

but

$$P_{bs} \leq 0.5 \times 1.0 \times 50 \times 10 \times \frac{460}{1000}\,kN$$

$$= 115\,kN$$

$$P_{bs} = 110.4\,kN$$

Intermediate bolts

The bearing capacity of the intermediate bolts is given by the following expression:

$$P_{bs} = k_{bs}\, d\, t_p\, p_{bs}$$　　　　　　6.3.3.5

$$k_{bs} = 1.0$$　　　　　　6.3.3.5

	Subject	Chapter ref.
BRE	**No. 1** **COLUMN SPLICE**	**26**
	Design code **BS 5950**	Made by **DBM** Checked by **BD**	Sheet no. **7**

d $= 24\,mm$

t_p $= 10\,mm$

p_{bs} $= 460\,N/mm^2$

$$P_{bs} = 1.0 \times 24 \times 10 \times \frac{460}{1000}\,kN$$

$$= 110.4\,kN$$

Table 32

Bearing capacity

$\Sigma P_{bs} = 2 \times 110.4 + 4 \times 110.4\,kN$

    $= 662.4\,kN$

     $662.4\,kN > 33.33\,kN$

Therefore the bearing capacity is adequate.

	Subject	Chapter ref.
BRE	**No. 2** **FIN PLATE CONNECTION**	**26**
	Design code	Made by *DBM*	Sheet no. *1*
	BS 5950	Checked by *BD*	

The connection is between a 610 × 229 × 101 UB (S275) supporting beam and a 457 × 152 × 52 UB (S275) supported beam.

The end reaction of the simply supported beam due to factored loads is 110 kN.

Refer to Table 26.8 for design checks.

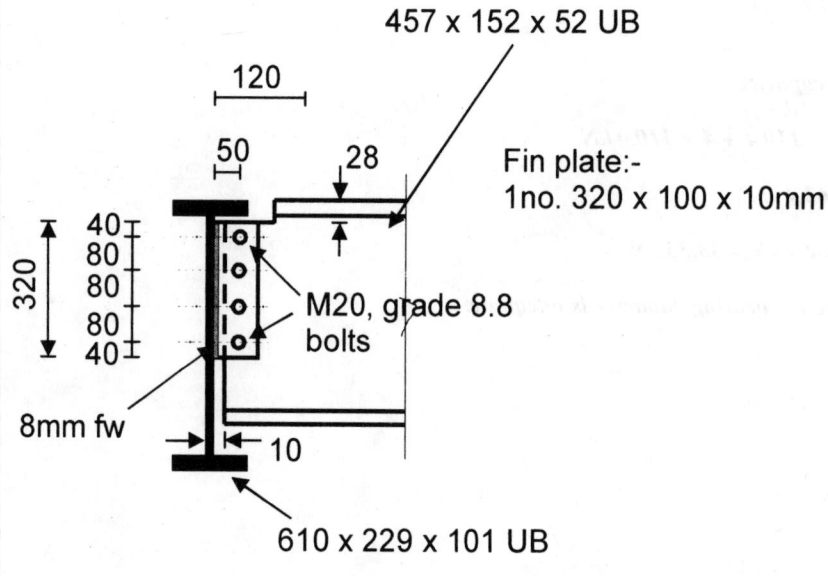

457 x 152 x 52 UB

120

50   28

Fin plate:-
1no. 320 x 100 x 10mm

320

40
80
80
80
40

M20, grade 8.8
bolts

8mm fw

10

610 x 229 x 101 UB

	Subject		Chapter ref.
BRE	**No. 2** **FIN PLATE CONNECTION**		**26**
	Design code **BS 5950**	Made by **DBM** Checked by **BD**	Sheet no. **2**

*Check 2 – Capacity of bolt group connecting the fin plate to the web
of supported beam*

*The basic requirement is that the bearing capacity per bolt (P_{bs}) must
be greater than the resultant force on the outermost bolt due to direct
shear and moment (F_s)*

$$P_{bs} \geq F_s$$

For a single line of bolts the resultant shear is given by:

$$F_s = \left(F_{sv}^2 + F_{sm}^2\right)^{1/2}$$

where F_{sv} is the vertical force on the bolt due to shear.

$$F_{sv} = \frac{F_v}{n}$$

F_{sm} is the force on the outmost bolt due to the moment

$$F_{sm} = \frac{F_v a}{Z_{bg}}$$

a is the eccentricity of the vertical force

Z_{bg} is the elastic section modulus of the bolt group.

$$Z_{bg} = \frac{n(n-1)}{6}P$$

$$= \frac{4 \times 3 \times 80}{6} = 160\,mm^3$$

$$F_s = \frac{F_v}{n}$$

$$= \frac{110}{4} = 27.5\,kN$$

$$F_{sm} = \frac{110 \times 50}{160} = 34.4\,kN$$

$$F_s = \left(27.5^2 + 34.4^2\right)^{1/2}$$

$$= 44.0\,kN$$

	Subject		Chapter ref.
BRE	*No. 2* *FIN PLATE CONNECTION*		**26**
	Design code BS 5950	Made by *DBM*	Sheet no. *3*
		Checked by *BD*	

Bearing capacity connected plate (fin plate)

The bearing capacity of the connected plate is given by the following expressions:

$$P_{bs} = k_{bs} d t_p p_{bs}$$

but

6.3.3.3

$$P_{bs} \leq 0.5 k_{bs} e t_p p_{bs}$$

For the fin plate

$k_{bs} = 1.0$
$d = 20\,mm$
$t_p = 10\,mm$
$e = 40\,mm$ (conservative edge distance)
$p_{bs} = 460\,N/mm^2$

6.3.3.3

Table 32

$$P_{bs} = \frac{1.0 \times 20 \times 10 \times 460}{1000}\,kN$$
$$= 92\,kN$$

but

$$P_{bs} \leq \frac{0.5 \times 1.0 \times 40 \times 10 \times 460}{1000}\,kN$$
$$= 92\,kN$$

Therefore $P_{bs} = 92\,kN$

Bearing capacity connected part (beam web)

The bearing capacity of the connected part is given by the following expression:

$$P_{bs} = k_{bs} d t_p p_{bs}$$

but

6.3.3.3

$$P_{bs} \leq 0.5 k_{bs} e t_p p_{bs}$$

	Subject		Chapter ref.
BRE	**No. 2** **FIN PLATE CONNECTION**		**26**
	Design code **BS 5950**	Made by **DBM**	Sheet no. **4**
		Checked by **BD**	

For the beam web

k_{bs} = 1.0
d = 20 mm
t_p = 7.6 mm   **6.3.3.3**
e = 40 mm (conservative edge distance)
p_{bs} = 460 N/mm²   **Table 32**

$$\therefore P_{bs} = \frac{1.0 \times 20 \times 7.6 \times 460}{1000} \, kN$$

$$= 69.9 \, kN$$

$$\therefore P_{bs} = 69.9 \, kN$$

The bearing capacity of the beam web is smaller than the bearing
capacity of the fin plate.

$$69.9 \, kN > 44 \, kN$$

Therefore the bearing capacity of the bolt group is adequate.

	Subject		Chapter ref.
BRE	*No. 2* *FIN PLATE CONNECTION*		**26**
	Design code	Made by **DBM**	Sheet no. **5**
	BS 5950	Checked by **BD**	

Check 3a – plain end beams – strength of beam at net section
This check is not necessary because the beam is notched.

Check 3b – Shear and bending capacity of the supported beam

For shear

*The basic requirement is that the shear capacity of the beam must
be greater than the applied shear.*

$P_v \geq F_v$

For shear there are three checks to consider. These are:

- *shear of the gross section*
- *shear of the net section*
- *block shear*

Shear of the gross section

The shear capacity of the gross section is given by:

$$
\begin{aligned}
P_v &= 0.6 p_y A_v \\
p_y &= 275\,N/mm^2 \\
A_v &= \text{Gross area of notched section at the bolt line.} \\
&= (D - d_c)t \\
&= (449.8 - 28) \times 7.6\,mm^2 \\
&= 3206\,mm^2
\end{aligned}
$$ *4.2.3*

$$
\therefore P_v = \frac{0.6 \times 275 \times 3206}{1000}\,kN
$$
$$
= 529\,kN
$$

Shear of the net section

The shear capacity of the net section is given by:

$$
\begin{aligned}
P_v &= 0.7 p_y K_e A_{v,net} \\
K_e &= 1.2 \\
A_{v,net} &= A_v - nD_h t \\
&= 3206 - 4 \times 22 \times 7.6\,mm^2 \\
&= 2537\,mm^2
\end{aligned}
$$ *6.2.3*

	Subject		Chapter ref.
BRE	**No. 2** **FIN PLATE CONNECTION**		**26**
	Design code	Made by **DBM**	Sheet no. **6**
	BS 5950	Checked by **BD**	

$$P_v = \frac{0.7 \times 275 \times 1.2 \times 2537}{1000} \, kN$$

$$= 586 \, kN$$

Block shear

The block shear capacity of a notched beam is given by:

$$P_r = 0.6 \, p_y t[L_v + K_e(L_t - kD_h)]$$ 6.2.4

$$\begin{aligned} L_v &= e + (n - 1)p \\ &= 40 + (4 - 1) \times 80 \\ &= 280 \, mm \end{aligned}$$

t = 7.6 mm   6.2.4
k = 0.5   3.4.3
K_e = 1.2
L_t = 40 mm
D_h = 22 mm

$$P_r = 0.6 \times 275 \times 7.6[289 + 1.2 \times (40 - 0.5 \times 22)]$$
$$= 394.8 \, kN$$

Block shear gives the minimum shear capacity

$394.8 \, kN > 110 \, kN$

Therefore the shear capacity is adequate.

	Subject		Chapter ref.
BRE	**No. 2** **FIN PLATE CONNECTION**		**26**
	Design code **BS 5950**	Made by **DBM** Checked by **BD**	Sheet no. **7**

Shear and bending interaction at the notch

The basic requirement is that the moment capacity at the notch in the presence of shear must be greater than the moment from the product of the end reaction and the distance to the end of the notch.

$$M_{cn} \geq F_v(t_1 + c)$$

t_1 *is the gap*
c *is the notch length*

<u>*Check for low shear*</u>

$$F_v \leq 0.75P_v$$

from previous calculation   *4.2.5.4*

$$P_v = 394.8\,kN$$

$$\therefore 0.75P_v = 296.1\,kN$$

$$110\,kN < 296.1\,kN$$

$\therefore$ *section is subject to low shear*

$$M_{CN} = p_y Z$$

$p_y = 275\,N/mm^2$ *4.2.5.4*
Z = *elastic section modulus of the residual tee-section at the notch*
$Z = 373\,mm^3$

$$\therefore M_{CN} = \frac{275 \times 373 \times 10^3}{10^6}\,kNm$$
$$= 102.6\,kNm$$

Applied moment $= \dfrac{110 \times (10 + 110)}{10^3}\,kNm$
$$= 13.2\,kNm$$

$102.6\,kNm > 13.2\,kNm$

Therefore the moment capacity of the notched beam is adequate.

	Subject	Chapter ref.
BRE	**No. 2** **FIN PLATE CONNECTION**	**26**
	Design code *BS 5950*	Made by **DBM** Sheet no. **8**
		Checked by **BD**

Check 3c – Supported beam – local stability of notched beam

*The beam is restrained against lateral torsional buckling, therefore
no account need be taken of notch stability provided for one flange
notched the following requirement is satisfied:*

c = 120 mm
D = 449.8 mm
t_w = 7.6 mm

$$for \ \frac{D}{t_w} = \frac{449.8}{7.6} = 59.2$$

$$c \leq \frac{160000D}{(D/t_w)^3} \quad for \quad \frac{D}{t_w} > 54.3 \ (S275)$$

$$\frac{160000D}{(D/t_w)^3} = \frac{160000 \times 449.8}{(449.8/7.6)^3} = 346.8 \ mm$$

$120 \ mm < 346.8 \ mm$

Check 3d – Unrestrained supported beam overall stability of notched beam

*The notched beam is restrained therefore the overall stability of the
beam need not be checked.*

	Subject		Chapter ref.
BRE	**No. 2** **FIN PLATE CONNECTION**		**26**
	Design code **BS 5950**	Made by **DBM** Checked by **BD**	Sheet no. **9**

Check 4 – Shear and bending capacity of fin plate connected to supported beam

For shear

The basic requirement is that the shear capacity of the fin plate must be greater than the shear force.

$P_v \geq F_v$

For shear the following checks must be made:

- *shear at the gross section*
- *shear at the net section*
- *block shear*

<u>Shear (gross section)</u>

$P_v = 0.6 \, p_y A_v$

$p_y = 275 \, N/mm^2$ 4.2.3

$A_v = 0.9 \times 3200 = 2880 \, mm^2$

$\therefore P_v = \dfrac{0.6 \times 275 \times 2880}{1000} \, kN$

$\quad\ = 475.2 \, kN$

<u>Shear (net section)</u>

The effect of bolt holes need not be allowed for in the shear area provided that:

$A_{v,net} > \dfrac{0.85 A_v}{K_e}$

$\qquad = \dfrac{0.85 \times 2880}{1.2} = 2040 \, mm$ 6.2.3

$A_{v,net} = A_v - n D_h t$

$\qquad = 2880 - 4 \times 22 \times 10 = 2000 \, mm^2$

$2000 \, mm < 2040 \, mm$

Therefore the effect of bolt holes needs to be taken into account.

	Subject	Chapter ref.
BRE	*No. 2* *FIN PLATE CONNECTION*	**26**
	Design code **BS 5950**	Made by **DBM** Sheet no. **10** Checked by **BD**

The shear capacity of the net section is given by:

$$P_v = 0.7 p_y K_e A_{v,net}$$ **6.2.3**

$$K_e = 1.2$$ **3.4.3**

$$A_{v,net} = 0.9 A_v - nDt$$
$$= 0.9 \times 3200 - 4 \times 22 \times 10 \, mm^2$$
$$= 2000 \, mm^2$$

$$\therefore P_v = \frac{0.7 \times 275 \times 1.2 \times 2000}{1000} \, kN$$
$$= 462 \, kN$$

Block shear

The block shear capacity is given by the following expression:

$$P_r = 0.6 \, p_y t[L_v + K_e(L_t - kD_t)]$$
$$L_v = e + (n-1)p = 40 + (4-1)80$$
$$= 280 \, mm$$ **6.2.4**

$$t = 10 \, mm$$
$$K_e = 1.2$$ **3.4.3**
$$L_t = 50 \, mm$$
$$k = 0.5$$
$$D_t = 22 \, mm$$

$$P_r = 0.6 \times 275 \times 10[280 + 1.2(50 - 0.5 \times 22)]$$ **6.2.4**
$$= 539.2 \, kN$$

The shear capacity is given by the net shear capacity at the section

$$P_v = 462 \, kN$$

$$462 \, kN > 110 \, kN$$

Therefore the shear capacity of the fin plate is adequate.

	Subject		Chapter ref.
BRE	**No. 2** **FIN PLATE CONNECTION**		**26**
	Design code **BS 5950**	Made by **DBM**	Sheet no. **11**
		Checked by **BD**	

For bending

The basic requirement is that the moment capacity of the fin plate must be greater than the applied moment

$M_c \geq F_v a$

For low shear

$F_v \leq 0.6P_v$ 4.2.5.2

$P_v = 462\,kN$ *(from previous page)*

$0.6P_v = 277.2\,kN$

__$110\,kN < 277.2\,kN$__

$\therefore$ *Section is subject to low shear*

$$M_c = \frac{p_y t}{6}\left[2e + (n-1)p\right]^2$$

$p_y = 275\,N/mm^2$
$t\ = 10\,mm$
$e\ = 40\,mm$
$p\ = 80\,mm$
$n\ = 4$

$$M_c = \frac{275 \times 10}{6}\left[2 \times 40 + (4-1) \times 80\right]^2/10^6\,kNm$$
$$= 46.9\,kNm$$
$$F_v a = \frac{110 \times 50}{1000}\,kNm$$
$$= 5.5\,kNm$$

$46.9\,kNm > 5.5\,kNm$

Therefore the fin plate is adequate in bending.

	Subject	Chapter ref.
BRE	**No. 2** **FIN PLATE CONNECTION**	**26**
	Design code *BS 5950*	Made by *DBM*	Sheet no. *12*
		Checked by *BD*	

Check 5 – Lateral torsional buckling resistance of long fin-plate

<u>*For long fin-plates*</u>

$$a \geq \frac{t}{0.15}$$

$a = 50\,mm$
$t = 10\,mm$

$t/0.15 = 66\,mm$

$50\,mm < 66\,mm$

Therefore this is a short fin plate and does not need to be checked for lateral torsional buckling.

Check 6 – Supporting beam welds

The basic requirement is that the leg length of fillet weld must be greater than or equal to 0.8 times the thickness of the fin plate.

leg length = 8 mm

$0.8\,t = 0.8 \times 10 = 8\,mm$

Therefore 8 mm fillet weld is adequate.

	Subject		Chapter ref.
BRE	**No. 2** **FIN PLATE CONNECTION**		**26**
	Design code **BS 5950**	Made by **DBM** Checked by **BD**	Sheet no. **13**

Check 7 – Supporting beam – local capacity (with one supported beam)

The basic requirement is that the local shear capacity of the supporting beam web should be greater than the end reaction.

$P_v \geq F_v$

The following two modes of failure should be considered:

- *Local shear failure of the supporting beam web*
- *Punching shear capacity*

<u>*Local shear failure (beam web)*</u>

The local shear failure is given by the following expression:

$$P_v = 0.6p_yA_v \qquad\qquad 4.2.3$$

$p_y = 275\,N/mm^2$
$A_v = 0.9A\,mm^2$

where A is local shear area

$A = lt_w$

l is the depth of fin plate
t_w *is web thickness of supporting beam*

$l\ \ = 320\,mm$
$t_w = 10.6\,mm$

$A_v = 0.9 \times 320 \times 10.6\,mm^2$
$\quad = 3052.8\,mm^2$

$$\therefore P_v = \frac{0.6 \times 275 \times 3052.8}{1000}\,kN$$
$$\quad = 503.7\,kN$$

$503.7\,kN > 55\,kN$

Therefore the local shear capacity is adequate.

	Subject		Chapter ref.
BRE	**No. 2** ***FIN PLATE CONNECTION***		**26**
	Design code **BS 5950**	Made by **DBM**	Sheet no. **14**
		Checked by **BD**	

Punching shear capacity

The punching shear capacity of the supporting web is satisfactory provided that

$$t_f \leq t_w \left\{ \frac{U_s}{p_y} \right\}$$

where t_f *is the fin plate thickness*
          t_w *is the thickness of the web of the supporting beam*
          U_s *is the ultimate tensile strength of the supporting beam* $= 410\,N/mm^2$
          p_y *is the design strength of the fin plate*
          $t_f = 10\,mm$

$$t_w \left\{ \frac{U_s}{p_y} \right\} = 10.6 \left\{ \frac{410}{275} \right\} = 15.8\,mm$$

$10\,mm < 15.8\,mm$

Therefore the punching shear capacity is adequate.

	Subject		Chapter ref.
BRE	**No. 2** **FIN PLATE CONNECTION**		**26**
	Design code **BS 5950**	Made by **DBM** Checked by **BD**	Sheet no. **15**

Check 8 – Structural integrity – connecting elements

The basic requirement is that the tension capacity of the fin plate must be greater than the tie force.

Tension capacity of fin plate ≥ Tie force

<u>*Tie force*</u>

The minimum tying force is 75 kN.   *2.4.5.3*

Note that in certain cases the tie force will be greater than 75 kN and it may be necessary to check for a tying force equal to the end reaction of the supported beam.

<u>*Tension capacity of fin plate*</u>

The following two checks should be considered:

- *capacity of gross section*
- *capacity of net section*

<u>*Capacity of the gross section*</u>

The tension capacity of the gross section is given by:

$P_t = p_y A_e$ *4.6.1*

$p_y = 275 \, N/mm^2$

At the gross section a_n is given by the following expression: *3.4.3*

$$a_n = a_g = L \times t_f$$
$$= 320 \times 10$$
$$= 3200 \, mm^2$$

$$P_t = \frac{275 \times 3200}{1000} \, kN$$
$$= 880 \, kN$$

<u>*Capacity of net section*</u>

The tension capacity of the net section is given by:

$P_t = p_y A_e$ *4.6.1*

$A_e = k_e a_n$ *3.4.3*
$k_e = 1.2$
$a_n = A - n D_h$ *3.4.4.1*
$$= 3200 - 4 \times 22$$
$$= 3112 \, mm^2$$

	Subject		Chapter ref.
BRE	**No. 2** **FIN PLATE CONNECTION**		**26**
	Design code **BS 5950**	Made by **DBM** Checked by **BD**	Sheet no. **16**

$$\therefore P_t = \frac{275 \times 1.2 \times 3112}{1000} \, kN$$
$$= 1026.9 \, kN$$

The tension capacity of the fin plate is *880 kN.*

$$880 \, kN > 75 \, kN$$

Therefore the tension capacity of the fin plate is adequate.

Bearing capacity

The bearing capacity of each bolt hole is given by:

$$P_{bs} = k_{bs}dt_f p_{bs} \qquad\qquad\qquad 6.3.3.3$$

but

$$P_{bs} \leq 0.5 k_{bs} e t_f p_{bs}$$

$k_{bs} = 1.0$
$d = 20 \, mm$
$t_f = 10 \, mm$
$p_{bs} = 460 \, N/mm^2 \ (S275)$ *Table 32*
$e = 50 \, mm$

$$\therefore P_{bs} \leq \frac{1.0 \times 20 \times 10 \times 460}{1000} \, kN$$
$$= 92 \, kN$$

but

$$P_{bs} \leq \frac{0.5 \times 1.0 \times 50 \times 10 \times 460}{1000} \, kN$$
$$= 115 \, kN$$

Capacity of fin plate $= np_{bs}$
$$= 4 \times 92 \, kN$$
$$= \underline{360 \, kN}$$

$$\underline{360 \, kN > 75 \, kN}$$

$\therefore$ *The bearing capacity of the fin plate is adequate.*

	Subject		Chapter ref.
BRE	**No. 2** **FIN PLATE CONNECTION**		**26**
	Design code **BS 5950**	Made by **DBM** Checked by **BD**	Sheet no. **17**

Check 9 – Structural integrity – supported beam

<u>*For tension*</u>

The basic requirement is that the tension capacity of the supported beam web must be greater than the tie force.

Tension capacity of the beam web ≥ Tie force

For tension the following checks should be considered:

- *net tension capacity*
- *local tension capacity*
- *bearing capacity*

<u>*Net tension capacity*</u>

The net tension capacity of the beam's web is given by the following expression:

$$P_t = p_y A_e \hspace{3cm} \text{4.6.1}$$

For the notched section

$A_e = K_e(A_g - n D_h t)$

where A_g is the gross area of the notched beam
$\quad K_e = 1.2 \ (for \ S275 \ steel)$ \hspace{2cm} 3.4.3
$\quad D_h = 22\,mm$
$\quad n = 4$
$\quad t = 7.6\,mm$
$\quad A_g = 4868.7\,mm^2$

$A_e = 1.2(4868.9 - 4 \times 22 \times 7.6)$
$\quad = 5802.4\,mm^2 > A_g$

$P_t = \dfrac{275 \times 4868.7}{1000}\,kN$

$\quad = 1338.9\,kN$

1338.9 kN > 75 kN

Therefore the net tension capacity of the web is adequate.

	Subject		Chapter ref.
BRE	**No. 2** **FIN PLATE CONNECTION**		**26**
	Design code	Made by **DBM**	Sheet no. **18**
	BS 5950	Checked by **BD**	

Local tension capacity

This is tension failure through a group of bolt holes at a free edge.
This consists of tension failure at the row of bolt holes accompanied
by shear failure along a line from the end bolts to the free edge.

The tension capacity is given by the following expression:

$$P_t = 2(0.6\,p_y A_v) + (n-1)(p-D_h)t\,p_y$$

where A_v is the shear area and is given by $0.9t\left(e - \dfrac{D_h}{2}\right)$

p is the bolt pitch
t is the web thickness

$$P_t = \frac{2 \times [0.6 \times 275 \times 0.9 \times 7.6 \times (40 - 22/2)] + (4-1)(80-22) \times 7.6 \times 275}{1000}\,kN$$

$$= 429.1\,kN$$

429.1 kN > 75 kN

Therefore the local tension capacity of the web is adequate.

Bearing

The basic requirement is that the bearing capacity of the beam web
must be greater than the tie force.

bearing capacity of the beam web ≥ tie force

For a single line of bolts the bearing capacity of the beam web is
given by the smaller of the following expressions:

$$P_{bs} = 1.5nk_{bs}dtp_{bs} \qquad\qquad\qquad\qquad 6.3.3.3$$

but

$$P_{bs} \leq 0.5nk_{bs}etp_{bs}$$

	Subject		Chapter ref.
BRE	**No. 2** **FIN PLATE CONNECTION**		**26**
	Design code **BS 5950**	Made by **DBM**	Sheet no. **19**
		Checked by **BD**	

where

n is the number of bolts

$k_{bs} = 1.0$

$p_{bs} = 460 \, N/mm^2$ *(for grade S275 steel)*   *Table 32*

$e = 40 \, mm$

$$P_{bs} = \frac{1.5 \times 4 \times 1.0 \times 20 \times 7.6 \times 460}{1000} \, kN$$
$$= 419.5 \, kN$$

but

$$P_{bs} \leq \frac{0.5 \times 4 \times 1.0 \times 40 \times 7.6 \times 460}{1000} \, kN$$
$$= 279.7 \, kN$$

$279.7 \, kN > 75 \, kN$

Therefore the bearing capacity of the web is adequate.

	Subject		Chapter ref.
BRE	**No. 3** **WEB CLEAT CONNECTION**		**26**
	Design code **BS 5950**	Made by **DBM**	Sheet no. **1**
		Checked by **BD**	

The connection is between a 356 × 171 × 45 UB (S275) beam and a 254 × 254 × 73 UC (S275) column.

The end reaction of the simply supported beam due to factored loads is 185 kN.

Refer to Table 26.3 for design checks.

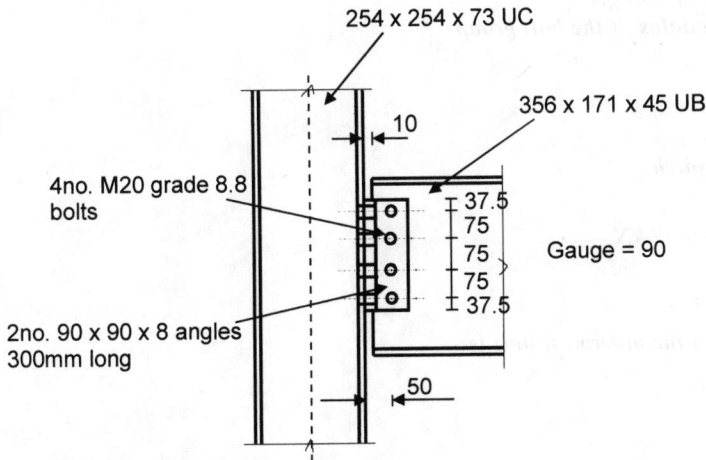

Check 2 – Shear capacity of bolt group connecting cleats to web of supported beam
Basic check is:

$$P_s \geq F_s/2$$

The resulting shear force on the outermost bolt due to direct shear and moment is given by the following expression:

$$F_s = \left(F_{sv}^2 + F_{sm}^2\right)^{1/2}$$

where F_{sv} is the force on the bolt due to direct shear
  F_{sm} is the force on the outermost bolt due to moment

	Subject		Chapter ref.
BRE	**No. 3** **WEB CLEAT CONNECTION**		**26**
	Design code	Made by **DBM**	Sheet no. **2**
	BS 5950	Checked by **BD**	

F_{sv} = end reaction/number of bolts

$$= \frac{185}{4} = 46.25 \, kN$$

$$F_{sm} = \frac{F_v a}{Z_{bg}}$$

a is the eccentricity of bolt group
Z_{bg} *is the elastic modulus of the bolt group*

$$= \frac{n(n+1)p}{6}$$

where p is the bolt pitch

$$F_{sm} = \frac{185 \times 50}{\left(\dfrac{4 \times 5 \times 75}{6} \right)} = 37 \, kN$$

∴ *resultant shear on the outermost bolt is*

$$F_s = (46.25^2 + 37^2)^{1/2}$$
$$= 59.2 \, kN$$

$F_s/2 = 29.6 \, kN$

The shear resistance of a single bolt in double shear is given by:

$$P_s = p_s A_s \qquad\qquad\qquad 6.3.2.1$$

where p_s is the bolt shear strength
       A_s is the shear area of the bolt   *Table 30*

$$F_v = 375 \times 245/10^3 \, kN$$
$$= 91.88 \, kN$$

$P_s > F_s/2$ ∴ *the shear capacity of the bolt group is adequate.*

	Subject	Chapter ref.
BRE	No. 3 **WEB CLEAT CONNECTION**	**26**
	Design code **BS 5950**	Made by **DBM** Checked by **BD**	Sheet no. **3**

Check 3 – Shear and bearing capacity of cleat connected to supported beam

Shear

The basic check for shear is that the shear capacity of the leg of the angle cleat (P_v) must be greater than half the end reaction $\left(\dfrac{F_v}{2}\right)$

$$P_v \geq \frac{F_v}{2}$$

For shear the following three checks must be made:

- *shear capacity of the gross section*
- *shear capacity of the net section*
- *block shear*

Shear capacity of the gross section

The shear capacity of the gross section is given by the following expression:

$$P_v = 0.6\, p_y A_v$$

where p_y is the design strength of the angle   **4.2.3**
* A_v is the shear area of the angle*

For an 8 mm thick angle

$$p_y = 275\, N/mm^2$$

$$A_v = 0.9A$$ **Table 9**
 4.2.3

where A is the gross area for a $90 \times 90 \times 8$ angle 300 mm long

$$A = 300 \times 8 = 2400\, mm^2$$
$$\therefore A_v = 0.9A = 0.9 \times 2400\, mm^2$$
$$= 2160\, mm^2$$

$$\therefore P_v = 0.6 \times 275 \times 2160/10^3\, kN$$
$$= 356.4\, kN$$

356.4 kN > 92.5 kN

Shear capacity of angle cleat at gross section is adequate.

	Subject		Chapter ref.
BRE	No. 3 **WEB CLEAT CONNECTION**		**26**
	Design code **BS 5950**	Made by **DBM** Checked by **BD**	Sheet no. **4**

Shear capacity of the net section

The effect of bolt holes on the shear capacity of the section need not be considered provided that:

$$A_{v,net} \geq \frac{0.85 A_v}{K_e}$$

where $A_{v,net}$ *is the net shear area after deducting bolt holes*   6.2.3
       K_e *is the effective net area coefficient*
       $A_{v,net} = 0.9A - nD_h t$

where n is the number of bolt holes
       D_h *is the diameter of the hole*
       t is the thickness of cleat

$$\therefore A_{v,net} = 0.9 \times 2400 - 4 \times 22 \times 8 \, mm^2$$
$$= 1456 \, mm^2$$

$$\frac{0.85 A_v}{K_e} = \frac{0.85 \times 2160}{1.2} = 1530 \, mm^2$$

$$1456 < 1530$$

Therefore the effect of bolt holes must be taken into account.

The shear capacity of the net section is given by the following expression:

$$P_v = 0.7 p_y K_e A_{v,net}$$ 6.2.3

for grade S275 steel $K_e = 1.2$

            3.4.3

$$\therefore P_v = \frac{0.7 \times 275 \times 1.2 \times 1456}{10^3} \, kN$$
$$= 336 \, kN$$

$$336 \, kN > 92.5 \, kN$$

Shear capacity of angle cleat at the net section is adequate.

	Subject		Chapter ref.
BRE	*No. 3* *WEB CLEAT CONNECTION*		**26**
	Design code *BS 5950*	Made by *DBM* Checked by *BD*	Sheet no. *5*

Block shear

The block shear capacity through a group of bolts at a free edge is given by the following expression:

$$P_r = 0.6\, p_y t [L_v + K_e(L_t - kD_t)]$$

6.2.4

where t is the thickness of the cleat
 L_v is the length of the shear face
 L_t is the length of the tension face
 k is a coefficient with the following values:

• *for a single line of bolts k = 0.5*
• *for a double line of bolts k = 2.5*

$$L_v = e_1 + (n-1)p$$

where e_1 is the end distance
 p is the bolt pitch

$$L_v = 37.5 + (4-1) \times 75\ mm$$
$$= 262.5\ mm$$
$$L_t = e_2$$

where e_2 is the horizontal end distance

$$L_t = 40\ mm$$
$$\therefore P_r = 0.6 \times 275 \times 8[262.5 + 1.2(40 - 22/2)]$$
$$= 392.4\ kN$$

392.4 kN > 92.5 kN

Block shear capacity of the angle is adequate.

	Subject		Chapter ref.
BRE	*No. 3* **WEB CLEAT CONNECTION**		**26**
	Design code	Made by **DBM**	Sheet no. **6**
	BS 5950	Checked by **BD**	

Bearing

The basic check for bearing is that the bearing capacity of the leg of the angle cleat per bolt (P_{bs}) must be greater than half the resultant force on the outermost bolt ($F_s/2$).

$$P_{bs} \geq \frac{F_s}{2}$$

For bearing the following checks should be made:

- *bearing capacity of bolt*
- *bearing capacity of the angle*

Bearing capacity of bolt

The bearing capacity of a bolt is given by:

$$P_{bb} = d t_p p_{bb} \qquad\qquad\qquad\qquad 6.3.3.2$$

where d is the nominal diameter of the bolt
  p_{bb} *is the bearing strength of the bolt = 1000 N/mm²*   *Table 31*
  t_p *is the thickness of the angle cleat*

$$\therefore P_{bs} = \frac{69 \times 8 \times 1000}{10^3} \, kN$$

$$= 160 \, kN$$

$$160 \, kN > 29.6 \, kN$$

The bearing capacity of the bolt is adequate.

Bearing capacity of the angle

The bearing capacity of the angle is given by the following expression:

$$P_{bs} = k_{bs} \, d \, t_p \, p_{bs} \qquad\qquad\qquad\qquad 6.3.3.3$$

but

$$P_{bs} \leq 0.5 k_{bs} \, e t_p \, p_{bs}$$

where e is the end distance and is usually taken as the minimum of the horizontal and vertical end distances

	Subject	Chapter ref.
BRE	**No. 3** **WEB CLEAT CONNECTION**	**26**
	Design code **BS 5950**	Made by **DBM** Checked by **BD**	Sheet no. **7**

p_{bs} is the bearing strength of the angle $= 460\,N/mm^2$ for S275 steel

Table 32

k_{bs} is a coefficient allowing for the type of hole $= 1.0$

6.3.3.3

$$\therefore P_{bs} = \frac{1.0 \times 20 \times 8 \times 460}{10^3}\,kN$$
$$= 73.6\,kN$$

$$P_{bs} \leq \frac{0.5 \times 1.0 \times 37.5 \times 8 \times 460}{10^3}\,kN$$
$$= 69\,kN$$

$$\therefore P_{bs} = 69\,kN$$

$$69\,kN > 29.6\,kN$$

The bearing capacity of the angle cleat is adequate.

	Subject		Chapter ref.
BRE	*No. 3* *WEB CLEAT CONNECTION*		**26**
	Design code	Made by **DBM**	Sheet no. **8**
	BS 5950	Checked by **BD**	

Check 4 – Shear and bearing capacity of supporting beam

Shear

The basic check is that the shear capacity of the beam (P_v) must be greater than the reaction at the end of the beam (F_v)

$$P_v \geq F_v$$

For shear the following checks should be considered:

- *shear capacity of the gross section*
- *shear capacity of the net section*
- *block shear (notched beams only)*

Shear capacity of the gross section

The shear capacity of the gross section is given by the following expression:

$$P_v = 0.6p_yA_v \hspace{4cm} \textbf{4.2.3}$$

where p_y is the design strength of the beam $= 275\,N/mm^2$
* A_v is the shear area*

The shear area of a rolled I-section is given by:

$$
\begin{aligned}
A_v \quad &= tD \\
&= 6.9 \times 352.0\,mm^2 \\
&= 24288\,mm^2
\end{aligned}
\hspace{3cm} \textbf{4.2.3}
$$

$$\therefore P_v = \frac{0.6 \times 275 \times 24288}{10^3}\,kN$$
$$= 4007.52\,kN$$

4007.52 kN > 185 kN

The shear capacity of the gross section of the beam is adequate.

Shear capacity of the net section

The effect of bolt holes on the shear capacity of the section need not be considered provided that:

$$A_{v,net} \geq \frac{0.85\,A_v}{K_e} \hspace{4cm} \textbf{6.2.3}$$

	Subject		Chapter ref.
BRE	*No. 3* *WEB CLEAT CONNECTION*		**26**
	Design code **BS 5950**	Made by **DBM** Checked by **BD**	Sheet no. **9**

where

K_e *is the effective net area coefficient = 1.2 for S275*

$A_{v,net}$ *is the shear area after deducting bolt holes*   3.4.3

$A_{v,net} = tD - nD_h t$

where n is the number of bolts

D_h *is the diameter of the bolt hole*

t is the thickness of the web

$$A_{v,net} = 6.9 \times 352 - 4 \times 22 \times 6.9$$
$$= 24288 - 607.2$$
$$= 23680.0 \, mm^2$$

$$\frac{0.85 A_v}{K_e} = \frac{0.85 \times 24288}{1.2} = 17204 \, mm$$

23680.0 > 17204

Therefore the effect of bolt holes does not need to be considered.

Bearing

The basic check for bearing is that the bearing capacity of the beam web per bolt (P_{bs}) must be greater than the resultant force on the outermost bolt (F_s)

$P_{bs} \geq F_s$

For bearing the following checks should be considered:

- *bearing capacity of the bolt*
- *bearing capacity of the web*

Bearing capacity of the bolt

The bearing capacity of the bolt is given by the following expression:

$P_{bb} = d \, t_p \, p_{bb}$ 6.3.3.2

where

	Subject	Chapter ref.
BRE	*No. 3* *WEB CLEAT CONNECTION*	**26**
	Design code	Made by **DBM**	Sheet no. **10**
	BS 5950	Checked by **BD**	

d is the nominal diameter of the bolt
p_{bb} is the bearing strength of the bolt = 1000 N/mm²
t_p is the web thickness   *Table 31*

$\therefore P_{bb} = 20 \times 6.9 \times 1000/10^3 \, kN$
$\quad = 138 \, kN$

$\quad 138 \, kN > 59.2 \, kN$

The bearing capacity of the bolt is adequate.

<u>*Bearing capacity of web*</u>

The bearing capacity of the web is given by the following expression:

$\quad P_{bs} = k_{bs} \, d \, t_p \, p_{bs}$

but

$\quad P_{bs} \leq 0.5 \, k_{bs} \, e \, t_p \, p_{bs}$

where e is the end distance in the direction of bearing and is con-
servatively taken as the horizontal end distance
p_{bs} is the bearing strength of the web = 460 N/mm² for S275   *Table 32*
steel   *6.3.3.3*
k_{bs} is the coefficient allowing for the type of hole = 1.0

$P_{bs} = \dfrac{1.0 \times 20 \times 6.9 \times 460}{10^3} \, kN$

$\quad = 63.5 \, kN$

but

$P_{bs} \leq \dfrac{0.5 \times 1.0 \times 40 \times 6.9 \times 460}{10^3} \, kN$

$\quad = 63.5 \, kN$

$\therefore P_{bs} = 63.5 \, kN$

$63.5 \, kN > 59.2 \, kN$

The bearing capacity of the web is adequate.

	Subject	Chapter ref.
BRE	*No. 3* *WEB CLEAT CONNECTION*	**26**

Design code	Made by *DBM*	Sheet no. *11*
BS 5950	Checked by *BD*	

Check 5 – Shear capacity of bolt group connecting cleats to column

<u>*Shear*</u>

The basic check is that the shear capacity of the bolt group connecting the cleats to the column (ΣP_s) must be greater than the reaction at the end of the beam (F_v)

$$\sum P_s \geq F_v$$

where P_s is the shear capacity of a single bolt

To calculate the shear capacity of the bolt group the shear capacity of the top and intermediate bolts should be calculated.

<u>*Top row of bolts*</u>

The shear capacity of the top bolts is given by the smaller of the following expressions:

$$P_s = p_s A_s \qquad\qquad 6.3.2.1$$

where p_s is the shear strength of the bolt. For grade 8.8 bolts
$p_s = 375\,N/mm$
A_s is the shear area of the bolt   *Table 30*

but

$$P_s \leq 0.5\, k_{bs}\, e\, t_p\, p_{bs} \qquad\qquad 6.3.3.3$$

where e is the vertical end distance
k_{bs} is a coefficient allowing for the type of hole = 1.0   *6.3.3.3*
t_p is the thickness of the angle
p_{bs} is the bearing strength of the angle. $p_{bs} = 460\,N/mm^2$ for S275 steel

$$\therefore P_s = \frac{375 \times 245}{10^3}\,kN$$
$$= 91.9\,kN \qquad\qquad \text{Table 32}$$

but

	Subject		Chapter ref.
BRE	*No. 3* *WEB CLEAT CONNECTION*		**26**
	Design code **BS 5950**	Made by **DBM** Checked by **BD**	Sheet no. **12**

$$P_s \leq \frac{0.5 \times 1.0 \times 37.5 \times 8 \times 460}{10^3} kN$$

$$= 69 \, kN$$

∴ *the bearing capacity of the top bolts is 69 kN*

Intermediate bolts

The shear capacity of the intermediate bolts is given by the following expression:

$$P_s = p_s A_s$$

$$P_s = \frac{375 \times 245}{10^3} kN$$

$$= 91.9 \, kN$$

6.3.2.1

Capacity of bolt group

The capacity of the bolt group is given by:

$$\sum P_s = 2 \times 69 + 6 \times 91.9 \, kN$$

$$= 689.4 \, kN$$

$$689.4 \, kN > 185 \, kN$$

∴ *The shear capacity of the bolt group is adequate.*

	Subject	Chapter ref.
BRE	**No. 3** **WEB CLEAT CONNECTION**	26
	Design code **BS 5950**	Made by **DBM** Checked by **BD**	Sheet no. 13

Check 6 – Shear and bearing capacity of cleats connected to column

Shear

The basic check is that the shear capacity of the leg of the angle cleat (P_v) must be greater than half the reaction at the end of the beam ($F_v/2$)

$$P_v \geq \frac{F_v}{2}$$

For shear the following three checks should be considered:

- *shear capacity of the gross section*
- *shear capacity of the net section*
- *block shear*

Shear capacity of the gross section

The shear capacity of the gross section is given by the following expression:

$$P_v = 0.6 p_y A_v \qquad\qquad 4.2.3$$

where p_y is the design strength of the angle
A_v is the shear area of the angle

For an 8 mm angle

$$p_y = 275 \, N/mm^2$$

$$A_v = 0.9A \qquad\qquad Table\ 9$$

where A is the gross area for a $90 \times 90 \times 8$ mm angle 300 mm long $\qquad 4.2.3$

$$A = 300 \times 8 \, mm^2$$
$$= 2400 \, mm^2$$

$$\therefore A_v = 2160 \, mm^2$$

$$\therefore P_v = \frac{0.6 \times 275 \times 2160}{10^3} \, kN$$
$$= 356.4 \, kN$$

356.4 kN > 92.5 kN

∴ The shear capacity of the gross section is adequate.

	Subject	Chapter ref.
BRE	*No. 3* **WEB CLEAT CONNECTION**	**26**
	Design code **BS 5950**	Made by **DBM**	Sheet no. **14**
		Checked by **BD**	

Shear capacity of the net section

The shear capacity of the net section is given by the following expression:

$$P_v = 0.7 p_y K_e A_{v,net}$$

where $A_{v,net}$ is the net shear area after deducting bolt holes   6.2.3
K_e is the effective net area coefficient. For S275 steel, $K_e = 1.2$
$A_{v,net} = 0.9A - nD_h t$ 3.4.3

where n is the number of bolt holes
D_h is the diameter of the hole
t is the thickness of the cleat

$$\therefore A_{v,net} = 2160 - 4 \times 22 \times 8$$
$$= 1456 \, mm^2$$

$$\therefore P_v = \frac{0.7 \times 275 \times 1.2 \times 1456}{10^3} \, kN$$
$$= 336.3 \, kN$$

336.3 kN > 92.5 kN

The shear capacity of the angle cleat is adequate.

Block shear

The block shear capacity through a group of bolts at a free edge is given by the following expression:

$$P_r = 0.6 \, p_y t [L_v + K_e (L_t - k D_t)]$$

where t is the thickness of the cleat   6.2.4
L_v is the length of the shear face
L_t is the length of the tension face

	Subject		Chapter ref.
BRE	*No. 3* *WEB CLEAT CONNECTION*		**26**
	Design code **BS 5950**	Made by **DBM**	Sheet no. **15**
		Checked by **BD**	

k is a coefficient for a single line of bolts = 0.5

$L_v = e_1 + (n - 1)p$

where e_1 is the end distance
p is the bolt pitch

$L_v = 37.5 + (4 - 1) \times 75 \, mm$

$\quad = 262.5 \, mm$

$L_t = e_2$

where e_2 is the horizontal end distance = 48 mm

$\therefore P_r = 0.6 \times 275 \times 8 \times \left[262.5 + 1.2 \times \left(48 - \frac{22}{2} \right) \right]$

$\quad = 405 \, kN$

$405 \, kN > 92.5 \, kN$

Block shear capacity of the angle is adequate.

Bearing

The basic check is that the bearing capacity of the leg of a single angle cleat (ΣP_{bs}) must be greater than half the reaction at the end of the beam ($F_v/2$).

$$\sum P_{bs} \geq \frac{F_v}{2}$$

where ΣP_{bs} is the bearing capacity of the single angle cleat (i.e. for n bolts)

To calculate the bearing capacity of the angle cleat, the bearing capacity of the top and intermediate bolts must be calculated.

	Subject		Chapter ref.
BRE	**No. 3** **WEB CLEAT CONNECTION**		**26**
	Design code	Made by **DBM**	Sheet no. **16**
	BS 5950	Checked by **BD**	

Bearing capacity of top bolts

The bearing capacity of the top bolts is given by the smaller of the following expressions:

$$P_{bs} = k_{bs} \, d \, t_p \, p_{bs}$$

but   6.3.3.3

$$P_{bs} \leq 0.5 \, k_{bs} \, e \, t_p \, p_{bs}$$

where *e is the end distance = 37.5 mm*
p_{bs} *is the bearing strength of the cleat. For S275*
$p_{bs} = 460 \, N/mm^2$
k_{bs} *is a coefficient allowing for the type of hole.* $k_{bs} = 1.0$ *Table 32*

$$\therefore P_{bs} = \frac{1.0 \times 20 \times 8 \times 460}{10^3} \, kN$$

$$= 73.6 \, kN$$
                                                                                              6.3.3.3
but

$$P_{bs} \leq \frac{0.5 \times 1.0 \times 37.5 \times 8 \times 460}{10^3} \, kN$$

$$= 69 \, kN$$

$$\therefore P_{bs} = 69 \, kN$$

Bearing capacity of intermediate bolts

The bearing capacity of the intermediate bolts is given by the following expression:

$$P_{bs} = k_{bs} \, d \, t_p \, p_{bs}$$ 6.3.3.3

where p_{bs} *is the bearing strength of the cleat*
k_{bs} *is a coefficient allowing for bolt type.*

$$\therefore P_{bs} = \frac{1.0 \times 20 \times 8 \times 460}{10^3} \, kN$$

$$= 73.6 \, kN$$

	Subject		Chapter ref.
BRE	*No. 3* *WEB CLEAT CONNECTION*		*26*
	Design code *BS 5950*	Made by *DBM* Checked by *BD*	Sheet no. *17*

Bearing capacity of the angle

The bearing capacity of the angle is given by:

$$\sum P_{bs} = 69 + 3 \times 73.6 \, kN$$
$$= 289.8 \, kN$$

289.8 kN > 92.5 kN

∴ The bearing capacity of the angle cleat is adequate.

Check 7 – Local shear and bearing capacity of column web

This check is not applicable in this example because the beam is connected to the column flange.

	Subject		Chapter ref.
BRE	**No. 3** **WEB CLEAT CONNECTION**		**26**
	Design code	Made by **DBM**	Sheet no. **18**
	BS 5950	Checked by **BD**	

Check 8 – Structural integrity – tension capacity of a pair of web angle cleats

The basic check is that the tying capacity of double angle web cleats must be greater that the tie force.

$$\text{Tying capacity} \geq \text{tie force} \qquad\qquad 2.4.5.3$$

The minimum tie force is 75 kN for either an internal or an edge tie. For certain structures the tie force may be as large as the end reaction.

The tying capacity of a double angle web cleat is given by the following expression:

Tying capacity = 0.6 L_e t_c p_y – for S275 steel

where $L_e = 2e_e + (n - 1)p_e - nD_h$
  e_e is end distance
  $p_e = p$ but $\leq 2e_2$
  e_2 is the edge distance = 48 mm
  D_h is the diameter of the bolt hole
  t_c is the thickness of the cleat

Check 11 in Joints in steel construction: Simple connections

$$\therefore L_e = 2 \times 37.5 + (4 - 1) \times 75 - 4 \times 22 \text{ mm}$$
$$= 212 \text{ mm}$$

$$\therefore \text{Tying capacity} = \frac{0.6 \times 212 \times 8 \times 275}{10^3} \text{ kN}$$
$$= 279.8 \text{ kN}$$

279.8 kN > 75 kN

The tying capacity of the angle cleats is adequate.

	Subject	Chapter ref.
BRE	*No. 3* *WEB CLEAT CONNECTION*	**26**
	Design code *BS 5950*	Made by *DBM* Checked by *BD*	Sheet no. *19*

Check 9 – Structural integrity – tension and bearing capacity of beam web

Tension

The basic check is that the tension capacity of the beam web must be greater than the tie force.

   Tension capacity of beam web ≥ tie force

For tension the following checks should be considered:

* **net tension capacity**
* **local tension capacity**
* **bearing capacity**

Net tension capacity

The net tension capacity of the beam's web is given by the following expression:

$$P_t = p_y A_e$$ **4.6.1**

where A_e is the sum of the effective net areas

For an I-section

$$A_e = K_e (A_g - n D_h t)$$

where A_g is the gross area of the I-section
K_e is a coefficient. For S275 steel $K_e = 1.2$ **3.4.3**

$$\therefore A_e = 1.2 \times (5700 - 4 \times 22 \times 6.9) \, mm^2$$
$$= 6111.4 \, mm^2$$

As $A_e > A_g \therefore A_e = A_g$

$$\therefore P_t = \frac{275 \times 5700}{10^3} \, kN$$
$$= 1567.5 \, kN$$

1567.5 kN > 75 kN

The net tension capacity of the web is adequate.

	Subject	Chapter ref.
BRE	*No. 3* *WEB CLEAT CONNECTION*	26
	Design code *BS 5950*	Made by *DBM* Checked by *BD*	Sheet no. *20*

Local tension capacity

This is tension failure through a group of bolt holes at a free edge. This consists of tension failure at the row of bolt holes accompanied by shear failure along a line from the end bolts to the free edge.

The tension capacity is given by the following expression:

$$P_t = 2 \times (0.6\, p_y\, A_v) + (n-1)(p - D_h)t p_y$$

where A_v is the shear area and is given by $0.9\,t\left(\dfrac{e-D_h}{2}\right)$

p is the bolt pitch
t is the web thickness

$$P_t = 2 \times (0.6 \times 275 \times 180) + (4-1)(75-22)6.9 \times 275$$
$$= 361\,kN$$

$361\,kN > 75\,kN$

Bearing capacity

Bearing capacity of beam web $= 1.5\,n\,d\,t\,p_{bs}$ *but* $\leq 0.5\,n\,e\,t\,p_{bs}$

p_{bs} *is the bearing strength of the web. For S275 steel*
$p_{bs} = 460\,N/mm^2$ *Table 32*
e is the edge distance

$$1.5\,n\,d\,t\,p_{bs} = \frac{1.5 \times 4 \times 20 \times 6.9 \times 460}{10^3} = 380.9\,kN$$

$$0.5\,n\,e\,t\,p_{bs} = \frac{0.5 \times 4 \times 40 \times 6.9 \times 460}{10^3} = 253.9\,kN$$

$253.9\,kN > 75\,kN$

∴ *The bearing capacity of the web is adequate.*

Note:
For this accidental limit state deformations can be ignored and a higher bearing strength can be used, i.e. $1.5\,p_{bs}$ rather than p_{bs} (see clause 6.4.4 of BS 5950: Part 1). This applies to friction grip connections designed to be non-slip in service but the principle may be extended to non-preloaded bolts.

	Subject		Chapter ref.
BRE	**No. 3** **WEB CLEAT CONNECTION**		**26**
	Design code *BS 5950*	Made by *DBM* Checked by *BD*	Sheet no. **21**

Check 10 – Structural integrity – tension capacity of bolts

The basic requirement is that the tension capacity of the bolt group must be greater than the tie force.

   Tension capacity ≥ tie force

The tension capacity of the bolt group is given by the following expression:

$$P_t = 2nA_t p^*_t$$

where n is the number of rows of bolts
    A_t is the tensile stress area of a bolt
    p^*_t is the reduced tension strength of a bolt in the presence of extreme prying

Check 13 in Joints in steel construction: Simple connections

    = 300 N/mm² for grade 8.8 bolts

$$\therefore P_t = \frac{2 \times 4 \times 245 \times 300}{10^3} \, kN$$
$$= 588 \, kN$$

588 kN > 75 kN

∴ the tension capacity of the bolt group is OK

Chapter 27

Foundations and holding-down systems

by HUBERT BARBER

27.1 Foundations

27.1.1 Types of foundation

Pad foundations are used primarily to support the major structural elements in either sheds or multi-storey buildings. The pad foundations to major elements may be either mass concrete or reinforced concrete, the latter when either heavy loads or very poor ground conditions are present. They may be used in the context of cladding to support intermediate posts carrying sheeting rails, in which case the load is almost all from wind forces and is horizontal.

Strip foundations are used in steel-framed buildings to support external masonry or brickwork cladding and masonry internal partitions. In some cases the ground floor is thickened at these locations to provide a foundation but care should be taken with respect to the appropriate depth for clay or frost heave and for compatibility between such foundations and those of the main frame.

Piled foundations, either driven, bored or cast in place, are used on sites where ground conditions are poor or for buildings or structures in which differential settlement is critical. They may also be required in circumstances where heavy concentrations of load occur. In general when piled foundations are used the whole of the construction should be supported on piles. The ground floor slab, ground floor cladding and internal partitions should be carried by ground beams between the pile cap locations. If it is necessary for reasons of economy to support the ground floor independently, provision should be made for differential settlement by the inclusion of suitable movement joints.

Ground improvement techniques are appropriate for some types of poor ground. The most usual techniques are vibro-compaction or vibro-replacement but dynamic compaction can also be useful for improvement of large isolated sites. Ground improvement specialists or specialist consultants should be approached as economy will be the most important factor in the decision.

Typical foundation layouts are shown in Fig. 27.1.

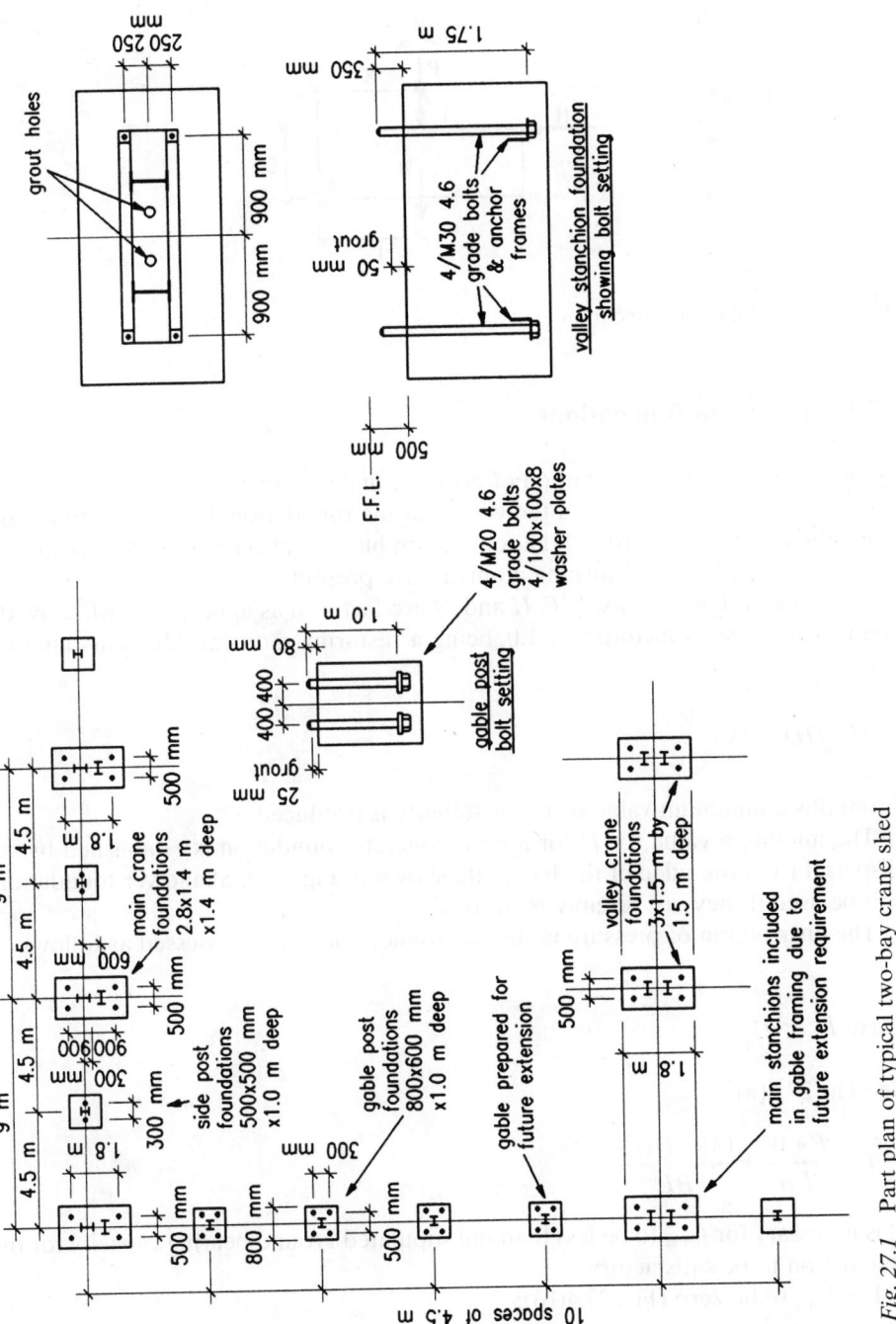

Fig. 27.1 Part plan of typical two-bay crane shed

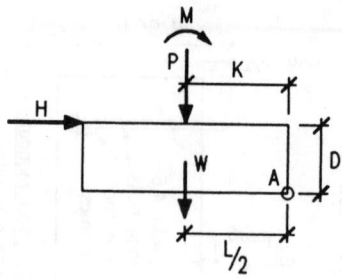

Fig. 27.2   Stability of foundation

27.1.2 Design of foundations

In order to assess the distribution of pressure under a foundation it is necessary to make a reasonable estimate of the weight of the foundation. In addition to distributing the forces to the ground the foundation block is also required to provide stability in cases where overturning moments are present.

Referring to Fig. 27.2, loads P, H and M are factored as appropriate while W, the foundation mass, is factored by 1.0, being a restoring moment. Moments about A give

$$M + HD - PK - \frac{WL}{2} \leq 0$$

From this a minimum value of W for stability is produced.

The minimum value for D for a mass concrete foundation is established by 45° dispersal from the edge of the baseplate shown in Fig. 27.3. Shallower foundations can be used if they are suitably reinforced.

The distribution of pressure under the foundation is then assessed as follows.

Case 1

See Fig. 27.4(a):

$$f_g = \frac{P + W}{LB} \pm \frac{(M + DH)6}{BL^2}$$

It is necessary for f_{gmax} to be less than the stipulated ground bearing capacity for the foundation to be satisfactory.

For f_{gmin} to be zero (Fig. 27.4(b)):

$$\frac{P + W}{LB} - \frac{(M + DH)6}{BL^2} = 0$$

Replacing the forces by the resultant acting at eccentricity x:

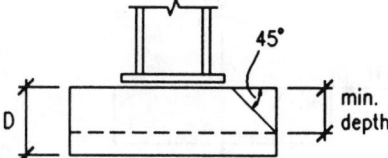

Fig. 27.3   Thickness of foundation

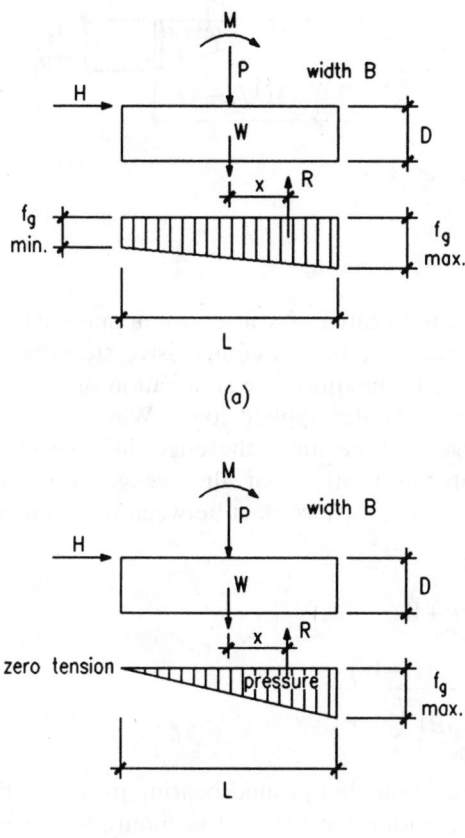

Fig. 27.4   Ground pressure – Case 1

$$\frac{R}{LB} - \frac{6Rx}{BL^2} = 0$$

from which

$$x = \frac{L}{6}$$

This is the limiting condition for the application of Case 1.

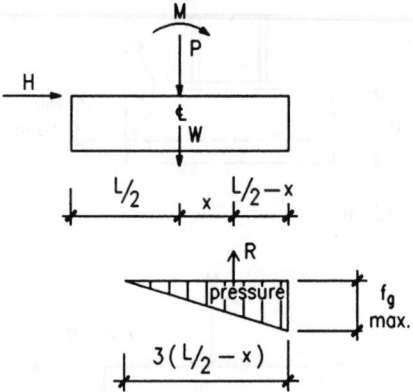

Fig. 27.5   Ground pressure – Case 2

Case 2

This occurs when f_{gmin} is negative. As no tension can exist between the soil and the underside of the concrete base a compressive stress wedge is formed at the compression side of the foundation. The summation of the stress under the block must equal the resultant of the applied loads. When $x > L/6$, the length of the triangular stress wedge is three times the edge distance $(L/2 - x)$ in order that the resultant acts at the centroid of the wedge. The theory proposes that $3(L/2 - x)$ is the length of surface contact between the foundation and the ground (Fig. 27.5).

$$\frac{f_{gmax}}{2}\left(\frac{L}{2} - x\right)3B = P + W$$

$$f_{gmax} = \frac{2(P + W)}{3B\left(\dfrac{L}{2} - x\right)}$$

When f_{gmax} exceeds the stipulated ground bearing pressure, dimensions B or L or both may be increased within the limits of economy, after which piling or ground improvement techniques can be investigated.

27.1.3 Sub-soil bearing pressure

Foundation bearing pressure should always be determined on the basis of experimental results and field assessments taken during the soil investigation. Almost all sites have considerable variation of strength and quality of the sub-strata and it is therefore necessary to undertake a comprehensive investigation producing a large number of test results in order to be satisfied that reasonable average values are

Table 27.1 Terzaghi's constants for clayey soils

ϕ	N_c	N_q	N_γ
0	5.7	1.0	0
10	10	3.0	2.0
20	18	8.0	6.5
30	36	21.0	20
35	60	50.0	45.0

obtained for the various parameters. A factor of safety of 3 is usually applied to the bearing strength to obtain the safe foundation bearing pressure.

27.1.3.1 Clayey soils – Terzaghi's method

(1) Foundation long in relation to width i.e. strip footing.

$$q = cN_c + \gamma z N_q + 0.5\gamma B N_\gamma$$

where q = bearing capacity (ultimate) (kN/m^2)
c = cohesion (kN/m^2)
γ = bulk density of the soil (kN/m^3)
z = depth of foundation (m)
B = breadth of foundation (m)
and N_c, N_q and N_γ are constants dependent upon the angle of cohesion. These constants are given in graph form in soil engineering references; Table 27.1 gives approximate values as guidance only.

(2) Square and circular foundations to isolated piers or columns.

The bearing capacity of foundations rectangular, square or circular is higher than that of strip footings. Terzaghi's expression is adjusted as follows:

$$q = 1.3cN_c + \gamma z N_q + 0.3\gamma B N_\gamma$$

while Skempton applies a factor $(1 + 0.2B/L)$ to the Terzaghi strip footing calculation where B is the breadth and L the length. In the case of a square the enhancement factor is then 1.2.

27.1.3.2 Sandy or cohesionless soils

The appropriate site test for cohesionless soils is the standard penetration test (SPT) in which the number of blows of a standard weight is recorded for unit penetration of a standard cylindrical implement. According to Meyerhof the ultimate bearing capacity is given, in kN/m^2, by

$$q = 10.7 \times NB\left(1 + \frac{z}{B}\right)$$

where N is the number of blows per metre and q, z and B are as before.

Cohesionless soils subject to flooding will suffer a reduction of capacity at water table level:

capacity when flooded $= K \times$ unflooded capacity

where $K = (\gamma - 9.8)/\gamma$ and γ is the soil bulk density given for convenience in kN/m³.

27.2 Connection of the steelwork

27.2.1 Fixed and pinned bases

The function of a column baseplate is to distribute the column forces to the concrete foundation. In general a plain or slab base is used for pinned conditions or when there is very little tension between the plate and the concrete. A gusseted base is used occasionally to spread very heavy loads but more generally for conditions of large moment in relation to the vertical applied loads, the principal function of the gusset being to allow the holding-down bolt lever arm to be increased to give maximum efficiency while keeping the baseplate thickness to an acceptable minimum. Gusseted or built-up bases give an ideal solution for compound or twin crane stanchions in industrial shed buildings.

Fixed bases are used primarily in low-rise construction either in portal buildings specifically designed as 'fixed base' or in industrial sheds in which the main columns cantilever from the foundations. They are also used, though less frequently, in multi-storey rigid-frame construction. In each of these cases it is assumed by definition that no angular rotation takes place, and although this is unlikely to be achieved it is generally accepted that sufficient rigidity can be obtained to justify the assumption.

Pinned bases are those in which it is assumed that there is no restraint against angular rotation. Although this is also difficult to achieve it is accepted that sufficient flexibility can be introduced by minimizing the size of the foundation and similarly reducing the anchorage system. Pinned bases are used in portal and in multi-storey construction.

Typical pinned and fixed bases are shown in Fig. 27.6.

27.2.2 Baseplate design

27.2.2.1 Plain bases

The empirical method for determining the size of baseplates in BS 5950-1: 1990 was not suitable for use with deep UBs or with bases that have very small outstand

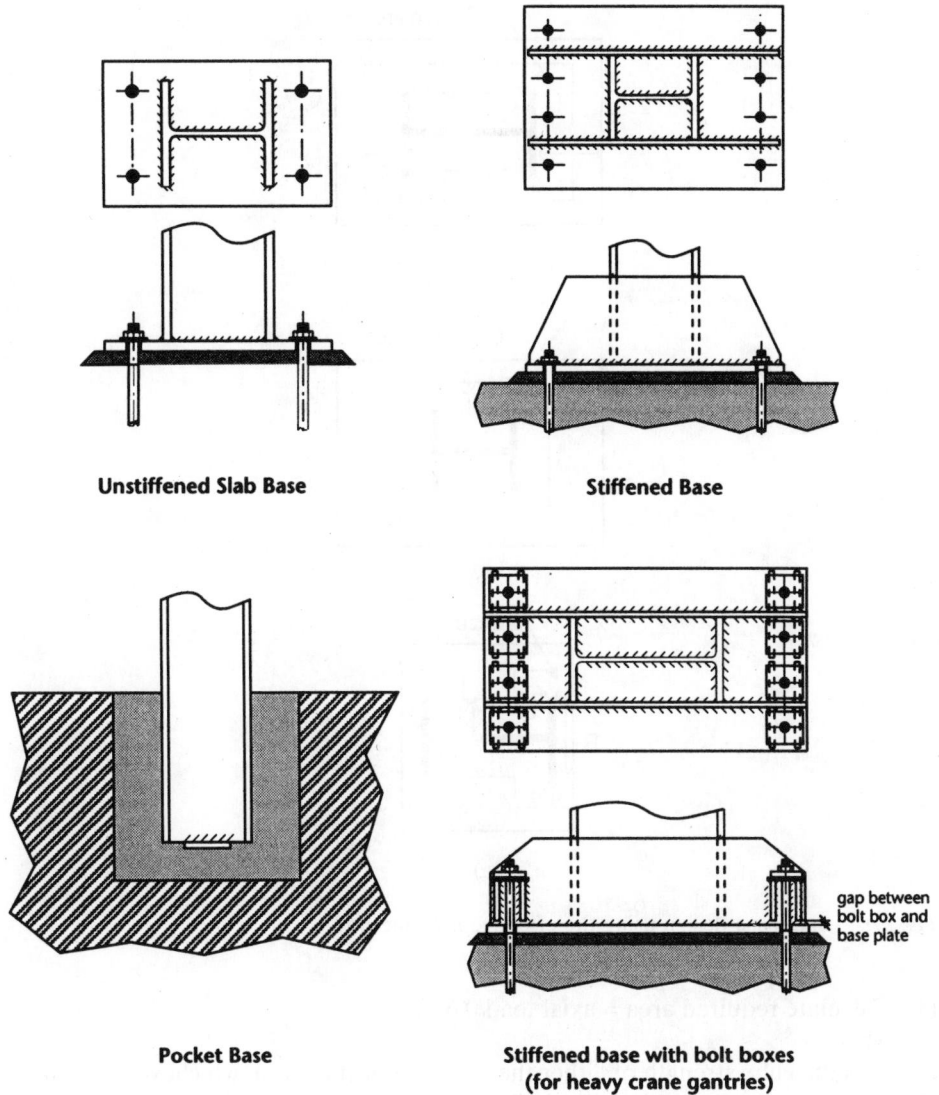

Unstiffened Slab Base

Stiffened Base

Pocket Base

**Stiffened base with bolt boxes
(for heavy crane gantries)**

gap between
bolt box and
base plate

Fig. 27.6 Typical column base connections

dimensions *a* and *b* (see Fig. 27.7). The empirical method was replaced in BS5950: 2000 by the effective area method, which offers more economy than the empirical method while still producing safe designs when compared to test results.

The effective area method for baseplate design may initially seem to be more complex than the empirical method given in BS 5950-1: 1990. However, the approach is much more reliable and can be used for all column sections.

The basic design procedure is set out below.

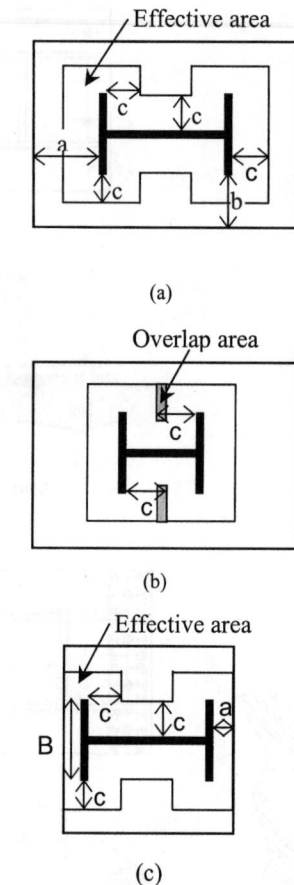

Fig. 27.7   Slab base design using the effective area method

(1) Calculate required area = axial load/0.6 f_{cu}.
    where:
    f_{cu} is the cube strength of either the concrete or the grout, whichever is weaker.
(2) Calculate outstand c (see Fig. 27.7(a)) by equating required area to actual area expressed as a function of c. The expression for the actual effective area of an I or H section may be approximated to $4c^2$ + (column perimeter) $\times c$ + column area.
(3) Check that there is no overlap of effective area between flanges (see Fig. 27.7(b)). This will occur if $2c >$ the distance between the inner faces of the flanges. If an overlap exists, modify the expression for effective area and recalculate c.
(4) Check the effective area fits on the size of baseplate selected (see Fig. 27.7(c)). If the effective area does not fit on the baseplate, modify the expression for effective area to allow for the limitations of the plate size and recalculate c, or

select a larger base plate. For the case shown in Figure 27.7(c), the modified expression for the effective area will be $4c^2 +$ (column perimeter) $\times c +$ column area $- 2 \times (B + 2c) \times (c - a)$.

(5) If c has been recalculated step 3 will need to be repeated.
(6) Calculate required plate thickness t_p using expression below (given in clause 4.13.2.2):

$$t_p = c\left(\frac{3w}{p_{yp}}\right)^{0.5}$$

where:

> w $= 0.6f_{cu}$
> p_{yp} is the plate design strength

The expression for the plate thickness can be derived from equating the moment produced by the uniform load w to the elastic moment capacity of the baseplate (both per unit length).

> Moment from uniform load on cantilever = Elastic moment capacity of plate
> $w\,c^2/2 = p_y Z$ (per unit length)
> $w\,c^2/2 = p_y t_p^2/6$

Rearranging gives

$$t_p = c(3\,w/p_y)^{0.5}$$

When the outstand of the effective area is equal either side of the flange (as in Fig. 27.7(a)) the cantilevers are balanced and there is no resultant moment induced in the flange. However, if the cantilevers do not balance either side of the flange, as would be the case in Fig. 27.7(c), then theoretically to satisfy equilibrium there is a resultant moment induced in the flange. However, it is important to remember that the method given in BS 5950-1 is a design model, and the remainder of the plate (not only the 'effective' area) does exist and does carry load. With this in mind, the moment induced in the column flange due to unbalanced cantilevers does not need to be explicitly considered in the design of either the column or the base-plate.

The method is also applied to tubular columns.

The dispersal dimension K taken radially on either side of the tube wall gives an annular contact area between the plate and the bedding material, as shown in Fig. 27.8(a). Then

$$A_e = (2K + t)(D - t)\pi$$

where D is the tube diameter and $(W/A_e) \not> $ the bearing strength of the bedding.

After solving for K, M and t are determined in the same way as for rolled sections (see the second worked example at the end of this chapter).

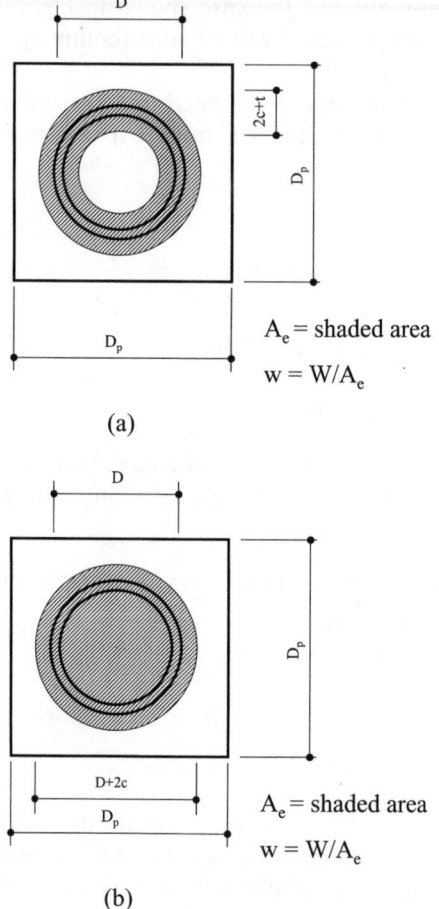

A_e = shaded area

$w = W/A_e$

(a)

A_e = shaded area

$w = W/A_e$

(b)

Fig. 27.8   Square base plate for CHS or solid column

If K is greater than $D/2$:

$$A_e = (D+2k)^2 \frac{\pi}{4}$$

as shown in Fig. 27.8(b).

Similarly after solving for K, M and t are obtained in the same way as for rolled sections (see the third worked example at the end of this chapter).

27.2.2.2 Gusseted bases

In a stiffened or gusseted base the moment in the gusset due to the bearing pressure under the effective area of the baseplate or due to the tensile forces in the

holding-down bolts should not exceed $p_{yg}Z_g$, where Z_g is the elastic modulus of the gusset and p_{yg} is the design strength of the gusset ($p_{gy} \not> 270\,\text{N/mm}^2$). When the effective area of the baseplate is less than its gross area, the connections of the gusset should be checked for the effects of a nominal distribution of bearing pressure on the gross area as well as for the distribution used in the design.

27.2.2.3 Beam bearing plates

Bearing plates at beam seatings are required to distribute the beam reaction to the masonry support at stress levels within the capacity of the masonry and to ensure that the web-crushing capacity of the beam is not exceeded. The distribution of bearing stresses under the plate is extremely complex although simplifying assumptions are usually made in appropriate cases.

The bending of the plate, shown in Fig. 27.9 in the direction transverse to the beam, will depend on the stiffness of the beam flange and the fixing of the flange to the plate. It is usual to assume that the position of maximum bending is the outside edge of the root of the web and that the plate carries the whole of the bending.

In the longitudinal direction, shown in Fig. 27.10, the deflection and rotation of the beam due to its loading will cause a concentration of bearing at the front edge and, depending upon the load from above the bearing, a possible lifting of the back edge of the plate. It is often assumed, therefore, that the distribution will be either trapezoidal or triangular; possibly the triangle may not reach the back of the bearing. If it is expected that the front edge concentration will be high the plate is set back from the front of the pier as shown in Fig. 27.11. This is to reduce the possibility of spalling at the front of the pier but also has the advantage of applying the beam reaction more centrally to the masonry.

A method of assessing the rotation of the bearing has been proposed by Lothers. From this a more accurate estimate of the stress distribution can be made. The method, however, can only be applied in cases of isolated masonry piers and is dependent on the homogeneity of the masonry. It may be justified in cases of very heavy beam reactions provided the workmanship in constructing the pier can be reasonably guaranteed.

27.3 Analysis

27.3.1 Bolt forces

The area required to transmit the compressive forces under the baseplate is calculated at the appropriate bearing strength of the concrete. The stress block may be assumed to be rectangular with a maximum stress of $0.6 f_{cu}$ where f_{cu} is the char-

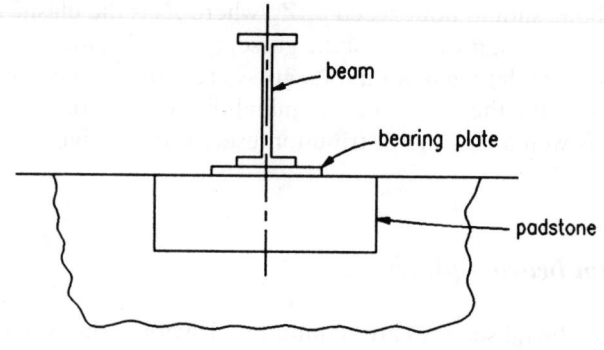

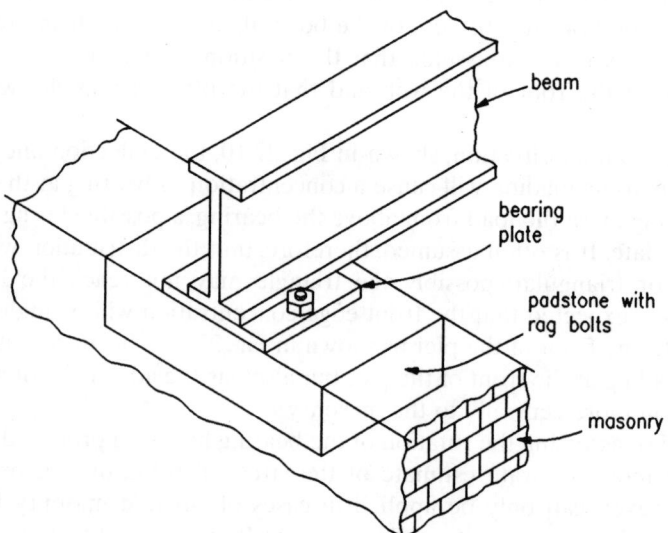

Fig. 27.9   Beam bearing

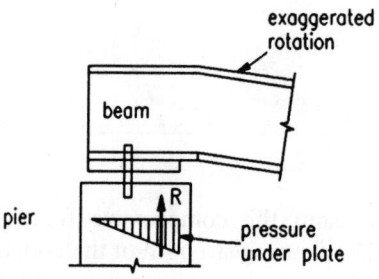

Fig. 27.10   Pressure under bearing plate (1)

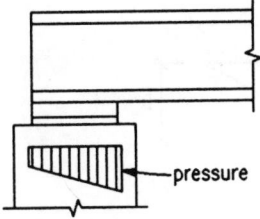

Fig. 27.11   Pressure under bearing plate (2)

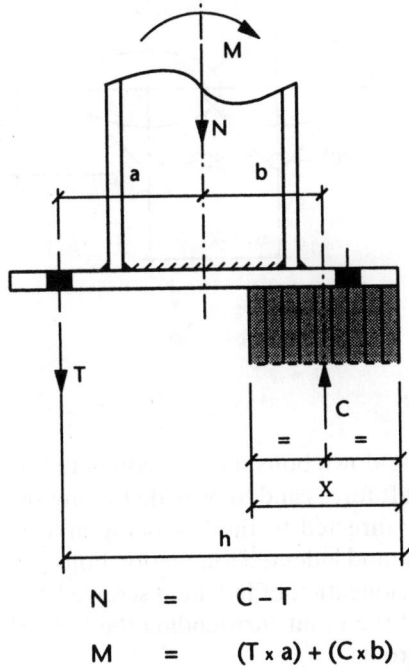

$$N \quad = \quad C - T$$

$$M \quad = \quad (T \times a) + (C \times b)$$

Fig. 27.12   Distribution of forces and equations of equilibrium

acteristic cube strength of the concrete base or the bedding material, whichever is less. The lever arm for the design of the bolts is then from the centroid of this stress block to the bolt position as shown in Fig. 27.12. The centroid of the stress block is often less than the edge distance from the compression edge of the plate to the holding-down bolt: it is therefore often assumed that the lever arm for the bolts is equal to the centres of the bolts. It is also very likely that the point of application of the compressive forces will be near to the holding-down bolts due to the extra stiffening that is often included in the vicinity of the bolts. This is illustrated in the typical design given as the fourth worked example at the end of this chapter.

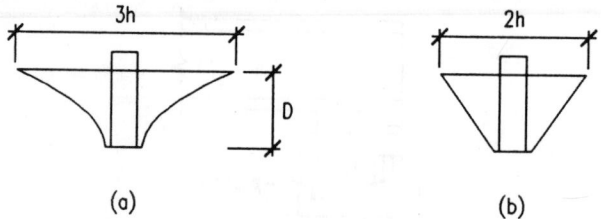

Fig. 27.13 Typical pull-out from concrete block

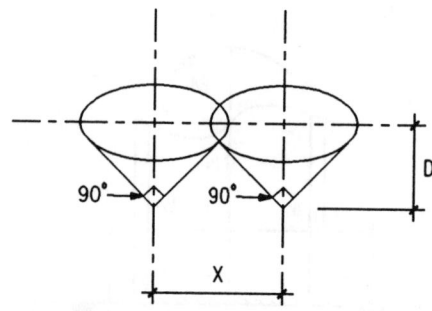

Fig. 27.14 Conical pull-out

27.3.2 Bolt anchorage

Anchorage of the holding-down bolts into the concrete foundation should be sufficient to cater for any uplift forces and to provide for any shears applied to the bolts. Attention is particularly directed to the last paragraph of clause 6.7 of BS 5950, which states that rag bolts and indented foundation bolts should not be used to resist uplift forces. The elastic elongation of indented screwed rods or bolts under tension causes the breakdown of the grout surrounding the bolt. This is even more critical in the case of resinous grouts.

The failure mode of bolts pulled from a concrete block is shown in Fig. 27.13(a); a reasonable approximation is shown in Fig. 27.13(b). The surface area of the conical pull-out (Fig. 27.14) is $4.44D^2$, where D is the depth of embedment. The factored tensile capacity of an M20 (4.6) bolt is $245 \times 195 \times 10^{-3} = 47.7\,\text{kN}$. For an M20 HD bolt 450 mm long with an embedment of say 350 mm

the conical surface is $4.44 \times 350^2 = 544 \times 10^3\,\text{mm}^2$

the surface stress is $\dfrac{47.7\,\text{kN}}{544 \times 10^3} = 0.09\,\text{N/mm}^2$

As holding-down bolts usually act in pairs the conical pull-outs often overlap depending on the depth of embedment. The BCSA have tabulated the surface areas including those which overlap (Table 27.2).

Table 27.2 Embedded lengths of holding-down bolts based on conical pull-out

Depth, D (mm)	Distance between centres, X (mm)										
	75	100	125	150	200	225	300	450	600	750	1000
	Effective conical surface area (allowing for overlap) (cm²)										
75	402.1	445.1	479.9	499.8	499.8	499.8	499.8	499.8	499.8	499.8	499.8
100	651.3	714.9	773.2	824.5	888.6	888.6	888.6	888.6	888.6	888.6	888.6
125	955.3	1038.1	1117	1191	1316	1362	1388	1388	1388	1388	1388
150	1315	1416	1514	1608	1780	1855	1999	1999	1999	1999	1999
200	2199	2337	2473	2605	2859	2979	3298	3554	3554	3554	3554
225	2724	2880	3034	3186	3479	3619	4006	4498	4498	4498	4498
300	4633	4843	5052	5258	5664	5862	6434	7420	7997	7997	7997
450	9950	10267	10583	10897	11521	11831	12743	14476	16022	17277	17994
600	17266	17689	18112	18533	19373	19790	21032	23448	25735	27837	30715
750	26581	27111	27640	28168	29221	29746	31312	34392	37371	40211	44507
1000	46550	47256	47962	48667	50076	50779	52882	57048	61141	65134	71486

Values to the right of the heavy zig-zag line are for two non-intersecting cones

27.4 Holding-down systems

27.4.1 Holding-down bolts

The most generally used holding-down bolts are of grade 4.6 although 8.8 grade are also available (see Table 27.3). They are usually supplied □□O×: square head, square shoulder, round shank, hexagon nut. Each bolt must be provided with an anchor washer (square hole to match the shoulder) or an appropriate anchor frame to embed in the concrete in circumstances of high uplift forces. In such cases the anchor frame may be composed of angles or channels. Typical frames are shown in Fig. 27.15. When long anchors are required a rod threaded at both ends may be used, and in exceptional circumstances when prestressing is required a high tensile rod

Table 27.3 Holding-down bolts; tension capacity per pair of bolts

Nominal diameter (mm)	Tensile stress area (mm²)[b]	Bolts grade 4.6 @ 192 N/mm² (kN)	Bolts grade 8.8 @ 448 N/mm² (kN)
M16	314	60.29	140.67
M20	490	94.08	219.52
M22[a]	606	116.35	271.49
M24	706	135.55	316.29
M27[a]	918	176.26	411.26
M30	1122	215.42	502.66

[a] Non-preferred size
[b] Tensile stress areas are taken from BS 4190 and BS 3692

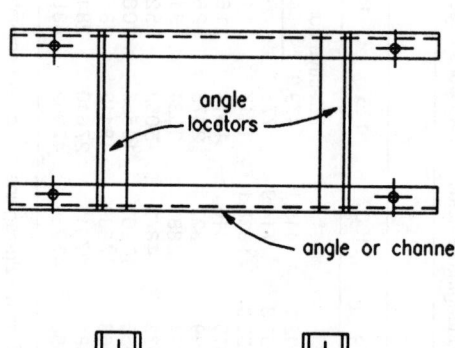

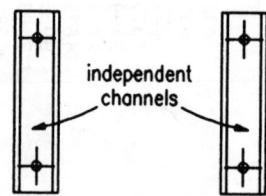

Fig. 27.15 Typical anchorages for holding-down bolts

(usually Macalloy bar) is adopted. In both these cases provision should be made to prevent rotation of the rod during tightening which may result in the embedded nut being slackened.

Corrosion of holding-down bolts to a significant extent has been reported in some instances. This usually occurs between the level of the concrete block and the underside of the steel baseplate, in aggressive chemical environments or at sites where moisture ingress to this level is recurrent. Fine concrete grout, well mixed, well placed and well compacted will provide the best protection against corrosion, but in cases where this is not adequate for the prevailing conditions, an allowance may be made in the sizing of the bolts or by specifying a higher grade bolt which provides a larger factor of safety against tensile failure in the event that some corrosion does occur.

27.4.2 Grouting

The casting-in of the holding-down bolts with adequate provision for adjustment requires that they are positioned in the concrete surrounded by a tube, conical or cylindrical, or a polystyrene former. After removal of the tube or former, which should be delayed until the last possible time before the erection of the columns, the available lateral movement of the bolts should be between three and four times the bolt diameter. In cases where open tubes are used they should be provided with a cap or cover to prevent the ingress of water, rubbish and mud. After erection, lining, levelling and plumbing of the frame the grout voids around the bolts should be cleaned out by compressed air immediately prior to grouting. The bolt grouting and baseplate filling should be done as two separate operations to allow shrinkage to take place. During the levelling and plumbing operations wedges and packings are driven into the grout space. Before final grouting these should be removed, otherwise, after shrinkage of the grout filling material, they will become hard spots, preventing the even distribution of the compressive forces to the concrete base.

27.4.3 Bedding

Bedding materials are required to perform a number of functions, one of which is the provision of the corrosion protection referred to earlier in section 27.4.1. In accordance with BS 5950, steel baseplates are designed for a compression under the plate of $0.6 f_{cu}$, where f_{cu} is the characteristic cube strength of the concrete base or the bedding material, whichever is less. The bedding material therefore transmits high vertical stresses including those resulting from the applied moment. The third function is to transmit the horizontal forces or shears resulting from wind or crane surge. It is clear therefore that the bedding material is a structural medium and should be specified, controlled and supervised accordingly.

For heavily-loaded columns or those carrying large moments resulting in high compressive forces the bedding should be fine concrete using a maximum aggregate of 10 mm size. The usual mix is $1:1\frac{1}{4}:2$ with a water–cement ratio of between 0.4 and 0.45 (this is not suitable for filling the bolt tubes as it is too stiff; a pure cement water mix has suitable flow properties and is usually used). It also has high shrinkage properties and should be allowed to set fully before continuing with the bedding. A cement mortar mix is often used for moderately-loaded columns. A suitable mix would be $1:2\frac{1}{2}$. Weaker filling than this should only be used for lightly-loaded columns where the erection packs are left in position and transfer all the load to the foundation.

In order to facilitate the compaction of the bedding material, holes are cut in the baseplate of the order of 50 mm diameter or more, near to the centre of the plate, in order to allow the escape of air pockets and to ensure that the bedding reaches to centre.

Further reading for Chapter 27

British Constructional Steelwork Association/The Concrete Society/Constructional Steel Research and Development Council (1980) *Holding-Down Systems for Steel Stanchions.*

British Standards Institution (1997) *Structural use of concrete.* Part 1: *Code of practice for design and construction.* BS 8110, BSI, London.

British Standards Institution (2000) *Structural use of steelwork in building.* Part 1: *Code of practice for design – Rolled and welded sections.* BS 5950, BSI, London.

Capper P.L. & Cassie W.F. (1976) *The Mechanics of Engineering Soils*, 6th edn. E. & F.N. Spon Ltd.

Capper P.L., Cassie W.F. & Geddes J.W. (1980) *Problems in Engineering Soils.*, 3rd edn. E. & F.N. Spon, London.

Lothers J.E. (1972) *Design in Structural Steel*, 3rd edn. Prentice Hall, Engleword Cliffs, NJ.

Pounder C.C. (1940) *The Design of Flat Plates.* Association of Engineering and Shipbuilding Draughtsmen.

Skempton A.W. & McDonald D.H. (1956) The allowable settlement of buildings. *Proc. Instn Civ. Engrs*, **5**, Part 3, 727–68, 5 Dec.

Skempton A.W. & Bjerrum L. (1957) A contribution to the settlement analysis of foundations on clay. *Géotechnique*, **7**, No. 4, 168–78.

Terzaghi K., Peck R.B. & Nesri G. (1996) *Soil Mechanics in Engineering Practice*, 3rd edn. Wiley, New York.

The Steel Construction Institute/British Constructional Steelwork Association (2002) *Joints in Steel Construction.* Simple Connections. SCI/BCSA.

Tomlinson M.J. (2001) *Foundation Design and Construction*, 7th edn. Prentice Hall, Harlow.

A series of worked examples follows which are relevant to Chapter 27.

The **Steel Construction Institute** Silwood Park, Ascot, Berks SL5 7QN	Subject ***FOUNDATION EXAMPLE 1***	Chapter ref. **27**
	Design code ***BS 5950: Part 1***	Made by ***HB***	Sheet no. ***1***
		Checked by ***GWO***	

Problem

Design a simple base plate for a 254 × 254 × 73 UC to carry a factored axial load of 1000 kN

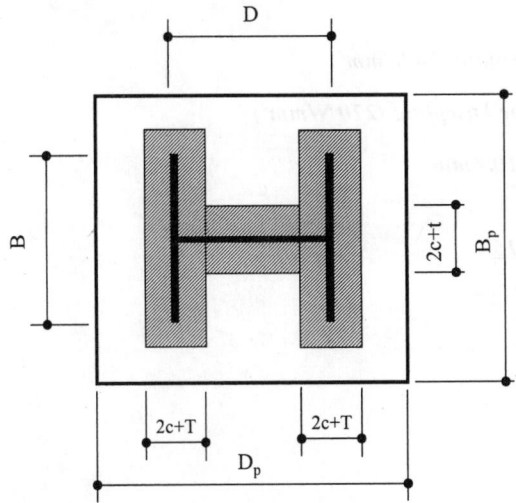

Design by the effective area method (4.13.2)

Bearing strength of concrete $= 0.6 f_{cu} -$ *take* f_{cu} *as* $40 N/mm^2$

Area required (in mm^2) $= \dfrac{1000 \times 10^3}{0.6 \times 40} = 41667\, mm^2$

Bearing area $=$ *hatched area*

$4c^2 + $ *(column perimeter)* $\times c +$ *column area*

$\therefore 4c^2 + [254.6 \times 4 + 2 \times (254.1 - 28.4)] \times c + 9310 = 41667$

$\therefore 4c^2 + 1470c + 9310 = 0$

The **Steel Construction Institute** Silwood Park, Ascot, Berks SL5 7QN	Subject ***FOUNDATION EXAMPLE 1***	Chapter ref. **27**
	Design code **BS 5950: Part 1** / Made by ***HB*** / Checked by ***GWO***	Sheet no. **2**

$$c = \frac{-1470 \pm \sqrt{1470^2 + 517712}}{8} = 20.8\,mm$$

$$t_p = c\left[\frac{3w}{p_{yp}}\right]^{0.5}$$

4.13.2.2

where w = pressure under baseplate (24 N/mm²)

and p_{yp} = design strength of the baseplate (270 N/mm²)

$\therefore t_p = 20.8\ (3 \times 24/270)^{0.5} = 10.7\,mm$

<u>Use a baseplate 300 × 300 × 15</u>

The **Steel Construction Institute** Silwood Park, Ascot, Berks SL5 7QN	Subject *FOUNDATION* *EXAMPLE 2*	Chapter ref. **27**
	Design code **BS 5950: Part 1**	Made by **HB** Checked by **GWO**	Sheet no. **1**

Problem

Design a simple base plate for a 219 × 6.3 CHS to carry a factored axial load of 1010 kN

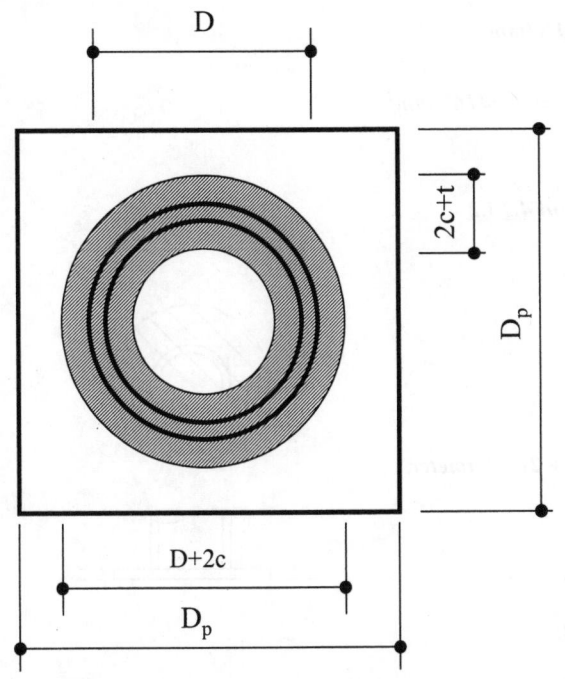

219 × 6.3 CHS

Factored axial load = 1010 kN

Assumed bedding material
 $f_{cu} = 40\,N/mm^2$

Design by the effective area method 4.13.2

Area required $\quad = \dfrac{1010 \times 10^3}{0.6 \times 40} \quad = \quad 42083\,mm^2$

Area of shaded annulus $= (2c + t)\,(D - t)\,\pi = 42083$

$(2c + 6.3)\,(219 - 6.3) \quad = \quad 13395,\ \text{hence } c \quad = \quad 28.3\,mm$

$t_p \quad = \quad 28.3 \times \left[\dfrac{3 \times 24}{270}\right]^{0.5} \quad = \quad 14.62\,mm \quad \therefore \underline{\textit{Use } 280 \times 280 \times 15 \textit{ plate}}$

The Steel Construction Institute Silwood Park, Ascot, Berks SL5 7QN	Subject *FOUNDATION EXAMPLE 3*	Chapter ref. 27
	Design code **BS 5950: Part 1**	Made by *HB*	Sheet no. *1*
		Checked by *GWO*	

Problem

Design a simple base plate for a 273 × 25 CHS to carry a factored axial load of 6340 kN

Strength of bedding material = 40 N/mm²

$$\text{Area required} = \frac{6340 \times 10^3}{0.6 \times 40} = 264167 \, mm^2$$

Design by the effective area method 4.13.2

$(2c + 25)(273 - 25)\,\pi = 264167$

from which $c = 157$ mm

When $c > \dfrac{D - 2t}{2}$

the bearing area is a circle of $(D + 2c)$ diameter.

Then:

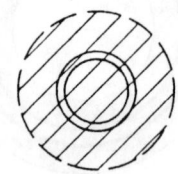

$$(D + 2c)^2 \frac{\pi}{4} = 264167$$

$$(273 + 2c)^2 = 336348$$

$$4c^2 + 1092c + 74529 = 336348$$

$$c = \frac{-1092 \pm [1092^2 - 16 \times (-261819)]^{0.5}}{8}$$

$$= 153 \, mm$$

$$t_p = 153\left[\frac{3 \times 24}{270}\right]^{0.5} \qquad\qquad\qquad 4.13.2.2$$

$$= 79 \, mm \qquad \therefore \underline{\textit{Use } 600 \times 600 \times 80 \textit{ plate}}$$

The Steel Construction Institute	Subject	Chapter ref.
Silwood Park, Ascot, Berks SL5 7QN	**FOUNDATION EXAMPLE 4**	**27**

Design code	Made by **HB**	Sheet no. **1**
BS 5950: Part 1	Checked by **GWO**	

Problem

Design a built-up base for the valley stanchion of a double bay crane shed that is shown below. The stanchion comprises twin 406 × 178 UB.

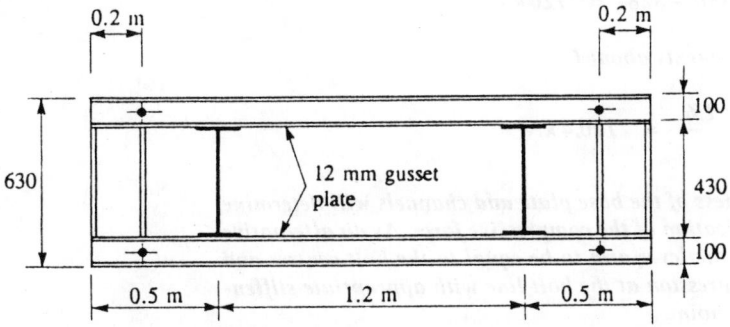

Taking moments about the tensile bolt, with n, the trial neutral axis as 0.4 m depth, and taking

$$f_c = 0.6 f_{cu} = 12 \, N/mm^2 \quad (assuming \ f_{cu} = 20 \, N/mm^2)$$

4.13.1

Loading

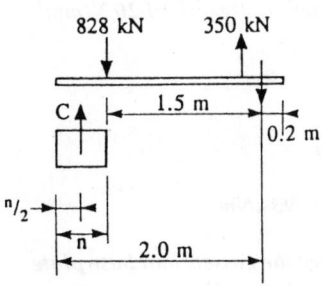

$$828 \times 1.5 - 350 \times 0.3 \ = \ C(2.0 - 0.2)$$

$$C \ = \ \frac{828 \times 1.5 - 350 \times 0.3}{1.8} \ = \ 632 \, kN$$

The Steel Construction Institute Silwood Park, Ascot, Berks SL5 7QN	Subject **FOUNDATION EXAMPLE 4**	Chapter ref. **27**
	Design code **BS 5950: Part 1**	Made by **HB** Checked by **GWO**	Sheet no. **2**

$$f_c = \frac{632 \times 10^3}{0.4 \times 0.6 \times 10^6} = 2.63 \, N/mm^2$$

Taking n as 0.2 m, C becomes 598 kN and f_c is 4.98 N/mm²

T is then: $598 + 350 - 828 = 120 \, kN$

To check, take moments about C

$$\frac{828 \times 0.4 - 350 \times 1.6}{1.9} = 120.4 \, kN$$

The relative stiffness of the base plate and channels will determine the point of application of the compressive force. As an alternative therefore assume the lever arm to be equal to the bolt centres and the centre of compression at the bolt line with appropriate stiffening added at this point.

$$C = \frac{828 \times 1.5 - 350 \times 0.3}{1.7} = 669 \, kN$$

$$T = 669 + 350 - 828 = 191 \, kN$$

$$n = \frac{669 \times 10^3}{630 \times 12 \, N/mm^2} = 88 \, mm$$

This is the minimum value of n for concrete strength of 20 N/mm².

<u>Design of channels & gusset</u>

$M = 669 \times 300/10^3 = 200.7 \, kNm$

Use 2/229 × 89 × 32.76 RSCs, $M_{cx} = 95 \, kNm$

These are satisfactory by inspection since the gussets and base plate acting compositely would also make a contribution.

The internal stiffener and base plate would similarly be designed as a composite member taking the maximum outstand given in Table 11 of BS 5950.

The Steel Construction Institute Silwood Park, Ascot, Berks SL5 7QN	Subject *FOUNDATION EXAMPLE 4*	Chapter ref. **27**
Design code **BS 5950: Part 1**	Made by **HB** Checked by **GWO**	Sheet no. **3**

The base plate panel between the stiffeners should be checked using the Pounder expressions given in Chapter 30 as follows – the panel is shown below.

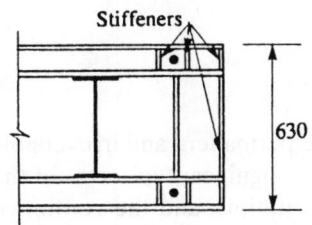

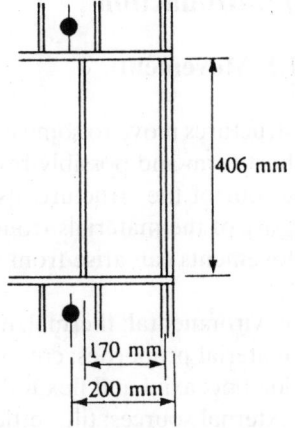

Stiffeners

630

406 mm

170 mm

200 mm

Base plate

$$\frac{L}{B} = \frac{406}{170} = 2.38$$

$$K = \frac{2.38^4}{2.38^4 + 1} = 0.97$$

K_{me}, *the Pounder expression for moment, in the centre of the long edge, when all four edges are encastre, is given below:*

$$K_{me} = K\left[1 + \frac{11}{35}(1-k) + \frac{79}{141}(1-K)^2\right]$$
$$= 0.9796$$

w_{me}, *the ultimate load intensity is given by:*

$$W_{me} = \frac{275 \times 12 \times 1.2 \times t^2}{6 \times 170^2\,[K_{me}]}$$

$$f_c = \frac{669 \times 10^3}{600 \times 200} = 5.58\,N/mm^2,\ cf.f_c = 4.98\,N/mm^2\ for\ n = 0.2\,m$$

The plate thickness of 16 mm is therefore satisfactory.

Chapter 28
Bearings and joints

by STEPHEN MATTHEWS

28.1 Introduction

28.1.1 Movement

All structures move to some extent. Movements may be permanent and irreversible or short-term and possibly reversible. The effects can be significant in terms of the behaviour of the structure, its performance during its lifetime, and the continued integrity of the materials from which it is built.

Movements can arise from a variety of sources:

(1) environmental: thermal, humidity, wind-induced.
(2) material properties: creep, shrinkage.
(3) loading: axial and flexural strains, impact, braking, traction, centrifugal forces.
(4) external sources: tilt, settlement, subsidence, seismic loads.
(5) use of the building: heating, cold storage.
(6) others: requirements for moving or lifting bridges, allowances for jacking procedures, during or after construction.

In general it is necessary to consider the behaviour of the structure at each point in terms of its possible movement in each of three principal directions, together with any associated rotations. The movements of a structure are not in themselves detrimental; the problems arise where movements are restrained, either by the way in which the structure is connected to the ground, or by surrounding elements such as claddings, adjacent buildings, or other fixed or more rigid items. If provision is not made for such movements and associated forces it is possible that they will lead to, or contribute towards, deterioration in one or more elements. Deterioration in this context can range from, for example, cracking or disturbance of the finishes on a building to buckling or failure of primary structural elements due to large forces developed through inadvertent restraint.

Note that for bridges with total lengths of up to 60m, it is possible to dispense with bearings and expansion joints through use of abutments and piers which are designed to be integral with the bridge deck. Further guidance on this topic can be found in Reference 1.

28.1.2 Design philosophies

In catering for movement of a structure, one of three methods can be adopted:

(1) Design the structure to withstand all the forces developed by restraint of movement. This is possible with smaller structures (small-span bridges) or structures which are comparatively flexible (portal frames, in the plane of the frame). The method will avoid joints but may require the use of additional material in construction.

(2) Subdivide the structure into smaller structurally stable units, each of which then becomes essentially a structure in its own right, able to move independently of the surrounding units. This principle is ideal for controlling those factors such as thermal movement which are related to the size of the overall structure. In many cases, the need for bearings as discrete elements can be eliminated. The disadvantage lies in the need to provide joints between the various units of the structure capable of accommodating all the anticipated relative movements between the units, while at the same time fulfilling all the other requirements, i.e. visual, practical, etc. It is, however, generally possible to achieve a balance by subdividing the structure so that the movements at the joints between units are kept relatively small, permitting the joints to be simple and economical (possibly at the expense of larger numbers of joints).

(3) Subdivide the structure into fewer but larger sections, and make provision for a smaller number of joints, each with larger movement capacity, and thus possibly more complex than those that would be used at (2). Examples are to be found in bridges where use of the least number of road deck joints is preferable both in terms of riding quality, and also in the minimization of long-term maintenance requirements.

The need to restrict strains on elements and thus to protect finishes will lead to the adoption of the second of the above methods for design of building structures. Bridges, for reasons cited above, are more frequently designed adopting the third method.

28.2 Bearings

28.2.1 Criteria for design and selection

28.2.1.1 Form of the unit

Choice of form depends on several criteria:

(1) *Physical size limitations.* The space available in the structure for the bearing. As bearings are subject to more wear than other parts of the structure they

may have a shorter life and consequently this space should include allowance for access, inspection, maintenance and possible replacement.

(2) *Bearing pressure.* The allowable bearing pressure on the materials above and below the bearing will dictate the minimum size of the top and bottom faces of the bearing unit.

(3) *Loading.* The magnitude of the design load to be withstood by the bearing in each of the three principal directions will govern the form and type of the bearing. For each direction the maximum and minimum load should be considered at ultimate limit state, serviceability limit state or working load depending on the requirements of the design. In each case co-existent load and movement effects should be considered, together with a check for the existence of any load combinations which would act so as to separate the components of the bearing (e.g. uplift). For bearings carrying both horizontal and vertical loads it is common that the design of the bearing requires a *minimum* vertical load to be present to ensure satisfactory performance under horizontal loads.

(4) *Rotations.* The magnitude of the maximum anticipated rotations in the three principal directions should be considered. For certain types of bearing (e.g. elastomeric bearings) there exists an interaction between maximum load-carrying capacity and rotation/translation capacity, so that it may be necessary to consider co-existent effects under loading (3) and movement (5).

(5) *Movements.* Provision for maximum calculated movements can affect the size of the moving parts of the bearing and thus the overall size of the unit. As with rotations, the design of certain types of bearing is sensitive to the interaction of movement and loading requirements.

(6) *Stiffness (vertical, rotational or translational).* Certain structures may be sensitive to the deformation which occurs within the bearing during its support of the loads. The various types of bearing have different stiffness characteristics so that an appropriate form can be selected.

(7) *Dynamic considerations.* Any particularly onerous dynamic loadings on the structure will have to be considered. Certain types of bearings (e.g. elastomeric bearings) have damping characteristics which may be desirable in particular instances, such as vibration of footbridges or machine foundations.

(8) *Connections to structure.* The form of connection of the bearing to the structure requires careful consideration of the materials involved and the need for installation, maintenance and replacement of the bearing. In addition, bearings are frequently at a position in the structure where different forms of construction meet, perhaps constructed by different contractors. In this case, it is necessary to ensure that surrounding construction is properly detailed so that design requirements for load transfer are achieved.

(9) *Use of proprietary bearings.* Many types of bearings are commercially available. These range from items which are available 'off the shelf' to more specialized units which may be designed and proven, but which are only produced to order. It is often appropriate for bearings to be individually designed to meet a particular need in situations where proprietary types may not be suit-

able. In these instances the engineer has the option of designing the units using available literature (see references to Chapter 28) and perhaps incorporating standard bearings from a manufacturer as components of a completed assembly or alternatively engaging a recognized manufacturer to design and produce the item as a special bearing. For straightforward applications such as may be required on a short single-span bridge, it may be worthwhile investigating the relative costs of a simple fabricated bearing compared with the equivalent proprietary unit. Bearings (particularly 'special' bearings) can prove to be a large item of expenditure in a structure and an estimate of the costs involved should be made early in the design stage.

(10) *Summary of design requirements.* Before selecting a particular bearing it is suggested that a summary of all relevant parameters is prepared. This can then be used if necessary for submission to the bearing manufacturers for examination and recommendations as to particular bearing types. A typical format for such a sheet is given in Table 9 of BS 5400: Section 9.1.[2]

28.2.1.2 Materials

Generally materials fall into three groups:

(1) those able to withstand high localized contact pressures e.g. steel.
(2) those able to withstand lower contact pressures but having a low coefficient of friction; these slide easily in a direction perpendicular to the direction of the pressure and thus accommodate translational movement, e.g. polytetrafluoroethylene (PTFE).
(3) those able to withstand contact pressure and also to accommodate translational or rotational movements by deformation of the material (e.g. elastomers). Certain of these materials may be confined within a steel cylinder in order to increase their compressive resistance.

(a) Mild or high-yield steel

The coefficient of friction of steel on steel is of the order of 0.3 to 0.5, unless continuously lubricated; in order to provide for movement alternative arrangements are usually necessary. Traditionally this has been through the use of single or multiple rollers or knuckles. Rollers will permit translation in one direction and, if a single roller is used, rotation about an axis perpendicular to that direction. Knuckles permit rotation about one axis only. Rotation in two directions may be achieved using spherical-shaped bearing surfaces.

The allowable pressures between surfaces for steel on steel contact depend upon the radii of the two surfaces and the hardness and ultimate tensile strength of the material used. BS 5400: Section 9.1[2] gives expressions for design load effects in such

cases. As load-carrying requirements increase, the use of steels with greater hardness is dictated. This can be achieved by use of high-grade alloy steels of various compositions. For design purposes, Table 2 of BS 5400: Section 9.1[2] gives indicative values of coefficients of friction of between 0.01 and 0.05 for steel roller bearings.

(b) Stainless steel

Stainless steel is frequently used in strip or plate form to provide a smooth path for sliding surfaces. It is important to utilize a material for the sliding surface which will not deteriorate and adversely affect the coefficient of friction assumed for design of the structure. A typical arrangement is a polished austenitic stainless steel surface sliding against dimpled PTFE.

(c) Polytetrafluoroethylene (PTFE)

PTFE has good chemical resistance and very low coefficients of static and dynamic friction. Unfortunately, pure PTFE has a low compressive strength, high thermal expansion and very low thermal conductivity. As a consequence it is frequently used in conjunction with 'filler' materials which improve these detrimental effects without significantly affecting the coefficient of friction.

The coefficient of friction varies with the bearing stress acting upon it. BS 5400: Section 9.1[2] gives the relationship shown in Fig. 28.1 for continuously lubricated pure PTFE sliding on stainless steel.

Lubrication of the pure PTFE is commonly achieved by means of silicone grease confined in dimples which are rolled on to the surface of the material. References 2 and 3 give further guidance on the restrictions on shape, thickness and containment on the PTFE and stainless steel components.

In preliminary design and assessment of forces on structures using PTFE sliding

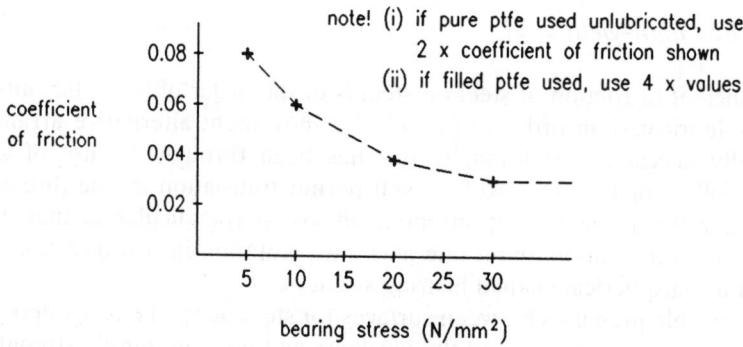

Fig. 28.1   Coefficient of friction for continuously lubricated pure PTFE

bearings, a figure of 0.06 is usually assumed for the coefficient of friction, and the value is checked later when the bearing selection is complete.

(d) Phosphor bronze

For particular applications, such as bearing guides, phosphor bronze may be used, BS 5400: Section 9.1 suggests a coefficient of friction of 0.35 for phosphor bronze sliding on steel or cast iron.

(e) Elastomers

An elastomer is either a natural rubber or a man-made material which has rubber-like characteristics. Elastomers are used frequently in bearings; they either constitute the bulk of the bearing itself or act as a medium for permitting rotation to take place (see sections 28.2.2.2 and 28.2.2.3(7)).

Elastomers are principally characterized by their hardness, which is measured in several ways, the most common of which is the international rubber hardness (IRHD). This ranges on a scale from very soft at 0 to very hard at 100. Those elastomers used in bearings which are to comply with BS 5400: Part 9 have hardnesses in the range 45 IRHD to 75 IRHD.

The tensile capacity of most elastomers is considerable. As an illustration BS 5400: Part 9 specifies a minimum tensile elongation at failure of between 300% and 450% depending on IRHD.

When considering the behaviour of a block of elastomer under vertical compression it is assumed that the material is securely bonded to top and bottom loading plates. In this case (which is representative of most bearing situations) the vertical behaviour is related to the material's ability to bulge on the four non-loaded faces and is expressed in terms of the *shape factor* for the block, which is the ratio of the loaded area to the force free surface area (see Fig. 28.2).

$$S = \frac{LB}{2t(L+B)} \quad \text{for a rectangle}$$

$$S = \frac{D}{4t} \quad \text{for a circle}$$

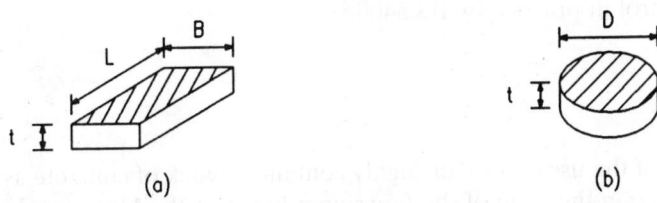

(a)   (b)

Fig. 28.2 Elastomeric bearing dimensions for (a) a rectangular block, (b) a circular block

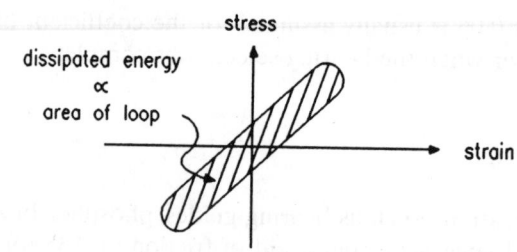

Fig. 28.3   Energy loss in elastomer

The vertical stiffness of the block is significantly reduced if the loaded faces can slip laterally.

The shear stiffness of a block of elastomer bonded to top and bottom plates is more or less linear and independent of shape factor. A detailed treatment of these effects is given in References 4 and 2.

An important property of elastomers is related to the fact that the strain in the material tends to lag behind the stress which causes it. As a consequence, some of the energy input during deformation is dissipated within the bearing as heat. A typical plot of stress versus strain for an elastomer is shown in Fig. 28.3. The energy lost as heat in one loading cycle is represented by the area of the loop.

This effect, known as *hysteresis*, has two implications:

(1) it can result in a build-up of heat in the bearing under dynamic loading conditions,
(2) if appropriately sized it can be used to act as a form of damping device to the structure.

The sensitivity of the elastomer to dynamic loading depends upon both the frequency of the applied stress and the temperature, as elastomers exhibit hardening at low temperatures.

Elastomers are prone to creep and are sensitive to attack by atmospheric oxygen and ozone, petrol and radiation from nuclear sources. They are not suitable for operation in temperatures above 120°C.

A more detailed discussion of these aspects is given by Long,[5] and limits are given for their control in practice by BS 5400.[2]

(f) Concrete

The concept of the use of a small, highly contained block of concrete as a hinge has been employed in the form of the Freysinnet hinge or the Mesnager hinge. Details of these are given in Reference 6.

28.2.2 Types of bearing

28.2.2.1 General

Structural bearings can be broadly divided into two types: elastomeric and mechanical. Elastomeric bearings comprise blocks of elastomeric material reinforced as necessary with other materials. They can be made to accommodate movement by shearing of the elastomer block. Mechanical bearings are generally formed from metal and employ sliding surfaces of PTFE to achieve any necessary movement capabilities. It should be noted that the above divisions are not exclusive as, for example, some bearings are commercially available which employ elastomer blocks with sliding surfaces which cater for larger movements than could be accommodated by shearing of the block, but which do not merit the larger expense of a mechanical bearing.

In the following sections the principal types of bearing are briefly described.

28.2.2.2 Elastomeric bearings

Elastomeric bearings rely for their operation on the interaction between vertical load, rotation and translation. As a consequence, design of most elastomeric bearings must be carefully checked. Large proprietary ranges are available, and although load tables of the various capacities are published by manufacturers, it is prudent to ask the supplier to confirm that the selected bearing is suitable for the load/movement conditions under which it will be used. The basis of design of these bearings is related to controlling strains and stresses in the elastomer and any reinforcing material and ensuring that the bearing does not deform excessively, become unstable, lift off, or slip under the anticipated design effects. Further guidance on this subject is given in BS 5400: Part 9.1[2] and by Long.[5] The latter reference also gives a detailed discussion of the properties of elastomers.

Bearing types are:

(1) *Rubber pad or strip bearings.* As the name implies, these bearings consist simply of a block or strip of elastomer. They have the advantage of being inexpensive and simple although their load-carrying and movement capability is limited.
(2) *Fabric-reinforced bearings.* In order to increase the capabilities of the simple pad bearing, use is made of fabric (e.g. compressed cotton duck) to reinforce the elastomer. Movement in these bearings is usually provided by use of a PTFE surface bonded to the top of the block and sliding against a stainless steel plate attached to the underside of the superstructure. In this manner, the elastomer is used to provide rotational capability only, rather than rotation and movement as in the case of other elastomeric bearings.
(3) *Elastomeric-laminated bearings.* This type of bearing consists of a block of elastomeric material reinforced with steel plates to which the elastomer is also

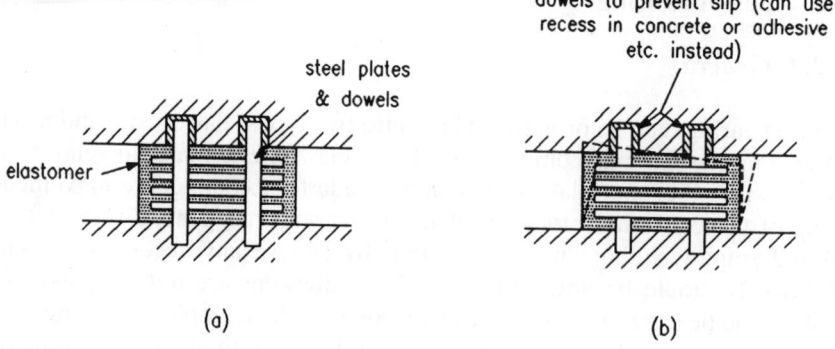

Fig. 28.4   Two types of elastomeric bearing: (a) fixed, (b) free

bonded. The characteristics of the bearings can be varied considerably by alteration of the size, shape and disposition of the layers as well as the usual parameters of bearing plan area and thickness. Generally, these bearings are either 'fixed' for translation by means of a steel dowel passing through the bearing layers (Fig. 28.4(a)) or 'free' bearings which permit translation and rotation by deformation of the bearing (see Fig. 28.4(b)). This type of bearing is capable of carrying quite substantial loadings and movements and has the benefit of being cheaper than mechanical bearings.

28.2.2.3 Mechanical bearings

(1) *Roller* (Fig. 28.5(a)). The earlier and more traditional forms of bearing comprised single or multiple steel rollers sandwiched between upper and lower steel plates. Single rollers will allow for longitudinal movement and rotation about the axis of the roller, while at the same time carrying comparatively high vertical loads, hut will not permit transverse rotation or movement. Bearings of very large capacity have been produced by use of special alloy steels to form the contact surfaces. Note that bearings which utilize multiple rollers will not allow rotation about an axis parallel to the axis of the rollers. Rollers are sometimes used enclosed in an oil bath or grease box to exclude deleterious matter. Other forms of bearing have, to a large extent, supplanted the use of rollers for the most common applications.

(2) *Rocker* (Fig. 28.5(b)). Rocker bearings will not permit translational movement. The bearings may be cylindrical or spherical on one surface with the other surface flat or curved. In the cylindrical form there is no provision for transverse rotation, which may have consequences for design of the structures above and below the unit. Rocker bearings usually incorporate a pin or shear key between the two surfaces to maintain relative position.

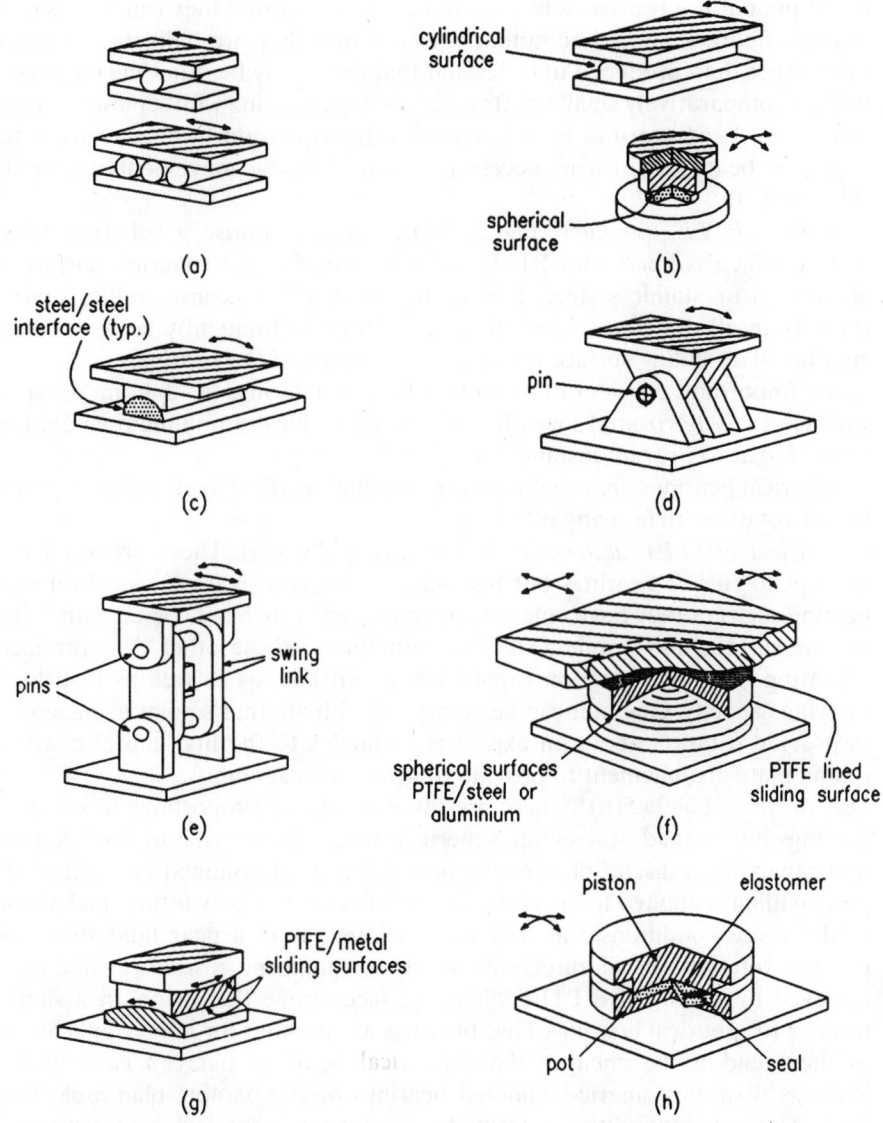

Fig. 28.5 Mechanical bearings: (a) single/multiple roller, (b) cylindrical/spherical rocker, (c) cylindrical knuckle, (d) knuckle leaf, (e) swing link, (f) spherical – with sliding top plate, (g) cylindrical PTFE bearings combined to form 'anticlastic' bearing, (h) 'pot' bearing which can have sliding top plate similar to (f)

(3) *Knuckle bearings* (Fig. 28.5(c)). These are similar to rocker bearings.

(4) *Leaf bearings* (Fig. 28.5(d)). These are formed of leaves of steel with a common pin. They will carry large vertical loads and permit large rotations about the axis

of the pin but not transversely. They have the benefit that they can be designed to resist uplift. It should be noted, however, that they are unlikely to be anything other than produced to order and that there may be other means of controlling comparatively small uplifts (e.g. 'pot' type bearing with separate vertical restraints). Leaf bearings have been used in suspension bridges to form the swing link bearings which are necessary to cater for large movements and uplifts (Fig. 28.5(e)).

(5) *Spherical (PTFE, circular)* (Fig. 28.5(f)). These comprise a spherical lower surface which is lined with PTFE and a matched upper spherical surface of aluminium or stainless steel. This arrangement allows considerable rotation capacity in all directions. Horizontal translation is frequently achieved using another (flat) sliding surface above the upper part of the bearing.

An important consideration with spherical bearings is that in order to withstand any horizontal loads it is necessary to have a minimum co-existent vertical load to prevent instability.

Spherical bearings are capable of carrying high vertical loads and also permit higher rotations than many other types.

(6) *Cylindrical (PTFE) 'anticlastic' bearings* (Fig. 28.5(g)). These are similar in concept to rocker bearings but instead of using (for example) steel on steel bearing surfaces they have enlarged bearing areas which are coated with PTFE on one surface and stainless steel or aluminium on the other. This produces a bearing with high rotation capabilities about an axis as well as high load-carrying capacity. One unit can be combined with another similar arrangement to provide rotation about an axis at right angles to the first and also with a sliding plate arrangement to provide translational capability.

(7) *Disc or 'pot'* (Fig 28.5(h)). These are often of similar proportions to spherical hearings but instead of a sliding spherical surface being used to provide rotation capability, a disc of elastomeric material is used, confined in a cylindrical pot. Loading is applied to the surface of the disc via a closely fitting steel piston. Under these conditions, the confined elastomer is in a near fluid state, and permits rotation in all directions without significant resistance. Sliding is achieved by means of a PTFE/sliding surface above the piston, in a similar manner to spherical bearings. Disc bearings are popular for many applications, as they tend to be cheaper than spherical bearings but can carry higher loadings than elastomeric-laminated bearings of comparable plan area. They have rotation capabilities intermediate between spherical and laminated bearings.

(8) *Fabricated.* Fabricated bearings have become less popular largely through the availability of a wide range of proprietary units. They are used for footbridges and temporary works applications. There is, however, no reason why properly designed fabricated bearings should not be used to support a structure, particularly, say, for a fixed bearing where there is no requirement for sliding surfaces. Guidance on design of bearings is given in References 2–9.

(9) *Special.* Special bearings will always be required for particular locations. Perhaps the most common demands are for:

(a) bearings which will resist horizontal loads only in order to restrain the structure in the horizontal plane, but without providing any vertical support (see section 28.2.4.2(3))

(b) bearings which will withstand uplift under certain loading conditions.

Uplift bearings can be special versions of normal proprietary bearing types, or can use a proprietary bearing set in a subframe which controls the tendency to uplift within prescribed limits adopted in consultation with the bearing manufacturer.

28.2.3 Use of bearings

28.2.3.1 General

The parameters which dictate the form of the bearing as a unit are discussed in section 28.2.1. It is also necessary to consider the action of the bearing in the broader concept of the behaviour of the two elements of structure which the unit connects.

28.2.3.2 Fixings

Various forms of fixings are utilized to connect bearings to the structure. These include:

(1) shear studs, usually in conjunction with a subsidiary plate which is tapped to receive the bearing fixing bolts,

(2) square or cylindrical dowels, tapped to receive the bearing fixing bolts,

(3) direct bolting of the bearing to the structure.

In all forms it is desirable to allow for tolerances in the processes of installation and possible need for replacement of the bearing. The system shown in Fig. 28.6 allows for support of the bearing during fine adjustment, but requires large jacking

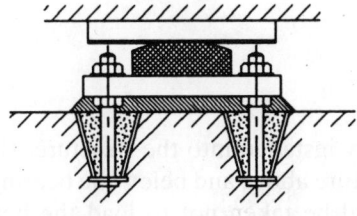

Fig. 28.6   Bearing fixing

capability (possibly more than a continuous structure could accommodate) to remove it.

In bearings subjected to dynamic loadings such as machine foundations, it is necessary to ensure that the fixings are vibration-proof.

28.2.3.3 Effect on the structure

The elements of the structure above and below the bearing are affected by the type of bearing, which can be classified as:

(1) *fixed* – not permitting movement in any horizontal direction,
(2) *guided* – movement, constrained by guides of some form, to be in one horizontal direction only,
(3) *free* – movement permitted in all horizontal directions,
(4) *elastomeric*, which may be laminated or not. These bearings can be 'fixed' by means of steel dowels passing through them but are more often used 'free' in all directions and their capability to generate forces when shearing takes place is utilized to withstand horizontal loadings. If the whole structure is supported on such bearings it effectively 'floats', with all horizontal loads shared by all bearings. (See also section 28.2.4.2.)

If the bearings are fixed or guided, the neighbouring structure must be designed for the forces arising from the restraints. Even when the bearing is free in a particular direction and movement is permitted, some forces are developed – either from friction effects at the movement interfaces of a sliding mechanical bearing, or from shearing deformation in the case of an elastomeric bearing (see Fig. 28.7(a)).

In addition to forces developed laterally, the effects of the eccentricities produced by the movement must be allowed for, and also the rotation capability of the bearing in the transverse direction (see Figs 28.7(b), (c) and (d)). It is possible to control the extent of the additional eccentricity effects on a steel superstructure by use of a sliding bearing inverted which transfers the eccentricity to the substructure, where it may be more easily accommodated. In this case however care should be taken to protect the sliding surfaces against falling dust, debris, etc. by use of a flexible skirt enclosure.

28.2.3.4 Installation

Bearings must be correctly installed into the structure. The procedure will depend upon the form of the structure above and below the bearing, and the type of bearing, but in general care should be taken not to load the bearing significantly before bedding materials between the bearing and the structure have fully cured, or to load

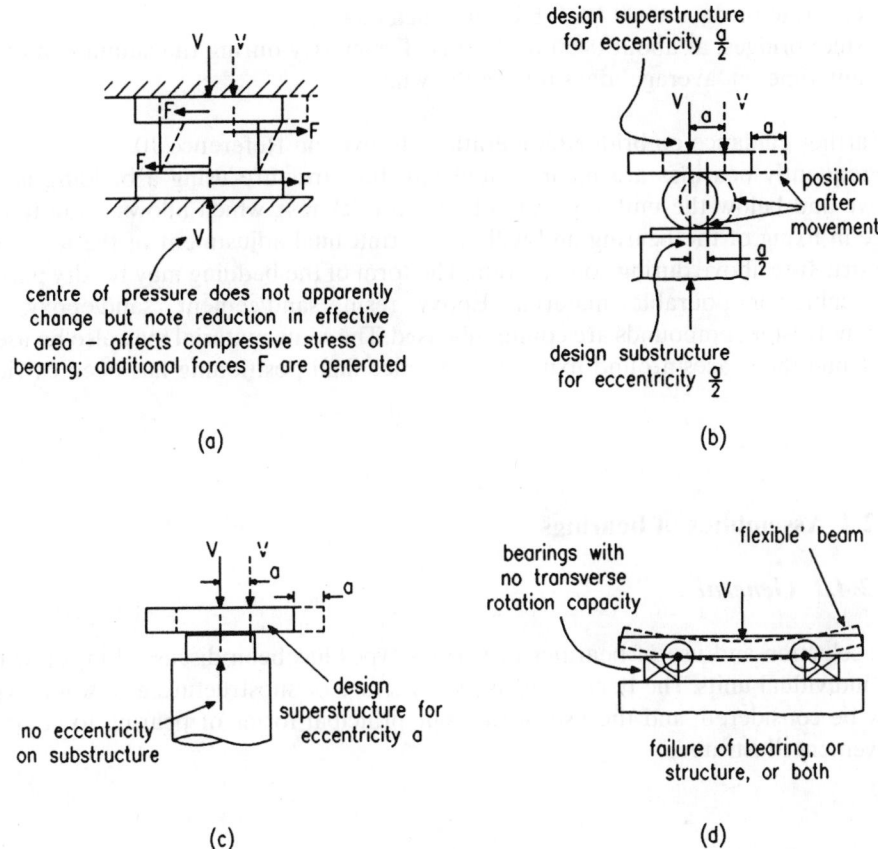

Fig. 28.7 Effect of bearing on structure: (a) elastomeric, (b) roller, (c) sliding, (d) need for transverse rotation capacity

the bearing in a manner for which it has not been designed. A common cause of the latter effect is the incorporation of temporary packs or shims into the bedding in such a manner that they subsequently form hard spots under the bearing. (Note that there is, however, no fault in principle with the concept of temporary packing provided that it is responsibly carried out.)

Factory-assembled bearings are usually provided with transit straps to prevent inadvertent dismantling of the unit. When a significant irreversible movement is anticipated at a bearing, due to shrinkage or prestressing for example, an allowance for this movement may be pre-set in the factory. To allow for departure of the actual structure temperature, when the bearings are set, from the mean temperature assumed in design, it may be necessary to make alterations of the relative positions of the fixed and moving elements of the bearing. For bridges, Lee[6] suggests times when this may be most conveniently carried out – when the bridge temperature is approximately equivalent to the air shade temperature. This can be taken as:

(1) concrete bridges: 0900 BST ± 1 hour each day,
(2) steel bridges: at about 0400 to 0600 BST each day during the summer, and at any time on 'average' days during the winter.

Further guidance on bridge temperatures is given in Reference 10.

Frequently bearings are incorporated into the structure using a bedding layer above and below the unit, typically of 25 mm thickness, which allows some tolerance in fixing of the bearing and will also permit final adjustment of the levels of the structure above during construction. The form of the bedding may be 'dry pack', trowelable or pourable material. Epoxy resin, sand/cement, sand/epoxy, or sand/polyester compounds are commonly used. The same material may also be used for filling the spaces around fixing devices once final positioning has been carried out.

28.2.4 Assemblies of bearings

28.2.4.1 General

The selection and use of bearings of various types has been discussed in terms of the individual units. The behaviour of the structure or substructure as a whole will now be considered, and the use of the four principal forms of bearing to control movement illustrated.

28.2.4.2 Structures straight in plan

As an example, the movement of a typical bridge deck will be considered in the horizontal plane, although the principles involved can equally be applied in other directions.

The four forms of bearing commonly available are given in section 28.2.3.3.

Consider the bridge deck shown in plan in Fig. 28.8.

The deck vertical loading arises from dead and live loads, from which maximum and minimum values of bearing loads can be derived at each position. Longitudinal loading on the deck will arise from wind loads, braking and traction of vehicles, and also from the manner in which the chosen restraint system accommodates

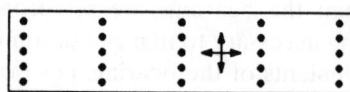

Fig. 28.8 Straight bridge deck

movements. Forces from similar effects are generated in the transverse direction also.

Thermal expansion and contraction of the deck is frequently the predominant reversible movement. This can normally be considered to act radially from a particular fixed point on the deck. The options for bearing arrangement are many, but three typical layouts are given in Fig. 28.9(a), (b) and (c). It should be remembered that whether a bearing accommodates horizontal movement through PTFE/sliding or by shearing of an elastomer block a horizontal force (due to friction or shear respectively) will be generated, and the bearing system should be arranged so that wherever possible these forces cancel one another out, and so minimize the net horizontal force to be resisted by the substructure.

(1) In Figure 28.9(a) all the bearings are elastomeric with *no* fixed bearings. The horizontal loads in both directions are shared between all bearings and the structure 'floats'.

   Thus all substructures will be loaded when horizontal loads or expansion/contraction occur. This system is economic, but is limited by the maximum capabilities of the bearings in rotation, load, and movement.

(2) In Fig. 28.9(b) all the bearings are mechanical (typically spherical or pot bearings). Line 'C' provides fixity in the transverse direction. Line 1 provides fixity in the longitudinal direction. All longitudinal forces from external sources are taken at abutment 1, together with longitudinal forces arising from friction at

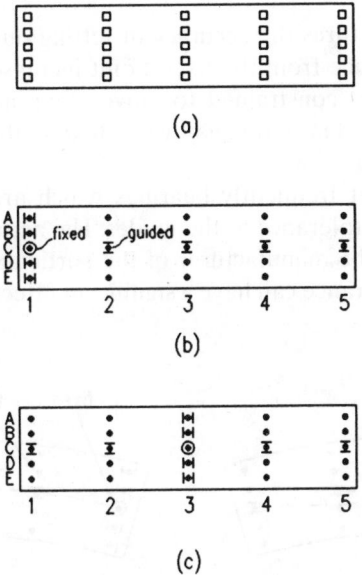

Fig. 28.9 Typical bearing layouts for straight bridge decks: (a) all elastomeric, (b) all mechanical (i), (c) all mechanical (ii)

bearings on piers 2 to 5. At each pier transversely lateral loads are taken by 'C' line bearings. Friction forces due to transverse expansion, etc. will tend to cancel one another out.

(3) The arrangement in Fig. 28.9(c) is better for longitudinal effects than that in Fig. 28.9(b) as friction forces in this direction tend to cancel one another out. It has the disadvantage that external loads are transmitted to an intermediate pier rather than an abutment. The forces due to movement of the deck are minimized in both horizontal directions. Occasionally, the line of fixed or guided bearings with both horizontal and vertical capability such as at 'C' may be replaced by two lines, one with bearings with vertical capability only, and one with bearings with horizontal capability only.

28.2.4.3 *Structures curved in plan*

If the structure shown in Fig. 28.9(b) is curved in plan, then any expansion or contraction movements longitudinally are accompanied by lateral movements also. This effect can be controlled in two ways:

(1) set the bearing guides to permit radial expansion from a fixed point on the structure,
(2) set the bearing guides tangential to the plan curvature, and so constrain the structure to follow this line when it moves (see Fig. 28.10).

In radially-guided structures the accuracy of setting out and alignment becomes more critical as the distance from the fixed point increases. In tangentially-guided structures, the structure is constrained to move along a particular path, and the horizontal forces developed in so doing must be taken into account in the design of the structure and supports.

It should be noted that frequently bearings which are nominally 'guided' are manufactured with a gap tolerance at the guides. The actual value of this tolerance should be checked with the manufacturer of the particular bearing, but a value of 0.5 mm is typical. This tolerance can have a significant effect on the permissible accu-

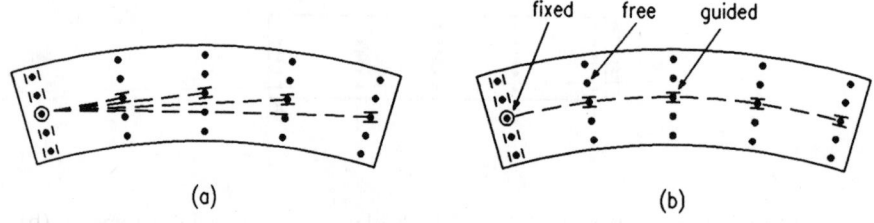

Fig. 28.10   Curved bridge deck: (a) radially-guided, (b) tangentially-guided

racy of setting out of radially-guided structures, and the magnitude of the forces developed in tangentially-guided structures.[7]

28.2.4.4 Structures with fixed bearings and flexible supports

An alternative to the use of systems of guided bearings is to provide fixed bearings at more than one (possibly all) supports. In this case the supporting structures (e.g. bridge piers) have to be designed to flex and accommodate the necessary movements. They also have to cater for the forces developed by these movements in addition to any other design loading effects. This arrangement may be appropriate when it is required to share horizontal load effects over several supports, but it should be noted that replacement of the bearings may be more difficult owing to horizontal loads which may be locked into the bearing/support arrangement.

28.2.4.5 Other considerations

(1) *Wedging action.* It is possible to utilize a form of 'wedging action' to resist horizontal loadings by setting two (usually elastomeric) bearings on planes inclined to one another as shown in Fig. 28.11. Equally, it is also possible to develop the action inadvertently by errors in bearing setting out, and thus attract more loading than that for which the unit is designed.[4,5,7]
(2) *Shock transmission units (STUs).* Although not strictly bearings, these units can be utilized in conjunction with bearings to distribute certain components of loading to other parts of the structure. The units typically consist of a cylinder filled with putty-like material which is acted on by a piston with a hole in it through which the putty can flow. Slow, steadily applied forces such as thermal expansion forces will cause the putty to flow from one side of the piston to the other, and allow dissipation of the force through movement. Rapidly applied forces such as seismic loads, braking loads, or wind gusts are too fast to allow the flow to occur, and the unit therefore effectively transmits this 'shock' loading without significant movement. A description of the use of these devices is given in Reference 9.

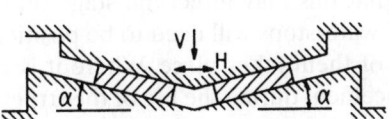

Fig. 28.11   Wedging action

28.3 Joints

28.3.1 General

The form of joints in a structure will vary to suit particular requirements at each position. The basic parameters to be considered in derivation of a joint detail are discussed below, although they are not all appropriate to every situation. Despite the fact that significant differences exist in the final application, many of the factors involved in joint design are common to both buildings and bridges. Joint detailing and construction is considerably facilitated by the many forms of proprietary sealants, gaskets, and fillers that are now commercially available for use as components, as well as complete prefabricated units which may be used in particular applications. The manufacturers of these products will generally be able to supply technical information on their products, and also to give guidance as to the suitability of items for use in particular applications.

28.3.2 Basic criteria

28.3.2.1 Form of the structure

The form of the structure, and the location and orientation of the joint within the structure, will dictate to a large extent the arrangement of the detail. The basic categories of joint are:

(1) *Wall joints.* These may be vertical (e.g. expansion joint in a building or a bridge substructure) or horizontal (e.g. joint between preformed cladding units on a building façade).
(2) *Floor/roof joints.* Examples are expansion joints in a building, or road deck joints in a bridge.
(3) *Internal/external joints.* This type of joint needs to be weather-proofed.

28.3.2.2 Material to be joined, and method of fixing

The material either side of the joint may be steel or aluminium cladding, concrete, brickwork, blockwork, or various forms of surfacing. The detail of the joint will vary considerably with the properties of the material and the method of fixing to be used. It is important to note that this may affect the stage of construction at which the joint is formed: e.g. PVC waterstops will need to be positioned before concreting of the walls on either side of them takes place. Where it is anticipated that the joint may need repair or replacement during the life of the structure (e.g. expansion joints on heavily trafficked bridges) the fixings of the joint should allow for easy removal and reinstatement.

28.3.2.3 *Weather-resistance*

It is important to consider the degree of weather-resistance required for a joint. In this respect, joints can be classified (in a somewhat over-simplified form) into three types (see Fig. 28.12(a), (b) and (c)):

(a) 'Closed' joint, with a filler material and exterior sealant,
(b) 'Closed' joint, with a compressible gasket and exterior sealant,
(c) 'Open' joint, with a flexible membrane seal, and arrangements to drain rain-water, etc. from the inside surfaces of the joint which are 'open' to the weather.

Where appropriate, arrangements should be made at joints in buildings for continuity or sealing of insulation and vapour barriers, etc. to prevent formation of condensation, or loss of heat.

In structures where there is likely to be water in contact with the structural envelope, e.g. structures buried in ground which has a high water table, or water-retaining structures, flexible waterbars are usually incorporated at construction and movement joints. These have the effect of interrupting the path along which any water present has to travel.

Figure 28.13(a), (b) and (c) are typical of wall details in reinforced concrete or brickwork construction. A typical joint detail for steel cladding in a building is

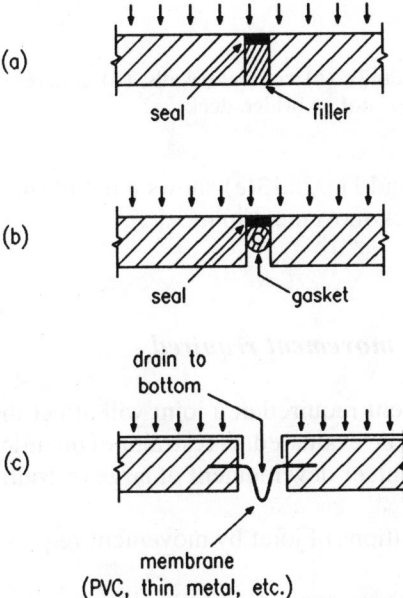

Fig. 28.12   Weather-resistance of joints: (a) closed, with filler, (b) closed, with gasket, (c) open, with membrane

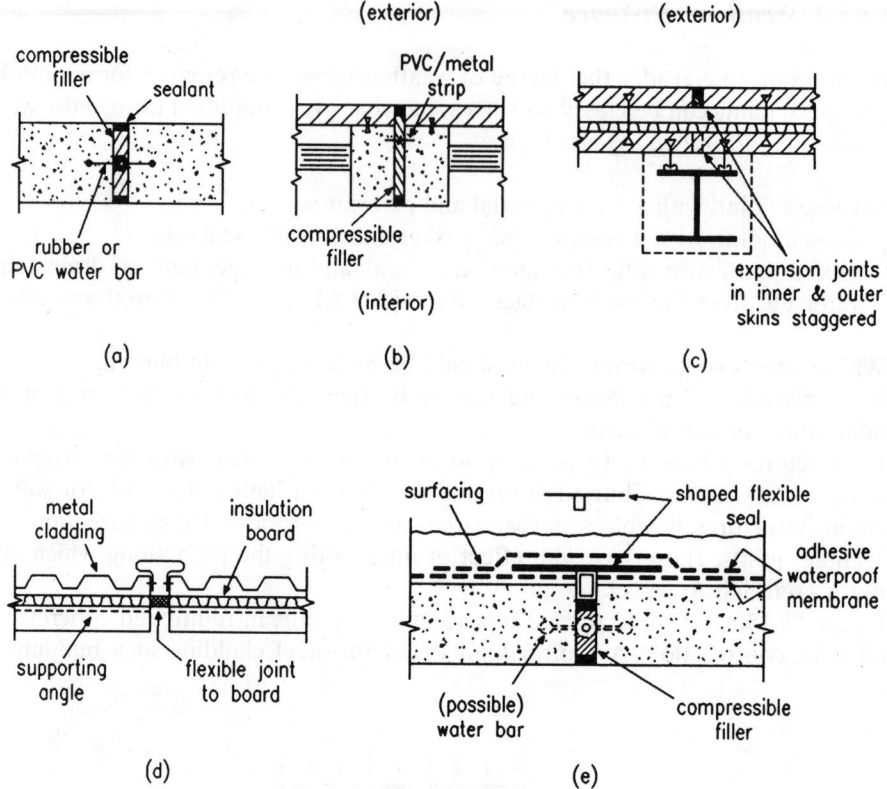

Fig. 28.13 Typical joint details: (a) concrete walls, (b) concrete columns, (c) brick walls, (d) cladding, (e) roof or bridge deck

shown in Fig. 28.13(d), and Fig. 28.13(e) shows a detail suitable for a building roof joint, or small bridge deck movement joint.

28.3.2.4 *Direction of movement required*

The direction of movement required at a joint will affect the form of the joint. All likely movements should be evaluated, as restraint of unanticipated movements may result in failure of the joint, or development of large restraint forces in the joint and the adjacent structure.

Typical broad classifications of joint by movement requirements are given in Fig. 28.14(a), (b), (c) and (d).

In order to illustrate the nature of additional movements which can occur in a structure, two particular cases relating to a bridge superstructure are considered (see Fig. 28.15(a) and (b)).

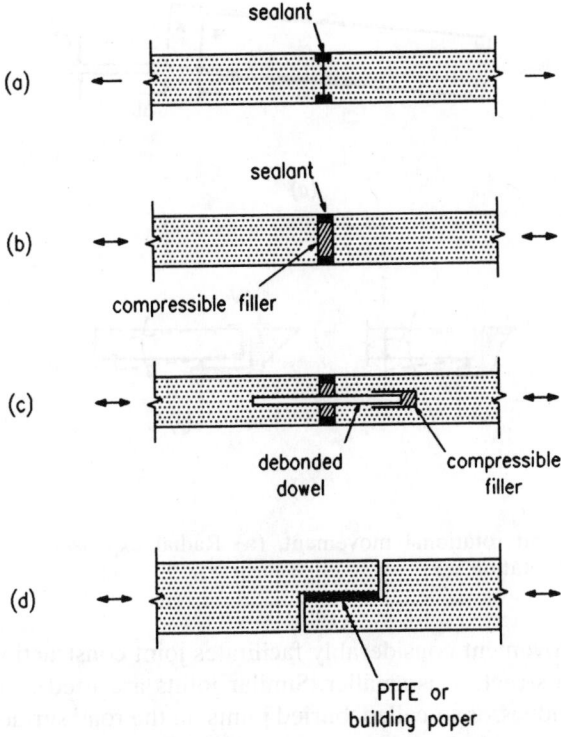

Fig. 28.14 Typical joint details allowing for moderate movements

In Fig. 28.15(a), the bridge deck is set to allow radial expansion from a fixed point. It will be observed that the expansion joint will require both longitudinal *and* lateral movement capability.

Figure 28.15(b) represents a cross-section throughout the end of the last span of a bridge. In (i) the distance from the actual point of rotation to the joint is small, and little vertical movement of the joint is necessary when flexure of the beam causes rotation. In (ii) the effect on the joint of a large overhang is demonstrated: a much larger vertical movement is induced. Further discussion of these and other related effects is given in References 4–7.

28.3.2.5 *Magnitude of movement required*

The magnitude of movement to be allowed for has a significant effect on the type of joint. Joints can be broadly classified into three groups on this basis:

(1) *0–25 mm movement*: Common for buildings, where there is a tendency towards the use of larger numbers of joints, each with a small movement. The restricted

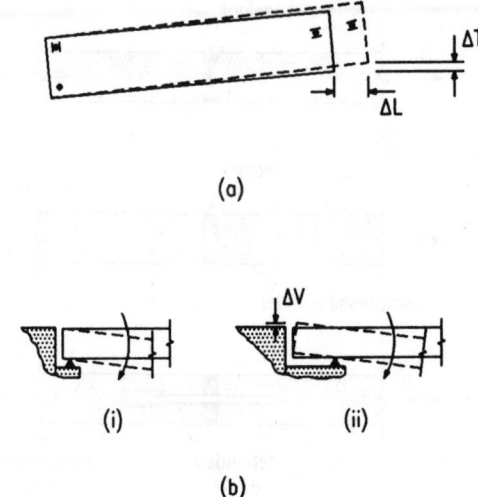

(a)

(i)   (ii)

(b)

Fig. 28.15   Radial and rotational movement. (a) Radial expansion from a fixed point,
(b) end rotation

amount of movement considerably facilitates joint construction as the discontinuity in the structure is smaller. Similar joints are used in bridge work for small-span bridges, or so-called 'buried joints' in the road surface. Typical examples of joints suitable for movements within this range are given in Fig. 28.14. They may be preformed units, but are more likely to be assemblies of proprietary components arranged to suit the particular requirements of the detail and its location.

(2) *25–300 mm movement*: Joints capable of movements of this extent are more common in bridge work, and are usually prefabricated units, although a simple plate arrangement can be used in some cases (Fig. 28.16(a)). Two basic types of prefabricated units are common. Figure 28.16(b) shows the first type, which uses an elastomeric material in its construction to accommodate the movement. Figure 28.16(c) shows the second type, which is suited to the larger movement ranges. This type uses an arrangement of mechanical components supported on stub beams to form a moving joint. Detailing of the structure must allow for the space required by these units.

(3) *>300 mm movements*: Joints for movements of this magnitude are unusual and will be tailored to particular requirements, e.g. expansion joints for suspension bridge decks.

28.3.2.6 *Required performance of the joint surface*

The nature of effects acting on the surface of the joint should be considered. In building structures, this will frequently be limited to environmental effects on

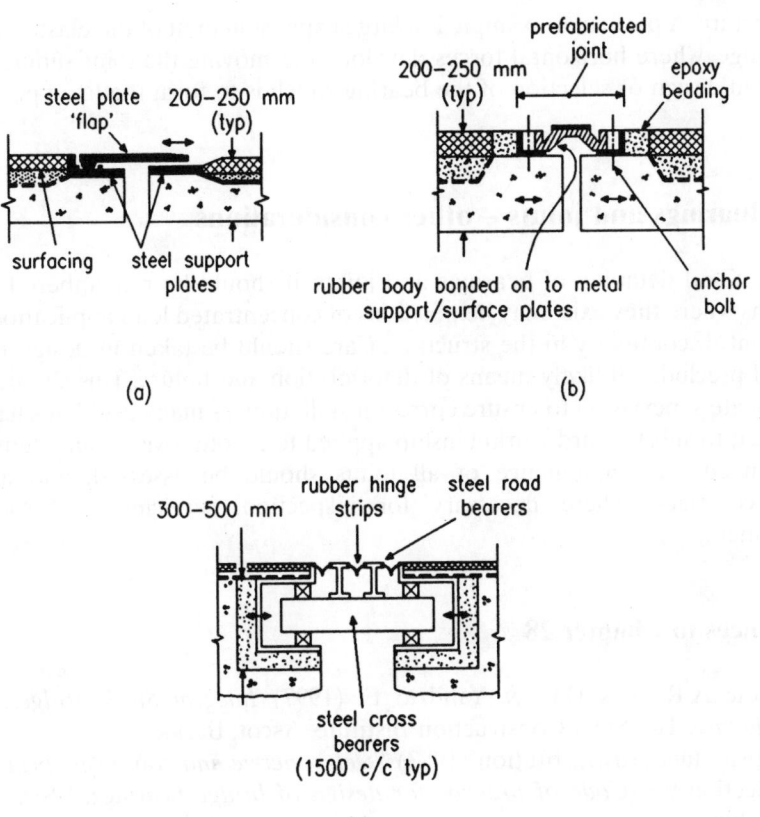

Fig. 28.16 Joints allowing for larger movements: (a) (b) 200–250 mm, (c) 300–500 mm

exterior walls or roof, but consideration of the size of the joint gap is necessary in, say, a floor or roof joint where excessive gaps could prove hazardous to pedestrians, or could lead to jamming of the joint through ingress of debris, etc.

Joints on bridge decks are subject to heavy localized effects from wheel loads and corrosive effects from de-icing salts and spilt chemicals. As a result, the practicalities of maintenance and perhaps eventual replacement of such joints should be considered. The resistance to skidding of the surface of such joints is also important.

28.3.2.7 Load generation at joints

Frequently the loads generated by compression or extension of movement joints are insignificant in terms of overall structural behaviour. In some cases, however, the loads developed may be large enough to affect the design of other elements of

the structure. A particular example is a large expansion joint of the elastomeric type in a bridge, where horizontal forces developed in moving the joint should be considered in design or selection of the bearing which is to form the fixed point of the deck.

28.4 Bearings and joints – other considerations

In design and detailing of bearings and joints, it should be remembered that the positions where they exist may be positions of concentrated load application and/or significant discontinuity in the structure. Care should be taken in design to establish and preclude all likely means of deterioration and failure. This should extend to adequate supervision to ensure correct installation, as many problems have been attributed to substandard workmanship applied to an otherwise competent design.

The need for maintenance of all joints should be assessed, and adequate allowance made where necessary for inspection, servicing, and facility of replacement.

References to Chapter 28

1. Biddle A.R., Iles D.C. & Yandzio E. (1997) *Integral Steel Bridges: Design Guidance.* The Steel Construction Institute, Ascot, Berks.
2. British Standards Institution (1983) *Steel, concrete and composite bridges.* Part 9: Section 9.1: *Code of practice for design of bridge bearings.* BS 5400, BSI, London.
3. Kaushke W. & Baigent M. (1986) *Improvements in the Long Term Durability of Bearings in Bridges.* ACI Congress, San Antonio, USA, Sept.
4. Baigent M. *The Design and Application of Structural Bearings in Bridges.* Glacier Metal Co. Ltd.
5. Long J. (1974) *Bearings in Structural Engineering.* Newnes-Butterworths, London.
6. Lee D.J. (1971) *The Theory and Practice of Bearings and Expansion Joints in Bridges.* Cement and Concrete Association.
7. Nicol T. & Baigent M. *The Importance of Accurate Installation of Structural Bearings and Expansion Joints.* Glacier Metal Co. Ltd.
8. Wallace A.A.C. (1988) *Design: Bearings and Deck Joints.* ECCS/BCSA International Symposium on Steel Bridges, Feb.
9. Pritchard B. & Hayward A.C.G. (1988) Upgrading of the viaducts for the Docklands Light Railway. *Symposium on Repair and Maintenance of Bridges*, June. Construction Marketing Ltd.
10. Emerson M. *Bridge Temperatures and Movements in the British Isles.* Transport and Road Research Laboratory Report LR 228. (*See also* TRRL reports LR 382 (W. Black); LR 491 & LR 532 (M. Taylor); LR 696, LR 744, LR 748, LR 765 (M. Emerson *et al.*))

Chapter 29
Steel piles

by TONY BIDDLE and ED YANDZIO

This chapter is intended to serve as an introduction to the subject of steel piling and a source of reference for further reading where detail design is required. It is divided into two main parts, i.e. 29.1 for bearing piles and 29.2 for sheet piles.

29.1 Bearing piles

29.1.1 Uses

Bearing piles are used mainly to support vertical loads for which the main design requirements are to restrict average settlement to a reasonable amount, eliminate differential settlement, and achieve an adequate factor of safety or load factor against overload.

Steel piling offers many advantages compared with other types including:

- Reduced foundation construction time and site occupation
- Reliable section properties without need for onsite pile integrity checking
- Inherently high strength to weight ratio
- Higher end bearing resistance in granular soils and rocks mobilized by pile driving as compared with boring
- Pile load capacity can be confirmed during driving by dynamic pile analysis (DPA) on each pile
- Low displacement of adjacent soils during driving and no arisings
- Easily removed for site reinstatement
- Reusable or recyclable following extraction
- Closer spacing possible.

The ability to be transported and handled without concern for overstressing due to self-weight is an important consideration where piles need to be lifted, pitched and driven in long lengths, e.g. maritime structures. Steel piles can withstand high driving stresses, thereby enabling them to be installed without significant damage through difficult strata such as brownfield sites and subsurface layers of boulders or weathered rock. Energy-absorbing structures such as jetties and dolphins can be particularly well constructed in steel by virtue of its ability to offer large elastic deformation

and bending strength without affecting its durability. On projects such as river or estuary crossings, steel piles offer clear advantages, as the soils are typically granular and waterlogged and unsuitable for satisfactory pile boring. New applications such as supports and abutments for integral bridges offer more economic construction methods and the property of compliance under lateral loading that reduces bending stresses and therefore section requirements.

29.1.2 Types of pile

Steel bearing piles are available in three basic profiles: universal column H-sections, tubular sections and box piles. In addition steel sheet piles, High Modulus Piles and Combi-piles can be used to support vertical load as well as a soil retaining function.

29.1.2.1 *Universal bearing piles*

Universal bearing piles are H-sections produced on a universal hot rolling mill – hence their name. They are essentially the same as universal column sections, except that they have uniform thickness throughout the section. See Fig. 29.1. Universal column H-piles are made in two basic qualities of steel, grade S275 (yield = $275 \, N/mm^2$) and grade S355 (yield = $355 \, N/mm^2$) to BS EN 10025.[1] Typical section sizes are presented in the Appendices.

Steel H-piles are very efficient in providing a large surface area for generating shaft friction resistance. In any given foundation plan area, a greater number of steel H-piles can be used in a group with a standard spacing of $2B$ (or $2 \times$ dia.)

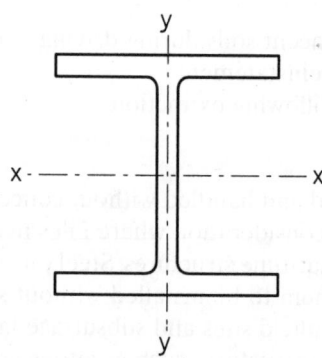

Fig. 29.1 Universal bearing pile

than concrete piles and either the load supported can be greater or the size of the group made smaller. The stiffness of H-piles is different about each orthogonal axis, allowing designers to select the orientation necessary to achieve the most efficient design.

H-piles are used principally for terrestrial structures where the full length of the pile is embedded, e.g. foundations for bridges and buildings, and therefore buckling is not a problem because there is good soil lateral support. They are used very effectively to transfer bearing loads into buried bedrock and to get around buried obstructions. It is sometimes advisable to use special cast steel pile shoes to strengthen the tip and prevent damage or buckling under hard driving conditions into bedrock where very high end bearing pressure can be achieved.

They are generally not the most suitable for conditions where long lengths of the pile shaft protrude above ground level and are unsupported by surrounding soil or through water, because buckling failure about the minor axis can occur at relatively low axial loads. In these cases consider the use of tubular or box piles that have more stiffness.

29.1.2.2 Tubular piles

Tubular piles were first used as foundations for offshore oil platforms in the oil fields of Lake Maracaibo in Venezuela in the 1920s. Initially, spare oil pipe was used out of convenience but, as the supporting structures became more sophisticated, the cold rolling of piles in structural plate to project-specific diameters and wall thicknesses became more common.

Purpose-rolled tubular piles can be used, but high quality steel line pipe is perfectly suitable for piling. Tubular piles are available in a large range of diameters and wall thicknesses. Typical sections used for piling purposes are produced as line pipe to API 5 L grades X52 (yield stress of 52 ksi, approximately equal to 355 N/mm^2) to X80 (yield stress of 80 ksi, approximately equal to 555 N/mm^2).[2] Tubular sections can be hot- or cold-rolled. The cold-rolling process produces consistently higher yield strengths than those of hot-rolled steel products and this can have significant benefits for highly loaded bearing pile and structural column-pile applications, and can also permit harder driving.

Steel tubular piles have a high stiffness and are therefore also suitable for sites where it is necessary to transfer bearing loads into buried bedrock. They are particularly suitable for marine structures, especially where sited in deep water, e.g. berthing jetties for deep draught vessels. Increasing use is being made of steel tubulars in composite columns for buildings and bridges where the tube is first driven as a pile before filling the upstand above ground level with a reinforced concrete core for added strength and rigidity. This permits significant savings in cost due to faster construction. More applications will become possible as the results of research into the effects of pile driving, dynamic load testing, corrosion, and new coatings are analysed and reported to design engineers.

29.1.2.3 Box sheet piling

Hot-rolled steel sheet piling is more extensively used nowadays in permanent as well as temporary retaining walls. Applications are not only above-ground earth-retaining structures but also composite basement wall construction, urban railway and road cuttings, bridge abutments, river bridge pier caissons, as well as the traditional river and coastal protection works. Economic design is enabled by better knowledge of soil–structure interaction and highly developed design and analysis programmes.

Box piles are particularly suitable for marine structures, such as jetties and dolphins, where part of the pile shaft is exposed above seabed level and the pile functions as a free-standing column or is connected at the head in clusters of columns.

Box piles are formed by welding two or more sheet pile sections together. Both Larssen and Frodingham steel sheet piles (see section 29.2.3) can be used. They can be introduced into a line of sheet piling at any point where local heavy loads are to be applied, for instance beneath bridge beams, or used separately. They are clutched together with adjacent sheet piles and can be positioned in a sheet pile abutment so that its appearance is unaffected.

Larssen box piles are formed by welding together two sheet pile sections with continuous welds, and Frodingham plated box piles are formed by continuously welding a plate to a pair of interlocked and intermittently welded sheet piles (see Fig. 29.2).

Special box piles can be formed using other combinations of sheet piles. Further information can be obtained from steel manufacturers.

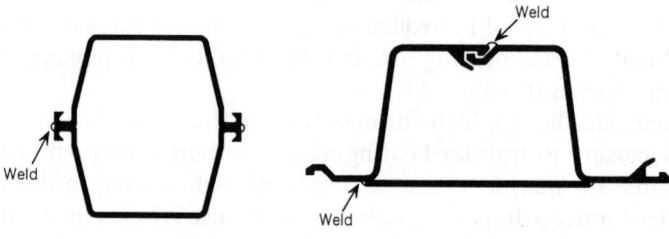

Larssen   Frodingham

Fig. 29.2 Box piles

29.1.2.4 High modulus piles

Buttressing a sheet pile wall with deep universal beams placed in 'soldier' fashion creates a *high modulus pile wall* as shown in Fig. 29.3. This section provides additional moment capacity and bearing capability. The UBs are welded to the back of a pair of Frodingham sheet piles and all three are driven together clutched to the next panel.

29.1.2.5 Combi-piles

Combi-piles are another form of part-fabricated piling used to form walls for deep excavations. The wall is formed from alternating tubular and sheet piles as shown in Fig. 29.4. The number of sheets between the tubes varies dependent on the required section modulus. The tubular piles take vertical loads and stiffen the wall.

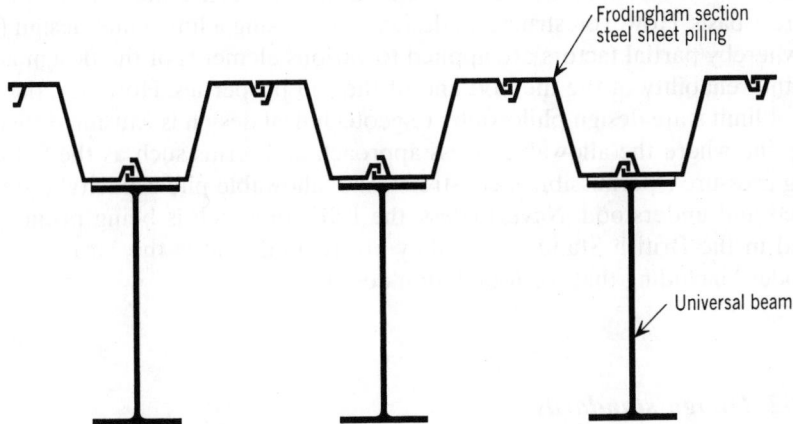

Fig. 29.3   High modulus piles

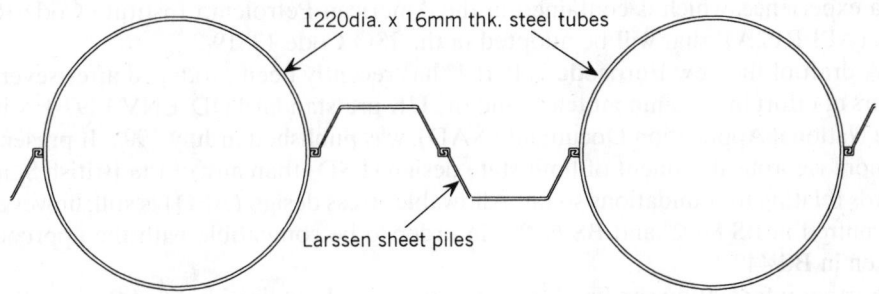

Fig. 29.4   Combi-piles

Tubular piles are available in a wide variety of sizes, and both Larssen and Frodingham sheet piles are suitable as infill piles. Appropriate clutches are welded to the walls of the tubes, and these provide the designer with simple corner details for many different wall angles. It is apparent that the clutch locking bar strips from U-section sheet piles (or Larssen sections) are more stable to weld onto the tubulars than those from Frodingham section sheet piles.

29.1.3 Design

29.1.3.1 Design basis

The basis of design of any bearing pile is its ultimate axial capacity in the particular soil conditions at the site where the structure is to be built. The design resistance is determined from the ultimate capacity, and the designer verifies that this is adequate to carry the required loads from the structure.

Until recently, design resistance of foundations has been evaluated on an allowable stress basis. However, structural design is now using a limit state design (LSD) basis, whereby partial factors are applied to various elements of the design according to the reliability of the method and of the soil properties. However, the application of limit state design philosophy to geotechnical design is causing difficulty in a discipline where the allowable stress approach and terms such as the 'allowable bearing pressure', 'permissible steel stress', and 'allowable pile capacity' are widely accepted and understood. Nevertheless, the LSD approach is being progressively adopted in the British Standards as they are revised, and is the basis for all the Eurocodes,[3] including that for foundation design.

29.1.3.2 Design standards

The common design standard used for the design of bearing piles is the offshore industry's recommended practice for steel tubular piles, based on US and UK North Sea experience, which is contained in the American Petroleum Institute Code RP 2A (API RP2A)[4] that will be adopted in the ISO Code 13819-2.[5]

A draft of the new Eurocode 7: Part 1[6] has recently been produced after several years of effort by an eminent team, and the UK pre-standard DD-ENV 1997-1[7] with the National Application Document (NAD), was published in July 1995. It presents a more rigorous treatment of limit state design (LSD) than any of the British Standards relating to foundations so far. Allowable stress design (ASD) is still, however, permitted in BS 8002[8] and BS 6349,[9] in order to be compatible with the approach taken in BS 449.[10]

Further information can be obtained from the SCI publication *Steel Bearing Piles Guide*.[11]

29.1.3.3 Loading

In most cases the structural loads are principally vertical, with relatively minor horizontal components. Small horizontal loads are commonly resisted by passive soil resistance on the side of pile caps and beams rather than by bending in the pile itself, because the structural movements are very small. Where the lateral loading is larger, it is better resisted by the inclusion of raking piles which provide greatly improved stiffness in the horizontal plane. Any induced uplift forces resulting from the use of raking piles can be resisted by the weight of the superstructure or by tension in the remaining associated piles in a group (see Fig. 29.5).

29.1.3.4 Axial capacity and load transfer

A pile subjected to a load parallel to its longitudinal axis will support that load partly by shear generated over its shaft length, due to the soil–pile skin friction or adhesion, and partly by normal stresses generated at the base or tip of the pile, due to end bearing resistance of the soil (see Fig. 29.6).

The basic relationship is given in BS 8004[12] for the ultimate capacity R_c of the pile. This relationship assumes that the ultimate capacity R_c is equal to the sum of the shaft friction capacity R_s and base capacity R_b, i.e.

$$R_c = R_s + R_b = q_s A_s + q_b A_b$$

where q_s is the unit shaft friction value*
      A_s is the surface area of the pile in contact with the soil[†]
      q_b is the unit base resistance value

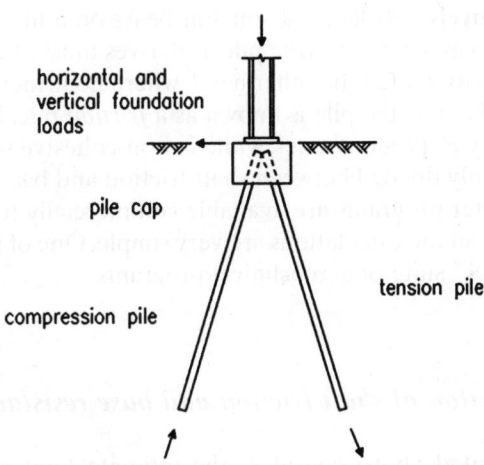

Fig. 29.5   Raking pile foundation

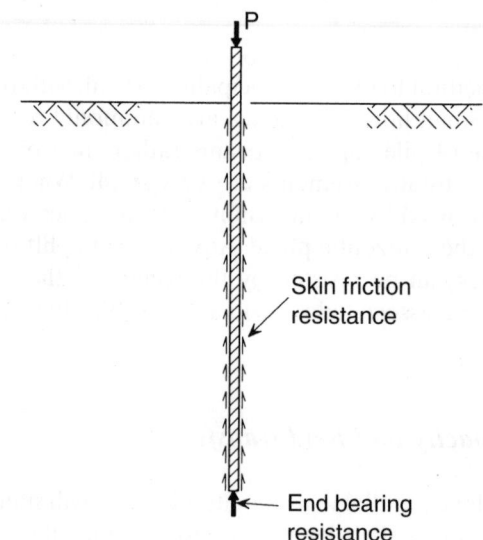

Fig. 29.6   Wall friction and end bearing resistance against vertical loads

A_b is the steel cross-section area of the base of the pile or plug cross-sectional
area.

*Where there is more than one soil type, the average value of q_s is taken over each
layer down the length of the pile.

The relative magnitudes of the ultimate shaft friction and ultimate end bearing
resistances depend on the geometry of the pile and the soil profile. Where a pile is
embedded in a relatively soft layer of soil, but bears on a firmer stratum beneath,
the pile is referred to as an *end bearing* pile. It derives most of its capacity from the
base resistance capacity R_b. On the other hand, where no firmer stratum is available
on which to found the pile, the pile is known as a *friction* pile. In cohesive soils the
shaft friction capacity R_s predominates, while in non-cohesive soils the overall axial
capacity is more evenly divided between shaft friction and base resistance capacity.

Numerous computer programs are available commercially to calculate the vertical capacity of piles, but the calculations are very simple. One of these is PILE, which
is part of the OASYS[13] suite of geotechnical programs.

29.1.3.5 Mobilization of shaft friction and base resistance

The equation presented above considers the ultimate limit state condition only
where the pile has been allowed to move sufficiently to allow both the ultimate shaft

friction and the ultimate base resistance capacities to be developed. Commonly, *load transfer curves* are produced, which are plots of load resistance versus vertical movement of the pile head for displacements ranging from zero to the ultimate limit or to a permissible maximum value (e.g. 40 mm). These plots include mobilized soil–pile shear transfer versus local pile movement and mobilized base resistance versus tip movement.

Axial load transfer curves for clays and sands have been obtained from full-scale research static pile load tests. These pile tests monitor both the load and strain at a number of positions along the pile. Tests that have been performed on clays and fine granular soils show the general behaviour (see Fig. 29.7).

It is found that, at any position along the pile, full shaft friction resistance is not developed until the pile has moved axially (relative displacement between pile and soil) to a magnitude in the range 7–10 mm. Once this movement is reached no further additional wall friction resistance develops, and apart from any small reduction in peak value of wall friction (due possibly to strain softening), the curve tends to a nearly constant resistance. This constant resistance value can be assumed to be the ultimate unit shaft friction of the soil at any level.

The behaviour of a pile in clay in base resistance is also shown in Fig. 29.7. It is seen from the base resistance load versus pile tip deflection (equal to pile head movement) curve that a greater pile tip deflection is required to achieve the ultimate base resistance capacity than is required to develop the shaft friction capacity, i.e. in excess of 40 mm.

The mobilization of the total axial resistance with increasing displacement for a pile is obtained from the summation of the mobilized shaft friction resistance and the mobilized base resistance (see Fig. 29.7).

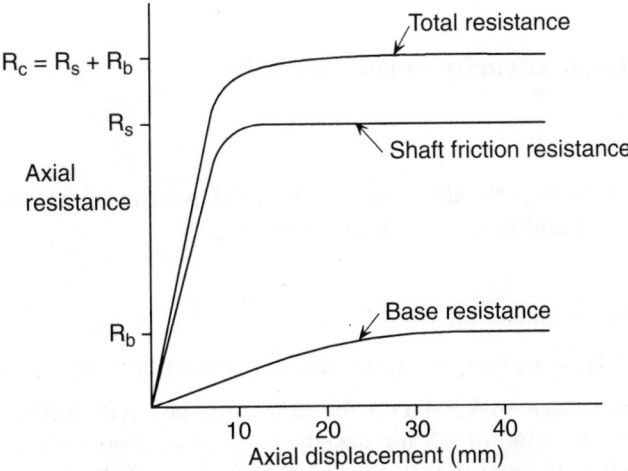

Fig. 29.7   Axial load transfer curves for soils

29.1.3.6 Vertical settlement and serviceability

The design ultimate capacity of a steel pile R_{cd} is given by:

$$R_{cd} = \frac{R_s}{\zeta\gamma_s} + \frac{R_b}{\xi\gamma_b}$$

where R_s is the ultimate shaft friction resistance

R_b is the ultimate base resistance

γ_s is the factor for shaft friction resistance

γ_b is the factor for base resistance

ξ is the material factor to take into account uncertainty of soil parameters determined on site or in the laboratory.

γ_s and γ_b are partial factors for the resistance side of the limit state equation, while ξ is a material partial factor. These factors are not provided by BS 8002[8] or BS 8004[12] but are given in DD-ENV 1997-1 Eurocode 7.[7] In Eurocode 7 for driven piles:

$\gamma_s = 1.3$

$\gamma_b = 1.3$

$\xi = 1.5$

The design vertical capacity of the sheet pile–soil interface is adequate provided that:

$$\frac{P_{des}}{R_{cd}} \leq 1$$

where P_{des} is the design magnitude of the axial load including all appropriate partial factors.

29.1.3.7 Ultimate capacity in cohesive soils

Shaft friction

The unit shaft friction qs for clays can be estimated in terms of the undrained shear strength of the soil and is given by the relationship

$$q_s = \alpha.c_u$$

where α is a dimensionless factor

c_u is the undrained triaxial test shear strength of the soil.

BS 8004[12] offers a range of 0.3–0.6 for the value of α; however, further research has verified an average value of 0.5 for design for most over-consolidated clays. Off-shore research has discovered that for soft clays of $c_u < 24\,\text{kPa}$, the α value should be taken as 1.0, and that the α value then reduces linearly to 0.5 when c_u becomes

72 kPa. Further information can be found in API RP 2A[4] or in the SCI *Steel Bearing Piles Guide.*[11]

Base resistance

It is common to calculate the base capacity of piles in clay in terms of the undrained shear strength c_u. The magnitude of the base resistance capacity q_b generated in cohesive soils is given in BS 8004[12] as:

$$q_b = 9c_u$$

where c_u is the undrained triaxial test shear strength of the soil beneath the toe.

As an indication, the undrained shear strength of cohesive soils is generally in the range from 20 kN/m² for soft clay to 400 kN/m² for very stiff or hard clays. For this range of c_u, the base resistance values are in the range 0.2–3.6 MPa.

29.1.3.8 Ultimate capacity in cohesionless soils

Shaft friction

The unit shaft friction q_s for a driven sheet pile in granular soil is given by:

$$q_s = 2N_b$$

where N_b is the average standard penetration test (SPT) value for each soil layer.

Base resistance

The magnitude of the base resistance capacity, q_b, generated by a driven sheet pile in cohesionless soils can be obtained from the relationship:

$$q_b = 400N_b$$

where N_b is the standard penetration test value at the level of the pile tip.

For sands, the base resistance values are an order of magnitude greater than for cohesive soils and range in value up to 40 MPa, which is experienced over the steel wall end bearing area. This value may seem high in relation to the 10 MPa quoted in BS 8004,[12] but that was based on bored concrete piles. The higher values have been verified by steel pile load tests, and even higher values (up to 70 MPa) have been measured in sands offshore. The wall endbearing resistance in granular soils is generally some 50% of the total load resistance for driven steel piles and therefore should always be calculated carefully.

29.1.3.9 Ultimate capacity in rock

Where piles are driven through clay/sand strata but are terminated at depth in a relatively incompressible rock stratum, the pile base resistance will be the main component. In these cases, only small axial movements of the pile will occur, owing to the high compressive stiffness of the rock, and it may not be possible to generate appreciable shaft friction resistance in any clay/sand layers higher up. In many cases, the maximum design load for such a pile will be governed by the stresses in the pile steel itself, rather than the ultimate base resistance of the rock because it is so high. A table of typical values for rocks measured by the Norwegian Geotechnical Institute is shown in Table 29.1, where compressive strength is given in MPa.

29.1.3.10 Lateral load resistance

Lateral loads on bearing piles range in importance from the major load component in such structures as transmission towers and mooring dolphins to a relatively insignificant force in the foundations of low-rise buildings.

The designer must first judge whether the imposed lateral load on the proposed foundation is significant enough to warrant special analysis. On buildings, for instance, the lateral loading is mainly due to wind pressures. For low-rise buildings not exceeding three storeys in height any foundation shear resistance required is normally accommodated by passive earth pressure acting on the buried pile caps

Table 29.1 End bearing resistance for rocks (MPa)

Rock type		Compressive strength
Hard	Leptite Diabase Basalt Granite Syenite Quartz porphyry Diorite, gabbro Quartzite Quartzitic phyllite	
Weak	Metamorphic phyllite Layered phyllite Hornblende Chalkstone Marble Dolerite Oil shale Mica shist Sandstone Lava	

Key :

⊢—•—⊣	Variation about mean
•	Cylinder sample H = D
◻	Cylinder sample H = 2D
▫	Cube test

and ground beams of a piled foundation, and on the frictional sliding resistance beneath the ground floor slab and the foundations. Therefore, the lateral loading on the bearing piles would be insignificant.

Lateral load resistance from vertical bearing piles is particularly dependent on soil type; in the extreme case of very soft soil, raking piles to an underlying more competent soil or rock would be required to provide any significant resistance.

Lateral load resistance from vertical bearing piles also requires significant pile displacement, of the order of many centimetres, so if this is unacceptable then the designer should think of alternative foundation elements to provide it, such as raking piles or embedded sheet pile perimeter walls.

Where the contribution to lateral loading resistance of vertical bearing piles is vital and is an acceptable solution, the designer is recommended to use CIRIA Report 103 *Design of laterally loaded piles*[14] and the textbooks by Poulos and Davis,[15] or Tomlinson.[16]

Methods of analysis

The two most extensively validated methods are the *P–Y curve method* and elastic continuum analysis FE programs. Both are explained in CIRIA Report 103.[14]

P–Y curves originate from instrumented lateral load tests carried out on 762 mm OD tubular piles in the USA in the 1960s for offshore design. The models of load resistance were derived from soil resistance distributions required to match the bending stresses measured by the pile shaft strain gauge instrumentation, i.e. curve-fitting to match bending moment diagrams. The *P–Y curve method* is the only one in which it is possible to allow for significant cyclic loading of piles. This is useful for the structural design of the pile section, but does not give accurate displacements because the single piles had no head restraint. It is explained in detail in the American Petroleum Institute Code RP2A[4] and in computer programs like 'ALP' in the OASYS suite.[13]

Assessment of soil properties

There are no clear guidelines for the assessment of soil property values to be adopted for the design of laterally-loaded piles. Many factors influence the actual mobilized values. In particular, the disturbing effect of pile installation is difficult to quantify. Most soils exhibit considerable loss of stiffness under the action of cyclic loading, which is virtually impossible to relate to laboratory test data. Generally, it is desirable to select upper and lower limits to the critical soil properties, and to check the sensitivity of the design to variation within the chosen range.

Structural designers would be wise to specify their design requirements for soils data carefully within a site investigation contract, so that the work includes soil tests appropriate to the type of loading to be applied to the pile. For example, if granular soils are found on site then in situ soil testing, such as static cone penetration

tests or pressuremeter tests, will be necessary to derive sensible values for lateral loading analysis parameters.

29.1.3.11 Pile group effects

Conceptual design – vertical load resistance

The first stage of a conceptual design is to estimate the number of piles required using the predicted individual axial load design capacity from the procedures given in section 29.1.3.4. These piles will then be laid out in a group within the plan area of the supported structure. However, depending on the relative positions of the piles there may be a group effect that will modify the capacity of each pile.

Assessing pile group effect

Piles in a group may be subject to the following effects:

(1) the vertical load resistance of a group of piles could be less than the sum of the resistances of all the piles in the group acting independently, and
(2) the pile head deflection of the group, or of its pile cap, may be different from that of a single pile.

There are some simple spacing rules that have been derived from experiment and experience, and piles will only interact to cause a 'group effect' if they are closer than a particular spacing to each other. This is about $3D$ (where D = pile diameter) for piles in clay or about $4D$ if the piles are in sands. When they are closer than those limits, the pile group behaves as a single block with shaft friction around its external periphery and a base resistance over the whole area of the block, because the individual soil resistance 'envelopes' overlap.

Therefore, in order to achieve best performance and economy of a pile group, the designer should adjust the layout and spacing to comply with these criteria and to ensure that piles act independently wherever possible. If this can be achieved, then no pile group effect will occur and the vertical load resistance of the group is the sum of the individual piles, and the vertical deflection at the pile head is the same as that of an individual pile under its share of the load.

It is worth adjusting the size of each pile, removing any conservatism in the assumptions on soil parameters involved in pile load resistance prediction, or waiting for the results of the trial pile load testing, before deciding on a pile group configuration that involves a pile group effect.

Where such adjustment of the spacing and arrangement of the piles in a group still infringes the spacing limits, the pile group may have to be designed as the enclosing large block.

Practical pile group design

Advice is given in section 2.2 of CIRIA Report 103[14] *The design of laterally loaded piles*, which provides a very good description of the design procedure for a pile group, including a flowchart. It is suggested that the three levels of appraisal detailed in the CIRIA report should be adopted, i.e.:

(1) Consideration of the ultimate failure mechanism of the foundation and incorporation of an overall reserve of strength for safety
(2) Computation of the lateral translation and rotation of the foundation at working loads, and consideration of the effect of this deformation on the whole structure
(3) Bending resistance of the piles.

29.1.4 Installation

Steel piles can be installed by a variety of methods and equipment. There are three types of installation equipment, which operate by impact, vibration or by jacking. Each has particular advantages and disadvantages, and the final choice is, in most cases, a balance between speed and economy of installation. A further deciding element is the increasing concern for noise and vibration control to which the industry has responded with the development of new installation equipment such as hydraulic hammers.

The installation of steel piles is a specialist activity, calling for considerable knowledge and experience of handling piles and hammers to achieve an acceptable placement within specified tolerances of position and level. Guidance on the practical limits that can be achieved in position and level for driven steel piles is available from the FPS (Federation of Piling Specialists)[17] and in the TESPA (Technical European Sheet Piling Association) publication *Installation of steel sheet piles* (see also section 29.3.1).[18] Guidance is also included at the back of the new ICE *Specification for Piling and Embedded Retaining Walls*[19] and in the *Corus Piling Handbook*.[20]

The designer should refer to these documents before carrying out a design, because the advice given will often affect the details at connections to the pile cap in the structure.

There is much to gain from matching the stiffness of the pile to the hammer and to the anticipated soil resistance at the site to achieve satisfactory driveability and to ensure achievement of the required design penetration.

In addition, following research work over the last 30 years, there is a developing understanding of the benefits of measuring the soil resistance during driving as a check on the designer's predicted compressive axial static capacity (see the ICE *Specification for Piling and Embedded Retaining Walls*[19]). Although there are some caveats to be applied to this practice, it is indisputable that both designer and installer gain from using dynamic analysis of pile driving and that this is of ultimate

benefit to their clients in many ways. For instance, it is a QA tool to permit the evaluation of piles that have an unexpectedly high resistance at a higher level; it can lead to fewer and shorter piles being acceptable; it can save construction time by avoiding delays; and it provides an equitable means of resolving disputes about specifications that turn out to be unrealistic due to inadequate site investigation or inaccuracy in design prediction methods.

The relationship between the static and dynamic capacity is now better understood, and in many soils and rocks there is very little difference between them. However, it is still sound practice to carry out static load testing to check the design penetration because in some soils pile capacity has been known to decrease after driving. Test piles should therefore have both static and dynamic measurements in order to relate the two, and then dynamic pile analysis can be used to check the project piles.

29.1.4.1 Determining bearing capacity

The safe load capacity of driven steel piles should be verified. The only certain way to do this is by static test loading. Unfortunately, to do this for every pile is neither practical nor economic for most projects.

Alternative ways of load testing piles are by stress wave analysis using the CAPWAP program[21] and a dynamic pile analyser.

Pile head settlement is most accurately determined by static test load. Dynamic formulae give no indication of this but stress wave analysis will give an indication the technique is rapidly improving.

Static test loading

Two forms of test loading are in general use, but the maintained load method is considered the more accurate. Details of both test methods are given in BS 8004 clauses 7.5.5 and 7.5.6.[12]

Excessive conservatism has been found in current practice and in the currently used specifications for load testing piles, which has been compounded by unrealistic design assumptions on the soil parameters that are used in pile resistance prediction methods to determine a 'working load'. This has led to costly overdesign of piling.

The amount of kentledge or tension resistance should always be in excess of the estimated ultimate bearing capacity of the pile. A factor of 1.5 should be a minimum, since loading to twice the 'working' or 'serviceability' load is the accepted requirement, and the working load is normally about half the ultimate capacity in both LSD and ASD procedures.

In the maintained increment load test (MLT), kentledge or adjacent tension piles or soil anchors are used to provide a reaction for the test load applied by jack(s)

placed over the pile being tested. Figure 29.8 shows a typical arrangement. The load is increased in defined steps, and is held at each level of loading until all settlement has either ceased or does not exceed a specified amount in a stated period of time.

A plot of load versus head settlement (Fig. 29.9) provides an understanding of pile head performance, which is an essential requirement for ensuring compatibility between superstructure and foundation, but it reveals only part of the picture of overall pile performance.

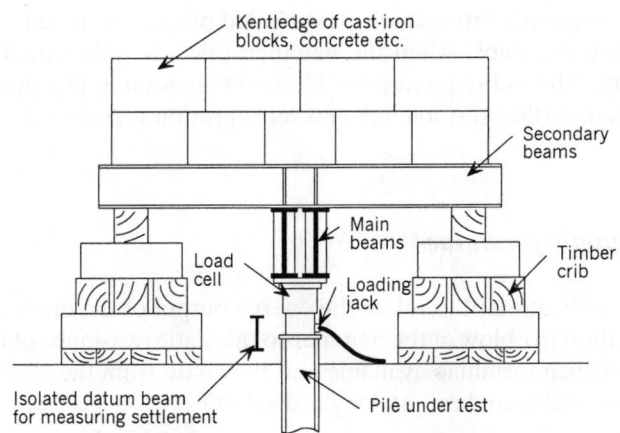

Fig. 29.8   Test load arrangement using kentledge

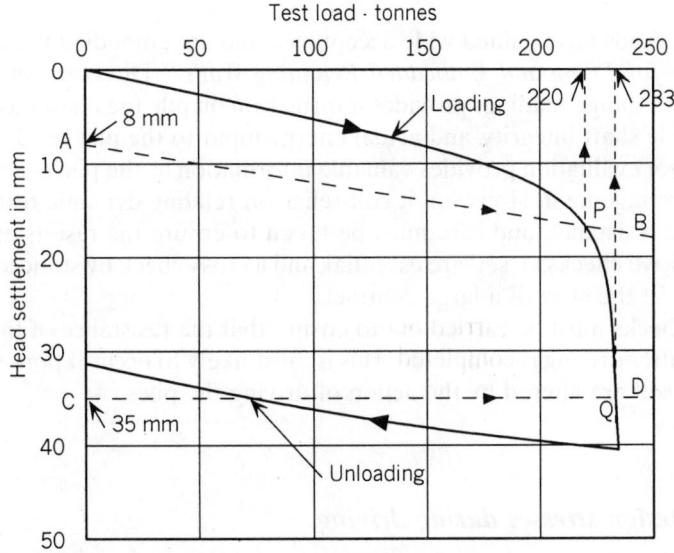

Fig. 29.9   Load–settlement plot

Elastic shortening can be significant in steel piles, especially those which are long and fully stressed in grade S355 quality. Such shortening is of no consequence provided it is anticipated and allowed for in design and evaluation.

29.1.4.2 Pile-driver calibration

To overcome the need to test load every pile, careful records taken during the driving of a subsequently satisfactorily test-loaded pile can be used to evaluate later piles. This system is reliable when the hammer, pile size, pile length, and geology remain unchanged for subsequent piles. However, as soon as any one of these elements is changed, further test loading and recalibration is required.

29.1.4.3 Dynamic formulae

These were an early attempt to relate the energy output of the impact hammer and the pile penetration per blow of the hammer to the static resistance of the pile. There are a number of such formulae available and they date from the 1920s but all have proved to be unreliable and are no longer used.

29.1.4.4 Dynamic stress wave analysis

Dynamic methods have gained wide acceptance and are embodied in the new ICE *Specification for Piling and Embedded Retaining Walls*.[19] Direct on-site computer analysis of pile gauge readings provides information on pile toe resistance, pile shaft resistance, pile shaft integrity, and actual energy input to the pile head.

This form of evaluation provides valuable information as the pile is driven, which is an ideal arrangement. However, it still relies on relating dynamic resistance and 'set' to static resistance, and care must be taken to ensure the results are not misleading. Redrive checks at 'set' are essential, and a cross-check by static loading may be advisable at the start of a large contract.

Redrive checks must be carried out to ensure that the resistance of the pile does not change after driving is completed. This is most likely to occur if pore water pressures in the soil are altered by the action of driving the piles.

29.1.4.5 Design stresses during driving

A driven pile is usually driven to a resistance of twice the working load or to ultimate bearing capacity to demonstrate an adequate factor of safety. BS EN 12699[22]

embodies offshore practice where it states that the maximum dynamic stress in the shaft during driving should be limited to be less than 90% of yield.

29.1.5 Worked examples

Insufficient space is available in this publication to provide comprehensive worked examples for the design of steel bearing piles but they may be found in other publications such as those listed below:

- SCI Publication P-250 *Integral Steel Bridges – Design of a multi-span bridge-Worked Example*[23]
- *Eurocode 7: A Commentary.*[24]

29.2 Sheet piles

29.2.1 Uses

There are various types of earth-retaining structures. Where a differential surface level is to be established with a vertical interface, a retaining wall is used. Steel sheet piling is one form of retaining wall construction. Sheet piles resist soil and water pressures by functioning as a beam spanning vertically between points of support, as shown in Fig. 29.10.

Steel sheet piles are used for temporary and permanent retaining walls. In temporary works and cofferdams the sheet piles enable deep excavations to be made to facilitate construction below ground and water level of other permanent works. On completion of construction the sheet piles are usually extracted for reuse on other projects.

More recent applications have seen the use of sheet piles as permanent retaining walls. Structures include basements, underground carparks and abutments for bridges including integral bridges.

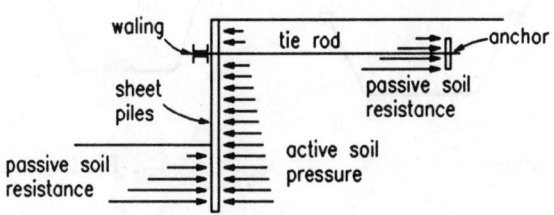

Fig. 29.10   Sheet pile retaining wall

29.2.2 Benefits

Benefits of the use of steel sheet piling include:

- Construction is faster than that for reinforced concrete walls
- Are a narrow form of construction, which can be installed close up to the boundary of the site, maximizing usable building space
- Suitable for all soil types
- No requirement to excavate for wall foundations
- No disturbance of existing ground or arisings unlike that for bored concrete piling
- The steel components are factory quality as opposed to site quality
- High ductility that can reduce bending stresses and soil reactions
- Can easily be made aesthetically pleasing
- Can be placed in advance of other works
- Immediate load-carrying capacity
- Forms a 'curtain walling' to contain the working site
- They are a 'sustainable' product as they can be extracted easily and minimize waste

29.2.3 Types of piles

In the UK, two profiles, designated as U and Z, are available. U profiles are also commonly referred to as *Larssen* sections and Z profiles as *Frodingham* sections. See Fig. 29.11. In addition, where the depth of excavation is small, trench sheet sections may be used as alternatives to sheet piles.

The essential difference between these sections and ordinary structural beams is their ability to interlock with each other to form a continuous membrane in the soil. (Sections and their properties are included in Appendices.)

Steel sheet piling, including sections for box piles, is produced in accordance with BS EN 10248,[25] the most typical grades being S270GP and S355GP with yield strengths of $270\,N/mm^2$ and $355\,N/mm^2$ respectively. In addition, higher-strength

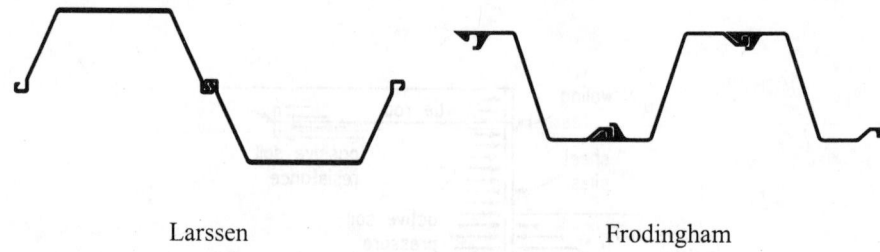

Larssen   Frodingham

Fig. 29.11   Larssen and Frodingham sheet piles

steel sheet piles can be obtained to grades S390GP and S430GP (yield strengths 390 N/mm^2 and 430 N/mm^2 respectively).

29.2.4 Design

29.2.4.1 Design standards

There is no single code of practice for retaining wall design, nor is there likely to be in the future. The principal national standards/guidance associated with the design of retaining walls are:

BS 8002 *Code of practice for earth retaining structures*[8]
CP2 Code of practice No 2: *Earth retaining structures*[26]
CIRIA Report 104 *Design of retaining walls embedded in stiff clays*[27]

BS 8002[8] is a complete revision of the Civil Engineering Code of Practice No 2, which was issued by the Institution of Structural Engineers in 1951 on behalf of the Civil Engineering Codes of Practice Joint Committee. However, in certain situations reference is still made to CP2.

CIRIA Report 104 (1984)[27] was adopted as an unofficial design standard before the publication of BS 8002 and is still widely used today due to familiarity. CIRIA 104 does not address multi-prop walls, but its principle of factoring soil strength has been used in analyses of such walls by deformation methods. CIRIA 104 is currently being updated and will be published as CIRIA Report 629 in 2003.[28]

Recently, numerous CEN 'pre-standard' Eurocodes have become available. The Design Eurocodes provide a set of design *principles* and *application rules* that are 'deemed to satisfy' the principles. They use a limit state design basis and a partial factor approach.

DD ENV 1997-1 *Eurocode 7: Geotechnical design – General rules 1995*[7]
ENV 1993 Eurocode 3: *Design of steel structures Part 5: Piling*[3]

To complement the Eurocodes, there are a number of CEN *'execution'* standards which relate to the installation and construction of the works. The standard most relevant to sheet piles is EN 12063: *Execution of special works – Sheet pile walls.*[29]

29.2.4.2 Limit state design philosophy

Limit state design philosophy is now generally accepted for structural design in the UK and Europe. The old working stress design methods are being replaced by limit state methods.

The design philosophy considers two principal limit states:

- *Ultimate limit state.* Collapse of all or part of the structure
- *Serviceability limit state.* A state, short of collapse, at which deformation, appearance or condition of the structure becomes unacceptable.
- *Accidental limit state.* Unintended loading is applied to the wall i.e. surcharge overload, burst water main etc.

To satisfy ultimate limit state requirements, it must be shown that there is an adequate margin of safety against collapse of any significant element of the soil–structure system, for the worst combination of loading and material properties that can occur. For a satisfactory design, the occurrence of each limit state must have an acceptably low probability. A selection of potential failure modes is shown in Fig. 29.12.

Geotechnical design has traditionally followed an essentially working stress philosophy that, in very many cases, is governed by limits on soil movement (wall deflections) rather than on strength.

Design of retaining walls according to limit state philosophy has to recognize that the geotechnical design may be governed by SLS considerations (deflections) at the same time as the structural design of the walls and props is governed by ULS considerations (for strength). This is a new concept that will require time and experience to come to terms with.

29.2.4.3 *Simplistic representation of soil pressures*

Active soil pressure and passive soil resistance at a given depth are both functions of the effective vertical pressure at that depth and of the strength of the stratum being considered.

For *long-term design*, drained effective stress analysis is used, the effective horizontal active and passive earth pressure equations being given in generalized form by:

$$\sigma_a' = k_a(\gamma z - u + q) - c'k_{ac}$$
$$\sigma_p' = k_p(\gamma z - u + q) + c'k_{pc}$$

where σ_a' is the effective active pressure acting at a depth in the soil
σ_p' is the effective passive pressure acting at a depth in the soil
γ is the bulk density (saturated density if below water level)
z is the depth below ground surface
u is the pore water pressure
q is any uniform surcharge at ground surface
c' is the effective shear strength of the soil.
k_a, k_p, k_{ac}, and k_{pc} are earth pressure coefficients.

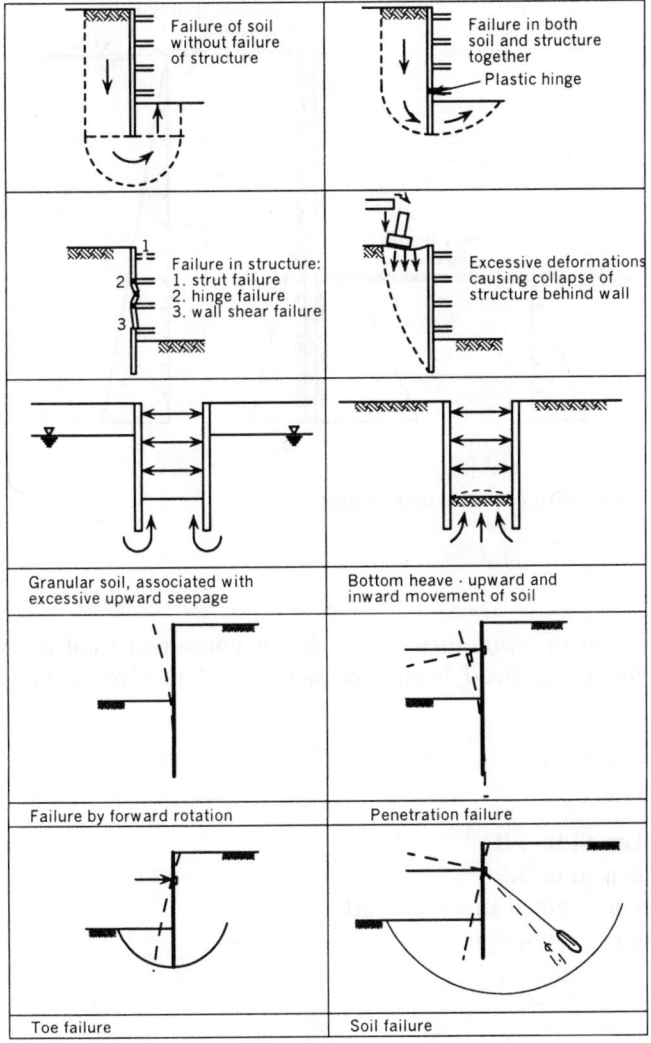

Fig. 29.12   Potential ultimate limit states

See Fig. 29.13 for diagrammatic representation of earth pressures.

The generalized form of k_{ac}, k_{pc} and δ can be obtained from BS 8002[8] or CIRIA 104.[27]

The total horizontal active and passive earth pressures acting against the wall are given by:

$$\sigma_a = \sigma'_a + u \qquad \text{and} \qquad \sigma_p = \sigma'_p + u$$

where u is the pore water pressure.

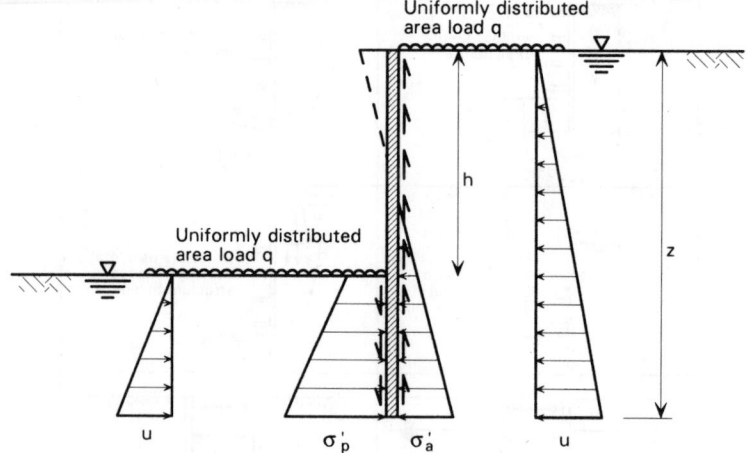

Fig. 29.13   General earth pressure distribution

For a *short-term* or *temporary works design* undrained total stress analysis is used, where the generalized horizontal active and passive earth pressures are reduced to:

$$\sigma_a = (q + \gamma z) - c_u k_{ac}$$
$$\sigma_p = (q + \gamma z) - c_u k_{pc}$$

where k_a is taken to be 1.0

k_p is taken to be 1.0

c_u is the undrained shear strength

c_w is the wall adhesion.

29.2.4.4 Water pressures

Groundwater forces exerted on a retaining wall can often be greater than those from the soil. Careful consideration therefore needs to be given to the variation of water levels and pressures acting on each side of the wall. This applies to both the temporary case (i.e. during construction) and the permanent case. Consideration should also be given to the effects of water seepage, where water flows around the base of the wall into the excavation. This water seepage through the ground will affect the value of pore water pressures and may reduce significantly the value of passive resistance of the soil in front of the toe of the wall and increase the active soil load.

29.2.4.5 Wall stability and depth of embedment for cantilever and propped retaining walls

For a retaining wall, the governing criterion for stability or security against overturning of the wall is one of moment equilibrium. Although other possible failures may occur, they are much less likely than that of overturning. In certain cases (particularly for waterfront structures or in sloping ground), a check should be made to ensure that a deep-seated slip plane (passing behind and below the wall) does not develop.

A minimum required depth of embedment for the wall may be obtained from the equation defining moment equilibrium; restoring moments should exceed overturning moments by a safety margin that is stated in the relevant codes of practice or standards.

The method used by CIRIA 104[27] and BS 8002[8] to determine the depth of embedment for a cantilever wall assumes that the toe of the wall is fixed.

For a propped wall it is assumed that there is sufficient embedment of the wall to prevent horizontal movement but rotation can still take place at the toe. Consequently, the wall is assumed to rotate as a rigid body about the prop. The wall/prop system is assumed to move far enough to develop active pressure in the retained soil (see Fig. 29.14).

The required depth of embedment is determined by equating moments about the prop, assuming fully mobilized active and passive earth pressures (expressed as resultant forces P_a and P_p), as shown in Fig. 29.14. Consideration of horizontal equilibrium allows the necessary prop force to be calculated.

For design, the moment equilibrium condition is used directly or indirectly to ensure that restoring moments exceed overturning moment by the required safety margin. This is achieved either by the use of partial factors (applied to the soil properties) in a limit state design method (often called the *factor on strength method*), or by use of a single or 'lumped' factor of safety which is applied to the bending moment (often called the *factor on moment method*). Both methods are applicable to single-prop walls.

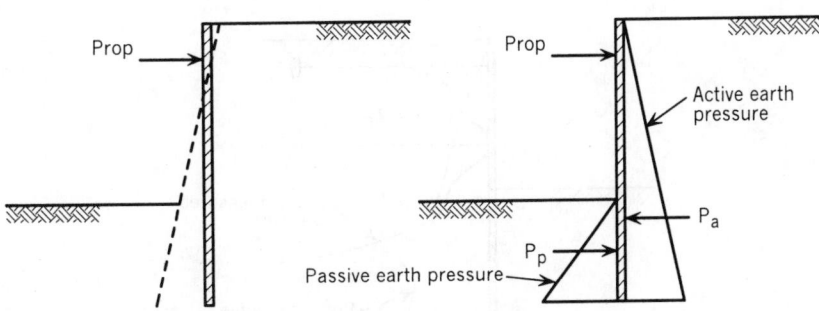

Fig. 29.14 Free-earth boundary condition for a single-prop wall

29.2.4.6 Vertical load resistance

For retaining walls that resist vertical loads from building or bridge superstructures, a method for predicting the axial capacity of sheet piles. Reference should be made to section 29.1 Bearing piles for information or to the SCI publication *Steel bearing piles guide*[10] for more detailed information.

29.2.4.7 Design for lateral loading

Structural forces acting on a retaining wall due to lateral forces are calculated using one of two methods. One method is to model soil–structure interaction effects using advanced analytical methods, and the second method is to use the simplistic distribution of earth pressures assumed in the limit equilibrium method. Both methods are adequate to determine conservative values of the structural forces in cantilever and singly propped retaining walls.

Bending moments and shear forces in the wall and forces in the prop due to pressures assumed in the limit equilibrium method can be calculated either manually or by computer software. Reference should be made to BS 8002[8] or CIRIA Report 104.[27] Two methods are presented to calculate structural forces acting on cantilever and propped retaining walls.

Soil–structure interaction analysis methods predict the earth pressure distribution that acts on the design configuration of the wall. As the relative stiffnesses of the wall and the soil are modelled, the earth pressure profiles predicted using these methods are much more realistic and compare favourably with actual earth pressures. Figure 29.15 shows a typical earth pressure profile for an anchored wall obtained from a soil–structure interaction analysis.

A soil–structure interaction method may be more appropriate where:

- soil movements need to be estimated
- the effect of construction on wall behaviour needs to be studied

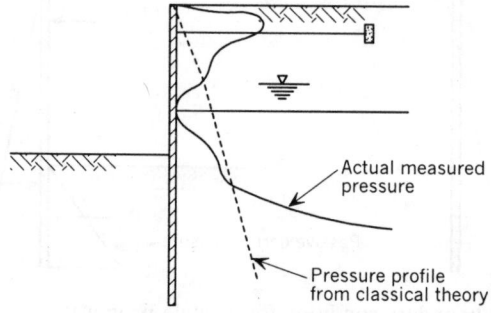

Fig. 29.15 Actual horizonal earth pressure distribution for a flexible sheet pile abutment

- the influence of high initial in-situ at-rest soil stresses needs to be analysed
- wall flexibility and soil stiffness effects are to be modelled.

These aspects cannot be analysed by using simplistic limiting equilibrium methods.

There are a number of commercially-available soil–structure interaction analysis software products that are commonly in use by design practices. Two of the most well known are FREW[13] and WALLAP.[30]

29.2.4.8 Design of multi-prop walls

Multi-prop walls are highly redundant structures and the degree of soil–structure interaction can have a very significant effect on the distribution of forces and moments. The method of construction can have a large influence on the earth pressures acting on the wall. While numerical methods are more appropriate for the analysis of this type of structure, the simple and empirical methods can and do allow approximate solutions to be obtained.

For an accurate analysis of multi-prop walls, deformation methods should be used. These methods consider soil–structure interaction and calculate forces acting on the wall and supports and calculate wall deflections. With the advent of powerful desktop computers, these more complex methods of analysis are now widely available to practising engineers.

Simple methods

The simplistic limiting pressure methods do not satisfy all of the fundamental theoretical requirements to simulate soil–structure behaviour. In particular they do not consider compatibility or the displacement boundary conditions, and hence the methods are only approximate. Although earth pressures acting against multi-propped walls are extremely difficult to predict, simple methods based on modified classical earth pressure distributions have been developed and used. Two methods that have been used are the *Hinge* method and the *continuous beam method*.

The Hinge method, see Fig. 29.16, is mentioned in both BS 8002[8] and the *British Steel Piling Handbook*[20] and is included in the ReWaRD analysis software.[31]

Deformation methods

Deformation methods using a soil–structure interaction approach can produce a much more realistic representation of the behaviour of a retaining wall by taking into account wall and soil stiffness, in situ soil stresses, and the load distribution capability of the soil and wall continuum.

Deformation methods predict the earth pressure distribution that acts on the design configuration of the wall. As the stiffnesses of the wall and the soil are

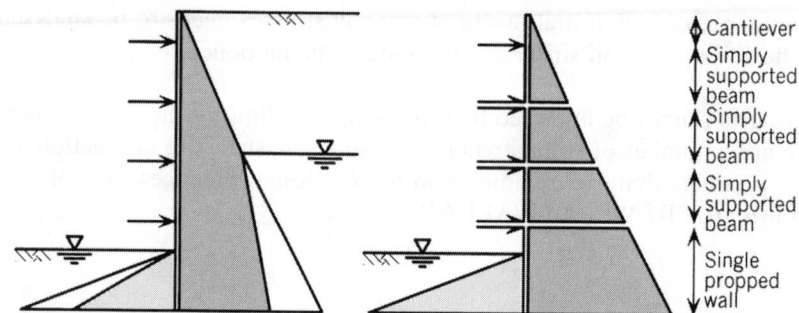

Fig. 29.16   Hinge method for multi-prop walls

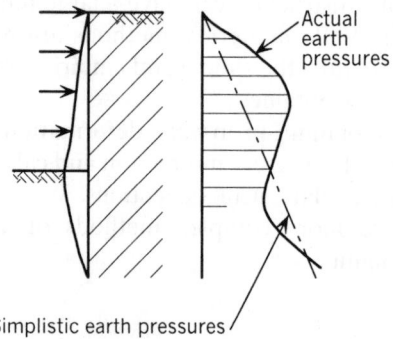

Fig. 29.17   Earth pressure distribution for a multi-propped sheet pile wall

modelled, the earth pressure profiles predicted using these methods compare favourably with actual earth pressures. Figure 29.17 shows a typical earth pressure profile for an multi-prop wall obtained from a soil–structure interaction analysis.

Commercially-available soil–structure interaction analysis software products that are commonly in use by design practices are FREW[13] and WALLAP.[30] In addition finite element and finite difference analyses can be performed. Software products include SAFE,[32] CRISP[33] and FLAC.[34]

29.2.4.9 Base stability

There are several possible modes of instability that can occur in supported excavations. These include:

- base heave
- hydraulic failures

Base heave

One form of base heave can arise as a result of higher pore water pressure in under-lying soil layers that is created by the excavation. The heave can occur in clay layers overlying a permeable sand or gravel that has a sufficiently high pore water pressure to force the clay up into the base of the excavation. A simple calculation comparing the weight of the thin clay layer with the pore water pressure beneath it will indicate the potential of failure in this manner. See Fig. 29.18.

Another form of base heave arises if the soil at the base is not strong enough to support the overburden stress imposed by the soil adjacent to the excavation. In this case the base of the excavation will fail and the soil will be forced up. See Fig. 29.19.

As this type of failure can occur during construction and before any base slab is installed, analysis is usually performed using the undrained shear strength, c_u. The most commonly used method is the Bjerrum and Eide method.[35]

Hydraulic instability in granular soils

Where groundwater exists above the base of the excavation, and where the toe of the wall does not penetrate into an impermeable layer, flow will occur under the

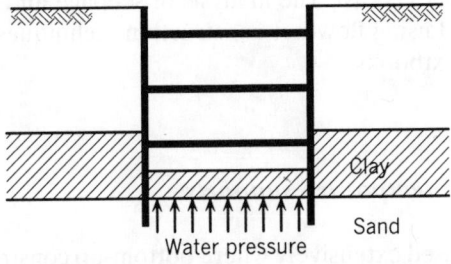

Fig. 29.18 Base heave resulting from excessive water pressures

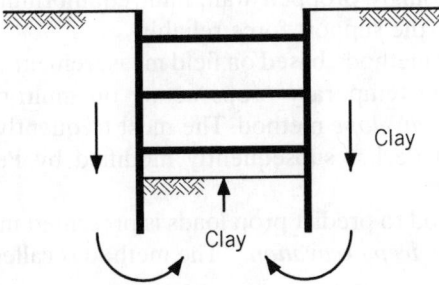

Fig. 29.19 Base heave due to weight of adjacent soil

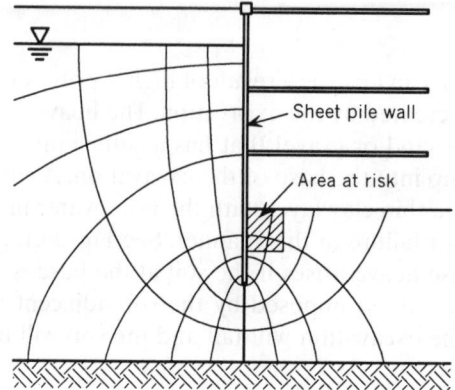

Fig. 29.20   Piping in soils

wall and upward through the base of the excavation. The result is loosening of the bottom soils, which may cause collapse of the wall and loss of the bearing capacity for building foundations.

The most effective control is dewatering, but in deep excavations it may be more economical to penetrate the wall to a depth sufficient to intercept the potential flow lines with high heads of water: see Fig. 29.20. It should be noted that corners of basements and cofferdams are at the greatest risk from piping failure during construction, before the base slab is cast. The analysis of seepage and the determination of flow lines are obtained using flow-net construction techniques, which can be found in many foundation textbooks.

29.2.4.10 *Design of temporary supports*

Temporary props are used extensively where bottom-up construction is undertaken. Steel sections, which include tubulars, box sections and universal columns, are the most common forms of temporary propping. In certain cases soil berms may be used either on their own or with temporary props.

For the design of a singly propped wall, limit equilibrium methods are usually adequate to determine the support force reliably.

Numerous empirical methods, based on field measurements, have been developed to determine forces on temporary props acting on multi-prop walls. One such method is the *pressure envelope* method. The most frequently used 'envelopes' are those of Terzaghi and Peck,[36] subsequently modified by Peck.[37] This method is included in BS 8002.[8]

A more recent method to predict prop loads is presented in CIRIA Report C517 *Temporary propping of deep excavation.*[38] The method is called the *distributed prop load* method.

An increase or decrease in the temperature of a prop from its installation temperature will cause the prop to expand or contract. If the prop is restricted or prevented from expanding freely, an additional load is generated in the prop. It is not usual for deformation methods of analysis to include temperature effects, although this is within the scope of most of the available methods. Temperature effects are normally added to the predicted prop loads after the analysis is complete. No common approach exists in the UK for the design of props loaded as a result of increases in prop temperature above the installation temperature.

29.2.4.11 Design for driveability

It is important to check that the pile section chosen is capable of withstanding the rigours of driving and will reach the desired penetration in a condition suited to the application for which it is intended. Relationship tables are presented in the *British Steel Piling Handbook*.[20] However, it is important to recognize that these relationships should only be treated as the preliminary process in hammer sizing and that they are not intended to be a substitute for engineering experience or local knowledge.

Generally, the driving capability of sheet piles increases with the section modulus and can also be improved by specification of a higher grade of steel.

When the soil to be penetrated is hard, the pile section required for structural purposes will be lighter than that required for driving. This does not need to be a penalty as the additional steel can be considered as sacrificial, enhancing the corrosion performance of the steel piles.

Driveability of sheet piles in granular soils

Vibratory driving is the most effective means of installing piles in granular soils but this method may not be efficient when standard penetration test N values exceed approximately 50 blows. In these conditions it will be necessary either to treat the ground with pre-boring or water jetting or to adopt impact driving. Care must be taken in the design as changes to the soil properties adjacent to the pile wall may result from ground treatment measures.

Driveability of sheet piles in cohesive soils

Impact driving is the most efficient means of installing sheet piles in cohesive materials. When noise and vibration are issues, pile jacking techniques are available. In either case, it is essential that an appropriate pile section is adopted for the ground conditions present. If the section selected is too light installation stresses may cause the pile to be damaged or become misaligned. A guide to selecting a pile section

Table 29.2 Selection of sheet piles in cohesive soils – *British Steel Piling Handbook*[20]

Clay description	Minimum wall modulus (cm³/m)	
	Grade S270GP	Grade S355GP
Soft to firm	450	
Firm	600–700	450–600
Firm to stiff	700–1600	600–1300
Stiff	1600–2500	1300–2000
Very stiff	2500–3000	2000–2500
Hard ($c_u > 200$)	Not recommended	4200–5000

Note. The ability of piles to penetrate any type of ground depends upon attention being given to good pile practice.

size in cohesive soil is given in BS 8002[8] and in the *British Steel Piling Handbook.*[24] Table 29.2 summarizes the guidance.

29.2.5 Worked examples

No space is available in this publication to provide comprehensive worked examples for the design of embedded sheet pile retaining walls. However, worked examples of retaining wall design and analysis can be found in various other publications. A selection of these publications are listed below for reference:

- CIRIA Report R104 *Design of retaining walls embedded in stiff clays*[27]
- CIRIA Report 629 *Embedded retaining walls in stiff clay: Guidance for more economic design* (to be published in 2002)[28]
- SCI Publication P-187 *Design guide for steel sheet pile bridge abutments*[39]
- SCI Publication P-180 *Integral Steel Bridges – Design of a single span bridge – Worked Example*[40]
- SCI Publication P-250 *Integral Steel Bridges – Design of a multi-span bridge – Worked Example*[23]
- Eurocode 7 *A commentary*[24]
- CIRIA Special Publication 95 *The design and construction of sheet-piled cofferdams*[41]

29.3 Pile driving and installation

29.3.1 Steel pile installation tolerances

Information on tolerances that are achievable using commonly available pile driving equipment and methods is quoted in the Institution of Civil Engineers publication

Table 29.3   Steel pile driving tolerances (TESPA)[18]

Type of pile and method of driving	For pitch and drive method or over water	For panel drive method
Deviation normal to the wall centre line at pile head	±50 mm Dependent on equipment used	±50 mm Dependent on equipment used

Finished level deviation from a specified level
  of pile head, after trim
  of pile toe
	
	±20 mm
	±120 mm
		
		±20 mm
		±120 mm

Deviation from specified inclination measured over the top 1 m of wall
  Normal to line of piles
  Along line of piles
	
	±1%
	±1%
		
		±1%
		±0.5%

Specification for piling[19] and specifications issued by the Federation of Piling Specialists,[17] the CEN Standard EN 12063 *Execution of special geotechnical works – Sheet-pile walls,*[29] and the TESPA publication *Installation of steel sheet piles.*[18]

Table 29.3 is included in the TESPA publication and represents tolerance levels for sheet piling which should not be too onerous to achieve but will give results which are visually acceptable – an important feature for permanent exposed sheet piling.

Accuracy of alignment will also be affected by pile stiffness, the driving equipment, and the experience of the workforce. Use of pile guide frames, which are often formed from universal beams aligned with their webs horizontal, will ensure that good alignment of the sheet piles is achieved.

29.3.2 Environmental factors: noise and vibration prediction

Increasing attention has been directed to environmental factors with regard to driven piles in recent years. Although the duration of the piling contract may be short in comparison with the whole contract period, noise and vibration perception may be more acute during the piling phase. Human perception is very intolerant of noise and vibration or shock transmitted through the ground, and tolerance requires careful prior education of the public. Efforts made to advise the public and to plan the precise times of driving carefully can reassure those likely to be affected in the vicinity of a pile installation and can result in the necessary cooperation.

In the UK, the Control of Pollution Act (1974) provides a legislative framework for, amongst other things, the control of construction site noise. The Act defines noise as including vibration and provides for the publication and approval of codes of practice, the approved code being BS 5228.[42] Part 4 of the code deals specifically with piling noise. This code was revised in 1992 to include guidance on vibration.

Two relevant documents include the TRRL Research Report RR53 *Ground vibration caused by civil engineering works*,[43] and the British Steel publication *Control of vibration and noise during piling*.[44] The SCI publication *Specifiers' Guide to Steel Piling*[45] also contains useful advice.

BS 6472 deals specifically with evaluation of human exposure to noise and vibration in buildings.[46]

29.3.2.1 Noise from piling operations

Pile driving is perceived to be an inherently noisy operation because impact-based methods of installation have historically been used. Typical data on noise levels produced by piling operations have been published by CIRIA Report No. 64 *Noise from construction and demolition sites – Measured levels and their prediction*.[47] These are discussed and interpreted in CIRIA Report PG9 *Noise and vibrations from piling operations*.[48] It is important to note that modern pile installation equipment such as the noise-free pile jacking machines will not be included as they post date these publications.

Impact driving of steel sheet piling is often noisy since the operation involves steel-to-steel contact. In areas where severe restrictions are placed on noise levels, pile vibratory or jacking equipment should be used. Such machines emit a different frequency and lower level of noise which may be acceptable, and recent advances in noise reduction technology can ensure that the noise from the auxiliary power pack is also lowered.

29.3.2.2 Ground vibrations caused by piling

It is widely recognized that noise and vibration, although related, are not amenable to similar curative treatment. In the main, noise from a site is airborne and consequently the prediction of noise levels is relatively straightforward, given the noise characteristics and mode of use of the equipment. On the other hand, the transmission of vibration is determined largely by site soil conditions and the particular nature of the structures involved. General guidance can be derived from the study of case histories of similar situations. Useful references on the subject of ground vibrations are provided by CIRIA Technical Note 142 *Ground-borne vibrations arising from piling*,[49] the publication *Dynamic ground movements – Man-made vibrations in ground movements and their effects on structures*,[50] and BRE Digest No. 403 *Damage to structures from ground-borne vibration*.[51]

29.4 Durability

It is important that the long-term performance of the structure is considered both in the choice of structural form and in the design of construction details. Failure to do so may result in maintenance problems requiring costly repair.

29.4.1 Corrosion allowances

The means for countering the effect of corrosion of steel piles are well developed. Guidance is given in the *British Steel Piling Handbook*.[20]

BS 8002[8] considers that the end of the effective life of a steel sheet pile occurs when the loss of section, due to corrosion, causes the stress to reach the specified minimum yield strength. A pile section chosen for the in-service condition has to be adequate at its end-of-design-life, at which time the effective pile section will have been reduced by corrosion.

As the corrosion loss allowance varies along the pile according to the corrosion environment, the designer needs to be aware that the maximum corrosion may occur at a different level to that of the maximum forces and moments, and should allow for this accordingly.

Also, since redistribution of earth pressures may occur as a result of increased flexure of a corroded section, the end-of-design-life condition may be a critical design load case in the selection of the sheet pile section.

29.4.2 Corrosion and protection of steel piles

The design life requirements for proposed buildings and individual components or assemblies are defined in BS 7543 *Guide to durability of buildings and building elements*,[52] where a building design life can range from 10 years for a building with a 'short' life to 120 years for civic and other high quality buildings. A retaining wall which is part of a building structure, i.e. a basement, must therefore comply with these requirements and be designed with sacrificial thicknesses applied to each surface, depending on the exposure conditions. The exposure conditions are based on the advice given in BS 8002 clause 4.4.4.4.3 and are shown in Table 29.4.[8]

The reduced (corroded) section properties can be obtained either by calculation or from the *British Steel Piling Handbook*.[20]

Another way of allowing for sacrificial thickness is to use a higher strength steel than would be required if no corrosion were assumed (i.e. use steel grade S355GP, to BS EN 10248,[25] in a wall designed for steel grade S270GP). This permits a greater loss of metal before stresses become critical.

Table 29.4 Sacrificial thicknesses for piling according to BS 8002[8]

Exposure zone	Sacrificial thickness (for one side of the pile only) (mm/year)
Atmospheric	0.035 (mean)
Continuous immersion in water or effluent	0.035 (mean)
In contact with natural soil	0.015 (max)
Splash and alternating wet/dry conditions	0.075 (mean)

It should be noted that the corrosion allowances apply to unprotected steel piles. Although it is generally cost effective to provide the sacrificial steel thickness, consideration can alternatively be given to the following corrosion protection options:

- Protective coatings, particularly in the exposed section of the pile.
- Cathodic protection in soil below the water table or in a marine environment.

Details of these options are given in the *British Steel Piling Handbook*.[20]

29.4.3 Corrosion in fill or industrial soils

Buildings can be constructed in areas of recent fill or industrial soils. Corrosion protection of the steel in contact with the fill material may be required, and this can be assessed by testing the material for pH and resistivity.

The nature of in situ fill soils can be variable, and a full soil analysis is required to assess the likely corrosion performance of steel in the environment. Soil tests to determine the pH of the soil should be in accordance with BS 1377-3[53] and as directed by the contract to determine resistivity. Other tests may be relevant, and most of these are reviewed in CIRIA' s series of reports on contaminated land (contact CIRIA for further details).

In a controlled fill, no special measures are required, and the same corrosion rates as in natural undisturbed soils can be assumed.

Corus (formerly British Steel) has undertaken significant research and development into corrosion of steel and corrosion protection. Further advice on corrosion assessment and protection can be obtained from Corus Construction Centre or from The Steel Construction Institute.

29.4.4 Corrosion and structural forces

It is not immediately obvious whether the start of in-service life case or the end-of-design-life case will be the most critical for the structural design of retaining walls. At the end-of-design-life, the reduced stiffness of the corroded steel pile will permit

increased deflection, which will in turn reduce the soil pressures acting upon it (and therefore the induced moments and shears).

As the corrosion loss allowance varies along the pile according to the corrosion environment, the designer should consider that the maximum corrosion may not occur at the same level as the maximum section stresses.

The reduced (corroded) section properties can be obtained either by calculation or from the *British Steel Piling Handbook*[20] or other Corus brochure.

References to Chapter 29

1. British Standards Institution (1998) BS EN 10025: *Hot rolled products of non-alloy structural steels. Technical delivery conditions.* BSI, London.
2. American Petroleum Institute (1995) API 5L: *Specifications for steel linepipe.* API.
3. European Committee for Standardization (1998) ENV 1993: Eurocode 3: *Design of steel structures* Part 5: *Piling.* CEN.
4. American Petroleum Institute (1993) API Recommended practice 2A-LRFD (RP2A-LRFD), 1st Edition, *Recommended practice for planning, designing and constructing fixed offshore platforms – load and resistance factor design.* API.
5. International Organization for Standardization (1995) 13819-2: *Petroleum and natural gas industries: offshore structures Part 2: Fixed steel structures.* ISO.
6. European Committee for Standardization (1994) ENV 1997: Eurocode 7 *Geotechnical design* Part 1: *General rules.* CEN.
7. British Standards Institution (1995) Draft for Development DD ENV 1997-1: 1995 Eurocode 7. *Geotechnical design. General rules* (together with United Kingdom National Application Document). BSI, London.
8. British Standards Institution (1994) BS 8002: 1994: *Code of practice for earth retaining structure.* BSI, London.
9. British Standards Institution (1988, 2000) BS 6349: *Maritime structures,* Part 1 2000 *Code of practice for general criteria,* Part 2 1988 *Design of quay walls, jetties and dolphins.* BSI, London.
10. British Standards Institution (1969) BS 449: *Specification for the use of structural steel in building.* BSI, London.
11. Biddle, A.R. (1997) *Steel bearing piles guide.* The Steel Construction Institute, Ascot.
12. British Standards Institution (1986) BS 8004: 1986. *Code of practice for foundations.* BSI, London.
13. OASYS (YEAR) Geotechnical suite of programs. ALP – *Laterally loaded pile analysis,* PILE: *Vertical pile capacity,* FREW *Flexible retaining wall analysis,* SAFE *Soil analysis using finite elements.* Ove Arup Oasys Ltd, London.
14. Elson W.K. (1984) *Design of laterally-loaded piles,* CIRIA Report 103. CIRIA, London.
15. Poulos H.J. & Davis E.H. (1980) *Pile foundation analysis and design.* John Wiley and Sons.

16. Tomlinson M.J. (1994) *Pile design and construction practice*, 4th edn. E & FN Spon.
17. Federation of Piling Specialists (1991) *Specification for steel sheet piling*. FPS, London.
18. Technical European Sheet Piling Association (1995) *Installation of steel sheet piles*. TESPA, L-2930 Luxembourg.
19. Institution of Civil Engineers (1996) *Specification for piling and embedded retaining walls*. ICE.
20. British Steel Sections, Plates & Commercial Steels (1997) *Piling handbook*, 7th edn. BS SP & CS.
21. CAPWAP (1966) Case Pile Wave Analysis Program. In: *CAPWAP Users Manual*. Goble Rausche Likins and Associates Inc.
22. British Standards Institution (2001) BS EN 12699: *Execution of special geotechnical work. Displacement piles*. BSI. London.
23. Way J.A. & Biddle A.R. (1998) P-250 *Integral Steel Bridges Design of a multi-span bridge – Worked Example.* The Steel Construction Institute, Ascot, Berks.
24. Simpson B. & Driscoll R. (1998) *Eurocode 7 A commentary*. DETR, BRE & ARUP.
25. British Standards Institution BS EN 10248: *Hot rolled sheet piling of non alloy steels*. Part 1: 1996 *Technical delivery conditions*. Part 2: 1996 *Tolerances on shape and dimensions*. BSI, London.
26. Institution of Structural Engineers (1951) Code of Practice No 2: *Earth retaining structures*. IStructE.
27. Construction Industry Research and Information Association (1984) *Design of retaining walls embedded in stiff clays*, CIRIA Report 104. CIRIA, London.
28. Construction Industry Research and Information Association (2003) *Embedded retaining walls in stiff clay: guidance for more economic design*, CIRIA Report 629. CIRIA, London.
29. British Standards Institution (2001) BS EN 12063: *Execution of special geotechnical work. Sheet pile walls*. BSI, London.
30. Borin D.L. (1997) *WALLAP anchored and cantilevered retaining wall analysis program: User's manual* (Version 4). Geosolve, London.
31. Bond A.J., & Potts D.M. (1995) *A program for advanced retaining wall design and analysis*, ReWaRD. Geotechnical Consulting Group.
32. SAFE – Oasys Geo 4 Package, *Soil Analysis using Finite Elements program*, Oasys suite of geotechnical analysis programs. Ove Arup Oasys Ltd, London.
33. Sage CRISP (1995) *Critical State soil mechanics Program*. Sage Engineering Ltd and University of Cambridge.
34. FLAC (*Fast Lagragian Analysis of Continua*), ITASCA Consulting Group Inc., USA.
35. Bjerrum L. & Eide O. (1956) Stability of strutted excavations in clay, *Géotechnique*, Vol. 6, No. 1, March, 32–47.
36. Terzaghi K. & Peck R.B. (1967) *Soil Mechanics in Engineering Practice*. pp. 729. John Wiley and Sons, New York.

37. Peck R.B. (1969) Deep excavations and tunnelling in soft ground, *Proc. 7th Int. Conf. SM & FE, Mexico.*
38. Construction Industry Research and Information Association (2000) *Temporary propping of deep excavations*, CIRIA Report C517. CIRIA, London.
39. Yandzio E. (1998) P-187 *Design guide for steel sheet pile bridge abutments.* The Steel Construction Institute, Ascot, Berks.
40. Way J.A. & Yandzio E. (1997) P-180 *Integral Steel Bridges – Design of a single span bridge – Worked Example.* The Steel Construction Institute, Ascot, Berks.
41. Williams B.P. & Waite D. (1993) *The design and construction of sheet-piled cofferdams.* CIRIA Special Publication 95, Construction Industry Research and Information Association.
42. British Standards Institution (1992) BS 5228: *Noise control on construction and open sites.* Part 4: 1992 *Code of practice for noise and vibration control applicable to piling operations.* BSI, London.
43. Transport and Road Research Laboratory (1986) *Ground vibration caused by civil engineering works.* TRRL Report RR53, Crowthorne.
44. British Steel (1997) *Control of vibration and noise during piling.*
45. Biddle A.R. & Yandzio E. (2002) *Specifiers' Guide to Steel Piling*, SCI Publication P-308. The Steel Construction Institute, Ascot, Berks.
46. British Standards Institution (1992) BS 6472: 1992 *Guide to evaluation of human exposure to vibration in buildings (1 Hz to 80 Hz).* BSI, London.
47. Construction Industry Research and Information Association (1977) *Noise from construction and demolition sites – Measured levels and their prediction.* CIRIA Report 64. CIRIA, London.
48. Weltman A.J. (1980) *Noise and vibrations from piling operations.* DOE and CIRIA Piling Development Group. CIRIA Report PG9. CIRIA, London.
49. Head J.M. & Jardine F.M. (1992) *Ground-borne vibrations arising from piling.* CIRIA Technical Note 142. CIRIA, London.
50. Skipp B.O. (1984) *Dynamic Ground Movements – Man-Made Vibrations in Ground Movements and their Effects on Structures.* pp. 381–434. Blackie, Glasgow and London.
51. Building Research Establishment (1995) *Damage to structures from ground-borne vibration.* BRE Digest No. 403.
52. British Standards Institution (1992) BS 7543: 1992 *Guide to durability of buildings and building elements, products and components.* BSI, London.
53. British Standards Institution (1990) BS 1377: 1990 *Methods of text for soils for civil engineering purposes.* Part 3: *Chemical and electro-chemical tests.* BSI, London.

Chapter 30
Floors and orthotropic decks

by DICK STAINSBY

30.1 Steel plate floors

Steel plate used for floors, walkways or for staircase treads normally has a raised pattern of the non-slip type such as Durbar plate supplied by Corus. Where a plated area will have only occasional use, such as walkways for maintenance access, then plain plate may be used.

If properly designed, both plain and non-slip plates have a dual function in carrying floor loads and in acting as a horizontal diaphragm in place of a separate bracing system. For example, a plate may be connected to the top flange of crane gantry girders to act as a surge girder and also as a means of crane or maintenance access; the surge girder elements should be designed as a horizontal plate girder.

When steel plate floors are in situations where moisture or high humidity is present, then the plates should be continuously welded to the support members to avoid corrosion problems. In dry areas connections may be fastened by bolting with countersunk bolts, intermittent welds, or clips. When clips are used they must be arranged so that they cannot move out of position.

Plain plate is commonly available in grades S275JR and S355JR to BS EN 10025.

Durbar plate

The pattern of Durbar plate is shown in Fig. 30.1. The pattern is raised 1.5 mm to 2.2 mm above the plate surface. When ordering, the thickness should be specified as 'thickness on plain'.

Durbar-pattern plate is free draining and can be matched width to length without affecting appearance.

Durbar plate is commonly available in grade S275JR and S355JR to BS EN 10025, also 'commercial quality' with a UTS not less than 355 N/mm².

30.1.1 Design of plates simply supported on four edges

This condition is for plates resting on four edges or attached by clips or tack welds only to supporting members. Plates supported on two edges, but with stiffeners in the other direction, can also be considered as simply supported on four edges.

Table 30.1   Plain plate

Thickness (mm)	3	4.5	6	8	10	12.5	15
Mass (kg/m²)	23.6	35.3	47.1	62.8	78.5	98.1	118

Table 30.2   Durbar plate

Thickness on plain (mm)	3	4.5	6	8	10	12.5
Mass (kg/m²)	24.9	36.9	48.7	64.4	80.1	99.7

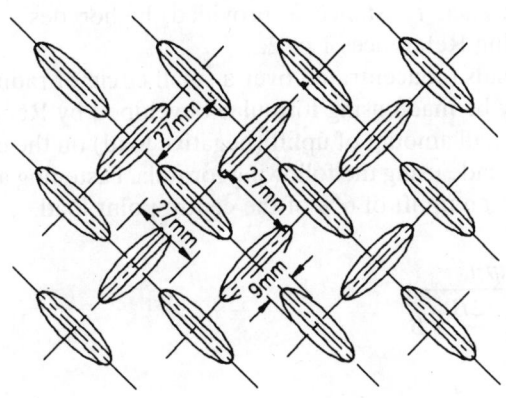

Fig. 30.1   Durbar plate

Table 30.3   Recommended standard sizes for Durbar plate

Width (mm)	600	1000	1000	1250	1500	1750	1830
Length (m)	2.0, 2.5, 3.0			Any between 1.8 and 12.0			

(1) Uniform distributed loading

The design may be made using Pounder's plate theory.

$$w = \frac{4p_y t^2}{3kB^2 \left[1 + \frac{14}{75}(1-k) + \frac{20}{57}(1-k)^2\right]}$$

$$d = \frac{m^2 - 1}{m^2} \times \frac{5kwB^4}{32\gamma_f Et^3}\left[1 + \frac{37}{175}(1-k) + \frac{79}{201}(1-k)^2\right]$$

where w = uniformly distributed load on plate (ultimate) (N/mm²)

d = maximum deflection (mm), occurring at serviceability

$\frac{1}{m}$ = Poisson's ratio ($m = 3$)

L = length of plate (mm) ($L > B$)

B = breadth of plate (mm)

t = thickness of plate (mm)

$k = \left(\dfrac{L^4}{L^4 + B^4}\right)$

p_y = yield stress of plate (N/mm²)

E = Young's modulus 205×10^3 (N/mm²)

γ_f = load factor

This formula, which is the basis of the first two load tables in the Appendix *Floor plate design tables*, assumes that there is no resistance to uplift in plate corners. Where such resistance is provided, higher design strength may be obtained by using References 1 or 2.

(2) Centre point loads (concentrated over a small circle of radius r)

The design may be made using formulae developed by Roark.[1]

NB There is a small amount of uplift (negative load) on the corners of the plate when design is made using the following formula. Fastening at each corner must be able to resist an uplift of 6% of the centre point load.

$$w = \frac{0.67 \pi p_y t^2}{\left(1 + \dfrac{1}{m}\right) \ln \dfrac{2B}{\pi r} + \beta}$$

$$d = \frac{\alpha W B^2}{\gamma_f E t^3}$$

where W = concentrated load (ultimate) (N)

d = maximum deflection (mm), occurring at serviceability

L = length of plate (mm) ($L > B$)

B = breadth of plate (mm)

t = thickness of plate (mm)

p_y = yield stress of plate (N/mm²)

E = Young's modulus 205×10^3 (N/mm²)

$\dfrac{1}{m}$ = Poisson's ratio ($m = 3$)

r = radius of contact for a load concentrated on a small area (mm)

$\ln$ = natural logarithm of $(2B/\pi r)$

γ_f = load factor

Table 30.4 Values of β and α for centre point loads on simply supported plates

L/B	1.0	1.2	1.4	1.6	1.8	2.0	∞
β	0.435	0.650	0.789	0.875	0.972	0.958	1.0
α	0.1267	0.1478	0.1621	0.1715	0.1770	0.1805	0.1851

30.1.2 Design of plates fixed on four edges

The plates must be secured by continuous bolting or welding, which will achieve the fixed condition and prevent uplift which would otherwise occur at the plate corners. The bolt or weld detail should be such as to develop the strength in bending of the chosen plate thickness.

(1) Uniformly distributed loading
   The design may be made using Pounder's plate theory.

$$w = \frac{2p_y t^2}{kB^2 \left[1 + \frac{11}{35}(1-k) + \frac{79}{141}(1-k)^2\right]}$$

$$d = \frac{m^2 - 1}{m^2} \times \frac{kwB^4}{32\gamma_f Et^3}\left(1 + \frac{47}{210}(1-k) + \frac{200}{517}(1-k)^2\right)$$

where L = length of plate (mm) $(L > B)$
   B = breadth of plate (mm)
   t = thickness of plate (mm)
   k = $\left(\dfrac{L^4}{L^4 + B^4}\right)$
   p_y = yield stress of plate (N/mm²)
   w = uniformly distributed load on plate (ultimate) (N/mm²)
   E = Young's modulus 205×10^3 (N/mm²)
   $\dfrac{1}{m}$ = Poisson's ratio $(m = 3)$
   d = maximum deflection (mm), occurring at serviceability
   γ_f = load factor

   This formula is the basis of the third load table in the Appendix *Floor plate design tables*.
(2) Centre point loads (concentrated over a small circle of radius r)
   The design may be made using formulae developed by Roark.[1]

$$W = \frac{0.67\pi p_y t^2}{\left(1 + \dfrac{1}{m}\right)\ln\dfrac{2B}{\pi r} + \beta_1} \quad \text{or} \quad \frac{p_y t^2}{\beta_2}$$

   whichever is the lesser, where

$$d = \frac{\alpha W B^2}{\gamma_f E t^3}$$
   W = concentrated load (ultimate) (N)
   d = maximum deflection (mm), occurring at serviceability
   L = length of plate (mm) $(L > B)$
   B = breadth of plate (mm)

t = thickness of plate (mm)

p_y = yield stress of plate (N/mm^2)

E = Young's modulus 205×10^3 (N/mm^2)

$\dfrac{1}{m}$ = Poisson's ratio ($m = 3$)

r = radius of contact for a load concentrated on a small area (mm)

ln = natural logarithm of $\dfrac{2B}{\pi r}$

γ_f = load factor

30.1.3 Design criteria

Imposed loading

Walkway loading is specified in British Standard BS 6399: Part 1 *Loading for buildings*. Table 30.6 is an extract from table 10 dealing with workshops and factories.

Horizontal loading applied to handrailing should be taken as the following applied at 1.1 m above the walkway (irrespective of the handrail height):

General duty – regular two-way pedestrian traffic	0.36 kN/m
Heavy duty – high-density pedestrian traffic: escape routes	0.74 kN/m
Potentially crowded areas over 3 m wide	3.00 kN/m

(See also BS 5395: Part 3: 1985, *Code of practice for the design of industrial type stairs, permanent ladders and walkways* and BS 6180: 1999 *Barriers in and about buildings – Code of practice*.)

Table 30.5 Values of β_1, β_2 and α for centre point loads on fixed plates

L/B	1.0	1.2	1.4	1.6	1.8	2.0	∞
β_1	−0.238	−0.078	0.011	0.053	0.068	0.067	0.067
β_2	0.7542	0.8940	0.9624	0.9906	1.000	1.004	1.008
α	0.0611	0.0706	0.0754	0.0777	0.0786	0.0788	0.0791

Table 30.6 Extract from table 10 of BS 6399: Part 1

Floor area usage	Intensity of distribution (kN/m^2)	Concentrated load (kN)
Factories, workshops and similar buildings	5.0	4.5
Access hallways, stairs and footbridges	4.0	4.5
Factories, workshops and similar buildings	5.0	4.5
Cat walks	–	1 kN @ 1 m centres

Load factors

When designing to BS 5950: Part 1: 2000, the load factors in Table 30.7 should be used.

Deflection

Deflection of floor plating under the action of dead and imposed loads should not exceed $B/100$, where B is the minimum spanning dimension of the plate.

Table 30.7 Load factors

Loading	Factor
Dead load	1.4
Vertical load	1.6
Horizontal load	1.6
Vertical & horizontal combined	1.4

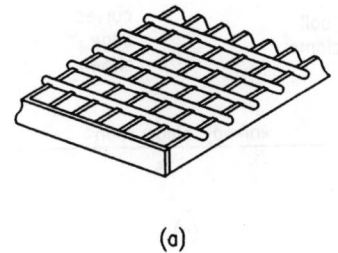

(a)

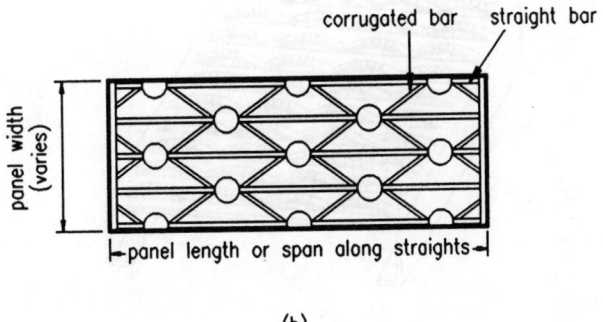

(b)

Fig. 30.2   Open grid flooring. (a) Open bar; (b) diamond pattern; (c) expanded metal; (d) Q-grating

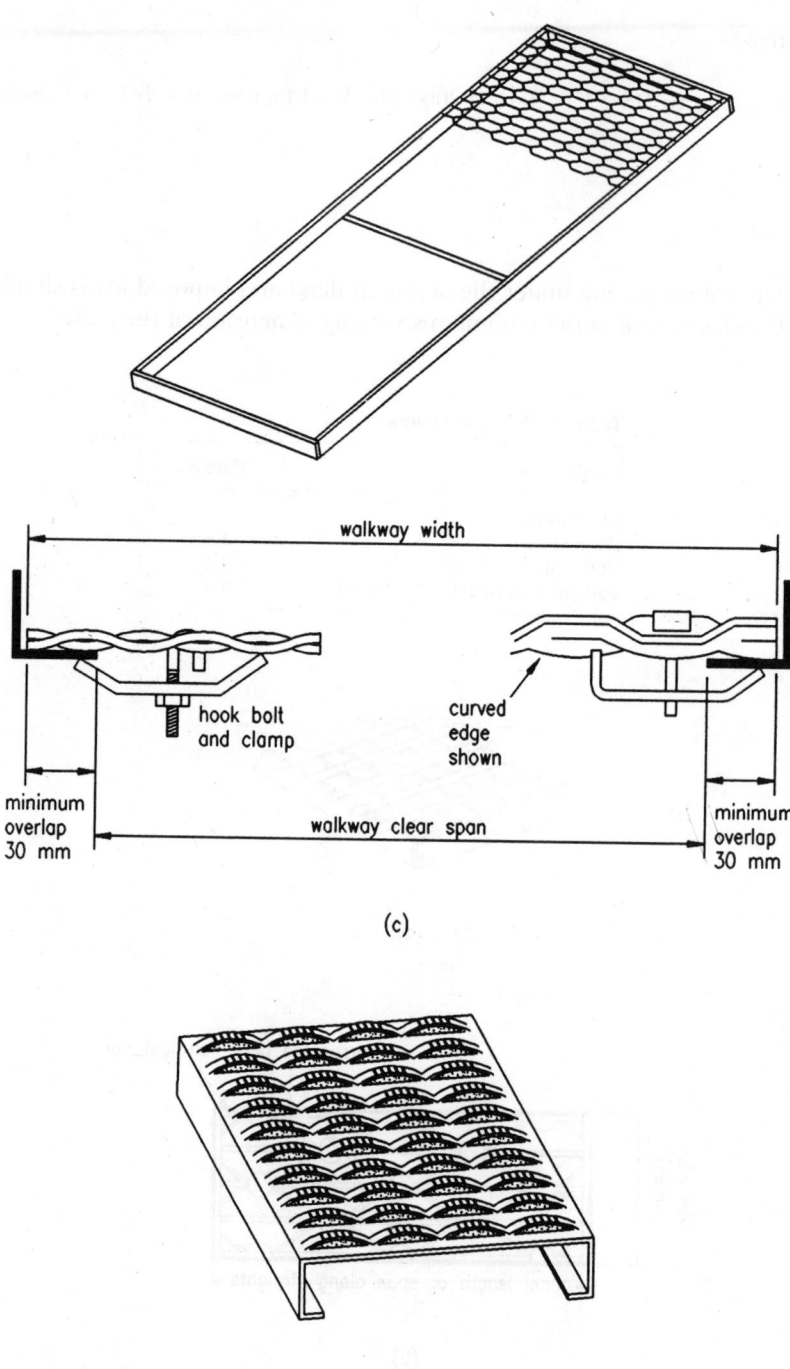

(c)

(d)

Fig. 30.2 continued

30.2 Open-grid flooring

Open-grid proprietary systems are often the economical solution for industrial flooring and walkways, particularly where ventilation or light must be available through the flooring. They are generally available in spans up to 2.0 metres, but can be supplied to span 4.0 metres. The design allows spanning in one direction only. Stair treads are also available. (See Fig. 30.2.)

Open-grid systems are usually manufactured in grade S275 steel, with painted or galvanized finish. Aluminium and stainless steel types are also available. Load-carrying capacities and fixing details are readily available from suppliers' catalogues.

Open-grid flooring is manufactured in accordance with BS 4592: *Industrial type metal flooring, walkways and stair treads*. It has the following relevant parts:

BS 4592: Part 1: 1995, *Specification for open bar gratings*
BS 4592: Part 2: 1987, *Specification for expanded metal grating panels*
BS 4592: Part 3: 1987, *Specification for cold formed planks.*

30.3 Orthotropic decks

Orthotropic decks first came into regular use for long-span bridges in the 1950s. They replaced the earlier battle deck system where steel plates were supported on an independent grillage of steel plates.

In the orthotropic system a stiffened plate deck is integral with the support members which together form the primary members of the bridge. A typical orthotropic bridge system is shown in Fig. 30.3.

All steel orthotropic decks are used when the dead weight of the bridge must be reduced to a minimum. They are therefore ideally suited to long-span bridges with single spans exceeding 120 metres, and are usually incorporated in suspension and cable-stayed bridges. Other applications are for bridges with moving or lifting spans.

It should be noted that for short-span bridges, steel/concrete composite decks are generally more economic than orthotropic decks. A composite deck of depth 250 mm will be three to four times the mass of an orthotropic deck but may be only one-third of the cost.

A typical orthotropic deck consists of a 12–15 mm deck plate stiffened with longitudinal stringers about 300 mm wide at 600 mm centres, intersecting with deeper transverse cross girders. Stringers are of the closed or open type.

Closed stringers

Closed, torsionally stiff, stringers are formed as a 'trough' and are rolled or pressed from plate of 6–10 mm thickness. They provide the lightest overall weight, since they

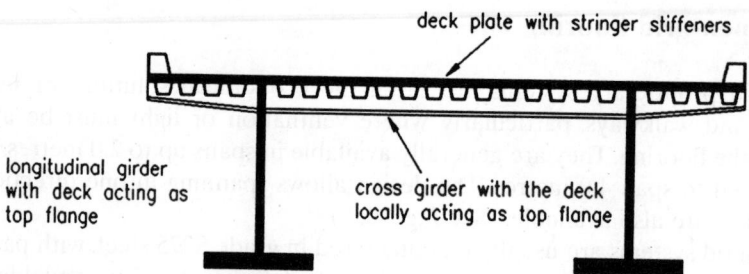

Fig. 30.3   Section through orthotropic deck

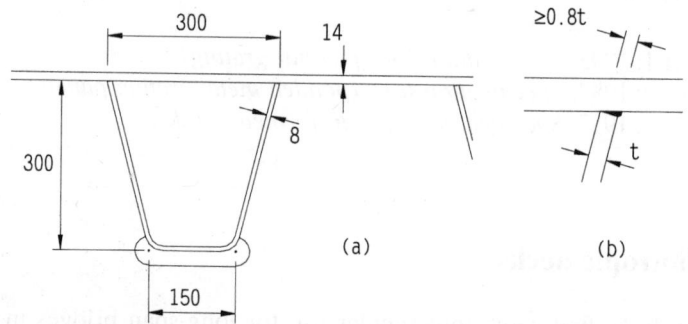

Fig. 30.4   Typical closed section stringers (a) trapezoidal trough (b) trough to deck weld

distribute heavy wheel loads most effectively between the members. The welding of the trough to deck plate has in the past suffered fatigue cracking, but this has been attributed to lack of penetration on the inaccessible side of the weld. Full, or very nearly full, penetration butt welds are now adopted, but they require careful fabrication. Closed stringers are preferably made in long lengths, slotted through cross girders, to achieve better fatigue life. No cope hole is provided at the trough to deck weld, again for better fatigue performance. Figure 30.4 shows the trough/crossgirder detail as presented in Eurocode 3: Part 2[3] (informative annex); the cut-out around the bottom of the trough may be used to ease fabrication, but, for decks with shallow crossgirders carrying high shear stresses, it should be omitted for best fatigue performance.

Open stringers

Open-section, torsionally weak stringers may be formed from flats or, for decks not subject to heavy fatigue loading, bulb flats or angles. A slightly greater dead weight results because the transverse distribution is less efficient than with closed stringers. The advantage of open stringers is that the welds to the floor plate can be made

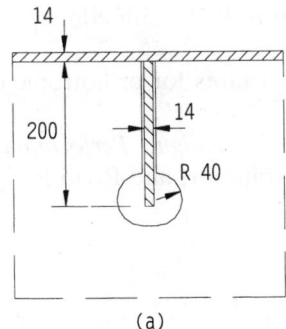
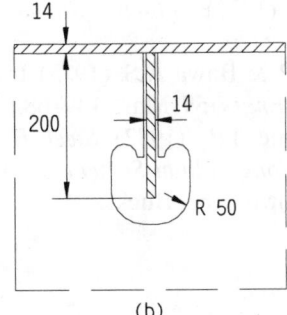

(a)   (b)

Fig. 30.5   Typical open section stringers

from both sides, and take the form of fillet welds. Open-section stringers are suitable for bridges where the individual wheel loads are not so great, especially railway bridges (with ballasted track) and footbridges. Again, best fatigue performance is achieved when the stringers are slotted through the cross girders; typical details given in Eurocode 3: Part 2 are shown in Fig. 30.5.

Loading, analysis and design

For loading to bridges reference should be made to BS 5400: Part 2: *Specification for loads*.

An analysis to take account of the distribution of point loads and partially distributed loads is best achieved with a bridge deck computer program, but hand methods of creating an 'influence surface' for moment and shears are available (see References 4 and 5).

Design should be made in accordance with BS 5400: Part 3: *Code of practice for design of steel bridges*. Fatigue considerations for bridgework are generally covered by BS 5400: Part 10: *Code of practice for fatigue*. It does not specifically deal with orthotropic decks, for which it is necessary to consult research papers.[6] Good detailing is essential to prevent premature fatigue cracking on heavily trafficked bridges.

References to Chapter 30

1. Young W.C. (2002) *Roark's Formulas for Stress and Strain*, 7th edn. McGraw-Hill, New York.
2. Timoshenko S. & Weinowsky-Krieger S. (1959) *Theory of Plates and Shells*, 2nd edn. McGraw-Hill, New York.
3. Eurocode 3 (1997) *Design of steel structures.* Part 2: *Steel bridges.* BSI, London.

4. Pucher A. (1977) *Influence Surfaces of Elastic Plates*, 5th edn. Springer-Verlag, Wien.
5. Dowling P. & Bawa A.S. (1975) Influence surfaces for orthotropic decks. *Proc. Instn Civ. Engrs*, **59**, Mar., 149–68.
6. Cuninghame J.R. (1982) *Steel Bridge Decks, Fatigue Performance of Joints Between Longitudinal Stiffeners*. LR 1066, Transport and Road Research Laboratory, Crowthorne, Bucks.

Further reading for Chapter 30

Beales C. (1990) *Assessment of Trough to Crossbeam Connections in Orthotropic Steel Bridge Decks*. TRL Report RR 276. Transport Research Laboratory, Crowthorne, Bucks.
Cuninghame J.R. (1990) *Fatigue Classification of Welded Joints in Orthotropic Steel Bridge Decks*. TRL Report RR 259. Transport Research Laboratory, Crowthorne, Bucks.
Gurney T.R. (1992) *Fatigue of Steel Bridge Decks*. HMSO, London.

Chapter 31
Tolerances

by COLIN TAYLOR

31.1 Introduction

31.1.1 Why set tolerances?

Compared to other structural materials, steel (and aluminium) structures can be made economically to much closer tolerances. Compared to mechanical parts, however, it is neither economic nor necessary to achieve extreme accuracy.

There are a number of distinct reasons why tolerances may need to be considered. It is important to be quite clear which actually apply in any given case, particularly when deciding the values to be specified, or when deciding the actions to be taken in cases of non-compliance.

The various reasons for specifying tolerances are outlined in Table 31.1. In all cases no closer tolerances than are actually needed should normally be specified, because while additional accuracy may be achievable, it generally increases the costs disproportionately.

31.1.2 Terminology

'Tolerance' as a general term means a permitted range of values. Other terms which need definition are given in Table 31.2.

31.1.3 Classes of tolerance

Table 31.3 defines the three classes of tolerances which are recognized in Euro-code 3.

It is important to draw attention to any particular or special tolerances when calling for tenders, as they usually have cost implications. Where nothing is stated, fabricators will automatically assume that only normal tolerances are required.

Table 31.1   Reasons for specifying tolerances

Structural safety	Dimensions (particularly of cross-sections, straightness, etc.) associated with structural resistance and safety of the structure.
Assembly requirements	Tolerances necessary to enable fabricated parts to be put together.
Fit-up	Requirements for fixing non-structural components, such as cladding panels, to the structure.
Interference	Tolerances to ensure that the structure does not foul with walls, door or window openings or service runs, etc.
Clearances	Clearances necessary between structures and moving parts, such as overhead travelling cranes, elevators, etc. or for rail tracks, and also between the structure and fixed or moving plant items.
Site boundaries	Boundaries of sites to be respected for legal reasons. Besides plan position, this can include limits on the inclination of outer faces of tall buildings.
Serviceability	Floors must be sufficiently flat and even, and crane gantry tracks etc. must be accurately aligned, to enable the structure to fulfil its function.
Appearance	The appearance of a building may impose limits on verticality, straightness, flatness and alignment, though generally the tolerance limits required for other reasons will already be sufficient.

Table. 31.2   Definitions – deviations and tolerances

Deviation	The difference between a specified value and the actual measured value, expressed vectorially (i.e. as a positive or negative value).
Permitted deviation	The vectorial limit specified for a particular deviation.
Tolerance range	The sum of the absolute values of the permitted deviations each side of a specified value.
Tolerance limits	The permitted deviations each side of a specified value, e.g. $\pm 3.5\,$mm or $+5\,$mm $-0\,$mm.

Table 31.3   Classes of tolerances

Normal tolerances	Those which are generally necessary for all buildings. They include those normally required for structural safety, together with normal structural assembly tolerances.
Particular tolerances	Tolerances which are closer than normal tolerances, but which apply only to *certain* components or only to *certain* dimensions. They may be necessary in specific cases for reasons of fit-up or interference or in order to respect clearances or boundaries.
Special tolerances	Tolerances which are closer than normal tolerances, and which apply to a *complete* structure or project. They may be necessary in specific cases for reasons of serviceability or appearance, or possibly for special structural reasons (such as dynamic or cyclic loading or critical design criteria), or for special assembly requirements (such as interchangeability or speed of assembly).

31.1.4 Types of tolerances

For structural steel there are three types of dimensional tolerance:

(1) *Manufacturing tolerances*, such as plate thickness and dimensions of sections.
(2) *Fabrication tolerances*, applicable in the workshops.
(3) *Erection tolerances*, relevant to work on site.

Manufacturing tolerances are specified in standards such as BS 4, BS 4848, BS EN 10024, BS EN 10029, BS EN 10034 and BS EN 10210. Only fabrication and erection tolerances will be covered here.

31.2 Standards

31.2.1 Relevant documents

The standards covering tolerances applicable to building steelwork are:

(1) BS 5950 *Structural use of steelwork in building.*
Part 2: *Specification for materials fabrication and erection: hot rolled sections.*
Part 7: *Specification for materials and workmanship: cold formed sections and sheeting.*
(2) *National structural steelwork specification for building construction* NSSS, 4th edition.
(3) ENV 1090-1 *Execution of steel structures: Part 1: General rules and rules for buildings.*
(4) ISO 10721-2: 1999 *Steel structures:* Part 2: *Fabrication and erection.*
(5) BS 5606 *Guide to accuracy in building.*

31.2.2 BS 5950 *Structural use of steelwork in building*

The specification of tolerances for building steelwork was first introduced into British Standards in BS 5950: Part 2: 1985. The current edition was issued in 2001. This revision of the 1992 edition updates cross-references to other standards, many of which are now European Standards (BS EN standards). In addition the opportunity was taken to align the code more closely with the industry standard document, the *National structural steelwork specification for building construction.*

31.2.3 *National structural steelwork specification* (NSSS)

The limitations of the tolerances specified in earlier versions of BS 5950: Part 2 have been extended by an extensive coverage of tolerances in the *National structural steelwork specification for building construction*. This is an industry standard based on established sound practice. The widely accepted document, promoted by the British Constructional Steelwork Association (BCSA), is now in its 4th edition.

31.2.4 ENV 1090-1 *Execution of steel structures*

As part of the harmonization of construction standards in Europe, CEN has issued ENV 1090: Part 1: *General rules and rules for buildings*, which is available through BSI as DD ENV 1090-1: 1998.

This document includes comprehensive recommendations for both erection and manufacturing tolerances. To a large extent these recommendations are consistent with BS 5950: Part 2 and the NSSS. However, some of them are more detailed.

31.2.5 ISO 1071-2 *Steel structures:* Part 2: *Fabrication and erection*

This is very similar to ENV 1090-1 and BS 5950: Part 2. It is unlikely to be issued as a BSI standard.

31.2.6 BS 5606 *Guide to accuracy in building*

BS 5606 is concerned with buildings generally and is not specific to steelwork. The 1990 version has been rewritten as a guide, following difficulties due to incorrect application of the previous (1978) version, which was in the form of a code.

BS 5606 is not intended as a document to be simply called up in a contract specification. It is primarily addressed to designers to explain the need for them to include means for adjustment, rather than to call for unattainable accuracy of construction. Provided that this advice is heeded, its tables of 'normal' accuracy can then be included in specifications, except where they conflict with overriding structural requirements. This can in fact happen, so it is important to remember that the requirements of BS 5950 must take precedence over BS 5606.

BS 5606 introduces the idea of *characteristic accuracy*, the concept that any construction process will inevitably lead to deviations from the target dimensions, and its objective is to advise designers on how to avoid resulting problems on site by appropriate detailing. The emphasis in BS 5606 is on the practical tolerances which will normally be achieved by good workmanship and proper site supervision. This can only be improved upon by adopting intrinsically more accurate techniques,

which are likely to incur greater costs. These affect the fit-up, the boundary dimensions, the finishes and the interference problems. Data are given on the normal tolerances (to be expected and catered for in detailed design) under two headings:

(1) Site construction (table 1 of BS 5606).
(2) Manufacture (table 2 of BS 5606).

Unfortunately many of the values for site construction of steelwork are only estimated. No specific consideration is given in BS 5606 to dimensional tolerances necessary to comply with the assumptions inherent in structural design procedures, which may in fact be more stringent. It does however recognize that special accuracy may be necessary for particular details, joints and interfaces.

Another important point mentioned in BS 5606 is the need to specify methods of monitoring compliance, including methods of measurement. It has to be recognized that methods of measurement are also subject to deviations; for the methods necessary for monitoring site dimensions, these measurement deviations may in fact be quite significant compared to the permitted deviations of the structure itself.

31.3 Implications of tolerances

31.3.1 Member sizes

31.3.1.1 Encasement

The tolerances on cross-sectional dimensions have to be allowed for when encasing steel columns or other members, whether for appearance, fire resistance or structural reasons. It should not be forgotten that the permitted deviations represent a further variation over and above the difference between the serial size and the nominal size.

For example, a $356 \times 406 \times 235$ UC has a nominal size of 381 mm deep by 395 mm wide, but with tolerances to BS 4 may actually measure 401 mm wide by 387 mm deep one side, and have a depth of 381 mm the other side. The same is true of continental sections. A $400 \times 400 \times 237$ HD also has a nominal size of 381 mm deep by 395 mm wide, but with tolerances to Euronorm 34 may actually measure 398 mm wide by 389 mm deep one side, and have a depth of 380 mm the other side.

31.3.1.2 Fabrication

Variations of cross-sectional dimensions (with permitted deviations) may also need to be allowed for, either in detailing the workmanship drawings or in the fabrication process itself, if problems are to be avoided during erection on site.

The most obvious case is a splice between two components of the same nominal size, where packs may be needed before the flange splice plates fit properly, unless the components are carefully matched. Similarly variations in the depths of adjacent crane girders or runway beams may necessitate the provision of packs, unless the members are carefully matched.

Less obviously, if the sizes of columns vary, the lengths of beams connected between them will need some form of adjustment, even if the columns are accurately located and the beams are exactly to length.

31.3.2 Attachment of non-structural components

It is good practice to ensure that all other items attached to the steel frame have adequate provision for adjustment in their fixings to cater for the effects of all steelwork tolerances, plus an allowance for deviations in their own dimensions. Where necessary, further allowances may be needed to cater for structural movements under load and for differential expansion due to temperature changes.

Where possible, the number of fixing points should be limited to three or four, only one of which should be positive with all the others having slotted holes or other means of adjustment.

31.3.3 Building envelope

It must be appreciated that erection tolerances, including variation in the position of the site grid lines, will affect the exact location of the external building envelope relative to other buildings or to site boundaries, and there may be legal constraints to be respected which will have to be taken into account at the planning and preliminary stages of design.

These effects also need to be taken into account where a building is intended to have provision for future extension or where the project is an extension of an existing building, in which case deviations in the actual dimensions have to be catered for at the interface.

In the case of tall multi-storey buildings, the building envelope deviates increasingly with height compared to the location at ground level, even though permitted deviations for column lean generally reduce with height. Unless there are step-backs or other features with a similar effect, it may be necessary to impose particular tolerance limits on the outward deviations of the columns.

31.3.4 Lift shafts for elevators

The deviations from verticality that can be tolerated in the construction of guides for 'lifts' or elevators are commonly more stringent than those for the construction of the building in which they operate. In low-rise buildings sufficient adjustment can be provided in association with the clearances, but in tall buildings it becomes necessary either to impose 'special' tolerances on column verticality or else to impose 'particular' tolerances on those columns bounding the lift shaft.

In agreeing the limits to be observed with the lift supplier, it should not be overlooked that the horizontal deflections of the building due to wind load also have implications for the verticality of the lift shafts.

31.4 Fabrication tolerances

31.4.1 Scope of fabrication tolerances

The description 'fabrication tolerances' is used here to include tolerances for all normal workshop operations except welding. It thus covers tolerances for:

(1) cross sections, other than rolled sections,
(2) member length, straightness and squareness,
(3) webs, stiffened plates and stiffeners,
(4) holes, edges and notches,
(5) bolted joints and splices,
(6) column baseplates and cap plates.

However, tolerances for cross sections of rolled sections and for thicknesses of plates and flats are treated as manufacturing tolerances. Welding tolerances (including tolerances on weld preparations and fit-up and sizes of permitted weld defects) are treated elsewhere.

31.4.2 Relation to erection tolerances

An overriding requirement for accuracy of fabrication must always be to ensure that it is possible to erect the steelwork within the specified erection tolerances.

Due to the wide variety of steel structures and the even wider variety of their components, any recommended tolerances must always be specified in a very general way. Even if it were possible to specify fabrication tolerances in such a way that their cumulative effect would always permit the specified erection tolerances to be satisfied, the resulting permitted deviations would be so small as to be unreasonably expensive, if not impossible, to achieve.

Fortunately in most cases it is possible to rely on the inherent improbability of all unfavourable extreme deviations occurring together. Also the usually accepted values for fabrication tolerances do make some limited allowances for the need to avoid cumulative effects developing on site. They are tolerances that have been shown by experience to be workable, provided that simple means of adjustment are incorporated where the effects of a number of deviations could otherwise become cumulative. For example, beams with bolted end cleats usually have sufficient adjustment available due to hole clearances, but where a line of beams all have end plate connections, provision for packing at intervals may be advisable, unless other measures are taken to ensure that the beams are not all systematically over-length or under-length by the normal permitted deviation. Other possible means for adjustment include threaded rods and slotted holes.

Where it can be seen from the drawings that the fabrication tolerances could easily accumulate in such a way as to create a serious problem in erection, either closer tolerances or means of adjustment should be considered; however, the coincident occurrence of all extreme deviations is highly improbable, and judgement should be exercised both on the need for providing means of adjustment and on the range of adjustment to be incorporated.

31.4.3 Full contact bearing

31.4.3.1 Application

The requirements for contact surfaces in joints which are required to transmit compression by 'full contact bearing' probably cause more trouble than any other item in a fabrication specification, largely due to misapprehension of what is actually intended to be achieved.

First it is necessary to be clear about the kind of joint to which the requirements for full contact bearing should be applied. Figure 31.1(a) shows the normal case, where the profile of a member is required to be in full contact bearing on a baseplate or cap plate or division plate. The stress on the contact area equals the stress in the member: thus full contact is needed to transmit this stress from the member into the plate. Only that part of the plate in contact with the member need satisfy the full contact bearing criteria, though it may be easier to prepare the whole plate.

Figure 31.1(b) shows two end plates in simple bearing. The potential contact area is substantially larger than the cross-sectional area of the member: thus full contact bearing is not necessary. All that is needed is for the end plates to be square to the axis of the member. Another common case of simple bearing is shown in Fig. 31.1(c).

By contrast, the case shown in Fig. 31.1(d) is one where, if full contact bearing is needed, it is also necessary to take special measures to ensure that the profiles of the two members align accurately, otherwise the area in contact may be significantly less than the area required to transmit the load. Particular tolerances should be specified in such cases, based on the maximum local reduction of area that can be

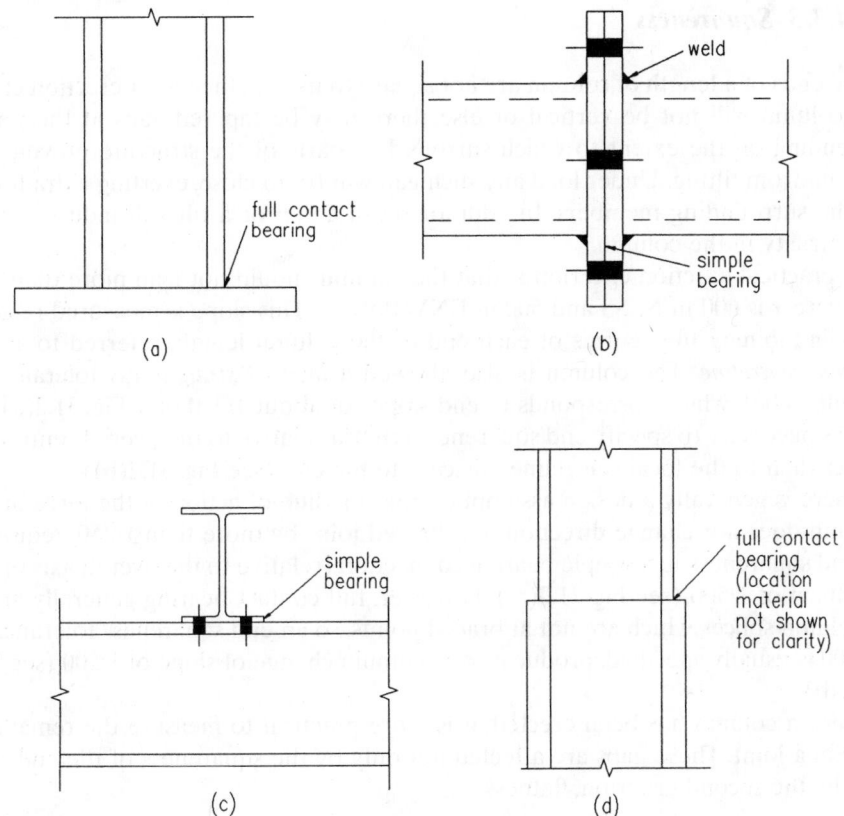

Fig. 31.1 Types of member-to-member bearing: (a) profile to plate, (b) plate to plate, (c) flange to flange, (d) profile to profile (accurate alignment necessary)

accepted according to the design calculations. Alternatively a division plate could be introduced; if the stresses are high this may well prove to be the most practical solution.

31.4.3.2 Requirements

Where full contact bearing is required, there are in fact three different criteria involved:

(1) Squareness.
(2) Flatness.
(3) Smoothness.

31.4.3.3 Squareness

If the ends of a length of column are not square to its axis, then after erection either the column will not be vertical or else there may be tapered gaps at the joints, depending on the extent to which surrounding parts of the structure prevent the column from tilting. Under load any such gap will try to close, exerting extra forces on the surrounding members. In addition, both a gap or a tilt will induce a local eccentricity in the column.

A practical erection criterion is that the column should not lean more than 1 in x (where x is 600 in NSSS and 500 in ENV 1090-1). This slope is measured relative to a line joining the centres of each end of the column length, referred to as the *overall centreline*. The column is also allowed a *lack of straightness* tolerance of (length/1000), which corresponds to end slopes of about 1/300 (see Fig. 31.2(a)). It is thus necessary to specify end squareness criteria relative to the overall centreline, rather than to the local centreline adjacent to the end (see Fig. 31.2(b)).

There is generally a design assumption that the line of action of the force in the column does not change direction at a braced joint by more than 1/250, requiring an end squareness in a simple bearing connection (relative to the overall axis of the member) of 1/500 (see Fig 31.2(c)). However, full contact bearing generally arises at column splices which are not at braced points, so an end squareness tolerance of 1/1000 is usually specified, producing a maximum change of slope of 1/500 (see Fig. 31.2(d)).

Once a column has been erected, it is more practical to measure the remaining gaps in a joint. These gaps are affected not only by the squareness of the ends but also by the second criterion, flatness.

31.4.3.4 Flatness

Ends have to be reasonably flat (as distinct from curved or grossly uneven) to enable the load to be transferred properly. Following a history of arguments over appropriate specifications, the American Institute of Steel Construction (AISC) commissioned some tests, which are the basis for their current specifications.

It was found that a surprisingly high tolerance was quite acceptable, and that beyond its limit (or to compensate for end squareness deviations) the use of localized packs or shims was acceptable. Basically similar rules are now beginning to appear in other specifications including the CEN standard (see section 31.5.6 in relation to erection tolerances). This is an essentially simple and effective method of correcting excessive gaps on site (see also section 31.5.6). However, inserting shims into column joints is not a matter to be undertaken lightly. It is normally more economic to avoid the need for shimming by working to close fabrication tolerances in joints where full contact bearing is required.

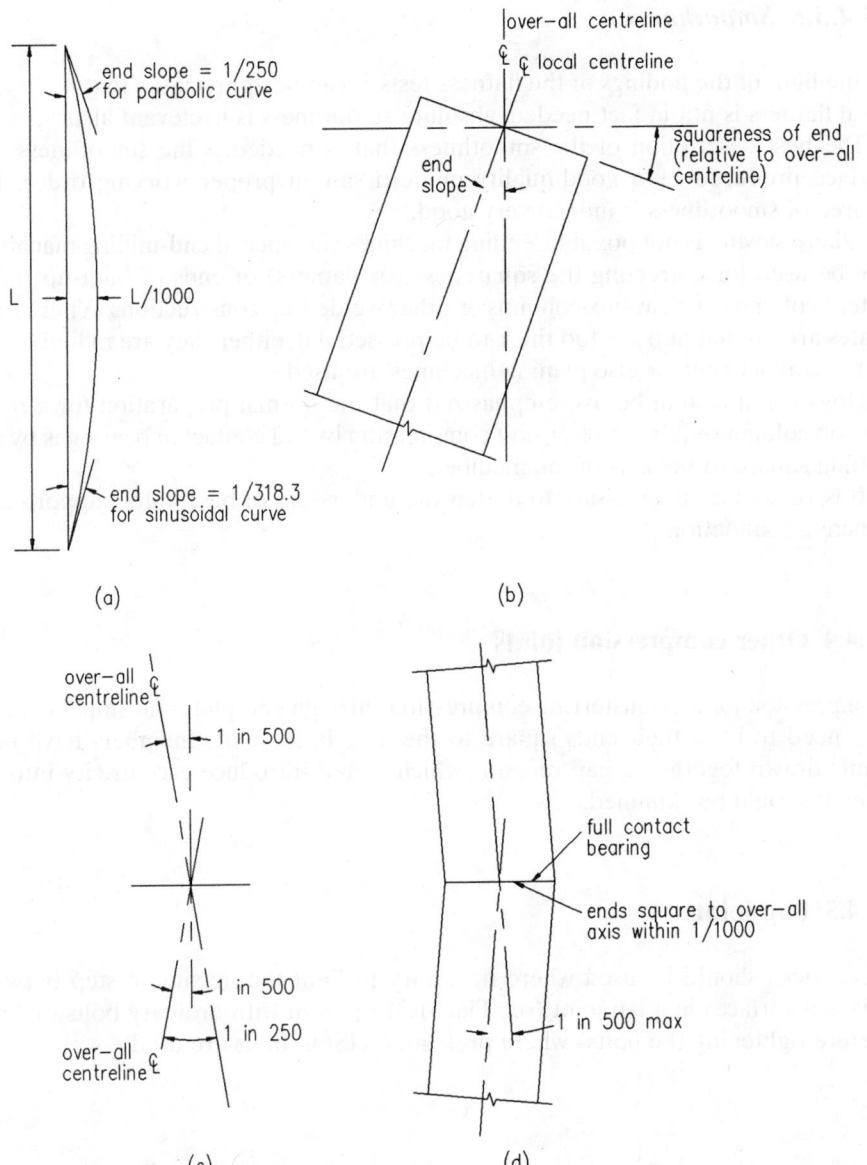

Fig. 31.2 Squareness of column ends. (a) Bow of 1/1000 giving end slopes of about 1/300. (b) Squareness of end measured relative to overall centreline. (c) Change of direction at a braced joint. (d) End squareness at full contact bearing splice

31.4.3.5 Smoothness

In the light of the findings of the flatness tests, it can be appreciated that if absolute local flatness is not in fact needed, absolute smoothness is irrelevant also.

The best description of the smoothness that is needed is the smoothness of a surface produced by a good-quality modern saw in proper working order. This degree of smoothness is indeed very good.

Where sawing is not possible, ending machines (i.e. special end-milling machines) can be used for correcting the squareness (or flatness) of ends of built-up (fabricated) columns, such as box columns or other welded-up constructions. Where baseplates are not flat and are too thick to be pressed flat, either they are milled locally in the contact zone or else planing machines are used.

However, it cannot be overemphasized that the normal preparation for a rolled section column required to transmit compression by full contact in bearing is by saw cutting square to the axis of the member.

It is, of course, unnecessary to flatten the undersides of baseplates supported on concrete foundations.

31.4.4 Other compression joints

Compression joints, transferring compression through end plates in simple bearing, also need to have their ends square to the axis. If, after the members have been firmly drawn together, a gap remains which would introduce eccentricity into the joint, it should be skimmed.

31.4.5 Lap joints

Steel packs should be used where necessary to limit the maximum step between adjacent surfaces in a lap joint (see Fig. 31.3) to 2 mm with ordinary bolts or 1 mm (before tightening the bolts) where preloaded HSFG bolts are used.

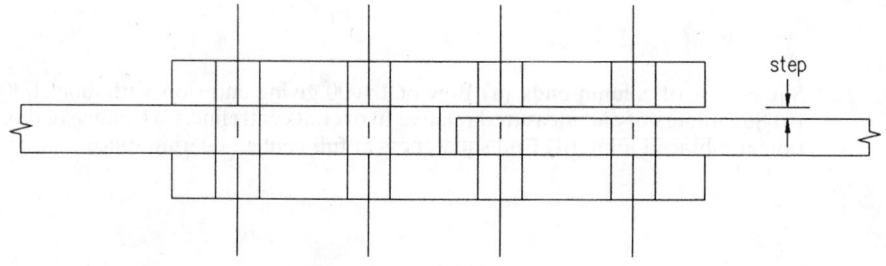

Fig. 31.3   Maximum step between adjacent surfaces

31.4.6 Beam end plates

Where the length of a beam with end plates is too short to fit between the supporting columns, or other supporting members, packs should be supplied to make up the difference.

Gaps arising from distortion caused by welding, as shown in Fig. 31.4, need not be packed if the members can be firmly drawn together. However, they may need to be filled or sealed to avoid corrosion where the steelwork is external or is exposed to an aggressive internal environment.

31.4.7 Values for fabrication tolerances

The values for fabrication tolerances currently given in the NSSS are reproduced for convenience in Table 31.4. Each of the specified criteria should be considered and satisfied separately. The cumulative effect of several permitted deviations should not be considered as overriding the specific criteria.

These values represent current practice and are taken from the fourth edition of the NSSS.

The clause numbers referred to in Table 31.4 are clause numbers in the NSSS, which should be referred to for further information.

31.5 Erection tolerances

31.5.1 Importance of erection tolerances

Erection tolerances potentially have a significant effect on structural behaviour. There are four matters to be considered:

(1) overall position,
(2) fixing bolts,
(3) internal accuracy,
(4) external envelope.

31.5.2 Erection – positional tolerance

31.5.2.1 Setting out

The position in plan, level and orientation can only be defined relative to some fixed references, such as the National Grid and the Ordnance datum level. From the

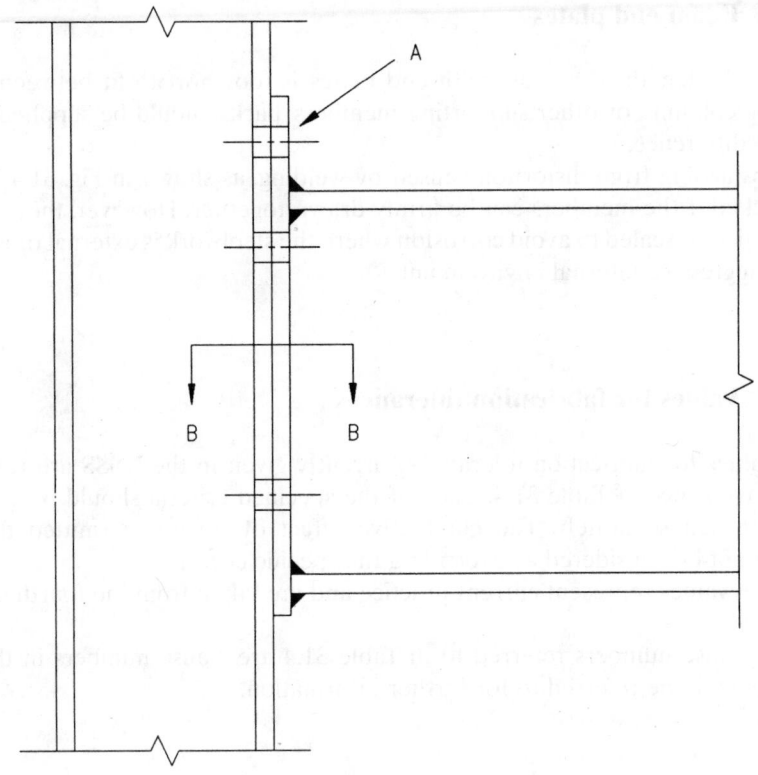

A

B   B

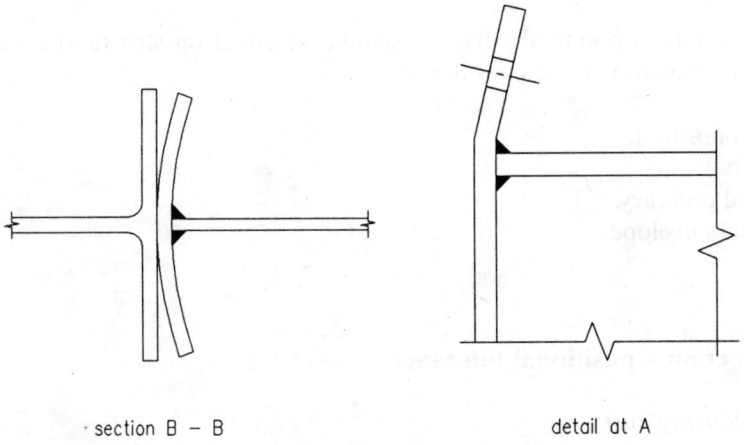

section B – B   detail at A

Fig. 31.4   End plate with welding (exaggerated)

Table 31.4 (Extract from National Structural Steelwork Specification 4th edn.)

SECTION 7
WORKMANSHIP - ACCURACY OF FABRICATION

7.1 PERMITTED DEVIATIONS

Permitted deviations in cross section, length, straightness, flatness, cutting, holing and position of fittings shall be as specified in 7.2 to 7.5 below.

7.2 PERMITTED DEVIATIONS IN ROLLED COMPONENTS AFTER FABRICATION
(Including Structural Hollow Sections)

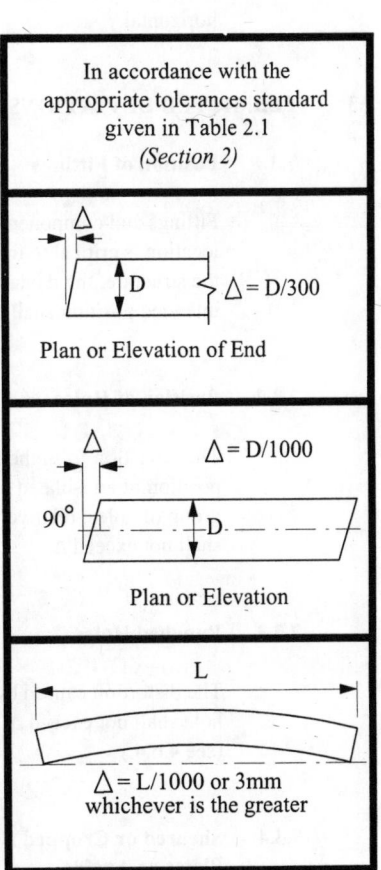

7.2.1 Cross Section after Fabrication

In accordance with the appropriate tolerances standard given in Table 2.1
(Section 2)

7.2.2 Squareness of Ends Not Prepared for Bearing

See also 4.3.3 (i).

$\Delta = D/300$

Plan or Elevation of End

7.2.3 Squareness of Ends Prepared for Bearing

Prepare ends with respect to the longitudinal axis of the member. See also 4.3.3 (ii) and (iii).

$\Delta = D/1000$

90°

Plan or Elevation

7.2.4 Straightness on Both Axes

L

$\Delta = L/1000$ or 3mm whichever is the greater

Table 31.4   *(contd)*

7.2.5   **Length**

Length after cutting, measured on the centre line of the section or on the corner of angles.

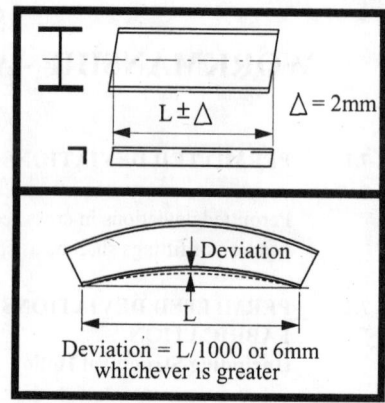

$L \pm \triangle$ $\triangle = 2mm$

7.2.6   **Curved or Cambered**

Deviation from intended curve or camber at mid-length of curved portion when measured with web horizontal.

Deviation
L
Deviation = L/1000 or 6mm whichever is greater

7.3   PERMITTED DEVIATIONS FOR ELEMENTS OF FABRICATED MEMBERS

7.3.1   **Position of Fittings**

Fittings and components whose location is critical to the force path in the structure, the deviation from the intended position shall not exceed Δ.

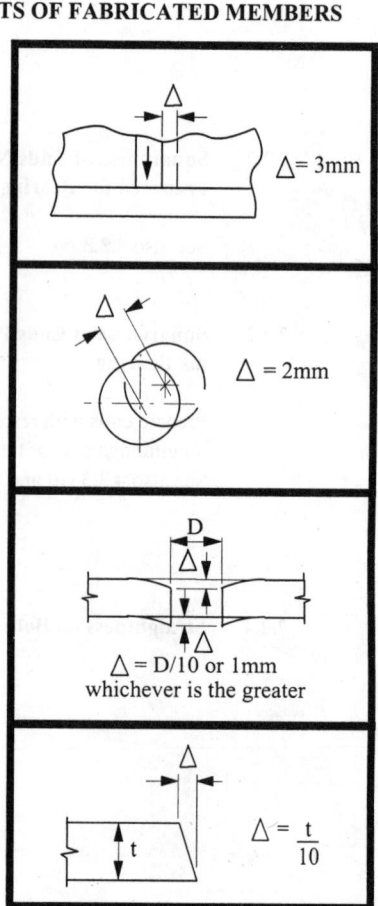

$\triangle = 3mm$

7.3.2   **Position of Holes**

The deviation from the intended position of an isolated hole, also a group of holes, relative to each other shall not exceed Δ.

$\triangle = 2mm$

7.3.3   **Punched Holes**

The distortion caused by a punched hole shall not exceed Δ.
(see 4.6.4.)

D
$\triangle$
$\triangle = D/10$ or 1mm whichever is the greater

7.3.4   **Sheared or Cropped Edges of Plates or Angles**

The deviation from a 90° edge shall not exceed Δ.

$\triangle$
t
$\triangle = \dfrac{t}{10}$

Table 31.4 *(contd)*

7.3.5 Flatness

Where full contact bearing is specified, the flatness shall be such that when measured against a straight edge not exceeding one metre long, which is laid against the full bearing surface in any direction, the gap does not exceed Δ.

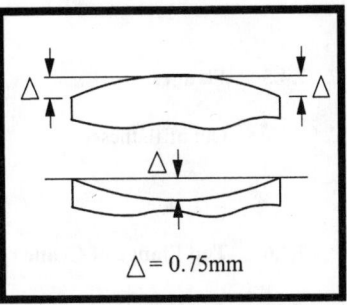

$\triangle = 0.75\text{mm}$

7.4   PERMITTED DEVIATIONS IN PLATE GIRDER SECTIONS

7.4.1 Depth

Depth on centre line.

7.4.2 Flange Width

Width of B_w or B_n

7.4.3 Squareness of Section

Out of Squareness of Flanges.

7.4.4 Web Eccentricity

Intended position of web from one edge of flange.

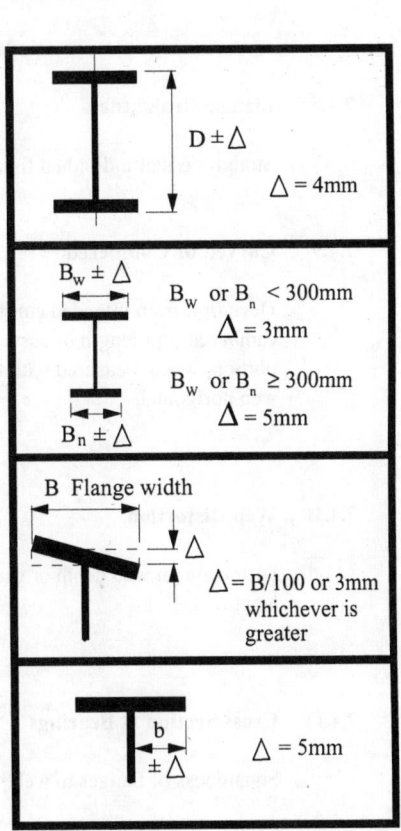

$D \pm \triangle$

$\triangle = 4\text{mm}$

$B_w \pm \triangle$

B_w or B_n $< 300\text{mm}$

$\triangle = 3\text{mm}$

B_w or B_n $\geq 300\text{mm}$

$\triangle = 5\text{mm}$

$B_n \pm \triangle$

B Flange width

$\triangle = B/100$ or 3mm whichever is greater

$\pm \triangle$ b

$\triangle = 5\text{mm}$

Table 31.4   (*contd*)

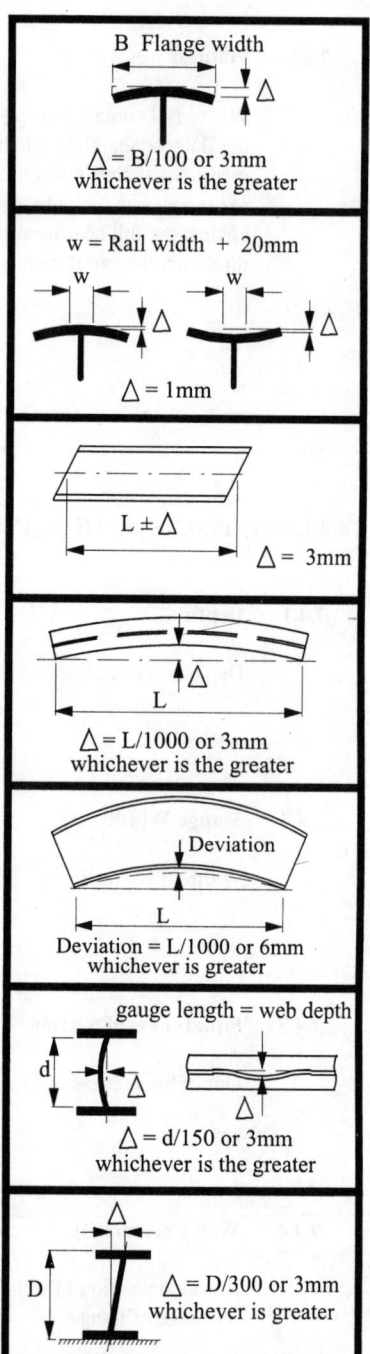

7.4.5   Flanges

Out of flatness.

7.4.6   Top Flange of Crane Girder

Out of flatness where the rail seats.

7.4.7   Length

Length on centre line.

7.4.8   Flange Straightness

Straightness of individual flanges.

7.4.9   Curved or Cambered

Deviation from intended curve or
camber at mid-length of curved
portion, when measured with the
web horizontal.

7.4.10   Web Distortion

Distortion on web depth or gauge
length.

7.4.11   Cross Section at Bearings

Squareness of flanges to web.

©BCSA

Table 31.4 (*contd*)

7.4.12   Web Stiffeners

Straightness of stiffener out of plane with web after welding.

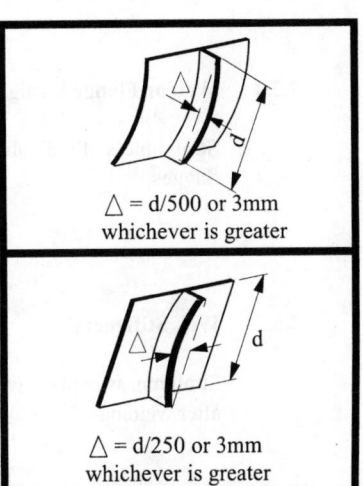

$\triangle$ = d/500 or 3mm
whichever is greater

7.4.13   Web Stiffeners

Straightness of stiffener in plane with web after welding.

$\triangle$ = d/250 or 3mm
whichever is greater

7.5   PERMITTED DEVIATIONS IN BOX SECTIONS

7.5.1   Plate Widths

Width of B_f or B_w

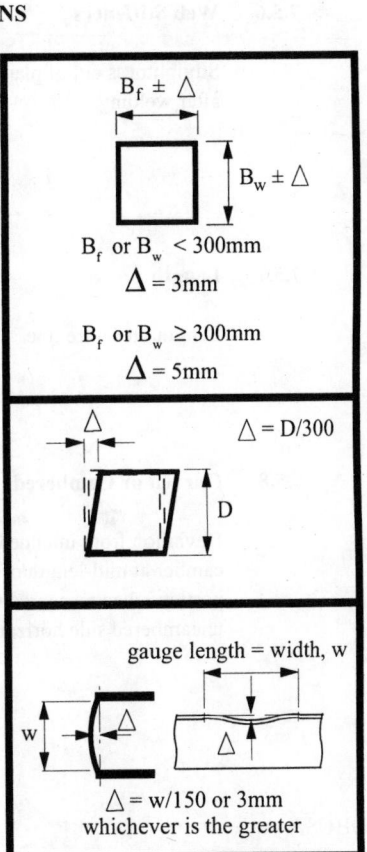

$B_f \pm \triangle$

$B_w \pm \triangle$

B_f or B_w < 300mm
$\triangle$ = 3mm

B_f or B_w ≥ 300mm
$\triangle$ = 5mm

7.5.2   Squareness

Squareness at diaphragm positions.

$\triangle$ = D/300

D

7.5.3   Plate Distortion

Distortion on width or gauge length.

gauge length = width, w

w

$\triangle$ = w/150 or 3mm
whichever is the greater

Table 31.4   (*contd*)

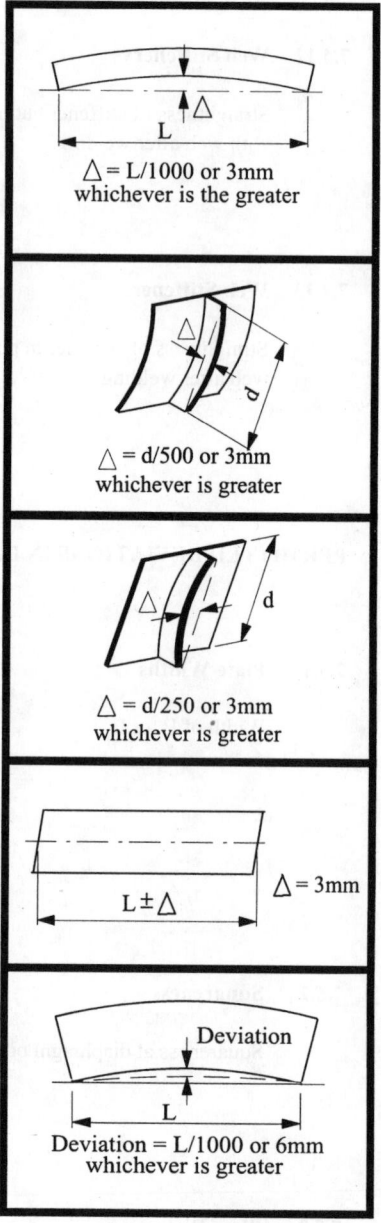

7.5.4   Web or Flange Straightness

Straightness of individual web or flanges.

$\triangle$ = L/1000 or 3mm
whichever is the greater

7.5.5   Web Stiffeners

Straightness in plane with plate after welding.

$\triangle$ = d/500 or 3mm
whichever is greater

7.5.6   Web Stiffeners

Straightness out of plane to plate after welding.

$\triangle$ = d/250 or 3mm
whichever is greater

7.5.7   Length

Length on centre line.

L $\pm$ $\triangle$ $\triangle$ = 3mm

7.5.8   Curved or Cambered

Deviation from intended curve or camber at mid-length of curved portion when measured with the uncambered side horizontal.

Deviation

Deviation = L/1000 or 6mm
whichever is greater

national system, it is usual to set subsidiary site datum points, and often a site datum level, and then refer the accuracy of the structure to these.

For any site the use of a grid of established column lines together with an established site level is strongly recommended. For a large site it is virtually indispensable. To help appreciate this, consider what happens on the site of a steel structure.

31.5.2.2 Site practice

Normal site practice is for the supporting concrete foundations, and other supporting structures, to be prepared in advance of steel erection, generally by an organization separate from the steel erector. Depending on the system of holding-down bolts or other fixings to be used, this may involve casting-in of holding-down bolts, preparation of pockets in the concrete, and preparation of surfaces to receive fixings to the steelwork.

Even with care, the standard of accuracy achievable is limited, and the concrete requires time to harden to a sufficient strength for steel erection to proceed. Once all the foundations etc. are available for steel erection (or at least a sufficient proportion of them on a large site), it is prudent to survey them to review their accuracy.

31.5.2.3 Established column lines and established site level

From this survey it is convenient to introduce a grid of established column lines (ECL) and an established site level (ESL) of the foundations and other supporting structures in such a way that the positions and levels of steel columns etc. can readily be related to the site grid and site level.

The established column lines are defined as that grid of site grid lines that best represents the actual mean positions of the installed foundations and fixings. Similarly the established site level is defined as that level which best represents the actual mean level of the installed foundations. Of course it should also be verified that the deviation of the ECL grid and the ESL from those specified are within the relevant permitted deviations.

31.5.3 Erection – fixing bolts

31.5.3.1 Types of fixing bolts

Fixing bolts include both holding-down bolts for columns and various types of fixing bolts used to locate or to support other members, such as beams or brackets carried by walls or concrete members.

Holding-down bolts and other fixing bolts are either:

(1) fixed in position, or
(2) adjustable, in sleeves or pockets.

31.5.3.2 Fixed bolts

Fixed bolts used to be solidly cast in, an operation requiring care and the use of jigs or templates to achieve accurately. However, they are now also commonly produced by placing resin-grouted bolts in holes drilled in the concrete after casting. It may also be possible to use expanding bolts.

In whatever way fixed bolts are achieved, they need to be positioned accurately, as the only adjustment possible is in the steelwork, so relatively close tolerances are normally specified.

31.5.3.3 Adjustable bolts

Adjustable bolts are placed in tubes or in tapered trapezoidal or conical holes cast in the concrete, so that a degree of movement of the threaded end of the bolt is possible, while the other end is held in place by a steel washer or other anchoring device embedded in the concrete.

This alternative permits the use of more easily achieved tolerances for the bolts, while using relatively simple details for the steelwork. Adjustment of the bolt necessitates its axis deviating from the vertical to some extent, and the holes in the steelwork need to be large enough to allow for this, particularly if the baseplate is thick. The use of loose plate washers is recommended to span oversize holes if necessary. If required they can be welded in place after the bolts are tightened, but this should not normally be necessary. 'Particular' tolerances need to be worked out for each case, depending on the details, including the length of the bolts, because this affects their slope.

31.5.3.4 Length of bolts

The level of the top of an HD bolt is also important to ensure that the nuts can be fitted properly after erection. To provide the necessary tolerances for the fixing of the bolts they should be longer than theoretically required, long threaded lengths should be provided, and the nominal level for the top should be above the theoretical position.

Similar considerations apply to the lengths of fixing bolts located horizontally.

31.5.4 Erection – internal accuracy

In terms of structural performance, the main erection tolerance is verticality of columns; positions of beams etc. on brackets may also be important. Levels of beams, particularly of one end relative to the other end and of one beam relative to the next one, are important in terms of serviceability.

Otherwise the internal accuracy of one part of the structure relative to another is largely a matter of assembly tolerances, provided that these do not cause any problem of fit-up, interference or clearances. Where the structural accuracy resulting from the assembly tolerances is liable to infringe any of these limits, 'particular' tolerances should be specified.

The necessary tolerances are specified in relation to readily identifiable points and levels. For columns and other vertical members, the reference points are conveniently defined as the actual centre of the member at each end of the fabricated piece. For beams and other horizontal members the reference points are more conveniently defined by the actual centre of the top surface at each end. Either the column system or the beam system should be used for any other cases, and the relevant system should be indicated on the erection drawings. The tolerances are then defined by the permitted deviations of these reference points from the established column lines ECL and established floor level EFL.

The concept of an ECL grid and an established site level ESL have already been explained in section 31.5.2.3. The established floor level EFL is defined as that level which best represents the actual mean level of the as-built floor levels. The EFL must not deviate from the specified floor level (relative to the ESL) by more than the permitted deviation for height of columns.

The reference points for each beam must then be within the permitted deviation from the EFL. In addition the difference in level of each end of a beam and the difference in level between adjacent beams must also be within their respective limits.

In the case of columns, the permitted deviations at each level form an 'envelope' within which the column must lie at all levels. In addition, the permitted inclination of each column within a storey height is limited, but except where columns are fabricated as individual storey-height pieces, the overall envelope normally governs.

31.5.5 Erection – external envelope

Generally the same erection tolerances for verticality apply to external columns as to internal columns. When the envelope of extreme permitted deviations is plotted from the extreme position of the base (allowing for the permitted deviation of the ECL from the theoretical position as well as the permitted deviation of the column base from the ECL), it may be found that this is unacceptable in terms of site boundaries or building lines, especially for a tall multi-storey building. If so, 'particular' tolerances should be specified.

Fit-up problems with cladding could also occur if alternate adjacent columns at the periphery were allowed too large a deviation alternately in and out from the theoretical line of the building face. Even if the fit-up problems could be overcome, the visual appearance might be affected. Again, 'particular' tolerances should be specified if necessary.

31.5.6 Shimming full contact bearing splices

As mentioned in section 31.4.3.4 in relation to fabrication tolerances, tests commissioned by AISC, and used as the basis for several modern standards, showed that shims can be used to reduce gaps in full contact bearings to within the specified tolerances. Shimmed gaps up to 6.35 mm were tested, so it is not prudent to permit shimming for gaps exceeding 6 mm; gaps larger than this should be corrected by other means.

Gaps which would otherwise remain over the specified tolerance when the members are in their final alignment should be shimmed. As the tests were on flat shims, it is acceptable to use flat shims in practice. In the tests the shims were of mild steel, and this is permitted in the AISC specification and ENV 1090-1.

The shims should be inserted such that no remaining gap exceeds the specified permitted deviation. Short lengths of shim are appropriate in a variety of thicknesses

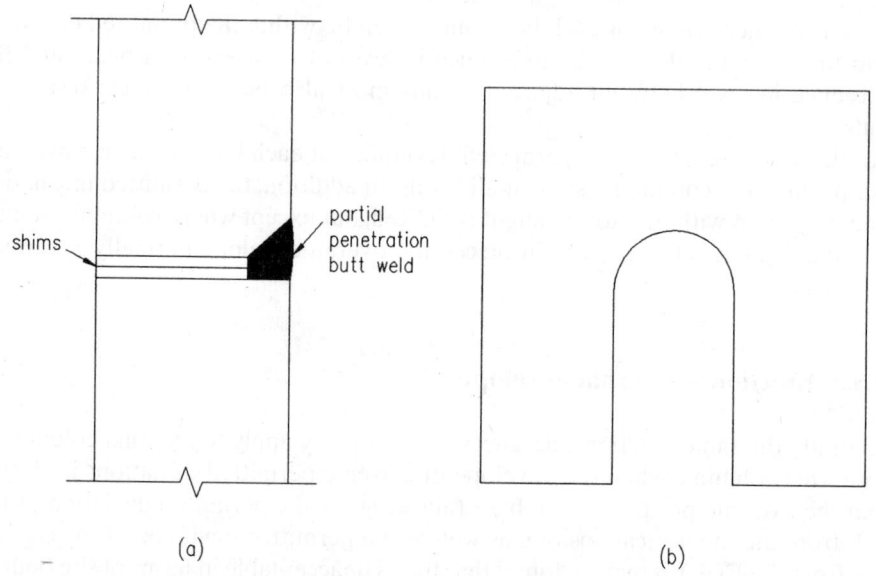

(a)   (b)

Fig. 31.5   Shims for full contact bearing. (a) Shims with partial penetration butt weld. (b) Finger shim

Table 31.5   (Extract from National Structural Steelwork Specification 4th edn.)

SECTION 9
WORKMANSHIP
ACCURACY OF ERECTED STEELWORK

9.1   **PERMITTED DEVIATIONS FOR FOUNDATIONS, WALLS AND FOUNDATION BOLTS**

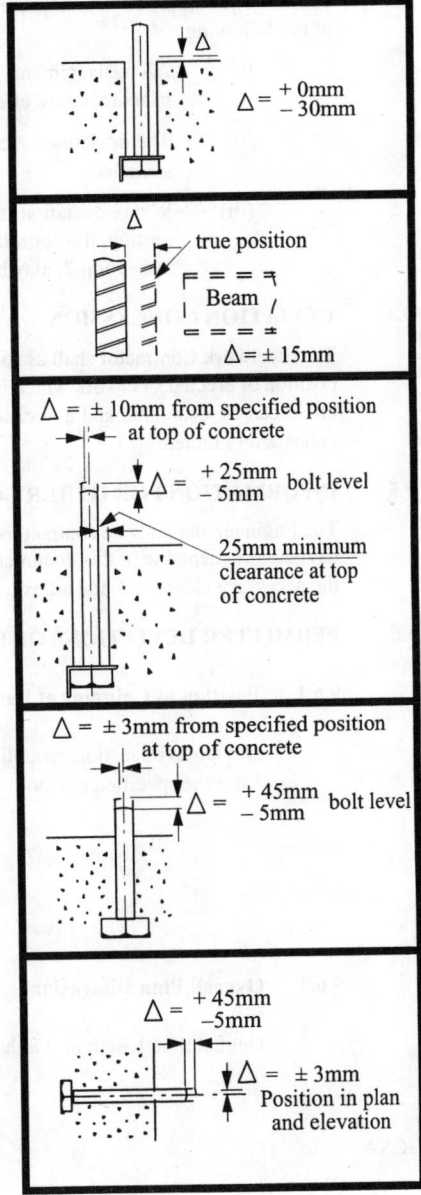

9.1.1   **Foundation Level**

Deviation from exact level.

$$\Delta = \begin{array}{l} + \text{0mm} \\ - \text{30mm} \end{array}$$

9.1.2   **Vertical Wall**
Deviation from exact position at steelwork support point.

true position

Beam

$\Delta = \pm 15\text{mm}$

9.1.3   **Pre-set Foundation Bolt or Bolt Groups when Prepared for Adjustment**

Deviation from specified position

$\Delta = \pm 10\text{mm}$ from specified position at top of concrete

$\Delta = \begin{array}{l} + \text{25mm} \\ - \text{5mm} \end{array}$ bolt level

25mm minimum clearance at top of concrete

9.1.4   **Pre-set Foundation Bolt or Bolt Groups when Not Prepared for Adjustment**

Deviation from specified position

$\Delta = \pm 3\text{mm}$ from specified position at top of concrete

$\Delta = \begin{array}{l} + \text{45mm} \\ - \text{5mm} \end{array}$ bolt level

9.1.5   **Pre-set Wall Bolt or Bolt Groups when Not Prepared for Adjustment**

Deviation from specified position

$\Delta = \begin{array}{l} + \text{45mm} \\ - \text{5mm} \end{array}$

$\Delta = \pm 3\text{mm}$
Position in plan and elevation

©BCSA

Table 31.5   *(contd)*

9.2   FOUNDATION INSPECTION

The Steelwork Contractor shall inspect the prepared foundations and holding down bolts for position and level not less than seven days before erection of steelwork starts. He shall then inform the Employer if he finds any discrepancies which are outside the deviations specified in clause 9.1 requesting that remedial work be carried out before erection commences.

9.3   STEELWORK

Permitted maximum deviations in erected steelwork shall be as specified in 9.6 taking account of the following:

(i)   All measurements be taken in calm weather, and due note is to be taken of temperature effects on the structure. (see 8.6.2.).

(ii)   The deviations shown for **I** sections apply also to box and tubular sections.

(iii) Where deviations are shown relative to nominal centrelines of the section, the permitted deviation on cross-section and straightness, given in Section 7, may be added.

9.4   DEVIATION CONCESSION

The Steelwork Contractor shall as soon as possible inform the Engineer of any deviation in position of erected steelwork which is greater than the permitted deviation in 9.6 so that the effect can be evaluated and a decision reached on whether remedial work is needed or a concession granted.

9.5   INFORMATION FOR OTHER CONTRACTORS

The Engineer shall advise contractors engaged in operations following steel erection of the deviations acceptable in this document in fabrication and erection, so that they can provide the necessary clearances and adjustments.

9.6   PERMITTED DEVIATIONS OF ERECTED COMPONENTS

9.6.1   **Position of Columns at Base**

Deviation of section centreline from the specified position.

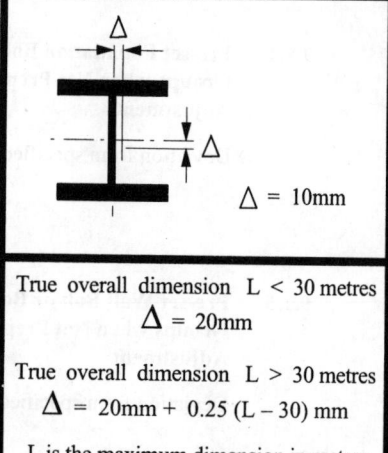

Δ = 10mm

9.6.2   **Overall Plan Dimensions**

Deviation in length or width.

True overall dimension L < 30 metres
$$\Delta = 20mm$$

True overall dimension L > 30 metres
$$\Delta = 20mm + 0.25 (L - 30) \ mm$$

L is the maximum dimension in metres

Table 31.5 (*contd*)

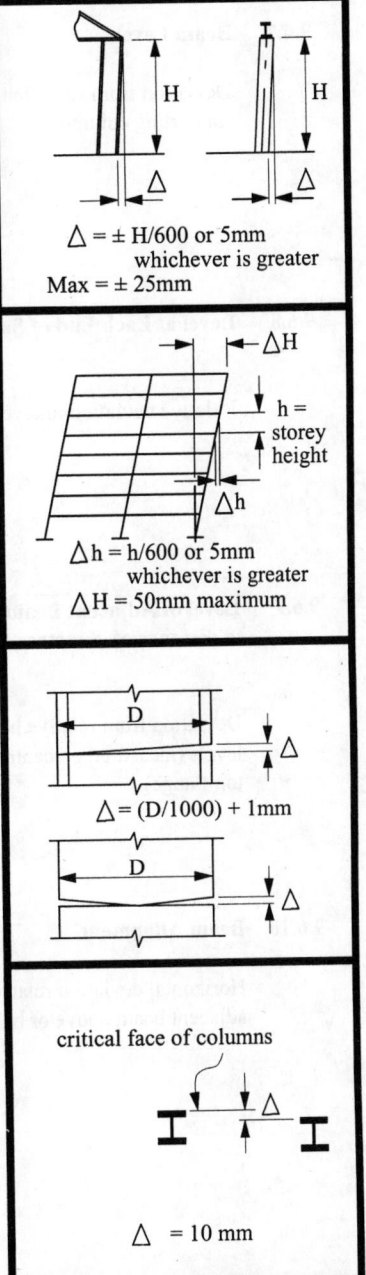

9.6.3   **Single Storey Columns Plumb**

Deviation of top relative to base, excluding portal frame columns, on main axes.
See clause 1.2A (xvii) and 3.4.4 (iii) regarding pre-setting portal frames.

$\triangle = \pm$ H/600 or 5mm whichever is greater
Max $= \pm$ 25mm

9.6.4   **Multi-storey Columns Plumb**

Deviation in each storey and maximum deviation relative to base.

$\triangle$h $=$ h/600 or 5mm whichever is greater
$\triangle$H $=$ 50mm maximum

9.6.5   **Gap Between Bearing Surfaces**

(See clauses 4.3.3 (iii), 6.2.1 and 7.2.3.)

$\triangle = $ (D/1000) $+$ 1mm

9.6.6   **Alignment of Adjacent Perimeter Columns**

Deviation relative to next column on a line parallel to the grid line when measured at base or splice level.

critical face of columns

$\triangle = 10$ mm

Table 31.5 (*contd*)

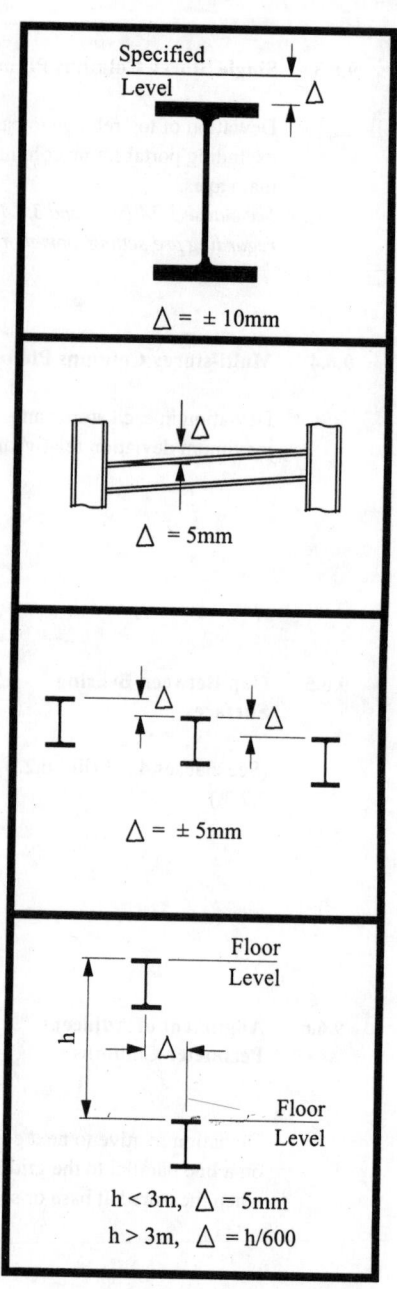

9.6.7 **Beam Level** Deviation from specified level at supporting column.	
9.6.8 **Level at Each End of Same Beam** Relative deviation in level at ends.	
9.6.9 **Level of Adjacent Beams within a distance of 5 metres** Deviation from relative horizontal levels (measured on centreline of top flange).	
9.6.10 **Beam Alignment** Horizontal deviation relative to an adjacent beam above or below.	

Table 31.5 (*contd*)

9.6.11 Crane Gantry Columns Plumb

Deviation of cap relative to base.

9.6.12 Crane Gantries Gauge of Rail Tracks

Deviation from true gauge.

9.6.13 Joints in Gantry Crane Rails Rail surface

9.6.14 Joints in Gantry Crane Rails Rail edge

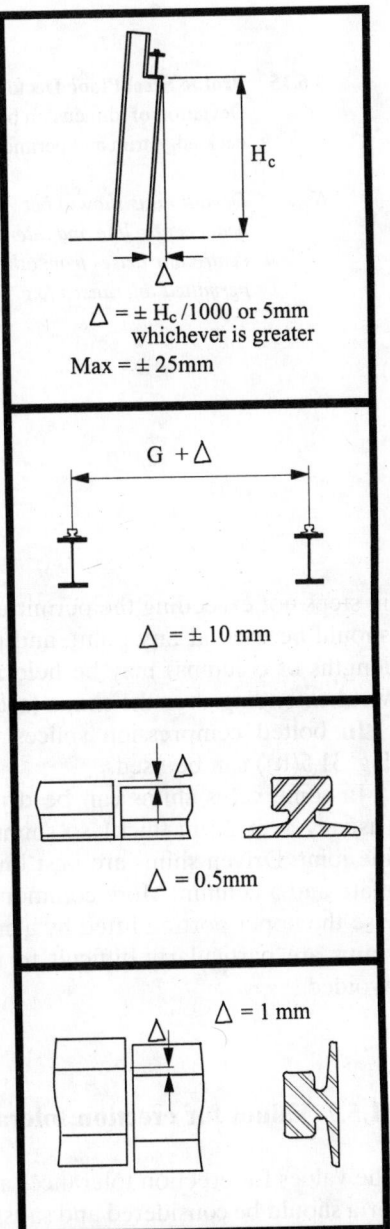

$\Delta = \pm H_c /1000$ or 5mm whichever is greater
Max $= \pm 25$mm

$G + \Delta$

$\Delta = \pm 10$ mm

$\Delta = 0.5$mm

$\Delta = 1$ mm

9.6.15 Profile Steel Floor Decking
Deviation of dimension between
deck edge trim and perimeter beam

*Note :   Deviation(as shown) between actual
beam centre line and intended beam
centre line arises from other
permitted tolerances (e.g. 9.6.4)*

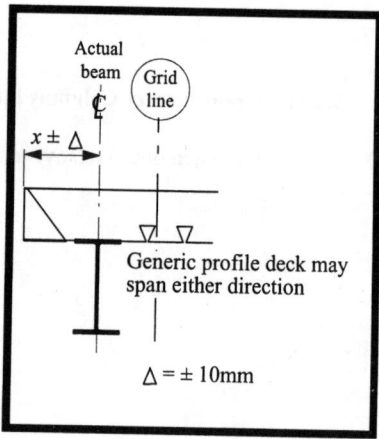

in steps not exceeding the permitted deviation. No more than three layers of shims should be used at any point, and preferably only one or two. The shims (and the lengths of columns) may be held in place by means of a partial penetration butt weld extending over the shims (see Fig. 31.5(a)).

In bolted compression splices, bolted 'finger' shims (shaped as indicated in Fig. 31.5(b)) can be used.

In some cases shims can be driven in, but if so they need to be fairly robust (usually over 2 mm thick), so shims of various thicknesses are needed throughout the joint. Driven shims are best limited to vertical joints e.g. between a beam end plate and a column. More commonly the joint must be jacked or wedged open (or else the upper portion lifted by a crane) so that the shims can be inserted. Tapered shims are particularly difficult to insert; as they are not necessary they are best avoided.

31.5.7 Values for erection tolerances

The values for erection tolerances are given in Table 31.5. Each of the specified criteria should be considered and satisfied separately. The permitted deviations should not be considered as cumulative, except to the extent that they are specified relative to points or lines that also have permitted deviation. These values represent current practice and are taken from the fourth edition of the NSSS.

The clause numbers referred to in Table 31.5 are clause numbers in the NSSS, which should be referred to for further information.

Further reading for Chapter 31

British Standards Institution (1993) *Specification for hot rolled sections.* BS 4: Part 1: 1993, BSI, London.

British Standards Institution (1993) *Specification for hot rolled products of non-alloy structural steels and their technical delivery conditions.* BS EN 10025: 1993, BSI, London.

British Standards Institution (1986) *Equal and unequal angles.* BS 4848: Part 4: 1986, BSI, London.

British Standards Institution (1990) *Guide to accuracy in building.* BS 5606: 1990, BSI, London.

British Standards Institution (2001) *Specification for materials fabrication and erection: hot rolled sections.* BS 5950: Part 2: 2001, BSI, London.

British Standards Institution (1991) *Specification for tolerances and dimensions, shape and mass for hot rolled steel plates 3 mm thick and above.* BS EN 10029: 1991, BSI, London.

British Standards Institution (1998) *Execution of steel structures* Part 1: *General rules and rules for buildings:* DD ENV 1090-1: 1998, BSI, London.

The British Constructional Steelwork Association (2002) *National structural steelwork specification*, 4th edition, BCSA/SCI.

Acknowledgement

Extracts from the *National structural steelwork specification* 4th Edition are reproduced with the kind permission of the British Constructional Steelwork Association.

Chapter 32
Fabrication

by DAVID DIBB-FULLER

32.1 Introduction

The steel-framed building derives most of its competitive advantage from the virtues of prefabricated components which can be assembled speedily on site. Additional economies can be significant provided the designer seeks through the design to minimize the value added costs of fabrication. This is proper 'value engineering' of the product and is applied to perhaps the most influential sector of the delivered-to-site cost of structural steelwork.

This chapter explains the processes of fabrication and links them with design decisions. It is increasingly important for the designer to understand the skills and techniques available from different fabricators so that the design can be tailored to keep overall costs down. The choice of fabricator can then be made on the basis of ability to conform in production engineering terms to design assumptions made much earlier. It is unacceptable to allow the fabricator or designer to undertake the production engineering element of design in perfect isolation; the dialogue between designers and fabricators must be a continuous one. The effects that fabrication and assembly have on design assumptions and vice versa and, in particular, the achievable fit-up of components and permissible limits of tolerance must be addressed.

Design must be viewed as a complete process, covering strength and stiffness as well as production engineering, to achieve the most economical structures.

32.2 Economy of fabrication

Structural form has a significant effect on the delivered-to-site cost of steelwork. This is due to a number of factors additional to the cost of raw steel from steel suppliers. Some forms will prove to be more costly from some fabricators than others; they tend to attract work by aligning their production facilities to specific market sectors. For example, the industrial building market was largely taken over by the introduction of the portal-framed structure. The fabricators in this sector adopted high-volume, low-cost production and concentrated primarily on this area of the market. By the use of pre-engineered standards they were able to maximize repetition and minimize input from both design and drawing activities: a classic example

948

of a combination of design and production engineering. Other steelwork contractors specialize in tubular structures, lightweight sections and heavy sections.

32.2.1 Fabrication as a cost consideration

Figures 32.1 to 32.4 give an indication of the proportional costs associated with the fabrication and erection of structural steelwork. Actual costs in monetary terms have not been included as they will change with the demand level of the market. The proportions of cost will vary a little from fabricator to fabricator; those shown represent a reasonable average.

The cost headings are:

- raw steel
- fabrication ⎫
- painting   ⎬ fabrication shop activities
- transport   ⎭
- erection
- site painting.

Raw steel cost covers the average cost of rolled steel in S275 delivered to the steelwork contractor. No allowance has been made for extra costs arising from the use of stockholders' steel or the extremes of section variation.

Fabrication cost covers a number of different sized jobs and incorporates cleaning of the raw steel by shot blasting, preparation of small parts (cleats and plates and connection components), and assembly of the components into complete structural members ready for shop-applied paint treatments. It also includes an allowance for consumables.

Paint cost covers the shop application of 75 μm of primer by spray immediately after fabrication. No allowance has been made for blast cleaning of areas affected by welding. A coverage allowance of 28 m²/tonne at the rate of 3 m²/litre has been made, which represents the likely consumption for rolled section beams and columns. This allowance has been adjusted for various specific work types as described in the accompanying text.

Transport cost is based on 20 tonne loads per trailer which delivers finished products to sites within a 50 mile radius of the fabrication shop. Transport costs will rise for loads of less than 20 tonnes or when oversize components need special police escort or permission.

The erection cost is average for the work type and includes preliminaries for high-rise multi-storey work.

The cost of site painting has been included but it only covers the average cost for touching-up damage to the primer coat. Other site-applied protection systems vary enormously in cost and so have not been considered.

Portal-framed industrial buildings

The breakdown shown in Fig. 32.1 follows the assumptions given above. There is no adjustment in either the shop painting or the transport cost as this type of work fits the basic parameters well. It can be seen that design economies come principally from the weight of the structure combined with efficient fabrication processes.

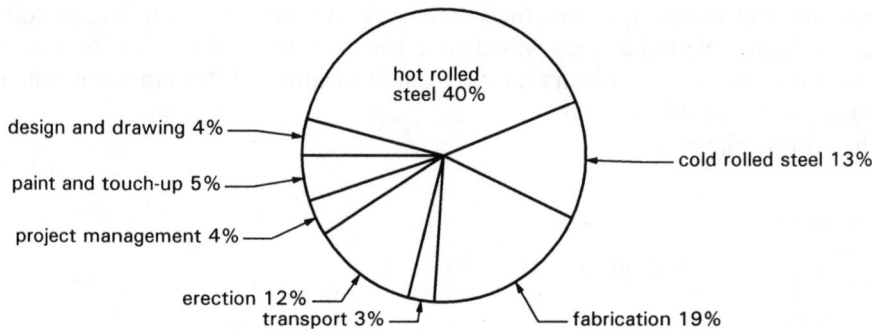

Fig. 32.1   Cost breakdown: portal-framed industrial buildings

Simple beam and column structures

The breakdown shown in Fig. 32.2 incorporates the following limitations:

(1) the maximum height of the building is three storeys,
(2) erection is carried out using mobile cranes on the ground floor slab.

It can be seen that again tonnage and fabrication efficiencies are the dominant criteria; 83% of the costs arise from these elements, with a slightly greater emphasis on tonnage than was the case with portal frames.

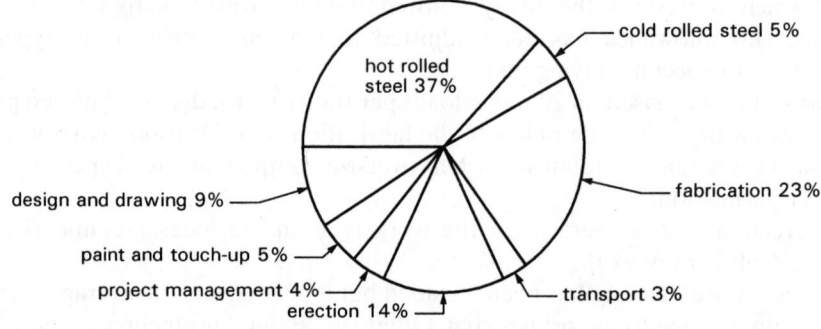

Fig. 32.2   Cost breakdown: simple beam and column structures

High-rise multi-storey

The breakdown shown in Fig. 32.3 incorporates the following adjustments:

(1) the steel grade has been taken as S355 with an allowance for cambering,
(2) the paint system generally is shot blast and 75 μm primer, 10% of steel coated with 100 μm primer,
(3) no allowance has been made for any concrete-encased beams or stanchions,
(4) transport includes for off-site stockpiling, bundling and out-of-hours delivery to site (city centre sites often incur these costs).

This sector of the market has a very different cost profile to those already shown. Raw steel still dominates but erection charges have now overtaken the fabrication element. This type of steelwork lends itself particularly well to automated fabrication techniques featuring drilling lines.

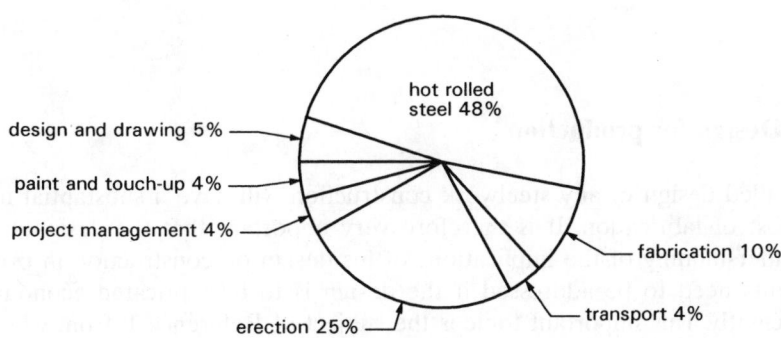

Fig. 32.3 Cost breakdown: high-rise multi-storey

Lattice structures

This is perhaps the most difficult of the sectors on which to carry out an analysis as lattice structures vary enormously in size, complexity and element make-up. The costs shown in Fig. 32.4 are indicative and incorporate the following:

(1) angle booms and lacings,
(2) welded joints without gusset plates,
(3) transportable lengths and widths with full-depth splices only.

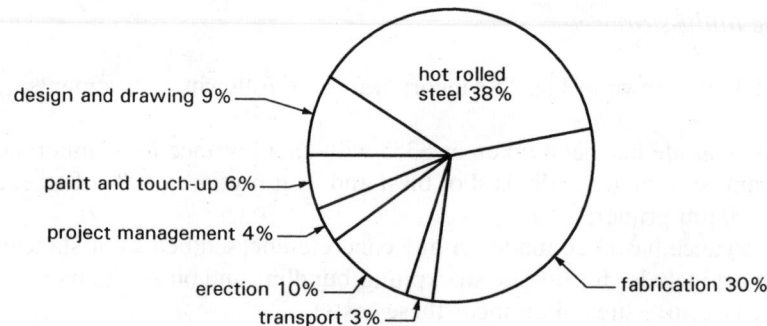

Fig. 32.4   Cost breakdown: lattice structures

Here fabrication is virtually the same cost as the raw steel. The designer should work very closely with the steelwork contractor to ensure simplicity of assembly for this form of structure, in particular checking joint capacities at the design stage.

32.2.2 Design for production

The detailed design of any steelwork construction will have a substantial impact on its cost of fabrication. It is therefore very important that the designer has a basic understanding of the implications of his design on construction in practice. Key points need to be addressed if the design is to be fabricated economically and efficiently. This important topic is the subject of Reference 1, from which the following key points are extracted.

Fabrication processes

- Modern computer numerically controlled (CNC) fabrication equipment is more effective with:
    (a) Single end cuts, arranged square to the member length
    (b) One hole diameter on any one piece, avoids drill bit changes
    (c) Alignment of holes on an axis square to the member length, holes in webs and flanges aligned not staggered to reduce piece moves between drill times
    (d) Web holes having adequate side clearance to the flanges.
- To allow efficient production of fittings:
    (a) Rationalize on the range of fittings sizes – use a limited range of flats and angles
    (b) Allow punching and cropping wherever possible.

- If possible select connections which avoid mixing welding and drilling in any one piece. This avoids double handling of the member during fabrication.

Materials grade and section selection

- The designer should rationalize the range of sections and grades used in any one structure. This will lead to benefits in purchasing and handling during all fabrication, transportation and erection phases of manufacture.
- Make maximum use of S355 material for main sections. This is typically 8% more expensive but up to 30% stronger than S275 steel. The exception is where deflection governs section selection.
- The specification of small quantities of S355 or other 'special' grade material should be avoided, particularly if the proposed material has poorer welding qualities.
- Choice of fittings material grade should be left with the fabricator wherever possible.
- Structural hollow sections are approximately 60–80% more expensive than equivalent weight open sections and have additional problems associated with the connection requirements. Limitations on mill lengths should also be remembered.

Connection design considerations

- Connections directly influence 40–60% of the total frame cost. They must therefore be taken into account during the frame design.
- Least-weight design solutions are rarely the cheapest. Increasing member thickness to eliminate stiffening at connections will often be an economic solution.
- The cost benefits from an integrated approach to frame and connection design will only be realized if the fabricator is given a full package of information at tender stage. Connection styles and design philosophy must be clearly marked on drawings.

Bolts and bolting

- Non-preloaded bolting is the preferred method for site connections. Preloaded (friction grip) bolts should only be used where joint slip is unacceptable or where there is a danger of fatigue.
- The use of different grade bolts of the same diameter on any one contract should be avoided.
- Threads should be permitted in the shear plane and in bearing.
- Direct and indirect cost savings can accrue by using only a small range of 'standard' bolts.

Recommended standards are:
  M20 grade 8.8 for shear connections
  M24 grade 8.8 for moment connections
  Mechanical properties to BS 3692, dimensions to BS 4190
  Fully threaded for shanks up to 70 mm long.
- The use of fully threaded bolts generally means additional thread protrusion is visible; specifiers should be aware of this and state at tender stage where this is *not* acceptable.
- Washers are *not* required for strength when using non-preloaded bolts in normal clearance holes; they may still be specified to provide a degree of protection to surface finishes.
- When used with corrosion-protected steelwork, bolts, nuts and washers should be supplied with a coating which does not require further protection applications.

Welding and inspection

- The welding content of a fabrication has a significant influence on the total cost of fabrication.
- In designing welded connections consideration should be given to the weldability of materials, access for welding and inspection, and the effects of distortion. Access is of primary importance – good welds cannot be formed without adequate access.
- Fillet welds up to 12 mm leg length are preferred to the equivalent-strength butt weld. Generally two fillet welds whose combined throat thicknesses equal the thickness of the plate to be connected are considered as equivalent in strength to a full penetration butt weld.
- Weld defect inspection and defect acceptance criteria should be defined; the use of the National Structural Steelwork Specification criteria is strongly recommended.

Corrosion protection

- In selecting a corrosion protection system the designer must consider the environment in which the steelwork will be placed and the design life of the corrosion protection system.
  If the environment does not require a corrosion protection system don't specify one.
- If a protection system is required, significant advantages are gained by use of a single-coat protection system applied during fabrication. These should be specified where possible.
- Wherever possible avoid using 'named' product specifications; allow the fabricator to use his preferred supplier or even alternative preferred coating system of equal capability.
  Specification of surface conditions should relate to the condition immediately prior to painting, not bound by any time-limit from shot blasting operations.

Example coating specifications for a range of environments are given, together with an indication of relative costs.

Trusses and lattice girders

- Lattice girders and trusses are effective for medium to long spans where deflection is a major criterion and are able to accommodate services within their depth, but always consider the use of a plain rolled section beam first.
- Most lattice frames are joint critical. Never select a section for the chords or internals without first checking whether it can be effectively joined – preferably without recourse to stiffening.
- Always check the limits on transport before starting the design.
- Be aware that SHS are only available in limited standard lengths, normally from stockists. Long lengths may therefore need additional butt welding.
- For internal members try to detail single bevel end cuts; for angles square-cut ends are better to allow use of an automatic cropping process.
- In tubular construction use of RHS chords leads to simpler end preparation for internals than that required if CHS chords are used.
- Think about access provisions for welding of internals to chords.
- Access for painting is difficult for double angle or double channel members; use of SHS reduces paint area and provides fewer locations for corrosion traps to be formed.

Transportation

- Police notification with associated programme and cost penalties will occur for road transport loads greater than 18.3 m long, or 2.9 m wide or 3.175 m high.

32.3 Welding

Fusion welding processes are used to join structural steel components together. These processes can be carried out either in the workshop or on site, though it is generally accepted that site welding should not be used as a primary source of fabrication. Welding should be undertaken only by welders who are certificated to the appropriate level required by British Standards or other recognized authority. Welding is covered in detail in Chapter 24.

32.4 Bolting

Shop bolting may form part of the fabrication process. There are implications arising from the use of shop bolting which need to be appreciated by the designer in order to ensure that a cost penalty does not occur.

At one time it was common practice to assemble components in the workshop using bolts or rivets. With the increased implementation of welding, this practice has declined due to the costs associated with bolted fabrications. Instead of a simple run of fillet weld, holes need to be drilled and bolts introduced, increasing total labour hours and cost. In many respects the ease with which welding can be undertaken has diverted designers' attention from the use of shop bolting. Today, steelwork contractors are looking at increased automation to keep costs down, and machines have been developed which considerably speed up hole drilling.

32.4.1 Shop bolting

There is still a demand for structural members to be bolted arising from a requirement to avoid welding because of the service conditions of the member under consideration. These may be low temperature criteria, the need to avoid welding stresses or the requirement for the component to be taken apart during service (e.g. bolted-on crane rails). For lattice structures, the designer should specify the bolting, bearing in mind the effect of hole clearances around bolt shanks. HSFG bolts will not give problems but other bolts in clearance holes will allow a 'shake-out' which can cause significant additional displacement at joints. Typically, a truss with bolted connections may deflect due to the take-up of lack of fit in clearance holes to such an extent that it loses its theoretical camber. The use of HSFG assemblies avoids this risk.

Large and complex assemblies which are to be bolted together on site may be trial assembled in the fabrication shop. This increases fabrication costs but may pay for itself many times over by ensuring that the steel delivered to site will fit. Restricting trial assembly to highly repetitive items or items critical to the site programme is to be recommended.

32.4.2 Types of bolt

The choice of which type of bolt to use may not necessarily be made on the basis of strength alone but may be influenced by the actual situation in which the bolt is used, e.g. in non-slip connections.

There are four basic types of bolts. They are structural bolts, friction-grip bolts and close-tolerance bolts.

Structural bolts

Bolts with low material strength and wide manufacturing tolerance were until recently known as 'black bolts' because of their appearance. Now they are called

structural bolts, normally have a protective coating which gives them a bright appearance, and are available in a range of tensile strengths.

Friction-grip bolts

Friction-grip bolts are normally supplied with a protective coating and are differentiated by material grade. Friction-grip bolts are used in connections which resist shear by clamping action, in contrast to structural bolts which resist shear and bearing. When considering the use of friction-grip bolts during the fabrication process, adequate means of access must be provided so that the bolt can be properly tensioned.

Close-tolerance bolts

This type of bolt differs from a structural bolt in that it is manufactured to closer tolerances. To gain the full benefit of close-tolerance bolts, they should be used in close-fitting holes produced by reaming, which adds considerably to the expense of fabrication. Where a limited slip connection is required, close-tolerance bolts can be used in holes of the same nominal diameter as the bolt but not reamed; this gives a connection which is subject to far less slip than would be the case for structural bolts in clearance holes.

32.4.3 Hole forming

To gain the best output, computer numerically controlled (CNC) machines are incorporated in conveyor lines. There is an infeed line which takes the unprocessed raw steel and an outfeed line which distributes the finished product either to the despatch area or further along the fabrication cycle and into an assembly area. The infeed conveyor for punching and drilling machines that handle angles, flats, small channels and joists will normally be configured to handle 12 m long bars. There will be a marking unit which carries a set of marking dyes that stamp an identification mark on to the steel if required. There may be a number of hydraulic punch presses each suitable for accepting up to three punching tools, and it is normal that the tool holders are quick-change units. Typically, hydraulic presses have 1000 kN capacity. The punch presses can form differently-shaped holes so that angles can be produced with slots. Some machines produce the slot by a series of circular punching operations, others have a single slot-shaped dye.

Once the material has passed through the punch presses, it is then positioned for shearing. The hydraulic shear has a capacity of between 3000 and 5000 kN. Machines which incorporate drilling facilities can have these positioned next to the punch presses prior to the hydraulic shear. The output of the machine can be as high as a

thousand holes per hour. Obviously, these machines can be obtained with different capacities so that if the work of the fabricator is at the light end of the section range, it is not necessary to purchase machines with large capacity.

For larger sections, where drilling and cutting are required, there are two basic machines available. One is a drilling and plate-cutting system which is used for heavy plates; the other machine is of a larger capacity and is used for drilling rolled sections of all sizes. Modern drills normally have three-axis numeric control and air-cooled drills. These machines have sensors which can detect the position of the web to ensure that hole patterns are symmetrical about the web centreline.

Where components only need a small number of holes, mobile drills are used. These are normally magnetic limpet drills hand operated by the fabricator. The holes are formed with rotary-broach drill bits, which have a central guide drill surrounded by a cylindrical cutter.

The fabricator will always carry out hole-forming operations prior to any further fabrication as the presence of any stiffeners or cleats on a bar would severely disrupt the input of NC machines.

32.5 Cutting

The fabrication process of cutting has become highly automated. Universal beams, columns and the larger angles, tees and channels are normally cut to length by saws. Hollow sections are also treated in this way. Small sections of joist, channel, angle and flat are cut to length by shearing, either as a separate operation or as part of a punching and cropping operation which is computer controlled. Large plate sections may be sheared but this involves specialist plant and equipment which is not available to all fabricators.

32.5.1 Cutting and shaping techniques

Flame cutting or burning

This technique produces a cut by the use of a cutting torch. The process may be either manual or machine controlled. Manual cutting produces a rough edge profile, which can be very jagged and may need further treatment to improve its appearance. Edges of plates cut manually require greater edge distances to holes for this reason. Notches or holes with square corners should not be hand cut unless the corner is first radiused by a drilled hole, Fig. 32.5.

When flame cutting is carried out under machine control, the cut edge is smooth and therefore the restrictions on edge distances to holes are relaxed. Notches which are machine cut need radiusing either by a predrilled hole or by the control of the cutting machine. Unradiused notches are to be avoided due to their stress-raising

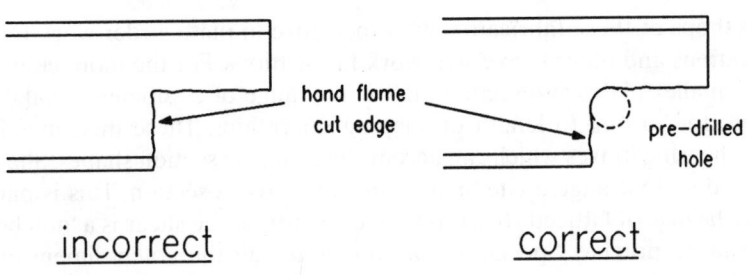

Fig. 32.5   Hand flame cutting

characteristics. Notching or coping of beam ends can now be carried out by computer-controlled machines which burn the webs first and create the radiused corner then cut the flanges with a cut which coincides with the web cut. Numerically controlled coping machines can cut top and bottom notches simultaneously, and the notches can be of different dimensions.

Castellated beams were traditionally formed by cutting a multi-linear, castellated pattern along a UB or UC section followed by realignment and rewelding. The web cutting is always machine controlled to ensure accurate fit-up and an edge which is suitable for subsequent welding.

More recently, a two-cut process has been developed that creates beams with circular openings, frequently called *Cellform* beams. These beams have proved to be very popular both for long span roof structures where the steelwork is exposed and as primary and secondary beams in composite floors. In the latter case, further economy may be achieved by cutting the top tee sections from a lighter rolling than the bottom.

Arc plasma cutting

Plates may be cut by arc plasma techniques. In this case the cutting energy is produced electrically by heating a gas in an electric arc produced between the tungsten electrode and the workpiece which ionizes the gas, enabling it to conduct an electric current. The high-velocity plasma jet melts the metal of the workpiece and blows it away. The cut so produced is very clean, and its quality can be improved by using a water injection arc plasma torch. Plasma cutting can be used on thicknesses up to about 150 mm but the process is then substantially reduced in speed.

Shearing and cropping

Sections can be cut to length or width by cropping or shearing using hydraulic shears. Heavy sections or long plates can be shaped and cut to length by specialist plate shears. These are large and very expensive machines normally to be found in

the workshops of those fabricators who specialize in plate girder work for bridges, power stations and other heavy steelwork fabrications. For the more commonplace range of smaller plates and sections, there is a range of equipment available which is suitable for cutting to length or shaping operations. These machines feature a range of shearing knives which can accept the differing section shapes. Shearing can be adjusted so that angled cuts may be made across a section. This is particularly useful for lacings of latticed structures. One version of the shear is a 'notcher' which can cut shaped notches. Special dies are made to suit the notch dimensions, and it is possible to obtain dies to cut the ends of hollow sections in preparation for welding together.

Cold sawing

When, because of either specification or size, a section cannot be cut to length by cropping or shearing, then it is normally sawn. All saws for structural applications are mechanical and feature some degree of computer control. Sawing is normally carried out after steel is shot blasted as the saw can be easily incorporated within the conveyor systems associated with shot blast plants.

There are three forms of mechanical saw: circular, band and hack. The circular saw has the blade rotating in a vertical plane which can cut either downwards or upwards, though the former is the more common. The blade is a large milling wheel approximately 5 mm thick. The diameter of the saw blade determines its capacity in terms of the maximum size of section which can be cut. Normally, steelwork contractors will have saws capable of dealing with the largest sections produced by Corus. For increased productivity, sections may be nested or stacked together and cut simultaneously. Some circular saws allow the blade to move transversely across the workpiece, which is useful for wide plates. Most circular saws can make raking cuts across a section of any angle, though some saws are restricted to single side movement. The preferred axis of cut is across the $Y–Y$ axis of the section. In the case of beams, channels and columns, this is with the web horizontal and the flange toes upwards. Depending on the control exercised, a circular saw can make a cut within the flatness and squareness tolerances necessary for end bearing of members.

Band saws have less capacity in terms of section size which can be cut. The saw blade is a continuous metal band edged with cutting teeth which is driven by an electric motor. The speed of cut is adjustable to suit the workpiece. Band saws can make mitre cuts and can cut through stacked sections. Cutting accuracy is dependent on machine set-up but produces results similar to circular saws.

Hack saws are as the name implies mechanically driven reciprocating saws. They have normal format blades carried in a heavy duty hack saw frame. Hack saws have more limited cutting capacity than band saws and have the capability to produce mitre cuts.

All saws feature computer-controlled positioning carriages for accurate length set-up; most also have computer sensing for angle cuts.

Gouging

The gouging process is the removal of metal at the underside of butt welds. Gouging techniques are also used for the removal of defective material or welds. The various forms of gouging are flame gouging, air-arc gouging, oxygen-arc gouging and metal-arc gouging.

Flame gouging is an oxy fuel gas cutting process and uses the same torch but with a different nozzle. It is important that a proper gouging nozzle is adopted so that the gouging profile is correct.

Air-arc gouging uses the same equipment as manual metal arc welding. The process differs in that the electrode is made of a bonded mixture of carbon and graphite encased in a layer of copper, and jets of compressed air are emitted from the specially designed electrode holder. These air jets blow away the parent metal from beneath the arc.

A special electrode holder is used in the oxygen-arc gouging process. A special tubular coated steel electrode controls the release of a supply of oxygen. Oxygen is released once the arc is established, and when the gouge is being made the oxygen flow is increased to a maximum.

Standard manual metal arc welding equipment is used for metal-arc gouging. The electrodes, however, are specially designed for cutting or gouging. This process relies on the metal being forced out of the cut by the arc and not blown away as in the other processes.

32.5.2 Surface preparation

Structural sections from the rolling mills may require surface cleaning prior to fabrication and painting. Hand preparation, such as wire brushing, does not normally conform to the requirements of modern paint or surface protection systems.

Blast cleaning is the accepted way of carrying out surface preparation. It involves blasting dry steelwork with either shot or grit at high velocity to remove rust, oil, paint, mill scale and any other surface contaminants. The most productive form of blast cleaning plant has special equipment comprising infeed conveyor, drying oven, blast chamber, spray chamber and outfeed conveyor.

Steel is loaded on to the infeed conveyor either as separate bars or side by side depending on the blast chamber passage opening. It then travels through the drying oven, which ensures that any surface moisture is removed prior to blasting. The blast chamber receives the steel and passes it over racks and between the blasting turbines, which are impellers fed centrally with either shot or grit. The material is thrown out from the edge of the turbine and impacts against the steelwork. The speed of the turbine, its location and aperture determine the velocity and direction of the blasting medium.

The blasting medium is retrieved from within the blast chamber and recycled until it is exhausted. Shot provides a good medium for steel which is to be painted; grit

can cause excessive wear of the blast chamber and results in a surface which is more pitted, giving a reduced paint thickness over high points.

The steel leaves the blast chamber and immediately enters the spray chamber. Depending upon the requirements of the specification, the steel is then sprayed with a primer paint which protects it from flash rusting. The use of prefabrication primers should be discussed with the fabricator and paint system supplier as requirements may vary. For most structural applications in building work, there is little need to specify a prefabrication primer. Once the steel has passed through the spray chamber, it is fed on to the outfeed conveyor and continues to its next process centre.

The other form of blast cleaning is called *vacu blasting*, a manual method where the blasting medium is blown out of a hand-held nozzle under the pressure of compressed air. This process is normally performed in a special sealed cabinet with the operator wearing protective clothing and breathing equipment. It relies heavily on the skill of the operator and does not give production rates approaching those of mechanical blasting plants.

Local areas of damage caused by welding can be cleaned by vacu blasting or by needle gunning. The needle gun has a set of hardened needles which are propelled and withdrawn rapidly. The action of the needles on the steel effectively removes weld slag and provides a good surface for painting. Needle gunning is not suitable for large areas or for heavily contaminated surfaces.

Surface preparation is intended to provide a uniform substrate for even paint application. It is important to consider the edges of plates, where paint thickness is often low.

32.5.3 Cambering, straightening and bending

Each of these operations has a different purpose. Cambering is done to compensate for anticipated deflections of beams or trusses under permanent loads, dead loads or superimposed dead loads such as finishes. Straightening is part of the fabrication process aimed at bringing sections back within straightness tolerances. Bending is to form the section to a shape which is outside normal cambering limits.

Rolled sections are normally bought in by the fabricator and then sent to specialist firms for cambering. The steel section is cold cambered by being passed between rolls. The rolls are adjusted on each pass until the required camber is achieved. The following cambers can be produced:

- circular profile (specify radius)
- parabolic profile (specify equation)
- specified offsets (tabulate co-ordinates)
- reverse cambers (circular, parabolic, offset or composite).

Bending is also carried out by specialists, the process being identical to that for cambering. Curves can be formed on either the X–X or Y–Y axes.

Straightening can be carried out by the fabricator or by the cambering specialists. Sections are often straightened by the fabricator by applying heat local to the flange or web. This is a skilled operation which achieves excellent results.

32.6 Handling and routeing of steel

The fabrication process involves the movement of steel around the workshop from one process activity to the next. Clearly, all these movements should be planned so that the maximum throughput of work is achieved and costs are minimized. Even with the highly automated facilities available today, planning is essential as work centres may become overloaded and cause delays to other activities. In this respect, the work flow has to be balanced between work centres.

The routeing of steel through a fabrication workshop is planned on the basis of the work content required on the workpiece. The workpiece will consist of one or more main components which may have smaller items attached (cleats, tabs and brackets). The main components are normally obtained from the steel mills though some fabricators obtain steel from stockholders. The steel is either stored in a stock-yard local to the workshop or preferably delivered from the stockist just in time for immediate fabrication. The control of steel stock is an important management function which can balance the cash flow of the fabricator and provide clear identification of the steel for traceability purposes. Stock control systems are commonly computerized and should allow the fabricator to identify:

(1) steel on order but not yet received,
(2) steel which is in the stockyard, and its location,
(3) allocation of steel to particular contracts,
(4) levels of unallocated steel (free stock).

Prior to the fabrication cycle, an estimate is made of the work content of a work-piece. This assessment may be crudely based upon tonnage or more commonly based upon time study data of previous similar work. The work content assessment is a production control function which will identify:

(1) weight of finished product,
(2) content of workpiece (parts list),
(3) work centre activities required (sawing, drilling, welding, etc.),
(4) labour content,
(5) machine centre utilization.

Production control systems are now often computerized and can produce a detailed map of the movement of the workpiece and its component parts within the fabrication workshop. Computer systems can identify any potential overloadings of work centres so that the production controller can specify an alternative path.

Table 32.1 Sequence of activities in fabricating shops

Cleaning and cutting to length	Shot blasting Sawing
Manufacture of attachments	Preparation of: cleats brackets stiffening plates end plates baseplates
Preparation of main components	Pre-assembly: drilling punching coping cropping notching cutting of openings
Assembly of workpiece	Marshalling of components Setting-out of components Shop bolting Welding Cleaning up

Quality control (assembly)	Check assembly – dimensional
	                   – NDT
	                   – visual inspection
	Sign-off

Surface treatment	Painting Metal spraying } often by other specialists Galvanizing
Quality control (final product)	Conformity with specification Sign-off
Transportation	Loading Despatch

Generally, the cycle of work centre activity is as shown in Table 32.1 (the exact sequence of events may, however, vary between fabricators).

32.6.1 Lifting equipment in fabrication workshops

Not all steel can be moved around a workshop on conveyor lines. It is not practical or cost effective to do so and would restrict the flexibility for planning workshop activities. Much of the steel will be moved by cranes of one form or another. Craneage capacity can be a restriction on the work which a fabricator can undertake. Restricted headroom is another problem, particularly when considering the fabrication of deep roof trusses.

There are various types of crane in common use in most workshops:

- electric overhead travelling (EOT)
- goliath

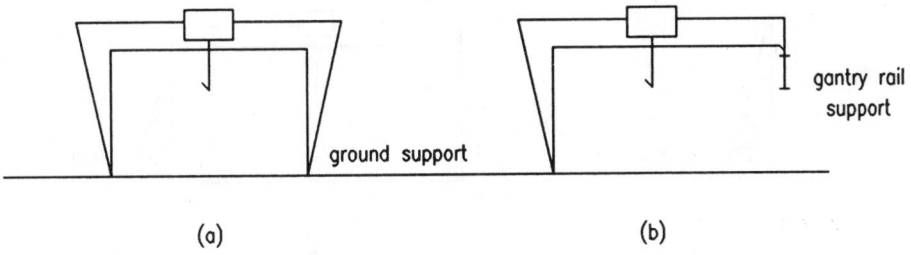

Fig. 32.6 Goliath cranes: (a) goliath, (b) semi-goliath

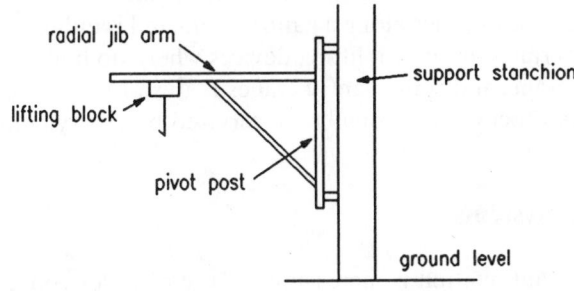

Fig. 32.7 Jib crane

- semi-goliath
- jib
- gantry.

EOT cranes are the main lifting vehicles in the larger fabrication workshops. Their capacity varies from fabricator to fabricator but rarely exceeds 30 tonnes SWL. EOT cranes run on rails supported on crane beams carried off either the main building stanchions or a separate gantry and are used for moving main components and finished workpieces between work centres. Heavy components need turning over for welding and this is achieved by the use of EOT cranes.

Goliath and semi-goliath cranes are heavy-duty lifting devices which tend to be used on a more local basis within the workshop than EOT cranes. The goliath is a free-standing gantry supported on rails laid along the workshop floor. Semi-goliath cranes are ground rail supported on one side and gantry rail supported on the other. The lifting capacity of goliath-type cranes is normally less than that of an EOT crane though greater in some workshops where there are EOT crane support restrictions (Fig. 32.6).

Jib cranes are light-duty (1–3 tonne capacity) cranes mounted from building side stanchions and operating on a radial arm with a reach of up to 5 m. They are useful for lifting or turning the lighter components (Fig. 32.7).

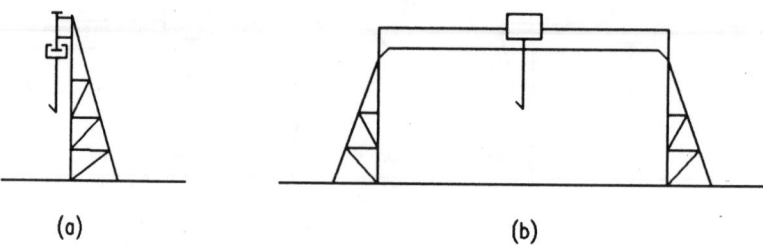

Fig. 32.8   Gantry cranes: (a) single underslung, (b) EOT

Gantry cranes are purpose-built structures with limited lifting capacity (3–10 tonne). The lifting block travels along a gantry beam and is either underslung or top mounted. Gantry cranes are useful lifting devices where no building can be used to support an EOT crane, and some gantry cranes feature EOT cranes riding on rails alongside gantries. Stockyards normally are serviced by gantry cranes (Fig. 32.8).

32.6.2 Conveyor systems

The use of factory automation is increasing the use of roller conveyor systems for moving main components. The roller conveyors can be designed so that they move steel between automated work centres either laterally across the workshop or longitudinally along it. The modern conveyor systems are computer controlled and bring steel accurately into position for the work centre activity to be performed. An example of a roller conveyor system is shown in Fig. 32.9.

Roller conveyors are suitable for steel which does not need turning and does not have many cleats or brackets attached to it. Assembly work is not performed on conveyors but the assembly work benches are fed with components from conveyors or cranes. Rolling buggies are also used, in which case fabrication operations are carried out with the steel section on the buggy.

32.6.3 Handling aids

Most steelwork assemblies can be picked up easily using chains or strops but the designer should be alert to the possibility of having to provide for temporary lifting brackets suitably stiffened. This may prove necessary where it is important that a component is lifted in a particular way either for stability reasons or for reasons of strength.

The designer should also be cautious in the provision of welded-on brackets. Flimsy outstands *will* get damaged, whether in the workshop, during transport or on site. Flimsy brackets should be supplied separately for bolting on at site.

Some components by nature of their design or geometry may require to be lifted in a specific way using strongbacks or lifting beams. The designer must make the

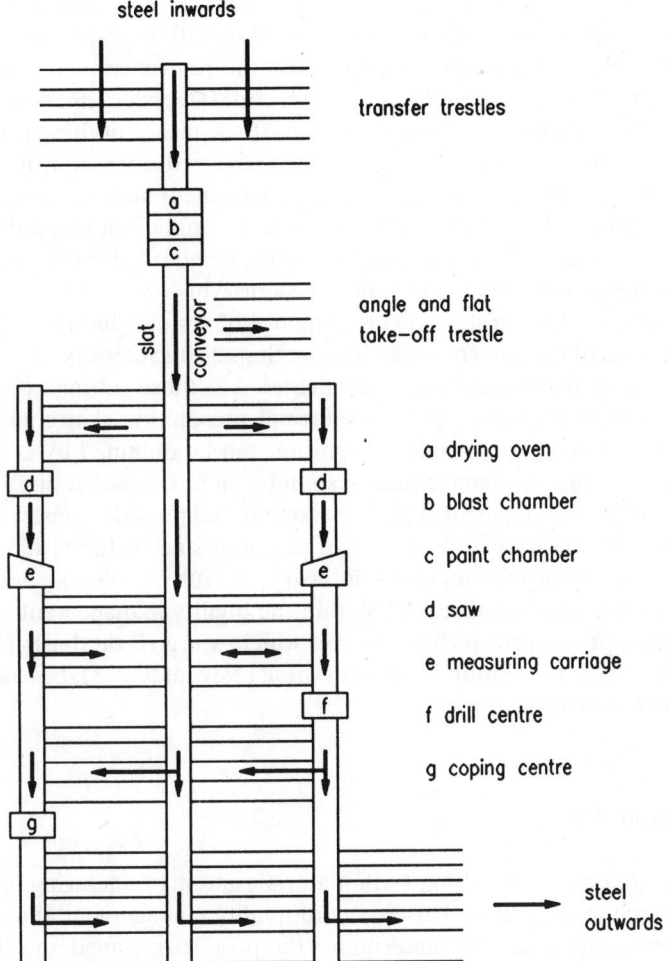

Fig. 32.9   Roller conveyor system

fabricator aware of this requirement to avoid possible accidental damage or injury to workshop staff; steel is a very unforgiving material.

The fabricator will normally organize transport of steel from the workshop to site. There are loading restrictions on vehicles in terms of weight, width, height and length.

32.7 Quality management

The key activities within any quality management system are quality assurance and quality control. These activities are fully embodied in many industry-wide quality

standards. While it is not a prerequisite that a manufacturer is registered under a standard-based scheme for appropriate levels of quality management to prevail, it does, however, allow specifiers to easily assess the relevance of the system under review. As far as design and manufacture of structural steelwork are concerned, there are a number of standards which apply. The most significant of these is the National Structural Steelwork Specification. These standards are the basis for the formulation and implementation of appropriate quality management systems. British Standard 5750 provides an excellent basis for a system to be applied in structural steelwork manufacturing and design. The various parts of the standard give the opportunity for fabricators to develop quality management systems which cover design, manufacture and the erection of structural steelwork. The use of systems based on BS 5750 will allow progression to the International System Standard ISO 9000.

While the use of these standards is commended to those setting up quality management systems, it does not imply any formal recognition that compliance with the standard has been met. Formal recognition can be obtained by application for assessment of the quality management system by an independent body. The British Standards Institution will perform this assessment and provide registration to those companies which can successfully demonstrate compliance with the standard. Other registration schemes in the steelwork industry are run by Lloyd's and Steelwork Construction Quality Assurance Ltd. Within the quality management system there must be written procedures to be followed which will give the basis for adequate QA and OC. A brief description of some of these relevant to the fabrication of structural steelwork is given below.

32.7.1 Traceability

It is necessary to demonstrate that both materials and manufacturing processes can be traced by a clear audit trail. This aspect of quality management allows the specifier to determine the source of material for the product received and the origin of workmanship. This is particularly important when a defect is discovered. In terms of fabrication, the materials should have the support of documentation from the steelmaker in the form of mill certificates. Suppliers of other materials or consumable items such as bolts, welding electrodes, etc. will also be required to give details which show appropriate manufacturing data. The fabrication process involves workmanship, and evidence should be maintained which traces the source of workmanship on each item.

32.7.2 Inspection

This is a prime area of quality control. It is important that quality inspectors are independent of workshop management: that is, they must be made responsible to a higher level of management in the company which is not concerned with the pro-

duction of fabricated steel in the workshop. They may report to the quality manager of the company, who in turn may report to the managing director. It is also important that all inspection activities are part of a predefined quality plan for the job in question. There should be no opportunity for the level or extent of inspection to be decided upon an *ad hoc* basis. The quality plan will define levels of inspection required and when the inspection is to take place, typically:

(1) inspection of incoming materials or components,
(2) inspection of fabricated assemblies during and/or after fabrication,
(3) inspection of finished products prior to despatch,
(4) calibration of measuring equipment,
(5) certification of welders.

32.7.3 Defect feedback

At some stage or another, there will be occasions when defects are discovered in the fabricated item. This may be during inspection or on site. It is important that the defect is reported, and the quality management system must cater for this. The report should be formal and it must identify the defect and possible cause. This report is then acted upon to prevent recurrence of the defect. The audit trail of traceability and inspection should highlight the actual source of the defect, and this may be corrected by defined and planned actions.

32.7.4 Corrective action

All defects will cause corrective action to be taken. This action may take the form of revised procedures but more commonly will involve a process change. The most important corrective action is training as many defects stem from a lack of understanding on the part of someone in the production cycle. An important part of the prevention of recurrence of defects is trend analysis. Defects are recorded and trends studied which may highlight underlying problems which can be tackled. Occasionally, a finished product does not fully comply with the client's requirements and an approach may be made to determine if the product is suitable. While this process is going on the offending items must be put into quarantine areas which clearly distinguish them from others which do conform to requirements.

Further reading for Chapter 32

British Constructional Steelworks Association (2002) *National Structural Steelwork Specification*, 4th edn. BCSA/SCI.

CIM Steel (1995) *Design for Manufacture Guidelines.* The Steel Construction Institute, Ascot, Berks.

CIM Steel (1997) *Design for Construction.* The Steel Construction Institute, Ascot, Berks.

Dewsnap H. (1987) Submerged-arc welding of plate girders. *Metal Construction,* Oct.

Gibbons C. (1995) Economic steelwork design. *The Structural Engineer,* **73,** No. 15 Aug., 250–3.

Hicks J.G. (1999) *Welded Joint Design,* 3rd edn. Abington, Cambridge.

Pratt J.L. (1989) *Introduction to the Welding of Structural Steelwork,* 3rd edn. SCI Publication 014, the Steel Construction Institute, Ascot, Berks.

Taggart R. (1986) Structural steelwork fabrication. *The Structural Engineer,* **64A,** No. 8, Aug.

Chapter 33
Erection

By ALAN ROGAN

33.1 Introduction

Planning for the erection of any structure should commence at the design phase. The incorporation of buildability into the design phase may allow significant additional benefits. If a structure is to be erected quickly, to programme, and to the lowest possible cost, consideration must be given to the planning of the site works.

Work at a height and on site should be minimized, especially if this work could be carried out at the factory in ideal conditions. Where possible, frames and components should be designed to be built in the factory or assembled at low level for subsequent incorporation into the works. This will save time and cost as the work is not so weather dependent and expensive temporary works are not required. Planners should:

- plan for repetition and standardization
- plan for simplicity of assembly
- plan for ease of erection – keep it simple
- agree information handover dates and design sign-off dates
- make allowance for trade interfaces
- allow realistic programmes for manufacture and erection
- recognize the complexity of the design process
- hold regular co-ordination meetings
- identify and fairly allocate risk
- consider long-term stable relations with teamwork.

Attention to site construction methods are now considered an essential part of any design. So too is observance of the Construction (Design and Management) (CDM) Regulations[1] aimed at enforcing the safe construction and use of structures. The designer of the structure should take into account site access, materials handling, the construction sequence, and any limitations these may impose on the construction project.

One of the key factors to improvement in construction is the specialist trade contractors (STCs) as they carry out as much as 90% of the contract value. Utilizing the expertise of specialist contractors at an early stage will maximize the speed of construction, reduce conflict, and give the opportunity of adding real value to the building.

33.2 The method statement

It is now customary for the steelwork specialist to provide an erection method statement for the site work. The purpose of this is to set out a safe system of work for the delivery, erection and completion of the intended structure, thereby allowing the design team the opportunity to appraise this plan and make any appropriate observations or changes.

The erection method statement will set out procedures for delivery and assembly including the phasing of the erection, bolting, welding, and methods of erection and safety of the steelwork package. Guidance note GS28 Parts 1 & 2,[2] produced by the Health and Safety Executive (HSE), contains full details of the contents of a method statement. The main items described include:

(1) Arrangements for scheme management, including co-ordination and responsibility allocation of supervisory personnel at all levels

(2) Delivery and off-loading of the steelwork and other materials, including lifting methods and weights of each load.

(3) Erection sequences, notification of the scheduled starting position or positions if phased construction is required.

(4) Methods of ensuring continual stability of individual members (including columns) and sub-assemblies, as well as of the partially erected structure.

(5) The detailed method of erecting the structure or erection scheme devised to ensure that activities such as lifting, un-slinging, initial connecting, alignment and final connecting can be carried out safely.

(6) Provisions to prevent falls from height, including safe means of access and safe places of work, special platforms and walkways, and arrangements for the early completion of permanent walkways, mobile towers, aerial platforms, slung, suspended or other scaffolds, secured ladders, safety harnesses and safety nets.

(7) Protection from falls of materials, tools and debris by the provision of barriers such as screens, fans and nets.

(8) The provision of suitable plant (including cranes), tools and equipment of sufficient strength and quantity.

(9) Contingency arrangements to guard against the unexpected, such as a breakdown of essential plant, or the delivery of components out of sequence.

(10) Arrangements for delivery, stacking, storing, on-site movement, fabrication or pre-assembly and the siting of offices and mess rooms.

(11) Details of site features, layout and access, with notes on how they may affect proposed arrangements and methods of working.

(12) A detailed risk assessment to ensure all safety precautions have been taken.

(13) Handover procedures, including a method of checking the alignment of the structure and methods of determining bolting and welding completion.

These items enable the requirements of the Health and Safety at Work Act 1974 to be satisfied.

A complex job may require a series of complex method statements, each covering a particular aspect of the work.

33.3 Planning

33.3.1 Design information

Construction planning should start at the design phase since decisions at this stage will dictate the performance of the fabrication and erection teams. Site erection performance depends on many external and internal factors, and it is essential that these areas are properly managed. The site operation process is a complex inter-relationship of other disciplines which influence performance, as shown in Fig. 33.1.

The use of value engineering has produced spectacular improvements in on-site performance by identifying the cost of alternative methods and highlighting the need to minimize change in the site erection programme. The precise methodology

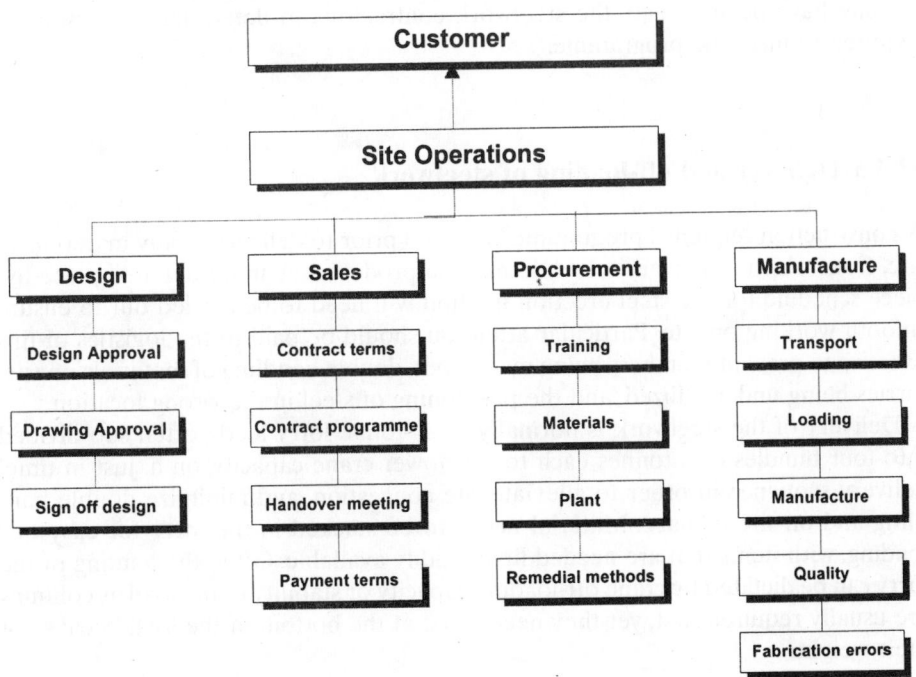

Fig. 33.1   Chart showing the dependence of the erection process on other disciplines

it offers provides a unique vehicle for the incorporation of a wide range of construction expertise into the design process, with the ultimate objective of achieving building designs and construction which offer clients better value for money.

33.3.2 Programming

Resources can be best utilized when the delivery and erection operations are well planned to follow a logical sequence. Typically, civil and other trades need to be sufficiently ahead of the steel erection (i.e. one week minimum) to allow for a reasonable flow of uninterrupted work. This will allow the steelwork contractor time to survey the foundation bolts prior to accepting them. It is not unusual for the civil contractor to have to carry out minor remedial work due to poor alignment of the holding-down bolts, and to this end it is advisable to allow the maximum time possible for handovers. Accurate positioning of foundations is essential, and tolerances defined in the specification and the National Structural Steelwork Specification (NSSS)[3] should be strictly adhered to.

Access restrictions or phasing of the works often govern the sequence of erection, which normally follows a grid pattern, split into zones. The siting of the erection crane will also dictate the construction pattern. Once the sequence and phasing have been agreed, the steelwork contractor can determine the resources required to meet the programme.

33.3.3 Delivery and off-loading of steelwork

A construction sequence programme is agreed prior to delivery of any materials to site, from which a delivery schedule may be produced. A more detailed piece-by-piece schedule for the steel erection method will need to be carried out to ensure smooth working on site. Particular attention should be paid to the logistics of this exercise to prevent members being missed out, double handling of materials on site, lorries being under-utilized, and the positioning of steel in the wrong location.

Delivery of the steelwork is normally in 20-tonne lorry loads, often sub-divided into four bundles of 5 tonnes each to suit tower crane capacity, on a 'just in time' delivery sequence in order to alleviate site congestion, and minimize double handling and on-site damage. Material needs to be stacked in the lorry for easy offloading, with items that are needed first, readily available. Often the loading of the lorry can be dictated by crane off-loading capacity or stability of the load as columns are usually required first, yet they need to be at the bottom of the load because of their weight.

Shakedown areas (areas for separating and selecting steel bundles) are required on site to allow the sorting of the steelwork. The area should be firm and level with a plentiful supply of wooden sleepers or other suitable material. The steel members

will have been allocated a unique code related to section, level and piece number: this is usually stamped on the top surface to identify its orientation. In multi-storey high-rise buildings it is usual to create temporary shakeout platforms at various stages of the construction to minimize crane hook time.

Cold-formed roof members and steel decking packs also need to be placed in location as the work proceeds since crane access is often restricted later. Each bundle should be numbered and placed in convenient locations as subsequent relocation may prove difficult and costly.

33.3.4 Sub-assemblies on site

The designer may have been limited by the size and weight of components for transportation: thus it may prove advantageous to erect canopies, roof trusses, lattice girders and the like 'piece small' and then lift them as completed components.

If the decision to sub-assemble has been made there will remain the further question of deciding whether an area in the stockyard is to be dedicated for this purpose or whether the work will be done behind the cranes on the erection front in order that the assembly can be lifted straight off the ground and into position as the crane is moved back.

The most common components to be assembled behind the crane are roof trusses or lattice girders, which, because of their size, are almost always delivered to site 'piece small'. However, there is not always space for this to be done and other locations, such as the stockyard, have to be made available.

Where the potential sub-assembly area, and even the stockyard itself, is remote from the erection site, very careful investigations are needed before a decision to sub-assemble can be agreed. Local transportation size restrictions on the route to the erection cranes may rule this option out.

There are three factors which affect the practicability and economy of sub-assembling a unit on the ground:

(1) The weight of the eventual assembly, including any lifting beams required
(2) The degree to which the unit is capable of being temporarily stiffened without unduly increasing its weight
(3) The bulk of the unit, i.e. will it be possible to lift it to the height needed without fouling the crane jib?

Sub-assembly is only worthwhile if the unit can be lifted and bolted into place almost as easily as a single beam. In other words, the object of the exercise is to avoid carrying out operations at height which can easily be done at ground level.

The speed of construction is dependent on the number and type of crane lifts and connection. The rate of erection will also vary according to the weight, size and location of the piece being erected. Erection speed is often dictated by the distance

between the crane and hook, thereby emphasizing the need to consider lifting bundles closer to the erection face and providing shake-out platforms.

It is impossible to provide a chart showing piece counts per hour or day. However, as a broad rule of thumb, an experienced erection crew working in ideal conditions at low level with average weight, standard-sized components should be able to erect 40 to 60 pieces per day per hook.

33.3.5 Interface management

In order to minimize on-site disruptiòn, the specialist/trade interface must be managed successfully. Failure to understand the requirements of other trades through unclear specifications and uncertain allocation of responsibility may result in conflict and disagreement. Thus the development of open communication between all parties and the establishment of clear objectives during regular site meetings between overlapping trades are necessary measures to ensure proper component, plant, and work area management. Unfortunately, these measures constitute some of the most neglected areas in the construction process. Failure to adequately address these issues can result in cladding fixings being left off or located in the wrong place, holding-down bolts being incorrectly positioned, lift shaft tolerances proving incompatible with the main frame, and so on.

Thus co-operation between all parties is necessary to ensure enhanced efficiency. This may be achieved via:

(1) An improved communication process
(2) A clear contractual definition
(3) The use of single-source suppliers.

33.3.6 Surveying and aligning the structure

Surveying should be carried out in accordance with BS 5964[4] (1990) Parts 1 to 3 *Building setting out and measurement*, and adjusted for temperature outside the range of 5°C to 15°C. Accurate setting-out is crucial for maintaining control of tolerances and achieving a structure which is acceptable to all the subsequent trades.

The site should be provided with a dumpy or quick-set level and with a simple theodolite. On larger sites, it is common practice to use an electronic digital measurement (EDM) instrument. The advantage of a theodolite is that the transit of the instrument can later be used for checking the plumb of the columns. The telescope can be employed for the reading of offsets to check their alignment. Time spent on laying down and marking the cross-centre lines of the column bases on individual boards at each base and subsequently on to the concrete of the base is

time very well spent. The base can then be quickly and accurately set in both directions and to the correct level on the pre-set pads. The value of this work is that it enables errors in the concrete work to be identified at a time when corrective measures can be undertaken. The accurate setting of grouted packer plates, on which to land the column, serves the same purpose, i.e. it will facilitate the identification of concrete which is high and holding-down bolts which are low.

33.4 Site practices

33.4.1 Erection sequence

Columns are placed in grids to suit the grid/zone areas as previously agreed. The columns are lifted with a Dawson Ratchet (see Fig. 33.2), nylon slings or chains.

Care must be taken to minimize damage to painted surfaces as repairs can often be difficult. Once placed into position the column is roughly positioned and plumbed. In extreme cases the columns may need guying off to ensure stability during the erection sequence. This procedure is repeated until a grid is formed. Beams are then placed in location and secured by two bolts at each end connection throughout the designated area. Once all of the beams are located, the remaining bolts are placed. Since many structures employ composite slab construction, an intricate part of the site erection is the placing and fixing of the decking bundles. A grid system should have been agreed for the placing of bundles which the erection crew situates in convenient locations to ensure that connections are not obstructed, nor the structure eccentrically loaded. Heavy wires are then placed and left in position. When this is achieved the erection crew can move on to the next area. The engineer with the line and levelling crew then follows on behind, pulling the structure to its correct alignment. Once this task is completed, the bolting-up crew can tighten all of the bolts and complete. The decking is then placed, followed by the stud welding. A typical flow chart of the process is shown in Fig. 33.3.

33.4.2 Lining, levelling and plumbing

If, in spite of having taken care over the initial levelling of packers and carefully positioning the column bases to the pre-set lines on the foundations, the structure still needs adjustment during lining, levelling and plumbing, then something may be wrong. The first thing to check is that no error has been made in the erection. If nothing is out of place then the drawings and the fabrication need to be checked. If there is an error it will be necessary to take careful note of all the circumstances and advise all concerned immediately.

If the steelwork has just been thrown together with no great regard for accurate positioning it will be necessary to provide equipment and manpower to carry out

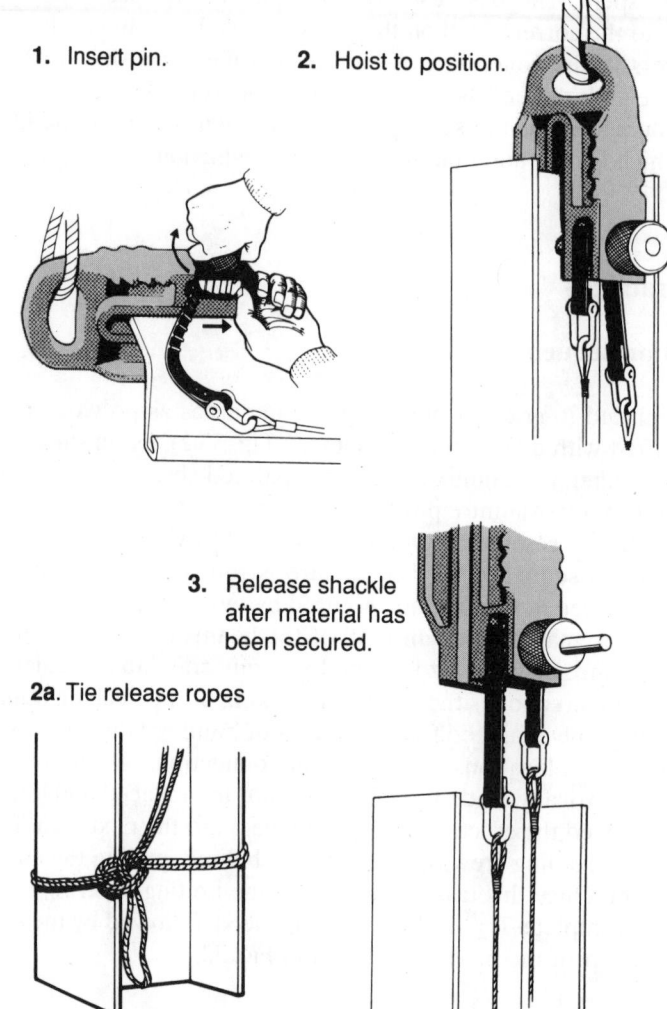

1. Insert pin.

2. Hoist to position.

3. Release shackle after material has been secured.

2a. Tie release ropes

Fig. 33.2   Dawson Ratchet (Courtesy of Dawson Construction Plant Ltd.)

the lining, levelling and plumbing. The equipment will include jacks, tirfor-type wire pullers with wires, straining screws, wedges, piano wires, and heavy plumb bobs with damping arrangements, all in addition to the level and simple theodolite which are needed for the basic check.

It is necessary to do the lining and levelling check before final bolting up is done. In practice this means that it must be done immediately following erection since it is inefficient to slacken off bolts already tightened in order move the steelwork about. There is no way that erection can go ahead until the braced bay is bolted up, and that cannot be done until it is level and plumb.

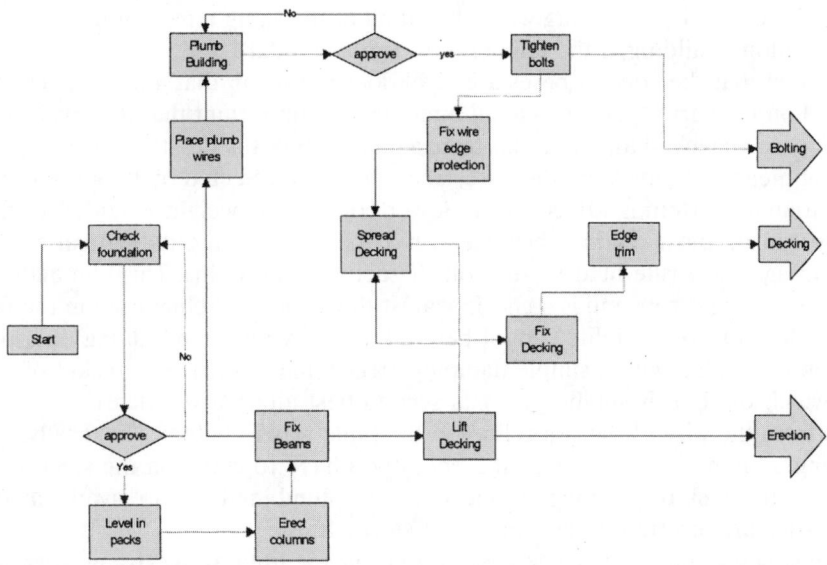

Fig. 33.3 Flow chart of the erection process

A supply of steel landing wedges and slip plates is needed to adjust the levels. They must be positioned in pairs opposite to each other on two sides of the base plate of the columns to be levelled. If the wedge is placed on one side only the column will be supported eccentrically and can in fact be brought down, especially if more than one column has been lifted on a series of wedges all driven in from the same direction. Alternatively toe jacks (jacks that can be used in tight areas) can be used in pairs to lift the columns. A temporary bench-mark should be established, and agreed with the client's representative, in the vicinity of the columns in a position where it cannot be disturbed. The level will then be used to check the final setting with the seating packers inserted, and the column landed back on them and bolted down with the holding-down bolts. The column can then be moved about in plan on these packers to bring it into line and bay length from its neighbours.

The position of a line, offset from the column centreline in order to clear the columns, should be marked and agreed. This line is then used either to string and strain a piano wire or to set and sight a theodolite telescope. The readings from it will then give the amounts by which movements must be made to keep the structure in tolerance.

Running dimensions from the previous column will give required longitudinal movement. However, care should be taken to watch the plumb of the columns in this direction in view of the tendency of steelwork to 'grow' due to cumulative tolerances. Regard should be paid to column-to-column dimensions rather than to running dimensions from the end of the building.

Having levelled the bases of the columns in their correct positions, it is possible to check the plumb. Attention to temperature effects is necessary as it is futile to

check the plumb of a building which is not all at the same temperature and, in the case of a long building, if the temperature is not standard.

The fact that the column bases are level does not mean that a level check is not needed on the various floors in a tall building. It is important that the levels of any one floor are checked for variations from a plane rather than on the basis of running vertical measurements from the base, since these are affected by temperature and the variable shortening effect on the lower columns as weight is added to them. Plumb can be most readily checked with a theodolite using its vertical axis and reading against a rule held to zero on the column centreline. This eliminates the effect of rolling errors and is a check against the same centreline used in the fabrication shops. If a theodolite cannot be used, a heavy plumb bob hung on a piano wire and provided with a simple damping arrangement, such as a bucket of water into which the bob is submerged, is a second-best alternative. Measurements are made from the wire to the centrelines in the same way as before. The disadvantage of using a plumb wire is that all the operatives have to climb on the steel to take and then to check the readings. Optical or laser plumbing units are available; these are particularly useful for checking multi-storey frames.

Adjustments, and in extreme cases provision for holding the framework in its correct position, are normally necessary only if the frame is not self-stable. For instance, if structural integrity depends on concrete diaphragm panels for stability, consideration should have been given at the design stage to one of two alternatives: either the concrete panels should be erected with the steel frame, or temporary bracing should have been provided as part of the original planning. To ensure proper function, any bracing must be positioned so that it can be left in place until after the concrete panel has been placed and fixed.

If diagonal wires have to be used they should be tied off to the frame at node points rather than in mid-beam, in order to avoid bending members. Also, their ends should be fixed using timber packers in the same way as is done when the pieces are slung for lifting. Turnbuckles or tirfor-type pullers provide the effort necessary to tension the wires. The sequence of placing, tensioning and ultimately removing these temporary arrangements should form part of the method statement prepared for each particular situation. Once the components of the building have been pulled into position the final bolting-up can be completed.

Portal frame construction in its temporary state often has the legs pre-set out of vertical, and becomes vertical only after the final loading has taken place. It is the responsibility of the designer to calculate the extent of pre-set. This often proves a difficult process to assess. Further, allowance must be made if the final position is needed for aesthetic purposes.

It is essential to ensure that the lining and levelling keep pace with the erection, since when additional floors are added it may become increasingly difficult to adjust, pull, or move the structure. As a rule of thumb, the lining and levelling of heavy sections should be within two to three bays of the erection face, whereas the lining and levelling of lighter sections should be within four to six bays of the erection face.

33.4.3 Tolerances

Prior to determining tolerances it is important to understand why they are needed. Rolling tolerances of the steel and manufacturing tolerances need to be accommodated. Tolerances must accommodate preceding work and subsequent building components. Tolerances must meet design requirements and architectural/ aesthetic requirements, and maximize buildability.

Tolerances can be categorized as structural (necessary for the integrity of the intended structure), architectural (necessary for the aesthetics of the structure), and buildability (necessary to construct all of the components in a building). For a detailed coverage of tolerances in steel construction, refer to Chapter 31.

It is important to inform the designer of any deviation from specified tolerances, as failure to comply can often adversely affect both the structure and subsequent operations.

33.4.4 Holding-down bolts

To facilitate the erection of the structure, holding-down bolts are normally positioned by the civils contractor. Movement tolerances must be accommodated in the foundation to ensure that correct alignment can take place (see Fig. 33.4).

Conical sleeves maximize movement without reducing the area of anchorage. It is important to have the bolts set both vertical and loose in the sleeve. The key benefit of using sleeved cast-in bolts is to allow movement for the alignment of the structure to meet tolerances. After checking and confirming that the holding-down bolts are acceptable, typical 100 mm × 100 mm steel packs (1–20 mm thick) or similar are placed on the base to the required level.

33.4.5 Site bolting

Wherever possible, site splices should be designed to be bolted. This process is less affected by adverse weather than welding, uses simple equipment, and presents much less complex work access and subsequent inspection problems.

Bolts should be chosen to minimize size variation, and ease of repetition and selection. *Design for Construction*[5] makes the following recommendations when selecting bolts:

(1) Preloaded bolts should only be used where relative movement of connected parts (slip) is unacceptable, or where there is a possibility of dynamic loading.
(2) The use of different-grade bolts of the same diameter on the same project should be avoided.

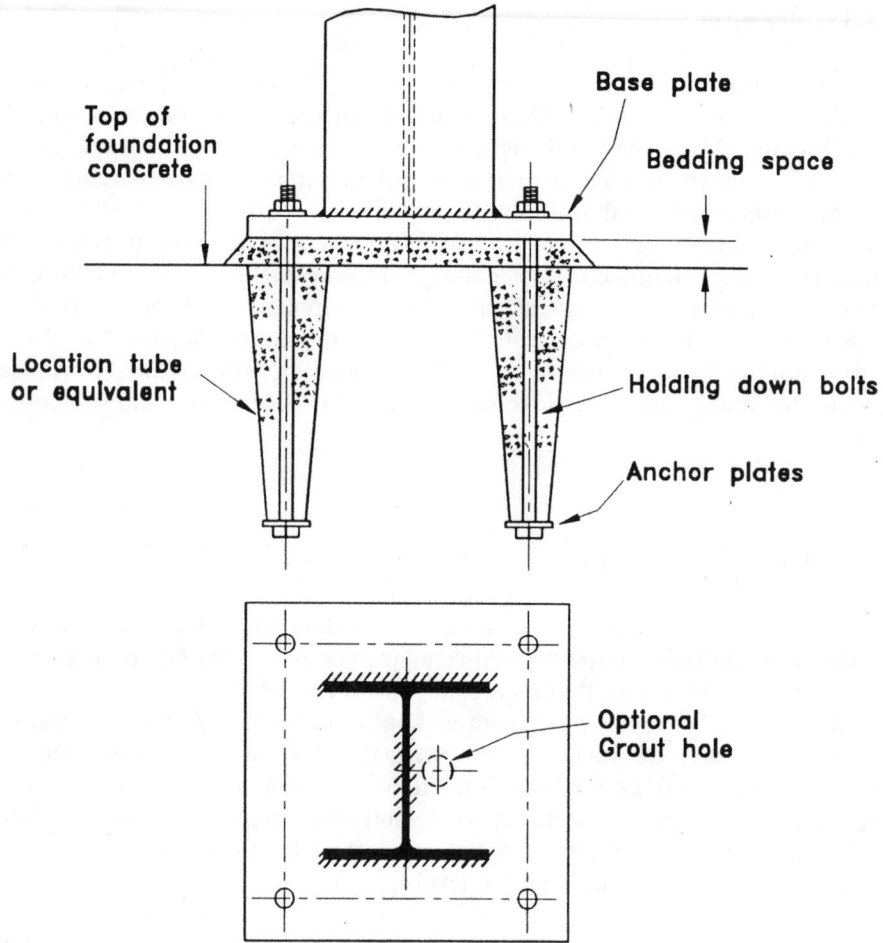

Fig. 33.4 Holding-down bolts

(3) Where appropriate, bolts, nuts and washers should be supplied with a corrosion protection coating that does not require further on-site protection.
(4) Bolt lengths should be rationalized.
(5) Bolts should be threaded full length where possible.
(6) Connections should be standardized where possible.
(7) 'Just-in-time' delivery should be used, based on simple, easy-to-follow bolt lists.
(8) Approximately 90% of simple connections could be made using M20 60 mm long bolts. With a choice of three lengths, 95% of connections are possible.

Impact wrenches and powered nut runners speed the tightening of bolts. When high strength friction-grip bolts (HSFG) are necessary the Construction Industry Research and Information Association technical note for friction grip bolted

connections[6] should be used. If small quantities of HSFG bolts are to be used, manual torque wrenches are available. Where large quantities of friction-grip bolts have to be tightened, an impact wrench is an essential tool. Where access is difficult, wrenches are made which run the nut up and are then used as manual ratchet wrenches to finally tighten the bolt.

Friction-grip bolts must be tightened first to bring the plates together and thereafter be brought up to the required preload either by applying a further part turn of the nut or by using a wrench calibrated to indicate that the required torque has been reached. The manual wrench has a break action which gives the indication by means of a spring without endangering the erector by a sudden complete collapse. Alternatively, load-indicating washers may be used to demonstrate that the bolt has been tightened to the required tension.

33.5 Site fabrication and modifications

Special consideration must be given to the planning of the erection of a structural steel frame where site welding has been specified. Site welding has to be carried out in appropriate weather conditions, and in positions which make it more difficult and expensive than welding carried out in workshop conditions. In a workshop the work can be positioned to provide optimum conditions for the deposition of the weld metal, whereas most welds on-site are positional. Most site welding is carried out by manual metal arc and electrodes, which is more flexible in use and is more able to lay good weld fillets in the vertical up-and-overhead positions.

Some means must be provided for temporarily aligning adjacent components which are to be welded together, and of holding them in position until they are welded (see Fig. 33.5).

The methods adopted to cope with the need for alignment may have to carry the weight of the components, and in some cases a substantial load from the growing structure.

Safe means of access and safe working conditions must be provided for the welder and his equipment. The working platform may also have to incorporate weather protection since wind, rain and cold can all adversely affect the quality of the weld.

The design of joint component weld preparation must take into account the position of these components in the structure. The erection method and weld procedure statements for each joint must take all these factors into account (see Chapter 24).

Provision must be made for the necessary run-off coupon plates for butt welds and for their subsequent removal. Run-off plates are required in order to ensure that the full quality of the weld metal being deposited is maintained along the full design length of the weld (see Fig. 33.6).

Attention must be paid to the initial setting and positioning of the components in order that the inevitable weld shrinkage resulting from the cooling of joints does not result in loss of the required dimensional tolerances across the joints.

Fig. 33.5   Temporary web splice plates supporting and aligning joint for site welding

Health and safety issues are important when using welding:

- Cutting and welding create sparks and hot metal
- Welding sets are very noisy
- Harmful substances are generated by the process
- Welding is often carried out in confined space, high up
- Access platforms are difficult to get to
- Vision can be obscured by eye protection
- Personal protection from heat is cumbersome and difficult to move about in
- Welding sets need earthing.

Where possible, prefabrication should be carried out at ground level; failing this, bolted connections should be used.

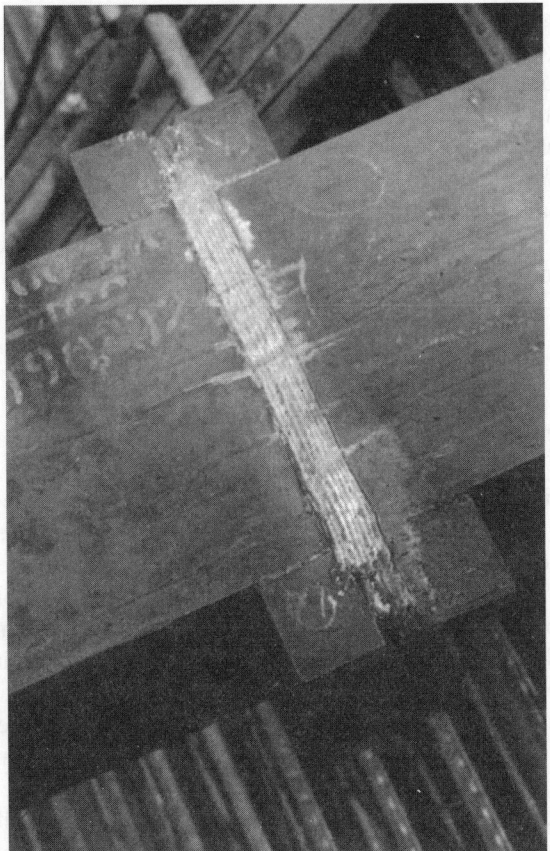

Fig. 33.6   Run-off plates for a butt weld

33.6 Steel decking and shear connectors

33.6.1 Introduction

Many steel-framed commercial buildings in the UK use composite slab construction, and so steel decking[7] and shear connectors now play an integral part in the steel erection programme. A key advantage of light-weight steel decking is that it can act as a working platform at each level. This eliminates the need for temporary platforms, and reduces the height an erector can fall – a critical safety issue in multi-storey construction.

Once the building or section has been lined, levelled and bolted up, then steel decking bundles previously positioned on the steelwork can be spread. It is important to remember that the sheets are very light and should be protected from the wind. Since decking operatives are not normally steel erectors, they need safety wires to hook on to as they work. As the decking is spread, exposed edges must be protected since these are the vulnerable points where operatives can fall. The

decking is then shot-fired to the steel using Hilti nails or similar. Once this is complete the edge trim is placed using shot-fired nails, and the joints are taped to minimize grout loss and then handed over to the welder for the shear connection installation. Shear connectors are usually placed and welded using automatic machines. Thus provision should be made for a high current supply or adequate space for a portable generator. Most shear studs are 100 mm × 19 mm diameter and are delivered in barrels weighing about 100 kg. It is therefore important to allow for the lifting of studs and equipment to the various levels. The welder will then work methodically in lines, placing and testing the studs until each area is ready for inspection and handing over.

33.6.2 Cold-formed sections

Secondary items in many structures are cold-formed sections, used as roofing rails or cladding rails. Since roof and wall material is very light and easy to supply in large volumes, cold-formed sections are often delivered by 20-tonne lorry. Care must be taken in the factory to ensure clear marking and banding of components to prevent wastage of labour hours, days and even weeks of subsequent on-site sorting. Once the material has been off-loaded in bundles it will be placed in the rough location for installation. The material will need a large shakeout area to facilitate speed of erection. In situations of multi storey high rise construction, the material can be hoisted to the floor below the erection face, then shaken out and subsequently pulled up into position. This can often eliminate crucial tower crane hook time. Most manufactures offer a total package of rails, sag rods and fastenings. Selection of quick-fit materials will expedite site erection, thereby reducing construction time and man hours.

33.7 Quality control

33.7.1 Introduction

The extent and detailed content of quality systems are recorded in British Standards publications BS 5750[8] and BS EN ISO 9001.[9] Any planned system should be recorded and documented and should form the basis of a quality manual which describes the way the organization conducts its business.

The selection of a quality-assured steelwork contractor with BS 5750 (EN ISO 9001) approval will help in monitoring quality. A quality control system should be in place identifying all materials and workmanship from purchase to delivery and subsequent incorporation into the works. It is essential that monitoring is carried out to ensure that the performance of materials and workmanship incorporated into the final works are not inferior to those specified by the designer. All

materials delivered to site should be stored and stacked carefully in designated areas to ensure that damage to those materials does not occur.

Within the method statement, an approved testing and inspection plan will have been agreed to ensure that the as-built frame conforms to specification and design intention. BS DD ENV 1090–1[10] *Execution of steel structures, General rules for buildings* defines the plan fully. However, as a guide, the following should be monitored:

(1) The building position is in accordance with the tolerances laid down.
(2) The methods of construction have allowed for process quality control, and the tracing of components from supply source to completion.
(3) The construction has not deviated from the design requirements or specification, without all changes being approved by the appropriate body, and a full record kept of those changes.

In the case of the alignment of the structure, the dimensional survey should include the location, zone or area tested. This should include the tolerance the survey has been carried out to. Survey results should be transposed onto a drawing, noting the location surveyed, and any deviations from line and level. Any out-of-tolerance areas should be highlighted and a non-conformance report (NCR) instigated to enable the design engineer to approve or reject the change.

Bolting or welding should be carried out to plan to enable ease of monitoring and signing off. In multi-bolt connections, as the bolt is tightened, it should be marked to ensure all bolts are adequately secured. It is often a good idea to require the operative who has tightened complex connections to initial each connection, and for management to record those connections. A drawing showing progress and areas completed should be kept and handed over to the customer as and when appropriate.

33.7.2 Non-conformance procedures

During the construction of a building, modifications to the existing structure or repairs to the new members may be necessary. All variations after the work has started must be accompanied by an NCR, traceable throughout the entire process and signed off by the design engineer.

• Any damage or discrepancy should be reported using the NCR system and repaired or replaced following this approved system.

On completion of the project a register with a copy of all NCRs should be given to the customer. This should include site location drawings, sections, sketches, test certificates, etc., to help complete the building's log and ensure that all parties are aware of the changes made. A typical NCR form is shown in Fig. 33.7.

Non Conformance Report		Date..................
Project	DWG. No	NCR No

The following item(s) do not conform to specification

Option
Inspected by...............................   Section ....

Suggested Remedial Action
Site Eng. ....

Clients comments   Accept/Reject...............................
Authorization   .... Position ....

Signed off Design Eng............................	Date closed ....

Fig. 33.7 A typical NCR form

33.8 Cranes and craneage

33.8.1 Introduction

The choice of cranes and positioning may be dependent on many factors. Care needs to be taken to ensure good progress of the work. Often, cranes that work at maximum capacity in restricted areas actually slow production and are not efficient. A well-located crane with spare capacity can speed erection and more than compensate for the additional cost of hire.

Principal items to be considered when selecting a crane are:[5]

- site location – access and adjacent features
- duration of construction

- the weight of the pieces to be erected and their position relative to potential crane standing positions
- size of piece to be erected
- the need for tandem lifts
- maximum height of lifts
- number of pieces to be erected per week
- ground conditions
- the need to travel with loads
- the need for craneage to be spread over a number of locations
- organization of off-loading and stockyard areas
- dismantling.

On a large job, cranes of varying size and capacity perform various tasks, i.e. a heavy one for the main columns and a smaller one for the side framing posts and angles. The number of pieces to be lifted in any one working period determines the number of cranes required to achieve the desired work-rate. In practice, the choice of crane type and capacity is a compromise, with efforts being made to get as near to the optimum as possible.

In order to decide on a crane layout it is necessary to prepare a sketch or drawing of each critical lifting position showing, on a large scale, the position of the crane on the ground; the location of the hook on plan; the clearance between the jib and its load, and between the jib and the existing structure as the piece is being landed in its final position. These drawings will enable a check to be made of the match of each crane to its load; of the ability of the ground beneath the crane to sustain the crane and its load; and of the clearance to lift the component and place it in position. It is important to allow adequate clearance for the tail of the crane as it slews. A series of these drawings will rapidly confirm whether a particular crane is suitable. If there is an anomalous heavy lift in the series, the designer can be asked whether a splice position can be moved to reduce weight.

Reports on overturned cranes and crippled jibs bear ample proof of the need to plan and to the extent to which it is neglected. The Code of Practice CP 3010[11] on the *Safe Use of Cranes* clearly describes the dangers of improper practice.

33.8.2 Types of crane

Mobile cranes

The mobile cranes include truck or wheel-mounted, and tracked cranes. Truck-mounted cranes are able to travel on the public roads under their own power and with their jibs shortened. Mobile cranes mounted on crawler tracks (see Fig. 33.8) are not permitted to travel on the road under their own power.

A truck-mounted crane may prove a correct choice if a crane is only required on site for a short time since it can come and go relatively easily. However, if a crane is to be on site for a longer period, the high transportation costs associated with

Fig. 33.8 Crawler crane (courtesy of Baldwin Industrial Services/Chapman Brown Photography)

crawler crane use can be justified. The real advantage of a crawler crane is that both its weight and the reaction from the load-lift are spread more widely by its tracks. However, it does not have the added stability of a truck-mounted crane with its outriggers set (Fig. 33.9).

Most types of mobile crane form part of the erection fleet owned by steel erection and/or plant hire companies. The decision to hire from a plant hire company is often determined by the geography of the site, i.e. is the site close to a plant hire company's yard, or is it close to the contractor's own plant depot? Is the crane to be used for short or long duration? Is the lift for which the crane is required one of a series or simply a one-off? Mobile crane capacity normally ranges between 15 and 800 tonnes. Figures 33.9 and 33.10 provide performance data for a typical 25 and 800 tonne crane. It is important to note *that the rated capacity of a mobile crane and the load which can be safely lifted are two very different things*. The maker's safe loading diagram for the crane must always be carefully consulted to see what load can be lifted with the jib length and the radius proposed. It is essential that a mobile crane is used on level ground so that no side loads are imposed on the jib structure.

25 Tonne DEMAG AC 75 "CITY CLASS"
Mobile Crane

Lifting Capacities Main Boom Extension

Telescopic Boom 20,7 - 25,0 m. On Fully Extended Outriggers 5,9 m. Working Range 360°

Radius m	Main Boom 20,7 m				Main Boom 25,0 m				Radius m
	7,1 m		13,0 m		7,1 m		13,0 m		
	0°	30°	0°	30°	0°	30°	0°	30°	
5	5,6	-	-	-	-	-	-	-	5
6	5,6	-	-	-	-	-	-	-	6
7	5,3	-	2,2	-	4,5	-	-	-	7
8	5,0	-	2,2	-	4,4	-	2,0	-	8
9	4,7	3,7	2,2	-	4,2	-	2,0	-	9
10	4,4	3,6	2,1	-	4,0	3,4	2,0	-	10
11	4,2	3,5	2,1	-	3,9	3,3	2,0	-	11
12	4,0	3,4	2,0	-	3,7	3,2	1,9	-	12
13	3,5	3,3	1,9	1,7	3,5	3,1	1,9	-	13
14	3,1	3,2	1,9	1,6	3,0	3,0	1,8	1,5	14
15	2,7	3,0	1,8	1,6	2,7	2,9	1,8	1,5	15
16	2,4	2,6	1,7	1,5	2,4	2,6	1,7	1,5	16
17	2,2	2,3	1,6	1,5	2,1	2,3	1,7	1,4	17
18	1,9	2,1	1,6	1,4	1,8	2,0	1,6	1,4	18
19	1,7	1,8	1,5	1,4	1,6	1,8	1,6	1,4	19
20	1,5	1,5	1,4	1,4	1,4	1,6	1,5	1,3	20
21	1,3	1,3	1,4	1,3	1,3	1,4	1,3	1,3	21
22	1,1	-	1,3	1,3	1,1	1,2	1,2	1,3	22
23	0,9	-	1,1	1,3	0,9	1,0	1,0	1,3	23
24	0,8	-	1,0	1,2	0,8	0,9	0,9	1,2	24
25	0,7	-	0,9	1,0	0,7	0,7	0,8	1,0	25
26	-	-	0,8	0,9	0,6	-	0,7	0,9	26
27	-	-	0,7	0,7	0,5	-	0,6	0,8	27
28	-	-	0,6	-	-	-	0,5	0,7	28
29	-	-	0,5	-	-	-	-	0,6	29

25 Tonne DEMAG AC 75 "CITY CLASS"
Mobile Crane

Dimensions

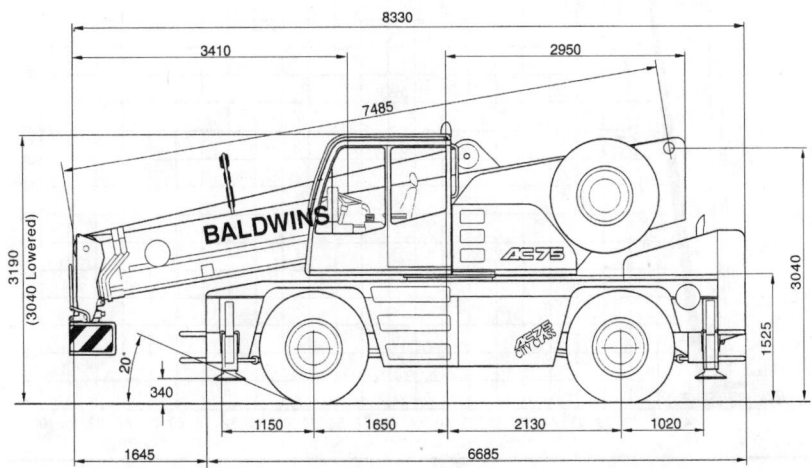

Fig. 33.9   Typical mobile crane (courtesy of Baldwin Industrial Services)

800 Tonne LIEBHERR LTM 1800
Mobile Crane

Working Ranges Luffing Lattice Jib

Main Boom 83°. Luffing Lattice Jib 21,0 - 91,0 m. On Outriggers 13 m x 13 m. Working Range 360° Counterweight 153 t.

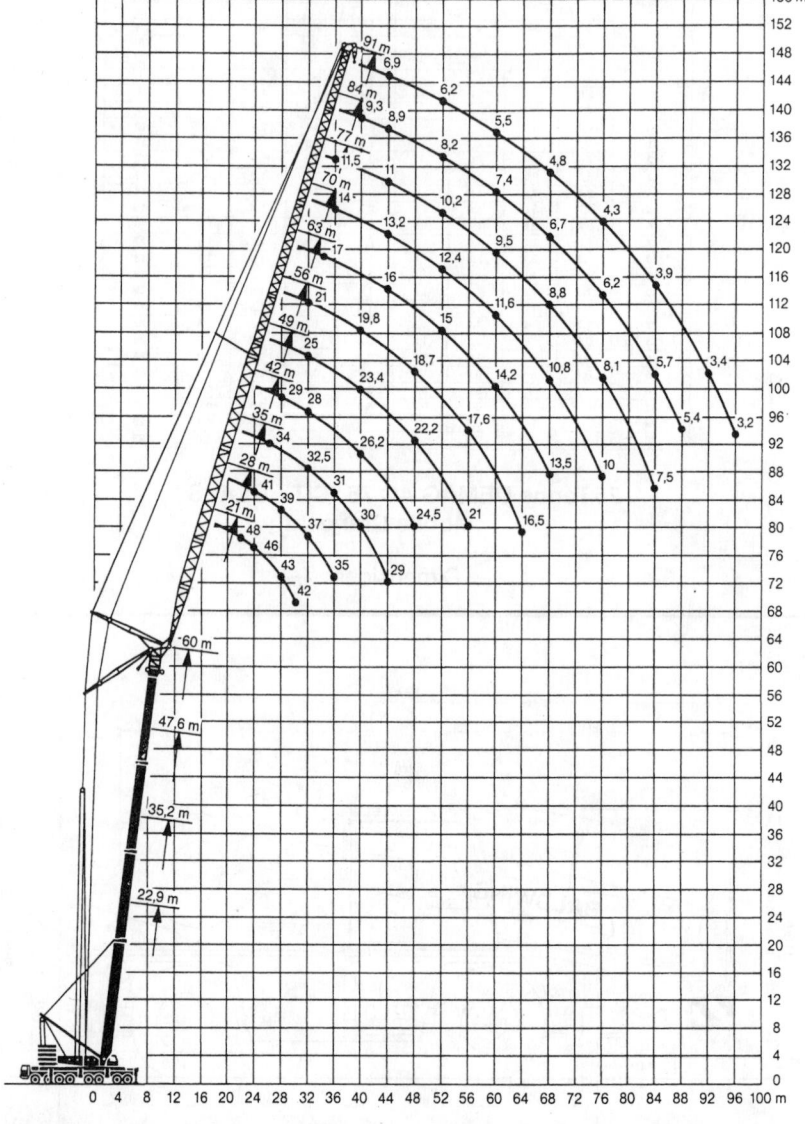

While every care has been taken in the preparation of the ratings given in this leaflet, and all reasonable steps have been taken to check the accuracy of the information, Baldwins cannot accept responsibility for any inaccuracy in respect of any matter arising out of, or in connection with the use of these tables.

Fig. 33.10 Special 800 tonne mobile crane (courtesy of Baldwin Industrial Services)

Mobility for non-mobile cranes

Non-mobile cranes can be made mobile by mounting them on rails. This has two advantages: the positioning of the crane can be more easily dictated and controlled, and the loads transmitted by the rails to the ground act in a precisely known location. Many cranes have collapsed because of insufficient support underneath. However, most rail-mounted crane failures have occurred from overloading. Where the crane is to work over complex plant foundations the rail can be carried on a beam supported, if necessary, on piles especially driven for the purpose. If the rail is supported only by a beam on sleepers in direct contact with the ground, the load can be properly distributed by suitable spreaders. In either case conditions must be properly considered and designed for. Problems often occur when too much faith is invested in the capability of the ground to support a mobile crane and its outriggers.

Non-mobile cranes

Non-mobile cranes are generally larger than their mobile counterparts. They can reach a greater height, and are able to lift their rated loads at a greater radius.

There are two main types of non-mobile crane: the tower crane and the (now rare) derrick. Due to their great size, the cranes must arrive on site in pieces. Thus the disadvantage of a non-mobile crane is that it has to be assembled on site. Having been assembled, the crane must receive structural, winch and stability tests before being put into service.

A tower crane with sufficient height and lifting capacity (see Fig. 33.11) has several advantages:

(1) It requires only two rails for it to be 'mobile'. These two rails, although at a wide gauge, take up less ground space than a derrick.
(2) It carries most of its ballast at the top of the tower on the sluing jib/counter balance structure, and so very much less ballast is needed at the bottom. Indeed, in some cases, there is no need for any ballast at the tower base or portal.
(3) Because the jib of a tower crane is often horizontal, with the luffing of a derrick jib replaced by a travelling crab, the crane can work much closer to the structure and can reach over to positions inaccessible to a luffing jib crane.
(4) A tower crane is 'self-erecting' in the sense that, after initial assembly at or near ground level, the telescoping tower eliminates the need for secondary cranes.
(5) As shown in Fig. 33.12 a tower crane can be tied into the structure it is erecting, thus permitting its use at heights beyond its free-standing capacity.

There exist several types of tower crane, e.g. articulated jib, luffing and saddle cranes as illustrated in Fig. 33.13. It is essential that manufacturers or plant hirers are consulted in order to make the most appropriate choice of crane.

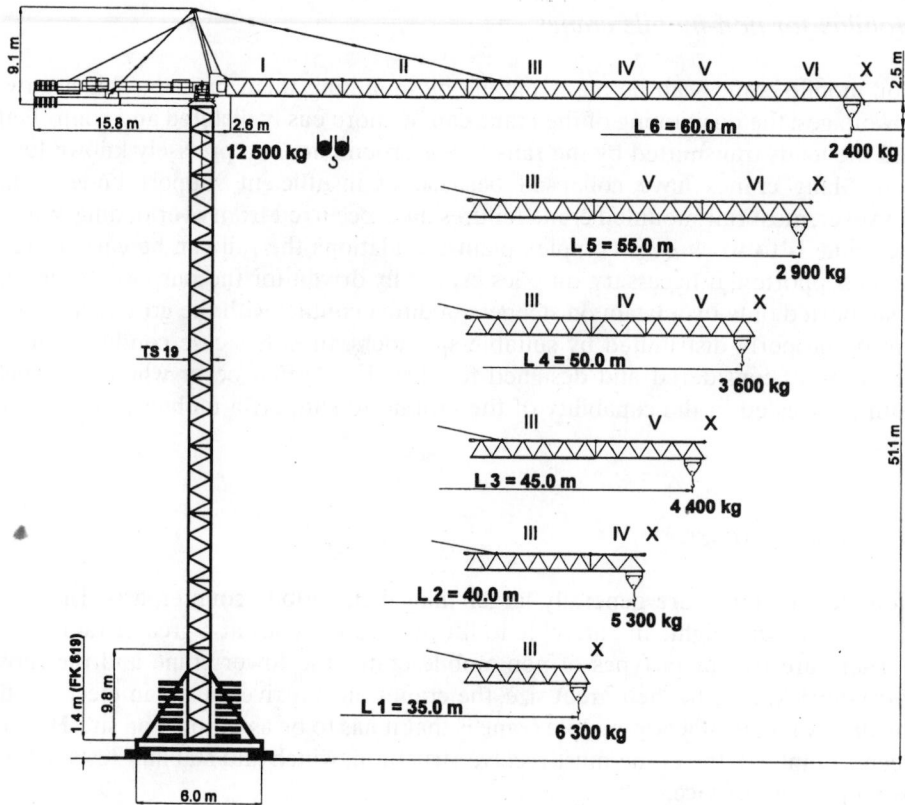

Fig. 33.11   Tower crane (courtesy of Delta Tower Cranes)

Cranes for the stockyard

Stockyard cranes have to work hard. The tonnage per job has to be handled twice in the same period of time, often with many fewer cranes. It is therefore important that cranes be selected and cited carefully to ensure maximum efficiency.

33.8.3 Other solutions

If there is no suitable crane, or if there is no working place around or inside the building where a crane may be placed, then consideration must be given to a special mounting device for a standard crane, or even a special lifting device to do the work of a crane, designed to be supported on the growing structure under construction. In either event, close collaboration between the designer and erector members of

Fig. 33.12   Citigroup Tower, London showing tower cranes tied into the building (courtesy of Victor-Buyck Hollandia)

the management team is of paramount importance. Conversely, inadequate communication may prove problematic. Once the decision to consider the use of a special lifting device has been made, a new range of options becomes available. The most important of these is the possibility of sub-assembling larger and heavier components, thereby reducing the number of labour hours worked at height. This is particularly true in the case of bridgework, where substantial sums would otherwise have to be expended on other temporary supports and stiffening. A major disadvantage of special lifting devices is that the apparatus being considered is often so specialized that is unlikely to be of use on another job. Thus

Fig. 33.13   Tower cranes used in the construction of Citigroup Tower, London (courtesy of Victor-Buyck Hollandia)

the whole cost is targeted at the one job for which it has been initially designed (see Fig. 33.14).

Where the frame is single-storey, and at the cost of only a slight increase in time and labour, it is possible to do without an on-site crane. With the help of a winch (powered by either compressed air or an internal combustion engine, and some blocks and tackle), a light lattice-guyed pole can be used to give very economic erection (see Fig. 33.15). In this instance the pole is carried in a cradle of wires attached at points on the tower. These connection points need to be carefully designed to ensure that they will carry the load without crippling the tower structure.

It is vital that all poles are used in as near a vertical position as possible, since capacity drops off severely as the droop increases. This requires careful planning and the employment of a gang of men experienced in the use of the method.

In a different context, pairs of heavier poles provided with a cat head to support the top block of the tackle can be used inside existing buildings to erect the components of, for example, an overhead travelling crane, or to lift in a replacement girder. The arrangements for a pole and its appurtenances take up much less floor space in a working bay than a mobile crane. This is because a mobile crane needs a wide access route and adequate space to manoeuvre itself into position – particularly useful where headroom is restricted.

Fig. 33.14 A purpose-made lifting beam for cantilever erection

33.8.4 Crane layout

It is important to decide on the type, size and number of cranes that are required to carry out the work, since each has a designated range of positions relative to the work it is to perform. These positions are then co-ordinated into an overall plan which enables each crane to work without interfering with its neighbours, and at the same time enables each to work in a position where adequate support can safely be provided (see Fig. 33.16). This plan will then form the basis of the erection method statement documentation.

A major factor in planning craneage is to ensure that access is both available and adequate to enable the necessary quantity and size of components to be moved. On large greenfield developments these movements may often have to take place along common access roads used by all contractors and along routes which may be subject to weight or size restrictions. On a tight urban site the access may be no more than a narrow one-way street subject to major traffic congestion.

33.8.5 The safe use of cranes

Mention has already been made of the UK Statutory Regulations. These lay down not only requirements for safe access and safe working but also a series of test requirements for cranes and other lifting appliances.

Fig. 33.15   An erection pole used to build a transmission tower

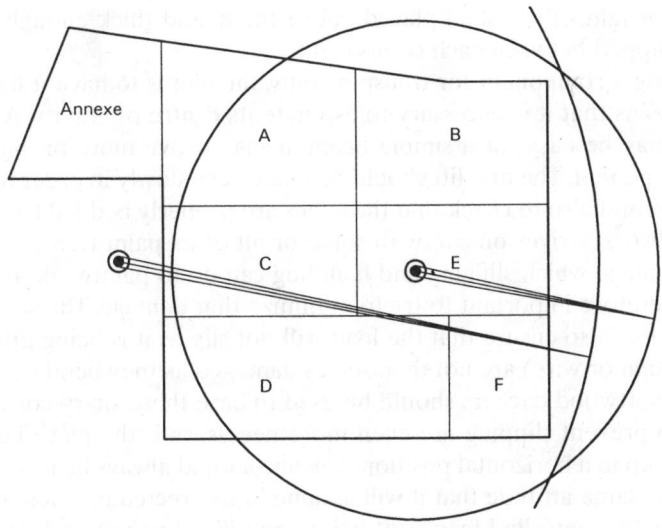

Fig. 33.16   Typical crane layout

It is the responsibility of management to ensure that plant put on to a site has a sufficient capacity to do the job for which it is intended, and that it remains in good condition during the course of the project. Shackles and slings must have test certificates showing when they were last tested. Cranes must be tested to an overload after they have been assembled. The crane test is to ensure that the winch capacity, as well as the resistance to overturning and the integrity of the structure, is adequate.

British Standards lay down the various requirements for safe working. Lists of those standards, and the necessary forms to enable each of the tests to be recorded, must be provided by management, often in the form of a 'site pack' which the site agent must then display and bring into use as each test is carried out. It is the site agent's responsibility to ensure that these requirements are fulfilled. The site agent may also be required to produce them from time to time for inspection by the factory inspector during one of his periodic visits to the site.

A crane which has been tested and used safely in many locations might overturn at its next location. Failure is often caused by inadequate foundation provision under the tracks or outriggers of the crane. In other words adequate support under the tracks or outriggers is an essential requirement. It is equally important that the crane should work on level ground, since an overload can easily be imposed, either directly or as a sideways twist to the jib, if the ground is not level.

33.8.6 Slinging and lifting

Components, whether they are on transport or are lying in the stockyard, should always be landed on timber packers. The packers should be strong enough to

support the weight of the steel placed above them, and thick enough to enable a sling to be slipped between each component.

When lifting a component for transport only, the aim is to have it hang horizontally. This means that it is necessary to estimate its centre of gravity. Although this calculation may be easy for a simple beam, it may prove more problematic for a complex component. The first lift should be made very slowly in order to check how it will behave, and also to check that the slings are properly bedded (see Fig. 33.17).

Most steelwork arrives on site with some or all of its paint treatment. Since the inevitable damage which slinging and handling can do to paintwork must be made good, it is therefore important to try to minimize that damage. The same measures that achieve this also ensure that the load will not slip as it is being lifted, and that the slings (chain or wire) are not themselves damaged as they bend sharply around the corners. Softwood packers should be used to ease these sharp corners.

Packers to prevent slipping are even more necessary if the piece being erected does not end up in a horizontal position. The aim should always be to sling the piece to hang at the same attitude that it will assume in its erected position. Pieces being lifted are usually controlled by a light hand line affixed to one end. This hand line is there to control the swing of the piece in the wind, and not to pull it into level. Wherever possible non-metallic slings should be used. They will reduce damage to paintwork and are less likely to slip than chain or wire slings.

In extreme cases two pieces may have to be erected simultaneously using two cranes. Staff, working back at the office, should account for this in the site erection method statement. It is too late to discover this omission when the erection is attempted with only one crane, or with no contingency plan to pull back the head of the column.

As discussed above, it is important to consider both the stiffness of large assemblies such as roof trusses as they are lifted from a horizontal position on the ground,

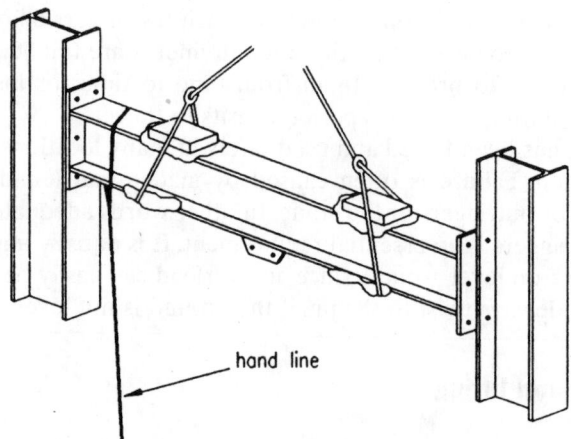

Fig. 33.17   Typical slinging of a piece of steelwork

and the need to build assemblies in a jig to represent the various points at which connection has to be made in the main framework. An additional jig for lifting can be particularly useful if there are many similar lifts to be made. This can be made to combine the need to stiffen with the need to connect to stiff points in the sub-frame, and the need to have the sub-frame hang in the correct attitude on the crane hook. The weight of any such stiffening and of any jig must of course be taken into account in the choice of crane.

Some temporary stiffening may be left in position after the initial erection until the permanent connections are made. This eventuality should also have been fore-seen, and sufficient stiffeners and lifting devices should be provided to avoid an unnecessary bottleneck caused by a shortage of a device for erection of the next sub-frame.

Where a particularly awkward or heavy lift has to be made, slinging and lifting can be made both quicker and safer if cleats for the slings have been incorporated in the fabrication. Each trial lift made after the first one wastes time until the piece hangs true. The drawing office should determine exactly where the centre of gravity is.

A chart giving details of standard hand signals is illustrated in Fig. 33.18. Their use is essential when a banksman is employed to control the rear end of the trans-port, thereby bringing the component to the hook as it is reversed. The banksman is needed to relay the signal from the man directing the movements of the crane if he is out of sight of the crane driver. A clear system of signals should be agreed for the handover of crane control from the man on the ground to the man up on the steel who controls the actual landing of the component. A banksman may also be needed up on the steel if the crane driver cannot clearly see the top man who is giving the control instructions: *it is vital that there is no confusion over who is giving instructions to the crane driver.*

33.9 Safety

33.9.1 The safety of the workforce

The health and safety regulations require a project safety plan to be drawn up, which should include a detailed assessment of anticipated risks.

There are a number of standards, regulations and guidance notes for the safety of the workforce during construction, as referenced in the further reading.

Site safety of the workforce is subject to statutory regulation and inspection by the Health and Safety Executive. Regulations lay down minimum acceptable standards for the width of working platforms; the height of guard-rails; the fixing of ladders; and so on. They refer to the use of safety belts and safety nets. They lay down the frequency with which a shackle or chain sling must be tested and the records that must be kept to show that this was done. Reference should be made to the appropriate regulation for the details of these requirements.

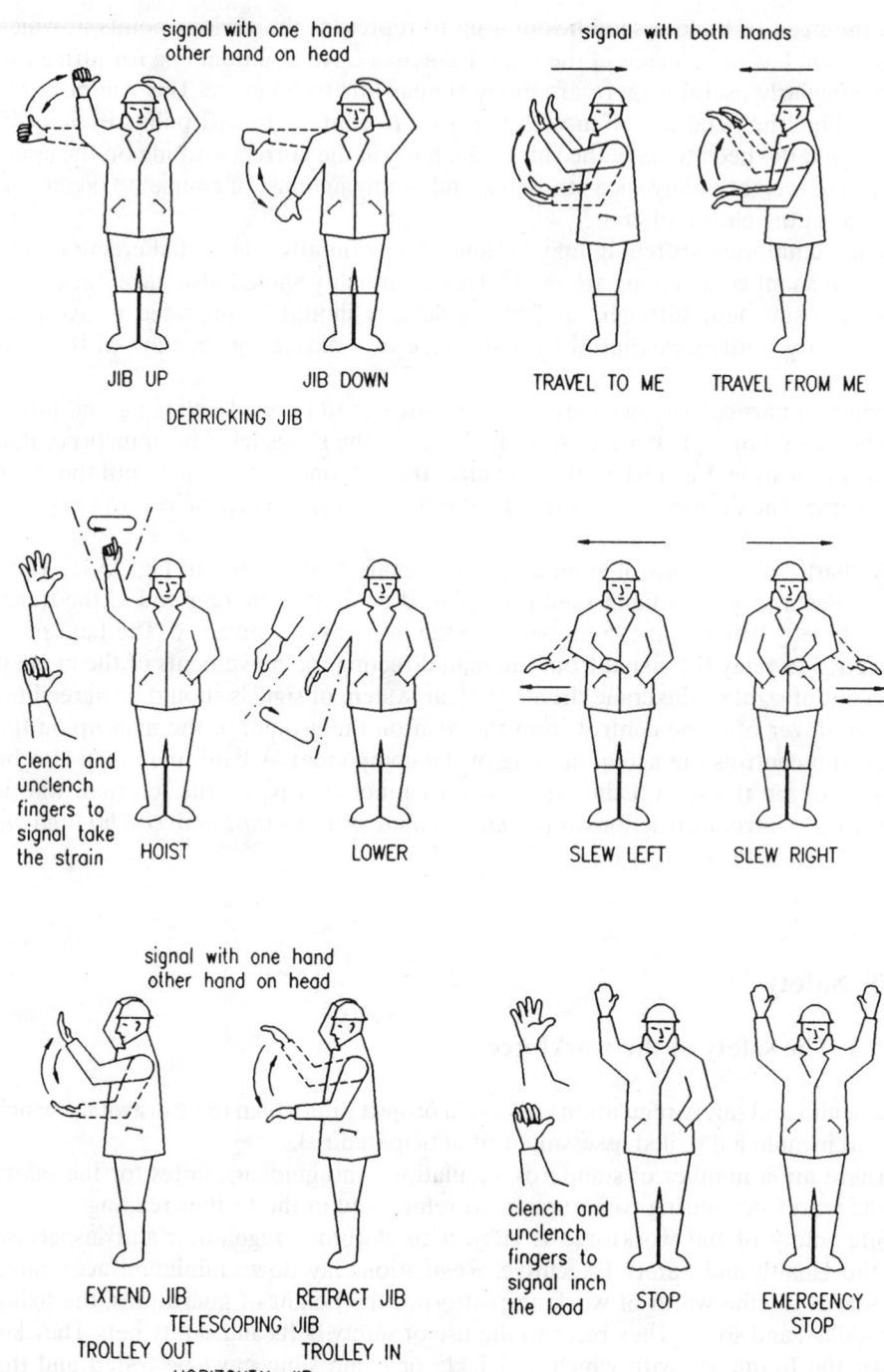

Fig. 33.18   Standard hand signals for lifting

33.9.2 Risk assessment

Identification of foreseeable risk should be carried out[5], as shown in Fig. 33.19, required before the start of work on site. These should be categorized into likelihood of occurrence (probability) and severity of occurrence (impact).

By carrying out a risk assessment, the risks can be identified and, where appropriate, avoided and reduced.

However, by its very nature, the erection of a structural frame is a process involving a certain amount of risk. The work is carried out at height, and until it has progressed to a certain point there is nothing to which a safe working platform can be attached. The process of erecting a safe platform can be as hazardous as the erection process itself. One solution is to provide mobile access equipment if ground conditions permit.

Different access platforms are appropriate in different circumstances. One advantage of modern composite floor construction is that the decking can quickly provide a safe working platform, requiring only the addition of a handrail. Figure 33.20 shows a safe platform for the erection of bare steelwork – a prefabricated platform slung over a convenient beam. In this case weather protection may be added for site welding.

It is the responsibility of the designers and planners to ensure that no platforms are erected in order to carry out work that ought to have been done either in the fabrication shops or on the ground before the component concerned was lifted into place.

A key planning-stage consideration is to see if the need for a working platform can be eliminated altogether, i.e. can the operation be carried out at ground level before the component is erected? If not, can the platform be designed so that it is assembled on the component while it is still on the ground? It is impractical to have to consider the provision of a safe working platform in order to be able to safely erect the main safe working platform.

The object of safety procedures is to ensure that everything possible is done to eliminate the risk of an on-site accidents. Methods of achieving safety include:

(1) An enhanced communication process
   Communication of the details of safety procedures to all concerned, the display of abstracts of the regulations themselves, the issuing of safety procedure documents, and the running of training courses all contribute to safe working practice. Individuals must be aware of the location of particularly hazardous

Risk	Likelihood of occurrence	Likely severity
High	Certain or near certain to occur	Fatality, major injury, long-term disability
Medium	Reasonably likely to occur	Injury or illness causing short-term disability
Low	Rarely or never occur	Other injury or illness

Fig. 33.19 Risk chart[5]

areas and the available protection, the types of protective clothing and equipment that are available and how to obtain them, the restrictions in force on the site regarding the use of scaffolding or certain items of plant, and any access restrictions to certain areas. They should be encouraged to tell someone in authority if they see a potential hazard developing before it causes an accident.

(2) Adequate equipment provision

It is important to make the necessary equipment available on the site and maintain it in good order. Equipment includes safety helmets, ladders and working platforms, safety belts and properly selected tools.

(3) Avoidance of working at height

Tasks should be organized to minimize work at height by: (a) the use of sub-assembly techniques; (b) the fixing of ladders and working platforms to the steelwork on the ground before it is lifted into place; (c) the early provision of horizontal access walkways; (d) the provision of temporary staircases or hoists where appropriate.

The above measures enable some of the hazards of working at height to be reduced by conferring on that work some of the advantages of ground-level working.

(4) Appropriate fixing of portable equipment

It is important to ensure that portable equipment such as gas bottles and welding plant is firmly anchored while it is being used. The horizontal pull on a gas pipe or a welding cable being used at height is considerable, and can dislodge plant from a working platform, thereby endangering the operator. Care should also be taken to ensure that there are no flammable materials below the working area, on which sparks could fall.

(5) Good design

A well thought-out design can make an important contribution to on-site safety. The positioning of a splice so that it is just above, rather than just below, a floor level will reduce the risks associated with the completion of an on-site splice. The arrangement of the splice so that the entry of the next component can be simply and readily completed will reduce the need to complete the splice up in the air.

Lifting cleats and connections for heavy and complex components should be designed and incorporated in the shop fabrications, as should fixing cleats, brackets or holes for working platforms and for safety belts or safety net anchorages. They can then be incorporated as part of the off-site fabrication, rather than having to be provided by the erector at a height. Access to a level should be provided by attaching a ladder and working platform to the member at ground level prior to lifting. Ideally these connections should be designed so that they can be dismantled after the erector has left the platform and descended the ladder. The erector should not have to come down an unfixed ladder or stand on an unfixed platform while removing these items after use.

Fig. 33.20 A prefabricated working platform slung over a convenient beam

Proper consideration of all of the above issues at the drawing-board stage will preclude resort to risk-laden, hastily improvised on-site solutions by unqualified personnel.

33.9.4 Employees' first visit to site

A tour of the site should be made during the induction process to aid in the identification and location of key personnel and citing of equipment, fire and first aid points, etc. Procedures should be laid down and constantly reviewed for any employee joining a company.

33.9.5 The safety of the structure

The safety of those working on a structure is prescribed by statutory regulations. However, the stability of the structure itself is not prescribed by any regulation. Where a collapse of a partly built structure occurs, the loss of life is generally heavy. Post-collapse investigation and inquest often show gaps in the understanding of the behaviour of the incomplete structure; lapses in the detailed consideration of each and every temporary condition; and, most important of all, blocks in the flow of communication of information to all involved. In the bridge collapse shown in Fig. 33.21 the temporary loading condition was not considered properly. At the point of failure of the bottom flange, a splice similar to that shown in Fig. 33.22 was unable to take the compression imposed during erection.

A designer must communicate the plan for building the structure to those who will actually have to do the building. In order to realize the design, the designer must be able to successfully translate the planning stage into the construction phase.

Columns

The end restraints of a column can change during the construction of a building; each condition must be checked to ensure that column capacity is not compromised. The risks inherent at each stage must be assessed and provided for, taking into account location, height, loads, temporary conditions, etc.

Plate girders and box girders

The checks which are applied to the webs and stiffeners of a plate girder during its design normally take account only of conditions at points where stiffeners are located and at points where loads are applied to the girder. It may not be obvious to the designer of a bridge girder that it may be subjected to a rolling load when

Fig. 33.21   Bridge failure during cantilever erection

Fig. 33.22   Typical splice detail that is adequate in tension but would fail if subject to significant compression

the bottom flange is rolled out over the piers, thus subjecting the girder to a compressive load which it is not required to carry in its permanent position.

Splices

The effects of stress reversals are most severe on splice details. These almost inevitably involve some degree of eccentricity, which can trigger a collapse if the condition, transient though it is, has not been considered in design.

Bracing

Bracing is built into all types of structures to give them the capacity to withstand horizontal forces produced by wind, temperature and the movements of cranes and other plant in and on the building.

Erection cranes carried on the structure produce vibration and load. These loads may not have been adequately accounted for post building completion. Movements resulting from cranes slewing, luffing and hoisting are carried by the framework supporting the crane to ground level. These loads and vibrations must be considered, and the structure's ability to carry them assessed at the outset.

Temporary bracings, which may be required at some stages of the work, must have properly designed connections and be specifically referred to in the erection method statement. Early or unauthorized removal of temporary bracings is a common cause of collapse in a partially completed frame.

Having considered the need for installing temporary bracings and the need to postpone fixing permanent bracings, consideration should be given to the overall economy of retaining the temporary bracings and perhaps leaving out the permanent bracings. It is a costly and potentially dangerous business to go back into a structure solely in order to take out temporary members, or to insert components that have had to be left out temporarily.

Effects of temperature and wind

On a partly erected and unclad building frame the effects of temperature on the framework can exceed the effects of the wind. A tall framework will lean away from the sun as the sun moves round from east to west: thus checking the plumb of a building should be done only on a cloudy day or after the whole structure has been allowed to reach a uniform temperature (e.g. at night), and then only when the temperature is at or near the design mean figure. Tightening the bolts in the bracing when a building is at non-uniform temperature can lock in an error which may prove difficult to correct later.

Wind effects can bring a building down if it is not adequately braced and guyed. The wind can have two effects, via the pressure exerted on anything in its path, or vibrations in a member obstructing its path.

The problem is compounded by the variability of the direction and speed of a wind, and by the variability of the aerodynamic shape of the structure as each new piece is added. Care must be taken to ensure that these issues have been properly addressed at each stage of the erection of potentially problematic structures, e.g. bridges erected by cantilevering. Bracings, guy ropes and damping weights may all have to be considered as methods of changing critical frequencies of vibration and of limiting movements as the job progresses.

33.9.6 Temporary supports and temporary conditions

Much time and effort is invested in the design of the structure. However, the design of the temporary works on which that structure may have to depend while it is being built may not have been given adequate attention. The number of recorded collapses that take place after an initial failure in the temporary supports bears testimony to this omission. For example, a temporary support may be designed only to take a vertical load. In practice, the structure it is intended to support may move due to changes in temperature and wind loading, thereby imposing significant additional horizontal loads.

Sufficient consideration should be given to the foundations. Settlement in a trestle foundation can profoundly affect the stress distribution in the girder work that it supports. Settlement under a crane outrigger from a load applied only momentarily can lead to the collapse of the crane and its load. The Code of Practice BS 5975[12] for falsework (which includes all temporary works, trestling, guy wires, etc., as well as temporary works associated with earthworks) deals with a wide range of falsework types and should be carefully read and observed. Particular attention should be paid to the paragraphs dealing with communication, co-ordination and supervision since failure in any of these areas can lead to a failure of the falsework itself.

Re-used steelwork showing signs of severe corrosion must not be used for temporary falsework carrying critical loading. In other situations re-used steel should be measured to ensure adequate performance.

During construction a structure will move as its parts take up their design load. Connections to temporary supports have to be capable of absorbing these movements.

Unless the design allows for these movements, eccentricities can result which may trigger a collapse. The cross-heads at the tops of bridge trestles have been known to fail from this cause since they are often called upon to resist wind-induced loads, vibration and temperature-induced movements in the structure, in addition to their more obvious direct loading burden. For these reasons they must receive a special design study.

Very tall buildings and chimneys as well as bridges can be affected by wind-induced vibrations, as can working platforms and those who have to work with them. The force of the wind can make welding impossible without adequate shelter: therefore the fixings for a working platform must be able to take the load of the wind blowing on shelter area.

Too many examples exist of a collapse following the removal of guy wires before the bracing was fitted, or before column bases designed to be 'fixed' had been actually grouted and fixed. What is needed here is a clear flow of communication from the designer to the foreman and the workforce of exactly which sequence of working must be followed. Supervision alone may not suffice. The only way to ensure that safe practice is adhered to is to issue a clear directive coupled with an explanation of why the instruction is being given. It also helps to have employed a skilled workforce who know what they are doing!

The need for provision of an organization chart has already been discussed. However, a second chart showing who needs to know what, why, and when should also be produced. If the lines of communication and the patterns of responsibility between various management levels and organizations are to be effective, there

Fig. 33.23   Humber Bridge during erection

must be a commitment made by all concerned to understand why the links are there, and how best to enhance speedy information exchange.

33.10 Special structures

All structures are, to some extent, special. However, there exist particular structures which, by their complexity, require special consideration when designing and planning their erection. The length, height and relative mobility of a completed structure, or the depth of individual members, may bring forward particular design problems.

Temperature differentials over the depth of bridge box girders will produce changes in the camber of the girder. Temperature differentials over the width of a structure will produce changes in verticality. Temperature changes will affect the vertical orientation of the columns at each end of a long single-storey factory building or bridge. Some of these effects can be, and commonly are, accommodated by the provision of expansion joints. Others must be addressed in the planning and execution phases.

The construction of suspension (see Fig. 33.23) or cable-stayed bridges provides good examples where movement and change to the shape of the structure become increasingly apparent as the construction process progresses. A radio telescope is the best example of a special structure which is designed to move and yet must maintain very close tolerances as the extremities of the structure are reached. Other structures move as they grow, and their temporary supports can fail as a result. These failures are too often the result of a lack of appreciation of construction movements, vibrations from wind, or local loads from erection plant.

References to Chapter 33

1. HMSO (1995) The Construction (Design & Management) Regulations 1994.
2. Health and Safety Executive (1984) *Guidance Note 28 (Parts 1–4)*, HMSO.
3. The British Constructional Steelwork Association (2002) *National Structural Steelwork Specification for Building Structures*, 4th edition, BCSA/SCI.
4. British Standards Institution (1990) *Building setting out and measurements*. Part 1: *Methods of measuring, planning and organisation and acceptance criteria*. Part 2: *Measuring stations and targets*, Part 3: *Check-lists for the procurement of surveys and measurement surveys*. BS 5964, BSI, London.
5. CIMSteel (1997) *Design for Construction*. The Steel Construction Institute, Ascot, Berks.
6. Cheal B.D. (1980) *Design Guidance Notes for Friction Grip Bolted Connections*. CIRIA Technical Note 98. Construction Industry Research and Information Association, London.

7. Couchman G.H., Mullett D.L. & Rackham L.W. (2000) *Composite slabs and beams using steel decking: best practice for design and construction.* The Metal Cladding & Roofing Manufacturers Association/The Steel Construction Institute, Ascot, Berks.
8. British Standards Institution (1993) *Quality systems. Part 14: Guide to dependability programme management.* BS 5750, BSI, London.
9. British Standards Institution (1994) *Quality systems. Model for quality assurance in design, development, production, installation and servicing.* BS EN ISO 9001, BSI, London.
10. British Standards Institution (1998) *Execution of steel structures.* Part 1: *General rules and rules for buildings.* DD ENV 1090, BSI, London.
11. British Standards Institution (1972) *Code of practice for safe use of cranes (mobile cranes, tower cranes and derrick cranes).* CP 3010, BSI, London.
12. British Standards Institution (1996) *Code of practice for falsework.* BS 5975, BSI, London.

Further reading for Chapter 33

British Standards Institution (1997) *Safety nets.* Part 1: *Safety requirements, test methods.* BS EN 1263, BSI, London.

British Standards Institution (1998) *Safety nets.* Part 2: *Safety requirements for the erection of safety nets.* BS EN 1263, BSI, London.

British Standards Institution (1988) *Code of practice for safety in erecting structural frames.* BS 5531, BSI, London.

British Standards Institution (1993) *Code of practice for access and working scaffolds and special scaffold structures in steel.* BS 5973, BSI, London.

British Standards Institution (1990) *Code of practice for temporarily installed suspended scaffolds and access equipment.* BS 5974, BSI, London.

British Standards Institution (1989) *Code of practice for safe use of cranes.* Part 1: *General.* BS 7121, BSI, London.

British Standards Institution (1991) *Code of practice for safe use of cranes.* Part 2: *Inspection, testing and examination.* BS 7121, BSI, London.

British Standards Institution (2000) *Code of practice for safe use of cranes.* Part 3: *Mobile cranes.* BS 7121, BSI, London.

British Standards Institution (1997) *Code of practice for safe use of cranes.* Part 5: *Tower cranes.* BS 7121, BSI, London.

British Standards Institution (1998) *Code of practice for safe use of cranes.* Part 11: *Offshore cranes,* BS 7121, BSI, London.

The Steel Construction Institute (1993) *A Case Study of the Steel Frame Erection at Senator House, London.* SCI Publication 136, Ascot, Berks.

The Steel Construction Institute (1994) *The Construction (Design and Management) Regulations 1994: Advice for Designers in Steel.* SCI Publication 162, Ascot, Berks.

Chapter 34
Fire protection and fire engineering

by JEF ROBINSON

34.1 Introduction

Fire safety must be regarded as a major priority at the earliest stage as it can have a major impact on the design of a building and its structural form. Nevertheless, it should not stifle aesthetic or functional freedom; *fire engineering* techniques are now available which permit a more rational treatment of fire development and fire protection in buildings.

The strength of all materials reduces as their temperature increases. Steel is no exception. It is essential that the structure should not weaken in fire to the extent that collapse occurs prematurely, while the occupants are seeking to make their way to safety. For this reason it is necessary to provide a minimum degree of *fire resistance* to the building structure. Additionally, a measure of *property protection* is implied in the current Approved Document B of the Building Regulations, although in principle the only concern of the Building Regulations is safety of life.

There are two basic ways to provide fire resistance: first, to design the structure using the ordinary temperature properties of the material and then to insulate the members so that the temperature of the structure remains sufficiently low, or secondly, to take into account the high-temperature properties of the material, in which case no insulation may be necessary.

34.2 Standards and building regulations

34.2.1 Building regulations

All buildings in the UK are required to comply with the Building Regulations,[1] which are concerned with safety of life. The provisions of Approved Document B, for England and Wales, are aimed at reducing the danger to people who are in or around a building when a fire occurs, by containing the fire and ensuring the stability of the structure for sufficient time to allow the occupants to reach safety. Generally, in Scotland and Northern Ireland the provisions are similar but not identical. Approved Document B requires that adequate provision for fire safety be provided either by fulfilling its recommendations given in Appendix A or by suitable alternative methods.

The fire-resistance requirements of Document B apply only to structural elements used in:

(a) buildings, or parts of buildings, of more than one storey,
(b) single-storey buildings that are built close to a property boundary.

The degree of fire resistance required of a structural member is governed by the building function (office, shop, factory, etc.), by the building height, by the compartment size in which the member is located, and by whether or not sprinklers are installed.

Fire resistance provisions are expressed in units of time: ½, 1, 1½ and 2 hours. It is important to realize that these times are not allowable escape times for building occupants or even survival times for the structure. They are simply a convenient way of grading different categories of buildings by fire load, from those in which a fire is likely to be relatively small, such as low-rise offices, to those in which a fire might result in a major conflagration, such as a library. Fire-resistance recommendations for structural elements are given in Reference 2.

34.2.2 BS 5950: Part 8

BS 5950: Part 8[2] permits two methods of assessing the fire resistance of bare steel members. The first, the *load ratio* method, consists of comparing the *design temperature*, which is defined as the temperature reached by an unprotected member in the required fire-resistance time, with the *limiting temperature*, which is the temperature at which it will fail. The load ratio is defined as:

$$\text{load ratio} = \frac{\text{load carried at the fire limit state}}{\text{load capacity at } 20°\text{C}}$$

If the limiting temperature exceeds the design temperature no protection is necessary. The method permits designers to make use of reduced loads and higher-strength steels to achieve improved fire-resistance times in unprotected sections.

The second method, which is applicable to beams only, gives benefits when members are partially exposed and when the temperature distribution is known. It consists of comparing the calculated moment capacity at the required fire-resistance time with the applied moment. When the moment capacity exceeds the applied moment no protection is necessary. This method of design is used for unusual structural forms such as 'shelf-angle' floor beams. Some examples of the use of the moment capacity method are given in the handbook to BS 5950: Part 8.

Limiting temperatures for various structural members are presented in the Appendix *Limiting temperatures*. These 'failure' temperatures are independent of the form or amount of fire protection. Beams supporting concrete floors fail at a much higher limiting temperature than columns, for example.

The rate of heating of a given section is related to its *section factor* which is the ratio of the surface perimeter exposed to radiation and convection and the mass, which is directly related to cross-sectional area:

$$\text{section factor} = \frac{A_m}{V} = \frac{\text{exposed surface area of section per unit length (m)}}{\text{cross-sectional area of the section per unit length (m}^2)}$$

A member with a low A_m/V value will heat up at a slower rate than one with a high A_m/V value and will require less insulation (fire protection) to achieve the same fire-resistance rating. Standard tables are available listing A_m/V ratios for structural sections (see the Appendix *Section factors for UBs, UCs, CHSs and RHSs*). These factors are calculated as indicated in Fig. 34.1.

Sections at the heavy end of the structural range have such low A_m/V ratios, and therefore such slow heating rates, that failure does not occur within ½ hour under standard BS 476 heating conditions even when they are unprotected.

Limiting section factors for various structural elements are given in Fig. 34.2 (corresponding to a load ratio of 0.6).

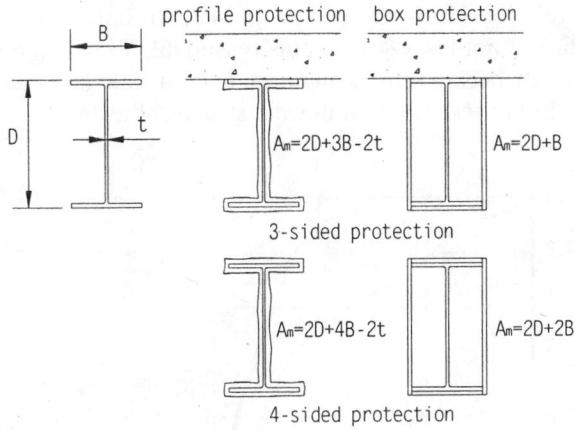

Fig. 34.1 Some different forms of fire protection to I-section members

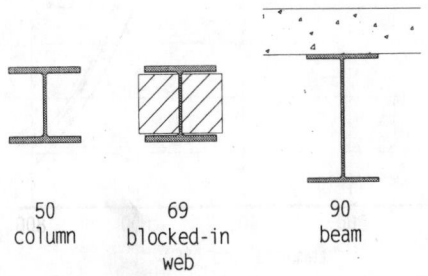

Fig. 34.2 Maximum ratios of A_m/V (m^{-1}) of exposed steel to give 30 minutes fire resistance

Manufacturers of fire-protection products now give guidance on the required thickness of fire protection depending on the section factor of the member. The example of a table for a typical spray applied (profile protection) shown in the Appendix *Minimum thickness of spray protection* is taken from Reference 3.

Some fire protection materials are assessed at the single limiting temperature of 550°C. However, there is an increasing trend for manufacturers to provide thickness recommendations for a range of temperatures. This allows designers to minimize cost by tailoring protection thickness in an individual project to the limiting temperature derived from BS 5950 Part 8.

34.3 Structural performance in fire

34.3.1 Strength of steel at elevated temperatures

Steel begins to lose strength at about 200°C and continues to lose strength at an increasing rate up to a temperature of about 750°C, when the rate of strength loss flattens off. This relationship is shown in Fig. 34.3. An important parameter is the strain at which the strength is assessed. It is reasonable to take a higher strain limit than in normal design, because fire is an ultimate limit state and much higher deflections are allowed in fire tests than in normal structural tests.

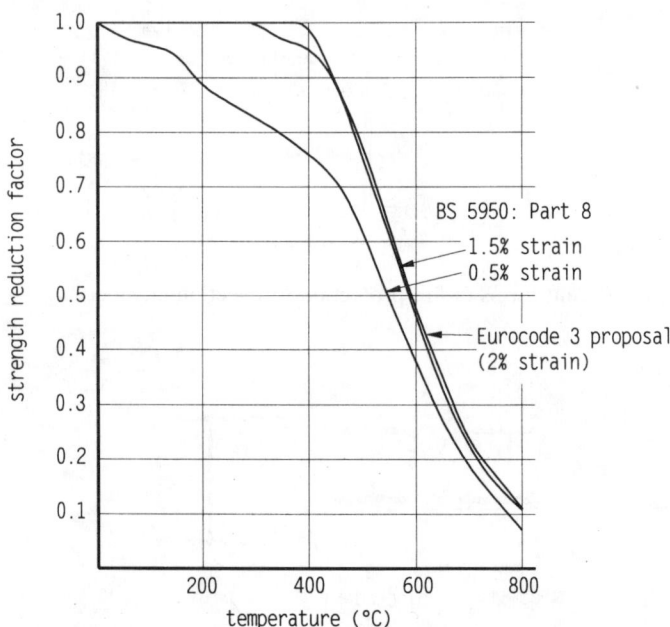

Fig. 34.3   Strength retention factor for grade 43 steel at elevated temperatures

BS 5950 Part 8 specifies the use of 2.0% strain values for design of composite beams and 1.5% strain for non-composite beams that are unprotected or protected with 'robust' (i.e. deformable) fire protection materials. For columns and tensile members or beams protected with 'brittle' material the use of 0.5% strain is specified for design calculations.

Design to Eurocode 3 (as proposed in the 2001 draft prEN 1993-1-2) is slightly less conservative, with 2.0% strain being specified for beams and 1.0% when consideration of deformation is required.

34.3.2 Performance of beams

Beams supporting concrete slabs behave better than uniformly heated sections, for which the material performance is the dominant factor. The concrete slab causes the top flange to be significantly cooler than the bottom flange and thus, as the section is heated, the plastic neutral axis of the section rises towards the top flange (see Fig. 34.4). The section resistance is determined by the strength of steel at 1.5% strain, but in this case more of the web is effective in resisting tension.

The limiting temperatures of beams supporting concrete slabs are shown in Fig. 34.5 with the test results for a range of beam sizes and load ratios. These limiting temperatures are increased by about 60°C relative to a uniformly heated section.

This temperature differential effect, and its beneficial influence on moving the neutral axis, can be enhanced by partially or fully embedding steel beams into the floors that they support. The shelf-angle beam and the Slimdek beam (shown in Fig. 34.6) are methods by which 30 minutes or 1 hour fire resistance can be achieved by design, without the need for subsequent protection.

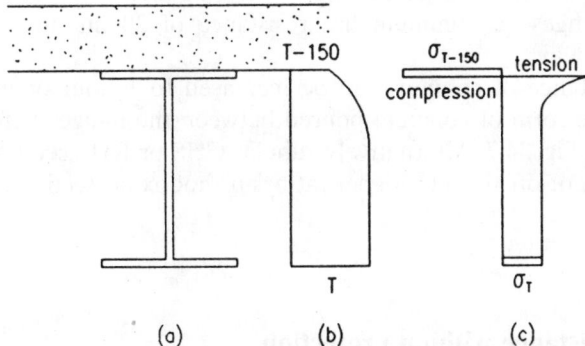

Fig. 34.4 Temperature and stress variation in I-beam supporting concrete slab when limiting temperature is reached. (a) Section through beam and slab. (b) Temperature variation. (c) Stress variation

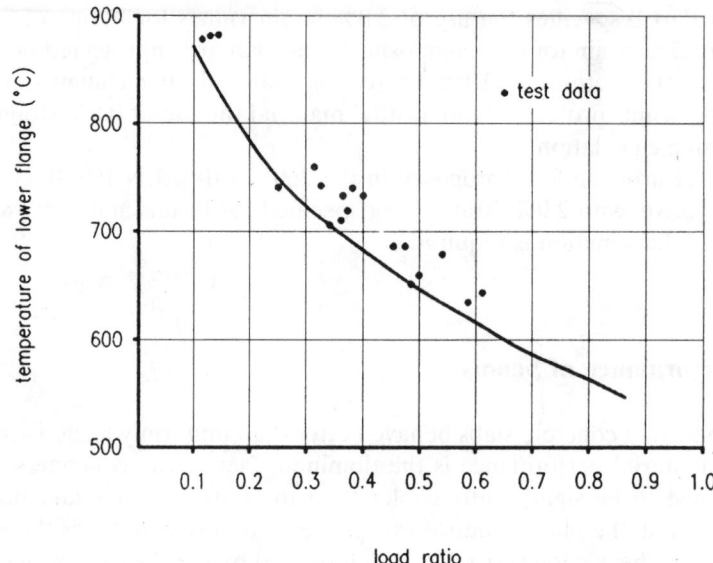

Fig. 34.5   Limiting temperatures for I-section beams supporting concrete floors on the upper flange

34.3.3 Performance of columns

Steel columns are considered to fail at a cross-section strain of less than 1%, and the 0.5% strain limit is taken as being appropriate when determining the limiting temperature of the members. This is applicable for columns up to the slenderness of 70 normally encountered in buildings.

The fire resistance of columns can also be increased by partial protection in the form of concrete blocks or bricks, either by building into a wall or by fitting blocks between the flanges. A minimum fire resistance of 30 minutes can usually be achieved in this way.

The fire resistance of columns can be increased to 1 hour or more by partial protection in the form of concrete poured between the flanges before delivery to site, as shown in Fig. 34.7. Alternatively, tubular CHS or RHS sections can be filled with concrete on or off site. For higher ratings orthodox protection methods can be used.

34.3.4 Fire resistance without protection

Table 34.1 summarizes the methods of attaining fire resistance given in the SCI publication *Design of steel framed buildings without applied fire protection*.[4] Partial

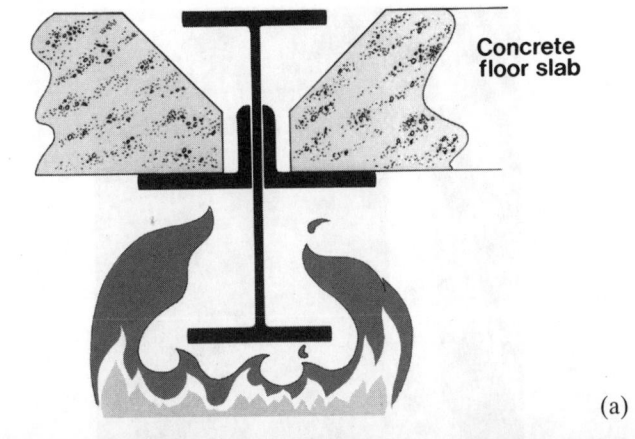

(a)

(b)

Fig. 34.6 (a) Universal beam with shelf angle to support precast concrete slab
(b) Asymmetric beam with deep metal deck

exposure of members reduces heat input and creates a temperature gradient that allows stress redistribution from hot to cold regions of the section. Ratings of up to 1 hour can be achieved without the application of fire-protecting materials.

34.3.5 Performance of composite slabs

Modern steel frames often involve the use of composite slabs comprising steel decking acting compositely with the concrete floor (see Chapter 21). A large number

Fig. 34.7   H-section columns in-filled with concrete have 1 hour fire resistance

of fire tests have now been carried out to justify the use of 90 minutes' fire resist-ance for composite slabs with standard mesh reinforcement and no additional fire protection. Guidance is given in *Handbook to RS 5950: Part 8.*[5]

In addition, it has been determined from tests that it is unnecessary to fill the voids between the beam top flange and the floor decking for ratings up to 90 minutes, although some increase in protection thickness on the beam may be required when a trapezoidal deck is used. (See Appendix – *Steelwork in fire infor-mation sheet* no. 6.)

34.3.6 Eurocodes

All of the methods given in this chapter are compatible with current (2001) drafts of Eurocodes:

Table 34.1 Table 2.1 from SCI publication 186 *Design of steel framed buildings without applied fire protection*

Beam type:	Fire resistance (mins)					
	Column type:					
	unprotected column	blocked-in column	partially encased unreinforced	partially encased reinforced	concrete filled hollow sections	protected column
unprotected beam	15	15	15	15	15	15
slim floor systems	15	30	60	60	60	60
shelf angle floor	15	30	60	60	60	60
partially encased	15	30	60	>60	>60	>60
protected beam	15	30	60	>60	>60	>60

prEN 1991-1-2: *Actions on structures exposed to fires*
prEN 1993-1-2: *Design of steel structures: Structural fire design*
prEN 1994-1-2: *Design of composite structures: Structural fire design.*

34.4 Developments in fire-safe design

In the mid 1990s, a series of realistic fire tests were carried out on a full-sized eight-storey steel-framed building at the Building Research Establishment's large-scale test facility at Cardington, Bedfordshire (see Fig. 34.8). Full analysis of the test results has shown that the behaviour of steel members in a whole building frame with all of the restraint, continuity and interaction that can occur is very different from the behaviour of single members in unrestrained standard fire tests. Columns will need protection because deformation of columns can cause damage beyond the compartment of fire origin, but unprotected composite beams were able to survive a temperature of 1100°C without collapse (see Fig. 34.9). It became clear that membrane action in the composite slab was responsible for the observed high performance and that it would be possible to design steel-framed buildings in such a way as to allow the beams to remain unprotected.

A further floor test was conducted at BRE's Garston laboratory to quantify the effect of tensile membrane action, and Bailey and Moore[6] showed that the strength

Fig. 34.8 Realistic fire tests in a modern steel-framed office building showed that stability requirements can be maintained without beam protection

Fig. 34.9   Unprotected beams were heated to 1100°C without collapse

and location of the reinforcement in a composite slab and the aspect ratio of the slab panel itself largely govern its load capacity.

This analysis has been incorporated into simple design guidance by the UK Steel Construction Institute (Newman *et al.*[7]), which allows secondary beams to remain unprotected in composite slab panels up to 9 metre span for fire ratings up to 60 minutes. These limitations may be extended by use of the Bailey and Moore[6] calculation method from first principles or by application of a specialized finite element program such as VULCAN,[8] which is adapted to deal with fire conditions.

34.5 Methods of protection

Basic information on methods of protection is summarized in the Appendix.

34.5.1 Spray-applied protection

Sprays are the cheapest method, with costs commonly in the range of £4 to £12/square metre applied (2001 prices), depending on the fire rating required and the size of the job. As implied, spray protection is applied around the exposed perimeter of the member, and therefore the relevant section factors are for profile protection (see the Appendix *Limiting temperatures*). Application is fast, and it is easy to protect complex shapes or connections. However, sprays are applied wet,

which can create problems in winter conditions, they can be messy, and the appearance is often poor. For this reason they are generally used in hidden areas such as on beams above suspended ceilings, or in plant rooms.

34.5.2 Board protection

Boards tend to be more expensive, commonly in the range £6 to £20/square metre without a decorative finish (2001 prices), because of the higher labour content in fixing. The price depends on the rating required and the surface finish chosen but tends to be less sensitive to job size. Board systems form a box around the section and therefore have a reduced heated perimeter in comparison to spray systems (see the Appendix *Limiting temperatures* and Fig. 34.1). They are dry fixed by gluing, stapling or screwing, so there is less interference with other trades on site, and the box appearance is often more suitable for frame elements, such as free-standing columns, which will be in view.

34.5.3 Intumescent coatings

Intumescent coatings have become more widely used in recent years. Unlike traditional protection materials their insulating layer is formed only by the action of heat when the fire breaks out.

The coating is applied as a thin layer, perhaps as thin as 1 mm, but it contains a compound in its formulation which releases a gas when heat is applied. This gas inflates the coating into a thick carbonaceous foam, which provides heat insulation to the steel underneath. The coatings are available in a range of colours and may be used for aesthetic reasons on visible steelwork.

Two types of intumescent coating are currently available. The first is commonly used for ratings up to 1½ hours used in dry interiors and is not recommended for wet applications such as swimming pools or in exterior conditions. Costs range from about £5 to £25/square metre (2001 prices) according to fire rating. The second type, which is water resistant, has a maximum rating of 2 hours but is expensive, with a 1 hour coating costing around £40/square metre (2001 prices).

34.5.4 Pre-delivery protection

Recent developments in intumescent coating formulation have made it practicable to merge two operations, fabrication and fire protection, into a single off-site contract. It is a method of improving efficiency that is rapidly gaining ground.

Fig. 34.10   Pre fire-protected sections awaiting erection

Application of intumescent coatings in the fabricator's works before delivery means that steelwork can arrive on site finished (see Fig. 34.10), eliminating a whole trade on site. This reduces construction time and can cut overall construction costs. Further information can be found in the Corus publication *Off-site fire protection using intumescent coatings.*[9]

34.6 Fire testing

Fire resistance requirements are based largely on the *Fire Grading of Buildings* report of 1946.[10] The fire-resistance periods refer to the time in a standard fire, defined by BS 476: Parts 20 and 21,[11] that an element of structure (a column, beam, compartment wall, etc.) should maintain:

(1) stability – it should not collapse under load at the fire limit state,
(2) integrity – it should not crack or otherwise allow the passage of flame to an adjoining compartment,
(3) insulation – it should not allow passage of heat by conduction which might induce ignition in an adjoining compartment.

The 'standard' fire time–temperature relationship defined in BS 476 is presented in Fig. 34.11. Most attention (and cost) is directed at satisfying the stability (or strength) criterion.

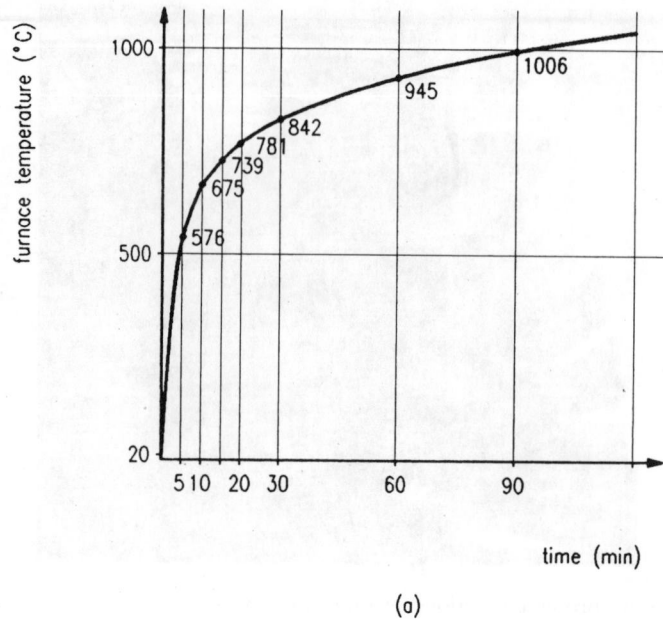

(a)

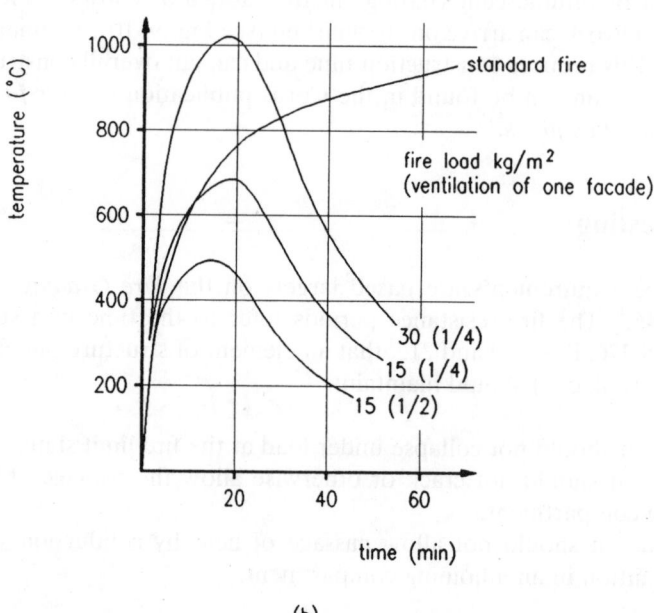

(b)

Fig. 34.11 Temperature–time curves: (a) standard ISO fire test, (b) standard and natural fires in small compartments

In a fire test columns are exposed to fire on all four sides and axially loaded vertically, whereas beams are loaded horizontally in bending and are exposed on three sides, the upper flange being in contact with the floor slab, which also acts as the furnace roof.

The stability limit is deemed to have been reached for columns when 'run-away' deflection occurs. For beams this is more accurately specified when the deflection rate reaches $L^2/9000d$ (where L = beam span, d = beam depth), within a deflection range of $L/30$ min to $L/20$ max.

34.7 Fire engineering

There are many special building forms which can take advantage of a rational approach, called *fire engineering*,[12,13] which takes account of the temperatures developed in a real fire, as opposed to a standard fire (see Fig. 34.11).

Essentially the fire engineering design method can be divided into two main steps:

(1) Determination of the fire load

The fire load of a compartment is the maximum heat that can theoretically be generated by the combustible items of contents and structure i.e. weight × calorific value per unit weight.

Fire load is usually expressed in relation to floor area, sometimes as MJ/m^2 or $Mcal/m^2$ but often converted to an equivalent weight of wood and expressed as 'kg wood/m^2' (1 kg wood $\equiv$ 18 MJ). Standard data tables giving fire loads of different materials are available.[12,13]

Examples of typical fire loads (wood equivalent) are:

schools	$15 \, kg/m^2$
hospital wards	$20 \, kg/m^2$
hotels	$25 \, kg/m^2$
offices	$35 \, kg/m^2$
department stores	$35 \, kg/m^2$
textile warehouses	$>200 \, kg/m^2$

(2) Prediction of maximum compartment temperature

The heat that is retained in the burning compartment depends upon the thermal characteristics of the wall, floor and ceiling materials and the degree of ventilation. Sheet steel walls will dissipate heat by conduction and radiation while blockwork will retain heat in the compartment and lead to higher temperatures.

It is assumed that window glass breaks in fire conditions, and calculations take into account the size and position of such ventilation. Openings close to the ceiling level of a compartment (or disintegrating roofing materials) will tend to dissipate heat whereas openings close to the floor will provide oxygen to feed the fire.

Since a great deal of data have been gathered over the years on the performance of materials in the standard fire test, methods have been sought to relate real fire conditions to standard fire performance in order that the existing data can be used in fire engineering. The *time equivalent* or equivalent required fire resistance is given by:

$$T_{eq} = CWQ_f$$

where Q_f = fire load density in MJ/m, i.e. the amount of combustible material per unit area of compartment floor,

W = ventilation factor relating to the area and height of door and window openings,

C = a constant relating to the thermal properties of the walls, floor and ceiling.

Detailed methods for calculation of temperatures in natural fires are given in Eurocode 1 (prEN1991-1-2[13]).

Fire engineering is not a concept that can be recommended for buildings that are subject to frequent change of use, such as advance factory units, but many buildings are 'fixed' in terms of their occupancy (car parks, hospitals, swimming pools, etc.) and in such cases fire engineering is a valid approach. It is most appropriate for buildings of large volume with low fire load.

In the UK a number of buildings have been built using unprotected steel on fire engineering principles, one example being the north stand at Ibrox football ground in Glasgow.

Other examples of the use of calculation methods in determining structural response in fire are external steel in framed buildings,[14] and portal frames with fire-resistant boundary wall.[15]

References to Chapter 34

1. The Building Regulations (a) Approved document B (1991), Fire safety, Department of the Environment and the Welsh Office. HMSO.
   (b) Technical standards for compliance with the Building Standards (Scotland) Regulations (1990) HMSO
   (c) The Building Regulations (Northern Ireland) (1990) Department of the Environment, HMSO.
2. British Standards Institution (1990) *Structural use of steelwork in building.* Part 8: *Code of practice for fire resistant design.* BS 5950, BSI, London.
3. The Steel Construction Institute/Association of Structural Fire Protection Contractors and Manufacturers (1989) *Fire Protection of Structural Steel in Buildings,* 2nd edn. SCI/ASFPCM.
4. Bailey C.G., Newman G.M. & Simms W.I. (1999) *Design of Steel Framed Build-*

ings without Applied Fire Protection. The Steel Construction Institute, Ascot, Berks.

5. Lawson R.M. & Newman G.M. (1990) *Fire Resistant Design of Steel Structures – A Handbook to BS 5950: Part 8.* The Steel Construction Institute, Ascot, Berks.

6. Bailey C.G. & Moore D.B. The structural behaviour of steel frames with composite floor slabs subject to fire. *The Structural Engineer*, **78** No. 11, 19–26 & 28–33, 6 June 2000.

7. Newman G.M., Robinson J.T. & Bailey C.G. (2000) *Fire Safe Design – A New Approach to Multi-Storey Buildings.* The Steel Construction Institute, Ascot, Berks.

8. Papers on VULCAN are published on the website http://www.shef.ac.uk/fire-research

9. Dowling J.J. *Off-Site Protection Using Intumescent Coatings.* Corus Construction Centre, PO Box 1, Brigg Road Scunthorpe. http://www.corusconstruction.com

10. Ministry of Public Buildings and Works (1946) *Fire Grading of Buildings, Part 1, General Principles and Structural Precautions.* Post War Building Studies No. 20. HMSO.

11. British Standards Institution (1987) *Fire tests on building materials and structures.* Part 20: *Method of determination of the fire resistance of element of construction (general principles).* Part 21: *Method for determination of the fire resistance of load-bearing elements of construction.* BS 476, BSI, London.

12. Report of CIB Workshop 14 (1983) Design guide: structural fire safety. *Fire Safety Journal*, **6**. No. 1, 1–79.

13. Eurocode 1 – *Actions on structures.* Part 1.2 *Actions on structures exposed to fire.* CEN, Central secretariat, rue de Stassart 36, Brussels.

14. Law M. & O'Brien T.P. (1981) *Fire and Steel Construction – Fire Safety of Bare External Structural Steel.* Constrado.

15. Newman G.M. (1990) *Fire and Steel Construction – The Behaviour of Steel Portal Frames in Boundary Conditions*, 2nd edn. The Steel Construction Institute, Ascot, Berks.

Further reading for Chapter 34

References 4, 7 and 9 are recommended further reading.
EUROFER (1990) *Steel and Fire Safety – A Global Approach.* Eurofer, Brussels (available from The Steel Construction Institute, Ascot, Berks).

Chapter 35

Corrosion and corrosion prevention

by ROGER HUDSON and KEN JOHNSON

35.1 The corrosion process

35.1.1 Introduction

The specification of cost-effective protective treatments for structural steelwork should not present a major problem for most common applications if the factors that affect durability are appreciated. Primarily it is important to recognize and define the corrosivity of the environment to which the structure is to be exposed to enable the specification of an appropriate protective system. Many structures are in relatively low risk category environments and therefore require minimal treatment. Conversely, a steel structure exposed to an aggressive environment needs to be protected with a durable system that may require maintenance for extended life.

The optimum protection treatment combines good surface preparation with suitable coating materials for a required durability at a minimum cost. Modern practices applied according to the relevant industry standards provide an opportunity to achieve the desired protection requirements for specific structures.

Much guidance for the protection of steel structures has, over the years, been sought from BS 5493: 1977 *Code of Practice for Protective Coating of Iron and Steel Structures Against Corrosion*. This document has now been superseded by a series of new ISO standards. One of the most important is ISO 12944 *Paints and Varnishes – Corrosion Protection of Steel Structures by Protective Paint Systems*. This standard, which is published in eight parts, should be referred to when drafting protection specifications for steel structures. Part 5 of the series *Protective paint systems* contains a range of paint coatings and systems for different environmental categories that are defined in Part 2 *Classification of environments*. However, specifiers concerned with UK projects should be aware that not all of the paints listed are 'compliant' with current national environmental legislation, and further advice should be sought from the paint manufacturer.

35.1.2 General corrosion

Most corrosion of steel can be considered as an electrochemical process which occurs in stages. Initial attack occurs at anodic areas on the surface, where ferrous ions go into solution. Electrons are released from the anode and move through the

1030

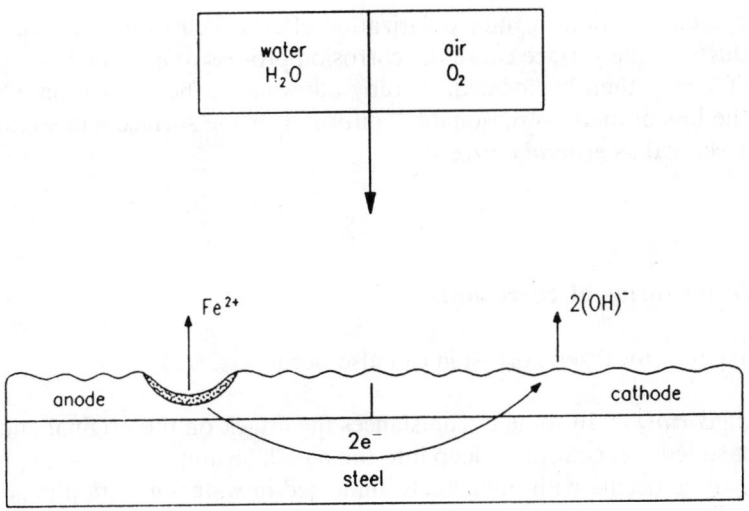

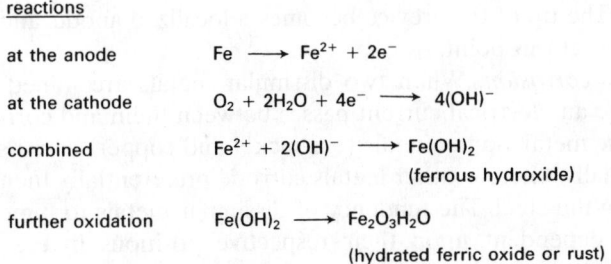

reactions

at the anode $Fe \longrightarrow Fe^{2+} + 2e^-$

at the cathode $O_2 + 2H_2O + 4e^- \longrightarrow 4(OH)^-$

combined $Fe^{2+} + 2(OH)^- \longrightarrow Fe(OH)_2$
                        (ferrous hydroxide)

further oxidation $Fe(OH)_2 \longrightarrow Fe_2O_3H_2O$
                        (hydrated ferric oxide or rust)

Fig. 35.1 Diagrammatic representation of the corrosion of steel

metallic structure to the adjacent catholic sites on the surface, where they combine with oxygen and water to form hydroxyl ions. These react with the ferrous ions from the anode to produce ferrous hydroxide, which itself is further oxidized in air to produce hydrated ferric oxide: red rust (Fig. 35.1).

The sum of these reactions is described by the following equation:

$$4Fe + 3O_2 + 2H_2O = 2Fe_2O_3H_2O$$
(iron/steel) + (oxygen) + (water) = rust

Two important points emerge:

(1) For iron or steel to corrode it is necessary to have the *simultaneous* presence of water and oxygen; in the absence of either, corrosion does not occur.
(2) All corrosion occurs at the anode; no corrosion occurs at the cathode.

However, after a period of time, polarization effects such as the growth of corrosion products on the surface cause the corrosion process to be stifled. New, reactive anodic sites may then be formed, thereby allowing further corrosion. Over long periods the loss of metal is reasonably uniform over the surface and so this case is usually described as *general corrosion*.

35.1.3 Other forms of corrosion

Various types of localized corrosion can also occur:

(1) *Pitting corrosion.* In some circumstances the attack on the original anodic area is not stifled and continues deep into the metal, forming a corrosion pit. Pitting more often occurs with mild steels immersed in water or buried in soil rather than those exposed in air.
(2) *Crevice corrosion.* Crevices can be formed by design-detailing, welding, surface debris, etc. Available oxygen in the crevice is quickly used by the corrosion process and, because of limited access, cannot be replaced. The entrance to the crevice becomes catholic, since it can satisfy the oxygen-demanding cathode reaction. The tip of the crevice becomes a localized anode, and high corrosion rates occur at this point.
(3) *Bimetallic corrosion.* When two dissimilar metals are joined together in an electrolyte an electrical current passes between them and corrosion occurs on the anodic metal. Some metals (e.g. nickel and copper) cause steel to corrode preferentially whereas other metals corrode preferentially themselves, thereby protecting the steel. The tendency of dissimilar metals to bimetallic corrosion is partly dependent upon their respective positions in the galvanic series (Table 35.1): the further apart the two metals are in the series the greater the tendency. Other aspects which influence bimetallic corrosion are the nature of the electrolyte and the respective surface areas of the anodic and cathodic metals. Bimetallic corrosion is most serious for immersed or buried structures but in less aggressive environments, e.g. stainless steel brick support angles attached to mild steel structural sections, the effect on the mild steel sections is practically minimal and no special precautions are required.

Further guidance for the avoidance of bimetallic corrosion can be found in BS PD 6484, *Commentary on corrosion at bimetallic contacts and its alleviation*.

35.1.4 Corrosion rates

The principal factors that determine the rate of corrosion of steel in air are:

Table 35.1   Bimetallic corrosion and structural steelwork

	Anodic end
	(more prone to corrosion)
Magnesium	Tendency to inhibit
Zinc	corrosion of structural
Aluminium	steels
Cadmium	
Carbon and low alloy (structural) steels	
Cast irons	Tendency to accelerate
Lead	corrosion of structural
Tin	steels
Copper	
Brasses	
Bronzes	
Nickel (passive)	
Titanium	
Stainless steels 430/304/316 (passive)	
	Cathodic end
	(less prone to corrosion)

(1) *Time of wetness.* This is the proportion of total time during which the surface is wet, due to rainfall condensation, etc. It follows, therefore, that for unprotected steel in dry environments, e.g. inside heated buildings, corrosion will be negligible due to the low availability of water.

(2) *Atmospheric pollution.* The type and amount of atmospheric pollution and contaminants, e.g. sulphur dioxide chlorides, dust, etc.

(3) *Sulphates.* These originate from sulphur dioxide gas, which is produced during the combustion of fossil fuels, e.g. sulphur-bearing oils and coal. The sulphur dioxide gas reacts with water or moisture in the atmosphere to form sulphurous and sulphuric acids. Industrial environments are a prime source of sulphur dioxide.

(4) *Chlorides.* These are mainly present in marine environments. The highest concentrations of chlorides are to be found in coastal regions, and there is a rapid reduction moving inland.

Within a given local environment corrosion rates can vary markedly due to the effects of sheltering and prevailing winds. It is therefore the 'microclimate' immediately surrounding the structure which determines corrosion rates for practical purposes.

35.2   Effect of the environment

Corrosion rate data cannot be generalized; however, environments can be broadly classified and corresponding corrosion rates provide a useful indication. A range of

UK environments and steel corrosion rates are considered as follows: (Note: Corrosion rates are usually expressed as μm/year; 1 μm = 0.001 mm.)

(1) *Rural atmospheric* – essentially inland, unpolluted environments; steel corrosion rates tend to be low, usually less than 50 μm/year
(2) *Industrial atmospheric* – inland, polluted environments; corrosion rates are usually between 40 and 80 μm/year, depending upon level of SO_2
(3) *Marine atmospheric* – in the UK a 2 km strip around the coast is broadly considered as being in a marine environment; corrosion rates are usually between 50 and 100 μm/year, largely dependent upon proximity to the sea
(4) *Marine/industrial atmospheric* – polluted coastal environments which produce the highest corrosion rates e.g. between 50 and 150 μm/year
(5) *Sea-water immersion* – in tidal waters four vertical zones are usually encountered:
   (a) the splash zone, immediately above the high-tide level, is usually the most corrosive zone with a mean corrosion rate of about 75 μm/year
   (b) the tidal zone, between high-tide and low-tide levels, is often covered with marine growths and exhibits low corrosion rates e.g. 35 μm/year
   (c) the low-water zone, a narrow band just below the low-water level, exhibits corrosion rates similar to the splash zone
   (d) the permanent immersion zone, from the low-water level down to bed level, exhibits low corrosion rates e.g. 35 μm/year
(6) *Fresh-water immersion* – corrosion rates are lower in fresh water than in salt water e.g. 30–50 μm/year
(7) *Soils* – the corrosion process is complex and very variable; various methods are used to assess the corrosivity of soils:
   (a) resistivity; generally high-resistance soils are least corrosive
   (b) redox potential; to assess the soil's capability of anaerobic bacterial corrosion
   (c) pH; highly acidic soils (e.g. pH less than 4.0) can be corrosive
   (d) water content; corrosion depends upon the presence of moisture in the soil, the position of the water-table has an important bearing.

Long buried steel structures, e.g. pipelines, are most susceptible to corrosion. Steel piles driven into undisturbed soils are much less susceptible due to the low availability of oxygen.

35.3 Design and corrosion

In external or wet environments design can have an important bearing on the corrosion of steel structures. The prevention of corrosion should therefore be taken into account during the design stage of a project. The main points to be considered are:

(1) Entrapment of moisture and dirt:
   (a) avoid the creation of cavities, crevices, etc.
   (b) welded joints are preferable to bolted joints
   (c) avoid or seal lap joints
   (d) edge-seal HSFG faying surfaces
   (e) provide drainage holes for water, where necessary
   (f) seal box sections except where they are to be hot-dip galvanized
   (g) provide free circulation of air around the structure.
(2) Contact with other materials:
   (a) avoid bimetallic connections or insulate the contact surfaces (see BS PD 6484)
   (b) provide adequate depth of cover and quality of concrete (see BS 8110)
   (c) separate steel and timber by the use of coatings or sheet plastics.
(3) Coating application; design should ensure that the selected protective coatings can be applied efficiently:
   (a) provide vent-holes and drain-holes for items to be hot-dip galvanized (see BS 4479: Part 6)
   (b) provide adequate access for paint spraying, thermal (metal) spraying, etc. (see BS 4479: Part 7)
(4) General factors.
   (a) large flat surfaces are easier to protect than more complicated shapes
   (b) provide access for subsequent maintenance
   (c) provide lifting lugs or brackets where possible to reduce damage during handling and erection.

BS 5493, Appendix A,[1] provides a detailed account of designing for the prevention of corrosion.

35.4 Surface preparation

Structural steel is a hot-rolled product. Sections leave the last rolling pass at about 1000°C and as they cool the steel surface reacts with oxygen in the atmosphere to produce mill-scale, a complex oxide which appears as a blue-grey tenacious scale completely covering the surface of the as-rolled steel section. Unfortunately, mill-scale is unstable. On weathering, water penetrates fissures in the scale, and rusting of the steel surface occurs. The mill-scale loses adhesion and begins to shed. Mill-scale is therefore an unsatisfactory base and needs to be removed before protective coatings are applied.

As mill-scale sheds, further rusting occurs. Rust is a hydrated oxide of iron which forms at ambient temperatures, producing a layer on the surface which is itself an unsatisfactory base and also needs to be removed before protective coatings are applied.

Surface preparation of steel is therefore principally concerned with the removal of mill-scale and rust and is an essential process in corrosion protection treatments.

Various methods of surface preparation are presented in BS 7079: Part A1: 1989 or the equivalent ISO 8501–1: 1988 standards and are summarized as follows:

(1) *Hand and power tool cleaning (St)*
   St. 2: Thorough hand and power tool cleaning.
   St. 3: Very thorough hand and power tool cleaning.

Both manual and mechanical methods using scrapers, wire brushes, etc., can remove about 30–50% of rust and scale.

(2) *Blast cleaning (Sa)*
   Sa. 1: Light blast cleaning
   Sa. 2: Thorough blast cleaning
   Sa. $2\frac{1}{2}$: Very thorough blast cleaning
   Sa. 3: Blast cleaning to visually clean steel

The blast cleaning process involves the projection of abrasive particles (shot or grit) in a jet of compressed air or by centrifugal impellers at high velocities on to the steel surface. This process can be 100% efficient in the removal of rust and scale. The profile of the surface produced is dependent upon the size and shape of the abrasive used; angular grits produce angular surface profiles and round shot produce a rounded profile. The effectiveness of the surface preparation methods above are compared with the relevant photograph contained in the standard.

Grit-blast abrasive can be either metallic (e.g. chilled iron grit) or non-metallic (e.g. slag grit). The latter are used only once and are referred to as *expendable*. They are used exclusively for site work. Metallic grits are expensive and are only used where they can be recycled. Grit blasting is always used for thermal (metal) sprayed coatings, where adhesion is at least partly dependent upon mechanical keying. It is also used for some paint coatings, particularly on site and for primers where adhesion may be a problem (e.g. zinc silicates and high-build solvent-free paint coatings).

Shot-blast abrasives are always metallic, usually cast steel shot, and are used particularly in shot-blast plants, utilizing impeller wheels and abrasive recycling. They are the preferred abrasive for most paints, particularly for thin film coatings (e.g. prefabrication primers).

Blast-cleaned surfaces should be specified in terms of both surface cleanliness (BS 7079/ISO 8501 Group A) and surface roughness (BS 7079/ISO 8503 Group C).

(3) *Wet (abrasive) blast cleaning*
   A further variation on the blast cleaning process is described as *wet blasting*. In this process a small amount of water is entrained in the abrasive/compressed air stream. This is particularly useful in washing from the surface soluble iron salts which are formed in the rust by atmospheric pollutants (e.g. chlorides and sulphates) during weathering. These are often located deep in corrosion pits on

the steel surface and cannot be removed by conventional dry blast cleaning methods. Wet abrasive blasting has proved to be particularly useful on offshore structures and prior to maintenance painting for structures in heavily polluted environments.

(4) *Acid pickling*

This process involves immersing the steel in a bath of suitable inhibited acids which dissolve or remove mill-scale and rust but do not appreciably attack the exposed steel surface. It can be 100% effective. Acid pickling is normally used on structural steel intended for hot-dip galvanizing but is now rarely used as pre-treatment before painting.

35.5 Metallic coatings

There are four commonly used methods of applying metal coating to steel surfaces: hot-dip galvanizing, thermal (metal) spraying, electroplating and sherardizing. The latter two processes are not used in structural steelwork but are used for fittings, fasteners and other small items.

In general the corrosion protection afforded by metallic coatings is largely dependent upon the choice of coating metal and its thickness and is not greatly influenced by the method of application.

35.5.1 Hot-dip galvanizing

The most common method of applying a metal coating to structural steel is by galvanizing.

The galvanizing process involves the following stages:

(1) Any surface oil or grease is removed by suitable degreasing agents.
(2) The steel is cleaned of all rust and scale by acid pickling. This may be preceded by blast-cleaning to remove scale and roughen the surface but such surfaces are always subsequently pickled in inhibited hydrochloric acid.
(3) The cleaned steel is then immersed in a fluxing agent to ensure good contact between the zinc and steel during immersion
(4) The cleaned and fluxed steel is dipped into a bath of molten zinc at a temperature of about 450°C at which the steel reacts with the molten zinc to form a series of zinc/iron alloys on its surface.
(5) As the steel workpiece is removed from the bath a layer of relatively pure zinc is deposited on top of the alloy layers.

As the zinc solidifies it assumes a crystalline metallic lustre, usually referred to as *spangling*. The thickness of the galvanized coating is influenced by various factors

including the size and thickness of the workpiece and the surface preparation of the steel. Thick steels and steels which have been abrasive blast cleaned tend to produce heavier coatings. Additionally, the steel composition has an effect on the coating produced. Steels containing silicon and phosphorus can have a marked effect on the thickness, structure and appearance of the coating. The thickness of the coating varies mainly with the silicon content of the steel and the bath immersion time. These thick coatings sometimes have a dull dark grey appearance and can be susceptible to mechanical damage.

Since hot-dip galvanizing is a dipping process, there is obviously some limitation on the size of components which can be galvanized. Double dipping can often be used when the length or width of the workpiece exceeds the size of the bath.

Some aspects of design need to take the galvanizing process into account, particularly filling, venting, draining and distortion. To enable a satisfactory coating, suitable holes must be provided in hollow articles (e.g. tubes and rectangular hollow sections) to allow access for the molten zinc, venting of hot gases to prevent explosions and the subsequent draining of zinc. Distortion of fabricated steel-work can be caused by differential thermal expansion and contraction and by the relief of unbalanced residual stresses during the galvanizing process. Further guidance on the design of articles to be hot-dip galvanized can be found in BS 4479: Part 6.

EN ISO 1461, which replaces BS 729, is the specification of hot-dip galvanized coating for structural steelwork. This requires, for sections not less than 6 mm thick, a minimum zinc coating weight of $610 \, g/m^2$, equivalent to a minimum average coating thickness of 85 μm. EN ISO 14713 provides additional guidance on the performance of metallic coatings applied by dipping and thermal spray.

For many applications, hot-dip galvanizing is used without further protection. However, to provide extra durability, or where there is a decorative requirement, paint coatings are applied. The combination of metal and paint coatings is usually referred to as a *duplex* coating. When applying paints to galvanized coatings, special surface preparation treatments should be used to ensure good adhesion. These include light blast cleaning to roughen the surface and provide a mechanical key, and the application of special etch primers or 'T' wash, which is an acidified solution designed to react with the surface and provide a visual indication of effectiveness.

35.5.2 Thermal (metal) spray coatings

An alternative method of applying a metallic coating to structural steelwork is by thermal (metal) spraying of either zinc or aluminium. The metal, in powder or wire form, is fed through a special spray-gun containing a heat source which can be either an oxy-gas flame or an electric arc. Molten globules of the metal are blown by a compressed air jet on to the previously blast-cleaned steel surface. No alloying occurs and the coating which is produced consists of overlapping platelets of metal

and is porous. The pores are subsequently sealed, either by applying a thin organic coating which soaks into the surface, or by allowing the metal coating to weather, when corrosion products block the pores.

The adhesion of sprayed metal coatings to steel surfaces is considered to be essentially mechanical in nature. It is therefore necessary to apply the coating to a clean roughened surface for which blast-cleaning with a coarse grit abrasive is normally specified, usually chilled-iron grit, but for steels with a hardness exceeding 360HV, alumina or silicon carbide grits may be necessary.

Typical specified coating thicknesses vary between 150–200 µm for aluminium and 100–150 µm for zinc.

Thermal (metal) spray coatings can be applied in the shops or at site, and there is no limitation on the size of the workpiece as there is with hot-dip galvanizing. Since the steel surface remains cool there are no distortion problems. Guidance on the design of articles to be thermally sprayed can be found in BS 4479: Part 7 and EN ISO 14713. However, thermal spraying is considerably more expensive than hot-dip galvanizing.

For many applications thermal spray coatings are further protected by the subsequent application of paint coatings. A sealer is first applied, which fills the pores in the thermal spray coating and provides a smooth surface for application of the paint coating.

The protection of structural steelwork against atmospheric corrosion by thermal sprayed aluminium or zinc coatings is covered in BS EN 22063: 1994.

35.6 Paint coatings

Painting is the principal method of protecting structural steelwork from corrosion.

35.6.1 Composition of paints and film formation

Paints are made by mixing and blending three main components:

(1) Pigments: finely ground inorganic or organic powders which provide colour, opacity, film cohesion and sometimes corrosion inhibition.
(2) Binders: usually resins or oils but can be inorganic compounds such as soluble silicates. The binder is the film-forming component in the paint.
(3) Solvents: used to dissolve the binder and to facilitate application of the paint. Solvents are usually organic liquids or water.

Paints are applied to steel surfaces by many methods but in all cases they produce a *wet film*. The thickness of the wet film can be measured, before the solvent evaporates, using a comb-gauge.

As the solvent evaporates, film-formation occurs, leaving the binder and pigments on the surface as a *dry film*. The thickness of the dry film can be measured, usually with a magnetic induction gauge.

The relationship between the applied wet film thickness and the final dry film thickness (d.f.t.) is determined by the percentage volume solids of the paint, i.e.

d.f.t. = wet film thickness × % vol. solids

In general the corrosion protection afforded by a paint film is directly proportional to its dry film thickness.

35.6.2 Classification of paints

Since, in the broadest terms, a paint consists of a particular pigment, dispersed in a particular binder, dissolved in a particular solvent, the number of generic types of paint is limited. The most common methods of classifying paints are either by their pigmentation or by their binder-type.

Primers for steel are usually classified according to the main corrosion-inhibitive pigments used in their formulation, e.g. zinc phosphate, metallic zinc. Each of these inhibitive pigments can be incorporated into a range of binder resins, e.g. zinc phosphate alkyd primers, zinc phosphate epoxy primers, zinc phosphate acrylated-rubber primers.

Intermediate coats and finishing coats are usually classified according to their binders, e.g. vinyl finishes, urethane finishes.

35.6.3 Painting systems

Paints are usually applied one coat on top of another, each coat having a specific function or purpose.

The primer is applied directly on to the cleaned steel surface. Its purpose is to wet the surface and to provide good adhesion for subsequently applied coats. Primers for steel surfaces are also usually required to provide corrosion inhibition.

The intermediate coats (or undercoats) are applied to build the total film thickness of the system. This may involve the application of several coats.

The finishing coats provide the first-line defence against the environment and also determine the final appearance in terms of gloss, colour, etc.

The various superimposed coats within a painting system have, of course, to be compatible with one another. They may be all of the same generic type or may be different, e.g. acrylated-rubber-based intermediate coats may be applied on to an epoxy primer. However, as a first precaution, all paints within a system should normally be obtained from the same manufacturer.

35.6.4 Main generic types of paint and their properties

(1) *Air-drying paints*, e.g. oil-based, alkyds. These materials dry and form a film by an oxidative process which involves absorption of oxygen from the atmosphere. They are therefore limited to relatively thin films. Once the film has formed it has limited solvent resistance and usually poor chemical resistance.

(2) *One-pack chemical-resistant paints*, e.g. acrylated rubbers, vinyls. For these materials, film formation is by solvent evaporation and no oxidative process is involved. They can be applied as moderately thick films, although retention of solvent in the film can be a problem at the upper end of the range. The film formed remains relatively soft, and has poor solvent resistance but good chemical resistance.

   Bituminous paints also dry by solvent evaporation. They are essentially solutions of either asphaltic bitumen or coal-tar pitch in organic solvents.

(3) *Two-pack chemical-resistant paints*, e.g. epoxy, urethane. These materials are supplied as two separate components, usually referred to as the base and the curing agent. When the two components are mixed, immediately before use, a chemical reaction begins. These materials therefore have a limited 'pot-life' by which the mixed coating must be applied. The polymerization reaction continues after the paint has been applied and after the solvent has evaporated to produce a densely cross-linked film which can be very hard and has good solvent and chemical resistance.

   Liquid resins of low viscosity can be used in the formulation thereby avoiding the need for a solvent. Such coatings are referred to as solventless or solvent-free and can be applied as very thick films.

A summary of the main generic types of paint and their properties is shown in Table 35.2.

35.6.5 Prefabrication primers (also referred to as blast primers, shop-primers, weldable primers, temporary primers, holding primers, etc.)

These primers are used on structural steelwork, immediately after blast-cleaning, to hold the reactive blast-cleaned surface in a rust-free condition until final painting can be undertaken. They are mainly applied to steel plates and sections before fabrication. The main requirements of a blast primer are as follows:

(1) The primer should be capable of airless-spray application to produce a very thin even coating. Dry-film thickness is usually limited to 15–25 μm. Below 15 μm the peaks of the blast profile are not protected and 'rust-rashing' occurs on weathering. Above 25 μm the primer affects the quality of the weld and produces excessive weld-fume.

Table 35.2 Main generic types of paint and their properties

	Cost	Tolerance of poor surface preparation	Chemical resistance	Solvent resistance	Over-coatability after ageing	Other comments
Bituminous	Low	Good	Moderate	Poor	Good with coatings of same type	Limited to black and dark colours Thermoplastic
Alkyds	Low–medium	Moderate	Poor	Poor–moderate	Good	Good decorative properties
Acrylated-rubber	Medium	Poor	Good	Poor	Good	High-build films remain soft and are susceptible to 'sticking'
Vinyl	High	Poor	Good	Poor	Good	
Epoxy	Medium–high	V. poor	V. good	Good	Poor	Very susceptible to chalking in UV
Urethane	High	V. poor	V. good	Good	Poor	Can be more decorative than epoxies
Inorganic or organic silicate	High	V. poor	Moderate	Good	Moderate	May require special surface preparation

(2) The primer must dry very quickly. Priming is often done in-line with automatic blast-cleaning plant, which may be handling plates or sections at a pass-rate of 1–3 metres/minute. The interval between priming and handling is usually of the order of 1–10 minutes and hence the primer film must dry within this time.

(3) Normal fabrication procedures e.g. welding, gas-cutting, must not be significantly impeded by the coating and the primer should not cause excessive weld porosity.

(4) Weld fumes emitted by the primer must not exceed the appropriate occupational exposure limits. Proprietary primers are tested and certificated by the Newcastle Occupational Health Agency.

(5) The primer coating should provide adequate protection. It should be noted that manufacturers may claim extended durability for their prefabrication primers, and suggested exposure periods of 6–12 months are not uncommon. In practice, such claims are rarely met except in the least arduous conditions, e.g. indoor storage. In aggressive conditions, durability can be measured in weeks rather than months.

Many proprietary blast-primers are available but they can be classified under the following main generic types:

(1) *Etch primers* are based on polyvinyl butyral resin reinforced with a phenolic resin to uprate water resistance. These primers can be supplied in a single-pack or two-pack form.

(2) *Epoxy primers* are two-pack materials utilizing epoxy resins and usually either polyamide or polyamine curing agents. They are pigmented with a variety of inhibitive and non-inhibitive pigments. Zinc phosphate epoxy primers are the most frequently encountered and give the best durability within the group.

(3) *Zinc epoxy primers* can be subdivided into zinc-rich and reduced-zinc types. Zinc-rich primers produce films which contain about 80% by weight of metallic zinc powder and the reduced-zinc as low as 55% by weight.

When exposed in either marine or highly industrial environments, zinc epoxy primers are prone to the formation of insoluble white zinc corrosion products, which must be removed from the surface before subsequent overcoating. All zinc epoxy primers produce zinc oxide fumes during welding and gas cutting and may be a health hazard.

(4) *Zinc silicate primers* produce a level of protection which is comparable to the zinc-rich epoxy types and they suffer from the same drawbacks, e.g. formation of zinc salts and production of zinc oxide fumes during welding. They are however more expensive and usually are less convenient to use.

There are currently different categories of silicate primer based upon the binder (organic or inorganic) and the zinc content. Low-zinc primers in this group have been developed to improve weldability and minimize weld porosity. However, their durability is reduced. The organic silicate primers are the most suitable as prefabrication primers.

35.7 Application of paints

35.7.1 Methods of application

The method of application and the conditions under which paints are applied have a significant effect on the quality and durability of the coating.

The standard methods used for applying paints to structural steelwork are brush, roller, conventional air-spray, and airless spray, although other methods, e.g. dipping, can be used.

(1) *Brush*. This is the simplest and also the slowest and therefore most expensive method. Nevertheless it has certain advantages over the other methods, e.g. better wetting of the surface; can be used in restricted spaces; useful for small areas; less wastage and less contamination of surroundings.
(2) *Roller*. This process is much quicker than brushing and is useful for large flat areas but demands suitable rheological properties of the paint.
(3) *Air-spray*. The paint is atomized at the gun-nozzle by jets of compressed air; application rates are quicker than for brushing or rolling; paint wastage by over-spray is high.
(4) *Airless spray*. The paint is atomized at the gun-nozzle by very high hydraulic pressures; application rates are higher than for air-spray and overspray wastage is greatly reduced.

Airless spraying has become the most commonly used method of applying paint coatings to structural steelwork under controlled shop-conditions. Brush and roller application are more commonly used for site application, though spraying methods are also used.

35.7.2 Conditions for application

The principal conditions which affect the application of paint coatings are tempera-ture and humidity. These can be more easily controlled under shop conditions than on site.

(1) *Temperature*. Air temperature and steel temperature affect solvent evaporation, brushing and spraying properties, drying and curing times, pot-life of two-pack materials, etc. Heating, if required, should only be by indirect methods.
(2) *Humidity*. Paints should not be applied when there is condensation present on the steel surface or the relative humidity of the atmosphere is such that it will affect the application or drying of the coating. Normal practice is to measure the steel temperature with a contact thermometer and to ensure that it is main-tained at at least 3°C above dew-point.

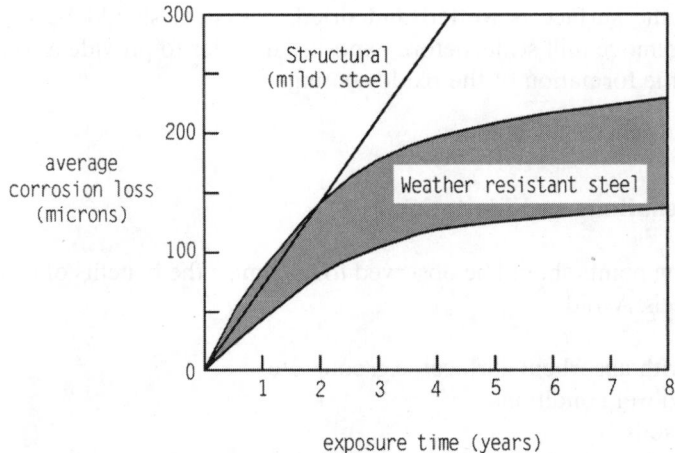

average
corrosion loss
(microns)

exposure time (years)

Fig. 35.2 Typical corrosion losses of structural (mild) steel and weather-resistant steels in the UK

35.8 Weather-resistant steels

Weather-resistant steels are high strength, low alloy weldable structural steels which possess good weathering resistance in many atmospheric conditions without the need for protective coatings. They contain up to 2.5% of alloying elements, e.g. chromium, copper, nickel and phosphorus. On exposure to air, under suitable conditions, they form an adherent protective rust layer. This acts as a protective film, which with time causes the corrosion rate to reduce until it reaches a low terminal level, usually between 2 and 3 years.

Conventional structural steels form rust layers that eventually become non-adherent and detach from the steel surface. The rate of corrosion progresses as a series of incremental curves approximating to a straight line, the slope of which is related to the aggressiveness of the environment. With weather-resistant steels, the rusting process is initiated in the same way but the alloying elements react with the environment to form an adherent, less porous rust layer. With time, this rust layer becomes protective and reduces the corrosion rate (see Fig. 35.2).

Weather-resistant steels are specified in BS EN 10155: 1993, and within this category Corten is one of the best known proprietary weather-resistant steels. These steels have mechanical properties comparable to those of grade S355 steels to BS EN 10025: 1993.

35.8.1 Formation of the protective oxide layer

The time required for a weather-resistant steel to form a stable protective rust layer depends upon its orientation, the degree of atmospheric pollution and the frequency

with which the surface is wetted and dried. The steel should be abrasive blast cleaned, to remove mill-scale, before exposure in order to provide a sound uniform surface for the formation of the oxide coatings.

35.8.2 Precautions and limitations

The following points should be observed to maximize the benefits of using weather-resistant steels. Avoid:

- contact with absorbent surfaces, e.g. concrete
- prolonged wet conditions
- burial in soils
- contact with dissimilar metals
- aggressive environments, e.g. marine atmospheres.

Drainage of corrosion products can be expected during the first years of exposure and can stain or streak adjacent materials, e.g. concrete piers. Provision should be made to divert corrosion products from vulnerable surfaces. Often the north faces of buildings experience long periods of wetness and do not favour the formation of a protective rust patina.

35.8.3 Welding and bolted connections

Weather-resistant steels can be welded by all the usual methods, e.g. manual metal arc, gas shielded, submerged arc, and electrical resistance, including spot welding. Welding electrodes should be compatible with the welding process. For structural joints where high strength bolts are required, ASTM A325, Type 3 bolts (Corten X) must be used. Where lower strength bolts are satisfactory these may be in Corten A or stainless steel. Galvanized, sherardized or electroplated nuts and bolts are not suitable for use in weather resistant-steel structures since, in time, the coatings will be consumed leaving an unprotected fastener that is less corrosion resistant than the surrounding weather-resistant steel.

35.8.4 Painting of weather-resistant steels

The practive of painting weather-resistant steels, should this be required, does not differ from those practices employed for conventional structural steels. They require the same surface preparation, and the same painting systems may be used.

35.9 The protective treatment specification

35.9.1 Factors affecting choice

For a given structure the following will be largely predetermined:

(1) The expected life of the structure and the feasibility of maintenance
(2) The environment/s to which the steelwork will be subjected
(3) The size and shape of the structural members
(4) The shop-treatment facilities which are available to the fabricator and/or the coatings sub-contractor
(5) The site conditions, which will determine whether the steelwork can be treated after erection
(6) The money which is available to provide protection.

These facts, and possibly others, have to be considered before making decisions on:

- the types of coating to be used,
- the method of surface preparation,
- the method/s of application,
- the number of coats and the thickness of each coat.

In general, each case has to be decided on its own merits. However, the following points may be of assistance in making these decisions:

(1) Protection requirements are minimal inside dry, heated buildings. Hidden steelwork in such situations requires no protection at all.
(2) The durability of painting systems is increased several times over by using abrasive blast-cleaning rather than manual surface preparation.
(3) Shot-blasting is preferred for most painting systems.
(4) Grit-blasting is essential for thermal (metal) spraying and some primers, e.g. zinc silicates.
(5) If blast-cleaning is to be used, two alternative process routes are available, i.e.
  (a) blast/prime/fabricate/repair damage
  (b) fabricate/blast/prime
  The former is usually cheaper but requires the use of a weldable prefabrication primer.
(6) Prefabrication primers have to be applied to blast-cleaned surfaces as thin films, usually 25 μm maximum. Their durability is therefore limited and further shop-coating is often desirable.
(7) Manual preparation methods are dependent upon weathering to loosen the mill-scale. These methods are therefore not usually appropriate for shop treatments. On site an adequate weathering period, usually several months, must be allowed.

(8) Many modern primers based on synthetic resins are not compatible with manually prepared steel surfaces since they have a low tolerance to rust and scale.

(9) Many oil-based and alkyd-based primers cannot be overcoated with finishing coats which contain strong solvents, e.g. acrylated rubbers, epoxies. bituminous coatings, etc.

(10) Two-pack epoxies have poor resistance to UV radiation and are highly susceptible to 'chalking'. Overcoating problems can arise with two-pack epoxies unless they are overcoated before the prior coat is fully cured. This is particularly relevant when an epoxy system is to be applied partly in the shops and partly on site.

(11) Steelwork which is to be encased in concrete does not normally require any other protection, given an adequate depth of concrete cover (British Standard BS 8110).

(12) Perimeter steelwork hidden in cavity walls falls into two categories:
   (a) Where an air gap (40 mm min.) exists between the steelwork and the outer brick or stone leaf, then adequate protection can be achieved by applying relatively simple painting systems.
   (b) Where the steelwork is in direct contact with the outer leaf, or is embedded in it, then the steel should be hot-dip galvanized and painted with a water-resistant coating, e.g. bitumen.

(13) Where fire-protection systems are to be applied to the steelwork, consideration must be given to the question of compatibility between the corrosion-protection and the fire-protection systems.

(14) New hot-dip galvanized surfaces can be difficult to paint and, unless special primers are used, adhesion problems can arise. Weathering the zinc surface before painting reduces this problem.

(15) Thermal spraying produces a porous coating which should be sealed by applying a thin low-viscosity sealant. Further painting is then optional.

(16) Particular attention should be paid to the treatment of weld areas. Flux residues and weld-spatter should be removed before application of coatings. In general the objective should be to achieve the same standard of surface preparation and coating on the weld area as on the general surface.

(17) Black-bolted joints require protection of the contact surfaces. This is normally restricted to the priming coat, which can be applied either in the shops or on site before the joint is assembled.

(18) For high-strength friction-grip bolted joints, the faying surfaces must be free of any contaminant or coating which would reduce the slip-factor required on the joint. Some metal-spray coatings and some inorganic zinc silicate primers can be used but virtually all organic coatings adversely affect the slip-factor.

35.9.2 Writing the specification

The specification is intended to provide clear and precise instructions to the contractor on what is to be done and how it is to be done. It should be written in a

logical sequence, starting with surface preparation, going through each paint or metal coat to be applied and finally dealing with specific areas e.g. welds. It should also be as brief as possible, consistent with providing all the necessary information. The most important items of a specification are as follows:

(1) The method of surface preparation and the standard required, which can often be specified by reference to an appropriate standard, e.g. BS 7079: Part A1 (ISO 8501–1), Sa2–3 qualities.
(2) The maximum interval between surface preparation and subsequent priming.
(3) The types of paint to be used, supported by standards where these exist.
(4) The method/s of application to be used.
(5) The number of coats to be applied and the interval between coats.
(6) The wet and dry film thickness for each coat.
(7) Where each coat is to be applied (e.g. shop or site) and the application conditions that are required, in terms of temperature, humidity, etc.
(8) Details for treatment of welds, connections, etc.
(9) Rectification procedures for damage, etc.

Possibly the most convenient method of presenting this information is in tabular form, and an example is shown in Table 35.3.

35.9.3 Inspection

Inspection must be carried out to ensure that the requirements of the specification are being met. Ideally this inspection should be carried out throughout the course of the contract at each separate phase of the work, i.e. surface preparation, first coat, second coat, etc. Instruments are available which can be used to assess surface roughness and cleanliness on blast-cleaned steel, wet-film thickness on paint coatings, dry-film thickness on paint coatings and metal coatings. It is suggested that an independent coating inspector qualified through the Institute of Corrosion's training and certification scheme be employed.

35.9.4 Environmental protection

In addition to the requirements for corrosion protection, there is increasing pressure being introduced by legislation to use paints and coatings that are 'environmentally friendly' and thereby minimize damage to the atmosphere. Only coatings that do not contain high quantities of organic solvents and toxic or harmful substances may be used. Following the introduction of the Environmental Protection Act (1990) the Secretary of State's Process Guidance Note PG6/23 *Coating of Metal and Plastic* was produced which has tables of maximum limits of volatile organic

Table 35.3 Protective coating specification for structural steelwork (example)

Contract no. – 1234/56/R
Client – J. Bloggs
Project title – Docklands Warehouse
Date – 27.6.90

Surface preparation
Shot-blast
Grit-blast
Other
Standard 8a 2½

Before fabrication or After fabrication
Maximum interval before overcoating – 4 hours
Level of inspection – Periodic
Other comments –

	1st coat	2nd coat	3rd coat	4th coat	
	Shop	Shop	Site	Site	
Brand name of paint Generic type	Excote 2 Zinc-rich epoxy	Excote HB ZP zinc phosphate epoxy	Excote HB CR HB chlor rubber	Excote CR Chlor rubber finish	Treatment of welds: Blast clean Sa 2½ and full system.
Reference number	1/597/P	5/643/u	3/124/u	4/510/F	Treatment of bolted connections: HSFG faying surface left bare. Edges sealed on site.
Specification	BS 4652 Type 3	N/A	N/A	N/A	
Supplier Method of application	A.N. Other Airless spray	A.N. Other Airless spray	A.N. Other Airless spray or brush	A.N. Other Airless spray or brush	
Wet film thickness	N/A	150 µm	250 µm	75 µm	Treatment of damaged areas: Abrade smooth or local blast cleaning. Full system.
Dry film thickness	20 µm	75 µm	75 µm	25 µm	
Spreading rate	10 m²/l	5 m²/l	2.5 m²/l	10 m²/l	
Minimum temperature	10°C	5°C	5°C	5°C	
Maximum relative humidity	95%	95%	95%	95%	Other comments: Site finish colour – BS 4800, 12 B 15.
Overcoating period, minimum	4 hours	24 hours	48 hours	N/A	
Overcoating period, maximum	7 days	7 days	N/A	N/A	
Inspection required	Full	Full	Random	Random	

compounds (VOCs). 'Compliant' coatings are those which contain high solids or water as the primary solvent or are solvent free.

Corus, in conjunction with paint and coatings manufacturers, has progressively removed from specifications products with identified harmful substances and moved towards the specification of EPA compliant coatings. Recent editions of the Corus Protection Guides for structural steelwork take due consideration of the changes brought about by the EPA and have included compliant coatings in the protection specifications where practical.

Further reading for Chapter 35

See Appendix *Basic data on corrosion* for extracts from corus *Corrosion Protection Guides.*

Corus Corrosion Protection Guides

The prevention of corrosion on structural steelwork
Steelwork exposed to exterior environments
Steelwork in building interiors and perimeter walls
Steelwork in indoor swimming pool buildings
Weather-resistant steels – uses and applications

Relevant standards

Surface preparation

BS 7079: Part 0: 1990 *Preparation of steel substrates before application of paints and related products.*
BS 7079: Part Al: 1989 (ISO 8501-1) *Preparation of steel substrates before application of paints and related products – visual assessment of surface cleanliness. Specification for rust and preparation grades of uncoated steel substrates after overall removal of previous coatings.*
BS 7079: Part Al: 1996 (ISO 8501-1) Supplement *Visual assessment of surface cleanliness – representative photographic examples of change of appearance imparted to steel when blast cleaned with different abrasives.*
BS 7079: Part A2: 1996 (ISO 8501-2) *Preparation of grades of previously coated steel substrates after localised removal of previous coatings.*
BS 7079: Parts B2, B3–B4 (ISO 8502/2-3-4) *Surface cleanliness, methods of assessment of surface cleanliness.*
BS 7079: Parts C1–C4 (ISO 8503/1 to 4) *Surface roughness characteristics of blast cleaned steel substrates.*
BS 7079: Parts Dl–D3 (ISO 8504/1 to 3) *Methods of surface preparation.*
BS 7079: Parts E1–E12 (ISO 11124/1 to 4 and 11125/1 to 7) *Metallic blast cleaning abrasives.*

BS 7079: Parts F1–F17: 1994 (ISO 11126/1 to 8 and 11127/1 to 7) *Non-metallic blast cleaning abrasives.*

Paints

BS 1070: 1993 *Black paint (tar based).*
BS 2015: 1992 *Glossary of paint and related terms.*
BS 3416: 1991 *Bitumen based coatings for cold application, suitable for use in contact with potable water.*
BS 3900: *Methods of tests for paints (46 parts).*
BS 4147: 1980 *Bitumen based hot applied coating materials for protecting iron and steel products, including suitable primers where required.*
BS 4164: 1987 (ISO 5256) *Coal tar based hot applied coating materials for protecting iron and steel products, including suitable primers where required.*
BS 4652: 1995 *Zinc rich priming paint (organic media).*
BS 6949: 1991 *Bitumen based coatings for cold application excluding use in contact with potable water.*

Metallic coatings

BS EN ISO 1461 (1999) *Hot dip galvanized coatings on fabricated iron and steel articles – Specifications and test methods.*
BS EN ISO 14713 (1999) *Protection against corrosion of iron and steel structures – zinc and aluminium coatings – Guidelines.*
BS EN 12329 (2000) *Corrosion protection of metals. Electroplated deposited coatings of zinc on iron and steel.*
BS EN 12330 (2000) *Corrosion protection of metals. Electroplated deposited coatings of cadmium on iron and steel.*
BS 3083: 1990 *Hot dip zinc coated, hot dip aluminium/zinc coated corrugated steel sheets for general purposes.*
BS 4921: 1988 *Specification for sherardized coatings on iron or steel.*
BS EN 10147: 2000 *Continuously hot dip metal coated steel sheet and strip. Technical delivery conditions.*
BS EN 22063: 1994 (ISO 2063) *Metallic and other inorganic coatings – thermal spraying – zinc, aluminium and their alloys.*
BS 3382: 1961 Parts 1 and 2 *Electroplated coatings on threaded components – cadmium and zinc on steel components.*
ISO 12944 Parts 1–8, 1998 *Paints and varnishes – Corrosion protection of steel structures by protective paint systems.*
BS PD 6484: 1979 *Commentary on corrosion at bimetallic contacts and its alleviation.*
BS 8110: Part 1: 1985 *Structural use of concrete. Code of practice for design and construction.*
BS EN ISO 14713 1999 *Design of articles that are to be coated. Design recommendations for hot-dip metal coatings.*
BS 4479: Part 7: 1990 *Design of articles that are to be coated. Recommendations for thermally sprayed coatings.*

Chapter 36
The Eurocodes

by COLIN TAYLOR and MIKE BANFI

36.1 The Eurocodes – background and timescales

In November 2001 the main parts of both Eurocode 3 on Steel Structures and Eurocode 4 on Composite Construction received a positive vote from their relevant CEN committees. There will be a period before they are published but this is a bureaucratic process and there are no more political or technical obstacles to their final publication. After a process that started over 25 years ago, the first fruits of the programme are being delivered. A total of five parts will appear in the next few years. They will cover all the main materials, i.e. concrete, timber masonry and aluminium as well as steel. They will also cover loading: dead, imposed, wind, snow, thermal, traffic together with accidental impact and explosion. Steel accounts for the largest number of parts. There are 20 that, as well as more typical subjects, cover stainless steel, cold-formed sections, shells, cables, silos, piling, and towers. There are also three parts for composite structures.

The aim of the Eurocodes is to 'establish a set of common technical rules for the design of buildings and civil engineering works which will ultimately replace the differing rules in Member States'. They are being produced by the European standards organization CEN. The actual work is being done by standards organizations in various countries and project teams set up for each part. Each country has a National Technical Contact for each part. The role of the NTC is to comment on drafts and, as far as the UK is concerned, liase with the relevant BSI committee. For most parts there has been an 'ENV' version, a 'pre-standard', which could be used for design with the relevant National Application Document. The recent phase has been conversion of the ENV into a full EN standard.

The process of getting 15 countries to agree on such a large number of technical issues has been long and hard. There is the rational argument that, as the laws of physics don't change, common codes should be possible. Against this is the practical argument, especially in countries with well developed codes, not to change a system that works. Not surprisingly there has not been total agreement. The codes identify areas that are subject to national determination. For these areas and these alone the member states can input their own values, which will be in National Annexes to be published with the code. The programme for each part is that, after the formal vote and process by CEN, the part is made available in English, French and German. Each member country then has up to two years to fix the Nationally Determined Parameters and adapt national provisions to allow the part to be used.

The part is then published by the national standards body together with the National Annex. Some countries will also have to translate the code during this period. After publication there is a formal coexistence period of up to 6 or 7 years, at the end of which conflicting national standards must be withdrawn. Any nation can decide to implement the parts in a shorter period than mentioned above. CEN plan to make all National Annexes available on a central website.

One important factor about the Eurocodes is that there will not be repetition. Values and properties will only be given in one code. This means that, for the design of composite structures, you will need EC0 for load factors, EC1 for loads, EC2 for concrete properties, EC3 for steel properties and EC4 for the design information. This will affect the date when the codes can be used, as all the relevant parts must be available. Another important factor is that the Eurocodes tend to give the rules for working out capacities but do not give precise equations. An example is the capacity of composite beams, where the stresses that can taken by the steel and concrete are given but there are no equations to work out the capacity. The latter is considered to be 'text book' material.

The Eurocodes include Principles that set out the basic requirements and Application Rules that give guidance on how to satisfy the Principles. The Principles must be complied with; in theory, alternative Application Rules are possible and can even be included in the National Annex, if they do not conflict with the published rules. However, if alternative application rules are used, the design cannot be said to be in accordance with the Eurocode.

There has been coordination between the project teams working on various parts. For the material codes, e.g. EC2, EC3 and EC4, there is a common layout for the first seven sections and, where possible, common wording has been used.

The Eurocodes will present opportunities by making it easier to work throughout Europe. There will need to be guidance and manuals to fill the gap between the new codes and the more detailed guidance to which we are accustomed in the UK. There may be some well-known methods that are not acceptable without further validation. This may be a chance to provide a solid background to certain methods. The recent work on the wind moment method for the design of frames shows what sort of thing can be done. Although the current UK codes will not be developed further, they will still be around and it is likely they will be acceptable for the design of most buildings for many years to come.

36.2 Conformity with EN 1990 – basis of design (EC O)

In principle all the structural Eurocodes have to conform with EN 1990. This is generally done by using direct references to it, rather than by repeating its provisions. Thus users need to use EN 1990 in defining the design actions and load combinations to be considered, including the values of the partial factors for actions.

In addition, the validity of as many as possible of the design expressions given in EC3 and EC4 have been established in conformity with the principles of structural

reliability defined in EN 1990, and the necessary values of the partial factors for the various resistances have been determined accordingly. This is intended to give a consistent, uniform level of structural reliability for designs in conformity with EC3 and EC4.

This level of reliability will depend upon the values of the nationally determined parameters for use in EN 1990 that will be required by the authorities of the country concerned for the relevant type of structure.

36.3 EC3 Design of steel structures

36.3.1 Scope

Steel has several key features that influenced the nature of a comprehensive standard for structural design:

- The versatility of steel requires that a large number of different type of structure have to be covered.
- The ductility of steel requires the inclusion of both elastic and plastic design.
- The key role of joints is reflected by including design methods for bolted and welded joints.
- Slender components commonly occur and require the inclusion of comprehensive approaches to the stability of structures, members, flat plates and shells.

These key features combine to produce a need for a significantly longer document to cover all aspects of the design of steel structures to the same level of detail as the other structural materials.

For example, the scope of EC3 has to cover that of BS 5950: Parts 1, 5, 6 and 9 for buildings; BS 5400, Parts 3, 9 and 10 for bridges; BS 2573 Part 1 for cranes; BS 2853 for runway beams; BS 4076 for steel chimneys; BS 4604 for HSFG bolts; and BS 8100, Part 3 for lattice towers and masts. It also has to cover many other topics that have not until now featured in a British Standard, such as design of cables, semi-rigid joints, tubular joints, bearing piles, sheet piling, tanks, hoppers, silos, pipelines and shell structures, plus design of structures using stainless steel. Several of these topics have not previously been codified. In addition, advances in knowledge of such topics as resistance to fatigue, low temperature brittle fracture and design for through-thickness stresses call for more comprehensive treatment than before.

36.3.2 Contents

Like most of the other structural Eurocodes, EC3 has developed in stages. The earliest documents seeking to harmonize design rules between European

countries were the various Recommendations published by ECCS. From these were developed the initial draft Eurocode 3 published by the European Commission, followed by the various parts of ENV 1993 issued by CEN. The best known is Part 1.1 *General rules and rules for buildings*; this is what is often thought of as EC3 but is only the first of a total of 18 documents issued as ENVs.

However, attention must now turn to the forthcoming EN version, which will have full status as British Standard BS EN 1993. This is also being developed in a number of parts. Part 1 will probably be called *Generic rules* and will be sub-divided into 11 sub-parts dealing with different types of steel components. Each of the other parts will then cover the application of the relevant generic rules to various types of structure.

The various parts are listed in Tables 36.1 and 36.2. At present only a minority of drafts are nearing completion, so the titles should be understood as indicating the intended scope, rather than precisely defining the final titles. The position may change, but it seems probable that the total length of the nine application documents comprising Parts 2 to 7 will be far less than that of Part 1.

Table 36.1 Components of EC3 Design of Steel Structures Part 1 – generic rules

Part no.	Generic rules
1.1	Common rules
1.2	Fire design
1.3	Cold-formed thin gauge members and sheeting
1.4	Stainless steels
1.5	Stiffened plating subject to in-plane loading
1.6	Steel shells
1.7	Stiffened plating subject to out-of-plane bending
1.8	Joints
1.9	Fatigue
1.10	Fracture toughness and through-thickness properties
1.11	Cables

Table 36.2 Application of the generic rules from EC3 to various structure types

Parts	Application
2	Bridges
3	Buildings
4.1	Tanks
4.2	Silos
4.3	Pipelines
5	Piling
6	Crane supporting structures
7.1	Towers and masts
7.2	Chimneys

36.3.3 Design rules

Material resistance

The design resistances of steel members and other steel components is generally related to the yield strength of the material. However, in the case of both bolted connections and welded connections, the design strength is related to the tensile strength (i.e. the UTS). This is also relevant when considering net cross-section resistance at bolt holes.

Thus generally the design resistance R_d of a cross-section needs to be related to a characteristic value of the yield strength f_y using:

$$R_d = Af_y/\gamma_M$$

in which A is the cross-section area.

For convenience, the characteristic yield strength f_y is taken as equal to the specified 'guaranteed' minimum value for yield strength R_{eH} given in the relevant product standard for structural steel, generally EN 10025. This is not the 'true' characteristic strength (based on 95% or any other statistically determined level of probability), which clearly will be higher. More importantly, the value of γ_M needs to cover the variability of the section properties represented by A as well as that of the yield strength f_y.

An ECSC funded study of the variability of section properties, yield strength and cross-section resistance using large numbers of samples of current production at a variety of European (including British) steel mills found that the mean value of A is approximately equal to the nominal value, but that the mean value of f_y is about 1.2 times the nominal value. In addition, the shape of the distribution curve showing its variability is markedly skewed, being noticeably truncated on the negative side, presumably due to the process control procedure.

Reliability analysis related to the EN 1990 target reliability β, and taking account of the actual distribution of yield strength and section properties, leads to a value of R_d related to the EN 1990 target reliability $\beta = 3.8$ that corresponds to a required value of γ_M of slightly less than 1.0.

In simplified terms, the mean yield strength is about 1.2 times the nominal value: thus the mean resistance is also about 1.2 times the nominal resistance Af_y (see Fig. 36.1). The precise value of the characteristic resistance is not important, because the design resistance has been determined directly from the data in conformity with EN 1990. This design resistance turns out to be equal to the nominal resistance, hence $\gamma_M = 1.0$.

It is important to note that this is only possible because this value of $\gamma_M = 1.0$ is used exclusively with the nominal 'guaranteed minimum' value of the yield strength. For steel reinforcement the yield strength used in design is a characteristic value, and a larger value of γ_M is needed to obtain the design value.

Table 36.3 The differences in axes between BS 5950-1 and the Eurocodes

Axes	BS 5950-1	Eurocodes
Along the member		X
Major axis	X	Y
Minor axis	Y	Z

Table 36.4 Comparison of frequently used symbols in BS 5950-1 and the Eurocodes

	BS 5950-1	Eurocodes
Area	A	A
Elastic modulus	Z	W_{el}
Plastic modulus	S	W_{pl}
Inertia about major axis	I_x	I_y
Inertia about minor axis	I_y	I_z
Warping constant	H	I_w
Torsion constant	J	I_t
Radius of gyration	r	i
Applied axial force	F	N
Resistance to axial force	P	N_{Rd}
Bending moment	M	M
Applied shear force	F_v	V
Shear resistance	P_v	V_{Rd}
Yield stress	p_y	f_y
Bending strength	p_b	$\chi_{LT}f_y$
Compressive strength	p_c	χf_y

Axes and notation

Tables 36.3 and 36.4 summarize the differences in axes and notation that exist between BS 5950 and the Eurocodes.

Compression members

As with BS5950-1, multiple strut curves are defined, depending on the type of cross-section and the axes of buckling. The formulations for these differ only slightly from current UK practice. However, the design process differs in detail. The member slenderness is first converted to a relative slenderness, as in BS 5400-3, thereby eliminating the need for the numerous pages of tables used in BS 5950-1. This leads to a reduction factor χ which is applied to the cross-section resistance to obtain the member buckling resistance.

Beams

For restrained beams, the cross-section resistance is based on the plastic modulus, provided that the cross-section is classified as class 1 (plastic) or class 2 (composite), just as in BS 5950-1, except for detailed differences in the limiting values for the different classes.

For laterally unrestrained beams, there are alternative methods. A method based on the column curve may be used for all members; this comprehensive approach gives lower resistances than those in BS 5950-1 for rolled sections.

There is a specific approach for rolled beams that gives higher resistances than those in BS 5950-1. The procedure is more similar to that of BS 5400-3 than to BS 5950-1. It also introduces the intermediate step of using the relative slenderness rather than the resistance to cover design cases with varying moments. (In terms familiar to users of the 1990 version of BS 5950-1, it uses n-factors not m-factors.)

Members subject to combined axial force and bending moments

There are two methods. A general method examines the overall lateral slenderness of the frame and then determines the resistance as a function of the ultimate resistance without buckling (e.g. the plastic collapse load) reduced by a buckling factor.

A more complex interaction method is also available. This considers the design of individual members separately. It differentiates between members that are not susceptible to torsional deformations and those that are. For each of these classes there are less complicated and more complicated methods. Even for the simplest case, i.e. members not susceptible to torsional deformation and treated more simply, the resulting equations are too complex for manual design and will require carefully prepared design aids, or software, before they can be implemented in practice without a significant increase in design time.

Connections

Connections are treated in a separate part of EC3 and in considerably more detail than is given in BS 5950-1.

In addition to conventional information on basic strengths, connection analysis is treated in considerable detail, at least for certain classes of connection. There is explicit consideration of both the strength and stiffness of the connections and also a recognition of their effect on overall structural behaviour.

Frame stability

Frame stability is checked by considering the effects of imperfection on both the global analysis and the performance of any bracing system. For the former, the sway

imperfections at height/200 (or an equivalent horizontal force) are similar to BS 5950-1. Imperfections of the bracing system are both more rigorous and require more design effort than current UK practice.

Frame analysis

The analysis is required to take account of the effects of the deformed geometry on the structure. First order analysis may be used where the increases in internal forces and moments are less than 10%. This implies a critical load ratio (λ_{cr}), under factored loading, of more than 10; it is equivalent to the BS 5950-1: 2000 'non-sway' frame approach.

A range of second order approaches may be used for structures for which γ_{cr} is less than 10. These methods include:

- effective length approaches
- amplification factors
- energy methods
- formal second order analysis.

36.3.4 Supporting standards

Relevant supporting standards include those for:

- Steel properties
- Tolerances on steel sections and plates
- Ordinary bolts
- HSFG bolts
- Welding electrodes
- Fabrication and erection.

Steel properties

Properties of structural steels are currently given in EN 10025, EN 10113, EN 10137, EN 10155, EN 10210 and EN 10219, the last two covering structural hollow sections. A new edition of EN 10025 is planned that will combine all of these except steel for hollow sections (and may include these also). All of these have been adopted as BS EN standards.

Further BS EN standards are available covering steel sheets and strip, both plain and galvanized, for use in cold-formed sections and sheeting. They are already in use in connection with BS 5950 Parts 5, 6, 7 and 9.

Tolerances on steel sections and plates

Tolerances for plates and for each type of section are similarly given in numerous EN standards adopted as BS ENs. Details are given in BS 5950 Part 2.

Ordinary bolts

Bolts, nuts and washers are covered in ISO standards in three series commencing with ISO 4014, ISO 4032 and ISO 7089, that have been adopted as BS EN ISO standards.

HSFG bolts

A new series of standards for high-strength bolts suitable for using as preloaded fasteners in friction grip applications has been under development for some years. It is hoped these may be finalized and agreed by 2002, but this remains to be seen.

Welding electrodes

A range of EN standards have been developed covering all types of steel welded electrodes, using common strength grades, and adopted as BS ENs. Details are given in BS 5950-2.

Fabrication and erection

The rules for execution (fabrication and erection) of steel structures originally developed as part of EC3 under the aegis of the European Commission were separated from the ENV version of the design rules and became ENV 1090-1. Parts 2 to 6 of ENV 1090 were subsequently added to cover cold-formed thin gauge sections and sheeting, higher strength steels, structural hollow sections, bridges and stainless steels. These are now being converted to EN status as a single EN 1090.

36.4 EC4 Design of composite steel and concrete structures

36.4.1 Scope and contents

Eurocode 4 will come in three parts: 1.1 *General rules for buildings*, 1.2 *Structural fire design* and 2 *Bridges*. This section covers Part 1.1 and compares it with BS 5950 Parts 3.1 and 4. It is based on the latest available draft so some details may change,

but the overall concepts are now well established. For certain parameters recommended values will be given, and it will be left to national bodies to decide whether to accept these or choose something different. Where necessary, recommended values have been used in this section. The plan is for Parts 1.1 and 1.2 to be made available by CEN in the first half of 2003, with Part 2 following a year later.

Eurocode 4 covers more than BS 5950 Part 3.1. It includes rules for partially encased sections and, more importantly, includes guidance on composite columns. As well as individual elements, the EC3 rules for moment connections are extended to composite joints. The basis for the work was the ENV version published in 1994, but there have been significant changes since then. Guidance on composite slabs equivalent to that in BS 5950 Part 4 is included in EC4 Part 1.1.

Taking an overall view EC4 will give very similar results to BS 5950 for composite beams and slabs. For composite columns the resistance will be larger than that determined by current UK guidance. One area where EC4 is more conservative than BS 5950 is in the capacity of stud shear connections.

36.4.2 Design rules

Materials

The Eurocode covers a wider range of properties than BS 5950. Steel grades S420 and S460 are included, as are concrete grades up to a cube strength of 75 MPa.

For concrete, the limiting stress used in EC4 is 85% of the design strength. According to EC2 the concrete design strength (f_{cd}) is $\alpha f_{ck}/\gamma_c$, where f_{ck} is the cylinder strength. The recommended value for the material safety factor γ_c is 1.5. To compare EC4 and BS 5950 you also need to account for the relationship between cube and cylinder strength. Typical values, given in Table 36.5, show that the limiting stresses are very similar.

For steel, both for sections and reinforcement, the limiting stress in EC4 is the design strength. For structural steel, EC3 recommends that this is equal to the nominal yield stress, but for reinforcement EC2 recommends a partial factor of 1.15

Table 36.5 Comparison of concrete strength in BS 5950 and EC4

Cube strength (MPa)	BS 5950 limiting stress	EC4					
		Normal weight concrete			Lightweight concrete		
		Cylinder strength	Limiting stress	EC4/BS 5950	Cylinder strength	Limiting stress	EC4/BS 5950
25	11.25	20.0	11.33	1.01	22.5	10.84	0.96
30	13.50	25.0	14.17	1.05	27.0	13.00	0.96
35	15.75	28.6	16.21	1.03	32.0	15.41	0.98
40	18.00	31.9	18.10	1.00	36.7	17.68	0.98

on the characteristic strength. The limiting stresses are therefore identical to those in BS 5950.

Beam design

For simply-supported secondary beams in buildings, the effective width limit of span/8 is the same as BS 5950. For primary beams, EC4 does not include the '0.8*b*' limit in the BS. For continuous beams the equivalent spans are slightly different from those in the BS, and there is no simplified method for calculating moments. Redistribution of moments in continuous beams is allowed in EC4, with similar limits depending on the class of steel section.

The plastic capacity of beams is based on rectangular stress blocks with the limiting stresses given above. For high-strength steels, i.e. S420 and S460, there are additional requirements. The capacities will therefore be very similar to those calculated to BS 5950. There is no requirement to check stresses under working loads for beams in buildings.

When calculating deflections, partial interaction only needs to be taken into account if it is less than 50%. However, shrinkage must be considered, unless the ratio of the span to overall depth is less than 20 and the free shrinkage strain of the

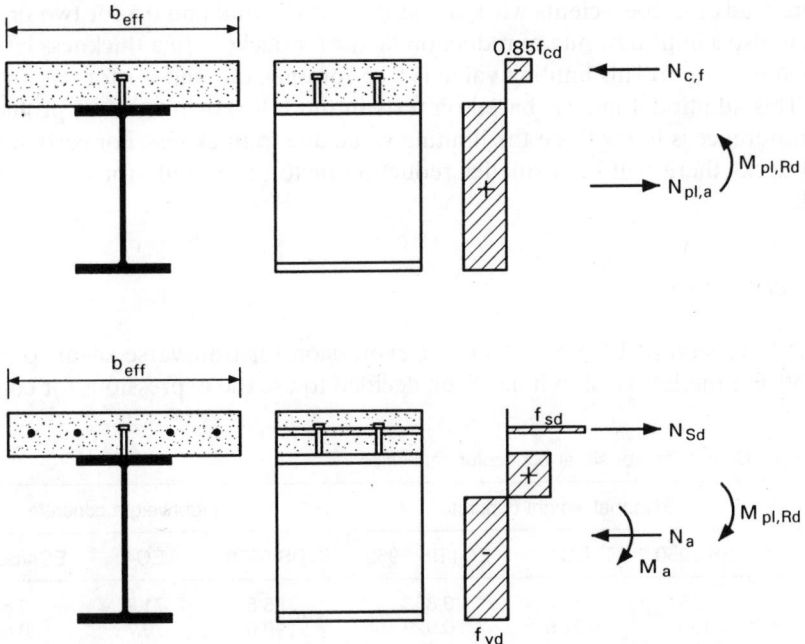

Fig. 36.1   Rectangular stress blocks for simply-supported composite beams

concrete is less than 400×16^{-6}. This will affect a lot of beams, particularly those with lightweight concrete.

Shear connection

The ENV version and initial drafts included many types of shear connections. It was decided to limit the standard to headed studs. Other connectors are allowed but no application rules are given.

Expressions are given for the capacity of headed studs. The governing expression is usually that based on the concrete but there is a limiting value based on the strength of the steel in the stud. For lightweight concrete, the capacity is reduced in the ratio of the density of the concrete compared with the 10% reduction given in BS 5950. For the typical 19/100 mm stud, characteristic capacities in kN are shown in Table 36.6.

To calculate the design capacity, the recommended partial safety factor is 1.25. The relationship between design strengths to EC4 and BS 5950 will therefore be exactly the same for positive moment regions. For negative moment regions, the 0.6 factor in BS 5950 means that the design capacity to EC4 will be relatively higher.

Like BS 5950, EC4 includes reduction factors for the stud capacity based on the geometry of the decking. For decking parallel to the beam the factor is identical to that in the BS. For transverse decking the factor is generally similar but there are two differences. Instead of the coefficients 0.85, 0.6 and 0.5 for one, two and three or more studs, the coefficients work out as 0.7 for one stud and 0.5 for two or more. There is also a limiting value of reduction factor for decks with a thickness less than or equal to 1.0 mm. This limiting value is 0.85 for one stud and 0.7 for two or more studs. This additional limit is based on tests in the UK. For re-entrant profiles the main difference is likely to be the limiting value due to thickness. For certain trapezoidal decks there will be a smaller reduction factor, especially for two studs per trough.

Transverse shear

The ENV version of EC4 had a similar expression for transverse shear to that in BS 5950. For the EN version it has been decided to use the expressions for concrete

Table 36.6 Comparison of shear connector capacities

Cube strength	Normal weight concrete			Lightweight concrete		
	BS 5950	EC4	EC4/BS 5950	BS 5950	EC4	EC4/BS 5950
25	95	81.0	0.853	85.5	71.2	0.833
30	100	92.9	0.929	90.0	79.7	0.885
35	104	100.9	0.970	93.6	88.5	0.945
40	109	102.1	0.936	98.1	96.3	0.982

tee-beams in EC2. The transverse shear capacity is typically dependent on the amount of reinforcement, though there is a cut off value that varies with concrete strength. The decking can contribute to the transverse shear capacity when it is perpendicular to the beam. The contribution is similar to that in BS 5950 but it only reduces the amount of reinforcement required and does not increase the cut-off strength based on the concrete strength.

The variation of the transverse shear capacity with percentage reinforcement for normal and lightweight concrete is shown in Figures 36.2 and 36.3.

It can be seen that the transverse shear capacity to EC4 is typically greater than that to BS 5950.

Composite slabs

EC4 Part 1.1 also covers composite slabs with profiled sheeting for buildings. The guidance is very similar to that in BS 5950 Part 4 but there are some differences in the details. There is no specific mention of testing as an alternative to the application rules as a design procedure.

The bearing distance at end supports for the slab should not be less than 70 mm for bearings on steel or concrete and 100 mm for other materials. This is for the slab; the bearing for the decking can be less, i.e. 70 and 50 mm respectively.

The bending capacity of the slab is based on similar limiting stresses to those used for composite beams, $0.85f_{ed}$ for concrete and $f_{yp,d}$ for the decking.

A continuous slab may be designed as a series of simply supported spans. In this case the minimum reinforcement is 0.2% of the concrete area above the ribs rather

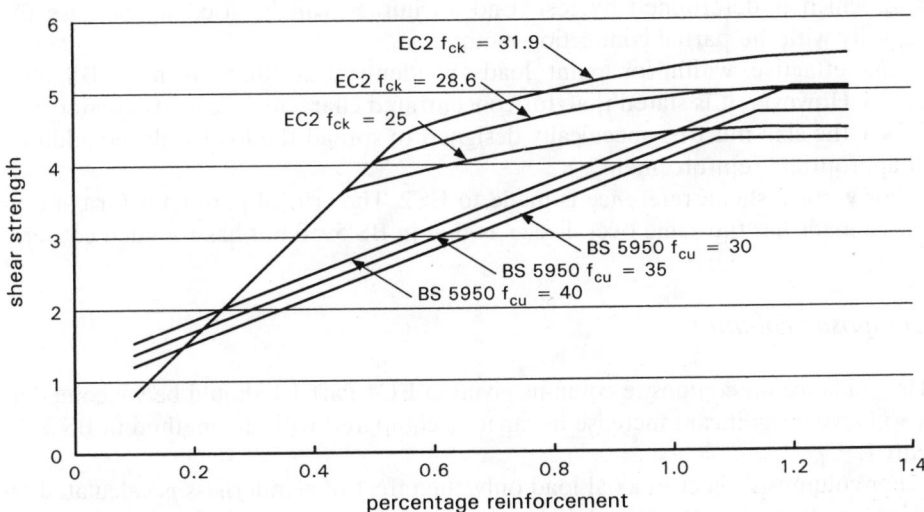

Fig. 36.2   Treatment of transverse shear for normal weight concrete to EC4 and BS 5950 Part 3.1

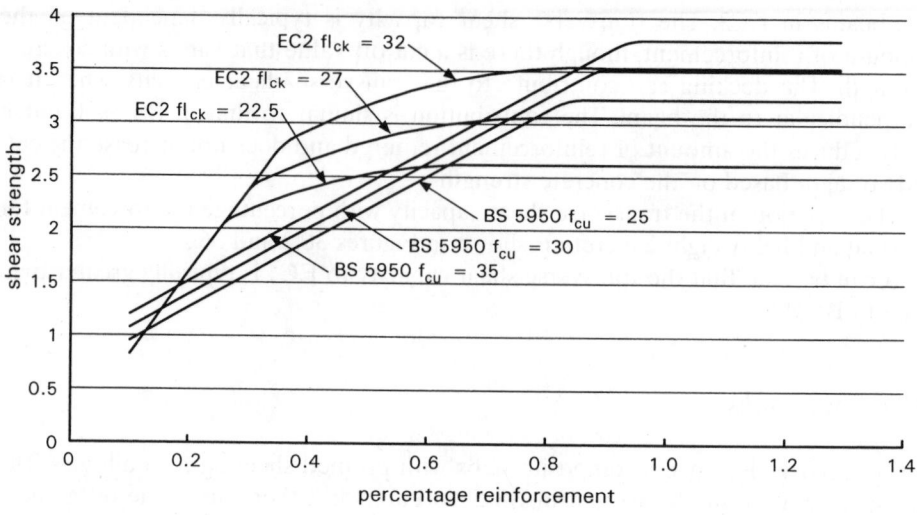

Fig. 36.3 Treatment of transverse shear for lightweight concrete to EC4 and BS 5950 Part 3.1

than 0.1% of the gross concrete area as given in BS 5950 Part 4. Where the decking is propped, the minimum reinforcement is 0.4%.

To calculate the shear bond capacity the m–k method can be used in an identical manner to BS 5950 but no increase in capacity is allowed due to end anchorage. An alternative 'partial connection method' is given which is similar to the design of beams with partial interaction. This method relies on a design shear bond strength $\tau_{u,Rd}$, which is determined by test. End anchorage can be used to increase the capacity with the partial connection method.

The effective width for point loads is identical to that given in BS 5950 Part 4. However, it is stated that, for concentrated characteristic loads greater than 7.5 kN, the slab must be specifically designed to spread the load with the addition of appropriate reinforcement.

For vertical shear, reference is made to EC2. The critical perimeter for concentrated loads has the same overall size as that in BS 5950 but has rounded corners.

Composite columns

The guidance on composite columns given in EC4 Part 1.1 should be welcomed as it will give a significant increase in capacity compared with the method in BS 5950 Part 1.

For columns subject to axial load only, the effect of slenderness is calculated by using a reduction factor from EC3, i.e. it is designed as an equivalent steel column. This is a similar method to BS 5950. The calculation of the appropriate slenderness is more accurate but more complicated.

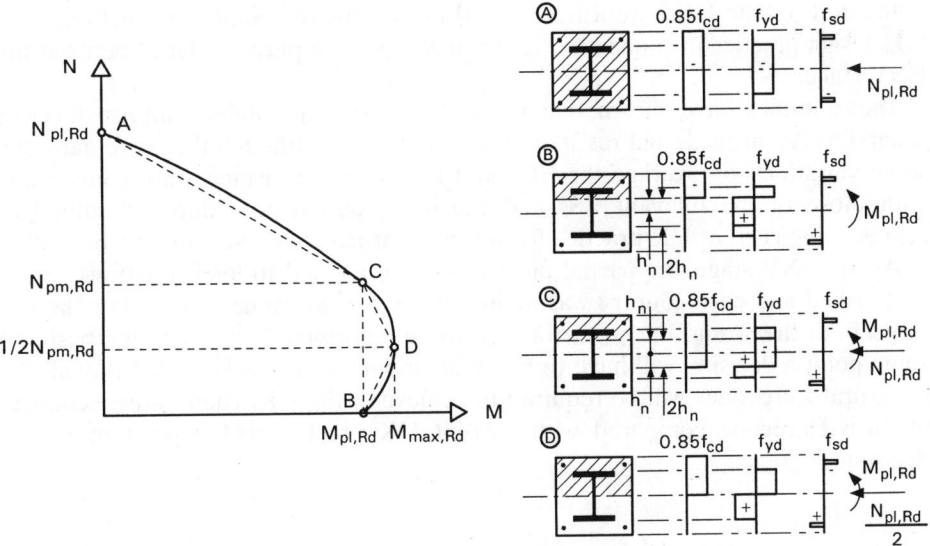

Fig. 36.4   Strength of composite column cross-sections to EC4

For members in combined compression and bending, a 'local capacity' check is carried out for axial force and moment where the moment includes effects due to slenderness as shown in Fig. 36.4. The local check is based on rectangular stress blocks with the normal limiting stresses and produces an interaction curve similar to that for reinforced concrete columns.

Composite joints

The section on joints in EC4 Part 1.1 covers moment connections where there is some composite action. This can be by the use of reinforcement in the slab to take tension or by encasement of the column to increase the shear and compression capacity of the column web. Rules are given for the strength and stiffness of these joint components. The stiffness values are in an 'informative' annex to reflect the fact that these are estimates and not hard and fast values.

36.5 Implications of the Eurocodes for practice in the UK

Extensive calibrations of the design rules in the various parts of EC3 and EC4 have been carried out in connection with their drafting, as part of the process of determining the values of γ_M needed to provide the uniform level of reliability represented by the target reliability β given in EN 1990.

In addition, the UK authorities carried out extensive comparisons with current UK design practice in connection with the National Application Documents at the ENV stage.

These studies have shown that the rules provide a suitable uniform level of reliability. As far as design resistances are concerned, although there are naturally some variations, on average the EC3 and EC4 design resistances are comparable with those of the relevant BSs and the other design procedures identified as representing current UK practice for forms of structure not yet covered by a BS.

At the ENV stage, the partial factors γ_F to be applied to loads ('actions') were established and each country was invited to fix its own values for γ_M. For the EN stage, both the γ_F and the γ_M factors are to be determined on a national basis. At some point a decision will have to be made about which level of reliability all the structural Eurocodes will be required to achieve in the UK. The relative economy of each Eurocode compared with current UK practice will depend upon this decision.

36.6 Conclusions

The scopes of Eurocode 3 and 4 are very wide in terms of types of structure, forms of construction, methods of design and materials. Much of their content has not been covered previously, particularly in Eurocode 3.

In relation to common types of building using rolled steel sections, the differences in results are only minor, though the design procedures are sometimes more tedious.

The real differences compared with existing British Standards lie in the new methods. For buildings, detailed methods for semi-rigid joints are given. For cold-formed steelwork, more advanced methods of design are included. Rules for stainless steels appear for the first time in a standard. The rules for shells and for the design of piles, sheet piling and silos are all new.

By presenting all the rules for as wide a variety of steel and composite designs as possible in a consistent format, the specialist designer will be given a greater versatility than ever before.

Appendix

Elastic properties of steel

Modulus of elasticity (Young's modulus)	$E = 205\,\text{kN/mm}^2$
Poisson's ratio	$v = 0.30$
Coefficient of linear thermal expansion	$\alpha = 12 \times 10^{-6}$ per °C

European standards for structural steels

Introduction

As part of the exercise towards the removal of technical barriers to trade, the European Committee for Iron and Steel Standardization (ECISS) has prepared a series of European Standards (ENs) for structural steels. The first of these standards, EN 10025, was published in the UK by BSI as BS EN 10025: 1990, partly superseding BS 4360: 1986, which was re-issued as BS 4360: 1990. In 1993, a second edition of BS EN 10025 was made available together with BS EN 10113: Parts 1, 2 and 3 and BS EN 10155. In June 1994, BS EN 10210: Part 1 was published and at the same time BS 4360 was officially withdrawn. The balance of the BS 4360 steels not affected by these ENs were re-issued in new British Standards BS 7613 and BS 7668. In 1996, with the publication of BS EN 10137, BS 7613 was withdrawn. BS 7668 will remain until an EN for atmospheric corrosion resistant hollow sections is available.

Designation systems

The designation systems used in the EN are in accordance with EN 10027: Parts 1 and 2, together with ECISS Information Circular IC 10 (published by BSI as DD 214). These designations are totally different from the familiar BS 4360 designations: therefore, the following is intended to help users understand them.

Table 1　European and British Standards which have superseded BS 4360

Standard	Superseded BS 4360 grades
BS EN 10025: 1993	40 A, B, C, D; 43 A, B, C, D; 50 A, B, C, D, DD
BS EN 10113: Parts 1, 2 & 3: 1993	40 DD, E, EE; 43 DD, E, EE; 50 E, EE; 55 C, EE
BS EN 10137: Parts 1, 2 & 3: 1996	50 F and 55 F
BS EN 10155: 1993	WR 50 A, B, C
BS EN 10210: Part 1: 1994	Hot-finished structural hollow section grades – excluding weather resistant grades
BS 7668: 1994	Hot-finished weather resistant hollow section grades

Table 2a   Symbols used in EN 10025

S....	Structural steel
E....	Engineering steel
.235...	Minimum yield strength (R.) in N/mm² @ 16 mm
...JR..	Longitudinal Charpy V-notch impacts 27J @ +20°C
...J0..	Longitudinal Charpy V-notch impacts 27J @ 0°C
...J2..	Longitudinal Charpy V-notch impacts 27J @ –20°C
...K2..	Longitudinal Charpy V-notch impacts 40J @ –20°C
....G1	Rimming steel (FU)
....G2	Rimming steel not permitted (FN)
....G3	FLAT products: Supply condition 'N', i.e. normalized or normalized rolled. LONG products: Supply condition at manufacturer's discretion
....G4	ALL products: Supply condition at manufacturer's discretion

Examples: S235JRG1, S355K2G4

Table 2b   Symbols used in EN 10155

S....	Structural steel
.235...	Minimum yield strength (R.) in N/mm² @ 16 mm
...J0..	Longitudinal Charpy V-notch impacts 27J @ 0°C
...J2..	Longitudinal Charpy V-notch impacts 27J @ –20°C
...K2..	Longitudinal Charpy V-notch impacts 40J @ –20°C
....G1	FLAT products: Supply condition 'N', i.e. normalized or normalized rolled. LONG products: Supply condition at manufacturer's discretion
....G2	All products: Supply condition at manufacturer's discretion
....W	Weather resistant steel
....P	High phosphorus grade

Examples: S235J0WP, S355K2G2W

Table 2c   Symbols used in EN 10113

S....	Structural steel
.275..	Minimum yield strength (R.) in N/mm² @ 16 mm
....N.	Normalized or normalized rolled
....M.	Thermomechanically rolled
.....L	Charpy V-notch impacts down to –50°C

Examples: S275N, S355ML

Table 2d   Symbols used in EN 10137

S....	Structural steel
.460..	Minimum yield strength (R.) in N/mm² @ 16 mm
....Q.	Quenched and tempered
.....L	Charpy V-notch impacts down to –40°C
.....L1	Charpy V-notch impacts down to –60°C

Examples: S460QL, S620QL1

Table 3 Comparison between grades in EN 10025: 1993 and BS 4360: 1986

	EN 10025: 1993						BS 4360: 1986					
Grade	Tensile strength (R_m) at $t=16$ mm (N/mm²)	Min yield strength (R_e) at $t=16$ mm (N/mm²)	Max thickness for specified yield strength (R_e) (mm)	Temp (°C)	Charpy V-notch impacts (longitudinal) Energy (J) ≤150 mm (1)	Energy (J) >150 mm ≤250 mm	Grade	Tensile strength (R_m) at $t=16$ mm (N/mm²)	Min yield strength (R_e) at $t=16$ mm (N/mm²)	Max thickness for specified yield strength (R_e) (mm) (2)	Temp (°C)	Charpy V-notch impacts (longitudinal) Energy (J) ≤100 mm (3)
S185 (4)	290/510	185	25	–	–	–	–	–	–	–	–	–
S235 (5)	340/470	235	250	–	–	–	40A	340/500	235	150	–	–
S235JR (4)	340/470	235	25	+20 (6)	27	–	–	–	–	–	–	–
S235JRG1 (4)	340/470	235	25	+20 (6)	27	–	–	–	–	–	–	–
S235JRG2	340/470	235	250	+20 (6)	27	–	40B	340/500	235	150	+20 (6)	27
S235J0	340/470	235	250	0	27	23	40C	340/500	235	150	0	27
S235J2G3	340/470	235	250	–20	27	23	40D	340/500	235	150	–20	27
S235J2G4	340/470	235	250	–20	27	23	40D	340/500	235	150	–20	27
S275 (5)	410/560	275	250	–	–	–	43A	430/580	275	150	–	–
S275JR	410/560	275	250	+20 (6)	27	23	43B	430/580	275	150	+20 (6)	27
S275J0	410/560	275	250	0	27	23	43C	430/580	275	150	0	27
S275J2G3	410/560	275	250	–20	27	23	43D	430/580	275	150	–20	27
S275J2G4	410/560	275	250	–20	27	23	43D	430/580	275	150	–20	27
S355 (5)	490/630	355	250	–	–	–	50A	490/640	355	150	–	–
S355JR	490/630	355	250	+20 (6)	27	23	50B	490/640	355	150	+20 (6)	27
S355J0	490/630	355	250	0	27	23	50C	490/640	355	150	0	27
S355J2G3	490/630	355	250	–20	27	23	50D	490/640	355	150	–20	27
S355J2G4	490/630	355	250	–20	27	23	50D	490/640	355	150	–20	27
S355K2G3	490/630	355	250	–20	40	33	50DD	490/640	355	150	–30	27
S355K2G4	490/630	355	250	–20	40	33	50DD	490/640	355	150	–30	27
E295	470/610	295	250	–	–	–	–	–	–	–	–	–
E335	570/710	335	250	–	–	–	–	–	–	–	–	–
E360	650/830	360	250	–	–	–	–	–	–	–	–	–

(1) For sections up to and including 100 mm only

(2) For wide flats and sections up to and including 63 mm and 100 mm respectively.

(3) For wide flats up to and including 63 mm and for sections no limit stated.

(4) Only available up to and including 25 mm

(5) The steel grades S235, S275 and S355 appear only in the English language version (BS EN 10025) as non-conflicting additions, and do not appear in other European versions

(6) Verification of the specified impact value is only carried out when agreed at the time of enquiry and order.

Table 4 Comparison between grades in EN 10113: Part 2 (Part 3): 1993 and BS 4360: 1990

	EN 10113: Part 2 (Part 3): 1993								BS 4360: 1990								
Grade	Tensile strength (Rm) at t=16mm (N/mm²)	Min yield Strength (Re) at t=16mm (N/mm²)	Max thickness for specified yield strength (Re) (mm) (1)	(2)	Charpy V-notch impacts (longitudinal) – see Table 4a — Temp (°C)	Energy (J)	Max thickness (mm) (1)	(2)	Grade	Tensile strength (Rm) at t=16mm (N/mm²)	Min yield strength (Re) at t=16mm (N/mm²)	Max thickness for specified yield strength (Re) (mm) (1)	(2)	Charpy V-notch impacts (longitudinal) — Temp (°C)	Energy (J)	Max thickness (mm) (1)	(2)
S275N (M)	370 (360) to 510 (510)	275	150 (63)	150 (150)	-20	40	150 (63)	150 (150)	43DD (3)	430 to 580	275	–	100	-30	27	–	(9)
S275NL (ML)		275	150 (63)	150 (150)	-50	27	150 (63)	150 (150)	43EE (4)		275	150 (5)	–	-50	27	75 (7)	–
S355N (M)	470 (450) to 630 (610)	355	150 (63)	150 (150)	-20	40	150 (63)	150 (150)	50E (3)	490 to 640	355	–	100	-40	27	–	(9)
S355NL (ML)		355	150 (63)	150 (150)	-50	27	150 (63)	150 (150)	50EE (4)		355	150 (5)	–	-50	27	75 (8)	–
S420N (M)	520 (500) to 680 (660)	420	150 (63)	150 (150)	-20	40	150 (63)	150 (150)	–		–	–	–	–	–	–	–
S420NL (ML)		420	150 (63)	150 (150)	-50	27	150 (63)	150 (150)	55C (3)	550 to 700	450	25	40	0	27	25	19
S460N (M)	550 (530) to 720 (720)	460	100 (63)	100 (150)	-20	40	100 (63)	100 (150)	–		–	–	–	–	–	–	–
S460NL (ML)	720 (720)	460	100 (63)	100 (150)	-50	27	100 (63)	100 (150)	55EE (4)		450	63 (6)	–	-50	27	63 (6)	–

Table 4a Longitudinal Charpy V-notch impacts

Grade	Min ave energy (J) at test temp (°C)						
	+20	0	-10	-20	-30	-40	-50
S_ _N (M)	55	47	43	40			
S_ _NL (ML)	63	55	51	47	40	31	27

(1) Applies to plate and wide flats
(2) Applies to sections
(3) Supply condition M by agreement
(4) Supply condition M not permitted
(5) For wide flats max thickness is 63 mm
(6) Not available as wide flats
(7) For wide flats max thickness is 50 mm
(8) For wide flats max thickness is 30 mm
(9) For sections no thickness limit is given

Table 5 Comparison between grades in EN 10137: Part 2: 1996 and BS 4360: 1990

	EN 10137: Part 2: 1996						BS 4360: 1990						
Grade	Tensile strength (R_m) at t = 16 mm (N/mm²)	Min yield strength (R_e) at t = 16 mm (N/mm²)	Max thickness for specified yield strength (R_e) (mm)	Charpy V-notch impacts (longitudinal) Temp (°C)	Energy (J)	Max thickness (mm)	Grade	Tensile strength (R_m) at t = 16 mm (N/mm²)	Min yield strength (R_e)	Max thickness for specified yield strength (R_e) (mm)	Charpy V-notch impacts (longitudinal) Temp (°C)	Energy (J)	Max thickness (mm)
S460Q	550 to 720	460	150	−20	27	150	50 F	490–640	390	40	−60	27	40
460QL		460	150	−40	27	150	–	–	–	–	–	–	–
S460QL1		460	150	−60	27	150	55F	550–700	450	40	−60	27	40

EN 10137: Part 2 also contains Grades S500Q/QL/QL1, S550Q/QL/QL1, S620Q/QL/QL1, S690Q/QL/QL1, S890Q/QL/QL1 and S960Q/QL

Table 6 Comparison between grades in EN 10155: 1993 and BS 4360: 1990

	EN 10155: 1993							BS 4360: 1990						
Grade	Product nominal thickness (mm) Flat	Long Sections Shapes	Long Bars	Tensile strength (R_m) at t = 16 mm (N/mm²)	Min yield strength (R_e) at t = 16 mm (N/mm²)	Charpy V-notch impacts (longitudinal) Temp (°C)	Energy (J)	Grade	Tensile strength (R_m) at t = 16 mm (N/mm²)	Min yield strength (R_e) at t = 16 mm (N/mm²)	Max thickness for specified yield strength (R_e) (mm)	Charpy V-notch impacts (longitudinal) Temp (°C)	Energy (J)	Max thickness (mm)
S235J0W	100	40	100	340–470	235	0	27	–	–	–	–	–	–	–
S235J2W	100	40	100	340–470	235	−20	27	–	–	–	–	–	–	–
S355J0WP	12	40	–	490–630	355	0	27	WR50A	≥480	345 (1)	40	0	27	12 (2)
S355J2WP	12	40	–	490–630	355	−20	27	–	–	–	–	–	–	–
S355J0W	100	40	100	490–630	355	0	27	WR50B	≥480	345	345	0	27	50
S355J2G1W	100	40	100	490–630	355	−20	27	WR50C	≥480	345	345	−15	27	50
S355J2G2W	100	40	100	490–630	355	−20	27	WR50C	≥480	345	345	−15	27	50
S355K2G1W	100	40	100	490–630	355	−20	40	–	–	–	–	–	–	–
S355K2G2W	100	40	100	490–630	355	−20	40	–	–	–	–	–	–	–

(1) Up to and incl. 12 mm thick. Over 12 mm min yield strength of 325 N/mm² applies. (2) For round and square bar max thickness is 25 mm

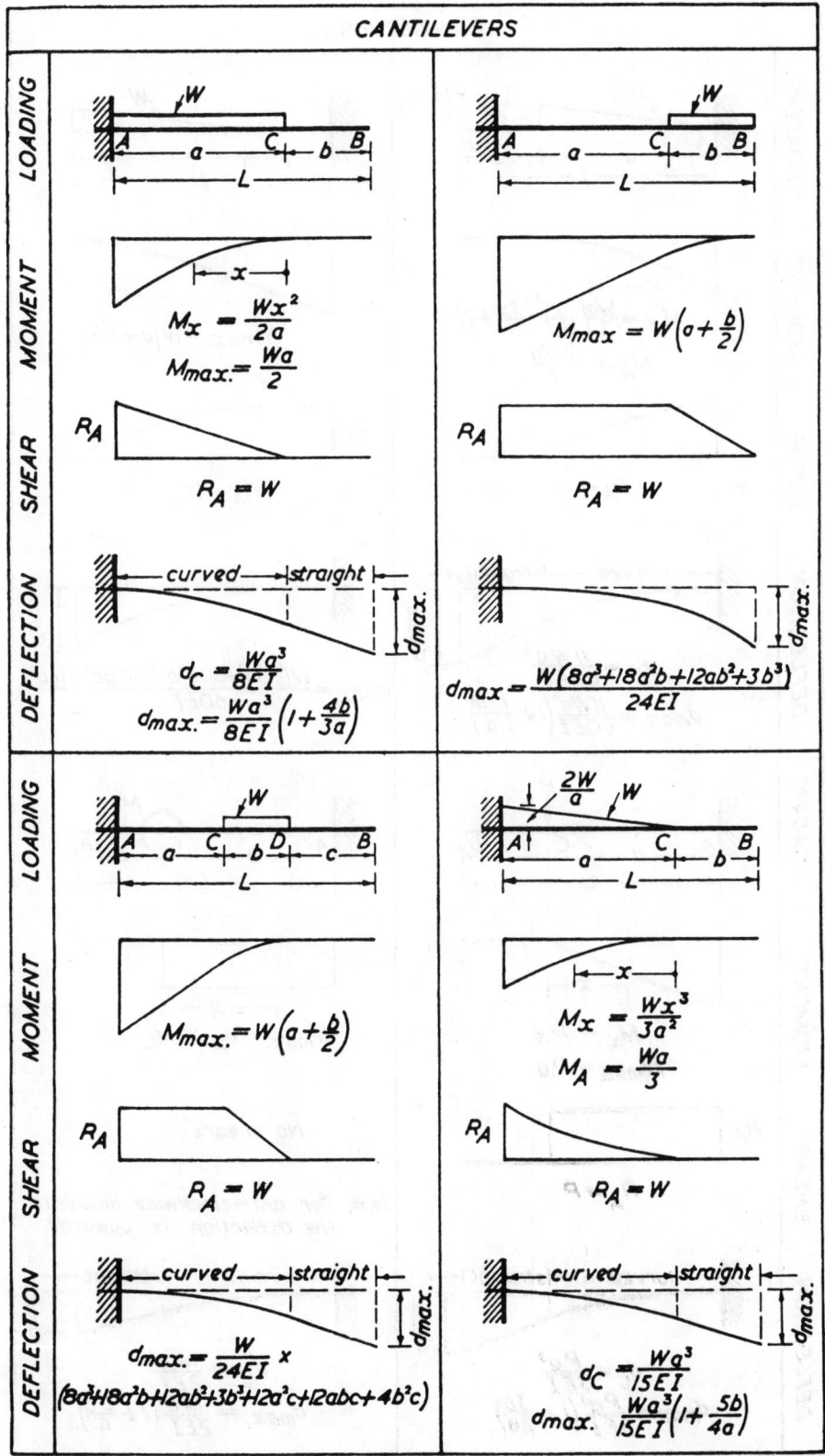

CANTILEVERS

LOADING

MOMENT

$$M_x = \frac{Wx^2}{2a}$$

$$M_{max.} = \frac{Wa}{2}$$

$$M_{max} = W\left(a + \frac{b}{2}\right)$$

SHEAR

$$R_A = W$$

$$R_A = W$$

DEFLECTION

curved | *straight*

$$d_C = \frac{Wa^3}{8EI}$$

$$d_{max.} = \frac{Wa^3}{8EI}\left(1 + \frac{4b}{3a}\right)$$

$$d_{max} = \frac{W(8a^3 + 18a^2b + 12ab^2 + 3b^3)}{24EI}$$

LOADING

MOMENT

$$M_{max.} = W\left(a + \frac{b}{2}\right)$$

$$M_x = \frac{Wx^3}{3a^2}$$

$$M_A = \frac{Wa}{3}$$

SHEAR

$$R_A = W$$

$$R_A = W$$

DEFLECTION

curved | *straight*

$$d_{max.} = \frac{W}{24EI} \times$$

$$(8a^3 + 18a^2b + 12ab^2 + 3b^3 + 12a^2c + 12abc + 4b^2c)$$

$$d_C = \frac{Wa^3}{15EI}$$

$$d_{max.} = \frac{Wa^3}{15EI}\left(1 + \frac{5b}{4a}\right)$$

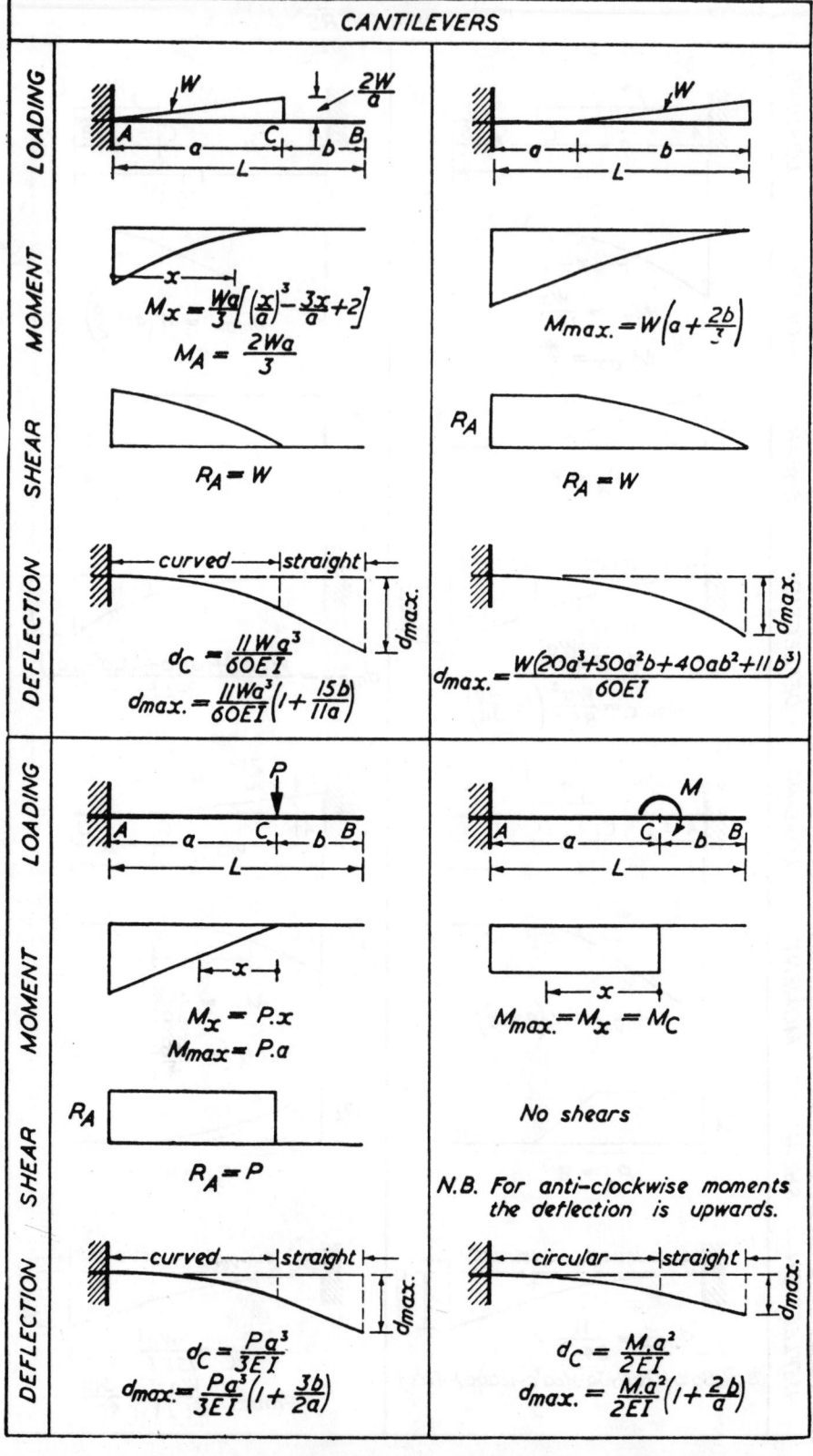

CANTILEVERS

LOADING

MOMENT

$$M_x = \frac{Wa}{3}\left[\left(\frac{x}{a}\right)^3 - \frac{3x}{a} + 2\right]$$

$$M_A = \frac{2Wa}{3}$$

$$M_{max.} = W\left(a + \frac{2b}{3}\right)$$

SHEAR

$$R_A = W$$

$$R_A = W$$

DEFLECTION

curved — straight

$$d_C = \frac{11Wa^3}{60EI}$$

$$d_{max.} = \frac{11Wa^3}{60EI}\left(1 + \frac{15b}{11a}\right)$$

$$d_{max.} = \frac{W(20a^3 + 50a^2b + 40ab^2 + 11b^3)}{60EI}$$

LOADING

MOMENT

$$M_x = P.x$$

$$M_{max} = P.a$$

$$M_{max.} = M_x = M_C$$

SHEAR

$$R_A = P$$

No shears

N.B. For anti-clockwise moments the deflection is upwards.

DEFLECTION

curved — straight

$$d_C = \frac{Pa^3}{3EI}$$

$$d_{max.} = \frac{Pa^3}{3EI}\left(1 + \frac{3b}{2a}\right)$$

circular — straight

$$d_C = \frac{M.a^2}{2EI}$$

$$d_{max.} = \frac{M.a^2}{2EI}\left(1 + \frac{2b}{a}\right)$$

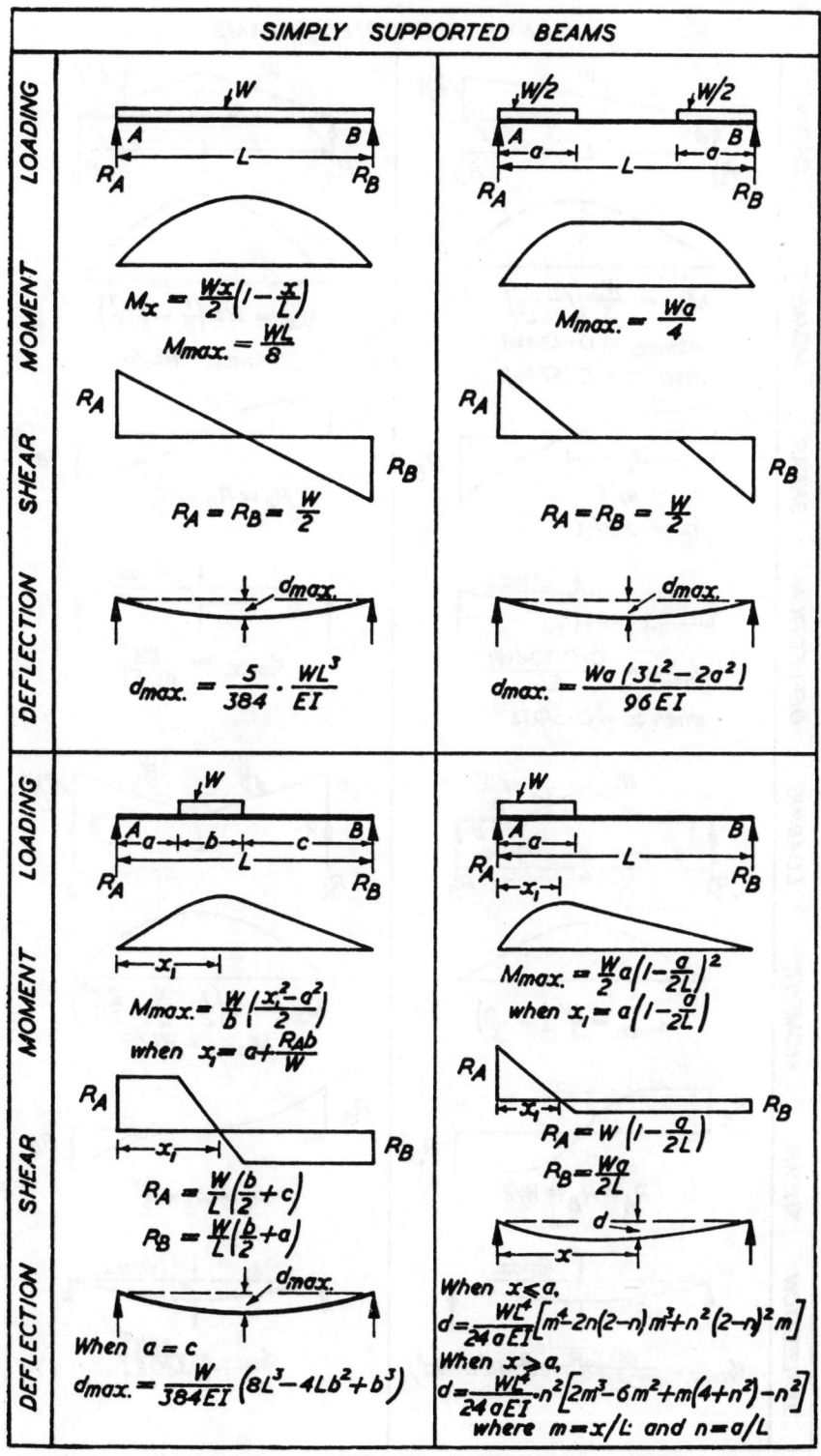

SIMPLY SUPPORTED BEAMS

LOADING — MOMENT — SHEAR — DEFLECTION

$$M_x = \frac{Wx}{2}\left(1 - \frac{x}{L}\right)$$
$$M_{max.} = \frac{WL}{8}$$
$$R_A = R_B = \frac{W}{2}$$
$$d_{max.} = \frac{5}{384} \cdot \frac{WL^3}{EI}$$

$$M_{max.} = \frac{Wa}{4}$$
$$R_A = R_B = \frac{W}{2}$$
$$d_{max.} = \frac{Wa(3L^2 - 2a^2)}{96EI}$$

$$M_{max.} = \frac{W}{b}\left(\frac{x_i^2 - a^2}{2}\right)$$
$$\text{when } x_i = a + \frac{Rab}{W}$$
$$R_A = \frac{W}{L}\left(\frac{b}{2} + c\right)$$
$$R_B = \frac{W}{L}\left(\frac{b}{2} + a\right)$$
$$\text{When } a = c$$
$$d_{max.} = \frac{W}{384EI}\left(8L^3 - 4Lb^2 + b^3\right)$$

$$M_{max.} = \frac{W}{2}a\left(1 - \frac{a}{2L}\right)^2$$
$$\text{when } x_i = a\left(1 - \frac{a}{2L}\right)$$
$$R_A = W\left(1 - \frac{a}{2L}\right)$$
$$R_B = \frac{Wa}{2L}$$

When $x \leqslant a$,
$$d = \frac{WL^4}{24aEI}\left[m^4 - 2n(2-n)m^3 + n^2(2-n)^2 m\right]$$
When $x \geqslant a$,
$$d = \frac{WL^4}{24aEI}n^2\left[2m^3 - 6m^2 + m(4+n^2) - n^2\right]$$
where $m = x/L$ and $n = a/L$

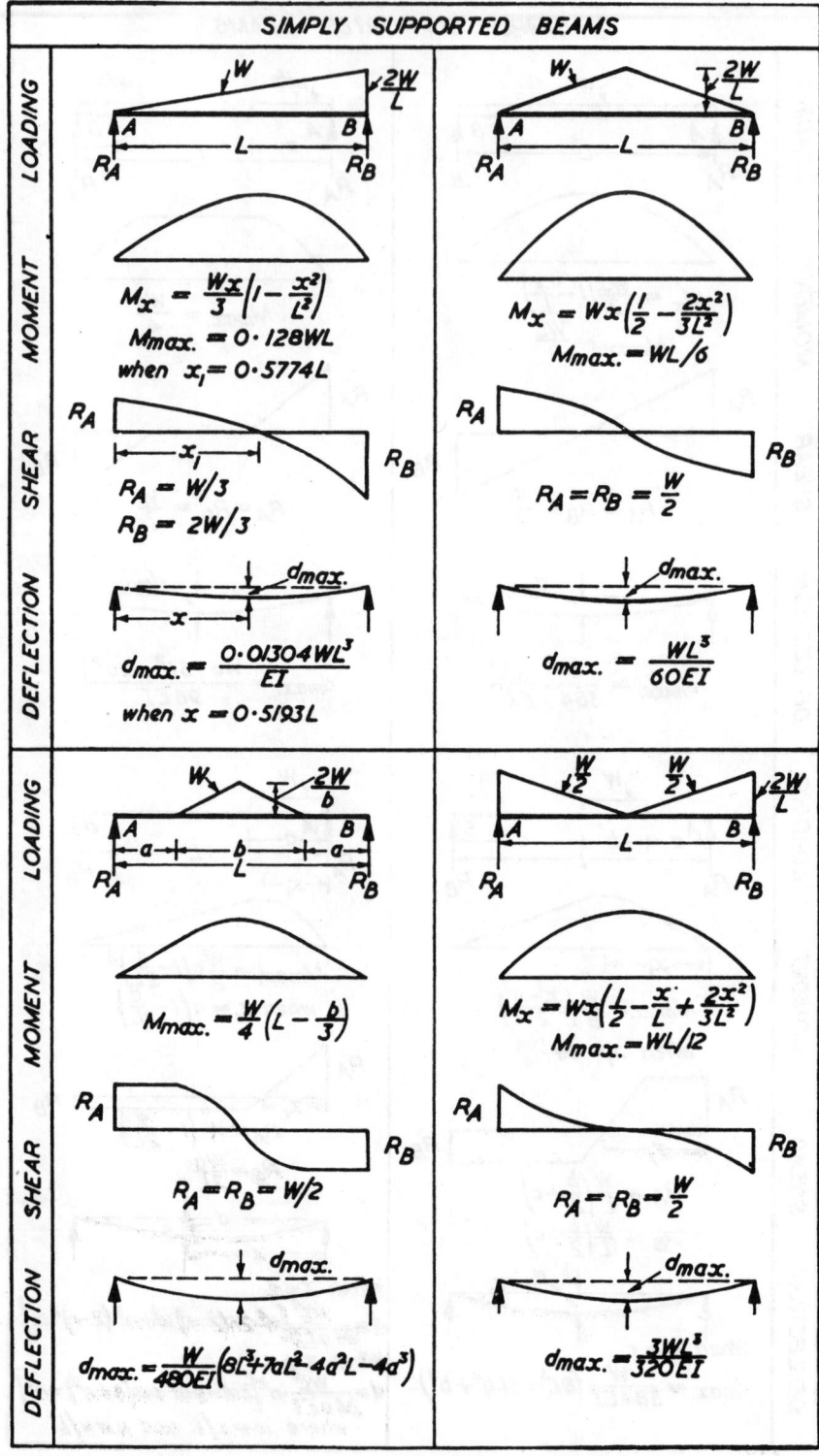

SIMPLY SUPPORTED BEAMS

LOADING

MOMENT

$$M_x = \frac{Wx}{3}\left(1 - \frac{x^2}{L^2}\right)$$

$M_{max.} = 0 \cdot 128WL$

when $x_1 = 0 \cdot 5774L$

$$M_x = Wx\left(\frac{1}{2} - \frac{2x^2}{3L^2}\right)$$

$M_{max.} = WL/6$

SHEAR

$R_A = W/3$

$R_B = 2W/3$

$R_A = R_B = \dfrac{W}{2}$

DEFLECTION

$$d_{max.} = \frac{0 \cdot 01304WL^3}{EI}$$

when $x = 0 \cdot 5193L$

$$d_{max.} = \frac{WL^3}{60EI}$$

LOADING

MOMENT

$$M_{max.} = \frac{W}{4}\left(L - \frac{b}{3}\right)$$

$$M_x = Wx\left(\frac{1}{2} - \frac{x}{L} + \frac{2x^2}{3L^2}\right)$$

$M_{max.} = WL/12$

SHEAR

$R_A = R_B = W/2$

$R_A = R_B = \dfrac{W}{2}$

DEFLECTION

$$d_{max.} = \frac{W}{480EI}\left(8L^3 + 7aL^2 - 4a^2L - 4a^3\right)$$

$$d_{max.} = \frac{3WL^3}{320EI}$$

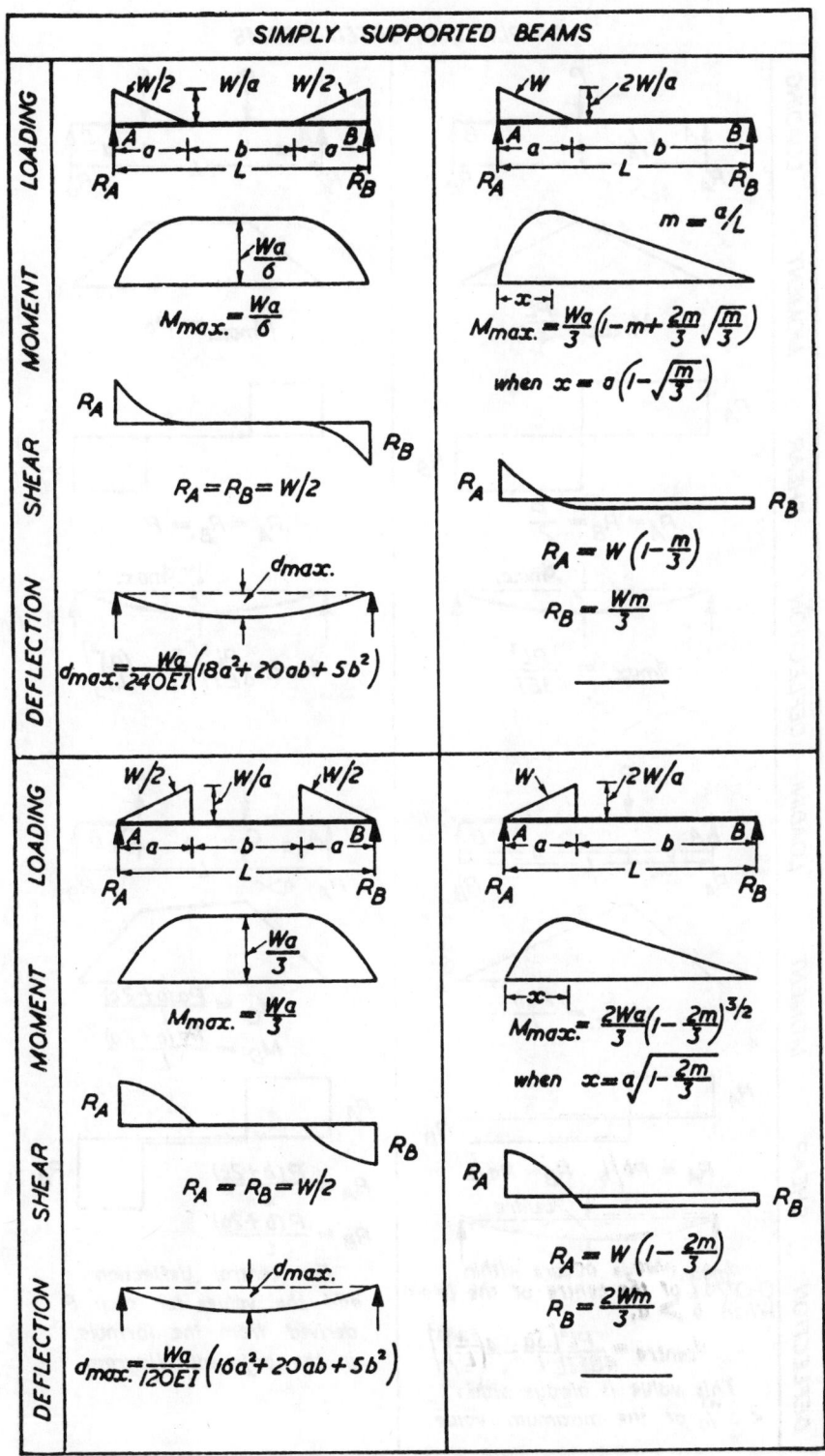

SIMPLY SUPPORTED BEAMS

Top left — LOADING / MOMENT / SHEAR / DEFLECTION

$M_{max.} = \dfrac{Wa}{6}$

$R_A = R_B = W/2$

$d_{max.} = \dfrac{Wa}{240EI}\left(18a^2 + 20ab + 5b^2\right)$

Top right

$m = {a}/{L}$

$M_{max.} = \dfrac{Wa}{3}\left(1 - m + \dfrac{2m}{3}\sqrt{\dfrac{m}{3}}\right)$

when $x = a\left(1 - \sqrt{\dfrac{m}{3}}\right)$

$R_A = W\left(1 - \dfrac{m}{3}\right)$

$R_B = \dfrac{Wm}{3}$

Bottom left

$M_{max.} = \dfrac{Wa}{3}$

$R_A = R_B = W/2$

$d_{max.} = \dfrac{Wa}{120EI}\left(16a^2 + 20ab + 5b^2\right)$

Bottom right

$M_{max.} = \dfrac{2Wa}{3}\left(1 - \dfrac{2m}{3}\right)^{3/2}$

when $x = a\sqrt{1 - \dfrac{2m}{3}}$

$R_A = W\left(1 - \dfrac{2m}{3}\right)$

$R_B = \dfrac{2Wm}{3}$

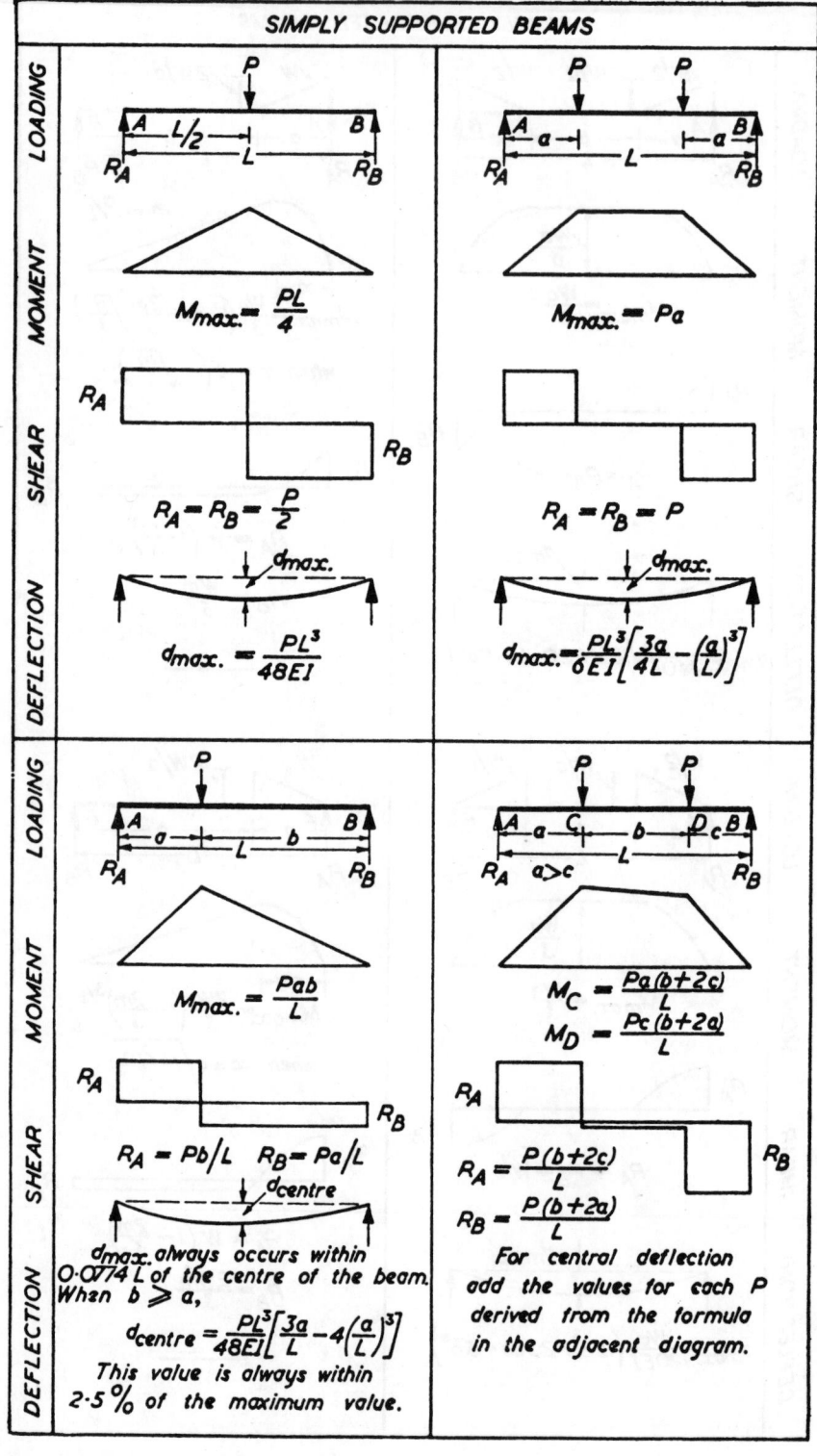

SIMPLY SUPPORTED BEAMS

LOADING · MOMENT · SHEAR · DEFLECTION

$$M_{max.} = \frac{PL}{4}$$

$$R_A = R_B = \frac{P}{2}$$

$$d_{max.} = \frac{PL^3}{48EI}$$

$$M_{max.} = Pa$$

$$R_A = R_B = P$$

$$d_{max.} = \frac{PL^3}{6EI}\left[\frac{3a}{4L} - \left(\frac{a}{L}\right)^3\right]$$

$$M_{max.} = \frac{Pab}{L}$$

$$R_A = Pb/L \quad R_B = Pa/L$$

$d_{max.}$ always occurs within
$0.0774\,L$ of the centre of the beam.
When $b \geqslant a$,

$$d_{centre} = \frac{PL^3}{48EI}\left[\frac{3a}{L} - 4\left(\frac{a}{L}\right)^3\right]$$

This value is always within
2.5% of the maximum value.

$$M_C = \frac{Pa(b+2c)}{L}$$
$$M_D = \frac{Pc(b+2a)}{L}$$

$$R_A = \frac{P(b+2c)}{L}$$
$$R_B = \frac{P(b+2a)}{L}$$

For central deflection
add the values for each P
derived from the formula
in the adjacent diagram.

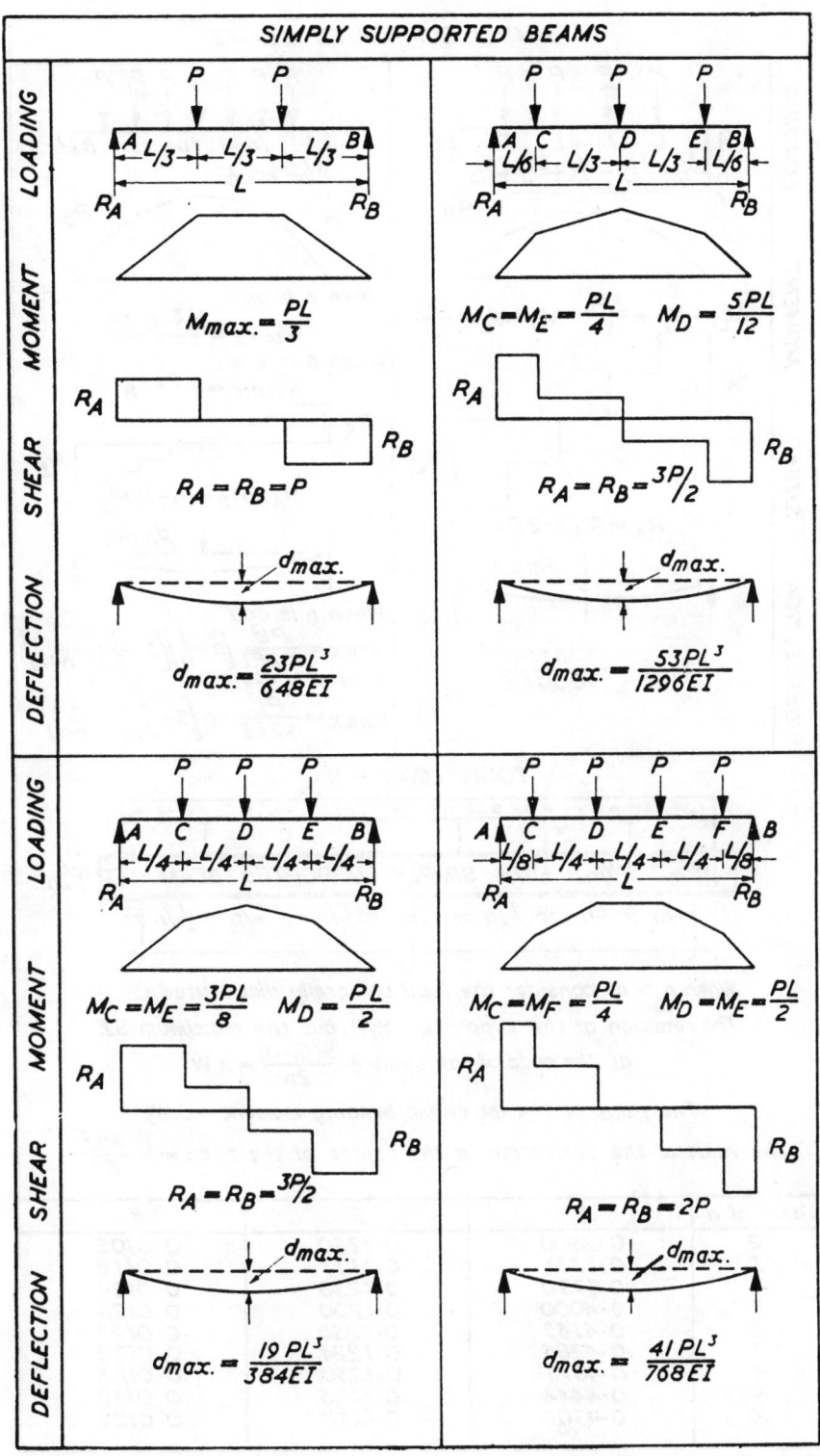

SIMPLY SUPPORTED BEAMS

LOADING

MOMENT

$$M_{max.} = \frac{PL}{3}$$

$$M_C = M_E = \frac{PL}{4} \qquad M_D = \frac{5PL}{12}$$

SHEAR

$$R_A = R_B = P$$

$$R_A = R_B = \frac{3P}{2}$$

DEFLECTION

$$d_{max.} = \frac{23PL^3}{648EI}$$

$$d_{max.} = \frac{53PL^3}{1296EI}$$

LOADING

MOMENT

$$M_C = M_E = \frac{3PL}{8} \qquad M_D = \frac{PL}{2}$$

$$M_C = M_F = \frac{PL}{4} \qquad M_D = M_E = \frac{PL}{2}$$

SHEAR

$$R_A = R_B = \frac{3P}{2}$$

$$R_A = R_B = 2P$$

DEFLECTION

$$d_{max.} = \frac{19PL^3}{384EI}$$

$$d_{max.} = \frac{41PL^3}{768EI}$$

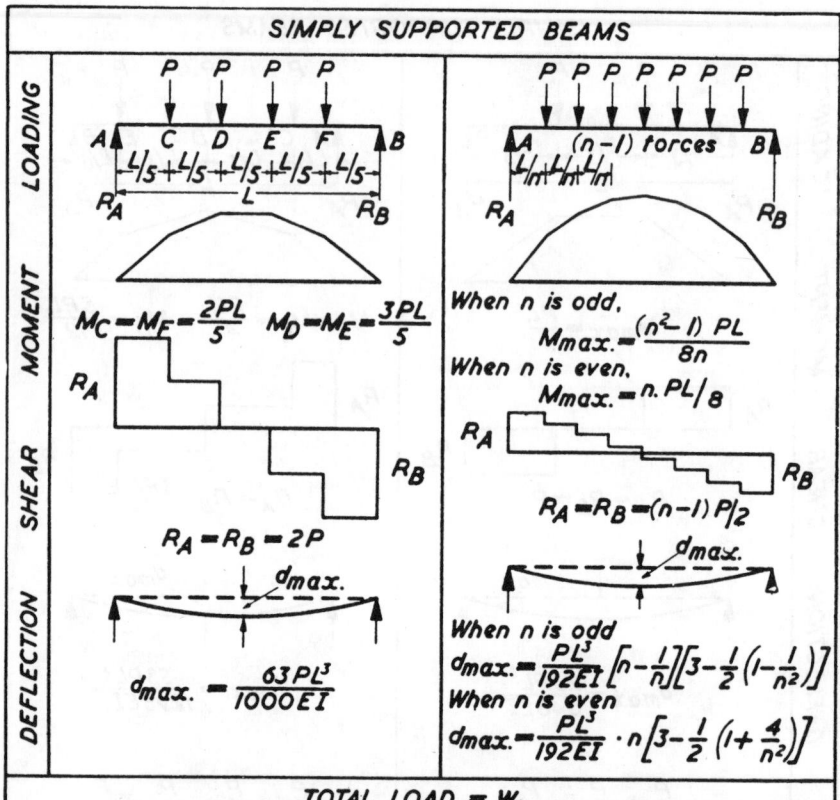

SIMPLY SUPPORTED BEAMS

LOADING

MOMENT

$$M_C = M_F = \frac{2PL}{5} \quad M_D = M_E = \frac{3PL}{5}$$

When n is odd,
$$M_{max.} = \frac{(n^2 - 1)\,PL}{8n}$$
When n is even,
$$M_{max.} = n.\,PL/8$$

SHEAR

$$R_A = R_B = 2P$$

$$R_A = R_B = (n-1)P/2$$

DEFLECTION

$$d_{max.} = \frac{63\,PL^3}{1000\,EI}$$

When n is odd
$$d_{max.} = \frac{PL^3}{192EI}\left[n - \frac{1}{n}\right]\left[3 - \frac{1}{2}\left(1 - \frac{1}{n^2}\right)\right]$$
When n is even
$$d_{max.} = \frac{PL^3}{192EI}\cdot n\left[3 - \frac{1}{2}\left(1 + \frac{4}{n^2}\right)\right]$$

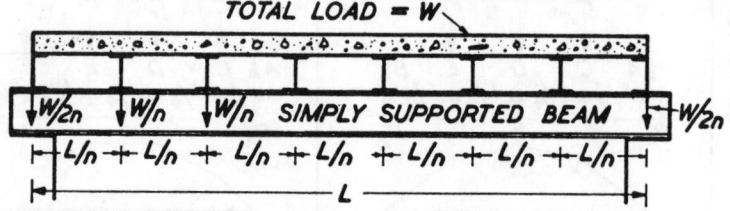

TOTAL LOAD = W

SIMPLY SUPPORTED BEAM

When n > 10, consider the load uniformly distributed

The reaction at the supports = W/2, but the maximum S.F.
at the ends of the beam $= \dfrac{W(n-1)}{2n} = A.W$

The value of the maximum bending moment $= C.WL$

The value of the deflection at the centre of the span $= k.\dfrac{WL^3}{EI}$

Value of n	A	C	k
2	0·2500	0·1250	0·0105
3	0·3333	0·1111	0·0118
4	0·3750	0·1250	0·0124
5	0·4000	0·1200	0·0126
6	0·4167	0·1250	0·0127
7	0·4286	0·1224	0·0128
8	0·4375	0·1250	0·0128
9	0·4444	0·1236	0·0129
10	0·4500	0·1250	0·0129

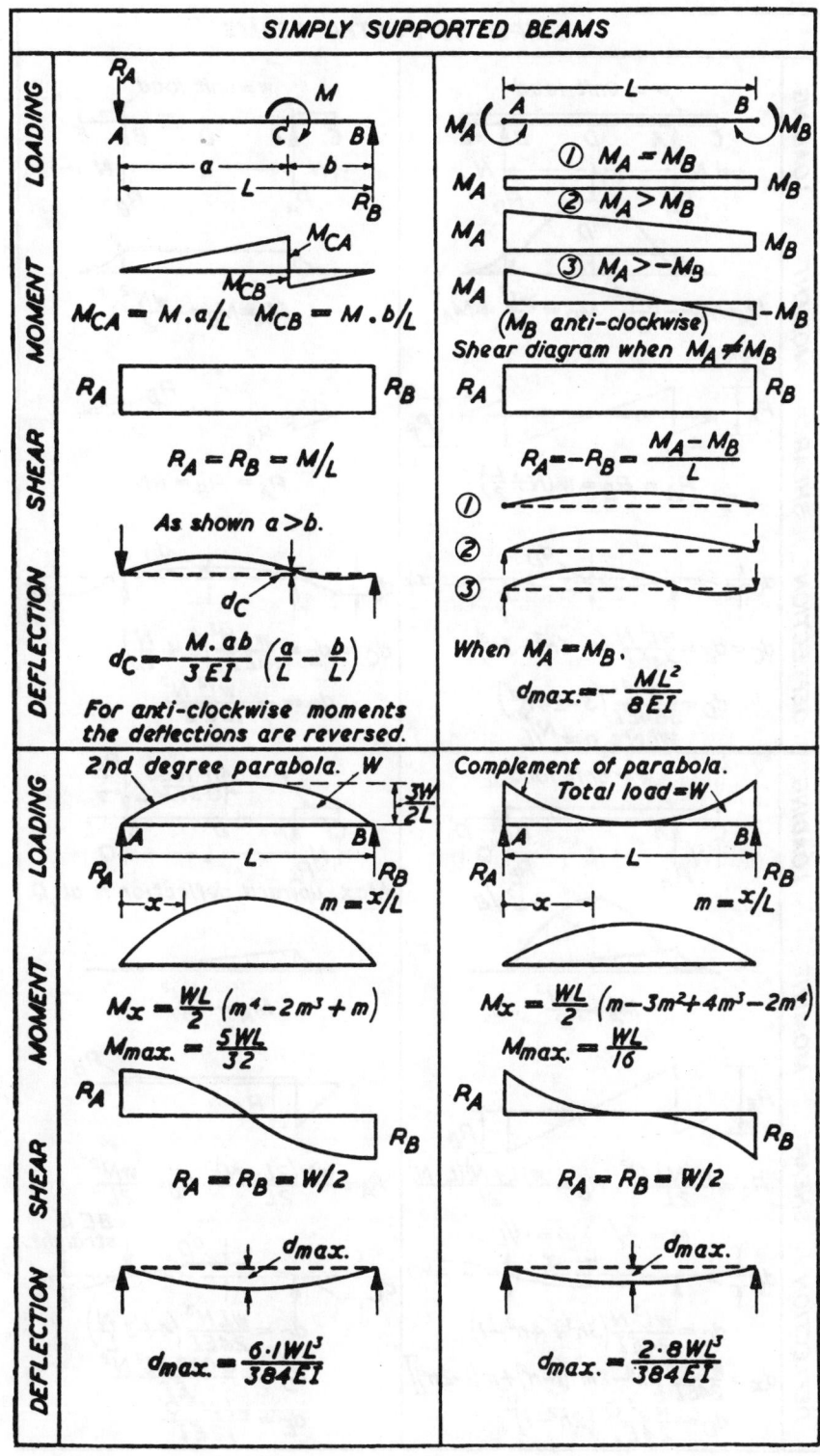

SIMPLY SUPPORTED BEAMS

LOADING / MOMENT / SHEAR / DEFLECTION

$M_{CA} = M \cdot a/L$ $M_{CB} = M \cdot b/L$

$R_A = R_B = M/L$

As shown $a > b$.

$d_C = \dfrac{M \cdot ab}{3EI}\left(\dfrac{a}{L} - \dfrac{b}{L}\right)$

For anti-clockwise moments the deflections are reversed.

① $M_A = M_B$
② $M_A > M_B$
③ $M_A > -M_B$

(M_B anti-clockwise)
Shear diagram when $M_A \neq M_B$

$R_A = -R_B = \dfrac{M_A - M_B}{L}$

When $M_A = M_B$.

$d_{max.} = -\dfrac{ML^2}{8EI}$

2nd degree parabola. W $\dfrac{3W}{2L}$

$m = x/L$

$M_x = \dfrac{WL}{2}(m^4 - 2m^3 + m)$

$M_{max.} = \dfrac{5WL}{32}$

$R_A = R_B = W/2$

$d_{max.} = \dfrac{6 \cdot 1 WL^3}{384EI}$

Complement of parabola.
Total load = W

$m = x/L$

$M_x = \dfrac{WL}{2}(m - 3m^2 + 4m^3 - 2m^4)$

$M_{max.} = \dfrac{WL}{16}$

$R_A = R_B = W/2$

$d_{max.} = \dfrac{2 \cdot 8 WL^3}{384EI}$

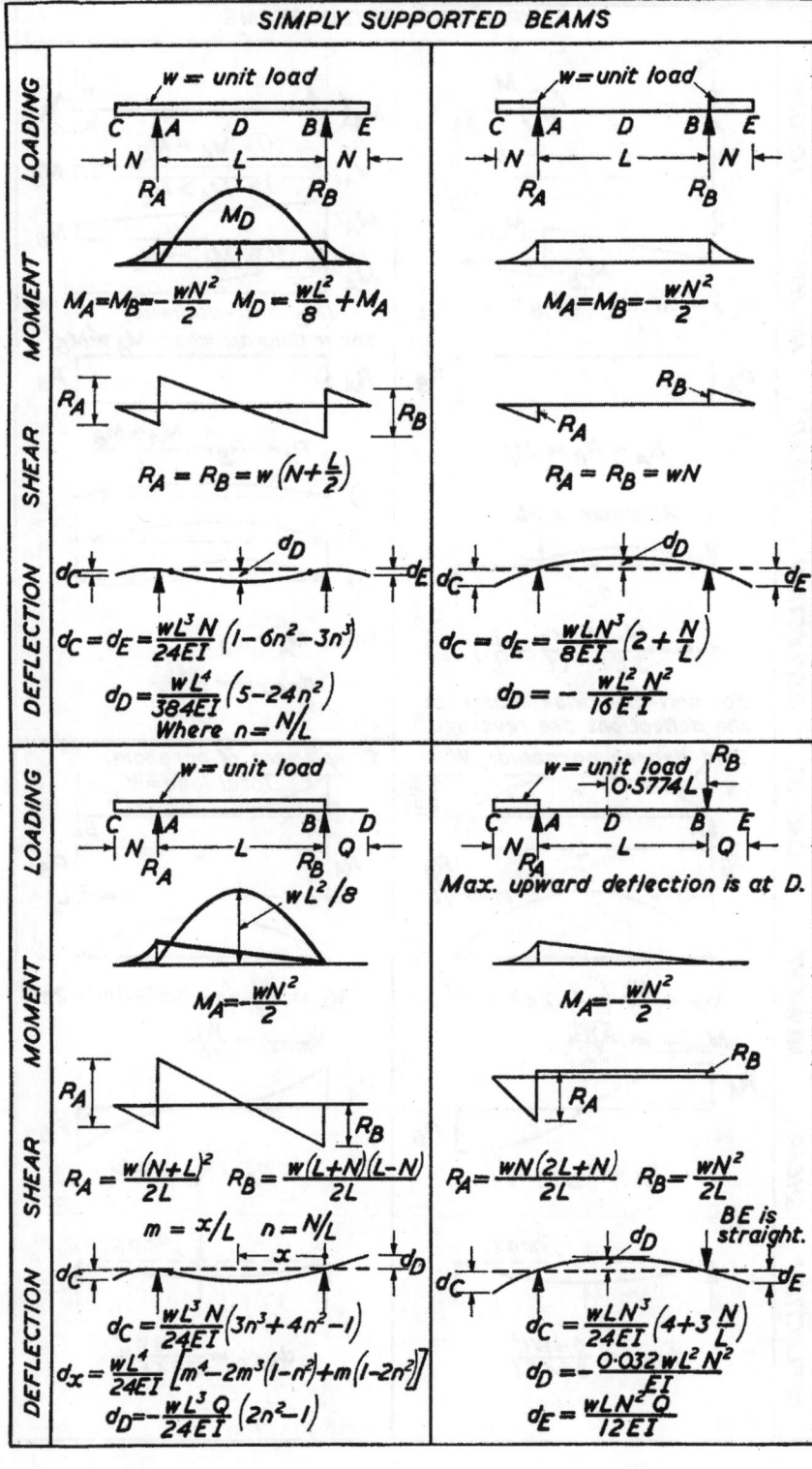

SIMPLY SUPPORTED BEAMS

LOADING

w = unit load

$M_A = M_B = -\dfrac{wN^2}{2}$ $M_D = \dfrac{wL^2}{8} + M_A$

$R_A = R_B = w\left(N + \dfrac{L}{2}\right)$

$d_C = d_E = \dfrac{wL^3 N}{24EI}\left(1 - 6n^2 - 3n^3\right)$

$d_D = \dfrac{wL^4}{384EI}\left(5 - 24n^2\right)$

Where $n = N/L$

w = unit load

$M_A = M_B = -\dfrac{wN^2}{2}$

$R_A = R_B = wN$

$d_C = d_E = \dfrac{wLN^3}{8EI}\left(2 + \dfrac{N}{L}\right)$

$d_D = -\dfrac{wL^2 N^2}{16EI}$

w = unit load

$M_A = -\dfrac{wN^2}{2}$

$R_A = \dfrac{w(N+L)^2}{2L}$ $R_B = \dfrac{w(L+N)(L-N)}{2L}$

$m = x/L$ $n = N/L$

$d_C = \dfrac{wL^3 N}{24EI}\left(3n^3 + 4n^2 - 1\right)$

$d_x = \dfrac{wL^4}{24EI}\left[m^4 - 2m^3(1-n^2) + m(1-2n^2)\right]$

$d_D = -\dfrac{wL^3 Q}{24EI}\left(2n^2 - 1\right)$

w = unit load
$\rightarrow\!|0.5774L$

Max. upward deflection is at D.

$M_A = -\dfrac{wN^2}{2}$

$R_A = \dfrac{wN(2L+N)}{2L}$ $R_B = \dfrac{wN^2}{2L}$

BE is straight.

$d_C = \dfrac{wLN^3}{24EI}\left(4 + 3\dfrac{N}{L}\right)$

$d_D = -\dfrac{0.032\,wL^2 N^2}{EI}$

$d_E = \dfrac{wLN^2 Q}{12EI}$

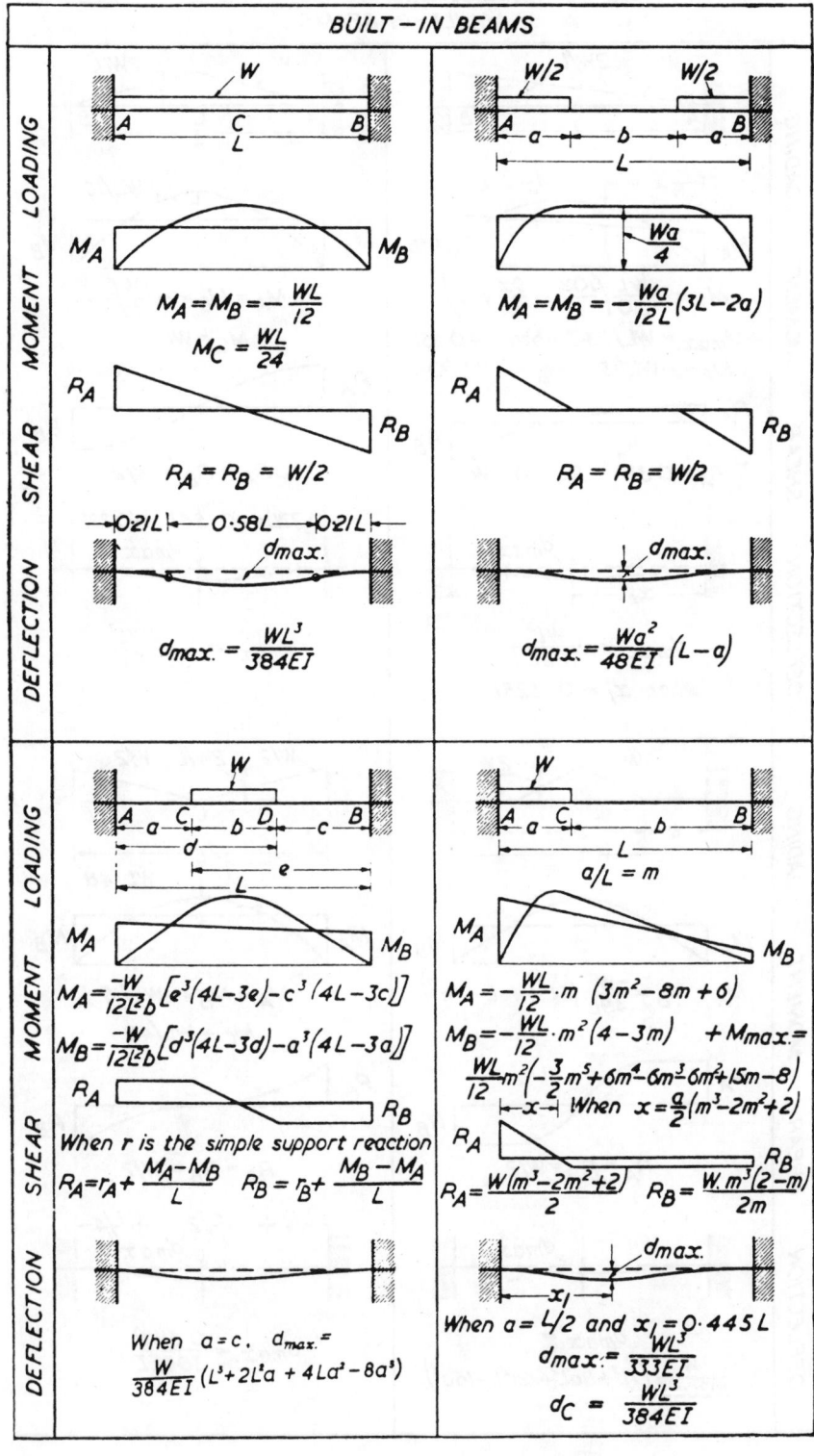

BUILT—IN BEAMS

$$M_A = M_B = -\frac{WL}{12}$$

$$M_C = \frac{WL}{24}$$

$$R_A = R_B = W/2$$

$$d_{max.} = \frac{WL^3}{384EI}$$

$$M_A = M_B = -\frac{Wa}{12L}(3L-2a)$$

$$R_A = R_B = W/2$$

$$d_{max.} = \frac{Wa^2}{48EI}(L-a)$$

$$M_A = \frac{-W}{12L^2 b}\left[e^3(4L-3e) - c^3(4L-3c)\right]$$

$$M_B = \frac{-W}{12L^2 b}\left[d^3(4L-3d) - a^3(4L-3a)\right]$$

When r is the simple support reaction

$$R_A = r_A + \frac{M_A - M_B}{L} \qquad R_B = r_B + \frac{M_B - M_A}{L}$$

When a = c. $d_{max.} =$

$$\frac{W}{384EI}(L^3 + 2L^2 a + 4La^2 - 8a^3)$$

$a/L = m$

$$M_A = -\frac{WL}{12}\cdot m\ (3m^2 - 8m + 6)$$

$$M_B = -\frac{WL}{12}\cdot m^2(4 - 3m) \qquad + M_{max.} =$$

$$\frac{WL}{12}m^2\left(-\frac{3}{2}m^5 + 6m^4 - 6m^3 + 6m^2 + 15m - 8\right)$$

When $x = \frac{a}{2}(m^3 - 2m^2 + 2)$

$$R_A = \frac{W(m^3 - 2m^2 + 2)}{2} \qquad R_B = \frac{W.m^3(2-m)}{2m}$$

When a = L/2 and $x_1 = 0.445L$

$$d_{max.} = \frac{WL^3}{333EI}$$

$$d_C = \frac{WL^3}{384EI}$$

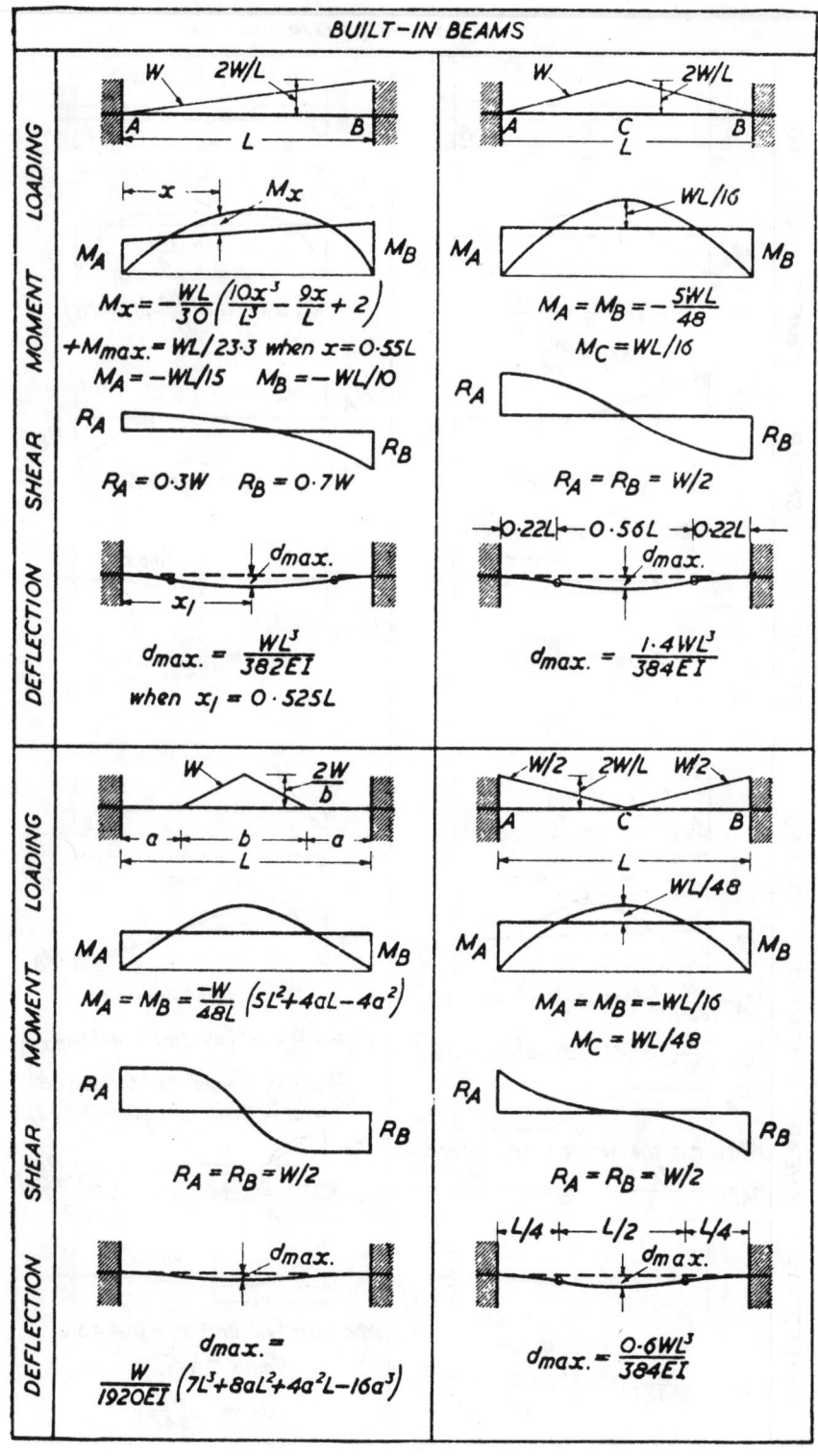

BUILT-IN BEAMS

First beam (top left):

$$M_x = -\frac{WL}{30}\left(\frac{10x^3}{L^3} - \frac{9x}{L} + 2\right)$$

$$+M_{max.} = WL/23.3 \text{ when } x = 0.55L$$

$$M_A = -WL/15 \qquad M_B = -WL/10$$

$$R_A = 0.3W \qquad R_B = 0.7W$$

$$d_{max.} = \frac{WL^3}{382EI}$$

$$\text{when } x_1 = 0.525L$$

Second beam (top right):

WL/16

$$M_A = M_B = -\frac{5WL}{48}$$

$$M_C = WL/16$$

$$R_A = R_B = W/2$$

$$-0.22L-0.56L-0.22L-$$

$$d_{max.} = \frac{1.4WL^3}{384EI}$$

Third beam (bottom left):

$$\frac{2W}{b}$$

$$M_A = M_B = \frac{-W}{48L}\left(5L^2 + 4aL - 4a^2\right)$$

$$R_A = R_B = W/2$$

$$d_{max.} = \frac{W}{1920EI}\left(7L^3 + 8aL^2 + 4a^2L - 16a^3\right)$$

Fourth beam (bottom right):

$$W/2 \quad 2W/L \quad W/2$$

WL/48

$$M_A = M_B = -WL/16$$

$$M_C = WL/48$$

$$R_A = R_B = W/2$$

$$-L/4-L/2-L/4-$$

$$d_{max.} = \frac{0.6WL^3}{384EI}$$

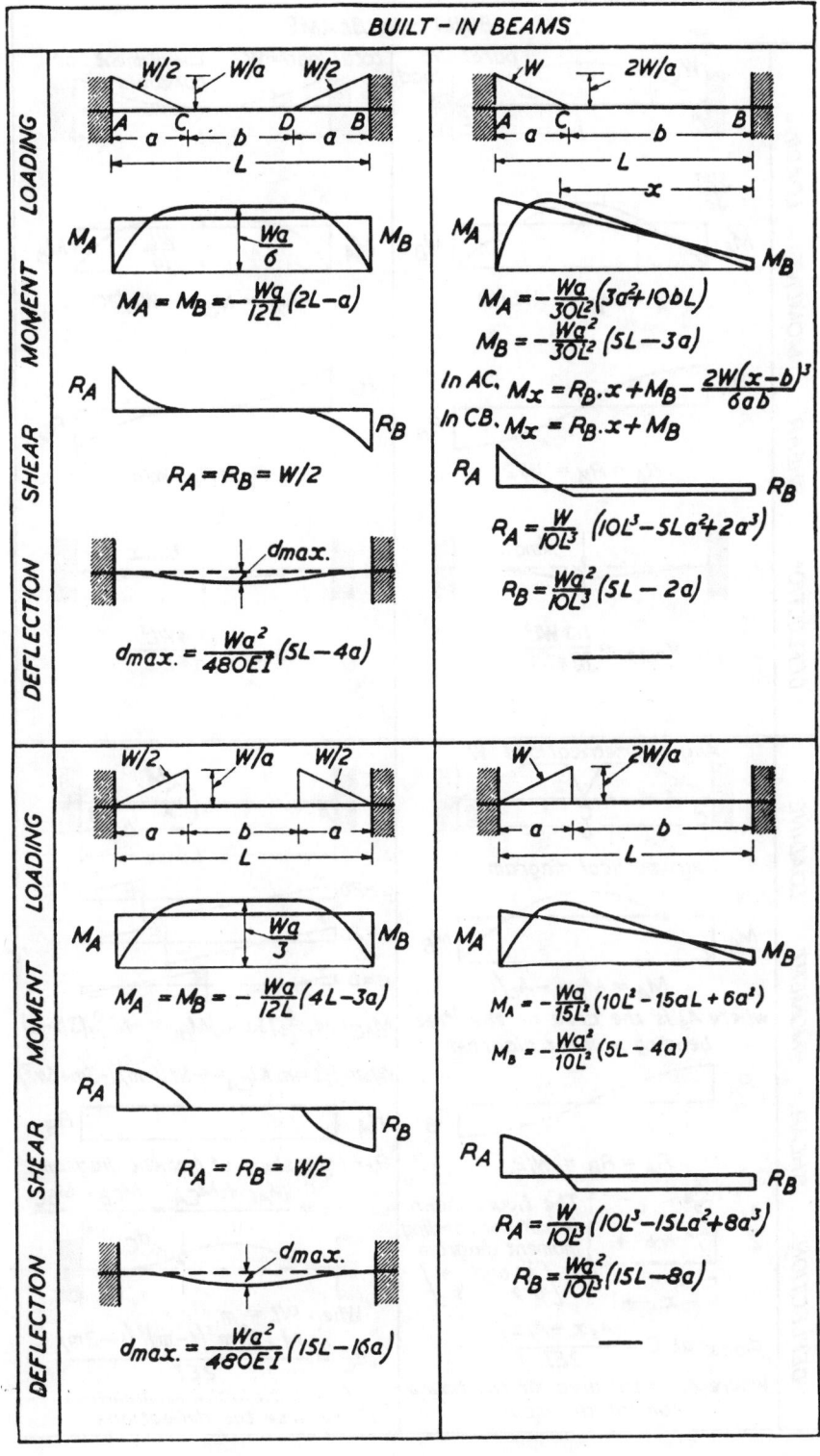

BUILT-IN BEAMS
LOADING
MOMENT
SHEAR
DEFLECTION
W/2
W/a
W/2
A
C
D
B
a
b
a
L
MA
Wa/6
MB
MA = MB = −Wa/12L (2L−a)
RA
RB
RA = RB = W/2
dmax.
dmax. = Wa²/480EI (5L−4a)
W
2W/a
A
C
B
a
b
L
x
MA
MB
MA = −Wa/30L² (3a²+10bL)
MB = −Wa²/30L² (5L−3a)
In AC. Mx = RB.x + MB − 2W(x−b)³/6ab
In CB. Mx = RB.x + MB
RA
RB
RA = W/10L³ (10L³−5La²+2a³)
RB = Wa²/10L³ (5L − 2a)
W/2
W/a
W/2
a
b
a
L
MA
Wa/3
MB
MA = MB = − Wa/12L (4L−3a)
RA
RB
RA = RB = W/2
dmax.
dmax. = Wa²/480EI (15L−16a)
W
2W/a
a
b
L
MA
MB
MA = −Wa/15L² (10L² − 15aL + 6a²)
MB = −Wa²/10L² (5L − 4a)
RA
RB
RA = W/10L³ (10L³−15La²+8a³)
RB = Wa²/10L³ (15L−8a)

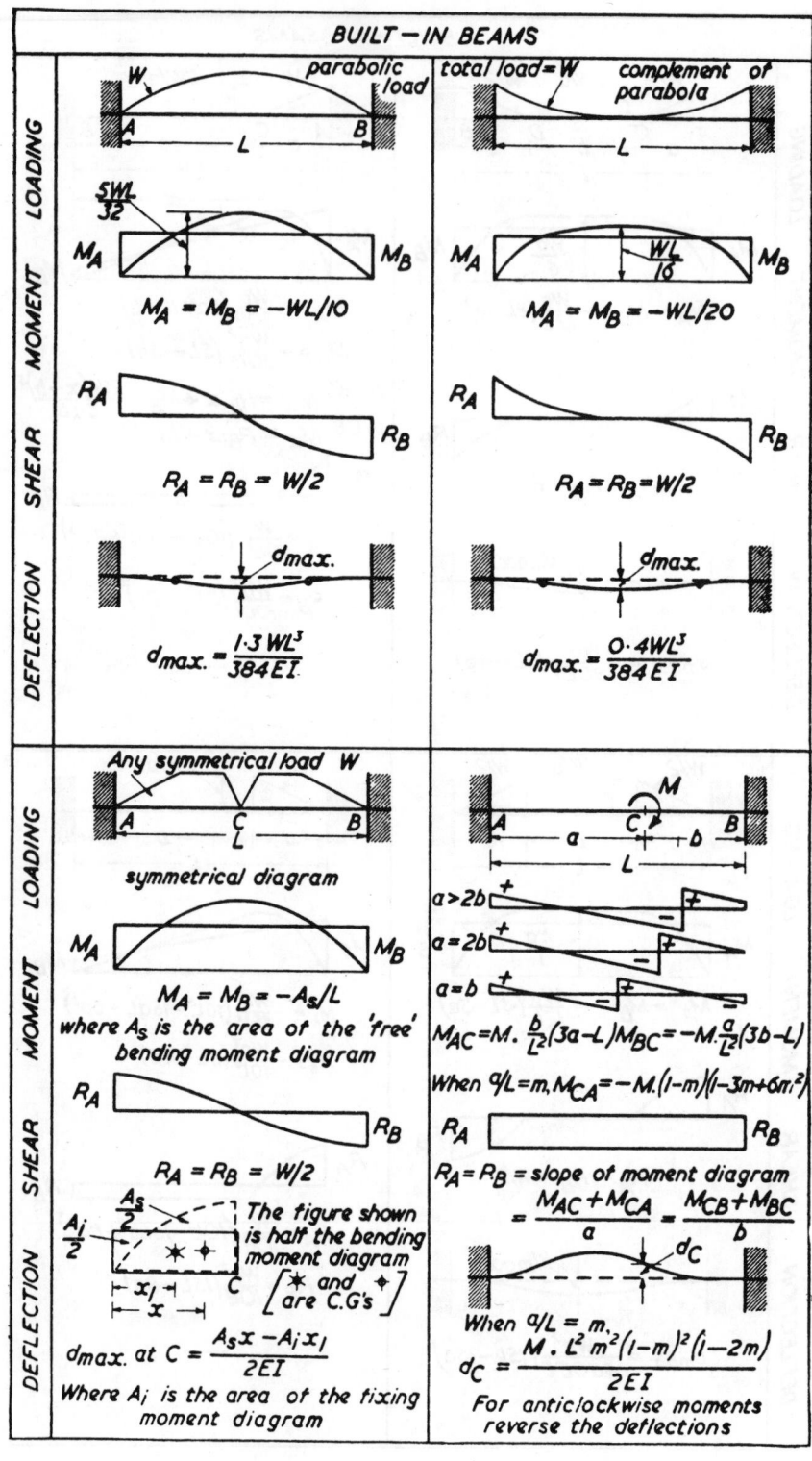

BUILT—IN BEAMS

LOADING — parabolic load / total load = W, complement of parabola

MOMENT:
$$M_A = M_B = -WL/10$$
$$M_A = M_B = -WL/20$$

SHEAR:
$$R_A = R_B = W/2$$
$$R_A = R_B = W/2$$

DEFLECTION:
$$d_{max.} = \frac{1.3\,WL^3}{384\,EI}$$
$$d_{max.} = \frac{0.4\,WL^3}{384\,EI}$$

LOADING — Any symmetrical load W

symmetrical diagram

MOMENT:
$$M_A = M_B = -A_s/L$$
where A_s is the area of the 'free' bending moment diagram

SHEAR:
$$R_A = R_B = W/2$$

The figure shown is half the bending moment diagram
[✱ and ◆ are C.G's]

DEFLECTION:
$$d_{max.} \text{ at } C = \frac{A_s x - A_i x_i}{2EI}$$
Where A_i is the area of the fixing moment diagram

$$M_{AC} = M.\frac{b}{L^2}(3a-L) \quad M_{BC} = -M.\frac{a}{L^2}(3b-L)$$

When $a/L = m, \; M_{CA} = -M.(1-m)(1-3m+6m^2)$

$$R_A = R_B = \text{slope of moment diagram}$$
$$= \frac{M_{AC} + M_{CA}}{a} = \frac{M_{CB} + M_{BC}}{b}$$

When $a/L = m$,
$$d_C = \frac{M.L^2 m^2(1-m)^2(1-2m)}{2EI}$$
For anticlockwise moments reverse the deflections

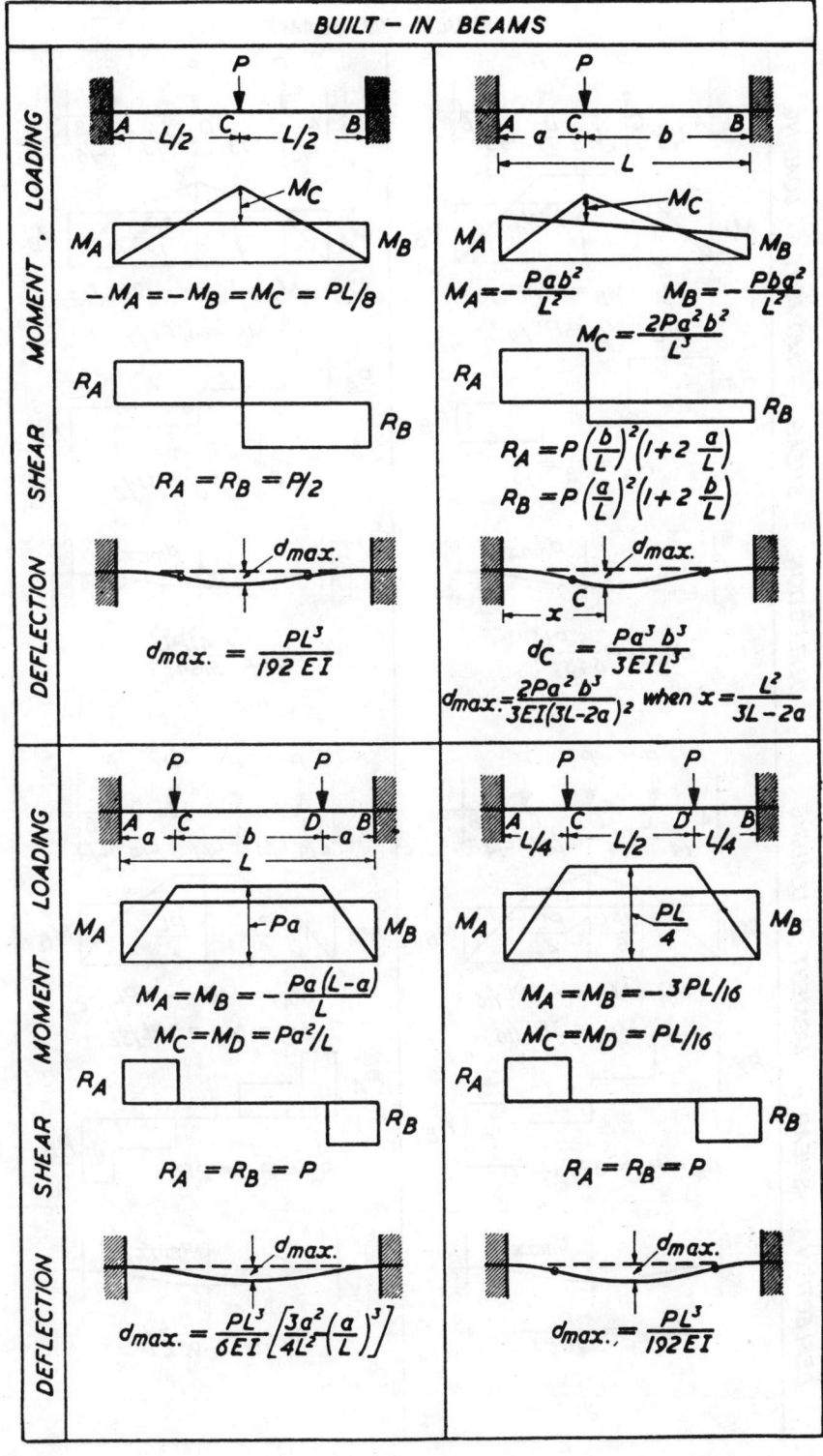

BUILT — IN BEAMS

MOMENT . LOADING

$$-M_A = -M_B = M_C = PL/8$$

$$M_A = -\frac{Pab^2}{L^2} \qquad M_B = -\frac{Pba^2}{L^2}$$

$$M_C = \frac{2Pa^2b^2}{L^3}$$

SHEAR

$$R_A = R_B = P/2$$

$$R_A = P\left(\frac{b}{L}\right)^2\left(1 + 2\frac{a}{L}\right)$$

$$R_B = P\left(\frac{a}{L}\right)^2\left(1 + 2\frac{b}{L}\right)$$

DEFLECTION

$$d_{max.} = \frac{PL^3}{192\,EI}$$

$$d_C = \frac{Pa^3 b^3}{3EIL^3}$$

$$d_{max} = \frac{2Pa^2 b^3}{3EI(3L-2a)^2} \quad \text{when } x = \frac{L^2}{3L-2a}$$

MOMENT LOADING

$$M_A = M_B = -\frac{Pa(L-a)}{L}$$

$$M_C = M_D = Pa^2/L$$

$$M_A = M_B = -3PL/16$$

$$M_C = M_D = PL/16$$

SHEAR

$$R_A = R_B = P$$

$$R_A = R_B = P$$

DEFLECTION

$$d_{max.} = \frac{PL^3}{6EI}\left[\frac{3a^2}{4L^2}\left(\frac{a}{L}\right)^3\right]$$

$$d_{max.} = \frac{PL^3}{192\,EI}$$

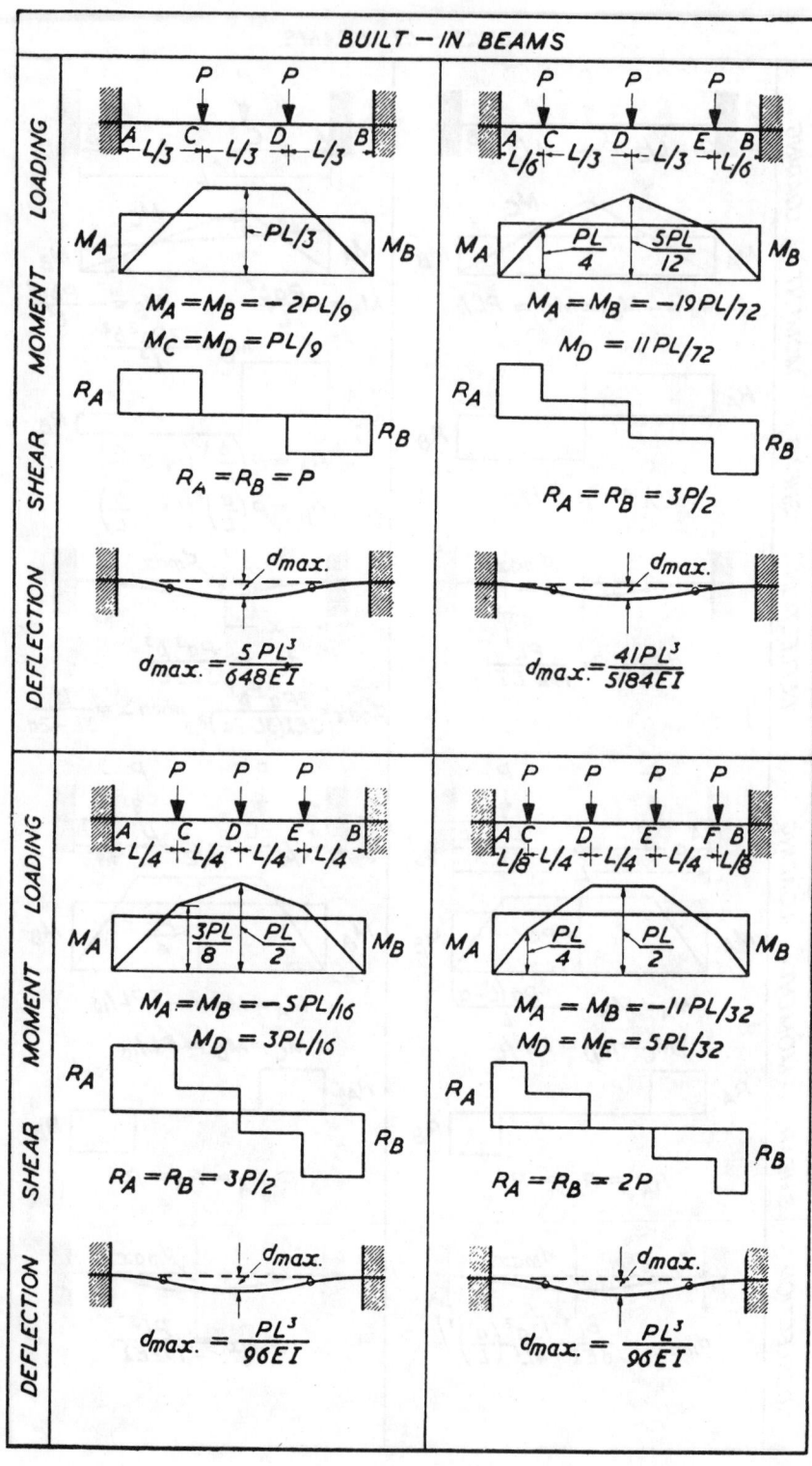

BUILT — IN BEAMS

(Upper left panel)

$M_A = M_B = -2PL/9$

$M_C = M_D = PL/9$

$R_A = R_B = P$

$d_{max.} = \dfrac{5 PL^3}{648EI}$

(Upper right panel)

$M_A = M_B = -19PL/72$

$M_D = 11PL/72$

$R_A = R_B = 3P/2$

$d_{max.} = \dfrac{41PL^3}{5184EI}$

(Lower left panel)

$M_A = M_B = -5PL/16$

$M_D = 3PL/16$

$R_A = R_B = 3P/2$

$d_{max.} = \dfrac{PL^3}{96EI}$

(Lower right panel)

$M_A = M_B = -11PL/32$

$M_D = M_E = 5PL/32$

$R_A = R_B = 2P$

$d_{max.} = \dfrac{PL^3}{96EI}$

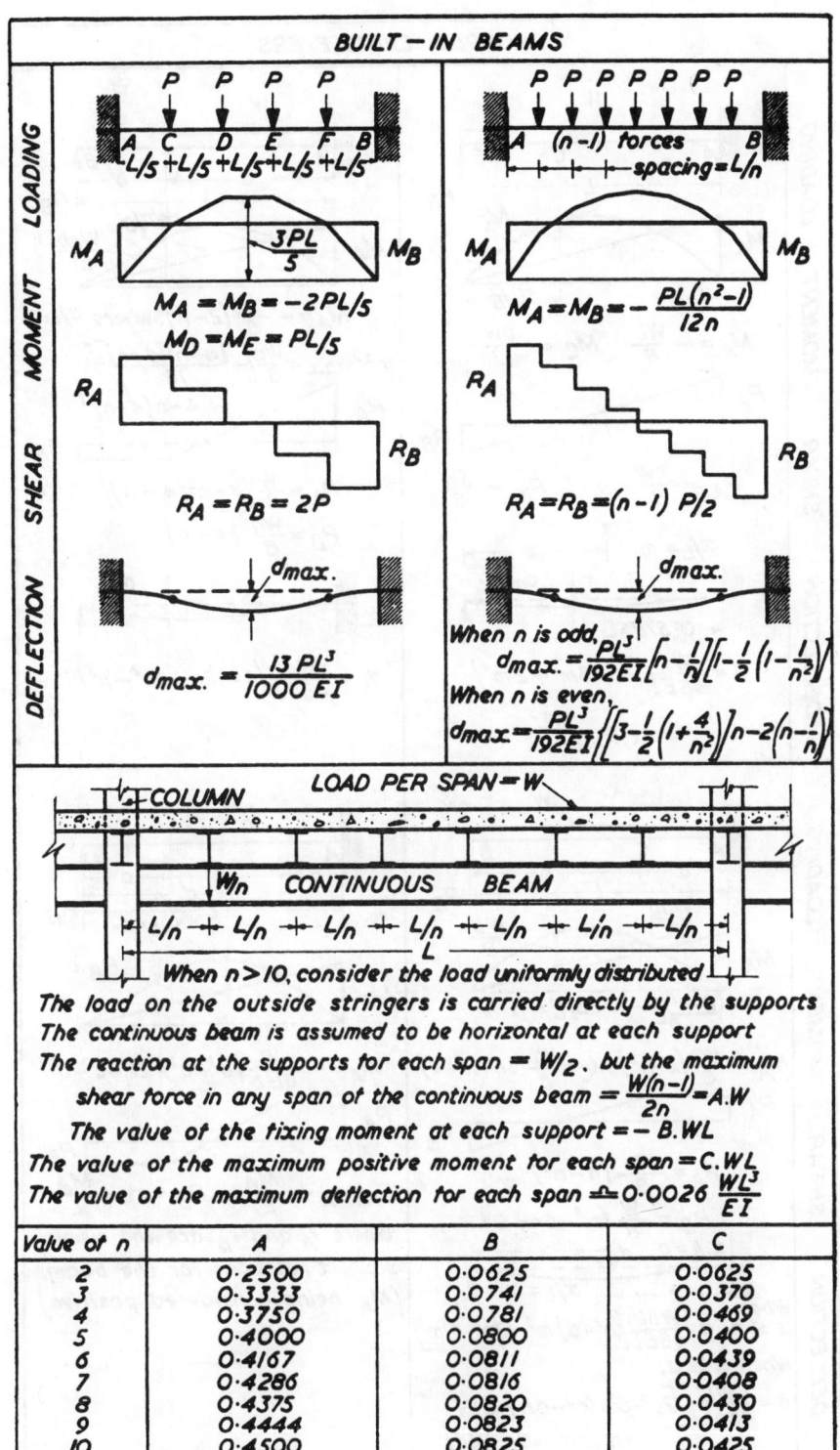

BUILT — IN BEAMS

LOADING

MOMENT

$M_A = M_B = -2PL/5$

$M_D = M_E = PL/5$

$M_A = M_B = -\dfrac{PL(n^2-1)}{12n}$

SHEER

$R_A = R_B = 2P$

$R_A = R_B = (n-1)\,P/2$

DEFLECTION

$d_{max.} = \dfrac{13\,PL^3}{1000\,EI}$

When n is odd,
$$d_{max.} = \dfrac{PL^3}{192EI}\left[n-\dfrac{1}{n}\right]\left[1-\dfrac{1}{2}\left(1-\dfrac{1}{n^2}\right)\right]$$

When n is even,
$$d_{max.} = \dfrac{PL^3}{192EI}\left\{\left[3-\dfrac{1}{2}\left(1+\dfrac{4}{n^2}\right)\right]n-2\left(n-\dfrac{1}{n}\right)\right\}$$

COLUMN   **LOAD PER SPAN = W**

W/n   **CONTINUOUS**   **BEAM**

When n > 10, consider the load uniformly distributed

The load on the outside stringers is carried directly by the supports

The continuous beam is assumed to be horizontal at each support

The reaction at the supports for each span $= W/2$, but the maximum
shear force in any span of the continuous beam $= \dfrac{W(n-1)}{2n} = A.W$

The value of the fixing moment at each support $= -B.WL$

The value of the maximum positive moment for each span $= C.WL$

The value of the maximum deflection for each span $\triangleq 0.0026\,\dfrac{WL^3}{EI}$

Value of n	A	B	C
2	0·2500	0·0625	0·0625
3	0·3333	0·0741	0·0370
4	0·3750	0·0781	0·0469
5	0·4000	0·0800	0·0400
6	0·4167	0·0811	0·0439
7	0·4286	0·0816	0·0408
8	0·4375	0·0820	0·0430
9	0·4444	0·0823	0·0413
10	0·4500	0·0825	0·0425

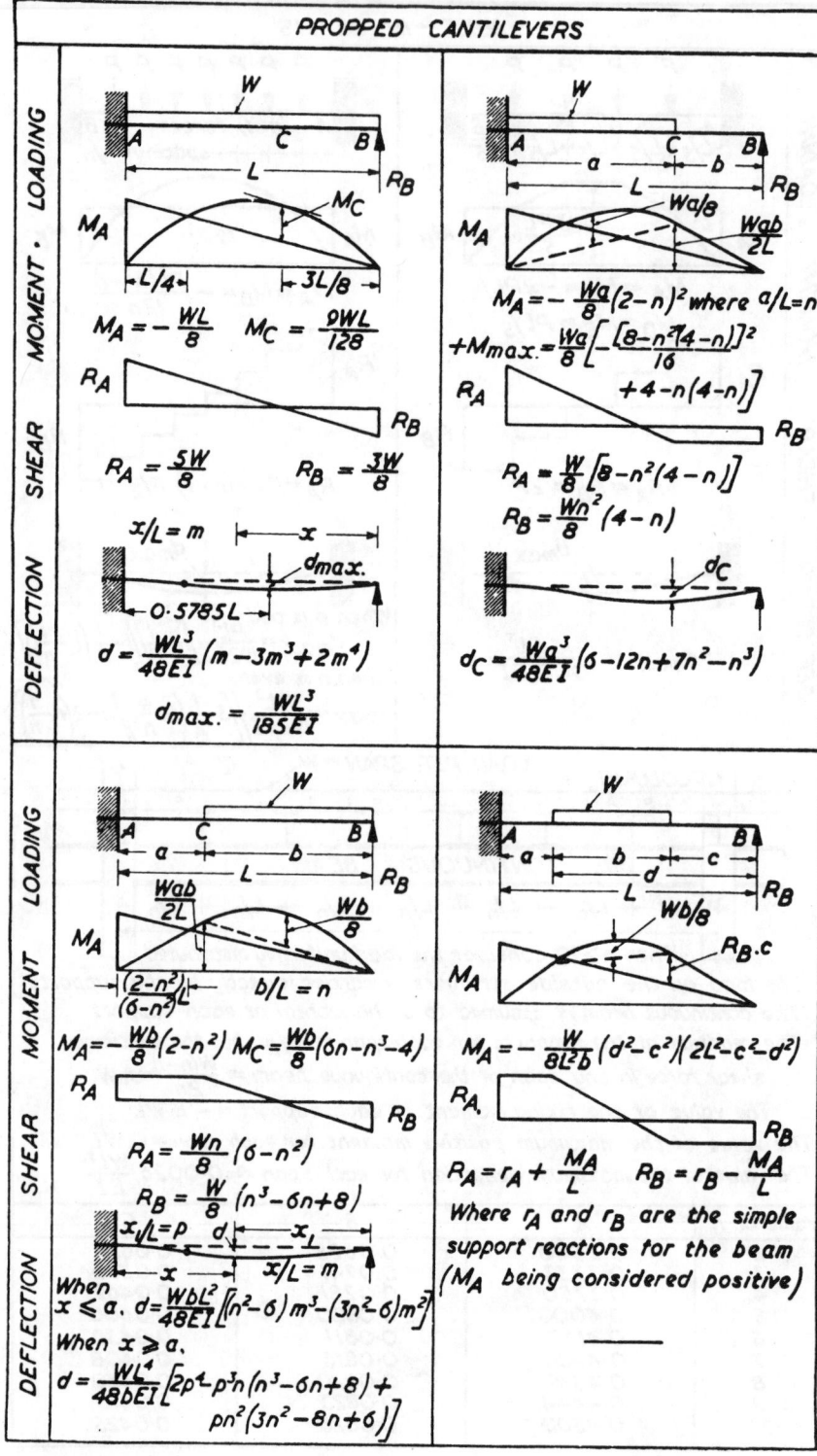

PROPPED CANTILEVERS

Top-left panel

$$M_A = -\frac{WL}{8} \qquad M_C = \frac{9WL}{128}$$

$$R_A = \frac{5W}{8} \qquad R_B = \frac{3W}{8}$$

$x/L = m$

$$d = \frac{WL^3}{48EI}(m - 3m^3 + 2m^4)$$

$$d_{max.} = \frac{WL^3}{185EI}$$

at $0.5785L$

Top-right panel

$$M_A = -\frac{Wa}{8}(2-n)^2 \text{ where } a/L = n$$

$$+M_{max.} = \frac{Wa}{8}\left[-\frac{[8-n^2(4-n)]^2}{16} + 4 - n(4-n)\right]$$

$$R_A = \frac{W}{8}\left[8 - n^2(4-n)\right]$$

$$R_B = \frac{Wn^2}{8}(4-n)$$

$$d_C = \frac{Wa^3}{48EI}(6 - 12n + 7n^2 - n^3)$$

Bottom-left panel

$b/L = n$

$$M_A = -\frac{Wb}{8}(2-n^2) \qquad M_C = \frac{Wb}{8}(6n - n^3 - 4)$$

$$R_A = \frac{Wn}{8}(6 - n^2)$$

$$R_B = \frac{W}{8}(n^3 - 6n + 8)$$

$x/L = p$, $x_L/L = m$

When $x \leqslant a$, $d = \frac{Wbl^2}{48EI}\left[(n^2 - 6)m^3 - (3n^2 - 6)m^2\right]$

When $x \geqslant a$,

$$d = \frac{WL^4}{48bEI}\left[2p^4 - p^3n(n^3 - 6n + 8) + pn^2(3n^2 - 8n + 6)\right]$$

Bottom-right panel

$$M_A = -\frac{W}{8L^2b}(d^2 - c^2)(2L^2 - c^2 - d^2)$$

$$R_A = r_A + \frac{M_A}{L} \qquad R_B = r_B - \frac{M_A}{L}$$

Where r_A and r_B are the simple support reactions for the beam (M_A being considered positive)

PROPPED CANTILEVERS

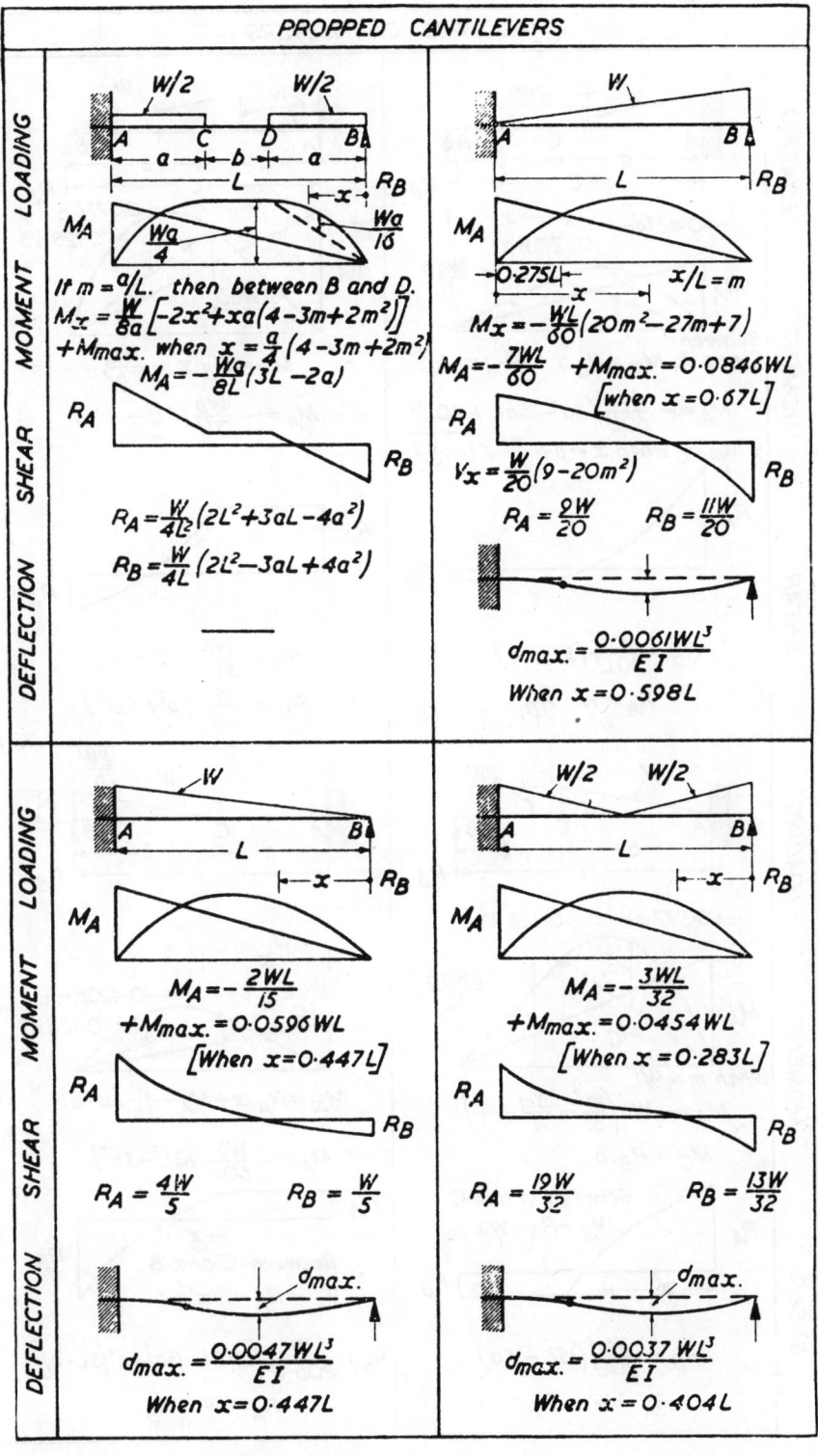

Top left — MOMENT / LOADING / SHEAR / DEFLECTION

If $m = a/L$, then between B and D.

$$M_x = \frac{W}{8a}\left[-2x^2+xa(4-3m+2m^2)\right]$$

$+M_{max.}$ when $x = \frac{a}{4}(4-3m+2m^2)$

$$M_A = -\frac{Wa}{8L}(3L-2a)$$

$$R_A = \frac{W}{4L^2}(2L^2+3aL-4a^2)$$

$$R_B = \frac{W}{4L}(2L^2-3aL+4a^2)$$

Top right

$$M_x = -\frac{WL}{60}(20m^2-27m+7)$$

$$M_A = -\frac{7WL}{60} \qquad +M_{max.} = 0.0846WL$$

$$[when\ x=0.67L]$$

$$V_x = \frac{W}{20}(9-20m^2)$$

$$R_A = \frac{9W}{20} \qquad R_B = \frac{11W}{20}$$

$$d_{max.} = \frac{0.00611WL^3}{EI}$$

When $x = 0.598L$

Bottom left

$$M_A = -\frac{2WL}{15}$$

$$+M_{max.} = 0.0596WL$$

$$[When\ x = 0.447L]$$

$$R_A = \frac{4W}{5} \qquad R_B = \frac{W}{5}$$

$$d_{max.} = \frac{0.0047WL^3}{EI}$$

When $x = 0.447L$

Bottom right

$$M_A = -\frac{3WL}{32}$$

$$+M_{max.} = 0.0454WL$$

$$[When\ x = 0.283L]$$

$$R_A = \frac{19W}{32} \qquad R_B = \frac{13W}{32}$$

$$d_{max.} = \frac{0.0037WL^3}{EI}$$

When $x = 0.404L$

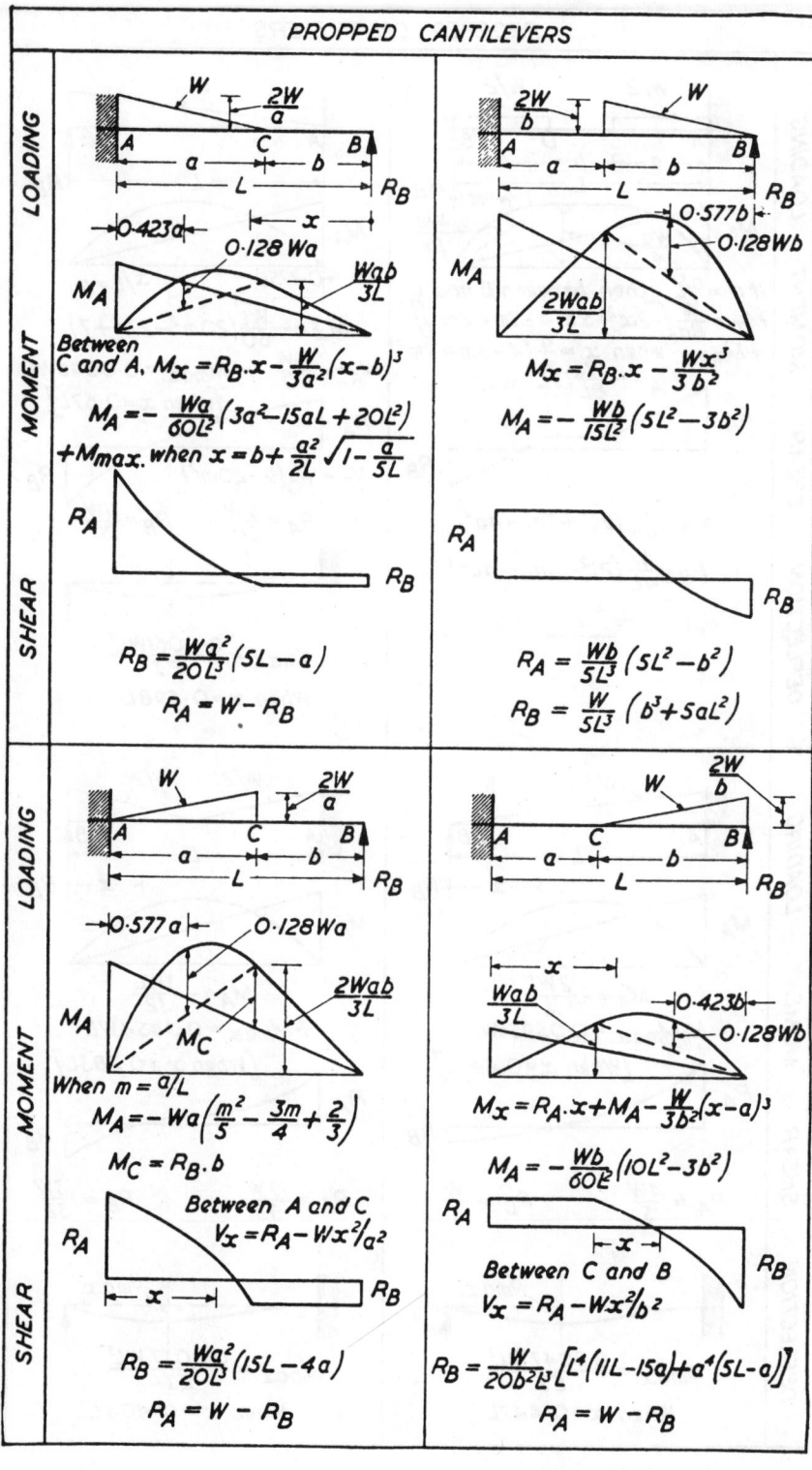

PROPPED CANTILEVERS

Between C and A, $M_x = R_B \cdot x - \dfrac{W}{3a^2}(x-b)^3$

$M_A = -\dfrac{Wa}{60L^2}(3a^2 - 15aL + 20L^2)$

$+M_{max.}$ when $x = b + \dfrac{a^2}{2L}\sqrt{1-\dfrac{a}{5L}}$

$R_B = \dfrac{Wa^2}{20L^3}(5L-a)$

$R_A = W - R_B$

$M_x = R_B \cdot x - \dfrac{Wx^3}{3b^2}$

$M_A = -\dfrac{Wb}{15L^2}(5L^2 - 3b^2)$

$R_A = \dfrac{Wb}{5L^3}(5L^2 - b^2)$

$R_B = \dfrac{W}{5L^3}(b^3 + 5aL^2)$

When $m = \dfrac{a}{L}$

$M_A = -Wa\left(\dfrac{m^2}{5} - \dfrac{3m}{4} + \dfrac{2}{3}\right)$

$M_C = R_B \cdot b$

Between A and C

$V_x = R_A - Wx^2/a^2$

$R_B = \dfrac{Wa^2}{20L^3}(15L - 4a)$

$R_A = W - R_B$

$M_x = R_A \cdot x + M_A - \dfrac{W}{3b^2}(x-a)^3$

$M_A = -\dfrac{Wb}{60L^2}(10L^2 - 3b^2)$

Between C and B

$V_x = R_A - Wx^2/b^2$

$R_B = \dfrac{W}{20b^2L^3}\left[L^4(11L - 15a) + a^4(5L - a)\right]$

$R_A = W - R_B$

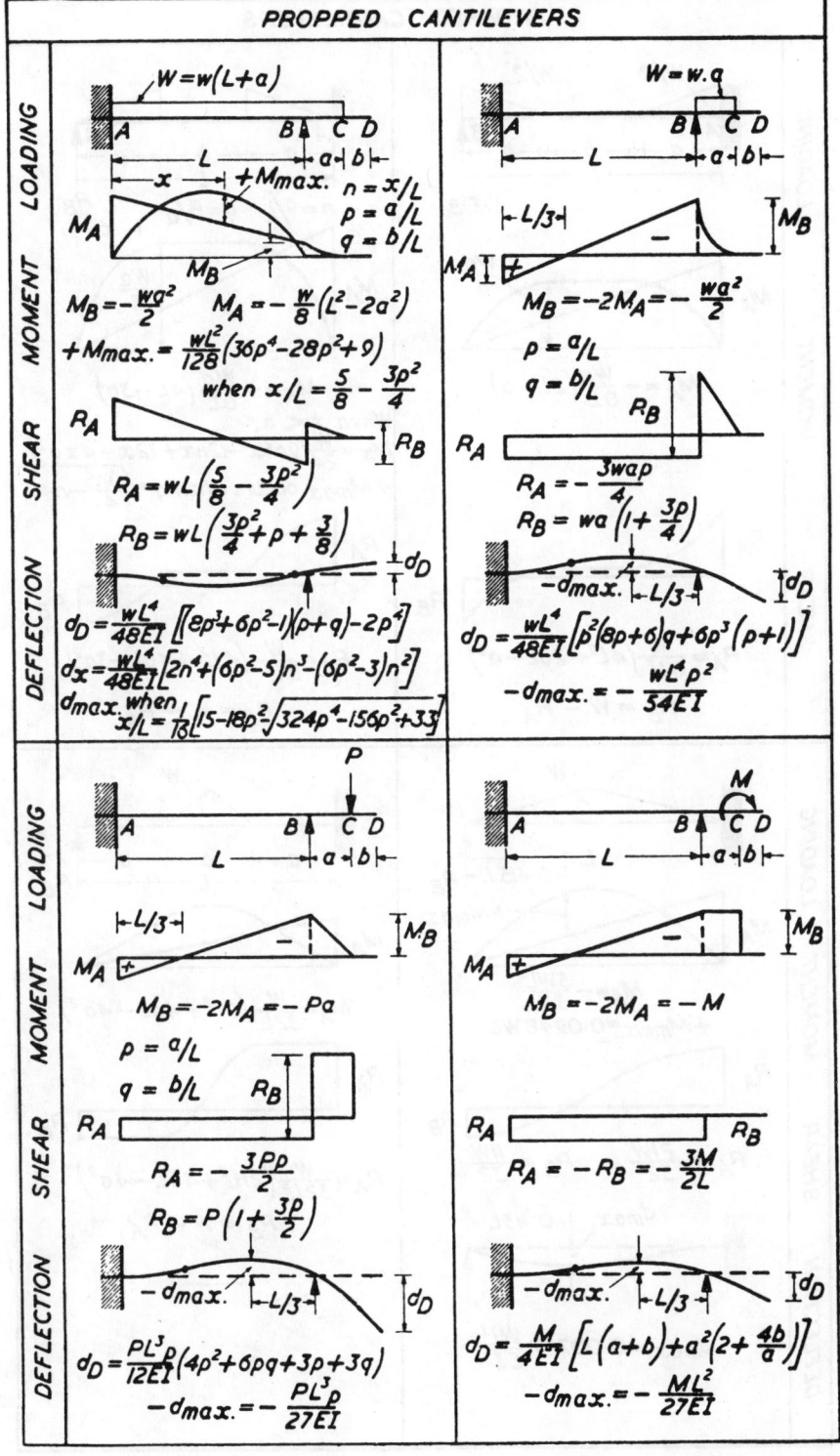

PROPPED CANTILEVERS

Top-left panel:

$W = w(L+a)$

$n = x/L$
$p = a/L$
$q = b/L$

$M_B = -\dfrac{wa^2}{2} \qquad M_A = -\dfrac{W}{8}(L^2 - 2a^2)$

$+Mmax. = \dfrac{wL^2}{128}(36p^4 - 28p^2 + 9)$

when $x/L = \dfrac{5}{8} - \dfrac{3p^2}{4}$

$R_A = wL\left(\dfrac{5}{8} - \dfrac{3p^2}{4}\right)$

$R_B = wL\left(\dfrac{3p^2}{4} + p + \dfrac{3}{8}\right)$

$d_D = \dfrac{wL^4}{48EI}\left[(8p^3 + 6p^2 - 1)(p+q) - 2p^4\right]$

$d_x = \dfrac{wL^4}{48EI}\left[2n^4 + (6p^2 - 5)n^3 - (6p^2 - 3)n^2\right]$

$d_{max.}$ when $\dfrac{x}{L} = \dfrac{1}{18}\left[15 - 18p^2 - \sqrt{324p^4 - 156p^2 + 33}\right]$

Top-right panel:

$W = w.a$

$|{-}L/3{-}|$

$M_B = -2M_A = -\dfrac{wa^2}{2}$

$p = a/L$
$q = b/L$

$R_A = -\dfrac{3wap}{4}$

$R_B = wa\left(1 + \dfrac{3p}{4}\right)$

$d_D = \dfrac{wL^4}{48EI}\left[p^2(8p+6)q + 6p^3(p+1)\right]$

$-d_{max.} = -\dfrac{wL^4 p^2}{54EI}$

Bottom-left panel:

P

$|{-}L/3{-}|$

$M_B = -2M_A = -Pa$

$p = a/L$
$q = b/L$

$R_A = -\dfrac{3Pp}{2}$

$R_B = P\left(1 + \dfrac{3p}{2}\right)$

$d_D = \dfrac{PL^3}{12EI}p(4p^2 + 6pq + 3p + 3q)$

$-d_{max.} = -\dfrac{PL^3 p}{27EI}$

Bottom-right panel:

M

$M_B = -2M_A = -M$

$R_A = -R_B = -\dfrac{3M}{2L}$

$d_D = \dfrac{M}{4EI}\left[L(a+b) + a^2\left(2 + \dfrac{4b}{a}\right)\right]$

$-d_{max.} = -\dfrac{ML^2}{27EI}$

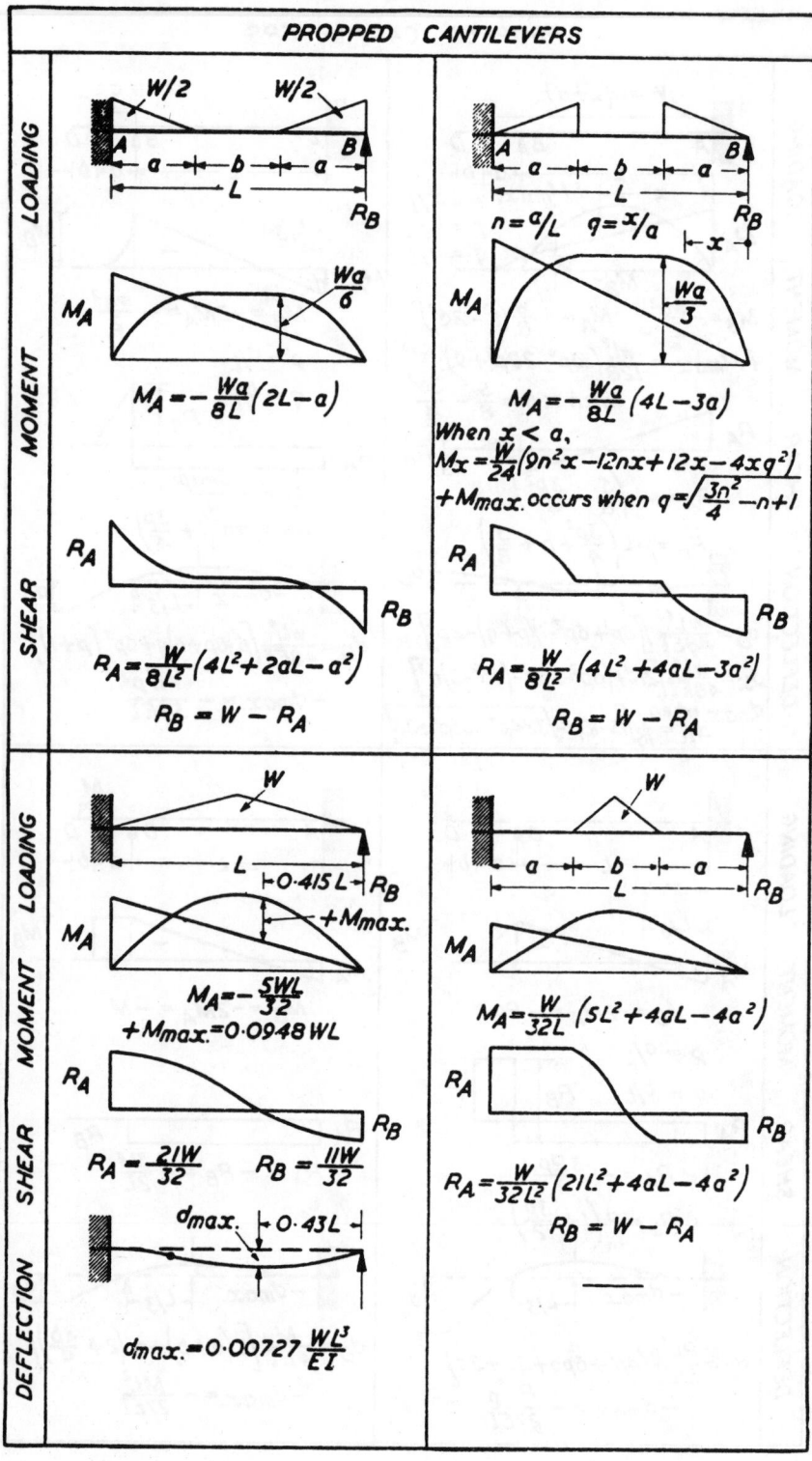

PROPPED CANTILEVERS

LOADING / MOMENT / SHEAR / DEFLECTION

Left upper:

$$M_A = -\frac{Wa}{8L}(2L-a)$$

$$R_A = \frac{W}{8L^2}(4L^2+2aL-a^2)$$

$$R_B = W - R_A$$

Right upper:

$$n = a/L \qquad q = x/a$$

$$M_A = -\frac{Wa}{8L}(4L-3a)$$

When $x < a$,

$$M_x = \frac{W}{24}(9n^2x-12nx+12x-4xq^2)$$

$+M_{max.}$ occurs when $q = \sqrt{\dfrac{3n^2}{4}-n+1}$

$$R_A = \frac{W}{8L^2}(4L^2+4aL-3a^2)$$

$$R_B = W - R_A$$

Left lower:

$$M_A = -\frac{5WL}{32}$$

$$+M_{max.} = 0.0948\,WL$$

$$R_A = \frac{21W}{32} \qquad R_B = \frac{11W}{32}$$

$$d_{max.} = 0.00727\,\frac{WL^3}{EI}$$

Right lower:

$$M_A = \frac{W}{32L}(5L^2+4aL-4a^2)$$

$$R_A = \frac{W}{32L^2}(21L^2+4aL-4a^2)$$

$$R_B = W - R_A$$

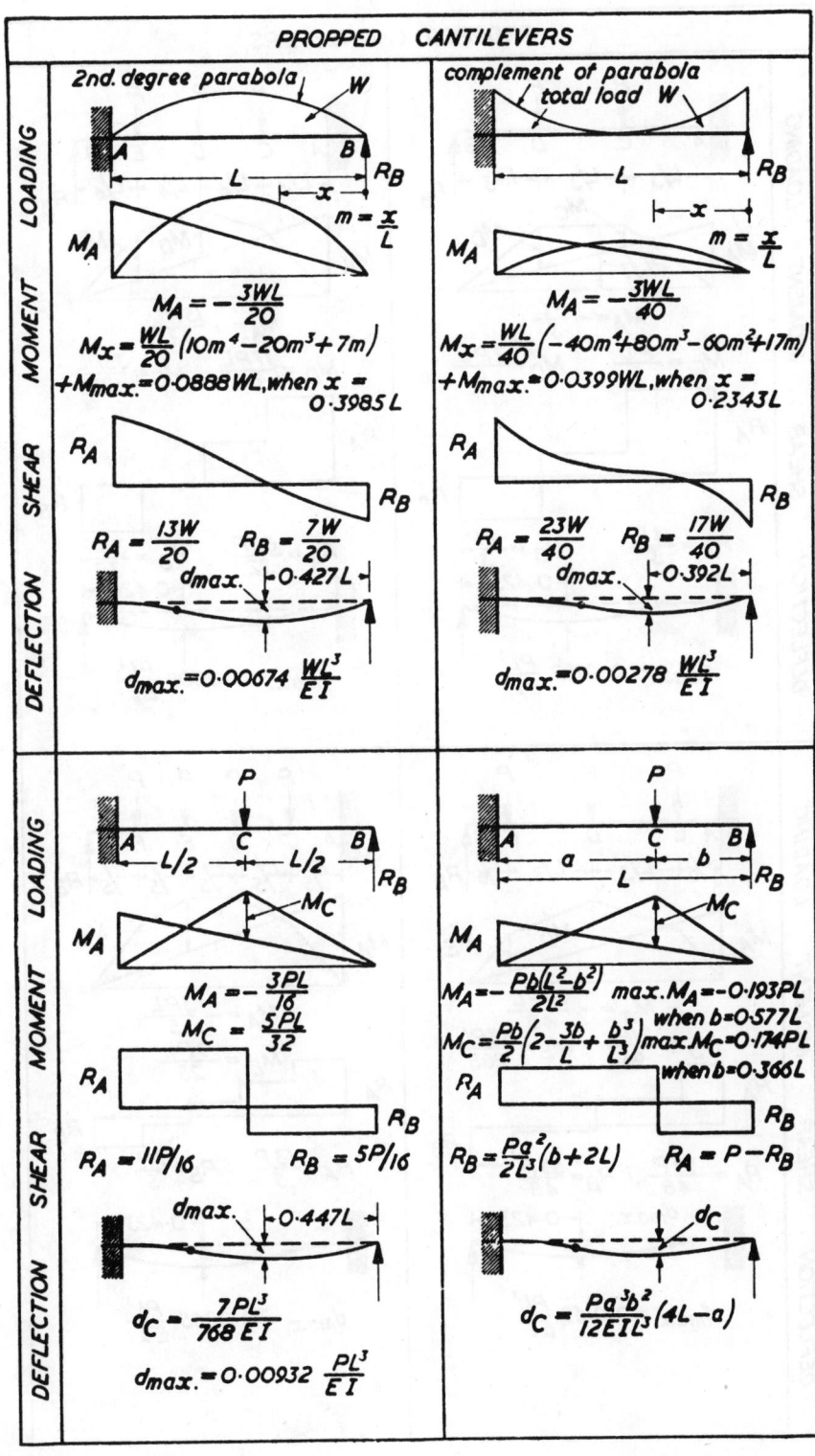

PROPPED CANTILEVERS

LOADING — 2nd. degree parabola, W — A, B, R_B — L, x, $m = \dfrac{x}{L}$

MOMENT

$M_A = -\dfrac{3WL}{20}$

$M_x = \dfrac{WL}{20}(10m^4 - 20m^3 + 7m)$

$+M_{max.} = 0.0888WL$, when $x = 0.3985L$

SHEAR

$R_A = \dfrac{13W}{20}$ $R_B = \dfrac{7W}{20}$

DEFLECTION

$d_{max.}$ — $0.427L$

$d_{max.} = 0.00674\,\dfrac{WL^3}{EI}$

LOADING — complement of parabola, total load W — R_B — L, x, $m = \dfrac{x}{L}$

MOMENT

$M_A = -\dfrac{3WL}{40}$

$M_x = \dfrac{WL}{40}(-40m^4 + 80m^3 - 60m^2 + 17m)$

$+M_{max.} = 0.0399WL$, when $x = 0.2343L$

SHEAR

$R_A = \dfrac{23W}{40}$ $R_B = \dfrac{17W}{40}$

DEFLECTION

$d_{max.}$ — $0.392L$

$d_{max.} = 0.00278\,\dfrac{WL^3}{EI}$

LOADING — P — A, C, B — $\dfrac{L}{2}$, $\dfrac{L}{2}$, R_B

MOMENT

M_C

$M_A = -\dfrac{3PL}{16}$

$M_C = \dfrac{5PL}{32}$

SHEAR

$R_A = 11P/16$ $R_B = 5P/16$

DEFLECTION

$d_{max.}$ — $0.447L$

$d_C = \dfrac{7PL^3}{768\,EI}$

$d_{max.} = 0.00932\,\dfrac{PL^3}{EI}$

LOADING — P — A, C, B — a, b, L, R_B

MOMENT

M_C

$M_A = -\dfrac{Pb(L^2 - b^2)}{2L^2}$ $max.\,M_A = -0.193PL$ when $b = 0.577L$

$M_C = \dfrac{Pb}{2}\left(2 - \dfrac{3b}{L} + \dfrac{b^3}{L^3}\right)$ $max.\,M_C = 0.174PL$ when $b = 0.366L$

SHEAR

$R_B = \dfrac{Pa^2}{2L^3}(b + 2L)$ $R_A = P - R_B$

DEFLECTION

d_C

$d_C = \dfrac{Pa^3b^2}{12EIL^3}(4L - a)$

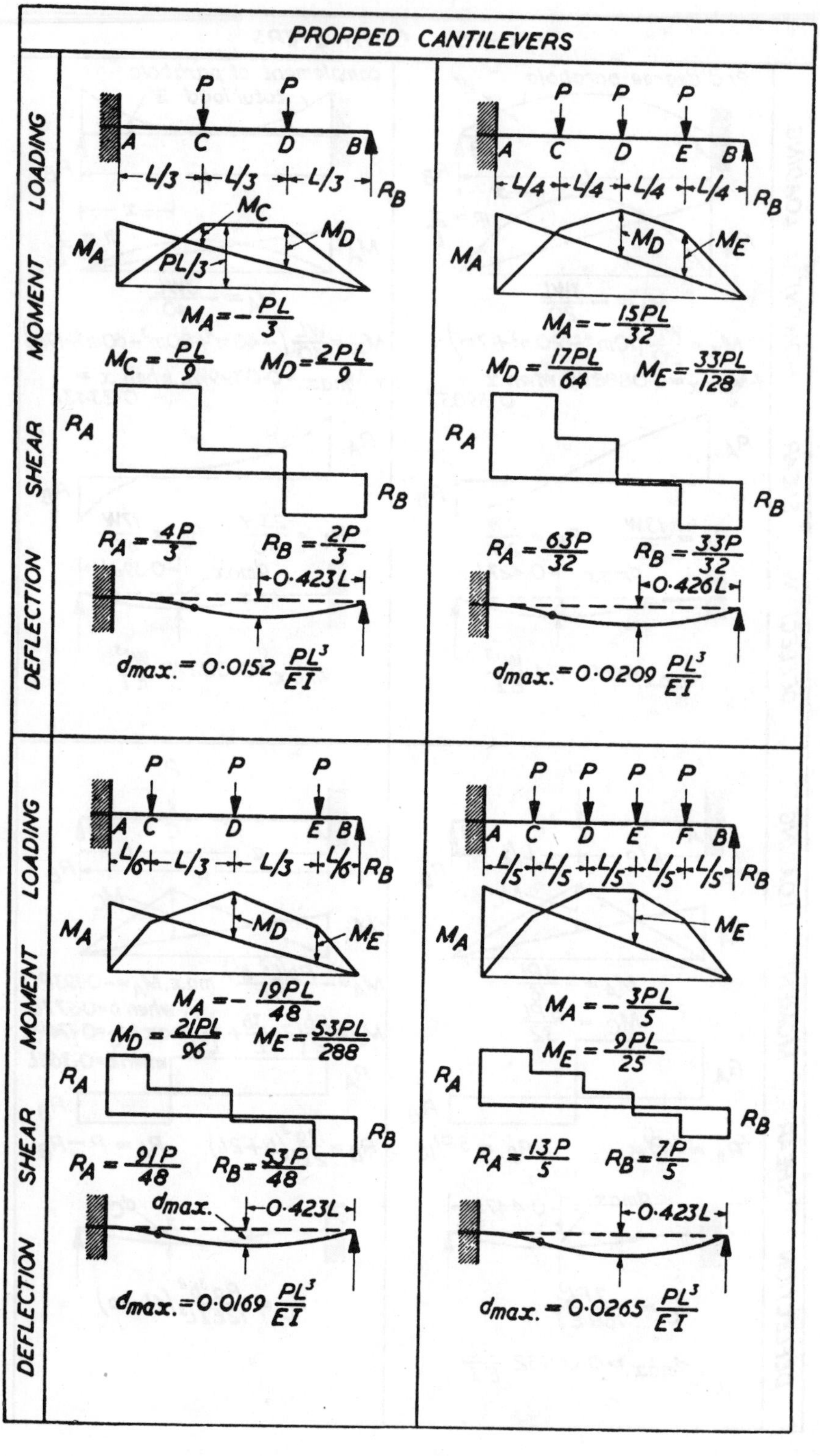
PROPPED CANTILEVERS
LOADING
MOMENT
SHEAR
DEFLECTION
P P
A C D B
L/3 L/3 L/3
R_B
M_C
M_A
PL/3
M_D
$M_A = -\frac{PL}{3}$
$M_C = \frac{PL}{9}$ $M_D = \frac{2PL}{9}$
R_A
R_B
$R_A = \frac{4P}{3}$ $R_B = \frac{2P}{3}$
0·423L
$d_{max.} = 0{\cdot}0152 \frac{PL^3}{EI}$
P P P
A C D E B
L/4 L/4 L/4 L/4
R_B
M_A
M_D
M_E
$M_A = -\frac{15PL}{32}$
$M_D = \frac{17PL}{64}$ $M_E = \frac{33PL}{128}$
R_A
R_B
$R_A = \frac{63P}{32}$ $R_B = \frac{33P}{32}$
0·426L
$d_{max.} = 0{\cdot}0209 \frac{PL^3}{EI}$
P P P
A C D E B
L/6 L/3 L/3 L/6
R_B
M_A
M_D
M_E
$M_A = -\frac{19PL}{48}$
$M_D = \frac{21PL}{96}$ $M_E = \frac{53PL}{288}$
R_A
R_B
$R_A = \frac{91P}{48}$ $R_B = \frac{53P}{48}$
dmax.
0·423L
$d_{max.} = 0{\cdot}0169 \frac{PL^3}{EI}$
P P P P
A C D E F B
L/5 L/5 L/5 L/5 L/5
R_B
M_A
M_E
$M_A = -\frac{3PL}{5}$
$M_E = \frac{9PL}{25}$
R_A
R_B
$R_A = \frac{13P}{5}$ $R_B = \frac{7P}{5}$
0·423L
$d_{max.} = 0{\cdot}0265 \frac{PL^3}{EI}$

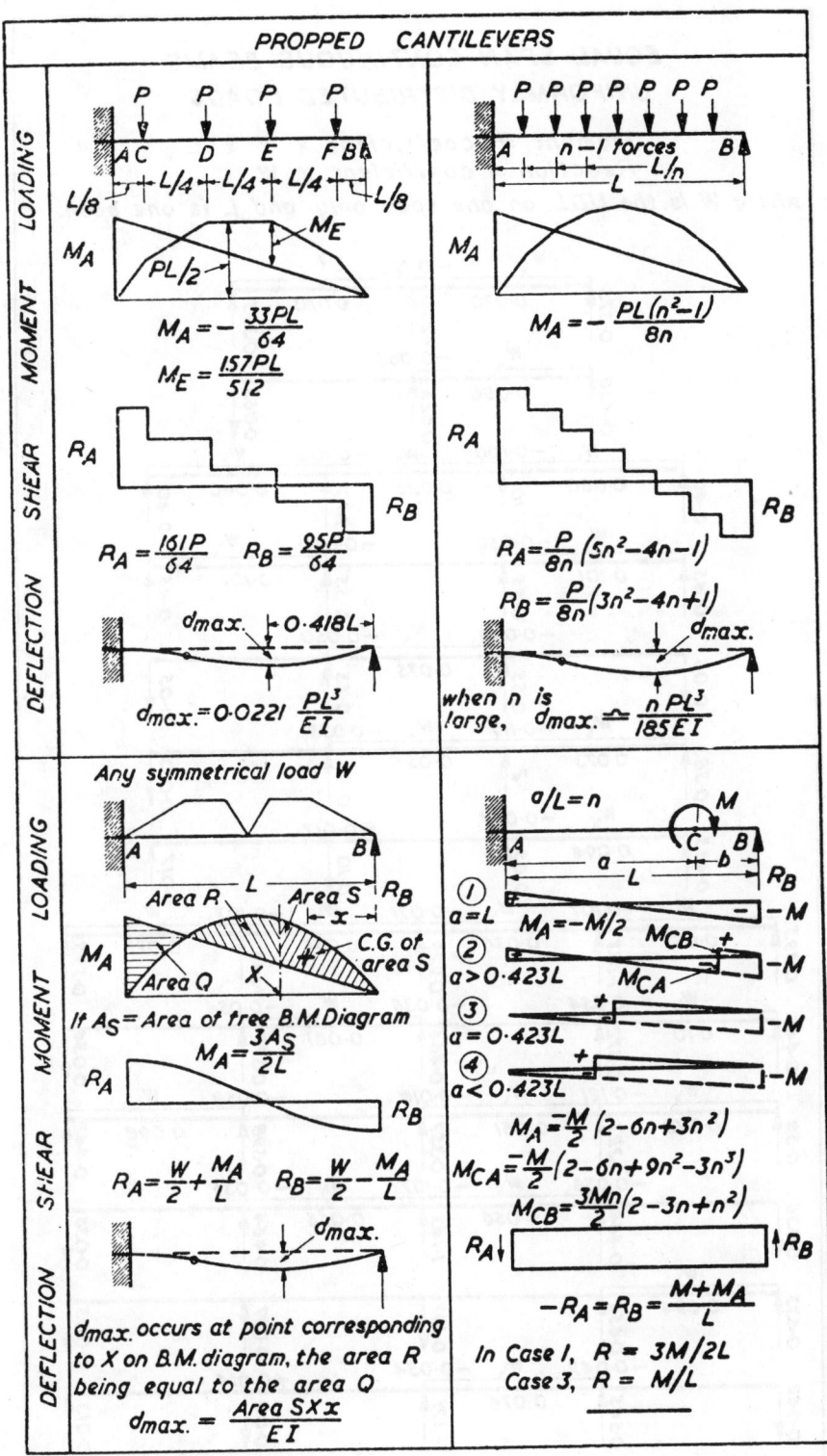

PROPPED CANTILEVERS

LOADING · MOMENT · SHEAR · DEFLECTION

$$M_A = -\frac{33PL}{64}$$

$$M_E = \frac{157PL}{512}$$

$$R_A = \frac{161P}{64} \qquad R_B = \frac{95P}{64}$$

$$d_{max.} = 0.0221 \frac{PL^3}{EI}$$

$d_{max.}$ — $0.418L$

$$M_A = -\frac{PL(n^2-1)}{8n}$$

$n-1$ forces, L/n

$$R_A = \frac{P}{8n}(5n^2-4n-1)$$

$$R_B = \frac{P}{8n}(3n^2-4n+1)$$

when n is large, $d_{max.} \simeq \frac{nPL^3}{185EI}$

Any symmetrical load W

If A_S = Area of free B.M. Diagram

$$M_A = \frac{3A_S}{2L}$$

$$R_A = \frac{W}{2} + \frac{M_A}{L} \qquad R_B = \frac{W}{2} - \frac{M_A}{L}$$

Area R   Area S   C.G. of area S   Area Q   X

$d_{max.}$ occurs at point corresponding to X on B.M. diagram, the area R being equal to the area Q

$$d_{max.} = \frac{Area\ S \times x}{EI}$$

$a/L = n$

① $a = L$ $\quad M_A = -M/2 \quad M_{CB}$ $\quad -M$

② $a > 0.423L$ $\quad M_{CA}$ $\quad -M$

③ $a = 0.423L$ $\quad -M$

④ $a < 0.423L$ $\quad -M$

$$M_A = \frac{-M}{2}(2-6n+3n^2)$$

$$M_{CA} = \frac{-M}{2}(2-6n+9n^2-3n^3)$$

$$M_{CB} = \frac{3Mn}{2}(2-3n+n^2)$$

$$-R_A = R_B = \frac{M+M_A}{L}$$

In Case 1, $R = 3M/2L$

Case 3, $R = M/L$

EQUAL SPAN CONTINUOUS BEAMS
UNIFORMLY DISTRIBUTED LOADS

Moment = coefficient x W x L
Reaction = coefficient x W
where W is the U.D.L. on one span only and L is one span

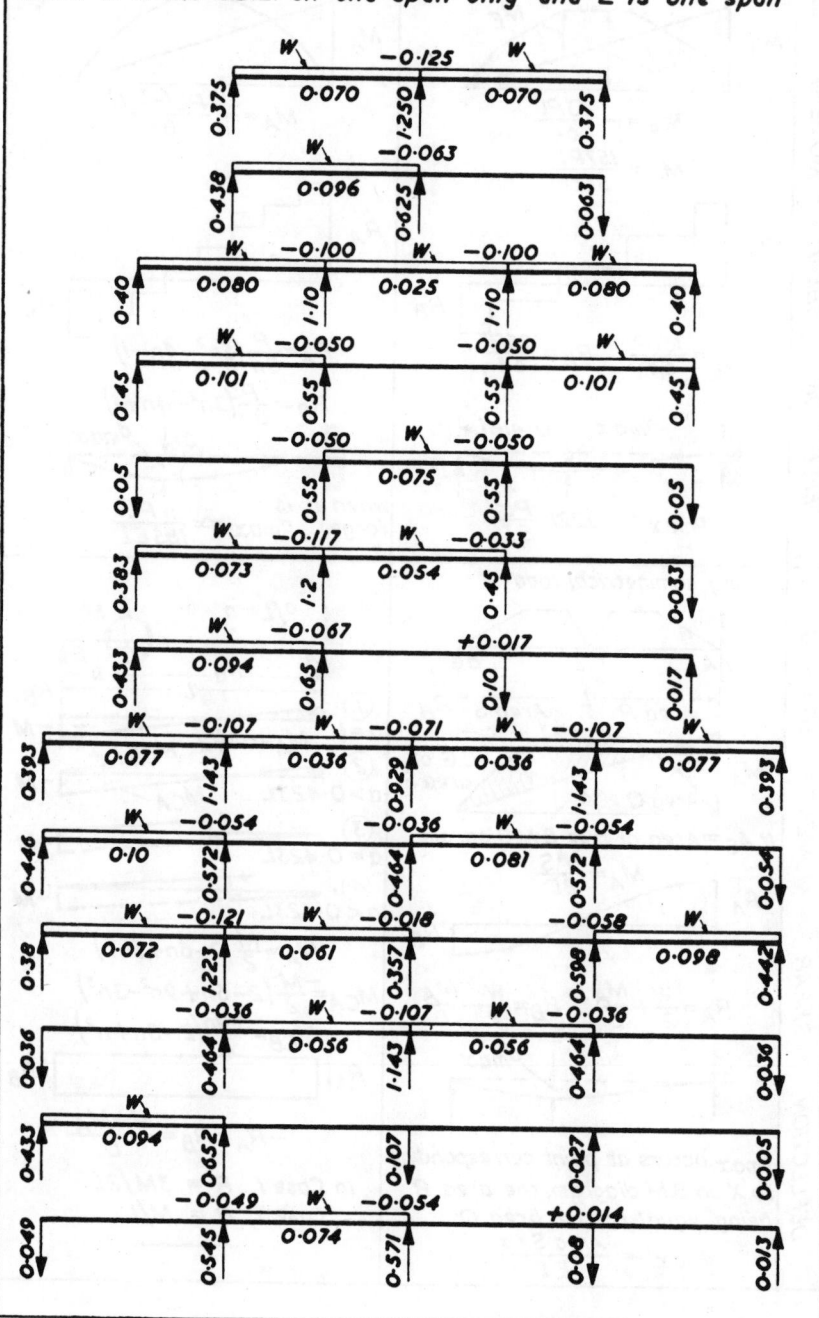

EQUAL SPAN CONTINUOUS BEAMS
CENTRAL POINT LOADS

Moment = coefficient x W x L
Reaction = coefficient x W
where W is the Load on one span only and L is one span

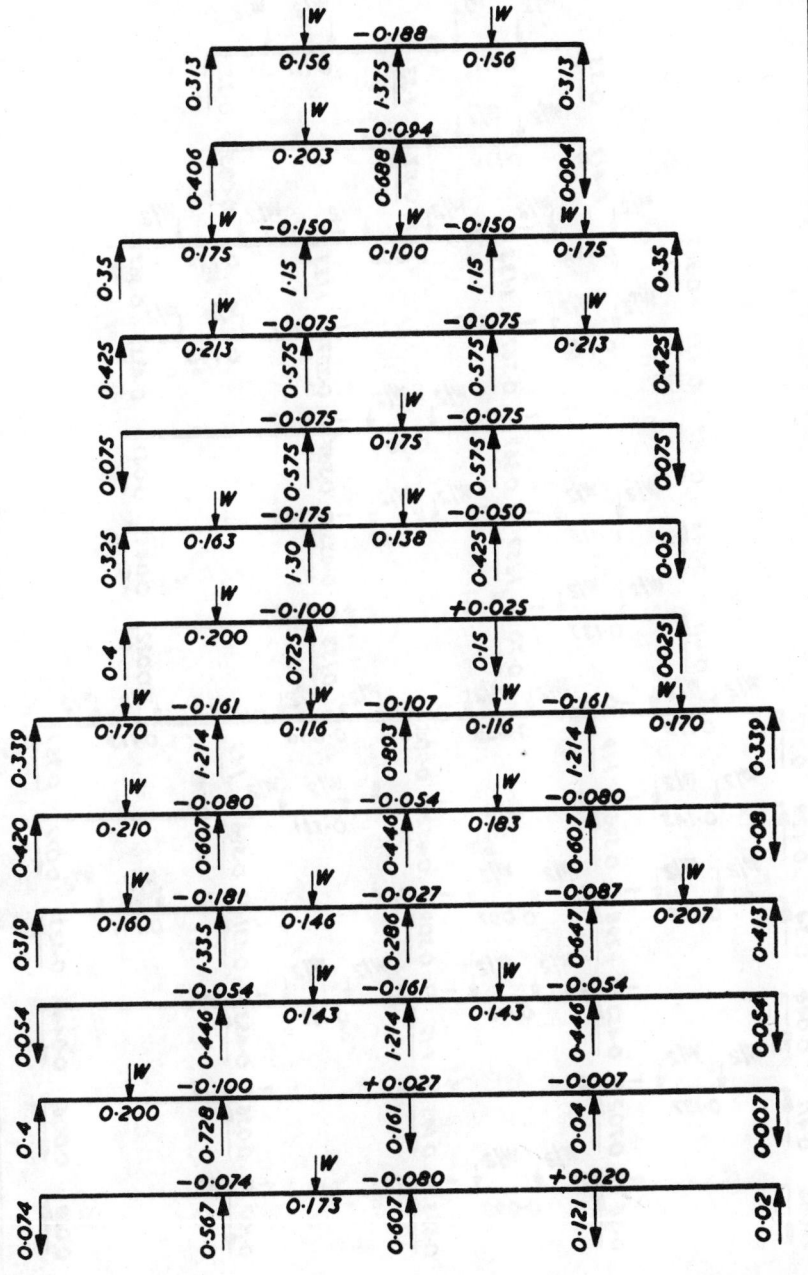

EQUAL SPAN CONTINUOUS BEAMS
POINT LOADS AT THIRD POINTS OF SPANS

Moment = coefficient x W x L
Reaction = coefficient x W
where W is the <u>total</u> load on one span only & L is one span

W/2 W/2   W/2 W/2
-0.167
0.33   0.111   1.33   0.111   0.33

W/2 W/2
-0.083
0.417   0.139   1.67   0.083

W/2 W/2 W/2 W/2 W/2 W/2
-0.133 -0.133
0.367   0.122   1.33   0.033   1.33   0.122   0.367

W/2 W/2   W/2 W/2
-0.067 -0.067
0.433   0.145   0.567   0.567   0.145   0.433

W/2 W/2
-0.067 -0.067
0.067   0.567   0.100   0.567   0.067

W/2 W/2 W/2 W/2
-0.156 -0.045
0.345   0.115   1.267   0.085   0.433   0.045

W/2 W/2
-0.089 $+0.022$
0.411   0.137   0.70   0.33   0.022

W/2 W/2 W/2 W/2 W/2 W/2 W/2 W/2
-0.143 -0.096 -0.143
0.357   0.119   1.19   0.056   0.905   0.056   1.19   0.119   0.357

W/2 W/2   W/2 W/2
-0.072 -0.048 -0.072
0.428   0.143   0.596   0.452   0.111   0.596   0.072

W/2 W/2   W/2 W/2
-0.161 -0.024 -0.078
0.34   0.113   1.298   0.097   0.309   0.631   0.141   0.423

W/2 W/2
-0.048 -0.143 -0.048
0.048   0.453   0.111   1.19   0.111   0.453   0.048

W/2 W/2
-0.089 $+0.024$ -0.006
0.411   0.137   0.702   0.143   0.036   0.006

W/2 W/2
-0.066 -0.071 $+0.018$
0.065   0.56   0.099   0.595   0.107   0.018

Influence lines for bending moments − two-span beam

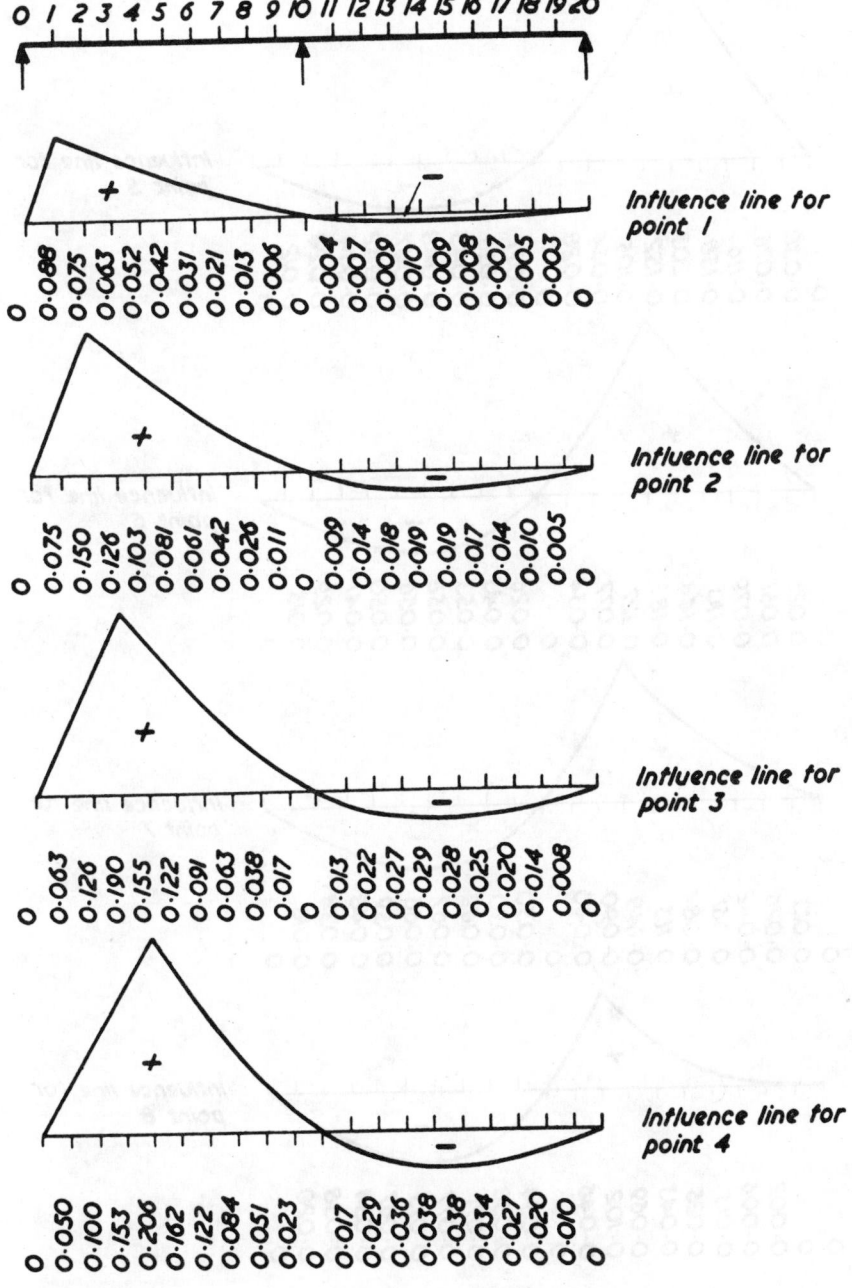

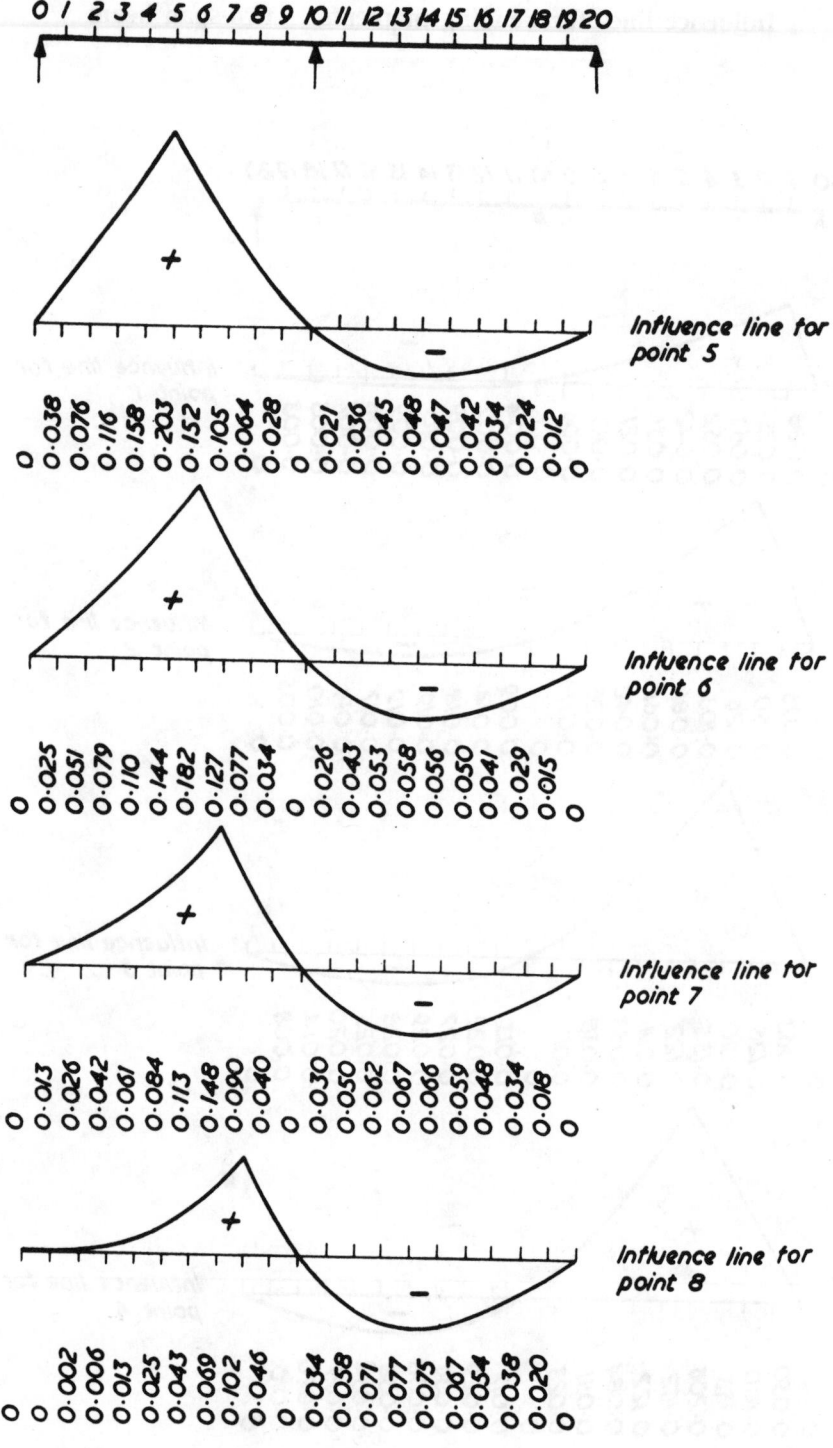

Influence line for point 5

Influence line for point 6

Influence line for point 7

Influence line for point 8

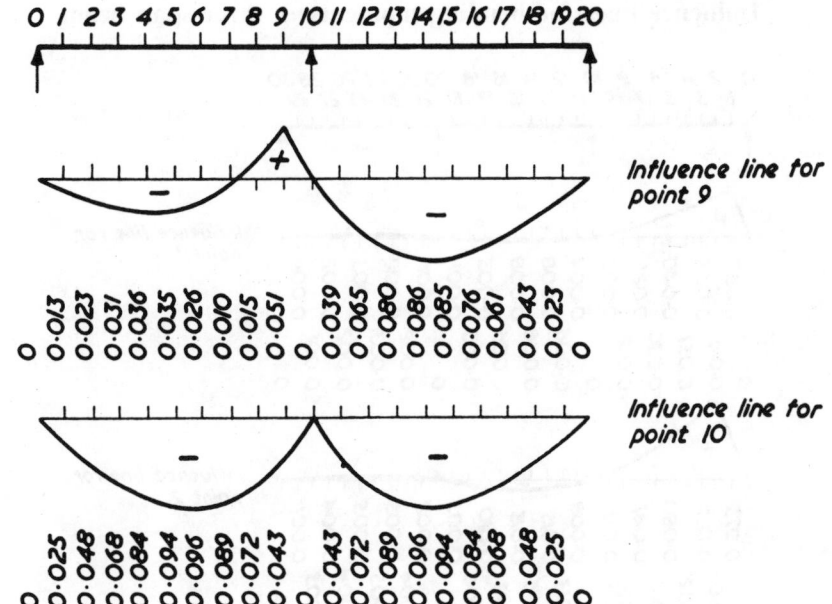

Influence line for point 9

Influence line for point 10

Influence lines for reactions and shear forces — two-span beam

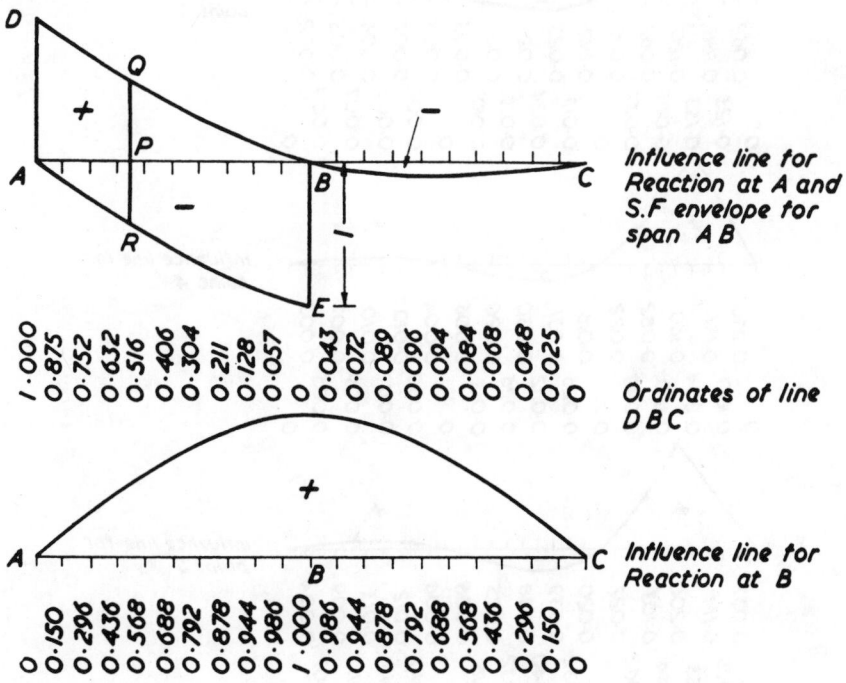

Influence line for Reaction at A and S.F envelope for span A B

Ordinates of line D B C

Influence line for Reaction at B

Influence lines for bending moments – three-span beam

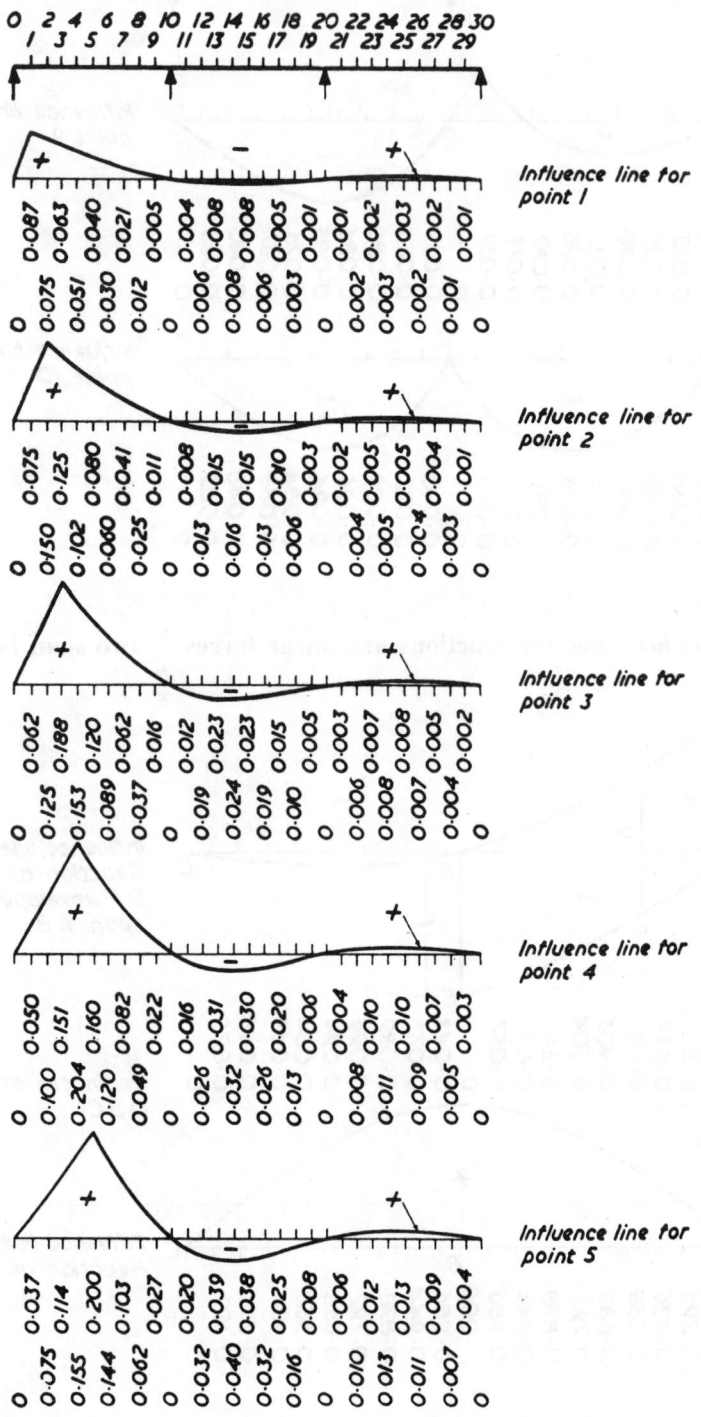

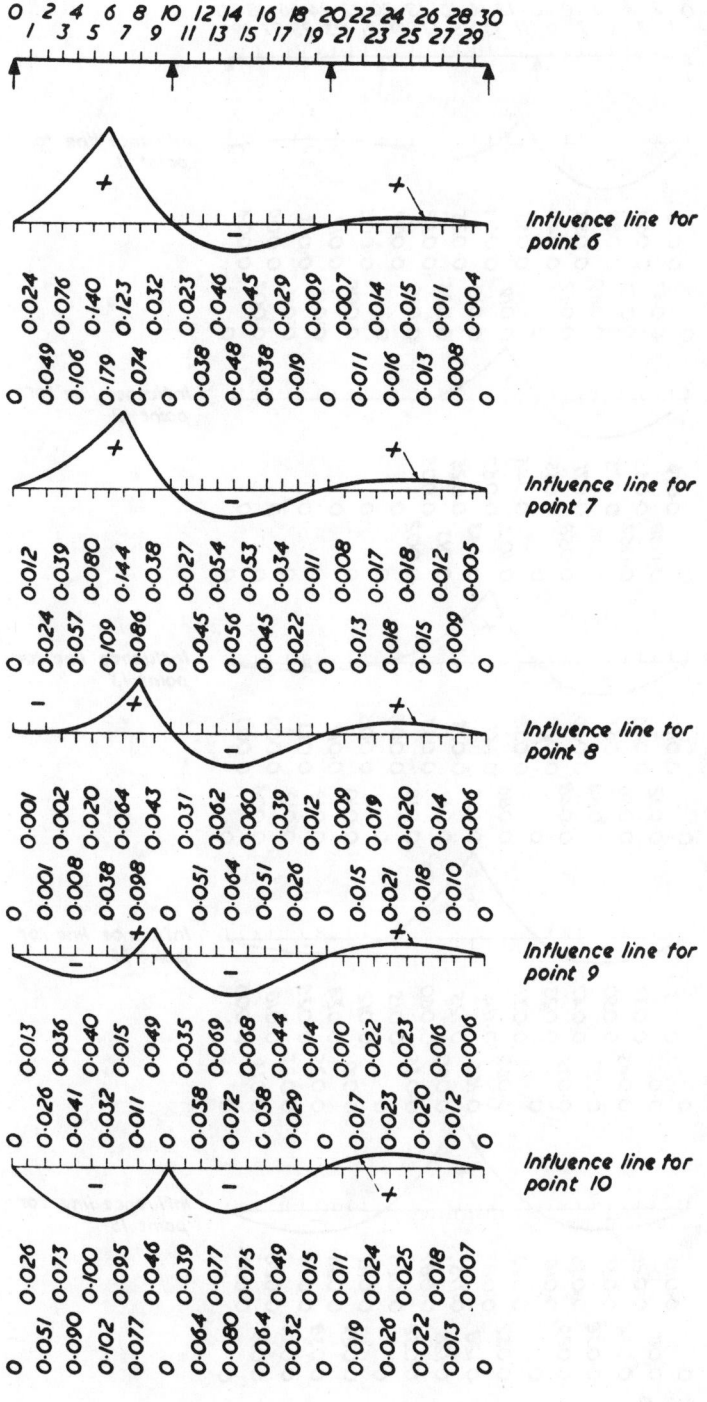

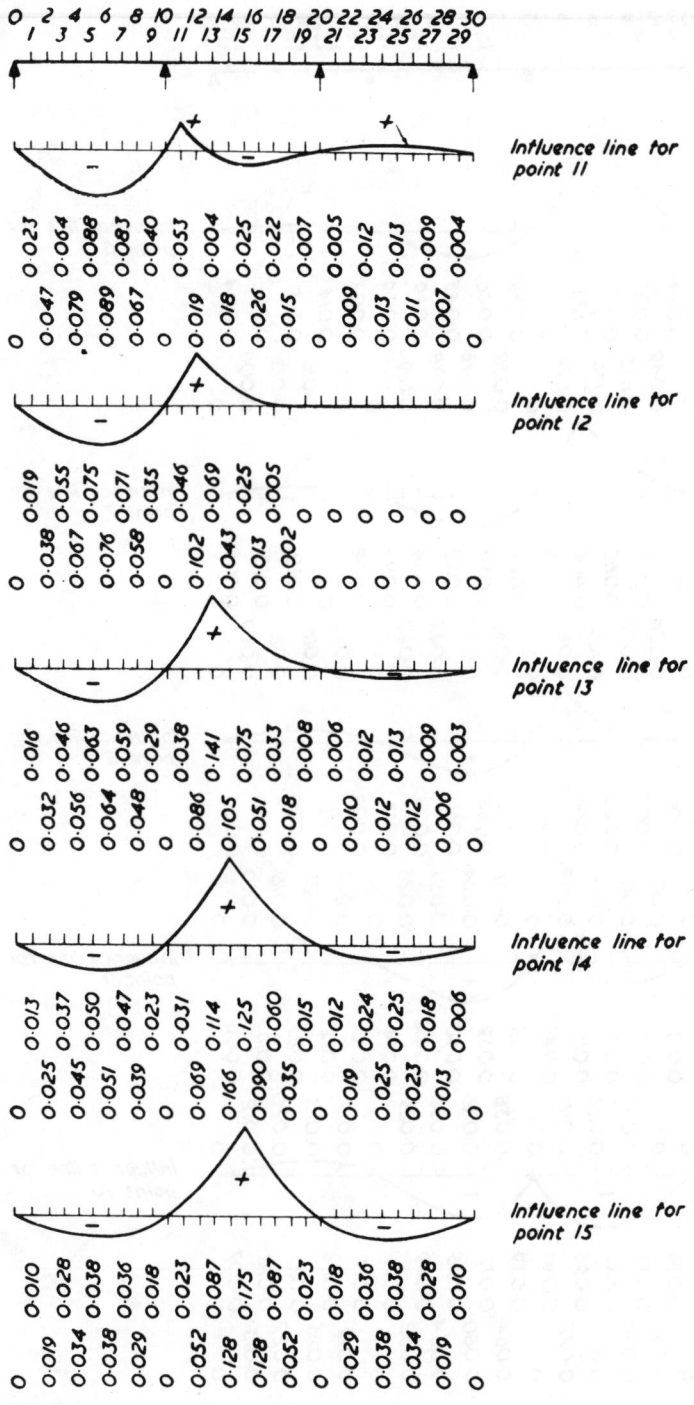

Influence line for point 11

Influence line for point 12

Influence line for point 13

Influence line for point 14

Influence line for point 15

Influence lines for reactions and shear forces — three-span beam

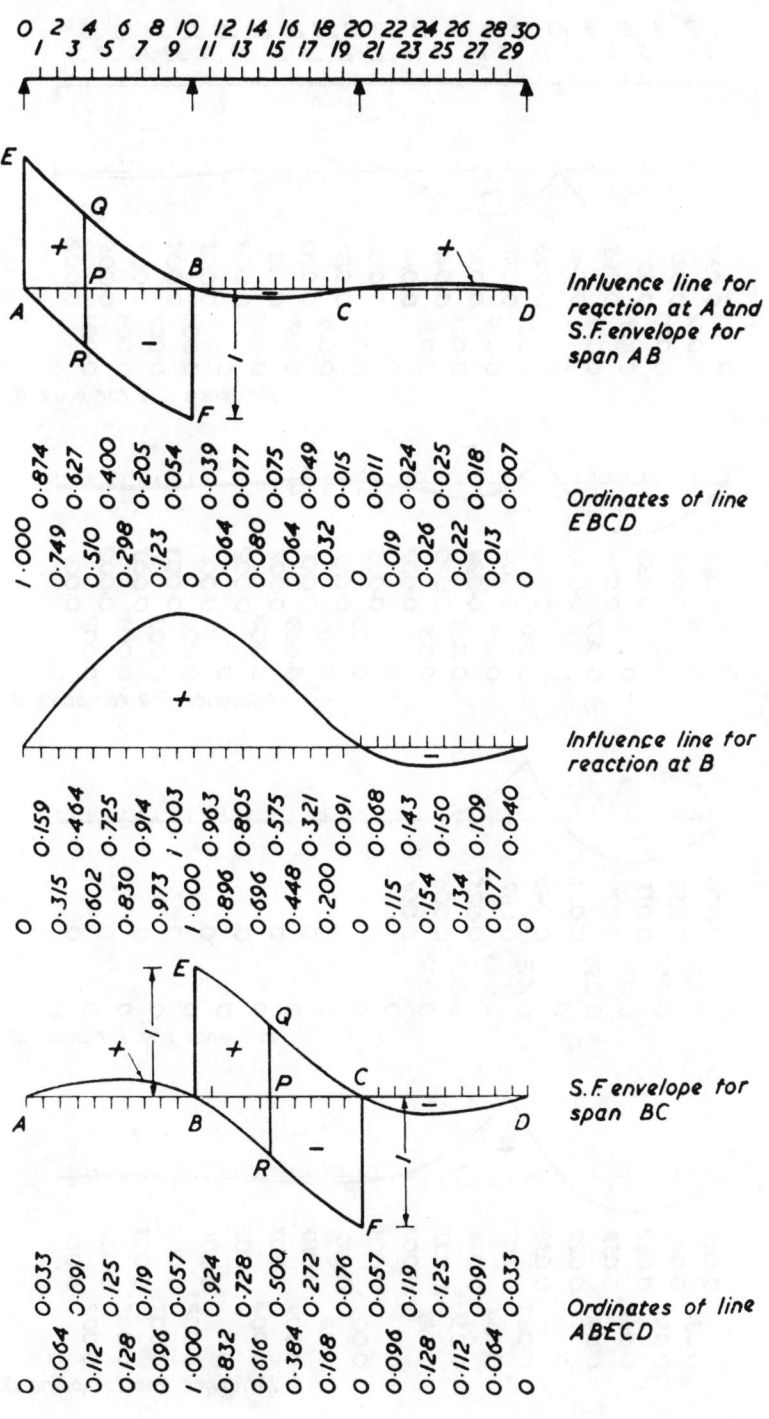

Influence line for
reaction at A and
S.F. envelope for
span AB

Ordinates of line
EBCD

Influence line for
reaction at B

S.F. envelope for
span BC

Ordinates of line
ABECD

Influence lines for bending moments — four-span beam

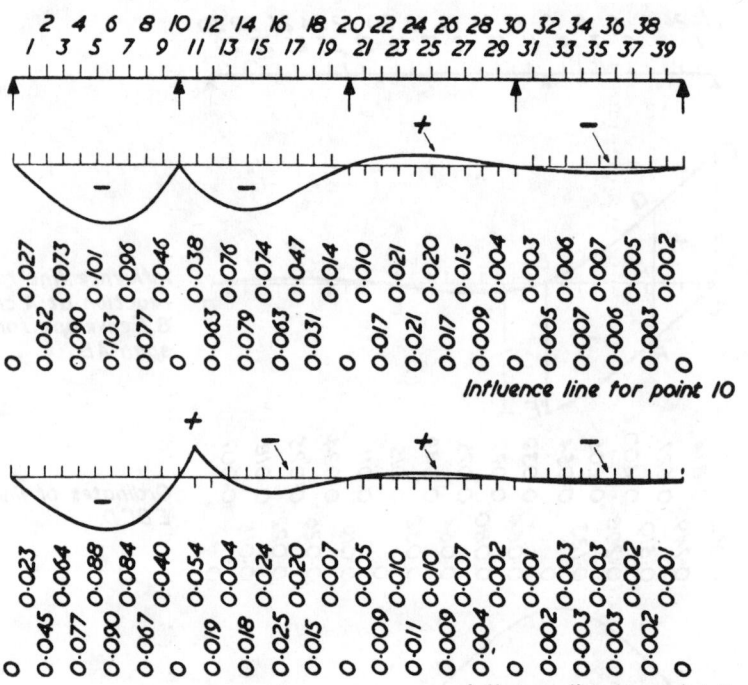

Influence line for point 10

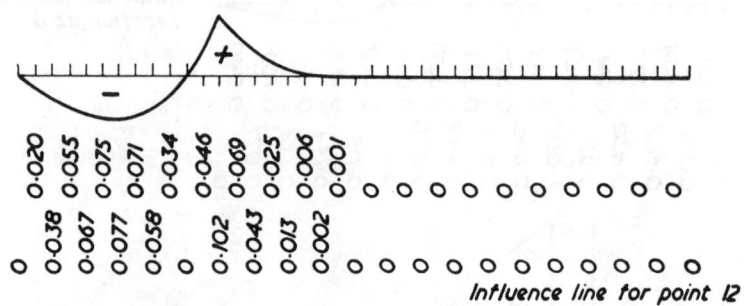

Influence line for point 11

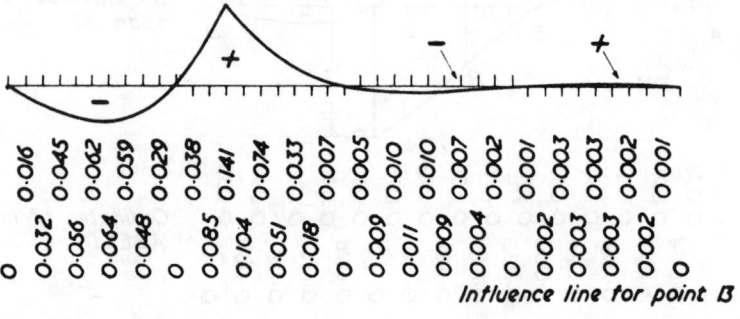

Influence line for point 12

Influence line for point 13

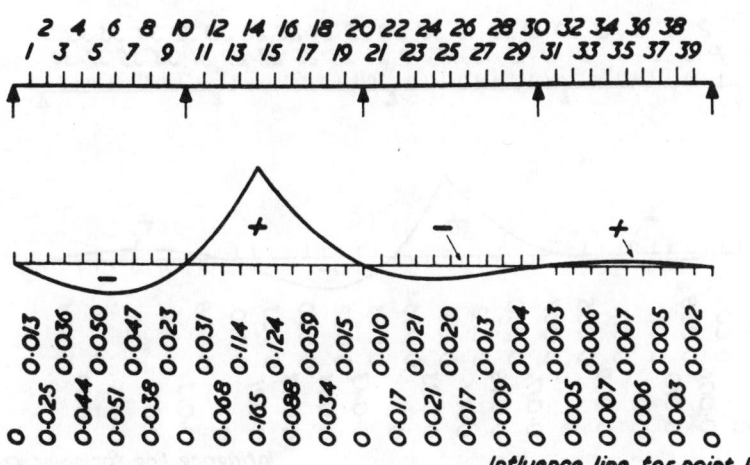

Influence line for point 14

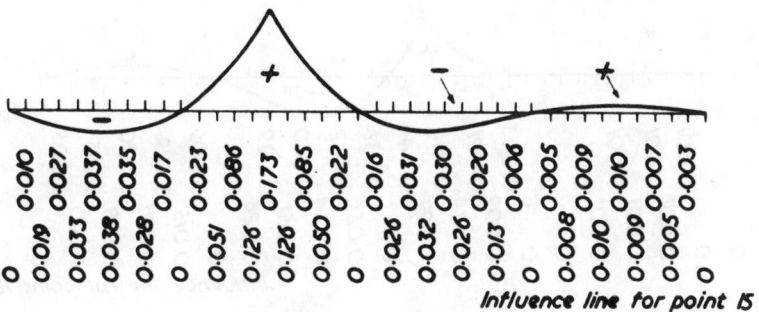

Influence line for point 15

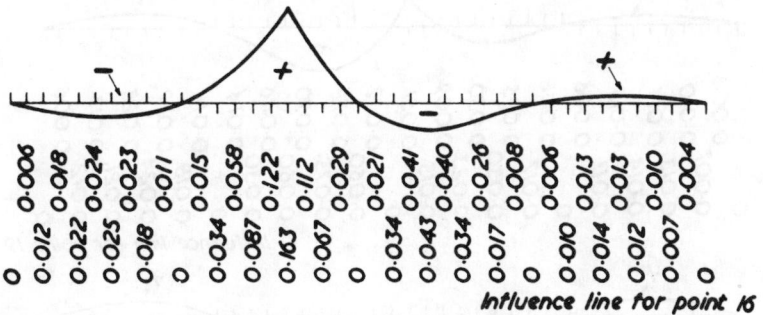

Influence line for point 16

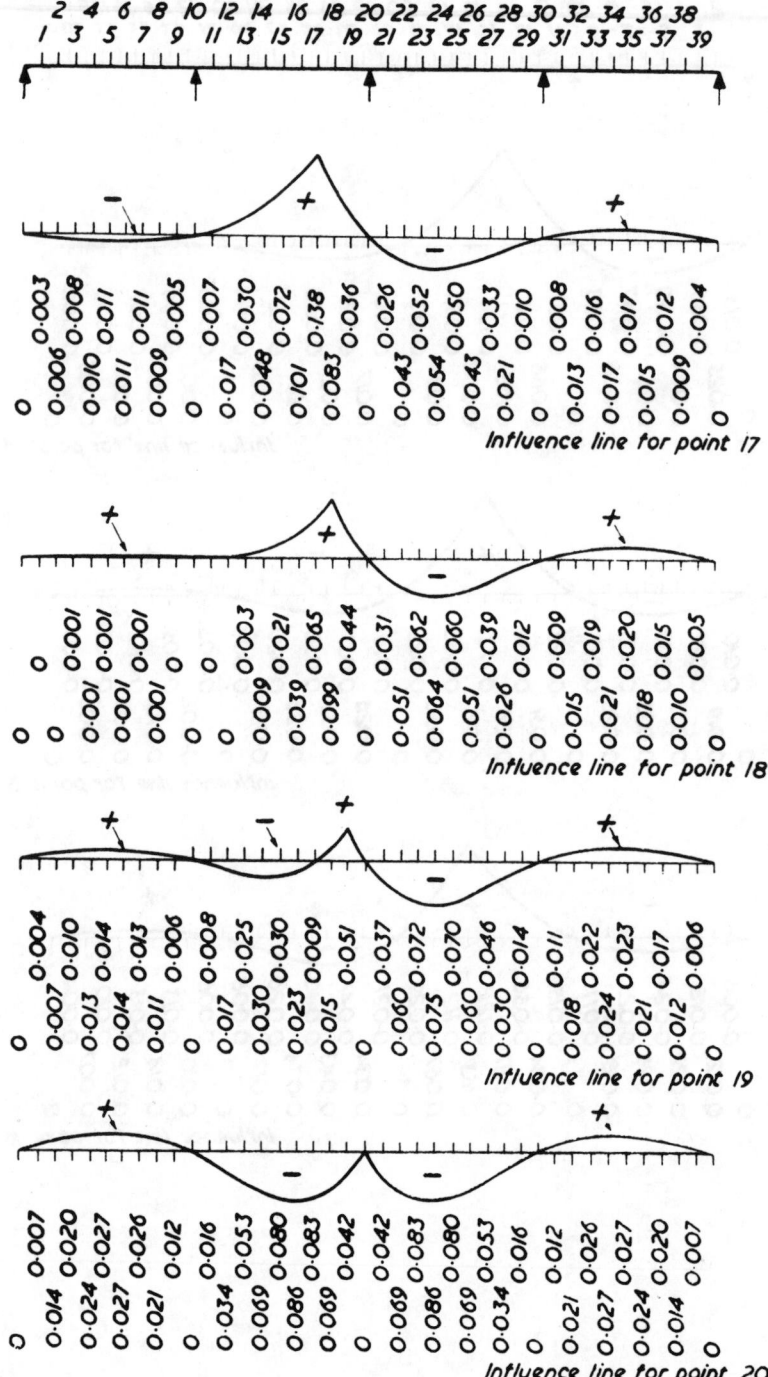

Influence line for point 17

Influence line for point 18

Influence line for point 19

Influence line for point 20

Influence lines for reactions and shear forces — four-span beam

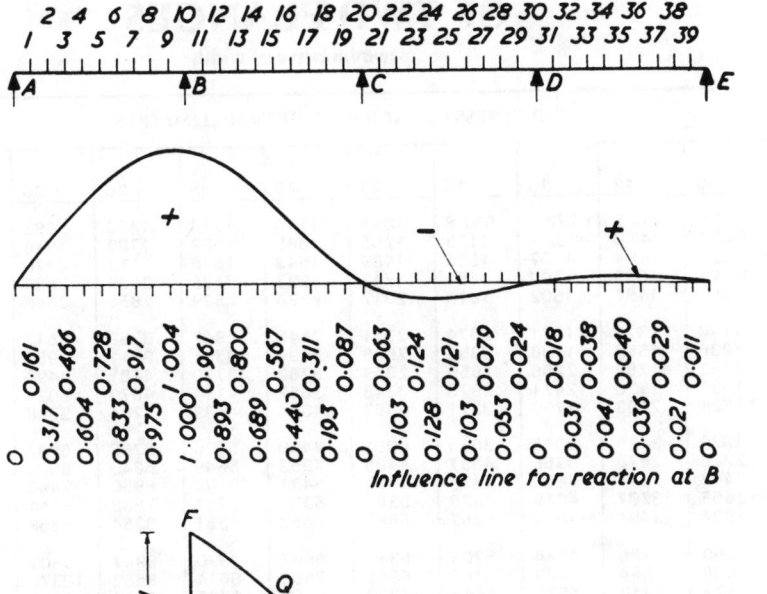

Influence line for reaction at B

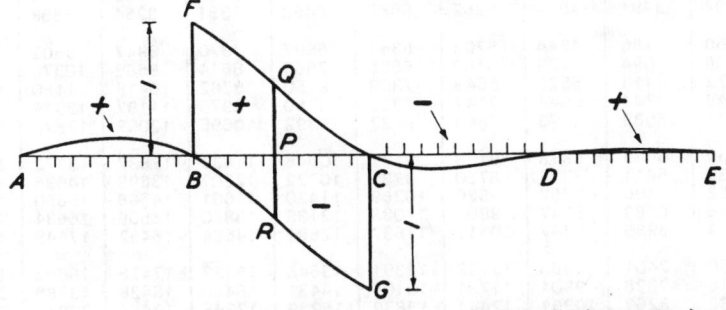

S.F. Envelope for span BC. Ordinates are for line ABFCDE

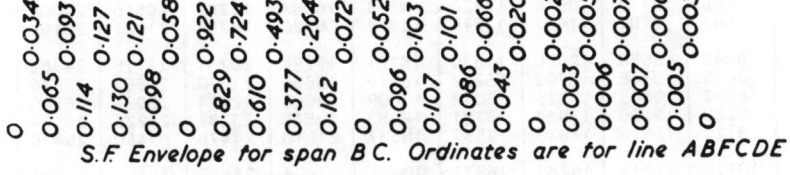

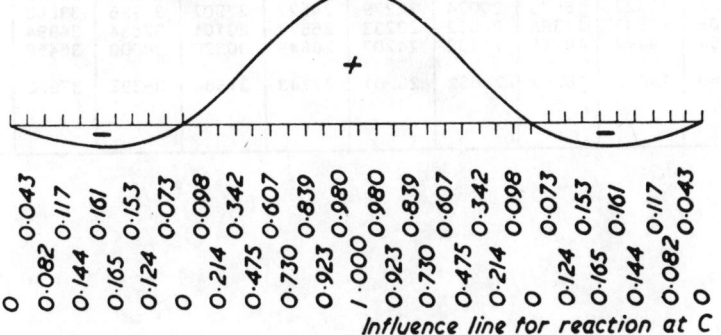

Influence line for reaction at C

SECOND MOMENTS OF AREA (cm⁴)
OF TWO FLANGES
per millimetre of width

Distance d_w mm	THICKNESS OF EACH FLANGE IN MILLIMETRES									
	10	12	15	18	20	22	25	28	30	32
1000	510.1	614.5	772.7	932.8	1041	1149	1314	1480	1592	1705
1100	616.1	742.0	932.5	1125	1255	1385	1582	1782	1916	2051
1200	732.1	881.4	1107	1335	1489	1643	1876	2112	2270	2429
1300	858.1	1033	1297	1564	1743	1923	2195	2469	2654	2839
1400	994.1	1196	1502	1810	2017	2224	2539	2855	3068	3282
1500	1140	1372	1721	2074	2311	2548	2907	3269	3512	3756
1600	1296	1559	1956	2356	2625	2894	3301	3711	3986	4262
1700	1462	1759	2206	2656	2959	3262	3720	4181	4490	4800
1800	1638	1970	2471	2975	3313	3652	4164	4679	5024	5371
1900	1824	2193	2750	3311	3687	4064	4632	5204	5588	5973
2000	2020	2429	3045	3665	4081	4498	5126	5758	6182	6607
2100	2226	2676	3355	4037	4495	4953	5645	6340	6806	7273
2200	2442	2936	3680	4428	4929	5431	6189	6950	7460	7971
2300	2668	3207	4019	4836	5383	5931	6757	7588	8144	8702
2400	2904	3491	4374	5262	5857	6453	7351	8254	8858	9464
2500	3150	3786	4744	5706	6351	6997	7970	8947	9602	10258
2600	3406	4094	5129	6169	6865	7563	8614	9669	10376	11084
2700	3672	4413	5528	6649	7399	8150	9282	10419	11180	11943
2800	3948	4744	5943	7147	7953	8760	9976	11197	12014	12833
2900	4234	5088	6373	7663	8527	9392	10695	12003	12878	13755
3000	4530	5443	6818	8198	9121	10046	11439	12837	13772	14709
3100	4836	5811	7277	8750	9735	10722	12207	13699	14696	15696
3200	5152	6190	7752	9320	10369	11420	13001	14588	15650	16714
3300	5478	6582	8242	9908	11023	12139	13820	15506	16634	17764
3400	5814	6985	8747	10515	11697	12881	14664	16452	17648	18846
3500	6160	7401	9266	11139	12391	13645	15532	17426	18692	19961
3600	6516	7828	9801	11781	13105	14431	16426	18428	19766	21107
3700	6882	8267	10351	12441	13839	15239	17345	19458	20870	22285
3800	7258	8719	10916	13120	14593	16069	18289	20515	22004	23495
3900	7644	9182	11495	13816	15367	16920	19257	21601	23168	24738
4000	8040	9658	12090	14530	16161	17794	20251	22715	24362	26012
4100	8446	10145	12700	15262	16975	18690	21270	23857	25586	27318
4200	8862	10645	13325	16012	17809	19608	22314	25027	26840	28656
4300	9288	11156	13964	16781	18663	20548	23382	26225	28124	30027
4400	9724	11679	14619	17567	19537	21510	24476	27450	29438	31429
4500	10170	12215	15289	18371	20431	22494	25595	28704	30782	32863
4600	10626	12762	15974	19193	21345	23499	26739	29986	32156	34329
4700	11092	13322	16673	20034	22279	24527	27907	31296	33560	35827
4800	11568	13893	17388	20892	23233	25577	29101	32634	34994	37358
4900	12054	14477	18118	21768	24207	26649	30320	34000	36458	38920
5000	12550	15072	18863	22662	25201	27743	31564	35393	37952	40514

SECOND MOMENTS OF AREA (cm⁴) OF TWO FLANGES

per millimetre of width

THICKNESS OF EACH FLANGE IN MILLIMETRES										Distance
35	**38**	**40**	**45**	**50**	**55**	**60**	**65**	**70**	**75**	d_w mm
1875	2048	2164	2459	2758	3064	3374	3691	4013	4341	1000
2255	2461	2600	2951	3308	3671	4040	4416	4797	5184	1100
2670	2913	3076	3489	3908	4334	4766	5205	5651	6103	1200
3120	3402	3592	4072	4558	5052	5552	6060	6575	7097	1300
3604	3930	4148	4700	5258	5825	6398	6980	7569	8166	1400
4124	4495	4744	5372	6008	6652	7304	7965	8633	9309	1500
4679	5099	5380	6090	6808	7535	8270	9014	9767	10528	1600
5269	5740	6056	6853	7658	8473	9296	10129	10971	11822	1700
5893	6420	6772	7661	8558	9466	10382	11309	12245	13191	1800
6553	7137	7528	8513	9508	10513	11528	12554	13589	14634	1900
7248	7892	8324	9411	10508	11616	12734	13863	15003	16153	2000
7978	8686	9160	10354	11558	12774	14000	15238	16487	17747	2100
8742	9517	10036	11342	12658	13987	15326	16678	18041	19416	2200
9542	10387	10952	12374	13808	15254	16712	18183	19665	21159	2300
10377	11294	11908	13452	15008	16577	18158	19752	21359	22978	2400
11247	12240	12904	14575	16258	17955	19664	21387	23123	24872	2500
12151	13223	13940	15743	17558	19388	21230	23087	24957	26841	2600
13091	14245	15016	16955	18908	20875	22856	24852	26861	28884	2700
14066	15304	16132	18213	20308	22418	24542	26681	28835	31003	2800
15076	16401	17288	19516	21758	24016	26288	28576	30879	33197	2900
16120	17537	18484	20864	23258	25669	28094	30536	32993	35466	3000
17200	18710	19720	22256	24808	27376	29960	32561	35177	37809	3100
18315	19922	20996	23694	26408	29139	31886	34650	37431	40228	3200
19465	21171	22312	25177	28058	30957	33872	36805	39755	42722	3300
20649	22459	23668	26705	29758	32830	35918	39025	42149	45291	3400
21869	23784	25064	28277	31508	34757	38024	41310	44613	47934	3500
23124	25147	26500	29895	33308	36740	40190	43659	47147	50653	3600
24414	26549	27976	31558	35158	38778	42416	46074	49751	53447	3700
25738	27988	29492	33266	37058	40871	44702	48554	52425	56316	3800
27098	29466	31048	35018	39008	43018	47048	51099	55169	59259	3900
28493	30981	32644	36816	41008	45221	49454	53708	57983	62278	4000
29923	32535	34280	38659	43058	47479	51920	56383	60867	65372	4100
31387	34126	35956	40547	45158	49792	54446	59123	63821	68541	4200
32887	35756	37672	42479	47308	52159	57032	61928	66845	71784	4300
34422	37423	39428	44457	49508	54582	59678	64797	69939	75103	4400
35992	39128	41224	46480	51758	57060	62384	67732	73103	78497	4500
37596	40872	43060	48548	54058	59593	65150	70732	76337	81966	4600
39236	42653	44936	50660	56408	62180	67976	73797	79641	85509	4700
40911	44473	46852	52818	58808	64823	70862	76926	83015	89128	4800
42621	46330	48808	55021	61258	67521	73808	80121	86459	92822	4900
44365	48226	50804	57269	63758	70274	76814	83381	89973	96591	5000

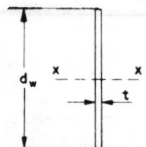

SECOND MOMENTS OF AREA (cm⁴)
OF RECTANGULAR PLATES
about axis x − x

Depth d_w mm	THICKNESS t MILLIMETRES					
	3	**4**	**5**	**6**	**8**	**10**
25	.391	.521	.651	.781	1.04	1.30
50	3.13	4.17	5.21	6.25	8.33	10.4
75	10.5	14.1	17.6	21.1	28.1	10.4
100	25.0	33.3	41.7	50.0	66.7	83.3
125	48.8	65.1	81.4	97.7	130	163
150	84.4	113	141	169	225	281
175	134	179	223	268	357	447
200	200	267	333	400	533	667
225	285	380	475	570	759	949
250	391	521	651	781	1042	1302
275	520	693	867	1040	1386	1733
300	675	900	1125	1350	1800	2250
325	858	1144	1430	1716	2289	2861
350	1072	1429	1786	2144	2858	3573
375	1318	1758	2197	2637	3516	4395
400	1600	2133	2667	3200	4267	5333
425	1919	2559	3199	3838	5118	6397
450	2278	3038	3797	4556	6075	7594
475	2679	3572	4465	5359	7145	8931
500	3125	4167	5208	6250	8333	10417
525	3618	4823	6029	7235	9647	12059
550	4159	5546	6932	8319	11092	13865
575	4753	6337	7921	9505	12674	15842
600	5400	7200	9000	10800	14400	18000
625	6104	8138	10173	12207	16276	20345
650	6866	9154	11443	13731	18308	22885
675	7689	10252	12814	15377	20503	25629
700	8575	11433	14292	17150	22867	28583
725	9527	12703	15878	19054	25405	31757
750	10547	14063	17578	21094	28125	35156
775	11637	15516	19395	23274	31032	38790
800	12800	17067	21333	25600	34133	42667
825	14038	18717	23396	28076	37434	46793
850	15353	20471	25589	30706	40942	51177
875	16748	22331	27913	33496	44661	55827
900	18225	24300	30375	36450	48600	60750

SECOND MOMENTS OF AREA (cm⁴) OF RECTANGULAR PLATES

about axis x–x

THICKNESS t MILLIMETRES						Depth
12	**15**	**18**	**20**	**22**	**25**	d_n mm
1.56	1.95	2.34	2.60	2.86	3.26	25
12.5	15.6	18.8	20.8	22.9	26.0	50
42.2	52.7	63 3	70.3	77.3	87.9	75
100	125	150	167	183	208	100
195	244	293	326	358	407	125
338	422	506	563	619	703	150
536	670	804	893	983	1117	175
800	1000	1200	1333	1467	1667	200
1139	1424	1709	1898	2088	2373	225
1563	1953	2344	2604	2865	3255	250
2080	2600	3120	3466	3813	4333	275
2700	3375	4050	4500	4950	5625	300
3433	4291	5149	5721	6293	7152	325
4288	5359	6431	7146	7860	8932	350
5273	6592	7910	8789	9668	10986	375
6400	8000	9600	10667	11733	13333	400
7677	9596	11515	12794	14074	15993	425
9113	11391	13669	15188	16706	18984	450
10717	13396	16076	17862	19648	22327	475
12500	15625	18750	20833	22917	26042	500
14470	18088	21705	24117	26529	30146	525
16638	20797	24956	27729	30502	34661	550
19011	23764	28516	31685	34853	39606	575
21600	27000	32400	36000	39600	45000	600
24414	30518	36621	40690	44759	50863	625
27463	34328	41194	45771	50348	57214	650
30755	38443	46132	51258	56384	64072	675
34300	42875	51450	57167	62883	71458	700
38108	47635	57162	63513	69864	79391	725
42188	52734	63281	70313	77344	87891	750
46548	58186	69823	77581	85339	96976	775
51200	64000	76800	85333	93867	106667	800
56152	70189	84227	93586	102945	116982	825
61413	76766	92119	102354	112590	127943	850
66992	83740	100488	111654	122819	139567	875
72900	91125	109350	121500	133650	151875	900

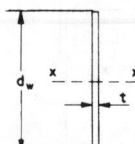

SECOND MOMENTS OF AREA (cm⁴)
OF RECTANGULAR PLATES

about axis x—x

Depth dₓ mm	THICKNESS t MILLIMETRES					
	3	**4**	**5**	**6**	**8**	**10**
1000	25000	33333	41667	50000	66667	83333
1100	33275	44367	55458	66550	88733	110917
1200	43200	57600	72000	86400	115200	144000
1300	54925	73233	91542	109850	146467	183083
1400	68600	91467	114333	137200	182933	228667
1500	84375	112500	140625	168750	225000	281250
1600	102400	136533	170667	204800	273067	341333
1700	122825	163767	204708	245650	327533	409417
1800	145800	194400	243000	291600	388800	486000
1900	171475	228633	285792	342950	457267	571583
2000	200000	266667	333333	400000	533333	666667
2100	231525	308700	385875	463050	617400	771750
2200	266200	354933	443667	532400	709867	887333
2300	304175	405567	506958	608350	811133	1013917
2400	345600	460800	576000	691200	921600	1152000
2500	390625	520833	651042	781250	1041667	1302083
2600	439400	585867	732333	878800	1171733	1464667
2700	492075	656100	820125	984150	1312200	1640250
2800	548800	731733	914667	1097600	1463467	1829333
2900	609725	812967	1016208	1219450	1625933	2032417
3000	675000	900000	1125000	1350000	1800000	2250000
3100	744775	993033	1241292	1489550	1986067	2482583
3200	819200	1092267	1365333	1638400	2184533	2730667
3300	898425	1197900	1497375	1796850	2395800	2994750
3400	982600	1310133	1637667	1965200	2620267	3275333
3500	1071875	1429167	1786458	2143750	2858333	3572917
3600	1166400	1555200	1944000	2332800	3110400	3888000
3700	1266325	1688433	2110542	2532650	3376867	4221083
3800	1371800	1829067	2286333	2743600	3658133	4572667
3900	1482975	1977300	2471625	2965950	3954600	4943250
4000	1600000	2133333	2666667	3200000	4266667	5333333
4100	1723025	2297367	2871708	3446050	4594733	5743417
4200	1852200	2469600	3087000	3704400	4939200	6174000
4300	1987675	2650233	3312792	3975350	5300467	6625583
4400	2129600	2839467	3549333	4259200	5678933	7098667
4500	2278125	3037500	3796875	4556250	6075000	7593750
4600	2433400	3244533	4055667	4866800	6489067	8111333
4700	2595575	3460767	4325958	5191150	6921533	8651917
4800	2764800	3686400	4608000	5529600	7372800	9216000
4900	2941225	3921633	4902042	5882450	7843267	9804083
5000	3125000	4166667	5208333	6250000	8333333	10416667

SECOND MOMENTS OF AREA (cm⁴) OF RECTANGULAR PLATES

about axis x—x

THICKNESS t MILLIMETRES						Depth
12	**15**	**18**	**20**	**22**	**25**	**d_w mm**
100000	125000	150000	166667	183333	208333	1000
133100	166375	199650	221833	244017	277292	1100
172800	216000	259200	288000	316800	360000	1200
219700	274625	329550	366167	402783	457708	1300
274400	343000	411600	457333	503067	571667	1400
337500	421875	506250	562500	618750	703125	1500
409600	512000	614400	682667	750933	853333	1600
491300	614125	736950	818833	900717	1023542	1700
583200	729000	874800	972000	1069200	1215000	1800
685900	857375	1028850	1143167	1257483	1428958	1900
800000	1000000	1200000	1333333	1466667	1666667	2000
926100	1157625	1389150	1543500	1697850	1929375	2100
1064800	1331000	1597200	1774667	1952133	2218333	2200
1216700	1520875	1825050	2027833	2230617	2534792	2300
1382400	1728000	2073600	2304000	2534400	2880000	2400
1562500	1953125	2343750	2604167	2864583	3255208	2500
1757600	2197000	2636400	2929333	3222267	3661667	2600
1968300	2460375	2952450	3280500	3608550	4100625	2700
2195200	2744000	3292800	3658667	4024533	4573333	2800
2438900	3048625	3658350	4064833	4471317	5081042	2900
2700000	3375000	4050000	4500000	4950000	5625000	3000
2979100	3723875	4468650	4965167	5461683	6206458	3100
3276800	4096000	4915200	5461333	6007467	6826667	3200
3593700	4492125	5390550	5989500	6588450	7486875	3300
3930400	4913000	5895600	6550667	7205733	8188333	3400
4287500	5359375	6431250	7145833	7860417	8932292	3500
4665600	5832000	6998400	7776000	8553600	9720000	3600
5065300	6331625	7597950	8442167	9286383	10552708	3700
5487200	6859000	8230800	9145333	10059867	11431667	3800
5931900	7414875	8897850	9886500	10875150	12358125	3900
6400000	8000000	9600000	10666667	11733333	13333333	4000
6892100	8615125	10338150	11486833	12635517	14358542	4100
7408800	9261000	11113200	12348000	13582800	15435000	4200
7950700	9938375	11926050	13251167	14576283	16563958	4300
8518400	10648000	12777600	14197333	15617067	17746667	4400
9112500	11390625	13668750	15187500	16706250	18984375	4500
9733600	12167000	14600400	16222667	17844933	20278333	4600
10382300	12977875	15573450	17303833	19034217	21629792	4700
11059200	13824000	16588800	18432000	20275200	23040000	4800
11764900	14706125	17647350	19608167	21568983	24510208	4900
12500000	15625000	18750000	20833333	22916667	26041667	5000

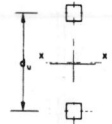

SECOND MOMENT
OF A PAIR OF UNIT AREAS

about axis x—x

Distance d_u mm	0	5	10	15	20	25	30	35	40	45
500	1250	1275	1301	1326	1352	1378	1405	1431	1458	1485
550	1513	1540	1568	1596	1625	1653	1682	1711	1741	1770
600	1800	1830	1861	1891	1922	1953	1985	2016	2048	2080
650	2113	2145	2178	2211	2245	2278	2312	2346	2381	2415
700	2450	2485	2521	2556	2592	2628	2665	2701	2738	2775
750	2813	2850	2888	2926	2965	3003	3042	3081	3121	3160
800	3200	3240	3281	3321	3362	3403	3445	3486	3528	3570
850	3613	3655	3698	3741	3785	3828	3872	3916	3961	4005
900	4050	4095	4141	4186	4232	4278	4325	4371	4418	4465
950	4513	4560	4608	4656	4705	4753	4802	4851	4901	4950
1000	5000	5050	5101	5151	5202	5253	5305	5356	5408	5460
1050	5513	5565	5618	5671	5725	5778	5832	5886	5941	5995
1100	6050	6105	6161	6216	6272	6328	6385	6441	6498	6555
1150	6613	6670	6728	6786	6845	6903	6962	7021	7081	7140
1200	7200	7260	7321	7381	7442	7503	7565	7626	7688	7750
1250	7813	7875	7938	8001	8065	8128	8192	8256	8321	8385
1300	8450	8515	8581	8646	8712	8778	8845	8911	8978	9045
1350	9113	9180	9248	9316	9385	9453	9522	9591	9661	9730
1400	9800	9870	9941	10011	10082	10153	10225	10296	10368	10440
1450	10513	10585	10658	10731	10805	10878	10952	11026	11101	11175
1500	11250	11325	11401	11476	11552	11628	11705	11781	11858	11935
1550	12013	12090	12168	12246	12325	12403	12482	12561	12641	12720
1600	12800	12880	12961	13041	13122	13203	13285	13366	13448	13530
1650	13613	13695	13778	13861	13945	14028	14112	14196	14281	14365
1700	14450	14535	14621	14706	14792	14878	14965	15051	15138	15225
1750	15313	15400	15488	15576	15665	15753	15842	15931	16021	16110
1800	16200	16290	16381	16471	16562	16653	16745	16836	16928	17020
1850	17113	17205	17298	17391	17485	17578	17672	17766	17861	17955
1900	18050	18145	18241	18336	18432	18528	18625	18721	18818	18915
1950	19013	19110	19208	19306	19405	19503	19602	19701	19801	19900
2000	20000	20100	20201	20301	20402	20503	20605	20706	20808	20910
2050	21013	21115	21218	21321	21425	21528	21632	21736	21841	21945
2100	22050	22155	22261	22366	22472	22578	22685	22791	22898	23005
2150	23113	23220	23328	23436	23545	23653	23762	23871	23981	24090
2200	24200	24310	24421	24531	24642	24753	24865	24976	25088	25200
2250	25313	25425	25538	25651	25765	25878	25992	26106	26221	26335
2300	26450	26565	26681	26796	26912	27028	27145	27261	27378	27495
2350	27613	27730	27848	27966	28085	28203	28322	28441	28561	28680
2400	28800	28920	29041	29161	29282	29403	29525	29646	29768	29890
2450	30013	30135	30258	30381	30505	30628	30752	30876	31001	31125
2500	31250	31375	31501	31626	31752	31878	32005	32131	32258	32385
2550	32513	32640	32768	32896	33025	33153	33282	33411	33541	33670
2600	33800	33930	34061	34191	34322	34453	34585	34716	34848	34980
2650	35113	35245	35378	35511	35645	35778	35912	36046	36181	36315
2700	36450	36585	36721	36856	36992	37128	37265	37401	37538	37675

Second moments are tabulated in cm^4 and are for unit areas of 1 cm^2 each.

SECOND MOMENT
OF A PAIR OF UNIT AREAS

about axis x—x

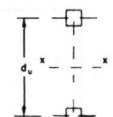

Distance d_u mm	0	5	10	15	20	25	30	35	40	45
2750	37813	37950	38088	38226	38365	38503	38642	38781	38921	39060
2800	39200	39340	39481	39621	39762	39903	40045	40186	40328	40470
2850	40613	40755	40898	41041	41185	41328	41472	41616	41761	41905
2900	42050	42195	42341	42486	42632	42778	42925	43071	43218	43365
2950	43513	43660	43808	43956	44105	44253	44402	44551	44701	44850
3000	45000	45150	45301	45451	45602	45753	45905	46056	46208	46360
3050	46513	46665	46818	46971	47125	47278	47432	47586	47741	47895
3100	48050	48205	48361	48516	48672	48828	48985	49141	49298	49455
3150	49613	49770	49928	50086	50245	50403	50562	50721	50881	51040
3200	51200	51360	51521	51681	51842	52003	52165	52326	52488	52650
3250	52813	52975	53138	53301	53465	53628	53792	53956	54121	54285
3300	54450	54615	54781	54946	55112	55278	55445	55611	55778	55945
3350	56113	56280	56448	56616	56785	56953	57122	57291	57461	57630
3400	57800	57970	58141	58311	58482	58653	58825	58996	59168	59340
3450	59513	59685	59858	60031	60205	60378	60552	60726	60901	61075
3500	61250	61425	61601	61776	61952	62128	62305	62481	62658	62835
3550	63013	63190	63368	63546	63725	63903	64082	64261	64441	64620
3600	64800	64980	65161	65341	65522	65703	65885	66066	66248	66430
3650	66613	66795	66978	67161	67345	67528	67712	67896	68081	68265
3700	68450	68635	68821	69006	69192	69378	69565	69751	69938	70125
3750	70313	70500	70688	70876	71065	71253	71442	71631	71821	72010
3800	72200	72390	72581	72771	72962	73153	73345	73536	73728	73920
3850	74113	74305	74498	74691	74885	75078	75272	75466	75661	75855
3900	76050	76245	76441	76636	76832	77028	77225	77421	77618	77815
3950	78013	78210	78408	78606	78805	79003	79202	79401	79601	79800
4000	80000	80200	80401	80601	80802	81003	81205	81406	81608	81810
4050	82013	82215	82418	82621	82825	83028	83232	83436	83641	83845
4100	84050	84255	84461	84666	84872	85078	85285	85491	85698	85905
4150	86113	86320	86528	86736	86945	87153	87362	87571	87781	87990
4200	88200	88410	88621	88831	89042	89253	89465	89676	89888	90100
4250	90313	90525	90738	90951	91165	91378	91592	91806	92021	92235
4300	92450	92665	92881	93096	93312	93528	93745	93961	94178	94395
4350	94613	94830	95048	95266	95485	95703	95922	96141	96361	96580
4400	96800	97020	97241	97461	97682	97903	98125	98346	98568	98790
4450	99013	99235	99458	99681	99905	100128	100352	100576	100801	101025
4500	101250	101475	101701	101926	102152	102378	102605	102831	103058	103285
4550	103513	103740	103968	104196	104425	104653	104882	105111	105341	105570
4600	105800	106030	106261	106491	106722	106953	107185	107416	107648	107880
4650	108113	108345	108578	108811	109045	109278	109512	109746	109981	110215
4700	110450	110685	110921	111156	111392	111628	111865	112101	112338	112575
4750	112813	113050	113288	113526	113765	114003	114242	114481	114721	114960
4800	115200	115440	115680	115921	116162	116403	116645	116886	117128	117370
4850	117613	117855	118098	118341	118585	118828	119072	119316	119561	119805
4900	120050	120295	120541	120786	121032	121278	121525	121771	122018	122265
4950	122513	122760	123008	123256	123505	123753	124002	124251	124501	124750

Second moments are tabulated in cm⁴ and are for unit areas of 1 cm² each.

GEOMETRICAL PROPERTIES OF PLANE SECTIONS

Section	Area	Position of Centroid	Moments of Inertia	Section Moduli
TRIANGLE	$A = \dfrac{bh}{2}$	$e_x = \dfrac{h}{3}$	$I_{XX} = bh^3/36$ $I_{YY} = hb^3/48$ $I_{aa} = bh^3/4$ $I_{bb} = bh^3/12$	Z_{XX} $base = bh^2/12$ $apex = bh^2/24$ $Z_{YY} = bh^2/24$
RECTANGLE	$A = bd$	$e_x = \dfrac{h}{2}$	$I_{XX} = bd^3/12$ $I_{YY} = db^3/12$ $I_{bb} = bd^3/3$	$Z_{XX} = bd^2/6$ $Z_{YY} = db^2/6$
RECTANGLE *axis on diagonal*	$A = bd$	$e_x = \dfrac{bd}{\sqrt{b^2+d^2}}$	$I_{XX} = \dfrac{b^3 d^3}{6(b^2+d^2)}$	$Z_{XX} = \dfrac{b^2 d^2}{6\sqrt{b^2+d^2}}$
RECTANGLE *axis through C.G.*	$A = bd$	$e_x = \dfrac{b.\sin\theta + d.\cos\theta}{2}$	$I_{XX} = \dfrac{bd(b^2\sin^2\theta + d^2\cos^2\theta)}{12}$	$Z_{XX} = \dfrac{bd(b^2\sin^2\theta + d^2\cos^2\theta)}{6(b.\sin\theta + d.\cos\theta)}$
SQUARE	$A = s^2$	$e_x = \dfrac{s}{2}$ $e_v = \dfrac{s}{\sqrt{2}}$	$I_{XX} = I_{YY} = s^4/12$ $I_{bb} = s^4/3$ $I_{VV} = s^4/12$	$Z_{XX} = Z_{YY} = \dfrac{s^3}{6}$ $Z_{VV} = \dfrac{s^3}{6\sqrt{2}}$
TRAPEZIUM	$A = \dfrac{d(a+b)}{2}$	$e_{x_I} = \dfrac{d(2a+b)}{3(a+b)}$	$I_{XX} = \dfrac{d^3(a^2+4ab+b^2)}{36(a+b)}$ $I_{YY} = \dfrac{d(a^3+a^2b+ab^2+b^3)}{48}$	$Z_{XX} = \dfrac{I_{XX}}{d-e_x}$ (two values) $Z_{YY} = \dfrac{2I_{YY}}{b}$
DIAMOND	$A = \dfrac{bd}{2}$	$e_x = \dfrac{d}{2}$	$I_{XX} = \dfrac{bd^3}{48}$ $I_{YY} = \dfrac{db^3}{48}$	$Z_{XX} = \dfrac{bd^2}{24}$ $Z_{YY} = \dfrac{db^2}{24}$
HEXAGON	$A = 0.866d^2$	$e_x = 0.866s$ $= d/2$	$I_{XX} = I_{YY} = I_{VV}$ $= 0.060Id^4$	$Z_{XX} = 0.1203d^3$ $Z_{YY} = Z_{VV}$ $= 0.1042d^3$

GEOMETRICAL PROPERTIES OF PLANE SECTIONS

Section		Area	Position of Centroid	Moments of Inertia	Section Moduli
OCTAGON		$A = 0.8284 d^2$ $s = 0.4142 d$	$e_x = \dfrac{d}{2}$ $e_v = 0.541 d$	$I_{XX} = I_{YY} = I_{VV}$ $= 0.0547 d^4$	$Z_{XX} = Z_{YY}$ $= 0.1095 d^3$ $Z_{VV} = 0.1011 d^3$
POLYGON	 *Regular figure*	$A = \dfrac{n s^2 \cot\theta}{4}$ $A = n r^2 \tan\theta$ $A = \dfrac{n R^2 \sin 2\theta}{2}$	$e = r$ or R depending on the axis and value of n	$I_1 = I_2$ $= \dfrac{A(6R^2 - s^2)}{24}$ $= \dfrac{A(12 r^2 + s^2)}{48}$	$Z = \dfrac{I}{e}$
CIRCLE		$A = \pi r^2$ $A = 0.7854 d^2$	$e = r = \dfrac{d}{2}$	$I = \dfrac{\pi d^4}{64}$ $I = 0.7854 r^4$	$Z = \dfrac{\pi d^3}{32}$ $Z = 0.7854 r^3$
SEMI-CIRCLE		$A = 1.5708 r^2$	$e_x = 0.424 r$	$I_{XX} = 0.1098 r^4$ $I_{YY} = 0.3927 r^4$	Z_{XX} base $= 0.2587 r^3$ crown $= 0.1907 r^3$ $Z_{YY} = 0.3927 r^3$
SEGMENT		$A =$ $\dfrac{r^2}{2}\left(\dfrac{\pi\theta^\circ}{180^\circ} - \sin\theta\right)$	$e_0 = \dfrac{c^3}{12A}$ $e_x = e_0 - r \cdot \cos\dfrac{\theta}{2}$	$I_{XX} = \dfrac{r^4}{16}\left(\dfrac{\pi\theta^\circ}{90^\circ} - \sin 2\theta\right)$ $- \dfrac{2 \theta r^4 (1 - \cos\theta)^3}{\pi\theta^\circ - 180^\circ \sin\theta}$ $I_{YY} = \dfrac{r^4}{48}\left(\dfrac{\pi\theta^\circ}{30^\circ} - 8\sin\theta + \sin 2\theta\right)$	Z_{XX} base $= I_{XX}/e_x$ crown $= \dfrac{I_{XX}}{b - e_x}$ $Z_{YY} = \dfrac{2 I_{YY}}{c}$
SECTOR		$A = \dfrac{\theta^\circ}{360^\circ}\pi r^2$	$e_x = \dfrac{2}{3}r\dfrac{c}{a}$ $e_x = \dfrac{r^2 c}{3A}$	$I_{XX} = I_0 - \dfrac{360^\circ}{\theta^\circ\pi}\sin\dfrac{2\theta}{2}\cdot\dfrac{4 r^4}{3}$ $I_{YY} = \dfrac{r^4}{8}\left(\dfrac{\pi\theta^\circ}{180^\circ} - \sin\theta\right)$ $I_0 = \dfrac{r^4}{8}\left(\dfrac{\pi\theta^\circ}{180^\circ} + \sin\theta\right)$	Z_{XX} centre $= I_{XX}/e_x$ crown $= \dfrac{I_{XX}}{r - e_x}$ $Z_{YY} = \dfrac{2 I_{YY}}{c}$
QUADRANT		$A = \dfrac{\pi r^2}{4}$	$e_x = 0.424 r$ $e_v = 0.6 r$ $e_u = 0.707 r$	$I_{XX} = I_{YY} = 0.0549 r^4$ $I_{bb} = 0.1963 r^4$ $I_{UU} = 0.0714 r^4$ $I_{VV} = 0.0384 r^4$	Minimum Values $Z_{XX} = Z_{YY}$ $= 0.0953 r^3$ $Z_{UU} = 0.1009 r^3$ $Z_{VV} = 0.064 r^3$
COMPLEMENT		$A = 0.2146 r^2$	$e_x = 0.777 r$ $e_v = 1.098 r$ $e_u = 0.707 r$ $e_a = 0.316 r$ $e_b = 0.391 r$	$I_{XX} = I_{YY} = 0.0076 r^4$ $I_{UU} = 0.012 r^4$ $I_{VV} = 0.0031 r^4$	Minimum Values $Z_{XX} = Z_{YY}$ $= 0.0097 r^3$ $Z_{UU} = 0.017 r^3$ $Z_{VV} = 0.0079 r^3$

GEOMETRICAL PROPERTIES OF PLANE SECTIONS

Section	Area	Position of Centroid	Moments of Inertia	Section Moduli
ELLIPSE	$A = \pi ab$	$e_x = a$ $e_y = b$	$I_{XX} = 0.7854ba^3$ $I_{YY} = 0.7854ab^3$	$Z_{XX} = 0.7854ba^2$ $Z_{YY} = 0.7854ab^2$
SEMI-ELLIPSE	$A = \dfrac{\pi ab}{2}$	$e_x = 0.424a$ $e_y = b$	$I_{XX} = 0.1098ba^3$ $I_{YY} = 0.3927ab^3$ $I_{base} = 0.3927ba^3$	Z_{XX}-base $= 0.2587ba^2$ Z_{XX}-crown $= 0.1907ba^2$ $Z_{YY} = 0.3927ab^2$
1/4 ELLIPSE	$A = 0.7854ab$	$e_x = 0.424a$ $e_y = 0.424b$	$I_{XX} = 0.0549ba^3$ $I_{YY} = 0.0549ab^3$ $I_{b_1a_1} = 0.1963ba^3$ $I_{b_1c_1} = 0.1963ab^3$	Z_{XX}-base $= 0.1293ba^2$ Z_{XX}-crown $= 0.0953ba^2$ Z_{YY}-base $= 0.1293ab^2$ Z_{YY}-crown $= 0.0953ab^2$
COMPLEMENT	$A = 0.2146ab$	$e_x = 0.777a$ $e_y = 0.777b$	$I_{XX} = 0.0076ba^3$ $I_{YY} = 0.0076ab^3$	Z_{XX}-base $= 0.0338ba^2$ Z_{XX}-apex $= 0.0097ba^2$ Z_{YY}-base $= 0.0338ab^2$ Z_{YY}-apex $= 0.0097ab^2$
FULL PARABOLA	$A = \dfrac{4ab}{3}$	$e_x = \dfrac{2a}{5}$ $e_y = b$	$I_{XX} = 0.0914ba^3$ $I_{YY} = 0.2666ab^3$ $I_{base} = 0.3048ba^3$	Z_{XX}-base $= 0.2286ba^2$ Z_{XX}-crown $= 0.1524ba^2$ $Z_{YY} = 0.2666ab^2$
SEMI-PARABOLA	$A = \dfrac{2ab}{3}$	$e_x = \dfrac{2a}{5}$ $e_y = \dfrac{3b}{8}$	$I_{XX} = 0.0457ba^3$ $I_{YY} = 0.0396ab^3$ $I_{b_1a_1} = 0.1524ba^3$ $I_{b_1c_1} = 0.1333ab^3$	Z_{XX}-base $= 0.1143ba^2$ Z_{XX}-crown $= 0.076ba^2$ Z_{YY}-base $= 0.1055ab^2$ Z_{YY}-crown $= 0.0633ab^2$
COMPLEMENT	$A = \dfrac{ab}{3}$	$e_x = \dfrac{7a}{10}$ $e_y = \dfrac{3b}{4}$	$I_{XX} = 0.0176ba^3$ $I_{YY} = 0.0125ab^3$ $I_{a_1b_1} = 0.181ba^3$ $I_{b_1c_1} = 0.2ab^3$	Z_{XX}-base $= 0.0587ba^2$ Z_{XX}-apex $= 0.0252ba^2$ Z_{YY}-base $= 0.05ab^2$ Z_{YY}-apex $= 0.0167ab^2$
FILLET (parabola)	$A = \dfrac{s^2}{6}$	$e_u = e_v = \dfrac{4s}{5}$	$I_{UU} = I_{VV} = 0.00524s^4$ $I_{ab} = 0.1119a^4$	$Z_{UU} = Z_{VV}$ base $= 0.0262s^3$ apex $= 0.0066s^3$

PLASTIC MODULUS OF TWO FLANGES

Dist d mm	Plastic Modulus Sxx(cm³) For Thickness t(mm)												
	15	20	25	30	35	40	45	50	55	60	65	70	75
1000	15.2	20.4	25.6	30.9	36.2	41.6	47.0	52.5	58.0	63.6	69.2	74.9	80.6
1100	16.7	22.4	28.1	33.9	39.7	45.6	51.5	57.5	63.5	69.6	75.7	81.9	88.1
1200	18.2	24.4	30.6	36.9	43.2	49.6	56.0	62.5	69.0	75.6	82.2	88.9	95.6
1300	19.7	26.4	33.1	39.9	46.7	53.6	60.5	67.5	74.5	81.6	88.7	95.9	103
1400	21.2	28.4	35.6	42.9	50.2	57.6	65.0	72.5	80.0	87.6	95.2	103	111
1500	22.7	30.4	38.1	45.9	53.7	61.6	69.5	77.5	85.5	93.6	102	110	118
1600	24.2	32.4	40.6	48.9	57.2	65.6	74.0	82.5	91.0	99.6	108	117	126
1700	25.7	34.4	43.1	51.9	60.7	69.6	78.5	87.5	96.5	106	115	124	133
1800	27.2	36.4	45.6	54.9	64.2	73.6	83.0	92.5	102	112	121	131	141
1900	28.7	38.4	48.1	57.9	67.7	77.6	87.5	97.5	108	118	128	138	148
2000	30.2	40.4	50.6	60.9	71.2	81.6	92.0	102	113	124	134	145	156
2100	31.7	42.4	53.1	63.9	74.7	85.6	96.5	107	119	130	141	152	163
2200	33.2	44.4	55.6	66.9	78.2	89.6	101	112	124	136	147	159	171
2300	34.7	46.4	58.1	69.9	81.7	93.6	106	117	130	142	154	166	178
2400	36.2	48.4	60.6	72.9	85.2	97.6	110	122	135	148	160	173	186
2500	37.7	50.4	63.1	75.9	88.7	102	115	127	141	154	167	180	193
2600	39.2	52.4	65.6	78.9	92.2	106	119	132	146	160	173	187	201
2700	40.7	54.4	68.1	81.9	95.7	110	124	137	152	166	180	194	208
2800	42.2	56.4	70.6	84.9	99.2	114	128	142	157	172	186	201	216
2900	43.7	58.4	73.1	87.9	103	118	133	147	163	178	193	208	223
3000	45.2	60.4	75.6	90.9	106	122	137	152	168	184	199	215	231
3100	46.7	62.4	78.1	93.9	110	126	142	157	174	190	206	222	238
3200	48.2	64.4	80.6	96.9	113	130	146	162	179	196	212	229	246
3300	49.7	66.4	83.1	99.9	117	134	151	167	185	202	219	236	253
3400	51.2	68.4	85.6	103	120	138	155	172	190	208	225	243	261
3500	52.7	70.4	88.1	106	124	142	160	177	196	214	232	250	268
3600	54.2	72.4	90.6	109	127	146	164	182	201	220	238	257	276
3700	55.7	74.4	93.1	112	131	150	169	187	207	226	245	264	283
3800	57.2	76.4	95.6	115	134	154	173	192	212	232	251	271	291
3900	58.7	78.4	98.1	118	138	158	178	197	218	238	258	278	298
4000	60.2	80.4	101	121	141	162	182	202	223	244	264	285	306
4100	61.7	82.4	103	124	145	166	187	207	229	250	271	292	313
4200	63.2	84.4	106	127	148	170	191	212	234	256	277	299	321
4300	64.7	86.4	108	130	152	174	196	217	240	262	284	306	328
4400	66.2	88.4	111	133	155	178	200	222	245	268	290	313	336
4500	67.7	90.4	113	136	159	182	205	227	251	274	297	320	343
4600	69.2	92.4	116	139	162	186	209	232	256	280	303	327	351
4700	70.7	94.4	118	142	166	190	214	237	262	286	310	334	358
4800	72.2	96.4	121	145	169	194	218	242	267	292	316	341	366
4900	73.7	98.4	123	148	173	198	223	247	273	298	323	348	373
5000	75.2	100.0	126	151	176	202	227	252	278	304	329	355	381

PLASTIC MODULUS OF RECTANGLES

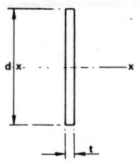

Depth d mm	Plastic Modulus Sxx(cm³) For Thickness t(mm)									
	5	6	7	8	9	10	12.5	15	20	25
25	0.78	0.93	1.09	1.25	1.41	1.56	1.95	2.34	3.13	3.91
50	3.13	3.75	4.37	5.00	5.62	6.25	7.81	9.37	12.5	15.6
75	7.03	8.44	9.84	11.3	12.7	14.1	17.6	21.1	28.1	35.2
100	12.5	15.0	17.5	20.0	22.5	25.0	31.2	37.5	50.0	62.5
125	19.5	23.4	27.3	31.2	35.2	39.1	48.8	58.6	78.1	97.7
150	28.1	33.8	39.4	45.0	50.6	56.2	70.3	84.4	112	141
175	38.3	45.9	53.6	61.2	68.9	76.6	95.7	115	153	191
200	50.0	60.0	70.0	80.0	90.0	100.0	125	150	200	250
225	63.3	75.9	88.6	101	114	127	158	190	253	316
250	78.1	93.7	109	125	141	156	195	234	312	391
275	94.5	113	132	151	170	189	236	284	378	473
300	112	135	158	180	203	225	281	338	450	563
325	132	158	185	211	238	264	330	396	528	660
350	153	184	214	245	276	306	383	459	613	766
375	176	211	246	281	316	352	439	527	703	879
400	200	240	280	320	360	400	500	600	800	1000
425	226	271	316	361	406	452	564	677	903	1130
450	253	304	354	405	456	506	633	759	1010	1270
475	282	338	395	451	508	564	705	846	1130	1410
500	312	375	437	500	562	625	781	937	1250	1560
525	345	413	482	551	620	689	861	1030	1380	1720
550	378	454	529	605	681	756	945	1130	1510	1890
575	413	496	579	661	744	827	1030	1240	1650	2070
600	450	540	630	720	810	900	1130	1350	1800	2250
625	488	586	684	781	879	977	1220	1460	1950	2440
650	528	634	739	845	951	1060	1320	1580	2110	2640
675	570	683	797	911	1030	1140	1420	1710	2280	2850
700	613	735	858	980	1100	1230	1530	1840	2450	3060
725	657	788	920	1050	1180	1310	1640	1970	2630	3290
750	703	844	984	1120	1270	1410	1760	2110	2810	3520
775	751	901	1050	1200	1350	1500	1880	2250	3000	3750
800	800	960	1120	1280	1440	1600	2000	2400	3200	4000
825	851	1020	1190	1360	1530	1700	2130	2550	3400	4250
850	903	1080	1260	1440	1630	1810	2260	2710	3610	4520
875	957	1150	1340	1530	1720	1910	2390	2870	3830	4790
900	1010	1210	1420	1620	1820	2020	2530	3040	4050	5060

PLASTIC MODULUS OF RECTANGLES

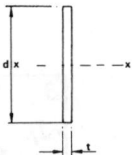

Depth d mm	Plastic Modulus Sxx(cm³) For Thickness t(mm)									
	5	6	7	8	9	10	12.5	15	20	25
1000	1250	1500	1750	2000	2250	2500	3120	3750	5000	6250
1100	1510	1810	2120	2420	2720	3020	3780	4540	6050	7560
1200	1800	2160	2520	2880	3240	3600	4500	5400	7200	9000
1300	2110	2530	2960	3380	3800	4220	5280	6340	8450	10600
1400	2450	2940	3430	3920	4410	4900	6130	7350	9800	12300
1500	2810	3370	3940	4500	5060	5620	7030	8440	11200	14100
1600	3200	3840	4480	5120	5760	6400	8000	9600	12800	16000
1700	3610	4330	5060	5780	6500	7220	9030	10800	14400	18100
1800	4050	4860	5670	6480	7290	8100	10100	12100	16200	20200
1900	4510	5410	6320	7220	8120	9020	11300	13500	18000	22600
2000	5000	6000	7000	8000	9000	10000	12500	15000	20000	25000
2100	5510	6620	7720	8820	9920	11000	13800	16500	22100	27600
2200	6050	7260	8470	9680	10900	12100	15100	18100	24200	30200
2300	6610	7930	9260	10600	11900	13200	16500	19800	26500	33100
2400	7200	8640	10100	11500	13000	14400	18000	21600	28800	36000
2500	7810	9370	10900	12500	14100	15600	19500	23400	31200	39100
2600	8450	10100	11800	13500	15200	16900	21100	25400	33800	42200
2700	9110	10900	12800	14600	16400	18200	22800	27300	36400	45600
2800	9800	11800	13700	15700	17600	19600	24500	29400	39200	49000
2900	10500	12600	14700	16800	18900	21000	26300	31500	42000	52600
3000	11200	13500	15700	18000	20200	22500	28100	33700	45000	56200
3100	12000	14400	16800	19200	21600	24000	30000	36000	48000	60100
3200	12800	15400	17900	20500	23000	25600	32000	38400	51200	64000
3300	13600	16300	19100	21800	24500	27200	34000	40800	54400	68100
3400	14400	17300	20200	23100	26000	28900	36100	43300	57800	72200
3500	15300	18400	21400	24500	27600	30600	38300	45900	61200	76600
3600	16200	19400	22700	25900	29200	32400	40500	48600	64800	81000
3700	17100	20500	24000	27400	30800	34200	42800	51300	68400	85600
3800	18000	21700	25300	28900	32500	36100	45100	54100	72200	90200
3900	19000	22800	26600	30400	34200	38000	47500	57000	76000	95100
4000	20000	24000	28000	32000	36000	40000	50000	60000	80000	100000
4100	21000	25200	29400	33600	37800	42000	52500	63000	84000	105000
4200	22100	26500	30900	35300	39700	44100	55100	66200	88200	110000
4300	23100	27700	32400	37000	41600	46200	57800	69300	92400	116000
4400	24200	29000	33900	38700	43600	48400	60500	72600	96800	121000
4500	25300	30400	35400	40500	45600	50600	63300	75900	101000	127000
4600	26500	31700	37000	42300	47600	52900	66100	79300	106000	132000
4700	27600	33100	38700	44200	49700	55200	69000	82800	110000	138000
4800	28800	34600	40300	46100	51800	57600	72000	86400	115000	144000
4900	30000	36000	42000	48000	54000	60000	75000	90000	120000	150000
5000	31200	37500	43700	50000	56200	62500	78100	93700	125000	156000

Frame I

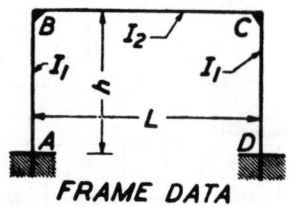

FRAME DATA

Coefficients:
$$k = \frac{I_2}{I_1} \cdot \frac{h}{L}$$
$$N_1 = k + 2 \qquad N_2 = 6k + 1$$

---

w per unit length

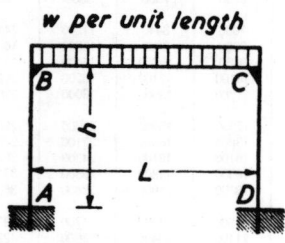

$$M_A = M_D = \frac{wL^2}{12N_1} \qquad M_B = M_C = -\frac{wL^2}{6N_1} = -2M_A$$

$$M_{max} = \frac{wL^2}{8} + M_B \qquad V_A = V_D = \frac{wL}{2} \qquad H_A = H_D = \frac{3M_A}{h}$$

---

w per unit length

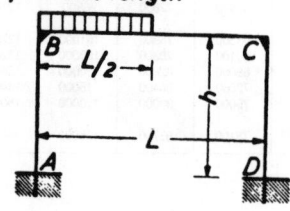

$$M_A = \frac{wL^2}{8}\left[\frac{1}{3N_1} - \frac{1}{8N_2}\right] \qquad M_B = -\frac{wL^2}{8}\left[\frac{2}{3N_1} + \frac{1}{8N_2}\right]$$

$$M_D = \frac{wL^2}{8}\left[\frac{1}{3N_1} + \frac{1}{8N_2}\right] \qquad M_C = -\frac{wL^2}{8}\left[\frac{2}{3N_1} - \frac{1}{8N_2}\right]$$

$$V_D = \frac{wL}{8}\left[1 - \frac{1}{4N_2}\right] \qquad V_A = \frac{wL}{2} - V_D \qquad H_A = H_D = \frac{wL^2}{8hN_1}$$

Extract: 'Kleinlogel, Rahmenformeln' 11. Auflage Berlin—Verlag von Wilhelm Ernst & Sohn.

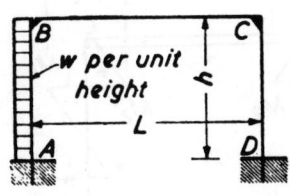

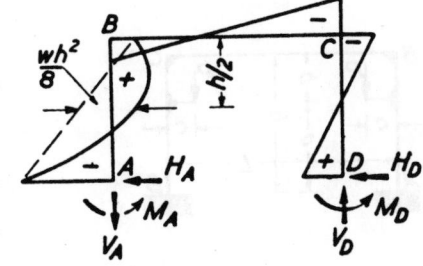

$$M_A = \frac{wh^2}{4}\left[-\frac{k+3}{6N_1} - \frac{4k+1}{N_2}\right] \qquad M_B = \frac{wh^2}{4}\left[-\frac{k}{6N_1} + \frac{2k}{N_2}\right]$$

$$M_D = \frac{wh^2}{4}\left[-\frac{k+3}{6N_1} + \frac{4k+1}{N_2}\right] \qquad M_C = \frac{wh^2}{4}\left[-\frac{k}{6N_1} - \frac{2k}{N_2}\right]$$

$$H_D = \frac{wh(2k+3)}{8N_1} \qquad H_A = -(wh - H_D) \qquad V_A = -V_D = -\frac{wh^2 k}{LN_2}$$

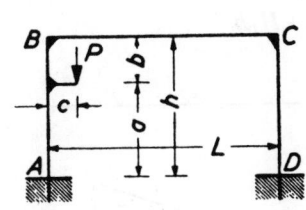

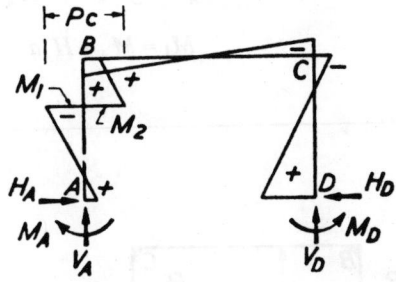

$$\text{Constants: } a_1 = \frac{a}{h} \qquad b_1 = \frac{b}{h}$$

$$X_1 = \frac{Pc}{2N_1}[1 + 2b_1 k - 3b_1^2(k+1)] \qquad X_2 = \frac{Pcka_1(3a_1 - 2)}{2N_1}$$

$$X_3 = \frac{3Pcka_1}{N_2}$$

$$M_A = +X_1 - \left(\frac{Pc}{2} - X_3\right) \qquad M_B = +X_2 + X_3$$

$$M_D = +X_1 + \left(\frac{Pc}{2} - X_3\right) \qquad M_C = +X_2 - X_3$$

$$H_A = H_D = \frac{Pc}{2h} + \frac{X_1 - X_2}{h} \qquad V_D = \frac{2X_3}{L} \qquad V_A = P - V_D$$

$$M_1 = M_A - H_A a \qquad M_2 = M_B + H_D b$$

Extract: 'Kleinlogel, Rahmenformeln' 11. Auflage Berlin—Verlag von Wilhelm Ernst & Sohn.

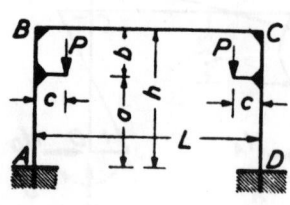
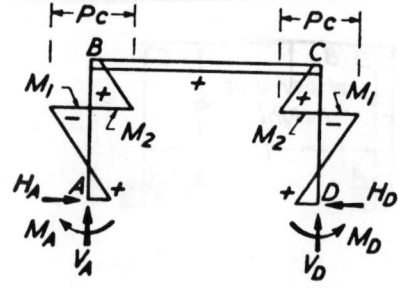

$$\text{Constants: } a_1 = \frac{a}{h} \qquad b_1 = \frac{b}{h}$$

$$X_1 = \frac{Pc}{2N_1}[1 + 2b_1k - 3b_1^2(k+1)] \qquad X_2 = \frac{Pcka_1(3a_1 - 2)}{2N_1}$$

$$M_A = M_D = \frac{Pc}{N_1}[1 + 2b_1k - 3b_1^2(k+1)] = 2X_1$$

$$M_B = M_C = \frac{Pcka_1(3a_1 - 2)}{N_1} = 2X_2$$

$$V_A = V_D = P \qquad H_A = H_D = \frac{Pc + M_A - M_B}{h}$$

$$M_1 = M_A - H_A a \qquad M_2 = M_B + H_D b$$

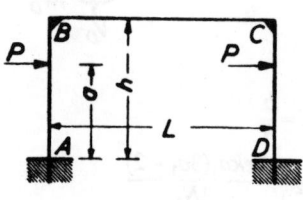
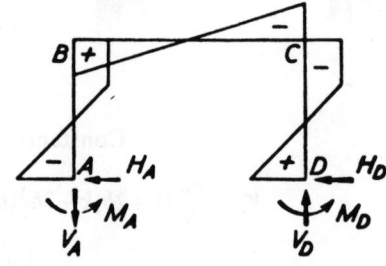

$$\text{Constants: } a_1 = \frac{a}{h} \qquad X_1 = \frac{3Paa_1k}{N_2}$$

$$M_A = -Pa + X_1 \qquad\qquad M_B = X_1$$

$$M_D = +Pa - X_1 \qquad\qquad M_C = -X_1$$

$$V_A = -V_D = -\frac{2X_1}{L} \qquad H_A = -H_D = -P$$

Extract: 'Kleinlogel, Rahmenformeln' 11. Auflage Berlin—Verlag von Wilhelm Ernst & Sohn.

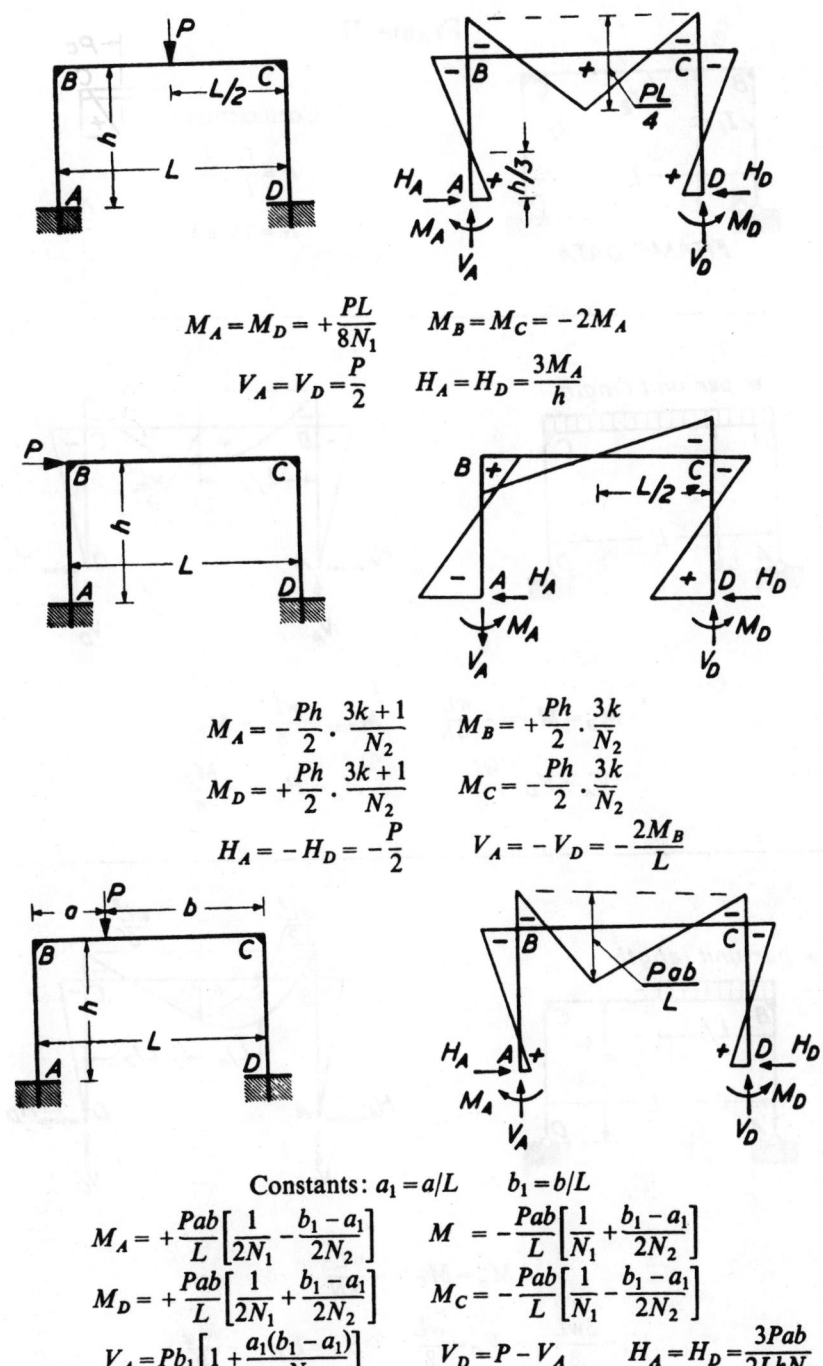

$$M_A = M_D = + \frac{PL}{8N_1} \qquad M_B = M_C = -2M_A$$

$$V_A = V_D = \frac{P}{2} \qquad H_A = H_D = \frac{3M_A}{h}$$

$$M_A = -\frac{Ph}{2} \cdot \frac{3k+1}{N_2} \qquad M_B = +\frac{Ph}{2} \cdot \frac{3k}{N_2}$$

$$M_D = +\frac{Ph}{2} \cdot \frac{3k+1}{N_2} \qquad M_C = -\frac{Ph}{2} \cdot \frac{3k}{N_2}$$

$$H_A = -H_D = -\frac{P}{2} \qquad V_A = -V_D = -\frac{2M_B}{L}$$

Constants: $a_1 = a/L \qquad b_1 = b/L$

$$M_A = +\frac{Pab}{L}\left[\frac{1}{2N_1} - \frac{b_1 - a_1}{2N_2}\right] \qquad M = -\frac{Pab}{L}\left[\frac{1}{N_1} + \frac{b_1 - a_1}{2N_2}\right]$$

$$M_D = +\frac{Pab}{L}\left[\frac{1}{2N_1} + \frac{b_1 - a_1}{2N_2}\right] \qquad M_C = -\frac{Pab}{L}\left[\frac{1}{N_1} - \frac{b_1 - a_1}{2N_2}\right]$$

$$V_A = Pb_1\left[1 + \frac{a_1(b_1 - a_1)}{N_2}\right] \qquad V_D = P - V_A \qquad H_A = H_D = \frac{3Pab}{2LhN_1}$$

Extract: 'Kleinlogel, Rahmenformeln' 11. Auflage Berlin—Verlag von Wilhelm Ernst & Sohn.

Frame II

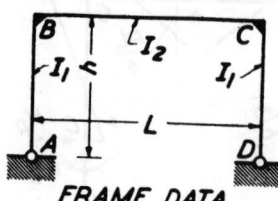

FRAME DATA

Coefficients:

$$k = \frac{I_2}{I_1} \cdot \frac{h}{L}$$

$$N = 2k + 3$$

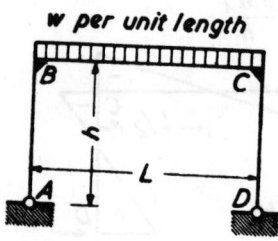

w per unit length

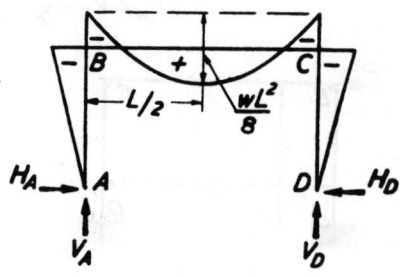

$$M_B = M_C = -\frac{wL^2}{4N} \qquad M_{\max} = \frac{wL^2}{8} + M_B$$

$$V_A = V_D = \frac{wL}{2} \qquad H_A = H_D = -\frac{M_B}{h}$$

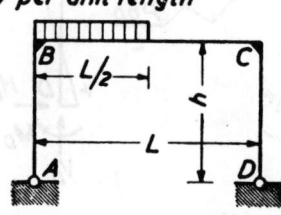

w per unit length

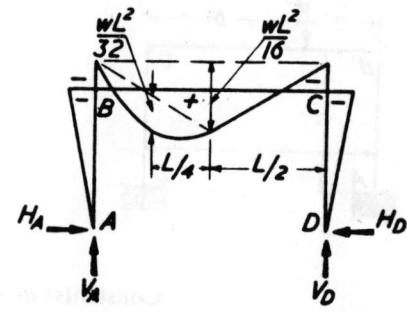

$$M_B = M_C = -\frac{wL^2}{8N}$$

$$V_A = \frac{3wL}{8} \qquad V_D = \frac{wL}{8} \qquad H_A = H_D = -\frac{M_B}{h}$$

Extract: '*Kleinlogel, Rahmenformeln*' 11. *Auflage Berlin—Verlag von Wilhelm Ernst & Sohn.*

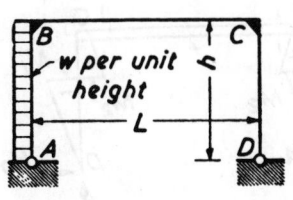
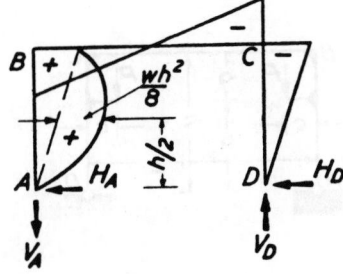

$$M_B = \frac{wh^2}{4}\left[-\frac{k}{2N}+1\right] \qquad H_D = -\frac{M_C}{h}$$

$$M_C = \frac{wh^2}{4}\left[-\frac{k}{2N}-1\right] \qquad H_A = -(wh - H_D)$$

$$V_A = -V_D = -\frac{wh^2}{2L}$$

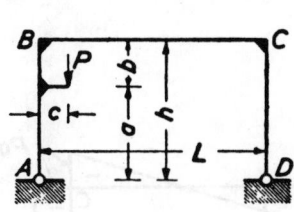
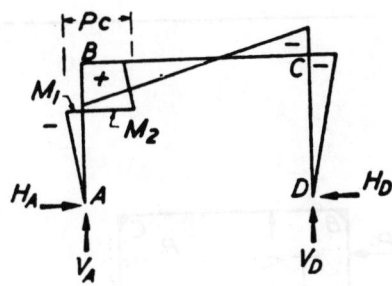

$$\text{Constant: } a_1 = \frac{a}{h}$$

$$M_B = \frac{Pc}{2}\left[\frac{(3a_1^2 - 1)k}{N} + 1\right]$$

$$H_A = H_D = -\frac{M_C}{h}$$

$$M_C = \frac{Pc}{2}\left[\frac{(3a_1^2 - 1)k}{N} - 1\right]$$

$$V_D = \frac{Pc}{L} \qquad V_A = P - V_D$$

$$M_1 = -H_A a \qquad M_2 = Pc - H_A a$$

Extract: 'Kleinlogel, Rahmenformeln' 11. Auflage Berlin—Verlag von Wilhelm Ernst & Sohn.

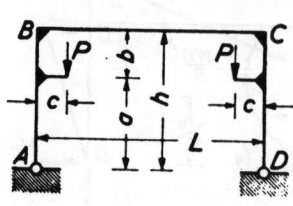
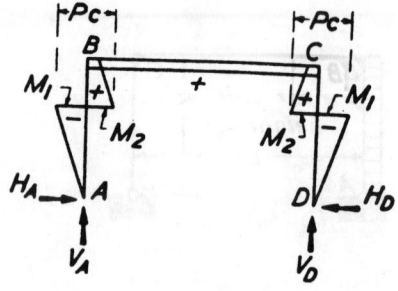

$$\text{Constant: } a_1 = \frac{a}{h}$$

$$M_B = M_C = \frac{Pc(3a_1^2 - 1)k}{N}$$

$$H_A = H_D = \frac{Pc - M_B}{h} \qquad V_A = V_D = P$$

$$M_1 = -H_A a \qquad M_2 = Pc - H_A a$$

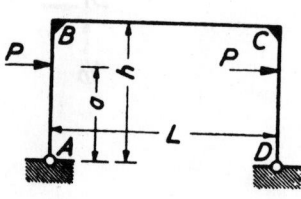
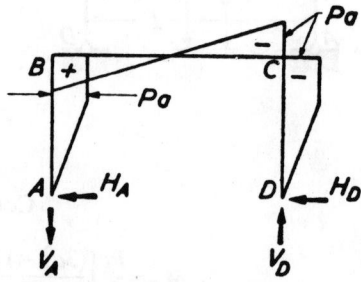

$$M_B = -M_C = Pa \qquad H_A = H_D = P$$

$$V_A = -V_D = -\frac{2Pa}{L}$$

$$\text{Moment at loads} = \pm Pa$$

Extract: 'Kleinlogel, Rahmenformeln' 11. Auflage Berlin—Verlag von Wilhelm Ernst & Sohn.

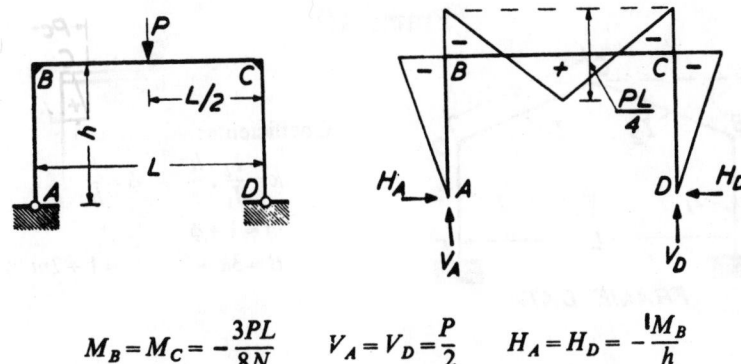

$$M_B = M_C = -\frac{3PL}{8N} \qquad V_A = V_D = \frac{P}{2} \qquad H_A = H_D = -\frac{^1M_B}{h}$$

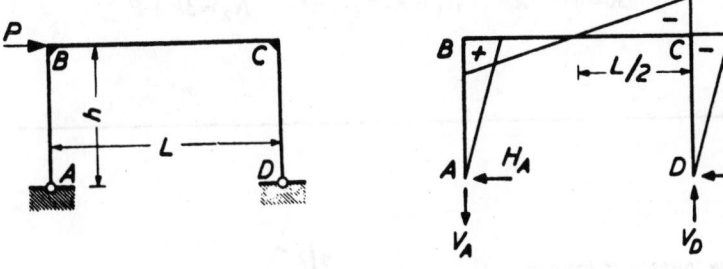

$$M_B = -M_C = +\frac{Ph}{2}$$

$$V_A = -V_D = -\frac{Ph}{L} \qquad H_A = -H_D = -\frac{P}{2}$$

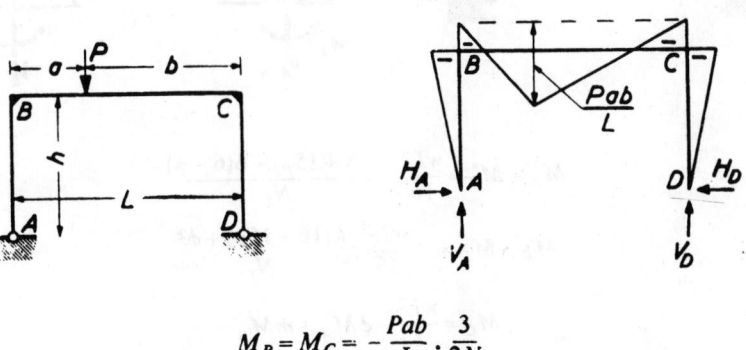

$$M_B = M_C = -\frac{Pab}{L} \cdot \frac{3}{2N}$$

$$V_A = \frac{Pb}{L} \qquad V_D = \frac{Pa}{L} \qquad H_A = H_D = -\frac{M_B}{h}$$

Extract: 'Kleinlogel, Rahmenformeln' 11. Auflage Berlin—Verlag von Wilhelm Ernst & Sohn.

Frame III

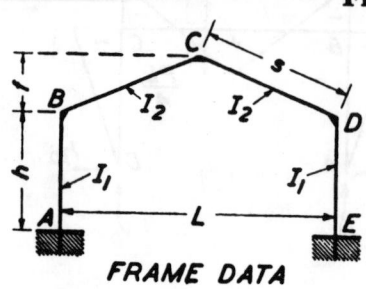

FRAME DATA

Coefficients:

$$k = \frac{I_2}{I_1} \cdot \frac{h}{s} \qquad \phi = \frac{f}{h}$$

$$m = 1 + \phi$$

$$B = 3k + 2 \qquad C = 1 + 2m$$

$$K_1 = 2(k + 1 + m + m^2) \qquad K_2 = 2(k + \phi^2)$$

$$R = \phi C - k \qquad N_1 = K_1 K_2 - R^2 \qquad N_2 = 3k + B$$

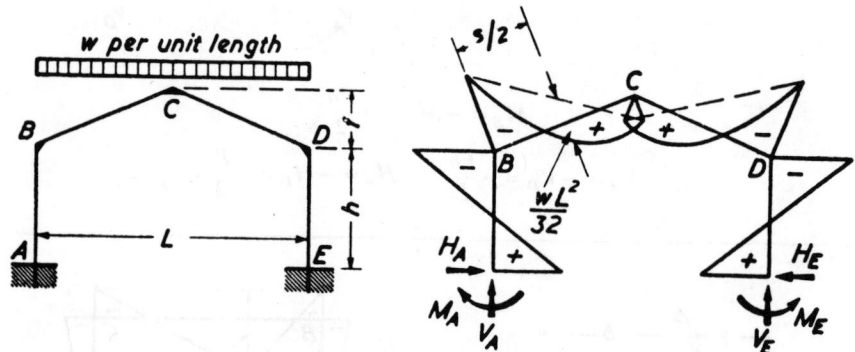

$$M_A = M_E = \frac{wL^2}{16} \cdot \frac{k(8 + 15\phi) + \phi(6 - \phi)}{N_1}$$

$$M_B = M_D = -\frac{wL^2}{16} \cdot \frac{k(16 + 15\phi) + \phi^2}{N_1}$$

$$M_C = \frac{wL^2}{8} - \phi M_A + m M_B$$

$$V_A = V_E = \frac{wL}{2} \qquad H_A = H_E = \frac{M_A - M_B}{h}$$

Extract: 'Kleinlogel, Rahmenformeln' 11. *Auflage Berlin—Verlag von Wilhelm Ernst & Sohn.*

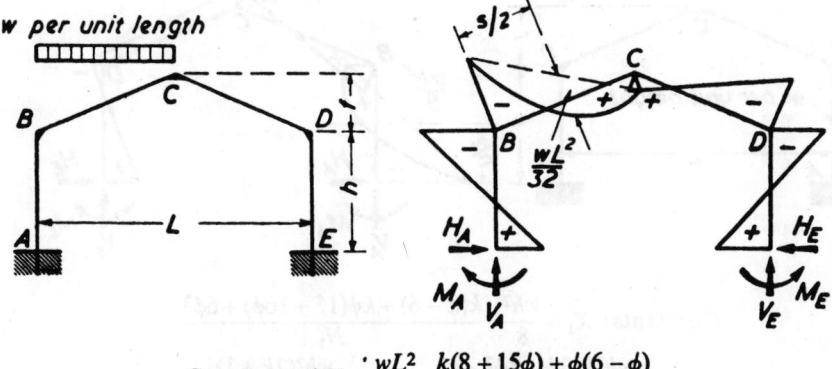

$$\text{Constants: } {}^*X_1 = \frac{wL^2}{32} \cdot \frac{k(8+15\phi)+\phi(6-\phi)}{N_1}$$

$$^*X_2 = \frac{wL^2}{32} \cdot \frac{k(16+15\phi)+\phi^2}{N_1} \qquad X_3 = \frac{wL^2}{32N_2}$$

$$M_A = +X_1 - X_3 \qquad M_B = -X_2 - X_3 \qquad M_E = +X_1 + X_3 \qquad M_D = -X_2 + X_3$$

$$^*M_C = \frac{wL^2}{16} - \phi X_1 - m X_2$$

$$V_E = \frac{wL}{8} - \frac{2X_3}{L} \qquad V_A = \frac{wL}{2} - V_E \qquad H_A = H_E = \frac{X_1 + X_2}{h}$$

* Note that X_1, $-X_2$ and M_C are respectively half the values of $M_A (=M_E)$, $M_B (=M_D)$ and M_C from the previous set of formulæ where the whole span was loaded.

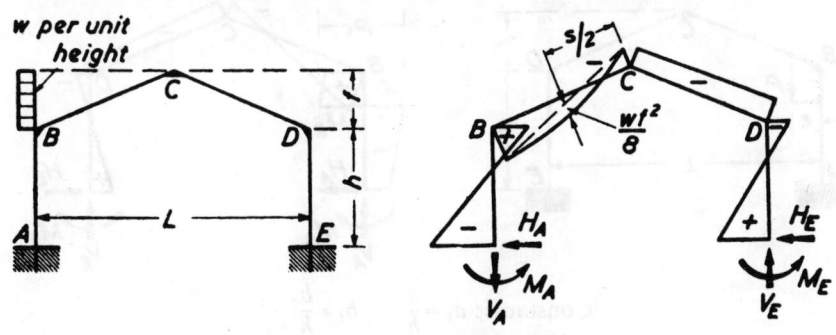

$$\text{Constants: } X_1 = \frac{wf^2}{8} \cdot \frac{k(9\phi+4)+\phi(6+\phi)}{N_1}$$

$$X_2 = \frac{wf^2}{8} \cdot \frac{k(8+9\phi)-\phi^2}{N_1} \qquad X_3 = \frac{wfh}{8} \cdot \frac{4B+\phi}{N_2}$$

$$M_A = -X_1 - X_3 \qquad M_B = +X_2 + \left(\frac{wfh}{2} - X_3\right)$$

$$M_E = -X_1 + X_3 \qquad M_D = +X_2 - \left(\frac{wfh}{2} - X_3\right)$$

$$M_C = -\frac{wf^2}{4} + \phi X_1 + m X_2$$

$$V_A = -V_E = -\frac{wfh(2+\phi)}{2L} + \frac{2X_3}{L} \qquad H_E = \frac{wf}{2} - \frac{X_1 + X_2}{h} \qquad H_A = -(wf - H_E)$$

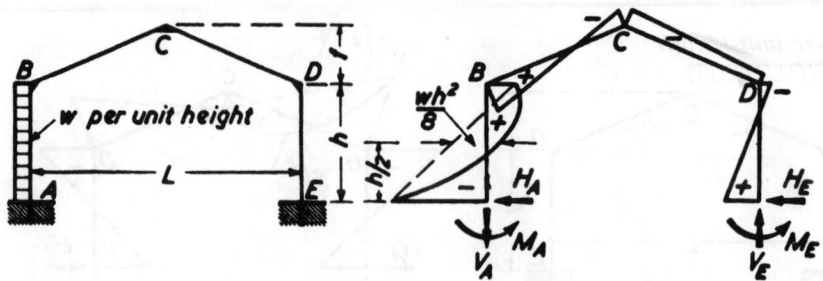

Constants: $X_1 = \dfrac{wh^2}{8} \cdot \dfrac{k(k+6) + k\phi(15 + 16\phi) + 6\phi^2}{N_1}$

$X_2 = \dfrac{wh^2 k(9\phi + 8\phi^2 - k)}{8N_1}$ $X_3 = \dfrac{wh^2(2k+1)}{2N_2}$

$M_A = -X_1 - X_3$ $M_B = +X_2 + \left(\dfrac{wh^2}{4} - X_3\right)$

$M_E = -X_1 + X_3$ $M_D = +X_2 - \left(\dfrac{wh^2}{4} - X_3\right)$

$M_C = -\dfrac{whf}{4} + \phi X_1 + m X_2$

$V_A = -V_E = -\dfrac{wh^2}{2L} + \dfrac{2X_3}{L}$ $H_E = \dfrac{wh}{4} - \dfrac{X_1 + X_2}{h}$ $H_A = -(wh - H_E)$

Constants: $a_1 = \dfrac{a}{h}$ $b_1 = \dfrac{b}{h}$

$Y_1 = Pc[2\phi^2 - (1 - 3b_1^2)k]$ $Y_2 = Pc[\phi C - (3a_1^2 - 1)k]$

$X_1 = \dfrac{Y_1 K_1 - Y_2 R}{2N_1}$ $X_2 = \dfrac{Y_2 K_2 - Y_1 R}{2N_1}$ $X_3 = \dfrac{Pc}{2} \cdot \dfrac{B - 3(a_1 - b_1)k}{N_2}$

$M_A = -X_1 - X_3$ $M_B = +X_2 + \left(\dfrac{Pc}{2} - X_3\right)$

$M_E = -X_1 + X_3$ $M_D = +X_2 - \left(\dfrac{Pc}{2} - X_3\right)$ $M_C = -\dfrac{\phi Pc}{2} + \phi X_1 + m X_2$

$M_1 = M_A - H_A a$ $M_2 = M_B + H_E b$

$V_E = \dfrac{Pc - 2X_3}{L}$ $V_A = P - V_E$ $H_A = H_E = \dfrac{Pc}{2h} - \dfrac{X_1 + X_2}{h}$

Extract: 'Kleinlogel, Rahmenformeln' 11. Auflage Berlin—Verlag von Wilhelm Ernst & Sohn.

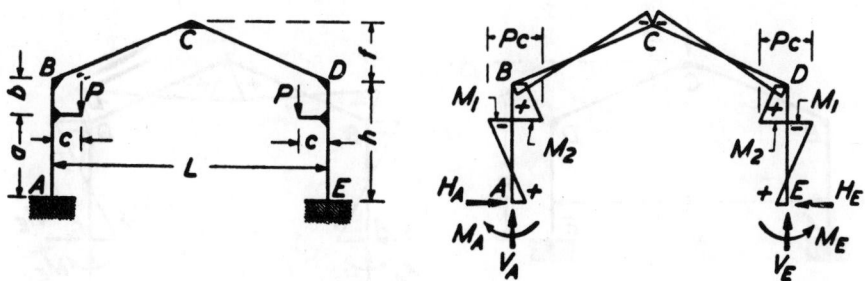

$$\text{Constants: } a_1 = \frac{a}{h} \qquad b_1 = \frac{b}{h}$$

$$Y_1 = Pc[2\phi^2 - (1 - 3b_1^2)k]$$

$$Y_2 = Pc[\phi C + (3a_1^2 - 1)k]$$

$$M_A = M_E = \frac{Y_2 R - Y_1 K_1}{N_1} \qquad M_B = M_D = \frac{Y_2 K_2 - Y_1 R}{N_1}$$

$$M_C = -\phi(Pc + M_A) + mM_B$$

$$V_A = V_D = P \qquad H_A = H_E = \frac{Pc + M_A - M_B}{h}$$

$$M_1 = M_A - H_A a \qquad M_2 = M_B + H_E b$$

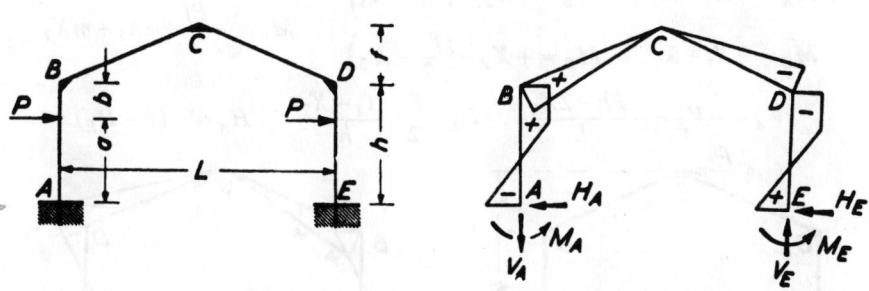

$$\text{Constant: } X_1 = \frac{Pa(B + 3b_1 k)}{N_2}$$

$$M_A = -M_E = -X_1 \qquad M_B = -M_D = Pa - X_1 \qquad M_C = 0$$

$$V_A = -V_E = -2\left[\frac{Pa - X_1}{L}\right] \qquad H_A = -H_E = -P$$

Extract: '*Kleinlogel, Rahmenformeln*' 11. *Auflage Berlin—Verlag von Wilhelm Ernst & Sohn.*

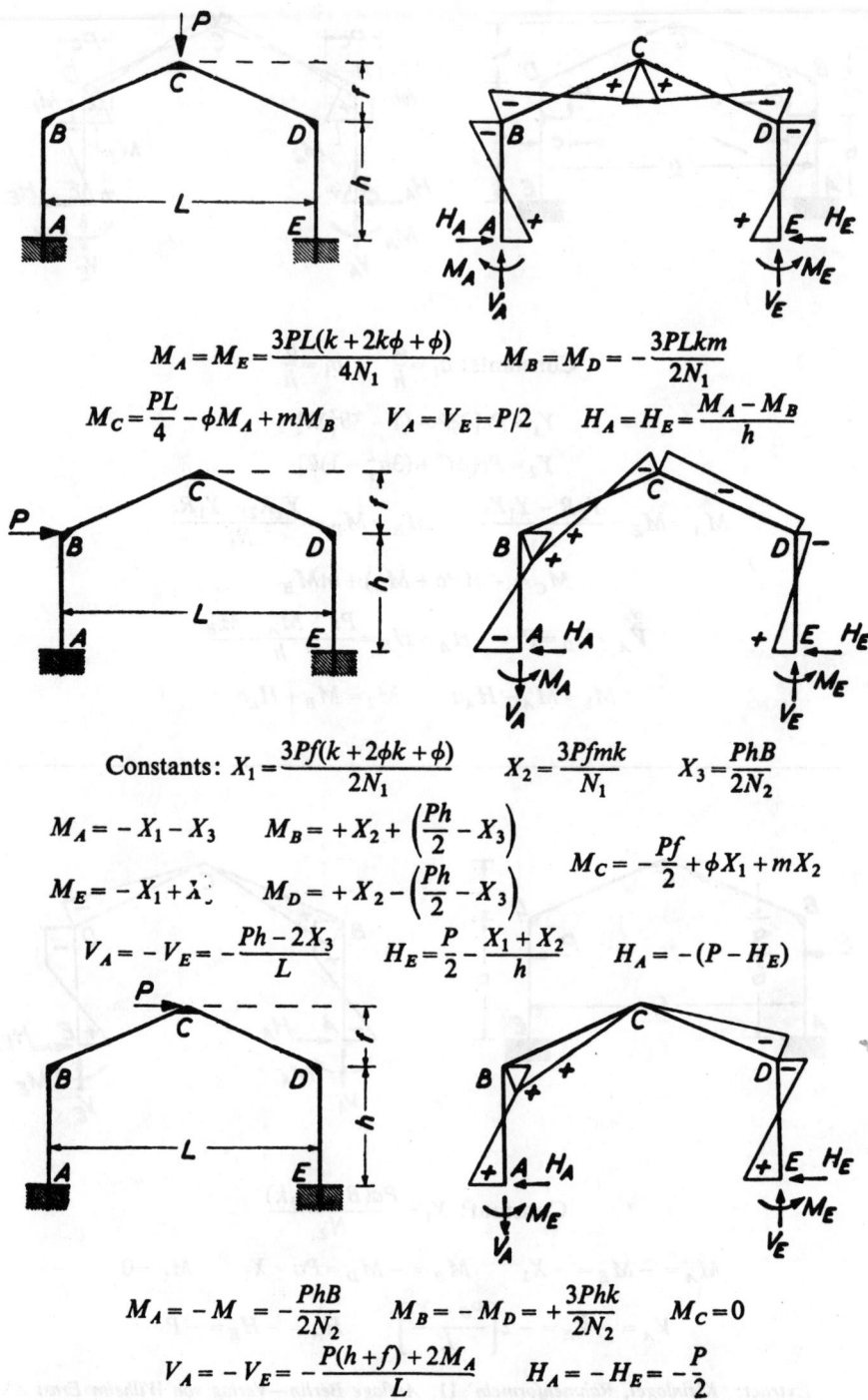

$$M_A = M_E = \frac{3PL(k + 2k\phi + \phi)}{4N_1} \qquad M_B = M_D = -\frac{3PLkm}{2N_1}$$

$$M_C = \frac{PL}{4} - \phi M_A + m M_B \qquad V_A = V_E = P/2 \qquad H_A = H_E = \frac{M_A - M_B}{h}$$

$$\text{Constants: } X_1 = \frac{3Pf(k + 2\phi k + \phi)}{2N_1} \qquad X_2 = \frac{3Pfmk}{N_1} \qquad X_3 = \frac{PhB}{2N_2}$$

$$M_A = -X_1 - X_3 \qquad M_B = +X_2 + \left(\frac{Ph}{2} - X_3\right)$$

$$M_E = -X_1 + X_3 \qquad M_D = +X_2 - \left(\frac{Ph}{2} - X_3\right) \qquad M_C = -\frac{Pf}{2} + \phi X_1 + m X_2$$

$$V_A = -V_E = -\frac{Ph - 2X_3}{L} \qquad H_E = \frac{P}{2} - \frac{X_1 + X_2}{h} \qquad H_A = -(P - H_E)$$

$$M_A = -M = -\frac{PhB}{2N_2} \qquad M_B = -M_D = +\frac{3Phk}{2N_2} \qquad M_C = 0$$

$$V_A = -V_E = -\frac{P(h + f) + 2M_A}{L} \qquad H_A = -H_E = -\frac{P}{2}$$

Extract: 'Kleinlogel, Rahmenformeln' 11. Auflage Berlin—Verlag von Wilhelm Ernst & Sohn.

Frame IV

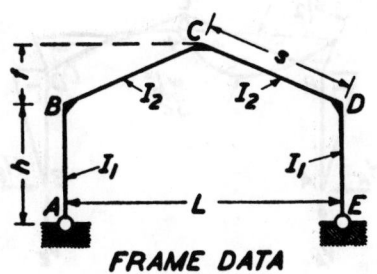

FRAME DATA

Coefficients:

$$k = \frac{I_2}{I_1} \cdot \frac{h}{s}$$

$$\phi = \frac{f}{h}$$

$$m = 1 + \phi$$

$$B = 2(k+1) + m \qquad C = 1 + 2m \qquad N = B + mC$$

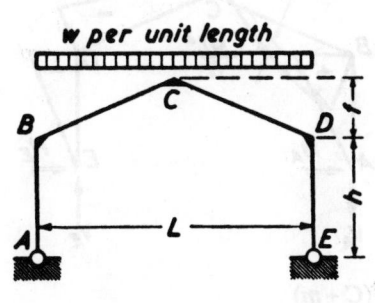

w per unit length

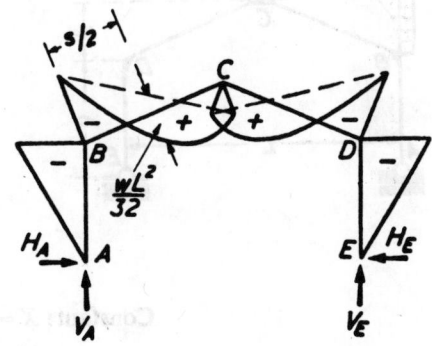

$$M_B = M_D = -\frac{wL^2(3 + 5m)}{16N} \qquad M_C = \frac{wL^2}{8} + mM_B$$

$$H_A = H_E = -\frac{M_B}{h} \qquad V_A = V_E = \frac{wL}{2}$$

Extract: 'Kleinlogel, Rahmenformeln' 11. Auflage Berlin—Verlag von Wilhelm Ernst & Sohn.

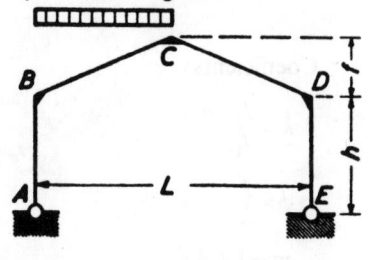

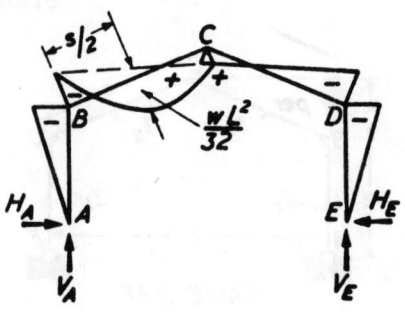

$$M_B = M_D = -\frac{wL^2(3+5m)}{32N} \qquad M_C = \frac{wL^2}{16} + mM_B$$

$$H_A = H_E = -\frac{M_B}{h} \qquad V_A = \frac{3wL}{8} \qquad V_E = \frac{wL}{8}$$

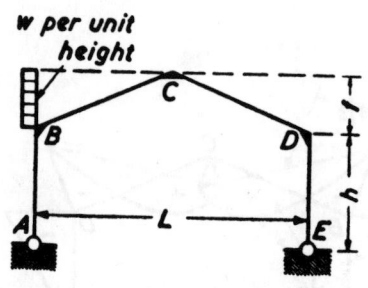

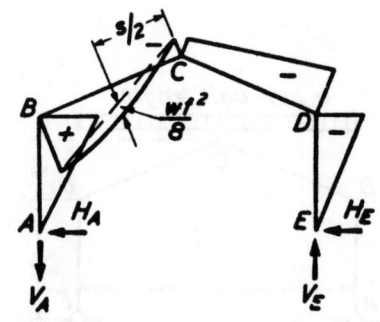

$$\text{Constant:} \ X = \frac{wf^2(C+m)}{8N}$$

$$M_B = +X + \frac{wfh}{2} \qquad M_C = -\frac{wf^2}{4} + mX$$

$$M_D = +X - \frac{wfh}{2} \qquad V_A = -V_E = -\frac{wfh(1+m)}{2L}$$

$$H_A = -\frac{X}{h} - \frac{wf}{2} \qquad H_E = -\frac{X}{h} + \frac{wf}{2}$$

Extract: 'Kleinlogel, Rahmenformeln' 11. *Auflage Berlin—Verlag von Wilhelm Ernst & Sohn.*

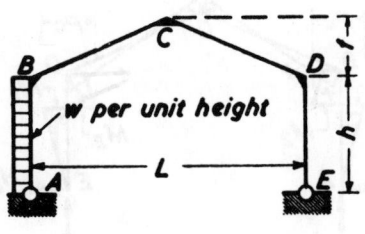
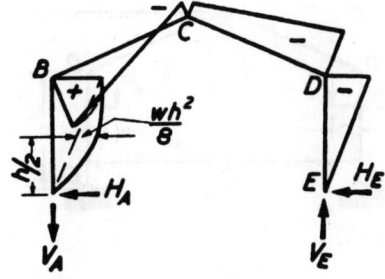

$$M_D = -\frac{wh^2}{8} \cdot \frac{2(B+C)+k}{N} \qquad M_B = \frac{wh^2}{2} + M_D$$

$$M_C = \frac{wh^2}{4} + mM_D$$

$$V_A = -V_E = -\frac{wh^2}{2L} \qquad H_E = -\frac{M_D}{h} \qquad H_A = -(wh - H_E)$$

Constants: $a_1 = \dfrac{a}{h}$ $\qquad X = \dfrac{Pc}{2} \cdot \dfrac{B+C-k(3a_1^2 - 1)}{N}$

$$M_B = Pc - X \qquad M_D = -X \qquad M_C = \frac{Pc}{2} - mX$$

$$M_1 = -a_1 X \qquad M_2 = Pc - a_1 X$$

$$V_E = \frac{Pc}{L} \qquad V_A = P - V_E \qquad H_A = H_E = \frac{X}{h}$$

Extract: 'Kleinlogel, Rahmenformeln' 11. *Auflage Berlin—Verlag von Wilhelm Ernst & Sohn.*

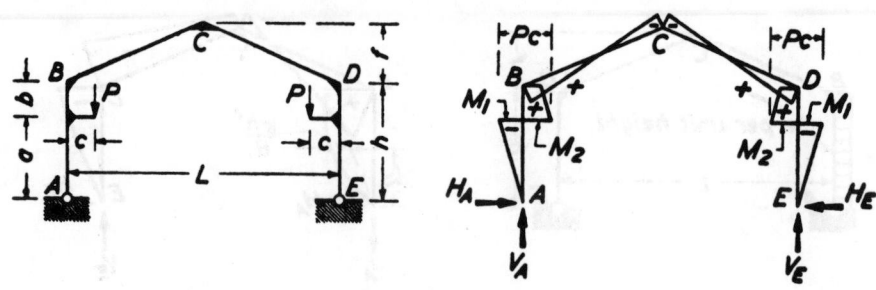

$$\text{Constant: } a_1 = \frac{a}{h}$$

$$M_B = M_D = Pc \cdot \frac{\phi C + k(3a_1^2 - 1)}{N} \qquad M_C = -\phi Pc + mM_B$$

$$H_A = H_E = \frac{Pc - M_B}{h} \qquad V_A = V_E = P$$

$$M_1 = -a_1(Pc - M_B) \qquad M_2 = (1 - a_1)Pc + a_1 M_B$$

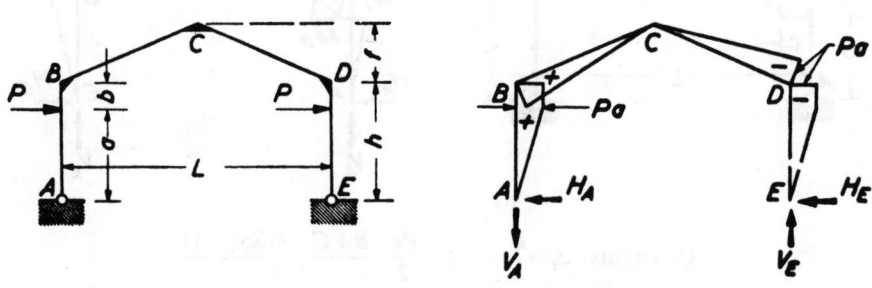

$$M_B = -M_D = Pa \qquad M_C = 0$$

$$H_A = -H_E = -P \qquad V_A = -V_E = -\frac{2Pa}{L}$$

$$\text{Moment at loads} = \pm Pa$$

Extract: 'Kleinlogel, Rahmenformeln' 11. *Auflage Berlin—Verlag von Wilhelm Ernst & Sohn.*

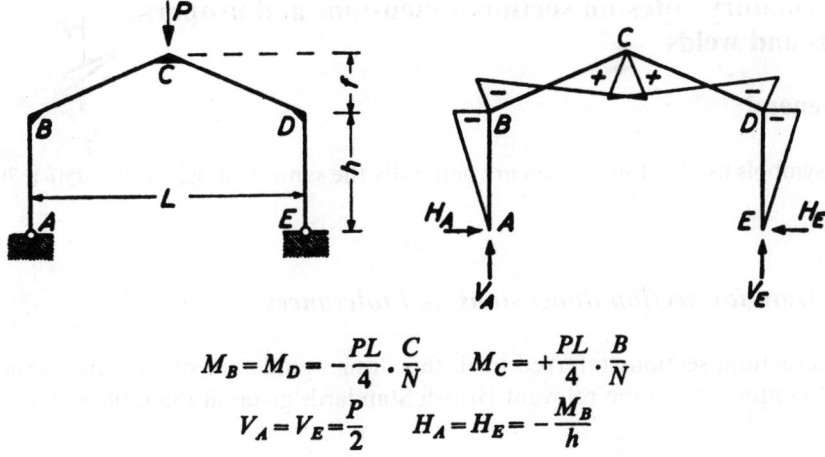

$$M_B = M_D = -\frac{PL}{4} \cdot \frac{C}{N} \qquad M_C = +\frac{PL}{4} \cdot \frac{B}{N}$$

$$V_A = V_E = \frac{P}{2} \qquad H_A = H_E = -\frac{M_B}{h}$$

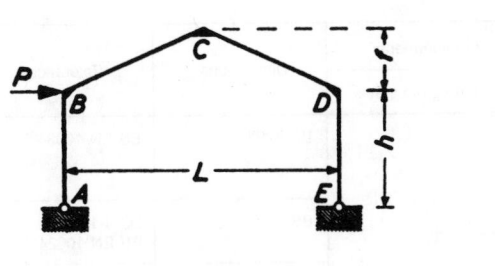
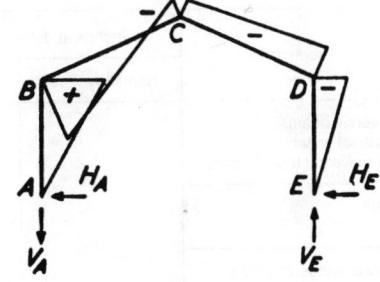

$$M_D = -\frac{Ph(B+C)}{2N} \qquad M_B = Ph + M_D \qquad M_C = \frac{Ph}{2} + mM_D$$

$$V_A = -V_E = -\frac{Ph}{L} \qquad H_E = -\frac{M_D}{h} \qquad H_A = -(P - H_E)$$

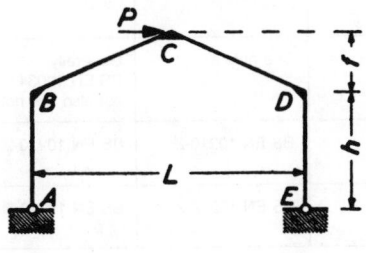
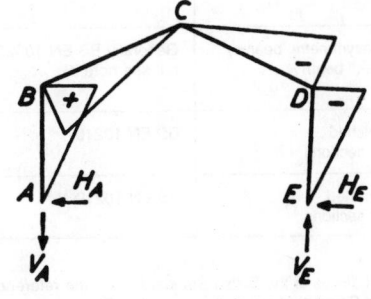

$$M_B = -M_D = +\frac{Ph}{2} \qquad M_C = 0 \qquad V_A = -V_E = -\frac{Phm}{L} \qquad H_A = -H_E = -\frac{P}{2}$$

Extract: '*Kleinlogel, Rahmenformeln*' 11. *Auflage Berlin—Verlag von Wilhelm Ernst & Sohn.*

Explanatory notes on section dimensions and properties, bolts and welds

1 General

The symbols used in this section are generally the same as those in BS 5950-1:2000.[1]

1.1 Material, section dimensions and tolerances

The structural sections referred to in this design guide are of weldable structural steels conforming to the relevant British Standards given in the table below:

Table – Structural steel products

Product	Technical delivery requirements		Dimensions	Tolerances
	Non-alloy steels	Fine grain steels		
Universal beams, universal columns, and universal bearing piles	BS EN 10025[2]	BS EN 10113-1[3]	BS 4-1[4]	BS EN 10034[5]
Joists			BS 4-1[4]	BS 4-1[4] BS EN 10024[6]
Parallel flange channels			BS 4-1[4]	BS EN 10279[7]
Angles			BS EN 10056-1[8]	BS EN 10056-2[8]
Structural tees cut from universal beams and universal columns			Bs 4-1[4]	–
Castellated universal beams Castellated universal columns			–	–
ASB (asymmetric beams) Slimdek® beam	Generally BS EN 10025[2], but see note b)		See note a)	Generally BS EN 10034[5], but also see note b)
Hot finished hollow sections	BS EN 10210-1[9]		BS EN 10210-2[9]	BS EN 10210-2[9]
Cold formed hollow sections	BS EN 10219-1[10]		BS EN 10219-2[10]	BS EN 10219-2[10]

Notes:
For full details of the British Standards, see the reference list at the end of the Explanatory Notes.
a) See Corus publication.[11]
b) For further details consult Corus.

1.2 Dimensional units

The dimensions of sections are given in millimetres (mm).

1.3 Property units

Generally, the centimetre (cm) is used for the calculated properties but for surface areas and for the warping constant (H), the metre (m) and the decimetre (dm) respectively are used.

Note: 1 dm   = 0.1 m   =   100 mm
       1 dm^6 = 1 × 10^{-6} m^6 =   1 × 10^{12} mm^6

1.4 Mass and force units

The units used are the kilogram (kg), the newton (N) and the metre per second^2 (m/s^2) so that $1\,N = 1\,kg \times 1\,m/s^2$. For convenience, a standard value of the acceleration due to gravity has been generally accepted as 9.80665 m/s^2. Thus, the force exerted by 1 kg under the action of gravity is 9.80665 N and the force exerted by 1 tonne (1000 kg) is 9.80665 kilonewtons (kN).

2 Dimensions of sections

2.1 Masses

The masses per metre have been calculated assuming that the density of steel is 7850 kg/m^3.

In all cases, including compound sections, the tabulated masses are for the steel section alone and no allowance has been made for connecting material or fittings.

2.2 Ratios for local buckling

The ratios of the flange outstand to thickness (b/T) and the web depth to thickness (d/t) are given for I, H and channel sections. The ratios of the outside diameter to thickness (D/t) are given for circular hollow sections. The ratios d/t and b/t are also given for square and rectangular hollow sections. All the ratios for local buckling have been calculated using the dimensional notation given in Figure 5 of

BS 5950-1: 2000 and are for use when element and section class are being checked to the limits given in Tables 11 and 12 of BS 5950-1: 2000.

2.3 Dimensions for detailing

The dimensions C, N and n have the meanings given in the figures at the heads of the tables and have been calculated according to the formulae below. The formulae for N and C make allowance for rolling tolerances, whereas the formulae for n make no such allowance.

2.3.1 Universal beams, universal columns and bearing piles

$$N = \frac{(B-t)}{2} + 10\,\text{mm} \qquad \text{(rounded to the nearest 2 mm above)}$$

$$n = \frac{(D-d)}{2} \qquad \text{(rounded to the nearest 2 mm above)}$$

$$C = \frac{t}{2} + 2\,\text{mm} \qquad \text{(rounded to the nearest mm)}$$

2.3.2 Joists

$$N = \frac{(B-t)}{2} + 6\,\text{mm} \qquad \text{(rounded to the nearest 2 mm above)}$$

$$n = \frac{(D-d)}{2} \qquad \text{(rounded to the nearest 2 mm above)}$$

$$C = \frac{t}{2} + 2\,\text{mm} \qquad \text{(rounded to the nearest mm)}$$

Note: Flanges of BS 4-1 joists have an 8° taper.

2.3.3 Parallel flange channels

$$N = (B-t) + 6\,\text{mm} \qquad \text{(rounded up to the nearest 2 mm above)}$$

$$n = \frac{(D-d)}{2} \qquad \text{(taken to the next higher multiple of 2 mm)}$$

$$C = t + 2\,\text{mm} \qquad \text{(rounded up to the nearest mm)}$$

2.3.4 Castellated sections

The depth of the castellated section. D_c, is given by:

$$D_c = D + D_s/2$$

where D is the actual depth of the original section

D_s is the serial depth of the original section, except that $D_s = 381\,\text{mm}$ for 356×406 UCs.

3 Section properties

3.1 General

All section properties have been accurately calculated and rounded to three signifi-cant figures. They have been calculated from the metric dimensions given in the appropriate standards (see section 1.2). For angles, BS EN 10056-1 assumes that the toe radius equals half the root radius.

3.2 Sections other than hollow sections

3.2.1 Second moment of area (I)

The second moment of area of the section, often referred to as moment of inertia, has been calculated taking into account all tapers, radii and fillets of the sections.

3.2.2 Radius of gyration (r)

The radius of gyration is a parameter used in buckling calculation and is derived as follows:

$$r = \left[\frac{I}{A}\right]^{1/2}$$

where A is the cross-sectional area.

For castellated sections, the radius of gyration given is calculated at the net section as required in design to BS 5950-1: 2000.

3.2.3 Elastic modulus (Z)

The elastic modulus is used to calculate the elastic moment capacity based on the design strength of the section or the stress at the extreme fibre of the section from a known moment. It is derived as follows:

$$Z = \frac{1}{y}$$

where y is the distance to the extreme fibre of the section from the elastic neutral axis.

For castellated sections, the elastic moduli given are those at the net section. The elastic moduli of the tee are calculated at the outer face of the flange and toe of the tee formed at the net section.

For parallel flange channels, the elastic modulus about the minor $(y–y)$ axis is given at the toe of the section, i.e.

$$y = B - c_y$$

where B is the width of the section

c_y is the distance from the back of the web to the centroidal axis.

For angles, the elastic moduli about both axes are given at the toes of the section, i.e.

$$y_x = A - c_x$$

$$y_y = B - c_y$$

Where A is the leg length perpendicular to the $x–x$ axis

B is the leg length perpendicular to the $y–y$ axis

C_x is the distance from the back of the angle to the centre of gravity, referred to as the $x–x$ axis

C_y is the distance from the back of the angle to the centre of gravity, referred to as the $y–y$ axis.

3.2.4 Buckling parameter (u) and torsional index (x)

The buckling parameter and torsional index used in buckling calculations are derived as follows:

(1) For bi-symmetric flanged sections and flanged sections symmetrical about the minor axis only:

$$u = [(4S_x^2 \gamma)/(A^2 h^2)]^{1/4}$$

$$x = 0.566h[A/J]^{1/2}$$

(2) For flanged sections symmetric about the major axis only:

$$u = [(I_y S_x^2 \gamma)/(A^2 H)]^{1/4}$$

$$x = 1.132[(A H)/(I_y J)]^{1/2}$$

where S_x is the plastic modulus about the major axis

$$\gamma = \left[1 - \frac{I_y}{I_x} \right]$$

I_x is the second moment of area about the major axis
I_y is the second moment of area about the minor axis
A is the cross-sectional area
h is the distance between shear centres of flanges (for T sections, h is the distance between the shear centre of the flange and the toe of the web)
H is the warping constant
J is the torsion constant.

3.2.5 Warping constant (H) and torsion constant (J)

(1) *I and H sections*

The warping constant and torsion constant for I and H sections are calculated using the formulae given in the SCI publication P057 *Design of members subject to combined bending and torsion.*[12]

(2) *Tee-sections*

For tee-sections cut from UB and UC sections, the warping constant (H) and torsion constant (J) have been derived as given below.

$$H = \frac{1}{144} T^3 B^3 + \frac{1}{36} \left(d - \frac{T}{2} \right)^3 t^3$$

$$J = \frac{1}{3} B T^3 + \frac{1}{3} (d - T) t^3 + \alpha_1 D_1^4 - 0.21\, T^4 - 0.105\, t^3$$

where $\alpha_1 = -0.042 + 0.2204 \dfrac{t}{T} + 0.1355 \dfrac{r}{T} - 0.0865 \dfrac{t\, r}{T^2} - 0.0725 \dfrac{t^2}{T^2}$

$$D_1 = \frac{(T + r)^2 + (r + 0.25\, t)t}{2r + T}$$

Note: These formulae do not apply to tee-sections cut from joists which have tapered flanges. For such sections, details are given in SCI publication 057.[12]

(3) *Parallel flange channels*

For parallel flange channels, the warping constant (H) and torsion constant (J) are calculated as follows:

$$H = \frac{h^2}{4} \left[I_y - A \left(c_y - \frac{t}{2} \right)^2 \left(\frac{h^2 A}{4 I_x} - 1 \right) \right]$$

$$J = \frac{2}{3} B T^3 + \frac{1}{3} (D - 2T) t^3 + 2 \alpha_3 D_3^4 - 0.42 T^4$$

where c_y = is the distance from the back of the web to the centroidal axis

$$\alpha_3 = -0.0908 + 0.2621\frac{t}{T} + 0.1231\frac{r}{T} - 0.0752\frac{t\,r}{T^2} - 0.0945\left(\frac{t}{T}\right)^2$$

$$D_3 = 2[(3r + t + T) - \sqrt{2(2r + t)(2r + T)}]$$

Note: The formula for the torsion constant (J) is applicable to parallel flange channels only and does not apply to tapered flange channels.

(4) *Angles*

For angles, the torsion constant (J) is calculated as follows:

$$J = \frac{1}{3}bt^3 + \frac{1}{3}(d - t)t^3 + \alpha_3 D_3^4 - 0.21t^4$$

where $\alpha_3 = 0.0768 + 0.0479\frac{r}{t}$

$$D_3 = 2\left[(3r + 2t) - \sqrt{2(2r + t)^2}\right]$$

(5) *ASB sections*

For ASB (asymmetric beams) Slimdek® beam, the warping constant (H) and torsion constant (J) are as given in Corus brochure, *Structural sections*.[11]

3.2.6 Plastic modulus (S)

The full plastic moduli about both principal axes are tabulated for all sections except angle sections. For angle sections, BS 5950-1: 2000 requires design using the elastic modulus.

The reduced plastic moduli under axial load are tabulated for both principal axes for all sections except asymmetric beams and angle sections. For angle sections, BS 5950-1: 2000 requires design using the elastic modulus.

When a section is loaded to full plasticity by a combination of bending and axial compression about the major axis, the plastic neutral axis shifts and may be located either in the web or in the tension flange (or in the taper part of the flange for a joist) depending on the relative values of bending and axial compression. Formulae giving the reduced plastic modulus under combined loading have to be used, which use a parameter n as follows:

$$n = \frac{F}{A\,p_y}$$ (This is shown in the member capacity tables as F/P_z)

where F is the factored axial load

A is the cross-sectional area

p_y is the design strength of the steel.

For each section, there is a 'change' value of n. Formulae for reduced plastic modulus and the 'change' value are given below.

(1) *Universal beams, universal columns and bearing piles*
   If the value of n calculated is less than the change value, the plastic neutral axis is in the web and the formula for lower values of n must be used. If n is greater than the change value, the plastic neutral axis lies in the tension flange and the formula for higher values of n must be used. The same principles apply when the sections are loaded axially and bent about the minor axis, lower and higher values of n indicating that the plastic neutral axis lies inside or outside the web respectively.

Major axis bending:

Reduced plastic modulus:

$$S_{rx} = K_1 - K_2 n^2$$

$$S_{rx} = K_3(1-n)(K_4+n)$$

Change value:

$$\text{for} \quad n < \frac{(D-2T)t}{A}$$

$$\text{for} \quad n \geq \frac{(D-2T)t}{A}$$

where $\quad K_1 = S_x$

$$K_3 = \frac{A^2}{4B}$$

$$K_2 = \frac{A^2}{4t}$$

$$K_4 = \frac{2DB}{A} - 1$$

Minor axis bending:

Reduced plastic modulus:

$$S_{ry} = K_1 - K_2 n^2$$

$$S_{ry} = K_3(1-n)(K_4+n)$$

Change value:

$$\text{for} \quad n < \frac{tD}{A}$$

$$\text{for} \quad n \geq \frac{tD}{A}$$

where $\quad K_1 = S_y$

$$K_3 = \frac{A^2}{8T}$$

$$K_2 = \frac{A^2}{4D}$$

$$K_4 = \frac{4BT}{A} - 1$$

(2) *Joists*

Major axis bending:

If the value of n calculated is less than the lower change value (n_1), the plastic neutral axis is in the web and the formula for lower values of n must be used. If n is greater than the higher change value (n_2), the plastic neutral axis lies in

the part of the tension flange that is not tapered and the formula for higher values of n must be used. If the value of n calculated lies between the lower change value (n_1) and the higher change value (n_2), the plastic neutral axis lies in the tapered part of the flange and then a linear interpolation between the two formulae is used to calculate the reduced plastic modulus.

Reduced plastic modulus   Change value

$$S_{rx} = S_{rx1} = K_1 - K_2 n^2 \qquad \text{for } n \leq n_1 = \left\{ \frac{D}{A} - \frac{2}{A}\left(T + \frac{B-t}{4}\tan(\theta)\right) \right\} t$$

$$S_{rx} = S_{rx2} = K_3(1-n)(K_4+n) \qquad \text{for } n \geq n_2 = 1 - \frac{2B}{A}\left(T - \frac{B-t}{4}\tan(\theta)\right)$$

$$S_{rx} = S_{rx1} + (S_{rx2} - S_{rx1})\left(\frac{n-n_1}{n_2-n_1}\right) \qquad \text{for } n_1 < n < n_2$$

where $K_1 = S_x$ $\qquad\qquad\qquad\qquad K_2 = \dfrac{A^2}{4t}$

$\qquad\quad K_3 = \dfrac{A^2}{4B}$ $\qquad\qquad\qquad K_4 = \dfrac{2DB}{A} - 1$

$\qquad\quad \theta = 8°$ (flange taper)

Minor axis bending:

The same principles apply when the sections are loaded axially and bent about the minor axis, lower and higher values of n indicating that the plastic neutral axis lies inside or outside the web respectively.

Reduced plastic modulus   Change value

$$S_{ry} = K_1 - K_2 n^2 \qquad\qquad \text{for } n < \frac{tD}{A}$$

$$S_{ry} = K_3(1-n)(K_4+n) \qquad \text{for } n \geq \frac{tD}{A}$$

where $K_1 = S_y$ $\qquad\qquad\qquad\qquad K_2 = \dfrac{A^2}{4D}$

$\qquad\quad K_3 = 0.87\dfrac{A^2}{8T}$ $\qquad\qquad K_4 = \dfrac{4BT}{A} - 1$

(3) *Parallel flange channels*

Major axis bending:

If the value of n calculated is less than the change value, the plastic neutral axis is in the web and the formula for lower values of n must be used. If n is greater

than the change value, the plastic neutral axis lies in the flange and the formula for higher values of n must be used.

Reduced plastic modulus   Change value

$$S_{rx} = K_1 - K_2 n^2 \qquad\qquad \text{for} \quad n < \frac{(D-2T)t}{A}$$

$$S_{rx} = K_3(1-n)(K_4 + n) \qquad\qquad \text{for} \quad n \geq \frac{(D-2T)t}{A}$$

where $K_1 = S_x$ $K_2 = \dfrac{A^2}{4t}$

$$K_3 = \frac{A^2}{4B} \qquad\qquad\qquad K_4 = \frac{2DB}{A} - 1$$

Minor axis bending:

In calculating the reduced plastic modulus of a channel for axial force combined with bending about the minor axis, the axial force is considered as acting at the centroidal axis of the cross-section whereas it is considered to be resisted at the plastic neutral axis. The value of the reduced plastic modulus takes account of the resulting moment due to eccentricity relative to the net centroidal axis.

The reduced plastic modulus of a parallel flange channel bending about the minor axis depends on whether the stresses induced by the axial force and applied moment are the same or of opposite kind towards the back of the channel. Where the stresses are of the same kind, an initial increase in axial force may cause a small initial rise of the 'reduced' plastic modulus, due to the eccentricity of the axial force.

For each section there is again a change value of n. For minor axis bending the position of the plastic neutral axis when there is no axial load may be either in the web or in the flanges. When the value of n is less than the change value, the formula for lower values of n must be used. If n is greater than the change value, the formula for higher values of n must be used.

The formulae concerned are complex and are therefore not quoted here.

3.2.7 Equivalent slenderness coefficient (ϕ_a) and monosymmetry index (ψ_a)

The equivalent slenderness coefficient (ϕ_a) is tabulated for both equal and unequal angles. Two values of the equivalent slenderness coefficient are given for each unequal angle. The larger value is based on the major axis elastic modulus (Z_u) to the toe of the short leg and the lower value is based on the major axis elastic modulus to the toe of the long leg.

The equivalent slenderness coefficient (ϕ_a) is calculated as follows:

$$\phi_{\mathrm{a}} = \left[\frac{Z_{\mathrm{u}}^2 \gamma_{\mathrm{a}}}{AJ}\right]^{0.5}$$

Definitions of all the individual terms are given in BS 5950-1[1], clause B.2.9.

The monosymmetry index (ψ_{a}) is only applicable for unequal angles and is calculated as follows:

$$\psi_{\mathrm{a}} = \left[2v_0 - \frac{\int v_i(u_{\mathrm{t}}^2 + v_i^2)\mathrm{d}A}{I_{\mathrm{u}}}\right]\frac{1}{t}$$

Definitions of all the individual terms are given in BS 5950-1[1], Clause B.2.9.

3.3 Hollow sections

Section properties are given for both hot-finished and cold-formed hollow sections. The ranges of hot-finished and cold-formed sections covered are different. The section ranges listed are in line with sections that are readily available from the major section manufacturers. For the same overall dimensions and wall thickness, the section properties for hot-finished and cold-formed sections are different because the corner radii are different.

3.3.1 Common properties

For comment on second moment of area, radius of gyration and elastic modulus, see sections 3.2.1, 3.2.2 and 3.2.3.

For hot-finished square and rectangular hollow sections, the sectional properties have been calculated, using corner radii of $1.5t$ externally and $1.0t$ internally, as specified by BS EN 10210-2.[9]

For cold-formed square and rectangular hollow sections, the sectional properties have been calculated, using the external corner radii of $2t$ if $t \le 6\,\mathrm{mm}$, $2.5t$ if $6\,\mathrm{mm} < t \le 10\,\mathrm{mm}$ and $3t$ if $t > 10\,\mathrm{mm}$ as specified by BS EN 10219–2.[10] The internal corner radii used are $1.0t$ if $t \le 6\,\mathrm{mm}$, $1.5t$ if $6\,\mathrm{mm} < t \le 10\,\mathrm{mm}$ and $2t$ if $t > 10\,\mathrm{mm}$, as specified by BS EN 10219-2.[10]

3.3.2 Torsion constant (J)

For circular hollow sections:

$$J = 2I$$

For square and rectangular hollow sections:

$$J = \frac{4A_{\mathrm{h}}^2 t}{h} + \frac{t^3 h}{3}$$

where I is the second moment of area

         t is the thickness of section

        h is the mean perimeter $= 2\,[(B - t) + (D - t)] - 2\,R_c\,(4 - \pi)$

       A_h is the area enclosed by mean perimeter $= (B - t)\,(D - t) - R_c^2\,(4 - \pi)$

        B is the breadth of section

        D is the depth of section

       R_c is the average of internal and external corner radii.

3.3.3 Torsion modulus constant (C)

For circular hollow sections

$$C = 2Z$$

For square and rectangular hollow sections

$$C = J \Big/ \left(t + \frac{2A_h}{h} \right)$$

where Z is the elastic modulus and J, t, A_h and h are as defined in section 3.3.2.

3.3.4 Plastic modulus of hollow sections (S)

The full plastic modulus (S) is given in the tables. When a member is subject to a combination of bending and axial load the plastic neutral axis shifts. Formulae giving the reduced plastic modulus under combined loading have to be used, which use the parameter n as defined below.

$$n = \frac{F}{A\,p_y} \quad \text{(This is shown in the member capacity tables as } F/P_z)$$

where F is the factored axial load

        A is the cross-sectional area

       p_y is the design strength of the steel.

For square and rectangular hollow sections there is a 'change' value of n. Formulae for reduced plastic modulus and 'change' value are given below.

(1) *Circular hollow sections*

$$S_r = S \cos\left(\frac{n\pi}{2} \right)$$

(2) *Square and rectangular hollow sections*

If the value of n calculated is less than the change value, the plastic neutral axis is in the webs and the formula for lower values of n must be used. If n is greater than the change value, the plastic neutral axis lies in the flange and the formula for higher values of n must be used.

Major axis bending:

Reduced plastic modulus   Change value

$$S_{rx} = S_x - \frac{A^2 n^2}{8t}$$ for $n \le \frac{2t(D-2t)}{A}$

$$S_{rx} = \frac{A^2}{4(B-t)}(1-n)\left[\frac{2D(B-t)}{A} + n - 1\right]$$ for $n > \frac{2t(D-2t)}{A}$

Minor axis bending:

Reduced plastic modulus   Change value

$$S_{ry} = S_y - \frac{A^2 n^2}{8t}$$ for $n \le \frac{2t(B-2t)}{A}$

$$S_{ry} = \frac{A^2}{4(D-t)}(1-n)\left[\frac{2B(D-t)}{A} + n - 1\right]$$ for $n > \frac{2t(B-2t)}{A}$

where S, S_x, S_y are the full plastic moduli about the relevant axes

A is the gross cross-sectional area

D, B and t are as defined in section 3.3.2.

4 Bolts and welds

4.1 Bolt capacities

The types of bolts covered are:

- Grades 4.6, 8.8 and 10.9, as specified in BS 4190:[13] *ISO metric black hexagon bolts, screws and nuts.*
- Non-preloaded and preloaded HSFG bolts as specified in BS 4395:[14] *High strength friction grip bolts and associated nuts and washers for structural engineering.* Part 1: *General grade* and Part 2: *Higher grade.*
  Preloaded HSFG bolts should be tightened to minimum shank tension (P_o) as specified in BS 4604.[15]
- Countersunk bolts as specified in BS 4933:[16] *ISO metric black cup and countersunk bolts and screws with hexagon nuts.*

Information on assemblies of matching bolts, nuts and washers is given is BS 5950-2.[1]

(1) *Non-preloaded bolts*, Ordinary (Grades 4.6, 8.8 and 10.9) and HSFG (General and Higher Grade):
  (a) The tensile stress area (A_t) is obtained from the above standards.
  (b) The tension capacity of the bolt is given by:

$$P_{nom} = 0.8p_tA_t \quad \text{Nominal}$$ 6.3.4.2

$$P_t \quad = p_tA_t \quad\quad \text{Exact}$$ 6.3.4.3

  where p_t is the tension strength of the bolt.   Table 34

  (c) The shear capacity of the bolt is given by:

$$P_s = p_sA_s$$ 6.3.2.1

  where p_s is the shear strength of the bolt   Table 30

  A_s is the shear area of the bolt.

In the tables, A_s has been taken as equal to A_t.
  The shear capacity given in the tables must be reduced for large packings, large grip lengths, kidney shaped slots or long joints when applicable.   6.3.2.2
6.3.2.3
6.3.2.4
6.3.2.5

  (d) The effective bearing capacity given is the lesser of the bearing capacity of the bolt given by:

$$P_{bb} = dt_pp_{bb}$$ 6.3.3.2

  and the bearing capacity of the connected ply given by:

$$P_{bs} = k_{bs}dt_pp_{bs}$$ 6.3.3.3

  assuming that the end distance is greater than or equal to twice the bolt diameter to meet the requirement that $P_{bs} \leq 0.5\, k_{bs}et_pp_{bs}$

  where d is the nominal diameter of the bolt

  t_p is the thickness of the ply.

For countersunk bolts, t_p is taken as the ply thickness minus half the depth of countersinking. Depth of countersinking is taken as half the bolt diameter based on a 90° countersink.   6.3.3.2

  p_{bb} is the bearing strength of the bolt   Table 31
  p_{bs} is the bearing strength of the ply   Table 32

e is the end distance

k_{bs} is a coefficient to allow for hole type.   6.3.3.3

Tables assume standard clearance holes, therefore k_{bs} is taken as 1.0. For oversize holes and short slots, $k_{bs} = 0.7$. For long slots and kidney shaped slots, $k_{bs} = 0.5$.

(2) *Preloaded HSFG bolts* (general grade and higher grade):   6.4

(a) The proof load of the bolt (P_o) is obtained from BS 4604.[19] The same proof load is used for countersunk bolts as for non-countersunk bolts. For this to be acceptable the head dimensions must be as specified in BS 4933.[20]

(b) The tension capacity (P_t) of the bolt is taken as: 6.4.5

1.1 P_o for non-slip in service

0.9 P_o for non-slip under factored load

(c) The slip resistance of the bolt is given by: 6.4.2

$P_{SL} = 1.1\ K_s\mu P_o$ for non-slip in service

$P_{SL} = 0.9\ K_s\mu P_o$ for non-slip under factored load

where K_s is taken as 1.0 for fasteners in standard clearance holes   6.4.2

μ is the slip factor.   Table 35

(d) The bearing resistance is only applicable for non-slip in service and is taken as:

$P_{bg} = 1.5\ dt_p p_{bs}$ 6.4.4

assuming that the end distance is greater than or equal to three times the bolt diameter, to meet the requirement that $P_{bg} \le 0.5\ et_p p_{bs}$.

where d is the nominal diameter of the bolt

t_p is the thickness of the ply

p_{bs} is the bearing strength of the ply.   Table 32

(e) The shear capacity of the bolt is given by: 6.4.1(a)

$P_s = p_s A_s$ 6.3.2.1

where p_s is the shear strength of the bolt   Table 30

A_s is the shear area of the bolt

In the tables, A_s has been taken as equal to A_t.

4.2 Welds

Capacities of longitudinal and transverse fillet welds per unit length are tabulated. The weld capacities are given by:

Longitudinal shear capacity, $P_L = p_w a$ 6.8.7.3

Transverse capacity, $P_T = K p_w a$

where p_w is the weld design strength   Table 37

      a is the throat thickness, taken as $0.7 \times$ the leg length

      K is the enhancement factor for transverse welds.   6.8.7.3

The plates are assumed to be at $90°$ and therefore $K = 1.25$. Electrode classifications of E35 and E42 are assumed for steel grade S275 and S355 respectively. Welding consumables are in accordance with BS EN 440,[17] BS EN 449,[18] BS EN 756,[19] BS EN 758,[20] or BS EN 1668[21] as appropriate.   Table 37

References to explanatory notes

1. British Standards Institution
   BS 5950 *Structural use of steelwork in building.*
   BS 5950-1: 2000 *Code of Practice for design – Rolled and welded sections.*
   BS 5950-2: 2000 *Specification for materials, fabrication and erection: Rolled and welded sections.*
2. British Standards Institution
   BS EN 10025: 1993 *Hot-rolled products of non-alloy structural steels. Technical delivery conditions* (including amendment 1995).
3. British Standards Institution
   BS EN 10113 *Hot-rolled products in weldable fine grain structural steels.*
   BS EN 10113-1: 1993 *General delivery conditions* (replaces BS 4360: 1990).
4. British Standards Institution
   BS 4 *Structural steel sections.*
   BS 4-1: 1993 *Specification for hot rolled sections* (including amendment 2001).
5. British Standards Institution
   BS EN 10034: 1993 *Structural steel I and H sections. Tolerances on shape and dimensions* (replaces BS 4-1: 1980).
6. British Standards Institution
   BS EN 10024: 1995 *Hot rolled taper flange I sections. Tolerances on shape and dimensions.*
7. British Standards Institution
   BS EN 10279: 2002 *Hot-rolled steel channels. Tolerances on shape, dimension and mass* (including amendment 1, amendment 2: 200).

8. British Standards Institution
   BS EN 10056 *Specification for structural steel equal and unequal angles.*
   BS EN 10056-1: 1999 *Dimensions* (replaces BS 4848-4: 1972).
   BS EN 10056-2: 1999 *Tolerances on shape and dimensions* (replaces BS 4848-4: 1972).
9. British Standards Institution
   BS EN 10210 *Hot-finished structural hollow sections of non-alloy and fine grain structural steels.*
   BS EN 10210-1: 1994 *Technical delivery requirements* (replaces BS 4360: 1990).
   BS EN 10210-2: 1997 *Tolerances, dimensions and sectional properties* (replaces BS 4848-2: 1991).
10. British Standards Institution
    BS EN 10219 *Cold-formed welded structural sections of non-alloy and fine grain steels.*
    BS EN 10219-1: 1997 *Technical delivery requirements.*
    BS EN 10219-2: 1997 *Tolerances and sectional properties* (replaces BS 6363: 1983).
11. Structural sections to BS 4: Part 1: 1963 and BS EN 10056: 1999
    Corus Construction and Industrial Sections, 03/2001
12. Nethercot D.A., Salter P.R. & Malik A.S. (1989)
    Design of members subject to combined bending and torsion (SCI-P057)
    The Steel Construction Institute, Ascot, Berks.
13. British Standards Institution
    BS 4190: 2001 *ISO metric black hexagon bolts, screws and nuts – Specification.*
14. British Standards Institution
    BS 4395 *Specification for high strength friction grip bolts and associated nuts and washers for structural engineering.*
    BS 4395-1: 1969 *General grade* (including amendments 1, amendments 2: 1997).
    BS 4395-2: 1969 *Higher grade bolts and nuts and general grade washers* (including amendment 1, amendment 2: 1976).
15. British Standards Institution
    BS 4604 *Specification for the use of high strength friction grip bolts in structural steelwork.* Metric series.
    BS 4604-1: 1970 *General grade* (including amendment 1, amendment 2, and amendment 3: 1982).
    BS 4604-2: 1970 – *High grade* (parallel shank) (including amendment 1, amendment 2: 1972).
16. British Standards Institution
    BS 4933: 1973 *Specification for ISO metric black cup and countersunk head bolts and screws with hexagon nuts.*
17. British Standards Institution
    BS EN 440: 1995 *Welding consumables. Wire electrodes and deposits for gas shielded metal are welding of non-alloy and fine grain steels. Classification.*
18. British Standards Institution
    BS EN 499: 1995 *Welding consumables. Covered electrodes for manual metal are welding of non-alloy and fine grain. Classification.*

19. British Standards Institution
BS EN 756: 1996 *Welding consumables. Wire electrodes and wire-flux combinations for submerged arc welding of non-alloy and fine grain steels. Classification.*
20. British Standards Institution
BS EN 758: 1997 *Welding consumables. Tubular cored electrodes for metal arc welding with and without a gas shield of non-alloy and fine grain steels. Classification.*
21. British Standards Institution
BS EN 1668: 1997 *Welding consumables. Rods, wires and deposits for tungsten inert gas welding of non-alloy and fine grain steels. Classification.*

Tables of dimensions and gross section properties

UNIVERSAL BEAMS

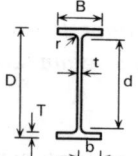

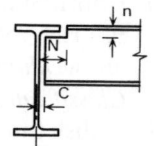

DIMENSIONS

Section Designation	Mass per Metre	Depth of Section	Width of Section	Thickness Web	Thickness Flange	Root Radius	Depth between Fillets	Ratios for Local Buckling Flange	Ratios for Local Buckling Web	Dimensions for Detailing End Clearance	Dimensions for Detailing Notch N	Dimensions for Detailing Notch n	Surface Area Per Metre	Surface Area Per Tonne
	kg/m	D mm	B mm	t mm	T mm	r mm	d mm	b/T	d/t	C mm	N mm	n mm	m²	m²
1016 × 305 × 487 # †	486.6	1036.1	308.5	30.0	54.1	30.0	867.9	2.85	28.9	17	150	86	3.19	6.57
1016 × 305 × 437 # †	436.9	1025.9	305.4	26.9	49.0	30.0	867.9	3.12	32.3	16	150	80	3.17	7.25
1016 × 305 × 393 # †	392.7	1016.0	303.0	24.4	43.9	30.0	868.2	3.45	35.6	14	150	74	3.14	8.01
1016 × 305 × 349 # †	349.4	1008.1	302.0	21.1	40.0	30.0	868.1	3.77	41.1	13	150	70	3.13	8.96
1016 × 305 × 314 # †	314.3	1000.0	300.0	19.1	35.9	30.0	868.2	4.18	45.5	12	150	66	3.11	9.90
1016 × 305 × 272 # †	272.3	990.1	300.0	16.5	31.0	30.0	868.1	4.84	52.6	10	152	63	3.10	11.4
1016 × 305 × 249 # †	248.7	980.2	300.0	16.5	26.0	30.0	868.2	5.77	52.6	10	152	56	3.08	12.4
1016 × 305 × 222 # †	222.0	970.3	300.0	16.0	21.1	30.0	868.1	7.11	54.3	10	152	52	3.06	13.8
914 × 419 × 388 #	388.0	921.0	420.5	21.4	36.6	24.1	799.6	5.74	37.4	13	210	62	3.44	8.87
914 × 419 × 343 #	343.3	911.8	418.5	19.4	32.0	24.1	799.6	6.54	41.2	12	210	58	3.42	9.95
914 × 305 × 289 #	289.1	926.6	307.7	19.5	32.0	19.1	824.4	4.81	42.3	12	156	52	3.01	10.4
914 × 305 × 253 #	253.4	918.4	305.5	17.3	27.9	19.1	824.4	5.47	47.7	11	156	48	2.99	11.8
914 × 305 × 224 #	224.2	910.4	304.1	15.9	23.9	19.1	824.4	6.36	51.8	10	156	44	2.97	13.3
914 × 305 × 201 #	200.9	903.0	303.3	15.1	20.2	19.1	824.4	7.51	54.6	10	156	40	2.96	14.7
838 × 292 × 226 #	226.5	850.9	293.8	16.1	26.8	17.8	761.7	5.48	47.3	10	150	46	2.81	12.4
838 × 292 × 194 #	193.8	840.7	292.4	14.7	21.7	17.8	761.7	6.74	51.8	9	150	40	2.79	14.4
838 × 292 × 176 #	175.9	834.9	291.7	14.0	18.8	17.8	761.7	7.76	54.4	9	150	38	2.78	15.8
762 × 267 × 197	196.8	769.8	268.0	15.6	25.4	16.5	686.0	5.28	44.0	10	138	42	2.55	13.0
762 × 267 × 173	173.0	762.2	266.7	14.3	21.6	16.5	686.0	6.17	48.0	9	138	40	2.53	14.6
762 × 267 × 147	146.9	754.0	265.2	12.8	17.5	16.5	686.0	7.58	53.6	8	138	34	2.51	17.1
762 × 267 × 134	133.9	750.0	264.4	12.0	15.5	16.5	686.0	8.53	57.2	8	138	32	2.51	18.7
686 × 254 × 170	170.2	692.9	255.8	14.5	23.7	15.2	615.1	5.40	42.4	9	132	40	2.35	13.8
686 × 254 × 152	152.4	687.5	254.5	13.2	21.0	15.2	615.1	6.06	46.6	9	132	38	2.34	15.4
686 × 254 × 140	140.1	683.5	253.7	12.4	19.0	15.2	615.1	6.68	49.6	8	132	36	2.33	16.6
686 × 254 × 125	125.2	677.9	253.0	11.7	16.2	15.2	615.1	7.81	52.6	8	132	32	2.32	18.5
610 × 305 × 238	238.1	635.8	311.4	18.4	31.4	16.5	540.0	4.96	29.3	11	158	48	2.45	10.3
610 × 305 × 179	179.0	620.2	307.1	14.1	23.6	16.5	540.0	6.51	38.3	9	158	42	2.41	13.5
610 × 305 × 149	149.2	612.4	304.8	11.8	19.7	16.5	540.0	7.74	45.8	8	158	38	2.39	16.0
610 × 229 × 140	139.9	617.2	230.2	13.1	22.1	12.7	547.6	5.21	41.8	9	120	36	2.11	15.1
610 × 229 × 125	125.1	612.2	229.0	11.9	19.6	12.7	547.6	5.84	46.0	8	120	34	2.09	16.7
610 × 229 × 113	113.0	607.6	228.2	11.1	17.3	12.7	547.6	6.60	49.3	8	120	30	2.08	18.4
610 × 229 × 101	101.2	602.6	227.6	10.5	14.8	12.7	547.6	7.69	52.2	7	120	28	2.07	20.5
533 × 210 × 122	122.0	544.5	211.9	12.7	21.3	12.7	476.5	4.97	37.5	8	110	34	1.89	15.5
533 × 210 × 109	109.0	539.5	210.8	11.6	18.8	12.7	476.5	5.61	41.1	8	110	32	1.88	17.2
533 × 210 × 101	101.0	536.7	210.0	10.8	17.4	12.7	476.5	6.03	44.1	7	110	32	1.87	18.5
533 × 210 × 92	92.1	533.1	209.3	10.1	15.6	12.7	476.5	6.71	47.2	7	110	30	1.86	20.2
533 × 210 × 82	82.2	528.3	208.8	9.6	13.2	12.7	476.5	7.91	49.6	7	110	26	1.85	22.5

† Section is not given in BS 4-1: 1993.
# Check availability.

UNIVERSAL BEAMS

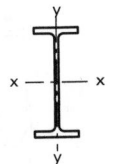

PROPERTIES

Section Designation	Second Moment of Area		Radius of Gyration		Elastic Modulus		Plastic Modulus		Buckling Parameter	Torsional Index	Warping Constant	Torsional Constant	Area of Section
	Axis x-x cm^4	Axis y-y cm^4	Axis x-x cm	Axis y-y cm	Axis x-x cm^3	Axis y-y cm^3	Axis x-x cm^3	Axis y-y cm^3	u	x	H dm^6	J cm^4	A cm^2
1016 × 305 × 487 # †	1020000	26700	40.6	6.57	19700	1730	23200	2800	0.867	21.1	64.4	4300	620
1016 × 305 × 437 # †	910000	23500	40.4	6.49	17700	1540	20800	2470	0.868	23.1	55.9	3190	557
1016 × 305 × 393 # †	808000	20500	40.2	6.40	15900	1350	18500	2170	0.868	25.5	48.4	2330	500
1016 × 305 × 349 # †	723000	18500	40.3	6.44	14400	1220	16600	1940	0.872	27.9	43.3	1720	445
1016 × 305 × 314 # †	644000	16200	40.1	6.37	12900	1080	14900	1710	0.872	30.7	37.7	1260	400
1016 × 305 × 272 # †	554000	14000	40.0	6.35	11200	934	12800	1470	0.872	35.0	32.2	835	347
1016 × 305 × 249 # †	481000	11800	39.0	6.09	9820	784	11400	1250	0.861	39.9	26.8	582	317
1016 × 305 × 222 # †	408000	9550	38.0	5.81	8410	636	9810	1020	0.849	45.8	21.5	390	283
914 × 419 × 388 #	720000	45400	38.2	9.59	15600	2160	17700	3340	0.885	26.7	88.9	1730	494
914 × 419 × 343 #	626000	39200	37.8	9.46	13700	1870	15500	2890	0.883	30.1	75.8	1190	437
914 × 305 × 289 #	504000	15600	37.0	6.51	10900	1010	12600	1600	0.867	31.9	31.2	926	368
914 × 305 × 253 #	436000	13300	36.8	6.42	9500	871	10900	1370	0.865	36.2	26.4	626	323
914 × 305 × 224 #	376000	11200	36.3	6.27	8270	739	9540	1160	0.861	41.3	22.1	422	286
914 × 305 × 201 #	325000	9420	35.7	6.07	7200	621	8350	982	0.853	46.9	18.4	291	256
838 × 292 × 226 #	340000	11400	34.3	6.27	7990	773	9160	1210	0.869	35.0	19.3	514	289
838 × 292 × 194 #	279000	9070	33.6	6.06	6640	620	7640	974	0.862	41.6	15.2	306	247
838 × 292 × 176 #	246000	7800	33.1	5.90	5890	535	6810	842	0.856	46.5	13.0	221	224
762 × 267 × 197	240000	8180	30.9	5.71	6230	610	7170	959	0.868	33.2	11.3	404	251
762 × 267 × 173	205000	6850	30.5	5.58	5390	514	6200	807	0.865	38.1	9.39	267	220
762 × 267 × 147	169000	5460	30.0	5.40	4470	411	5160	647	0.858	45.2	7.40	159	187
762 × 267 × 134	151000	4790	29.7	5.30	4020	362	4640	570	0.853	49.8	6.46	119	171
686 × 254 × 170	170000	6630	28.0	5.53	4920	518	5630	811	0.872	31.8	7.42	308	217
686 × 254 × 152	150000	5780	27.8	5.46	4370	455	5000	710	0.871	35.4	6.42	220	194
686 × 254 × 140	136000	5180	27.6	5.39	3990	409	4560	638	0.869	38.6	5.72	169	178
686 × 254 × 125	118000	4380	27.2	5.24	3480	346	3990	542	0.863	43.8	4.80	116	159
610 × 305 × 238	210000	15800	26.3	7.23	6590	1020	7490	1570	0.887	21.3	14.5	785	303
610 × 305 × 179	153000	11400	25.9	7.07	4940	743	5550	1140	0.886	27.7	10.2	340	228
610 × 305 × 149	126000	9310	25.7	7.00	4110	611	4590	937	0.886	32.7	8.17	200	190
610 × 229 × 140	112000	4510	25.0	5.03	3620	391	4140	611	0.875	30.6	3.99	216	178
610 × 229 × 125	98600	3930	24.9	4.97	3220	343	3680	535	0.874	34.1	3.45	154	159
610 × 229 × 113	87300	3430	24.6	4.88	2870	301	3280	469	0.870	38.1	2.99	111	144
610 × 229 × 101	75800	2920	24.2	4.75	2520	256	2880	400	0.863	43.1	2.52	77.0	129
533 × 210 × 122	76000	3390	22.1	4.67	2790	320	3200	500	0.878	27.6	2.32	178	155
533 × 210 × 109	66800	2940	21.9	4.60	2480	279	2830	436	0.874	31.0	1.99	126	139
533 × 210 × 101	61500	2690	21.9	4.57	2290	256	2610	399	0.873	33.2	1.81	101	129
533 × 210 × 92	55200	2390	21.7	4.51	2070	228	2360	356	0.873	36.4	1.60	75.7	117
533 × 210 × 82	47500	2010	21.3	4.38	1800	192	2060	300	0.863	41.6	1.33	51.5	105

† Section is not given in BS 4-1: 1993.
# Check availability.

UNIVERSAL BEAMS

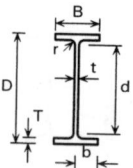

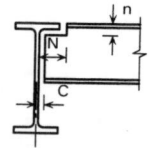

DIMENSIONS

Section Designation	Mass per Metre	Depth of Section	Width of Section	Thickness Web	Thickness Flange	Root Radius	Depth between Fillets	Ratios for Local Buckling Flange	Ratios for Local Buckling Web	Dimensions for Detailing End Clearance	Dimensions for Detailing Notch N	Dimensions for Detailing Notch n	Surface Area Per Metre	Surface Area Per Tonne
	kg/m	D mm	B mm	t mm	T mm	r mm	d mm	b/T	d/t	C mm	N mm	n mm	m²	m²
457 × 191 × 98	98.3	467.2	192.8	11.4	19.6	10.2	407.6	4.92	35.8	8	102	30	1.67	16.9
457 × 191 × 89	89.3	463.4	191.9	10.5	17.7	10.2	407.6	5.42	38.8	7	102	28	1.66	18.5
457 × 191 × 82	82.0	460.0	191.3	9.9	16.0	10.2	407.6	5.98	41.2	7	102	28	1.65	20.1
457 × 191 × 74	74.3	457.0	190.4	9.0	14.5	10.2	407.6	6.57	45.3	7	102	26	1.64	22.1
457 × 191 × 67	67.1	453.4	189.9	8.5	12.7	10.2	407.6	7.48	48.0	6	102	24	1.63	24.3
457 × 152 × 82	82.1	465.8	155.3	10.5	18.9	10.2	407.6	4.11	38.8	7	84	30	1.51	18.4
457 × 152 × 74	74.2	462.0	154.4	9.6	17.0	10.2	407.6	4.54	42.5	7	84	28	1.50	20.3
457 × 152 × 67	67.2	458.0	153.8	9.0	15.0	10.2	407.6	5.13	45.3	7	84	26	1.50	22.3
457 × 152 × 60	59.8	454.6	152.9	8.1	13.3	10.2	407.6	5.75	50.3	6	84	24	1.49	24.9
457 × 152 × 52	52.3	449.8	152.4	7.6	10.9	10.2	407.6	6.99	53.6	6	84	22	1.48	28.2
406 × 178 × 74	74.2	412.8	179.5	9.5	16.0	10.2	360.4	5.61	37.9	7	96	28	1.51	20.3
406 × 178 × 67	67.1	409.4	178.8	8.8	14.3	10.2	360.4	6.25	41.0	6	96	26	1.50	22.3
406 × 178 × 60	60.1	406.4	177.9	7.9	12.8	10.2	360.4	6.95	45.6	6	96	24	1.49	24.8
406 × 178 × 54	54.1	402.6	177.7	7.7	10.9	10.2	360.4	8.15	46.8	6	96	22	1.48	27.4
406 × 140 × 46	46.0	403.2	142.2	6.8	11.2	10.2	360.4	6.35	53.0	5	78	22	1.34	29.2
406 × 140 × 39	39.0	398.0	141.8	6.4	8.6	10.2	360.4	8.24	56.3	5	78	20	1.33	34.2
356 × 171 × 67	67.1	363.4	173.2	9.1	15.7	10.2	311.6	5.52	34.2	7	94	26	1.38	20.6
356 × 171 × 57	57.0	358.0	172.2	8.1	13.0	10.2	311.6	6.62	38.5	6	94	24	1.37	24.1
356 × 171 × 51	51.0	355.0	171.5	7.4	11.5	10.2	311.6	7.46	42.1	6	94	22	1.36	26.7
356 × 171 × 45	45.0	351.4	171.1	7.0	9.7	10.2	311.6	8.82	44.5	6	94	20	1.36	30.1
356 × 127 × 39	39.1	353.4	126.0	6.6	10.7	10.2	311.6	5.89	47.2	5	70	22	1.18	30.2
356 × 127 × 33	33.1	349.0	125.4	6.0	8.5	10.2	311.6	7.38	51.9	5	70	20	1.17	35.4
305 × 165 × 54	54.0	310.4	166.9	7.9	13.7	8.9	265.2	6.09	33.6	6	90	24	1.26	23.3
305 × 165 × 46	46.1	306.6	165.7	6.7	11.8	8.9	265.2	7.02	39.6	5	90	22	1.25	27.1
305 × 165 × 40	40.3	303.4	165.0	6.0	10.2	8.9	265.2	8.09	44.2	5	90	20	1.24	30.8
305 × 127 × 48	48.1	311.0	125.3	9.0	14.0	8.9	265.2	4.47	29.5	7	70	24	1.09	22.7
305 × 127 × 42	41.9	307.2	124.3	8.0	12.1	8.9	265.2	5.14	33.1	6	70	22	1.08	25.8
305 × 127 × 37	37.0	304.4	123.4	7.1	10.7	8.9	265.2	5.77	37.4	6	70	20	1.07	29.0
305 × 102 × 33	32.8	312.7	102.4	6.6	10.8	7.6	275.9	4.74	41.8	5	58	20	1.01	30.8
305 × 102 × 28	28.2	308.7	101.8	6.0	8.8	7.6	275.9	5.78	46.0	5	58	18	1.00	35.4
305 × 102 × 25	24.8	305.1	101.6	5.8	7.0	7.6	275.9	7.26	47.6	5	58	16	0.992	40.0
254 × 146 × 43	43.0	259.6	147.3	7.2	12.7	7.6	219.0	5.80	30.4	6	82	22	1.08	25.1
254 × 146 × 37	37.0	256.0	146.4	6.3	10.9	7.6	219.0	6.72	34.8	5	82	20	1.07	29.0
254 × 146 × 31	31.1	251.4	146.1	6.0	8.6	7.6	219.0	8.49	36.5	5	82	18	1.06	34.2
254 × 102 × 28	28.3	260.4	102.2	6.3	10.0	7.6	225.2	5.11	35.7	5	58	18	0.904	31.9
254 × 102 × 25	25.2	257.2	101.9	6.0	8.4	7.6	225.2	6.07	37.5	5	58	16	0.897	35.6
254 × 102 × 22	22.0	254.0	101.6	5.7	6.8	7.6	225.2	7.47	39.5	5	58	16	0.890	40.5
203 × 133 × 30	30.0	206.8	133.9	6.4	9.6	7.6	172.4	6.97	26.9	5	74	18	0.923	30.8
203 × 133 × 25	25.1	203.2	133.2	5.7	7.8	7.6	172.4	8.54	30.2	5	74	16	0.915	36.4
203 × 102 × 23	23.1	203.2	101.8	5.4	9.3	7.6	169.4	5.47	31.4	5	60	18	0.790	34.2
178 × 102 × 19	19.0	177.8	101.2	4.8	7.9	7.6	146.8	6.41	30.6	4	60	16	0.738	38.8
152 × 89 × 16	16.0	152.4	88.7	4.5	7.7	7.6	121.8	5.76	27.1	4	54	16	0.638	39.8
127 × 76 × 13	13.0	127.0	76.0	4.0	7.6	7.6	96.6	5.00	24.1	4	46	16	0.537	41.3

UNIVERSAL BEAMS

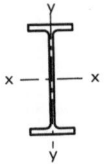

PROPERTIES

Section Designation	Second Moment of Area		Radius of Gyration		Elastic Modulus		Plastic Modulus		Buckling Parameter	Torsional Index	Warping Constant	Torsional Constant	Area of Section
	Axis x-x cm⁴	Axis y-y cm⁴	Axis x-x cm	Axis y-y cm	Axis x-x cm³	Axis y-y cm³	Axis x-x cm³	Axis y-y cm³	u	x	H dm⁶	J cm⁴	A cm²
457 × 191 × 98	45700	2350	19.1	4.33	1960	243	2230	379	0.882	25.7	1.18	121	125
457 × 191 × 89	41000	2090	19.0	4.29	1770	218	2010	338	0.879	28.3	1.04	90.7	114
457 × 191 × 82	37100	1870	18.8	4.23	1610	196	1830	304	0.879	30.8	0.922	69.2	104
457 × 191 × 74	33300	1670	18.8	4.20	1460	176	1650	272	0.877	33.8	0.818	51.8	94.6
457 × 191 × 67	29400	1450	18.5	4.12	1300	153	1470	237	0.872	37.9	0.705	37.1	85.5
457 × 152 × 82	36600	1190	18.7	3.37	1570	153	1810	240	0.871	27.4	0.591	89.2	105
457 × 152 × 74	32700	1050	18.6	3.33	1410	136	1630	213	0.873	30.2	0.518	65.9	94.5
457 × 152 × 67	28900	913	18.4	3.27	1260	119	1450	187	0.868	33.6	0.448	47.7	85.6
457 × 152 × 60	25500	795	18.3	3.23	1120	104	1290	163	0.868	37.5	0.387	33.8	76.2
457 × 152 × 52	21400	645	17.9	3.11	950	84.6	1100	133	0.859	43.8	0.311	21.4	66.6
406 × 178 × 74	27300	1550	17.0	4.04	1320	172	1500	267	0.882	27.6	0.608	62.8	94.5
406 × 178 × 67	24300	1370	16.9	3.99	1190	153	1350	237	0.880	30.5	0.533	46.1	85.5
406 × 178 × 60	21600	1200	16.8	3.97	1060	135	1200	209	0.880	33.8	0.466	33.3	76.5
406 × 178 × 54	18700	1020	16.5	3.85	930	115	1060	178	0.871	38.3	0.392	23.1	69.0
406 × 140 × 46	15700	538	16.4	3.03	778	75.7	888	118	0.872	39.0	0.207	19.0	58.6
406 × 140 × 39	12500	410	15.9	2.87	629	57.8	724	90.8	0.858	47.5	0.155	10.7	49.7
356 × 171 × 67	19500	1360	15.1	3.99	1070	157	1210	243	0.886	24.4	0.412	55.7	85.5
356 × 171 × 57	16000	1110	14.9	3.91	896	129	1010	199	0.882	28.8	0.330	33.4	72.6
356 × 171 × 51	14100	968	14.8	3.86	796	113	896	174	0.881	32.1	0.286	23.8	64.9
356 × 171 × 45	12100	811	14.5	3.76	687	94.8	775	147	0.874	36.8	0.237	15.8	57.3
356 × 127 × 39	10200	358	14.3	2.68	576	56.8	659	89.1	0.871	35.2	0.105	15.1	49.8
356 × 127 × 33	8250	280	14.0	2.58	473	44.7	543	70.3	0.863	42.2	0.081	8.79	42.1
305 × 165 × 54	11700	1060	13.0	3.93	754	127	846	196	0.889	23.6	0.234	34.8	68.8
305 × 165 × 46	9900	896	13.0	3.90	646	108	720	166	0.891	27.1	0.195	22.2	58.7
305 × 165 × 40	8500	764	12.9	3.86	560	92.6	623	142	0.889	31.0	0.164	14.7	51.3
305 × 127 × 48	9580	461	12.5	2.74	616	73.6	711	116	0.874	23.3	0.102	31.8	61.2
305 × 127 × 42	8200	389	12.4	2.70	534	62.6	614	98.4	0.872	26.6	0.0846	21.1	53.4
305 × 127 × 37	7170	336	12.3	2.67	471	54.5	539	85.4	0.871	29.7	0.0725	14.8	47.2
305 × 102 × 33	6500	194	12.5	2.15	416	37.9	481	60.0	0.867	31.6	0.0442	12.2	41.8
305 × 102 × 28	5370	155	12.2	2.08	348	30.5	403	48.5	0.859	37.4	0.0349	7.40	35.9
305 × 102 × 25	4460	123	11.9	1.97	292	24.2	342	38.8	0.846	43.4	0.0273	4.77	31.6
254 × 146 × 43	6540	677	10.9	3.52	504	92.0	566	141	0.890	21.2	0.103	23.9	54.8
254 × 146 × 37	5540	571	10.8	3.48	433	78.0	483	119	0.889	24.4	0.0857	15.3	47.2
254 × 146 × 31	4410	448	10.5	3.36	351	61.3	393	94.1	0.879	29.6	0.0660	8.55	39.7
254 × 102 × 28	4010	179	10.5	2.22	308	34.9	353	54.8	0.874	27.5	0.0280	9.57	36.1
254 × 102 × 25	3420	149	10.3	2.15	266	29.2	306	46.0	0.867	31.4	0.0230	6.42	32.0
254 × 102 × 22	2840	119	10.1	2.06	224	23.5	259	37.3	0.856	36.3	0.0182	4.15	28.0
203 × 133 × 30	2900	385	8.71	3.17	280	57.5	314	88.2	0.881	21.5	0.0374	10.3	38.2
203 × 133 × 25	2340	308	8.56	3.10	230	46.2	258	70.9	0.877	25.6	0.0294	5.96	32.0
203 × 102 × 23	2110	164	8.46	2.36	207	32.2	234	49.8	0.888	22.5	0.0154	7.02	29.4
178 × 102 × 19	1360	137	7.48	2.37	153	27.0	171	41.6	0.886	22.6	0.00987	4.41	24.3
152 × 89 × 16	834	89.8	6.41	2.10	109	20.2	123	31.2	0.889	19.6	0.00470	3.56	20.3
127 × 76 × 13	473	55.7	5.35	1.84	74.6	14.7	84.2	22.6	0.896	16.3	0.00199	2.85	16.5

UNIVERSAL BEAMS

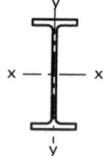

REDUCED PLASTIC MODULUS UNDER AXIAL LOAD

Section Designation	Plastic Modulus Axis x-x	Major Axis Reduced Modulus					Plastic Modulus Axis y-y	Minor Axis Reduced Modulus				
		Lower Values of n		Change Formula At n =	Higher Values of n			Lower Values of n		Change Formula At n =	Higher Values of n	
	cm³	K1	K2		K3	K4	cm³	K1	K2		K3	K4
1016 × 305 × 487 # †	23200	23200	32000	0.449	3120	9.31	2800	2800	928	0.501	8880	0.0768
1016 × 305 × 437 # †	20800	20800	28800	0.448	2540	10.2	2470	2470	756	0.495	7910	0.0747
1016 × 305 × 393 # †	18500	18500	25600	0.453	2060	11.3	2170	2170	615	0.496	7120	0.0641
1016 × 305 × 349 # †	16600	16600	23500	0.440	1640	12.7	1940	1940	491	0.478	6190	0.0858
1016 × 305 × 314 # †	14900	14900	20900	0.443	1330	14.0	1710	1710	400	0.478	5570	0.0770
1016 × 305 × 272 # †	12800	12800	18200	0.441	1000	16.1	1470	1470	304	0.471	4860	0.0720
1016 × 305 × 249 # †	11400	11400	15200	0.483	837	17.6	1250	1250	256	0.510	4830	−0.0158
1016 × 305 × 222 # †	9810	9810	12500	0.525	667	19.6	1020	1020	206	0.549	4740	−0.105
914 × 419 × 388 #	17700	17700	28500	0.367	1450	14.7	3340	3340	663	0.399	8340	0.246
914 × 419 × 343 #	15500	15500	24600	0.376	1140	16.5	2890	2890	524	0.405	7470	0.225
914 × 305 × 289 #	12600	12600	17400	0.457	1100	14.5	1600	1600	366	0.491	5300	0.0695
914 × 305 × 253 #	10900	10900	15100	0.462	853	16.4	1370	1370	284	0.492	4670	0.0561
914 × 305 × 224 #	9540	9540	12800	0.480	671	18.4	1160	1160	224	0.507	4270	0.0178
914 × 305 × 201 #	8350	8350	10800	0.509	540	20.4	982	982	181	0.533	4050	−0.0424
838 × 292 × 226 #	9160	9160	12900	0.445	709	16.3	1210	1210	245	0.475	3880	0.0915
838 × 292 × 194 #	7640	7640	10400	0.475	521	18.9	974	974	181	0.501	3510	0.0283
838 × 292 × 176 #	6810	6810	8960	0.498	430	20.7	842	842	150	0.522	3340	−0.0208
762 × 267 × 197	7170	7170	10100	0.448	586	15.5	959	959	204	0.479	3090	0.0863
762 × 267 × 173	6200	6200	8490	0.467	455	17.4	807	807	159	0.495	2810	0.0457
762 × 267 × 147	5160	5160	6840	0.492	330	20.4	647	647	116	0.516	2500	−0.00820
762 × 267 × 134	4640	4640	6060	0.506	275	22.2	570	570	97.0	0.528	2350	−0.0390
686 × 254 × 170	5630	5630	8110	0.432	459	15.3	811	811	170	0.463	2480	0.118
686 × 254 × 152	5000	5000	7130	0.439	370	17.0	710	710	137	0.468	2240	0.102
686 × 254 × 140	4560	4560	6420	0.449	314	18.4	638	638	116	0.475	2100	0.0806
686 × 254 × 125	3990	3990	5440	0.474	251	20.5	542	542	93.8	0.497	1960	0.0280
610 × 305 × 238	7490	7490	12500	0.348	739	12.1	1570	1570	362	0.386	3660	0.289
610 × 305 × 179	5550	5550	9220	0.354	423	15.7	1140	1140	210	0.383	2760	0.271
610 × 305 × 149	4590	4590	7650	0.356	296	18.6	937	937	147	0.380	2290	0.264
610 × 229 × 140	4140	4140	6060	0.421	345	14.9	611	611	129	0.454	1800	0.142
610 × 229 × 125	3680	3680	5330	0.428	277	16.6	535	535	104	0.457	1620	0.127
610 × 229 × 113	3280	3280	4670	0.442	227	18.3	469	469	85.3	0.469	1500	0.0970
610 × 229 × 101	2880	2880	3960	0.467	183	20.3	400	400	69.0	0.491	1400	0.0451
533 × 210 × 122	3200	3200	4750	0.410	285	13.8	500	500	111	0.445	1420	0.162
533 × 210 × 109	2830	2830	4160	0.419	229	15.4	436	436	89.4	0.451	1280	0.142
533 × 210 × 101	2610	2610	3830	0.421	197	16.5	399	399	77.1	0.450	1190	0.136
533 × 210 × 92	2360	2360	3410	0.432	165	18.0	356	356	64.6	0.459	1100	0.113
533 × 210 × 82	2060	2060	2850	0.460	131	20.1	300	300	51.9	0.484	1040	0.0531

† Section is not given in BS 4-1: 1993.
\# Check availability.
n = F/(A pᵧ), where F is the factored axial load, A is the gross cross sectional area and pᵧ is the design strength of the section.
For lower values of n, the reduced plastic modulus, $S_r = K1 - K2.n^2$, for both major and minor axis bending.
For higher values of n, the reduced plastic modulus, $S_r = K3(1 - n)(K4 + n)$, for both major and minor axis bending.

UNIVERSAL BEAMS

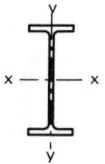

REDUCED PLASTIC MODULUS UNDER AXIAL LOAD

Section Designation	Plastic Modulus Axis x-x	Major Axis Reduced Modulus					Plastic Modulus Axis y-y	Minor Axis Reduced Modulus				
		Lower Values of n		Change Formula At n =	Higher Values of n			Lower Values of n		Change Formula At n =	Higher Values of n	
	cm³	K1	K2		K3	K4	cm³	K1	K2		K3	K4
457 × 191 × 98	2230	2230	3440	0.390	203	13.4	379	379	84.0	0.425	1000	0.207
457 × 191 × 89	2010	2010	3080	0.395	169	14.6	338	338	69.8	0.428	914	0.194
457 × 191 × 82	1830	1830	2760	0.406	143	15.8	304	304	59.3	0.436	853	0.172
457 × 191 × 74	1650	1650	2490	0.407	118	17.4	272	272	49.0	0.435	772	0.167
457 × 191 × 67	1470	1470	2150	0.425	96.3	19.1	237	237	40.3	0.451	720	0.128
457 × 152 × 82	1810	1810	2600	0.430	176	12.8	240	240	58.7	0.468	723	0.123
457 × 152 × 74	1630	1630	2320	0.435	145	14.1	213	213	48.3	0.469	656	0.111
457 × 152 × 67	1450	1450	2030	0.450	119	15.5	187	187	40.0	0.482	610	0.0786
457 × 152 × 60	1290	1290	1790	0.455	95.0	17.2	163	163	32.0	0.483	546	0.0670
457 × 152 × 52	1100	1100	1460	0.488	72.9	19.6	133	133	24.7	0.513	509	-0.00290
406 × 178 × 74	1500	1500	2350	0.383	124	14.7	267	267	54.1	0.415	698	0.216
406 × 178 × 67	1350	1350	2080	0.392	102	16.1	237	237	44.7	0.421	640	0.196
406 × 178 × 60	1200	1200	1850	0.393	82.3	17.9	209	209	36.0	0.420	572	0.190
406 × 178 × 54	1060	1060	1540	0.425	66.9	19.8	178	178	29.5	0.450	545	0.124
406 × 140 × 46	888	888	1260	0.442	60.5	18.6	118	118	21.3	0.468	384	0.0864
406 × 140 × 39	724	724	963	0.491	43.5	21.7	90.8	90.8	15.5	0.513	358	-0.0176
356 × 171 × 67	1210	1210	2010	0.353	105	13.7	243	243	50.3	0.387	582	0.272
356 × 171 × 57	1010	1010	1630	0.371	76.4	16.0	199	199	36.8	0.400	506	0.234
356 × 171 × 51	896	896	1420	0.379	61.4	17.8	174	174	29.7	0.405	458	0.215
356 × 171 × 45	775	775	1170	0.405	48.0	20.0	147	147	23.4	0.429	423	0.158
356 × 127 × 39	659	659	938	0.440	49.1	16.9	89.1	89.1	17.5	0.469	289	0.0836
356 × 127 × 33	543	543	740	0.473	35.4	19.8	70.3	70.3	12.7	0.497	261	0.0120
305 × 165 × 54	846	846	1500	0.325	70.8	14.1	196	196	38.1	0.357	431	0.330
305 × 165 × 46	720	720	1290	0.323	52.1	16.3	166	166	28.1	0.350	366	0.331
305 × 165 × 40	623	623	1100	0.331	39.9	18.5	142	142	21.7	0.355	323	0.312
305 × 127 × 48	711	711	1040	0.416	74.8	11.7	116	116	30.1	0.457	335	0.146
305 × 127 × 42	614	614	891	0.424	57.4	13.3	98.4	98.4	23.2	0.460	295	0.127
305 × 127 × 37	539	539	784	0.426	45.1	14.9	85.4	85.4	18.3	0.458	260	0.119
305 × 102 × 33	481	481	663	0.459	42.7	14.3	60.0	60.0	14.0	0.493	202	0.0576
305 × 102 × 28	403	403	536	0.487	31.6	16.5	48.5	48.5	10.4	0.516	183	-0.00120
305 × 102 × 25	342	342	431	0.534	24.6	18.6	38.8	38.8	8.18	0.560	178	-0.100
254 × 146 × 43	566	566	1040	0.308	50.9	13.0	141	141	28.9	0.341	295	0.366
254 × 146 × 37	483	483	883	0.313	38.0	14.9	119	119	21.7	0.342	255	0.353
254 × 146 × 31	393	393	656	0.354	26.9	17.5	94.1	94.1	15.7	0.380	229	0.267
254 × 102 × 28	353	353	517	0.420	31.8	13.8	54.8	54.8	12.5	0.455	163	0.133
254 × 102 × 25	306	306	428	0.450	25.2	15.4	46.0	46.0	9.98	0.482	153	0.0686
254 × 102 × 22	259	259	344	0.489	19.3	17.4	37.3	37.3	7.73	0.517	144	-0.0136
203 × 133 × 30	314	314	570	0.314	27.3	13.5	88.2	88.2	17.7	0.346	190	0.346
203 × 133 × 25	258	258	448	0.334	19.2	15.9	70.9	70.9	12.6	0.362	164	0.300
203 × 102 × 23	234	234	400	0.339	21.2	13.1	49.8	49.8	10.6	0.373	116	0.288
178 × 102 × 19	171	171	307	0.321	14.5	13.8	41.6	41.6	8.28	0.352	93.1	0.318
152 × 89 × 16	123	123	229	0.303	11.6	12.3	31.2	31.2	6.77	0.337	67.0	0.344
127 × 76 × 13	84.2	84.2	171	0.271	8.98	10.7	22.6	22.6	5.37	0.308	44.9	0.399

n = F/(A p$_y$), where F is the factored axial load, A is the gross cross sectional area and p$_y$ is the design strength of the section.
For lower values of n, the reduced plastic modulus, S$_r$ = K1 − K2.n^2, for both major and minor axis bending.
For higher values of n, the reduced plastic modulus, S$_r$ = K3(1 − n)(K4 + n), for both major and minor axis bending.

UNIVERSAL COLUMNS

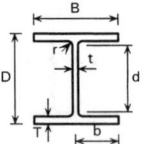

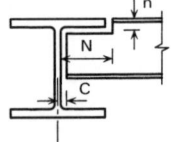

DIMENSIONS

Section Designation	Mass per Metre	Depth of Section	Width of Section	Thickness Web	Thickness Flange	Root Radius	Depth between Fillets	Ratios for Local Buckling Flange	Ratios for Local Buckling Web	Dimensions for Detailing End Clearance	Dimensions for Detailing Notch N	Dimensions for Detailing Notch n	Surface Area Per Metre	Surface Area Per Tonne
	kg/m	D mm	B mm	t mm	T mm	r mm	d mm	b/T	d/t	C mm	N mm	n mm	m²	m²
356 × 406 × 634 #	633.9	474.6	424.0	47.6	77.0	15.2	290.2	2.75	6.10	26	200	94	2.52	3.98
356 × 406 × 551 #	551.0	455.6	418.5	42.1	67.5	15.2	290.2	3.10	6.89	23	200	84	2.47	4.49
356 × 406 × 467 #	467.0	436.6	412.2	35.8	58.0	15.2	290.2	3.55	8.11	20	200	74	2.42	5.19
356 × 406 × 393 #	393.0	419.0	407.0	30.6	49.2	15.2	290.2	4.14	9.48	17	200	66	2.38	6.05
356 × 406 × 340 #	339.9	406.4	403.0	26.6	42.9	15.2	290.2	4.70	10.9	15	200	60	2.35	6.90
356 × 406 × 287 #	287.1	393.6	399.0	22.6	36.5	15.2	290.2	5.47	12.8	13	200	52	2.31	8.05
356 × 406 × 235 #	235.1	381.0	394.8	18.4	30.2	15.2	290.2	6.54	15.8	11	200	46	2.28	9.69
356 × 368 × 202 #	201.9	374.6	374.7	16.5	27.0	15.2	290.2	6.94	17.6	10	190	44	2.19	10.8
356 × 368 × 177 #	177.0	368.2	372.6	14.4	23.8	15.2	290.2	7.83	20.2	9	190	40	2.17	12.3
356 × 368 × 153 #	152.9	362.0	370.5	12.3	20.7	15.2	290.2	8.95	23.6	8	190	36	2.16	14.1
356 × 368 × 129 #	129.0	355.6	368.6	10.4	17.5	15.2	290.2	10.50	27.9	7	190	34	2.14	16.6
305 × 305 × 283	282.9	365.3	322.2	26.8	44.1	15.2	246.7	3.65	9.21	15	158	60	1.94	6.86
305 × 305 × 240	240.0	352.5	318.4	23.0	37.7	15.2	246.7	4.22	10.7	14	158	54	1.91	7.94
305 × 305 × 198	198.1	339.9	314.5	19.1	31.4	15.2	246.7	5.01	12.9	12	158	48	1.87	9.46
305 × 305 × 158	158.1	327.1	311.2	15.8	25.0	15.2	246.7	6.22	15.6	10	158	42	1.84	11.6
305 × 305 × 137	136.9	320.5	309.2	13.8	21.7	15.2	246.7	7.12	17.9	9	158	38	1.82	13.3
305 × 305 × 118	117.9	314.5	307.4	12.0	18.7	15.2	246.7	8.22	20.6	8	158	34	1.81	15.3
305 × 305 × 97	96.9	307.9	305.3	9.9	15.4	15.2	246.7	9.91	24.9	7	158	32	1.79	18.5
254 × 254 × 167	167.1	289.1	265.2	19.2	31.7	12.7	200.3	4.18	10.4	12	134	46	1.58	9.45
254 × 254 × 132	132.0	276.3	261.3	15.3	25.3	12.7	200.3	5.16	13.1	10	134	38	1.55	11.7
254 × 254 × 107	107.1	266.7	258.8	12.8	20.5	12.7	200.3	6.31	15.6	8	134	34	1.52	14.2
254 × 254 × 89	88.9	260.3	256.3	10.3	17.3	12.7	200.3	7.41	19.4	7	134	30	1.50	16.9
254 × 254 × 73	73.1	254.1	254.6	8.6	14.2	12.7	200.3	8.96	23.3	6	134	28	1.49	20.4
203 × 203 × 86	86.1	222.2	209.1	12.7	20.5	10.2	160.8	5.10	12.7	8	110	32	1.24	14.4
203 × 203 × 71	71.0	215.8	206.4	10.0	17.3	10.2	160.8	5.97	16.1	7	110	28	1.22	17.2
203 × 203 × 60	60.0	209.6	205.8	9.4	14.2	10.2	160.8	7.25	17.1	7	110	26	1.21	20.1
203 × 203 × 52	52.0	206.2	204.3	7.9	12.5	10.2	160.8	8.17	20.4	6	110	24	1.20	23.0
203 × 203 × 46	46.1	203.2	203.6	7.2	11.0	10.2	160.8	9.25	22.3	6	110	22	1.19	25.8
152 × 152 × 37	37.0	161.8	154.4	8.0	11.5	7.6	123.6	6.71	15.5	6	84	20	0.912	24.7
152 × 152 × 30	30.0	157.6	152.9	6.5	9.4	7.6	123.6	8.13	19.0	5	84	18	0.901	30.0
152 × 152 × 23	23.0	152.4	152.2	5.8	6.8	7.6	123.6	11.2	21.3	5	84	16	0.889	38.7

Check availability.

UNIVERSAL COLUMNS

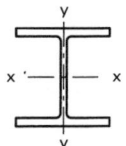

PROPERTIES

Section Designation	Second Moment of Area		Radius of Gyration		Elastic Modulus		Plastic Modulus		Buckling Parameter	Torsional Index	Warping Constant	Torsional Constant	Area of Section
	Axis x-x cm⁴	Axis y-y cm⁴	Axis x-x cm	Axis y-y cm	Axis x-x cm³	Axis y-y cm³	Axis x-x cm³	Axis y-y cm³	u	x	H dm⁶	J cm⁴	A cm²
356 × 406 × 634 #	275000	98100	18.4	11.0	11600	4630	14200	7110	0.843	5.46	38.8	13700	808
356 × 406 × 551 #	227000	82700	18.0	10.9	9960	3950	12100	6060	0.841	6.05	31.1	9240	702
356 × 406 × 467 #	183000	67800	17.5	10.7	8380	3290	10000	5030	0.839	6.86	24.3	5810	595
356 × 406 × 393 #	147000	55400	17.1	10.5	7000	2720	8220	4150	0.837	7.87	18.9	3550	501
356 × 406 × 340 #	123000	46900	16.8	10.4	6030	2330	7000	3540	0.836	8.84	15.5	2340	433
356 × 406 × 287 #	99900	38700	16.5	10.3	5080	1940	5810	2950	0.834	10.2	12.3	1440	366
356 × 406 × 235 #	79100	31000	16.3	10.2	4150	1570	4690	2380	0.835	12.0	9.54	812	299
356 × 368 × 202 #	66300	23700	16.1	9.60	3540	1260	3970	1920	0.844	13.4	7.16	558	257
356 × 368 × 177 #	57100	20500	15.9	9.54	3100	1100	3460	1670	0.843	15.0	6.09	381	226
356 × 368 × 153 #	48600	17600	15.8	9.49	2680	948	2970	1440	0.844	17.0	5.11	251	195
356 × 368 × 129 #	40300	14600	15.6	9.43	2260	793	2480	1200	0.845	19.8	4.18	153	164
305 × 305 × 283	78900	24600	14.8	8.27	4320	1530	5110	2340	0.856	7.65	6.35	2030	360
305 × 305 × 240	64200	20300	14.5	8.15	3640	1280	4250	1950	0.854	8.74	5.03	1270	306
305 × 305 × 198	50900	16300	14.2	8.04	3000	1040	3440	1580	0.854	10.2	3.88	734	252
305 × 305 × 158	38800	12600	13.9	7.90	2370	808	2680	1230	0.852	12.5	2.87	378	201
305 × 305 × 137	32800	10700	13.7	7.83	2050	692	2300	1050	0.852	14.1	2.39	249	174
305 × 305 × 118	27700	9060	13.6	7.77	1760	589	1960	895	0.851	16.2	1.98	161	150
305 × 305 × 97	22300	7310	13.4	7.69	1450	479	1590	726	0.852	19.2	1.56	91.2	123
254 × 254 × 167	30000	9870	11.9	6.81	2080	744	2420	1140	0.851	8.50	1.63	626	213
254 × 254 × 132	22500	7530	11.6	6.69	1630	576	1870	878	0.850	10.3	1.19	319	168
254 × 254 × 107	17500	5930	11.3	6.59	1310	458	1480	697	0.849	12.4	0.898	172	136
254 × 254 × 89	14300	4860	11.2	6.55	1100	379	1220	575	0.851	14.5	0.717	102	113
254 × 254 × 73	11400	3910	11.1	6.48	898	307	992	465	0.849	17.3	0.562	57.6	93.1
203 × 203 × 86	9450	3130	9.28	5.34	850	299	977	456	0.849	10.2	0.318	137	110
203 × 203 × 71	7620	2540	9.18	5.30	706	246	799	374	0.853	11.9	0.250	80.2	90.4
203 × 203 × 60	6130	2070	8.96	5.20	584	201	656	305	0.846	14.1	0.197	47.2	76.4
203 × 203 × 52	5260	1780	8.91	5.18	510	174	567	264	0.848	15.8	0.167	31.8	66.3
203 × 203 × 46	4570	1550	8.82	5.13	450	152	497	231	0.846	17.7	0.143	22.2	58.7
152 × 152 × 37	2210	706	6.85	3.87	273	91.5	309	140	0.849	13.3	0.0399	19.2	47.1
152 × 152 × 30	1750	560	6.76	3.83	222	73.3	248	112	0.849	16.0	0.0308	10.5	38.3
152 × 152 × 23	1250	400	6.54	3.70	164	52.6	182	80.2	0.840	20.7	0.0212	4.63	29.2

Check availability.

UNIVERSAL COLUMNS

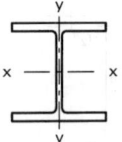

REDUCED PLASTIC MODULUS UNDER AXIAL LOAD

Section Designation	Plastic Modulus Axis x-x	Major Axis Reduced Modulus					Plastic Modulus Axis y-y	Minor Axis Reduced Modulus				
		Lower Values Of n		Change Formula At n =	Higher Values Of n			Lower Values Of n		Change Formula At n =	Higher Values Of n	
	cm³	K1	K2		K3	K4	cm³	K1	K2		K3	K4
356 × 406 × 634 #	14200	14200	34300	0.189	3850	3.98	7110	7110	3440	0.280	10600	0.617
356 × 406 × 551 #	12100	12100	29300	0.192	2940	4.43	6060	6060	2700	0.273	9120	0.610
356 × 406 × 467 #	10000	10000	24700	0.193	2150	5.05	5030	5030	2030	0.263	7630	0.607
356 × 406 × 393 #	8220	8220	20500	0.196	1540	5.81	4150	4150	1500	0.256	6370	0.600
356 × 406 × 340 #	7000	7000	17600	0.197	1160	6.56	3540	3540	1150	0.250	5460	0.597
356 × 406 × 287 #	5810	5810	14800	0.198	838	7.59	2950	2950	849	0.243	4580	0.593
356 × 406 × 235 #	4690	4690	12200	0.197	568	9.05	2380	2380	588	0.234	3710	0.593
356 × 368 × 202 #	3970	3970	10000	0.206	441	9.91	1920	1920	442	0.240	3060	0.573
356 × 368 × 177 #	3460	3460	8830	0.205	341	11.2	1670	1670	345	0.235	2670	0.573
356 × 368 × 153 #	2970	2970	7710	0.202	256	12.8	1440	1440	262	0.229	2290	0.575
356 × 368 × 129 #	2480	2480	6490	0.203	183	15.0	1200	1200	190	0.225	1930	0.570
305 × 305 × 283	5110	5110	12100	0.206	1010	5.53	2340	2340	889	0.272	3680	0.577
305 × 305 × 240	4250	4250	10200	0.208	734	6.34	1950	1950	663	0.265	3100	0.570
305 × 305 × 198	3440	3440	8340	0.210	506	7.47	1580	1580	469	0.257	2540	0.565
305 × 305 × 158	2680	2680	6420	0.217	326	9.11	1230	1230	310	0.257	2030	0.545
305 × 305 × 137	2300	2300	5510	0.219	246	10.4	1050	1050	237	0.254	1750	0.539
305 × 305 × 118	1960	1960	4700	0.221	183	11.9	895	895	179	0.251	1510	0.531
305 × 305 × 97	1590	1590	3850	0.222	125	14.2	726	726	124	0.247	1240	0.523
254 × 254 × 167	2420	2420	5900	0.204	427	6.20	1140	1140	392	0.261	1790	0.580
254 × 254 × 132	1870	1870	4620	0.205	270	7.59	878	878	256	0.251	1400	0.573
254 × 254 × 107	1480	1480	3630	0.212	180	9.12	697	697	174	0.250	1130	0.556
254 × 254 × 89	1220	1220	3120	0.205	125	10.8	575	575	123	0.237	928	0.565
254 × 254 × 73	992	992	2520	0.208	85.1	12.9	465	465	85.3	0.235	763	0.553
203 × 203 × 86	977	977	2370	0.210	144	7.48	456	456	135	0.257	733	0.564
203 × 203 × 71	799	799	2040	0.200	99.0	8.85	374	374	94.7	0.239	591	0.579
203 × 203 × 60	656	656	1550	0.223	70.9	10.3	305	305	69.6	0.258	513	0.531
203 × 203 × 52	567	567	1390	0.216	53.8	11.7	264	264	53.3	0.246	439	0.541
203 × 203 × 46	497	497	1200	0.222	42.4	13.1	231	231	42.4	0.249	392	0.525
152 × 152 × 37	309	309	694	0.236	35.9	9.61	140	140	34.3	0.275	241	0.508
152 × 152 × 30	248	248	563	0.236	23.9	11.6	112	112	23.2	0.268	195	0.503
152 × 152 × 23	182	182	369	0.275	14.0	14.9	80.2	80.2	14.0	0.302	157	0.416

\# Check availability.

n = F/(A p$_y$), where F is the factored axial load, A is the gross cross sectional area and p$_y$ is the design strength of the section.

For lower values of n, the reduced plastic modulus, S$_r$ = K1 − K2.n², for both major and minor axis bending.

For higher values of n, the reduced plastic modulus, S$_r$ = K3(1 − n)(K4 + n), for both major and minor axis bending.

JOISTS

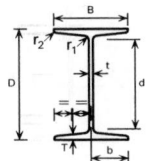

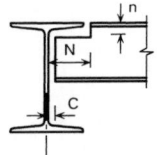

DIMENSIONS

Section Designation	Mass per Metre	Depth of Section	Width of Section	Thickness		Radii		Depth between Fillets	Ratios for Local Buckling		Dimensions for Detailing			Surface Area	
				Web	Flange	Root	Toe		Flange	Web	End Clearance	Notch		Per Metre	Per Tonne
		D	B	t	T	r_1	r_2	d	b/T	d/t	C	N	n		
	kg/m	mm	mm	mm	mm	mm	mm	mm			mm	mm	mm	m²	m²
254 × 203 × 82 #	82.0	254.0	203.2	10.2	19.9	19.6	9.7	166.6	5.11	16.3	7	104	44	1.21	14.8
254 × 114 × 37 ‡	37.2	254.0	114.3	7.6	12.8	12.4	6.1	199.3	4.46	26.2	6	60	28	0.899	24.2
203 × 152 × 52 #	52.3	203.2	152.4	8.9	16.5	15.5	7.6	133.2	4.62	15.0	6	78	36	0.932	17.8
152 × 127 × 37 #	37.3	152.4	127.0	10.4	13.2	13.5	6.6	94.3	4.81	9.07	7	66	30	0.737	19.8
127 × 114 × 29 #	29.3	127.0	114.3	10.2	11.5	9.9	4.8	79.5	4.97	7.79	7	60	24	0.646	22.0
127 × 114 × 27 #	26.9	127.0	114.3	7.4	11.4	9.9	5.0	79.5	5.01	10.7	6	60	24	0.650	24.2
127 × 76 × 16 ‡	16.5	127.0	76.2	5.6	9.6	9.4	4.6	86.5	3.97	15.4	5	42	22	0.512	31.0
114 × 114 × 27 ‡	27.1	114.3	114.3	9.5	10.7	14.2	3.2	60.8	5.34	6.40	7	60	28	0.618	22.8
102 × 102 × 23 #	23.0	101.6	101.6	9.5	10.3	11.1	3.2	55.2	4.93	5.81	7	54	24	0.549	23.9
102 × 44 × 7 #	7.5	101.6	44.5	4.3	6.1	6.9	3.3	74.6	3.65	17.3	4	28	14	0.350	46.6
89 × 89 × 19 #	19.5	88.9	88.9	9.5	9.9	11.1	3.2	44.2	4.49	4.65	7	46	24	0.476	24.4
76 × 76 × 15 ‡	15.0	76.2	80.0	8.9	8.4	9.4	4.6	38.1	4.76	4.28	6	42	20	0.419	27.9
76 × 76 × 13 #	12.8	76.2	76.2	5.1	8.4	9.4	4.6	38.1	4.54	7.47	5	42	20	0.411	32.1

‡ Not available from some leading producers. Check availability.
# Check availability.

JOISTS

PROPERTIES

Section Designation	Second Moment of Area		Radius of Gyration		Elastic Modulus		Plastic Modulus		Buckling Parameter	Torsional Index	Warping Constant	Torsional Constant	Area of Section
	Axis x-x cm⁴	Axis y-y cm⁴	Axis x-x cm	Axis y-y cm	Axis x-x cm³	Axis y-y cm³	Axis x-x cm³	Axis y-y cm³	u	x	H dm⁶	J cm⁴	A cm²
254 × 203 × 82 #	12000	2280	10.7	4.67	947	224	1080	371	0.888	11.0	0.312	152	105
254 × 114 × 37 ‡	5080	269	10.4	2.39	400	47.1	459	79.1	0.885	18.7	0.0392	25.2	47.3
203 × 152 × 52 #	4800	816	8.49	3.50	472	107	541	176	0.890	10.7	0.0711	64.8	66.6
152 × 127 × 37 #	1820	378	6.19	2.82	239	59.6	279	99.8	0.867	9.33	0.0183	33.9	47.5
127 × 114 × 29 #	979	242	5.12	2.54	154	42.3	181	70.8	0.853	8.77	0.00807	20.8	37.4
127 × 114 × 27 #	946	236	5.26	2.63	149	41.3	172	68.2	0.868	9.31	0.00788	16.9	34.2
127 × 76 × 16 ‡	571	60.8	5.21	1.70	90.0	16.0	104	26.4	0.891	11.8	0.00210	6.72	21.1
114 × 114 × 27 ‡	736	224	4.62	2.55	129	39.2	151	65.8	0.839	7.92	0.00601	18.9	34.5
102 × 102 × 23 #	486	154	4.07	2.29	95.6	30.3	113	50.6	0.836	7.42	0.00321	14.2	29.3
102 × 44 × 7 #	153	7.82	4.01	0.907	30.1	3.51	35.4	6.03	0.872	14.9	0.000178	1.25	9.50
89 × 89 × 19 #	307	101	3.51	2.02	69.0	22.8	82.7	38.0	0.830	6.58	0.00158	11.5	24.9
76 × 76 × 15 ‡	172	60.9	3.00	1.78	45.2	15.2	54.2	25.8	0.820	6.42	0.000700	6.83	19.1
76 × 76 × 13 #	158	51.8	3.12	1.79	41.5	13.6	48.7	22.4	0.853	7.21	0.000595	4.59	16.2

‡ Not available from some leading producers. Check availability.
# Check availability.

JOISTS

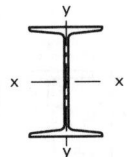

REDUCED PLASTIC MODULUS UNDER AXIAL LOAD

Section Designation	Plastic Modulus Axis x-x	Major Axis Reduced Modulus						Plastic Modulus Axis y-y	Minor Axis Reduced Modulus					
		Lower Values Of n		Change Formula		Higher Values Of n			Lower Values Of n		Change Formula At n =	Higher Values Of n		
	cm³	K1	K2	At n₁ =	At n₂ =	K3	K4	cm	K1	K2		K3	K4
254 × 203 × 82 #	1080	1080	2680	0.195	0.504	134	8.88	371	371	107	0.248	597	0.548
254 × 114 × 37 ‡	459	459	737	0.354	0.576	49.0	11.3	79.1	79.1	22.1	0.408	190	0.236
203 × 152 × 52 #	541	541	1250	0.213	0.490	72.7	8.30	176	176	54.5	0.272	292	0.511
152 × 127 × 37 #	279	279	542	0.256	0.533	44.4	7.15	99.8	99.8	37.0	0.334	186	0.412
127 × 114 × 29 #	181	181	342	0.262	0.542	30.5	6.77	70.8	70.8	27.5	0.347	132	0.407
127 × 114 × 27 #	172	172	396	0.208	0.507	25.6	7.48	68.2	68.2	23.1	0.275	112	0.523
127 × 76 × 16 ‡	104	104	198	0.272	0.500	14.6	8.18	26.4	26.4	8.75	0.337	50.3	0.388
114 × 114 × 27 ‡	151	151	313	0.234	0.557	26.0	6.58	65.8	65.8	26.0	0.315	121	0.420
102 × 102 × 23 #	113	113	226	0.239	0.534	21.2	6.04	50.6	50.6	21.2	0.329	90.8	0.428
102 × 44 × 7 #	35.4	35.4	52.5	0.391	0.575	5.07	8.52	6.03	6.03	2.22	0.460	16.1	0.143
89 × 89 × 19 #	82.7	82.7	163	0.240	0.515	17.4	5.36	38.0	38.0	17.4	0.340	67.9	0.416
76 × 76 × 15 ‡	54.2	54.2	103	0.250	0.533	11.4	5.37	25.8	25.8	12.0	0.354	47.4	0.404
76 × 76 × 13 #	48.7	48.7	129	0.170	0.463	8.66	6.15	22.4	22.4	8.66	0.239	34.2	0.576

‡ Not available from some leading producers. Check availability.
\# Check availability.
n = F/(A pᵧ), where F is the factored axial load, A is the gross cross sectional area and pᵧ is the design strength of the section.
For values of n lower than n₁, the reduced plastic modulus, $S_{rx} = S_{rx1} = K1 - K2.n^2$, for major axis bending.
For values of n higher than n₂, the reduced plastic modulus, $S_{rx} = S_{rx2} = K3(1 - n)(K4 + n)$, for major axis bending.
For values of n between n₁ and n₂, the reduced plastic modulus, $S_{rx} = S_{rx1} + (S_{rx2} - S_{rx1})(n - n_1)/(n_2 - n_1)$, for major axis bending.
For lower values of n, the reduced plastic modulus, $S_{ry} = K1 - K2.n^2$, for minor axis bending.
For higher values of n, the reduced plastic modulus, $S_{ry} = K3(1 - n)(K4 + n)$, for minor axis bending.

UNIVERSAL BEARING PILES

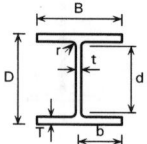

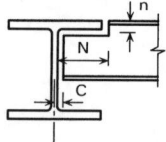

DIMENSIONS

Section Designation	Mass per Metre	Depth of Section	Width of Section	Thickness		Root Radius	Depth between Fillets	Ratios for Local Buckling		Dimensions for Detailing			Surface Area	
				Web	Flange			Flange	Web	End Clearance	Notch		Per Metre	Per Tonne
	kg/m	D mm	B mm	t mm	T mm	r mm	d mm	b/T	d/t	C mm	N mm	n mm	m²	m²
356 × 368 × 174 #	173.9	361.4	378.5	20.3	20.4	15.2	290.2	9.28	14.3	12	190	36	2.17	12.5
356 × 368 × 152 #	152.0	356.4	376.0	17.8	17.9	15.2	290.2	10.5	16.3	11	190	34	2.16	14.2
356 × 368 × 133 #	133.0	352.0	373.8	15.6	15.7	15.2	290.2	11.9	18.6	10	190	32	2.14	16.1
356 × 368 × 109 #	108.9	346.4	371.0	12.8	12.9	15.2	290.2	14.4	22.7	8	190	30	2.13	19.5
305 × 305 × 223 #	222.9	337.9	325.7	30.3	30.4	15.2	246.7	5.36	8.14	17	158	46	1.89	8.49
305 × 305 × 186 #	186.0	328.3	320.9	25.5	25.6	15.2	246.7	6.27	9.67	15	158	42	1.86	10.0
305 × 305 × 149 #	149.1	318.5	316.0	20.6	20.7	15.2	246.7	7.63	12.0	12	158	36	1.83	12.3
305 × 305 × 126 #	126.1	312.3	312.9	17.5	17.6	15.2	246.7	8.89	14.1	11	158	34	1.82	14.4
305 × 305 × 110 #	110.0	307.9	310.7	15.3	15.4	15.2	246.7	10.1	16.1	10	158	32	1.80	16.4
305 × 305 × 95 #	94.9	303.7	308.7	13.3	13.3	15.2	246.7	11.6	18.5	9	158	30	1.79	18.9
305 × 305 × 88 #	88.0	301.7	307.8	12.4	12.3	15.2	246.7	12.5	19.9	8	158	28	1.78	20.3
305 × 305 × 79 #	78.9	299.3	306.4	11.0	11.1	15.2	246.7	13.8	22.4	8	158	28	1.78	22.5
254 × 254 × 85 #	85.1	254.3	260.4	14.4	14.3	12.7	200.3	9.10	13.9	9	134	28	1.50	17.6
254 × 254 × 71 #	71.0	249.7	258.0	12.0	12.0	12.7	200.3	10.8	16.7	8	134	26	1.49	20.9
254 × 254 × 63 #	63.0	247.1	256.6	10.6	10.7	12.7	200.3	12.0	18.9	7	134	24	1.48	23.5
203 × 203 × 54 #	53.9	204.0	207.7	11.3	11.4	10.2	160.8	9.11	14.2	8	110	22	1.20	22.2
203 × 203 × 45 #	44.9	200.2	205.9	9.5	9.5	10.2	160.8	10.8	16.9	7	110	20	1.19	26.4

Check availability.

UNIVERSAL BEARING PILES

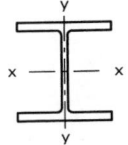

PROPERTIES

Section Designation	Second Moment of Area		Radius of Gyration		Elastic Modulus		Plastic Modulus		Buckling Parameter	Torsional Index	Warping Constant	Torsional Constant	Area of Section
	Axis x-x cm⁴	Axis y-y cm⁴	Axis x-x cm	Axis y-y cm	Axis x-x cm³	Axis y-y cm³	Axis x-x cm³	Axis y-y cm³	u	x	H dm⁶	J cm⁴	A cm²
356 × 368 × 174 #	51000	18500	15.2	9.13	2820	976	3190	1500	0.822	15.8	5.37	330	221
356 × 368 × 152 #	44000	15900	15.1	9.05	2470	845	2770	1290	0.821	17.9	4.55	223	194
356 × 368 × 133 #	38000	13700	15.0	8.99	2160	732	2410	1120	0.823	20.1	3.87	151	169
356 × 368 × 109 #	30600	11000	14.9	8.90	1770	592	1960	903	0.822	24.2	3.05	84.6	139
305 × 305 × 223 #	52700	17600	13.6	7.87	3120	1080	3650	1680	0.826	9.55	4.15	943	284
305 × 305 × 186 #	42600	14100	13.4	7.73	2600	881	3000	1370	0.827	11.1	3.24	560	237
305 × 305 × 149 #	33100	10900	13.2	7.58	2080	691	2370	1070	0.828	13.5	2.42	295	190
305 × 305 × 126 #	27400	9000	13.1	7.49	1760	575	1990	885	0.828	15.7	1.95	182	161
305 × 305 × 110 #	23600	7710	13.0	7.42	1530	496	1720	762	0.830	17.7	1.65	122	140
305 × 305 × 95 #	20000	6530	12.9	7.35	1320	423	1470	648	0.830	20.2	1.38	80.0	121
305 × 305 × 88 #	18400	5980	12.8	7.31	1220	389	1360	595	0.830	21.6	1.25	64.2	112
305 × 305 × 79 #	16400	5330	12.8	7.28	1100	348	1220	531	0.834	23.8	1.11	46.9	100
254 × 254 × 85 #	12300	4220	10.6	6.24	966	324	1090	498	0.826	15.6	0.607	81.8	108
254 × 254 × 71 #	10100	3440	10.6	6.17	807	267	904	409	0.826	18.4	0.486	48.4	90.4
254 × 254 × 63 #	8860	3020	10.5	6.13	717	235	799	360	0.827	20.5	0.421	34.3	80.2
203 × 203 × 54 #	5030	1710	8.55	4.98	493	164	557	252	0.827	15.8	0.158	32.7	68.7
203 × 203 × 45 #	4100	1380	8.46	4.92	410	134	459	206	0.828	18.6	0.126	19.2	57.2

Check availability.

UNIVERSAL BEARING PILES

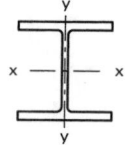

REDUCED PLASTIC MODULUS UNDER AXIAL LOAD

Section Designation	Plastic Modulus Axis x-x	Major Axis Reduced Modulus					Plastic Modulus Axis y-y	Minor Axis Reduced Modulus				
		Lower Values Of n		Change Formula At n =	Higher Values Of n			Lower Values Of n		Change Formula At n =	Higher Values Of n	
	cm³	K1	K2		K3	K4	cm³	K1	K2		K3	K4
356 × 368 × 174 #	3190	3190	6040	0.294	324	11.4	1500	1500	339	0.331	3010	0.394
356 × 368 × 152 #	2770	2770	5270	0.295	249	12.8	1290	1290	263	0.328	2620	0.390
356 × 368 × 133 #	2410	2410	4600	0.295	192	14.5	1120	1120	204	0.324	2280	0.386
356 × 368 × 109 #	1960	1960	3760	0.296	130	17.5	903	903	139	0.320	1870	0.380
305 × 305 × 223 #	3650	3650	6650	0.296	619	6.75	1680	1680	597	0.361	3320	0.395
305 × 305 × 186 #	3000	3000	5500	0.298	437	7.89	1370	1370	428	0.353	2740	0.387
305 × 305 × 149 #	2370	2370	4380	0.301	285	9.60	1070	1070	283	0.346	2180	0.378
305 × 305 × 126 #	1990	1990	3690	0.302	206	11.2	885	885	207	0.340	1830	0.371
305 × 305 × 110 #	1720	1720	3210	0.303	158	12.7	762	762	159	0.336	1590	0.366
305 × 305 × 95 #	1470	1470	2750	0.305	118	14.5	648	648	120	0.334	1380	0.358
305 × 305 × 88 #	1360	1360	2530	0.307	102	15.6	595	595	104	0.334	1280	0.351
305 × 305 × 79 #	1220	1220	2300	0.303	82.4	17.3	531	531	84.3	0.328	1140	0.354
254 × 254 × 85 #	1090	1090	2040	0.300	113	11.2	498	498	115	0.338	1030	0.375
254 × 254 × 71 #	904	904	1700	0.300	79.2	13.3	409	409	81.8	0.332	851	0.370
254 × 254 × 63 #	799	799	1520	0.298	62.7	14.8	360	360	65.1	0.327	752	0.369
203 × 203 × 54 #	557	557	1050	0.298	56.8	11.3	252	252	57.9	0.335	518	0.378
203 × 203 × 45 #	459	459	862	0.301	39.8	13.4	206	206	40.9	0.332	431	0.367

Check availability.

$n = F/(A\, p_Y)$, where F is the factored axial load, A is the gross cross sectional area and p_Y is the design strength of the section.

For lower values of n, the reduced plastic modulus, $S_r = K1 - K2.n^2$, for both major and minor axis bending.

For higher values of n, the reduced plastic modulus, $S_r = K3(1 - n)(K4 + n)$, for both major and minor axis bending.

HOT-FINISHED
CIRCULAR HOLLOW SECTIONS

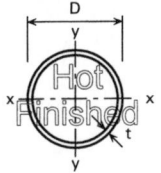

DIMENSIONS AND PROPERTIES

Section Designation		Mass per Metre	Area of Section	Ratio for Local Buckling	Second Moment of Area	Radius of Gyration	Elastic Modulus	Plastic Modulus	Torsional Constants		Surface Area	
Outside Diameter	Thickness										Per Metre	Per Tonne
D mm	t mm	kg/m	A cm²	D/t	I cm⁴	r cm	Z cm³	S cm³	J cm⁴	C cm³	m²	m²
26.9	3.2 ~	1.87	2.38	8.41	1.70	0.846	1.27	1.81	3.41	2.53	0.0845	45.2
42.4	3.2 ~	3.09	3.94	13.3	7.62	1.39	3.59	4.93	15.2	7.19	0.133	43.0
48.3	3.2 ~	3.56	4.53	15.1	11.6	1.60	4.80	6.52	23.2	9.59	0.152	42.7
	4.0 ~	4.37	5.57	12.1	13.8	1.57	5.70	7.87	27.5	11.4	0.152	34.8
	5.0 ~	5.34	6.80	9.66	16.2	1.54	6.69	9.42	32.3	13.4	0.152	28.5
60.3	3.2 ~	4.51	5.74	18.8	23.5	2.02	7.78	10.4	46.9	15.6	0.189	41.9
	5.0 ~	6.82	8.69	12.1	33.5	1.96	11.1	15.3	67.0	22.2	0.189	27.7
76.1	2.9 ^	5.24	6.67	26.2	44.7	2.59	11.8	15.5	89.0	23.5	0.239	45.6
	3.2 ~	5.75	7.33	23.8	48.8	2.58	12.8	17.0	97.6	25.6	0.239	41.6
	4.0 ~	7.11	9.06	19.0	59.1	2.55	15.5	20.8	118	31.0	0.239	33.6
	5.0 ~	8.77	11.2	15.2	70.9	2.52	18.6	25.3	142	37.3	0.239	27.3
88.9	3.2 ~	6.76	8.62	27.8	79.2	3.03	17.8	23.5	158	35.6	0.279	41.3
	4.0 ~	8.38	10.7	22.2	96.3	3.00	21.7	28.9	193	43.3	0.279	33.3
	5.0 ~	10.4	13.2	17.8	116	2.97	26.2	35.2	233	52.4	0.279	27.0
	6.3 ~	12.8	16.3	14.1	140	2.93	31.5	43.1	280	63.1	0.279	21.7
114.3	3.2 ~	8.77	11.2	35.7	172	3.93	30.2	39.5	345	60.4	0.359	40.9
	3.6	9.83	12.5	31.8	192	3.92	33.6	44.1	384	67.2	0.359	36.5
	5.0	13.5	17.2	22.9	257	3.87	45.0	59.8	514	89.9	0.359	26.6
	6.3	16.8	21.4	18.1	313	3.82	54.7	73.6	625	109	0.359	21.4
139.7	5.0	16.6	21.2	27.9	481	4.77	68.8	90.8	961	138	0.439	26.4
	6.3	20.7	26.4	22.2	589	4.72	84.3	112	1180	169	0.439	21.2
	8.0 ~	26.0	33.1	17.5	720	4.66	103	139	1440	206	0.439	16.9
	10.0 ~	32.0	40.7	14.0	862	4.60	123	169	1720	247	0.439	13.7
168.3	5.0 ~	20.1	25.7	33.7	856	5.78	102	133	1710	203	0.529	26.3
	6.3 ~	25.2	32.1	26.7	1050	5.73	125	165	2110	250	0.529	21.0
	8.0 ~	31.6	40.3	21.0	1300	5.67	154	206	2600	308	0.529	16.7
	10.0 ~	39.0	49.7	16.8	1560	5.61	186	251	3130	372	0.529	13.6
193.7	5.0 ~	23.3	29.6	38.7	1320	6.67	136	178	2640	273	0.609	26.1
	6.3 ~	29.1	37.1	30.7	1630	6.63	168	221	3260	337	0.609	20.9
	8.0 ~	36.6	46.7	24.2	2020	6.57	208	276	4030	416	0.609	16.6
	10.0 ~	45.3	57.7	19.4	2440	6.50	252	338	4880	504	0.609	13.4

~ Check availability in S275.
^ Check availability in S355.

HOT-FINISHED
CIRCULAR HOLLOW SECTIONS

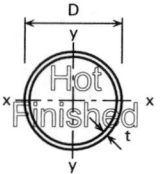

DIMENSIONS AND PROPERTIES

Section Designation		Mass per Metre	Area of Section	Ratio for Local Buckling	Second Moment of Area	Radius of Gyration	Elastic Modulus	Plastic Modulus	Torsional Constants		Surface Area	
Outside Diameter	Thickness										Per Metre	Per Tonne
D mm	t mm	kg/m	A cm²	D/t	I cm⁴	r cm	Z cm³	S cm³	J cm⁴	C cm³	m²	m²
219.1	5.0 ~	26.4	33.6	43.8	1930	7.57	176	229	3860	352	0.688	26.1
	6.3 ~	33.1	42.1	34.8	2390	7.53	218	285	4770	436	0.688	20.8
	8.0 ~	41.6	53.1	27.4	2960	7.47	270	357	5920	540	0.688	16.5
	10.0 ~	51.6	65.7	21.9	3600	7.40	328	438	7200	657	0.688	13.3
	12.5 ~	63.7	81.1	17.5	4350	7.32	397	534	8690	793	0.688	10.8
244.5	12.0 ~	68.8	87.7	20.4	5940	8.23	486	649	11900	972	0.768	11.2
273.0	5.0 ~	33.0	42.1	54.6	3780	9.48	277	359	7560	554	0.858	26.0
	6.3	41.4	52.8	43.3	4700	9.43	344	448	9390	688	0.858	20.1
	8.0 ~	52.3	66.6	34.1	5850	9.37	429	562	11700	857	0.858	16.4
	10.0	64.9	82.6	27.3	7150	9.31	524	692	14300	1050	0.858	13.2
	12.5	80.3	102	21.8	8700	9.22	637	849	17400	1270	0.858	10.1
	16.0 ~	101	129	17.1	10700	9.10	784	1060	21400	1570	0.858	8.46
323.9	6.3 ~	49.3	62.9	51.4	7930	11.2	490	636	15900	979	1.02	20.7
	8.0 ~	62.3	79.4	40.5	9910	11.2	612	799	19800	1220	1.02	16.4
	10.0 ~	77.4	98.6	32.4	12200	11.1	751	986	24300	1500	1.02	13.2
	12.5 ~	96.0	122	25.9	14800	11.0	917	1210	29700	1830	1.02	10.6
	16.0	122	155	20.2	18400	10.9	1140	1520	36800	2270	1.02	8.40
406.4	6.3 ~	62.2	79.2	64.5	15900	14.1	780	1010	31700	1560	1.28	20.6
	8.0 ~	78.6	100	50.8	19900	14.1	978	1270	39700	1960	1.28	16.3
	10.0 ~	97.8	125	40.6	24500	14.0	1210	1570	49000	2410	1.28	13.1
	12.5 ~	121	155	32.5	30000	13.9	1480	1940	60100	2960	1.28	10.5
	16.0	154	196	25.4	37500	13.8	1840	2440	74900	3690	1.28	8.31
457.0	8.0 ~	88.6	113	57.1	28500	15.9	1250	1610	56900	2490	1.44	16.3
	10.0 ~	110	140	45.7	35100	15.8	1540	2000	70200	3070	1.44	13.1
	12.5 ~	137	175	36.6	43100	15.7	1890	2470	86300	3780	1.44	10.5
	16.0	174	222	28.6	54000	15.6	2360	3110	108000	4730	1.44	8.28
508.0	8.0 ~	98.6	126	63.5	39300	17.7	1550	2000	78600	3090	1.60	16.2
	10.0 ~	123	156	50.8	48500	17.6	1910	2480	97000	3820	1.60	13.0
	12.5	153	195	40.6	59800	17.5	2350	3070	120000	4710	1.60	10.5
	16.0 ~	194	247	31.8	74900	17.4	2950	3870	150000	5900	1.60	8.25
	20.0 ~	241	307	25.4	91400	17.3	3600	4770	183000	7200	1.60	6.64

~ Check availability in S275.

HOT-FINISHED
SQUARE HOLLOW SECTIONS

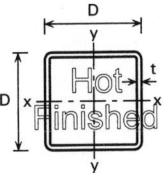

DIMENSIONS AND PROPERTIES

Section Designation		Mass per Metre	Area of Section	Ratio for Local Buckling	Second Moment of Area	Radius of Gyration	Elastic Modulus	Plastic Modulus	Torsional Constants		Surface Area	
											Per Metre	Per Tonne
Size D × D mm	Thickness t mm	kg/m	A cm²	d/t⁽¹⁾	I cm⁴	r cm	Z cm³	S cm³	J cm⁴	C cm³	m²	m²
40 × 40	3.0 ~	3.41	4.34	10.3	9.78	1.50	4.89	5.97	15.7	7.10	0.152	44.6
	3.2 ~	3.61	4.60	9.50	10.2	1.49	5.11	6.28	16.5	7.42	0.152	42.1
	4.0	4.39	5.59	7.00	11.8	1.45	5.91	7.44	19.5	8.54	0.150	34.2
	5.0 ~	5.28	6.73	5.00	13.4	1.41	6.68	8.66	22.5	9.60	0.147	27.8
50 × 50	3.0 ~	4.35	5.54	13.7	20.2	1.91	8.08	9.70	32.1	11.8	0.192	44.1
	3.2 ~	4.62	5.88	12.6	21.2	1.90	8.49	10.2	33.8	12.4	0.192	41.6
	4.0	5.64	7.19	9.50	25.0	1.86	9.99	12.3	40.4	14.5	0.190	33.7
	5.0	6.85	8.73	7.00	28.9	1.82	11.6	14.5	47.6	16.7	0.187	27.3
	6.3 ~	8.31	10.6	4.94	32.8	1.76	13.1	17.0	55.2	18.8	0.184	22.1
60 × 60	3.0 ~	5.29	6.74	17.0	36.2	2.32	12.1	14.3	56.9	17.7	0.232	43.9
	3.2 ~	5.62	7.16	15.8	38.2	2.31	12.7	15.2	60.2	18.6	0.232	41.3
	4.0 ~	6.90	8.79	12.0	45.4	2.27	15.1	18.3	72.5	22.0	0.230	33.3
	5.0	8.42	10.7	9.00	53.3	2.23	17.8	21.9	86.4	25.7	0.227	27.0
	6.3 ~	10.3	13.1	6.52	61.6	2.17	20.5	26.0	102	29.6	0.224	21.7
	8.0 ~	12.5	16.0	4.50	69.7	2.09	23.2	30.4	118	33.4	0.219	17.5
70 × 70	3.6 ~	7.40	9.42	16.4	68.6	2.70	19.6	23.3	108	28.7	0.271	36.6
	5.0 ~	9.99	12.7	11.0	88.5	2.64	25.3	30.8	142	36.8	0.267	26.7
	6.3 ~	12.3	15.6	8.11	104	2.58	29.7	36.9	169	42.9	0.264	21.5
	8.0 ~	15.0	19.2	5.75	120	2.50	34.2	43.8	200	49.2	0.259	17.3
80 × 80	3.6 ~	8.53	10.9	19.2	105	3.11	26.2	31.0	164	38.5	0.311	36.5
	4.0 ~	9.41	12.0	17.0	114	3.09	28.6	34.0	180	41.9	0.310	32.9
	5.0 ~	11.6	14.7	13.0	137	3.05	34.2	41.1	217	49.8	0.307	26.6
	6.3 ~	14.2	18.1	9.70	162	2.99	40.5	49.7	262	58.7	0.304	21.4
	8.0 ~	17.5	22.4	7.00	189	2.91	47.3	59.5	312	68.3	0.299	17.1
90 × 90	3.6 ~	9.66	12.3	22.0	152	3.52	33.8	39.7	237	49.7	0.351	36.3
	4.0 ~	10.7	13.6	19.5	166	3.50	37.0	43.6	260	54.2	0.350	32.7
	5.0 ~	13.1	16.7	15.0	200	3.45	44.4	53.0	316	64.8	0.347	26.5
	6.3 ~	16.2	20.7	11.3	238	3.40	53.0	64.3	382	77.0	0.344	21.2
	8.0 ~	20.1	25.6	8.25	281	3.32	62.6	77.6	459	90.5	0.339	16.9
100 × 100	4.0	11.9	15.2	22.0	232	3.91	46.4	54.4	361	68.2	0.390	32.8
	5.0	14.7	18.7	17.0	279	3.86	55.9	66.4	439	81.8	0.387	26.3
	6.3	18.2	23.2	12.9	336	3.80	67.1	80.9	534	97.8	0.384	21.1
	8.0 ~	22.6	28.8	9.50	400	3.73	79.9	98.2	646	116	0.379	16.8
	10.0	27.4	34.9	7.00	462	3.64	92.4	116	761	133	0.374	13.6
120 × 120	5.0 ~	17.8	22.7	21.0	498	4.68	83.0	97.6	777	122	0.467	26.2
	6.3	22.2	28.2	16.0	603	4.62	100	120	950	147	0.464	20.9
	8.0 ~	27.6	35.2	12.0	726	4.55	121	147	1160	176	0.459	16.6
	10.0 ~	33.7	42.9	9.00	852	4.46	142	175	1380	206	0.454	13.5
	12.5 ~	40.9	52.1	6.60	982	4.34	164	207	1620	236	0.448	11.0

~ Check availability in S275.
⁽¹⁾ For local buckling calculation d = D − 3t.

HOT-FINISHED
SQUARE HOLLOW SECTIONS

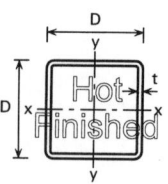

DIMENSIONS AND PROPERTIES

Section Designation		Mass per Metre	Area of Section	Ratio for Local Buckling	Second Moment of Area	Radius of Gyration	Elastic Modulus	Plastic Modulus	Torsional Constants		Surface Area	
Size D × D mm	Thickness t mm										Per Metre	Per Tonne
		kg/m	A cm²	$d/t^{(1)}$	I cm⁴	r cm	Z cm³	S cm³	J cm⁴	C cm³	m²	m²
140 × 140	5.0 ~	21.0	26.7	25.0	807	5.50	115	135	1250	170	0.547	26.0
	6.3 ~	26.1	33.3	19.2	984	5.44	141	166	1540	206	0.544	20.8
	8.0 ~	32.6	41.6	14.5	1200	5.36	171	204	1890	249	0.539	16.5
	10.0 ~	40.0	50.9	11.0	1420	5.27	202	246	2270	294	0.534	13.4
	12.5 ~	48.7	62.1	8.20	1650	5.16	236	293	2700	342	0.528	10.8
150 × 150	5.0 ~	22.6	28.7	27.0	1000	5.90	134	156	1550	197	0.587	26.0
	6.3	28.1	35.8	20.8	1220	5.85	163	192	1910	240	0.584	20.8
	8.0 ~	35.1	44.8	15.8	1490	5.77	199	237	2350	291	0.579	16.5
	10.0	43.1	54.9	12.0	1770	5.68	236	286	2830	344	0.574	13.3
	12.5 ~	52.7	67.1	9.00	2080	5.57	277	342	3370	402	0.568	10.8
	16.0	65.2	33.0	6.38	2430	5.41	324	411	4030	467	0.559	8.57
160 × 160	5.0 ~	24.1	30.7	29.0	1230	6.31	153	178	1890	226	0.627	26.0
	6.3 ~	30.1	38.3	22.4	1500	6.26	187	220	2330	275	0.624	20.7
	8.0 ~	37.6	48.0	17.0	1830	6.18	229	272	2880	335	0.619	16.5
	10.0 ~	46.3	58.9	13.0	2190	6.09	273	329	3480	398	0.614	13.3
	12.5 ~	56.6	72.1	9.80	2580	5.98	322	395	4160	467	0.608	10.7
180 × 180	6.3 ~	34.0	43.3	25.6	2170	7.07	241	281	3360	355	0.704	20.7
	8.0 ~	42.7	54.4	19.5	2660	7.00	296	349	4160	434	0.699	16.4
	10.0 ~	52.5	66.9	15.0	3190	6.91	355	424	5050	518	0.694	13.2
	12.5 ~	64.4	82.1	11.4	3790	6.80	421	511	6070	613	0.688	10.7
	16.0 ~	80.2	102	8.25	4500	6.64	500	621	7340	724	0.679	8.47
200 × 200	5.0 ~	30.4	38.7	37.0	2450	7.95	245	283	3760	362	0.787	25.9
	6.3 ~	38.0	48.4	28.7	3010	7.89	301	350	4650	444	0.784	20.6
	8.0 ~	47.7	60.8	22.0	3710	7.81	371	436	5780	545	0.779	16.3
	10.0	58.8	74.9	17.0	4470	7.72	447	531	7030	655	0.774	13.2
	12.5 ~	72.3	92.1	13.0	5340	7.61	534	643	8490	778	0.768	10.6
	16.0 ~	90.3	115	9.50	6390	7.46	639	785	10300	927	0.759	8.41
250 × 250	6.3 ~	47.9	61.0	36.7	6010	9.93	481	556	9240	712	0.984	20.5
	8.0 ~	60.3	76.8	28.3	7460	9.86	596	694	11500	880	0.979	16.2
	10.0 ~	74.5	94.9	22.0	9060	9.77	724	851	14100	1070	0.974	13.1
	12.5 ~	91.9	117	17.0	10900	9.66	873	1040	17200	1280	0.968	10.5
	16.0 ~	115	147	12.6	13300	9.50	1060	1280	21100	1550	0.959	8.31
300 × 300	6.3 ~	57.8	73.6	44.6	10500	12.0	703	809	16100	1040	1.18	20.4
	8.0 ~	72.8	92.8	34.5	13100	11.9	875	1010	20200	1290	1.18	16.2
	10.0 ~	90.2	115	27.0	16000	11.8	1070	1250	24800	1580	1.17	13.0
	12.5 ~	112	142	21.0	19400	11.7	1300	1530	30300	1900	1.17	10.5
	16.0 ~	141	179	15.8	23900	11.5	1590	1900	37600	2330	1.16	8.26
350 × 350	8.0 ~	85.4	109	40.8	21100	13.9	1210	1390	32400	1790	1.38	16.2
	10.0 ~	106	135	32.0	25900	13.9	1480	1720	39900	2190	1.37	12.9
	12.5 ~	131	167	25.0	31500	13.7	1800	2110	48900	2650	1.37	10.5
	16.0 ~	166	211	18.9	38900	13.6	2230	2630	61000	3260	1.36	8.19
400 × 400	10.0 ~	122	155	37.0	39100	15.9	1960	2260	60100	2900	1.57	12.9
	12.5	151	192	29.0	48700	15.8	2390	2780	73900	3530	1.57	10.4
	16.0 ~	191	243	22.0	59300	15.6	2970	3480	92400	4360	1.56	8.17
	20.0 ~	235	300	17.0	71500	15.4	3580	4250	113000	5240	1.55	6.60

~ Check availability in S275.
(1) For local buckling calculation d = D − 3t.

HOT-FINISHED
RECTANGULAR HOLLOW SECTIONS

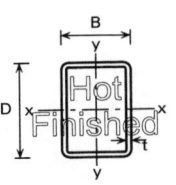

DIMENSIONS AND PROPERTIES

Section Designation		Mass per Metre	Area of Section	Ratios for Local Buckling		Second Moment of Area		Radius of Gyration		Elastic Modulus		Plastic Modulus		Torsional Constants		Surface Area	
Size	Thickness					Axis x-x	Axis y-y	Axis x-x	Axis y-y	Axis x-x	Axis y-y	Axis x-x	Axis y-y			Per Metre	Per Tonne
D × B mm	t mm	kg/m	A cm²	d/t[1]	b/t[1]	cm⁴	cm⁴	cm	cm	cm³	cm³	cm³	cm³	J cm⁴	C cm³	m²	m²
50 × 30	3.2 ~	3.61	4.60	12.6	6.38	14.2	6.20	1.76	1.16	5.68	4.13	7.25	5.00	14.2	6.80	0.152	42.1
60 × 40	3.0 ~	4.35	5.54	17.0	10.3	26.5	13.9	2.18	1.58	8.82	6.95	10.9	8.19	29.2	11.2	0.192	44.1
	4.0 ~	5.64	7.19	12.0	7.00	32.8	17.0	2.14	1.54	10.9	8.52	13.8	10.3	36.7	13.7	0.190	33.7
	5.0 ~	6.85	8.73	9.00	5.00	38.1	19.5	2.09	1.50	12.7	9.77	16.4	12.2	43.0	15.7	0.187	27.3
80 × 40	3.2 ~	5.62	7.16	22.0	9.50	57.2	18.9	2.83	1.63	14.3	9.46	18.0	11.0	46.2	16.1	0.232	41.3
	4.0 ~	6.90	8.79	17.0	7.00	68.2	22.2	2.79	1.59	17.1	11.1	21.8	13.2	55.2	18.9	0.230	33.3
	5.0 ~	8.42	10.7	13.0	5.00	80.3	25.7	2.74	1.55	20.1	12.9	26.1	15.7	65.1	21.9	0.227	27.0
	6.3 ~	10.3	13.1	9.70	3.35	93.3	29.2	2.67	1.49	23.3	14.6	31.1	18.4	75.6	24.8	0.224	21.7
	8.0 ~	12.5	16.0	7.00	2.00	106	32.1	2.58	1.42	26.5	16.1	36.5	21.2	85.8	27.4	0.219	17.5
90 × 50	3.6 ~	7.40	9.42	22.0	10.9	98.3	38.7	3.23	2.03	21.8	15.5	27.2	18.0	89.4	25.9	0.271	36.6
	5.0 ~	9.99	12.7	15.0	7.00	127	49.2	3.16	1.97	28.3	19.7	36.0	23.5	116	32.9	0.267	26.7
	6.3 ~	12.3	15.6	11.3	4.94	150	57.0	3.10	1.91	33.3	22.8	43.2	28.0	138	38.1	0.264	21.5
100 × 50	3.0 ~	6.71	8.54	30.3	13.7	110	36.8	3.58	2.08	21.9	14.7	27.3	16.8	88.4	25.0	0.292	43.5
	3.2 ~	7.13	9.08	28.3	12.6	116	38.8	3.57	2.07	23.2	15.5	28.9	17.7	93.4	26.4	0.292	41.0
	5.0 ~	10.8	13.7	17.0	7.00	167	54.3	3.48	1.99	33.3	21.7	42.6	25.8	135	36.9	0.287	26.6
	6.3 ~	13.3	16.9	12.9	4.94	197	63.0	3.42	1.93	39.4	25.2	51.3	30.8	160	42.9	0.284	21.4
	8.0 ~	16.3	20.8	9.50	3.25	230	71.7	3.33	1.86	46.0	28.7	61.4	36.3	186	48.9	0.279	17.1
	10.0 ^	19.6	24.9	7.00	2.00	259	78.4	3.22	1.77	51.8	31.4	71.2	41.4	209	53.6	0.274	14.0
100 × 60	3.6 ~	8.53	10.9	24.8	13.7	145	64.8	3.65	2.44	28.9	21.6	35.6	24.9	142	35.6	0.311	36.5
	5.0 ~	11.6	14.7	17.0	9.00	189	83.6	3.58	2.38	37.8	27.9	47.4	32.9	188	45.9	0.307	26.5
	6.3 ~	14.2	18.1	12.9	6.52	225	98.1	3.52	2.33	45.0	32.7	57.3	39.5	224	53.8	0.304	21.4
	8.0 ~	17.5	22.4	9.50	4.50	264	113	3.44	2.25	52.8	37.8	68.7	47.1	265	62.2	0.299	17.1
120 × 60	3.6 ~	9.70	12.3	30.3	13.7	227	76.3	4.30	2.49	37.9	25.4	47.2	28.9	183	43.3	0.351	36.2
	5.0 ~	13.1	16.7	21.0	9.00	299	98.8	4.23	2.43	49.9	32.9	63.1	38.4	242	56.0	0.347	26.5
	6.3 ~	16.2	20.7	16.0	6.52	358	116	4.16	2.37	59.7	38.8	76.7	46.3	290	65.9	0.344	21.2
	8.0 ~	20.1	25.6	12.0	4.50	425	135	4.08	2.30	70.8	45.0	92.7	55.4	344	76.6	0.339	16.9
120 × 80	5.0 ~	14.7	18.7	21.0	13.0	365	193	4.42	3.21	60.9	48.2	74.6	56.1	401	77.9	0.387	26.3
	6.3	18.2	23.2	16.0	9.70	440	230	4.36	3.15	73.3	57.6	91.0	68.2	487	92.9	0.384	21.1
	8.0	22.6	28.8	12.0	7.00	525	273	4.27	3.08	87.5	68.1	111	82.6	587	110	0.379	16.8
	10.0 ~	27.4	34.9	9.00	5.00	609	313	4.18	2.99	102	78.1	131	97.3	688	126	0.374	13.6
150 × 100	5.0 ~	18.6	23.7	27.0	17.0	739	392	5.58	4.07	98.5	78.5	119	90.1	807	127	0.487	26.2
	6.3	23.1	29.5	20.8	12.9	898	474	5.52	4.01	120	94.8	147	110	986	153	0.484	21.0
	8.0 ~	28.9	36.8	15.8	9.50	1090	569	5.44	3.94	145	114	180	135	1200	183	0.479	16.6
	10.0 ~	35.3	44.9	12.0	7.00	1280	665	5.34	3.85	171	133	216	161	1430	214	0.474	13.4
	12.5 ~	42.8	54.6	9.00	5.00	1490	763	5.22	3.74	198	153	256	190	1680	246	0.468	10.9
160 × 80	4.0 ~	14.4	18.4	37.0	17.0	612	207	5.77	3.35	76.5	51.7	94.7	58.3	493	88.0	0.470	32.6
	5.0 ~	17.8	22.7	29.0	13.0	744	249	5.72	3.31	93.0	62.3	116	71.1	600	106	0.467	26.2
	6.3	22.2	28.2	22.4	9.70	903	299	5.66	3.26	113	74.8	142	86.8	730	127	0.464	20.9
	8.0	27.6	35.2	17.0	7.00	1090	356	5.57	3.18	136	89.0	175	106	883	151	0.459	16.6
	10.0 ~	33.7	42.9	13.0	5.00	1280	411	5.47	3.10	161	103	209	125	1040	175	0.454	13.5

~ Check availability in S275.
^ Check availability in S355.
[1] For local buckling calculation d = D − 3t and b = B − 3t.

HOT-FINISHED
RECTANGULAR HOLLOW SECTIONS

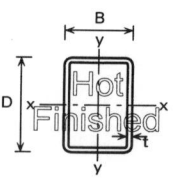

DIMENSIONS AND PROPERTIES

Section Designation		Mass per Metre	Area of Section	Ratios for Local Buckling		Second Moment of Area		Radius of Gyration		Elastic Modulus		Plastic Modulus		Torsional Constants		Surface Area	
Size	Thickness															Per Metre	Per Tonne
$D \times B$ mm	t mm	kg/m	A cm²	$d/t^{(1)}$	$b/t^{(1)}$	Axis x-x cm⁴	Axis y-y cm⁴	Axis x-x cm	Axis y-y cm	Axis x-x cm³	Axis y-y cm³	Axis x-x cm³	Axis y-y cm³	J cm⁴	C cm³	m²	m²
200 × 100	5.0 ~	22.6	28.7	37.0	17.0	1500	505	7.21	4.19	149	101	185	114	1200	172	0.587	26.0
	6.3 ~	28.1	35.8	28.7	12.9	1830	613	7.15	4.14	183	123	228	140	1470	208	0.584	20.8
	8.0	35.1	44.8	22.0	9.50	2230	739	7.06	4.06	223	148	282	172	1800	251	0.579	16.5
	10.0	43.1	54.9	17.0	7.00	2660	869	6.96	3.98	266	174	341	206	2160	295	0.574	13.3
	12.5	52.7	67.1	13.0	5.00	3140	1000	6.84	3.87	314	201	408	245	2540	341	0.568	10.8
200 × 120	5.0 ~	24.1	30.7	37.0	21.0	1690	762	7.40	4.98	168	127	205	144	1650	210	0.627	26.0
	6.3 ~	30.1	38.3	28.7	16.0	2070	929	7.34	4.92	207	155	253	177	2030	255	0.624	20.7
	8.0 ~	37.6	48.0	22.0	12.0	2530	1130	7.26	4.85	253	188	313	218	2490	310	0.619	16.5
	10.0 ~	46.3	58.9	17.0	9.00	3030	1340	7.17	4.76	303	223	379	263	3000	367	0.614	13.3
200 × 150	8.0 ~	41.4	52.8	22.0	15.8	2970	1890	7.50	5.99	297	253	359	294	3640	398	0.679	16.4
	10.0 ~	51.0	64.9	17.0	12.0	3570	2260	7.41	5.91	357	302	436	356	4410	475	0.674	13.2
250 × 100	10.0 ~	51.0	64.9	22.0	7.00	4730	1070	8.54	4.06	379	214	491	251	2910	376	0.674	13.2
	12.5 ~	62.5	79.6	17.0	5.00	5620	1250	8.41	3.96	450	249	592	299	3440	438	0.668	10.7
250 × 150	5.0 ~	30.4	38.7	47.0	27.0	3360	1530	9.31	6.28	269	204	324	228	3280	337	0.787	25.9
	6.3	38.0	48.4	36.7	20.8	4140	1870	9.25	6.22	331	250	402	283	4050	413	0.784	20.6
	8.0 ~	47.7	60.8	28.3	15.8	5110	2300	9.17	6.15	409	306	501	350	5020	506	0.779	16.3
	10.0 ~	58.8	74.9	22.0	12.0	6170	2760	9.08	6.06	494	367	611	426	6090	605	0.774	13.2
	12.5	72.3	92.1	17.0	9.00	7390	3270	8.96	5.96	591	435	740	514	7330	717	0.768	10.6
	16.0 ~	90.3	115	12.6	6.38	8880	3870	8.79	5.80	710	516	906	625	8870	849	0.759	8.41
300 × 100	8.0 ~	47.7	60.8	34.5	9.50	6310	1080	10.2	4.21	420	216	546	245	3070	387	0.779	16.3
	10.0 ~	58.8	74.9	27.0	7.00	7610	1280	10.1	4.13	508	255	666	296	3680	458	0.774	13.2
300 × 200	6.3 ~	47.9	61.0	44.6	28.7	7830	4190	11.3	8.29	522	419	624	472	8480	681	0.984	20.5
	8.0 ~	60.3	76.8	34.5	22.0	9720	5180	11.3	8.22	648	518	779	589	10600	840	0.979	16.2
	10.0 ~	74.5	94.9	27.0	17.0	11800	6280	11.2	8.13	788	628	956	721	12900	1020	0.974	13.1
	12.5 ~	91.9	117	21.0	13.0	14300	7540	11.0	8.02	952	754	1170	877	15700	1220	0.968	10.5
	16.0	115	147	15.8	9.50	17400	9110	10.9	7.87	1160	911	1440	1080	19300	1470	0.959	8.34
400 × 200	8.0 ~	72.8	92.8	47.0	22.0	19600	6660	14.5	8.47	978	666	1200	743	15700	1140	1.18	16.2
	10.0 ~	90.2	115	37.0	17.0	23900	8080	14.4	8.39	1200	808	1480	911	19300	1380	1.17	13.0
	12.5 ~	112	142	29.0	13.0	29100	9740	14.3	8.28	1450	974	1810	1110	23400	1660	1.17	10.5
	16.0 ~	141	179	22.0	9.50	35700	11800	14.1	8.13	1790	1180	2260	1370	28900	2010	1.16	8.26
450 × 250	8.0 ~	85.4	109	53.3	28.3	30100	12100	16.6	10.6	1340	971	1620	1080	27100	1630	1.38	16.2
	10.0 ~	106	135	42.0	22.0	36900	14800	16.5	10.5	1640	1190	2000	1330	33300	1990	1.37	12.9
	12.5 ~	131	167	33.0	17.0	45000	18000	16.4	10.4	2000	1440	2460	1630	40700	2410	1.37	10.5
	16.0	166	211	25.1	12.6	55700	22000	16.2	10.2	2480	1760	3070	2030	50500	2950	1.36	8.19
500 × 300	8.0 ~	98.0	125	59.5	34.5	43700	20000	18.7	12.6	1750	1330	2100	1480	42600	2200	1.58	16.1
	10.0 ~	122	155	47.0	27.0	53800	24400	18.6	12.6	2150	1630	2600	1830	52400	2700	1.57	12.9
	12.5 ~	151	192	37.0	21.0	65800	29800	18.5	12.5	2630	1990	3200	2240	64400	3280	1.57	10.4
	16.0 ~	191	243	28.3	15.8	81800	36800	18.3	12.3	3270	2450	4010	2800	80300	4040	1.56	8.17
	20.0 ~	235	300	22.0	12.0	98800	44100	18.2	12.1	3950	2940	4890	3410	97400	4840	1.55	6.60

~ Check availability in S275.
(1) For local buckling calculation d′= D – 3t and b = B – 3t.

COLD-FORMED
CIRCULAR HOLLOW SECTIONS

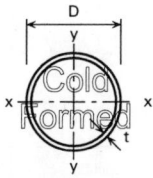

DIMENSIONS AND PROPERTIES

Section Designation		Mass per Metre	Area of Section	Ratio for Local Buckling	Second Moment of Area	Radius of Gyration	Elastic Modulus	Plastic Modulus	Torsional Constants		Surface Area	
Outside Diameter	Thickness										Per Metre	Per Tonne
D mm	t mm	kg/m	A cm²	D/t	I cm	r cm	Z cm³	S cm³	J cm⁴	C cm³	m²	m²
26.9	2.0 ‡	1.23	1.56	13.5	1.22	0.883	0.907	1.24	2.44	1.81	0.0845	68.7
	2.5 ‡	1.50	1.92	10.8	1.44	0.867	1.07	1.49	2.88	2.14	0.0845	56.3
	3.0 ‡	1.77	2.25	8.97	1.63	0.852	1.21	1.72	3.27	2.43	0.0845	47.7
33.7	2.0 ‡	1.56	1.99	16.9	2.51	1.12	1.49	2.01	5.02	2.98	0.106	67.9
	2.5 ‡	1.92	2.45	13.5	3.00	1.11	1.78	2.44	6.00	3.56	0.106	55.2
	3.0 ‡	2.27	2.89	11.2	3.44	1.09	2.04	2.84	6.88	4.08	0.106	46.7
	4.0 ‡	2.93	3.73	8.43	4.19	1.06	2.49	3.55	8.38	4.97	0.106	36.2
	4.5 ‡	3.24	4.13	7.49	4.50	1.04	2.67	3.87	9.01	5.35	0.106	32.7
42.4	2.5 ‡	2.46	3.13	17.0	6.26	1.41	2.95	3.99	12.5	5.91	0.133	54.1
	3.0 ‡	2.91	3.71	14.1	7.25	1.40	3.42	4.67	14.5	6.84	0.133	45.7
	3.5 ‡	3.36	4.28	12.1	8.16	1.38	3.85	5.31	16.3	7.69	0.133	39.6
	4.0 ‡	3.79	4.83	10.6	8.99	1.36	4.24	5.92	18.0	8.48	0.133	35.1
48.3	2.5 ‡	2.82	3.60	19.3	9.46	1.62	3.92	5.25	18.9	7.83	0.152	53.9
	3.0 ‡	3.35	4.27	16.1	11.0	1.61	4.55	6.17	22.0	9.11	0.152	45.4
	3.5 ‡	3.87	4.93	13.8	12.4	1.59	5.15	7.04	24.9	10.3	0.152	39.3
	4.0 ‡	4.37	5.57	12.1	13.8	1.57	5.70	7.87	27.5	11.4	0.152	34.8
	5.0 ‡	5.34	6.80	9.66	16.2	1.54	6.69	9.42	32.3	13.4	0.152	28.5
60.3	2.5 ‡	3.56	4.54	24.1	19.0	2.05	6.30	8.36	38.0	12.6	0.189	53.1
	3.0 ‡	4.24	5.40	20.1	22.2	2.03	7.37	9.86	44.4	14.7	0.189	44.6
	3.5 ‡	4.90	6.25	17.2	25.3	2.01	8.39	11.3	50.6	16.8	0.189	38.6
	4.0 ‡	5.55	7.07	15.1	28.2	2.00	9.34	12.7	56.3	18.7	0.189	34.1
	5.0 ‡	6.82	8.69	12.1	33.5	1.96	11.1	15.3	67.0	22.2	0.189	27.7
76.1	2.5 ‡	4.54	5.78	30.4	39.2	2.60	10.3	13.5	78.4	20.6	0.239	52.6
	3.0 ‡	5.41	6.89	25.4	46.1	2.59	12.1	16.0	92.2	24.2	0.239	44.2
	3.5 ‡	6.27	7.98	21.7	52.7	2.57	13.9	18.5	105	27.7	0.239	38.1
	4.0 ‡	7.11	9.06	19.0	59.1	2.55	15.5	20.8	118	31.0	0.239	33.6
	5.0 ‡	8.77	11.2	15.2	70.9	2.52	18.6	25.3	142	37.3	0.239	27.3
88.9	3.0 ‡	6.36	8.10	29.6	74.8	3.04	16.8	22.1	150	33.6	0.279	43.9
	3.5 ‡	7.37	9.39	25.4	85.7	3.02	19.3	25.5	171	38.6	0.279	37.9
	4.0 ‡	8.38	10.7	22.2	96.3	3.00	21.7	28.9	193	43.3	0.279	33.1
	5.0 ‡	10.3	13.2	17.8	116	2.97	26.2	35.2	233	52.4	0.279	27.1
114.3	3.0 ‡	8.23	10.5	38.1	163	3.94	28.4	37.2	325	56.9	0.359	43.6
	3.5 ‡	9.56	12.2	32.7	187	3.92	32.7	43.0	374	65.5	0.359	37.6
	4.0 ‡	10.9	13.9	28.6	211	3.90	36.9	48.7	422	73.9	0.359	32.9
	5.0 ‡	13.5	17.2	22.9	257	3.87	45.0	59.8	514	89.9	0.359	26.6
	6.0 ‡	16.0	20.4	19.1	300	3.83	52.5	70.4	600	105	0.359	22.4
139.7	4.0 ‡	13.4	17.1	34.9	393	4.80	56.2	73.7	786	112	0.439	32.8
	5.0 ‡	16.6	21.2	27.9	481	4.77	68.8	90.8	961	138	0.439	26.4
	6.0 ‡	19.8	25.2	23.3	564	4.73	80.8	107	1130	162	0.439	22.2
	8.0 ‡	26.0	33.1	17.5	720	4.66	103	139	1440	206	0.439	16.9
	10.0 ‡	32.0	40.7	14.0	862	4.60	123	169	1720	247	0.439	13.7
	12.5 ‡	39.2	50.0	11.2	1020	4.52	146	203	2040	292	0.439	11.2

‡ Grade S275 not available from some leading producers. Check availability.

COLD-FORMED
CIRCULAR HOLLOW SECTIONS

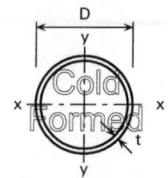

DIMENSIONS AND PROPERTIES

Section Designation		Mass per Metre	Area of Section	Ratio for Local Buckling	Second Moment of Area	Radius of Gyration	Elastic Modulus	Plastic Modulus	Torsional Constants		Surface Area	
Outside Diameter	Thickness										Per Metre	Per Tonne
D mm	t mm	kg/m	A cm²	D/t	I cm⁴	r cm	Z cm³	S cm³	J cm⁴	C cm³	m²	m²
168.3	4.0 ‡	16.2	20.6	42.1	697	5.81	82.8	108	1390	166	0.529	32.7
	5.0 ‡	20.1	25.7	33.7	856	5.78	102	133	1710	203	0.529	26.3
	6.0 ‡	24.0	30.6	28.1	1010	5.74	120	158	2020	240	0.529	22.0
	8.0 ‡	31.6	40.3	21.0	1300	5.67	154	206	2600	308	0.529	16.7
	10.0 ‡	39.0	49.7	16.8	1560	5.61	186	251	3130	372	0.529	13.6
	12.5 ‡	48.0	61.2	13.5	1870	5.53	222	304	3740	444	0.529	11.0
193.7	4.0 ‡	18.7	23.8	48.4	1070	6.71	111	144	2150	222	0.609	32.6
	4.5 ‡	21.0	26.7	43.0	1200	6.69	124	161	2400	247	0.609	29.0
	5.0 ‡	23.3	29.6	38.7	1320	6.67	136	178	2640	273	0.609	26.1
	6.0 ‡	27.8	35.4	32.3	1560	6.64	161	211	3120	322	0.609	21.9
	8.0 ‡	36.6	46.7	24.2	2020	6.57	208	276	4030	416	0.609	16.6
	10.0 ‡	45.3	57.7	19.4	2440	6.50	252	338	4880	504	0.609	13.4
	12.5 ‡	55.9	71.2	15.5	2930	6.42	303	411	5870	606	0.609	10.9
219.1	4.0 ‡	21.2	27.0	54.8	1560	7.61	143	185	3130	286	0.688	32.5
	4.5 ‡	23.8	30.3	48.7	1750	7.59	159	207	3490	319	0.688	28.9
	5.0 ‡	26.4	33.6	43.8	1930	7.57	176	229	3860	352	0.688	26.1
	6.0 ‡	31.5	40.2	36.5	2280	7.54	208	273	4560	417	0.688	21.8
	8.0 ‡	41.6	53.1	27.4	2960	7.47	270	357	5920	540	0.688	16.5
	10.0 ‡	51.6	65.7	21.9	3600	7.40	328	438	7200	657	0.688	13.3
	12.0 ‡	61.3	78.1	18.3	4200	7.33	383	515	8400	767	0.688	11.2
	12.5 ‡	63.7	81.1	17.5	4350	7.32	397	534	8690	793	0.688	10.8
	16.0 ‡	80.1	102	13.7	5300	7.20	483	661	10600	967	0.688	8.59
244.5	4.5 ‡	26.6	33.9	54.3	2440	8.49	200	259	4890	400	0.768	28.9
	5.0 ‡	29.5	37.6	48.9	2700	8.47	221	287	5400	441	0.768	26.0
	6.0 ‡	35.3	45.0	40.8	3200	8.43	262	341	6400	523	0.768	21.8
	8.0 ‡	46.7	59.4	30.6	4160	8.37	340	448	8320	681	0.768	16.4
	10.0 ‡	57.8	73.7	24.5	5070	8.30	415	550	10200	830	0.768	13.3
	12.0 ‡	68.8	87.7	20.4	5940	8.23	486	649	11900	972	0.768	11.2
	12.5 ‡	71.5	91.1	19.6	6150	8.21	503	673	12300	1010	0.768	10.7
	16.0 ‡	90.2	115	15.3	7530	8.10	616	837	15100	1230	0.768	8.51
273.0	4.0 ‡	26.5	33.8	68.3	3060	9.51	224	289	6120	448	0.858	32.4
	4.5 ‡	29.8	38.0	60.7	3420	9.49	251	324	6840	501	0.858	28.8
	5.0 ‡	33.0	42.1	54.6	3780	9.48	277	359	7560	554	0.858	26.0
	6.0 ‡	39.5	50.3	45.5	4490	9.44	329	428	8970	657	0.858	21.7
	8.0 ‡	52.3	66.6	34.1	5850	9.37	429	562	11700	857	0.858	16.4
	10.0 ‡	64.9	82.6	27.3	7150	9.31	524	692	14300	1050	0.858	13.2
	12.0 ‡	77.2	98.4	22.8	8400	9.24	615	818	16800	1230	0.858	11.1
	12.5 ‡	80.3	102	21.8	8700	9.22	637	849	17400	1270	0.858	10.7
	16.0 ‡	101	129	17.1	10700	9.10	784	1060	21400	1570	0.858	8.50

‡ Grade S275 not available from some leading producers. Check availability.

COLD-FORMED
CIRCULAR HOLLOW SECTIONS

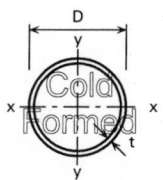

DIMENSIONS AND PROPERTIES

Section Designation		Mass per Metre	Area of Section	Ratio for Local Buckling	Second Moment of Area	Radius of Gyration	Elastic Modulus	Plastic Modulus	Torsional Constants		Surface Area	
Outside Diameter D	Thickness t		A	D/t	I	r	Z	S	J	C	Per Metre	Per Tonne
mm	mm	kg/m	cm²		cm⁴	cm	cm³	cm³	cm⁴	cm³	m²	m²
323.9	5.0 ‡	39.3	50.1	64.8	6370	11.3	393	509	12700	787	1.02	26.0
	6.0 ‡	47.0	59.9	54.0	7570	11.2	468	606	15100	935	1.02	21.7
	8.0 ‡	62.3	79.4	40.5	9910	11.2	612	799	19800	1220	1.02	16.4
	10.0 ‡	77.4	98.6	32.4	12200	11.1	751	986	24300	1500	1.02	13.2
	12.0 ‡	92.3	118	27.0	14300	11.0	884	1170	28600	1770	1.02	11.1
	12.5 ‡	96.0	122	25.9	14800	11.0	917	1210	29700	1830	1.02	10.6
	16.0 ‡	121	155	20.2	18400	10.9	1140	1520	36800	2270	1.02	8.43
355.6	5.0 ‡	43.2	55.1	71.1	8460	12.4	476	615	16900	952	1.12	25.9
	6.0 ‡	51.7	65.9	59.3	10100	12.4	566	733	20100	1130	1.12	21.7
	8.0 ‡	68.6	87.4	44.5	13200	12.3	742	967	26400	1490	1.12	16.3
	10.0 ‡	85.2	109	35.6	16200	12.2	912	1200	32400	1830	1.12	13.1
	12.0 ‡	102	130	29.6	19100	12.2	1080	1420	38300	2150	1.12	11.0
	12.5 ‡	106	135	28.4	19900	12.1	1120	1470	39700	2230	1.12	10.6
	16.0 ‡	134	171	22.2	24700	12.0	1390	1850	49300	2770	1.12	8.36
406.4	6.0 ‡	59.2	75.5	67.7	15100	14.2	745	962	30300	1490	1.28	21.6
	8.0 ‡	78.6	100	50.8	19900	14.1	978	1270	39700	1960	1.28	16.3
	10.0 ‡	97.8	125	40.6	24500	14.0	1210	1570	49000	2410	1.28	13.1
	12.0 ‡	117	149	33.9	28900	14.0	1420	1870	57900	2850	1.28	10.9
	12.5 ‡	121	155	32.5	30000	13.9	1480	1940	60100	2960	1.28	10.6
	16.0 ‡	154	196	25.4	37400	13.8	1840	2440	74900	3690	1.28	8.31
457.0	8.0 ‡	88.6	113	57.1	28400	15.9	1250	1610	56900	2490	1.44	16.3
	10.0 ‡	110	140	45.7	35100	15.8	1540	2000	70200	3070	1.44	13.1
	12.0 ‡	132	168	38.1	41600	15.7	1820	2380	83100	3640	1.44	10.9
	12.5 ‡	137	175	36.6	43100	15.7	1890	2470	86300	3780	1.44	10.5
	16.0 ‡	174	222	28.6	54000	15.6	2360	3110	108000	4720	1.44	8.28
508.0	8.0 ‡	98.6	126	63.5	39300	17.7	1550	2000	78600	3090	1.60	16.2
	10.0 ‡	123	156	50.8	48500	17.6	1910	2480	97000	3820	1.60	13.0
	12.0 ‡	147	187	42.3	57500	17.5	2270	2950	115000	4530	1.60	10.9
	12.5 ‡	153	195	40.6	59800	17.5	2350	3070	120000	4710	1.60	10.5
	16.0 ‡	194	247	31.8	74900	17.4	2950	3870	150000	5900	1.60	8.25

‡ Grade S275 not available from some leading producers. Check availability.

COLD-FORMED
SQUARE HOLLOW SECTIONS

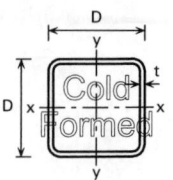

DIMENSIONS AND PROPERTIES

Section Designation		Mass per Metre	Area of Section	Ratio for Local Buckling	Second Moment of Area	Radius of Gyration	Elastic Modulus	Plastic Modulus	Torsional Constants		Surface Area	
Size	Thickness										Per Metre	Per Tonne
D × D mm	t mm	kg/m	A cm²	d/t[1]	I cm⁴	r cm	Z cm³	S cm³	J cm⁴	C cm³	m²	m²
25 × 25	2.0 ‡	1.36	1.74	7.50	1.48	0.924	1.19	1.47	2.53	1.80	0.0931	68.5
	2.5 ‡	1.64	2.09	5.00	1.69	0.899	1.35	1.71	2.97	2.07	0.0914	55.7
	3.0 ‡	1.89	2.41	3.33	1.84	0.874	1.47	1.91	3.33	2.27	0.0897	47.5
30 × 30	2.0 ‡	1.68	2.14	10.0	2.72	1.13	1.81	2.21	4.54	2.75	0.113	67.3
	2.5 ‡	2.03	2.59	7.00	3.16	1.10	2.10	2.61	5.40	3.20	0.111	54.7
	3.0 ‡	2.36	3.01	5.00	3.50	1.08	2.34	2.96	6.15	3.58	0.110	46.6
40 × 40	2.0 ‡	2.31	2.94	15.0	6.94	1.54	3.47	4.13	11.3	5.23	0.153	66.2
	2.5 ‡	2.82	3.59	11.0	8.22	1.51	4.11	4.97	13.6	6.21	0.151	53.5
	3.0 ‡	3.30	4.21	8.33	9.32	1.49	4.66	5.72	15.8	7.07	0.150	45.5
	4.0 ‡	4.20	5.35	5.00	11.1	1.44	5.54	7.01	19.4	8.48	0.146	34.8
50 × 50	2.0 ‡	2.93	3.74	20.0	14.1	1.95	5.66	6.66	22.6	8.51	0.193	65.9
	2.5 ‡	3.60	4.59	15.0	16.9	1.92	6.78	8.07	27.5	10.2	0.191	53.1
	3.0 ‡	4.25	5.41	11.7	19.5	1.90	7.79	9.39	32.1	11.8	0.190	44.7
	4.0 ‡	5.45	6.95	7.50	23.7	1.85	9.49	11.7	40.4	14.4	0.186	34.1
	5.0 ‡	6.56	8.36	5.00	27.0	1.80	10.8	13.7	47.5	16.6	0.183	27.9
60 × 60	3.0 ‡	5.19	6.61	15.0	35.1	2.31	11.7	14.0	57.1	17.7	0.230	44.3
	4.0 ‡	6.71	8.55	10.0	43.6	2.26	14.5	17.6	72.6	22.0	0.226	33.7
	5.0 ‡	8.13	10.4	7.00	50.5	2.21	16.8	20.9	86.4	25.6	0.223	27.4
70 × 70	2.5 ‡	5.17	6.59	23.0	49.4	2.74	14.1	16.5	78.5	21.2	0.271	52.4
	3.0 ‡	6.13	7.81	18.3	57.5	2.71	16.4	19.4	92.4	24.7	0.270	44.0
	3.5 ‡	7.06	8.99	15.0	65.1	2.69	18.6	22.2	106	28.0	0.268	38.0
	4.0 ‡	7.97	10.1	12.5	72.1	2.67	20.6	24.8	119	31.1	0.266	33.4
	5.0 ‡	9.70	12.4	9.00	84.6	2.62	24.2	29.6	142	36.7	0.263	27.1
80 × 80	3.0 ‡	7.07	9.01	21.7	87.8	3.12	22.0	25.8	140	33.0	0.310	43.8
	3.5 ‡	8.16	10.4	17.9	99.8	3.10	25.0	29.5	161	37.6	0.308	37.7
	4.0 ‡	9.22	11.7	15.0	111	3.07	27.8	33.1	180	41.8	0.306	33.2
	5.0 ‡	11.3	14.4	11.0	131	3.03	32.9	39.7	218	49.7	0.303	26.8
	6.0 ‡	13.2	16.8	8.33	149	2.98	37.3	45.8	252	56.6	0.299	22.7
90 × 90	3.0 ‡	8.01	10.2	25.0	127	3.53	28.3	33.0	201	42.5	0.350	43.7
	3.5 ‡	9.26	11.8	20.7	145	3.51	32.2	37.9	232	48.5	0.348	37.6
	4.0 ‡	10.5	13.3	17.5	162	3.48	36.0	42.6	261	54.2	0.346	33.0
	5.0 ‡	12.8	16.4	13.0	193	3.43	42.9	51.4	316	64.7	0.343	26.8
	6.0 ‡	15.1	19.2	10.0	220	3.39	49.0	59.5	368	74.2	0.339	22.5
100 × 100	3.0 ‡	8.96	11.4	28.3	177	3.94	35.4	41.2	279	53.2	0.390	43.5
	4.0 ‡	11.7	14.9	20.0	226	3.89	45.3	53.3	362	68.1	0.386	33.0
	5.0 ‡	14.4	18.4	15.0	271	3.84	54.2	64.6	441	81.7	0.383	26.6
	6.0 ‡	17.0	21.6	11.7	311	3.79	62.3	75.1	514	94.1	0.379	22.3
	8.0 ‡	21.4	27.2	7.50	366	3.67	73.2	91.1	645	114	0.366	17.1
120 × 120	4.0 ‡	14.2	18.1	25.0	402	4.71	67.0	78.3	637	101	0.466	32.8
	5.0 ‡	17.5	22.4	19.0	485	4.66	80.9	95.4	778	122	0.463	26.5
	6.0 ‡	20.7	26.4	15.0	562	4.61	93.7	112	913	141	0.459	22.2
	8.0 ‡	26.4	33.6	10.0	677	4.49	113	138	1160	175	0.446	16.9
	10.0 ‡	31.8	40.6	7.00	777	4.38	129	162	1380	203	0.437	13.7

‡ Grade S275 not available from some leading producers. Check availability.
[1] For local buckling calculation d = D − 5t.

COLD-FORMED
SQUARE HOLLOW SECTIONS

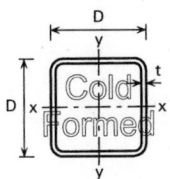

DIMENSIONS AND PROPERTIES

Section Designation		Mass per Metre	Area of Section	Ratio for Local Buckling	Second Moment of Area	Radius of Gyration	Elastic Modulus	Plastic Modulus	Torsional Constants		Surface Area	
Size	Thickness										Per Metre	Per Tonne
$D \times D$ mm	t mm	kg/m	A cm²	d/t⁽¹⁾	I cm⁴	r cm	Z cm³	S cm³	J cm⁴	C cm³	m²	m²
140 × 140	4.0 ‡	16.8	21.3	30.0	652	5.52	93.1	108	1020	140	0.546	32.5
	5.0 ‡	20.7	26.4	23.0	791	5.48	113	132	1260	170	0.543	26.2
	6.0 ‡	24.5	31.2	18.3	920	5.43	131	155	1480	198	0.539	22.0
	8.0 ‡	31.4	40.0	12.5	1130	5.30	161	194	1900	248	0.526	16.8
	10.0 ‡	38.1	48.6	9.00	1310	5.20	187	230	2270	291	0.517	13.6
150 × 150	4.0 ‡	18.0	22.9	32.5	808	5.93	108	125	1260	162	0.586	32.6
	5.0 ‡	22.3	28.4	25.0	982	5.89	131	153	1550	197	0.583	26.1
	6.0 ‡	26.4	33.6	20.0	1150	5.84	153	180	1830	230	0.579	21.9
	8.0 ‡	33.9	43.2	13.8	1410	5.71	188	226	2360	289	0.566	16.7
	10.0 ‡	41.3	52.6	10.0	1650	5.61	220	269	2840	341	0.557	13.5
160 × 160	4.0 ‡	19.3	24.5	35.0	987	6.34	123	143	1540	185	0.626	32.4
	5.0 ‡	23.8	30.4	27.0	1200	6.29	150	175	1900	226	0.623	26.2
	6.0 ‡	28.3	36.0	21.7	1410	6.25	176	206	2240	264	0.619	21.9
	8.0 ‡	36.5	46.4	15.0	1740	6.12	218	260	2900	334	0.606	16.6
	10.0 ‡	44.4	56.6	11.0	2050	6.02	256	311	3490	395	0.597	13.4
180 × 180	5.0 ‡	27.0	34.4	31.0	1740	7.11	193	224	2720	290	0.703	26.0
	6.0 ‡	32.1	40.8	25.0	2040	7.06	226	264	3220	340	0.699	21.8
	8.0 ‡	41.5	52.8	17.5	2550	6.94	283	336	4190	432	0.686	16.5
	10.0 ‡	50.7	64.6	13.0	3020	6.84	335	404	5070	515	0.677	13.4
	12.0 ‡	58.5	74.5	10.0	3320	6.68	369	454	5870	584	0.658	11.2
	12.5 ‡	60.5	77.0	9.40	3410	6.65	378	467	6050	600	0.656	10.8
200 × 200	5.0 ‡	30.1	38.4	35.0	2410	7.93	241	279	3760	362	0.783	26.0
	6.0 ‡	35.8	45.6	28.3	2830	7.88	283	330	4460	426	0.779	21.8
	8.0 ‡	46.5	59.2	20.0	3570	7.76	357	421	5820	544	0.766	16.5
	10.0 ‡	57.0	72.6	15.0	4250	7.65	425	508	7070	651	0.757	13.3
	12.0 ‡	66.0	84.1	11.7	4730	7.50	473	576	8230	743	0.738	11.2
	12.5 ‡	68.3	87.0	11.0	4860	7.47	486	594	8500	765	0.736	10.8
250 × 250	6.0 ‡	45.2	57.6	36.7	5670	9.92	454	524	8840	681	0.979	21.7
	8.0 ‡	59.1	75.2	26.3	7230	9.80	578	676	11600	878	0.966	16.3
	10.0 ‡	72.7	92.6	20.0	8710	9.70	697	822	14200	1060	0.957	13.2
	12.0 ‡	84.8	108	15.8	9860	9.55	789	944	16700	1230	0.938	11.1
	12.5 ‡	88.0	112	15.0	10200	9.52	813	975	17300	1270	0.936	10.6
300 × 300	8.0 ‡	71.6	91.2	32.5	12800	11.8	853	991	20300	1290	1.17	16.3
	10.0 ‡	88.4	113	25.0	15500	11.7	1030	1210	25000	1570	1.16	13.1
	12.0 ‡	104	132	20.0	17800	11.6	1180	1400	29500	1830	1.14	11.0
	12.5 ‡	108	137	19.0	18300	11.6	1220	1450	30600	1890	1.14	10.6

‡ Grade S275 not available from some leading producers. Check availability.
⁽¹⁾ For local buckling calculation d = D – 5t.

COLD-FORMED RECTANGULAR HOLLOW SECTIONS

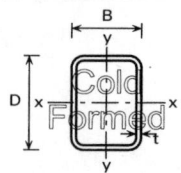

DIMENSIONS AND PROPERTIES

Section Designation		Mass per Metre	Area of Section	Ratios for Local Buckling		Second Moment of Area		Radius of Gyration		Elastic Modulus		Plastic Modulus		Torsional Constants		Surface Area	
Size D × B mm	Thickness t mm	kg/m	A cm²	$d/t^{(1)}$	$b/t^{(1)}$	Axis x-x cm⁴	Axis y-y cm⁴	Axis x-x cm	Axis y-y cm	Axis x-x cm³	Axis y-y cm³	Axis x-x cm³	Axis y-y cm³	J cm⁴	C cm³	Per Metre m²	Per Tonne m²
50 × 25	2.0 ‡	2.15	2.74	20.0	7.50	8.38	2.81	1.75	1.01	3.35	2.25	4.26	2.62	7.06	3.92	0.143	66.5
	2.5 ‡	2.62	3.34	15.0	5.00	9.89	3.28	1.72	0.991	3.95	2.62	5.11	3.12	8.43	4.60	0.141	53.8
	3.0 ‡	3.07	3.91	11.7	3.33	11.2	3.67	1.69	0.969	4.47	2.93	5.86	3.56	9.64	5.18	0.140	45.6
50 × 30	2.0 ‡	2.31	2.94	20.0	10.0	9.54	4.29	1.80	1.21	3.81	2.86	4.74	3.33	9.77	4.84	0.153	66.2
	2.5 ‡	2.82	3.59	15.0	7.00	11.3	5.05	1.77	1.19	4.52	3.37	5.70	3.98	11.7	5.72	0.151	53.5
	3.0 ‡	3.30	4.21	11.7	5.00	12.8	5.70	1.75	1.16	5.13	3.80	6.57	4.58	13.5	6.49	0.150	45.5
	4.0 ‡	4.20	5.35	7.50	2.50	15.3	6.69	1.69	1.12	6.10	4.46	8.05	5.58	16.5	7.71	0.146	34.8
60 × 30	3.0 ‡	3.77	4.81	15.0	5.00	20.5	6.80	2.06	1.19	6.83	4.53	8.82	5.39	17.5	7.95	0.170	45.1
	4.0 ‡	4.83	6.15	10.0	2.50	24.7	8.06	2.00	1.14	8.23	5.37	10.9	6.62	21.5	9.52	0.166	34.4
60 × 40	2.5 ‡	3.60	4.59	19.0	11.0	22.1	11.7	2.19	1.60	7.36	5.87	9.06	6.84	25.1	9.72	0.191	53.1
	3.0 ‡	4.25	5.41	15.0	8.33	25.4	13.4	2.17	1.58	8.46	6.72	10.5	7.94	29.3	11.2	0.190	44.7
	4.0 ‡	5.45	6.95	10.0	5.00	31.0	16.3	2.11	1.53	10.3	8.14	13.2	9.89	36.7	13.7	0.186	34.1
	5.0 ‡	6.56	8.36	7.00	3.00	35.3	18.4	2.06	1.48	11.8	9.21	15.4	11.5	42.8	15.6	0.183	27.9
70 × 40	3.0 ‡	4.72	6.01	18.3	8.33	37.3	15.5	2.49	1.61	10.7	7.75	13.4	9.05	36.5	13.2	0.210	44.5
	4.0 ‡	6.08	7.75	12.5	5.00	46.0	18.9	2.44	1.56	13.1	9.44	16.8	11.3	45.8	16.2	0.206	33.9
70 × 50	3.0 ‡	5.19	6.61	18.3	11.7	44.1	26.1	2.58	1.99	12.6	10.4	15.4	12.2	53.6	17.1	0.230	44.3
	4.0 ‡	6.71	8.55	12.5	7.50	54.7	32.2	2.53	1.94	15.6	12.9	19.5	15.4	68.1	21.2	0.226	33.7
80 × 40	3.0 ‡	5.19	6.61	21.7	8.33	52.3	17.6	2.81	1.63	13.1	8.78	16.5	10.2	43.9	15.3	0.230	44.3
	4.0 ‡	6.71	8.55	15.0	5.00	64.8	21.5	2.75	1.59	16.2	10.7	20.9	12.8	55.2	18.8	0.226	33.7
	5.0 ‡	8.13	10.4	11.0	3.00	75.1	24.6	2.69	1.54	18.8	12.3	24.7	15.0	65.0	21.7	0.223	27.4
80 × 50	3.0 ‡	5.66	7.21	21.7	11.7	61.1	29.4	2.91	2.02	15.3	11.8	18.8	13.6	65.0	19.7	0.250	44.2
	4.0 ‡	7.34	9.35	15.0	7.50	76.4	36.5	2.86	1.98	19.1	14.6	24.0	17.2	82.7	24.6	0.246	33.5
	5.0 ‡	8.91	11.4	11.0	5.00	89.2	42.3	2.80	1.93	22.3	16.9	28.5	20.5	98.4	28.7	0.243	27.3
80 × 60	3.0 ‡	6.13	7.81	21.7	15.0	70.0	44.9	3.00	2.40	17.5	15.0	21.2	17.4	88.3	24.1	0.270	44.0
	4.0 ‡	7.97	10.1	15.0	10.0	87.9	56.1	2.94	2.35	22.0	18.7	27.0	22.1	113	30.3	0.266	33.4
	5.0 ‡	9.70	12.4	11.0	7.00	103	65.7	2.89	2.31	25.8	21.9	32.2	26.4	136	35.7	0.263	27.1
90 × 50	3.0 ‡	6.13	7.81	25.0	11.7	81.9	32.7	3.24	2.05	18.2	13.1	22.6	15.0	76.7	22.4	0.270	44.0
	3.5 ‡	7.06	8.99	20.7	9.29	92.7	36.9	3.21	2.03	20.6	14.8	25.8	17.1	87.5	25.3	0.268	38.0
	4.0 ‡	7.97	10.1	17.5	7.50	103	40.7	3.18	2.00	22.8	16.3	28.8	19.1	97.7	28.0	0.266	33.4
	5.0 ‡	9.70	12.4	13.0	5.00	121	47.4	3.12	1.96	26.8	18.9	34.4	22.7	116	32.7	0.263	27.1
100 × 40	3.0 ‡	6.13	7.81	28.3	8.33	92.3	21.7	3.44	1.67	18.5	10.8	23.7	12.4	59.0	19.4	0.270	44.0
	4.0 ‡	7.97	10.1	20.0	5.00	116	26.7	3.38	1.62	23.1	13.3	30.3	15.7	74.5	24.0	0.266	33.4
	5.0 ‡	9.70	12.4	15.0	3.00	136	30.8	3.31	1.58	27.1	15.4	36.1	18.5	87.9	27.9	0.263	27.1
100 × 50	3.0 ‡	6.60	8.41	28.3	11.7	106	36.1	3.56	2.07	21.3	14.4	26.7	16.4	88.6	25.0	0.290	43.9
	4.0 ‡	8.59	10.9	20.0	7.50	134	44.9	3.50	2.03	26.8	18.0	34.1	20.9	113	31.3	0.286	33.3
	5.0 ‡	10.5	13.4	15.0	5.00	158	52.5	3.44	1.98	31.6	21.0	40.8	25.0	135	36.8	0.283	27.0
	6.0 ‡	12.3	15.6	11.7	3.33	179	58.7	3.38	1.94	35.8	23.5	46.9	28.5	154	41.4	0.279	22.7
100 × 60	3.0 ‡	7.07	9.01	28.3	15.0	121	54.6	3.66	2.46	24.1	18.2	29.6	20.8	122	30.6	0.310	43.8
	3.5 ‡	8.16	10.4	23.6	12.1	137	61.9	3.63	2.44	27.4	20.6	33.8	23.8	139	34.8	0.308	37.7
	4.0 ‡	9.22	11.7	20.0	10.0	153	68.7	3.60	2.42	30.5	22.9	37.9	26.6	156	38.7	0.306	33.2
	5.0 ‡	11.3	14.4	15.0	7.00	181	80.8	3.55	2.37	36.2	26.9	45.6	31.9	188	45.8	0.303	26.8
	6.0 ‡	13.2	16.8	11.7	5.00	205	91.2	3.49	2.33	41.1	30.4	52.5	36.6	216	51.9	0.299	22.7

‡ Grade S275 not available from some leading producers. Check availability.
$^{(1)}$ For local buckling calculation d = D − 5t and b = B − 5t.

COLD-FORMED
RECTANGULAR HOLLOW SECTIONS

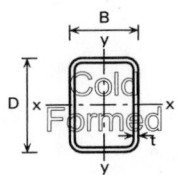

DIMENSIONS AND PROPERTIES

Section Designation		Mass per Metre	Area of Section	Ratios for Local Buckling		Second Moment of Area		Radius of Gyration		Elastic Modulus		Plastic Modulus		Torsional Constants		Surface Area	
Size	Thickness					Axis x-x	Axis y-y	Axis x-x	Axis y-y	Axis x-x	Axis y-y	Axis x-x	Axis y-y			Per Metre	Per Tonne
$D \times B$ mm	t mm	kg/m	A cm²	$d/t^{(1)}$	$b/t^{(1)}$	cm⁴	cm⁴	cm	cm	cm³	cm³	cm³	cm³	J cm⁴	C cm³	m²	m²
100 × 80	3.0 ‡	8.01	10.2	28.3	21.7	149	106	3.82	3.22	29.8	26.4	35.4	30.4	196	41.9	0.350	43.7
	4.0 ‡	10.5	13.3	20.0	15.0	189	134	3.77	3.17	37.9	33.5	45.6	39.2	254	53.4	0.346	33.0
	5.0 ‡	12.8	16.4	15.0	11.0	226	160	3.72	3.12	45.2	39.9	55.1	47.2	308	63.7	0.343	26.8
120 × 40	3.0 ‡	7.07	9.01	35.0	8.33	148	25.8	4.05	1.69	24.7	12.9	32.2	14.6	74.6	23.5	0.310	43.8
	4.0 ‡	9.22	11.7	25.0	5.00	187	31.9	3.99	1.65	31.1	15.9	41.2	18.5	94.2	29.2	0.306	33.2
	5.0 ‡	11.3	14.4	19.0	3.00	221	36.9	3.92	1.60	36.8	18.5	49.4	22.0	111	34.1	0.303	26.8
120 × 60	3.0 ‡	8.01	10.2	35.0	15.0	189	64.4	4.30	2.51	31.5	21.5	39.2	24.2	156	37.1	0.350	43.7
	3.5 ‡	9.26	11.8	29.3	12.1	216	73.1	4.28	2.49	35.9	24.4	44.9	27.7	179	42.2	0.348	37.6
	4.0 ‡	10.5	13.3	25.0	10.0	241	81.2	4.25	2.47	40.1	27.1	50.5	31.1	201	47.0	0.346	33.0
	5.0 ‡	12.8	16.4	19.0	7.00	287	96.0	4.19	2.42	47.8	32.0	60.9	37.4	242	55.8	0.343	26.8
	6.0 ‡	15.1	19.2	15.0	5.00	328	109	4.13	2.38	54.7	36.3	70.6	43.1	280	63.6	0.339	22.5
120 × 80	3.0 ‡	8.96	11.4	35.0	21.7	230	123	4.49	3.29	38.4	30.9	46.2	35.0	255	50.8	0.390	43.5
	4.0 ‡	11.7	14.9	25.0	15.0	295	157	4.44	3.24	49.1	39.3	59.8	45.2	331	64.9	0.386	33.0
	5.0 ‡	14.4	18.4	19.0	11.0	353	188	4.39	3.20	58.9	46.9	72.4	54.7	402	77.8	0.383	26.6
	6.0 ‡	17.0	21.6	15.0	8.33	406	215	4.33	3.15	67.7	53.8	84.3	63.5	469	89.4	0.379	22.3
	8.0 ‡	21.4	27.2	10.0	5.00	476	252	4.18	3.04	79.3	62.9	102	76.9	584	108	0.366	17.1
140 × 80	3.0 ‡	9.90	12.6	41.7	21.7	334	141	5.15	3.35	47.8	35.3	58.2	39.6	317	59.7	0.430	43.4
	4.0 ‡	13.0	16.5	30.0	15.0	430	180	5.10	3.30	61.4	45.1	75.5	51.3	412	76.5	0.426	32.8
	5.0 ‡	16.0	20.4	23.0	11.0	517	216	5.04	3.26	73.9	54.0	91.8	62.2	501	91.8	0.423	26.4
	6.0 ‡	18.9	24.0	18.3	8.33	597	248	4.98	3.21	85.3	62.0	107	72.4	584	106	0.419	22.2
	8.0 ‡	23.9	30.4	12.5	5.00	708	293	4.82	3.10	101	73.3	131	88.4	731	129	0.406	17.0
	10.0 ‡	28.7	36.6	9.00	3.00	804	330	4.69	3.01	115	82.6	152	103	851	147	0.397	13.8
150 × 100	4.0 ‡	14.9	18.9	32.5	20.0	595	319	5.60	4.10	79.3	63.7	95.7	72.5	662	105	0.486	32.6
	5.0 ‡	18.3	23.4	25.0	15.0	719	384	5.55	4.05	95.9	76.8	117	88.3	809	127	0.483	26.4
	6.0 ‡	21.7	27.6	20.0	11.7	835	444	5.50	4.01	111	88.8	137	103	948	147	0.479	22.1
	8.0 ‡	27.7	35.2	13.8	7.50	1010	536	5.35	3.90	134	107	169	128	1210	182	0.466	16.8
	10.0 ‡	33.4	42.6	10.0	5.00	1160	614	5.22	3.80	155	123	199	150	1430	211	0.457	13.7
160 × 80	5.0 ‡	17.5	22.4	27.0	11.0	722	244	5.68	3.30	90.2	61.0	113	69.7	601	106	0.463	26.5
	6.0 ‡	20.7	26.4	21.7	8.33	836	281	5.62	3.26	105	70.2	132	81.3	702	122	0.459	22.2
	8.0 ‡	26.4	33.6	15.0	5.00	1000	335	5.46	3.16	125	83.7	163	100	882	150	0.446	16.9
180 × 80	4.0 ‡	15.5	19.7	40.0	15.0	802	227	6.37	3.39	89.1	56.7	112	63.5	578	99.6	0.506	32.6
	5.0 ‡	19.1	24.4	31.0	11.0	971	272	6.31	3.34	108	68.1	137	77.2	704	120	0.503	26.3
	6.0 ‡	22.6	28.8	25.0	8.33	1130	314	6.25	3.30	125	78.5	160	90.2	823	139	0.499	22.1
	8.0 ‡	28.9	36.8	17.5	5.00	1360	377	6.08	3.20	151	94.1	198	111	1040	170	0.486	16.8
	10.0 ‡	35.0	44.6	13.0	3.00	1570	429	5.94	3.10	174	107	234	131	1210	196	0.477	13.6
180 × 100	4.0 ‡	16.8	21.3	40.0	20.0	926	374	6.59	4.18	103	74.8	126	84.0	854	127	0.546	32.5
	5.0 ‡	20.7	26.4	31.0	15.0	1120	452	6.53	4.14	125	90.4	154	103	1040	154	0.543	26.2
	6.0 ‡	24.5	31.2	25.0	11.7	1310	524	6.48	4.10	146	105	181	120	1230	179	0.539	22.0
	8.0 ‡	31.4	40.0	17.5	7.50	1600	637	6.32	3.99	178	127	226	150	1570	222	0.526	16.8
	10.0 ‡	38.1	48.6	13.0	5.00	1860	736	6.19	3.89	207	147	268	177	1860	260	0.517	13.6

‡ Grade S275 not available from some leading producers. Check availability.
$^{(1)}$ For local buckling calculation $d = D - 5t$ and $b = B - 5t$.

COLD-FORMED
RECTANGULAR HOLLOW SECTIONS

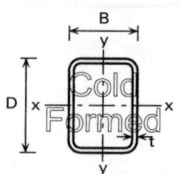

DIMENSIONS AND PROPERTIES

Section Designation		Mass per Metre	Area of Section	Ratios for Local Buckling		Second Moment of Area		Radius of Gyration		Elastic Modulus		Plastic Modulus		Torsional Constants		Surface Area	
Size	Thickness					Axis x-x	Axis y-y	Axis x-x	Axis y-y	Axis x-x	Axis y-y	Axis x-x	Axis y-y			Per Metre	Per Tonne
D × B mm	t mm	kg/m	A cm²	d/t[1]	b/t[1]	cm⁴	cm⁴	cm	cm	cm³	cm³	cm³	cm³	J cm⁴	C cm³	m²	m²
200 × 100	4.0 ‡	18.0	22.9	45.0	20.0	1200	411	7.23	4.23	120	82.2	148	91.7	985	142	0.586	32.6
	5.0 ‡	22.3	28.4	35.0	15.0	1460	497	7.17	4.19	146	99.4	181	112	1210	172	0.583	26.1
	6.0 ‡	26.4	33.6	28.3	11.7	1700	577	7.12	4.14	170	115	213	132	1420	200	0.579	21.9
	8.0 ‡	33.9	43.2	20.0	7.50	2090	705	6.95	4.04	209	141	267	165	1810	250	0.566	16.7
	10.0 ‡	41.3	52.6	15.0	5.00	2440	818	6.82	3.94	244	164	318	195	2150	292	0.557	13.5
200 × 120	4.0 ‡	19.3	24.5	45.0	25.0	1350	618	7.43	5.02	135	103	164	115	1350	172	0.626	32.4
	5.0 ‡	23.8	30.4	35.0	19.0	1650	750	7.37	4.97	165	125	201	141	1650	210	0.623	26.2
	6.0 ‡	28.3	36.0	28.3	15.0	1930	874	7.32	4.93	193	146	237	166	1950	245	0.619	21.9
	8.0 ‡	36.5	46.4	20.0	10.0	2390	1080	7.17	4.82	239	180	298	209	2510	308	0.606	16.6
	10.0 ‡	44.4	56.6	15.0	7.00	2810	1260	7.04	4.72	281	210	356	250	3010	364	0.597	13.4
200 × 150	4.0 ‡	21.2	26.9	45.0	32.5	1580	1020	7.67	6.16	158	136	187	154	1940	219	0.686	32.4
	5.0 ‡	26.2	33.4	35.0	25.0	1930	1250	7.62	6.11	193	166	230	189	2390	267	0.683	26.1
	6.0 ‡	31.1	39.6	28.3	20.0	2270	1460	7.56	6.06	227	194	271	223	2830	313	0.679	21.8
	8.0 ‡	40.2	51.2	20.0	13.8	2830	1820	7.43	5.95	283	242	344	283	3660	396	0.666	16.6
	10.0 ‡	49.1	62.6	15.0	10.0	3350	2140	7.31	5.85	335	286	413	339	4430	471	0.657	13.4
250 × 150	5.0 ‡	30.1	38.4	45.0	25.0	3300	1510	9.28	6.27	264	201	320	225	3280	337	0.783	26.0
	6.0 ‡	35.8	45.6	36.7	20.0	3890	1770	9.23	6.23	311	236	378	266	3890	396	0.779	21.8
	8.0 ‡	46.5	59.2	26.3	13.8	4890	2220	9.08	6.12	391	296	482	340	5050	504	0.766	16.5
	10.0 ‡	57.0	72.6	20.0	10.0	5830	2630	8.96	6.02	466	351	582	409	6120	602	0.757	13.3
	12.0 ‡	66.0	84.1	15.8	7.50	6460	2930	8.77	5.90	517	390	658	463	7090	684	0.738	11.2
	12.5 ‡	68.3	87.0	15.0	7.00	6630	3000	8.73	5.87	531	400	678	477	7310	704	0.736	10.8
300 × 100	6.0 ‡	35.8	45.6	45.0	11.7	4780	842	10.2	4.30	318	168	411	188	2400	306	0.779	21.8
	8.0 ‡	46.5	59.2	32.5	7.50	5980	1040	10.0	4.20	399	209	523	238	3080	385	0.766	16.5
	10.0 ‡	57.0	72.6	25.0	5.00	7110	1220	9.90	4.11	474	245	631	285	3680	455	0.757	13.3
	12.0 ‡	66.0	84.1	20.0	3.33	7810	1340	9.64	4.00	521	269	710	321	4180	508	0.738	11.2
	12.5 ‡	68.3	87.0	19.0	3.00	8010	1370	9.59	3.97	534	275	732	330	4290	521	0.736	10.8
300 × 200	6.0 ‡	45.2	57.6	45.0	28.3	7370	3960	11.3	8.29	491	396	588	446	8120	651	0.979	21.7
	8.0 ‡	59.1	75.2	32.5	20.0	9390	5040	11.2	8.19	626	504	757	574	10600	838	0.966	16.3
	10.0 ‡	72.7	92.6	25.0	15.0	11300	6060	11.1	8.09	754	606	921	698	13000	1010	0.957	13.2
	12.0 ‡	84.8	108	20.0	11.7	12800	6850	10.9	7.96	853	685	1060	801	15200	1170	0.938	11.1
	12.5 ‡	88.0	112	19.0	11.0	13200	7060	10.8	7.94	879	706	1090	828	15800	1200	0.936	10.6
400 × 200	8.0 ‡	71.6	91.2	45.0	20.0	19000	6520	14.4	8.45	949	652	1170	728	15800	1130	1.17	16.3
	10.0 ‡	88.4	113	35.0	15.0	23000	7860	14.3	8.36	1150	786	1430	888	19400	1370	1.16	13.1
	12.0 ‡	104	132	28.3	11.7	26200	8980	14.1	8.24	1310	898	1660	1030	22800	1590	1.14	11.0
	12.5 ‡	108	137	27.0	11.0	27100	9260	14.1	8.22	1360	926	1710	1060	23600	1640	1.14	10.6

‡ Grade S275 not available from some leading producers. Check availability.
[1] For local buckling calculation d = D − 5t and b = B − 5t.

ASB (ASYMMETRIC BEAMS)

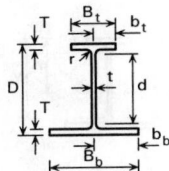

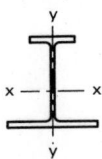

DIMENSIONS AND PROPERTIES

Section Designation	Mass per Metre	Depth of Section	Width of Flange		Thickness		Root Radius	Depth between Fillets	Ratios for Local Buckling			Second Moment of Area		Surface Area	
			Top	Bottom	Web	Flange			Flanges		Web	Axis x-x	Axis y-y	Per Metre	Per Tonne
		D	B_t	B_b	t	T	r	d	b_t/T	b_b/T	d/t				
	kg/m	mm	mm	mm	mm	mm	mm	mm				cm⁴	cm⁴	m²	m²
300 ASB 249†	249	342	203	313	40.0	40.0	27.0	208	2.54	3.91	5.20	52900	13200	1.59	6.38
300 ASB 196	196	342	183	293	20.0	40.0	27.0	208	2.29	3.66	10.4	45900	10500	1.55	7.93
300 ASB 185†	185	320	195	305	32.0	29.0	27.0	208	3.36	5.26	6.50	35700	8750	1.53	8.29
300 ASB 155	155	326	179	289	16.0	32.0	27.0	208	2.80	4.52	13.0	34500	7990	1.51	9.71
300 ASB 153†	153	310	190	300	27.0	24.0	27.0	208	3.96	6.25	7.70	28400	6840	1.50	9.81
280 ASB 136†	136	288	190	300	25.0	22.0	24.0	196	4.32	6.82	7.84	22200	6260	1.46	10.7
280 ASB 124	124	296	178	288	13.0	26.0	24.0	196	3.42	5.54	15.1	23500	6410	1.46	11.8
280 ASB 105	105	288	176	286	11.0	22.0	24.0	196	4.00	6.50	17.8	19200	5300	1.44	13.7
280 ASB 100†	100	276	184	294	19.0	16.0	24.0	196	5.75	9.19	10.3	15500	4250	1.43	14.2
280 ASB 74	73.6	272	175	285	10.0	14.0	24.0	196	6.25	10.2	19.6	12200	3330	1.40	19.1

† Sections are fire engineered with thick webs.
ASB sections are only available in S355.

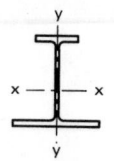

ASB (ASYMMETRIC BEAMS)

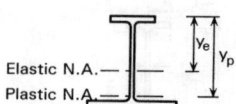

Elastic N.A.
Plastic N.A.

PROPERTIES (CONTINUED)

Section Designation	Radius of Gyration		Elastic Modulus			Neutral Axis Position		Plastic Modulus		Buckling Parameter	Torsional Index	Warping Constant	Torsional Constant	Area of Section
	Axis x-x	Axis y-y	Axis x-x Top	Axis x-x Bottom	Axis y-y	Elastic	Plastic	Axis x-x	Axis y-y					
	cm	cm	cm^3	cm^3	cm^3	y_e cm	y_p cm	cm^3	cm^3	u	x	H dm^6	J cm^4	A cm^2
300 ASB 249†	12.9	6.40	2760	3530	843	19.2	22.6	3760	1510	0.820	6.80	2.00	2000	318
300 ASB 196	13.6	6.48	2320	3180	714	19.8	28.1	3060	1230	0.840	7.86	1.50	1180	249
300 ASB 185†	12.3	6.10	1980	2540	574	18.0	21.0	2660	1030	0.820	8.56	1.20	871	235
300 ASB 155	13.2	6.35	1830	2520	553	18.9	27.3	2360	950	0.840	9.40	1.07	620	198
300 ASB 153†	12.1	5.93	1630	2090	456	17.4	20.4	2160	817	0.820	9.97	0.895	513	195
280 ASB 136†	11.3	6.00	1370	1770	417	16.3	19.2	1810	741	0.810	10.2	0.710	379	174
280 ASB 124	12.2	6.37	1360	1900	445	17.3	25.7	1730	761	0.830	10.5	0.721	332	158
280 ASB 105	12.0	6.30	1150	1610	370	16.8	25.3	1440	633	0.830	12.1	0.574	207	133
280 ASB 100†	11.0	5.76	995	1290	289	15.6	18.4	1290	511	0.810	13.2	0.451	160	128
280 ASB 74	11.4	5.96	776	1060	234	15.7	21.3	978	403	0.830	16.7	0.338	72.0	93.7

† Sections are fire engineered with thick webs.
ASB sections are only available in S355.

PARALLEL FLANGE CHANNELS

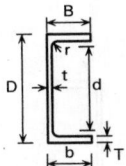

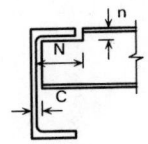

DIMENSIONS

Section Designation	Mass per Metre	Depth of Section	Width of Section	Thickness		Root Radius	Depth between Fillets	Ratios for Local Buckling		Dimensions for Detailing			Surface Area	
				Web	Flange			Flange	Web	End Clearance	Notch		Per Metre	Per Tonne
		D	B	t	T	r	d	b/T	d/t	C	N	n		
	kg/m	mm	mm	mm	mm	mm	mm			mm	mm	mm	m²	m²
430 × 100 × 64	64.4	430	100	11.0	19.0	15	362	5.26	32.9	13	96	36	1.23	19.0
380 × 100 × 54	54.0	380	100	9.5	17.5	15	315	5.71	33.2	12	98	34	1.13	20.9
300 × 100 × 46	45.5	300	100	9.0	16.5	15	237	6.06	26.3	11	98	32	0.969	21.3
300 × 90 × 41	41.4	300	90	9.0	15.5	12	245	5.81	27.2	11	88	28	0.932	22.5
260 × 90 × 35	34.8	260	90	8.0	14.0	12	208	6.43	26.0	10	88	28	0.854	24.5
260 × 75 × 28	27.6	260	75	7.0	12.0	12	212	6.25	30.3	9	74	26	0.796	28.8
230 × 90 × 32	32.2	230	90	7.5	14.0	12	178	6.43	23.7	10	90	28	0.795	24.7
230 × 75 × 26	25.7	230	75	6.5	12.5	12	181	6.00	27.8	9	76	26	0.737	28.7
200 × 90 × 30	29.7	200	90	7.0	14.0	12	148	6.43	21.1	9	90	28	0.736	24.8
200 × 75 × 23	23.4	200	75	6.0	12.5	12	151	6.00	25.2	8	76	26	0.678	28.9
180 × 90 × 26	26.1	180	90	6.5	12.5	12	131	7.20	20.2	9	90	26	0.697	26.7
180 × 75 × 20	20.3	180	75	6.0	10.5	12	135	7.14	22.5	8	76	24	0.638	31.4
150 × 90 × 24	23.9	150	90	6.5	12.0	12	102	7.50	15.7	9	90	26	0.637	26.7
150 × 75 × 18	17.9	150	75	5.5	10.0	12	106	7.50	19.3	8	76	24	0.579	32.4
125 × 65 × 15 #	14.8	125	65	5.5	9.5	12	82.0	6.84	14.9	8	66	22	0.489	33.1
100 × 50 × 10 #	10.2	100	50	5.0	8.5	9	65.0	5.88	13.0	7	52	18	0.382	37.5

Check availability.

PARALLEL FLANGE CHANNELS

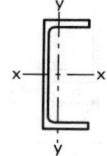

PROPERTIES

Section Designation	Second Moment of Area		Radius of Gyration		Elastic Modulus		Plastic Modulus		Buckling Parameter	Torsional Index	Warping Constant	Torsional Constant	Area of Section
	Axis x-x cm⁴	Axis y-y cm⁴	Axis x-x cm	Axis y-y cm	Axis x-x cm³	Axis y-y cm³	Axis x-x cm³	Axis y-y cm³	u	x	H dm⁶	J cm⁴	A cm²
430 × 100 × 64	21900	722	16.3	2.97	1020	97.9	1220	176	0.917	22.5	0.219	63.0	82.1
380 × 100 × 54	15000	643	14.8	3.06	791	89.2	933	161	0.933	21.2	0.150	45.7	68.7
300 × 100 × 46	8230	568	11.9	3.13	549	81.7	641	148	0.944	17.0	0.0813	36.8	58.0
300 × 90 × 41	7220	404	11.7	2.77	481	63.1	568	114	0.934	18.4	0.0581	28.8	52.7
260 × 90 × 35	4730	353	10.3	2.82	364	56.3	425	102	0.943	17.2	0.0379	20.6	44.4
260 × 75 × 28	3620	185	10.1	2.30	278	34.4	328	62.0	0.932	20.5	0.0203	11.7	35.1
230 × 90 × 32	3520	334	9.27	2.86	306	55.0	355	98.9	0.949	15.1	0.0279	19.3	41.0
230 × 75 × 26	2750	181	9.17	2.35	239	34.8	278	63.2	0.945	17.3	0.0153	11.8	32.7
200 × 90 × 30	2520	314	8.16	2.88	252	53.4	291	94.5	0.952	12.9	0.0197	18.3	37.9
200 × 75 × 23	1960	170	8.11	2.39	196	33.8	227	60.6	0.956	14.7	0.0107	11.1	29.9
180 × 90 × 26	1820	277	7.40	2.89	202	47.4	232	83.5	0.950	12.8	0.0141	13.3	33.2
180 × 75 × 20	1370	146	7.27	2.38	152	28.8	176	51.8	0.945	15.3	0.00754	7.34	25.9
150 × 90 × 24	1160	253	6.18	2.89	155	44.4	179	76.9	0.937	10.8	0.00890	11.8	30.4
150 × 75 × 18	861	131	6.15	2.40	115	26.6	132	47.2	0.945	13.1	0.00467	6.10	22.8
125 × 65 × 15 #	483	80.0	5.07	2.06	77.3	18.8	89.9	33.2	0.942	11.1	0.00194	4.72	18.8
100 × 50 × 10 #	208	32.3	4.00	1.58	41.5	9.89	48.9	17.5	0.942	10.0	0.000491	2.53	13.0

Check availability.

PARALLEL FLANGE CHANNELS

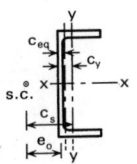

MAJOR AXIS REDUCED PLASTIC MODULUS UNDER AXIAL LOAD

Section Designation	Area of Section	Dimension				Plastic Modulus Axis x-x	Major Axis Reduced Modulus				
							Lower Values of n		Change Formula at n =	Higher Values of n	
	A cm²	e₀ cm	C_s cm	C_y cm	C_eq cm	cm³	K1	K2		K3	K4
430 × 100 × 64	82.1	3.27	5.34	2.62	0.954	1220	1220	1530	0.525	168	9.48
380 × 100 × 54	68.7	3.48	5.79	2.79	0.904	933	933	1240	0.477	118	10.1
300 × 100 × 46	58.0	3.68	6.29	3.05	1.31	641	641	934	0.414	84.1	9.35
300 × 90 × 41	52.7	3.18	5.33	2.60	0.879	568	568	772	0.459	77.2	9.24
260 × 90 × 35	44.4	3.32	5.66	2.74	1.14	425	425	615	0.418	54.7	9.55
260 × 75 × 28	35.1	2.62	4.37	2.10	0.676	328	328	441	0.470	41.2	10.1
230 × 90 × 32	41.0	3.46	6.01	2.92	1.69	355	355	559	0.370	46.6	9.11
230 × 75 × 26	32.7	2.78	4.75	2.30	1.03	278	278	411	0.408	35.6	9.55
200 × 90 × 30	37.9	3.60	6.37	3.12	2.24	291	291	512	0.318	39.8	8.51
200 × 75 × 23	29.9	2.91	5.09	2.48	1.53	227	227	372	0.352	29.7	9.04
180 × 90 × 26	33.2	3.64	6.48	3.17	2.36	232	232	424	0.304	30.6	8.76
180 × 75 × 20	25.9	2.87	4.98	2.41	1.34	176	176	280	0.368	22.4	9.42
150 × 90 × 24	30.4	3.71	6.69	3.30	2.66	179	179	356	0.269	25.7	7.88
150 × 75 × 18	22.8	2.99	5.29	2.58	1.81	132	132	236	0.314	17.3	8.88
125 × 65 × 15 #	18.8	2.56	4.53	2.25	1.55	89.9	89.9	161	0.310	13.6	7.64
100 × 50 × 10 #	13.0	1.94	3.43	1.73	1.18	48.9	48.9	84.5	0.319	8.45	6.69

\# Check availability.
e_0 is the distance from the centre of the web to the shear centre.
C_s is the distance from the centroidal axis to the shear centre.
C_y is the distance from the back of the web to the centroidal axis.
C_{eq} is the distance from the back of the web to the equal area axis.
$n = F/(A_g \cdot p_y)$, where F is the factored axial load, A_g is the gross cross sectional area and p_y is the design strength of the section.
For lower values of n, the reduced plastic modulus, $S_r = K1 - K2.n^2$
For higher values of n, the reduced plastic modulus, $S_r = K3(1 - n)(K4 + n)$

PARALLEL FLANGE CHANNELS

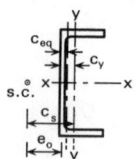

MINOR AXIS REDUCED PLASTIC MODULUS UNDER AXIAL LOAD

Section Designation	Dimension	Plastic Modulus Axis y-y	Minor Axis reduced Modulus under axial load about centroidal axis												
			Axial load and moment inducing stresses of the same kind towards back of web						Change Formula at n =	Axial load and moment inducing stresses of the opposite kind towards back of web					
			Lower Values of n			Higher Values of n				Lower Values of n			Higher Values of n		
	C_y cm	cm³	K1	K2	K3	K1	K2	K3		K1	K2	K3	K1	K2	K3
430 × 100 × 64	2.62	176	176	39.2	3.49	162	443	0.634	0.152	176	39.2	3.49	176	39.2	3.49
380 × 100 × 54	2.79	161	161	31.1	4.17	158	338	0.532	0.0503	161	31.1	4.17	161	31.1	4.17
300 × 100 × 46	3.05	148	148	255	0.419	148	255	0.419	0.0689	148	255	0.419	149	28.0	4.32
300 × 90 × 41	2.60	114	114	23.2	3.92	113	224	0.495	0.0241	114	23.2	3.92	114	23.2	3.92
260 × 90 × 35	2.74	102	102	176	0.419	102	176	0.419	0.0626	102	176	0.419	103	18.9	4.42
260 × 75 × 28	2.10	62.0	62.0	11.9	4.22	61.0	129	0.525	0.0359	62.0	11.9	4.22	62.0	11.9	4.22
230 × 90 × 32	2.92	98.9	98.9	150	0.338	99.1	150	0.338	0.158	98.9	150	0.338	101	18.2	4.56
230 × 75 × 26	2.30	63.2	63.2	107	0.410	63.1	107	0.410	0.0854	63.2	107	0.410	63.6	11.6	4.47
200 × 90 × 30	3.12	94.5	94.5	128	0.261	94.5	128	0.261	0.260	94.5	128	0.261	100	17.9	4.60
200 × 75 × 23	2.48	60.6	60.6	89.2	0.318	60.6	89.2	0.318	0.196	60.6	89.2	0.318	62.8	11.2	4.64
180 × 90 × 26	3.17	83.5	83.5	110	0.242	83.5	110	0.242	0.295	83.5	110	0.242	89.8	15.3	4.87
180 × 75 × 20	2.41	51.8	51.8	79.9	0.350	51.9	79.9	0.350	0.166	51.8	79.9	0.350	53.1	9.32	4.70
150 × 90 × 24	3.30	76.9	76.9	96.3	0.201	76.9	96.3	0.201	0.359	76.9	96.3	0.201	85.0	15.4	4.52
150 × 75 × 18	2.58	47.2	47.2	64.8	0.271	47.2	64.8	0.271	0.275	47.2	64.8	0.271	50.1	8.64	4.80
125 × 65 × 15 #	2.25	33.2	33.2	46.5	0.281	33.4	46.5	0.281	0.269	33.2	46.5	0.281	35.2	7.07	3.98
100 × 50 × 10 #	1.73	17.5	17.5	24.8	0.291	17.6	24.8	0.291	0.231	17.5	24.8	0.291	18.3	4.22	3.33

\# Check availability.
C_y is the distance from the back of the web to the centroidal axis.
$n = F/(A_g \cdot p_y)$, where F is the factored axial load, A_g is the gross cross sectional area and p_y is the design strength of the section.
For axial load and moment inducing stresses of the same kind towards back of web, the reduced plastic modulus, $S_r = K1 + K2.n.(K3 - n)$
For axial load and moment inducing stresses of the opposite kind towards back of web, the reduced plastic modulus, $S_r = K1 - K2.n.(K3 + n)$

TWO PARALLEL FLANGE CHANNELS LACED

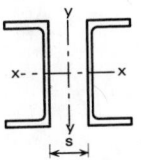

DIMENSIONS AND PROPERTIES

Composed of Two Channels	Total Mass per Metre kg/m	Total Area cm²	Space between Webs s mm	Second Moment of Area		Radius of Gyration		Elastic Modulus		Plastic Modulus	
				Axis x-x cm⁴	Axis y-y cm⁴	Axis x-x cm	Axis y-y cm	Axis x-x cm³	Axis y-y cm³	Axis x-x cm³	Axis y-y cm³
430 × 100 × 64	129	164	270	43900	44100	16.3	16.4	2040	1880	2440	2650
380 × 100 × 54	108	137	235	30100	30400	14.8	14.9	1580	1400	1870	2000
300 × 100 × 46	91.1	116	170	16500	16600	11.9	12.0	1100	898	1280	1340
300 × 90 × 41	82.8	105	175	14400	14400	11.7	11.7	962	811	1140	1200
260 × 90 × 35	69.7	88.8	145	9460	9560	10.3	10.4	727	588	849	886
260 × 75 × 28	55.2	70.3	155	7240	7190	10.1	10.1	557	472	656	692
230 × 90 × 32	64.3	81.9	120	7040	7190	9.27	9.37	612	479	709	731
230 × 75 × 26	51.3	65.4	135	5500	5720	9.17	9.35	478	401	557	592
200 × 90 × 30	59.4	75.7	90.0	5050	5030	8.16	8.15	505	372	583	577
200 × 75 × 23	46.9	59.7	105	3930	3910	8.11	8.09	393	306	454	462
180 × 90 × 26	52.1	66.4	75.0	3640	3730	7.40	7.49	404	292	464	459
180 × 75 × 20	40.7	51.8	90.0	2740	2770	7.27	7.31	304	231	352	358
150 × 90 × 24	47.7	60.8	45.0	2320	2380	6.18	6.26	310	212	357	338
150 × 75 × 18	35.7	45.5	65.0	1720	1810	6.15	6.30	230	168	264	265
125 × 65 × 15 #	29.5	37.6	50.0	966	1010	5.07	5.18	155	112	180	178
100 × 50 × 10 #	20.4	26.0	40.0	415	427	4.00	4.05	83.1	61.0	97.7	97.1

Check availability.

TWO PARALLEL FLANGE CHANNELS BACK TO BACK

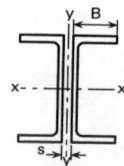

DIMENSIONS AND PROPERTIES

Composed of Two Channels	Total Mass per Metre kg/m	Total Area cm²	Properties about Axis x-x				Radius of Gyration r_y about Axis y-y (cm)				
			I_x cm⁴	r_x cm	Z_x cm³	S_x cm³	Space between webs, s (mm)				
							0	8	10	12	15
430 × 100 × 64	129	164	43900	16.3	2040	2440	3.96	4.23	4.31	4.38	4.49
380 × 100 × 54	108	137	30100	14.8	1580	1870	4.14	4.42	4.49	4.57	4.68
300 × 100 × 46	91.1	116	16500	11.9	1100	1280	4.37	4.66	4.73	4.81	4.92
300 × 90 × 41	82.8	105	14400	11.7	962	1140	3.80	4.08	4.16	4.23	4.35
260 × 90 × 35	69.7	88.8	9460	10.3	727	849	3.93	4.22	4.29	4.37	4.48
260 × 75 × 28	55.2	70.3	7240	10.1	557	656	3.11	3.40	3.47	3.55	3.66
230 × 90 × 32	64.3	81.9	7040	9.27	612	709	4.09	4.38	4.46	4.53	4.65
230 × 75 × 26	51.3	65.4	5500	9.17	478	557	3.29	3.58	3.66	3.73	3.85
200 × 90 × 30	59.4	75.7	5050	8.16	505	583	4.25	4.55	4.63	4.71	4.83
200 × 75 × 23	46.9	59.7	3930	8.11	393	454	3.44	3.74	3.82	3.89	4.01
180 × 90 × 26	52.1	66.4	3640	7.40	404	464	4.29	4.59	4.67	4.75	4.87
180 × 75 × 20	40.7	51.8	2740	7.27	304	352	3.39	3.68	3.76	3.84	3.95
150 × 90 × 24	47.7	60.8	2320	6.18	310	357	4.39	4.69	4.77	4.85	4.98
150 × 75 × 18	35.7	45.5	1720	6.15	230	264	3.52	3.82	3.90	3.98	4.10
125 × 65 × 15 #	29.5	37.6	966	5.07	155	180	3.05	3.36	3.44	3.52	3.64
100 × 50 × 10 #	20.4	26.0	415	4.00	83.1	97.7	2.34	2.65	2.73	2.82	2.94

\# Check availability.
Properties about y axis
$I_y = (\text{Total Area}) \cdot (r_y)^2$
$Z_y = I_y/(B + 0.5s)$
where s is the space between webs.

EQUAL ANGLES

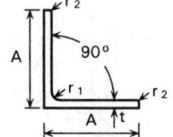

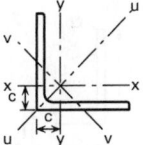

DIMENSIONS AND PROPERTIES

Section Designation		Mass per Metre	Radius		Area of Section	Dimension	Second Moment of Area			Radius of Gyration			Elastic Modulus	Torsional Constant	Equivalent Slenderness Coefficient
			Root	Toe			Axis x-x, y-y	Axis u-u	Axis v-v	Axis x-x, y-y	Axis u-u	Axis v-v	Axis x-x, y-y		
Size A × A mm	Thickness t mm	kg/m	r_1 mm	r_2 mm	cm²	c cm	cm^4	cm^4	cm^4	cm	cm	cm	cm^3	J cm^4	ϕ_a
200 × 200	24 #	71.1	18.0	9.00	90.6	5.84	3330	5280	1380	6.06	7.64	3.90	235	182	2.50
	20	59.9	18.0	9.00	76.3	5.68	2850	4530	1170	6.11	7.70	3.92	199	107	3.05
	18	54.3	18.0	9.00	69.1	5.60	2600	4150	1050	6.13	7.75	3.90	181	78.9	3.43
	16	48.5	18.0	9.00	61.8	5.52	2340	3720	960	6.16	7.76	3.94	162	56.1	3.85
150 × 150	18 #	40.1	16.0	8.00	51.2	4.38	1060	1680	440	4.55	5.73	2.93	99.8	58.6	2.48
	15	33.8	16.0	8.00	43.0	4.25	898	1430	370	4.57	5.76	2.93	83.5	34.6	3.01
	12	27.3	16.0	8.00	34.8	4.12	737	1170	303	4.60	5.80	2.95	67.7	18.2	3.77
	10	23.0	16.0	8.00	29.3	4.03	624	990	258	4.62	5.82	2.97	56.9	10.80	4.51
120 × 120	15 #	26.6	13.0	6.50	34.0	3.52	448	710	186	3.63	4.57	2.34	52.8	27.0	2.37
	12	21.6	13.0	6.50	27.5	3.40	368	584	152	3.65	4.60	2.35	42.7	14.2	2.99
	10	18.2	13.0	6.50	23.2	3.31	313	497	129	3.67	4.63	2.36	36.0	8.41	3.61
	8 #	14.7	13.0	6.50	18.8	3.24	259	411	107	3.71	4.67	2.38	29.5	4.44	4.56
100 × 100	15 #	21.9	12.0	6.00	28.0	3.02	250	395	105	2.99	3.76	1.94	35.8	22.3	1.92
	12	17.8	12.0	6.00	22.7	2.90	207	328	85.7	3.02	3.80	1.94	29.1	11.8	2.44
	10	15.0	12.0	6.00	19.2	2.82	177	280	73.0	3.04	3.83	1.95	24.6	6.97	2.94
	8	12.2	12.0	6.00	15.5	2.74	145	230	59.9	3.06	3.85	1.96	19.9	3.68	3.70
90 × 90	12 #	15.9	11.0	5.50	20.3	2.66	149	235	62.0	2.71	3.40	1.75	23.5	10.46	2.17
	10	13.4	11.0	5.50	17.1	2.58	127	201	52.6	2.72	3.42	1.75	19.8	6.20	2.64
	8	10.9	11.0	5.50	13.9	2.50	104	166	43.1	2.74	3.45	1.76	16.1	3.28	3.33
	7 #	9.61	11.0	5.50	12.2	2.45	92.6	147	38.3	2.75	3.46	1.77	14.1	2.24	3.80
80 × 80	10 ‡	11.9	10.0	5.00	15.1	2.34	87.5	139	36.4	2.41	3.03	1.55	15.4	5.45	2.33
	8 ‡	9.63	10.0	5.00	12.3	2.26	72.2	115	29.9	2.43	3.06	1.56	12.6	2.88	2.94
75 × 75	8 ‡	8.99	9.00	4.50	11.4	2.14	59.1	93.8	24.5	2.27	2.86	1.46	11.0	2.65	2.76
	6 ‡	6.85	9.00	4.50	8.73	2.05	45.8	72.7	18.9	2.29	2.89	1.47	8.41	1.17	3.70
70 × 70	7 ‡	7.38	9.00	4.50	9.40	1.97	42.3	67.1	17.5	2.12	2.67	1.36	8.41	1.69	2.92
	6 ‡	6.38	9.00	4.50	8.13	1.93	36.9	58.5	15.3	2.13	2.68	1.37	7.27	1.093	3.41
65 × 65	7 ‡	6.83	9.00	4.50	8.73	2.05	33.4	53.0	13.8	1.96	2.47	1.26	7.18	1.58	2.67
60 × 60	8 ‡	7.09	8.00	4.00	9.03	1.77	29.2	46.1	12.2	1.80	2.26	1.16	6.89	2.09	2.14
	6 ‡	5.42	8.00	4.00	6.91	1.69	22.8	36.1	9.44	1.82	2.29	1.17	5.29	0.922	2.90
	5 ‡	4.57	8.00	4.00	5.82	1.64	19.4	30.7	8.03	1.82	2.30	1.17	4.45	0.550	3.48
50 × 50	6 ‡	4.47	7.00	3.50	5.69	1.45	12.8	20.3	5.34	1.50	1.89	0.968	3.61	0.755	2.38
	5 ‡	3.77	7.00	3.50	4.80	1.40	11.0	17.4	4.55	1.51	1.90	0.973	3.05	0.450	2.88
	4 ‡	3.06	7.00	3.50	3.89	1.36	8.97	14.2	3.73	1.52	1.91	0.979	2.46	0.240	3.57
45 × 45	4.5 ‡	3.06	7.00	3.50	3.90	1.25	7.14	11.4	2.94	1.35	1.71	0.870	2.20	0.304	2.84
40 × 40	5 ‡	2.97	6.00	3.00	3.79	1.16	5.43	8.60	2.26	1.20	1.51	0.773	1.91	0.352	2.26
	4 ‡	2.42	6.00	3.00	3.08	1.12	4.47	7.09	1.86	1.21	1.52	0.777	1.55	0.188	2.83
35 × 35	4 ‡	2.09	5.00	2.50	2.67	1.00	2.95	4.68	1.23	1.05	1.32	0.678	1.18	0.158	2.50
30 × 30	4 ‡	1.78	5.00	2.50	2.27	0.878	1.80	2.85	0.754	0.892	1.12	0.577	0.850	0.137	2.07
	3 ‡	1.36	5.00	2.50	1.74	0.835	1.40	2.22	0.585	0.899	1.13	0.581	0.649	0.0613	2.75
25 × 25	4 ‡	1.45	3.50	1.75	1.85	0.762	1.02	1.61	0.430	0.741	0.931	0.482	0.586	0.1070	1.75
	3 ‡	1.12	3.50	1.75	1.42	0.723	0.803	1.27	0.334	0.751	0.945	0.484	0.452	0.0472	2.38
20 × 20	3 ‡	0.882	3.50	1.75	1.12	0.598	0.392	0.618	0.165	0.590	0.742	0.383	0.279	0.0382	1.81

‡ Not available from some leading producers. Check availability.
# Check availability.
c is the distance from the back of the leg to the centre of gravity.

UNEQUAL ANGLES

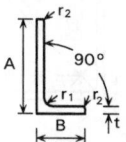

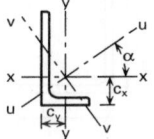

DIMENSIONS AND PROPERTIES

Section Designation		Mass per Metre	Radius		Dimension		Second Moment of Area				Radius of Gyration			
			Root	Toe			Axis x-x	Axis y-y	Axis u-u	Axis v-v	Axis x-x	Axis y-y	Axis u-u	Axis v-v
Size A × B mm	Thickness t mm	kg/m	r_1 mm	r_2 mm	c_x cm	c_y cm	cm⁴	cm⁴	cm⁴	cm⁴	cm	cm	cm	cm
200 × 150	18 #	47.1	15.0	7.50	6.34	3.86	2390	1160	2920	623	6.30	4.38	6.97	3.22
	15	39.6	15.0	7.50	6.21	3.73	2020	979	2480	526	6.33	4.40	7.00	3.23
	12	32.0	15.0	7.50	6.08	3.61	1650	803	2030	430	6.36	4.44	7.04	3.25
200 × 100	15	33.8	15.0	7.50	7.16	2.22	1760	299	1860	193	6.40	2.64	6.59	2.12
	12	27.3	15.0	7.50	7.03	2.10	1440	247	1530	159	6.43	2.67	6.63	2.14
	10	23.0	15.0	7.50	6.93	2.01	1220	210	1290	135	6.46	2.68	6.65	2.15
150 × 90	15	33.9	12.0	6.00	5.21	2.23	761	205	841	126	4.74	2.46	4.98	1.93
	12	21.6	12.0	6.00	5.08	2.12	627	171	694	104	4.77	2.49	5.02	1.94
	10	18.2	12.0	6.00	5.00	2.04	533	146	591	88.3	4.80	2.51	5.05	1.95
150 × 75	15	24.8	12.0	6.00	5.52	1.81	713	119	753	78.6	4.75	1.94	4.88	1.58
	12	20.2	12.0	6.00	5.40	1.69	588	99.6	623	64.7	4.78	1.97	4.92	1.59
	10	17.0	12.0	6.00	5.31	1.61	501	85.6	531	55.1	4.81	1.99	4.95	1.60
125 × 75	12	17.8	11.0	5.50	4.31	1.84	354	95.5	391	58.5	3.95	2.05	4.15	1.61
	10	15.0	11.0	5.50	4.23	1.76	302	82.1	334	49.9	3.97	2.07	4.18	1.61
	8	12.2	11.0	5.50	4.14	1.68	247	67.6	274	40.9	4.00	2.09	4.21	1.63
100 × 75	12	15.4	10.0	5.00	3.27	2.03	189	90.2	230	49.5	3.10	2.14	3.42	1.59
	10	13.0	10.0	5.00	3.19	1.95	162	77.6	197	42.2	3.12	2.16	3.45	1.59
	8	10.6	10.0	5.00	3.10	1.87	133	64.1	162	34.6	3.14	2.18	3.47	1.60
100 × 65	10 #	12.3	10.0	5.00	3.36	1.63	154	51.0	175	30.1	3.14	1.81	3.35	1.39
	8 #	9.94	10.0	5.00	3.27	1.55	127	42.2	144	24.8	3.16	1.83	3.37	1.40
	7 #	8.77	10.0	5.00	3.23	1.51	113	37.6	128	22.0	3.17	1.83	3.39	1.40
100 × 50	8 ‡	8.97	8.00	4.00	3.60	1.13	116	19.7	123	12.8	3.19	1.31	3.28	1.06
	6 ‡	6.84	8.00	4.00	3.51	1.05	89.9	15.4	95.4	9.92	3.21	1.33	3.31	1.07
80 × 60	7 ‡	7.36	8.00	4.00	2.51	1.52	59.0	28.4	72.0	15.4	2.51	1.74	2.77	1.28
80 × 40	8 ‡	7.07	7.00	3.50	2.94	0.963	57.6	9.61	60.9	6.34	2.53	1.03	2.60	0.838
	6 ‡	5.41	7.00	3.50	2.85	0.884	44.9	7.59	47.6	4.93	2.55	1.05	2.63	0.845
75 × 50	8 ‡	7.39	7.00	3.50	2.52	1.29	52.0	18.4	59.6	10.8	2.35	1.40	2.52	1.07
	6 ‡	5.65	7.00	3.50	2.44	1.21	40.5	14.4	46.6	8.36	2.37	1.42	2.55	1.08
70 × 50	6 ‡	5.41	7.00	3.50	2.23	1.25	33.4	14.2	39.7	7.92	2.20	1.43	2.40	1.07
65 × 50	5 ‡	4.35	6.00	3.00	1.99	1.25	23.2	11.9	28.8	6.32	2.05	1.47	2.28	1.07
60 × 40	6 ‡	4.46	6.00	3.00	2.00	1.01	20.1	7.12	23.1	4.16	1.88	1.12	2.02	0.855
	5 ‡	3.76	6.00	3.00	1.96	0.972	17.2	6.11	19.7	3.54	1.89	1.13	2.03	0.860
60 × 30	5 ‡	3.36	5.00	2.50	2.17	0.684	15.6	2.63	16.5	1.71	1.91	0.784	1.97	0.633
50 × 30	5 ‡	2.96	5.00	2.50	1.73	0.741	9.36	2.51	10.3	1.54	1.57	0.816	1.65	0.639
45 × 30	4 ‡	2.25	4.50	2.25	1.48	0.740	5.78	2.05	6.65	1.18	1.42	0.850	1.52	0.640
40 × 25	4 ‡	1.93	4.00	2.00	1.36	0.623	3.89	1.16	4.35	0.700	1.26	0.687	1.33	0.534
40 × 20	4 ‡	1.77	4.00	2.00	1.47	0.480	3.59	0.600	3.80	0.393	1.26	0.514	1.30	0.417
30 × 20	4 ‡	1.46	4.00	2.00	1.03	0.541	1.59	0.553	1.81	0.330	0.925	0.546	0.988	0.421
	3 ‡	1.12	4.00	2.00	0.990	0.502	1.25	0.437	1.43	0.256	0.935	0.553	1.00	0.424

‡ Not available from some leading producers. Check availability.
# Check availability.
c_x is the distance from the back of the short leg to the centre of gravity.
c_y is the distance from the back of the long leg to the centre of gravity.

UNEQUAL ANGLES

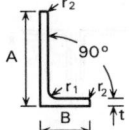

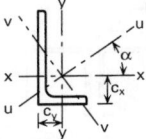

DIMENSIONS AND PROPERTIES (CONTINUED)

Section Designation		Elastic Modulus		Angle Axis x-x to Axis u-u	Torsional Constant	Equivalent Slenderness Coefficient		Mono-symmetry index	Area of Section
Size A × B mm	Thickness t mm	Axis x-x cm³	Axis y-y cm³	Tan α	J cm⁴	Min ϕ_a	Max ϕ_a	ψ_a	cm²
200 × 150	18 #	175	104	0.549	67.9	2.93	3.72	4.60	60.1
	15	147	86.9	0.551	39.9	3.53	4.50	5.55	50.5
	12	119	70.5	0.552	20.9	4.43	5.70	6.97	40.8
200 × 100	15	137	38.5	0.260	34.3	3.54	5.17	9.19	43.0
	12	111	31.3	0.262	18.0	4.42	6.57	11.5	34.8
	10	93.2	26.3	0.263	10.66	5.26	7.92	13.9	29.2
150 × 90	15	77.7	30.4	0.354	26.8	2.58	3.59	5.96	33.9
	12	63.3	24.8	0.358	14.1	3.24	4.58	7.50	27.5
	10	53.3	21.0	0.360	8.30	3.89	5.56	9.03	23.2
150 × 75	15	75.2	21.0	0.253	25.1	2.62	3.74	6.84	31.7
	12	61.3	17.1	0.258	13.2	3.30	4.79	8.60	25.7
	10	51.6	14.5	0.261	7.80	3.95	5.83	10.4	21.7
125 × 75	12	43.2	16.9	0.354	11.6	2.66	3.73	6.23	22.7
	10	36.5	14.3	0.357	6.87	3.21	4.55	7.50	19.1
	8	29.6	11.6	0.360	3.62	4.00	5.75	9.43	15.5
100 × 75	12	28.0	16.5	0.540	10.05	2.10	2.64	3.46	19.7
	10	23.8	14.0	0.544	5.95	2.54	3.22	4.17	16.6
	8	19.3	11.4	0.547	3.13	3.18	4.08	5.24	13.5
100 × 65	10 #	23.2	10.5	0.410	5.61	2.52	3.43	5.45	15.6
	8 #	18.9	8.54	0.413	2.96	3.14	4.35	6.86	12.7
	7 #	16.6	7.53	0.415	2.02	3.58	5.00	7.85	11.2
100 × 50	8 ‡	18.2	5.08	0.258	2.61	3.30	4.80	8.61	11.4
	6 ‡	13.8	3.89	0.262	1.14	4.38	6.52	11.6	8.71
80 × 60	7 ‡	10.7	6.34	0.546	1.66	2.92	3.72	4.78	9.38
80 × 40	8 ‡	11.4	3.16	0.253	2.05	2.61	3.73	6.85	9.01
	6 ‡	8.73	2.44	0.258	0.899	3.48	5.12	9.22	6.89
75 × 50	8 ‡	10.4	4.95	0.430	2.14	2.36	3.18	4.92	9.41
	6 ‡	8.01	3.81	0.435	0.935	3.18	4.34	6.60	7.19
70 × 50	6 ‡	7.01	3.78	0.500	0.899	2.96	3.89	5.44	6.89
65 × 50	5 ‡	5.14	3.19	0.577	0.498	3.38	4.26	5.08	5.54
60 × 40	6 ‡	5.03	2.38	0.431	0.735	2.51	3.39	5.26	5.68
	5 ‡	4.25	2.02	0.434	0.435	3.02	4.11	6.34	4.79
60 × 30	5 ‡	4.07	1.14	0.257	0.382	3.15	4.56	8.26	4.28
50 × 30	5 ‡	2.86	1.11	0.352	0.340	2.51	3.52	5.99	3.78
45 × 30	4 ‡	1.91	0.910	0.436	0.166	2.85	3.87	5.92	2.87
40 × 25	4 ‡	1.47	0.619	0.380	0.142	2.51	3.48	5.75	2.46
40 × 20	4 ‡	1.42	0.393	0.252	0.131	2.57	3.68	6.86	2.26
30 × 20	4 ‡	0.807	0.379	0.421	0.1096	1.79	2.39	3.95	1.86
	3 ‡	0.621	0.292	0.427	0.0486	2.40	3.28	5.31	1.43

‡ Not available from some leading producers. Check availability.
# Check availability.

EQUAL ANGLES BACK TO BACK

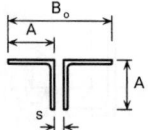

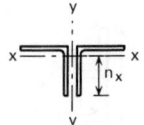

DIMENSIONS AND PROPERTIES

Composed of Two Angles		Total Mass per Metre	Distance	Total Area	Properties about Axis x-x			Radius of Gyration r_y about Axis y-y (cm)				
								Space between angles, s, (mm)				
$A \times A$ mm	t mm	kg/m	n_x cm	cm^2	I_x cm^4	r_x cm	Z_x cm^3	0	8	10	12	15
200×200	24 #	142	14.2	181	6660	6.06	470	8.42	8.70	8.77	8.84	8.95
	20	120	14.3	153	5700	6.11	398	8.34	8.62	8.69	8.76	8.87
	18	109	14.4	138	5200	6.13	362	8.31	8.58	8.65	8.72	8.83
	16	97.0	14.5	124	4680	6.16	324	8.27	8.54	8.61	8.68	8.79
150×150	18 #	80.2	10.6	102	2120	4.55	200	6.32	6.60	6.67	6.75	6.86
	15	67.6	10.8	86.0	1800	4.57	167	6.24	6.52	6.59	6.66	6.77
	12	54.6	10.9	69.6	1470	4.60	135	6.18	6.45	6.52	6.59	6.70
	10	46.0	11.0	58.6	1250	4.62	114	6.13	6.40	6.47	6.54	6.64
120×120	15 #	53.2	8.48	68.0	896	3.63	106	5.06	5.34	5.42	5.49	5.60
	12	43.2	8.60	55.0	736	3.65	85.4	4.99	5.27	5.35	5.42	5.53
	10	36.4	8.69	46.4	626	3.67	72.0	4.94	5.22	5.29	5.36	5.47
	8 #	29.4	8.76	37.6	518	3.71	59.0	4.93	5.20	5.27	5.34	5.45
100×100	15 #	43.8	6.98	56.0	500	2.99	71.6	4.25	4.54	4.62	4.69	4.81
	12	35.6	7.10	45.4	414	3.02	58.2	4.19	4.47	4.55	4.62	4.74
	10	30.0	7.18	38.4	354	3.04	49.2	4.14	4.43	4.50	4.57	4.69
	8	24.4	7.26	31.0	290	3.06	39.8	4.11	4.38	4.46	4.53	4.64
90×90	12 #	31.8	6.34	40.6	298	2.71	47.0	3.80	4.09	4.16	4.24	4.36
	10	26.8	6.42	34.2	254	2.72	39.6	3.75	4.04	4.11	4.19	4.30
	8	21.8	6.50	27.8	208	2.74	32.2	3.71	3.99	4.06	4.13	4.25
	7 #	19.2	6.55	24.4	185	2.75	28.2	3.69	3.96	4.04	4.11	4.22
80×80	10 ‡	23.8	5.66	30.2	175	2.41	30.8	3.36	3.65	3.72	3.80	3.92
	8 ‡	19.3	5.74	24.6	144	2.43	25.2	3.31	3.60	3.67	3.75	3.86
75×75	8 ‡	18.0	5.36	22.8	118	2.27	22.0	3.12	3.41	3.49	3.56	3.68
	6 ‡	13.7	5.45	17.5	91.6	2.29	16.8	3.07	3.35	3.43	3.50	3.62
70×70	7 ‡	14.8	5.03	18.8	84.6	2.12	16.8	2.89	3.18	3.26	3.33	3.45
	6 ‡	12.8	5.07	16.3	73.8	2.13	14.5	2.87	3.16	3.23	3.31	3.42
65×65	7 ‡	13.7	4.45	17.5	66.8	1.96	14.4	2.83	3.14	3.21	3.29	3.42
60×60	8 ‡	14.2	4.23	18.1	58.4	1.80	13.8	2.52	2.82	2.90	2.97	3.10
	6 ‡	10.8	4.31	13.8	45.6	1.82	10.6	2.48	2.77	2.85	2.92	3.04
	5 ‡	9.14	4.36	11.6	38.8	1.82	8.90	2.45	2.74	2.81	2.89	3.01
50×50	6 ‡	8.94	3.55	11.4	25.6	1.50	7.22	2.09	2.38	2.46	2.54	2.66
	5 ‡	7.54	3.60	9.60	22.0	1.51	6.10	2.06	2.35	2.43	2.51	2.63
	4 ‡	6.12	3.64	7.78	17.9	1.52	4.92	2.04	2.32	2.40	2.48	2.60

‡ Not available from some leading producers. Check availability.
# Check availability.
Properties about y-y axis:
$I_y = (\text{Total Area}) \cdot (r_y)^2$
$Z_y = I_y/(0.5B_o)$

UNEQUAL ANGLES LONG LEGS BACK TO BACK

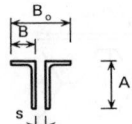

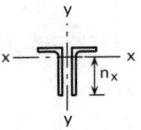

DIMENSIONS AND PROPERTIES

Composed of Two Angles		Total Mass per Metre kg/m	Distance n_x cm	Total Area cm^2	Properties about Axis x-x			Radius of Gyration r_y about Axis y-y (cm)				
								Space between angles, s, (mm)				
$A \times B$ mm	t mm				I_x cm^4	r_x cm	Z_x cm^3	0	8	10	12	15
200 × 150	18 #	94.2	13.7	120	4780	6.30	350	5.84	6.11	6.18	6.25	6.36
	15	79.2	13.8	101	4040	6.33	294	5.77	6.04	6.11	6.18	6.28
	12	64.0	13.9	81.6	3300	6.36	238	5.72	5.98	6.05	6.12	6.22
200 × 100	15	67.5	12.8	86.0	3520	6.40	274	3.45	3.72	3.79	3.86	3.97
	12	54.6	13.0	69.6	2880	6.43	222	3.39	3.65	3.72	3.79	3.90
	10	46.0	13.1	58.4	2440	6.46	186	3.35	3.61	3.67	3.74	3.85
150 × 90	15	53.2	9.79	67.8	1522	4.74	155	3.32	3.60	3.67	3.75	3.86
	12	43.2	9.92	55.0	1250	4.77	127	3.27	3.55	3.62	3.69	3.80
	10	36.4	10.0	46.4	1070	4.80	107	3.23	3.50	3.57	3.64	3.75
150 × 75	15	49.6	9.48	63.4	1430	4.75	150	2.65	2.94	3.01	3.09	3.21
	12	40.4	9.60	51.4	1180	4.78	123	2.59	2.87	2.94	3.02	3.14
	10	34.0	9.69	43.4	1000	4.81	103	2.56	2.83	2.90	2.97	3.08
125 × 75	12	35.6	8.19	45.4	708	3.95	86.4	2.76	3.04	3.11	3.19	3.30
	10	30.0	8.27	38.2	604	3.97	73.0	2.72	2.99	3.07	3.14	3.26
	8	24.4	8.36	31.0	494	4.00	59.2	2.68	2.95	3.02	3.09	3.20
100 × 75	12	30.8	6.73	39.4	378	3.10	56.0	2.95	3.24	3.31	3.39	3.51
	10	26.0	6.81	33.2	324	3.12	47.6	2.91	3.19	3.27	3.34	3.46
	8	21.2	6.90	27.0	266	3.14	38.6	2.87	3.15	3.22	3.29	3.41
100 × 65	10 #	24.6	6.64	31.2	308	3.14	46.4	2.43	2.72	2.79	2.87	2.99
	8 #	19.9	6.73	25.4	254	3.16	37.8	2.39	2.67	2.74	2.82	2.93
	7 #	17.5	6.77	22.4	226	3.17	33.2	2.37	2.65	2.72	2.79	2.91
100 × 50	8 ‡	17.9	6.40	22.8	232	3.19	36.4	1.73	2.02	2.09	2.17	2.29
	6 ‡	13.7	6.49	17.4	180	3.21	27.6	1.69	1.97	2.04	2.12	2.24
80 × 60	7 ‡	14.7	5.49	18.8	118	2.51	21.4	2.31	2.59	2.67	2.74	2.86
80 × 40	8 ‡	14.1	5.06	18.0	115	2.53	22.8	1.41	1.71	1.79	1.87	2.00
	6 ‡	10.8	5.15	13.8	89.8	2.55	17.5	1.37	1.66	1.74	1.82	1.97
75 × 50	8 ‡	14.8	4.98	18.8	104	2.35	20.8	1.90	2.19	2.27	2.35	2.47
	6 ‡	11.3	5.06	14.4	81.0	2.37	16.0	1.86	2.14	2.22	2.30	2.42
70 × 50	6 ‡	10.8	4.77	13.8	66.8	2.20	14.0	1.90	2.19	2.26	2.34	2.46
65 × 50	5 ‡	8.70	4.51	11.1	46.4	2.05	10.3	1.93	2.21	2.28	2.36	2.48
60 × 40	6 ‡	8.92	4.00	11.4	40.2	1.88	10.1	1.51	1.80	1.88	1.96	2.09
	5 ‡	7.52	4.04	9.58	34.4	1.89	8.50	1.49	1.78	1.86	1.94	2.06

‡ Not available from some leading producers. Check availability.
\# Check availability.
Properties about y-y axis:
$I_y = (\text{Total Area}) \cdot (r_y)^2$
$Z_y = I_y/(0.5B_o)$

CASTELLATED UNIVERSAL BEAMS

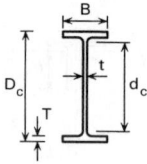

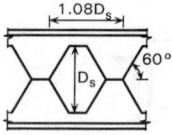

DIMENSIONS AND PROPERTIES

Section Designation		Mass per Metre	Depth of Section	Width of Section	Thickness		Depth between Fillets	Pitch 1.08 × D$_s$	Net Second Moment of Area		Net Radius of Gyration	
Original	Castellated		D$_c$	B	Web t	Flange T	d$_c$		Axis x-x	Axis y-y	Axis x-x	Axis y-y
		kg/m	mm	mm	mm	mm	mm	mm	cm^4	cm^4	cm	cm
914 × 419 × 388	1371 × 419 × 388	388.0	1378.0	420.5	21.4	36.6	1256.6	987.1	1670000	45400	64.8	10.7
914 × 419 × 343	1371 × 419 × 343	343.3	1368.8	418.5	19.4	32.0	1256.6	987.1	1450000	39100	64.6	10.6
914 × 305 × 289	1371 × 305 × 289	289.1	1383.6	307.7	19.5	32.0	1281.4	987.1	1160000	15600	64.5	7.47
914 × 305 × 253	1371 × 305 × 253	253.4	1375.4	305.5	17.3	27.9	1281.4	987.1	1010000	13300	64.3	7.38
914 × 305 × 224	1371 × 305 × 224	224.2	1367.4	304.1	15.9	23.9	1281.4	987.1	873000	11200	64.0	7.26
914 × 305 × 201	1371 × 305 × 201	200.9	1360.0	303.3	15.1	20.2	1281.4	987.1	757000	9410	63.6	7.10
838 × 292 × 226	1257 × 292 × 226	226.5	1269.9	293.8	16.1	26.8	1180.7	905.0	781000	11400	59.4	7.16
838 × 292 × 194	1257 × 292 × 194	193.8	1259.7	292.4	14.7	21.7	1180.7	905.0	645000	9060	59.0	6.99
838 × 292 × 176	1257 × 292 × 176	175.9	1253.9	291.7	14.0	18.8	1180.7	905.0	570000	7790	58.7	6.86
762 × 267 × 197	1143 × 267 × 197	196.8	1150.8	268.0	15.6	25.4	1067.0	823.0	554000	8160	53.8	6.53
762 × 267 × 173	1143 × 267 × 173	173.0	1143.2	266.7	14.3	21.6	1067.0	823.0	475000	6840	53.5	6.42
762 × 267 × 147	1143 × 267 × 147	146.9	1135.0	265.2	12.8	17.5	1067.0	823.0	392000	5450	53.2	6.27
762 × 267 × 134	1143 × 267 × 134	133.9	1131.0	264.4	12.0	15.5	1067.0	823.0	351000	4780	53.0	6.19
686 × 254 × 170	1029 × 254 × 170	170.2	1035.9	255.8	14.5	23.7	958.1	740.9	393000	6620	48.5	6.30
686 × 254 × 152	1029 × 254 × 152	152.4	1030.5	254.5	13.2	21.0	958.1	740.9	348000	5780	48.4	6.23
686 × 254 × 140	1029 × 254 × 140	140.1	1026.5	253.7	12.4	19.0	958.1	740.9	316000	5180	48.2	6.17
686 × 254 × 125	1029 × 254 × 125	125.2	1020.9	253.0	11.7	16.2	958.1	740.9	274000	4380	47.9	6.06
610 × 305 × 238	915 × 305 × 238	238.1	940.8	311.4	18.4	31.4	845.0	658.8	478000	15800	44.0	8.00
610 × 305 × 179	915 × 305 × 179	179.0	925.2	307.1	14.1	23.6	845.0	658.8	352000	11400	43.6	7.85
610 × 305 × 149	915 × 305 × 149	149.2	917.4	304.8	11.8	19.7	845.0	658.8	291000	9300	43.4	7.77
610 × 229 × 140	915 × 229 × 140	139.9	922.2	230.2	13.1	22.1	852.6	658.8	258000	4500	43.2	5.71
610 × 229 × 125	915 × 229 × 125	125.1	917.2	229.0	11.9	19.6	852.6	658.8	228000	3930	43.1	5.65
610 × 229 × 113	915 × 229 × 113	113.0	912.6	228.2	11.1	17.3	852.6	658.8	203000	3430	42.9	5.58
610 × 229 × 101	915 × 229 × 101	101.2	907.6	227.6	10.5	14.8	852.6	658.8	176000	2910	42.7	5.48
533 × 210 × 122	800 × 210 × 122	122.0	811.0	211.9	12.7	21.3	743.0	575.6	175000	3380	37.9	5.28
533 × 210 × 109	800 × 210 × 109	109.0	806.0	210.8	11.6	18.8	743.0	575.6	154000	2940	37.8	5.22
533 × 210 × 101	800 × 210 × 101	101.0	803.2	210.0	10.8	17.4	743.0	575.6	142000	2690	37.7	5.19
533 × 210 × 92	800 × 210 × 92	92.1	799.6	209.3	10.1	15.3	743.0	575.6	128000	2390	37.6	5.14
533 × 210 × 82	800 × 210 × 82	82.2	794.8	208.8	9.6	13.2	743.0	575.6	110000	2010	37.4	5.03
457 × 191 × 98	686 × 191 × 98	98.3	695.7	192.8	11.4	19.6	636.1	493.6	105000	2340	32.6	4.86
457 × 191 × 89	686 × 191 × 89	89.3	691.9	191.9	10.5	17.7	636.1	493.6	94600	2090	32.5	4.82
457 × 191 × 82	686 × 191 × 82	82.0	688.5	191.3	9.9	16.0	636.1	493.6	85600	1870	32.3	4.78
457 × 191 × 74	686 × 191 × 74	74.3	685.5	190.4	9.0	14.5	636.1	493.6	77200	1670	32.3	4.75
457 × 191 × 67	686 × 191 × 67	67.1	681.9	189.9	8.5	12.7	636.1	493.6	68200	1450	32.1	4.69
457 × 152 × 82	686 × 152 × 82	82.1	694.3	155.3	10.5	18.9	636.1	493.6	84300	1180	32.4	3.83
457 × 152 × 74	686 × 152 × 74	74.2	690.5	154.4	9.6	17.0	636.1	493.6	75500	1050	32.3	3.80
457 × 152 × 67	686 × 152 × 67	67.2	686.5	153.8	9.0	15.0	636.1	493.6	67000	911	32.1	3.74
457 × 152 × 60	686 × 152 × 60	59.8	683.1	152.9	8.1	13.3	636.1	493.6	59200	794	32.0	3.71
457 × 152 × 52	686 × 152 × 52	52.3	678.3	152.4	7.6	10.9	636.1	493.6	49800	644	31.8	3.62

CASTELLATED UNIVERSAL BEAMS

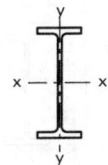

PROPERTIES (CONTINUED)

Section Designation		Net Elastic Modulus		Elastic Modulus of Tee		Net Plastic Modulus		Net Buckling Parameter	Net Torsional Index	Net Warping Constant	Net Torsional Constant	Net Area
Original	Castellated	Axis x-x cm³	Axis y-y cm³	Flange x-x cm³	Toe x-x cm³	Axis x-x cm³	Axis y-y cm³	u	x	H dm⁶	J cm⁴	A_n cm²
914 × 419 × 388	1371 × 419 × 388	24200	2160	1380	314	25600	3290	0.975	37.9	204	1590	396
914 × 419 × 343	1371 × 419 × 343	21200	1870	1250	276	22400	2850	0.974	43.0	175	1080	349
914 × 305 × 289	1371 × 305 × 289	16800	1010	1080	288	17900	1560	0.972	44.8	71.1	813	279
914 × 305 × 253	1371 × 305 × 253	14700	869	964	249	15600	1340	0.971	50.9	60.3	547	244
914 × 305 × 224	1371 × 305 × 224	12800	738	863	223	13600	1140	0.971	58.4	50.6	361	213
914 × 305 × 201	1371 × 305 × 201	11100	620	772	205	11800	956	0.969	67.1	42.2	239	187
838 × 292 × 226	1257 × 292 × 226	12300	772	811	204	13100	1180	0.972	49.0	43.8	455	221
838 × 292 × 194	1257 × 292 × 194	10200	619	704	179	10900	951	0.971	59.0	34.7	261	185
838 × 292 × 176	1257 × 292 × 176	9080	534	639	166	9660	821	0.970	66.4	29.7	183	165
762 × 267 × 197	1143 × 267 × 197	9620	609	620	159	10200	935	0.972	46.7	25.8	356	191
762 × 267 × 173	1143 × 267 × 173	8310	513	554	141	8840	788	0.971	53.9	21.5	230	166
762 × 267 × 147	1143 × 267 × 147	6900	411	476	122	7330	632	0.971	64.7	17.0	132	138
762 × 267 × 134	1143 × 267 × 134	6200	362	435	113	6590	556	0.969	71.7	14.9	97.0	125
686 × 254 × 170	1029 × 254 × 170	7590	518	480	120	8070	793	0.973	44.8	17.0	273	167
686 × 254 × 152	1029 × 254 × 152	6750	454	438	107	7160	695	0.972	50.2	14.7	193	149
686 × 254 × 140	1029 × 254 × 140	6160	408	407	98.6	6520	625	0.972	54.8	13.1	147	136
686 × 254 × 125	1029 × 254 × 125	5380	346	367	90.4	5700	531	0.972	62.7	11.0	97.9	119
610 × 305 × 238	915 × 305 × 238	10200	1020	555	135	10800	1550	0.974	30.1	32.7	722	247
610 × 305 × 179	915 × 305 × 179	7610	743	444	96.3	8040	1130	0.974	39.3	23.2	312	185
610 × 305 × 149	915 × 305 × 149	6340	610	385	77.8	6670	927	0.974	46.6	18.7	183	154
610 × 229 × 140	915 × 229 × 140	5590	391	348	86.2	5950	598	0.974	43.0	9.11	194	138
610 × 229 × 125	915 × 229 × 125	4970	343	317	76.6	5280	525	0.973	48.1	7.91	137	123
610 × 229 × 113	915 × 229 × 113	4440	301	291	69.7	4700	460	0.973	53.9	6.87	97.4	110
610 × 229 × 101	915 × 229 × 101	3880	256	263	64.1	4120	392	0.971	61.6	5.80	65.2	96.9
533 × 210 × 122	800 × 210 × 122	4310	319	262	65.9	4590	489	0.971	39.0	5.27	160	122
533 × 210 × 109	800 × 210 × 109	3820	279	240	58.6	4060	427	0.973	43.8	4.55	112	108
533 × 210 × 101	800 × 210 × 101	3540	256	225	53.8	3750	392	0.973	46.9	4.15	89.8	99.9
533 × 210 × 92	800 × 210 × 92	3200	228	209	49.3	3390	349	0.972	51.8	3.67	66.5	90.5
533 × 210 × 82	800 × 210 × 82	2780	192	188	45.4	2940	294	0.971	59.5	3.06	43.7	79.1
457 × 191 × 98	686 × 191 × 98	3020	243	177	43.6	3220	371	0.974	36.3	2.68	110	99.2
457 × 191 × 89	686 × 191 × 89	2730	218	164	39.2	2900	332	0.974	40.0	2.37	81.9	89.8
457 × 191 × 82	686 × 191 × 82	2490	195	153	36.1	2640	298	0.973	43.8	2.11	61.8	81.9
457 × 191 × 74	686 × 191 × 74	2250	175	142	32.3	2380	267	0.973	48.1	1.88	46.2	74.1
457 × 191 × 67	686 × 191 × 67	2000	153	130	29.7	2120	233	0.973	54.0	1.62	32.5	66.1
457 × 152 × 82	686 × 152 × 82	2430	152	147	39.2	2600	234	0.974	38.3	1.35	80.4	80.5
457 × 152 × 74	686 × 152 × 74	2190	135	135	35.0	2330	208	0.974	42.2	1.18	59.2	72.5
457 × 152 × 67	686 × 152 × 67	1950	118	124	32.0	2080	182	0.972	47.2	1.03	42.1	65.0
457 × 152 × 60	686 × 152 × 60	1730	104	113	28.2	1840	159	0.973	52.8	0.890	29.8	57.7
457 × 152 × 52	686 × 152 × 52	1470	84.5	99.2	25.5	1560	130	0.971	62.5	0.717	18.0	49.3

The values of the elastic modulus of the Tee are the elastic modulus at the flange and at the toe of the Tee formed at the net section.

CASTELLATED UNIVERSAL BEAMS

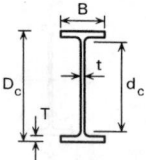

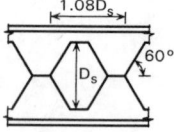

DIMENSIONS AND PROPERTIES

Section Designation		Mass per Metre	Depth of Section	Width of Section	Thickness		Depth between Fillets	Pitch 1.08 × D_s	Net Second Moment of Area		Net Radius of Gyration	
Original	Castellated				Web	Flange			Axis x-x	Axis y-y	Axis x-x	Axis y-y
		kg/m	D_c mm	B mm	t mm	T mm	d_c mm	mm	cm^4	cm^4	cm	cm
406 × 178 × 74	609 × 178 × 74	74.2	615.8	179.5	9.5	16.0	563.4	438.5	62900	1540	28.9	4.53
406 × 178 × 67	609 × 178 × 67	67.1	612.4	178.8	8.8	14.3	563.4	438.5	56200	1360	28.8	4.49
406 × 178 × 60	609 × 178 × 60	60.1	609.4	177.9	7.9	12.8	563.4	438.5	50000	1200	28.7	4.46
406 × 178 × 54	609 × 178 × 54	54.1	605.6	177.7	7.7	10.9	563.4	438.5	43500	1020	28.6	4.37
406 × 140 × 46	609 × 140 × 46	46.0	606.2	142.2	6.8	11.2	563.4	438.5	36400	538	28.5	3.46
406 × 140 × 39	609 × 140 × 39	39.0	601.0	141.8	6.4	8.6	563.4	438.5	29200	409	28.2	3.34
356 × 171 × 67	534 × 171 × 67	67.1	541.4	173.2	9.1	15.7	489.6	384.5	44800	1360	25.4	4.43
356 × 171 × 57	534 × 171 × 57	57.0	536.0	172.2	8.1	13.0	489.6	384.5	37100	1110	25.3	4.36
356 × 171 × 51	534 × 171 × 51	51.0	533.0	171.5	7.4	11.5	489.6	384.5	32800	968	25.2	4.32
356 × 171 × 45	534 × 171 × 45	45.0	529.4	171.1	7.0	9.7	489.6	384.5	28100	811	25.0	4.25
356 × 127 × 39	534 × 127 × 39	39.1	531.4	126.0	6.6	10.7	489.6	384.5	23700	357	24.9	3.07
356 × 127 × 33	534 × 127 × 33	33.1	527.0	125.4	6.0	8.5	489.6	384.5	19300	280	24.8	2.98
305 × 165 × 54	458 × 165 × 54	54.0	462.9	166.9	7.9	13.7	417.7	329.4	27000	1060	21.8	4.33
305 × 165 × 46	458 × 165 × 46	46.1	459.1	165.7	6.7	11.8	417.7	329.4	22900	895	21.7	4.30
305 × 165 × 40	458 × 165 × 40	40.3	455.9	165.0	6.0	10.2	417.7	329.4	19800	764	21.6	4.26
305 × 127 × 48	458 × 127 × 48	48.1	463.5	125.3	9.0	14.0	417.7	329.4	22100	460	21.6	3.11
305 × 127 × 42	458 × 127 × 42	41.9	459.7	124.3	8.0	12.1	417.7	329.4	19000	388	21.5	3.07
305 × 127 × 37	458 × 127 × 37	37.0	456.9	123.4	7.1	10.7	417.7	329.4	16700	336	21.4	3.04
305 × 102 × 33	458 × 102 × 33	32.8	465.2	102.4	6.6	10.8	428.4	329.4	14900	194	21.7	2.47
305 × 102 × 28	458 × 102 × 28	28.2	461.2	101.8	6.0	8.8	428.4	329.4	12400	155	21.5	2.41
305 × 102 × 25	458 × 102 × 25	24.8	457.6	101.6	5.8	7.0	428.4	329.4	10300	123	21.3	2.32
254 × 146 × 43	381 × 146 × 43	43.0	386.6	147.3	7.2	12.7	346.0	274.3	15100	677	18.2	3.85
254 × 146 × 37	381 × 146 × 37	37.0	383.0	146.4	6.3	10.9	346.0	274.3	12800	570	18.1	3.82
254 × 146 × 31	381 × 146 × 31	31.1	378.4	146.1	6.0	8.6	346.0	274.3	10300	447	17.9	3.74
254 × 102 × 28	381 × 102 × 28	28.3	387.4	102.2	6.3	10.0	352.2	274.3	9190	178	18.1	2.52
254 × 102 × 25	381 × 102 × 27	25.2	384.2	101.9	6.0	8.4	352.2	274.3	7870	148	18.0	2.47
254 × 102 × 22	381 × 102 × 22	22.0	381.0	101.6	5.7	6.8	352.2	274.3	6580	119	17.8	2.39
203 × 133 × 30	305 × 133 × 30	30.0	308.3	133.9	6.4	9.6	273.9	219.2	6680	384	14.5	3.48
203 × 133 × 25	305 × 133 × 25	25.1	304.7	133.2	5.7	7.8	273.9	219.2	5430	307	14.4	3.43
203 × 102 × 23	305 × 102 × 23	23.1	304.7	101.8	5.4	9.3	270.9	219.2	4910	164	14.3	2.62
178 × 102 × 19	267 × 102 × 19	19.0	266.8	101.2	4.8	7.9	235.8	192.2	3160	137	12.6	2.61
152 × 89 × 16	228 × 89 × 16	16.0	228.4	88.7	4.5	7.7	197.8	164.2	1950	89.7	10.7	2.30
127 × 76 × 13	191 × 76 × 13	13.0	190.5	76.0	4.0	7.6	160.1	137.2	1120	55.7	8.93	2.00

CASTELLATED UNIVERSAL BEAMS

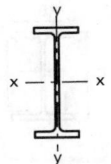

PROPERTIES (CONTINUED)

Section Designation		Net Elastic Modulus		Elastic Modulus of Tee		Net Plastic Modulus		Net Buckling Parameter	Net Torsional Index	Net Warping Constant	Net Torsional Constant	Net Area
Original	Castellated	Axis x-x cm³	Axis y-y cm³	Flange x-x cm³	Toe x-x cm³	Axis x-x cm³	Axis y-y cm³	u	x	H dm⁶	J cm⁴	A_n cm²
406 × 178 × 74	609 × 178 × 74	2040	172	121	28.4	2170	262	0.974	39.0	1.39	57.0	75.2
406 × 178 × 67	609 × 178 × 67	1840	153	112	25.6	1940	233	0.973	43.2	1.22	41.5	67.7
406 × 178 × 60	609 × 178 × 60	1640	135	103	22.6	1730	206	0.974	48.0	1.07	30.0	60.5
406 × 178 × 54	609 × 178 × 54	1440	115	94.1	21.3	1520	175	0.973	54.9	0.902	20.0	53.3
406 × 140 × 46	609 × 140 × 46	1200	75.6	78.6	18.7	1270	116	0.974	54.8	0.476	16.9	44.8
406 × 140 × 39	609 × 140 × 39	971	57.7	66.8	16.8	1030	88.8	0.970	68.0	0.359	8.93	36.7
356 × 171 × 67	534 × 171 × 67	1660	157	92.1	21.2	1760	239	0.974	34.6	0.940	51.2	69.3
356 × 171 × 57	534 × 171 × 57	1380	129	81.5	18.0	1460	196	0.974	41.1	0.757	30.2	58.1
356 × 171 × 51	534 × 171 × 51	1230	113	74.8	16.0	1300	172	0.974	45.9	0.658	21.4	51.7
356 × 171 × 45	534 × 171 × 45	1060	94.7	67.8	14.7	1120	144	0.972	53.1	0.547	13.8	44.9
356 × 127 × 39	534 × 127 × 39	891	56.7	56.5	13.8	945	87.1	0.974	49.6	0.242	13.4	38.0
356 × 127 × 33	534 × 127 × 33	732	44.6	48.6	12.1	775	68.7	0.971	60.1	0.188	7.51	31.5
305 × 165 × 54	458 × 165 × 54	1170	127	61.9	13.5	1230	193	0.974	33.7	0.536	32.3	56.7
305 × 165 × 46	458 × 165 × 46	998	108	54.7	11.0	1050	164	0.975	38.8	0.448	20.7	48.5
305 × 165 × 40	458 × 165 × 40	866	92.6	49.6	9.58	910	140	0.974	44.4	0.379	13.6	42.2
305 × 127 × 48	458 × 127 × 48	954	73.4	54.8	14.9	1020	113	0.973	33.1	0.232	28.1	47.5
305 × 127 × 42	458 × 127 × 42	827	62.4	49.0	12.8	881	96.0	0.972	37.8	0.194	18.5	41.2
305 × 127 × 37	458 × 127 × 37	730	54.4	44.3	11.1	775	83.5	0.972	42.4	0.167	12.9	36.4
305 × 102 × 33	458 × 102 × 33	641	37.8	41.9	11.3	685	58.4	0.971	44.3	0.100	10.7	31.8
305 × 102 × 28	458 × 102 × 28	536	30.5	36.4	9.87	572	47.1	0.970	52.7	0.0794	6.30	26.7
305 × 102 × 25	458 × 102 × 25	451	24.2	31.6	9.11	482	37.5	0.966	62.6	0.0623	3.78	22.8
254 × 146 × 43	381 × 146 × 43	780	91.9	39.0	8.65	827	139	0.974	30.3	0.237	22.3	45.6
254 × 146 × 37	381 × 146 × 37	670	77.9	34.8	7.24	707	118	0.973	34.9	0.197	14.3	39.2
254 × 146 × 31	381 × 146 × 31	544	61.2	31.3	6.45	572	93.0	0.971	42.9	0.153	7.64	32.1
254 × 102 × 28	381 × 102 × 28	474	34.9	29.3	7.51	506	53.6	0.972	38.8	0.0635	8.51	28.1
254 × 102 × 25	381 × 102 × 25	410	29.1	26.5	6.85	436	44.9	0.971	44.8	0.0524	5.50	24.4
254 × 102 × 22	381 × 102 × 22	345	23.5	23.3	6.22	368	36.2	0.968	52.7	0.0417	3.36	20.8
203 × 133 × 30	305 × 133 × 30	433	57.4	22.1	4.86	459	87.2	0.970	31.0	0.0857	9.42	31.7
203 × 133 × 25	305 × 133 × 25	357	46.2	19.6	4.08	376	70.1	0.969	37.2	0.0678	5.34	26.2
203 × 102 × 23	305 × 102 × 23	322	32.2	16.4	3.79	342	49.0	0.976	32.1	0.0357	6.49	23.9
178 × 102 × 19	267 × 102 × 19	237	27.0	11.8	2.60	251	41.1	0.974	32.4	0.0229	4.08	20.0
152 × 89 × 16	228 × 89 × 16	171	20.2	7.86	1.80	181	30.8	0.974	28.1	0.0109	3.33	16.9
127 × 76 × 13	191 × 76 × 13	117	14.7	4.63	1.12	125	22.3	0.976	23.5	0.00466	2.72	14.0

The values of the elastic modulus of the Tee are the elastic modulus at the flange and at the toe of the Tee formed at the net section.

CASTELLATED UNIVERSAL BEAMS

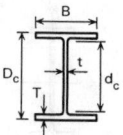

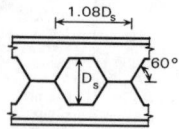

DIMENSIONS AND PROPERTIES

Section Designation		Mass per Metre	Depth of Section	Width of Section	Thickness		Depth between Fillets	Pitch $1.08 \times D_s$	Net Second Moment of Area		Net Radius of Gyration	
Original	Castellated				Web	Flange			Axis x-x	Axis y-y	Axis x-x	Axis y-y
			D_c	B	t	T	d_c					
		kg/m	mm	mm	mm	mm	mm	mm	cm⁴	cm⁴	cm	cm
356 × 406 × 634	534 × 406 × 634	633.9	665.1	424.0	47.6	77.0	480.7	411.5	600000	98000	28.9	11.7
356 × 406 × 551	534 × 406 × 551	551.0	646.1	418.5	42.1	67.5	480.7	411.5	504000	82600	28.5	11.5
356 × 406 × 467	534 × 406 × 467	467.0	627.1	412.2	35.8	58.0	480.7	411.5	413000	67800	28.0	11.3
356 × 406 × 393	534 × 406 × 393	393.0	609.5	407.0	30.6	49.2	480.7	411.5	336000	55300	27.6	11.2
356 × 406 × 340	534 × 406 × 340	339.9	596.9	403.0	26.6	42.9	480.7	411.5	284000	46800	27.3	11.1
356 × 406 × 287	534 × 406 × 287	287.1	584.1	399.0	22.6	36.5	480.7	411.5	235000	38700	27.0	10.9
356 × 406 × 235	534 × 406 × 235	235.1	571.5	394.8	18.4	30.2	480.7	411.5	188000	31000	26.7	10.8
356 × 368 × 202	534 × 368 × 202	201.9	552.6	374.7	16.5	27.0	468.2	384.5	152000	23700	25.8	10.2
356 × 368 × 177	534 × 368 × 177	177.0	546.2	372.6	14.4	23.8	468.2	384.5	132000	20500	25.7	10.1
356 × 368 × 153	534 × 368 × 153	152.9	540.0	370.5	12.3	20.7	468.2	384.5	113000	17600	25.5	10.1
356 × 368 × 129	534 × 368 × 129	129.0	533.6	368.6	10.4	17.5	468.2	384.5	94000	14600	25.4	10.0
305 × 305 × 283	458 × 305 × 283	282.9	517.8	322.2	26.8	44.1	399.2	329.4	172000	24600	23.2	8.78
305 × 305 × 240	458 × 305 × 240	240.0	505.0	318.4	23.0	37.7	399.2	329.4	142000	20300	22.9	8.66
305 × 305 × 198	458 × 305 × 198	198.1	492.4	314.5	19.1	31.4	399.2	329.4	114000	16300	22.6	8.54
305 × 305 × 158	458 × 305 × 158	158.1	479.6	311.2	15.8	25.0	399.2	329.4	88100	12600	22.3	8.42
305 × 305 × 137	458 × 305 × 137	136.9	473.0	309.2	13.8	21.7	399.2	329.4	75100	10700	22.1	8.35
305 × 305 × 118	458 × 305 × 118	117.9	467.0	307.4	12.0	18.7	399.2	329.4	63800	9060	22.0	8.29
305 × 305 × 97	458 × 305 × 97	96.9	460.4	305.3	9.9	15.4	399.2	329.4	51700	7310	21.8	8.21
254 × 254 × 167	381 × 254 × 167	167.1	416.1	265.2	19.2	31.7	327.3	274.3	67100	9860	18.9	7.23
254 × 254 × 132	381 × 254 × 132	132.0	403.3	261.3	15.3	25.3	327.3	274.3	51200	7530	18.6	7.11
254 × 254 × 107	381 × 254 × 107	107.1	393.7	258.8	12.8	20.5	327.3	274.3	40300	5930	18.3	7.02
254 × 254 × 89	381 × 254 × 89	88.9	387.3	256.3	10.3	17.3	327.3	274.3	33200	4860	18.2	6.96
254 × 254 × 73	381 × 254 × 73	73.1	381.1	254.6	8.6	14.2	327.3	274.3	26700	3910	18.0	6.90
203 × 203 × 86	305 × 203 × 86	86.1	323.7	209.1	12.7	20.5	262.3	219.2	21400	3130	14.9	5.68
203 × 203 × 71	305 × 203 × 71	71.0	317.3	206.4	10.0	17.3	262.3	219.2	17400	2540	14.7	5.62
203 × 203 × 60	305 × 203 × 60	60.0	311.1	205.8	9.4	14.2	262.3	219.2	14200	2060	14.6	5.56
203 × 203 × 52	305 × 203 × 52	52.0	307.7	204.3	7.9	12.5	262.3	219.2	12200	1780	14.5	5.52
203 × 203 × 46	305 × 203 × 46	46.1	304.7	203.6	7.2	11.0	262.3	219.2	10700	1550	14.4	5.49
152 × 152 × 37	228 × 152 × 37	37.0	237.8	154.4	8.0	11.5	199.6	164.2	5030	706	11.1	4.15
152 × 152 × 30	228 × 152 × 30	30.0	233.6	152.9	6.5	9.4	199.6	164.2	4020	560	11.0	4.10
152 × 152 × 23	228 × 152 × 23	23.0	228.4	152.2	5.8	6.8	199.6	164.2	2910	400	10.8	4.01

CASTELLATED UNIVERSAL BEAMS

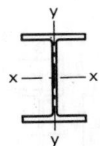

PROPERTIES (CONTINUED)

Section Designation		Net Elastic Modulus		Elastic Modulus of Tee		Net Plastic Modulus		Net Bucking Parameter	Net Torsional Index	Net Warping Constant	Net Torsional Constant	Net Area
Original	Castellated	Axis x-x cm³	Axis y-y cm³	Flange x-x cm³	Toe x-x cm³	Axis x-x cm³	Axis y-y cm³	u	x	H dm⁶	J cm⁴	Aₙ cm²
356 × 406 × 634	534 × 406 × 634	18100	4620	706	325	20600	7000	0.946	7.81	84.7	13000	717
356 × 406 × 551	534 × 406 × 551	15600	3950	573	245	17600	5970	0.946	8.72	69.1	8770	622
356 × 406 × 467	534 × 406 × 467	13200	3290	452	177	14700	4970	0.946	9.96	54.9	5520	527
356 × 406 × 393	534 × 406 × 393	11000	2720	358	127	12200	4110	0.947	11.5	43.4	3360	442
356 × 406 × 340	534 × 406 × 340	9530	2320	298	96.8	10400	3510	0.948	13.0	35.9	2220	382
356 × 406 × 287	534 × 406 × 287	8040	1940	244	71.4	8680	2930	0.947	15.1	29.0	1370	323
356 × 406 × 235	534 × 406 × 235	6580	1570	195	50.4	7040	2370	0.949	17.9	22.7	772	264
356 × 368 × 202	534 × 368 × 202	5500	1260	196	46.2	5870	1910	0.949	19.5	16.4	531	228
356 × 368 × 177	534 × 368 × 177	4820	1100	174	37.7	5120	1660	0.949	21.9	14.0	364	200
356 × 368 × 153	534 × 368 × 153	4180	947	153	30.2	4410	1430	0.949	25.0	11.8	239	173
356 × 368 × 129	534 × 368 × 129	3520	793	134	23.8	3700	1190	0.949	29.2	9.73	146	146
305 × 305 × 283	458 × 305 × 283	6650	1530	256	90.4	7390	2320	0.950	10.9	13.8	1940	320
305 × 305 × 240	458 × 305 × 240	5620	1280	210	67.6	6180	1930	0.950	12.5	11.1	1210	271
305 × 305 × 198	458 × 305 × 198	4630	1040	170	48.7	5030	1570	0.952	14.7	8.66	699	223
305 × 305 × 158	458 × 305 × 158	3670	807	137	34.4	3940	1220	0.952	18.1	6.49	358	177
305 × 305 × 137	458 × 305 × 137	3180	692	120	27.8	3390	1050	0.953	20.6	5.45	236	153
305 × 305 × 118	458 × 305 × 118	2730	589	106	22.5	2890	890	0.952	23.6	4.55	152	132
305 × 305 × 97	458 × 305 × 97	2240	479	90.9	17.1	2360	722	0.954	28.2	3.62	86.3	108
254 × 254 × 167	381 × 254 × 167	3220	744	115	37.9	3540	1130	0.952	12.2	3.64	596	188
254 × 254 × 132	381 × 254 × 132	2540	576	88.0	25.1	2750	871	0.950	15.0	2.69	303	149
254 × 254 × 107	381 × 254 × 107	2050	458	71.7	18.1	2200	692	0.952	18.1	2.06	164	120
254 × 254 × 89	381 × 254 × 89	1710	379	59.5	13.3	1820	572	0.953	21.2	1.66	97.7	100
254 × 254 × 73	381 × 254 × 73	1400	307	50.6	10.1	1480	463	0.952	25.4	1.31	54.9	82.2
203 × 203 × 86	305 × 203 × 86	1320	299	46.8	13.5	1440	452	0.951	14.8	0.718	130	96.7
203 × 203 × 71	305 × 203 × 71	1100	246	37.7	9.61	1180	371	0.952	17.4	0.571	76.9	80.3
203 × 203 × 60	305 × 203 × 60	911	201	33.5	7.83	971	303	0.951	20.6	0.455	44.4	66.8
203 × 203 × 52	305 × 203 × 52	796	174	29.2	6.22	843	263	0.952	23.3	0.387	30.1	58.3
203 × 203 × 46	305 × 203 × 46	702	152	26.7	5.34	740	230	0.952	26.1	0.334	20.9	51.4
152 × 152 × 37	228 × 152 × 37	423	91.4	16.8	4.13	453	138	0.952	19.4	0.0904	17.9	41.0
152 × 152 × 30	228 × 152 × 30	344	73.3	14.0	3.05	365	111	0.952	23.4	0.0704	9.82	33.3
152 × 152 × 23	228 × 152 × 23	255	52.5	12.0	2.37	268	79.5	0.952	30.7	0.0491	4.14	24.8

The values of the elastic modulus of the Tee are the elastic modulus at the flange and at the toe of the Tee formed at the net section.

STRUCTURAL TEES
CUT FROM UNIVERSAL BEAMS

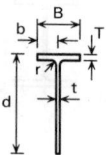

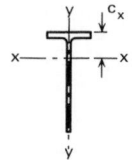

DIMENSIONS AND PROPERTIES

Section Designation	Cut from Universal Beam Section Designation	Mass per Metre kg/m	Width of Section B mm	Depth of Section d mm	Thickness Web t mm	Thickness Flange T mm	Root Radius r mm	Ratios for Local Buckling Flange b/T	Ratios for Local Buckling Web d/t	Dimension c_x cm	Second Moment of Area Axis x-x cm⁴	Second Moment of Area Axis y-y cm⁴
305 × 457 × 127	914 × 305 × 253	126.7	305.5	459.1	17.3	27.9	19.1	5.47	26.5	12.0	32700	6650
305 × 457 × 112	914 × 305 × 224	112.1	304.1	455.1	15.9	23.9	19.1	6.36	28.6	12.1	29100	5620
305 × 457 × 101	914 × 305 × 201	100.4	303.3	451.4	15.1	20.2	19.1	7.51	29.9	12.5	26400	4710
292 × 419 × 113	838 × 292 × 226	113.3	293.8	425.4	16.1	26.8	17.8	5.48	26.4	10.8	24600	5680
292 × 419 × 97	838 × 292 × 194	96.9	292.4	420.3	14.7	21.7	17.8	6.74	28.6	11.1	21300	4530
292 × 419 × 88	838 × 292 × 176	87.9	291.7	417.4	14.0	18.8	17.8	7.76	29.8	11.4	19600	3900
267 × 381 × 99	762 × 267 × 197	98.4	268.0	384.8	15.6	25.4	16.5	5.28	24.7	9.89	17500	4090
267 × 381 × 87	762 × 267 × 173	86.5	266.7	381.0	14.3	21.6	16.5	6.17	26.6	9.98	15500	3430
267 × 381 × 74	762 × 267 × 147	73.5	265.2	376.9	12.8	17.5	16.5	7.58	29.4	10.2	13200	2730
267 × 381 × 67	762 × 267 × 134	66.9	264.4	374.9	12.0	15.5	16.5	8.53	31.2	10.3	12100	2390
254 × 343 × 85	686 × 254 × 170	85.1	255.8	346.4	14.5	23.7	15.2	5.40	23.9	8.67	12100	3320
254 × 343 × 76	686 × 254 × 152	76.2	254.5	343.7	13.2	21.0	15.2	6.06	26.0	8.61	10800	2890
254 × 343 × 70	686 × 254 × 140	70.0	253.7	341.7	12.4	19.0	15.2	6.68	27.6	8.63	9910	2590
254 × 343 × 63	686 × 254 × 125	62.6	253.0	338.9	11.7	16.2	15.2	7.81	29.0	8.85	8980	2190
305 × 305 × 119	610 × 305 × 238	119.0	311.4	317.8	18.4	31.4	16.5	4.96	17.3	7.11	12300	7920
305 × 305 × 90	610 × 305 × 179	89.5	307.1	310.0	14.1	23.6	16.5	6.51	22.0	6.69	9040	5700
305 × 305 × 75	610 × 305 × 149	74.6	304.8	306.1	11.8	19.7	16.5	7.74	25.9	6.45	7420	4650
229 × 305 × 70	610 × 229 × 140	69.9	230.2	308.5	13.1	22.1	12.7	5.21	23.5	7.61	7740	2250
229 × 305 × 63	610 × 229 × 125	62.5	229.0	306.0	11.9	19.6	12.7	5.84	25.7	7.54	6900	1970
229 × 305 × 57	610 × 229 × 113	56.5	228.2	303.7	11.1	17.3	12.7	6.60	27.4	7.58	6270	1720
229 × 305 × 51	610 × 229 × 101	50.6	227.6	301.2	10.5	14.8	12.7	7.69	28.7	7.78	5690	1460
210 × 267 × 61	533 × 210 × 122	61.0	211.9	272.2	12.7	21.3	12.7	4.97	21.4	6.66	5160	1690
210 × 267 × 55	533 × 210 × 109	54.5	210.8	269.7	11.6	18.8	12.7	5.61	23.3	6.61	4600	1470
210 × 267 × 51	533 × 210 × 101	50.5	210.0	268.3	10.8	17.4	12.7	6.03	24.8	6.53	4250	1350
210 × 267 × 46	533 × 210 × 92	46.1	209.3	266.5	10.1	15.6	12.7	6.71	26.4	6.55	3890	1200
210 × 267 × 41	533 × 210 × 82	41.1	208.8	264.1	9.6	13.2	12.7	7.91	27.5	6.75	3530	1000
191 × 229 × 49	457 × 191 × 98	49.2	192.8	233.5	11.4	19.6	10.2	4.92	20.5	5.53	2970	1170
191 × 229 × 45	457 × 191 × 89	44.6	191.9	231.6	10.5	17.7	10.2	5.42	22.1	5.47	2680	1050
191 × 229 × 41	457 × 191 × 82	41.0	191.3	229.9	9.9	16.0	10.2	5.98	23.2	5.47	2470	935
191 × 229 × 37	457 × 191 × 74	37.1	190.4	228.4	9.0	14.5	10.2	6.57	25.4	5.38	2220	836
191 × 229 × 34	457 × 191 × 67	33.6	189.9	226.6	8.5	12.7	10.2	7.48	26.7	5.46	2030	726
152 × 229 × 41	457 × 152 × 82	41.0	155.3	232.8	10.5	18.9	10.2	4.11	22.2	5.96	2600	592
152 × 229 × 37	457 × 152 × 74	37.1	154.4	230.9	9.6	17.0	10.2	4.54	24.1	5.88	2330	523
152 × 229 × 34	457 × 152 × 67	33.6	153.8	228.9	9.0	15.0	10.2	5.13	25.4	5.91	2120	456
152 × 229 × 30	457 × 152 × 60	29.9	152.9	227.2	8.1	13.3	10.2	5.75	28.0	5.84	1880	397
152 × 229 × 26	457 × 152 × 52	26.2	152.4	224.8	7.6	10.9	10.2	6.99	29.6	6.04	1670	322

STRUCTURAL TEES
CUT FROM UNIVERSAL BEAMS

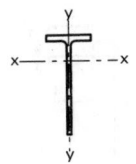

PROPERTIES (CONTINUED)

Section Designation	Radius of Gyration		Elastic Modulus			Plastic Modulus		Buckling Parameter	Torsional Index	Mono-symmetry Index	Warping Constant (*)	Torsional Constant	Area of Section
	Axis x-x	Axis y-y	Axis x-x		Axis y-y	Axis x-x	Axis y-y						
			Flange	Toe				u	x	ψ	H	J	A
	cm	cm	cm³	cm³	cm³	cm³	cm³				cm⁶	cm⁴	cm²
305 × 457 × 127	14.2	6.42	2720	965	435	1730	685	0.656	18.1	0.749	17000	313	161
305 × 457 × 112	14.3	6.27	2400	871	369	1570	582	0.666	20.6	0.753	12400	211	143
305 × 457 × 101	14.4	6.07	2110	808	311	1460	491	0.685	23.4	0.759	9820	146	128
292 × 419 × 113	13.1	6.27	2280	776	387	1380	606	0.640	17.5	0.742	11500	257	144
292 × 419 × 97	13.1	6.06	1930	689	310	1240	487	0.660	20.8	0.747	7830	153	123
292 × 419 × 88	13.2	5.90	1720	644	267	1160	421	0.675	23.2	0.751	6320	111	112
267 × 381 × 99	11.8	5.71	1770	613	305	1090	479	0.641	16.6	0.741	7620	202	125
267 × 381 × 87	11.9	5.58	1550	550	257	986	404	0.654	19.0	0.745	5450	134	110
267 × 381 × 74	11.9	5.40	1300	481	206	867	324	0.670	22.6	0.749	3600	79.5	93.6
267 × 381 × 67	11.9	5.30	1180	445	181	806	285	0.679	24.9	0.753	2850	59.2	85.3
254 × 343 × 85	10.5	5.53	1390	464	259	826	406	0.624	15.9	0.731	4720	154	108
254 × 343 × 76	10.5	5.46	1250	417	227	743	355	0.627	17.7	0.732	3420	110	97.0
254 × 343 × 70	10.5	5.39	1150	388	204	691	319	0.633	19.3	0.734	2720	84.3	89.2
254 × 343 × 63	10.6	5.24	1010	358	173	643	271	0.651	21.9	0.740	2090	58.1	79.7
305 × 305 × 119	9.02	7.23	1740	500	509	894	787	0.483	10.6	0.661	11300	393	152
305 × 305 × 90	8.91	7.07	1350	372	371	657	572	0.485	13.8	0.664	4710	170	114
305 × 305 × 75	8.83	7.00	1150	307	305	539	469	0.483	16.3	0.666	2690	100	95.0
229 × 305 × 70	9.32	5.03	1020	333	196	592	306	0.613	15.3	0.727	2560	108	89.1
229 × 305 × 63	9.31	4.97	915	299	172	531	268	0.617	17.0	0.728	1840	77.1	79.7
229 × 305 × 57	9.33	4.88	826	275	150	489	235	0.626	19.0	0.731	1400	55.7	72.0
229 × 305 × 51	9.40	4.76	732	255	128	456	200	0.645	21.5	0.736	1080	38.5	64.4
210 × 267 × 61	8.15	4.67	775	251	160	446	250	0.600	13.8	0.719	1660	89.2	77.7
210 × 267 × 55	8.14	4.60	697	226	140	401	218	0.605	15.4	0.721	1200	63.2	69.4
210 × 267 × 51	8.12	4.57	650	209	128	371	200	0.606	16.6	0.722	951	50.5	64.3
210 × 267 × 46	8.14	4.51	593	193	114	343	178	0.613	18.2	0.724	737	37.8	58.7
210 × 267 × 41	8.21	4.38	523	179	96.1	320	150	0.634	20.8	0.730	565	25.8	52.3
191 × 229 × 49	6.88	4.33	536	167	122	296	189	0.573	12.9	0.705	835	60.6	62.6
191 × 229 × 45	6.87	4.29	491	152	109	269	169	0.576	14.1	0.706	628	45.3	56.9
191 × 229 × 41	6.88	4.23	452	141	97.8	250	152	0.583	15.4	0.708	494	34.6	52.2
191 × 229 × 37	6.86	4.20	413	127	87.8	225	136	0.583	16.9	0.709	365	25.9	47.3
191 × 229 × 34	6.90	4.12	372	118	76.5	209	119	0.597	18.9	0.713	280	18.6	42.7
152 × 229 × 41	7.05	3.37	436	150	76.3	267	120	0.634	13.7	0.740	534	44.6	52.3
152 × 229 × 37	7.03	3.33	397	135	67.8	242	107	0.637	15.1	0.742	396	33.0	47.2
152 × 229 × 34	7.04	3.27	359	125	59.3	223	93.3	0.646	16.8	0.745	305	23.8	42.8
152 × 229 × 30	7.02	3.23	322	111	52.0	199	81.5	0.649	18.7	0.746	217	16.9	38.1
152 × 229 × 26	7.08	3.11	276	102	42.3	183	66.7	0.671	21.9	0.753	161	10.7	33.3

(*) Note units are cm⁶ and not dm⁶.

STRUCTURAL TEES
CUT FROM UNIVERSAL BEAMS

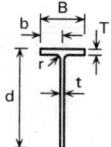

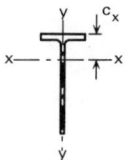

DIMENSIONS AND PROPERTIES

Section Designation	Cut from Universal Beam Section Designation	Mass per Metre	Width of Section	Depth of Section	Thickness		Root Radius	Ratios for Local Buckling		Dimension	Second Moment of Area	
					Web	Flange		Flange	Web		Axis x-x	Axis y-y
			B	d	t	T	r	b/T	d/t	c_x		
		kg/m	mm	mm	mm	mm	mm			cm	cm⁴	cm⁴
178 × 203 × 37	406 × 178 × 74	37.1	179.5	206.3	9.5	16.0	10.2	5.61	21.7	4.76	1740	773
178 × 203 × 34	406 × 178 × 67	33.6	178.8	204.6	8.8	14.3	10.2	6.25	23.2	4.73	1570	682
178 × 203 × 30	406 × 178 × 60	30.0	177.9	203.1	7.9	12.8	10.2	6.95	25.7	4.64	1400	602
178 × 203 × 27	406 × 178 × 54	27.1	177.7	201.2	7.7	10.9	10.2	8.15	26.1	4.83	1290	511
140 × 203 × 23	406 × 140 × 46	23.0	142.2	201.5	6.8	11.2	10.2	6.35	29.6	5.02	1120	269
140 × 203 × 20	406 × 140 × 39	19.5	141.8	198.9	6.4	8.6	10.2	8.24	31.1	5.32	979	205
171 × 178 × 34	356 × 171 × 67	33.5	173.2	181.6	9.1	15.7	10.2	5.52	20.0	4.00	1150	681
171 × 178 × 29	356 × 171 × 57	28.5	172.2	178.9	8.1	13.0	10.2	6.62	22.1	3.97	986	554
171 × 178 × 26	356 × 171 × 51	25.5	171.5	177.4	7.4	11.5	10.2	7.46	24.0	3.94	882	484
171 × 178 × 23	356 × 171 × 45	22.5	171.1	175.6	7.0	9.7	10.2	8.82	25.1	4.05	798	406
127 × 178 × 20	356 × 127 × 39	19.5	126.0	176.6	6.6	10.7	10.2	5.89	26.8	4.43	728	179
127 × 178 × 17	356 × 127 × 33	16.5	125.4	174.4	6.0	8.5	10.2	7.38	29.1	4.56	626	140
165 × 152 × 27	305 × 165 × 54	27.0	166.9	155.1	7.9	13.7	8.9	6.09	19.6	3.21	642	531
165 × 152 × 23	305 × 165 × 46	23.1	165.7	153.2	6.7	11.8	8.9	7.02	22.9	3.07	536	448
165 × 152 × 20	305 × 165 × 40	20.1	165.0	151.6	6.0	10.2	8.9	8.09	25.3	3.03	468	382
127 × 152 × 24	305 × 127 × 48	24.0	125.3	155.4	9.0	14.0	8.9	4.47	17.3	3.94	662	231
127 × 152 × 21	305 × 127 × 42	21.0	124.3	153.5	8.0	12.1	8.9	5.14	19.2	3.87	573	194
127 × 152 × 19	305 × 127 × 37	18.5	123.4	152.1	7.1	10.7	8.9	5.77	21.4	3.78	501	168
102 × 152 × 17	305 × 102 × 33	16.4	102.4	156.3	6.6	10.8	7.6	4.74	23.7	4.14	487	97.0
102 × 152 × 14	305 × 102 × 28	14.1	101.8	154.3	6.0	8.8	7.6	5.78	25.7	4.20	420	77.7
102 × 152 × 13	305 × 102 × 25	12.4	101.6	152.5	5.8	7.0	7.6	7.26	26.3	4.43	377	61.5
146 × 127 × 22	254 × 146 × 43	21.5	147.3	129.7	7.2	12.7	7.6	5.80	18.0	2.64	343	339
146 × 127 × 19	254 × 146 × 37	18.5	146.4	127.9	6.3	10.9	7.6	6.72	20.3	2.55	292	285
146 × 127 × 16	254 × 146 × 31	15.6	146.1	125.6	6.0	8.6	7.6	8.49	20.9	2.66	259	224
102 × 127 × 14	254 × 102 × 28	14.2	102.2	130.1	6.3	10.0	7.6	5.11	20.7	3.24	277	89.3
102 × 127 × 13	254 × 102 × 25	12.6	101.9	128.5	6.0	8.4	7.6	6.07	21.4	3.32	250	74.3
102 × 127 × 11	254 × 102 × 22	11.0	101.6	126.9	5.7	6.8	7.6	7.47	22.3	3.45	223	59.7
133 × 102 × 15	203 × 133 × 30	15.0	133.9	103.3	6.4	9.6	7.6	6.97	16.1	2.11	154	192
133 × 102 × 13	203 × 133 × 25	12.5	133.2	101.5	5.7	7.8	7.6	8.54	17.8	2.10	131	154

STRUCTURAL TEES
CUT FROM UNIVERSAL BEAMS

PROPERTIES (CONTINUED)

Section Designation	Radius of Gyration		Elastic Modulus			Plastic Modulus		Buckling Parameter	Torsional Index	Mono-symmetry Index	Warping Constant (*)	Torsional Constant	Area of Section
	Axis x-x	Axis y-y	Axis x-x		Axis y-y	Axis x-x	Axis y-y						
			Flange	Toe				u	x	ψ	H	J	A
	cm	cm	cm³	cm³	cm³	cm³	cm³				cm⁶	cm⁴	cm²
178 × 203 × 37	6.06	4.04	365	109	86.1	194	133	0.556	13.8	0.696	350	31.4	47.2
178 × 203 × 34	6.07	3.99	332	100	76.3	177	118	0.561	15.2	0.698	262	23.1	42.8
178 × 203 × 30	6.04	3.97	301	89.0	67.6	157	105	0.561	16.9	0.699	186	16.7	38.3
178 × 203 × 27	6.13	3.85	268	84.6	57.5	150	89.1	0.588	19.1	0.705	146	11.6	34.5
140 × 203 × 23	6.19	3.03	224	74.2	37.8	132	59.1	0.633	19.5	0.740	93.7	9.51	29.3
140 × 203 × 20	6.28	2.87	184	67.2	28.9	121	45.4	0.668	23.7	0.750	66.3	5.35	24.8
171 × 178 × 34	5.20	3.99	288	81.5	78.6	145	121	0.500	12.2	0.672	249	27.8	42.7
171 × 178 × 29	5.21	3.91	248	70.9	64.4	125	99.4	0.514	14.4	0.676	154	16.7	36.3
171 × 178 × 26	5.21	3.86	224	63.9	56.5	113	87.1	0.522	16.0	0.677	110	11.9	32.4
171 × 178 × 23	5.28	3.76	197	59.1	47.4	104	73.3	0.545	18.4	0.683	79.2	7.92	28.7
127 × 178 × 20	5.41	2.68	164	55.0	28.4	98.1	44.5	0.632	17.6	0.739	57.1	7.55	24.9
127 × 178 × 17	5.45	2.58	137	48.6	22.3	87.2	35.1	0.654	21.1	0.746	38.0	4.40	21.1
165 × 152 × 27	4.32	3.93	200	52.2	63.7	92.9	97.8	0.389	11.8	0.636	128	17.4	34.4
165 × 152 × 23	4.27	3.91	174	43.7	54.1	77.2	82.8	0.380	13.6	0.636	78.6	11.1	29.4
165 × 152 × 20	4.27	3.86	155	38.6	46.3	67.7	70.9	0.393	15.5	0.638	52.0	7.37	25.7
127 × 152 × 24	4.65	2.74	168	57.1	36.8	102	58.0	0.602	11.7	0.714	104	15.9	30.6
127 × 152 × 21	4.63	2.70	148	49.9	31.3	88.9	49.2	0.606	13.2	0.716	69.2	10.6	26.7
127 × 152 × 19	4.61	2.67	132	43.8	27.2	78.0	42.7	0.606	14.9	0.718	47.4	7.38	23.6
102 × 152 × 17	4.82	2.15	118	42.3	19.0	75.8	30.0	0.656	15.8	0.749	36.8	6.10	20.9
102 × 152 × 14	4.84	2.08	100	37.4	15.3	67.4	24.2	0.673	18.7	0.756	25.2	3.70	17.9
102 × 152 × 13	4.88	1.97	85.0	34.8	12.1	63.4	19.4	0.702	21.7	0.766	20.4	2.39	15.8
146 × 127 × 22	3.54	3.52	130	33.2	46.0	59.6	70.5	0.195	10.6	0.613	64.9	11.9	27.4
146 × 127 × 19	3.52	3.48	115	28.5	39.0	50.7	59.7	0.233	12.2	0.616	41.0	7.67	23.6
146 × 127 × 16	3.61	3.36	97.4	26.2	30.6	46.1	47.1	0.376	14.8	0.623	24.5	4.28	19.8
102 × 127 × 14	3.92	2.22	85.5	28.3	17.5	50.5	27.4	0.608	13.7	0.720	21.0	4.78	18.0
102 × 127 × 13	3.95	2.15	75.3	26.2	14.6	46.9	23.0	0.629	15.7	0.727	15.9	3.21	16.0
102 × 127 × 11	3.99	2.06	64.5	24.1	11.7	43.4	18.6	0.655	18.2	0.735	12.0	2.07	14.0
133 × 102 × 15	2.84	3.17	73.1	18.8	28.7	33.5	44.1	–	10.7	0.569	21.7	5.15	19.1
133 × 102 × 13	2.86	3.10	62.4	16.2	23.1	28.7	35.5	–	12.8	0.572	12.6	2.98	16.0

(*) Note units are cm⁶ and not dm⁶.
– Indicates that no values of u and x are given, as lateral torsional buckling due to bending about the x-x axis is not possible, because the second moment of area about the y-y axis exceeds the second moment of area about the x-x axis.

STRUCTURAL TEES
CUT FROM UNIVERSAL COLUMNS

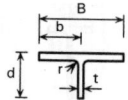

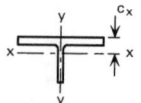

DIMENSIONS

Section Designation	Cut from Universal Column Section Designation	Mass per Metre kg/m	Width of Section B mm	Depth of Section d mm	Thickness		Root Radius r mm	Ratios for Local Buckling		Dimension c_x cm
					Web t mm	Flange T mm		Flange b/T	Web d/t	
406 × 178 × 118	356 × 406 × 235	117.5	394.8	190.4	18.4	30.2	15.2	6.54	10.3	3.40
368 × 178 × 101	356 × 368 × 202	100.9	374.7	187.2	16.5	27.0	15.2	6.94	11.3	3.29
368 × 178 × 89	356 × 368 × 177	88.5	372.6	184.0	14.4	23.8	15.2	7.83	12.8	3.09
368 × 178 × 77	356 × 368 × 153	76.5	370.5	180.9	12.3	20.7	15.2	8.95	14.7	2.88
368 × 178 × 65	356 × 368 × 129	64.5	368.6	177.7	10.4	17.5	15.2	10.5	17.1	2.69
305 × 152 × 79	305 × 305 × 158	79.0	311.2	163.5	15.8	25.0	15.2	6.22	10.3	3.04
305 × 152 × 69	305 × 305 × 137	68.5	309.2	160.2	13.8	21.7	15.2	7.12	11.6	2.86
305 × 152 × 59	305 × 305 × 118	58.9	307.4	157.2	12.0	18.7	15.2	8.22	13.1	2.69
305 × 152 × 49	305 × 305 × 97	48.4	305.3	153.9	9.9	15.4	15.2	9.91	15.5	2.50
254 × 127 × 66	254 × 254 × 132	66.0	261.3	138.1	15.3	25.3	12.7	5.16	9.03	2.70
254 × 127 × 54	254 × 254 × 107	53.5	258.8	133.3	12.8	20.5	12.7	6.31	10.4	2.45
254 × 127 × 45	254 × 254 × 89	44.5	256.3	130.1	10.3	17.3	12.7	7.41	12.6	2.21
254 × 127 × 37	254 × 254 × 73	36.5	254.6	127.0	8.6	14.2	12.7	8.96	14.8	2.05
203 × 102 × 43	203 × 203 × 86	43.0	209.1	111.0	12.7	20.5	10.2	5.10	8.74	2.20
203 × 102 × 36	203 × 203 × 71	35.5	206.4	107.8	10.0	17.3	10.2	5.97	10.8	1.95
203 × 102 × 30	203 × 203 × 60	30.0	205.8	104.7	9.4	14.2	10.2	7.25	11.1	1.89
203 × 102 × 26	203 × 203 × 52	26.0	204.3	103.0	7.9	12.5	10.2	8.17	13.0	1.75
203 × 102 × 23	203 × 203 × 46	23.0	203.6	101.5	7.2	11.0	10.2	9.25	14.1	1.69
152 × 76 × 19	152 × 152 × 37	18.5	154.4	80.8	8.0	11.5	7.6	6.71	10.1	1.53
152 × 76 × 15	152 × 152 × 30	15.0	152.9	78.7	6.5	9.4	7.6	8.13	12.1	1.41
152 × 76 × 12	152 × 152 × 23	11.5	152.2	76.1	5.8	6.8	7.6	11.2	13.1	1.39

STRUCTURAL TEES
CUT FROM UNIVERSAL COLUMNS

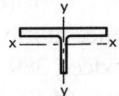

PROPERTIES

Section Designation	Second Moment of Area		Radius of Gyration		Elastic Modulus			Plastic Modulus		Mono-symmetry Index	Warping Constant (*)	Torsional Constant	Area of Section
	Axis x-x	Axis y-y	Axis x-x	Axis y-y	Axis x-x		Axis y-y	Axis x-x	Axis y-y				
					Flange	Toe							
	cm⁴	cm⁴	cm	cm	cm³	cm³	cm³	cm³	cm³	ψ	H cm⁶	J cm⁴	A cm²
406 × 178 × 118	2860	15500	4.37	10.2	843	183	785	367	1190	0.165	12700	405	150
368 × 178 × 101	2460	11800	4.38	9.60	749	160	632	312	960	0.216	7840	278	129
368 × 178 × 89	2090	10300	4.30	9.54	676	136	551	263	835	0.212	5270	190	113
368 × 178 × 77	1730	8780	4.22	9.49	601	114	474	216	717	0.209	3390	125	97.4
368 × 178 × 65	1420	7310	4.16	9.43	527	94.1	396	175	600	0.207	2010	76.2	82.2
305 × 152 × 79	1530	6290	3.90	7.90	503	115	404	225	615	0.268	3650	188	101
305 × 152 × 69	1290	5350	3.84	7.83	450	97.7	346	188	526	0.263	2340	124	87.2
305 × 152 × 59	1080	4530	3.79	7.77	401	82.8	295	156	448	0.262	1470	80.3	75.1
305 × 152 × 49	858	3650	3.73	7.69	343	66.5	239	123	363	0.258	806	45.5	61.7
254 × 127 × 66	871	3770	3.22	6.69	323	78.3	288	159	439	0.250	2200	159	84.1
254 × 127 × 54	676	2960	3.15	6.59	276	62.1	229	122	349	0.245	1150	85.9	68.2
254 × 127 × 45	524	2430	3.04	6.55	237	48.5	190	94.1	288	0.242	660	51.1	56.7
254 × 127 × 37	417	1950	2.99	6.48	204	39.2	153	74.1	233	0.236	359	28.8	46.5
203 × 102 × 43	373	1560	2.61	5.34	169	41.9	150	84.6	228	0.257	605	68.1	54.8
203 × 102 × 36	280	1270	2.49	5.30	143	31.8	123	63.6	187	0.254	343	40.0	45.2
203 × 102 × 30	244	1030	2.53	5.20	129	28.4	100	54.4	153	0.245	195	23.5	38.2
203 × 102 × 26	200	889	2.46	5.18	115	23.4	87.0	44.5	132	0.243	128	15.8	33.1
203 × 102 × 23	177	774	2.45	5.13	105	20.9	76.0	39.0	115	0.242	87.2	11.0	29.4
152 × 76 × 19	93.1	353	1.99	3.87	60.7	14.2	45.7	27.1	69.8	0.277	44.9	9.54	23.5
152 × 76 × 15	72.2	280	1.94	3.83	51.4	11.2	36.7	20.9	55.8	0.269	23.7	5.24	19.1
152 × 76 × 12	58.5	200	2.00	3.70	41.9	9.41	26.3	16.9	40.1	0.278	9.78	2.30	14.6

(*) Note units are cm⁶ and not dm⁶.
Values of u and x are not given, as lateral torsional buckling due to bending about the x-x axis is not possible, because the second moment of area about the y-y axis exceeds the second moment of area about the x-x axis.

Extracts from BS 5950: Part 1: 2000

Acknowledgement

Extracts from BS 5950 Part 1: 2000 reproduced with the permission of BSI under licence number 2002SK/0070. British Standards can be obtained from BSI Customer Services, 389 Chiswick High Road, London W4 4AL, United Kingdom. (Tel + 44 (0) 20 8996 9001).

BS 5950: Part 1: 2000: Section two

Table 8.   Suggested limits for calculated deflections

(a) *Vertical deflection of beams due to imposed load*	
Cantilevers	Length/180
Beams carrying plaster or other brittle finish	Span/360
Other beams (except purlins and sheeting rails)	Span/200
Purlins and sheeting rails	See clause **4.12.2**
(b) *Horizontal deflection of columns due to imposed load and wind load*	
Tops of columns in single-storey buildings, except portal frames	Height/300
Columns in portal frame buildings, not supporting crane runways	To suit cladding
Columns supporting crane runways	To suit crane runway
In each storey of a building with more than one storey	Height of that storey/300
(c) *Crane girders*	
Vertical deflection due to static vertical wheel loads from overhead travelling cranes	Span/600
Horizontal deflection (calculated on the top flange properties alone) due to horizontal crane loads	Span/500

BS 5950: Part 1: 2000: Section three

Table 9. Design strength p_y

Steel grade	Thickness[a] less than or equal to mm	Design strength p_y N/mm^2
S 275	16	275
	40	265
	63	255
	80	245
	100	235
	150	225
S 355	16	355
	40	345
	63	335
	80	325
	100	315
	150	295
S 460	16	460
	40	440
	63	430
	80	410
	100	400

[a] For rolled sections, use the specified thickness of the thickest element of the cross-section.

BS 5950: Part 1: 2000: Section three

Table 11. Limiting width-to-thickness ratios for sections other than CHS and RHS

Compression element		Ratio[a]	Limiting value[b]		
			Class 1 plastic	Class 2 compact	Class 3 semi-compact
Outstand element of compression flange	Rolled section	b/T	9ε	10ε	15ε
	Welded section	b/T	8ε	9ε	13ε
Internal element of compression flange	Compression due to bending	b/T	28ε	32ε	40ε
	Axial compression	b/T	Not applicable		
Web of an I-, H- or box section[c]	Neutral axis at mid-depth	d/t	80ε	100ε	120ε
	Generally[d] If r_1 is negative:	d/t		$\dfrac{100\varepsilon}{1+r_1}$	
	If r_1 is positive:	d/t	$\dfrac{80\varepsilon}{1+r_1}$ but $\geq 40\varepsilon$	$\dfrac{100\varepsilon}{1+1.5r_1}$ but $\geq 40\varepsilon$	$\dfrac{120\varepsilon}{1+2r_2}$ but $\geq 40\varepsilon$
	Axial compression[d]	d/t	Not applicable		
Web of a channel		d/t	40ε	40ε	40ε
Angle, compression due to bending (Both criteria should be satisfied)		b/t	9ε	10ε	15ε
		d/t	9ε	9ε	$10\varepsilon\ 15\varepsilon$
Single angle, or double angles with the components separated, axial compression (All three criteria should be satisfied)		b/t	Not applicable		15ε
		d/t			15ε
		$(b+d)/t$			24ε
Outstand leg of an angle in contact back-to-back in a double angle member Outstand leg of an angle with its back in continuous contact with another component		b/t	9ε	10ε	15ε
Stem of a T-section, rolled or cut from a rolled I- or H-section		D/t	8ε	9ε	18ε

[a] Dimensions b, D, d, T and t are as defined in Figure 5 of BS 5950-1. For a box section b and T are flange dimensions and d and t are web dimensions, where the distinction between webs and flanges depends upon whether the box section is bent about its major axis or its minor axis: see clause **3.5.1**.
[b] The parameter $\varepsilon = (275/p_y)^{0.5}$.
[c] For the web of a hybrid section ε should be based on the design strength p_{yf} of the flanges.
[d] The stress ratios r_1 and r_2 are defined in clause **3.5.5**.

BS 5950: Part 1: 2000: Section three

Table 12. Limiting width-to-thickness ratios for CHS and RHS

Compression element			Ratio[a]	Limiting value[b]		
				Class 1 plastic	Class 2 compact	Class 3 semi-compact
CHS	Compression due to bending		D/t	$40\varepsilon^2$	$50\varepsilon^2$	$140\varepsilon^2$
	Axial compression		D/t	Not applicable	Not applicable	$80\varepsilon^2$
HF RHS	Flange	Compression due to bending	b/t	28ε but $\leq 80\varepsilon - d/t$	32ε but $\leq 62\varepsilon - 0.5d/t$	40ε
		Axial compression	b/t	Not applicable	Not applicable	
	Web	Neutral axis at mid-depth	d/t	64ε	80ε	120ε
		Generally[cd]	d/t	$\dfrac{64\varepsilon}{1+0.6r_1}$ but $\geq 40\varepsilon$	$\dfrac{80\varepsilon}{1+r_1}$ but $\geq 40\varepsilon$	$\dfrac{120\varepsilon}{1+2r_2}$ but $\geq 40\varepsilon$
		Axial compression[d]	d/t	Not applicable	Not applicable	
CF RHS	Flange	Compression due to bending	b/t	26ε but $\leq 72\varepsilon - d/t$	28ε but $\leq 54\varepsilon - 0.5d/t$	35ε
		Axial compression[d]	b/t	Not applicable	Not applicable	
		Neutral axis at mid-depth	d/t	56ε	70ε	105ε
		Generally[cd]	d/t	$\dfrac{56\varepsilon}{1+0.6r_1}$ but $\geq 35\varepsilon$	$\dfrac{70\varepsilon}{1+r_1}$ but $\geq 35\varepsilon$	$\dfrac{105\varepsilon}{1+2r_2}$ but $\geq 35\varepsilon$
		Axial compression[d]	d/t	Not applicable	Not applicable	

Abbreviations

CF   Cold-formed;
CHS Circular hollow section – including welded tube;
HF   Hot-finished;
RHS Rectangular hollow section – including square hollow section.

[a] For an RHS, the dimensions b and d should be taken as follows:
 – for HF RHS to BS EN 10210: $b = B - 3t$. $d = D - 3t$
 – for CF RHS to BS EN 10219: $b = B - 5t$. $d = D - 5t$
and B, D and t are as defined in Figure 5 of BS 5950-1. For an RHS subject to bending B and b are always flange dimensions and D and d are always web dimensions, but the definition of which sides of the RHS are webs and which are flanges changes according to the axis of bending: see clause **3.5.1**.
[b] The parameter $\varepsilon = (275/p_y)^{0.5}$.
[c] For RHS subject to moments about both axes see **H.3**.
[d] The stress rations r_1 and r_2 are defined in clause **3.5.5**.

BS 5950: Part 1: 2000: Section four

Table 16. Bending strength p_b (N/mm^2) for rolled sections

λ_{LT}	Steel grade and design strength p_y (N/mm^2)														
	S 275					S 355					S 460				
	235	245	255	265	275	315	325	335	345	355	400	410	430	440	460
25	235	245	255	265	275	315	325	335	345	355	400	410	430	440	460
30	235	245	255	265	275	315	325	335	345	355	395	403	421	429	446
35	235	245	255	265	273	307	316	324	332	341	378	386	402	410	426
40	229	238	246	254	262	294	302	309	317	325	359	367	382	389	404
45	219	227	235	242	250	280	287	294	302	309	340	347	361	367	381
50	210	217	224	231	238	265	272	279	285	292	320	326	338	344	356
55	199	206	213	219	226	251	257	263	268	274	299	305	315	320	330
60	189	195	201	207	213	236	241	246	251	257	278	283	292	296	304
65	179	185	190	196	201	221	225	230	234	239	257	261	269	272	279
70	169	174	179	184	188	206	210	214	218	222	237	241	247	250	256
75	159	164	168	172	176	192	195	199	202	205	219	221	226	229	234
80	150	154	158	161	165	178	181	184	187	190	201	203	208	210	214
85	140	144	147	151	154	165	168	170	173	175	185	187	190	192	195
90	132	135	138	141	144	153	156	158	160	162	170	172	175	176	179
95	124	126	129	131	134	143	144	146	148	150	157	158	161	162	164
100	116	118	121	123	125	132	134	136	137	139	145	146	148	149	151
105	109	111	113	115	117	123	125	126	128	129	134	135	137	138	140
110	102	104	106	107	109	115	116	117	119	120	124	125	127	128	129
115	96	97	99	101	102	107	108	109	110	111	115	116	118	118	120
120	90	91	93	94	96	100	101	102	103	104	107	108	109	110	111
125	85	86	87	89	90	94	95	96	96	97	100	101	102	103	104
130	80	81	82	83	84	88	89	90	90	91	94	94	95	96	97
135	75	76	77	78	79	83	83	84	85	85	88	88	89	90	90
140	71	72	73	74	75	78	78	79	80	80	82	83	84	84	85
145	67	68	69	70	71	73	74	74	75	75	77	78	79	79	80
150	64	64	65	66	67	69	70	70	71	71	73	73	74	74	75
155	60	61	62	62	63	65	66	66	67	67	69	69	70	70	71
160	57	58	59	59	60	62	62	63	63	63	65	65	66	66	67
165	54	55	56	56	57	59	59	59	60	60	61	62	62	62	63
170	52	52	53	53	54	56	56	56	57	57	58	58	59	59	60
175	49	50	50	51	51	53	53	53	54	54	55	55	56	56	56
180	47	47	48	48	49	50	51	51	51	51	52	53	53	53	54
185	45	45	46	46	46	48	48	48	49	49	50	50	50	51	51
190	43	43	44	44	44	46	46	46	46	47	48	48	48	48	48
195	41	41	42	42	42	43	44	44	44	44	45	45	46	46	46
200	39	39	40	40	40	42	42	42	42	42	43	43	44	44	44
210	36	36	37	37	37	38	38	38	39	39	39	40	40	40	40
220	33	33	34	34	34	35	35	35	35	36	36	36	37	37	37
230	31	31	31	31	31	32	32	33	33	33	33	33	34	34	34
240	28	29	29	29	29	30	30	30	30	30	31	31	31	31	31
250	26	27	27	27	27	28	28	28	28	28	29	29	29	29	29
λ_{L0}	37.1	36.3	35.6	35.0	34.3	32.1	31.6	31.1	30.6	30.2	28.4	28.1	27.4	27.1	26.5

Table 17. Bending strength p_b (N/mm²) for welded sections

λ_{LT}	Steel grade and design strength p_y (N/mm²)														
	S 275					S 355					S 460				
	235	245	255	265	275	315	325	335	345	355	400	410	430	440	460
25	235	245	255	265	275	315	325	335	345	355	400	410	430	440	460
30	235	245	255	265	275	315	325	335	345	355	390	397	412	419	434
35	235	245	255	265	272	300	307	314	321	328	358	365	378	385	398
40	224	231	237	244	250	276	282	288	295	301	328	334	346	352	364
45	206	212	218	224	230	253	259	265	270	276	300	306	316	321	332
50	190	196	201	207	212	233	238	243	248	253	275	279	288	293	302
55	175	180	185	190	195	214	219	223	227	232	251	255	263	269	281
60	162	167	171	176	180	197	201	205	209	212	237	242	253	258	269
65	150	154	158	162	166	183	188	194	199	204	227	232	242	247	256
70	139	142	146	150	155	177	182	187	192	196	217	222	230	234	242
75	130	135	140	145	151	170	175	179	184	188	207	210	218	221	228
80	126	131	136	141	146	163	168	172	176	179	196	199	205	208	214
85	122	127	131	136	140	156	160	164	167	171	185	187	190	192	195
90	118	123	127	131	135	149	152	156	159	162	170	172	175	176	179
95	114	118	122	125	129	142	144	146	148	150	157	158	161	162	164
100	110	113	117	120	123	132	134	136	137	139	145	146	148	149	151
105	106	109	112	115	117	123	125	126	128	129	134	135	137	138	140
110	101	104	106	107	109	115	116	117	119	120	124	125	127	128	129
115	96	97	99	101	102	107	108	109	110	111	115	116	118	118	120
120	90	91	93	94	96	100	101	102	103	104	107	108	109	110	111
125	85	86	87	89	90	94	95	96	96	97	100	101	102	103	104
130	80	81	82	83	84	88	89	90	90	91	94	94	95	96	97
135	75	76	77	78	79	83	83	84	85	85	88	88	89	90	90
140	71	72	73	74	75	78	78	79	80	80	82	83	84	84	85
145	67	68	69	70	71	73	74	74	75	75	77	78	79	79	80
150	64	64	65	66	67	69	70	70	71	71	73	73	74	74	75
155	60	61	62	62	63	65	66	66	67	67	69	69	70	70	71
160	57	58	59	59	60	62	62	63	63	63	65	65	66	66	67
165	54	55	56	56	57	59	59	59	60	60	61	62	62	62	63
170	52	52	53	53	54	56	56	56	57	57	58	58	59	59	60
175	49	50	50	51	51	53	53	53	54	54	55	55	56	56	56
180	47	47	48	48	49	50	51	51	51	51	52	53	53	53	54
185	45	45	46	46	46	48	48	48	49	49	50	50	50	51	51
190	43	43	44	44	44	46	46	46	46	47	48	48	48	48	48
195	41	41	42	42	42	43	44	44	44	44	45	45	46	46	46
200	39	39	40	40	40	42	42	42	42	42	43	43	44	44	44
210	36	36	37	37	37	38	38	38	39	39	39	40	40	40	40
220	33	33	34	34	34	35	35	35	35	36	36	36	37	37	37
230	31	31	31	31	31	32	32	33	33	33	33	33	34	34	34
240	28	29	29	29	29	30	30	30	30	30	31	31	31	31	31
250	26	27	27	27	27	28	28	28	28	28	29	29	29	29	29
λ_{L0}	37.1	36.3	35.6	35.0	34.3	32.1	31.6	31.1	30.6	30.2	28.4	28.1	27.4	27.1	26.5

BS 5950: Part 1: 2000: Section four

Table 23. Allocation of strut curve

Type of section	Maximum thickness (see note 1)	Axis of buckling	
		x–x	y–y
Hot-finished structural hollow section		(a)	(a)
Cold-formed structural hollow section		(c)	(c)
Rolled I-section	≤40 mm	(a)	(b)
	>40 mm	(b)	(c)
Rolled H-section	≤40 mm	(b)	(c)
	>40 mm	(c)	(d)
Welded I- or H-section (see note 2 and clause **4.7.5**)	≤40 mm	(b)	(c)
	>40 mm	(b)	(d)
Rolled I-section with welded flange cover plates with $0.25 < U/B < 0.8$ as shown in Figure 14(a) of BS 5950-1	≤40 mm	(a)	(b)
	>40 mm	(b)	(c)
Rolled H-section with welded flange cover plates with $0.25 < U/B < 0.8$ as shown in Figure 14(a) of BS 5950-1	≤40 mm	(b)	(c)
	>40 mm	(c)	(d)
Rolled I- or H-section with welded flange cover plates with $U/B \geq 0.8$ as shown in Figure 14(b) of BS 5950-1	≤40 mm	(b)	(a)
	>40 mm	(c)	(b)
Rolled I- or H-section with welded flange cover plates with $U/B \leq 0.25$ as shown in Figure 14(c) of BS 5950-1	≤40 mm	(b)	(c)
	>40 mm	(b)	(d)
Welded box section (see note 3 and clause **4.7.5**)	≤40 mm	(b)	(b)
	>40 mm	(c)	(c)
Round, square or flat bar	≤40 mm	(b)	(b)
	>40 mm	(c)	(c)
Rolled angle, channel or T-section		Any axis: (c)	
Two rolled sections laced, battened or back-to-back			
Compound rolled sections			

NOTE 1   For thicknesses between 40 mm and 50 mm the value of p_c may be taken as the average of the values for thicknesses up to 40 mm and over 40 mm for the relevant value of p_y.

NOTE 2   For welded I- or H-sections with their flanges thermally cut by machine without subsequent edge grinding or machining, for buckling about the y–y axis, strut curve (b) may be used for flanges up to 40 mm thick and strut curve (c) for flanges over 40 mm thick. Table 24 gives values for p_c for strut curves (a)–(d).

NOTE 3   The category 'welded box section' includes any box section fabricated from plates or rolled sections, provided that all of the longitudinal welds are near the corners of the cross-section. Box sections with longitudinal stiffeners are *not* included in this category.

BS 5950: Part 1: 2000: Section four

Table 24. Compressive strength p_c (N/mm²)

(1) Values of p_c in N/mm² with λ < 110 for strut curve a

	Steel grade and design strength p_y (N/mm²)													
	S 275					S 355					S 460			
λ	235	245	255	265	275	315	325	335	345	355	400	410	430	440	460
15	235	245	255	265	275	315	325	335	345	355	399	409	429	439	458
20	234	244	254	264	273	312	322	332	342	351	395	405	424	434	453
25	232	241	251	261	270	309	318	328	338	347	390	400	419	429	448
30	229	239	248	258	267	305	315	324	333	343	385	395	414	423	442
35	226	236	245	254	264	301	310	320	329	338	380	389	407	416	434
40	223	233	242	251	260	296	305	315	324	333	373	382	399	408	426
42	222	231	240	249	258	294	303	312	321	330	370	378	396	404	422
44	221	230	239	248	257	292	301	310	319	327	366	375	392	400	417
46	219	228	237	246	255	290	299	307	316	325	363	371	388	396	413
48	218	227	236	244	253	288	296	305	313	322	359	367	383	391	407
50	216	225	234	242	251	285	293	302	310	318	355	363	378	386	401
52	215	223	232	241	249	282	291	299	307	315	350	358	373	380	395
54	213	222	230	238	247	279	287	295	303	311	345	353	367	374	388
56	211	220	228	236	244	276	284	292	300	307	340	347	361	368	381
58	210	218	226	234	242	273	281	288	295	303	334	341	354	360	372
60	208	216	224	232	239	269	277	284	291	298	328	334	346	352	364
62	206	214	221	229	236	266	273	280	286	293	321	327	338	344	354
64	204	211	219	226	234	262	268	275	281	288	314	320	330	335	344
66	201	209	216	223	230	257	264	270	276	282	307	312	321	326	334
68	199	206	213	220	227	253	259	265	270	276	299	303	312	316	324
70	196	203	210	217	224	248	254	259	265	270	291	295	303	306	313
72	194	201	207	214	220	243	248	253	258	263	282	286	293	296	302
74	191	198	204	210	216	238	243	247	252	256	274	277	283	286	292
76	188	194	200	206	212	232	237	241	245	249	265	268	274	276	281
78	185	191	197	202	208	227	231	235	239	242	257	259	264	267	271
80	182	188	193	198	203	221	225	229	232	235	248	251	255	257	261
82	179	184	189	194	199	215	219	222	225	228	240	242	246	248	251
84	176	181	185	190	194	209	213	216	219	221	232	234	237	239	242
86	172	177	181	186	190	204	207	209	212	214	224	225	229	230	233
88	169	173	177	181	185	198	200	203	205	208	216	218	220	222	224
90	165	169	173	177	180	192	195	197	199	201	209	210	213	214	216
92	162	166	169	173	176	186	189	191	193	194	201	203	205	206	208
94	158	162	165	168	171	181	183	185	187	188	194	196	198	199	200
96	154	158	161	164	166	175	177	179	181	182	188	189	191	192	193
98	151	154	157	159	162	170	172	173	175	176	181	182	184	185	186
100	147	150	153	155	157	165	167	168	169	171	175	176	178	178	180
102	144	146	149	151	153	160	161	163	164	165	169	170	172	172	174
104	140	142	145	147	149	155	156	158	159	160	164	165	166	166	168
106	136	139	141	143	145	150	152	153	154	155	158	159	160	161	162
108	133	135	137	139	141	146	147	148	149	150	153	154	155	156	157

Table 24. Compressive strength p_c (N/mm²) (*continued*)

(2) Values of p_c (N/mm²) with $\lambda \geq 110$ for strut curve a

| | Steel grade and design strength p_y (N/mm²) | | | | | | | | | | | | | |
| | S 275 | | | | | S 355 | | | | | S 460 | | | | |
λ	235	245	255	265	275	315	325	335	345	355	400	410	430	440	460
110	130	132	133	135	137	142	143	144	144	145	148	149	150	150	151
112	126	128	130	131	133	137	138	139	140	141	144	144	145	146	146
114	123	125	126	128	129	133	134	135	136	136	139	140	141	141	142
116	120	121	123	124	125	129	130	131	132	132	135	135	136	137	137
118	117	118	120	121	122	126	126	127	128	128	131	131	132	132	133
120	114	115	116	118	119	122	123	123	124	125	127	127	128	128	129
122	111	112	113	114	115	119	119	120	120	121	123	123	124	124	125
124	108	109	110	111	112	115	116	116	117	117	119	120	120	121	121
126	105	106	107	108	109	112	113	113	114	114	116	116	117	117	118
128	103	104	105	105	106	109	109	110	110	111	112	113	113	114	114
130	100	101	102	103	103	106	106	107	107	108	109	110	110	110	111
135	94	95	95	96	97	99	99	100	100	101	102	102	103	103	103
140	88	89	90	90	91	93	93	93	94	94	95	95	96	96	96
145	83	84	84	85	85	87	87	87	88	88	89	89	90	90	90
150	78	79	79	80	80	82	82	82	82	83	83	84	84	84	84
155	74	74	75	75	75	77	77	77	77	78	78	79	79	79	79
160	70	70	70	71	71	72	72	73	73	73	74	74	74	74	75
165	66	66	67	67	67	68	68	69	69	69	70	70	70	70	70
170	62	63	63	63	64	64	65	65	65	65	66	66	66	66	66
175	59	59	60	60	60	61	61	61	61	62	62	62	62	63	63
180	56	56	57	57	57	58	58	58	58	58	59	59	59	59	59
185	53	54	54	54	54	55	55	55	55	55	56	56	56	56	56
190	51	51	51	51	52	52	52	52	53	53	53	53	53	53	53
195	48	49	49	49	49	50	50	50	50	50	50	51	51	51	51
200	46	46	46	47	47	47	47	47	48	48	48	48	48	48	48
210	42	42	42	43	43	43	43	43	43	43	44	44	44	44	44
220	39	39	39	39	39	39	39	40	40	40	40	40	40	40	40
230	35	36	36	36	36	36	36	36	36	36	37	37	37	37	37
240	33	33	33	33	33	33	33	33	33	33	34	34	34	34	34
250	30	30	30	30	30	31	31	31	31	31	31	31	31	31	31
260	28	28	28	28	28	28	29	29	29	29	29	29	29	29	29
270	26	26	26	26	26	26	27	27	27	27	27	27	27	27	27
280	24	24	24	24	24	25	25	25	25	25	25	25	25	25	25
290	23	23	23	23	23	23	23	23	23	23	23	23	23	23	23
300	21	21	21	21	21	22	22	22	22	22	22	22	22	22	22
310	20	20	20	20	20	20	20	20	20	20	20	20	20	20	20
320	19	19	19	19	19	19	19	19	19	19	19	19	19	19	19
330	18	18	18	18	18	18	18	18	18	18	18	18	18	18	18
340	17	17	17	17	17	17	17	17	17	17	17	17	17	17	17
350	16	16	16	16	16	16	16	16	16	16	16	16	16	16	16

Table 24. Compressive strength p_c (N/mm²) (*continued*)

(3) Values of p_c (N/mm²) with $\lambda < 110$ for strut curve b

	Steel grade and design strength p_y (N/mm²)													
	S 275					S 355					S 460			
λ	235	245	255	265	275	315	325	335	345	355	400	410	430	440	460
15	235	245	255	265	275	315	325	335	345	355	399	409	428	438	457
20	234	243	253	263	272	310	320	330	339	349	391	401	420	429	448
25	229	239	248	258	267	304	314	323	332	342	384	393	411	421	439
30	225	234	243	253	262	298	307	316	325	335	375	384	402	411	429
35	220	229	238	247	256	291	300	309	318	327	366	374	392	400	417
40	216	224	233	241	250	284	293	301	310	318	355	364	380	388	404
42	213	222	231	239	248	281	289	298	306	314	351	359	375	383	399
44	211	220	228	237	245	278	286	294	302	310	346	354	369	377	392
46	209	218	226	234	242	275	283	291	298	306	341	349	364	371	386
48	207	215	223	231	239	271	279	287	294	302	336	343	358	365	379
50	205	213	221	229	237	267	275	283	290	298	330	337	351	358	372
52	203	210	218	226	234	264	271	278	286	293	324	331	344	351	364
54	200	208	215	223	230	260	267	274	281	288	318	325	337	344	356
56	198	205	213	220	227	256	263	269	276	283	312	318	330	336	347
58	195	202	210	217	224	252	258	265	271	278	305	311	322	328	339
60	193	200	207	214	221	247	254	260	266	272	298	304	314	320	330
62	190	197	204	210	217	243	249	255	261	266	291	296	306	311	320
64	187	194	200	207	213	238	244	249	255	261	284	289	298	302	311
66	184	191	197	203	210	233	239	244	249	255	276	281	289	294	301
68	181	188	194	200	206	228	233	239	244	249	269	273	281	285	292
70	178	185	190	196	202	223	228	233	238	242	261	265	272	276	282
72	175	181	187	193	198	218	223	227	232	236	254	257	264	267	273
74	172	178	183	189	194	213	217	222	226	230	246	249	255	258	264
76	169	175	180	185	190	208	212	216	220	223	238	241	247	250	255
78	166	171	176	181	186	203	206	210	214	217	231	234	239	241	246
80	163	168	172	177	181	197	201	204	208	211	224	226	231	233	237
82	160	164	169	173	177	192	196	199	202	205	217	219	223	225	229
84	156	161	165	169	173	187	190	193	196	199	210	212	216	218	221
86	153	157	161	165	169	182	185	188	190	193	203	205	208	210	213
88	150	154	158	161	165	177	180	182	185	187	196	198	201	203	206
90	146	150	154	157	161	172	175	177	179	181	190	192	195	196	199
92	143	147	150	153	156	167	170	172	174	176	184	185	188	189	192
94	140	143	147	150	152	162	165	167	169	171	178	179	182	183	185
96	137	140	143	146	148	158	160	162	164	165	172	173	176	177	179
98	134	137	139	142	145	153	155	157	159	160	167	168	170	171	173
100	130	133	136	138	141	149	151	152	154	155	161	162	164	165	167
102	127	130	132	135	137	145	146	148	149	151	156	157	159	160	162
104	124	127	129	131	133	141	142	144	145	146	151	152	154	155	156
106	121	124	126	128	130	137	138	139	141	142	147	148	149	150	151
108	118	121	123	125	126	133	134	135	137	138	142	143	144	145	147

Table 24. Compressive strength p_c (N/mm²) (*continued*)

(4) Values of p_c (N/mm²) with $\lambda \geq 110$ for strut curve b

	Steel grade and design strength p_y (N/mm²)														
	S 275					S 355					S 460				
λ	235	245	255	265	275	315	325	335	345	355	400	410	430	440	460
110	115	118	120	121	123	129	130	131	133	134	138	139	140	141	142
112	113	115	117	118	120	125	127	128	129	130	134	134	136	136	138
114	110	112	114	115	117	122	123	124	125	126	130	130	132	132	133
116	107	109	111	112	114	119	120	121	122	122	126	126	128	128	129
118	105	106	108	109	111	115	116	117	118	119	122	123	124	124	125
120	102	104	105	107	108	112	113	114	115	116	119	119	120	121	122
122	100	101	103	104	105	109	110	111	112	112	115	116	117	117	118
124	97	99	100	101	102	106	107	108	109	109	112	112	113	114	115
126	95	96	98	99	100	103	104	105	106	106	109	109	110	111	111
128	93	94	95	96	97	101	101	102	103	103	106	106	107	107	108
130	90	92	93	94	95	98	99	99	100	101	103	103	104	105	105
135	85	86	87	88	89	92	93	93	94	94	96	97	97	98	98
140	80	81	82	83	84	86	87	87	88	88	90	90	91	91	92
145	76	77	78	78	79	81	82	82	83	83	84	85	85	86	86
150	72	72	73	74	74	76	77	77	78	78	79	80	80	80	81
155	68	69	69	70	70	72	72	73	73	73	75	75	75	76	76
160	64	65	65	66	66	68	68	69	69	69	70	71	71	71	72
165	61	62	62	62	63	64	65	65	65	65	66	67	67	67	68
170	58	58	59	59	60	61	61	61	62	62	63	63	63	64	64
175	55	55	56	56	57	58	58	58	59	59	60	60	60	60	60
180	52	53	53	53	54	55	55	55	56	56	56	57	57	57	57
185	50	50	51	51	51	52	52	53	53	53	54	54	54	54	54
190	48	48	48	48	49	50	50	50	50	50	51	51	51	51	52
195	45	46	46	46	46	47	47	48	48	48	49	49	49	49	49
200	43	44	44	44	44	45	45	45	46	46	46	46	47	47	47
210	40	40	40	40	41	41	41	41	42	42	42	42	42	43	43
220	36	37	37	37	37	38	38	38	38	38	39	39	39	39	39
230	34	34	34	34	34	35	35	35	35	35	35	36	36	36	36
240	31	31	31	31	32	32	32	32	32	32	33	33	33	33	33
250	29	29	29	29	29	30	30	30	30	30	30	30	30	30	30
260	27	27	27	27	27	27	28	28	28	28	28	28	28	28	28
270	25	25	25	25	25	26	26	26	26	26	26	26	26	26	26
280	23	23	23	23	24	24	24	24	24	24	24	24	24	24	24
290	22	22	22	22	22	22	22	22	22	22	23	23	23	23	23
300	20	20	21	21	21	21	21	21	21	21	21	21	21	21	21
310	19	19	19	19	19	20	20	20	20	20	20	20	20	20	20
320	18	18	18	18	18	18	18	19	19	19	19	19	19	19	19
330	17	17	17	17	17	17	17	17	17	18	18	18	18	18	18
340	16	16	16	16	16	16	16	16	17	17	17	17	17	17	17
350	15	15	15	15	15	16	16	16	16	16	16	16	16	16	16

Table 24. Compressive strength p_c (N/mm²) *(continued)*

(5) Values of p_c (N/mm²) with $\lambda < 110$ for strut curve c

| | Steel grade and design strength p_y (N/mm²) | | | | | | | | | | | | | | |
| | S 275 | | | | | S 355 | | | | | S 460 | | | | |
λ	235	245	255	265	275	315	325	335	345	355	400	410	430	440	460
15	235	245	255	265	275	315	325	335	345	355	398	408	427	436	455
20	233	242	252	261	271	308	317	326	336	345	387	396	414	424	442
25	226	235	245	254	263	299	308	317	326	335	375	384	402	410	428
30	220	228	237	246	255	289	298	307	315	324	363	371	388	396	413
35	213	221	230	238	247	280	288	296	305	313	349	357	374	382	397
40	206	214	222	230	238	270	278	285	293	301	335	343	358	365	380
42	203	211	219	227	235	266	273	281	288	296	329	337	351	358	373
44	200	208	216	224	231	261	269	276	284	291	323	330	344	351	365
46	197	205	213	220	228	257	264	271	279	286	317	324	337	344	357
48	195	202	209	217	224	253	260	267	274	280	311	317	330	337	349
50	192	199	206	213	220	248	255	262	268	275	304	310	323	329	341
52	189	196	203	210	217	244	250	257	263	270	297	303	315	321	333
54	186	193	199	206	213	239	245	252	258	264	291	296	308	313	324
56	183	189	196	202	209	234	240	246	252	258	284	289	300	305	315
58	179	186	192	199	205	229	235	241	247	252	277	282	292	297	306
60	176	183	189	195	201	225	230	236	241	247	270	274	284	289	298
62	173	179	185	191	197	220	225	230	236	241	262	267	276	280	289
64	170	176	182	188	193	215	220	225	230	235	255	260	268	272	280
66	167	173	178	184	189	210	215	220	224	229	248	252	260	264	271
68	164	169	175	180	185	205	210	214	219	223	241	245	252	256	262
70	161	166	171	176	181	200	204	209	213	217	234	238	244	248	254
72	157	163	168	172	177	195	199	203	207	211	227	231	237	240	246
74	154	159	164	169	173	190	194	198	202	205	220	223	229	232	238
76	151	156	160	165	169	185	189	193	196	200	214	217	222	225	230
78	148	152	157	161	165	180	184	187	191	194	207	210	215	217	222
80	145	149	153	157	161	176	179	182	185	188	201	203	208	210	215
82	142	146	150	154	157	171	174	177	180	183	195	197	201	203	207
84	139	142	146	150	154	167	169	172	175	178	189	191	195	197	201
86	135	139	143	146	150	162	165	168	170	173	183	185	189	190	194
88	132	136	139	143	146	158	160	163	165	168	177	179	183	184	187
90	129	133	136	139	142	153	156	158	161	163	172	173	177	178	181
92	126	130	133	136	139	149	152	154	156	158	166	168	171	173	175
94	124	127	130	133	135	145	147	149	151	153	161	163	166	167	170
96	121	124	127	129	132	141	143	145	147	149	156	158	160	162	164
98	118	121	123	126	129	137	139	141	143	145	151	153	155	157	159
100	115	118	120	123	125	134	135	137	139	140	147	148	151	152	154
102	113	115	118	120	122	130	132	133	135	136	143	144	146	147	149
104	110	112	115	117	119	126	128	130	131	133	138	139	142	142	144
106	107	110	112	114	116	123	125	126	127	129	134	135	137	138	140
108	105	107	109	111	113	120	121	123	124	125	130	131	133	134	136

Table 24. Compressive strength p_c (N/mm²) (*continued*)

(6) Values of p_c (N/mm²) with $\lambda \geq 110$ for strut curve c

Steel grade and design strength p_y (N/mm²)

	S 275					S 355					S 460				
λ	235	245	255	265	275	315	325	335	345	355	400	410	430	440	460
110	102	104	106	108	110	116	118	119	120	122	126	127	129	130	132
112	100	102	104	106	107	113	115	116	117	118	123	124	125	126	128
114	98	100	101	103	105	110	112	113	114	115	119	120	122	123	124
116	95	97	99	101	102	108	109	110	111	112	116	117	118	119	120
118	93	95	97	98	100	105	106	107	108	109	113	114	115	116	117
120	91	93	94	96	97	102	103	104	105	106	110	110	112	112	113
122	89	90	92	93	95	99	100	101	102	103	107	107	109	109	110
124	87	88	90	91	92	97	98	99	100	100	104	104	106	106	107
126	85	86	88	89	90	94	95	96	97	98	101	102	103	103	104
128	83	84	86	87	88	92	93	94	95	95	98	99	100	100	101
130	81	82	84	85	86	90	91	91	92	93	96	96	97	98	99
135	77	78	79	80	81	84	85	86	87	87	90	90	91	92	92
140	72	74	75	76	76	79	80	81	81	82	84	85	85	86	87
145	69	70	71	71	72	75	76	76	77	77	79	80	80	81	81
150	65	66	67	68	68	71	71	72	72	73	75	75	76	76	76
155	62	63	63	64	65	67	67	68	68	69	70	71	71	72	72
160	59	59	60	61	61	63	64	64	65	65	66	67	67	67	68
165	56	56	57	58	58	60	60	61	61	61	63	63	64	64	64
170	53	54	54	55	55	57	57	58	58	58	60	60	60	60	61
175	51	51	52	52	53	54	54	55	55	55	56	57	57	57	58
180	48	49	49	50	50	51	52	52	52	53	54	54	54	54	55
185	46	46	47	47	48	49	49	50	50	50	51	51	52	52	52
190	44	44	45	45	45	47	47	47	47	48	49	49	49	49	49
195	42	42	43	43	43	45	45	45	45	45	46	46	47	47	47
200	40	41	41	41	42	43	43	43	43	43	44	44	45	45	45
210	37	37	38	38	38	39	39	39	40	40	40	40	41	41	41
220	34	34	35	35	35	36	36	36	36	36	37	37	37	37	38
230	31	32	32	32	32	33	33	33	33	34	34	34	34	34	35
240	29	29	30	30	30	30	31	31	31	31	31	31	32	32	32
250	27	27	27	28	28	28	28	28	29	29	29	29	29	29	29
260	25	25	26	26	26	26	26	26	27	27	27	27	27	27	27
270	23	24	24	24	24	24	25	25	25	25	25	25	25	25	25
280	22	22	22	22	22	23	23	23	23	23	23	24	24	24	24
290	21	21	21	21	21	21	21	22	22	22	22	22	22	22	22
300	19	19	20	20	20	20	20	20	20	20	21	21	21	21	21
310	18	18	18	19	19	19	19	19	19	19	19	19	19	19	20
320	17	17	17	17	18	18	18	18	18	18	18	18	18	18	18
330	16	16	16	16	17	17	17	17	17	17	17	17	17	17	17
340	15	15	15	16	16	16	16	16	16	16	16	16	16	16	16
350	15	15	15	15	15	15	15	15	15	15	15	15	15	15	15

Table 24. Compressive strength p_c (N/mm²) (*continued*)

(7) Values of p_c (N/mm²) with $\lambda < 110$ for strut curve d

| | Steel grade and design strength p_y (N/mm²) | | | | | | | | | | | | | | |
| | S 275 | | | | | S 355 | | | | | S 460 | | | | |
λ	235	245	255	265	275	315	325	335	345	355	400	410	430	440	460
15	235	245	255	265	275	315	325	335	345	355	397	407	425	435	453
20	232	241	250	259	269	305	314	323	332	341	381	390	408	417	434
25	223	231	240	249	257	292	301	309	318	326	365	373	390	398	415
30	213	222	230	238	247	279	287	296	304	312	348	356	372	380	396
35	204	212	220	228	236	267	274	282	290	297	331	339	353	361	375
40	195	203	210	218	225	254	261	268	275	283	314	321	334	341	355
42	192	199	206	214	221	249	256	263	270	277	307	314	327	333	346
44	188	195	202	209	216	244	251	257	264	271	300	306	319	325	337
46	185	192	199	205	212	239	245	252	258	265	293	299	311	317	329
48	181	188	195	201	208	234	240	246	252	259	286	291	303	309	320
50	178	184	191	197	204	228	235	241	247	253	278	284	295	301	311
52	174	181	187	193	199	223	229	235	241	246	271	277	287	292	303
54	171	177	183	189	195	218	224	229	235	240	264	269	279	284	294
56	167	173	179	185	191	213	219	224	229	234	257	262	271	276	285
58	164	170	175	181	187	208	213	218	224	229	250	255	264	268	277
60	161	166	172	177	182	203	208	213	218	223	243	247	256	260	268
62	157	163	168	173	178	198	203	208	212	217	236	240	248	252	260
64	154	159	164	169	174	193	198	202	207	211	229	233	241	245	252
66	150	156	160	165	170	188	193	197	201	205	223	226	234	237	244
68	147	152	157	162	166	184	188	192	196	200	216	220	226	230	236
70	144	149	153	158	162	179	183	187	190	194	210	213	219	222	228
72	141	145	150	154	158	174	178	182	185	189	203	207	213	215	221
74	138	142	146	150	154	170	173	177	180	183	197	200	206	209	214
76	135	139	143	147	151	165	169	172	175	178	191	194	199	202	207
78	132	136	139	143	147	161	164	167	170	173	186	188	193	195	200
80	129	132	136	140	143	156	160	163	165	168	180	182	187	189	194
82	126	129	133	136	140	152	155	158	161	163	175	177	181	183	187
84	123	126	130	133	136	148	151	154	156	159	169	171	176	177	181
86	120	123	127	130	133	144	147	149	152	154	164	166	170	172	175
88	117	120	123	127	129	140	143	145	148	150	159	161	165	167	170
90	114	118	121	123	126	137	139	141	144	146	154	156	160	161	164
92	112	115	118	120	123	133	135	137	139	142	150	152	155	156	159
94	109	112	115	117	120	129	132	134	136	138	145	147	150	152	154
96	107	109	112	115	117	126	128	130	132	134	141	143	146	147	150
98	104	107	109	112	114	123	125	126	128	130	137	138	141	143	145
100	102	104	107	109	111	119	121	123	125	126	133	134	137	138	141
102	99	102	104	106	108	116	118	120	121	123	129	131	133	134	136
104	97	99	102	104	106	113	115	116	118	120	126	127	129	130	132
106	95	97	99	101	103	110	112	113	115	116	122	123	125	126	128
108	93	95	97	99	101	107	109	110	112	113	119	120	122	123	125

Table 24. Compressive strength p_c (N/mm²) (*continued*)

(8) Values of p_c (N/mm²) with $\lambda \geq 110$ for strut curve d

| | Steel grade and design strength p_y (N/mm²) | | | | | | | | | | | | | | |
| | S 275 | | | | | S 355 | | | | | S 460 | | | | |
λ	235	245	255	265	275	315	325	335	345	355	400	410	430	440	460
110	91	93	95	96	98	105	106	108	109	110	115	116	118	119	121
112	88	90	92	94	96	102	103	105	106	107	112	113	115	116	118
114	86	88	90	92	94	99	101	102	103	104	109	110	112	113	114
116	85	86	88	90	91	97	98	99	101	102	106	107	109	110	111
118	83	84	86	88	89	95	96	97	98	99	103	104	106	107	108
120	81	82	84	86	87	92	93	94	95	96	101	101	103	104	105
122	79	81	82	84	85	90	91	92	93	94	98	99	100	101	102
124	77	79	80	82	83	88	89	90	91	92	95	96	98	98	99
126	76	77	78	80	81	86	87	88	89	89	93	94	95	96	97
128	74	75	77	78	79	84	85	85	86	87	91	91	93	93	94
130	72	74	75	76	77	82	83	83	84	85	88	89	90	91	92
135	68	70	71	72	73	77	78	79	79	80	83	84	85	85	86
140	65	66	67	68	69	73	73	74	75	75	78	79	80	80	81
145	62	63	64	65	65	69	69	70	71	71	74	74	75	75	76
150	59	60	60	61	62	65	66	66	67	67	69	70	71	71	72
155	56	57	57	58	59	62	62	63	63	64	66	66	67	67	68
160	53	54	55	55	56	58	59	59	60	60	62	62	63	63	64
165	50	51	52	53	53	55	56	56	57	57	59	59	60	60	61
170	48	49	49	50	51	53	53	54	54	54	56	56	57	57	57
175	46	47	47	48	48	50	51	51	51	52	53	53	54	54	55
180	44	45	45	46	46	48	48	49	49	49	50	51	51	51	52
185	42	43	43	44	44	46	46	46	47	47	48	48	49	49	49
190	40	41	41	42	42	44	44	44	44	45	46	46	46	47	47
195	38	39	39	40	40	42	42	42	42	43	44	44	44	45	45
200	37	37	38	38	39	40	40	40	41	41	42	42	42	43	43
210	34	34	35	35	35	37	37	37	37	37	38	38	39	39	39
220	31	32	32	32	33	34	34	34	34	34	35	35	36	36	36
230	29	29	30	30	30	31	31	31	32	32	32	33	33	33	33
240	27	27	28	28	28	29	29	29	29	29	30	30	30	30	31
250	25	25	26	26	26	27	27	27	27	27	28	28	28	28	28
260	24	24	24	24	24	25	25	25	25	25	26	26	26	26	26
270	22	22	22	23	23	23	23	23	24	24	24	24	24	24	25
280	21	21	21	21	21	22	22	22	22	22	23	23	23	23	23
290	19	20	20	20	20	20	21	21	21	21	21	21	21	21	21
300	18	18	19	19	19	19	19	19	19	20	20	20	20	20	20
310	17	17	17	18	18	18	18	18	18	18	19	19	19	19	19
320	16	16	16	17	17	17	17	17	17	17	18	18	18	18	18
330	15	15	16	16	16	16	16	16	16	16	17	17	17	17	17
340	15	15	15	15	15	15	15	15	15	15	16	16	16	16	16
350	14	14	14	14	14	14	14	15	15	15	15	15	15	15	15

Bolt data

Hole sizes – for ordinary bolts and friction grip connections

Nominal diameter (mm)	Clearance hole diameter[b] (mm)	Oversize hole diameter[a] (mm)	Short slotted holes[a] (mm)		Long slotted holes[a] (mm)	
			Narrow dimension	Slot dimension	Narrow dimension	Maximum dimension
M12[a]	14	17	14	18	14	30
M16	18	21	18	22	18	40
M20	22	25	22	26	22	50
M22	24	27	24	28	24	55
M24	26	30	26	32	26	60
M27	30	35	30	37	30	67
M30	33	38	33	40	33	75

[a]Hardened washers to be used
[b]In cases where there are more than three plies in joint the holes in the inner plies should be one millimetre larger than those in the outer plies

Bolt strengths

	Bolt grade		Steel grade		
	4.6	8.8	S275	S355	S460
Shear strength, p_s (N/mm^2)	160	375			
Bearing strength, p_{bb} (N/mm^2)	460	1000[a]	460	550	670
Tension strength, p_t (N/mm^2)	240	560			

[a]The bearing value of the connected part is critical

Spacing, end and edge distances – minimum values (see Fig. 23.1)

Nominal diameter of fastener (mm)	Diameter of clearance hole (mm)	Minimum spacing (mm)	Edge distance to rolled, sawn, planed, or machine flame cut edge (mm)	Edge distance to sheared edge or hand flame cut edge and end distance (mm)
M12	14	30	18	20
M16	18	40	23	26
M20	22	50	28	31
M22[a]	24	55	30	34
M24	26	60	33	37
M27[a]	30	68	38	42
M30	33	75	42	47

[a] Non-preferred size

Maximum centres of fasteners

Thickness of element (mm)	Spacing in the direction of stress (mm)	Spacing in any direction in corrosive environments (mm)
5	70	80
6	84	96
7	98	112
8	112	128
9	126	144
10	140	160
11	154	176
12	168	192
13	182	200
14	196	200
15	210	200

Maximum edge distances (1)

BS 4360 grade	Thickness less than or equal to (mm)	p_y (N/mm²)	$\varepsilon = \left(\dfrac{275}{p_y}\right)^{1/2}$	$11 t\varepsilon^a$
S275	16	275	1.0	$11t$
	40	265	1.02	$11.21t$
	63	255	1.04	$11.44t$
	80	245	1.06	$11.65t$
	100	235	1.08	$11.90t$
S355	16	355	0.88	$9.68t$
	40	345	0.89	$9.79t$
	63	335	0.91	$10.0t$
	80	325	0.92	$10.12t$
	100	315	0.93	$10.28t$
S460	16	460	0.77	$8.47t$
	40	440	0.79	$8.69t$
	63	430	0.80	$8.80t$
	80	410	0.82	$9.02t$
	100	400	0.83	$9.13t$

[a]This rule does not apply to fasteners interconnecting the components of back-to-back tension members
This table is expanded in the next table (Maximum edge distances (2))

Maximum edge distances (2)

Thickness of element t (mm)	Corrosive environment 40 mm + 4t (mm)	Steel grade S275 $\varepsilon = 11t\varepsilon$ (mm)	Steel grade S355 $\varepsilon = 11t\varepsilon$ (mm)	Steel grade S460 $\varepsilon = 11t\varepsilon$ (mm)
5	60[a]	55[a]	48[a]	42[a]
6	64[a]	66	58[a]	51[a]
7	68[a]	77	67[a]	59
8	72[a]	88	77	68[a]
9	76	99	87	76
10	80	110	96	85
11	84	121	106	93
12	88	132	116	102
13	92	143	125	110
14	96	154	135	119
15	100	165	145	127
16	104	176	154	136
20	120	224	196	174
25	140	280	245	217
30	160	336	294	261
35	180	392	343	304
40	200	448	392	347
45	220	515	445	396
50	240	572	501	440
55	260	629	551	484
60	280	686	601	528
65	300	757	657	586
70	320	816	708	631
75	340	874	759	677

[a] Use the lesser values for the appropriate grade of steel

Back marks in channel flanges

RSC	Nominal flange width (mm)	Back mark (mm)	Edge dist. (mm)	Recommended diameter (mm)
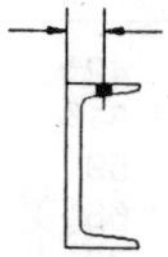	102	55	47	24
	89	55	34	20
	76	45	31	20
	64	35	29	16
	51	30	21	10
	38	22	–	–

Back marks in angles

Nominal leg (mm)	S_1 (mm)	S_2 (mm)	S_3 (mm)	S_4 (mm)	S_5 (mm)	S_6 (mm)	Nominal leg (mm)	S_1 (mm)
250							75	45 (20)
200		75 (30)	75 (30)	55 (20)	55 (20)	55 (20)	70	40 (20)
150		55 (20)	55 (20)				65	35 (20)
125		45 (20)	50 (20)				60	35 (16)
120		45 (16)	50 (16)				50	28 (12)
100	55 (24)						45	25
90	50 (24)						40	23
80	45 (20)						30	20
							25	15

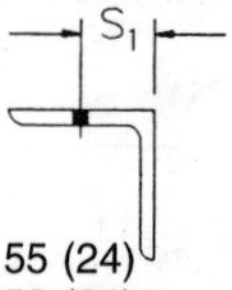

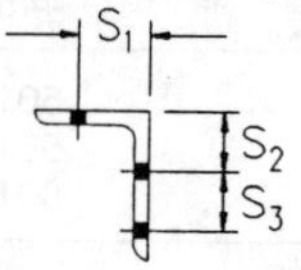

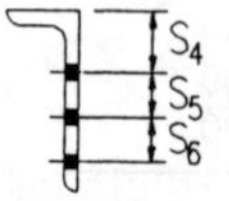

Maximum recommended bolt sizes are given in brackets

This table is reproduced from BCSA Publication No. 5/79, *Metric Practice for Structural Steelworks*, 3rd edn, 1979.

Cross centres through flanges

Flange width (mm)	Minimum for accessibility (mm)	Maximum for edge dist. (mm)	S_1 (mm)	S_2 (mm)	S_3 (mm)	S_4 (mm)
Joists						
44	27 (5)	30	30			
64	38 (10)	39	40			
76	48 (10)	51	48			
89	54 (12)	59	56			
102	60 (16)	62	60			
114	66 (16)	74	70			
127	72 (20)	77	75			
152	75 (20)	102	90			
203	91 (24)	143	140			
UCs						
152	65 (24)	92	90			
203	75 (24)	143	140			
254	87 (24)	194	140			
305	100 (24)	245	140	120 (24)	60 (24)	240 (24)
368	88 (24)	308	140	140 (24)	75 (24)	290 (24)
406	120 (24)	346	140	140 (24)	75 (24)	290 (24)
UBs						
102	50 (16)	62	54			
127	62 (20)	77	70			
133	57 (20)	83	70			
140	69 (24)	80	70			
146	64 (24)	86	70			
152	73 (24)	92	90			
165	67 (24)	105	90			
171	72 (24)	111	90			
178	72 (24)	118	90			
191	74 (24)	131	90			
210	80 (24)	150	140			
229	80 (24)	169	140			
254	87 (24)	194	140			
267	91 (24)	207	140	90 (20)	50 (20)	190 (20)
292	94 (24)	232	140	100 (24)	60 (24)	220 (24)
305	100 (24)	245	140	120 (24)	60 (24)	240 (24)
419	112 (24)	359	140	140 (24)	75 (24)	290 (24)

Maximum bolt diameters for dimensions shown are given in brackets

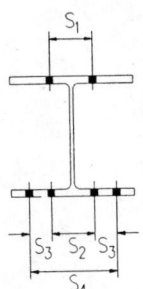

BS 5950-1: 2000
BS 4190: 2001

BOLT CAPACITIES

NON-PRELOADED ORDINARY BOLTS

GRADE 4.6 BOLTS IN S275

Diameter of Bolt	Tensile Stress Area	Tension Capacity		Shear Capacity		Bearing Capacity in kN (Minimum of P_{bb} and P_{bs}) End distance equal to 2 x bolt diameter. Thickness in mm of ply passed through.										
		Nominal $0.8A_t p_t$	Exact $A_t p_t$	Single Shear	Double Shear											
	A_t	P_{nom}	P_t	P_s	$2P_s$											
mm	mm^2	kN	kN	kN	kN	5	6	7	8	9	10	12	15	20	25	30
12	84.3	16.2	20.2	13.5	27.0	27.6	33.1	38.6	44.2	49.7	55.2	66.2	82.8	110	138	166
16	157	30.1	37.7	25.1	50.2	36.8	44.2	51.5	58.9	66.2	73.6	88.3	110	147	184	221
20	245	47.0	58.8	39.2	78.4	46.0	55.2	64.4	73.6	82.8	92.0	110	138	184	230	276
22	303	58.2	72.7	48.5	97.0	50.6	60.7	70.8	81.0	91.1	101	121	152	202	253	304
24	353	67.8	84.7	56.5	113	55.2	66.2	77.3	88.3	99.4	110	132	166	221	276	331
27	459	88.1	110	73.4	147	62.1	74.5	86.9	99.4	112	124	149	186	248	311	373
30	561	108	135	89.8	180	69.0	82.8	96.6	110	124	138	166	207	276	345	414

Values in **bold** are less than the single shear capacity of the bolt.
Values in *italic* are greater than the double shear capacity of the bolt.
Bearing values assume standard clearance holes.
If oversize or short slotted holes are used, bearing values should be multiplied by 0.7.
If long slotted or kidney shaped holes are used, bearing values should be multiplied by 0.5.
If appropriate, shear capacity must be reduced for large packings, large grip lengths and long joints.

GRADE 8.8 BOLTS IN S275

Diameter of Bolt	Tensile Stress Area	Tension Capacity		Shear Capacity		Bearing Capacity in kN (Minimum of P_{bb} and P_{bs}) End distance equal to 2 x bolt diameter. Thickness in mm of ply passed through.										
		Nominal $0.8A_t p_t$	Exact $A_t p_t$	Single Shear	Double Shear											
	A_t	P_{nom}	P_t	P_s	$2P_s$											
mm	mm^2	kN	kN	kN	kN	5	6	7	8	9	10	12	15	20	25	30
12	84.3	37.8	47.2	31.6	63.2	27.6	33.1	38.6	44.2	49.7	55.2	66.2	82.8	110	138	166
16	157	70.3	87.9	58.9	118	36.8	44.2	51.5	58.9	66.2	73.6	88.3	110	147	184	221
20	245	110	137	91.9	184	46.0	55.2	64.4	73.6	82.8	92.0	110	138	184	230	276
22	303	136	170	114	227	50.6	60.7	70.8	81.0	91.1	101	121	152	202	253	304
24	353	158	198	132	265	55.2	66.2	77.3	88.3	99.4	110	132	166	221	276	331
27	459	206	257	172	344	62.1	74.5	86.9	99.4	112	124	149	186	248	311	373
30	561	251	314	210	421	69.0	82.8	96.6	110	124	138	166	207	276	345	414

Values in **bold** are less than the single shear capacity of the bolt.
Values in *italic* are greater than the double shear capacity of the bolt.
Bearing values assume standard clearance holes.
If oversize or short slotted holes are used, bearing values should be multiplied by 0.7.
If long slotted or kidney shaped holes are used, bearing values should be multiplied by 0.5.
If appropriate, shear capacity must be reduced for large packings, large grip lengths and long joints.

BOLT CAPACITIES

BS 5950-1: 2000
BS 4190: 2001

NON-PRELOADED ORDINARY BOLTS

GRADE 10.9 BOLTS IN S275

Diameter of Bolt	Tensile Stress Area	Tension Capacity		Shear Capacity		Bearing Capacity in kN (Minimum of P_{bb} and P_{bs}) End distance equal to 2 x bolt diameter. Thickness in mm of ply passed through.										
		Nominal $0.8A_t p_t$	Exact $A_t p_t$	Single Shear	Double Shear											
	A_t	P_{nom}	P_t	P_s	$2P_s$											
mm	mm²	kN	kN	kN	kN	5	6	7	8	9	10	12	15	20	25	30
12	84.3	47.2	59.0	33.7	67.4	27.6	33.1	38.6	44.2	49.7	55.2	66.2	82.8	*110*	*138*	*166*
16	157	87.9	110	62.8	126	36.8	44.2	51.5	58.9	66.2	73.6	88.3	110	*147*	*184*	*221*
20	245	137	172	98.0	196	46.0	55.2	64.4	73.6	82.8	92.0	110	138	184	*230*	*276*
22	303	170	212	121	242	50.6	60.7	70.8	81.0	91.1	101	121	152	202	*253*	*304*
24	353	198	247	141	282	55.2	66.2	77.3	88.3	99.4	110	132	166	221	276	*331*
27	459	257	321	184	367	62.1	74.5	86.9	99.4	112	124	149	186	248	311	*373*
30	561	314	393	224	449	69.0	82.8	96.6	110	124	138	166	207	276	345	414

Values in **bold** are less than the single shear capacity of the bolt.

Values in *italic* are greater than the double shear capacity of the bolt.

Bearing values assume standard clearance holes.

If oversize or short slotted holes are used, bearing values should be multiplied by 0.7.

If long slotted or kidney shaped holes are used, bearing values should be multiplied by 0.5.

If appropriate, shear capacity must be reduced for large packings, large grip lengths and long joints.

BS 5950-1: 2000
BS 4190: 2001
BS 4933: 1973

BOLT CAPACITIES

NON-PRELOADED COUNTERSUNK BOLTS

GRADE 4.6 COUNTERSUNK BOLTS IN S275

Diameter of Bolt	Tensile Stress Area	Tension Capacity		Shear Capacity		Bearing Capacity in kN (Minimum of P_{bb} and P_{bs}) End distance equal to 2 x bolt diameter. Thickness in mm of ply passed through.										
		Nominal $0.8A_t p_t$ P_{nom}	Exact $A_t p_t$ P_t	Single Shear P_s	Double Shear $2P_s$											
mm	mm^2	kN	kN	kN	kN	5	6	7	8	9	10	12	15	20	25	30
12	84.3	16.2	20.2	13.5	27.0	**11.0**	16.6	22.1	27.6	*33.1*	*38.6*	*49.7*	*66.2*	*93.8*	*121*	*149*
16	157	30.1	37.7	25.1	50.2	**7.36**	**14.7**	22.1	29.4	36.8	44.2	*58.9*	*81.0*	*118*	*155*	*191*
20	245	47.0	58.8	39.2	78.4	0	**9.20**	**18.4**	27.6	36.8	46.0	64.4	92.0	*138*	*184*	*230*
22	303	58.2	72.7	48.5	97.0	0	**5.06**	**15.2**	**25.3**	35.4	45.5	65.8	96.1	*147*	*197*	*248*
24	353	67.8	84.7	56.5	113	0	0	**11.0**	**22.1**	**33.1**	44.2	66.2	99.4	*155*	*210*	*265*
27	459	88.1	110	73.4	147	0	0	**3.11**	**15.5**	**27.9**	**40.4**	65.2	102	*165*	*227*	*289*
30	561	108	135	89.8	180	0	0	0	**6.90**	**20.7**	**34.5**	**62.1**	104	*173*	*242*	*311*

Values in **bold** are less than the single shear capacity of the bolt.
Values in *italic* are greater than the double shear capacity of the bolt.
Bearing values assume standard clearance holes.
If oversize or short slotted holes are used, bearing values should be multiplied by 0.7.
If long slotted or kidney shaped holes are used, bearing values should be multiplied by 0.5.
Depth of countersink is taken as half the bolt diameter.

GRADE 8.8 COUNTERSUNK BOLTS IN S 275

Diameter of Bolt	Tensile Stress Area	Tension Capacity		Shear Capacity		Bearing Capacity in kN (Minimum of P_{bb} and P_{bs}) End distance equal to 2 x bolt diameter. Thickness in mm of ply passed through.										
		Nominal $0.8A_t p_t$ P_{nom}	Exact $A_t p_t$ P_t	Single Shear P_s	Double Shear $2P_s$											
mm	mm^2	kN	kN	kN	kN	5	6	7	8	9	10	12	15	20	25	30
12	84.3	37.8	47.2	31.6	63.2	**11.0**	16.6	22.1	27.6	33.1	38.6	49.7	66.2	*93.8*	*121*	*149*
16	157	70.3	87.9	58.9	118	**7.36**	**14.7**	22.1	29.4	36.8	44.2	58.9	81.0	*118*	*155*	*191*
20	245	110	137	91.9	184	0	**9.20**	**18.4**	27.6	36.8	46.0	64.4	92.0	138	*184*	*230*
22	303	136	170	114	227	0	**5.06**	**15.2**	**25.3**	35.4	45.5	65.8	96.1	147	*197*	*248*
24	353	158	198	132	265	0	0	**11.0**	**22.1**	**33.1**	44.2	66.2	99.4	155	*210*	*265*
27	459	206	257	172	344	0	0	**3.11**	**15.5**	**27.9**	**40.4**	65.2	102	165	*227*	*289*
30	561	251	314	210	421	0	0	0	**6.90**	**20.7**	**34.5**	**62.1**	104	173	*242*	*311*

Values in **bold** are less than the single shear capacity of the bolt.
Values in *italic* are greater than the double shear capacity of the bolt.
Bearing values assume standard clearance holes.
If oversize or short slotted holes are used, bearing values should be multiplied by 0.7.
If long slotted or kidney shaped holes are used, bearing values should be multiplied by 0.5.
Depth of countersink is taken as half the bolt diameter.

BOLT CAPACITIES

NON-PRELOADED COUNTERSUNK BOLTS

BS 5950-1: 2000
BS 4190: 2001
BS 4933: 1973

GRADE 10.9 COUNTERSUNK BOLTS IN S 275

Diameter of Bolt	Tensile Stress Area	Tension Capacity		Shear Capacity		Bearing Capacity in kN (Minimum of P_{bb} and P_{bs}) End distance equal to 2 x bolt diameter. Thickness in mm of ply passed through.										
		Nominal $0.8A_tp_t$	Exact A_tp_t	Single Shear	Double Shear											
	A_t	P_{nom}	P_t	P_s	$2P_s$											
mm	mm²	kN	kN	kN	kN	5	6	7	8	9	10	12	15	20	25	30
12	84.3	47.2	59.0	33.7	67.4	**11.0**	**16.6**	**22.1**	**27.6**	**33.1**	38.6	49.7	66.2	*93.8*	*121*	*149*
16	157	87.9	110	62.8	126	**7.36**	**14.7**	**22.1**	**29.4**	**36.8**	**44.2**	**58.9**	81.0	118	*155*	*191*
20	245	137	172	98.0	196	**0**	**9.20**	**18.4**	**27.6**	**36.8**	**46.0**	**64.4**	**92.0**	138	184	*230*
22	303	170	212	121	242	**0**	**5.06**	**15.2**	**25.3**	**35.4**	**45.5**	**65.8**	**96.1**	147	197	*248*
24	353	198	247	141	282	**0**	**0**	**11.0**	**22.1**	**33.1**	**44.2**	**66.2**	**99.4**	155	210	265
27	459	257	321	184	367	**0**	**0**	**3.11**	**15.5**	**27.9**	**40.4**	**65.2**	**102**	165	227	289
30	561	314	393	224	449	**0**	**0**	**0**	**6.90**	**20.7**	**34.5**	**62.1**	**104**	173	242	311

Values in **bold** are less than the single shear capacity of the bolt.
Values in *italic* are greater than the double shear capacity of the bolt.
Bearing values assume standard clearance holes.
If oversize or short slotted holes are used, bearing values should be multiplied by 0.7.
If long slotted or kidney shaped holes are used, bearing values should be multiplied by 0.5.
Depth of countersink is taken as half the bolt diameter.

BS 5950-1: 2000
BS 4395: 1969

BOLT CAPACITIES

NON-PRELOADED HSFG BOLTS

GENERAL GRADE HSFG BOLTS IN S275

Diameter of Bolt	Tensile Stress Area	Tension Capacity		Shear Capacity		Bearing Capacity in kN (Minimum of P_{bb} and P_{bs}) End distance equal to 2 x bolt diameter. Thickness in mm of ply passed through.										
		Nominal $0.8A_t p_t$	Exact $A_t p_t$	Single Shear	Double Shear											
mm	A_t mm^2	P_{nom} kN	P_t kN	P_s kN	$2P_s$ kN	5	6	7	8	9	10	12	15	20	25	30
12	84.3	39.8	49.7	33.7	67.4	27.6	33.1	38.6	44.2	49.7	55.2	66.2	82.8	*110*	*138*	*166*
16	157	74.1	92.6	62.8	126	36.8	44.2	51.5	58.9	66.2	73.6	88.3	110	*147*	*184*	*221*
20	245	116	145	98.0	196	46.0	55.2	64.4	73.6	82.8	92.0	110	138	184	*230*	*276*
22	303	143	179	121	242	50.6	60.7	70.8	81.0	91.1	101	121	152	202	*253*	*304*
24	353	167	208	141	282	55.2	66.2	77.3	88.3	99.4	110	132	166	221	276	*331*
27	459	189	236	161	321	62.1	74.5	86.9	99.4	112	124	149	186	248	311	*373*
30	561	231	289	196	393	69.0	82.8	96.6	110	124	138	166	207	276	345	*414*

Values in **bold** are less than the single shear capacity of the bolt.
Values in *italic* are greater than the double shear capacity of the bolt.
Bearing values assume standard clearance holes.
If oversize or short slotted holes are used, bearing values should be multiplied by 0.7.
If long slotted or kidney shaped holes are used, bearing values should be multiplied by 0.5.
If appropriate, shear capacity must be reduced for large packings, large grip lengths and long joints.

HIGHER GRADE HSFG BOLTS IN S 275

Diameter of Bolt	Tensile Stress Area	Tension Capacity		Shear Capacity		Bearing Capacity in kN (Minimum of P_{bb} and P_{bs}) End distance equal to 2 x bolt diameter. Thickness in mm of ply passed through.										
		Nominal $0.8A_t p_t$	Exact $A_t p_t$	Single Shear	Double Shear											
mm	A_t mm^2	P_{nom} kN	P_t kN	P_s kN	$2P_s$ kN	5	6	7	8	9	10	12	15	20	25	30
16	157	87.9	110	62.8	126	36.8	44.2	51.5	58.9	66.2	73.6	88.3	110	*147*	*184*	*221*
20	245	137	172	98.0	196	46.0	55.2	64.4	73.6	82.8	92.0	110	138	184	*230*	*276*
22	303	170	212	121	242	50.6	60.7	70.8	81.0	91.1	101	121	152	202	*253*	*304*
24	353	198	247	141	282	55.2	66.2	77.3	88.3	99.4	110	132	166	221	276	*331*
27	459	257	321	184	367	62.1	74.5	86.9	99.4	112	124	149	186	248	311	*373*
30	561	314	393	224	449	69.0	82.8	96.6	110	124	138	166	207	276	345	*414*

Values in **bold** are less than the single shear capacity of the bolt.
Values in *italic* are greater than the double shear capacity of the bolt.
Bearing values assume standard clearance holes.
If oversize or short slotted holes are used, bearing values should be multiplied by 0.7.
If long slotted or kidney shaped holes are used, bearing values should be multiplied by 0.5.
If appropriate, shear capacity must be reduced for large packings, large grip lengths and long joints.

BS 5950-1: 2000
BS 4395: 1969
BS 4604: 1970

BOLT CAPACITIES

PRELOADED HSFG BOLTS: NON-SLIP IN SERVICE

GENERAL GRADE HSFG BOLTS IN S275

Diameter of Bolt	Min. Shank Tension	Tension		Shear Capacity		Slip Resistance for $\mu = 0.5$		Bearing Capacity, P_{bg} in kN — End distance equal to 3 x bolt diameter. Thickness in mm of ply passed through.										
				Single Shear	Double Shear	Single Shear	Double Shear												
	P_o	$1.1P_o$	$A_t p_t$																
mm	kN	kN	kN	kN	kN	kN	kN	5	6	7	8	9	10	12	15	20	25	30
12	49.4	54.3	49.7	33.7	67.4	27.2	54.3	41.4	49.7	58.0	66.2	74.5	82.8	99.4	124	166	207	248
16	92.1	101	92.6	62.8	126	50.7	101	55.2	66.2	77.3	88.3	99.4	110	132	166	221	276	331
20	144	158	145	98.0	196	79.2	158	69.0	82.8	96.6	110	124	138	166	207	276	345	414
22	177	195	179	121	242	97.4	195	75.9	91.1	106	121	137	152	182	228	304	380	455
24	207	228	208	141	282	114	228	82.8	99.4	116	132	149	166	199	248	331	414	497
27	234	257	236	161	321	129	257	93.2	112	130	149	168	186	224	279	373	466	559
30	286	315	289	196	393	157	315	104	124	145	166	186	207	248	311	414	518	621

Values in **bold** are less than the single shear capacity of the bolt.
Values in *italic* are greater than the double shear capacity of the bolt.
Shading indicates that the ply thickness is not suitable for an outer ply.

HIGHER GRADE HSFG BOLTS IN S275

Diameter of Bolt	Min. Shank Tension	Tension		Shear Capacity		Slip Resistance for $\mu = 0.5$		Bearing Capacity, P_{bg} in kN — End distance equal to 3 x bolt diameter. Thickness in mm of ply passed through.										
				Single Shear	Double Shear	Single Shear	Double Shear												
	P_o	$1.1P_o$	$A_t p_t$																
mm	kN	kN	kN	kN	kN	kN	kN	5	6	7	8	9	10	12	15	20	25	30
16	104	114	110	62.8	126	57.1	114	55.2	66.2	77.3	88.3	99.4	110	132	166	221	276	331
20	162	178	172	98.0	196	89.0	178	69.0	82.8	96.6	110	124	138	166	207	276	345	414
22	200	220	212	121	242	110	220	75.9	91.1	106	121	137	152	182	228	304	380	455
24	233	257	247	141	282	128	257	82.8	99.4	116	132	149	166	199	248	331	414	497
27	303	333	321	184	367	167	333	93.2	112	130	149	168	186	224	279	373	466	559
30	370	407	393	224	449	204	407	104	124	145	166	186	207	248	311	414	518	621

Values in **bold** are less than the single shear capacity of the bolt.
Values in *italic* are greater than the double shear capacity of the bolt.
Shading indicates that the ply thickness is not suitable for an outer ply.

BS 5950-1: 2000
BS 4395: 1969
BS 4604: 1970

BOLT CAPACITIES

PRELOADED HSFG BOLTS: NON-SLIP UNDER FACTORED LOADS

GENERAL GRADE HSFG BOLTS IN S275

Diameter of Bolt	Min. Shank Tension	Bolt Tension Capacity	Slip Resistance P_{sL}							
			$\mu = 0.2$		$\mu = 0.3$		$\mu = 0.4$		$\mu = 0.5$	
			Single Shear	Double Shear	Single Shear	Double Shear	Single Shear	Double Shear	Single Shear	Double Shear
	P_o	$0.9P_o$								
mm	kN	kN	kN	kN	kN	kN	kN	kN	kN	kN
12	49.4	44.5	8.89	17.8	13.3	26.7	17.8	35.6	22.2	44.5
16	92.1	82.9	16.6	33.2	24.9	49.7	33.2	66.3	41.4	82.9
20	144	130	25.9	51.8	38.9	77.8	51.8	104	64.8	130
22	177	159	31.9	63.7	47.8	95.6	63.7	127	79.7	159
24	207	186	37.3	74.5	55.9	112	74.5	149	93.2	186
27	234	211	42.1	84.2	63.2	126	84.2	168	105	211
30	286	257	51.5	103	77.2	154	103	206	129	257

HIGHER GRADE HSFG BOLTS IN S 275

Diameter of Bolt	Min. Shank Tension	Bolt Tension Capacity	Slip Resistance P_{sL}							
			$\mu = 0.2$		$\mu = 0.3$		$\mu = 0.4$		$\mu = 0.5$	
			Single Shear	Double Shear	Single Shear	Double Shear	Single Shear	Double Shear	Single Shear	Double Shear
	P_o	$0.9P_o$								
mm	kN	kN	kN	kN	kN	kN	kN	kN	kN	kN
16	104	93.5	18.7	37.4	28.1	56.1	37.4	74.8	46.8	93.5
20	162	146	29.1	58.2	43.7	87.4	58.2	116	72.8	146
22	200	180	36.0	72.1	54.1	108	72.1	144	90.1	180
24	233	210	42.0	84.0	63.0	126	84.0	168	105	210
27	303	273	54.5	109	81.8	164	109	218	136	273
30	370	333	66.6	133	99.9	200	133	266	167	333

BS 5950-1: 2000
BS 4395: 1969
BS 4604: 1970
BS 4933: 1973

BOLT CAPACITIES

PRELOADED HSFG BOLTS: NON-SLIP IN SERVICE

GENERAL GRADE COUNTERSUNK HSFG BOLTS IN S275

Diameter of Bolt	Min. Shank Tension	Tension		Shear Capacity		Slip Resistance for $\mu = 0.5$		Bearing Capacity, P_{bg} in kN — End distance equal to 3 x bolt diameter. Thickness in mm of ply passed through.										
				Single Shear	Double Shear	Single Shear	Double Shear	5	6	7	8	9	10	12	15	20	25	30
	P_o	$1.1P_o$	$A_t p_t$															
mm	kN	kN	kN	kN	kN	kN	kN											
12	49.4	54.3	49.7	33.7	67.4	27.2	54.3	**16.6**	**24.8**	**33.1**	41.4	49.7	58.0	74.5	*99.4*	*141*	*182*	*224*
16	92.1	101	92.6	62.8	126	50.7	101	**11.0**	**22.1**	**33.1**	**44.2**	**55.2**	66.2	88.3	121	*177*	*232*	*287*
20	144	158	145	98.0	196	79.2	158	**0**	**13.8**	**27.6**	**41.4**	**55.2**	69.0	96.6	138	*207*	*276*	*345*
22	177	195	179	121	242	97.4	195	**0**	**7.59**	**22.8**	**38.0**	**53.1**	**68.3**	98.7	144	220	*296*	*372*
24	207	228	208	141	282	114	228	**0**	**0**	**16.6**	**33.1**	**49.7**	**66.2**	**99.4**	149	232	*315*	*397*
27	234	257	236	161	321	129	257	**0**	**0**	**4.66**	**23.3**	**41.9**	**60.5**	**97.8**	154	247	*340*	*433*
30	286	315	289	196	393	157	315	**0**	**0**	**0**	**10.4**	**31.1**	**51.8**	**93.2**	155	259	362	*466*

Values in **bold** are less than the single shear capacity of the bolt.
Values in *italic* are greater than the double shear capacity of the bolt.
Shading indicates that the ply thickness is not suitable for an outer ply.

HIGHER GRADE COUNTERSUNK HSFG BOLTS IN S275

Diameter of Bolt	Min. Shank Tension	Tension		Shear Capacity		Slip Resistance for $\mu = 0.5$		Bearing Capacity, P_{bg} in kN — End distance equal to 3 x bolt diameter. Thickness in mm of ply passed through.										
				Single Shear	Double Shear	Single Shear	Double Shear	5	6	7	8	9	10	12	15	20	25	30
	P_o	$1.1P_o$	$A_t p_t$															
rnm	kN	kN	kN	kN	kN	kN	kN											
16	104	114	110	62.8	126	57.1	114	**11.0**	**22.1**	**33.1**	**44.2**	**55.2**	66.2	88.3	121	*177*	*232*	*287*
20	162	178	172	98.0	196	89.0	178	**0**	**13.8**	**27.6**	**41.4**	**55.2**	69.0	96.6	138	*207*	*276*	*345*
22	200	220	212	121	242	110	220	**0**	**7.59**	**22.8**	**38.0**	**53.1**	**68.3**	98.7	144	220	*296*	*372*
24	233	257	247	141	282	128	257	**0**	**0**	**16.6**	**33.1**	**49.7**	**66.2**	**99.4**	149	232	*315*	*397*
27	303	333	321	184	367	167	333	**0**	**0**	**4.66**	**23.3**	**41.9**	**60.5**	**97.8**	154	247	*340*	*433*
30	370	407	393	224	449	204	407	**0**	**0**	**0**	**10.4**	**31.1**	**51.8**	**93.2**	155	259	362	*466*

Values in **bold** are less than the single shear capacity of the bolt.
Values in *italic* are greater than the double shear capacity of the bolt.
Shading indicates that the ply thickness is not suitable for an outer ply.

BS 5950-1: 2000
BS 4395: 1969
BS 4604: 1970
BS 4933: 1973

BOLT CAPACITIES

PRELOADED HSFG BOLTS: NON-SLIP UNDER FACTORED LOADS

GENERAL GRADE COUNTERSUNK HSFG BOLTS IN S275

Diameter of Bolt	Min. Shank Tension	Bolt Tension Capacity	Slip Resistance P_{sL}							
			$\mu = 0.2$		$\mu = 0.3$		$\mu = 0.4$		$\mu = 0.5$	
			Single Shear	Double Shear	Single Shear	Double Shear	Single Shear	Double Shear	Single Shear	Double Shear
	P_o	$0.9P_o$								
mm	kN	kN	kN	kN	kN	kN	kN	kN	kN	kN
12	49.4	44.5	8.89	17.8	13.3	26.7	17.8	35.6	22.2	44.5
16	92.1	82.9	16.6	33.2	24.9	49.7	33.2	66.3	41.4	82.9
20	144	130	25.9	51.8	38.9	77.8	51.8	104	64.8	130
22	177	159	31.9	63.7	47.8	95.6	63.7	127	79.7	159
24	207	186	37.3	74.5	55.9	112	74.5	149	93.2	186
27	234	211	42.1	84.2	63.2	126	84.2	168	105	211
30	286	257	51.5	103	77.2	154	103	206	129	257

HIGHER GRADE COUNTERSUNK HSFG BOLTS IN S 275

Diameter of Bolt	Min. Shank Tension	Bolt Tension Capacity	Slip Resistance P_{sL}							
			$\mu = 0.2$		$\mu = 0.3$		$\mu = 0.4$		$\mu = 0.5$	
			Single Shear	Double Shear	Single Shear	Double Shear	Single Shear	Double Shear	Single Shear	Double Shear
	P_o	$0.9P_o$								
mm	kN	kN	kN	kN	kN	kN	kN	kN	kN	kN
16	104	93.5	18.7	37.4	28.1	56.1	37.4	74.8	46.8	93.5
20	162	146	29.1	58.2	43.7	87.4	58.2	116	72.8	146
22	200	180	36.0	72.1	54.1	108	72.1	144	90.1	180
24	233	210	42.0	84.0	63.0	126	84.0	168	105	210
27	303	273	54.5	109	81.8	164	109	218	136	273
30	370	333	66.6	133	99.9	200	133	266	167	333

BS 5950-1: 2000
BS 4190: 2001

BOLT CAPACITIES

NON-PRELOADED ORDINARY BOLTS

GRADE 4.6 BOLTS IN S355

Diameter of Bolt	Tensile Stress Area	Tension Capacity		Shear Capacity		Bearing Capacity in kN (Minimum of P_{bb} and P_{bs}) End distance equal to 2 x bolt diameter. Thickness in mm of ply passed through.										
		Nominal $0.8A_tp_t$	Exact A_tp_t	Single Shear P_s	Double Shear $2P_s$											
	A_t	P_{nom}	P_t													
mm	mm²	kN	kN	kN	kN	5	6	7	8	9	10	12	15	20	25	30
12	84.3	16.2	20.2	13.5	27.0	27.6	33.1	38.6	44.2	49.7	55.2	66.2	82.8	110	138	166
16	157	30.1	37.7	25.1	50.2	36.8	44.2	51.5	58.9	66.2	73.6	88.3	110	147	184	221
20	245	47.0	58.8	39.2	78.4	46.0	55.2	64.4	73.6	82.8	92.0	110	138	184	230	276
22	303	58.2	72.7	48.5	97.0	50.6	60.7	70.8	81.0	91.1	101	121	152	202	253	304
24	353	67.8	84.7	56.5	113	55.2	66.2	77.3	88.3	99.4	110	132	166	221	276	331
27	459	88.1	110	73.4	147	62.1	74.5	86.9	99.4	112	124	149	186	248	311	373
30	561	108	135	89.8	180	69.0	82.8	96.6	110	124	138	166	207	276	345	414

Values in **bold** are less than the single shear capacity of the bolt.
Values in *italic* are greater than the double shear capacity of the bolt.
Bearing values assume standard clearance holes.
If oversize or short slotted holes are used, bearing values should be multiplied by 0.7.
If long slotted or kidney shaped holes are used, bearing values should be multiplied by 0.5.
If appropriate, shear capacity must be reduced for large packings, large grip lengths and long joints.

GRADE 8.8 BOLTS IN S355

Diameter of Bolt	Tensile Stress Area	Tension Capacity		Shear Capacity		Bearing Capacity in kN (Minimum of P_{bb} and P_{bs}) End distance equal to 2 x bolt diameter. Thickness in mm of ply passed through.										
		Nominal $0.8A_tp_t$	Exact A_tp_t	Single Shear P_s	Double Shear $2P_s$											
	A_t	P_{nom}	P_t													
mm	mm²	kN	kN	kN	kN	5	6	7	8	9	10	12	15	20	25	30
12	84.3	37.8	47.2	31.6	63.2	33.0	39.6	46.2	52.8	59.4	66.0	79.2	99.0	132	165	198
16	157	70.3	87.9	58.9	118	44.0	52.8	61.6	70.4	79.2	88.0	106	132	176	220	264
20	245	110	137	91.9	184	55.0	66.0	77.0	88.0	99.0	110	132	165	220	275	330
22	303	136	170	114	227	60.5	72.6	84.7	96.8	109	121	145	182	242	303	363
24	353	158	198	132	265	66.0	79.2	92.4	106	119	132	158	198	264	330	396
27	459	206	257	172	344	74.3	89.1	104	119	134	149	178	223	297	371	446
30	561	251	314	210	421	82.5	99.0	116	132	149	165	198	248	330	413	495

Values in **bold** are less than the single shear capacity of the bolt.
Values in *italic* are greater than the double shear capacity of the bolt.
Bearing values assume standard clearance holes.
If oversize or short slotted holes are used, bearing values should be multiplied by 0.7.
If long slotted or kidney shaped holes are used, bearing values should be multiplied by 0.5.
If appropriate, shear capacity must be reduced for large packings, large grip lengths and long joints.

BS 5950-1: 2000
BS 4190: 2001

BOLT CAPACITIES

NON-PRELOADED ORDINARY BOLTS

GRADE 10.9 BOLTS IN S355

Diameter of Bolt	Tensile Stress Area	Tension Capacity		Shear Capacity		Bearing Capacity in kN (Minimum of P_{bb} and P_{bs}) End distance equal to 2 x bolt diameter. Thickness in mm of ply passed through.										
		Nominal $0.8A_tp_t$	Exact A_tp_t	Single Shear P_s	Double Shear $2P_s$											
mm	A_t mm^2	P_{nom} kN	P_t kN	P_s kN	kN	5	6	7	8	9	10	12	15	20	25	30
12	84.3	47.2	59.0	33.7	67.4	**33.0**	**39.6**	**46.2**	**52.8**	**59.4**	**66.0**	**79.2**	*99.0*	*132*	*165*	*198*
16	157	87.9	110	62.8	126	**44.0**	**52.8**	**61.6**	**70.4**	**79.2**	**88.0**	106	*132*	*176*	*220*	*264*
20	245	137	172	98.0	196	**55.0**	**66.0**	**77.0**	**88.0**	**99.0**	110	132	165	*220*	*275*	*330*
22	303	170	212	121	242	**60.5**	**72.6**	**84.7**	**96.8**	109	121	145	182	242	*303*	*363*
24	353	198	247	141	282	**66.0**	**79.2**	**92.4**	106	119	132	158	198	264	*330*	*396*
27	459	257	321	184	367	**74.3**	**89.1**	104	119	134	149	178	223	297	*371*	*446*
30	561	314	393	224	449	**82.5**	**99.0**	116	132	149	165	198	248	330	413	*495*

Values in **bold** are less than the single shear capacity of the bolt.

Values in *italic* are greater than the double shear capacity of the bolt.

Bearing values assume standard clearance holes.

If oversize or short slotted holes are used, bearing values should be multiplied by 0.7.

If long slotted or kidney shaped holes are used, bearing values should be multiplied by 0.5.

If appropriate, shear capacity must be reduced for large packings, large grip lengths and long joints.

BS 5950-1: 2000
BS 4190: 2001
BS 4933: 1973

BOLT CAPACITIES

NON-PRELOADED COUNTERSUNK BOLTS

GRADE 4.6 COUNTERSUNK BOLTS IN S355

Diameter of Bolt	Tensile Stress Area	Tension Capacity		Shear Capacity		Bearing Capacity in kN (Minimum of P_{bb} and P_{bs}) End distance equal to 2 x bolt diameter. Thickness in mm of ply passed through.										
		Nominal	Exact	Single Shear	Double Shear											
	A_t	$0.8A_t p_t$ P_{nom}	$A_t p_t$ P_t	P_s	$2P_s$											
mm	mm^2	kN	kN	kN	kN	5	6	7	8	9	10	12	15	20	25	30
12	84.3	16.2	20.2	13.5	27.0	**11.0**	*16.6*	*22.1*	*27.6*	*33.1*	*38.6*	*49.7*	*66.2*	*93.8*	*121*	*149*
16	157	30.1	37.7	25.1	50.2	**7.36**	**14.7**	**22.1**	**29.4**	*36.8*	*44.2*	*58.9*	*81.0*	*118*	*155*	*191*
20	245	47.0	58.8	39.2	78.4	**0**	**9.20**	**18.4**	**27.6**	**36.8**	*46.0*	*64.4*	*92.0*	*138*	*184*	*230*
22	303	58.2	72.7	48.5	97.0	**0**	**5.06**	**15.2**	**25.3**	**35.4**	**45.5**	*65.8*	*96.1*	*147*	*197*	*248*
24	353	67.8	84.7	56.5	113	**0**	**0**	**11.0**	**22.1**	**33.1**	**44.2**	*66.2*	*99.4*	*155*	*210*	*265*
27	459	88.1	110	73.4	147	**0**	**0**	**3.11**	**15.5**	**27.9**	**40.4**	*65.2*	*102*	*165*	*227*	*289*
30	561	108	135	89.8	180	**0**	**0**	**0**	**6.90**	**20.7**	**34.5**	**62.1**	*104*	*173*	*242*	*311*

Values in **bold** are less than the single shear capacity of the bolt.
Values in *italic* are greater than the double shear capacity of the bolt.
Bearing values assume standard clearance holes.
If oversize or short slotted holes are used, bearing values should be multiplied by 0.7.
If long slotted or kidney shaped holes are used, bearing values should be multiplied by 0.5.
Depth of countersink is taken as half the bolt diameter.

GRADE 8.8 COUNTERSUNK BOLTS IN S 355

Diameter of Bolt	Tensile Stress Area	Tension Capacity		Shear Capacity		Bearing Capacity in kN (Minimum of P_{bb} and P_{bs}) End distance equal to 2 x bolt diameter. Thickness in mm of ply passed through.										
		Nominal	Exact	Single Shear	Double Shear											
	A_t	$0.8A_t p_t$ P_{nom}	$A_t p_t$ P_t	P_s	$2P_s$											
mm	mm^2	kN	kN	kN	kN	5	6	7	8	9	10	12	15	20	25	30
12	84.3	37.8	47.2	31.6	63.2	**13.2**	**19.8**	**26.4**	*33.0*	*39.6*	*46.2*	*59.4*	*79.2*	*112*	*145*	*178*
16	157	70.3	87.9	58.9	118	**8.80**	**17.6**	**26.4**	**35.2**	**44.0**	*52.8*	*70.4*	*96.8*	*141*	*185*	*229*
20	245	110	137	91.9	184	**0**	**11.0**	**22.0**	**33.0**	**44.0**	**55.0**	*77.0*	*110*	*165*	*220*	*275*
22	303	136	170	114	227	**0**	**6.05**	**18.2**	**30.3**	**42.4**	**54.5**	*78.7*	*115*	*175*	*236*	*296*
24	353	158	198	132	265	**0**	**0**	**13.2**	**26.4**	**39.6**	**52.8**	*79.2*	*119*	*185*	*251*	*317*
27	459	206	257	172	344	**0**	**0**	**3.71**	**18.6**	**33.4**	**48.3**	*78.0*	*123*	*197*	*271*	*345*
30	561	251	314	210	421	**0**	**0**	**0**	**8.25**	**24.8**	**41.3**	**74.3**	*124*	*206*	*289*	*371*

Values in **bold** are less than the single shear capacity of the bolt.
Values in *italic* are greater than the double shear capacity of the bolt.
Bearing values assume standard clearance holes.
If oversize or short slotted holes are used, bearing values should be multiplied by 0.7.
If long slotted or kidney shaped holes are used, bearing values should be multiplied by 0.5.
Depth of countersink is taken as half the bolt diameter.

BS 5950-1: 2000
BS 4190: 2001
BS 4933: 1973

BOLT CAPACITIES

NON-PRELOADED COUNTERSUNK BOLTS

GRADE 10.9 COUNTERSUNK BOLTS IN S355

Diameter of Bolt	Tensile Stress Area	Tension Capacity		Shear Capacity		Bearing Capacity in kN (Minimum of P_{bb} and P_{bs}) End distance equal to 2 x bolt diameter. Thickness in mm of ply passed through.										
		Nominal $0.8A_t p_t$	Exact $A_t p_t$	Single Shear	Double Shear											
	A_t	P_{nom}	P_t	P_s	$2P_s$											
mm	mm^2	kN	kN	kN	kN	5	6	7	8	9	10	12	15	20	25	30
12	84.3	47.2	59.0	33.7	67.4	**13.2**	**19.8**	**26.4**	**33.0**	39.6	46.2	59.4	79.2	*112*	*145*	*178*
16	157	87.9	110	62.8	126	**8.80**	**17.6**	**26.4**	**35.2**	**44.0**	**52.8**	70.4	96.8	*141*	*185*	*229*
20	245	137	172	98.0	196	**0**	**11.0**	**22.0**	**33.0**	**44.0**	**55.0**	**77.0**	110	165	220	275
22	303	170	212	121	242	**0**	**6.05**	**18.2**	**30.3**	**42.4**	**54.5**	**78.7**	115	175	236	296
24	353	198	247	141	282	**0**	**0**	**13.2**	**26.4**	**39.6**	**52.8**	**79.2**	119	185	251	317
27	459	257	321	184	367	**0**	**0**	**3.71**	**18.6**	**33.4**	**48.3**	**78.0**	123	197	271	345
30	561	314	393	224	449	**0**	**0**	**0**	**8.25**	**24.8**	**41.3**	**74.3**	**124**	206	289	371

Values in **bold** are less than the single shear capacity of the bolt.

Values in *italic* are greater than the double shear capacity of the bolt.

Bearing values assume standard clearance holes.

If oversize or short slotted holes are used, bearing values should be multiplied by 0.7.

If long slotted or kidney shaped holes are used, bearing values should be multiplied by 0.5.

Depth of countersink is taken as half the bolt diameter.

BS 5950-1: 2000
BS 4395: 1969

BOLT CAPACITIES

NON-PRELOADED HSFG BOLTS

GENERAL GRADE HSFG BOLTS IN S355

Diameter of Bolt	Tensile Stress Area	Tension Capacity		Shear Capacity		Bearing Capacity in kN (Minimum of P_{bb} and P_{bs}) End distance equal to 2 x bolt diameter. Thickness in mm of ply passed through.										
		Nominal $0.8A_t p_t$	Exact $A_t p_t$	Single Shear	Double Shear											
	A_t	P_{nom}	P_t	P_s	$2P_s$											
mm	mm²	kN	kN	kN	kN	5	6	7	8	9	10	12	15	20	25	30
12	84.3	39.8	49.7	33.7	67.4	**33.0**	**39.6**	**46.2**	**52.8**	**59.4**	**66.0**	**79.2**	*99.0*	*132*	*165*	*198*
16	157	74.1	92.6	62.8	126	**44.0**	**52.8**	**61.6**	**70.4**	**79.2**	**88.0**	106	*132*	*176*	*220*	*264*
20	245	116	145	98.0	196	**55.0**	**66.0**	**77.0**	**88.0**	**99.0**	110	132	165	*220*	*275*	*330*
22	303	143	179	121	242	**60.5**	**72.6**	**84.7**	**96.8**	109	121	145	182	*242*	*303*	*363*
24	353	167	208	141	282	**66.0**	**79.2**	**92.4**	106	119	**132**	158	198	264	*330*	*396*
27	459	189	236	161	321	**74.3**	**89.1**	104	119	134	149	178	223	297	*371*	*446*
30	561	231	289	196	393	**82.5**	**99.0**	116	132	149	165	198	248	330	*413*	*495*

Values in **bold** are less than the single shear capacity of the bolt.

Values in *italic* are greater than the double shear capacity of the bolt.

Bearing values assume standard clearance holes.

If oversize or short slotted holes are used, bearing values should be multiplied by 0.7.

If long slotted or kidney shaped holes are used, bearing values should be multiplied by 0.5.

If appropriate, shear capacity must be reduced for large packings, large grip lengths and long joints.

HIGHER GRADE HSFG BOLTS IN S 355

Diameter of Bolt	Tensile Stress Area	Tension Capacity		Shear Capacity		Bearing Capacity in kN (Minimum of P_{bb} and P_{bs}) End distance equal to 2 x bolt diameter. Thickness in mm of ply passed through.										
		Nominal $0.8A_t p_t$	Exact $A_t p_t$	Single Shear	Double Shear											
	A_t	P_{nom}	P_t	P_s	$2P_s$											
mm	mm²	kN	kN	kN	kN	5	6	7	8	9	10	12	15	20	25	30
16	157	87.9	110	62.8	126	**44.0**	**52.8**	**61.6**	**70.4**	**79.2**	**88.0**	106	*132*	*176*	*220*	*264*
20	245	137	172	98.0	196	**55.0**	**66.0**	**77.0**	**88.0**	**99.0**	110	132	165	*220*	*275*	*330*
22	303	170	212	121	242	**60.5**	**72.6**	**84.7**	**96.8**	109	121	145	182	*242*	*303*	*363*
24	353	198	247	141	282	**66.0**	**79.2**	**92.4**	106	119	132	158	198	264	*330*	*396*
27	459	257	321	184	367	**74.3**	**89.1**	104	119	134	149	178	223	297	*371*	*446*
30	561	314	393	224	449	**82.5**	**99.0**	116	132	149	165	198	248	330	413	*495*

Values in **bold** are less than the single shear capacity of the bolt.

Values in *italic* are greater than the double shear capacity of the bolt.

Bearing values assume standard clearance holes.

If oversize or short slotted holes are used, bearing values should be multiplied by 0.7.

If long slotted or kidney shaped holes are used, bearing values should be multiplied by 0.5.

If appropriate, shear capacity must be reduced for large packings, large grip lengths and long joints.

BS 5950-1: 2000
BS 4395: 1969
BS 4604: 1970

BOLT CAPACITIES

PRELOADED HSFG BOLTS: NON-SLIP IN SERVICE

GENERAL GRADE HSFG BOLTS IN S355

Diameter of Bolt	Min. Shank Tension	Tension		Shear Capacity		Slip Resistance for $\mu = 0.5$		Bearing Capacity, P_{bg} in kN End distance equal to 3 x bolt diameter. Thickness in mm of ply passed through.										
				Single Shear	Double Shear	Single Shear	Double Shear												
	P_o	$1.1P_o$	$A_t p_t$					5	6	7	8	9	10	12	15	20	25	30
mm	kN	kN	kN	kN	kN	kN	kN											
12	49.4	54.3	49.7	33.7	67.4	27.2	54.3	49.5	59.4	69.3	79.2	89.1	99.0	*119*	*149*	*198*	*248*	*297*
16	92.1	101	92.6	62.8	126	50.7	101	66.0	79.2	92.4	106	119	*132*	*158*	*198*	*264*	*330*	*396*
20	144	158	145	98.0	196	79.2	158	**82.5**	99.0	116	132	149	165	*198*	*248*	*330*	*413*	*495*
22	177	195	179	121	242	97.4	195	**90.8**	**109**	127	145	163	182	*218*	*272*	*363*	*454*	*545*
24	207	228	208	141	282	114	228	**99.0**	**119**	**139**	158	178	198	238	*297*	*396*	*495*	*594*
27	234	257	236	161	321	129	257	**111**	**134**	**156**	178	200	223	267	*334*	*446*	*557*	*668*
30	286	315	289	196	393	157	315	**124**	**149**	**173**	198	223	248	297	371	*495*	*619*	*743*

Values in **bold** are less than the single shear capacity of the bolt.
Values in *italic* are greater than the double shear capacity of the bolt.
Shading indicates that the ply thickness is not suitable for an outer ply.

HIGHER GRADE HSFG BOLTS IN S355

Diameter of Bolt	Min. Shank Tension	Tension		Shear Capacity		Slip Resistance for $\mu = 0.5$		Bearing Capacity, P_{bg} in kN End distance equal to 3 x bolt diameter. Thickness in mm of ply passed through.										
				Single Shear	Double Shear	Single Shear	Double Shear												
	P_o	$1.1P_o$	$A_t p_t$					5	6	7	8	9	10	12	15	20	25	30
mm	kN	kN	kN	kN	kN	kN	kN											
16	104	114	110	62.8	126	57.1	114	66.0	79.2	92.4	106	119	*132*	*158*	*198*	*264*	*330*	*396*
20	162	178	172	98.0	196	89.0	178	**82.5**	99.0	116	132	149	165	*198*	*248*	*330*	*413*	*495*
22	200	220	212	121	242	110	220	**90.8**	**109**	127	145	163	182	*218*	*272*	*363*	*454*	*545*
24	233	257	247	141	282	128	257	**99.0**	**119**	**139**	158	178	198	238	*297*	*396*	*495*	*594*
27	303	333	321	184	367	167	333	**111**	**134**	**156**	178	200	223	267	*334*	*446*	*557*	*668*
30	370	407	393	224	449	204	407	**124**	**149**	**173**	**198**	223	248	297	371	*495*	*619*	*743*

Values in **bold** are less than the single shear capacity of the bolt.
Values in *italic* are greater than the double shear capacity of the bolt.
Shading indicates that the ply thickness is not suitable for an outer ply.

<table>
<tr><td>BS 5950-1: 2000
BS 4395: 1969
BS 4604: 1970</td></tr>
</table>

BOLT CAPACITIES

PRELOADED HSFG BOLTS: NON-SLIP UNDER FACTORED LOADS

GENERAL GRADE HSFG BOLTS IN S355

Diameter of Bolt	Min. Shank Tension	Bolt Tension Capacity	Slip Resistance P_{sL}							
			$\mu = 0.2$		$\mu = 0.3$		$\mu = 0.4$		$\mu = 0.5$	
			Single Shear	Double Shear	Single Shear	Double Shear	Single Shear	Double Shear	Single Shear	Double Shear
	P_o	$0.9P_o$								
mm	kN	kN	kN	kN	kN	kN	kN	kN	kN	kN
12	49.4	44.5	8.89	17.8	13.3	26.7	17.8	35.6	22.2	44.5
16	92.1	82.9	16.6	33.2	24.9	49.7	33.2	66.3	41.4	82.9
20	144	130	25.9	51.8	38.9	77.8	51.8	104	64.8	130
22	177	159	31.9	63.7	47.8	95.6	63.7	127	79.7	159
24	207	186	37.3	74.5	55.9	112	74.5	149	93.2	186
27	234	211	42.1	84.2	63.2	126	84.2	168	105	211
30	286	257	51.5	103	77.2	154	103	206	129	257

HIGHER GRADE HSFG BOLTS IN S355

Diameter of Bolt	Min. Shank Tension	Bolt Tension Capacity	Slip Resistance P_{sL}							
			$\mu = 0.2$		$\mu = 0.3$		$\mu = 0.4$		$\mu = 0.5$	
			Single Shear	Double Shear	Single Shear	Double Shear	Single Shear	Double Shear	Single Shear	Double Shear
	P_o	$0.9P_o$								
mm	kN	kN	kN	kN	kN	kN	kN	kN	kN	kN
16	104	93.5	18.7	37.4	28.1	56.1	37.4	74.8	46.8	93.5
20	162	146	29.1	58.2	43.7	87.4	58.2	116	72.8	146
22	200	180	36.0	72.1	54.1	108	72.1	144	90.1	180
24	233	210	42.0	84.0	63.0	126	84.0	168	105	210
27	303	273	54.5	109	81.8	164	109	218	136	273
30	370	333	66.6	133	99.9	200	133	266	167	333

BS 5950-1: 2000
BS 4395: 1969
BS 4604: 1970
BS 4933: 1973

BOLT CAPACITIES

PRELOADED HSFG BOLTS: NON-SLIP IN SERVICE

GENERAL GRADE COUNTERSUNK HSFG BOLTS IN S355

Diameter of Bolt	Min. Shank Tension	Tension		Shear Capacity		Slip Resistance for μ = 0.5		Bearing Capacity, P_{bg} in kN End distance equal to 3 x bolt diameter. Thickness in mm of ply passed through.										
				Single Shear	Double Shear	Single Shear	Double Shear												
mm	P_o kN	$1.1P_o$ kN	$A_t p_t$ kN	kN	kN	kN	kN	5	6	7	8	9	10	12	15	20	25	30
12	49.4	54.3	49.7	33.7	67.4	27.2	54.3	**19.8**	**29.7**	39.6	49.5	59.4	69.3	*89.1*	*119*	*168*	*218*	*267*
16	92.1	101	92.6	62.8	126	50.7	101	**13.2**	**26.4**	**39.6**	**52.8**	66.0	79.2	106	*145*	*211*	*277*	*343*
20	144	158	145	98.0	196	79.2	158	**0**	**16.5**	**33.0**	**49.5**	**66.0**	82.5	116	165	*248*	*330*	*413*
22	177	195	179	121	242	97.4	195	**0**	**9.08**	**27.2**	**45.4**	**63.5**	**81.7**	118	172	*263*	*354*	*445*
24	207	228	208	141	282	114	228	**0**	**19.8**	**39.6**	**59.4**	**79.2**	119	178	*277*	*376*	*475*	
27	234	257	236	161	321	129	257	**0**	**0**	**5.57**	**27.8**	**50.1**	**72.4**	117	184	295	*407*	*518*
30	286	315	289	196	393	157	315	**0**	**0**	**0**	**12.4**	**37.1**	**61.9**	**111**	186	309	*433*	*557*

Values in **bold** are less than the single shear capacity of the bolt.

Values in *italic* are greater than the double shear capacity of the bolt.

Shading indicates that the ply thickness is not suitable for an outer ply.

HIGHER GRADE COUNTERSUNK HSFG BOLTS IN S355

Diameter of Bolt	Min. Shank Tension	Tension		Shear Capacity		Slip Resistance for μ = 0.5		Bearing Capacity, P_{bg} in kN End distance equal to 3 x bolt diameter. Thickness in mm of ply passed through.										
				Single Shear	Double Shear	Single Shear	Double Shear												
mm	P_o kN	$1.1P_o$ kN	$A_t p_t$ kN	kN	kN	kN	kN	5	6	7	8	9	10	12	15	20	25	30
16	104	114	110	62.8	126	57.1	114	**13.2**	**26.4**	**39.6**	**52.8**	66.0	79.2	106	*145*	*211*	*277*	*343*
20	162	178	172	98.0	196	89.0	178	**0**	**16.5**	**33.0**	**49.5**	**66.0**	82.5	116	165	*248*	*330*	*413*
22	200	220	212	121	242	110	220	**0**	**9.08**	**27.2**	**45.4**	**63.5**	**81.7**	118	172	*263*	*354*	*445*
24	233	257	247	141	282	128	257	**0**	**0**	**19.8**	**39.6**	**59.4**	**79.2**	119	178	*277*	*376*	*475*
27	303	333	321	184	367	167	333	**0**	**0**	**5.57**	**27.8**	**50.1**	**72.4**	117	184	295	*407*	*518*
30	370	407	393	224	449	204	407	**0**	**0**	**0**	**12.4**	**37.1**	**61.9**	**111**	186	309	*433*	*557*

Values in **bold** are less than the single shear capacity of the bolt.

Values in *italic* are greater than the double shear capacity of the bolt.

Shading indicates that the ply thickness is not suitable for an outer ply.

BS 5950-1: 2000
BS 4395: 1969
BS 4604: 1970
BS 4933: 1973

BOLT CAPACITIES

PRELOADED HSFG BOLTS: NON-SLIP UNDER FACTORED LOADS

GENERAL GRADE COUNTERSUNK HSFG BOLTS IN S355

Diameter of Bolt	Min. Shank Tension	Bolt Tension Capacity	Slip Resistance P_{sL}							
			$\mu = 0.2$		$\mu = 0.3$		$\mu = 0.4$		$\mu = 0.5$	
			Single Shear	Double Shear	Single Shear	Double Shear	Single Shear	Double Shear	Single Shear	Double Shear
	P_o	$0.9P_o$								
mm	kN	kN	kN	kN	kN	kN	kN	kN	kN	kN
12	49.4	44.5	8.89	17.8	13.3	26.7	17.8	35.6	22.2	44.5
16	92.1	82.9	16.6	33.2	24.9	49.7	33.2	66.3	41.4	82.9
20	144	130	25.9	51.8	38.9	77.8	51.8	104	64.8	130
22	177	159	31.9	63.7	47.8	95.6	63.7	127	79.7	159
24	207	186	37.3	74.5	55.9	112	74.5	149	93.2	186
27	234	211	42.1	84.2	63.2	126	84.2	168	105	211
30	286	257	51.5	103	77.2	154	103	206	129	257

HIGHER GRADE COUNTERSUNK HSFG BOLTS IN S 355

Diameter of Bolt	Min. Shank Tension	Bolt Tension Capacity	Slip Resistance P_{sL}							
			$\mu = 0.2$		$\mu = 0.3$		$\mu = 0.4$		$\mu = 0.5$	
			Single Shear	Double Shear	Single Shear	Double Shear	Single Shear	Double Shear	Single Shear	Double Shear
	P_o	$0.9P_o$								
mm	kN	kN	kN	kN	kN	kN	kN	kN	kN	kN
16	104	93.5	18.7	37.4	28.1	56.1	37.4	74.8	46.8	93.5
20	162	146	29.1	58.2	43.7	87.4	58.2	116	72.8	146
22	200	180	36.0	72.1	54.1	108	72.1	144	90.1	180
24	233	210	42.0	84.0	63.0	126	84.0	168	105	210
27	303	273	54.5	109	81.8	164	109	218	136	273
30	370	333	66.6	133	99.9	200	133	266	167	333

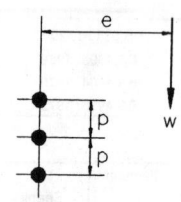

Bolt groups
One row of fasteners; fasteners in the plane of the force

No. of fasteners in vertical row	Pitch, p (mm)	Values of Z_{xx} (cm³) for diameter of bolt, D (mm)						
		12	16	20	22	24	27	30
2	70	6.0	11.4	18.0	22.5	26.6	35.1	43.8
3		11.9	22.3	35.0	43.4	50.8	66.5	81.9
4		19.7	36.9	57.8	71.6	83.6	109.1	133.9
5		29.6	55.2	86.3	106.9	124.7	162.5	199.2
6		41.4	77.5	120.6	149.3	174.1	226.7	277.6
7		55.1	102.8	160.6	198.7	231.7	301.6	369.2
8		70.9	132.1	206.3	255.3	297.6	387.3	473.8
9		88.6	165.1	257.8	318.9	371.7	483.6	591.6
2	100	8.5	16.0	25.1	31.2	36.6	48.0	59.3
3		16.9	31.6	49.5	61.3	71.6	93.4	114.6
4		28.2	52.5	82.1	101.6	118.5	154.4	189.1
5		42.2	78.7	122.9	152.1	177.3	230.8	282.5
6		59.1	110.1	171.9	212.7	247.9	322.6	394.6
7		78.7	146.7	229.0	283.3	330.2	429.6	525.5
8		101.2	188.5	294.4	364.1	424.3	552.0	675.0
9		126.5	235.6	367.9	455.0	530.2	689.7	843.3

Bolt groups
Two rows of fasteners; fasteners in the plane of the force

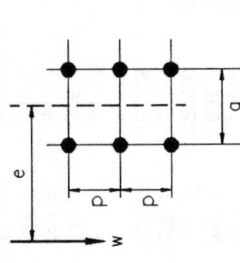

No. of fasteners in vertical row	Pitch, p (mm)	Gauge, g (mm)	Values of Z_{xx} (cm³) for diameter of bolt, D (mm)						
			12	16	20	22	24	27	30
2	70	60	15.7	29.5	46.5	57.9	67.9	89.2	110.3
3			27.8	52.0	81.7	101.3	118.4	154.8	190.4
4			43.5	81.3	127.2	157.6	184.1	240.0	294.5
5			63.1	117.8	184.2	228.0	266.1	346.7	424.8
6			86.7	161.6	252.6	312.7	364.6	474.8	581.4
7			114.2	212.8	332.5	411.5	479.7	624.4	764.2
8			145.6	271.4	423.9	524.4	611.3	795.6	973.4
9			181.0	337.3	526.7	651.6	759.5	988.2	1208.8
2	100	60	19.8	37.1	58.2	72.2	84.6	110.6	136.3
3			36.8	68.6	107.4	133.1	155.4	202.6	248.5
4			59.2	110.4	172.5	213.5	249.1	324.4	397.3
5			87.2	162.6	254.0	314.3	366.4	476.9	583.7
6			120.9	225.3	351.8	435.3	507.4	660.2	807.7
7			160.2	298.5	466.0	576.5	671.9	874.2	1069.1
8			205.1	382.2	596.6	738.0	860.1	1118.8	1368.1
9			255.7	476.3	743.5	919.7	1071.8	1394.0	1704.5

Bolt groups
Four rows of fasteners; fasteners in the plane of the force

No. of fasteners in vertical row	Pitch, p (mm)	S₁ (mm)	S₂ (mm)	S₃ (mm)	Values of Z_{xx} (cm³) for diameter of bolt, D (mm)						
					12	16	20	22	24	27	30
2	70	120	60	240	55.3	103.2	161.3	199.8	233.1	303.7	372.2
3					89.5	166.9	261.0	323.0	376.8	490.8	601.3
4					128.1	238.9	373.4	462.2	539.0	701.9	859.4
5					172.1	320.8	501.3	620.4	723.4	941.6	1152.6
6					222.3	414.3	647.2	800.8	933.6	1215.1	1486.9
7					279.3	520.6	813.0	1005.9	1172.6	1525.7	1866.6
8					343.6	640.2	999.8	1236.9	1441.7	1875.7	2294.4
9					415.3	773.8	1208.2	1494.7	1742.0	2266.3	2771.8
2	70	120	85	290	61.3	114.3	178.7	221.2	258.0	336.1	411.6
3					98.0	182.8	285.6	353.5	412.3	536.8	657.3
4					139.1	259.3	405.1	501.3	584.6	761.0	931.5
5					185.2	345.1	539.1	667.1	777.8	1012.3	1238.8
6					237.1	441.8	690.0	853.7	995.2	1295.1	1584.5
7					295.4	550.4	859.6	1063.5	1239.6	1612.9	1973.0
8					360.7	672.0	1049.3	1298.1	1512.9	1968.3	2407.5
9					433.1	807.0	1259.9	1558.6	1816.5	2363.1	2890.0
2	70	140	50	240	58.8	109.7	171.5	212.4	247.8	322.8	395.5
3					94.2	175.7	274.7	340.0	396.7	516.6	632.8
4					133.6	249.2	389.4	481.9	562.0	731.8	896.0
5					178.1	331.9	518.6	641.7	748.2	974.0	1192.1
6					228.5	425.9	665.2	823.1	959.6	1248.8	1528.1
7					285.7	532.4	831.4	1028.7	1199.1	1560.3	1908.8
8					350.0	652.2	1018.4	1260.0	1468.6	1910.7	2337.2
9					421.8	785.9	1227.1	1518.0	1769.1	2301.5	2814.8

	70	140	75	290							
2					64.3	119.8	187.3	231.8	270.3	352.1	431.1
3					102.1	190.4	297.5	368.2	429.4	559.1	684.5
4					144.0	268.4	419.4	518.9	605.1	787.7	964.1
5					190.6	355.3	554.9	686.7	800.5	1041.9	1275.0
6					242.8	452.5	706.8	874.5	1019.4	1326.6	1623.0
7					301.4	561.6	877.0	1085.1	1264.8	1645.6	2013.0
8					366.8	683.4	1067.2	1320.2	1538.7	2001.8	2448.5
9					439.4	818.6	1278.1	1581.1	1842.7	2397.1	2931.6

	100	120	60	240							
2					59.8	111.5	174.4	215.8	251.8	328.0	401.9
3					101.6	189.5	296.1	366.4	427.4	556.4	681.3
4					151.1	281.7	440.1	544.5	634.9	826.3	1011.3
5					209.8	391.0	610.6	755.5	880.6	1145.9	1401.9
6					278.6	519.2	810.6	1002.9	1168.8	1520.7	1860.0
7					358.1	667.2	1041.7	1288.6	1501.7	1953.5	2389.0
8					448.6	835.6	1304.5	1613.6	1880.4	2445.9	2990.8
9					550.0	1024.6	1599.4	1978.4	2305.4	2998.5	3666.1

	100	120	85	290							
2					65.2	121.6	190.1	235.3	274.4	357.3	437.6
3					109.1	203.4	317.9	393.3	458.7	597.0	730.8
4					160.6	299.3	467.4	578.3	674.2	877.4	1073.6
5					220.7	411.2	642.2	794.5	926.1	1205.0	1474.0
6					290.5	541.3	845.2	1045.6	1218.6	1585.4	1939.0
7					370.7	690.7	1078.3	1333.9	1554.5	2022.1	2472.8
8					461.7	860.0	1342.6	1660.7	1935.3	2517.2	3077.9
9					563.5	1049.7	1638.6	2026.8	2361.7	3071.7	3755.7

	100	140	50	240							
2					63.2	117.8	184.2	228.0	265.9	346.4	424.3
3					105.8	197.3	308.3	381.6	445.0	579.4	709.3
4					155.7	290.2	453.3	560.9	654.0	851.2	1041.6
5					214.5	399.7	624.3	772.3	900.3	1171.5	1433.1
6					283.4	528.0	824.4	1019.9	1188.7	1546.5	1891.5
7					362.9	676.1	1055.5	1305.7	1521.6	1979.4	2420.6
8					453.3	844.5	1318.3	1630.7	1900.2	2471.7	3022.3
9					554.8	1033.4	1613.2	1995.4	2325.1	3024.2	3697.6

	100	140	75	290							
2					68.1	127.0	198.4	245.6	286.4	372.9	456.6
3					112.9	210.4	328.7	406.8	474.3	617.4	755.6
4					164.8	307.1	479.6	593.5	691.8	900.3	1101.6
5					225.1	419.5	655.1	810.5	944.7	1229.1	1503.6
6					295.1	549.8	858.4	1062.0	1237.7	1610.2	1969.3
7					375.4	699.3	1091.7	1350.5	1573.8	2047.2	2503.5
8					466.3	868.7	1356.0	1677.4	1954.6	2542.4	3108.7
9					568.2	1058.3	1652.0	2043.4	2381.1	3097.0	3786.5

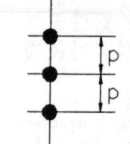

Bolt groups
One row of fasteners; fasteners not in the plane of the force

No. of fasteners in vertical row	Pitch, p (mm)	Values of Z_{xx} (cm³) for diameter of bolt, D (mm)					
		12	16	20	22	24	27	30
2	70	13.3	24.8	38.8	48.0	55.9	72.8	89.1
3		25.5	47.5	74.2	91.8	107.0	139.2	170.3
4		41.7	77.6	121.2	149.9	174.6	227.2	277.8
5		61.8	115.0	179.5	222.1	258.8	336.5	411.4
6		85.8	159.8	249.4	308.4	359.4	467.3	571.3
7		113.7	211.8	330.6	408.9	476.4	619.6	757.3
8		145.6	271.2	423.3	523.5	610.0	793.2	969.6
9		181.5	338.0	527.4	652.3	760.0	988.3	1208.0
2	100	15.4	28.7	44.8	55.4	64.6	84.1	102.9
3		31.4	58.5	91.3	112.9	131.6	171.2	209.3
4		53.0	98.8	154.2	190.7	222.2	288.9	353.2
5		80.3	149.5	233.4	288.6	336.3	437.3	534.6
6		113.2	210.8	328.9	406.8	474.0	616.3	753.4
7		151.6	282.4	440.8	545.2	635.1	825.9	1009.5
8		195.8	364.6	569.0	703.7	819.9	1066.1	1303.1
9		245.5	457.2	713.5	882.5	1028.1	1336.9	1634.1

Centre of rotation is assumed 60 mm below the bottom bolt
The tabulated values are conservative when the centre of rotation is located more than 60 mm below the bottom bolt. The tabulated values are unconservative when the centre of rotation is located less than 60 mm below the bottom line.

Bolt groups
Two rows of fasteners; fasteners not in the plane of the force

No. of fasteners in vertical row	Pitch, p (mm)	Values of Z_{xx} (cm³) for diameter of bolt, D (mm)					
		12	16	20	22	24	27	30
2	70	26.6	49.6	77.5	95.9	111.8	145.6	178.2
3		51.0	95.1	148.5	183.7	214.1	278.5	340.6
4		83.3	155.2	242.3	299.7	349.3	454.3	555.5
5		123.5	230.1	359.1	444.1	517.5	673.1	822.9
6		171.5	319.5	498.7	616.8	718.7	934.7	1142.6
7		227.5	423.7	661.2	817.8	952.8	1239.1	1514.7
8		291.2	542.5	846.6	1047.1	1219.9	1586.4	1939.2
9		362.9	675.9	1054.8	1304.6	1520.0	1976.5	2416.0
2	100	30.8	57.4	89.6	110.9	129.3	168.2	205.8
3		62.8	117.0	182.6	225.9	263.2	342.4	418.6
4		106.1	197.5	308.3	381.4	444.4	577.9	706.5
5		160.6	299.1	466.7	577.3	672.6	874.7	1069.2
6		226.3	421.5	657.8	813.6	947.9	1232.6	1506.7
7		303.3	564.9	881.6	1090.3	1270.3	1651.8	2019.1
8		391.5	729.2	1138.0	1407.4	1639.7	2132.2	2606.2
9		491.0	914.5	1427.1	1765.0	2056.3	2673.8	3268.2

Centre of rotation is assumed 60 mm below the bottom bolts
The tabulated values are conservative when the centre of rotation is located more than 60 mm below the bottom bolts. The tabulated values are unconservative when the centre of rotation is located less than 60 mm below the bottom bolts.

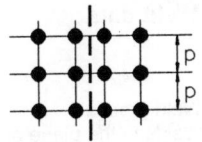

Bolt groups
Four rows of fasteners; fasteners not in the plane of the force

No. of fasteners in vertical row	Pitch, p (mm)	Values of Z_{xx} (cm³) for diameter of bolt, D (mm)						
		12	16	20	22	24	27	30
2	70	53.2	99.2	155.0	191.8	223.7	291.1	356.3
3		102.1	190.2	296.9	367.3	428.1	557.0	681.2
4		166.7	310.5	484.7	599.5	698.6	908.7	1111.1
5		247.0	460.1	718.2	888.3	1035.0	1346.1	1645.7
6		343.1	639.1	997.4	1233.7	1437.4	1869.3	2285.2
7		454.9	847.3	1322.4	1635.6	1905.7	2478.2	3029.4
8		582.5	1084.9	1693.2	2094.1	2439.9	3172.8	3878.3
9		725.8	1351.8	2109.7	2609.2	3039.9	3953.1	4832.0
2	100	61.6	114.8	179.2	221.8	258.5	336.4	411.5
3		125.6	234.0	365.2	451.8	526.5	684.8	837.3
4		212.1	395.1	616.7	762.7	888.7	1155.8	1413.0
5		321.1	598.1	933.5	1154.5	1345.2	1749.3	2138.4
6		452.6	843.0	1315.6	1627.2	1895.8	2465.3	3013.5
7		606.6	1129.8	1763.1	2180.6	2540.6	3303.7	4038.2
8		783.1	1458.4	2276.0	2814.9	3279.5	4264.5	5212.5
9		982.0	1828.9	2854.2	3529.9	4112.6	5347.7	6536.4

Centre of rotation is assumed 60 mm below the bottom bolts
The tabulated values are conservative when the centre of rotation is located more than 60 mm below the bottom bolts. The tabulated values are unconservative when the centre of rotation is located less than 60 mm below the bottom bolts.

Weld data

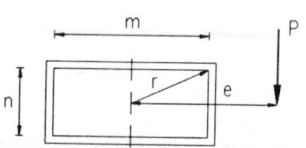

Weld groups
Welds in the plane of the force

Values of Z_p (cm³) for 1 mm throat thickness

Values of n (mm)	Values of m (mm)										
	50	75	100	125	150	175	200	225	250	275	300
50	4.7	7.2	10.1	13.3	16.9	20.9	25.3	30.1	35.3	40.9	47.0
75	7.2	10.6	14.3	18.3	22.6	27.4	32.5	37.9	43.8	50.1	56.8
100	10.1	14.3	18.9	23.7	28.9	34.4	40.2	46.5	53.1	60.1	67.5
125	13.3	18.3	23.7	29.5	35.5	41.8	48.5	55.5	62.9	70.6	78.7
150	16.9	22.6	28.9	35.5	42.4	49.6	57.2	65.0	73.2	81.7	90.6
175	20.9	27.4	34.4	41.8	49.6	57.7	66.1	74.8	83.9	93.2	102.9
200	25.3	32.5	40.2	48.5	57.2	66.1	75.4	85.0	94.9	105.1	115.6
225	30.1	37.9	46.5	55.5	65.0	74.8	85.0	95.5	106.2	117.3	128.6
250	35.3	43.8	53.1	62.9	73.2	83.9	94.9	106.2	117.9	129.8	142.0
275	40.9	50.1	60.1	70.6	81.7	93.2	105.1	117.3	129.8	142.6	155.7
300	47.0	56.8	67.5	78.7	90.6	102.9	115.6	128.6	142.0	155.7	169.7
325	53.5	64.0	75.3	87.2	99.8	112.9	126.4	140.3	154.5	169.1	184.0
350	60.3	71.5	83.4	96.1	109.4	123.3	137.6	152.3	167.4	182.8	198.6
375	67.6	79.4	92.0	105.4	119.4	134.0	149.1	164.6	180.6	196.9	213.5
400	75.4	87.8	101.1	115.1	129.8	145.1	161.0	177.3	194.1	211.2	228.7
425	83.5	96.5	110.5	125.2	140.6	156.7	173.3	190.4	207.9	225.9	244.2
450	92.0	105.7	120.3	135.7	151.8	168.5	185.9	203.8	222.1	240.9	260.0
475	101.0	115.3	130.5	146.6	163.4	180.8	198.9	217.5	236.6	256.2	276.2
500	110.4	125.3	141.2	157.9	175.4	193.5	212.3	231.7	251.6	271.9	292.7
525	120.2	135.8	152.3	169.6	187.8	206.6	226.1	246.2	266.8	288.0	309.5
550	130.4	146.6	163.8	181.8	200.6	220.1	240.3	261.1	282.5	304.4	326.8
575	141.0	157.9	175.7	194.3	213.8	234.0	254.9	276.4	298.5	321.2	344.3
600	152.0	169.5	188.0	207.3	227.4	248.3	269.8	292.1	314.9	338.3	362.2

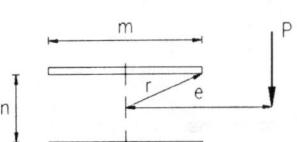

Weld groups
Welds in the plane of the force

Values of Z_P (cm³) for 1 mm throat thickness

Values of n (mm)	Values of m (mm)										
	50	75	100	125	150	175	200	225	250	275	300
50	2.4	3.6	5.2	7.2	9.5	12.2	15.4	18.9	22.9	27.3	32.1
75	3.6	5.3	7.2	9.3	11.7	14.6	17.8	21.3	25.3	29.7	34.6
100	4.8	7.1	9.4	11.9	14.6	17.5	20.9	24.6	28.6	33.1	37.9
125	6.1	9.0	11.8	14.7	17.8	21.0	24.6	28.4	32.6	37.2	42.1
150	7.4	10.9	14.3	17.7	21.2	24.8	28.7	32.8	37.2	41.9	47.0
175	8.6	12.8	16.8	20.8	24.8	28.9	33.1	37.5	42.2	47.1	52.4
200	9.9	14.7	19.4	24.0	28.5	33.1	37.7	42.5	47.5	52.7	58.2
225	11.2	16.6	21.9	27.1	32.2	37.3	42.5	47.7	53.1	58.7	64.5
250	12.4	18.5	24.5	30.3	36.0	41.7	47.4	53.1	58.9	64.9	71.1
275	13.7	20.4	27.0	33.4	39.8	46.1	52.3	58.6	64.9	71.3	77.9
300	14.9	22.3	29.5	36.6	43.6	50.5	57.3	64.1	71.0	77.8	84.9
325	16.2	24.2	32.0	39.8	47.4	54.9	62.3	69.7	77.1	84.5	92.0
350	17.4	26.1	34.6	43.0	51.2	59.3	67.4	75.4	83.3	91.3	99.2
375	18.7	27.9	37.1	46.1	55.0	63.8	72.5	81.0	89.6	98.1	106.6
400	19.9	29.8	39.6	49.3	58.8	68.2	77.5	86.7	95.8	104.9	114.0
425	21.2	31.7	42.1	52.4	62.6	72.7	82.6	92.4	102.1	111.8	121.5
450	22.5	33.6	44.7	55.6	66.4	77.1	87.7	98.1	108.5	118.7	129.0
475	23.7	35.5	47.2	58.7	70.2	81.5	92.7	103.8	114.8	125.7	136.5
500	25.0	37.4	49.7	61.9	74.0	86.0	97.8	109.5	121.1	132.6	144.1
525	26.2	39.2	52.2	65.0	77.8	90.4	102.9	115.2	127.5	139.6	151.6
550	27.5	41.1	54.7	68.2	81.6	94.8	107.9	120.9	133.8	146.6	159.2
575	28.7	43.0	57.2	71.3	85.4	99.2	113.0	126.6	140.1	153.5	166.8
600	30.0	44.9	59.7	74.5	89.1	103.7	118.1	132.3	146.5	160.5	174.4

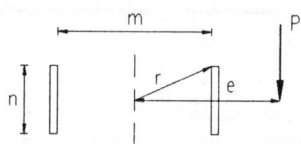

Weld groups
Welds in the plane of the force

Values of Z_p (cm³) for 1 mm throat thickness

Values of n (mm)	Values of m (mm)										
	50	75	100	125	150	175	200	225	250	275	300
50	2.4	3.6	4.8	6.1	7.4	8.6	9.9	11.2	12.4	13.7	14.9
75	3.6	5.3	7.1	9.0	10.9	12.8	14.7	16.6	18.5	20.4	22.3
100	5.2	7.2	9.4	11.8	14.3	16.8	19.4	21.9	24.5	27.0	29.5
125	7.2	9.3	11.9	14.7	17.7	20.8	24.0	27.1	30.3	33.4	36.6
150	9.5	11.7	14.6	17.8	21.2	24.8	28.5	32.2	36.0	39.8	43.6
175	12.2	14.6	17.5	21.0	24.8	28.9	33.1	37.3	41.7	46.1	50.5
200	15.4	17.8	20.9	24.6	28.7	33.1	37.7	42.5	47.4	52.3	57.3
225	18.9	21.3	24.6	28.4	32.8	37.5	42.5	47.7	53.1	58.6	64.1
250	22.9	25.3	28.6	32.6	37.2	42.2	47.5	53.1	58.9	64.9	71.0
275	27.3	29.7	33.1	37.2	41.9	47.1	52.7	58.7	64.9	71.3	77.8
300	32.1	34.6	37.9	42.1	47.0	52.4	58.2	64.5	71.1	77.9	84.9
325	37.3	39.8	43.2	47.4	52.4	58.0	64.1	70.6	77.4	84.6	92.0
350	42.9	45.4	48.9	53.2	58.2	63.9	70.2	76.9	84.1	91.6	99.3
375	48.9	51.5	55.0	59.3	64.4	70.2	76.7	83.6	91.0	98.8	106.9
400	55.4	57.9	61.4	65.8	71.0	76.9	83.5	90.6	98.2	106.3	114.7
425	62.3	64.8	68.3	72.8	78.0	84.0	90.7	98.0	105.8	114.0	122.7
450	69.6	72.1	75.7	80.1	85.4	91.5	98.2	105.7	113.6	122.1	131.0
475	77.3	79.8	83.4	87.8	93.2	99.3	106.2	113.7	121.9	130.5	139.7
500	85.4	88.0	91.5	96.0	101.4	107.6	114.5	122.2	130.4	139.3	148.6
525	94.0	96.5	100.1	104.6	110.0	116.2	123.2	131.0	139.4	148.4	157.9
550	102.9	105.5	109.0	113.6	119.0	125.3	132.4	140.2	148.7	157.8	167.5
575	112.3	114.9	118.4	123.0	128.4	134.7	141.9	149.8	158.4	167.6	177.5
600	122.1	124.7	128.2	132.8	138.2	144.6	151.8	159.8	168.5	177.8	187.8

Weld groups
Welds in the plane of the force

Values of Z_p (cm³) for 1 mm of throat thickness

m or n (mm)	Z_p (cm³)
50	0.4
75	0.9
100	1.7
125	2.6
150	3.8
175	5.1
200	6.7
225	8.4
250	10.4
275	12.6
300	15.0
325	17.6
350	20.4
375	23.4
400	26.7
425	30.1
450	33.8
475	37.6
500	41.7
525	45.9
550	50.4
575	55.1
600	60.0

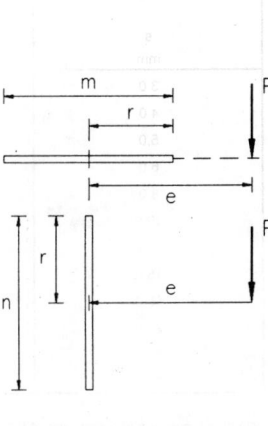

BS 5950-1 :2000
BS EN 440
BS EN 499
BS EN 756
BS EN 758
BS EN 1668

FILLET WELDS

WELD CAPACITIES WITH E35 ELECTRODE WITH S275

Leg Length	Throat Thickness	Longitudinal Capacity	Transverse Capacity
s	a	P_L	P_T
mm	mm	kN/mm	kN/mm
3.0	2.1	0.462	0.577
4.0	2.8	0.616	0.770
5.0	3.5	0.770	0.963
6.0	4.2	0.924	1.155
8.0	5.6	1.232	1.540
10.0	7.0	1.540	1.925
12.0	8.4	1.848	2.310
15.0	10.5	2.310	2.888
18.0	12.6	2.772	3.465
20.0	14.0	3.080	3.850
22.0	15.4	3.388	4.235
25.0	17.5	3.850	4.813

Welds are between two elements at 90° to each other.

$P_L = p_w a$

$P_T = K p_w a$

$p_w = 220 \text{ N/mm}^2$

K = 1.25 for elements at 90° to each other.

BS 5950-1 :2000
BS EN 440
BS EN 499
BS EN 756
BS EN 758
BS EN 1668

FILLET WELDS

WELD CAPACITIES WITH E42 ELECTRODE WITH S355

Leg Length	Throat Thickness	Longitudinal Capacity	Transverse Capacity
s	a	P_L	P_T
mm	mm	kN/mm	kN/mm
3.0	2.1	0.525	0.656
4.0	2.8	0.700	0.875
5.0	3.5	0.875	1.094
6.0	4.2	1.050	1.312
8.0	5.6	1.400	1.750
10.0	7.0	1.750	2.188
12.0	8.4	2.100	2.625
15.0	10.5	2.625	3.281
18.0	12.6	3.150	3.938
20.0	14.0	3.500	4.375
22.0	15.4	3.850	4.813
25.0	17.5	4.375	5.469

Welds are between two elements at 90° to each other.

$P_L = p_w a$

$P_T = K p_w a$

$p_w = 250 \text{ N/mm}^2$

K = 1.25 for elements at 90° to each other.

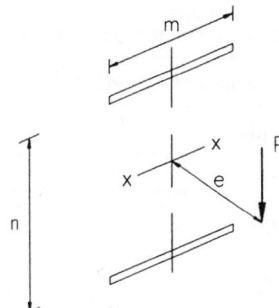

Weld groups
Welds not in the plane of the force

	Values of Z_{xx} (cm³) for 1 mm throat thickness										
Values of n (mm)	**Values of m (mm)**										
	50	75	100	125	150	175	200	225	250	275	300
50	2.5	3.8	5.0	6.3	7.5	8.8	10.0	11.3	12.5	13.8	15.0
75	3.8	5.6	7.5	9.4	11.3	13.1	15.0	16.9	18.8	20.6	22.5
100	5.0	7.5	10.0	12.5	15.0	17.5	20.0	22.5	25.0	27.5	30.0
125	6.3	9.4	12.5	15.6	18.8	21.9	25.0	28.1	31.3	34.4	37.5
150	7.5	11.3	15.0	18.8	22.5	26.3	30.0	33.8	37.5	41.3	45.0
175	8.8	13.1	17.5	21.9	26.3	30.6	35.0	39.4	43.8	48.1	52.5
200	10.0	15.0	20.0	25.0	30.0	35.0	40.0	45.0	50.0	55.0	60.0
225	11.3	16.9	22.5	28.1	33.8	39.4	45.0	50.6	56.3	61.9	67.5
250	12.5	18.8	25.0	31.3	37.5	43.8	50.0	56.3	62.5	68.8	75.0
275	13.8	20.6	27.5	34.4	41.3	48.1	55.0	61.9	68.8	75.6	82.5
300	15.0	22.5	30.0	37.5	45.0	52.5	60.0	67.5	75.0	82.5	90.0
325	16.3	24.4	32.5	40.6	48.8	56.9	65.0	73.1	81.3	89.4	97.5
350	17.5	26.3	35.0	43.8	52.5	61.3	70.0	78.8	87.5	96.3	105.0
375	18.8	28.1	37.5	46.9	56.3	65.6	75.0	84.4	93.8	103.1	112.5
400	20.0	30.0	40.0	50.0	60.0	70.0	80.0	90.0	100.0	110.0	120.0
425	21.3	31.9	42.5	53.1	63.8	74.4	85.0	95.6	106.3	116.9	127.5
450	22.5	33.8	45.0	56.3	67.5	78.8	90.0	101.3	112.5	123.8	135.0
475	23.8	35.6	47.5	59.4	71.3	83.1	95.0	106.9	118.8	130.6	142.5
500	25.0	37.5	50.0	62.5	75.0	87.5	100.0	112.5	125.0	137.5	150.0
525	26.3	39.4	52.5	65.6	78.8	91.9	105.0	118.1	131.3	144.4	157.5
550	27.5	41.3	55.0	68.8	82.5	96.3	110.0	123.8	137.5	151.3	165.0
575	28.8	43.1	57.5	71.9	86.3	100.6	115.0	129.4	143.8	158.1	172.5
600	30.0	45.0	60.0	75.0	90.0	105.0	120.0	135.0	150.0	165.0	180.0

Weld groups
Welds not in the plane of the force

Values of Z_{xx} (cm^3) for 1 mm throat thickness

n (mm)	Z_{xx} (cm^3)
50	0.8
75	1.9
100	3.3
125	5.2
150	7.5
175	10.2
200	13.3
225	16.9
250	20.8
275	25.2
300	30.0
325	35.2
350	40.8
375	46.9
400	53.3
425	60.2
450	67.5
475	75.2
500	83.3
525	91.9
550	100.8
575	110.2
600	120.0

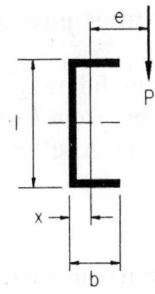

Weld groups
Welds in the plane of the force

	Values of Z_p (cm³) for 1 mm throat thickness									
Values of n (mm)	Values of m (mm)									
	50	75	100	125	150	175	200	225	250	275	300
50	2.8	4.3	6.2	8.5	11.2	14.3	17.9	21.9	26.2	31.0	36.2
75	4.3	6.2	8.4	11.0	13.9	17.3	21.1	25.2	29.8	34.8	40.3
100	6.2	8.5	11.0	13.9	17.1	20.7	24.7	29.2	34.0	39.2	44.9
125	8.3	11.0	14.0	17.2	20.7	24.6	28.9	33.6	38.7	44.1	50.0
150	10.6	13.9	17.3	20.9	24.8	29.0	33.5	38.5	43.8	49.6	55.7
175	13.2	17.0	20.8	24.9	29.1	33.7	38.6	43.8	49.4	55.4	61.8
200	16.0	20.3	24.7	29.1	33.8	38.8	44.0	49.6	55.5	61.8	68.4
225	19.0	23.9	28.8	33.7	38.8	44.2	49.8	55.7	61.9	68.5	75.5
250	22.2	27.7	33.1	38.5	44.1	49.9	55.9	62.2	68.8	75.7	82.9
275	25.6	31.7	37.7	43.6	49.7	55.9	62.3	69.0	75.9	83.2	90.8
300	29.2	35.9	42.5	49.0	55.5	62.2	69.0	76.1	83.4	91.1	99.0
325	33.1	40.4	47.5	54.5	61.6	68.7	76.0	83.5	91.3	99.3	107.6
350	37.1	45.0	52.7	60.3	67.9	75.5	83.3	91.3	99.4	107.8	116.5
375	41.4	49.9	58.2	66.3	74.5	82.6	90.9	99.3	107.8	116.7	125.7
400	45.9	55.0	63.8	72.6	81.2	89.9	98.7	107.5	116.6	125.8	135.3
425	50.6	60.3	69.7	79.0	88.3	97.5	106.7	116.1	125.6	135.3	145.1
450	55.4	65.8	75.8	85.7	95.5	105.3	115.0	124.9	134.8	145.0	155.3
475	60.6	71.5	82.2	92.6	103.0	113.3	123.6	133.9	144.4	155.0	165.7
500	65.9	77.4	88.7	99.8	110.7	121.5	132.4	143.2	154.2	165.2	176.5
525	71.4	83.6	95.4	107.1	118.6	130.0	141.4	152.8	164.2	175.8	187.5
550	77.1	89.9	102.4	114.6	126.7	138.7	150.6	162.5	174.5	186.5	198.7
575	83.0	96.5	109.6	122.4	135.1	147.6	160.1	172.6	185.0	197.6	210.2
600	89.2	103.2	116.9	130.4	143.7	156.8	169.8	182.8	195.8	208.9	222.0

Sheet pile sections

The full range of current Corus steel sheet pile and bearing pile sections are available from Corus Piling on 01724 404040 or from website www.corusconstruction.com. A selection of Corus piling products are shown below.

Larssen sections

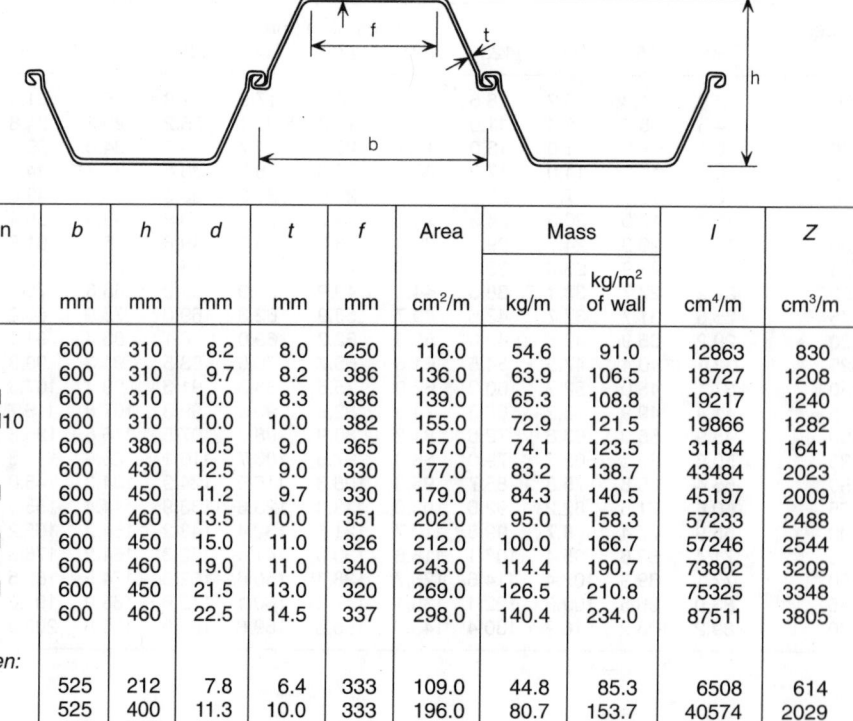

Section	b	h	d	t	f	Area	Mass		I	Z
								kg/m²		
	mm	mm	mm	mm	mm	cm²/m	kg/m	of wall	cm⁴/m	cm³/m
LX8	600	310	8.2	8.0	250	116.0	54.6	91.0	12863	830
LX12	600	310	9.7	8.2	386	136.0	63.9	106.5	18727	1208
LX12d	600	310	10.0	8.3	386	139.0	65.3	108.8	19217	1240
LX12d10	600	310	10.0	10.0	382	155.0	72.9	121.5	19866	1282
LX16	600	380	10.5	9.0	365	157.0	74.1	123.5	31184	1641
LX20	600	430	12.5	9.0	330	177.0	83.2	138.7	43484	2023
LX20d	600	450	11.2	9.7	330	179.0	84.3	140.5	45197	2009
LX25	600	460	13.5	10.0	351	202.0	95.0	158.3	57233	2488
LX25d	600	450	15.0	11.0	326	212.0	100.0	166.7	57246	2544
LX32	600	460	19.0	11.0	340	243.0	114.4	190.7	73802	3209
LX32d	600	450	21.5	13.0	320	269.0	126.5	210.8	75325	3348
LX38	600	460	22.5	14.5	337	298.0	140.4	234.0	87511	3805
Larssen:										
6W	525	212	7.8	6.4	333	109.0	44.8	85.3	6508	614
20Wd	525	400	11.3	10.0	333	196.0	80.7	153.7	40574	2029
GSP2	400	200	10.5	8.6	266	157.0	49.4	123.5	8756	876
GSP3	400	250	13.5	8.6	270	191.0	60.1	150.3	16316	1305
GSP4	400	340	15.5	9.7	259	242.0	76.1	190.3	38742	2279
6-42	500	450	20.5	14.0	329	339.0	133.0	266.0	94755	4211
6(122)	420	440	22.0	14.0	250	371.0	122.5	291.7	92115	4187
6(131)	420	440	25.4	14.0	250	396.0	130.7	311.2	101598	4618
6(138.7)	420	440	28.6	14.0	251	419.0	138.3	329.3	110109	5005

Frodingham sections

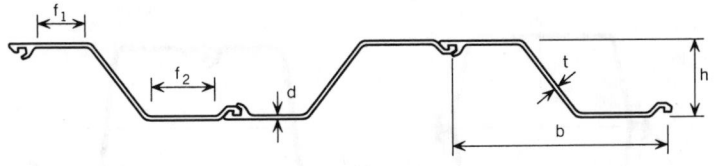

Section	b	h	d	t	f_1	f_2	Area	Mass		I	Z
									kg/m² of wall		
	mm	mm	mm	mm	mm	mm	cm²/m	kg/m		cm⁴/m	cm³/m
1BXN	476	143	12.7	12.7	77	122	170	63.4	133.2	4947	692
1N	483	170	9.0	9.0	107	142	126	48.0	99.4	6072	714
2N	483	235	9.7	8.4	91	146	145	54.8	113.5	13641	1161
3NA	483	305	9.7	9.5	90	148	166	62.7	129.8	25710	1687
4N	483	330	14.0	10.4	75	128	218	82.7	171.2	39869	2415
5	426	311	17.1	11.9	87	119	302	101.0	237.1	49329	3171

Frodingham straight web sheet piles

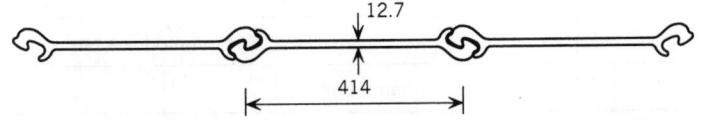

Section	b	t	Area Single Pile	Mass per m of pile	Mass per m² of pile	Mass per m of junction	Ultimate strength		interlock	Coating area per pile	Coating area per m wall	Max deviation angle
							S270GP	S355GP				
	mm	mm	cm²	kg/m	kg/m²	kg/m	t/m	t/m		m²/m	m²/m²	degrees
SW1A	414	12.7	81	63.5	153.7	95.2	285	384		1.00	2.41	6

Box sheet piles

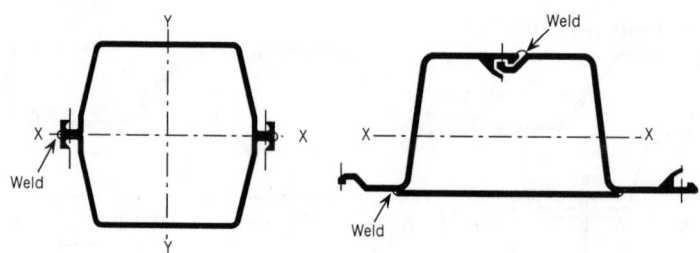

Larssen   Frodingham

Larssen box piles		
Section	Section modulus cm³	
	XX axis cm³	YY axis cm³
LX25	3424	3257
LX32	4377	3544
LX38	5271	4374
6-42	4920	3902

Frodingham 4N box piles		
Section	Section modulus cm³	
	Plated box	Double box
Frod 4N	2662	5805

High modulus piles

Frodingham high modulus piles

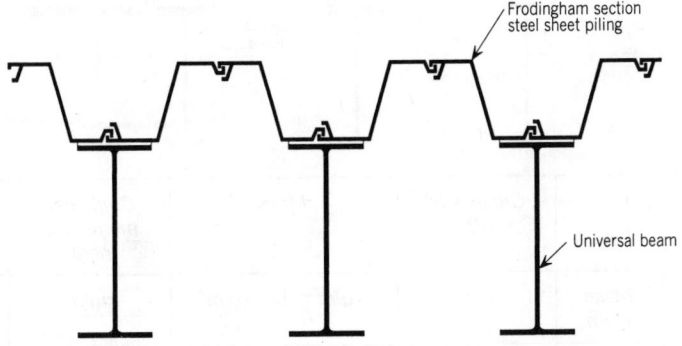

Universal beam		Centres of UBs	Mass		Combined moment of inertia	Elastic section modulus
Serial size mm	Mass kg/m	mm	kg/m	kg/m²	cm⁴/m	cm³/m
533 × 210	101	966	267	276	259478	4832
610 × 305	147	966	314	326	397108	7198
762 × 267	176	966	338	350	584576	9026
838 × 292	194	966	359	372	732365	10621
914 × 305	253	966	419	433	1005797	14254
914 × 419*	388	966	522	540	1353126	21435

* Denotes beam section with one flange reduced to 310 mm to facilitate fabrication.

Larssen LX20 high modulus piles

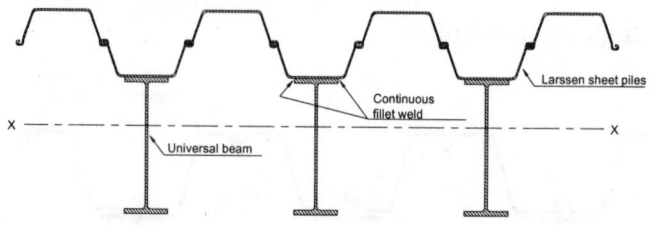

Universal beam		Centres of UBs	Mass		Combined moment of inertia	Elastic section modulus
Serial size mm	Mass kg/m	mm	kg/m	kg/m²	cm⁴/m	cm³/m
686 × 254	125	1200	208.4	243.0	200426	3945
762 × 267	147	1200	230.1	261.1	267505	4918
838 × 292	176	1200	259.1	285.3	363085	6273
914 × 305	253	1200	336.6	349.8	558248	9453
1016 × 305	222	1200	305.2	323.7	555491	8673
1016 × 305	487	1200	570.2	544.5	1132123	16938

H-piles

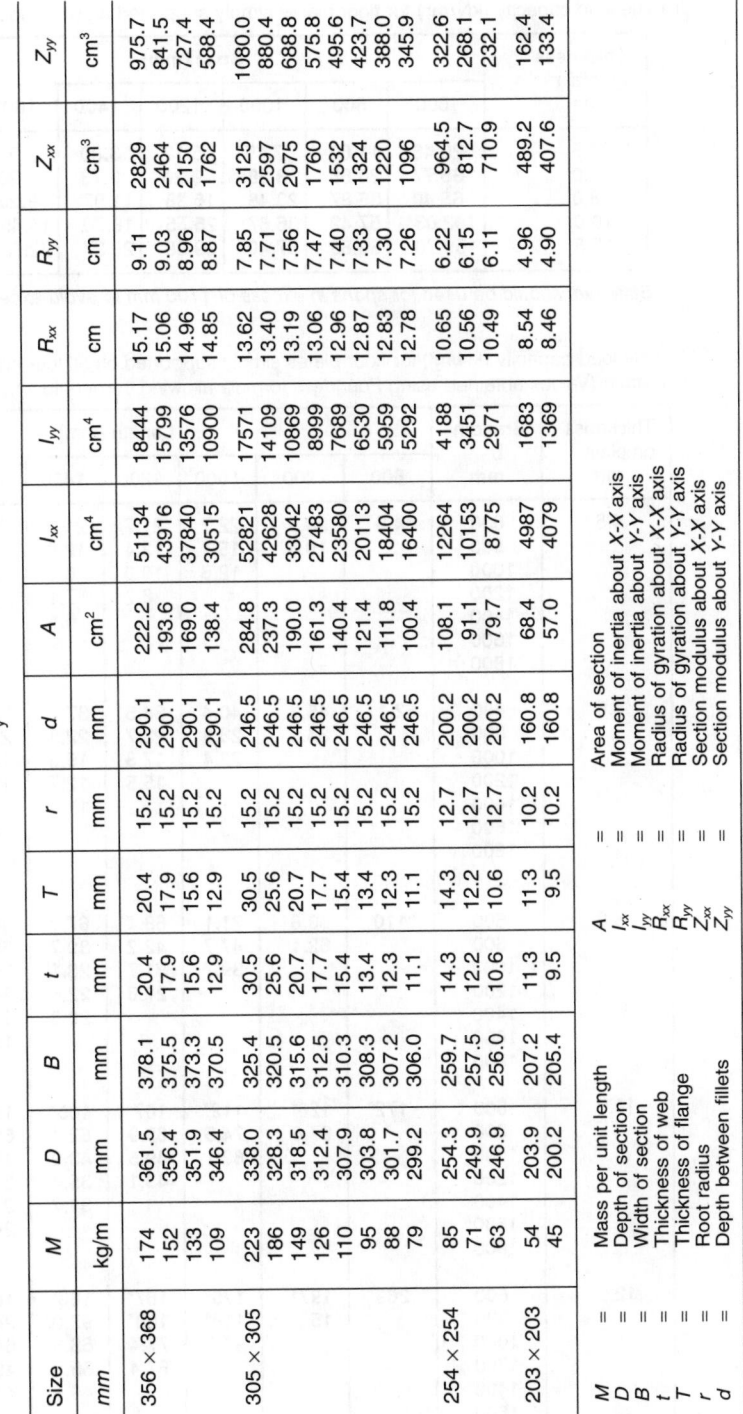

Size	M	D	B	t	T	r	d	A	I_{xx}	I_{yy}	R_{xx}	R_{yy}	Z_{xx}	Z_{yy}
mm	kg/m	mm	mm	mm	mm	mm	mm	cm²	cm⁴	cm⁴	cm	cm	cm³	cm³
356 × 368	174	361.5	378.1	20.4	20.4	15.2	290.1	222.2	51134	18444	15.17	9.11	2829	975.7
	152	356.4	375.5	17.9	17.9	15.2	290.1	193.6	43916	15799	15.06	9.03	2464	841.5
	133	351.9	373.3	15.6	15.6	15.2	290.1	169.0	37840	13576	14.96	8.96	2150	727.4
	109	346.4	370.5	12.9	12.9	15.2	290.1	138.4	30515	10900	14.85	8.87	1762	588.4
305 × 305	223	338.0	325.4	30.5	30.5	15.2	246.5	284.8	52821	17571	13.62	7.85	3125	1080.0
	186	328.3	320.5	25.6	25.6	15.2	246.5	237.3	42628	14109	13.40	7.71	2597	880.4
	149	318.5	315.6	20.7	20.7	15.2	246.5	190.0	33042	10869	13.19	7.56	2075	688.8
	126	312.4	312.5	17.7	17.7	15.2	246.5	161.3	27483	8999	13.06	7.47	1760	575.8
	110	307.9	310.3	15.4	15.4	15.2	246.5	140.4	23580	7689	12.96	7.40	1532	495.6
	95	303.8	308.3	13.4	13.4	15.2	246.5	121.4	20113	6530	12.87	7.33	1324	423.7
	88	301.7	307.2	12.3	12.3	15.2	246.5	111.8	18404	5959	12.83	7.30	1220	388.0
	79	299.2	306.0	11.1	11.1	15.2	246.5	100.4	16400	5292	12.78	7.26	1096	345.9
254 × 254	85	254.3	259.7	14.3	14.3	12.7	200.2	108.1	12264	4188	10.65	6.22	964.5	322.6
	71	249.9	257.5	12.2	12.2	12.7	200.2	91.1	10153	3451	10.56	6.15	812.7	268.1
	63	246.9	256.0	10.6	10.6	12.7	200.2	79.7	8775	2971	10.49	6.11	710.9	232.1
203 × 203	54	203.9	207.2	11.3	11.3	10.2	160.8	68.4	4987	1683	8.54	4.96	489.2	162.4
	45	200.2	205.4	9.5	9.5	10.2	160.8	57.0	4079	1369	8.46	4.90	407.6	133.4

M = Mass per unit length
D = Depth of section
B = Width of section
t = Thickness of web
T = Thickness of flange
r = Root radius
d = Depth between fillets

A = Area of section
I_{xx} = Moment of inertia about X-X axis
I_{yy} = Moment of inertia about Y-Y axis
R_{xx} = Radius of gyration about X-X axis
R_{yy} = Radius of gyration about Y-Y axis
Z_{xx} = Section modulus about X-X axis
Z_{yy} = Section modulus about Y-Y axis

Ultimate load capacity (kN/m²) for floor plates simply supported on two edges stressed to 275 N/mm²

Thickness on plain mm	Span (mm)							
	600	800	1000	1200	1400	1600	1800	2000
4.5	20.48	11.62	7.45	5.17	3.80	2.95	2.28	1.87
6.0	36.77	20.68	13.28	9.20	6.73	5.20	4.07	3.30
8.0	65.40	36.87	23.48	16.38	11.97	9.23	7.23	5.93
10.0	102.03	57.42	36.67	25.55	18.70	14.45	11.30	9.25
12.5	159.70	89.85	57.40	39.98	29.27	22.62	17.68	14.50

Stiffeners should be used for spans in excess of 1100 mm to avoid excessive deflections.

Ultimate load capacity (kN/m²) for floor plates simply supported on all four edges stressed to 275 N/mm² (Values obtained using Pounder's formula allowing corners to lift)

Thickness on plain mm	Breadth B mm	Length (mm)							
		600	800	1000	1200	1400	1600	1800	2000
4.5	600	34.9	25.5	22.7	21.7	21.2	21.0	20.8	20.8
	800		19.6	15.1	13.4	12.6	12.2	12.0	11.8
	1000			12.6	10.0	8.8	8.3	7.9	7.7
	1200				8.7	7.1	6.3	5.9	5.6
	1400					6.4	5.3	4.8	4.4
	1600						4.9	4.1	3.7
	1800							3.8	3.3
6.0	600	62.1	45.3	40.4	38.5	37.7	37.3	37.0	36.9
	800		34.9	26.8	23.7	22.3	21.7	21.3	21.1
	1000			22.4	17.8	15.8	14.8	14.2	13.9
	1200				15.5	12.7	11.3	10.6	10.1
	1400					11.4	9.5	8.5	7.9
	1600						8.7	7.4	6.7
	1800							6.9	5.9
8.0	600	110	80.6	71.1	68.4	67.0	66.2	65.8	65.6
	800		62.1	47.7	42.2	39.7	38.5	37.8	37.4
	1000			39.7	31.7	28.1	26.2	25.2	24.6
	1200				27.6	22.6	20.1	18.8	17.9
	1400					20.3	17.0	15.2	14.1
	1600						15.5	13.3	11.9
	1800							12.3	10.6
10.0	600	172*	126*	112*	107*	105*	103*	103*	103*
	800		97.0	74.5	65.9	62.1	60.1	59.1	58.5
	1000			62.1	49.5	43.9	41.0	39.4	38.5
	1200				43.1	35.4	31.5	29.3	28.0
	1400					31.7	26.6	23.8	22.1
	1600						24.3	20.7	18.6
	1800							19.2	16.6
12.5	600	269*	197*	175*	167*	163*	162*	161*	160*
	800		152	116*	103*	97.0*	94.0*	92.3*	91.4*
	1000			97.0	77.4	68.5	64.1	61.6	60.1
	1200				67.4	55.3	49.2	45.8	43.8
	1400					49.5	41.5	37.1	34.5
	1600						37.9	32.4	29.1
	1800							29.9	25.9

Ultimate load capacity (kN/m²) for floor plates fixed on all four edges stressed to 275 N/mm²

Thickness on plain mm	Breadth B mm	Length (mm)							
		600	800	1000	1200	1400	1600	1800	2000
4.5	600	47.7*	36.8*	33.5*	32.2*	31.6*	31.4*	31.2*	31.1*
	800		26.8	21.5*	19.5*	18.6*	18.1*	17.9*	17.7*
	1000			17.2*	14.2*	12.9*	12.2*	11.8*	11.6*
	1200				11.9	10.1	9.1	8.6	8.3
	1400					8.7	7.5	6.9	6.5
	1600						6.7	5.8	5.3
	1800							5.3	4.7
6.0	600	84.8*	65.4*	59.5*	57.3*	56.2*	55.7*	55.5*	55.3*
	800		47.7*	38.3*	34.7*	33.1*	32.2*	31.7*	31.5*
	1000			30.5*	25.3*	22.9*	21.7*	21.0*	20.6*
	1200				21.2*	18.0*	16.3*	15.4*	14.9*
	1400					15.6*	13.4*	12.3*	11.6
	1600						11.9	10.4	9.5
	1800							9.4	8.3
8.0	600	151*	116*	106*	102*	100*	99.1*	98.6*	98.3*
	800		68.1*	61.7*	58.8*	57.3*	56.4*	55.9*	
	1000			54.3*	44.9*	40.7*	38.6*	37.4*	36.7*
	1200				37.7*	31.9*	29.0*	27.4*	26.5*
	1400					27.7*	23.9*	21.8*	20.6*
	1600						21.2*	18.6*	17.0*
	1800							16.9*	14.8*
10.0	600	236*	182*	165*	159*	156*	155*	154*	154*
	800		132*	106*	96.4*	91.8*	89.5*	88.2*	87.4*
	1000			84.8*	70.2*	63.7*	60.3*	58.4*	57.3*
	1200				58.9*	49.9*	45.4*	42.9*	41.3*
	1400					43.3*	37.3*	34.1*	32.2*
	1600						33.1*	29.0*	26.6*
	1800							26.2*	23.2*
12.5	600	368*	284*	258*	249*	244*	242*	241*	240*
	800		207*	166*	151*	144*	140*	138*	137*
	1000			132*	110*	99.5*	94.2*	91.2*	89.5*
	1200				92.0*	77.9*	70.9*	67.0*	64.6*
	1400					67.6*	58.3*	53.3*	50.3*
	1600						51.8*	45.3*	41.6*
	1800							40.9*	36.2*

Note on tables:
Values without an asterisk cause deflection greater than B/100 at serviceability, assuming that the only dead load present is due to self-weight.

STEELWORK IN FIRE
INFORMATION SHEET

This series of information sheets is intended to illustrate methods of achieving fire resistance in steel structures. It should not be used for design without consulting detailed design guidance referenced below.

SPRAYED PROTECTION

UP TO 4 HRS

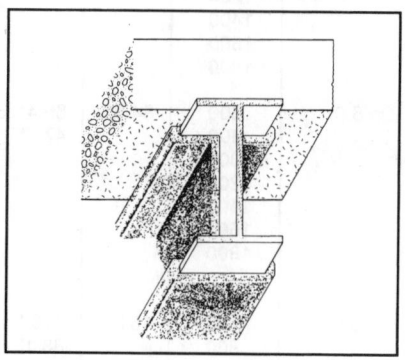

METHOD

Fire protective insulation can be applied by spraying to almost any type of steel member. Most products can achieve up to 4 hours rating.

PRINCIPLE

Insulation reduces the heating rate of a steel member so that its limiting temperature is not exceeded for the required fire resistance period. The protection material thickness necessary depends on the section factor (Hp/A) of the member and the fire rating required.

ADVANTAGES

a) Low cost
b) Rapid application
c) Easy to cover complex details
d) Often applied to non-primed steelwork
e) Some products may be suitable for external use

LIMITATIONS (check with manufacturer)

a) Appearance may be inadequate for visible members
b) Overspray may need masking or shielding
c) Primer, if used, must be compatible

FOR MORE DETAILED INFORMATION SEE:-
"Fire protection of Structural Steel in Building"
Published jointly by:
ASFP - (01252 336318) and
The Steel Construction Institute - (01344 23345)

Sheet Code
ISF/No.01
January 1997

PROTECTION THICKNESS

Thickness recommendations given in "Fire Protection of Structural Steel in Building" have normally been derived from fire tests on orthodox H or I rolled sections. For other sections the recommended thickness for a given section factor and fire rating should be modified as follows:

CASTELLATED SECTIONS

The thickness of fire protection material on a castellated section should be 20% greater than that required for the section from which it was cut.

HOLLOW SECTIONS

For spray applied fire protection materials the recommended thickness (t) should be increased as follows

For section factor (Hp/A) less than 250 modified thickness = t [1+(Hp/A) / 1000]

For section factor (Hp/A) 250 or over modified thickness = 1.25 x t

STEELWORK IN FIRE
INFORMATION SHEET

This series of information sheets is intended to illustrate methods of achieving fire resistance in steel structures. It should not be used for design without consulting detailed design guidance referenced below.

BOARD PROTECTION

UP TO 4 HRS

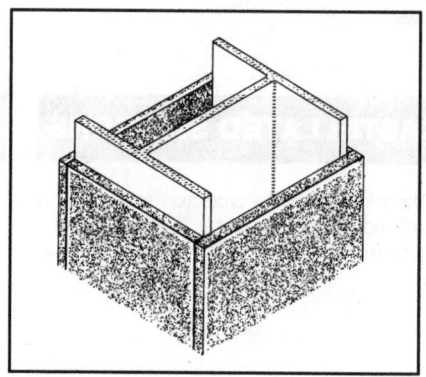

METHOD

Fire protective insulation can be applied by fixing boards to almost any type of steel member. Most products can achieve up to 4 hour rating. Fixing methods vary.

PRINCIPLE

Insulation reduces the heating rate of a steel member so that its limiting temperature is not exceeded during the required fire resistance period. The protection board thickness necessary depends on the section factor (Hp/A) of the member and the fire rating required.

ADVANTAGES

a) Boxed appearance suitable for visible members
b) Clean dry fixing
c) Factory manufactured, guaranteed thickness
d) Often applied to non-primed steelwork
e) Some products may be suitable for external use

LIMITATIONS (check with manufacturer)

a) Require fitting around complex details
b) May be more expensive and slower to fix than sprays

FOR MORE DETAILED INFORMATION SEE:-
"Fire protection of Structural Steel in Building"
Publication jointly by:
ASFP - (01252 336318) and
The Steel Construction Institute - (01344 23345)

Sheet Code
ISF/No.02
January 1997

PROTECTION THICKNESS

Thickness recommendations given in "Fire Protection of Structural Steel in Building" have normally been derived from fire tests on orthodox H or I rolled sections. For other sections the recommended thickness for a given section factor and fire rating should be modified as follows.

CASTELLATED SECTIONS

The thickness of fire protection material on a castellated section should be 20% greater than that required for the section from which it was cut.

STEELWORK IN FIRE
INFORMATION SHEET

This series of information sheets is intended to illustrate methods of achieving fire resistance in steel structures. It should not be used for design without consulting detailed design guidance referenced below.

THIN FILM INTUMESCENT COATINGS

UP TO 2 HRS

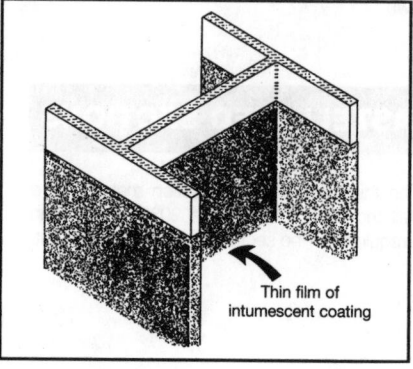

Thin film of
intumescent coating

METHOD

Most thin film intumescent coatings can be applied by spray, brush or roller and can achieve up to 1 hour fire resistance on fully exposed steel members. Some products can achieve up to 2 hours fire resistance on some section sizes.

PRINCIPLE

Insulation is created by swelling of the coating at elevated temperatures to generate a foam like char. This reduces the heating rate so that the limiting temperature of the steel member is not exceeded during the required fire resistance period. The coating thickness necessary depends on the section factor (Hp/A) and the fire rating required.

ADVANTAGES

a) Decorative finish
b) Rapid application
c) Easy to cover complex details
d) Easy post protection fixings to steelwork
   eg service hangers

LIMITATIONS (check with manufacturer)

a) May be suitable for dry internal environments only
b) May be more expensive than sprayed insulation
c) May require blast cleaned surface and compatible primer

FOR MORE DETAILED INFORMATION SEE:-
"Fire protection of Structural Steel in Building"
Publication jointly by:
ASFP - (01252 336318) and
The Steel Construction Institute - (01344 23345)

**Sheet Code
ISF/No.03
January 1997**

PROTECTION THICKNESS

Thickness recommendations given in "Fire Protection of Structural Steel in Building" have normally been derived from fire tests on orthodox H or I rolled sections. For other sections the recommended thickness for a given section factor and fire rating should be modified as follows:

CASTELLATED SECTIONS

The thickness of fire protection material on a castellated section should be 20% greater than that required for the section from which it was cut.

HOLLOW SECTIONS

For intumescent materials applied to hollow sections the manufacturers should have carried out separate tests and appraisal

STEELWORK IN FIRE
INFORMATION SHEET

This series of information sheets is intended to illustrate methods of achieving fire resistance in steel structures.
It should not be used for design without consulting detailed design guidance referenced below.

BLOCK - FILLED COLUMNS

30 MINUTES

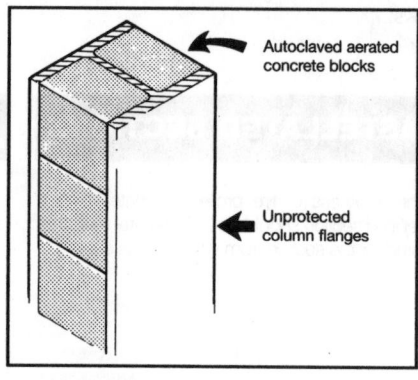

Autoclaved aerated
concrete blocks

Unprotected
column flanges

METHOD

Unprotected universal sections with section factors up to 69m^{-1} (see overleaf) can attain 30 minutes fire resistance by fitting autoclaved aerated concrete blocks between the flanges tied to the web at approximately 1m intervals

PRINCIPLE

Partial exposure of steel members affects fire resistance in two ways-

Firstly the reduction of exposed surface area reduces the rate of heating by radiation and thus increases the time to reach failure temperature.

Secondary, if the exposure creates both hot and cold regions in the cross section, plastic yielding occurs in the hot region and load is transferred to the stronger cooler region. Thus a non-uniformly heated section has a higher fire resistance than one heated evenly.

ADVANTAGES

a) Reduced cost - compared with total encasement with insulation
b) More slender finished columns occupy less floor space
c) Good durability - high resistance to impact and abrasion damage

LIMITATIONS

With unprotected steel the method is limited to 30 minutes fire rating.

When higher ratings are required exposed steel must be treated with the full insulation or intumescent coating thickness recommended for the higher rating.

This method should not be used when the blockwork also forms a separating wall. In this case the column will be heated on one side only and thermal bowing may cause the wall to crack or collapse. In such cases the flange(s) should be protected. Alternatively, if the limit of wall deformation is known, the bowing can be calculated to ensure no integrity failure.

FOR MORE DETAILED INFORMATION SEE:-
BRE digest 317, Building Research Establishment,
Garston, Watford WD2 7SR
Telephone 01923 894040

**Sheet Code
ISF/No.04
January 1997**

METHODS OF ACHIEVING 30 MINUTES FIRE RESISTANCE

COLUMN SECTION - AXIALLY LOADED [1] FREE STANDING

SERIAL SIZE mm	MASS/METRE kg	PROTECTION METHOD RECOMMENDED
305 x 406	393 and over	No fire protection required
356 x 406 305 x 305 254 x 254 203 x 203 203 x 203	340 and under All weights All weights 52 and over 46[2]	Block filling with autoclaved aerated concrete blocks
152 x 203	All weights	Apply fire protection material as per manufacturer's recommendations

BEAM SECTIONS ACTING AS PORTAL FRAME STANCHIONS [1]

914 x 419 914 x 305 *610 x 305	All weights 289 238	No fire protection required
*914 x 305 838 x 292 762 x 267 686 x 254 *610 x 305 610 x 229 533 x 210 457 x 191 457 x 152 406 x 178 356 x 171 305 x 165 305 x 127 254 x146	252 and under All weights All weights All weights 179 and under All weights All weights All weights 60 and over 60 and over 57 and over 54 48 43	Block filling with autoclaved aerated concrete blocks
Other beam sizes		Apply fire protection material as per manufacturer's recommendations

Notes:
1) This table applies to sections designed to BS 5950: Part 1:1990 provided the load factor (γf) does not exceed 1.5
2) To achieve 30 min fire resistance, a 203 x 203 x 46 kg/m column with blocked in webs should be loaded only up to 80% of the maximum allowable per BS 449:Part 2:1969 or BS 5950:Part 1:1990
*3) The table revises BRE Digest 317 (1986) in accordance with BS 5950:Part 8:1990

STEELWORK IN FIRE
INFORMATION SHEET

This series of information sheets is intended to illustrate methods of achieving fire resistance in steel structures.
It should not be used for design without consulting detailed design guidance referenced below.

CONCRETE FILLED HOLLOW COLUMNS

UP TO 2 HRS

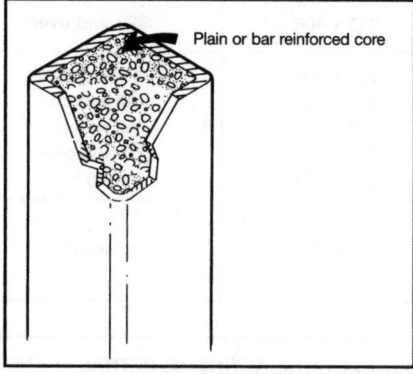

Plain or bar reinforced core

METHOD

Unprotected square or rectangular hollow sections can attain up to 120 minutes fire resistance by filling with plain, fibre reinforced or bar reinforced concrete.

PRINCIPLE

Heat flows through the steel wall into the concrete core which being a poor conductor heats up slowly.
As the temperature increases the steel yield strength reduces and the load is progressively transferred into the concrete core.
The steel acts as a restraint to the concrete preventing spalling and hence the rate of degradation of the concrete.

ADVANTAGES

a) Steel acts as a permanent shuttering
b) More slender finished columns occupy less floor space
c) Good durability - high resistance to impact and abrasion damage

LIMITATIONS

a) A minimum column size of 140mm x 140mm or 100mm x 200mm is required for plain or fibre reinforced sections.
b) A minimum column size of 200mm x 200mm. or 150mm x 250mm is required for bar reinforced sections.
c) CHS columns are not included due to insufficient data at present.

FOR MORE DETAILED INFORMATION SEE:-
BS 5950 Part 8
Concrete filled column design manual TD 296 from
British Steel Tubes & Pipes - (01536 404005)

Sheet Code
ISF/No.05
January 1997

CONCRETE FILLED RECTANGULAR HOLLOW SECTIONS

The fire resistance of externally unprotected concrete filled hollow sections is dependent on three main variables.

- The concrete strength selected
- The ratio of axial load and moment
- The addition of fibre or bar reinforcement

CONCRETE STRENGTH

The core capacity and hence its fire resistance is directly related to the concrete strength selected.

AXIAL LOAD AND MOMENT

Plain concrete does not perform well in tension and when subject to axial load and moment it is necessary to produce a resultant compressive stress in the core.

REINFORCEMENT

Fibre reinforcement will enhance the core axial capacity yet retain the advantage of filling into a section without obstructions.
Bar reinforcement will enhance the moment capacity.

COMBINED PROTECTION

As an alternative the concrete filled section can be deigned for full factored loads and provided with external fire protection.
The thickness of the external fire protection is assessed as far for an unfilled section, and, due to the effect of the core, the thickness can be reduced.

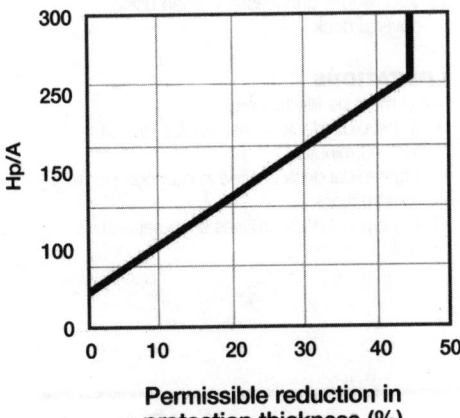

Permissible reduction in protection thickness (%)

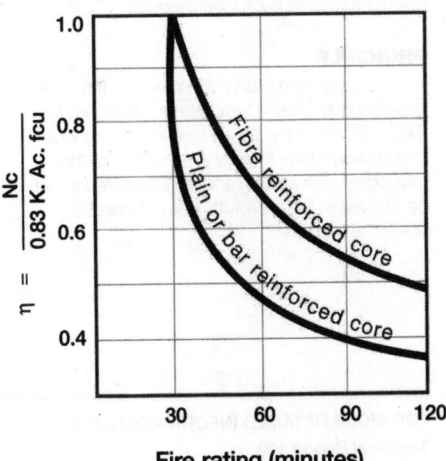

Fire rating (minutes)

STEELWORK IN FIRE
INFORMATION SHEET

This series of information sheets is intended to illustrate methods of achieving fire resistance in steel structures. It should not be used for design without consulting detailed design guidance referenced below.

COMPOSITE SLABS WITH PROFILED METAL DECK WITH UNFILLED VOIDS

| UP TO 2 HRS |

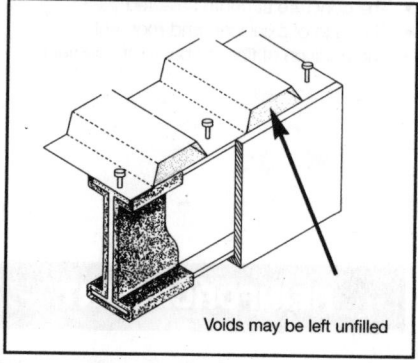

Voids may be left unfilled

METHOD

In composite construction using profiled metal deck floors it is unnecessary to fill the deck voids above the top flange for any fire resistance period using dovetail deck, or up to 90 minutes using trapezoidal deck (see overleaf)

PRINCIPLE

In a composite beam/slab member the neutral axis in bending lies in, or close to, the beam top flange. Thus the top flange makes little significant contribution to the structural behaviour of the total composite system and it's temperature can be allowed to increase with little detriment to performance in fire.

ADVANTAGES

a) Saving in time on site
b) Saving in cost for filling voids
c) It is unnecessary to build up the full thickness of protection on toes of upper flange
d) Void filling is unnecessary when using dovetail deck

LIMITATIONS

Voids must be filled where:-
a) Trapezoidal deck is used for fire ratings over 90 minutes
b) Trapezoidal deck is used in non-composite construction
c) Any type of deck crosses a fire separating wall

FOR MORE DETAILED INFORMATION SEE:-
Technical Report 109
"Fire resistance of composite beams"
The British Steel Construction Institute - (01344 23345)

Sheet Code
ISF/No.06
January 1997

COMPOSITE BEAMS - UNFILLED VOIDS

TRAPEZOIDAL DECK

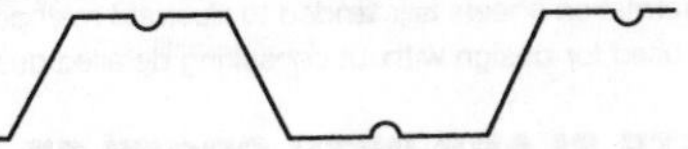

Construction	Fire Protection On Beam	Fire Resistance (minutes)		
		Up to 60	90	Over 90
Composite Beams	BOARD or SPRAY	No Increase in thickness*	Increase thickness* by 10% (or use thickness* appropriate to beam Hp/A + 15% whichever is less)	Fill voids
	INTUMESCENT	Increase thickness* by 20% (or use thickness* appropriate to beam Hp/A + 30% whichever is less)	Increase thickness* by 30% (or use thickness* appropriate to beam Hp/A + 50% whichever is less)	Fill voids
Non-Composite Beams	All types	Fill voids		

DOVETAIL DECK

Construction	Fire Protection On Beam	Fire Resistance (minutes)
Composite or Non-composite Beams	All Types	Voids may be left unfilled for all fire resistance periods.

* Thickness is the board, spray or intumescent thickness given for 30, 60 or 90 minutes rating in "Fire Protection for Structural Steel in Buildings" published by ASFP (01252 336318) and The Steel Construction Institute (01344 23345)

STEELWORK IN FIRE
INFORMATION SHEET

This series of information sheets is intended to illustrate methods of achieving fire resistance in steel structures. It should not be used for design without consulting detailed design guidance referenced below.

COMPOSITE SLABS WITH PROFILED METAL DECK | UP TO 2 HRS |

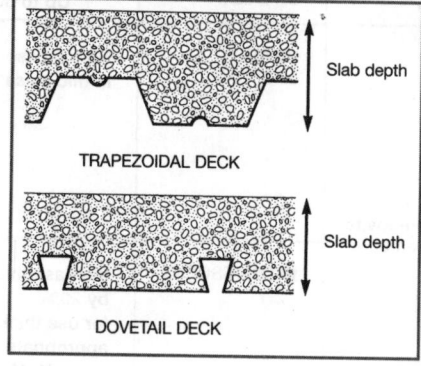

Slab depth

TRAPEZOIDAL DECK

Slab depth

DOVETAIL DECK

METHOD

Fire resistance of composite slabs up to 90 mins can be achieved using normal A142 mesh reinforcement. This can be increased to 120 mins if heavier mesh is used and the slab depth increased (see overleaf).

Other cases outside the limit overleaf can be evaluated by the "Fire Engineering Method" (See below)

PRINCIPLE

Mesh reinforcement, which is not designed to act structurally under normal conditions, makes a significant contribution to structural continuity in fire.

ADVANTAGES

a) Standard mesh, without additional reinforcing bars, may be used
b) No fire protection is required on the deck soffit

LIMITATIONS

a) Applies only to slabs designed to BS5950 Part 4
b) Mesh overlaps should exceed 50 times bar diameters
c) Mesh bar ductility should exceed 12% elongation in tension (to BS 4449)
d) Mesh should lie between 20 & 45mm from slab upper surface
e) Imposed load should not exceed 6.7kN/m^2 (including finishes)

FOR MORE DETAILED INFORMATION SEE:-
SCI Technical Report 056
"Fire resistance of composite floors with steel decking".
The Steel Construction Institute - (01344 23345) and
CIRIA Special publication 42 CIRIA (0171 222 8891)

Sheet Code
ISF/No.07
January 1997

FIRE RESISTANT COMPOSITE SLABS

TRAPEZOIDAL DECK

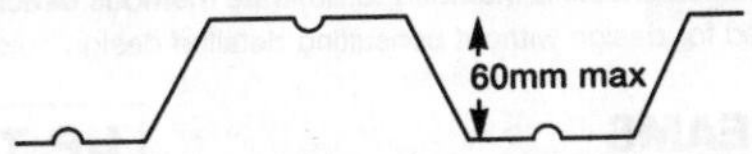
60mm max

Maximum Span (m)	Fire Rating (h)	Minimum Dimensions			Mesh Size
		Sheet thickness	Slab depth (mm)		
			NWC (2)	LWC (3)	
2.7	1	0.8	130	120	A142
3.0	1	0.9	130	120	A142
	1.5	0.9	140	130	A142
	2	0.9	155	140	A193
3.6	1	1.0	130	120	A193
	1.5	1.2	140	130	A193
	2	1.2	155	140	A252

DOVETAIL DECK

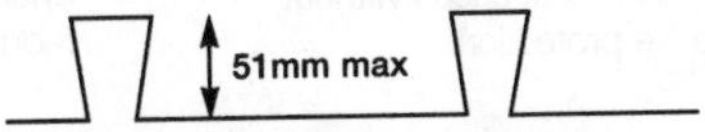
51mm max

Maximum Span (m)	Fire Rating (h)	Minimum Dimensions			Mesh Size
		Sheet thickness	Slab depth (mm)		
			NWC (2)	LWC (3)	
2.5	1	0.8	100	100	A142
	1.5	0.8	110	105	A142
3.0	1	0.9	120	110	A142
	1.5	0.9	130	120	A142
	2	0.9	140	130	A193
3.6	1	1.0	125	120	A193
	1.5	1.2	135	125	A193
	2	1.2	145	130	A252

1) Imposed load not exceeding $5kN/M^2$ (+ $1.7kN/m^2$ ceiling and services)
2) NWC = Normal weight concrete
3) LWC = Light weight concrete

NOTE: Minimum slab depths given in BS 5950 part 8 are to safety the insulation criterion only. Figures given in the table above incorporate a strength criterion also and thus may exceed the minimum depth given in the code.

STEELWORK IN FIRE
INFORMATION SHEET

This series of information sheets is intended to illustrate methods of achieving fire resistance in steel structures. It should not be used for design without consulting detailed design guidance referenced below.

SLIMDEK BEAMS
WITH DEEP DECK

UP TO 1 HOUR UNPROTECTED

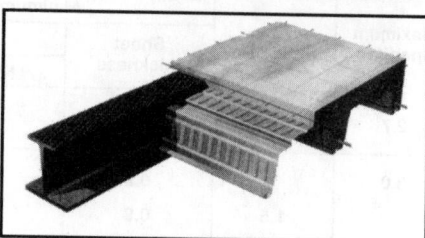

METHOD

The SLIMDEK system consists of an asymmetric beam, with a narrow upper flange and a 225mm deep deck positioned on the outstand of the lower flange. The floor is formed from in-situ applied concrete. This arrangement, shown above, can be designed to provide 60 minutes fire resistance in most cases without the need for applied fire protection.

PRINCIPLE

The section is protected form the effects of fire by the insulating concrete floor. Thus only the bottom flange is directly exposed in fire. Composite action, which develops as a consequence of the raised pattern on the upper flange compensates for much of the loss of strength in the steel at high temperatures

ADVANTAGES

- Fire resistance periods of 60 minutes can be achieved in most instances without any restrictions in loadings.

- Flat slab construction.

*SLIMDEK is a Registered Trade Mark of British Steel plc.

- Clear service runs.

- Reduced construction runs and building heights.

- Services can be passed through prepared openings in the rib of the decking to further reduce floor depth.

LIMITATIONS

a) For fire resistance periods greater than 60 minutes, the exposed lower flange requires fire protection.

b) Where holes are cut in the beam to allow services to pass through, the exposed bottom flange will generally require fire protection.

FOR MORE DETAILED INFORMATION SEE:-
SCI Publication 175 "Design of Asymmetric Slimflor Beams using Deep Composite Decking"
Ths Steel Construction Institute (01344 23345)

Sheet Code
ISF/No.9
April 1997

Table 1

Summary of recommendations

Fire Resistance (Minutes)	Design Type	
	Without holes or action	With service holes
30 Minutes	No protection required	No protection required
60 Minutes	No protection required in most circumstances (see table 2)	Protect bottom flange
Greater than 60	Protect bottom flange	

Table 2

Load table for ASB beams for 60 minutes fire resistance (Concrete grade 30, steel grade S355)

Section	Span of Beam (mm)	Effective width of slab (mm)	Moment resistance at ultimate limit state (kNm)	Moment resistance at fire resistance of 60 mins. (kNm)	Maximum Load Ratio
280 ASB 100	5500	688	554	257	0.48
	6000	750	562	258	0.47
	6500	813	570	260	0.47
280 ASB 136	5500	688	726	376	0.52
	6000	750	736	378	0.51
	6500	813	745	381	0.51
	7000	875	754	383	0.51
300 ASB 153	6000	750	870	472	0.51
	6500	813	880	475	0.54
	7000	875	890	478	0.54
	7500	938	900	481	0.54
300 ASB 153 (slab flush with top flange, and lightweight concrete used)	6000	750	835	440	0.53
	6500	813	842	442	0.53
	7000	875	849	443	0.52
	7500	938	856	445	0.52

STEELWORK IN FIRE
INFORMATION SHEET

This series of information sheets is intended to illustrate methods of achieving fire resistance in steel structures. It should not be used for design without consulting detailed design guidance referenced below.

SHELF ANGLE FLOOR BEAMS

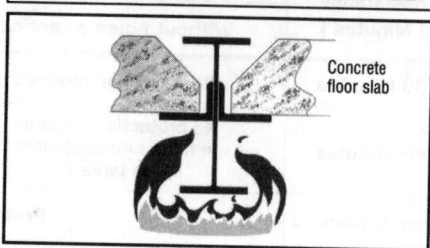

UP TO 1 HOUR UNPROTECTED

Concrete floor slab

METHOD

Shelf Angle Floor Beams consist of Universal Beams with angles bolted or welded to the web. The floor is formed from concrete floor slabs which sit on the outstand of the angle. The gap between the web and the floor slab is filled with grout or concrete to ensure that an effective heat sink is created around the section. This arrangement can be designed to provide 30 or 60 minutes fire resistance in many cases without the need for applied fire protection.

PRINCIPLE

The section is partly protected from the effects of fire by the insulating concrete floor and infill. Thus only part of the web and the bottom flange is exposed to the fire. The angles, which are ignored in cold design, are considered in fire and provide additional capacity. As the angles are placed further down the web, the insulated area, and thus the fire resistance is increased. Where fire protection is required, the reduced exposed perimeter leads in turn to reduced fire protection thicknesses.

ADVANTAGES

- Fire resistance periods of 30 minutes can be achieved in most cases without fire protection to the exposed web and bottom flange.

- Fire resistance periods of 60 minutes can be achieved in some instances depending on the load and the exposed area of the beam.

- Reduced construction runs and building height

- Where the exposed steelwork requires fire protection, reduced thicknesses are possible

LIMITATIONS

a) The angle must be 125 x 75 x 12 Grade S355, short leg attached vertically to the beam as shown overleaf.

b) For 60 minutes fire resistance at high loads, the required depth of floor slab may be unavailable or uneconomic.

c) For fire resistance periods over 60 minutes, fire protection to the exposed perimeter will always be necessary.

FOR MORE DETAILED INFORMATION SEE:-
SCI Publication 126 "The Fire Resistance of Shelf Angle Floor Beams to BS5950 Part 8" The Steel Construction Institute (01344 23345)

Sheet Code
ISF/No.10
April 1997

Typical recommendations for fire resistance of Shelf Angle Floor Beams, taken from the Steel Construction Institute publication, are given in Table 1. These can be expanded to take into account other grades of steel and fire resistance periods. The allowable angle connection force must also be considered.

Table 1

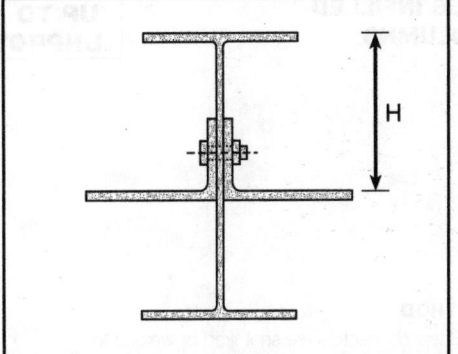

Fire Resistance	60 minutes

Beam Grade	S355
Angle Grade	S355

		H(mm), Position of angle below top of beam for load ratio[1] of						
Section Size	M_p (KNm)	0.4	0.45	0.5	0.55	0.6	0.65	0.7
305 x 102 x 25 UB	120	129	144	158	172	184	196	208
305 x 102 x 28 UB	145	137	152	167	180	193	205	217
305 x 102 x 33 UB	170	144	159	174	188	201	214	227
305 x 127 x 37 UB	192	145	160	174	188	202	209	222
305 x 127 x 42 UB	217	150	166	181	196	207	216	230
305 x 127 x 48 UB	251	158	174	190	206	212	227	235

[1]Load Ratio is defined in BS5950 Part 8 is the ratio of the load at the fire limit state to the cold capacity of the section.

STEELWORK IN FIRE
INFORMATION SHEET

This series of information sheets is intended to illustrate methods of achieving fire resistance in steel structures. It should not be used for design without consulting detailed design guidance referenced below.

WEB INFILLED COLUMNS

UP TO 1 HOUR UNPROTECTED

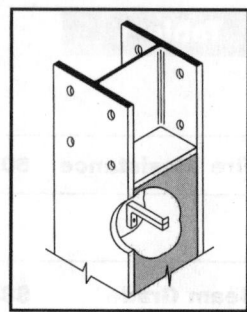

METHOD

Shear connectors are shot fired or welded to the web of the column. A web stiffening plate is welded to the column below the connection zone to contain the concrete. The area below the stiffener and between the flanges is then filled with concrete.

PRINCIPLE

In cold design any beneficial effects of the concrete are ignored. In fire however, as the steel becomes hot, the load is transferred to the concrete. Load transfer is accommodated both by the shear studs and the welded stiffener plate. The unconcreted part of the column and the connections are protected by the same system used to protect the steel beam.

ADVANTAGES

- Fire resistance periods of 60 minutes can be achieved in most cases without fire protection to the exposed flanges.

- For higher periods of fire resistance reduced fire protection thicknesses are required.

- The complete column can be constructed off-site.

- The complete section takes up no space not already occupied by the steel column.

LIMITATIONS

a) The method should not be used where a high specification is required and the column is exposed.

b) Although the system can be used in simple construction where moments are relatively small, it is not suitable for columns in moment resisting frames.

c) For fire resistance periods over 60 minutes, fire protection to the exposed perimeter will always be necessary.

d) The method can only be used for 203x203x46 UCs and above in size.

FOR MORE DETAILED INFORMATION SEE:-
SCI Publication 124 "The Fire Resistance of Web Infilled Steel Columns"
Ths Steel Construction Institute (01344 23345)

Sheet Code
ISF/No.11
April 1997

Typical recommendations for fire resistance of Web Infilled Columns, taken from the Steel Construction Institute publication, are given in Table 1

Table 1

Fire Resistance	60 minutes

Column Grade	S275

Section Size	Moment Capacity (kNm)		Load Ratio[1]	Compressive Capacity (kN) at effective lengths used for normal design				
	M_{fx}	M_{fy}		2500mm	3000mm	3500mm	4000mm	4500mm
203 x 203 x 46	40.3	22.9	0.57	744	682	616	551	488
203 x 203 x 52	47.0	25.4	0.54	803	738	669	600	533
203 x 203 x 60	56.7	29.0	0.52	882	881	735	659	586
203 x 203 x 71	69.9	34.6	0.49	974	899	820	740	662

[1]Load Ratio is defined in BS5950 Part 8 as the ratio of the load at the fire limit state to the cold capacity of the section.

M_{fx} = the moment capacity in fire conditions for bending about the x - x axis.

M_{fy} = the moment capacity in fire conditions for bending about the y - y axis.

Universal beams						Section factor $\frac{A_m}{V}$			
						Profile		Box	
						3 sides	4 sides	3 sides	4 sides
Designation		Depth of section D	Width of section B	Thickness		Area of section				
Serial size	Mass per metre			Web t	Flange T					
mm	kg	mm	mm	mm	mm	cm²	m⁻¹	m⁻¹	m⁻¹	m⁻¹
914×419	388	920.5	420.5	21.5	36.6	494.4	60	70	45	55
	343	911.4	418.5	19.4	32.0	437.4	70	80	50	60
914×305	289	926.6	307.8	19.6	32.0	368.8	75	80	60	65
	253	918.5	305.5	17.3	27.9	322.8	85	95	65	75
	224	910.3	304.1	15.9	23.9	285.2	95	105	75	85
	201	903	303.4	15.2	20.2	256.4	105	115	80	95
838×292	226	850.9	293.8	16.1	26.8	288.7	85	95	70	80
	194	840.7	292.4	14.7	21.7	247.1	100	115	80	90
	176	834.9	291.6	14	18.8	224.1	110	125	90	100
762×267	197	769.6	268	15.6	25.4	250.7	90	100	70	85
	173	762	266.7	14.3	21.6	220.4	105	115	80	95
	147	753.9	265.3	12.9	17.5	188.0	120	135	95	110
686×254	170	692.9	255.8	14.5	23.7	216.5	95	110	75	90
	152	687.6	254.5	13.2	21.0	193.8	110	120	85	95
	140	683.5	253.7	12.4	19.0	178.6	115	130	90	105
	125	677.9	253	11.7	16.2	159.6	130	145	100	115
610×305	238	633	311.5	18.6	31.4	303.7	70	80	50	60
	179	617.5	307	14.1	23.6	227.9	90	105	70	80
	149	609.6	304.8	11.9	19.7	190.1	110	125	80	95
610×229	140	617	230.1	13.1	22.1	178.3	105	120	80	95
	125	611.9	229	11.9	19.6	159.5	115	130	90	105
	113	607.3	228.2	11.2	17.3	144.4	130	145	100	115
	101	602.2	227.6	10.6	14.8	129.1	145	160	110	130
533×210	122	544.6	211.9	12.8	21.3	155.7	110	120	85	95
	109	539.5	210.7	11.6	18.8	138.5	120	135	95	110
	101	536.7	210.1	10.9	17.4	129.7	130	145	100	115
	92	533.1	209.3	10.2	15.6	117.7	140	160	110	125
	82	528.3	208.7	9.6	13.2	104.4	155	175	120	140
457×191	98	467.4	192.8	11.4	19.6	125.2	120	135	90	105
	89	463.6	192	10.6	17.7	113.9	130	145	100	115
	82	460.2	191.3	9.9	16.0	104.5	140	160	105	125
	74	457.2	190.5	9.1	14.5	94.98	155	175	115	135
	67	453.6	189.9	8.5	12.7	85.44	170	190	130	150
457×152	82	465.1	153.5	10.7	18.9	104.4	130	145	105	120
	74	461.3	152.7	9.9	17.0	94.99	140	155	115	130
	67	457.2	151.9	9.1	15.0	85.41	155	175	125	145
	60	454.7	152.9	8.0	13.3	75.93	175	195	140	160
	52	449.8	152.4	7.6	10.9	66.49	200	220	160	180
406×178	74	412.8	179.7	9.7	16.0	94.95	140	160	105	125
	67	409.4	178.8	8.8	14.3	85.49	155	175	115	140
	60	406.4	177.8	7.8	12.8	76.01	175	195	130	155
	54	402.6	177.6	7.6	10.9	68.42	190	215	145	170
406×140	46	402.3	142.4	6.9	11.2	58.96	205	230	160	185
	39	397.3	141.8	6.3	8.6	49.40	240	270	190	220
356×171	67	364	173.2	9.1	15.7	85.42	140	160	105	125
	57	358.6	172.1	8	13.0	72.18	165	190	125	145
	51	355.6	171.5	7.3	11.5	64.58	185	210	135	165
	45	352	171	6.9	9.7	56.96	210	240	155	185
356×127	39	352.8	126	6.5	10.7	49.40	215	240	170	195
	33	248.5	125.4	5.9	8.5	41.83	250	280	195	225
305×165	54	310.9	166.8	7.7	13.7	68.38	160	185	115	140
	46	307.1	165.7	6.7	11.8	58.90	185	210	130	160
	40	303.8	165.1	6.1	10.2	51.50	210	240	150	180
305×127	48	310.4	125.2	9.9	14.0	60.83	160	180	125	145
	42	306.6	124.3	8	12.1	53.18	180	205	140	160
	37	303.8	123.5	7.2	10.7	47.47	200	225	155	180
305×102	33	312.7	102.4	6.6	10.8	41.77	215	240	175	200
	28	308.9	101.9	6.1	8.9	36.30	245	275	200	225
	25	304.8	101.6	5.8	6.8	31.39	285	315	225	260
254×146	43	259.6	147.3	7.3	12.7	55.10	170	195	120	150
	37	256	146.4	6.4	10.9	47.45	195	225	140	170
	31	251.5	146.1	6.1	8.6	40.00	230	265	160	200
254×102	28	260.4	102.1	6.4	10.0	36.19	220	250	170	200
	25	257	101.9	6.1	8.4	32.17	245	280	190	225
	22	254	101.6	5.8	6.8	28.42	275	315	215	250
203×133	30	206.8	133.8	6.3	9.6	38.00	210	245	145	180
	25	203.2	133.4	5.8	7.8	32.31	240	285	165	210
203×102	23	203.2	101.6	5.2	9.3	29	235	270	175	210
178×102	19	177.8	101.6	4.7	7.9	24.2	265	305	190	230
152×89	16	152.4	88.9	4.6	7.7	20.5	270	310	190	235
127×76	13	127	76.2	4.2	7.6	16.8	275	320	195	240

Universal columns							Section factor $\frac{A_m}{V}$			
							Profile		Box	
							3 sides	4 sides	3 sides	4 sides
Designation		Depth of section D	Width of section B	Thickness		Area of section				
Serial size	Mass per metre			Web t	Flange T					
mm	kg	mm	mm	mm	mm	cm²	m⁻¹	m⁻¹	m⁻¹	m⁻¹
356 × 406	634	474.7	424.1	47.6	77.0	808.1	25	30	15	20
	551	455.7	418.5	42.0	67.5	701.8	30	35	20	25
	467	436.6	412.4	35.9	58.0	595.5	35	40	20	30
	393	419.1	407.0	30.6	49.2	500.9	40	45	25	35
	340	406.4	403.0	26.5	42.9	432.7	45	55	30	35
	287	393.7	399.0	22.6	36.5	366.0	50	65	30	45
	235	381.0	395.0	18.5	30.2	299.8	65	75	40	50
356 × 368	202	374.7	374.4	16.8	27.0	257.9	70	85	45	60
	177	368.3	372.1	14.5	23.8	225.7	80	95	50	65
	153	362.0	370.2	12.6	20.7	195.2	90	110	55	75
	129	355.6	368.3	10.7	17.5	164.9	105	130	65	90
305 × 305	283	365.3	321.8	26.9	44.1	360.4	45	55	30	40
	240	352.6	317.9	23.0	37.7	305.6	50	60	35	45
	198	339.9	314.1	19.2	31.4	252.3	60	75	40	50
	158	327.2	310.6	15.7	25.0	201.2	75	90	50	65
	137	320.5	308.7	13.8	21.7	174.6	85	105	55	70
	118	314.5	306.8	11.9	18.7	149.8	100	120	60	85
	97	307.8	304.8	9.9	15.4	123.3	120	145	75	100
254 × 254	167	289.1	264.5	19.2	31.7	212.4	60	75	40	50
	132	276.4	261.0	15.6	25.3	167.7	75	90	50	65
	107	266.7	258.3	13.0	20.5	136.6	90	110	60	75
	89	260.4	255.9	10.5	17.3	114.0	110	130	70	90
	73	254.0	254.0	8.6	14.2	92.9	130	160	80	110
203 × 203	86	222.3	208.8	13.0	20.5	110.1	95	110	60	80
	71	215.9	206.2	10.3	17.3	91.1	110	135	70	95
	60	209.6	205.2	9.3	14.2	75.8	130	160	80	110
	52	206.2	203.9	8.0	12.5	66.4	150	180	95	I25
	46	203.2	203.2	7.3	11.0	58.8	165	200	105	140
152 × 152	37	161.8	154.4	8.1	11.5	47.4	160	190	100	135
	30	157.5	152.9	6.6	9.4	38.2	195	235	120	160
	23	152.4	152.4	6.1	6.8	29.8	245	300	155	205

Circular hollow sections				Section factor $\frac{A_m}{V}$
				Profile or Box

Designation		Mass per metre	Area of section	Section factor $\frac{A_m}{V}$
Outside diameter D	Thickness t			Profile or Box
mm	mm	kg	cm²	m⁻¹
21.3	3.2	1.43	1.82	370
26.9	3.2	1.87	2.38	355
33.7	2.6	1.99	2.54	415
	3.2	2.41	3.07	345
	4.0	2.93	3.73	285
42.4	2.6	2.55	3.25	410
	3.2	3.09	3.94	340
	4.0	3.79	4.83	275
48.3	3.2	3.56	4.53	335
	4.0	4.37	5.57	270
	5.0	5.34	6.80	225
60.3	3.2	4.51	5.74	330
	4.0	5.55	7.07	270
	5.0	6.82	8.69	220
76.1	3.2	5.75	7.33	325
	4.0	7.11	9.06	265
	5.0	8.77	11.2	215
88.9	3.2	6.76	8.62	325
	4.0	8.38	10.70	260
	5.0	10.3	13.2	210
114.3	3.6	9.83	12.5	285
	5.0	13.5	17.2	210
	6.3	16.8	21.4	170
139.7	5.0	16.6	21.2	205
	6.3	20.7	26.4	165
	8.0	26.0	33.1	135
	10.0	32.0	40.7	110
168.3	5.0	20.1	25.7	205
	6.3	25.2	37.1	165
	8.0	31.6	40.3	130
	10.0	39.0	49.7	105
193.7	5.0	23.3	29.6	205
	6.3	29.1	37.1	165
	8.0	36.6	46.7	130
	10.0	45.3	57.7	105
	12.5	55.9	71.2	85
	16.0	70.1	89.3	70
219.1	5.0	26.4	33.6	205
	6.3	33.1	42.1	165
	8.0	41.6	53.1	130
	10.0	51.6	65.7	105
	12.5	63.7	81.1	85
	16.0	80.1	102	65
	20.0	98.2	125	55

Designation		Mass per metre	Area of section	Section factor $\frac{A_m}{V}$
Outside diameter D	Thickness t			Profile or Box
mm	mm	kg	cm²	m⁻¹
244.5	6.3	37.0	47.1	165
	8.0	46.7	59.4	130
	10.0	57.8	73.7	105
	12.5	71.5	91.1	85
	16.0	90.2	115	65
	20.0	111	141	55
273.0	6.3	41.4	52.8	160
	8.0	52.3	66.6	130
	10.0	64.9	82.6	105
	12.5	80.3	102	85
	16.0	101	129	65
	20.0	125	159	55
	25.0	153	195	45
323.9	6.3	49.3	62.9	160
	8.0	62.3	79.4	130
	10.0	77.4	98.6	105
	12.5	96.0	122	85
	16.0	121	155	65
	20.0	150	191	55
	25.0	184	235	45
355.6	8.0	68.6	87.4	130
	10.0	85.2	109	100
	12.5	106	135	85
	16.0	134	171	65
	20.0	166	211	55
	25.0	204	260	45
406.4	10.0	97.8	125	100
	12.5	121	155	80
	16.0	154	196	65
	20.0	191	243	55
	25.0	235	300	45
	32.0	295	376	35
457.0	10.0	110	140	105
	12.5	137	175	80
	16.0	174	222	65
	20.0	216	275	50
	25.0	266	339	40
	32.0	335	427	35
	40.0	411	524	25
508.0	10.0	123	156	100
	12.5	153	195	80
	16.0	194	247	65

				Section factor $\frac{A_m}{V}$		
Rectangular hollow sections				3 sides		4 sides
Designation		Mass per metre	Area of section			
Size D×B	Thickness t					
mm	mm	kg	cm²	m⁻¹	m⁻¹	m⁻¹
50×25	2.5	2.72	3.47	360	290	430
	3.0	3.22	4.10	305	245	365
	3.2	3.41	4.34	290	230	345
50×30	2.5	2.92	3.72	350	295	430
	3.0	3.45	4.40	295	250	365
	3.2	3.66	4.66	280	235	345
	4.0	4.46	5.68	230	195	280
	5.0	5.40	6.88	190	160	235
60×40	2.5	3.71	4.72	340	295	425
	3.0	4.39	5.60	285	250	355
	3.2	4.66	5.94	270	235	335
	4.0	5.72	7.28	220	190	275
	5.0	6.97	8.88	180	160	225
	6.3	8.49	10.8	150	130	185
80×40	3.0	5.34	6.80	295	235	355
	3.2	5.67	7.22	275	220	330
	4.0	6.97	8.88	225	180	270
	5.0	8.54	10.9	185	145	220
	6.3	10.5	13.3	150	120	180
	8.0	12.8	16.3	125	100	145
90×50	3.0	6.28	8.00	290	240	350
	3.6	7.46	9.50	240	200	295
	5.0	10.1	12.9	180	145	215
	6.3	12.5	15.9	145	120	175
	8.0	15.3	19.5	120	95	145
100×50	3.0	6.75	8.60	290	235	350
	3.2	7.18	9.14	275	220	330
	4.0	8.86	11.3	220	175	265
	5.0	10.9	13.9	180	145	215
	6.3	13.4	17.1	145	115	175
	8.0	16.6	21.1	120	95	140
100×60	3.0	7.22	9.20	285	240	350
	3.6	8.59	10.9	240	200	295
	5.0	11.7	14.9	175	150	215
	6.3	14.4	18.4	140	120	175
	8.0	17.8	22.7	115	95	140
120×60	3.6	9.72	12.4	240	195	290
	5.0	13.3	16.9	180	140	215
	6.3	16.4	20.9	145	115	170
	8.0	20.4	25.9	115	95	140
120×80	5.0	14.8	18.9	170	150	210
	6.3	18.4	23.4	135	120	170
	8.0	22.9	29.1	110	95	135
	10.0	27.9	35.5	90	80	115
150×100	5.0	18.7	23.9	165	145	210
	6.3	23.8	29.7	135	120	170
	8.0	29.1	37.1	110	95	135
	10.0	35.7	45.5	90	75	110
	12.5	43.6	55.5	70	65	90
160×80	5.0	18.0	22.9	175	140	210
	6.3	22.3	28.5	140	110	170
	8.0	27.9	35.5	115	90	135
	10.0	34.2	43.5	90	75	110
	12.5	41.6	53.0	75	60	90
200×100	5.0	22.7	28.9	175	140	210
	6.3	28.3	36.0	140	110	165
	8.0	35.4	45.1	110	90	135
	10.0	43.6	55.5	90	70	110
	12.5	53.4	68.0	75	60	90
	16.0	66.4	84.5	60	45	70
250×150	6.3	38.2	48.6	135	115	165
	8.0	48.0	61.1	105	90	130
	10.0	59.3	75.5	85	75	105
	12.5	73.0	93.0	70	60	85
	16.0	91.5	117	55	45	70
300×200	6.3	48.1	61.2	130	115	165
	8.0	60.5	77.1	105	90	130
	10.0	75.0	95.5	85	75	105
	12.5	92.6	118	70	60	85
	16.0	117	149	55	45	65
400×200	10.0	90.7	116	85	70	105
	12.5	112	143	70	55	85
	16.0	142	181	55	45	65
450×250	10.0	106	136	85	70	105
	12.5	132	168	70	55	85
	16.0	167	213	55	45	65

Rectangular hollow sections (square) — Section factor $\frac{A_m}{V}$

Designation		Mass per metre	Area of section	3 sides	4 sides
Size D × D (mm)	Thickness t (mm)	kg	cm²	m⁻¹	m⁻¹
20×20	2.0	1.12	1.42	425	565
	2.5	1.35	1.72	350	465
25×25	2.0	1.43	1.82	410	550
	2.5	1.74	2.22	340	450
	3.0	2.04	2.60	290	385
	3.2	2.15	2.74	275	365
30×30	2.5	2.14	2.72	330	440
	3.0	2.51	3.20	280	375
	3.2	2.65	3.38	265	355
40×40	2.5	2.92	3.72	325	430
	3.0	3.45	4.40	275	365
	3.2	3.66	4.66	260	345
	4.0	4.46	5.68	210	280
	5.0	5.40	6.88	175	235
50×50	2.5	3.71	4.72	320	425
	3.0	4.39	5.60	270	355
	3.2	4.66	5.94	255	335
	4.0	5.72	7.28	205	275
	5.0	6.97	8.88	170	225
	6.3	8.49	10.8	140	185
60×60	3.0	5.34	6.80	265	355
	3.2	5.67	7.22	250	330
	4.0	6.97	8.88	205	270
	5.0	8.54	10.9	165	220
	6.3	10.5	13.3	135	180
	8.0	12.8	16.3	110	145
70×70	3.0	6.28	8.00	260	350
	3.6	7.46	9.50	220	295
	5.0	10.1	12.9	165	215
	6.3	12.5	15.9	130	175
	8.0	15.3	19.5	110	145
80×80	3.0	7.22	9.20	260	350
	3.6	8.59	10.9	220	295
	5.0	11.7	14.9	160	215
	6.3	14.4	18.4	130	175
	8.0	17.8	22.7	105	140
90×90	3.6	9.72	12.4	220	290
	5.0	13.3	16.9	160	215
	6.3	16.4	20.9	130	170
	8.0	20.4	25.9	105	140
100×100	4.0	12.0	15.3	195	260
	5.0	14.8	18.9	160	210
	6.3	18.4	23.4	130	170
	8.0	22.9	29.1	105	135
	10.0	27.9	35.5	85	115

Designation		Mass per metre	Area of section	3 sides	4 sides
Size D × D (mm)	Thickness t (mm)	kg	cm²	m⁻¹	m⁻¹
120×120	5.0	18.0	22.9	155	210
	6.3	22.3	28.5	125	170
	8.0	27.9	35.5	100	135
	10.0	34.2	43.5	85	110
	12.5	41.6	53.0	70	90
140×140	5.0	21.1	26.9	155	210
	6.3	26.3	33.5	125	165
	8.0	32.9	41.9	100	135
	10.0	40.4	51.5	80	110
	12.5	49.5	63.0	65	90
150×150	5.0	22.7	28.9	155	210
	6.3	28.3	36.0	125	165
	8.0	35.4	45.1	100	135
	10.0	43.6	55.5	80	110
	12.5	53.4	68.0	65	90
	16.0	66.4	84.5	55	70
180×180	6.3	34.2	43.6	125	165
	8.0	43.0	54.7	100	130
	10.0	53.0	67.5	80	105
	12.5	65.2	83.0	65	85
	16.0	81.4	104	50	70
200×200	6.3	38.2	48.6	125	165
	8.0	48.0	61.1	100	130
	10.0	59.3	75.5	80	105
	12.5	73.0	93.0	65	85
	16.0	91.5	117	50	70
250×250	6.3	48.1	61.2	125	165
	8.0	60.5	77.1	95	130
	10.0	75.0	95.5	80	105
	12.5	92.6	118	65	85
	16.0	117	149	50	65
300×300	10.0	90.7	116	80	105
	12.5	112	143	65	85
	16.0	142	181	50	65
350×350	10.0	106	136	75	105
	12.5	132	168	65	85
	16.0	167	213	50	65
400×400	10.0	122	156	75	105
	12.5	152	193	60	85
	16.0	192	245	50	65

Minimum thickness of a typical spray protection to an I-section

A_m/V up to	Dry thickness (mm) to provide fire resistance of					
	$\frac{1}{2}$ hour	1 hour	$1\frac{1}{2}$ hours	2 hours	3 hours	4 hours
30	10	10	10	10	15	25
50	10	10	10	14	21	29
70	10	10	12	17	27	36
90	10	10	14	20	31	42
110	10	10	16	22	34	47
130	10	10	17	24	37	51
150	10	11	18	25	40	54
170	10	12	19	27	42	57
190	10	12	20	28	44	59
210	10	13	21	29	45	
230	10	13	21	30	47	
250	10	14	22	31	48	
270	10	14	23	31	49	
290	10	14	23	32	50	
310	10	14	23	33	51	

Linear interpolation is permissible between values of H_p/A

Basic data on corrosion

Atmospheric corrosivity categories and examples of typical environments (ISO 12944 Part 2)

Corrosivity category and risk	Mass loss per unit surface/thickness loss (see Note 1)		Examples of typical environments in a temperate climate (informative only)	
	Low-carbon steel Thickness loss μm	Exterior	Interior	
C1 very low	≤1.3	–	Heated buildings with clean atmospheres, e.g. offices, shops, schools, hotels	
C2 low	>1.3–25	Atmospheres with low level of pollution. Mostly rural areas	Unheated buildings where condensation may occur, e.g. depots, sports halls	
C3 medium	>25–60	Urban and industrial atmospheres, moderate sulphur dioxide pollution. Coastal area with low salinity	Production rooms with high humidity and some air pollution, e.g. food-processing plants, laundries, breweries, dairies	
C4 high	>50–80	Industrial areas and coastal areas with moderate salinity	Chemical plants, swimming pools, coastal, ship and boatyards	
C5-I very high (industrial)	>80–200	Industrial areas with high humidity and aggressive atmosphere	Buildings or areas with almost permanent condensation and high pollution	
C5-M very high (marine)	>80–200	Coastal and offshore areas with high salinity	Buildings or areas with almost permanent condensation and with high pollution	

1. The thickness loss values are after the first year of exposure. Losses may reduce over subsequent years.
2. The loss values used for the corrosivity categories are identical to those given in ISO 9223.
3. In coastal areas in hot, humid zones, the thickness losses can exceed the limits of category C5-M. Special precautions must therefore be taken when selecting protective paint systems for structures in such areas.

$1\,\mu m = 0.001\,mm$

Main generic types of paint and their properties

Binder	System cost	Tolerance of poor surface	Chemical resistance	Solvent resistance	Water resistance	Overcoating after ageing	Comments
Black coatings (based on tar products)	Low	Good	Moderate	Poor	Good	Very Good with coatings of same type	Limited to black or dark colours. May soften in hot conditions
Alkyds	Low – medium	Moderate	Poor	Poor – moderate	Moderate	Good	Good decorative properties. High solvent levels
Acrylated rubbers	Medium – high	Poor	Good	Poor	Good	Good	High build films that remain soft and are susceptible to sticking
Epoxy Surface tolerant	Medium – high	Good	Good	Good	Good	Good	Can be applied to a range of surfaces and coatings*
High performance	Medium – high	Very poor	Very good	Good	Very good	Poor	Susceptible to 'chalking' in U.V. light
Urethane and polyurethane	High	Very poor	Very good	Good	Very good	Poor	Can be more decorative than epoxies
Organic silicate and inorganic silicate	High	Very poor	Moderate	Good	Good	Moderate	May require special surface preparation

*Widely used for maintenance painting

SOME EXAMPLES OF DETAILING TO MINIMISE CORROSION

Details should be designed to enhance durability by avoiding water entrapment.

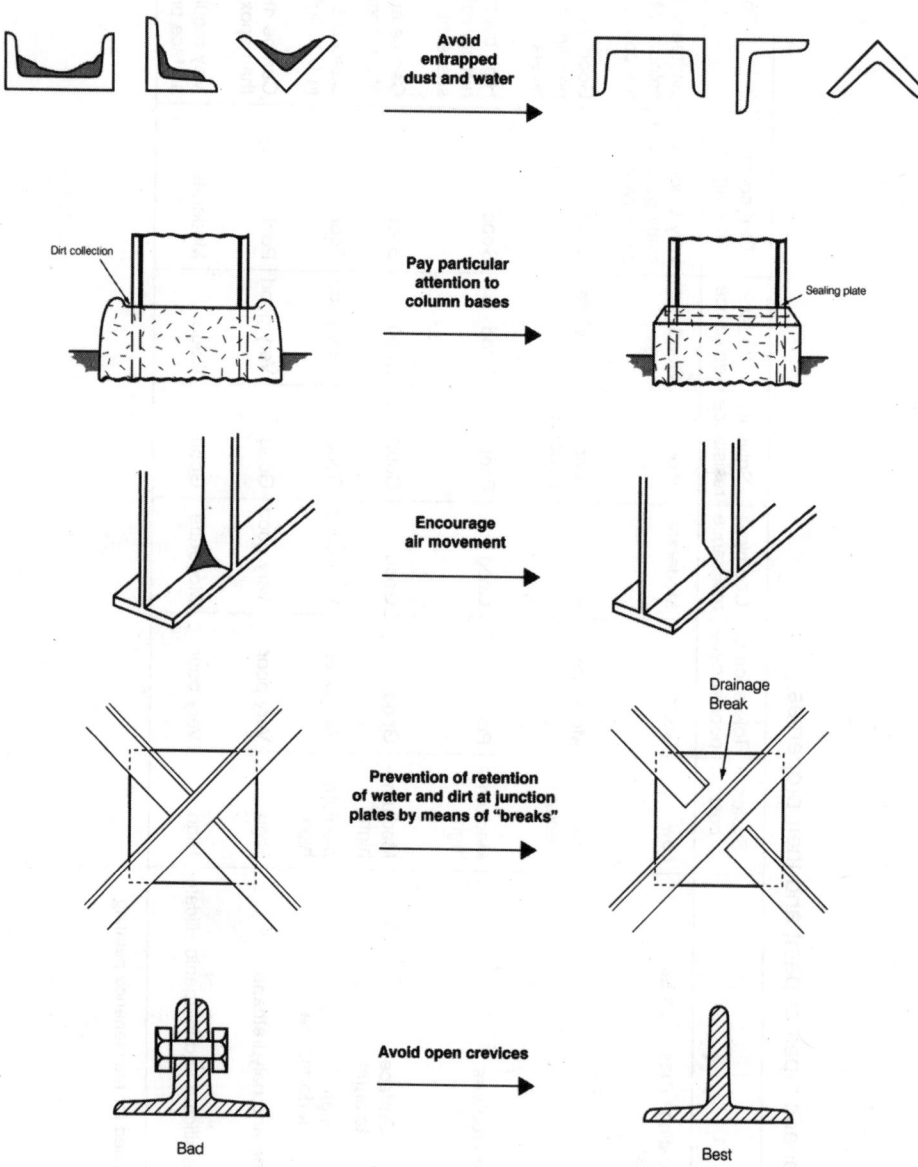

British and European Standards covering the design and construction of steelwork

A basic list of standards covering the design and construction of steelwork is given.

Loading

BS 5400   Steel concrete and composite bridges
          Part 2: 1978 Specification for loads
          Part 6: 1999 Specification for materials and workmanship, steel

BS 6399   Loading for buildings
          Part 1: 1997 Code of practice for dead and imposed loads
          Part 2: 1997 Code of practice for wind loads
          Part 3: 1998 Code of practice for imposed roof loads

DD ENV 1991   Eurocode 1 Basis of design and actions on structures
DD ENV 1991-1:   1996 Basis of design
DD ENV 1991-2-2:   1996 Actions on structures exposed to fire (together with UK NAD)
DD ENV 1991-2-3:   1996 Actions on structures – Snow loads
DD ENV 1991-2-4:   1997 Actions on structures – Wind actions (together with UK NAD)
DD ENV 1991-2-6:   2000 Actions on structures – Actions during execution
DD ENV 1991-3:   2000 Traffic loads on bridges (together with UK NAD)
DD ENV 1991-4:   1996 Actions in silos and tanks

Design

BS 5400   Steel concrete and composite bridges
          Part 3: 2000 Code of practice for design of steel bridges
          Part 5: 1979 Code of practice for design of composite bridges
          Part 9: 1983 Bridge bearings
          Part 10: 1980 Code of practice for fatigue
          Part 10c: 1999 Charts for classification of details for fatigue

BS 5427   Code of practice for the use of profiled sheet for roof and wall cladding on buildings
          Part 1: 1996 Design

BS 5950   Structural use of steelwork in building
          Part 1: 2000 Code of practice for design in simple and continuous construction – Hot rolled sections

Part 3: Section 3.1: 1990 Design of simple and continuous composite beams

Part 4: 1994 Code of practice for design of floors with profiled steel sheeting

Part 5: 1998 Code of practice for design of cold-formed sections

Part 6: 1995 Code of practice for design of light gauge profiled steel sheeting

Part 8: 1990 Code of practice for fire resistant design

Part 9: 1994 Code of practice for stressed skin design

ENV 1991   Eurocode 1 Basis of design and actions on structures

ENV 1991-2-7:   1998 Actions on structures – Accidental actions due to impact and explosions

ENV 1991-5:   1998 Actions induced by cranes and other machinery

ENV 1993   Eurocode 3 Design of steel structures

ENV 1993-1-1:   1992 General rules and rules for buildings (together with UK NAD)

ENV 1993-1-2:   2001 General rules – Structural fire design (together with UK NAD)

ENV 1993-1-3:   2001 General rules – supplementary rules for cold formed thin gauge members and sheeting (together with UK NAD)

ENV 1993   Eurocode 3 Design of steel structures

ENV 1993-1-6:   1999 General rules – Supplementary rules for shell structures

ENV 1993-1-7:   1999 General rules – Supplementary rules for planar plated structural elements with out of plane loading

ENV 1993-4-1:   1999 Silos, tanks and pipelines – Silos

ENV 1993-4-2:   1999 Silos, tanks and pipelines – Tanks

ENV 1993-4-3:   1999 Silos, tanks and pipelines – Pipelines

ENV 1994   Eurocode 4 Design of composite steel and concrete structures

ENV 1994-1-1:   1994 General rules and rules for buildings (together with UK NAD)

ENV 1998   Eurocode 8 Design provisions for earthquake resistance of structures

ENV 1998-1-1:   1996 General rules – Seismic actions and general requirements for structures

ENV 1998-1-2:   1996 General rules – General rules for buildings

ENV 1998-1-3:   1996 General rules – Specific rules for various materials and elements

ENV 1998-1-4:   1996 General rules – Strengthening and repair of buildings

ENV 1998-2:   1996 Bridges

ENV 1998-4:   1999 Silos, tanks and pipelines

ENV 1998-5:   1996 Foundations, retaining structures and geotechnical aspects

Steel fabrication and erection

BS 4604   Specification for the use of high strength friction grip bolts in structural
steelwork – Metric series
Part 1: 1970 General grade
Part 2: 1970 Higher grade (parallel shank)

BS 5400   Steel, concrete and composite bridges
Part 6: 1999 Specification for materials and workmanship, steel

BS 5950   Structural use of steelwork in building
Part 2: 1992 Specification for materials, fabrication and erection – Hot-
rolled sections
Part 7: 1992 Specification for materials and workmanship – Cold-formed
sections

ENV 1090   Execution of steel structures
ENV 1090-1:   1998 General rules and rules for buildings
ENV 1090-2:   1998 Supplementary rules for cold-formed thin gauge components
and sheeting
ENV 1090-3:   2000 Supplementary rules for high yield strength steels
ENV 1090-4:   2000 Supplementary rules for hollow section structures
ENV 1090-6:   2000 Supplementary rules for stainless steel

Foundations and piling

BS 4   Part 1: 1993 Structural steel sections – Specification for hot-rolled sections
BS 449   Specification for the use of structural steel in building
Part 2: 1969 Metric units

BS 5400:   Steel concrete and composite bridges
Part 1: 1988 General statement

BS 5493:   1977 Code of practice for protective coating of iron and steel structures
against corrosion

BS 5950   Structural use of steelwork in building
BS 5950-1:   2000 Code of practice for design in simple and continuous construction – Hot-rolled sections

BS 8002:   1994 Code of practice for earth-retaining systems

BS 8004:   1986 Code of practice for foundations

BS 8081:   1989 Code of practice for ground anchorages

BS EN 10248   Hot-rolled sheet piling of non alloy steels

BS EN 20149   Cold-formed sheet piling of non alloy steels

BS EN 12063:   1999 Execution of special geotechnical work – Sheet pile walls

Structural steel

BS 970   Specification for wrought steels for mechanical and allied engineering purposes
(NB. Some sections withdrawn and replaced by BS ENs)

BS 1449   Steel plate, sheet and strip (many sections)
(NB. Some sections withdrawn and replaced by BS ENs)

BS 4360   Specification for weldable structural steels
(Withdrawn. Replaced by BS 7668: 1994, BS EN 10029: 1991, Parts 1–3 of BS EN 10133: 1999, BS EN 10137: 1996, BS EN 10155: 1993 and BS EN 10210–1: 1994)

BS EN 10111:   1998 Continuously hot-rolled low carbon steel sheet and strip for cold forming – Technical delivery conditions

BS EN 10113   Hot rolled products in weldable fine grain structural steels

BS EN 10130:   1999 Cold-rolled low carbon steel flat products for cold forming – Technical delivery conditions

BS EN 10137:   Plates and wide flats made of high yield strength structural steels in the quenched and tempered or precipitation hardened conditions

BS EN 10139:   1998 Cold-rolled uncoated mild steel narrow strip for cold forming – Technical delivery conditions

BS EN 10164:   1993 Steel products with improved deformation properties perpendicular to the surface of the product – Technical delivery conditions

BS EN 10210-1:   1994 Hot-finished structural hollow sections of non-alloy and fine grain structural steels – Technical delivery requirements.

BS EN 10219-1:   1997 Cold-formed structural hollow sections of non-alloy and fine grain structural steels – Technical delivery requirements

BS EN 10268:   1999 Cold-rolled flat products made of high yield strength micro-alloyed steels for cold forming – General delivery conditions
(Supersedes BS 1449-1-1.5: 1991 and BS 1449-1-1.11: 1991)

BS EN 10273:   2000 Hot-rolled weldable steel bars for pressure purposes with specified elevated temperature properties

Steel products

BS 4   Part 1: 1993 Structural steel sections. Specification for hot-rolled sections
BS EN 10029:   1991 Tolerances on dimensions, shape and mass for hot-rolled steel plates 3 mm thick or above

BS EN 10051:   1992 + A1: 1997 Continuously hot-rolled uncoated plate, sheet and strip of non-alloy and alloy steels – Tolerances on dimensions and shape (including amendment A1: 1997)

BS EN 10055:   1996 Hot-rolled steel equal flange tees with radiused root and toes – Dimensions and tolerances on shape and dimensions

BS EN 10056   Specification for structural steel equal and unequal angles
BS EN 10056-1:   1998 Dimensions
BS EN 10056-2:   1993 Tolerances on shape and dimensions

BS EN 10067:   1997 Hot-rolled bulb flats – Dimensions and tolerances on shape, dimensions and mass

BS EN 10163   Specification for delivery requirements for surface conditions of hot-rolled steel plates, wide flats and sections

BS EN 10210-2:   1997 Hot-finished structural hollow sections of non-alloy and fine grain structural steels – Tolerances, dimensions and sectional properties

BS EN 10219-2:   1997 Cold-formed structural hollow sections of non-alloy and fine grain structural steels – Tolerances, dimensions and sectional properties

Euronorm 91:   1981 Hot-rolled wide flats – Tolerances on dimensions, shape and mass

Cold-rolled thin gauge sections and sheets

BS 1449   Steel plate, sheet and strip Section 1.9: 1991 Specification for cold-rolled narrow strip based on formability

BS 2994:   1976 (1987) Specification for cold-rolled steel sections

BS 5950   Structural use of steelwork in buildings
Part 5: 1987 Code of practice for the design of cold-rolled sections
Part 6: 1995 Code of practice for design of light gauge profiled steel sheeting
Part 7: 1992 Specification for materials and workmanship – Cold-formed sections
Part 9: 1994 Code of practice for stressed skin design

BS EN 10048:   1997 Hot-rolled narrow steel strip – Tolerances on dimensions and shape

BS EN 10031:   1991 Cold-rolled uncoated low carbon and high yield strength steel flat products for cold forming

BS EN 10140:   1997 Cold-rolled narrow steel strip – Tolerances on dimensions and shape

BS EN 10143:   1993 Continuously hot-dip metal coated steel sheet and strip – Tolerances on dimensions and shape

BS EN 10147:   2000 Continuously hot-dip zinc coated structural steels strip and sheet – Technical delivery conditions

BS EN 10149   Hot-rolled flat products made of high yield strength steels for cold forming

BS EN 10214:   1995 Continuously hot-dip zinc-aluminium (ZA) coated steel strip and sheet – Technical delivery conditions

BS EN 10215:   1995 Continuously hot-dip aluminium-zinc (AZ) coated steel strip and sheet – Technical delivery conditions

BS EN 10238:   1997 Automatically blast cleaned and automatically primed structural steel products

BS EN 10292:   2000 Continuously hot-dip coated strip and sheet of steels with higher yield strength for cold forming – Technical delivery conditions

DD ENV 10169   Continuously organic coated (coil coated) steel flat products
DD ENV 10169-2:   1999 Products for building exterior applications

ISO 4995:   1993 Hot-rolled steel sheet of structural quality

ISO 4997:   1991 Cold reduced steel sheet of structural quality

ISO 4999:   1999 Continuous hot-dip terne (lead alloy) coated cold-reduced steel sheet of commercial drawing and structural qualities

ISO 5951:   1993 Hot-rolled steel sheet of higher yield strength with improved formability

ISO 6316:   1993 Hot-rolled steel strip of structural quality

ISO 16162:   2000 Continuously cold-rolled products – Dimensional and shape tolerances

ISO 16163:   2000 Continuously hot-dipped coated steel products – Dimensional and shape tolerances

Stainless steels

BS 7475:   1991 Specification for fusion welding of austenitic stainless steels

BS EN 10088   Stainless steels

BS EN ISO 3506   Mechanical properties of corrosion resistant stainless steel fasteners

Castings and forgings

BS 3100:   1991 Specification for steel castings for general engineering purposes

BS EN 1560:   1997 Founding – Designation system for cast iron – Material symbols and material number

BS EN 1561:   1997 Founding – Grey cast irons

BS EN 1563:   1997 Founding – Spheroidal graphite cast irons

Steel construction components

BS 2994:   1976 (1987) Specification for cold-rolled steel sections

BS 5427   Code of practice for the use of profiled sheet for roof and wall cladding
on buildings
Part 1: 1996 Design

BS EN 1337   Structural bearings

BS EN 1462:   1997 Brackets for eaves gutters: Requirements and testing

Welding materials and processes

General

BS 499   Specification for symbols for welding (formerly Welding terms and
symbols)
Part 1: 1991 Glossary for welding, brazing and thermal cutting
BS 499-2C: 1999 European arc welding symbols in chart form

BS EN 29692:   1994 Specification for metal-arc welding with covered electrode, gas shielded metal arc-welding and gas welding. Joint preparation for steel

BS EN ISO 4063:   2000 Welding and allied processes – Nomenclature of processes and reference numbers

BS EN ISO 9692   Welding and allied processes – Joint preparation

BS EN ISO 9692-2:   1998 Submerged arc welding of steels

Processes and consumables

BS EN 440:   1995 Welding consumables: wire electrodes and deposits for gas shielded metal arc welding of non-alloy and fine grain steels: Classification

BS EN 499:   1995 Welding consumables – Covered electrodes for manual metal arc welding of non-alloy and fine grain steels: Classification

BS EN 756:   1996 Welding consumables – Wire electrodes and wire-flux combinations for submerged arc welding of non-alloy and fine grain steels – Classification

BS EN 757:   1997 Welding consumables – Covered electrodes for metal arc welding of high strength steels: Classification

BS EN 758:   1997 Welding consumables – Tubular cored electrodes for metal arc welding with and without a gas shield of non-alloy and fine grain steels – Classification

Testing and examination

BS 709:   1983 Methods of destructive testing fusion welded joints and weld metal in steel

BS 2600   Radiographic examination of fusion welded butt joints in steel
        Part 1: 1983 Methods for steel 2 mm up to and including 50 mm thick
        Part 2: 1973 Methods for steel over 50 mm up to and including 200 mm thick

BS 2910:   1986 Methods for radiographic examination of fusion welded circumferential butt joints in steel pipes

BS 3683   Glossary of terms used in non-destructive testing
        Part 2: 1985 Magnetic particle flaw detection
        Part 3: 1984 Radiological flaw detection

BS 4871   Specification for approval testing of welders working to approve welding procedures

BS 4872   Specification for approval testing of welders when welding procedure approval is not required

BS 4872-1:   1982 Fusion welding of steel

BS 6072:   1981 (1986) Method for magnetic particle flaw detection

BS EN 287   Approval testing of welders for fusion welding

BS EN 288   Specification and approval of welding procedures for metallic materials

BS EN 970:   1997 Non-destructive examination of fusion welds – Visual examination

BS EN 1289:   1998 Non-destructive examination of welds – Penetrant testing of welds – Acceptance testing of welds – Acceptance level

BS EN 1290:   1998 Non-destructive examination of welds – Magnetic particle examination of welds

BS EN 1291:   1998 Non-destructive examination of welds – Magnetic particle testing of welds – Acceptance levels

BS EN 1435:   1997 Non-destructive examination of welds – Radiographic examination of welded joints

BS EN 1713:   1998 Non-destructive examination of welds – ultrasonic examination – characterization of indication in welds

BS EN 1714:   1998 Non-destructive examination of welded joints – Ultrasonic examination of welded joints (*Supersedes BS 3923-1: 1986*)

BS EN 12062:   1998 Non-destructive examination of welds – General rules for metallic materials

BS EN 25817:   1992 Arc-welded joints in steel: Guidance on quality levels for imperfections

Bolts and fasteners

BS 4395   Specification for high strength friction grip bolts and associated nuts and washers for structural engineering
Part 1: 1969 General grade
Part 2: 1969 Higher grade bolts and nuts and general grade washers

BS 4604   Specification for the use of high strength friction grip bolts in structural steelwork – Metric series
Part 1: 1970 General grade
Part 2: 1970 Higher grade (parallel shank)

BS 7419:   1991 Specification for holding down bolts

BS 7644   Direct tension indicators
Part 1: 1993 Specification for compressible washers
Part 2: 1993 Specification for nut face and bolt face washers

Fire resistance

BS 476   Fire tests on building materials and structures
Part 20: 1987 Method for determination of the fire resistance of elements of construction (general principles)
Part 21: 1987 Methods for determination of the fire resistance of load-bearing elements of construction
Part 22: 1987 Methods for determination of the fire resistance of non-loadbearing elements of construction
Part 23: 1987 Methods for determination of the contribution of components to the fire resistance of a structure

BS 5588   Fire precautions in the design, construction and use of buildings

BS 5950   Structural use of steelwork in building
Part 8: 1990 Code of practice for fire resistant design

BS 8202   Coatings for fire protection of building elements
Part 1: 1995 Code of practice for the selection and installation of sprayed mineral coatings
Part 2: 1992 Code of practice for the use of intumescent coating systems to metallic substrates for providing fire resistance

Corrosion prevention and coatings

BS 2569   Specification for sprayed coatings
Part 2: 1965 (1997) Protection of iron and steel against corrosion and oxidation at elevated temperatures

BS 4652:   1995 Specification for metallic zinc-rich priming paints (organic media)

BS 4921:   1988 (1994) Specification for sherardized coatings on iron or steel

BS 5493:   1977 Code of practice for protective coating of iron and steel structures against corrosion

BS 7079   Preparation of steel substrates before application of paints and related products

Quality assurance

BS EN ISO 9000   Quality management and quality assurance standards

BS EN ISO 9001:   1994 Quality systems – Model for quality assurance in design, development, production, installation and servicing

Environmental

BS 6187:   2000 Code of practice for demolition

BS EN ISO 14001:   1996 Environmental management systems. Specification with guidance for use

BS EN ISO 14004:   1996 Environmental management systems. General guidelines on principles, systems and supporting techniques

BS EN ISO 14010:   1996 Guidelines for environmental auditing. General principles

Index